The Tectonic and Climatic Evolution
of the Arabian Sea Region

Geological Society Special Publications
Society Book Editors
A. J. FLEET (CHIEF EDITOR)
P. DOYLE
F. J. GREGORY
J. S. GRIFFITHS
A. J. HARTLEY
R. E. HOLDSWORTH
A. C. MORTON
N. S. ROBINS
M. S. STOKER
J. P. TURNER

Reviewing procedures

The Society makes every effort to ensure that the scientific and production quality of its books matches that of its journals. Since 1997, all book proposals have been refereed by specialist reviewers as well as by the Society's Books Editorial Committee. If the referees identify weaknesses in the proposal, these must be addressed before the proposal is accepted.

Once the book is accepted, the Society has a team of Book Editors (listed above) who ensure that the volume editors follow strict guidelines on refereeing and quality control. We insist that individual papers can only be accepted after satisfactory review by two independent referees. The questions on the review forms are similar to those for *Journal of the Geological Society*. The referees' forms and comments must be available to the Society's Book Editors on request.

Although many of the books result from meetings, the editors are expected to commission papers that were not presented at the meeting to ensure that the book provides a balanced coverage of the subject. Being accepted for presentation at the meeting does not guarantee inclusion in the book.

Geological Society Special Publications are included in the ISI Index of Scientific Book Contents, but they do not have an impact factor, the latter being applicable only to journals.

More information about submitting a proposal and producing a Special Publication can be found on the Society's web site: www.geolsoc.org.uk.

It is recommended that reference to all or part of this book should be made in one of the following ways:

CLIFT, P. D., KROON, D., GAEDICKE, C. & CRAIG, J. (eds) 2002. *The Tectonic and Climatic Evolution of the Arabian Sea Region*. Geological Society, London, Special Publications, **195**

UCHUPI, E., SWIFT, S. A. & ROSS, D. A. 2002. Tectonic geomorphology of the Gulf of Oman Basin. *In*: CLIFT, P. D., KROON, D., GAEDICKE, C. & CRAIG, J. (eds) *The Tectonic and Climatic Evolution of the Arabian Sea Region*. Geological Society, London, Special Publications, **195** 1–69.

GEOLOGICAL SOCIETY SPECIAL PUBLICATION NO. 195

The Tectonic and Climatic Evolution of the Arabian Sea Region

EDITED BY

P. D. CLIFT
Woods Hole Oceanographic Institution, USA

D. KROON
Free University Amsterdam, The Netherlands

C. GAEDICKE
Bundesanstalt für Geowissenschaften und Rohstoffe (BGR), Germany

and

J. CRAIG
Lasmo plc, UK

2002
Published by
The Geological Society
London

THE GEOLOGICAL SOCIETY

The Geological Society of London (GSL) was founded in 1807. It is the oldest national geological society in the world and the largest in Europe. It was incorporated under Royal Charter in 1825 and is Registered Charity 210161.

The Society is the UK national learned and professional society for geology with a worldwide Fellowship (FGS) of 9000. The Society has the power to confer Chartered status on suitably qualified Fellows, and about 2000 of the Fellowship carry the title (CGeol). Chartered Geologists may also obtain the equivalent European title, European Geologist (EurGeol). One fifth of the Society's fellowship resides outside the UK. To find out more about the Society, log on to www.geolsoc.org.uk.

The Geological Society Publishing House (Bath, UK) produces the Society's international journals and books, and acts as European distributor for selected publications of the American Association of Petroleum Geologists (AAPG), the American Geological Institute (AGI), the Indonesian Petroleum Association (IPA), the Geological Society of America (GSA), the Society for Sedimentary Geology (SEPM) and the Geologists' Association (GA). Joint marketing agreements ensure that GSL Fellows may purchase these societies' publications at a discount. The Society's online bookshop (accessible from www.geolsoc.org.uk) offers secure book purchasing with your credit or debit card.

To find out about joining the Society and benefiting from substantial discounts on publications of GSL and other societies worldwide, consult www.geolsoc.org.uk, or contact the Fellowship Department at: The Geological Society, Burlington House, Piccadilly, London W1J 0BG: Tel. +44 (0)20 7434 9944; Fax +44 (0)20 7439 8975; E-mail: enquiries@geolsoc.org.uk.

For information about the Society's meetings, consult *Events* on www.geolsoc.org.uk. To find out more about the Society's Corporate Affiliates Scheme, write to enquiries@geolsoc.org.uk.

Published by The Geological Society from:
The Geological Society Publishing House
Unit 7, Brassmill Enterprise Centre
Brassmill Lane
Bath BA1 3JN, UK
(*Orders*: Tel. +44 (0)1225 445046
       Fax +44 (0)1225 442836)
Online bookshop: http://bookshop.geolsoc.org.uk

The publishers make no representation, express or implied, with regard to the accuracy of the information contained in this book and cannot accept any legal responsibility for any errors or omissions that may be made.

© The Geological Society of London 2002. All rights reserved. No reproduction, copy or transmission of this publication may be made without written permission. No paragraph of this publication may be reproduced, copied or transmitted save with the provisions of the Copyright Licensing Agency, 90 Tottenham Court Road, London W1P 9HE. Users registered with the Copyright Clearance Center, 27 Congress Street, Salem, MA 01970, USA: the item-fee code for this publication is 0305-8719/02/$15.00.

British Library Cataloguing in Publication Data

A catalogue record for this book is available from the British Library.

ISBN 1-86239-111-4

Typeset by Keytec Typesetting Ltd., Bridport, Dorset, UK.

Printed by Cambrian Printers Ltd, Aberystwyth, UK

Distributors

USA
AAPG Bookstore
PO Box 979
Tulsa
OK 74101-0979
USA
Orders: Tel. +1 918 584-2555
         Fax +1 918 560-2652
         E-mail bookstore@aapg.org

India
Affiliated East-West Press PVT Ltd
G-1/16 Ansari Road, Daryaganj,
New Delhi 110 002
Orders: Tel. +91 11 327-9113
         Fax +91 11 326-0538
         E-mail affiliat@nda.vsnl.net.in

Japan
Kanda Book Trading Co.
Cityhouse Tama 204
Tsurumaki 1-3-10
Tama-shi
Tokyo 206-0034
Japan
Orders: Tel. +81 (0)423 57-7650
         Fax +81 (0)423 57-7651

Contents

CLIFT, P. D., KROON, D., GAEDICKE, C. & CRAIG, J. Tectonic and climatic evolution of the Arabian Sea region: an introduction — 1

ROYER, J.-Y., CHAUBEY, A. K., DYMENT, J., BHATTACHARYA, G. C., SRINIVAS, K., YATHEESH, V. & RAMPRASAD, T. Paleogene plate tectonic evolution of the Arabian and Eastern Somali basins — 7

GAEDICKE, C., PREXL, A., SCHLÜTER, H.-U., MEYER, H., ROESER, H. & CLIFT, P. D. Seismic stratigraphy and correlation of major regional unconformities in the northern Arabian Sea — 25

UCHUPI, E., SWIFT, S. A. & ROSS, D. A. Tectonic geomorphology of the Gulf of Oman Basin — 37

CHAUBEY, A. K., DYMENT, J., BHATTACHARYA, G. C., ROYER, J.-Y., SRINIVAS, K. & YATHEESH, V. Paleogene magnetic isochrons and palaeo-propagators in the Arabian and Eastern Somali basins, NW Indian Ocean — 71

VITA-FINZI, C. Neotectonics on the Arabian Sea coasts — 87

CLIFT, P. D., CARTER, A., KROL, M. & KIRBY, E. Constraints on India–Eurasia collision in the Arabian Sea region taken from the Indus Group, Ladakh Himalaya, India — 97

BURGATH, K.-P., VON RAD, U., VAN DER LINDEN, W., BLOCK, M., KHAN, A. A., ROESER, H. A. & WEISS, W. Basalt and peridotite recovered from Murray Ridge: are they of supra-subduction origin? — 117

DELISLE, G. & BERNER, U. Gas hydrates acting as cap rock to fluid discharge in the Makran accretionary prism? — 137

McCALL, G. J. H. A summary of the geology of the Iranian Makran — 147

SATTARZADEH, Y., COSGROVE, J. W. & VITA-FINZI, C. The geometry of structures in the Zagros cover rocks and its neotectonic implications — 205

STOW, D. A. V., TABREZ, A. R. & PRINS, M. A. Quaternary sedimentation on the Makran margin: turbidity current–hemipelagic interaction in an active slope-apron system — 219

CLIFT, P. D. A brief history of the Indus River — 237

DALEY, T. & ALAM, Z. Seismic stratigraphy of the offshore Indus Basin — 259

SMEWING, J. D., WARBUTON, J., DALEY, T., COPESTAKE, P. & UL-HAQ, N. Sequence stratigraphy of the southern Kirthar Fold Belt and Middle Indus Basin, Pakistan — 273

GLENNIE, K. W., SINGHVI, A. K., LANCASTER, N. & TELLER, J. T. Quaternary climatic changes over southern Arabia and the Thar Desert, India — 301

WENDLER, I., ZONNEVELD, K. A. F. & WILLEMS, H. Calcareous cyst-producing dinoflagellates: ecology and aspects of cyst preservation in a highly productive oceanic region — 317

JUNG, S. J. A., IVANOVA, E., REICHART, G. J., DAVIES, G. R., GANSSEN, G., KROON, D. & HINTE, J. E. V. Centennial–millennial-scale monsoon variations off Somalia over the last 35 ka — 341

BRUMMER, G. J. A., KLOOSTERHUIS, H. T. & HELDER, W. Monsoon-driven export fluxes and early diagenesis of particulate nitrogen and its $\delta^{15}N$ across the Somalia margin — 353

WILLIAMS, A. H. & WALKDEN, G. M. Late Quaternary highstand deposits of the southern Arabian Gulf: a record of sea-level and climate change — 371

VON RAD, U., KHAN, A. A., BERGER, W. H., RAMMLMAIR, D. & TREPPKE, U. Varves, turbidites and cycles in upper Holocene sediments (Makran Slope, northern Arabian Sea) — 387

REICHART, G. J., NORTIER, J., VERSTEEGH, G. & ZACHARIASSE, W. J. Periodical breakdown of the Arabian Sea oxygen minimum zone caused by deep convective mixing — 407

LUCKGE, A., REINHARDT, L., ANDRULEIT, H., DOSE-ROLINSKI, H., RAD, U. V., SCHULZ, H. & TREPPKE, U. Formation of varve-like laminae off Pakistan: decoding five years of sedimentation — 421

STAUBWASSER, M. & DULSKI, P. On the evolution of the oxygen minimum zone in the Arabian Sea during Holocene time and its relation to the South Asian monsoon — 433

VON RAD, U., BURGATH, K. P., PERVAZ, M. & SCHULZ, H. Discovery of the Toba Ash (c. 70 ka) in a high-resolution core recovering millennial monsoonal variability off Pakistan — 445

PEETERS, F. J. C. & BRUMMER, G. J. A. The seasonal and vertical distribution of living planktic foraminifera in the NW Arabian Sea — 463

SCHULZ, H., RAD, U. V. & ITTEKKOT, V. Planktic forminifera, particle flux and oceanic productivity off Pakistan, NE Arabian Sea: modern analogues and application to the paleoclimatic record — 499

Index — 517

Acknowledgements

The papers in the volume arise from a special meeting of the Tectonic Studies, Marine Studies and Petroleum Groups of the Geological Society of London, held at the Geological Society, Burlington House, Piccadilly, on 4–5th April 2001. The meeting was convened by Peter Clift, Christoph Gaedicke, Jonathan Craig and Dirk Kroon. Lasmo UK plc and the Geological Society are thanked for their partial financial support of the meeting, proceeds from which helped cover the costs of colour figures and fold-outs in this publication. Chryseis Fox is thanked for creating and maintaining the related web pages.

The editors wish to acknowledge reviews by the following geoscientists:

A. Beach	P. Hildebrand	J.J.G. Reijmer
D. Benn	M. Higginson	L.R. Sautter
C. Betzler	F. Jansen	B. Schreckenberger
J. P. Burg	G. Jones	D. Schelling,
S. Carey	C. Kendall	R. Schneider
S. Clemens	M. Khan	M.P. Searle
R. Coleman	E. Kirby	P. Sharland
P. D. Clift	N. Kukowski	A.D. Singh
T. Daley	F. Marret	F. Sirocko
K. Darling	K. McIntyre	J. Smewing
P. Degnan	J. McManus	S. Swift
P. DeMenocal	P. Miles	A.H.F. Robertson
R. Edwards	T. Minshull	E. Uchupi
P. Friend	G. Mountain	J. Warburton
R. Ganeshram	N. Nowaczyk	R. Whittington
E. Garzanti	F. Peeters	C. Vita-Finzi
E. Gnos	J. Pike	U. von Rad
R. Graham	M. Prins	
R. Harland	G.J. Reichart	

Tectonic and climatic evolution of the Arabian Sea region: an introduction

PETER D. CLIFT[1], DICK KROON[2], CHRISTOPH GAEDICKE[3] & JONATHAN CRAIG[4]

[1]*Woods Hole Oceanographic Institution, Woods Hole, MA 02543, USA*
(e-mail: pclift@whoi.edu)
[2]*Institute of Earth Sciences, Free University Amsterdam, de Boelelaan 1085, 1081 HV Amsterdam, The Netherlands*
[3]*Bundesanstalt für Geowissenschaften und Rohstoffe (BGR), Stilleweg 2, D-30655 Hannover, Germany*
[4]*Lasmo plc, 101 Bishops Gate, London EC2M 3XH, UK*

The evolution of the global oceanic and atmospheric circulation systems has been affected by several forcing processes, with orbital variations being dominant on shorter geological time scales. Over longer periods of time (>10 Ma) the tectonic evolution of the solid Earth has been recognized as the major control on the development of the global climate system. Tectonic activity acts in one of two different ways to influence regional and global climate. The earliest solid Earth–climatic interaction recognized was the effect that the opening and closure of oceanic gateways had on the circulation patterns in the global ocean. Major effects on regional and sometimes global climate have been attributed to such changes, e.g. closure of the Isthmus of Panama (Driscoll & Haug 1998). Since the late 1980s a second form of climate–tectonic interaction has been recognized, involving the growth and erosion of orogenic belts. In this second category the Arabian Sea region must be considered the global type example.

Growth of the Himalaya and associated Tibetan Plateau is now believed to have substantially altered Cenozoic climate. Raymo *et al.* (1988) suggested that chemical erosion of the uplifting orogen resulted in the draw-down of atmospheric CO_2, which was deposited as limestone, causing long-term global cooling, as a result of the reduction in this important greenhouse gas. Orogenic uplift is also believed to affect regional climate and in particular development of the monsoon. Summer heating of air above the Tibetan Plateau is known to have induced a strengthening of the monsoon (Manabe & Terpstra, 1974), and so tectonic uplift of the plateau has been linked to strengthening of the monsoon (Raymo & Ruddiman 1992; Molnar *et al.* 1993).

The Arabian Sea is the natural laboratory to study the interaction between orogenic growth and regional climate change. Ocean Drilling Program (ODP) sampling of pelagic sediments on the Oman margin has revealed a detailed history of monsoonal variability (Kroon *et al.* 1991; Prell *et al.* 1992), most notably an intensification of the monsoon at 8.5 Ma, as traced by the abundance of *Globigerina bulloides* and other eutrophic species, foraminifers associated with monsoon-induced coastal upwelling in the modern Arabian Sea. Initially this result appeared to correlate well with the start of extension on the Tibetan Plateau, an event that was linked to a period of rapid plateau uplift. Consequently, Harrison *et al.* (1992) and Molnar *et al.* (1993) proposed a direct correlation between rapid Tibetan uplift at that time and a strong SW monsoon and increased rainfall. Modelling studies support a shift of increased precipitation from Indochina toward the Himalayas and a strengthening of the Asian monsoon during late Tertiary time, separate from the strengthening at 8.5 Ma (Fluteau *et al.* 1999).

More recently, fieldwork in Tibet has complicated our ideas about the timing of uplift, with some workers indicating at least southern Tibet reaching maximum elevation at *c.* 19 Ma (Williams *et al.* 2001), whereas others suggest rapid uplift of the northern plateau during Pliocene time (Zheng *et al.* 2000). Work on the monsoon also continues to change our understanding of the climatic evolution of South Asia. Derry & France-Lanord (1997) proposed that the 8–7 Ma period in fact represented a weak monsoon, with greater regional aridity rather than the standard image of greater precipitation. Clearly much work remains in defining the timing of the SW monsoon and the timing of tectonism in Tibet.

Testing climate–tectonic models?

Despite the current debate, the Arabian Sea has several features that make it the best area globally to examine solid Earth–climatic interactions. The climatic signal is strong, especially with regard to the monsoon, and the rates of sedimentation are often high, allowing detailed palaeoceanographic records to be reconstructed. The Indus River has a large output, equivalent to that of the Mississippi River, and carries an erosional record of the high topography north of the High Himalaya, i.e. the Karakoram and western Tibet, which are the regions responsible for the monsoon. The modern climate of arid conditions, punctuated by heavy seasonal rains, contributes to enhancing erosion, the products of which can be found interbedded with dateable fossiliferous sediments in the Indus Fan. The western Himalaya and Karakoram have been the focus of much recent fieldwork, often involving high-resolution radiometric dating of basement units. This means that the thermal and tectonic development of these ranges can be accurately correlated with the marine records. Indeed, the very fact that the Arabian Sea is surrounded on three sides by land, often arid and with good exposure, makes this an ideal area to attempt linked marine and land-based field experiments.

Much work remains to be done in understanding the tectonic and climatic evolution of the Arabian Sea area. Efforts are now under way to expand our sampling on- and offshore so as to understand the temporal variations in climate, erosion and tectonism. Such efforts necessarily require international collaboration between researchers within the region and elsewhere. Additional marine geophysical surveys and deep scientific drilling are required in the Arabian Sea to construct the detailed records required to quantify the nature of solid Earth–climatic coupling. None the less, much has been achieved by a series of cruises and field campaigns over the last 10 years, since the last major review of the region. It was the desire to bring together workers spread across many countries and disciplines that prompted the special meeting of the Geological Society, where much of the work published here was presented.

The development of the Arabian Sea before India–Asia collision is reconstructed by three of the papers presented here. **Royer et al.** use a compilation of marine geophysical data to demonstrate that spreading in the Arabian Sea at 61–46 Ma has been affected by a series of successive ridge propagation events along the Carlsberg Ridge, and that the Arabia–India plate boundary was located west of the Owen Ridge along the Oman margin during most of Paleogene time. **Chaubey et al.** analyse marine magnetic anomaly patterns, and conclude that not only was Paleogene spreading very asymmetric, with the Arabian Sea accreting much faster than the Somali Basin, but also India–Somali motion slowed rapidly after 52 Ma. The evolution of the passive Indian margin before collision with Asia is described by **Smewing et al.** in a stratigraphic study of the southern Kirthar Fold Belt in Pakistan. Rift shoulder uplift and erosion of Middle Jurassic limestones dates the separation of India from East Africa. Subsequently, two phases of ophiolite emplacement are recognized, in Late Cretaceous and Early Tertiary time. Late Cretaceous (Campanian and Maastrichtian) submarine fans are interpreted to reflect thermal uplift of India as a result of the Deccan Plume. Subsequently, Middle Eocene northerly-derived clastic sediments from the early Himalaya constrain the age of collision in this region. The Kirthar Fold Belt is interpreted as an offshore equivalent of the Murray Ridge, which is underlain by thinned continental crust.

The slowing of India–Asia convergence after 52 Ma described by **Chaubey et al.** is broadly in accord with the evolving sedimentary character within the Asian forearc basin revealed by **Clift et al.** This study shows an influx of ophiolitic and carbonate platform material from a deformed Indian margin into the Jurutze forearc basin no later than Ypresian time (>49 Ma). The subsequent deposition of the alluvial Indus Group in the Indus Suture in Ladakh has been linked by **Clift** to the earliest activity of a palaeo-Indus River, which flowed into a delta in the Katawaz Basin of western Pakistan and Afghanistan by Early Eocene time, and must have reached the Arabian Sea by Mid-Eocene time. Some of that material has subsequently been accreted to the active margin of Asia. **McCall** shows that the Makran Accretionary Complex contains large volumes of Paleogene turbidites that were added to that margin, and may represent a palaeo-Indus Fan, in addition to the material still found in the Arabian Sea. **Clift et al.** show that the Indus Group Basin was inverted no later than 14 Ma, probably close to the time of tectonic uplift along the Murray Ridge (20 Ma), but that the Indus River has remained stationary in the suture since that time.

Clift presents a sediment budget for the Indus Fan that shows peak sedimentation rates in mid-Miocene time accompanied by channel–levee accumulation on the mid-fan, during a period of strong tectonic activity in the Karakoram, and possibly linked to an early initiation of the SW monsoon. Closer to the coast, **Daley & Alam** provide a review of the seismic stratigraphy of the offshore Indus Basin (Pakistan Shelf), highlighting the existence of two distinct phases of canyon development on the upper part of the Indus Fan: one in Early Miocene time and the other in Plio-Pleistocene time. Both phases are interpreted as relating to pulses of tectonic activity in the collision zone between the Indo-Pakistan and Eurasian plates.

The regional stratigraphy of the northern Arabian Sea is defined by **Gaedicke et al.**, who employ seismic data to construct a four-part stratigraphy to the Indus Fan and Gulf of Oman. They show that opening of the Gulf of Aden in mid- to late Miocene time caused a reorganization of the plates and subsequent tilting of the oceanic crust of the Arabian Plate toward the Makran subduction zone, thus generating the regional 'M-unconformity'. Uplift of the Murray Ridge during late Miocene to early Pliocene time has caused a separation in sedimentation between the Gulf of Oman and the Arabian Sea, with the modern Indus Fan prevented from flowing westward by the Murray Ridge. **Burgath et al.** provide sedimentary evidence that the Murray Ridge was uplifted close to sea level in late Miocene to early Pliocene time and subsequently subsided to modern depths. These workers also present geochemical data derived from samples for the volcanic and plutonic basement to the Murray Ridge demonstrating an origin within a Mesozoic Neotethyan active margin setting.

Pleistocene–Holocene sedimentation

Glacial sea-level lowstand periods are identified by **Williams & Walkden** as times when the Persian Gulf was subaerially exposed, allowing changes in sea level to be quantified. Sediments deposited during the last interglacial contain evidence for two highstands during this period, peaking at around 1.5 and 6 m above present sea level. Following the second highstand, sea level fell to more than 23 m below present level. Palaeocurrent directions indicate that during the last interglacial (125 ka) the prevailing wind blew from the NE. Evidence for a pluvial episode during this period is provided by palaeokarstic pits, believed to represent the former positions of trees or large plants.

Glennie et al. provide an overview of the development of the distribution of sand dunes over the southern half of Arabia during the Quaternary period. They discuss the expansion and contraction of deserts and their relationship to changes in climate. Two wind systems influenced desert formation: the northern 'Shamal' and the strong winds of the SW monsoon. The Shamal was particularly active during the cold parts of the Quaternary period and the SW monsoon was active during interglacial periods.

Upper Holocene sediments cored from the Pakistan margin within the oxygen minimum zone are used by **Von Rad et al.** to show a varving related to seasonal river flood events. A climate with reduced precipitation and river runoff, possibly 'winter monsoon' dominated, is recognized at 5600–4700 a bp on the basis of the thin varves, whereas other periods characterized by thicker varves document wetter periods, with summer-monsoon domination. These workers relate the cyclicity found in the varved sequence to lunar cycles, which result in variations in the amplitudes of tides, which in turn may have controlled sediment redeposition. Also, **Luckge et al.** describe Holocene varved sediments and their causes, and calibrate the variations in thickness of the varves to meteorological records such as rainfall.

In a separate contribution, **Von Rad et al.** discuss the discovery of a discrete ash layer derived from the Indonesian volcano Toba (70 ± 4 ka) and recovered from sediments deposited in the oxygen minimum zone offshore the Indus delta. In addition, two rhyolitic ash layers are dated at AD 1885–1900 and 1815–1830. The glass shards were probably derived from eruptions of Indonesian volcanoes, although it is not possible to correlate these two ashes with well-known historical eruptions. A complete, high-resolution stratigraphic record of the past 75 ka, with 21 interstadials or Dansgaard–Oeschger cycles and equivalents of Heinrich events H1–H6 is recognized. Rapid climate oscillations with periods around 1.5 ka can be tuned to the $\delta^{18}O$ record of a Bay of Bengal core and to the GISP-2 ice core from Greenland. Changes in the intensity of the Indian summer monsoon are tightly coupled with suborbital climate oscillations in the Northern Hemisphere.

The late glacial–Holocene evolution of the monsoon is reconstructed by **Jung et al.** using cores from the Arabian Sea offshore Somalia. Using a multiproxy approach they determine that bulk sediment chemistry, most notably Ba/Al, provides the best record of productivity associated with monsoon strength. Foraminifer-based records conflict with the sediment chemistry on centennial and millennial time scales, indicating that more than upwelling controls their productivity. Dissolution on the sea floor has affected foraminiferal abundances during the transition period. During glacial times upwelling and productivity were stronger during the winter NE monsoon, as opposed to the modern situation, where production is strongest during the summer SW monsoon.

Particle flux, sediment trap data, in the NE Arabian Sea are described by **Schulz et al.** Those workers document that the particle flux, including biota, in the NE Arabian Sea is determined by sediment resuspension and winter productivity rather than by summer upwelling. This is in contrast to numerous sediment trap observations in the western Arabian Sea. The new evidence is used to reconstruct the seasonal intensity of both monsoons for the past 25 ka. **Schultz et al.** use planktonic foraminiferal abundances for monsoon intensity in a core with a high temporal resolution. The NE monsoon peaked during the stadial phases and the SW monsoon during the Holocene period. Combined sediment trap and box core records of nitrogen isotopes ($\delta^{15}N$) in organic matter are studied by **Brummer et al.** Complex patterns of annual nitrogen fluxes and $\delta^{15}N$ are described, and their fate in the sediments. Upwelling processes have a strong effect on the $\delta^{15}N$ of settling nitrogen. Freshly upwelled waters show minima in $d^{15}N$, and maxima are found in stratified waters. Also, diagenesis is dicussed as a potential source for altering the primary signal.

Staubwasser & Dulski reconstruct the early–mid-Holocene evolution of the oxygen minimum zone in the Arabian Sea by using a trace metal proxy. Century-scale fluctuations are recorded in the laminated sediments off Pakistan. Complex ventilation patterns of the subsurface waters include lateral advection from Central Indian Water and ventilation by winter surface convection in the northern Arabian Sea.

Stow et al. describe shallow cores from the upper slope, mid-slope and the abyssal plain offshore the Makran. They employ an oxygen isotope stratigraphy to provide an age determination for statistical analyses. The upper 5–14 m of sediment is dominated by fine-grained, thin turbidites, averaging 5–10 turbidite events per metre of section, or one turbidite event per 200–300 a. The range of turbidite bed thicknesses is typical for beds triggered by seismicity on active margins. The lateral distribution of both turbidites and hemipelagites is influenced by sediment focusing along pathways between slope basins. At a larger scale, climate, sea level and tectonic effects have all controlled the margin sedimentation.

Two papers discuss the ecology of plankton groups in the Arabian Sea and their potential as 'proxies' in the sediments. **Wendler et al.** discuss abundances of calcareous dinoflagellate cyst species in surface sediment samples, and these are compared with environmental parameters in the upper water column. Certain assemblages can be clearly correlated with temperature and nutrient levels, particularly in the SW Arabian Sea. In the NE Arabian Sea diagenetic processes must be taken into account in interpreting sediment assemblages. **Peeters & Brummer** record the faunal composition and depth habitats of extant planktonic foraminifers by using plankton tows. The temporal distribution of assemblages of planktonic foraminifers is associated with the hydrography of the area, which is in turn controlled by the monsoon system. Highest abundances of eutrophic species occur during the upwelling phase triggered by the SW monsoon. Vertical abundance profiles in the water column are recorded to document depth habitats of individual species. This is important for interpretations of stable isotope studies using the shells of the planktonic foraminifers in sediments.

Neotectonics

The Arabian Sea region is still involved in the active deformation associated with India–Asia convergence and opening of the Red Sea and Gulf of Aden. In a review of this activity, **Vita Finzi** describes the Holocene record on the coasts of the Arabian Sea and interprets it in the context of the interaction between the Indian, Arabian and Eurasian plates. Marine terraces from the Makran margin show infrequent but vigorous coseismic uplift, similar to that seen in other accretionary subduction settings. Vita Finzi notes a landward rotation of the imbricate faults along which shortening is distributed. The lack of significant Holocene deformation on the SE coast of the Arabian peninsula is consistent with its position parallel to a transform, although large-scale buckling may be occurring at the Straits of Hormuz. Localized uplift on the SW of India represents compressional buckling related to India–Asia collision.

Sattarzadeh et al. examine the nature of folding in the Zagros Mountains as a result of the enhanced Pliocene deformation caused by faster Arabia–Asia convergence driven by opening of the Red Sea. The deformation in the Zagros is controlled by plate velocity and the regional stratigraphy, which is composed of four mechanically contrasting litho-structural units overlying a rigid basement. The sedimentary cover and the underlying metamorphic basement decouple along an important detachment horizon, the Hormuz Salt Formation. The uneven thickness and distribution of the Hormuz Salt plays a key role in determining the geometry of the deformation belt. The median surface of the fold belt has an asymmetric topography, its elevation ascending northeastward in a series of steps, consistent with neotectonic field evidence for serial folding and imbrication of the footwall.

Delisle & Berner present a numerical model of the geothermal field of the Makran accretionary prism, which suggests that conductive heat transport is dominant and that there is little contribution from fluid flow or frictional heat. They demonstrate that gas hydrate layers in the accretionary prism act as a very effective seal to the upward flow of fluids in water depths greater than 800 m, but that they break down when uplifted out of the gas hydrate stability field into shallower and warmer water.

The papers in the volume arise from a special meeting of the Tectonic Studies, Marine Studies and Petroleum Groups of the Geological Society of London, held at the Geological Society, Burlington House, Piccadilly, on 4–5 April 2001. The meeting was convened by Peter Clift, Christoph Gaedicke, Jonathan Craig and Dirk Kroon. Lasmo UK plc and the Geological Society are thanked for their partial financial support of the meeting, proceeds from which helped cover the costs of colour figures and fold-outs in this publication. Chryseis Fox is thanked for creating and maintaining the related web pages.

The editors wish to acknowledge reviews by the following geoscientists: A. Beach, D. Benn, C. Betzler, J. P. Burg, S. Carey, S. Clemens, R. Coleman, P. D. Clift, T. Daley, K. Darling, P. Degnan, P. DeMenocal, R. Edwards, P. Friend, R. Ganeshram, E. Garzanti, E. Gnos, R. Graham, R. Harland, P. Hildebrand, M. Higginson, F. Jansen, G. Jones, C. Kendall, M. Khan, E. Kirby, N. Kukowski, F. Marret, K. McIntyre, J. McManus, P. Miles, T. Minshull, G. Mountain, N. Nowaczyk, F. Peeters, J. Pike, M. Prins, G. J. Reichart, J. J. G. Reijmer, L. R. Sautter, B. Schreckenberger, D. Schelling, R. Schneider, M. P. Searle, P. Sharland, A. D. Singh, F. Sirocko, J. Smewing, S. Swift, A. H. F. Robertson, E. Uchupi, J. Warburton, R. Whittington, C. Vita Finzi, U. von Rad.

References

DERRY, L.A. & FRANCE-LANORD, C. 1997. Himalayan weathering and erosion fluxes; climate and tectonic controls. *In:* RUDDIMAN, W.F. (ed.) *Tectonic Uplift and Climate Change*. Plenum, New York, 289–312.

DRISCOLL, N.W. & HAUG, G.H. 1998. A short circuit in thermohaline circulation; a cause for Northern Hemisphere glaciation? *Science*, **282**, 436–438.

FLUTEAU, F., RAMSTEIN, G. & BESSE, J. 1999. Simulating the evolution of the Asian and African monsoons during the past 30 Myr using an atmospheric general circulation model. *Journal of Geophysical Research*, **104**, 11995–12018.

KROON, D., STEENS, T. & TROELSTRA, S.R. ET AL. 1991. Onset of monsoonal related upwelling in the western Arabian Sea as revealed by planktonic foraminifers. *In:* PRELL, W.L. & NIITSUMA, N. (eds) *Proceedings of the Ocean Drilling Program, Scientific Results, 117*. Ocean Drilling Program, College Station, TX, 257–263.

HARRISON, T.M., COPELAND, P., KIDD, W.S.F. & YIN, A. 1992. Raising Tibet. *Science*, **255**, 1663–1670.

MANABE, S. & TERPSTRA, T.B. 1974. The effects of mountains on the general circulation of the atmosphere as identified by numerical experiments. *Journal of Atmospheric Scences*, **31**, 3–42.

MOLNAR, P., ENGLAND, P. & MARTINOD, J. 1993. Mantle dynamics, the uplift of the Tibetan Plateau, and the Indian monsoon. *Reviews in Geophysics*, **31**, 357–396.

PRELL, W.L., MURRAY, D.W., CLEMENS, S.C. & ANDERSON, D.M. 1992. Evolution and variability of the Indian Ocean summer monsoon: evidence from the western Arabian Sea drilling program. *In:* DUNCAN, R.A., REA, D.K., KIDD, R.B., VON RAD, U. & WEISSEL, J.K. (eds) *Synthesis of Results from Scientific Drilling in the Indian Ocean*. Geophysical Monograph, American Geophysical Union, **70**, 447–469.

RAYMO, M.E., RUDDIMAN, W.F. & FROELICH, P. 1988. Influence of late Cenozoic mountain building on ocean geochemical cycles. *Geology*, **16**, 649–653.

RAYMO, M.E. & RUDDIMAN, W.F. 1992. Tectonic forcing of the late Cenozoic climate. *Nature*, **359**, 117–122.

WILLIAMS, H., TURNER, S., KELLEY, S. & HARRIS, N. 2001. Age and composition of dikes in southern Tibet: new constraints on the timing of east–west extension and its relations to post-collisional volcanism. *Geology*, **29**, 339–342.

ZHENG, H., POWELL, C.M., AN, Z., ZHOU, J. & DONG, G. 2000. Pliocene uplift of the northern Tibetan Plateau. *Geology*, **28**, 715–718.

Paleogene plate tectonic evolution of the Arabian and Eastern Somali basins

JEAN-YVES ROYER[1], A. K. CHAUBEY[2], JÉROME DYMENT[1],
G. C. BHATTACHARYA[2], K. SRINIVAS[2], V. YATHEESH[2]
& T. RAMPRASAD[2]

[1]*CNRS Domaines Océaniques, Institut Universitaire Européen de la Mer (IUEM), Place Copernic, 29280 Plouzané, France (e-mail: jyroyer@univ-brest.fr)*
[2]*Geological Oceanography Division, National Institute of Oceanography (NIO), Dona Paula, Goa 403 004, India*

Abstract: We review previous models for the Paleogene tectonic evolution of the Arabian and Eastern Somali basins and present a model based on a new compilation of magnetic and gravity data. Using plate reconstructions, we derive a self-consistent set of isochrons for Chron 27 to Chron 21 (61–46 Ma). The new isochrons account for the development of successive ridge propagation events along the Carlsberg Ridge, leading to an important spreading asymmetry between the conjugate basins. Our model predicts the growth of the outer and inner pseudo-faults associated with the ridge propagation events. The location of outer pseudo-faults appears to remain very stable despite a drastic change in the direction of ridge propagation before Chron 24 (c. 54 Ma). The motion of the Indian plate relative to the Somalian plate is stable in direction through Paleogene time; spreading velocities decrease from 6 to 3 cm a^{-1}. Our reconstructions also confirm that the Arabia–India plate boundary was located west of the Owen Ridge along the Oman margin during Paleogene time; some compression is predicted at about Chron 21 (47 Ma) between the Indian and Arabian plates.

The opening of the western Indian Ocean resulted from the break-up and dispersal of the African, Madagascar and Indian continental blocks. A common broad evolutionary proposal is that sea-floor spreading initiated in the western Somali Basin between Africa and a Madagascar–Seychelles–India block in Early Cretaceous time. In a second stage that started in mid-Cretaceous time, the Seychelles and India separated from Madagascar, leading to the opening of the Mascarene Basin. During Paleocene time, sea-floor spreading progressively stopped in the Mascarene Basin although resuming north of the Seychelles–Mascarene Plateau, creating the Eastern Somali and Arabian basins. Although this general evolution of the area appears to be fairly well understood, the detailed chronology and tectonic development of all the ocean basins of this area have not yet been fully unravelled. In this paper, we present a review of the various models that have been proposed as new data were collected, and a preliminary reconstruction model for Paleogene time (Chron 27–21, 61–46 Ma), that we derived from a new data compilation in this area (Chaubey *et al.* 2002).

Review of previous models

The first conjugate magnetic anomalies in the Arabian and Eastern Somali basins were observed in the late 1960s. From additional shipboard data, McKenzie & Sclater (1971) identified Paleogene magnetic anomalies 28–23. The Deep Sea Drilling Project (DSDP) in the Arabian Sea helped date the sea floor and define the structural trends of part of the basin (Whitmarsh 1974). The east–west-oriented magnetic lineations 28–18 were considered to be offset by four north–south-oriented fracture zones. A two-stage evolution was proposed with fast spreading stage between Chron 28 and 20 followed by a very slow spreading stage until the present day. The change in spreading rate and direction was supposed to be contemporaneous with similar changes in the Indian Ocean between Chron 18 and 20, related to the collision of India with Eurasia. Identifications of conjugate sequences of magnetic anomaly 27–23, and of additional fracture zones in the Eastern Somali Basin, further refined the picture (Norton & Sclater 1979; Schlich 1982). Extensive surveys by Russian vessels, in the early 1980s, led to a detailed tectonic chart of the Arabian Sea, which suggested an initiation of spreading at anomaly 29, or possibly 31, and a major change in rate and direction at about anomaly 11–12, thus much later than in the rest of the Indian Ocean (Karasik *et al.* 1986). Further studies suggested that the fast spreading rates prevailing during Chrons 27–23 (c. 6 cm a^{-1}) decreased at Chron 18 (40 Ma) when they became ultra-slow (<0.6 cm a^{-1}). The latest phase of spreading probably started shortly before anomaly 11 (c. 30 Ma) after a major change in spreading direction, and is continuing at present at about 1.2 cm a^{-1} (Chaubey *et al.* 1993; Mercuriev *et al.* 1996).

The fracture zone pattern, mostly inferred from discontinuities in the magnetic lineations, was still not consistent in the two conjugate basins. Satellite-derived gravity charts (Sandwell 1984; Haxby 1987; Sandwell & Ruiz 1992; Sandwell & Smith 1997) progressively unveiled, with increasing details, the structural grain of the region (Fig. 1): the old fast-spreading basins are fairly smooth with some structural trends oblique to the magnetic lineations; the young slow-spreading basins display a rough topography and, to the east, a series of well-defined and closely spaced fracture zones. Contour maps derived from dense magnetic surveys in the Eastern Somali and Arabian basins (Karasik *et al.* 1986; Mercuriev & Sochevanova

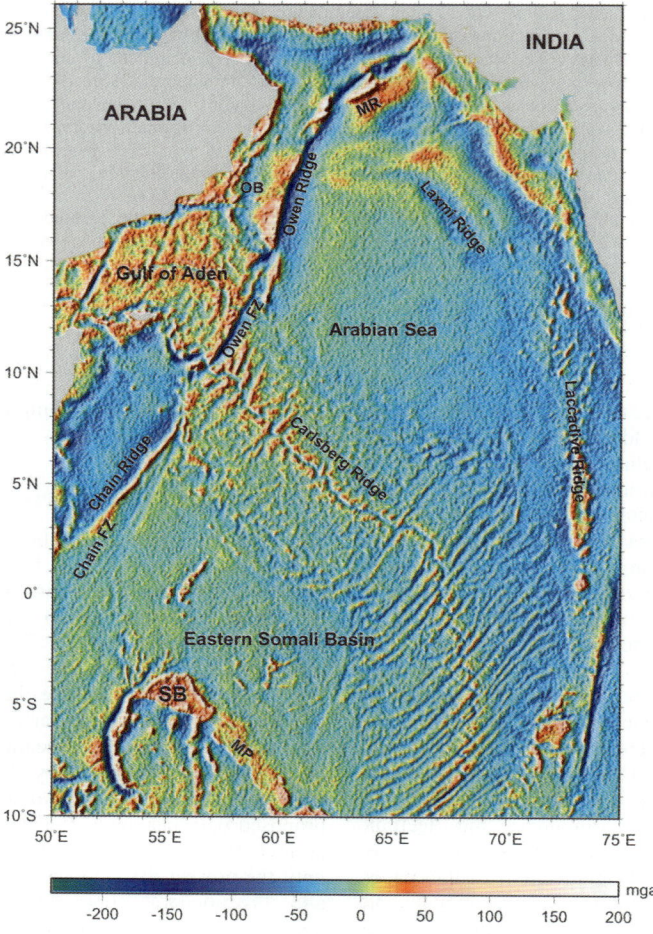

Fig. 1. Satellite-derived gravity chart of the northwestern Indian Ocean (1' grid; Sandwell & Smith 1997). SB, Seychelles Bank; MP, Mascarene Plateau; OB, Owen Basin; MR, Murray Ridge.

1990) showed incompatibility to most of the submeridional fracture zones inferred in earlier studies. Subsequent detailed reinterpretation of the Paleogene magnetic lineations revealed evidence of oblique pseudo-faults associated with systematic ridge propagation in both basins (Miles & Roest 1993; Chaubey et al. 1998; Dyment 1998). Ridge propagation explains the large spreading asymmetry between the Arabian and Eastern Somali basins. Between Chrons 26 and 25, c. 65% of the crust formed at the Carlsberg Ridge was accreted to the African plate, and after a change in the direction of ridge propagation at Chron 24r more than 75% of the crust was accreted to the Indian plate between Chrons 24 and 20 (Dyment 1998).

The early break-up between the Madagascar, Seychelles and Indian continental blocks is still a matter of debate. The break-up between Madagascar and India occurred during mid-Cretaceous time, according to the presence of anomaly 34 (83 Ma) off the Madagascar margin in the Mascarene Basin (Schlich 1982) and to basalt flows along the eastern coast of Madagascar (Storey et al. 1995). Evidence of early spreading (pre-Chron 28) between the Seychelles block and India is found in the Laxmi Basin and its adjacent areas (Bhattacharya et al. 1994; Malod et al. 1997). The oldest confidently identified magnetic lineations in the Arabian and Eastern Somali basins is Chron 27. Chron 28 is considered to be present in the Arabian Basin immediately south of Laxmi Ridge (Naini & Talwani 1983; Miles & Roest 1993; Chaubey et al. 1998); however, Miles et al. (1998) considered this anomaly as the signature of the ocean–continent transition. Magnetic lineations are also observed in the northwestern part of the Arabian Basin (Naini & Talwani 1983; Masson 1984; Karasik et al. 1986). They have been interpreted as sea-floor spreading anomaly 29–32n.2n (64–73 Ma; Malod et al. 1997; Shreider 1998; Todal & Edholm 1998), whereas Miles et al. (1998) considered them as magmatic intrusions within the ocean–continent boundary. No magnetic anomalies older than Chron 28 have been identified in the Somali Basin, north or west of the Seychelles Bank (McKenzie & Sclater 1971; Norton & Sclater 1979; Schlich 1982; de Ribet 1989). Conjugate sequence of anomalies 34–27, to the west, and 27–32, to the east, are found in the Mascarene Basin (Norton & Sclater 1979; Schlich 1982; Masson 1984; de Ribet 1989; Dyment 1991; Bernard & Munschy 2000); sea-floor spreading in the Mascarene Basin progressively stopped between Chron 30 and Chron 27 as sea-floor spreading started north of the Seychelles Bank and propagated southeastward along the Mascarene Plateau, leading to the development of a transient Seychelles microplate (Masson 1984; de Ribet 1989; Dyment 1991; Todal & Edholm 1998; Bernard & Munschy 2000).

Because of the complex ridge jumps from the Mascarene Basin to the Eastern Somali and Arabian basins, plate motions of the Indian plate relative to the Madagascar, African and Arabian plates have mainly been determined through a plate circuit passing by Antarctica (e.g. Norton & Sclater 1979; Patriat & Ségoufin 1988). The few attempts to derive the rotation parameters describing the India–Somalia motion directly from data in the Arabian and Eastern Somali basins (Molnar et al. 1988; Royer & Chang 1991) assume that Chain Ridge and Owen Ridge are conjugate features; we find that this hypothesis is not correct and hence introduces a bias in the reconstructions. Reconstructions of the Eastern Somali and Arabian basins are also more difficult because of the numerous propagators. Todal & Edholm (1998) derived reconstructions at Chron 26, 27 and 28 from a simplified magnetic lineation pattern that does not account for the propagators observed in the area (Chaubey et al. 1998; Dyment 1998).

A revised tectonic chart for Paleogene time

As part of a collaborative work between NIO, Goa, and IUEM, Brest, we attempted to compile all the digital magnetic data collected in the northwestern Indian Ocean and available to us. This compilation includes data from the National Geophysical Data Center (1998), data collected by French research vessels (R.V. *Galliéni*, R.V. *Marion Dufresne*, R.V. *Suroit*, R.V. *Jean Charcot*), and data collected by NIO (R.V. *Sagar Kanya*, DS.V. *Nand Rachit*). Magnetic anomaly data were automatically picked from the digital profiles, using the analytical signal method (Nabighian 1972, 1974; Miles & Roest 1993), and were further checked and digitally repicked when needed. The details of the data and their interpretation are given in a companion paper (Chaubey et al. 2002).

We compiled a total of 591 magnetic crossings (Fig. 2) spanning all reversals (young and old ends) between Chrons 27o and 20y. Ages and chron numbers refer to the Cande & Kent (1995) magnetic reversal time scale. In areas where data were sparse, we also used, as a guide, the magnetic contours drawn from the dense Russian magnetic surveys (Karasik et al. 1986; Mercuriev & Sochevanova 1990). As these contours were digitized from analogue figures, the locations of the magnetic reversals are not as accurate as those based on the digital profiles. For this reason, these contours were not used in the calculations of the rotation parameters.

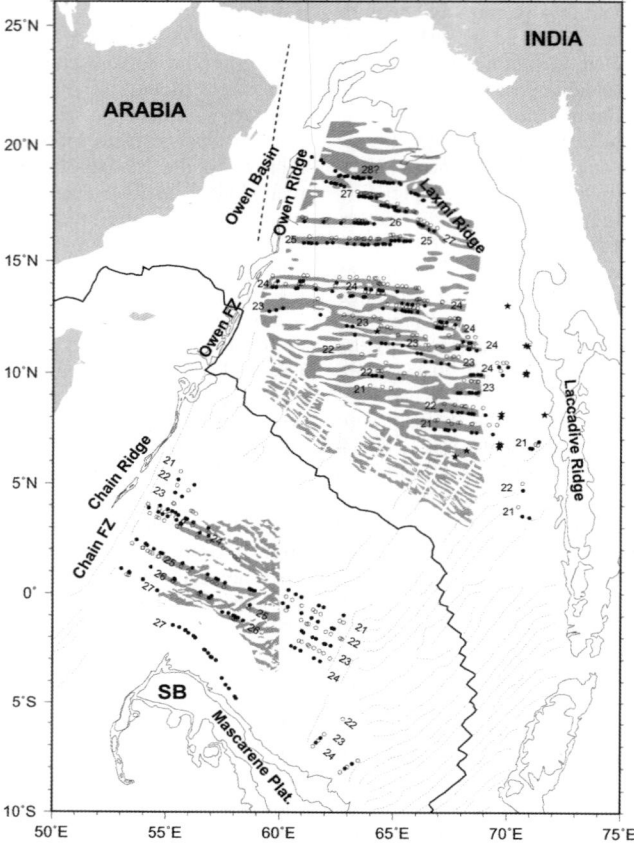

Fig. 2. Tectonic chart of the northwestern Indian Ocean. The shaded pattern corresponds to the magnetic anomaly contours from Karasik *et al.* (1986) and Mercuriev & Sochevanova (1990). Open symbols correspond to old ends of chrons; filled symbols correspond to young end of chrons. Magnetic crossings for anomaly 24 correspond to the ends of Chrons 24n1 and 24n3; anomaly 23 crossings correspond to 23n1y and 23n2o. Fracture zones are inferred from satellite-derived gravity data (dashed lines), from seismic profiles (stars), or from plate reconstructions (bold dashed line in the Owen Basin). SB, Seychelles Bank.

Plate reconstructions

Despite the occurrence of some clear magnetic lineations in the Eastern Somali and Arabian basins, deriving the Somalia–India relative motions by reconstructing these isochrons is a challenge. The first part of this section reviews the difficulties of this endeavour and how to circumvent them. The following paragraphs describe step by step the opening of the Arabian and Eastern Somali basins from Chron 27 (61 Ma) to Chron 21 (46 Ma).

Problems and method

First, plate reconstructions assume that the geometry of a spreading ridge is identically imprinted on the two conjugate basins that originated from this spreading ridge. Here, this assumption is not fully valid, as numerous propagators shaped these basins and left very dissymmetric magnetic patterns on each plate. During these processes, portions of Somalian or Indian crust were continuously transferred onto the conjugate plate. Luckily, it is possible to identify for each chron one or two symmetric lineations.

Second, conjugate magnetic crossings (i.e. lineation) do not fully constrain a reconstruction, as magnetic lineations can 'slide' on each other in the direction of their trend. To prevent this effect, fitting magnetic lineations generally requires some fracture zone crossings, which align perpendicularly to the magnetic lineations. As observed in the revised tectonic chart (Fig. 2), there are not many conjugate fracture zones. The fracture zone that we name Chain FZ in Fig. 1, immediately east of Chain Ridge, is perhaps the most obvious in the Somali Basin; however, its gravity signature displays a double trough and the actual offset along this fracture zone is not known as no magnetic anomaly has yet been identified in the basin bounded by Chain Ridge and the Chain FZ. In addition to these uncertainties, it is not clear at all where the conjugate fracture zone is in the Arabian Sea. The best candidate seems to be the eastern edge of Owen Ridge; previous studies even assumed that Chain Ridge and Owen Ridge were conjugate features (e.g. Whitmarsh 1979; Molnar et al. 1988; Royer & Chang 1991). However, there is evidence that the Owen Ridge represents only the Arabia–India plate boundary since the early opening of the Gulf of Aden in Oligocene–Early Miocene time, and that the conjugate fracture zone of Chain FZ may be located further west, between the Owen Ridge and the Arabian margin (see next section). Another fracture zone is clearly visible in the satellite-derived gravity data at 60°E in the Eastern Somali Basin. The conjugate fracture zone, at c. 68°E in the Arabian Basin, originally mapped by Whitmarsh (1974), is filled by sediments and hence has a subdued gravity signature that is difficult to trace precisely. We mostly rely for this part of the fracture zone on few seismic crossings (stars in Fig. 2). Finally, in some favourable cases, a good match of the propagating tips of the propagators against the pseudo-faults provides an additional verification of the quality of the fit.

The last difficulty, related to the previous problem, is that the location of the Paleogene boundary between the Indian and Arabian plates is not clearly assessed. The present-day plate boundary follows the Owen FZ, the Owen Ridge and the Murray Ridge complex, as indicated by the earthquake activity (e.g. Matthews 1966; Sykes 1968; Quittmeyer & Kafka 1984). This boundary has been active since the opening of the Gulf of Aden in Oligocene–Early Miocene time (Whitmarsh 1979). Before that time, some evidence suggests that the plate boundary was located further west, along the Oman margin. Depth to basement in the Owen Basin, sediment correlations from profiles in the Owen Basin with DSDP and Ocean Drilling Program drill holes on the Owen Ridge, and obliquity of (unidentified) magnetic lineations relative to the margin all suggest that the SE coast of Oman was a transform margin in Late Cretaceous time (Mountain & Prell 1989, 1990). Furthermore, the deep structure of the Oman margin shows large and abrupt contrasts in crustal thickness that could also be accounted for by strike-slip motion along the margin (Barton et al. 1990). As a first approach, we arbitrarily limited the Indian plate at the eastern edge of the modern Owen–Murray ridge complex. To examine the implications of this choice, we closed the Gulf of Aden by fitting the edges of the shelf-break along the Somalian and Arabian margins (rotation 38.5°E, 19.5°N, angle 17.38°).

The rotation parameters (Table 1) were computed with the method developed by Chang (1988), Royer & Chang (1991) and Kirkwood et al. (1999). This method allows calculation of the uncertainties in the rotation parameters from the uncertainties in the data. A discussion of the significance of these uncertainties is beyond the scope of this paper, which was to check the consistency of our interpretations for the magnetic data and the tectonic fabric.

Chron 27 (61 Ma)

Our oldest reconstruction corresponds to Chron 27ny (60.9 Ma; Figs. 3a and 4a). To the west, the young Carlsberg Ridge abuts the Chain FZ, whereas the magnetic lineation in the Arabian Basin is truncated by the Owen Ridge. The width of the missing part corresponds to the distance between the Owen Ridge and the Oman margin, which strongly suggest that the triangle of oceanic crust bounded by the Oman margin, the long-dashed line in Fig. 3 and the eastern side of Owen Ridge belonged to the Indian plate. To the east, the Carlsberg Ridge is at the foot of the Mascarene Plateau and possibly extended as far east as 59°E, where it coincides with structural changes on the Laxmi Ridge (Miles et al. 1998) and presumably connected through a transform fault, perpendicular to the plateau, to the spreading ridge in the Mascarene Basin. The magnetic anomalies at the foot of the plateau are difficult to identify, as are the magnetic anomalies SE of the Laxmi Ridge. North of Chron 27ny, a distinct

Table 1. Rotation parameters describing the relative motions between the (fixed) Indian and Somalian plates during Paleogene time (ages after Cande & Kent (1995) reversal time scale)

Chron	Age (Ma)	Latitude +(°N)	Longitude +(°E)	Angle (deg)
21ny	46.26	18.64	43.37	22.559
22ny	49.04	18.94	39.62	23.195
23n1y	50.78	18.55	38.73	26.157
24n1y	52.36	19.17	34.18	26.232
25ny	55.90	19.41	29.02	30.111
26ny	57.55	19.61	25.62	30.729
27ny*	60.92	18.83	24.86	35.411

*Sea-floor spreading is not fully stopped in the Mascarene Basin and some motions may still be occurring between the Seychelles Bank and Madagascar, so this rotation does not fully describe the India–Somalia relative position at this time.

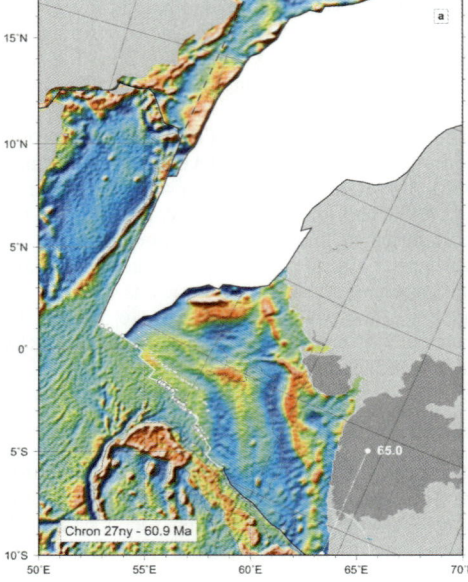
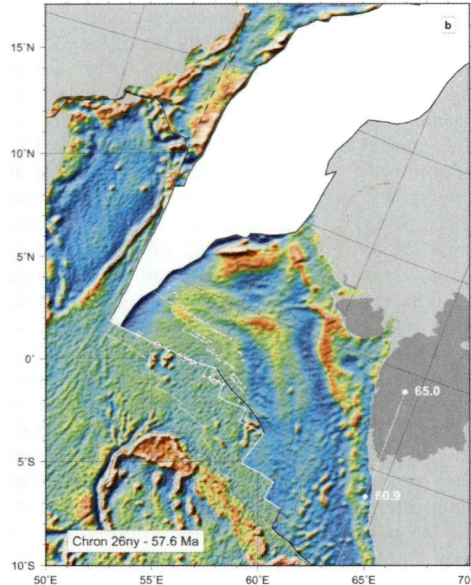
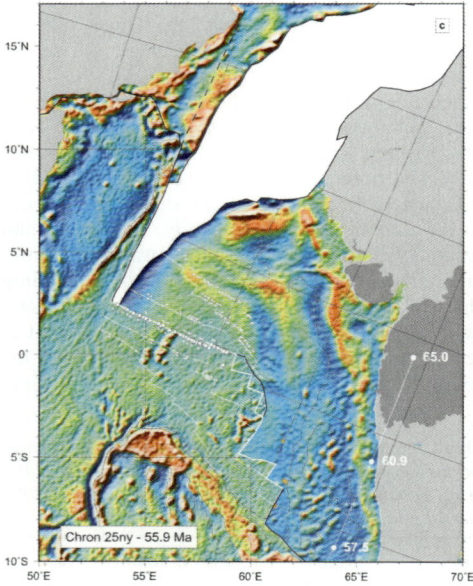

Fig. 3. Reconstructed satellite-derived gravity anomaly for (**a**) Chron 27ny (60.9 Ma), (**b**) Chron 26ny (57.6 Ma) and (**c**) Chron 25ny (55.9 Ma). All reconstructions are in the Somalia reference frame. The closure of the Gulf of Aden fits the shelf-breaks of Arabia and Somalia (rotation 38.5°E, 19.5°N, angle 17.38°). The bold dashed line, off Oman, would represent the conjugate of Chain FZ and the Paleogene India–Arabia plate boundary (see text). The dark area in India is the Deccan Trap; dots, ages in Ma and dotted lines correspond to the track of the Deccan–Réunion hotspot (after Müller *et al.* 1993). The circle around the hotspot location has a radius of 250 km. Open stars (in (**b**) and (**c**)) show fracture zone crossings inferred from seismic data. Large open circles correspond to magnetic crossings for the reconstructed chron; small open circles correspond to magnetic crossings from older chrons.

magnetic anomaly can be picked; it has been interpreted either as Chron 28ny (62.5 Ma) or as the magnetic contrast between the continent–ocean boundary and the oceanic crust. The presence of magnetic anomalies between isochron 27ny and the Seychelles Bank favours the former hypothesis, whereas the difficulty of identifying unambiguous sea-floor spreading anomalies favours the latter. Similarly on the Indian side, the cause of the magnetic anomaly pattern north of this lineation (shaded areas in Fig. 4) appears to remain equivocal as yet.

Chrons 26 (58 Ma) and 25 (56 Ma)

The next two steps, at Chrons 26ny (57.6 Ma) and 25ny (55.9 Ma), show a steady and fast eastward propagation of the Carlsberg Ridge, which was already initiated at Chron 27 (Figs. 3 and 4). The magnetic lineations east of Chain FZ almost tripled

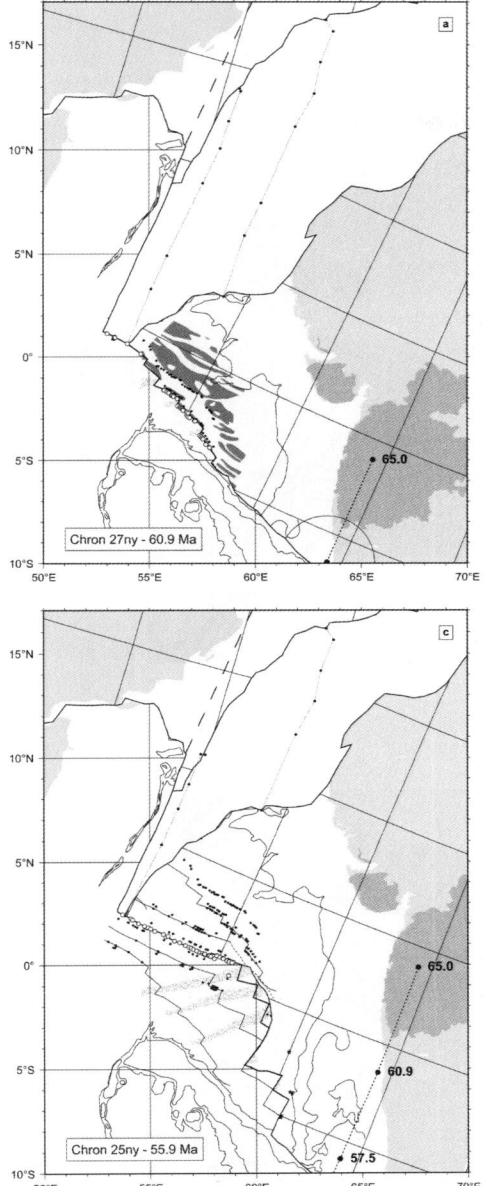

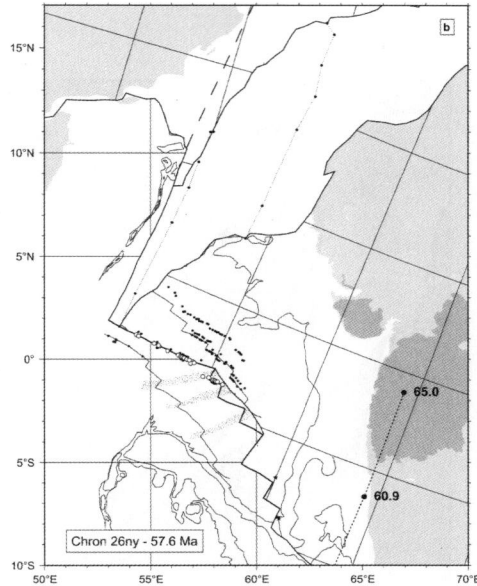

Fig. 4. Reconstructed tectonic chart for (**a**) Chron 27ny (60.9 Ma), (**b**) Chron 26ny (57.6 Ma) and (**c**) Chron 25ny (55.9 Ma). Stippled pattern represent the inner pseudo-faults abutting against the outer pseudo-fault (continuous line). Dotted lines show the trajectory of the Indian plate relative to the Arabian plate. Large open circles correspond to magnetic crossings for the reconstructed chron; small filled circles correspond to magnetic crossings from older chrons. Other symbols as in Fig. 3.

in length in 5 Ma. This rapid propagation follows the cessation of spreading in the Mascarene Basin, where the youngest fossil axis, located south of Réunion, is dated as Chron 27ny (Dyment 1991; Bernard & Munschy 2000). Another interesting observation is that all the inner pseudo-faults of the Eastern Somali Basin abut along the same outer pseudo-fault in the Arabian Basin, leaving three parallel wakes weakly visible in the gravity chart (Fig. 1 and stippled areas in Fig. 4). Because of lack of data, the eastern termination of the Carlsberg Ridge is less constrained; we believe, from the gravity data, that the outer pseudo-fault continued as far east as 61° or 62°E.

Chron 24 (52 Ma) to Chron 22 (49 Ma)

Chron 24 marked a major change in the direction of the ridge propagation from eastward to westward (Figs. 5a and 6a); this change had already

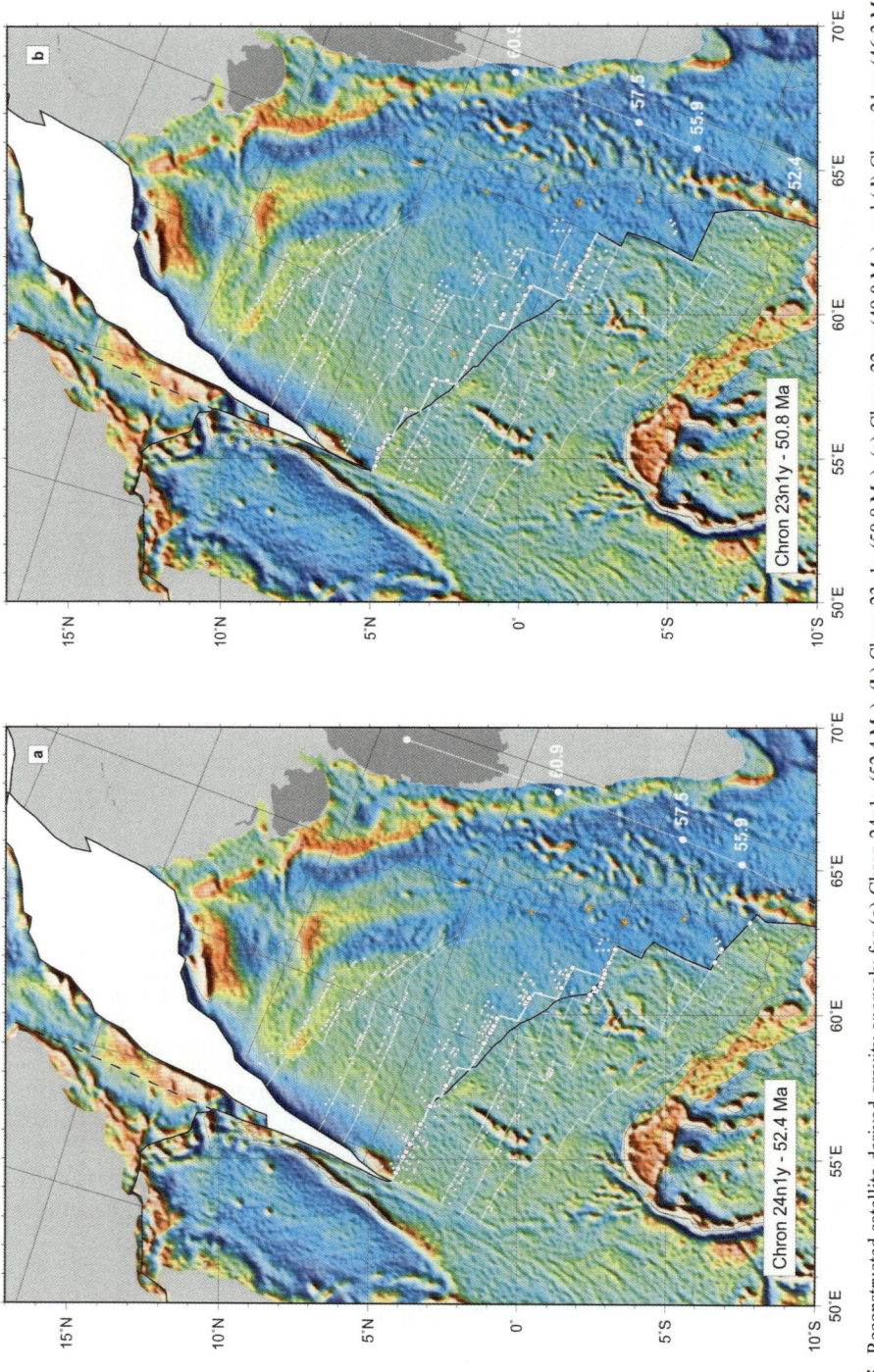

Fig. 5. Reconstructed satellite-derived gravity anomaly for (**a**) Chron 24n1y (52.4 Ma), (**b**) Chron 23n1y (50.8 Ma), (**c**) Chron 22ny (49.0 Ma), and (**d**) Chron 21ny (46.3 Ma). Same symbols as for Fig. 3.

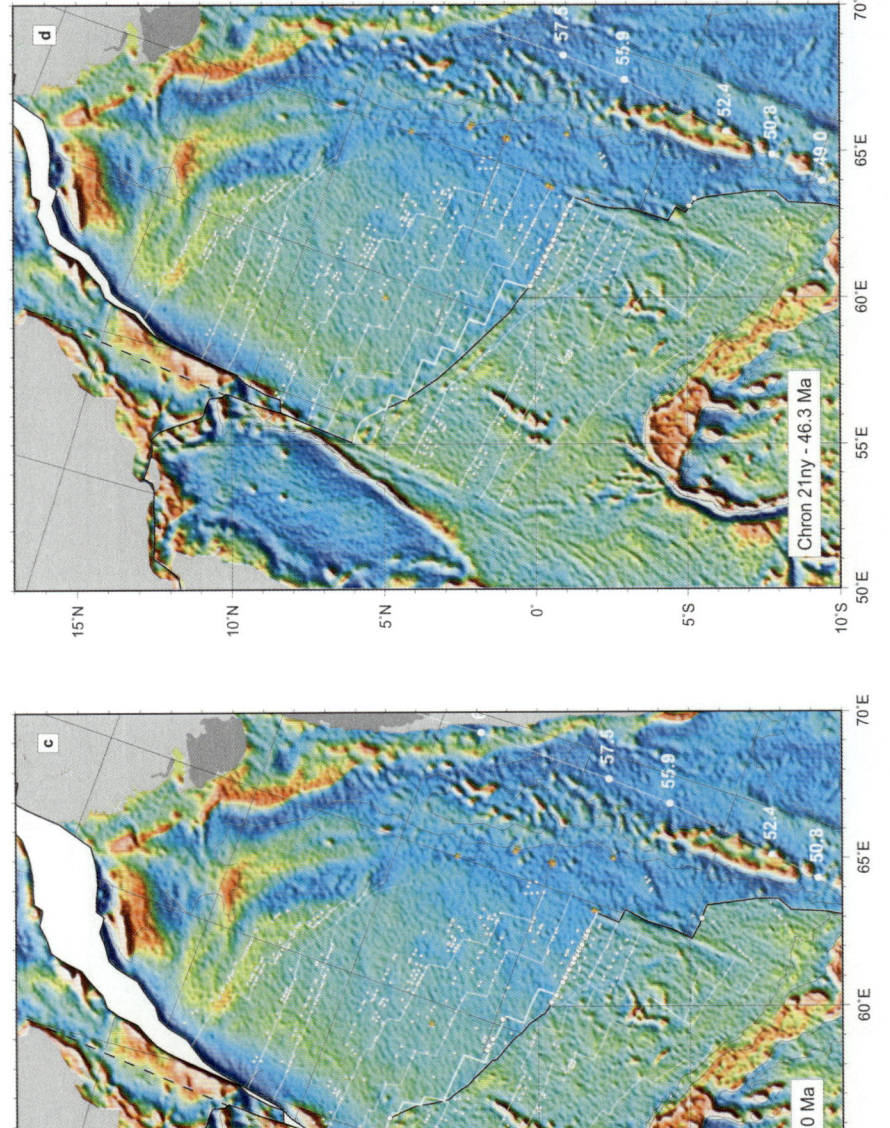

Fig. 5. (continued).

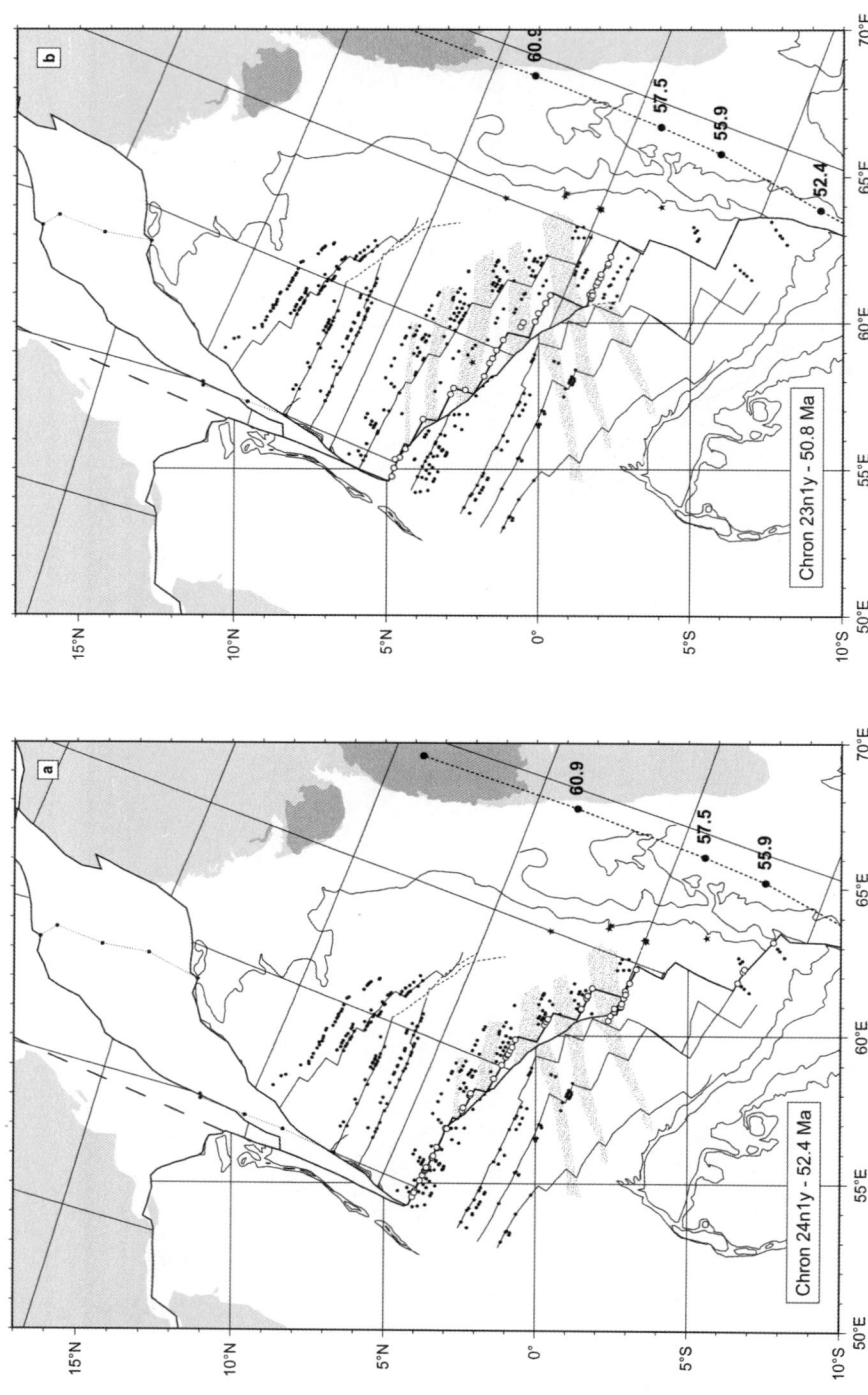

Fig. 6. Reconstructed tectonic chart for (**a**) Chron 24n1y (52.4 Ma), (**b**) Chron 23n1y (50.8 Ma), (**c**) Chron 22ny (49.0 Ma), and (**d**) Chron 21ny (46.3 Ma). Same symbols as for Fig. 4.

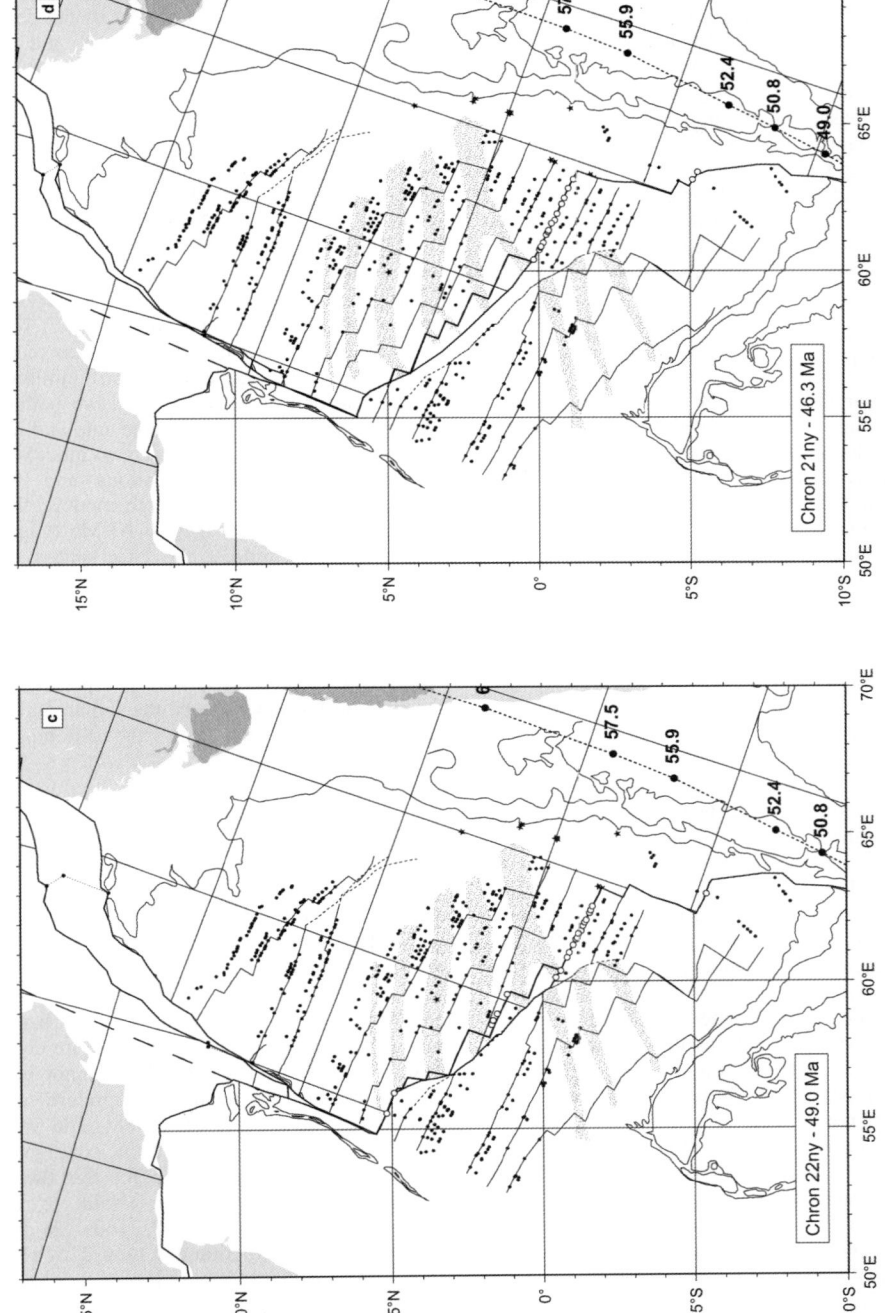

Fig. 6. (continued).

taken place at Chron 24n3o (53.3 Ma), which is the first reversal after Chron 25. The trace of the common outer pseudo-fault from the previous spreading stages remained in the Arabian Basin, whereas the new outer pseudo-fault started to develop in the Eastern Somali Basin. During the following stages (Chrons 23n1y and 22ny; Figs. 5 and 6), the easternmost segments of the Carlsberg Ridge increased in length whereas the westernmost segments progressively receded; the intermediate segments translated westward. From Chron 24 and younger chrons, the eastern side of Owen Ridge shears along the Chain FZ, suggesting that at this time the southern end of Owen Ridge did mark the western limit of the Indian plate.

Chron 21 (46 Ma)

Shortly after Chron 22, the speed of propagation decreased and as a result the geometry of the pseudo-faults changed (Figs. 5d and 6d). The outer pseudo-fault migrated to the north and the inner pseudo-faults bend anti-clockwise. All these spreading changes probably result from the collision of India with Eurasia. We note that this reconstruction implies some compression along the southern end of the Arabian–Indian plate boundary, which may have triggered or helped the opening of the Gulf of Aden as suggested by Whitmarsh (1979).

Discussion

These direct reconstructions of the India–Somalia relative motions provide new insights on several questions: the time of the initial India–Eurasia collision in the northwestern Indian Ocean, the amount of India–Capricorn motion, the relationship between the Laxmi Ridge and Seychelles Bank, the origin of the propagators, and the India–Arabia plate boundary in Paleogene time. Each of these questions would require some further work such as older and younger reconstructions, detailed analyses of the reconstruction uncertainties, further data compilation in the Owen and Mascarene basins, or some thermomechanical modelling.

The direction of motion (Fig. 7) of the Indian plate relative to the Somalian plate remained fairly stable between Chron 27 and Chron 21, as illustrated in Fig. 4a, which displays the path of the Indian plate relative to the welded Somalian–Arabian plate. The spreading (half) rates average 6 cm a^{-1} from Chron 26 to Chron 24, after when they decrease rapidly to 3 cm a^{-1} at Chron 21. A similar and synchronous drastic slow-down is observed in the Central Indian Basin and corresponds to the beginning of the collision of India with Eurasia (Patriat & Achache 1984).

In Fig. 8, we compare paths of the Indian plate relative to the Somalian plate derived from previous kinematic models. Because of the complexity of the western Indian Ocean, the India–Somalia relative motions are generally derived from magnetic anomalies in the Central Indian and Madagascar basins, with the additional constraint of closure of the Indian triple junction (e.g. Norton & Sclater 1979; Patriat 1983; Patriat & Ségoufin 1988; Royer & Sandwell 1989). The resulting paths (paths 2 in Fig. 8) are different from our model (paths 1) because the latter models do not account for the deformation of the Central Indian Basin; that is, for relative motions between the Indian and Capricorn (i.e. southern Central Indian Basin) plates. When corrected (paths 3) for the India–Capricorn motion since 20 Ma (Gordon et al. 1998), the two paths agree fairly well. This result has three implications. (1) The consistency between these two models, based on different magnetic lineations and fracture zones, gives confidence in both models; there is, however, some discrepancy at 61 Ma (Chron 27) which probably reflects the latest phase of seafloor spreading in the Mascarene Basin (Seychelles–Somalia motion), before extinction. (2) The applied correction, derived from reconstructions of the Carlsberg and Central Indian ridges at Chron 6 (Gordon et al. 1998), is representative of the integral motion between the Indian and Capricorn plates. (3) Conversely, the differences between the India–Somalia and Capricorn–Somalia motions in Paleogene time can be used to estimate and refine the Capricorn–India integral motion as suggested by Royer & Chang (1991). Such endeavour would require an accurate assessment of the rotation uncertainties for these two plate pairs. Finally, our model is broadly similar to, but different in details from the Molnar et al. (1988) model for the Arabian Sea (path 4 in Fig. 8). This is because the latter model assumes that the Chain and Owen ridges, at one end, and the Chagos and Mauritius trenches, at the other end, are conjugate fracture zones; the two trenches cannot be fitted directly because of the India–Capricorn motions, as noted by Molnar et al. (1988), and we show that the two other ridges cannot be conjugate.

Various studies appear to agree that the Laxmi Ridge is a continental sliver (Naini & Talwani 1983; Talwani & Reif 1998; Todal & Edholm 1998). The reconstruction at Chron 27ny (Fig. 3a) shows that a triangular wedge of oceanic crust separates the Seychelles Bank from the Laxmi Ridge, as if the Carlsberg Ridge split and propagated through these structures. A counter-clockwise rotation of Laxmi Ridge relative to

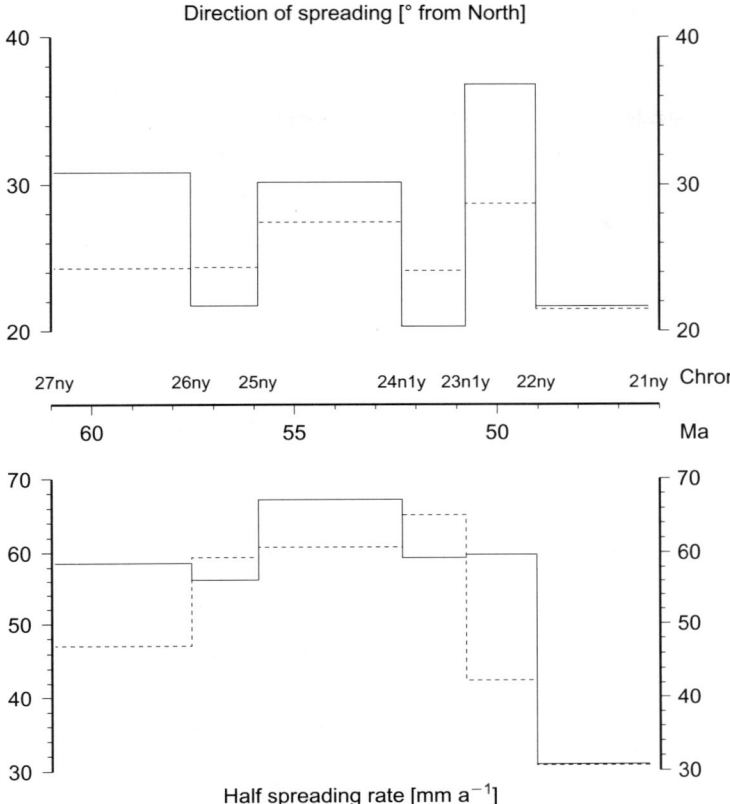

Fig. 7. Evolution of the predicted direction and half-rate of spreading along two flowlines in the Eastern Somali Basin (Somalia plate), between Chron 27ny (61 Ma) and Chron 21ny (46 Ma) assuming symmetric spreading (the flowlines are located at 56°E and 63°E; dashed and continuous lines, respectively).

Seychelles Bank about a pole located near 58°E and 5°S (in Fig. 3a) would fit the two features together; the gravity high, north of and perpendicular to the Laxmi Ridge, would then line up with the transform fault NW of Seychelles Bank.

The spreading history of the Arabian and Eastern Somali basins is dominated by propagating ridges. The two main propagating events, initially recognized from the data (Chaubey *et al.* 1998; Dyment 1998), are confirmed by these reconstructions: from west to east between Chrons 27 and 25y, the direction of propagation reversed itself before Chron 24o. Our step-by-step reconstructions show how the oceanic crust on one side developed at the expense of the conjugate side, leading to a very asymmetric set of isochrons (Fig. 6d). Overall, ridge propagation favoured the growth of the Arabian Basin (Indian plate) at the expense of the Eastern Somali Basin (Somalian plate). The isochrons drawn within the transferred crust zone in Figs. 3–6 correspond to idealized instantaneous isochrons; the actual isochrons here should be Z-shaped (Fig. 9), instead of stair-step shaped. On the conjugate side, isochrons stack along the outer pseudo-faults, such as Chrons 25ny and 26ny in the Arabian Basin, or Chrons 24n1y, 23n1y and 22ny in the Eastern Somali Basin. It is interesting to note that the length of the 'active' part of the outer pseudo-fault does not increase; but the 'active' part shifts as the magnetic lineations lengthen at one extremity and recede at the other (by active part, we mean part of the outer pseudo-fault bounded by the westernmost and easternmost propagator tips).

The origin of these continuously propagating ridges and of the abrupt shift in propagating direction is not clear. It has been noted that spreading ridges often propagate 'down-hill' along a gravity or bathymetry gradient (Morgan & Sandwell 1994). Although the Palaeogene bathymetry or gravity of the Carlsberg Ridge is not known, we can speculate that the Deccan–Réunion hotspot generated a regional swell large enough to affect the Carlsberg Ridge bathymetry.

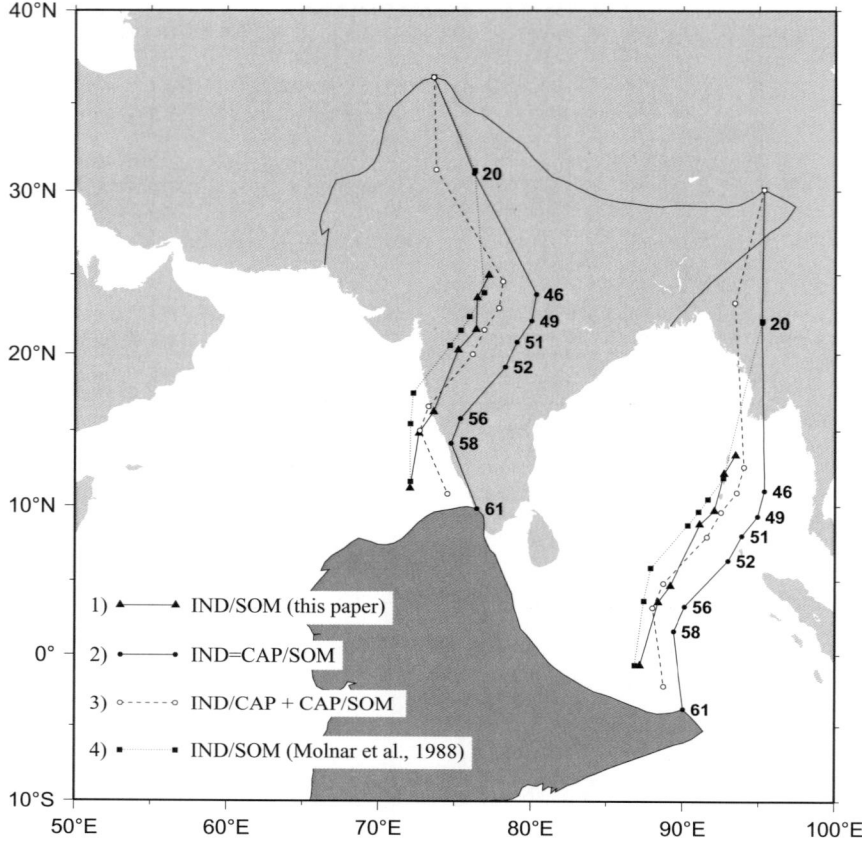

Fig. 8. Motions of the Indian plate relative to the Somalian plate: (1) reconstruction model based on magnetic crossings only from the Arabian and Eastern Somali basins (this paper, Table 1); (2) reconstruction model based on magnetic crossings from the Madagascar and Central Indian basins (Capricorn–Somalia motion, after Patriat & Ségoufin 1988; Royer & Sandwell 1989); (3) same model as (2) corrected for the relative motion between the Capricorn and Indian plates since Chron 6 (20 Ma; Gordon et al. 1998); (4) reconstruction model of Molnar et al. (1988) for the Arabian Basin assuming that Chain and Owen ridges and Mauritius and Chagos trenches are conjugate fracture zones. Except for Chron 27 (61 Ma), our solution is very close to model (3), confirming that the difference between model (1) and (2) is due to the deformation of the Central Indian Basin (i.e. to the relative motion between the Indian and Capricorn plates).

However, as pointed out by Dyment (1998), ridge segments initially propagate towards the Deccan–Réunion hotspot when the distance between the hotspot and the spreading ridge is the shortest (Figs. 3 and 4), and propagate away from the hotspot as the hotspot migrated southward (Figs. 5 and 6), so the attraction of propagator tips towards a bathymetry or gravity low does not seem to apply here. Asthenospheric flows or a thermal origin for these propagators seem also unlikely because of the abrupt shift (within a few million years) in the direction of propagation, and also because the effects would again be opposite to what is observed. Therefore, the ridge-propagation events in the Arabian and Eastern Somali basins, as documented in the present study, do not suggest an obvious link with the hotspot. The hypothesis that ridge propagations result from changes in the spreading direction (Hey et al. 1980) does not apply either, because the spreading direction is fairly stable, as seen both in the general orientation of the magnetic lineations and in our reconstructions. It is worth noting that the shift in propagating direction occurs at the same time (Chron 24r) as the spreading rates start to decrease, both in the Arabian–Eastern Somali and Central Indian–Crozet–Madagascar basins; this slow-down is interpreted as the result of the initial collision of the Greater India margin with Eurasian island arcs (Patriat & Achache 1984). This

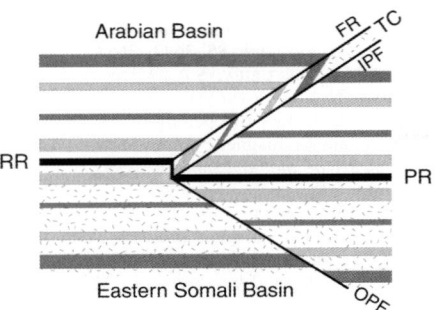

PR : Propagating Ridge   IPF : Inner Pseudofault
RR : Receding Ridge   OPF : Outer Pseudofault
FR : Failed Ridge trace TC : Transferred Crust

Fig. 9. Schematic diagram showing the main tectonic features associated with propagating ridges. Stippled areas show the oceanic crust formed on the Somalian side of the ridge; part of it (TC) is transferred onto the Indian plate as a result of the westward ridge propagation. This configuration corresponds to the direction of propagation at Chron 24 and younger times.

change in the plate boundary condition may explain the abrupt change in the direction of ridge propagation.

The oldest reconstructions (Chron 27–5) demonstrate that the Owen Ridge is neither the conjugate of Chain Ridge nor the conjugate of Chain Fracture Zone. The gap between the modelled Indian plate boundary and the Chain Fracture Zone has the same triangular shape as the Owen Basin that we left attached to the Arabian plate, suggesting that the missing portions of the magnetic lineations on the Indian plate are trapped in the Owen Basin. They may correspond to the WSW–NNE magnetic lineations observed in the Owen Basin (Whitmarsh 1979; Mountain & Prell 1989). This departure from the east–west orientation of the lineations in the Arabian Sea can be attributed to the SE–NW transpressional motion of the Indian plate relative to the Arabian plate between Chron 21 and 22, and younger times. This motion would be responsible for the uplift of the Owen Ridge in Miocene time, the breaking of the Indian plate along the Owen–Murray ridge complex, and the opening of the Gulf of Aden. The Chron 21–Chron 22 motion has a SE–NW component. Magnetic lineations in the Owen Basin are likely to be offset to the north, relative to the Arabian Sea lineations, as a result of the Arabia–India right-lateral motions (Gordon & DeMets 1989).

Conclusion

Despite the complexity of the tectonic fabric in the Arabian and Eastern Somali basins, the development of the northwestern Indian Ocean can be reconstructed step by step. We derived a consistent set of idealized isochrons which account for the large spreading asymmetry between the two basins, for the pattern of oblique pseudo-faults, and for the lack of 'normal' fracture zones in these basins. We are able to model in two main stages the growth of the outer pseudo-faults and the development of the inner pseudo-faults; the direction of ridge propagation reversed itself between Chrons 25 and 24. It is worth noting that the location of the outer pseudo-faults remained stable during these two stages. This stability and the repeated occurrence of ridge propagation have yet to be explained. Our reconstructions also confirm the proposition of Mountain & Prell (1990) that the India–Arabia plate boundary was located off the Oman margin during Paleogene time. The Chain Fracture Zone would be the conjugate fracture zone of the Oman transform margin (Barton et al. 1990). In this framework, Chain Ridge would be a scar of the break-up between the Indian and Madagascar plates in mid-Cretaceous time. The consistency of the Somalia–India motions (this model) with the Somalia–Capricorn motions (Patriat & Ségoufin 1988; Royer & Sandwell 1989) corrected for the post-Miocene India–Capricorn motions (Gordon et al. 1998) provides confidence in each of these links, based on entirely different datasets, and in their underlying hypotheses.

Our model requires some further work, such as reconstructing the intermediate chrons to improve our understanding of the complex ridge propagation, investigating more thoroughly the magnetic anomalies in the Owen Basin and eastern Gulf of Oman, and reconstructing the early opening of the northwestern Indian Ocean and the details of the simultaneous opening of the Mascarene Basin and Arabian Sea after the break-up of India and Madagascar in mid-Cretaceous time. A detailed comparison of the India–Somalia and Capricorn–Somalia motions would provide an independent estimate of the integral Capricorn–India motions responsible for the deformation of the Central Indian Basin.

This work was funded by grant 1911-1 from the Indo-French Centre for the Promotion of Advanced Research (IFCPAR), New Delhi. J.Y.R. and J.D. were supported by the Centre National de la Recherche Scientifique. This paper is published with the permission of E. Desa, Director, NIO, Goa. We thank P. Miles and B. Schreckenberger for their careful reviews of the manuscript. All figures were drafted with the GMT software (Wessel & Smith 1995). This paper is NIO contribution 3726.

References

BARTON, P.J., OWEN, T.R.E. & WHITE, R.S. 1990. The deep structure of the east Oman continental margin:

preliminary results and interpretation. *Tectonophysics*, **173**, 319-331.

BERNARD, A. & MUNSCHY, M. 2000. Le bassin des Mascareignes et le bassin de Laxmi (océan Indien occidental) se sont-ils formés à l'axe d'un même centre d'expansion? *Comptes Rendus de l'Académie des Sciences*, **330**, 777-783.

BHATTACHARYA, G.C., CHAUBEY, A.K., MURTY, G.P.S., SRINIVAS, K., SARMA, K.V.L.N.S., SUBRAHMANYAM, V. & KRISHNA, K.S. 1994. Evidence for seafloor spreading in the Laxmi Basin, northeastern Indian Ocean. *Earth and Planetary Science Letters*, **125**, 211-220.

CANDE, S.C. & KENT, D.V. 1995. Revised calibration of the geomagnetic polarity timescale for the Late Cretaceous and Cenozoic. *Journal of Geophysical Research*, **100**, 6093-6095.

CHANG, T. 1988. Estimating the relative rotation of two tectonic plates from boundary crossings. *Journal of the American Statistical Association*, **83**, 1178-1183.

CHAUBEY, A.K., BHATTACHARYA, G.C., MURTY, G.P.S. & DESA, M. 1993. Spreading history of the Arabian Sea: some new constraints. *Marine Geology*, **112**, 343-352.

CHAUBEY, A.K., BHATTACHARYA, G.C., MURTY, G.P.S., SRINIVAS, K., RAMPRASAD, T. & GOPALA RAO, D. 1998. Early Tertiary seafloor spreading magnetic anomalies and paleo-propagators in the northern Arabian Sea. *Earth and Planetary Science Letters*, **154**, 41-52.

CHAUBEY, A.K., DYMENT, J., BHATTACHARYA, G.C., ROYER, J.-Y., SRINIVAS, K. & YATHEESH, V. 2002. Palaeogene magnetic isochrons and paleo-propagators in the Arabian and Eastern Somali basins, north-west Indian Ocean. *In:* CLIFT, P.D. & GAEDICKE, C. (eds) *The Geological and Climatic Evolution of the Arabian Sea Region*. Geological Society, London, Special Publication, **195**, 71-86.

DE RIBET, B. 1989. *Étude géophysique du nord-ouest de l'océan Indien: cinématique Inde-Afrique*. Thèse de Doctorat, Université Denis Diderot (Paris 7), Paris.

DYMENT, J. 1991. *Structure et évolution de la lithosphère océanique dans l'océan Indien: apport des anomalies magnétiques*. Thèse de Doctorat, Université Louis Pasteur, Strasbourg.

DYMENT, J. 1998. Evolution of the Carlsberg ridge between 60 and 45 Ma: ridge propagation, spreading asymmetry, and the Deccan-Réunion hot spot. *Journal of Geophysical Research*, **103**, 24067-24084.

GORDON, R.G. & DEMETS, C. 1989. Present-day motion along the Owen fracture zone and Dalrymple trough in the Arabian Sea. *Journal of Geophysical Research*, **94**, 5560-5570.

GORDON, R.G., DEMETS, C. & ROYER, J.-Y. 1998. Evidence for long-term diffuse deformation in the equatorial Indian Ocean. *Nature*, **395**, 370-374.

HAXBY, W.F. 1987. *Gravity Field of the World's Oceans*. Office of Naval Research, United States Navy, Lamont-Doherty Geological Observatory, Boulder, CO.

HEY, R.N., DUENNEBIER, F.K. & MORGAN, J.P. 1980. Propagating rifts on midocean ridges. *Journal of Geophysical Research*, **85**, 3647-3658.

KARASIK, A.M., MERCURIEV, S.A., MITIN, L.I., SOCHEVANOVA, N.A. & YANOVSKY, V.N. 1986. Peculiarities in the history of opening of the Arabian Sea from systematic magnetic survey data (in Russian). *Documents of the Academy of Sciences of USSR*, **286**, 933-938.

KIRKWOOD, B.H., ROYER, J.-Y., CHANG, T.C. & GORDON, R.G. 1999. Statistical tools for estimating and combining finite rotations and their uncertainties. *Geophysical Journal International*, **137**, 408-428.

MALOD, J.A., DROZ, L., MUSTAFA KAMAL, B. & PATRIAT, P. 1997. Early spreading and continental to oceanic basement transition beneath the Indus deep-sea fan: northeastern Arabian Sea. *Marine Geology*, **141**, 221-235.

MASSON, D.G. 1984. Evolution of the Mascarene Basin, western Indian Ocean, and the significance of the Amirante Arc. *Marine Geophysical Research*, **6**, 365-382.

MATTHEWS, D.H. 1966. The Owen fracture zone and the northern end of the Carlsberg Ridge. *Philosophical Transactions of the Royal Society of London*, **259**, 172-186.

MCKENZIE, D.P. & SCLATER, J.G. 1971. The evolution of the Indian Ocean since the Late Cretaceous. *Geophysical Journal of the Royal Astronomical Society*, **25**, 437-528.

MERCURIEV, S., PATRIAT, P. & SOCHEVANOVA, N. 1996. Évolution de la dorsale de carlsberg: évidence pour une phase d'expansion très lente entre 40 et 25 Ma (A18 à A7). *Oceanologica Acta*, **19**, 1-13.

MERCURIEV, S.A. & SOCHEVANOVA, N.A. 1990. Complex patterns of the magnetic field as a consequence of an ancient triple junction on the Carlsberg Ridge? (in Russian). *In:* JDANOV, M. (ed.) *Electromagnetic Induction in the World Ocean*. USSR Academy of Sciences, Moscow, 48-56.

MILES, P.R., MUNSCHY, M. & SÉGOUFIN, J. 1998. Structure and early evolution of the Arabian Sea and East Somali Basin. *Geophysical Journal International*, **15**, 876-888.

MILES, P.R. & ROEST, W.R. 1993. Earliest seafloor spreading magnetic anomalies in the north Arabian Sea and the ocean-continent transition. *Geophysical Journal International*, **15**, 1025-1031.

MOLNAR, P., PARDO-CASAS, F. & STOCK, J. 1988. The Cenozoic and Late Cretaceous evolution of the Indian Ocean basin: uncertainties in the reconstructed positions of the Indian. *African and Antarctic plates, Basin Research*, **1**, 23-40.

MORGAN, J.P. & SANDWELL, D.T. 1994. Systematics of ridge propagation south of 30°S. *Earth and Planetary Science Letters*, **121**, 245-258.

MOUNTAIN, G.S. & PRELL, W.L. ET AL. 1989. Geophysical reconnaissance survey for ODP Leg 117 in the northwest Indian Ocean. *In:* PRELL, W.L. & NITSUMA, N. (eds) *Proceedings of the Ocean Drilling Program, Initial Reports, 117*. Ocean Drilling Program, College Station, TX, 51-64.

MOUNTAIN, G.S. & PRELL, W.L. 1990. A multiphase plate tectonic history of the southeast continental margin of Oman. *In:* ROBERTSON, A.H.F., SEARLE,

R. & RIES, A.C. (eds) *The Geology and Tectonics of the Oman Region*. Geological Society, London, Special Publications, **49**, 725–743.

MÜLLER, R.D., ROYER, J.-Y. & LAWVER, L.A. 1993. Revised plate motions relative to the hotspots from combined Atlantic and Indian Ocean hotspot tracks. *Geology*, **21**, 275–278.

NABIGHIAN, M.N. 1972. The analytic signal of two-diemnsional magnetic bodies with polygonal cross-section: its properties and use for automated anomaly interpretation. *Geophysics*, **37**, 507–517.

NABIGHIAN, M.N. 1974. Additional comments on the analytic signal of two-dimensional magnetic bodies with polygonal cross-section. *Geophysics*, **37**, 85–92.

NAINI, B.R. & TALWANI, M. 1983. Structural framework and the evolutionary history of the continental margin of Western India. *In:* WATKINS, J.S. & DRAKE, C.L. (eds) *Studies in Continental Margin Geology*. American Association of Petroleum Geologists, Memoir, **34**, 167–191.

NATIONAL GEOPHYSICAL DATA CENTER 1998. *GEODAS CD-ROM, worldwide marine geophysical data,98-MGG-04, updated 4th edn*. National Oceanic and Atmospheric Administration, US Department of Commerce, Boulder, CO.

NORTON, I.O. & SCLATER, J.G. 1979. A model for the evolution of the Indian Ocean and the breakup of Gondwanaland. *Journal of Geophysical Research*, **84**, 6803–6830.

PATRIAT, P. 1983. *Évolution du système de dorsales de l'Océan Indien*. Doctorat d'État, Université Pierre et Marie Curie (Paris 6), Paris.

PATRIAT, P. & ACHACHE, J. 1984. India–Eurasia collision chronology has implications for crustal shortening and driving mechanisms of plates. *Nature*, **311**, 615–621.

PATRIAT, P. & SÉGOUFIN, J. 1988. Reconstruction of the central Indian Ocean. *Tectonophysics*, **155**, 211–234.

QUITTMEYER, R.C. & KAFKA, A.L. 1984. Constraints on plate motions in Southern Pakistan and the northern Arabian Sea from the focal mechanisms of small earthquakes. *Journal of Geophysical Research*, **89**, 2444–2458.

ROYER, J.-Y. & CHANG, T. 1991. Evidence for relative motions between the Indian and Australian plates during the last 20 m.y. from plate tectonic reconstructions: implications for the deformation of the Indo-Australian plate. *Journal of Geophysical Research*, **96**, 11779–11802.

ROYER, J.-Y. & SANDWELL, D.T. 1989. Evolution of the Eastern Indian Ocean since the Late Cretaceous: constraints from Geosat altimetry. *Journal of Geophysical Research*, **94**, 13755–13782.

SANDWELL, D. T. 1984. *Along-track deflection of the vertical from SEASAT: GEBCO overlays*. National Oceanic and Atmospheric Administration, Rockville, MD, NOAA Technical Memorandum, **NOS NGS-40**.

SANDWELL, D.T. & RUIZ, M.B. 1992. Along-track gravity anomalies from Geosat and Seasat altimetry: GEBCO overlays. *Marine Geophysical Research*, **14**, 165–205.

SANDWELL, D.T. & SMITH, W.H.F. 1997. Marine gravity anomaly from Geosat and ERS-1 satellite altimetry. *Journal of Geophysical Research*, **102**, 10039–10054.

SCHLICH, R. 1982. The Indian Ocean: aseismic ridges, spreading centers and basins. *In:* NAIRN, A.E.M. & STEHLI, F.G. (eds) *The Ocean Basins and Margins, 6, The Indian Ocean*. Plenum, New York, 51–147.

SHREIDER, A.A. 1998. The oldest paleomagnetic anomalies in the Arabian Sea. *Okeanologiya*, **38**, 911–918.

STOREY, M., MAHONEY, J.J., SAUNDERS, A.D., DUNCAN, R.A., KELLEY, S.P. & COFFIN, M.F. 1995. Timing of hot-spot related volcanism and the breakup of Madagascar and India. *Science*, **267**, 852–855.

SYKES, L.R. 1968. Seismological evidence for transform faults, seafloor spreading, and continental drift. *In:* PHINNEY, R.A. (ed.) *The History of Earth's Crust*. Princeton University Press, Princeton, NJ, 120–150.

TALWANI, M. & REIF, C. 1998. Laxmi Ridge—a continental sliver in the Arabian Sea. *Marine Geophysical Research*, **20**, 259–271.

TODAL, A. & EDHOLM, O. 1998. Continental margin off western India and Deccan large igneous province. *Marine Geophysical Research*, **20**, 273–291.

WESSEL, P. & SMITH, W.H.F. 1995. New version of the Generic Mapping Tools released. *EOS Transactions, American Geophysical Union*, **76**, 329.

WHITMARSH, R.B. 1979. The Owen Basin off the southeast margin of Arabia and the evolution of the Owen Fracture Zone. *Geophysical Journal of the Royal Astronomical Society*, **58**, 441–470.

WHITMARSH, R.B. ET AL. 1974. Some aspects of plate tectonics in the Arabian Sea, in Leg XXXIII. *In:* WHITMARSH, R.B., WESER, O.E. & ROSS, D.A. (eds) *Initial Reports of the Deep Sea Drilling Program, 23*. US Government Printing Office, Washington, DC, 527–535.

Seismic stratigraphy and correlation of major regional unconformities in the northern Arabian Sea

CHRISTOPH GAEDICKE[1], ALEXANDER PREXL[1], HANS-ULRICH SCHLÜTER[1], HEINRICH MEYER[1], HANS ROESER[1] & PETER CLIFT[2]

[1]*Federal Institute for Geosciences and Natural Resources (BGR), Stilleweg 2, 30655 Hannover, Germany (e-mail: gaedicke@bgr.de)*
[2]*Department of Geology and Geophysics, Woods Hole Oceanographic Institution, Woods Hole, MA 02543, USA*

In the northern Arabian Sea the Arabian, Eurasian and Indian Plates are in tectonic interaction with one another. We present interpretations of multi-channel seismic profiles across the Makran subduction zone (which is part of the Eurasian–Arabian Plate boundary) and the transtensional Murray Ridge and Dalrymple Trough (which are part of the Arabian–Indian Plate boundary). We distinguish four megasequences in the sedimentary succession, which we correlate over the entire study area. Regional unconformities separate the megasequences and enable us to establish a common history of the region before Late Miocene time (c. 20 Ma). The Early Pliocene (c. 4.5 Ma) reopening of the Gulf of Aden caused a reorganization of the plates and subsequent tilting of the oceanic crust of the Arabian Plate toward the Makran subduction zone. This event is documented by the regional M-unconformity. Since that time, sedimentation on the Oman Abyssal Plain has been permanently separated from the Indus Fan by the Murray Ridge, on the northern end of which there has been no significant sedimentation.

The northern Arabian Sea can be divided into four units, on the basis of physiographic and structural characteristics: (1) the Makran Accretionary Wedge; (2) the Oman Abyssal Plain; (3) the Murray Ridge system; (4) the Indus Fan (Fig. 1). The units meet at the triple junction of the Eurasian, Arabian and Indian Plates west of Karachi.

The Makran Accretionary Wedge extends some 1000 km from Iran to central Pakistan. Its topography ranges from more than 3000 m below sea level to heights of 1500 m above sea level. The wedge was formed by the subduction of oceanic crust of the Arabian Plate under the Eurasian Plate. Subduction has continued since at least 45–55 Ma (Fruehn *et al.* 1997). Two features make this accretionary wedge especially interesting: first, the sediment thickness on the oceanic crust of the subducting plate is extremely high (more than c. 6 km); second, the angle of subduction is extremely low (Fruehn *et al.* 1997; Kukowski *et al.* 2000).

The Oman Abyssal Plain extends between the Makran deformation front in the north and the Murray Ridge system in the south. The SW–NE-trending Little Murray Ridge divides the abyssal plain into two parts. The sediments below the Oman Abyssal Plain reach a maximum thickness of 7 km at the toe of the Makran Accretionary Wedge. Because of the lack of magnetic anomalies the age of the underlying oceanic crust could not be determined.

The Murray Ridge system forms the northernmost extension of the Owen Fracture Zone and forms part of the boundary between the Indian and Arabian Plates (Fig. 1). Earthquake focal mechanism and seismic reflection profiles show present-day extensional structures (Gordon & DeMets 1989; Edwards *et al.* 2000) along the Murray Ridge and Dalrymple Trough, which contains more than 8 km of Cenozoic sediments. Transtensional strain has resulted in local volcanism close to the Murray Ridge crest (Fig. 1). Uplift during Early Miocene time means that the Murray Ridge system now divides the northern Arabian Sea into a northern and a southern depocentre.

The Indus Fan is substantially smaller than the Bengal Fan but none the less covers 1.1×10^6 km^2, stretching 1500 km into the Indian Ocean from the present delta front. The fan sediments are up to 9 km thick under the Pakistan Shelf (Clift *et al.* 2000). The present river drains an area of c. 1×10^6 km^2 with peak discharge during the summer months as a result of the seasonal melting of Himalayan glaciers, in combination with the increased runoff generated by the summer monsoon. Therefore, the sediments of the Indus Fan exhibit changes in climate and varia-

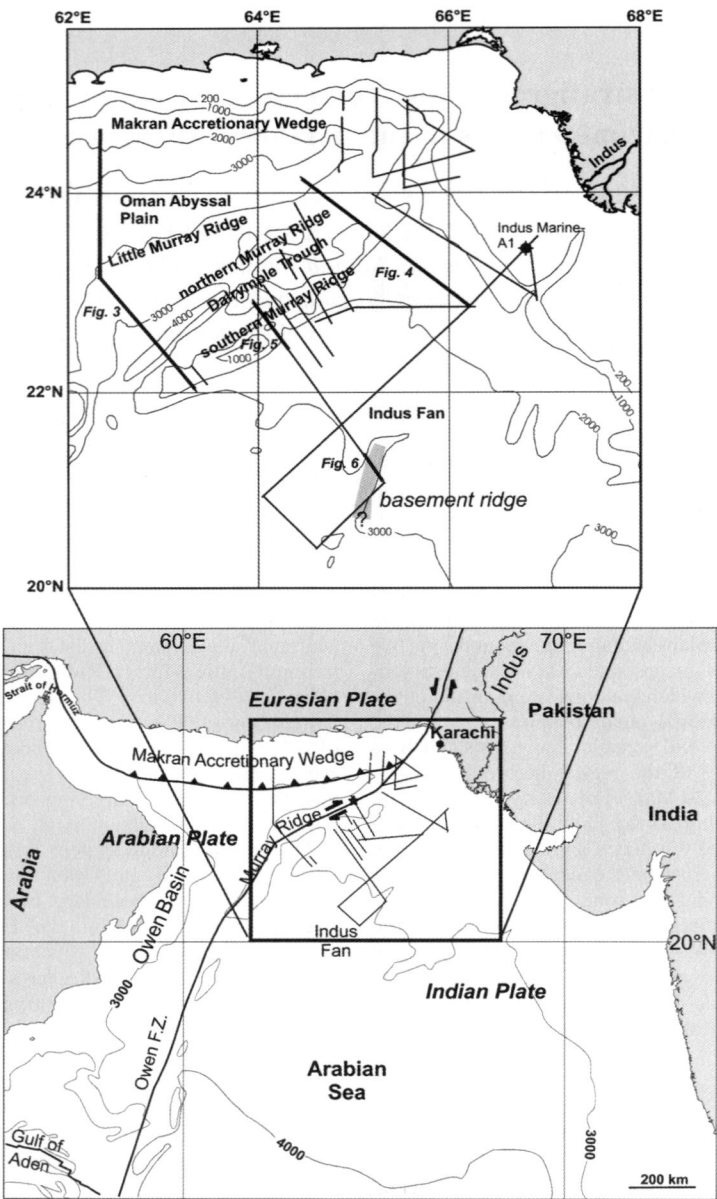

Fig. 1. Multi-channel seismic (MCS) survey of R.V. *Sonne* expedition SO122. The major plate boundaries are shown (below). A young submarine volcanic cone is marked with an asterisk. The occurrence of Sequence P (basement ridge) is shown as a shaded area (above; see also Figs. 2 and 6). Bold lines indicate MCS lines shown as figures. Industrial well Indus Marine A1 (Shuaib 1982) is marked. Water depth (contour lines) in metres.

bility of the monsoon, but also contain signatures of the collisional history of the Indian and Eurasian Plates, as well as of the uplift and erosion of the Tibetan Plateau and the Himalayas.

The four physiographic regions mentioned above belong to three plates and have been affected by the evolving nature of the interfering plate boundaries. Complex structures are caused by stresses driven by plate tectonic processes and their interpretation reveals the geological history of the northern Arabian Sea. The aim of this study is to establish a Cenozoic evolutionary history for the northern Arabian Sea, and to unravel the individual development of each region. To do this we present interpretations of multi-channel seismic reflection (MCS) profiles, which allow the correlation of principal sedimentary megasequences from the Indus Fan across the Murray Ridge to the Oman Abyssal Plain. The megasequences are separated by regional unconformities. From our correlation we propose a common geological history for these three provinces of the northern Arabian Sea, which seems to have been uniform until the Murray Ridge was uplifted to its present height, and the plate vector of the Arabian Plate changed significantly at the Miocene–Pliocene boundary.

Methods, data acquisition and interpretation

This study is based on 25 MCS lines (Fig. 1) with a total length of 2927 km that were acquired during R.V. *Sonne* cruise SO122 in 1997 (Roeser et al. 1997). The digital streamer length varied between 600 and 3000 m. Standard processing, including dip-moveout (DMO) corrections and a post-stack frequency wavenumber (FK)–time migration was applied. After calculation of the shotpoint–receiver geometry, the field data were sorted into common mid-point (CMP) gathers with a distance of 12.5 m, which were summed after processing into 25 m intervals for display and further interpretation. The seismic sequences and horizons were interpreted by application of the various tools of GeoQuest-IESX. Additional map data (i.e. satellite altimetry, high-resolution bathymetry) were imported and added to the basemap, allowing correlation with bathymetric and subsurface basement features. Published well data (Shuaib 1982) were integrated after conversion into the time domain using a simple velocity model as determined from stacking velocities.

Seismic stratigraphy and megasequences

In this study we focus on the sedimentary cover above the acoustic basement. In general, sediment cover is thin to absent on basement highs, such as the crest of the southern Murray Ridge, whereas it reaches a maximum depth of about 8 km in the centre of the northern Dalrymple Trough.

Droz & Bellaiche (1991) distinguished two megasequences in the middle and upper Indus Fan. The lower sedimentary megasequence is overlying and draping the acoustic basement. The megasequence consists of two units separated by an unconformity. Both units predate the onset of fan sedimentation. The upper megasequence is dominated by channel–levee complexes. Therefore this megasequence documents the evolution of the Indus Fan. McHargue (1991) focused on the channel–levee complexes and distinguished the characteristic seismic facies pattern of an upper fan.

Age estimations of the observed megasequences are crucial to establishing a stratigraphic and tectonic framework of the northern Arabian Sea and to constraining the evolution of the sheared Indian–Arabian Plate margin. Because precise age determinations of sedimentary sequences are rare and restricted to single well sites (Whitmarsh et al. 1974; Shuaib 1982; Quadri & Quadri 1986; Raza et al. 1990) we extrapolated age picks within the wells to our MCS profiles to define and correlate seismic sequences over long distances. It was possible to trace the identified megasequences and their regional unconformities throughout the area and to estimate their ages. We distinguish four sedimentary megasequences by their seismic reflectivity, internal geometry and external boundaries (Fig. 2).

Sequence M3 and M-(Makran-)unconformity

The sediments of the Oman Abyssal Plain are divided by the distinct M-(Makran-)unconformity (Fig. 3; Fruehn et al. 1997; Schlüter et al. 2002). The M-unconformity is parallel to the top of the oceanic crust and dips to the north, below the Makran Accretionary Wedge. In the south the M-unconformity crops out along the northern flank of Little Murray Ridge (Fig. 3). In the northeastern part of the Oman Abyssal Plain the M-unconformity crops out along the northern Murray Ridge (MCS line SO122-17, Fig. 4).

In front of the Makran Accretionary Wedge, Sequence M3, which is up to 2.4 s (two-way travel time; TWT) thick, shows an onlapping time-transgressive contact with the underlying Sequence M2, from which it is separated by the M-unconformity. Sequence M3 is characterized by closely spaced internal reflections with strong parallel bedding of very broad lateral continuity, high amplitudes and frequencies (Fig. 2). Sequence M3 is assumed to consist largely of turbidites from the growing Makran Accretionary Wedge. The onlap termination of Sequence M3 (Figs. 3 and 4) clearly shows a hiatus on the Little Murray Ridge and Murray Ridge, respectively indicating a phase of uplift and erosion or non-deposition. Consequently, these ridges operate as a barrier separating sediments supplied from the Makran coast and from the Indus River, and therefore divide the northern Arabian Sea into limited depocentres. Because of the similar reflec-

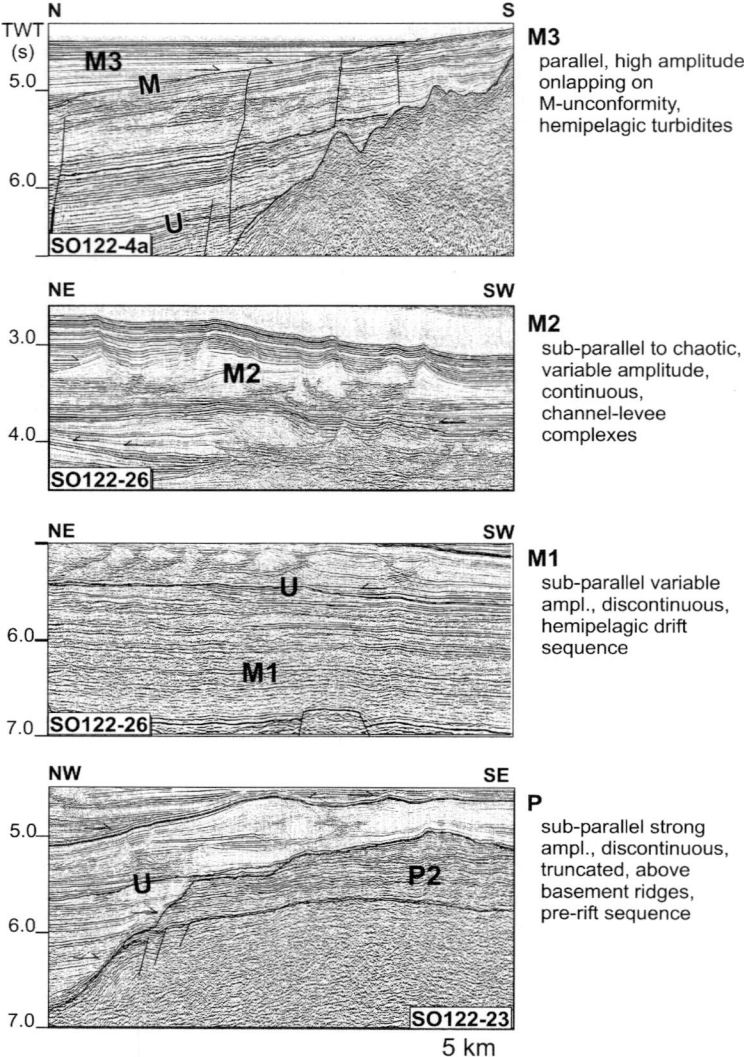

Fig. 2. MCS examples and main features of megasequences distinguished in the study area. All examples are shown at the same scale.

tion pattern of the uppermost deposits in the Dalrymple Trough we attribute these sediments to Sequence M3 (Fig. 4). However, the terrigenous sediments in this setting are probably derived from a source on the continental shelf offshore Karachi because the topography separates the Dalrymple Trough both from the Indus and Makran sources.

Sequence M2

The acoustic basement of the entire area between the Owen and Murray Ridges in the NW and west and the Indian continental margin in the NE and east, as far south as c. 5°S, is covered by the Indus Fan, a series of terrigenous deposits supplied by the Indus River from the Himalaya, Karakoram, Hindu Kush, Transhimalaya and the Tibetan Plateau (McHargue & Webb 1986; Droz & Bellaiche 1991; McHargue 1991). This Indus Fan Sequence M2 (Fig. 2) is composed of a heterogeneous unit of at least 21 individual channel–levee complexes (Clift et al. 2000, 2001). At the base of the Indian continental margin and adjacent to the recent Indus submarine canyon the megasequence reaches a

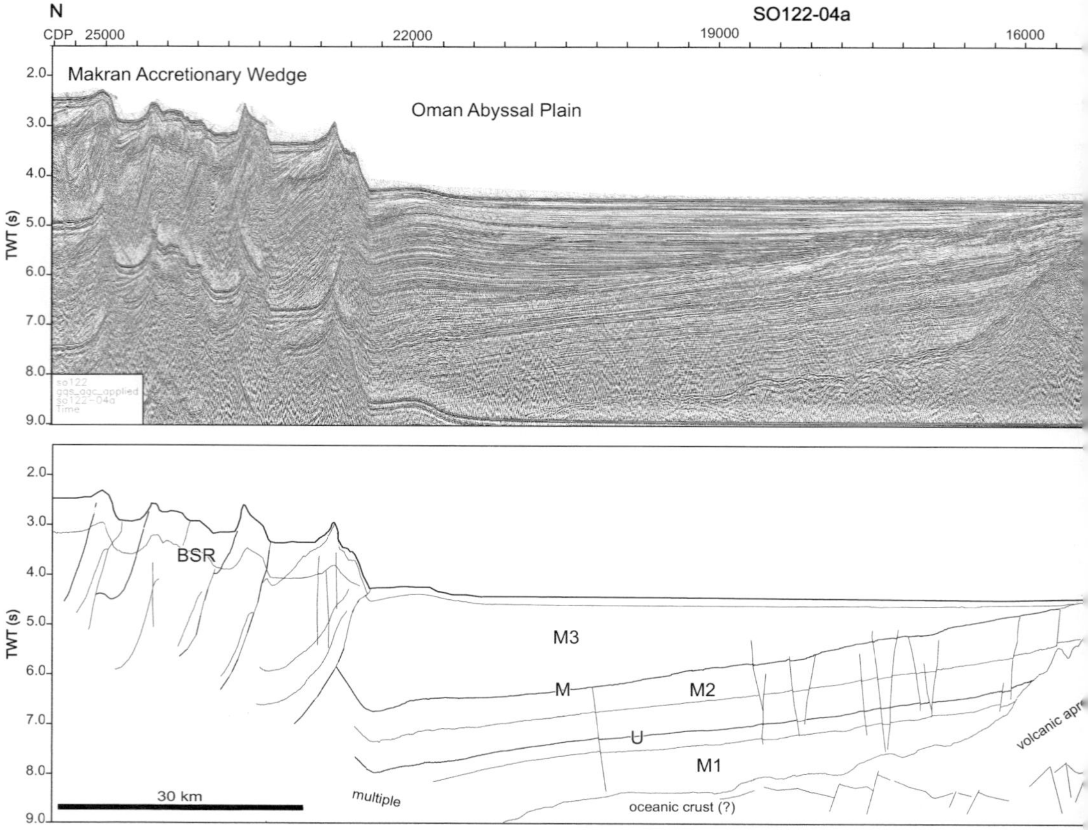

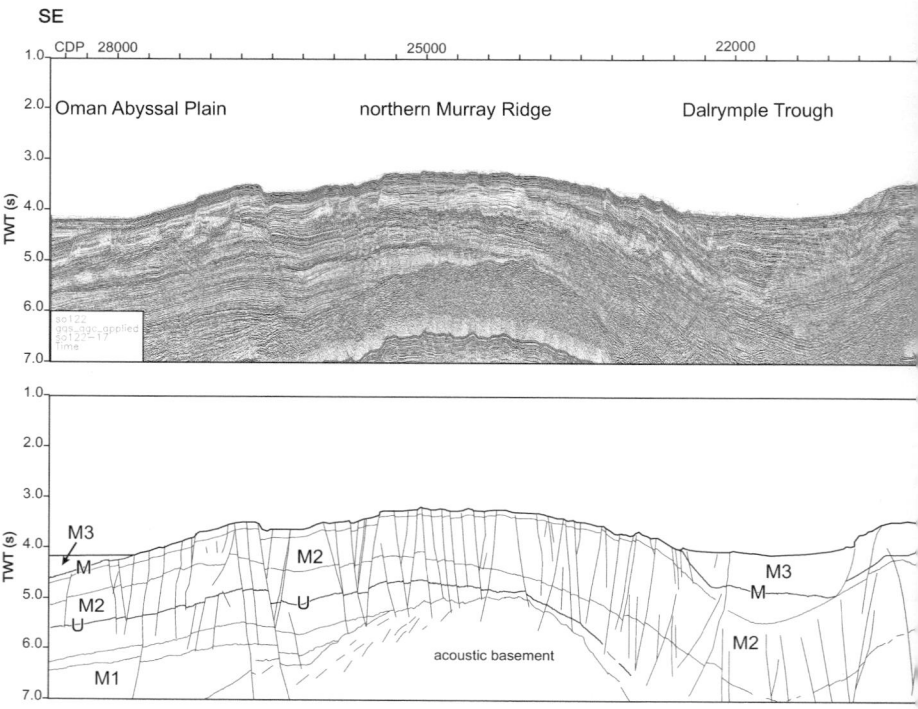

Fig. 3. Part of MCS profile SO122-4 and -4a from the Makran Accretionary Wedge across the Oman Abyssal Plain to the Lit The acoustic basement, and Sequences M1 and M2 are tilted and dip towards the Makran Accretionary Wedge. Sequence M3

Fig. 4. MCS profile SO122-17 (above). Sequence M2 is normal faulted and crops out on the northern Mur the central part of the northern Dalrymple Trough. (See Fig. 1 for location.)

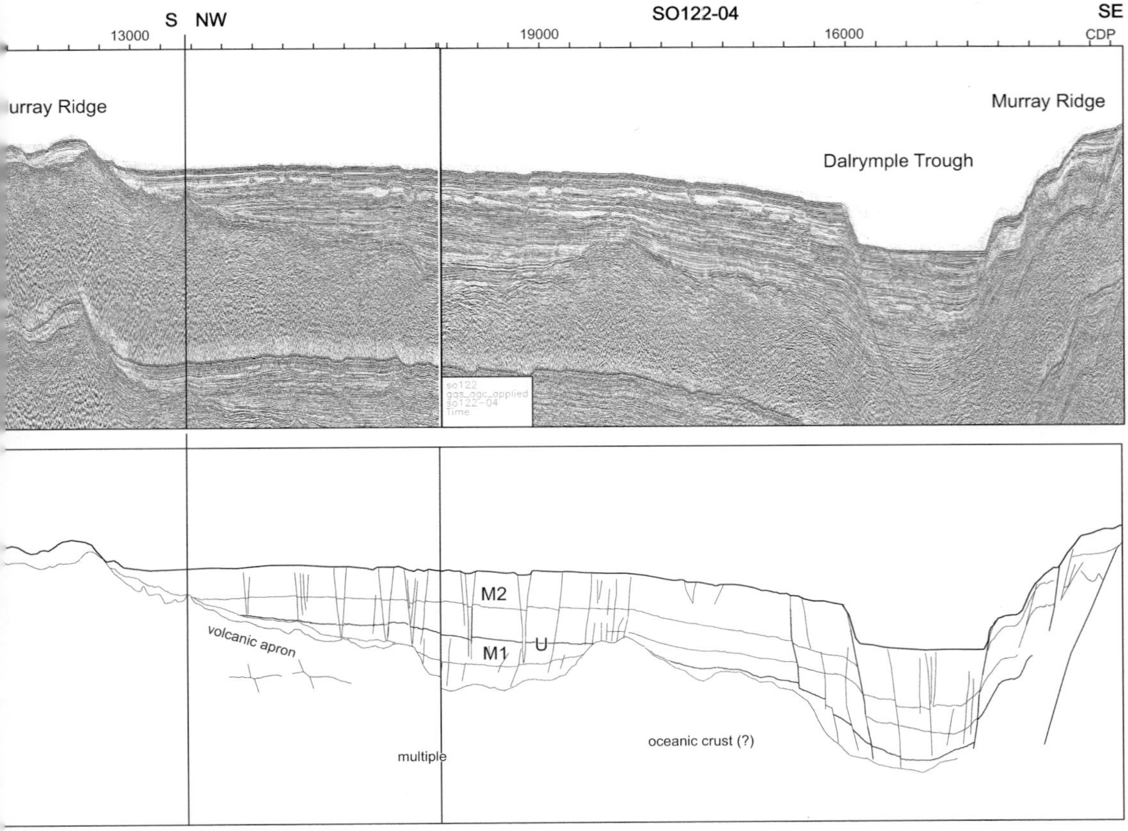

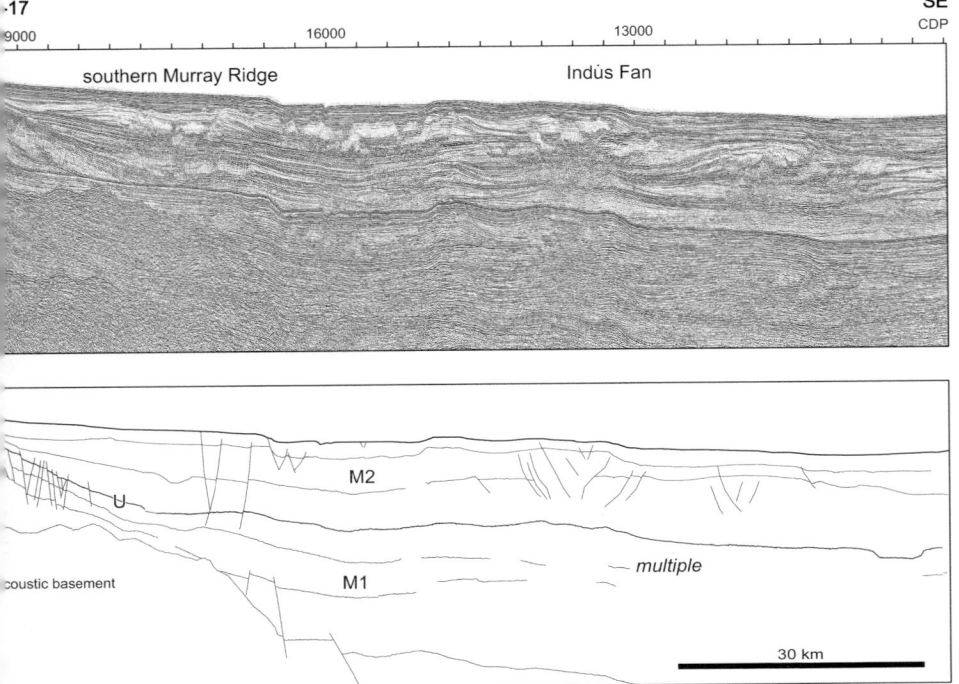

...dge (above). Sequence M2 is correlated across the Little Murray Ridge as a result of similarities in the reflection pattern (below). ...nce M2 with the M-unconformity. CDP, common depth point; BSR, bottom simulating reflector. (See Fig. 1 for location.)

...elow). Sequence M3 onlaps the ridge along the M-unconformity. Occurrence of Sequence M3 is restricted to

maximum thickness of 2.2 s (TWT) which decreases to the SW to 1.8 s (TWT) in our survey.

Most channel–levee complexes exhibit a typical seismic facies pattern (Kolla & Coumes 1987, 1990; Droz & Bellaiche 1991; Flood et al. 1991; McHargue 1991; Posamentier et al. 1991). A channel is characterized by discontinuous, high-amplitude reflections. The internal pattern is irregular and hummocky, the channel fill may onlap the channel flanks (Fig. 2). The channel packages exhibit a lens-like architecture with a concave lower boundary. Upper concordant boundaries are common. Migration of the channel axis through time is indicated by lateral shifts of the channel fill facies. Channels are up to 7 km wide and reach a thickness of 0.4 s (TWT) (c. 380 m).

The wedge-shaped levee on either side of a channel has a seismic signature of high-amplitude continuous reflections. Semi-transparent sections commonly separate bands of high-amplitude reflections. The reflectors (thought to be isochrons) converge with distance from the channel axis. Hence, the sedimentation rate decreases dramatically from the break-out point in the channel to the distal parts of the levee, leading to the mound-like architecture of channel–levee complexes. Semi-transparent sections may represent times of monotonous hemipelagic(?) sedimentation, predominantly mud. Depositional events of coarse sands create the impedance contrasts necessary to obtain high-amplitude reflections. The levees commonly onlap underlying layers. In the Indus Fan individual channel–levee complexes are up to 60 km wide and up to 1 s (TWT) thick (c. 900 m).

The channel–levee complexes are commonly separated by small bands of readily traceable, continuous, high-amplitude reflections. Flood et al. (1991) and Weimer (1991) interpreted these bands as hemipelagic sediments at the top of each channel–levee sequence deposited throughout and following the switch of sediment supply from the main channel system to another part of the fan. Consequently, these thin layers are condensed sections and exhibit significantly lower sedimentation rates than a channel–levee sequence. The hemipelagic sediments are commonly capped by a following onlapping channel sequence, which fills local depressions and then continues, to form mound-like structures. Erosion by channel activity may cause gaps in the underlying hemipelagic layer, which makes correlation over long distances more difficult.

Sequence M2 is also found on the northern Murray Ridge and below the regional M-unconformity, which truncates Sequence M2 in the area of the Oman Abyssal Plain (Figs. 3 and 4) and in the Dalrymple Trough (Fig. 4). On the northern Murray Ridge single channel–levee complexes are displaced by steep normal faults and tilted as a result of updoming of the ridge (Fig. 4). Because sequence M2 does not show any erosional truncations nor any thickness changes, the faulting, uplift and tilting of the northern Murray Ridge took place before sedimentation of the Sequence M3. Outcrops of Sequence M2 on the Little Murray Ridge and Murray Ridge provide evidence for a hiatus, caused by a tectonic event, which also caused the M-unconformity.

South of the Makran Accretionary Wedge, Sequence M2 below the M-unconformity is up to 1.5 s (TWT) thick (c. 1.6 km; Fig. 3). There, single channel–levee complexes are less thick than in the Indus Fan area and well-stratified hemipelagic sediments dominate. This indicates that this part of the Oman Abyssal Plain was in a distal position relative to the sediment source during deposition of Sequence M2.

Hitherto, Sequence M2 has been variously interpreted either to consist of turbidite layers that derived from the Himalayas (Fruehn et al. 1997; Harms et al. 1984) and correlate with the (Oligocene?) or Early to Mid-Miocene Panjur Formation of onshore Makran (Platt et al. 1985) or to consist of pelagic sediments (Minshull et al. 1992). On the basis of our seismo-stratigraphic correlation we assume that Sequence M2 is largely made up of sediments supplied by the Indus River. However, in contrast to the Recent channel–levee complexes of the Indus Fan, the channel–levee architecture of the Oman Abyssal Plain area represents a distal depositional area.

Sequence M1

The U-unconformity separates the channel–levee Sequence M2 from the more nearly uniform character of Sequence M1. Sedimentary Sequence M1 (Fig. 2) covers the acoustic basement below the Oman Abyssal Plain, the Murray Ridge, and the Indus Fan. The sequence consists of parallel to subparallel continuous reflections. The seismic facies are uniform within this sequence: the amplitudes are low to moderate, and consist of many closely spaced internal reflectors. The lower sequence boundary appears to be concordant with the strong reflection at the top of the basement (Fig. 5). Sequence M1 laps against the flanks of basement ridges.

Elsewhere in basement deeps, Sequence M1 drapes acoustic basement, reaching a maximum thickness of 1.2 s (TWT) (Fig. 6). The concordant upper boundary of the younger Sequence M2 is clearly visible as a result of a significant change of the internal reflection pattern. The depth of the upper boundary ranges from 5 to 6 s (TWT) in the Indus Fan area, but reaches 8 s (TWT) at the toe of

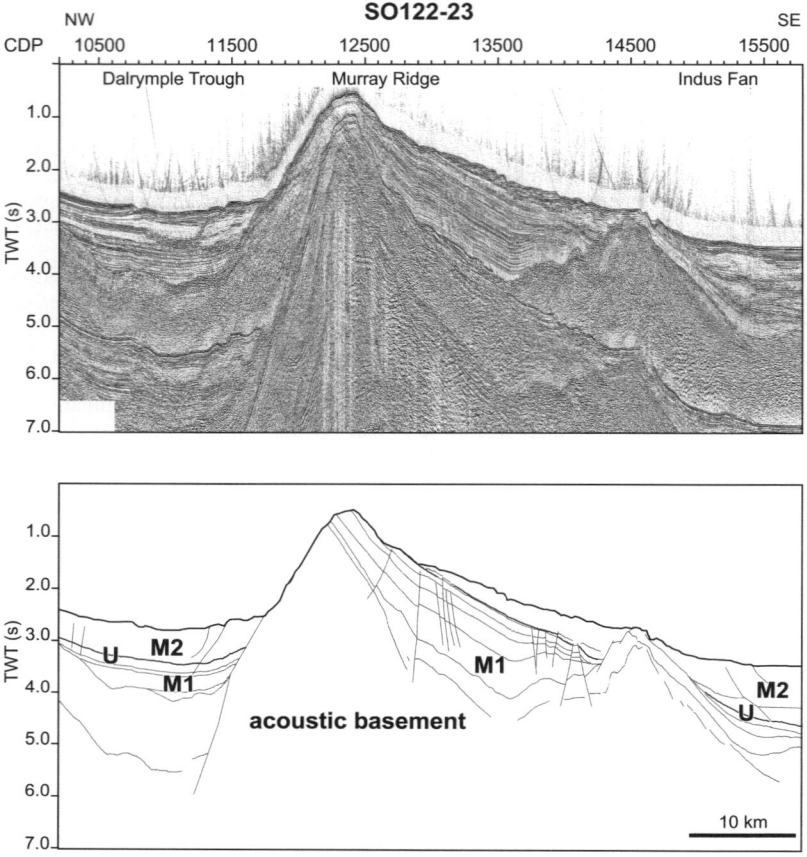

Fig. 5. Part of MCS profile SO122-23 across the Indus Fan and southern Murray Ridge (above). Timing of uplift is evident from tilting of Sequences M1 and M2 jointly with the acoustic basement (below). (See Fig. 1 for location.)

the Makran Accretionary Wedge. The seismic facies and geometry of Sequence M1 are typical for pelagic sediments, deep-sea carbonates or successions of volcanic ash layers intercalated with terrigenous clastic deposits that are transported by turbidity currents. Therefore, we interpret Sequence M1 to be deposited in a hemipelagic basin during Paleogene time following the Late Cretaceous break-up of India, the Seychelles and Africa. It also may represent initial detritus derived from the rapid erosion of the Karakoram, which was uplifted after the Eocene collision of India with Eurasia. Thus Sequence M1 comprises sediment deposited before the onset of the channel–levee complexes.

Sequence M1 conformably covers the SE flanks of the southern Murray Ridge (Fig. 5). The dip of the sequence is consistent with the angle of the ridge slope. Thus tilting and uplift of the southern Murray Ridge took place after deposition of Sequence M1. The sediments are truncated at the highest ridge crest where the whole sedimentary pile is exposed. Along the southeastern margin of the Murray Ridge the acoustic basement forms another small basement ridge, where the sedimentary cover is less than 0.1 s (TWT) thick. Sequence M1 has a 'downlap' contact to this basement ridge along its northwestern slope (Fig. 5). We interpret this downlap contact as caused by tectonic truncation and the contact being faulted, and therefore tilting and faulting took place after deposition of Sequence M1.

Pre-drift Sequence P

Sequence P (Figs. 1, 2 and 6) consists of continuous, subparallel reflections with moderate amplitudes. This sequence is up to 0.8 s (TWT) thick, and overlies an approximately north–south-trending buried acoustic basement ridge under the Indus Fan and extending over about 60 km width (Fig. 1). The north–south orientation of the ridge is supported by profiles published by Kolla &

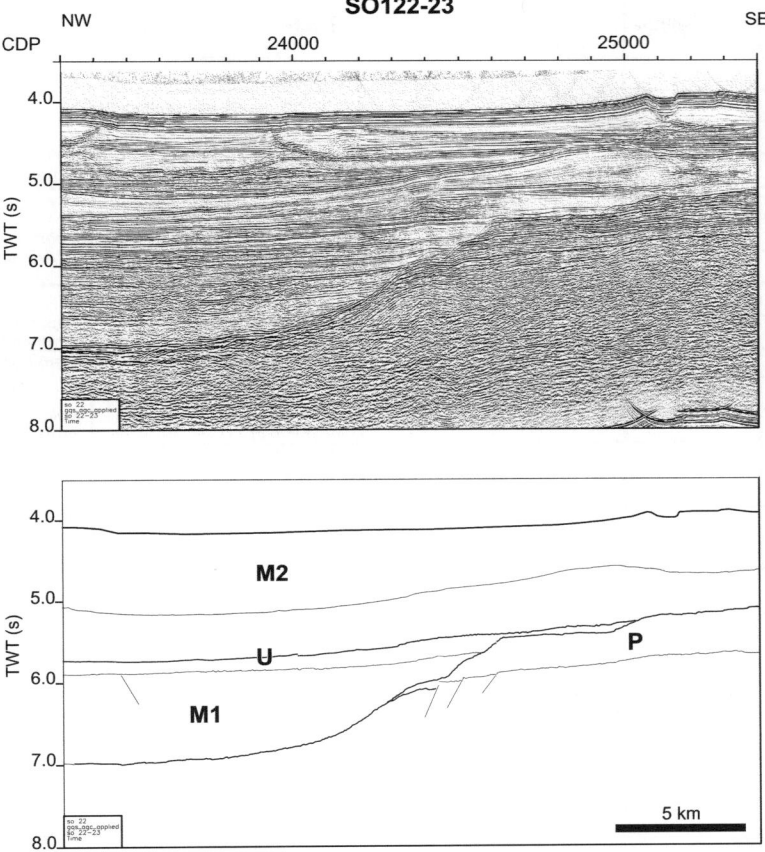

Fig. 6. Part of MCS profile SO122-23 across the Indus Fan (above). Sequences P, M1 and M2 are separated by major unconformities (below). A truncation caused by erosion of Sequence P and the underlying acoustic basement forms the boundary to younger deposits. The transition from hemipelagic sedimentation (Sequence M1) to deposition of channel–levee complexes (Sequence M2) is marked by the unconformity U. (See Fig. 1 for location.)

Coumes (1990) and Droz & Bellaiche (1991). There are no strong reflections at the transition from the sedimentary Sequence P to the underlying basement. The lack of a clear basement reflection might be evidence for the similarity of the acoustic properties of both the sedimentary and basement rocks. Both Sequence P and the basement ridge are truncated and exhibit pronounced slopes. Sequence P is interpreted as older, well-consolidated, stratified sediments, without stronger deformation but with considerable erosion before the deposition of younger sediments.

Because the overlying hemipelagic Sequence M1 onlaps the ridge slopes and because the fan deposits (Sequence M2) cover the older stratified Sequence P, a pre-Tertiary age for the ridge sequence seems reasonable. Therefore, we interpret the basement ridge to be overlain by remnants of well-stratified sediments of presumed Mesozoic age (Schreckenberger et al. 2001; Gaedicke et al. 2002).

Regional stratigraphy and tectonic evolution

The seismo-stratigraphic interpretation and the regional correlation of the sedimentary megasequences (Fig. 7) allow the following outline of a generalized geological history of the study area (Fig. 8).

Cretaceous time

The oldest Sequence P occurs only in restricted areas beneath the Indus Fan. The sequence was probably deposited before the break-up of India

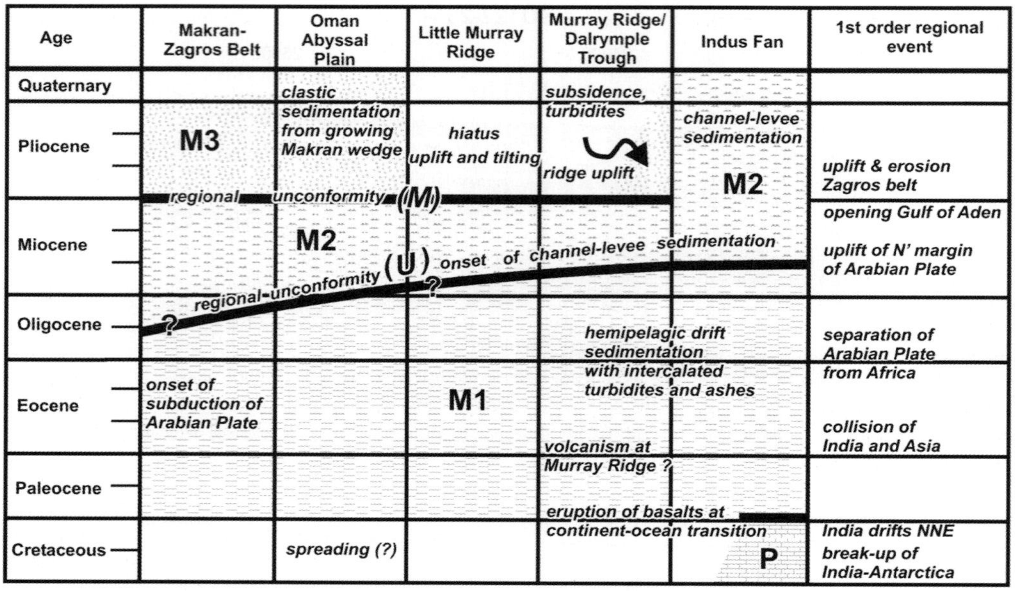

Age	Makran-Zagros Belt	Oman Abyssal Plain	Little Murray Ridge	Murray Ridge/Dalrymple Trough	Indus Fan	1st order regional event
Quaternary	M3	clastic sedimentation from growing Makran wedge	hiatus uplift and tilting	subsidence, turbidites ridge uplift	channel-levee sedimentation M2	uplift & erosion Zagros belt
Pliocene						
	regional unconformity (M)					opening Gulf of Aden
Miocene		M2	(U) onset of channel-levee sedimentation			uplift of N' margin of Arabian Plate
	regional-unconformity		?			
Oligocene	?			hemipelagic drift sedimentation with intercalated turbidites and ashes		separation of Arabian Plate from Africa
Eocene	onset of subduction of Arabian Plate		M1			collision of India and Asia
Paleocene				volcanism at Murray Ridge ?		
				eruption of basalts at continent-ocean transition		India drifts NNE
Cretaceous		spreading (?)			P	break-up of India-Antarctica

Fig. 7. Regional stratigraphy reconstructed from megasequences and correlated with first-order geological events.

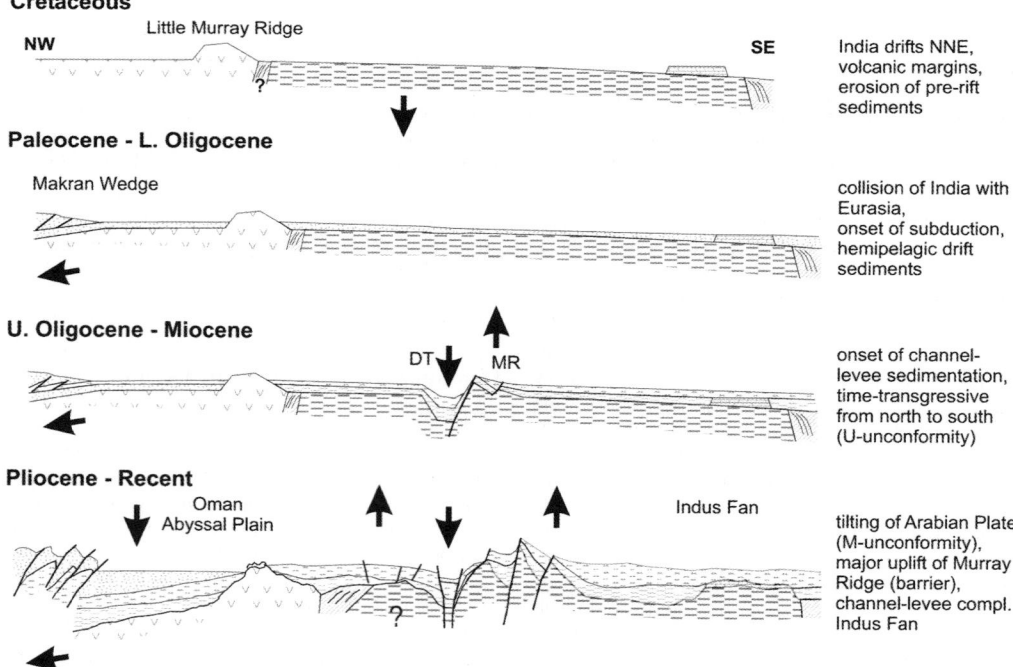

Fig. 8. Simplified tectonic evolution of the northern Arabian Sea area. DT, Dalrymple Trough; MR, Murray Ridge.

from the Seychelles, which started at the Cretaceous–Paleocene boundary and was accompanied by the extrusion of the Deccan Trap igneous rocks (Duncan & Pyle 1988; Vandamme et al. 1991; White & McKenzie 1995). It is believed that the Murray Ridge and the basement below the northern Indus Fan consists of extremely thinned continental crust (Edwards et al. 2000; Kopp et al. 2000; Gaedicke et al. 2002). For Sequence P we assume a Mesozoic age. Subsidence and erosion of Sequence P and presumably also parts of the acoustic basement occurred before the deposition of the younger Sequences M1–M3. The onlap contact of Sequence M1 with the acoustic basement of the Little Murray Ridge and the northern Murray Ridge, respectively (Figs. 3 and 4), provides evidence that these structures already existed during the drift phase.

Paleocene to Early Oligocene time

The drift phase of the Indian Plate is documented by Sequence M1, which, on the basis of seismic character, we interpret as a hemipelagic succession with intercalated clastic material. The sequence onlaps the acoustic basement of the entire study area and smooths pre-existing relief. The collision of India and Eurasia was initiated at 55 Ma (e.g Garzanti & van Haver 1988; Searle et al. 1989), and subduction of the Arabian Plate along the Makran subduction zone started in Eocene time (Fruehn et al. 1997). The uplift of the High Himalayas began in Early Miocene time (20–23 Ma; e.g. Hodges et al. 1992; Searle et al. 1997). As there are no significant thickness variations of Sequence M1 in the Dalrymple Trough we assume that this was not yet established at that time. However, as there are large thickness variations and onlap seismic patterns of Sequence M1 along the northern and southern Murray Ridges, we postulate that the Murray Ridge existed already at the time of Sequence M1, but that the main uplift along the ridge occurred later.

Late Oligocene–Miocene time

Uplift of the Karakoram combined with Late Oligocene to Early Miocene sea-level lowering and possibly higher precipitation (Clift et al. 2001) led to the initiation of unroofing, and deposition of detritus into the northern Arabian Sea. Qayyum et al. (1997) suggested that the fan development is a time-transgressive process starting in the Oman Abyssal Plain, which was closer to the palaeo-Indus River mouth during Late Oligocene time. The earliest fan deposits became accreted to the growing Makran Wedge. Plate convergence led to the subsequent southward migration of the delta until its present position was reached. Following this scenario channel–levee complexes were deposited starting in Mid-Miocene time in the south and on top of the Murray Ridge (Figs. 7 and 8). Our data support this idea because the sequence boundary of Sequences M1 and M2 occurs over the entire area but is better constrained in the Indus Fan area. There the channel–levee complexes are thicker and wider than in the north because the river mouth was closer.

Plate reorganization during Early Miocene time resulted in stresses that formed the transform margin of the Indian and Arabian Plates (Edwards et al. 2000). This caused the uplift of the Murray Ridge and subsidence of the Dalrymple Trough. Nevertheless, the thickness of Sequence M2 on the northern Murray Ridge (Fig. 4) provides evidence that this region was still under the influence of Indus Fan sedimentation during that time. This might indicate that the Murray Ridge was not then high enough to separate the Indus Fan and the Oman Abyssal Plain completely. No statement can be made on whether the uplift of the Murray Ridge was contemporaneous over its entire length, or uplift of only single segments occurred whereas others remained deep and linked of parts of the Dalrymple Trough to the Indus Fan.

Late Miocene–Recent time

The M-unconformity is of regional extent. It occurs over the entire Oman Abyssal Plain and in deep areas of the Dalrymple Trough (White 1983). The unconformity tapers off along the northern flanks of the Little Murray Ridge and the Murray Ridge, respectively. A major unconformity of Late Miocene or Early Pliocene age is also reported in the Strait of Hormuz (Ross et al. 1986). The opening of the Gulf of Aden in Mid- to Late Miocene time is a first-order plate tectonic event, which may have caused this unconformity (Hempton 1987; LePichon & Gaulier 1988). Subsequent reorganization of the Arabian Plate caused the uplift of the Zagros mountain belt, which is documented by the deposition of thick terrigenous successions into the Arabian Gulf (Hempton 1987; Alasharhan & Nairn 1997) since Early Pliocene time. We also correlate the tilting of the oceanic crust of the northern Arabian Sea toward the Makran subduction zone with this Miocene to Pliocene plate reorganization. Tilting of the Arabian Plate was accompanied by buckling of the northern Murray Ridge, which resulted in normal faulting of Sequence M2. As a result of decoupling of the transform margin of the Indian and Arabian Plates, the Indian Plate was not seriously affected by this event, but was uplifted along the southern Murray Ridge, as indicated by tilting of channel–levee complexes (Figs. 4 and 5). The common history of the Oman Abyssal Plain and the Indus Fan ceased at the Miocene–Pliocene boundary, and the northern Arabian Sea became divided by the Murray Ridge into two distinct depocentres, which were fed by different sediment sources (Fig. 9). The northern depocentre encompasses the Oman Abyssal Plain and Makran Accretionary Wedge, and the southern is covered by the Indus Fan south of the Murray Ridge. Today the higher sections of the Murray Ridge are areas of non-deposition.

Conclusion

The interpretations of seismic megasequences presented here show that what are now distinct tectonic and depositional regions of the northern Arabian Sea have a common geological history starting with the Late Mesozoic break-up of India from the Seychelles. Remnants of the early sedimentary cover overlying highly thinned continental crust are preserved on basement ridges in the Indus Fan area. Paleogene hemipelagic sediments onlap the Little Murray Ridge and Murray Ridge and indicate topography in this region since that time. The sediments cover the entire northern Arabian Sea, onlapping the acoustic basement. The onset of sedimentation as a result of erosion of the rapidly uplifting High Himalayas and Karakoram is documented by the U-unconformity. Proximal fan deposits occur in the Oman Abyssal Plain above the U-unconformity. Lower parts of this succession may be attached to the Makran Accretionary Wedge or are found in the terrestrial Indus foreland basins (Qayyum et al. 1997) but also form part of the older Sequence M2. The channel–levee sequences migrated to the south as a result of the plate convergence, which led to consumption of the oceanic crust of the Arabian Plate at the Makran subduction zone. Channel–levee deposits (Sequence M2) have dominated the Indus Fan area at least since Mid-Miocene time. Transtension along the Murray Ridge and subsidence of the Dalrymple Trough started in Early Miocene time with the reorganization of plates (Hempton 1987; Edwards et al. 2000), and is still active, as evidenced by seismicity and faults at the sea floor. The Murray Ridge experienced a significant phase of uplift at that time. Channel–levee sedimentation on the northern Murray Ridge points to pathways for sediments supplied by the Indus River to this northern region.

The opening of the Gulf of Aden and spreading in the Red Sea is the main tectonic driving mechanism for the reorganization of plates. Hempton (1987) distinguished a second phase of opening in Early Pliocene time, which led to enhanced westward extrusion of the Anatolian Plate, already under way because of continued convergence of Arabia and Eurasia. We correlate the tilting of the oceanic crust of the Arabian Plate toward the Makran subduction zone with this second opening phase at the Miocene–Pliocene boundary (c. 4.5 Ma). Tilting was accompanied by

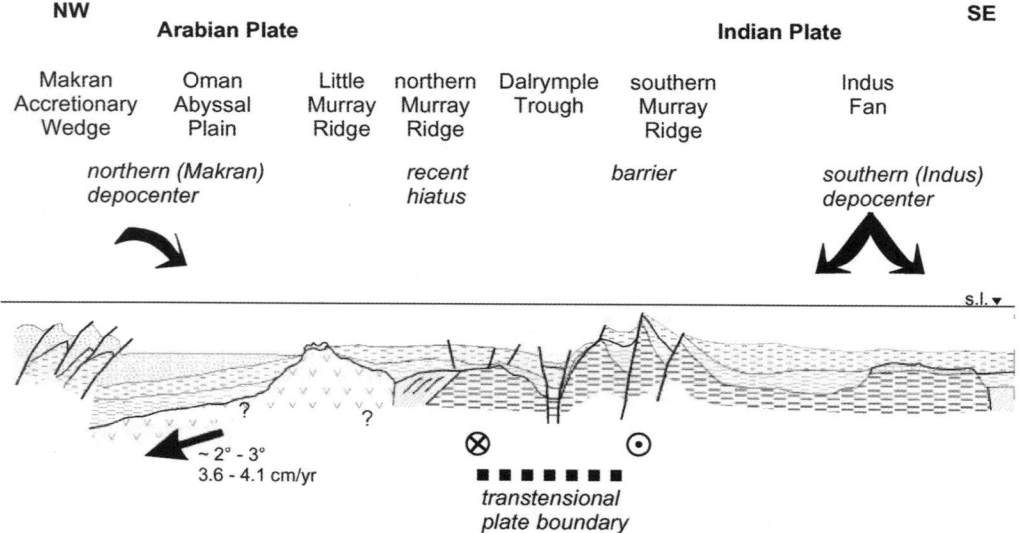

Fig. 9. Composite tectono-stratigraphic cross-section from the Makran Accretionary Wedge to the Indus Fan area. Recently two depocentres have been separated by the Murray Ridge. s.l., sea level.

buckling and renewed uplift of the Murray Ridge to its present topography, which finally led to the complete separation of the Oman Abyssal Plain from the Indus Fan (Fig. 9). Since that time the two depocentres have been divided by the physiographic barrier of the Murray Ridge. Additional Late Miocene to Early Pliocene uplift of the Murray Ridge resulted in a hiatus on the ridge. Whereas in the southern depocentre Himalaya- and Karakoram-derived channel–levee sedimentation continued until recent time, the Oman Abyssal Plain was filled with turbidites (conglomerates, sands and shales) of Pliocene to present age shed from the Zagros–Makran belt and the Oman Mountains. In the northeastern Dalrymple Trough coeval sediments were deposited that derived from the Murray Ridge or are of pelagic (biogenic) origin.

Special thanks go to the ship's master Andresen and the crew of R.V. *Sonne*. We are grateful to the government of the Republic of Pakistan for the permission to investigate in its territorial waters. We acknowledge the assistance of the National Institute of Oceanography, the Hydrocarbon Development Institute of Pakistan, and all scientists involved in this study. We gratefully acknowledge R. Edwards and G. Mountain for their critical review and constructive comments, which improved the manuscript. The research project was funded by the Bundesministerium für Forschung und Bildung, Germany (grant 03 G 122). P.C.'s involvement was supported by JOI–USSAC.

References

ALASHARHAN, A.S. & NAIRN, A.E.M. 1997. *Sedimentary Basins and Petroleum Geology of the Middle East*. Elsevier, Amsterdam.

CLIFT, P.D., LAYNE, G., SHIMIZU, N., GAEDICKE, C., SCHLÜTER, H.-U., CLARK, M. & AMJAD, S. 2000. 55 million years of Tibetan and Karakoram evolution recorded in the Indus Fan. *EOS Transactions, American Geophysical Union*, **81**, 277.

CLIFT, P.D., SHIMIZU, N., LAYNE, G., GAEDICKE, C., SCHLÜTER, H.-U., CLARK, M. & AMJAD, S. 2001. Development of the Indus Fan and its significance for the erosional history of the western Himalaya and Karakoram. *Geological Society of America Bulletin*, **113**, 1039–1051.

DROZ, L. & BELLAICHE, G. 1991. Seismic facies and geologic evolution of the central portion of the Indus Fan. *In:* WEIMER, P. & LINK, M.H. (eds) *Seismic Facies and Sedimentary Processes of Submarine Fans and Turbidite Systems*. Springer, New York, 383–402.

DUNCAN, R.A. & PYLE, D.G. 1988. Rapid eruption of the Deccan flood basalts at the Cretaceous/Tertiary boundary. *Nature*, **333**, 841–843.

EDWARDS, R.A., MINSHULL, T.A. & WHITE, R.S. 2000. Extension across the Indian–Arabian plate boundary: the Murray Ridge. *Geophysical Journal International*, **142**, 461–477.

FLOOD, R.D., MANLEY, P.L., KOWSMANN, R.O., APPI, C.J. & PIRMEZ, C. 1991. Seismic facies and Late Quaternary growth of Amazon submarine fan. *In:* WEIMER, P. & LINK, M.H. (eds) *Seismic Facies and Sedimentary Processes of Submarine Fans and Turbiditic Systems*. Springer, New York, 415–434.

FRUEHN, J., WHITE, R.S. & MINSHULL, T.A. 1997. Internal deformation and compaction of the Makran Accretionary Wedge. *Terra Nova*, **9**, 101–104.

GAEDICKE, C., SCHLÜTER, H.-U., ROESER, H. & 5 OTHERS 2002. Origin of the northern Indus Fan and Murray Ridge, northern Arabian Sea: interpretation from seismic and magnetic imaging. *Tectonophysics*, **355**, 127–143.

GARZANTI, E. & VAN HAVER, T. 1988. The Indus clastics: forearc basin sedimentation in the Ladakh Himalaya (India). *Sedimentary Geology*, **59**, 237–249.

GORDON, R.G. & DEMETS, C. 1989. Present-day motion along the Owen fracture zone and Dalrymple trough in the Arabian Sea. *Journal of Geophysical Research*, **94**(B5), 5560–5570.

HARMS, J.C., CAPPEL, H.N. & FRANCIS, D.C. 1984. The Makran coast of Pakistan: its stratigraphy and hydrocarbon potential. *In:* HAQ, B.U. & MILLIMAN, J.D. (eds) *Marine Geology and Oceanography of Arabian Sea and Coastal Pakistan*. Van Nostrand Reinhold, New York, 3–26.

HEMPTON, M.R. 1987. Constraints on Arabian Plate motion and extensional history of the Red Sea. *Tectonics*, **6**, 687–705.

HODGES, K.V., PARRISH, R.R., HOUSH, T.B., LUX, D.R., BURCHFIELD, B.C., ROYDEN, L.H. & CHEN, Z. 1992. Simultaneous Miocene extension and shortening in the Himalayan Orogen. *Science*, **258**, 1466–1470.

KOLLA, U. & COUMES, F. 1987. Morphology, internal structure, seismic stratigraphy, and sedimentation of Indus Fan. *AAPG Bulletin*, **71**, 650–677.

KOLLA, U. & COUMES, F. 1990. Extension of structural and tectonic trends from the Indian subcontinent into the eastern Arabian Sea. *Marine and Petroleum Geology*, **7**, 188–196.

KOPP, C., FRUEHN, J., FLUEH, E.R., REICHERT, C., KUKOWSKI, N., BIALAS, J. & KLAESCHEN, D. 2000. Structure of the Makran subduction zone from wide-angle and reflection seismic data. *Tectonophysics*, **329**, 171–191.

KUKOWSKI, N., SCHILLHORN, T., FLUEH, E. & HUHN, K. 2000. Newly identified strike-slip plate boundary in the northern Arabian Sea. *Geology*, **28**, 255–258.

LEPICHON, X. & GAULIER, J.-M. 1988. The rotation of Arabia and the Levant fault system. *Tectonophysics*, **153**, 271–294.

MCHARGUE, T.R. 1991. Seismic facies, processes, and evolution of Miocene inner fan channels, Indus submarine fan. *In:* WEIMER, P. & LINK, M.H. (eds) *Seismic Facies and Sedimentary Processes of Submarine Fans and Turbiditic Systems*. Springer, New York, 403–413.

MCHARGUE, T.R. & WEBB, J.E. 1986. Internal geometry, seismic facies, and petroleum potential of canyons and inner fan channels of the Indus submarine fan. *AAPG Bulletin*, **70**, 161–180.

MINSHULL, T.A., WHITE, R.S., BARTON, P.J. & COLL-

IER, J.S. 1992. Deformation at plate boundaries around the Gulf of Oman. *Marine Geology*, **104**, 265–277.

PLATT, J.P., LEGGETT, J.K., YOUNG, J., RAZA, H. & ALAM, S. 1985. Large-scale underplating in the Makran accretionary prism, southwest Pakistan. *Geology*, **13**, 507–511.

POSAMENTIER, H.W., ERSKINE, R.D. & MITCHUM, R.M. 1991. Models of submarine-fan deposition within a sequence-stratigraphic framework. *In:* WEIMER, P. & LINK, M.H. (eds) *Seismic Facies and Sedimentary Processes of Submarine Fans and Turbiditic Systems*. Springer, New York, 127–136.

QAYYUM, M., LAWRENCE, R.D. & NIEM, A.R. 1997. Discovery of the palaeo-Indus delta–fan complex. *Journal of the Geological Society, London*, **154**, 753–756.

QUADRI, V.N. & QUADRI, G.J. 1986. Indus Basin off Pakistan contains few wells. *Oil and Gas Journal*, **16**, 68–72.

RAZA, H.A., AHMED, R. & ALI, S.M. 1990. Pakistan offshore: an attractive frontier. *Pakistan Journal of Hydrocarbon Research*, **2**(2), 1–42.

ROESER, H., ADAM, J., BARGELOH, H. & 17 OTHERS 1997. *The Makran Accretionary Wedge off Pakistan: Tectonic Evolution and Fluid Migration—Part 1*. BGR Report, 116643.

ROSS, D.A., UCHUPI, E. & WHITE, R.S. 1986. The geology of the Persian Gulf–Gulf of Oman Region. *Reviews of Geophysics*, **24**, 537–556.

SCHLÜTER, H.U., PREXL, A., GAEDICKE, C., ROESER, H., REICHERT, C., MEYER, H. & VON DANILES, C. 2002. The Makran accretionary wedge: sediment thicknesses and ages, and the origin of mud volcanoes. *Marine Geology,*, **185**, 219–232.

SCHRECKENBERGER, B., GAEDICKE, C., SCHLÜTER, H.-U. & ROESER, H.A. 2001. Submerged continental crust and volcanism in the Indus Fan and Murray Ridge area: implications from gravity and magnetic modeling. *In: Workshop on The Geologic and Climatic Evolution of the Arabian Sea Region*. Geological Society, London, 9.

SEARLE, M.P., REX, A.J., TIRRUL, R., REX, D.C., BARNICOAT, A.C. & WINDLEY, B. 1989. Metamorphic, magmatic, and tectonic evolution of the central Karakoram in the Biafo–Baltoro–Hushe regions of northern Pakistan. *In:* MALINCONICO, L.L. & LILLIE, R.J. (eds) *Tectonics of the Western Himalayas*. Geological Society of America, Special Papers, **232**, 47–73.

SEARLE, M.P., PARRISH, R.R., HODGES, K.V., HURFORD, A., AYERS, M.W. & WHITEHOUSE, M.J. 1997. Shisha Pangma leucogranite, South Tibetan Himalaya: field relations, geochemistry, age, origin, and emplacement. *Journal of Geology*, **105**, 295–317.

SHUAIB, S.M. 1982. Geology and hydrocarbon potential of offshore Indus Basin, Pakistan. *AAPG Bulletin*, **66**, 940–946.

VANDAMME, D., COURTILLOT, V. & BESSE, J. 1991. Paleomagnetism and age determinations of the Deccan traps (India): results of a Nagpur–Bombay traverse and review of earlier work. *Reviews of Geophysics*, **29**(2), 159–190.

WEIMER, P. 1991. Seismic facies, characteristics, and variations in channel evolution, Mississippi Fan (Plio-Pleistocene), Gulf of Mexico. *In:* WEIMER, P. & LINK, M.H. (eds) *Seismic Facies and Sedimentary Processes of Submarine Fans and Turbiditic Systems*. Springer, New York, 323–348.

WHITE, R.S. 1983. The Makran accretionary prism. *In:* BALLY, A.W. (ed.) *Seismic Expression of Structural Styles*. American Association of Petroleum Geologist, Studies in Geology, **15**(3), 178–182.

WHITE, R.S. & MCKENZIE, D. 1995. Mantle plumes and flood basalts. *Journal of Geophysical Research*, **100**(B9), 17543–17585.

WHITMARSH, R.B., WESER, O.E., ROSS, D.A., *et al.* (eds) 1974. *Initial Reports of the Deep Sea Drilling Project, 23*. US Government Printing Office, Washington, DC.

Tectonic geomorphology of the Gulf of Oman Basin

ELAZAR UCHUPI, S.A. SWIFT & D.A. ROSS

Woods Hole Oceanographic Institution, Woods Hole, MA 02543, USA
(e-mail:euchupi@whoi.edu)

Abstract: The margins of the Gulf of Oman Basin range from convergent at the north to translation at the west and east, and passive at the south. The basin's northern margin has been a site of continuous subduction since Cretaceous time, which has led to the creation of an 800 km long and 650 km wide accretionary wedge, most of which is above sea level. Strata in the centre of the Gulf of Oman Basin display minor deformation resulting from the northward tilting of oceanic crust. A basin-wide unconformity dividing these strata in two was the result of erosion during Early Oligocene time when bottom water circulation was enhanced during a climatic deterioration. The morphology of the basin's south margin is due to Early Triassic rifting, deposition during Jurassic–Early Cretaceous time, early Late Cretaceous ophiolite obduction and Late Cretaceous–Cenozoic deposition. The western side of the accretionary wedge, along the north side of the Gulf of Oman Basin, is in sharp contact with the western translation margin. Structures along this margin are the result of post-Eocene convergence of the Lut and Central Iran microplates. The eastern end of the accretionary wedge, however, is not in contact with the eastern transform margin, but is separated from it by a north-trending trough. The landward extension of this trough is defined by the north-trending Las Bela Valley. The eastern side of the accretionary wedge turns northward at 65°30′N along the west side of the trough and becomes aligned with the north-trending Ornach–Nal Fault along the west side of the Las Bela Valley. Similarly, the Murray Ridge complex turns northward at 25°N and becomes aligned with the north-trending Surjan Fault on the Las Bela Valley's east side. The Ornach–Nal and Surjan faults merge at the apex of the Las Bela Valley with the north-trending Las Bela–Chaman Structural Axis. Differences between the eastern and western sides of the accretionary wedge may be due to the presence of the Ormara microplate on the eastern end of the wedge, a plate that is being pushed ahead of the Arabian plate. The morphology of the Murray Ridge complex is the result of transtension and secondary compression along the Indian–Arabian plate boundary. We infer that most of the relief of the Murray Ridge complex resulted from a change in plate geometry in Early Miocene time. Subsequent tectonic Pliocene–Quaternary events have enhanced this relief.

The Gulf of Oman Basin is a 300 km wide and 950 km long oceanic basin between Oman and southern Pakistan–southern Iran, bordered on the west by Arabia and on the east by the Murray Ridge along the Arabian–Indian plate boundary. West of the Gulf of Oman Basin is a shallow epicontinental sea, the Persian (Arabian) Gulf, which it is connected with it by the Strait of Hormuz (Fig. 1). The margins of the basin range from active convergent on the north to a passive margin on the south and active translation margins on the west and east. In the companion paper (Uchupi *et al.* 2002) we use seismic reflection and 3.5 and 12 kHz echosounding profiles to describe the morphology of the Gulf of Oman Basin. Surface sediment samples and gravity and piston core data also were used in that paper to define the Late Pleistocene–Holocene sedimentary regime of the basin. In the present paper we will use single- and multi-channel seismic reflection profiles (Fig. 2) augmented by a gravity map compiled from satellite altimetry data (Smith & Sandwell 1995) and onshore geological data to reconstruct the geological processes responsible for this morphology. We first describe the structure of the northern active convergent margin of the Gulf of Oman Basin. This will be followed by discussions on the structure of the centre of the basin, the passive southern margin, and the western and eastern active translation margins. We conclude with a brief summary of the geological evolution of the basin.

Morpho-tectonic setting

Convergent northern margin

Pre-Eocene northern accretionary wedge. Morphologically, the accretionary wedge between the Las Bela–Chaman left lateral structural axis in

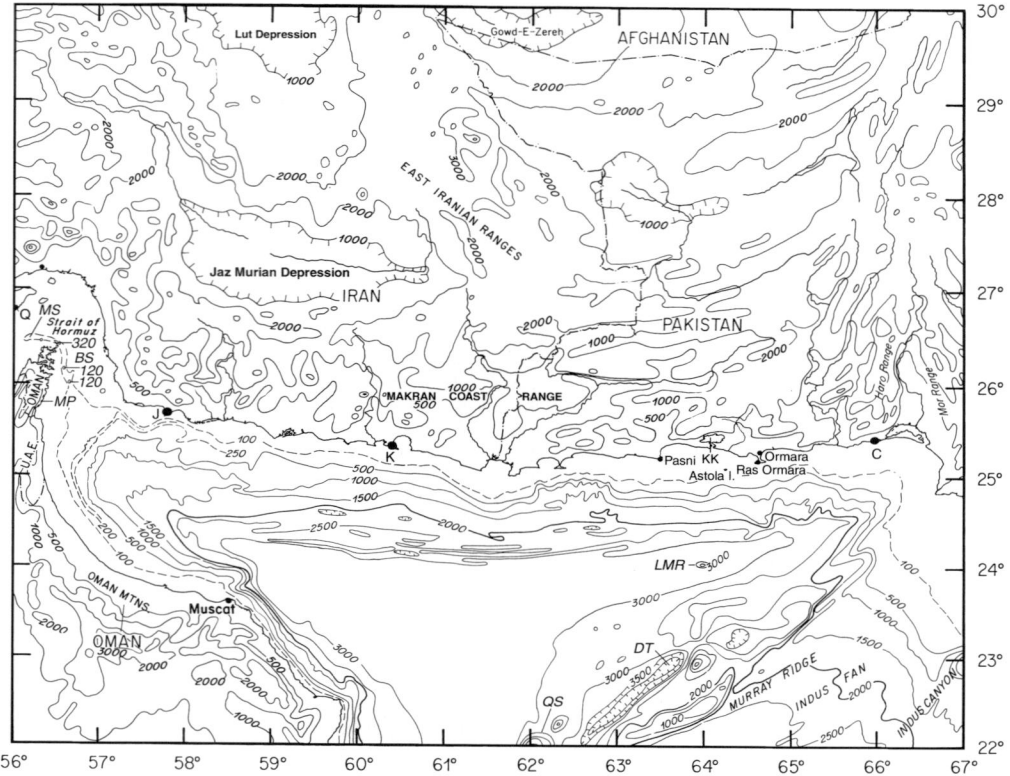

Fig. 1. Topographic map of the Gulf of Oman Basin. The map was compiled by Uchupi *et al.* (2002) from various sources, soundings recorded during R.V. *Atlantis II* Cruise 93, Legs 17 and 18 (1977), seismic reflection profiles described in this paper, General Bathymetric Chart of the Oceans Sheet 5.05, US Naval Oceanographic Office Chart 62028, Seibold & Ulrich (1970), White & Louden (1982), Coumes & Kolla (1984) and Collier & White (1990). BS, Baiban Shelf; C, Chandragup; DT, Dalrymple Trough; J, Jask; K, Konarak; LMR, Little Murray Ridge; MP, Musandam Peninsula; MS, Musandam Valley (Channel); Q, Qeshm Island; QS, Qualhat Seamount; U.A.E., United Arab Emirates. Bottom contours are in metres.

the east and the Zendan Fault–Oman Line right lateral transform in the west can be divided into two massive terranes, a 90–370 km wide Mesozoic–Paleocene northern terrane and a 180–240 km wide Eocene–Quaternary southern one (Figs. 3 and 4). The northern terrane is convex southward with the southern margin of the wedge turning gradually northward east of 62°E as it aligns itself with the Las Bela–Chaman Structural Axis. Interspersed within the wedge are the Afghan, Lut and Central Iranian microcontinents, the Sanandaj–Sirjan–Bajgan–Dur Kan continental sliver and obducted ophiolites (McCall & Kidd 1982; S, Fig. 3). The Afghan and Lut microcontinents are separated by the Sistan Suture (East Iranian Ranges; Fig. 3) and Lut and Central Iran by the north-trending dextral Oman Line (Naiband Fault of Glennie *et al.* 1990). The tectonic fabric was created during Late Cretaceous–Paleocene time by the north-directed subduction of the Neo-Tethys oceanic crust, and the collisions of the microcontinents with each other and against India to the east and Eurasia to the north.

Eocene–Holocene southern accretionary wedge. The terrane south of the pre-Eocene accretionary wedge, together with the Pleistocene andesitic volcanic arc and the Hamun-I-Mashkel and Jaz Muriel depressions to the north are creations of subduction processes since Eocene time (Figs. 3 and 4; Arthurton *et al.* 1982; McCall & Kidd 1982; Platt *et al.* 1985). This wedge, whose southern boundary is defined by the Makran Escarpment in the Gulf of Oman at a depth of 3000–3200 m and the frontal fold south of the scarp, is about 180–240 km wide, of which 70–200 km or 33–40% is below sea level (Figs. 3 and 4). The wedge consists of two segments, the coastal

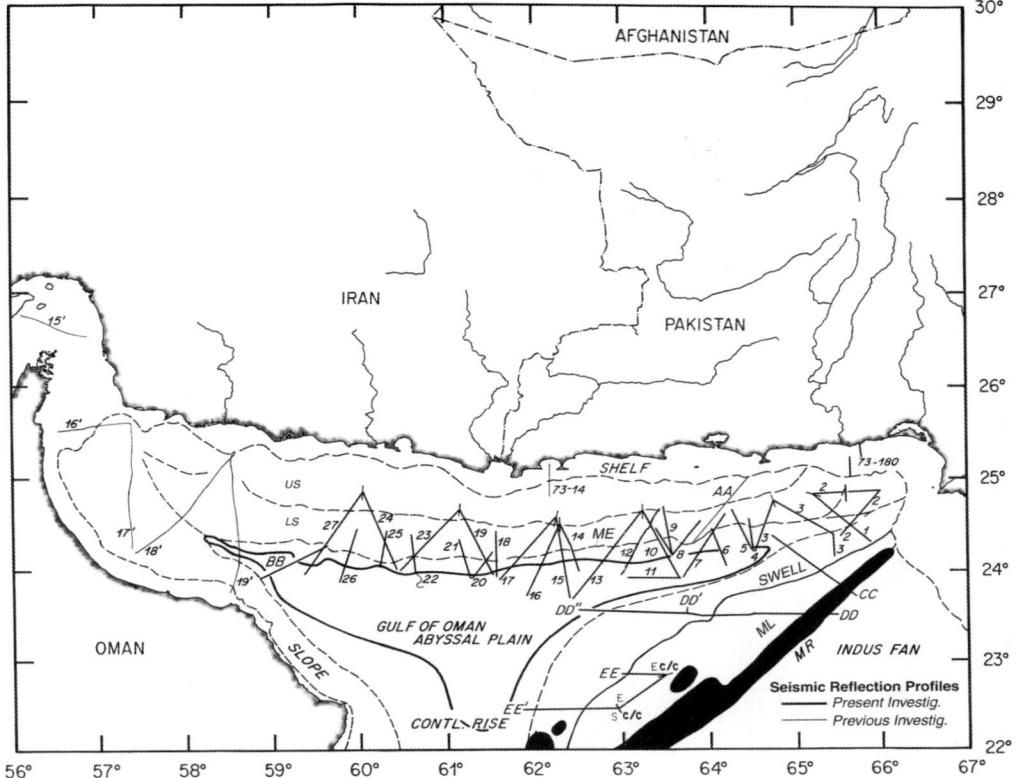

Fig. 2. Locations of seismic reflection profiles recorded during the present and previous investigations.

Makran and the offshore one. Intervening between the two segments is a southerly prograding sediment wedge that extends from south of the Makran Coast Fault to a depth of about 1500 m. This wedge displays limited tectonization that increases toward the west. Earthquake hypocentres and offshore seismic reflection and refraction measurements suggest that oceanic basement dips northward beneath the accretionary wedge at a constant gradient of about 3°, reaching a depth of about 15 km on the shelf's outer edge (Lehner et al. 1983) and 30 km in the Jaz Murian–Hamun-I-Maskhel depressions 300 km north of the coast. Such a subduction angle was recently confirmed by Kopp et al. (2000) using four wide-angle seismic lines and a 160 km multi-channel seismic dip line. In contrast to other convergent margins, the oceanic basement surface in the Makran margin does not display a hinge seaward of the accretionary wedge, along which its declivity increases to form a trench. From the Jaz Murian–Hamun-I-Mashkel depressions oceanic basement plunges northward beneath the Lut and Afghan microcontinents and the Pleistocene volcanic arc at a gradient of about 30° to a depth of 50 km (Fig. 4). From the distribution of the Pleistocene volcanoes, Dykstra & Birnie (1979) inferred that the subducted oceanic basement is divided by four north–south structures having the same hinge zone. They proposed that the two outer segments dip more gently than the inner two, with the major volcanoes being along steeper inner basements segments. More recently, Byrne et al. (1992) suggested that one of the north–south structures dividing the subducting oceanic basement may be in the area of the East Iranian Ranges (Sistan Suture). They also proposed that the Lut microcontinent west of the line is moving northward relative to the Afghan microcontinent east of the break.

Coastal Makran wedge. Earlier than mid-Oligocene time, c. 30 Ma, the 70–200 km coastal Makran segment of the Eocene–Holocene accretionary wedge was a site of distal muddy turbidite deposition containing graded sand beds displaying

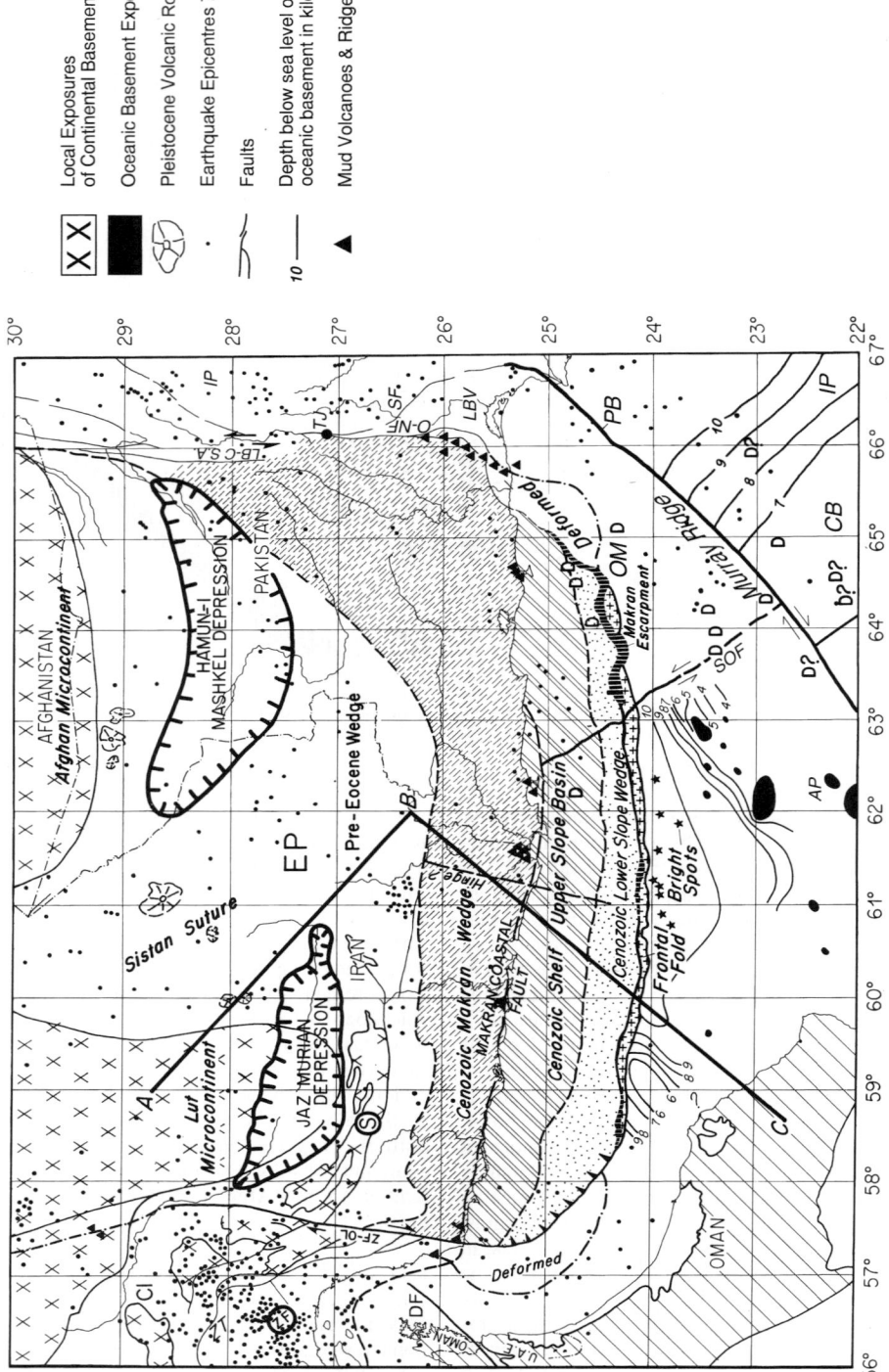

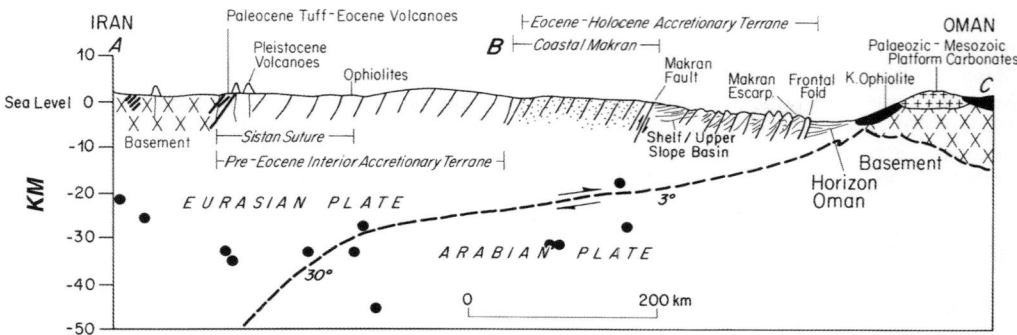

Fig. 4. Schematic geological cross-section of the Gulf of Oman Basin. (See Fig. 3 for location of cross-section and sources used to compiled section.)

features diagnostic of westward transport originating from the India–Eurasia collision zone (Arthurton et al. 1982; Harms et al. 1984). During a second depositional episode from mid-Oligocene(?) to Mid-Miocene time, westward transport of medium sands to coarse conglomerates and shale interbeds by turbidity currents in a slope environment formed a deep-sea fan estimated to be 400 m thick and about 1000 km long (Harms et al. 1984; Critelli et al. 1990; Garzanti et al. 1996). Construction of the Mid-Miocene fan coincided with an increase in fluvial sediment supply from the uplands formed by the collision of India and Eurasia (Harms et al. 1984; Platt et al. 1985). That the Indus drained westward during Mid-Miocene time also is supported by the location of its delta in the area of western Pakistan, in the Bugti area near Sibi at 28°N, 69°E (Kazmi 1984). This westward-directed Indus drainage during early to late Mid-Miocene time led to the construction of a massive fan along the Makran in southern Pakistan and southern Iran. During a third depositional phase, from Late Miocene to mid-Pleistocene time, several generations of slope–shelf and coastal plain sediment tens of kilometres wide and 90–120 km long with lobes as much as 7 km thick prograded southward over the fan sequence from uplifted northerly sources (Harms et al. 1984). The lobes migrated laterally in an east–west direction as the drainage direction shifted in response to tectonism to the north.

Structures in the coastal Makran were produced by north–south progressive deformation since at least Mid-Miocene time, which led to the creation of east–west-trending, southward overturned folds separated by broader synclines (Harms et al. 1984). The intensity of deformation on a regional scale decreases from the oldest (deepest) to the youngest (shallowest) strata, with the younger Neogene sandstones and mudstones near the coast displaying only broad open synclines. Deformation of these younger sediments, however, tends to increase eastward as the folds become aligned with the north–south-trending Las Bela–Chaman Structural Axis. Since mid-Pleistocene time the Makran part of the Eocene–Holocene accretionary wedge has undergone uplift and normal faulting. During this uplift mid-Pleistocene to Holocene shoreline calcareous and terrigenous sediments were deposited by westward littoral drift along the fringes of the progressive folded and raised older deposits (Harms et al. 1984). This uplift is documented by raised tilted and faulted terraces, flat-topped coastal hills, and the 10 m high sea

Fig. 3. Tectonic map of the Gulf of Oman Basin region. The map was compiled from the seismic reflection profiles described in this paper, Geological Survey of Iran (1963), Ahmed (1969), Woodward–Noorany Consulting Engineers Inc. & Woodward–Clyde Consultants, International (1975), Berberian (1976), UNESCO (1976), National Iranian Oil Company (1977), Arthurton et al. (1979, 1982), Asrarullah & Abbas (1979), Jacob & Quittmeyer (1979), Lawrence & Yeats (1979), Powell (1979), Sarwar & Dejong (1979), Coleman (1981), McCall & Kidd (1982), White & Louden (1982), Harms et al. (1984), Leggett & Platt (1984), Quittmeyer & Kafka (1984), Everett et al. (1986), Ross et al. (1986), Farah & Lawrence (1988), Glennie et al. (1990) and Byrne et al. (1992). Locations of earthquakes based on computer data provided by NOAA. AP, Arabian plate; CB, attenuated continental crust; CI, Central Iranian microcontinent; DF, Dibba Fault; EP, Eurasian plate; IP, Indian plate; LB-C S.A., Las Bela–Chaman Structural Axis; LBV, Las Bela Valley; OM, Ormara microplate; O-NF, Ornach–Nal Fault; S, Sanandaj–Sirjan–Bajdan–Dur Kan sliver; SF, Surjan Fault; SOF, Sonne Fault; TJ, triple junction; ZF, Zagros Foldbelt; ZF-OL, Zendan Fault–Oman Line.

cliffs truncating segments of the coastal plain on the Makran on its seaward side (Page *et al.* 1979). Many of these features are aligned along the Makran Coast Fault (Fig. 2; Ullah 1953; Snead 1967, 1970). The number of raised features also increases eastward along the Makran. Only one terrace with an elevation of a few metres occurs at the entrance of the Persian Gulf near 57° whereas there are nine terraces near 60°E. Quaternary uplift east of Makran, in the Indus Delta region, also is rather small, at 2–3 m (Snead 1967)

Parallel to the Ornach–Nal Fault, Las Bela–Chaman Structural Axis, Makran Coast Fault and Zendan Fault–Oman Line are numerous mud volcanoes formed by gas seeps composed of methane, ethane, carbon dioxide and traces of heavier hydrocarbons, and sulphur and thermal springs (Fig. 3; Hunting Survey Corporation, Limited 1960; Geological Survey of Iran 1963; Snead 1964, 1969; National Iranian Oil Company 1977; Harms *et al.* 1984). As these fault structures continue offshore, similar features also occur on the inner shelf and are responsible for the bubbling gas and turbid water offshore. Earthquake activity in 1935 and 1956 may have triggered some of the larger flows associated with these structures. During the 1945 earthquake four mud volcanoes rising 8–30 m above sea level were formed in 7–13 m water depth (Page *et al.* 1979). A recent investigation by Delisle *et al.* (2001) indicated that the volcanoes have existed for several centuries without major changes in their geometry, that their origin is due to major earthquakes or some other subsurface process, which resulted in the outpouring of mud along fracture zones, that the mud reservoir is at a minimum depth of 2–3 km, and that the methane gas associated with the volcanoes is of biological origin.

Offshore accretionary wedge. Shelf–upper slope prograding wedge. Morphologically the continental slope seaward of the Makran can be divided into two sections, an upper segment extending from the shelf's edge at a depth of 100 m to a depth of about 1000–1500 m, and a ridged surfaced lower segment extending from a depth of 1000–1500 to 3200 m. According to Kukowski *et al.* (2001), the upper slope, from the shelf edge to a depth of 700 m, has a gradient of 8° and displays evidence of mass wasting, with gullies and canyons being well developed. In contrast, the lower part of the upper slope from a depth of 700 to about 1000–1500 m is in the form of a smooth terrace. This terrace narrows eastward and disappears at 63°45'E. Huhn *et al.* (1999), Kopp *et al.* (2000) and Kukowski *et al.* (2001) stated that the seaward edge of the terrace agrees with the seaward edge of an underplated block adjacent to the ridged lower slope. The shelf and the upper slope are the surfaces of a 60–90 km wide clastic prism prograding southward over the ridged lower segment. This prograding prism extends southward from just south of the Makran Coast Fault to a depth of 1500 m and laterally from 61°E to 65°E (Fig. 3). Well data and facies patterns imaged by seismic sections indicate that the Late Miocene–Plio-Pleistocene sediments making up this wedge consist of bathyal mudstones with minor fine-grained turbidites, slope mudstone facies, and inner and outer shelf facies consisting of cyclic sequences of sandstones and siltstone interbedded with mudstone (Harms *et al.* 1984). The Holocene surface sediments above this sequence are mainly muds. The source of this sediment prism was to the north with its eastward thickness increase to more than 4 km reflecting the greater uplift and erosion of the coastal Makran in that direction (Fig. 5a). Reversal in the dip of the basin's lower strata along Profile 73-14 suggests that the wedge's lower part may be partially deformed (Fig. 6). West of 59°E this deformation is more intense, extending from the slope's lower segment to the Makran Coastal Fault (Profiles 16' and 17', 18' and 19', Figs. 6 and 7).

The ridged lower slope segment. The lower continental slope is in form of ridges and troughs with a regional gradient of 1.9° (Kukowski *et al.* 2001). The four to six ridges making up the lower slope range in length from <15 to >5 km, are generally 2–6 km wide, and are asymmetrical in cross-section, with the seaward slope being much steeper, displaying gradients of 8° to >20° (Kukowski *et al.* 2001). The most prominent of the ridges is the frontal one, rising over 1 km above the continental rise. We have named the southern side of this outer ridge the Makran Escarpment. As described by von Rad *et al.* (2000), the ridges are made up of imbricate thrust slices separated by perched basins filled with post-tectonic turbidites and hemipelagic sediments (Prins *et al.* 2000) displaying a maximum thickness of 0.5–1 km (Fig. 5a). Relief of the ridges decreases landward as the fill in the perched basin increases. Seaward of the Makran Escarpment is a frontal fold. The ridges and intervening lows, the Makran Escarpment along the front of the accretionary wedge and the frontal fold south of the scarp form a broad arc convex southward with the highs and lows trending east–west from 59°E to 64°E, NW west of 59°E, and NE east of 64°E. This arc is in the form of steps rising landward, which represent the uniformly spaced thrust packages that, according to Kopp *et al.* (2000), make up the wedge. Near 59°E, 63°E and 64°E the east–west trend of the arc is disrupted by the NE-

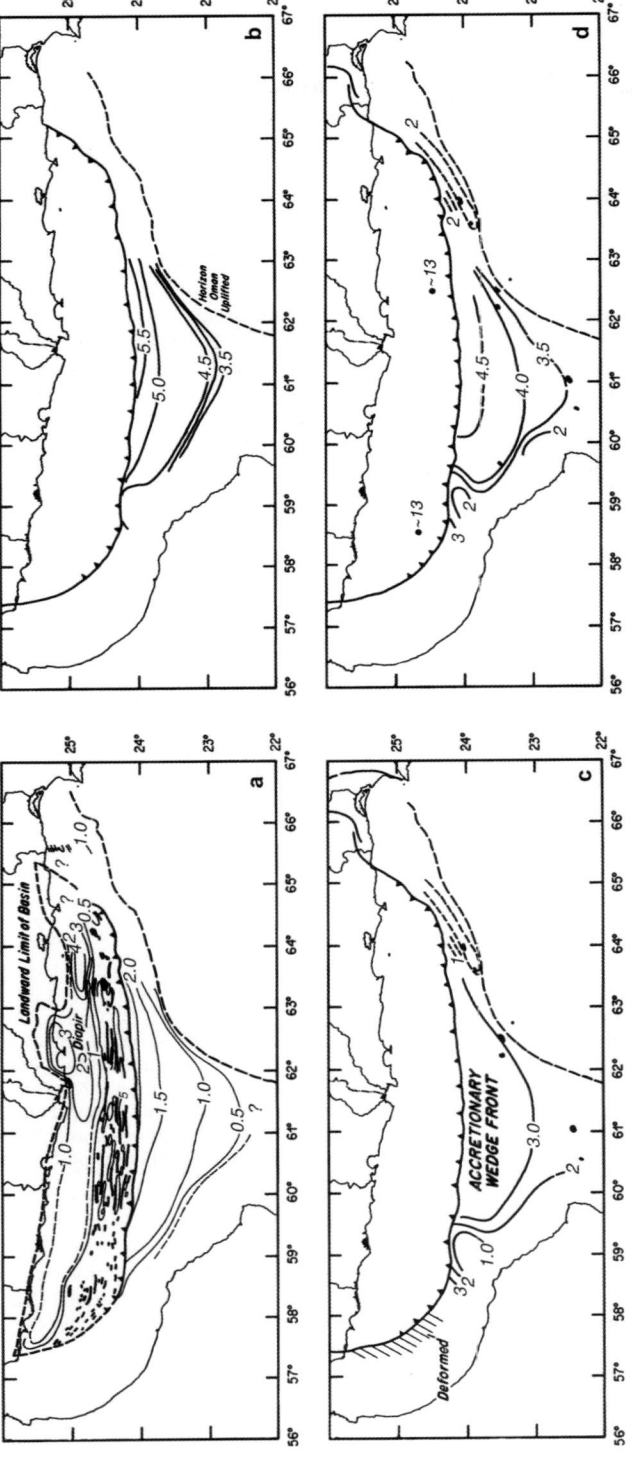

Fig. 5. (a) Isopach of the Plio-Quaternary sediments in the shelf–upper slope basin and the intra-ridge troughs on the lower slope. (b) Depth below sea level of the Oligocene Horizon Oman. (c) Isopach map of sediments below Horizon Oman. (d) Isopach map of sediment above the décollement–oceanic basement. Maps compiled from data described in this report supplemented with information from Bungenstock et al. (1966), White & Klitgord (1976), Niazi et al. (1980) and White & Louden (1982). Travel times were converted to kilometres assuming velocities of 1.5 km s^{-1} for water, 2.0 km s^{-1} for the sediments above Horizon Oman and 3.5 km s^{-1} for the layers beneath the horizon.

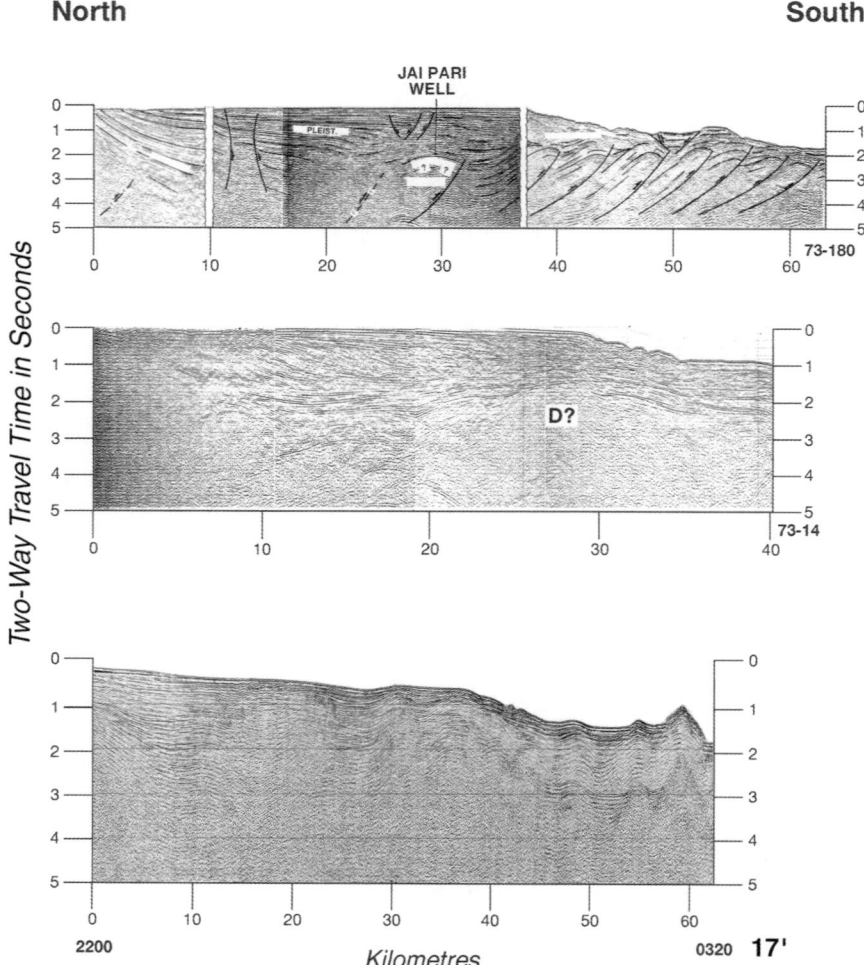

Fig. 6. Multi-channel seismic reflection profiles of the shelf–upper slope basin. (See Fig. 2 for locations of profiles.) Profile 73-180 (sea-floor vertical exaggeration 2×) is from Harms *et al.* (1983) and Profile 73-14 (sea-floor vertical exaggeration 7×) is from Harms *et al.* (1984). Profile 73-180 (Harms, J.C., Cappel, H.N., Francis, D.C. & Shackleford, T.J., ©1983) is reprinted by permission of the American Association of Petroleum Geologists, courtesy of Marathon Oil Company, and Profile 73-14 by permission of Van Nostrand Reinhold Company. Profile 17′ (sea-floor vertical exaggeration 7×) was recorded during R.V. *Atlantis II* Cruise 93, Leg 18, in 1977. Only the inner part of this profile is shown in this figure. The complete profile is shown in Fig. 7. It should be noted on 73-180 that strata on the shelf–upper slope basin rest on a tectonized unit whose deformation increases in a seaward direction. Along 17′ the folded fabric of the lower slope accretionary wedge extends across the length of the profile. On the shelf the ridge–trough texture of the lower slope wedge is buried by younger strata. D?, Diapir?

trending highs with a low gravity signature (Xs, Fig. 8). Another feature disrupting the accretionary wedge is the left lateral fault sinuous Sonne Fault near 63°E extending from the Gulf of Oman Abyssal Plain to the coastal Makran (Fig. 3). In the abyssal plain the fault offsets the Little Murray Ridge, with the offset of the ridges in the accretionary wedge increasing from <1 km on the seaward edge of the accretionary wedge to 10 km on the most landward high (Kukowski *et al.* 2000). This fault has been active during the last 2 Ma.

The easternmost high disrupting the accretionary wedge is the Little Murray Ridge, located at 64°E (Fig. 1). This offset by the Sonne Fault is largely buried by the rise and abyssal plain

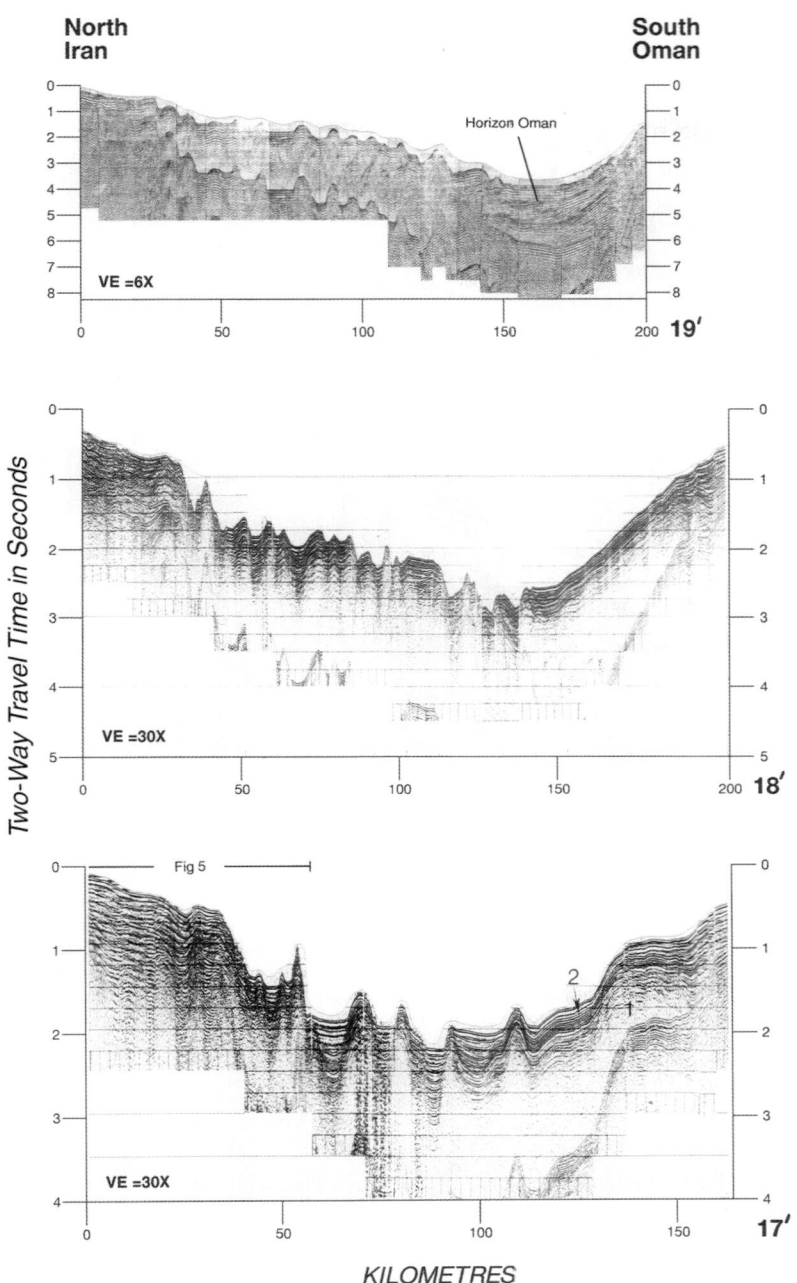

Fig. 7. Single-channel seismic reflection profiles of the western end of the slope accretionary wedge. Along 19' (seafloor vertical exaggeration 6×) the wedge is separated from the Oman margin by the Gulf of Oman Abyssal Plain, whereas on 18' and 17' (sea-floor vertical exaggerations 30×) the fold belt extends across the Gulf of Oman from Makran to Oman. As in Profile 17' the ridges on the uppermost continental slope on 19' and 18' are buried and do not have any topographic expression. Surface fault scarps on 17' and 18' are traces of deep-seated thrust faults. On line 17 unit 1 is of Late Cretaceous to mid-Miocene age and unit 2 is of post mid-Miocene age. All these lines were recorded during R.V. *Atlantis II* Cruise 93, Leg 18. Line 17' was published previously by White & Ross (1979) and Ross *et al.* (1986), and 18' by White & Ross (1979). Lines 17' and 18' are published with permission of the American Geophysical Union. (See Fig. 2 for locations of profiles.)

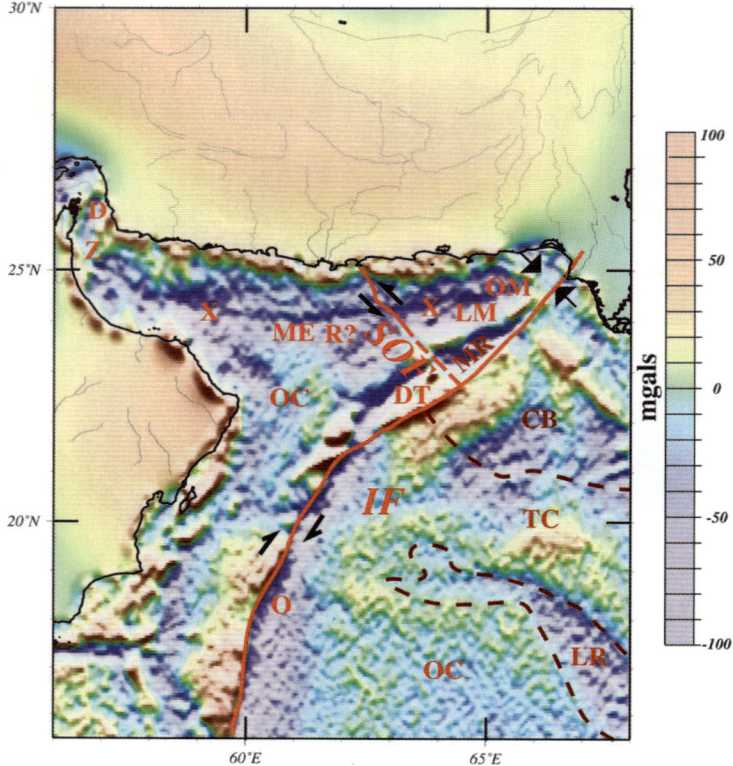

Fig. 8. Gravity of the Gulf of Oman from satellite altimetry data. From Smith & Sandwell (1995). Illumination from the north. CB, attenuated continental crust; D, Dibba Fault; DT, Dalrymple Trough; IF, Indus Fan; LM, Little Murray Ridge; LR, Laxmi Ridge; MR, Murray Ridge; ME, Makran Escarpment; O, Owen Ridge; OM, Ormara microplate; R?, Oceanic basement ridge?; SOF, Sonne Fault; TC, transitional crust; Z, Zendan Fault. It should be noted that R? with a gravity low signature trends northeastward north of about 22°N, parallel to the Murray Ridge. South of 22°N the gravity low changes trend sharply to east of north, parallel to the Owen Ridge and aligned along the seaward edge of the Oman margin. We infer that this is the former position of the plate boundary before Miocene time, when the boundary jumped to its present position along the Owen and Murray ridges. The displacement of the Makran Escarpment where Little Murray Ridge and oceanic basement ridges intersect the scarp (X) should be noted. Arrows indicate links between the Makran Escarpment and the Ornach–Nal Fault via the Sarpai Anticline (left arrow) along the west side of Las Bela Valley and the Murray Ridge and Surjan Fault (right arrow) along the east side of the valley. Red line indicates position of the Arabian–Indian plate boundary. Crustal types beneath the Indus Fan are from Malod *et al.* (1997) and Miles *et al.* (1998).

sediments south of the Makran Escarpment. Part of the Little Murray Ridge's crest protrudes through the sediments to form a line of seamounts, most of which have relief of about 500 m. Hutchinson *et al.* (1981) proposed that the ridge was created in Late Cretaceous time as a line of volcanic seamounts. White (1983a) reported that the ridge's NE subsurface trace in the accretionary wedge is marked by a large magnetic anomaly and that segments of the buried ridge are capped by sediment interpreted as lithified carbonate bank or reef complex. The gravity map computed from satellite altimetry data shows that the signature associated with the ridge extends across the slope accretionary wedge intersecting the coast east of 65°E (Fig. 7). As Little Murray Ridge is inserted into the wedge the sediments on the accretionary wedge become chaotic. White (1982) ascribed this deformation to uplift and imbricate thrusting of the wedge sediments as the shallow ridge was inserted into the wedge. We infer that the deformation results in the NE orientation of the ridge relative to the north direction of convergence. This difference in trend would lead to a sediment build-up on the SE side of the ridge, creating compressional highs parallel to the ridge.

Seismic reflection profiles of the offshore accretionary wedge. The single- and multi-channel seismic reflection data recorded during the present investigation and from the archives (these profiles are published for the first time in this paper) at Woods Hole Oceanographic Institution (Figs. 9–12; Profile BB, Fig. 13; Profile CD, Fig. 17 (see below)) and from previous investigations (White & Klitgord 1976; White 1979, 1982, 1983b, 1984; White & Ross 1979; White & Louden 1982;

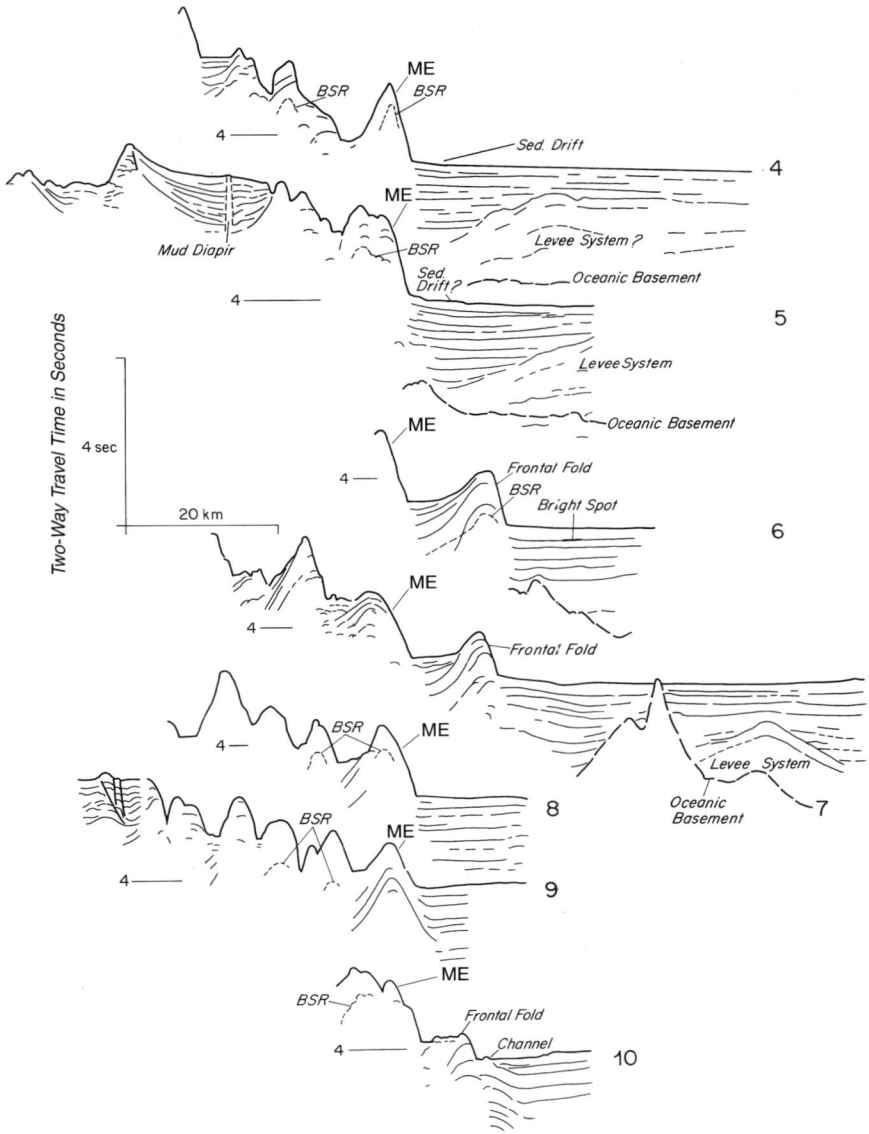

Fig. 9. Line interpretations of single-channel seismic reflection Profiles 4–10 of the Makran margin recorded during the present investigation, showing buried levee system, segments of Little Murray Ridge (oceanic basement highs on 6 and 7), frontal fold, mud diapir (5) and bottom current and turbidity current structures on the continental rise (4, 5 and 10). (For locations of profiles, see Fig. 2.) Vertical exaggeration of sea floor about 10×. BSR, bottom simulating reflector; ME, Makran Escarpment.

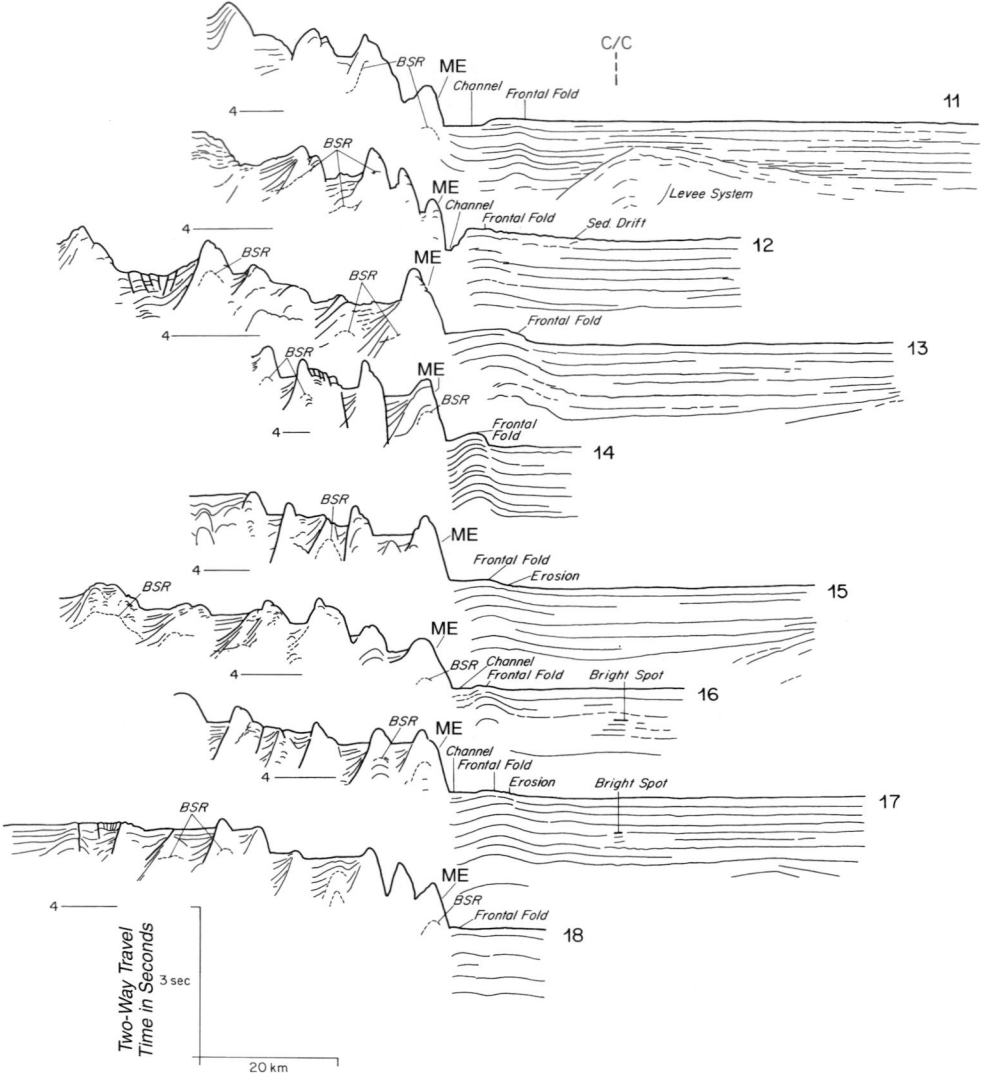

Fig. 10. Line interpretations of seismic reflection Profiles 11–18 of Makran margin recorded during the present investigation. Points of interest are 'fanning' reflectors associated with the slope accretionary wedge, frontal folds, BSR (bottom simulating reflector) on the wedge, bright spots and bottom current and turbidity current (channels on Profiles 11 and 12) features on continental rise. (See Fig. 2 for locations of profiles.) Vertical exaggeration of sea floor about 10×. ME, Makran Escarpment.

Harms et al. 1983, 1984; Lehner et al. 1983; Minshull et al. 1992) and the seismic refraction studies of White & Louden (1982) and Niazi et al. (1980) imaged the tectonic style of the accretionary wedge and the frontal fold south of the Makran Escarpment. As imaged by the multichannel seismic Profile AA off Pakistan (Figs. 2 and 13) the accretionary wedge is separated from the oceanic basement by a décollement slip surface. According to Kopp et al. (2000), this surface developed on a bright surface within the Himalayan Turbidites (see below), steps down to the surface creating considerable underplating. Kopp et al. inferred that more than 3 km of this

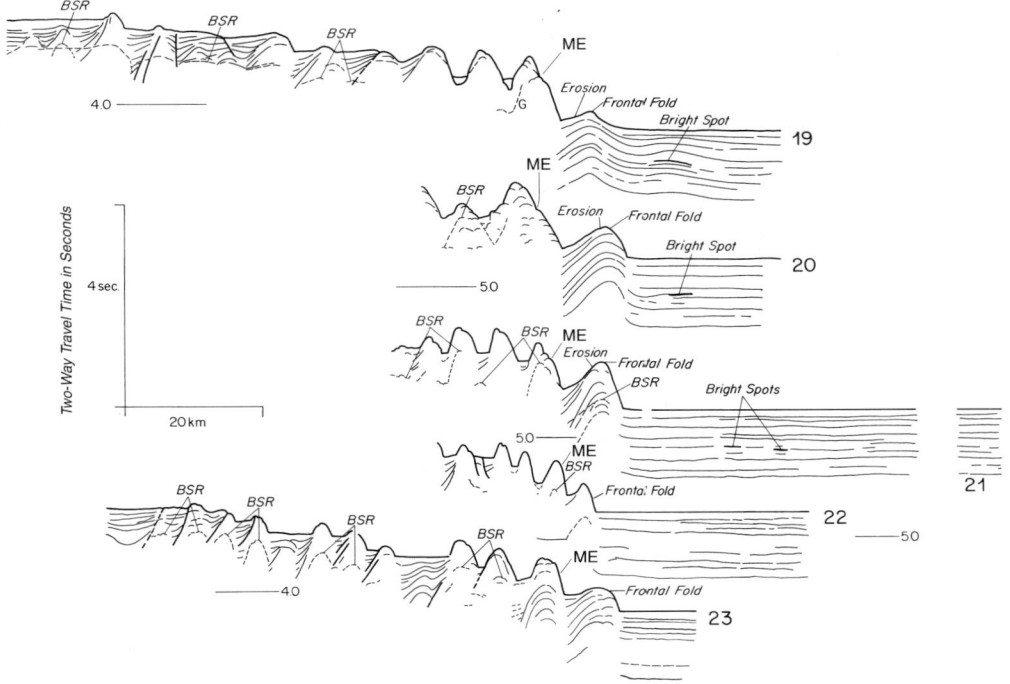

Fig. 11. Line interpretations of single-channel seismic reflection Profiles 19–23 of the Makran margin recorded during the present investigation. Points of interest are BSR and 'fanning' reflectors on slope, frontal folds and bright spots on rise. (See Fig. 2 for locations of profiles.) Vertical exaggeration of sea floor about 10×. ME, Makran Escarpment

underplated material bypasses the front of the accretionary wedge and is transported to greater depth. Lehner *et al.* (1983) stated that the tectonic style of the Makran accretionary wedge along this profile is one of a sequence of overthrust and thrust folds resembling the Canadian foothills. Anticlines become narrower and diapiric, and synclines widen and deepen in a landward direction. The synclinal basins separating the folds are 3–15 km wide and contain about 0.5–1 km of ponded sediment. The sediments in these ponds display 'fanning' of dips, indicating that the sediments are being progressively folded or tilted (Figs. 9–12; White & Louden 1982; Minshull *et al.* 1992).

Deformation of the strata in the area of the frontal fold in the upper rise south of the Makran Escarpment affects only the upper 3 km; the deeper strata pass undisturbed beneath the fold. As pointed out by White (1977a, 1982), this downward decrease in deformation is evidence that this deformation is not the result of mud diapirism. Whereas the folds within the continental accretionary wedge lack internal structure, the frontal fold south of the Makran Escarpment displays stratification (Figs. 10–12). The frontal fold, which defines the seaward edge of the offshore accretionary wedge, is not present in all the crossings of the Makran Escarpment (e.g. Profiles 4, 5, 8 and 9?, Fig. 9; Profiles 25 and 26, Fig. 12; Profile 3, Fig. 19 (see below)). Where the structure is present its amplitude varies along-strike; it is <100 m high on Profiles 16–18 (Fig. 10) and 1000 m on Profile 8 (Fig. 9). Along some profiles the frontal fold is asymmetric in cross-section, with its steeper limb facing seaward (Profiles 6 and 7, Fig. 8; Profiles 19 and 21, Fig. 10) and the low between the fold and Makran Escarpment being partially filled with sediment. White (1977a, 1982) estimated that a new frontal fold was created every million years and reached a relief of 400 m in only 0.4 Ma. From Parasound profiles, Kukowski *et al.* (2001) inferred that the 60–70 m high and 11–23 km wide frontal fold is disrupted by numerous thrusts spaced several hundred metres apart. Motion along these structures supposedly created a wave-shaped morphology with amplitudes of up to about 30 m. Such an inter-

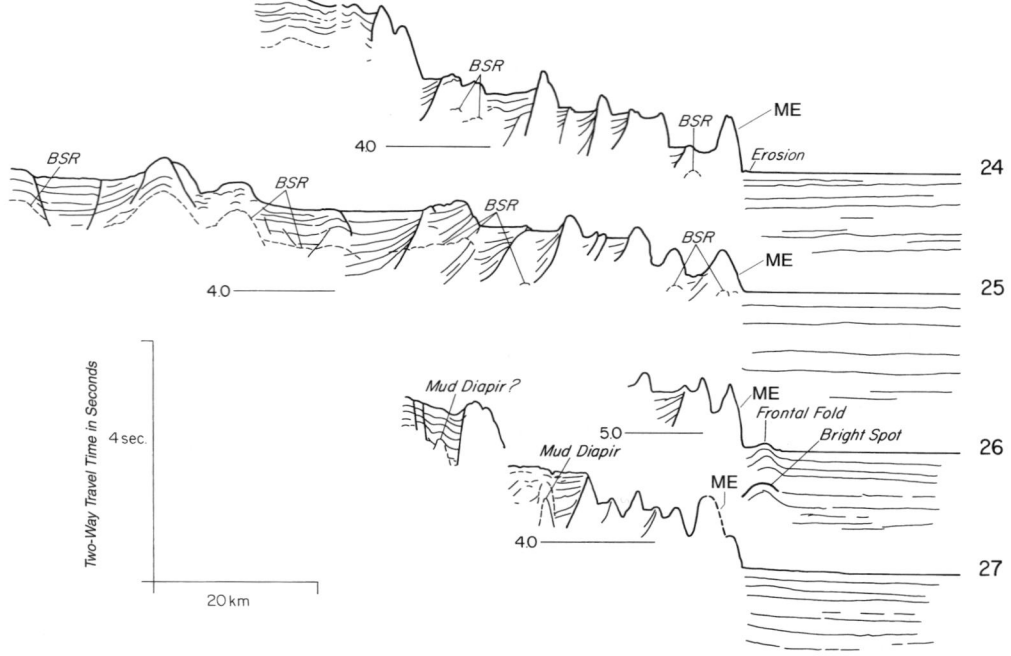

Fig. 12. Line interpretations of single-channel seismic reflection Profiles 24–27 of the Makran margin recorded during the present investigation. Points of interest are 'fanning' reflectors and BSR on the slope, frontal folds and bright spots on the rise. (See Fig. 2 for locations of profiles.) Vertical exaggeration of sea floor about 10×. ME, Makran Escarpment.

pretation appears reasonable in the presence of the tectonic activity in the region. However, in the absence of similar deformation in the subsurface layers of the frontal fold in the profiles recorded during the present investigation and the multi-channel seismic reflection profile in Grevemeyer et al. (2000), evidence of erosion along the crest of the frontal fold (Profiles 5 and 10, Fig. 9; Profiles 12, 13 and 15–18, Fig. 10; Grevemeyer et al. 2000), and the occurrence of similar structures on the tectonically inactive Oman margin and along the west side of the Murray Ridge complex makes us question such an origin. It is true that the seismic reflection profiles lack the resolution of the Parasound profiles, but the features attributed by Kukowski et al. (2001) to thrust faulting are so perverse on the frontal fold that they should be imaged in interval reflectors of the seismic profiles and they are not. Morphological similarity of the Parasound features to known current structures in the Atlantic Ocean, as seen in 3.5 kHz traces from the Atlantic Ocean (see Emery & Uchupi (1984) and Jacobi & Hayes (1992) and the numerous references quoted in those studies), leads us to infer that the Gulf of Oman Basin features are the result of either bottom or turbidity currents, not faulting (see Uchupi et al. 2002).

Some of the seismic profiles of the frontal fold and upper continental rise south of the fold recorded during the present investigation display regions of enhanced reflectivity (Figs. 3 and 9–12). These 'bright spots', which occur within the upper 2 km of the rise sediments and the frontal fold, represent local gas accumulations. A detailed seismic investigation of one these spots by White (1977b) indicated that the gas reservoir had a constant thickness of 35 m and that its gas saturation decreased to zero at its margins. South of the frontal fold also are 65 m high mud volcanoes 1–1.5 km in diameter (von Rad et al. 2000; Wiedicke et al. 2000; Kukowski et al. 2001). Sediments in the vicinity of the volcanoes are characterized by 100–300 m wide vertical transparent zones. Wiedicke et al. (2000) ascribed these zones and the mud volcanoes to the expulsion of mud and methane in front of the accretionary wedge.

Many of the seismic reflection profiles of the slope accretionary wedge recorded during the present investigation also show a bottom simulat-

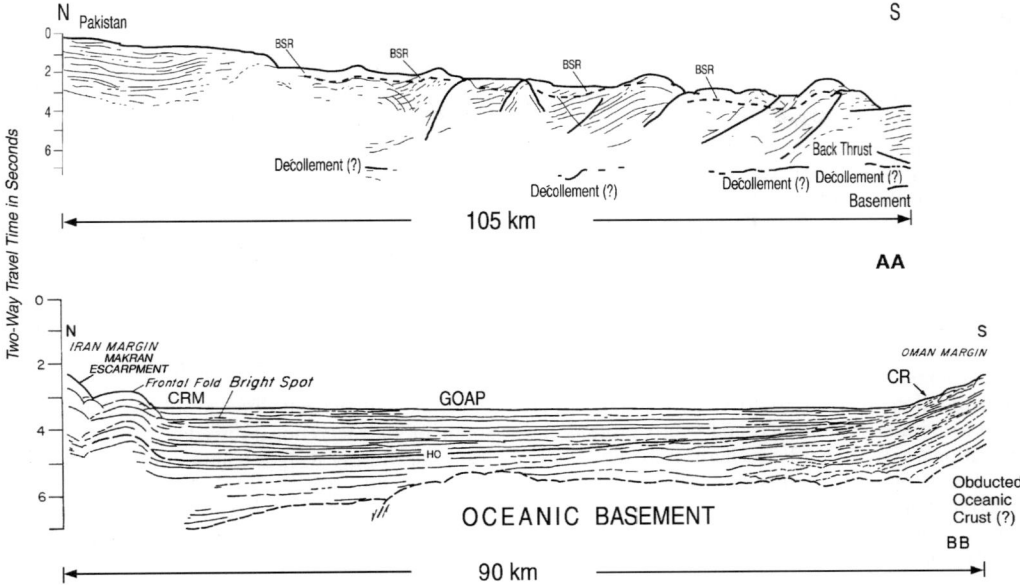

Fig. 13. Line interpretation of multi-channel seismic reflection profiles of the Makran margin (AA) and seaward edge of Makran margin, Gulf of Oman Abyssal Plain and Oman margin (BB). Vertical exaggeration of sea floor on AA recorded during the present investigation (redrawn from Lehner et al. (1983)) is about 3× and on BB (from the archives of the Woods Hole Oceanographic Institution) 6×. (For locations of profiles, see Fig. 2.) BSR, bottom simulating reflector; CR, continental rise off Oman; CRM, continental rise off Makran; GOAP, Gulf of Oman Abyssal Plain; HO, Horizon Oman.

ing reflector (BSR) marking the base of a frozen gas hydrate zone. This reflector occurs at a depth of about 500–800 m below the sea floor (White 1979; Minshull et al. 1992) and can be traced from the Makran Escarpment at a water depth of about 3000 m northward to a depth of about 1500 m. The reflector tends to be distorted in the region of the thrust faults, a result of enhanced advective heat flow being concentrated in the faults (Minshull et al. 1992). Grevemeyer et al. (2000) reported that the BSR characteristics vary across the accretionary wedge, variations they assumed are due to tectonic uplift and continuing sedimentation. According to von Rad et al. (2000), regional destabilization of the gas hydrate leading to focused flow occurs where active faults and deeply incised canyons reach the base of the gas hydrate. At one such location they discovered seeps of methane and H_2S-rich fluids associated with a chemoautrophic vent fauna and authigenic carbonates formed by the oxidation of the vented methane (von Rad et al. 2000).

Other features displayed by the surface of the rise strata are the result of external processes. Included among them are the box-shaped channels eroded by turbidity currents (Profiles 10–12, Figs. 10 and 11). Such features also occur within the sediments themselves. For example, incorporated within the rise strata NW of the Murray Ridge along Profile 3 (Fig. 19; see below) and Profiles 4, 5 and 7 (Fig. 9) is a southwestward draining subsurface levee network, probably part of the Indus Fan channel–levee complex described by Lafont-Petassou & Orszag-Sperber (1991). The flanks of the frontal fold and the continental rise immediately south of the frontal fold also display minor topographic irregularities, which, as stated above, are inferred to be the result of bottom current activity.

Age and composition of accretionary wedge. The age and lithology of the sediments making up the offshore accretionary wedge are not known. Extrapolation from well data on the shelf suggests that the strata are terrigenous, probably a mixture of distal turbidites and pelagic deposits of Cenozoic age. Fruehn et al. (1997) stated that the accretionary wedge was constructed of two sequences, a lower Himalayan Turbidite Series and an upper Makran Sands. They inferred that the lower sequence was derived from the Himalayas and is part of the east–west-trending proto-Indus Fan along the Makran. This would imply that these strata are of Early Miocene age, but as described below they may be older. The upper unit, Makran Sands, of Late Miocene age and younger, was derived from the north. As noted by Fruehn et al. (1997), these strata are incorporated into the accretionary wedge in a family of thrust sheets whose steep dips indicate a high degree of consolidation. In contrast to other wedges, which tend to be undercompacted and overpressured, the accretionary wedge off southern Pakistan–southern Iran is normally compacted. This is due to the Makran Sands being compacted before their incorporation into the wedge, a condition enhanced by its texture as sands tend to lose most of their fluids during normal compaction before accretion. Only the Himalayan Turbidite Series landward of the deformation front appear to be overpressured, as a result of the presence of mud layers in the sequence (Fruehn et al. 1997). This difference between the two units controls the location of the décollement. This surface is along the top of the turbidites, a position controlled by the low friction of the turbidites because of the fluids trapped in mud layers within the turbidites. More recently, Huhn et al. (1999), Kopp et al. (2000) and Kukowski et al. (2001) placed the décollement along a weak layer within the turbidite series.

Construction of the Makran and offshore wedge. Deformation of the coastal Makran was the result of folding along the front of the wedge and imbricate thrusting within the wedge in Miocene time. This shortened the wedge by about 1–2%. An additional 25–30% of shortening took place in Pliocene time. Later underplating and underthrusting raised the region about 500 m. Platt et al. (1985) attributed shortening in Pliocene time to a slowing down of subduction or stopping of the underthrusting by increasing friction along the décollement surface. White & Louden (1982) inferred that the 1–2% shortening of the offshore wedge is due to folding along its front and imbricate thrusting within the wedge. The mass balance between the sediment entering the subduction zone and the amount of sediment in the lower slope wedge indicates that deformation and thickening of the wedge occur throughout the prism (White & Louden 1982). Platt et al. (1985), however, estimated that the amount of shortening required to achieve mass balance in this manner is too large and proposed instead that mass balance is reached in large part by underplating. More recently, Huhn et al. (1999), Kopp et al. (2000) and Kukowski et al. (2001) have proposed that the underplating has played a major role in the formation of the offshore wedge. Kopp et al. (2000) inferred that the wedge consists of thrust packages scraped off the Arabian plate along a décollement within the Himalayan Turbidite Series. This surface is located at a depth of 8.5 km depth at the front of the wedge, stepping down northward to permit considerable underplating.

Centre of the Gulf of Oman Basin

From the seismic reflection profiles, oceanic crust south of the accretionary wedge can be mapped from the Murray Ridge to about 58°E (Profile 19', Fig. 8; Profile BB, Fig. 13). Farther west, in the vicinity of the Strait of Hormuz, oceanic basement has been consumed and the southern edge of the accretionary wedge is in contact with the Oman margin (Profiles 17' and 18', Fig. 7; White & Ross 1979). Oceanic basement in the central part of the Gulf of Oman Basin slowly plunges northward beneath the lower slope accretionary wedge from a depth of 7 km between the Oman margin and the Murray Ridge to a depth of about 10 km in the vicinity of the Makran Escarpment and 14 km immediately north of the scarp (Fig. 4). Rising above the surface of this northerly plunging oceanic basement surface are at least two NE-trending highs, one near 59°E and Little Murray Ridge near 64°E. The crest of the high near 59°, which has no topographic expression, is at a depth of less than 6 km below sea level, with its crest rising about 3 km above the general surface of oceanic basement. Limited seismic reflection coverage suggests that this basement high trends northeastward, plunging beneath the slope's accretionary wedge and possibly disrupting the Makran Escarpment (right X, Fig. 8). If it extends across the width of the lower slope wedge it intersects the coast near 61°E. A third oceanic basement high intersecting the Makran Escarpment may be documented by a poorly defined NE-trending gravity high west of Little Murray Ridge (R.?, Fig. 8). This high intersects the Makran Escarpment west of 63°E.

The >4.5 km to <3 km thick sediments resting on oceanic basement beneath the lower rise and the Gulf of Oman Abyssal Plain consist of two sequences separated by a prominent reflector (Profile BB, Fig. 13). We named the reflector 'Horizon Oman', whereas White & Klitgord (1976) referred to it as Horizon A. As this letter designation is used for a prominent horizon in the western Atlantic, the name Horizon Oman is more appropriate for the unconformity. This horizon ranges in depth from about 3 km below sea level at the southern end of the basin to 5.5 km south of the Makran Escarpment (Fig. 5b). In the vicinity of Little Murray Ridge, strata below the Horizon Oman form lobes, which tend to thicken away from where they onlap the ridge, and sediments above it form a scarp dipping northward where they overstep the high (White 1983a). These stratigraphic relationships were clearly illustrated by White & Klitgord (1976). Away from the Little Murray Ridge, sediments above the reflector dip gently southward, onlapping the Horizon Oman and the strata on the Oman continental rise. Strata below Horizon Oman rest on oceanic basement and dip northward. Our results and those of Fruehn et al. (1997) suggest that the lower unit may be up to 4 km thick (Fig. 5c). Whereas Fruehn et al. (1997) inferred that the Makran Sands above the unconformity are up to 3 km thick, our results suggest that the unit is >2 km thick near the Makran Escarpment (Fig. 5a). Fruehn et al. (1997) inferred that the lower unit, which they named the Himalayan Turbidite Series, originated in the Himalayan collisional zone and was transported southward along the Chamal–Ornach Nal lineation to the Makran and then westward to form the distal parts of proto-Indus fan along the Makran (see also Critelli et al. 1990; Garzanti et al. 1996). Fruehn et al. (1997) also proposed that the upper unit, which they named the Makran Sands, originated from the coastal Makran Mountains to the north.

Horizon Oman, separating the Himalayan Turbidite Series and Makran Sands, dips northward and can be traced from the western edge of the Murray Ridge to 58°30′E. West of this point the Gulf of Oman Abyssal Plain has been subducted beneath the Makran accretionary wedge and the wedge is in contact with the Oman passive margin. The age of Horizon Oman cannot be directly determined because the surface cannot be traced to Deep Sea Drilling Project (DSDP) boreholes to the south (Whitmarsh et al. 1974; Prell et al. 1989) or to boreholes on the Pakistani shelf (e.g. Harms et al. 1984). White (1983a) noted that Horizon Oman decreased in gradient away from Murray Ridge, and suggested that it was formed as a result of the uplift of the ridge. Minshull et al. (1992) ascribed the horizon to the uplift of the ridge, and suggested that the reflector marks the boundary between pelagic sediments below and turbidites above. Our Profile CC (see Murray Ridge section below) also demonstrates that a horizon, which we tentatively identified as Horizon Oman, is involved in the uplift of the Murray Ridge, but the extent of the horizon over much of the Gulf of Oman Basin makes its formation as a result of the uplift of the Murray Ridge unrealistic. Possibly the reflector is the creation of the acoustic impedence between the Makran Sands above and the finer-grained Himalayan Turbidite Series below. Another solution is that the horizon is an unconformity. Such an origin is supported by the onlapping of the Makran Sands on the horizon. If the horizon is an unconformity we propose that this erosion is due to enhanced bottom water circulation. If the Murray Ridge was uplifted subsequent to the erosion of Horizon Oman, then the unconformity must be present within the sediments forming the northwestern part of the Murray Ridge. However, we are able to detect it in only one of our profiles, CC (see Murray Ridge section below). The horizon may be present but unrecognized in other profiles or it may not be present. If it is not present, then at the time the horizon was eroded by bottom currents the Murray Ridge had some relief and the sediments were above the zone of bottom current circulation. Such a supposition is verified by the multi-channel seismic reflection profiles of Clift et al. (2001) that indicate that the Murray Ridge was uplifted in Early Miocene time and that two highs east of the ridge were positive elements before the uplift.

Although lack of adequate seismic coverage prevents us from tracing Horizon Oman outside the Gulf of Oman, we have tentatively correlated the horizon with a widespread unconformity in the Arabian Sea and western Indian Ocean eroded in Early Oligocene time (35 Ma) (Kidd et al. 1992). If this correlation is correct, the Himalayan Turbidite Series must be of Oligocene age and older, and the Makran Sands of post Early Oligocene age. The early Oligocene erosional event that may have formed Horizon Oman has been reported in the Somali, Madagascar and Mozambique basins, on the Chagos–Laccadive, Mozambique and Madagascar ridges, Broken Ridge, on the Kerguelen Plateau and the Australian margin (Fig. 14). Intensification of bottom current erosion during Early Oligocene time was the result of the glaciation in Antarctica and the climatic deterioration that began near the end of Eocene time or in Early Oligocene time (Kidd & Davies 1978; Kidd et al. 1992). It is this glaciation that triggered the production of excessive amounts of north-flowing cold bottom water and the establishment of vigorous circulation, which in turn led to the formation of extensive areas of erosion and non-deposition. Although sedimentation again resumed in most places in late Oligocene time once the circumpolar circulation was established (Kennett et al. 1974), bottom water entering the Gulf of Oman continues to influence sedimentation along the periphery of the Gulf of Oman Abyssal Plain.

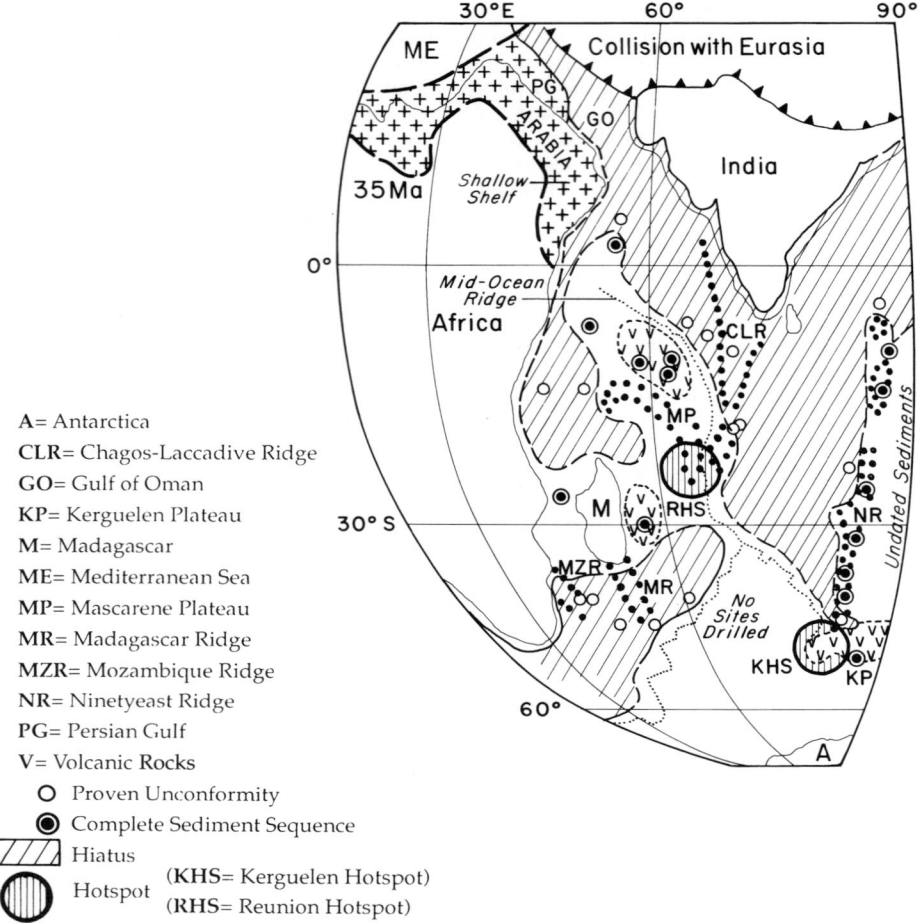

Fig. 14. Paleogeography of the western Indian Ocean, Arabian Sea, Gulf of Oman and Persian Gulf during Early Oligocene time, 35 Ma. Modified from Kidd *et al.* (1992), using geological–geophysical data compiled during the present investigation.

Southern passive rifted margin of the Gulf of Oman Basin

Except for Profile BB (Fig. 13), which covers the rise, no seismic reflection data were recorded on the southern margin of the Gulf of Oman Basin during the present investigation. Previous studies demonstrate that the tectonic events that led to the structural configuration of the Gulf of Oman Basin can be divided into four phases: rifting, sea-floor spreading, convergence and a divergence. The rifting phase probably began in mid- to Late Permian time (Koop & Stoneley 1982; Shirley 1991). White (1984) has proposed that the blocky topography of the Oman continental slope formed during this rifting phase by listric faulting. Rifting gave away to sea-floor spreading in Late Permian or Early Triassic time, and by Mid-Jurassic time an extensive southern Tethys developed in the region of the Gulf of Oman–Persian Gulf. At that time, deep- and shallow-water carbonates were deposited on the graben fill and overtopped the fault blocks Searle *et al.* 1983; Michaelis & Pauken 1990). In Aptian–Albian time, sea-floor spreading along the Oman margin was replaced by compression and the obduction of a melange of pelagic sediments, volcanic rocks and amphibolite blocks (the Hawasina Formation) and the Semail Ophiolite complex of Cenomanian–Turonian age (95 Ma) over the Mesozoic carbonates (Searle *et al.* 1983; Michaelis & Pauken 1990).

Western translation margin of the Gulf of Oman Basin

The Mesozoic–Cenozoic accretionary wedge is truncated on the western side by the right lateral fault, the Zendan Fault–Oman Line (Figs. 3 and 15). West of the transform are the Zagros Mountains of southern Iran, the Musandam Peninsula and Oman Mountains, and the Arabian Platform and the Dibba Fault. The Dibba Fault separates the ophiolites in the Oman Mountains from the Mesozoic carbonates in the Musandam Peninsula. The Zagros Mountains were created of the collision of the Arabian and Eurasian plates along the Zagros Crush Zone toward the end of Cenozoic time (White & Ross 1979; Koop & Stoneley 1982). The Musandam Peninsula and Oman Mountains are due to Mesozoic platform carbonate accretion, obduction of oceanic crust and ophiolites in Cretaceous time, carbonate deposition in Late Cretaceous–Cenozoic time, and recent neotectonism (Tilton *et al.* 1981; Searle *et al.* 1983).

Baiban Shelf

Sediments on the shelf (Baiban Shelf) south of the Strait of Hormuz, between the Zendan Fault–Oman Line right-lateral transform and the Dibba Fault, are argillaceous and detrital (Fig. 15). Ricateau & Riché (1980) speculated that the section below Horizon H_1 (Figs. 16 and 17), a Miocene unconformity east of the Dibba Fault, rests on Hawasina nappes or oceanic basement. That the region SE of the Dibba Fault may be underlain by oceanic crust is suggested by the investigation of Khattab (1993), who identified magnetic reversals M11 and M10N in the region. Others, however, claim (White & Ross 1979) that the region between the Zendan Fault–Oman Line and the Dibba Fault (Figs. 3 and 17) is one of continental collision and that oceanic crust is restricted to the area SE of the Zendan Fault–Oman Line.

The strata between Horizon H_1 (Figs. 16 and 17) and the Hawasina nappes or oceanic basement have been deformed into a series of north–south-trending ridges parallel to the Oman coast. Ricateau & Riché (1980) suggested that these structures may have been produced by mud diapirism either from the Hawasina Group of Permian to Senonian age or from Upper Cretaceous shales (Profile CD, Figs. 16 and 17). Between the north–south-trending ridges are Paleocene to Lower Miocene turbidites that prograde southward. Above H_1 are rhythmically alternating silts and shales or sandy shales and conglomerates of Late Miocene and younger age thickening southward toward the shelf's edge (Ricateau & Riché 1980). Strata beneath H_1 terminate eastward at the Zendan Fault–Oman Line transform, whereas the sediments above it extend across the transform, becoming folded as they approach the Makran coast (Profile 16', Fig. 16; Profile CD, Fig. 17). These folds, which dominate the lower continental slope east of the Strait of Hormuz, curve northward as they approach the north–south-trending Zendan Fault–Oman Line transform (Fig. 17).

Strait of Hormuz

Multi-channel seismic reflection profiles recorded in the Strait of Hormuz north of the Baiban Shelf (Profile 15', Fig. 16; Profiles 321 and 399, Fig.

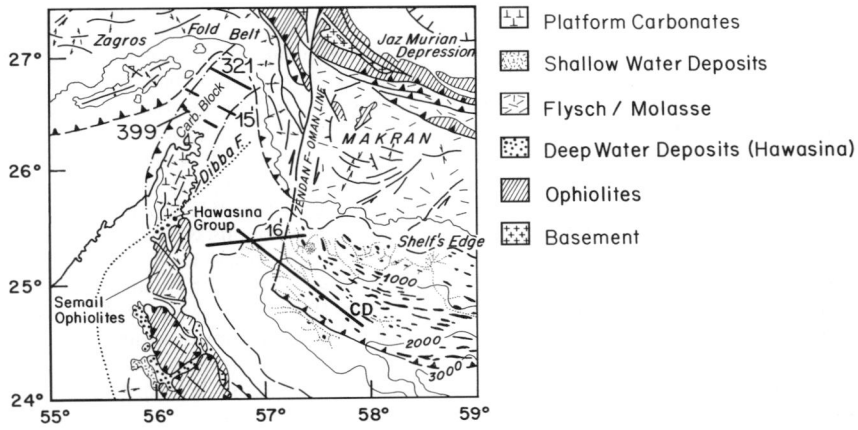

Fig. 15. Tectonic map of the Strait of Hormuz region adopted from Coleman (1981), showing locations of seismic reflection profiles displayed in Fig. 17.

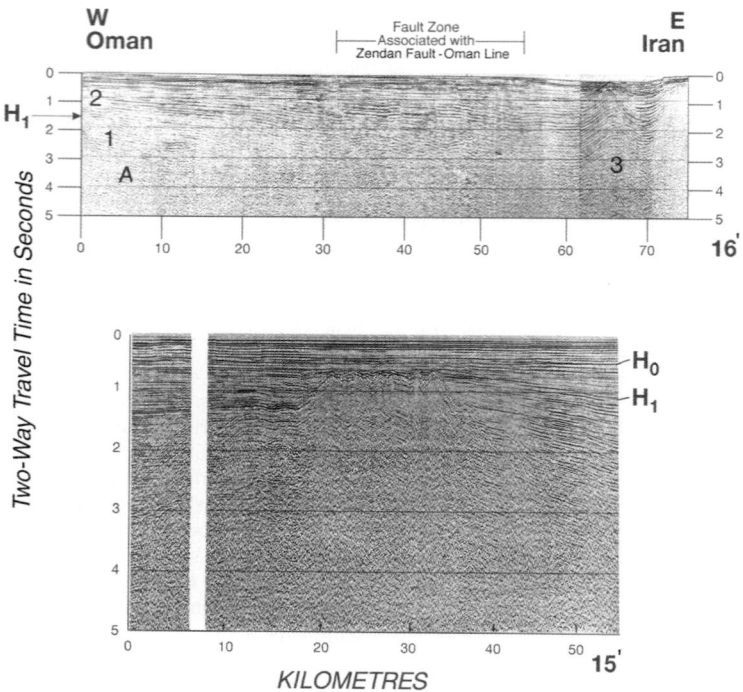

Fig. 16. Seismic reflection profiles of the Zendan Fault–Oman Line transform in the Strait of Hormuz (16′; sea-floor exaggeration 5×) and across the northern extension of the Musandam Peninsula high (15′; sea-floor exaggeration 9×). Horizons H_0 and H_1 represent Plio-Pleistocene and late Miocene unconformities (Ricateau & Riché 1980). Reflector A in 16′ represents the eastern edge of the Musandam peninsula high, unit 1 consists of diapiric ridges composed of Cretaceous muds and Paleocene–Miocene sediments between the ridges, and unit 2 is composed of undisturbed late Miocene and younger sediments. Units 1 and 2 are separated by Horizon H_1. Unit 1 terminates on the Zendan Fault–Oman Line transform, but the younger sediments extend across the fault, becoming more deformed eastward in the direction of the Makran. Both of these profiles were published originally by White & Ross (1979), and later by Ross *et al.* (1986). Published with permission of the American Geophysical Union. (See Fig. 2 for locations of profiles.)

17) display a NE-trending prominent block bounded on the SE side by the Dibba Fault (Fig. 15). The block separates two distinct sedimentary environments. Sediments on the Arabian Platform NW of the block are predominantly shallow-water carbonates whereas those SE of the block are neritic siliceous deposits. White & Ross (1979) argued that the block, a seaward extension of the Musandam Peninsula, extends across the Strait of Hormuz, plunging beneath the Zagros Mountains of Iran. They also suggested that this plunging block is the locus for the seismicity in southeastern Iran, events indicative of thrusting with a north–south slip vector (Jackson & McKenzie 1984). The surface of the block in the Strait of Hormuz, at a depth of 0.5–1.0 s below the sea floor (Horizon H_1, Profile 15′, Fig. 16), has a hummocky appearance, which has been interpreted as being of erosional origin. Ricateau & Riché (1980) have concluded that this erosion took place in Miocene time. Onlapping the erosional surface are Neogene sediments disrupted near the top by an erosional surface of Plio-Pleistocene age (Horizon H_0; Profile 15′, Fig. 15). According to Lees (1928), the Musandam Peninsula, and possibly its NE extension in the Strait of Hormuz, has experienced 450–900 m of subsidence since Pliocene time and more than 60 m in the last 10 ka (Vita-Finzi 1973). This subsidence would account for the drowning of Horizon H_0.

In summary, seismic reflection profiles suggest that the block in the Strait of Hormuz is composed of nappes of Musandam Limestone of the Arabian carbonate platform that were thrust northwestward (White & Ross 1979; Ricateau & Riché 1980; Michaelis & Pauken 1990). The age of this

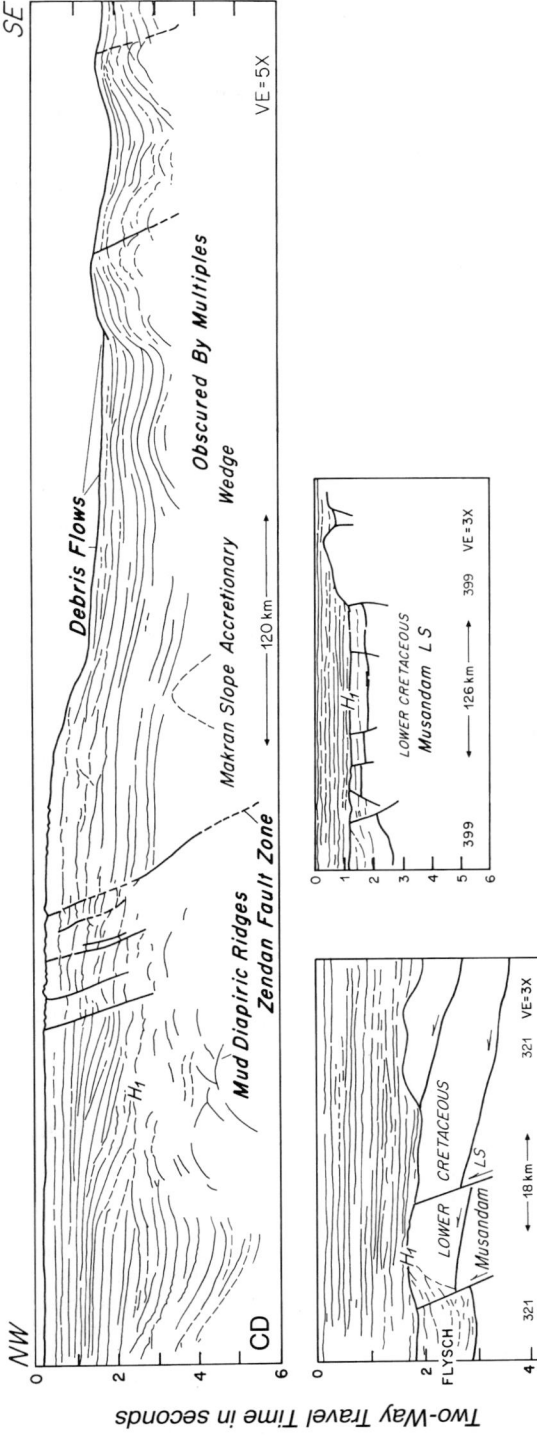

Fig. 17. Line interpretations of seismic reflection profiles in the Strait of Hormuz. Profile CD from the archives at Woods Hole Oceanographic Institution, and the other two profiles (321 and 399) are redrawn from Michaelis & Pauken (1990). (See Fig. 15 for locations of profiles.) Horizon H_1 is a late Miocene unconformity.

deformation is uncertain. Michaelis & Pauken (1990) concluded that thrusting took place during Late Cretaceous time, that syntectonic flysch-like sediments were deposited NW of the block during Late Cretaceous and/or Paleogene time and that the strata in the area of the block were refaulted during a Miocene compressional event. Glennie *et al.* (1990), however, argued that thrusting of the limestone took place much later, in post-Eocene time, as a result of the collision of Eurasia and Turkish–Iranian microcontinents.

Eastern side of the Makran subduction complex

Whereas the western side of the accretionary wedge on the north side of the Gulf of Oman Basin terminates abruptly against the Zendan Fault–Oman Line (Fig. 15), its eastern end turns gradually northward and merges with the Ornach–Nal Fault along the west side of the Las Bela Valley via the Sarpai anticline near 65°45'E (Figs. 3 and 18b; Snead 1969). This gradual change in trend is clearly imaged by the gravity map based on satellite altimetry (west arrow, Fig. 8). The Murray Ridge appears to merge northward with the Surjan Fault along the east side of the Las Bela Valley (east arrow, Fig. 8). Both of these faults converge with the Las Bela–Chaman Structural Axis at the northern apex of the triangular-shaped Las Bela Valley (Figs. 3 and 18b). Intervening between the north-trending segment of the Makran Escarpment and the Murray Ridge is a trough offshore and the Las Bela Valley onshore with the Eurasian–Indian–Arabian plates' triple junction located at the northern tip of the Las Bela Valley (Jacob & Quittmeyer 1979). Recently, Kukowski *et al.* (2000) has inferred that this low is the site of the Ormara microplate, bound on its SW side by the left-lateral Sonne Fault (Fig. 18b). The northern Murray Ridge forms the other margin of the triangular microplate offshore. We further speculate that the microplate extends onshore and that the Surjan and the Ornach–Nal faults on either side of the Las Bela Valley form the boundaries of the microplate. According to Kukowski *et al.* (2000), the presence of this microplate, which tore away from the Arabian plate along the Sonne Fault at 2 Ma, explains differences in the dip of the subduction plate in the eastern and western parts of the Makran accretionary wedge, and the different distances between the volcanic chain and the front deformation front.

Along the east side of the Las Bela Valley are the Mor and Pab Ranges, aligned along the Surjan Fault, and a thick sequence of stabilized dunes of probable Pleistocene age, which may been uplifted 30–60 m (Snead 1969). On the west side of the valley is the Haro Range along the Ornach–Nal Fault. This fault displays evidence of recent tectonism, including a raised a series of beach ridges and unweathered 23 ka oyster beds 10 m above sea level (Snead & Frishman 1968; Snead 1969). The floor of the Las Bela Valley also displays evidence of recent tectonism, including north-, NE- and NW-trending tectonic lineations disrupting the floor (Kazmi 1979; Sarwar & DeJong 1979) and nick points along the valley's entrenched dendritic drainage system (Snead 1969).

Like the Las Bela Valley floor, the west side of the offshore extension of the Las Bela Valley is folded and faulted (Profile 73-180, Fig. 6; Profiles 1–3, Fig. 19). Immediately west of the Las Bela Valley (Profile 73-180, Fig. 6) the Pliocene deposits on the prograding sediments on the shelf and upper continental slope are partially deformed, a tectonization that becomes more pronounced eastward and southward toward the axis of the trough, with the deformation extending into overlying Quaternary strata (Harms *et al.* 1984). This deformation takes the form of thrust faults, with the structures being aligned in a NE to north direction, parallel to the Makran Escarpment along the western side of the trough. Profiles 1–3 (Fig. 19) show that the strata in the offshore trough are deformed only next to the Makran accretionary wedge. This reverse faulting and folding may be related to transpression along the north-trending segment of the Makran Escarpment. Another feature displayed by the northern portion of Profile 3 is a lenticular-shaped acoustic transparent unit resting on an irregular strong reflecting horizon. We have identified the irregular reflector as the top of oceanic crust and the acoustic transparent unit as a levee system. This levee network is probably part of the Indus Fan channel–levee complex described by Lafont-Petassou & Orszag-Sperber (1991).

Eastern transform margin of the Gulf of Oman Basin

Morphology and structural setting

The active eastern translation margin of the Gulf of Oman Basin consists, from NW to SE, of a NE-trending oceanic basement swell, a Middle Low and the Murray Ridge. The Middle Low consists of a full graben to the NE and a half-graben, Dalrymple Trough, to the SW, separated by an east–west-trending high at 23°N. This morphology may indicate that the northern and southern Murray Ridge complex belong to two

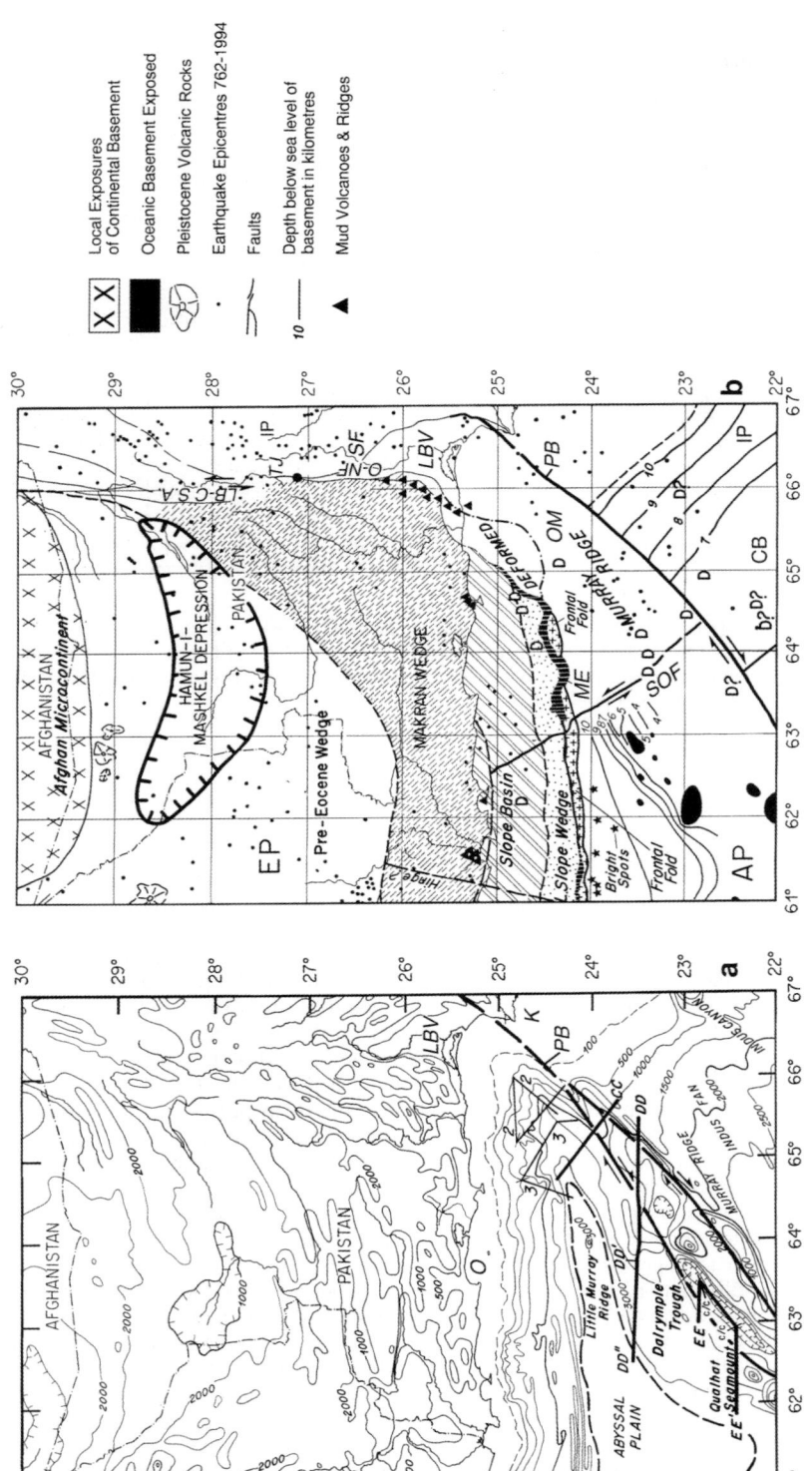

Fig. 18. (a) Bathymetry of the region of the Murray Ridge and vicinity, showing locations of the seismic reflection profiles in Figs. 19 and 20. K, Karachi; LBV, Las Bela Valley; PB, plate boundary. (b) Tectonic map of the Murray Ridge region and western side of the Indus Fan. AP, Arabian plate; CB, attenuated continental crust; D, mud diapir; EP, Eurasian plate; IP, Indian plate; LB-C S.A, Las Bela–Chaman Structural Axis; LBV, Las Bela Valley; O-NF, Ornach–Nal Fault; OM, Ornara microplate; PB, plate boundary; SF, Surjan Fault; SOF, Sonne Fault; TJ, triple juction. (See Figs. 1 and 3 for sources for these two maps.)

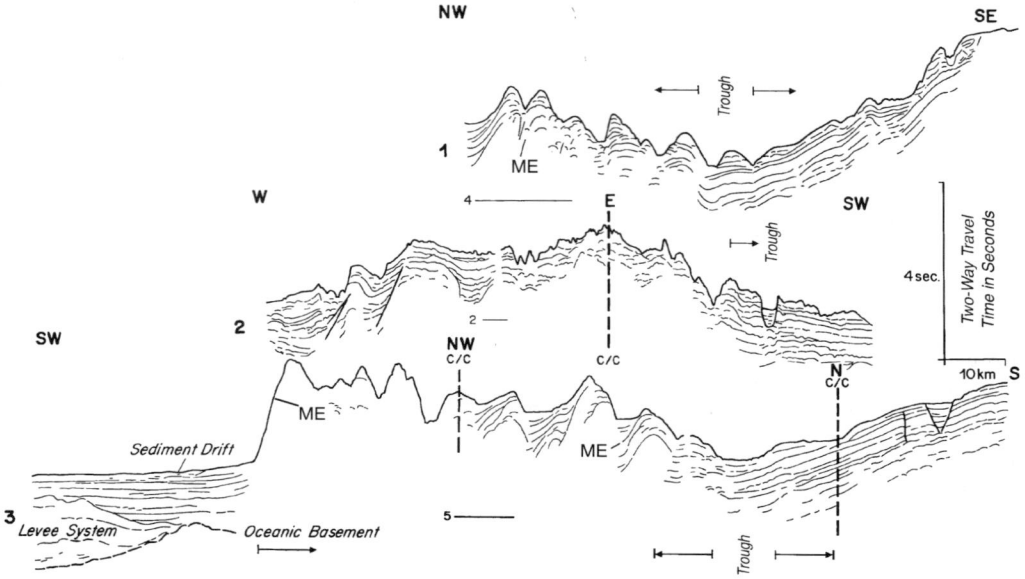

Fig. 19. Line interpretation of the single-channel seismic reflection Profiles 1–3 recorded during the present investigation of the trough SE of the Makran Escarpment. Along Profiles 1 and 3 the sediments on the NW side of the trough next to the Makran Escarpment are involved in the deformation of the accretionary wedge. ME, Makran Escarpment. (See Fig. 18a for locations of profiles.)

different plates with the east–west-trending high at 23°N marking the boundary between the two plates. Kukowski et al. (2000) proposed that oblique extension along the northern end of the Murray Ridge led to the formation of the NW-trending left-lateral Sonne Fault when the Arabian plate tore to accommodate oblique extension along the northern end of the Murray Ridge. The Ormara microplate was created from the Arabian plate as a result of of this tear.

The swell and Middle Low have topographic expressions to the shelf's edge, where they die out or plunge beneath the shelf's surface (Fig. 18a). The Murray Ridge has a topographic expression to the upper slope, where it too appears to die out (Fig. 18a). Earthquake distribution, however, indicates that the high may continue beneath the shelf, joining with the Surjan Fault along the east side of the Las Bela Valley (Fig. 18b). Farther north this fault and the Ornach–Nal Fault, along the west side of the valley, converge at the northern end of the Las Bela Valley with the Las Bela–Chaman Structural Axis (Figs. 3 and 18b). This axis, together with the Murray Ridge and the Owen Ridge south of the Murray Ridge, define the present-day Indian–Arabian plate boundary (Fig. 18b; Quittmeyer & Kafka 1984). Whitmarsh (1979) inferred that the plate boundary has been in its present position since India began to move northward at 90 Ma. In contrast, Mountain & Prell (1990) proposed that the plate boundary was initially located along the Oman margin on the SE side of the Arabian peninsula, which they postulated was of transform origin, and west of the Murray Ridge. This former boundary may be indicated the narrow linear gravity low (R?, Fig. 8) west of the Murray Ridge and along the seaward edge of the SE Oman margin and the line of seamounts west of the Murray and Little Murray Ridges (Fig. 3). Mountain & Prell (1990) further proposed that the boundary jumped eastward to its present position during Oligocene–Early Miocene time. Such a easterly jump may be supported by sediment from DSDP Sites 223 and 224 and Ocean Drilling Program (ODP) Site 731 (Shipboard Scientific Party 1989) on Owen Ridge, which display evidence that the ridge was uplifted in Early Miocene time, as a result of a plate reorganization following the opening of the Gulf of Aden. Geophysical data recorded by Barton *et al.* (1990) support the Mountain & Prell (1990) contention that the Oman margin is a former transform, indicating that the plate boundary was at one time located along this margin. According to Edwards *et al.* (2000), geophysical data from the Gulf of Oman, however, do not provide support for either model.

Crustal structure

Although Barker (1966) dredged spilitic basalt and tuff from the Murray Ridge near 22°30′N, the lack of any strong magnetic anomalies led Edwards *et al.* (2000) to infer that Murray Ridge is not of volcanic origin. From wide-angle seismic measurements and gravity and magnetic models, Edwards *et al.* (2000) proposed that that crustal thickness beneath the Gulf of Oman was 6 km (the thickness of normal oceanic crust), beneath Dalrymple Trough *c.* 13 km and beneath the Murray Ridge *c.* 17 km. From these crustal thicknesses and the presence of attenuated continental crust east of the ridge (Malod *et al.* 1997), Edwards *et al.* (2000) inferred that the Murray Ridge was of continental origin.

Proto-Indus and Indus fans

The Murray Ridge forms a divide between the north–south-trending Indus Fan to the east and an east–west-trending fan (Proto-Indus Fan) along the coastal Makran to the west. Both centres are constructed of detritus from the Himalayas, with the Indus River being the present depocentre (Coumes & Kolla 1984; Kolla & Coumes 1985). Construction of both of these fans was initiated by orogenic events resulting from the collision of India with Eurasia between 66 and 49 Ma and subsequent 1500–2000 km penetration of India into Eurasia (Molnar & Tapponnier 1975; Beck *et al.* 1995). These tectonic events, coupled with local climatic changes induced by them, increased the sediment supply to the coast. Harms *et al.* (1984), Critelli *et al.* (1990) and Garzanti *et al.* (1996) proposed that fan sedimentation from mid-Oligocene(?) to Mid-Miocene time from the Himalayas was westward. These proximal fan sediments were deposited over more distal Paleogene muddy turbidites, with graded sand beds displaying features diagnostic of westward transport. However, during Early Miocene time some of this deposition was diverted to the south, as suggested by the presence of muddy turbidites at the distal end of the Indus Fan (Davies *et al.* 1995). This diversion may document uplift of the Murray Ridge's landward extension, the Surjan Fault–Las Bela–Chaman Structural Axis. The present-day southerly sedimentary regime supposedly was established in late Mid-Miocene time (Harms *et al.* 1984), when uplift was renewed on the Murray Ridge and its landward extension. As the ridge experienced additional uplift, fan deposition shifted eastward, with the feed channels slowly migrating eastward in the manner described by Lafont-Petassou & Orszag-Sperber (1991). As the channels migrated eastward they created a series of overlapping fan lobes. The westernmost of these lobes was constructed in late Mid-Miocene time next to the Murray Ridge and the easternmost one in Pliocene–Pleistocene time along the present Indus Fan. This migration is a consequence of not only the uplift of the Murray Ridge, but also the tilting of the shelf eastward (Lafont-Petassou & Orszag-Sperber 1991).

On the basis of data from multi-channel seismic reflection profiles and a well on the Pakistani shelf, which penetrated to Middle Miocene units, Clift *et al.* (2001) have proposed a different scenario for construction of the Indus Fan. They stated that these data indicate that the Paleogene sediments in the Indus Fan are equal to these in the fan in the Makran, that the Early Miocene period was a time of limited deposition, and that the Mid-Miocene period, at the time the Makran fan was emplaced, was a time of significant deposition and the onset of strong channel–levee development on the Indus Fan. They further proposed that the Murray Ridge underwent two periods of uplift, one during the Eocene–Oligocene time and another during Early Miocene time. These observations are not compatible with one another. Studies of the coastal Makran and the offshore accretionary wedge suggest that the Makran was a site of distal fan turbidite deposition during Paleogene time. According to Clift *et al.* (2001), the Indus Fan was also a site of significant deposition at that time. Investigations of the coastal Makran and offshore accretionary wedge also indicate that the Makran was a site of massive fan proximal turbidite deposition during mid-Oligocene(?) and Mid-Miocene time. Clift *et al.* (2001) also stated that deposition in the Indus Fan also was significant at that time. In the absence of deep drill data coupled with multi-channel seismic reflection profiles from both sites, it is impossible to verify the validity of these observations. Only such a dataset would make it possible to determine the stratigraphy of these depocentres, how they relate to one another, and how their development was controlled by the tectonic history of the Himalayas and the plate boundary between India, Eurasia and Arabia. It is possible that both observations are correct, implying that Cenozoic fan deposition before the uplift of the Murray

Ridge migrated back and forth from west to south. Such a shift may have resulted in the creation of a fan of immense proportions covering the Makran, large expanses of the Gulf of Oman and the northern Arabian Sea. With uplift of the ridge, Miocene fan deposition was restricted to the northern Arabian Sea.

Seismic stratigraphy of the Murray Ridge complex

Murray Ridge. The seismic stratigraphy of the Murray Ridge, the Central Trough NW of the ridge and the oceanic basement swell NW of the low is imaged by reflection Profiles CC, DD–DD'– DD" and EE–EE' (Fig. 20). These profiles and those recorded by Edwards *et al.* (2000) demonstrate that the structural style of the Murray Ridge region is dominated by extensional tectonics, superimposed on which are secondary compressional features. Profiles CC and DD–DD'– DD", which extend across the width of the Murray Ridge region, from the Indus Fan to the Gulf of Oman, display the three topographic elements of the region, a broad swell on the NW whose strata dip northwestward (A, Profiles CC, DD–DD'– DD', Fig. 20), a Middle Low (B, Profiles CC, DD–DD'DD', Fig. 20), and a narrow basement ridge, which Edwards *et al.* (2000) tentatively identified as continental, towards the SE (Murray Ridge proper) (C, Profiles CC, DD–DD'DD', Fig. 19). Murray Ridge is asymmetric in cross-section, with its gentler slope facing southeastward. This side of the high is covered with sediments dipping to the SE. Along Profile CC, Indus Fan deposits onlapping onto the Murray Ridge completely buried the easterly dipping sequence resting on the continental basement ridge, whereas in Profile DD–DD'– DD" these are only partially buried by the fan deposits and form an easterly-facing sea floor. On both profiles the sediments on the Middle Low (B, Fig. 20) NW of the Murray Ridge are in faulted contact with the ridge. Along Profile CC the sediments in the low are tightly folded next to the ridge, but farther west they form a faulted sediment swell tilted to the NW and in faulted contact with Gulf of Oman sediments. Erosion of the deformed sediments on the Central Trough has created an irregular-floored

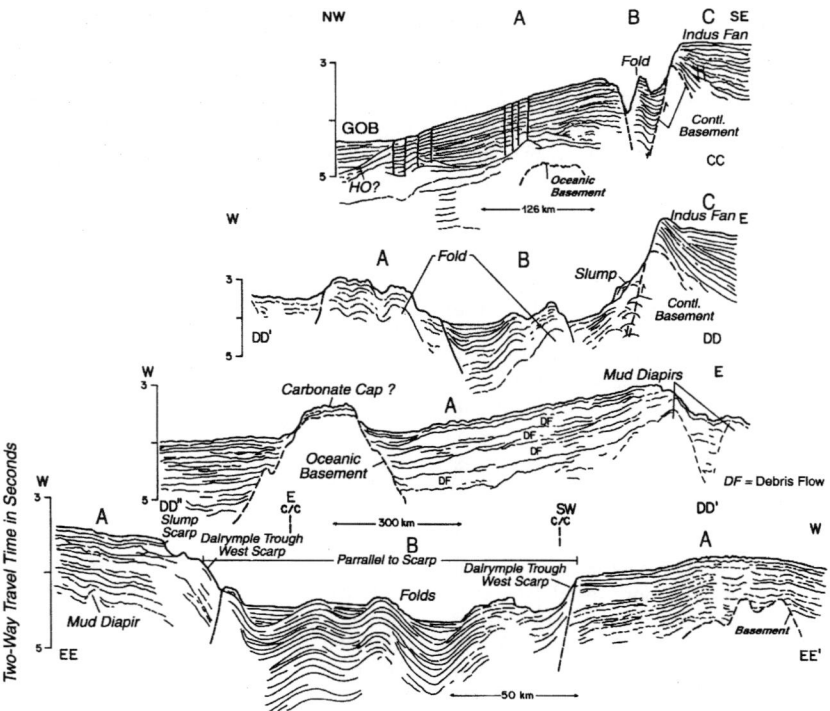

Fig. 20. Line interpretation of single-channel seismic reflection Profiles CC, DD–DD'– DD" and EE–EE' recorded during the present investigation of the Murray Ridge. A, NW swell; B, Middle Low; C, Murray Ridge. (For locations of profiles, see Fig. 18a.) DF, debris flows; GOB, Gulf of Oman Basin; HO?, Horizon Oman.

depression with a relief of about 590 m. The sediments of the oceanic basement swell NW of the Middle Low are tilted to the NW and may be in fault contact with the strata in the Middle Low. In the Gulf of Oman these sediments are partially underlain by a strongly reflective high that we identify as oceanic basement.

Middle Low–Dalrymple Trough. Farther south, along Profile DD–DD'– DD", sediments along the east flank of the Middle Low (B, Fig. 20) display a texture suggestive of slumping, a mass movement probably related to dislocation of the Murray Ridge. West of these slumps the sediments in the Middle Low have been folded, and the lows between the folds were filled with horizontal strata lacking any evidence of deformation; the crests of the structural highs display evidence of erosion. Strata of the swell next to the axial low are folded and in fault contract with the Middle Low sediments. On both crossings the NW scarp of Dalrymple Trough displays evidence of erosion. The unique feature displayed by Profile EE–EE' is the deformation of the sediments filling the trough, with the crests of two of the structures displaying evidence of erosion. Sediments filling the lows between the folds are undeformed, indicating that they are post-deformational. Fold spacing indicates that they trend at right angles or oblique to the axis of the trough and the trace of the Arabian–Indian plate boundary, although we lack other lines parallel to the axis of the trough to verify this. If they were parallel to the trough the fold axes would be much farther apart or would not be imaged at all along the trend of Profile EE–EE'.

A seismic reflection profile oriented approximately at right angles to the Murray Ridge in the area of Dalrymple Trough south of 23°N (Minshull *et al.* 1992) shows that the sediments on the NW oceanic basement swell are disrupted by normal faults with the fault planes dipping both away from and towards the trough. As imaged by this profile the trough is a graben with the master normal fault on its SE side. Minshull *et al.* observed that the sediment fill in the trough is comparable in thickness with sediments on either side of the graben, leading them to propose that the fill represents downfaulted deposits comparable with those on either side of the trough rather than post-tectonic sediments. Lack of a post-tectonic sediment cover in turn indicates that Dalrymple Trough is a recent feature. The strata in the trough dip SE and are warped and bent gently against the SE side of the Murray Ridge. Minshull *et al.* (1992) ascribed this bending to a combination of fault drag and sagging as a result of sediment compaction.

Oceanic basement swell. On Profile DD–DD'– DD" strata on the oceanic basement swell along the NW side of the Murray Ridge region are disrupted by highs that we have interpreted as oceanic basement hills or ridges. Lack of sediment deformation along the flanks of the highs indicates that they were constructed before the deposition of the sediments. Farther NW the oceanic basement swell sediments (A, Fig. 20) are disrupted by structures that we have inferred to be mud diapirs. As noted by Collier & White (1990) and Minshull *et al.* (1992), sediments of the Murray Ridge region, like those of the Indus Fan, are prone to diapirism. This suggests that they, like the fan deposits, were deposited rapidly, leading to the creation of overpressured muds. Collier & White (1990) proposed that the diapirs were formed at depths of less than 500 m below the sea floor, but Profile DD–DD'– DD" suggests that the mud structures may originate much deeper, at least at a depth of 2 s or about 1.6–2 km below the sea floor. The NW-dipping stratified sequence along Profile DD–DD'– DD" is disrupted by a basement high along the northwestern edge of the oceanic basement swell. Interbedded with the sediments onlapping the oceanic basement swell are three units with irregular tops lacking internal reflectors. We tentatively identify the sediments as debris flows (DF, Profile DD–DD'– DD', Fig. 20). The crest of the basement high itself is sediment covered. If this cap is made up of shallow carbonates then the high has subsided about 2400 m, the present water depth of the cap, since their deposition. The surface of this cap and the sediments on the swell NW of the Murray Ridge show evidence of recent erosion. Onlapping the western side of the high are the sediments of the Gulf of Oman Basin. These sediments also display evidence of erosion at their surface.

Profile EE–EE' (Fig. 20) begins on the NW swell (A, Fig. 20), descends to the Dalrymple Trough (Fig. 18a and b; Fig. 20), turns southwestward parallel to its west side, and then turns westward again to the NW swell (B, Fig. 20). The NW flank of the Dalrymple Trough consists of a sequence of reflectors dipping gently southeastward and terminating in faulted contact with the strata in the trough. From the geometry displayed by the reflectors, the sequence on the oceanic basement swell may consist of two units separated by an unconformity with the upper unit having a thickness of about 0.3 s; the thickness of the lower unit is not revealed by the profile. Along Profile EE–EE', sediments on the sediment rise NW of Dalrymple Trough are tilted southeastward in the direction of the trough. As imaged by this profile and Profiles CC and DD–DD'– DD", both the floor of the Middle Low and the surface of the oceanic basement swell NW of the low display evidence of bottom current erosion.

From the seismic reflection profiles described above it is clear that the Murray Ridge region is made of three structural elements. On the NW flank is a broad sediment-mantled basement swell tilted to the NW. On one profile (CC, Fig. 20) the basement swell has been uplifted above the floor of the Gulf of Oman Basin on its NW side. We have tentatively identified this basement as oceanic. Southeast of this swell is a basement high (Murray Ridge), which Edwards et al. (2000) have identified as having continental crust. Between these two positive elements is a Middle Low. The strata on the swell are faulted, intruded by mud diapirs and disrupted by oceanic basement highs. Some of the oceanic basement highs rise above their general surroundings, with at least one of them having a sediment cap. The Murray Ridge on the SE edge of the region is a faulted block tilted to the SE and overlapped by Indus Fan strata. Sediments in the Middle Low between these highs are partially folded and faulted, with the lows between the folds filled by undeformed deposits. Some of the crests of the structural highs display evidence of erosion. The presence of mud diapirs suggests that the strata in the area NW of the Murray Ridge represent a rapidly deposited sequence, a regime indicative of deposition by turbidity currents in a deep-sea fan environment rather than a hemipelagic–pelagic regime. It is this sedimentary style that makes us infer that the sediments associated with the Murray Ridge are part of a proto-Indus Fan that extended farther west than the present Indus Fan.

Origin of the topography of the Murray Ridge region

As Edwards et al. (2000) pointed out, wide-angle, gravity and magnetic modelling, as well as the ridge's proximity to the attenuated continental crust beneath the Indus Fan (Fig. 18b), indicate that the Murray Ridge may comprise continental crust. The presence of seamounts and its acoustic signature (chaotic–hyperbolic texture on its surface) suggests that the swell NW of the Murray Ridge has an oceanic crust. The Middle Low is along the contact of these two crustal types. Two models can explain the structures displayed by the seismic reflection profiles from the region, transpression and transtension. Two observations support a transpression model. These include the possible reverse faults and folds imaged by Profiles CC, DD–DD′– DD″ and EE–EE′. Such compressional features are common in transpression regimes (Wilcox et al. 1973; Harding 1985; Harding et al. 1985; Mount & Suppe 1987; Zoback et al. 1987), whereas they tend to be poorly developed or have low relief in the transtensional regime. Thus the structures in the region may have been produced by the westward thrusting of attenuated continental crust along the southeastern side of the Murray Ridge over the oceanic crust and its sediment cover. In this model the Middle Low results from the partial collapse of the face of the thrust continental crustal block and the emplacement of a load on the oceanic crust to the west. Thus the low is comparable with a foreland basin, with the broad swell NW of the Middle Low representing a peripheral bulge. This Middle Low was subsequently filled with sediments, strata that were locally deformed by a second cycle of thrusting. Such a style of deformation, but more subdued, also appears to have taken place along the Owen Ridge (Shipboard Scientific Party 1989). If such is the origin of the Murray Ridge then the sediments on top of the oceanic crustal swell over which the continental crust was tectonically transported should show some evidence of deformation. No such deformation is imaged in our single-channel seismic reflection profiles or the multi-channel profiles recorded by Edwards et al. (2000).

Edwards et al. (2000) inferred that the structure displayed by the seismic reflection profiles recorded in the Murray Ridge region by them can best be explained by transtension, with the compressional features being due to a small component of compression along the Arabian–Indian plate boundary. They assumed that the normal faulting was accompanied by subsidence of the hanging wall, uplift of the footwall and erosion. They also pointed out that, although earthquake activity indicates that episodic faulting is taking place in the region, the undisturbed nature of the sediment fill in the Central Low indicates that major vertical movements have ceased. Such an origin is supported by the model studies of Weissel & Karner (1989). These models suggested that rift flanks can experience extensive uplift by mechanical unloading of the lithosphere during extension. Such uplift may account for the magnitude of uplift displayed by the Murray Ridge. Although such a transtensional origin is not compatible with the broad warping of oceanic basement NW of the Murray Ridge–Middle Low, we tentatively support a transtensional origin with associated minor compression for the origin of the Murray Ridge.

Age of the Murray Ridge

Seismic stratigraphy of the western side of the Indus Fan (Clift et al. 2001) shows that initial uplift of the ridge probably took place sometime during the Eocene–Oligocene period. Seismic reflection data from the Gulf of Oman suggest that this uplift of the Murray Ridge post-dates the erosion of the Early Oligocene Horizon Oman in the Gulf of Oman. Thus the uplift is younger than Early Oligocene time. Regional tectonic events indicate that the major uplift of the Murray Ridge, together with the Owen Ridge, was probably initiated by changes in plate geometry in Early Miocene time (Shipboard Scientific Party 1989). During this uplift the Indus Fan's section east of the Murray Ridge was tilted eastward. Stratigraphic studies in the Makran (Harms et al. 1984) suggest that the landward extension of the Murray Ridge may have experienced initial uplift in late Mid-Miocene time during which fan deposition shifted eastward from the Makran. Before this uplift the Makran region was a zone of distal turbidite deposition before Oligocene time and proximal turbidite deposition from mid-Oligocene(?) to Mid-Miocene time. Whether the Murray Ridge also experienced uplift in late Mid-Miocene time cannot be verified from the offshore acoustic stratigraphy on either side of the ridge. The morphology of the Murray Ridge suggests that it may have experienced recent uplifts, reflected by the erosion of structural highs in the Middle Low of the ridge complex. This erosion is too extensive to have been created by turbidity current activity along the axis of the low and is too well preserved to be associated with Miocene tectonic events. Possibly this erosion is the result of the formation of the Ormara microplate from the Arabian plate at c. 2 Ma and recent activity along the left-lateral Sonne Fault on the SW side of this plate.

Conclusion

The 760 km long and 160 km wide Gulf of Oman Basin is the remnant of a more extensive Tethys Sea, which was reduced in size by the collision of India and Arabia with Eurasia, and rotations and collisions of microcontinents between Arabia and Eurasia. The tectonic style of the margins of the Gulf of Oman Basin margins is due to convergent, translation and divergent processes. Convergent processes created an accretionary wedge of 370–90 km width along the southern edge of Eurasia, 60–67% of which is above sea level. Morphologically this wedge can be divided into three segments: a northern, central and offshore segment. Included in the poorly developed northern segment are three microcontinents, Afghan, Lut and Central Iran, a continental sliver, Sanandaj–Sirjan–Bajgan–Dur–Kan, and obducted ophiolites. The northern part of the wedge, constructed in Cretaceous–Paleocene time, resulted from northerly directed subduction and collision of the microcontinents and continental sliver with Eurasia and with each other. The central segment along the Makran coast is made up of Eocene, Oligocene and middle Miocene distal and proximal turbidites derived from the east and upper Miocene–Holocene strata derived from the north from the wedge itself. It has undergone more than 30% shortening from mid-Miocene to Pliocene time, with frontal folding and imbricate thrusting accounting for 1–2% of the deformation and underplating the remainder. During Pleistocene–Holocene time this deformation was accompanied by uplift, with Holocene shoreline sediments being elevated as much as 500 m. The offshore lower slope Eocene–Holocene wedge offshore segment consists of Middle Miocene and and older Himalayan Turbidites derived from the east and Makran sands of Late Miocene age and younger derived from the north. The shortening experienced by these deposits was by frontal folding and imbricate thrusting within the wedge and underplating. This deformation has resulted in a sequence of thrust ridges separated by ponded basins filled by undisturbed Plio-Quaternary strata. Submarine canyons, some of which extend to the centre of the Gulf of Oman Basin, and the NW-trending left-lateral Sonne Fault near 63°E farther complicate the morphology and tectonic fabric of the accretionary wedge.

Deformation of the strata in the centre of the Gulf of Oman Basin is limited to minor faulting resulting from the northerly tilting of the oceanic basement. A basin-wide Early Oligocene unconformity divides the fill in two. The unconformity was eroded by north-flowing bottom currents before the uplift of the Murray Ridge. This bottom current erosion was enhanced by a climatic deterioration that began at the end of Eocene or in Early Oligocene time. The Oman margin along the southern side of the basin is the result of earliest Mesozoic rifting, Early Mesozoic carbonate deposition, early Late Cretaceous obduction of ophiolites and deposition since Late Cretaceous time.

Deformation along the Zendan Fault–Oman Line on the translation margin along the west side of the accretionary wedge is due to the east–west transpression. The geomorphology of the translation margin along the east side of the accretionary wedge involving a continental basement ridge on the SE (Murray Ridge), a Middle Low NW of the ridge and a oceanic basement swell NW of the low is due to transtension with secondary compression along the Arabian–Indian plate boundary. Whereas the western edge of the accretionary wedge terminates against the transform (Zendan Fault–Oman Line), the wedge's eastern side is separated from the transform by a north-trending trough. This trough extends onshore in the form of the triangular-shaped Las Bela Valley. Structural differences between the east and west sides of the accretionary wedge and its confining transforms are due to the presence of the Omara microplate on the east side. This microplate was torn from the Arabian plate along the left-lateral NW-trending Sonne Fault by oblique extension at the northern end of the Murray Ridge. The morphology of the Murray Ridge was probably initiated in Oligocene time after erosion of the Early Oligocene Horizon Oman in the centre of the Gulf of Oman Basin with the major uplift of the ridge taking placed in Early Miocene time. During recent tectonic episodes, probably associated with the formation of the Ormara microplate and motion along the Sonne Fault on the SW side of the microplate, strata in the Middle Low NW of the Murray Ridge were compressed and uplifted, an uplift that enhanced their erosion by bottom currents

This program was funded by Office of Naval Research Contract N00014-92-C-6028. Additional funds provided by Woods Hole Oceanographic Institution were used to complete drafting of the illustrations. S.T. Bolmer compiled the computer-generated gravity maps based on satellite altimetry data. The illustrations were drafted by P.E. Oberlander and photographed by D.L. Gray of Graphic Services at Woods Hole. We also wish to thank N. Driscoll for his comments during the preparation of this report. This paper is Contribution 10520 of the Woods Hole Oceanographic Institution.

References

AHMED, S.S. 1969. Tertiary geology of part of south Makran, Baluchistan, west Pakistan. *AAPG Bulletin*, **53**, 1480–1499.

ARTHURTON, R.S., ALAM, G.S., ANISUDDIN-AMAD, S. & IQBAL, S. 1979. Geological history of the Alamreg–Mashki Chad area, Chagai District, Baluchistan. *In:* FARAH, A. & DEJONG, K.A. (eds) *Geodynamics of Pakistan*. Geological Survey of Pakistan, Quetta, 325–331.

ARTHURTON, R.S., FARAH, A. & AHMED, W. 1982. The Late Cretaceous–Cenozoic history of western Baluchistan Pakistan—the northern margin of the Makran subduction complex. *In:* LEGGETT, J.K. (ed.) *Trench–Forearc Geology: Sedimentation and Tectonics of Modern and Ancient Plate Margins*. Geological Society, London, Special Publications, **10**, 373–385.

ASRARULLAH, A.Z. & ABBAS, S.G. 1979. Ophiolites in Pakistan: an introduction. *In:* FARAH, A. & DEJONG, K.A. (eds) *Geodynamics of Pakistan*. Geological Survey of Pakistan, Quetta, 181–192.

BARKER, P.F. 1966. A reconnaissance survey of the Murray Ridge. *Philosophical Tranactions of the Royal Society of London*, **A259**, 187–197.

BARTON, P.J., OWEN, T.R.E. & WHITE, R.S. 1990. The deep structure of the east Oman continental margin: preliminary results and interpretation. *Tectonophysics*, **173**, 319–331.

BECK, R.A., BURBANK, D.W., SERCOMBE, W.J. & 11 OTHERS 1995. Stratigraphic evidence for an early collision between northwest India and Asia. *Nature*, **373**, 55–58.

BERBERIAN, M. 1976. Contribution to the seismotectonics of Iran (Part II). *Geological Survey of Iran Reports*, **39**, 1–518.

BUNGENSTOCK, H., VON CLOSS, H. & HINZ, K. 1966. Seismische Untersuchungen im nördlichen Teil des Arrabischen Meeres (Golf von Oman). *Erdöl und Kohle, Erdgas, Petrochemie*, **19**, 237–243.

BYRNE, D.E., SYKES, L.R. & DAVIES, D.M. 1992. Great earthquakes from a seismic slip along the plate boundary of the Makran subduction zone. *Journal of Geophysical Research*, **97**, 449–478.

CLIFT, P.D., SHIMIZU, N., LAYNE, G.D., BLUSXTAJN, J.S., GAEDICKE, C., SCHÜLER, H.-U. & CLARK, M.K. 2001. Development of the Indus Fan and its significance for the erosional history of the western Himalayas and Karakoram. *Geological Society of America Bulletin*, **113**, 1039–1051.

COLEMAN, R.G. 1981. Tectonic setting for ophiolite obduction in Oman. *Journal of Geophysical Research*, **86**, 2497–2508.

COLLIER, J.S. & WHITE, R.S. 1990. Mud diapirism within Indus Fan sediments: Murray Ridge Baluchistan, Gulf of Oman. *Geophysical Journal International*, **101**, 345–353.

COUMES, F. & KOLLA, V. 1984. Indus Fan: seismic structure, channel migration and sediment-thickness in the upper fan. *In:* HAQ, B.U. & MILLIMAN, J.D. (eds) *Marine Geology and Oceanography of Arabian Sea and Coastal Pakistan*. Van Nostrand Reinhold, New York, 101–112.

CRITELLI, S., DE ROSA, R. & PLATT, J.P. 1990. Sandstone detrital modes in the Makran accretionary wedge, southwest Pakistan: implications for tectonic setting and long-distance turbidite transportation. *Sedimentary Geology*, **68**, 241–260.

DAVIES, T.A., KIDD, R.B. & RAMSEY, A.T.S. 1995. A time slice approach to the history of Cenozoic sedimentation in the Indian Ocean. *Sedimentary Geology*, **68**, 241–260.

DELISLE, G., VON RAD, U., ANDRULEIT, H., VON DANIELS, C.H., TABREZ, A.R. & INAM, A. 2001. Active mud volcanoes on- and offshore eastern Makran, Pakistan. *International Journal of Earth Sciences*, 90.

DYKSTRA, J.D. & BIRNIE, R.W. 1979. Segmentation of the Quaternary subduction zone under the Baluchistan region of Pakistan and Iran. *In:* FARAH, A. & DEJONG, K.A. (eds) *Geodynamics of Pakistan*. Geological Survey of Pakistan, Quetta, 319–323.

EDWARDS, R.A., MINSHULL, T.A. & WHITE, R.S. 2000. Extension across the Indian–Arabian boundary; the Murray Ridge. *Geophysical Journal International*, **142**, 461–477.

EMERY, K.O. & UCHUPI, E. 1984. *The Geology of the Atlantic Ocean*. Springer, New York.

EVERETT, J.R., MORISAWA, M. & SHORT, N.M. 1986.

Tectonic landforms. *In:* SHORT, N.M. & BLAIR, R.W. JR (eds) *Geomorphology from Space. A Global Overview of Regional Landforms*. National Aeronautics and Space Administration Special Publications, **486**, 27–184.

FARAH, A. & LAWRENCE, R.D. 1988. Plate geology and resource potential of off-shore region of Pakistan. *Acta Mineralogica Pakistanica*, **4**, 113–133.

FRUEHN, J., WHITE, R.S. & MINSHULL, T.A. 1997. Internal deformation and compaction of the Makran accretionary wedge. *Terra Nova*, **9**, 101–104.

GARZANTI, E., CRITELLI, S. & INGERSOLL, R.V. 1996. Paleogeographic and paleotectonic evolution of the Himalayan Range as reflected by detrital modes of Tertiary sandstones and modern sands (Indus transect, India and Pakistan). *Geological Society of America Bulletin*, **108**, 631–642.

GEOLOGICAL SURVEY OF IRAN 1963. *Geologic Map of the Strait of Hormuz*, 1: 250 000, plus notes.

GLENNIE, K.W., HUGHES CLARKE, M.W., BOEUF, M.G.A., PILAAR, W.F.H. & REINHARDT, B.M. 1990. Inter-relationship of Makran–Oman mountains belts of convergence. *In:* ROBERTSON, A.H.F., SEARLE, M.P. & RIES, A.C. (eds) *The Geology and Tectonics of the Oman Region*. Geological Society, London, Special Publications, **49**, 773–786.

GREVEMEYER, I., ROSENBERGER, A. & VILLINGER, H. 2000. Natural gas hydrates on the continental slope off Pakistan: constraints from seismic techniques. *Geophyical Journal International*, **140**, 295–310.

HARDING, T.P. 1985. Seismic characteristics and identification of negative flower structures, positive flower structures, and positive structural inversion. *AAPG Bulletin*, **69**, 582–600.

HARDING, T.P., VERBUCHEN, R.C. & CHRISTIE-BLICK, N. 1985. Structural style, plate-tectonic settings, and hydrocarbon traps of divergent (transtensional) wrench faults. *In:* BIDDLE, K.T. & CHRISTIE-BLICK, N. (eds) *Strike-slip Deformation, Basin Formation, and Sedimentation*. Society of Economic Paleontologists and Mineralogists, Special Publications, **37**, 51–77.

HARMS, J.C., CAPPEL, N. & FRANCIS, D.C. 1984. Geology of the Makran coast and offshore petroleum potential. *In:* HAQ, B.U. & MILLIMAN, J.D. (eds) *Marine Geology and Oceanography of Arabian Sea and Coastal Pakistan*. Van Nostrand Reinhold, New York, 3–26.

HARMS, J.C., CAPPEL, H.N., FRANCIS, D.C. & SHACKELFORD, J. 1983. Summary of the geology of the Makran coast. *In:* BALLY, A.W. (ed.) *Seismic Expression of Structural Styles, a Picture and Work Atlas*. American Association of Petroleum Geologists, Studies Series in Geology, **15-3**, 3.4.2-173–3.4.2-177.

HUHN, K., KUKOWSKI, H.N. & KOPP, C. 1999. Numerical finite element modeling of the deformation and mechanical behavior of the Makran accretionary wedge offshore Pakistan. *EOS Transactions, American Geophysical Union*, **80**, F572–F573.

HUNTING SURVEY CORPORATION LIMITED 1960. *Reconnaissance Geology of Part of West Pakistan. A Colombo Plain Co-operative Project*. Maracle Press, Toronto, OntÒ

HUTCHINSON, I., LOUDEN, K.E., WHITE, R.S. & VON HERZEN, R.P. 1981. Heat flow measurements and age in the Gulf of Oman. *Earth and Planetary Science Letters*, **56**, 252–262.

JACKSON, J.A. & MCKENZIE, D.P. 1984. Active tectonics in the Alpine Himalayan belt between western Turkey and Pakistan. *Geophysical Journal of the Royal Society of London*, **77**, 185–264.

JACOB, K.H. & QUITTMEYER, R.L. 1979. The Makran region of Pakistan and Iran: trench–arc system with active plate subduction. *In:* FARAH, A. & DEJONG, K.A. (eds) *Geodynamics of Pakistan*. Geological Survey of Pakistan, Quetta, 305–317.

JACOBI, R.D. & HAYES, D.E. 1992. Northwest African continental rise: effects of near-bottom processes inferred from high-resolution seismic data. *In:* POAG, C.W. & DE GRACIANSKY, P.C. (eds) *Geologic Evolution of Atlantic Continental Rises*. Van Nostrand Reinhold, New York, 293–326.

KAZMI, A.H. 1979. Active fault system in Pakistan. *In:* FARAH, A. & DEJONG, K.A. (eds) *Geodynamics of Pakistan*. Geological Survey of Pakistan, Quetta, 285–294.

KAZMI, A.H. 1984. Geology of the Indus Fan. *In:* HAQ, B.U. & MILLIMAN, J.D. (eds) *Marine Geology and Oceanography of Arabian Sea and Coastal Pakistan*. Van Nostrand Reinhold, New York, 71–84.

KENNETT, J.P., HOUTZ, R.E., ANDREWS, P.B. & 8 OTHERS 1974. Development of the Circum Antarctic Current. *Science*, **186**, 144–147.

KHATTAB, M.M. 1993. Identification of Early Cretaceous magnetic lineations in the northwestern part of the Gulf of Oman. *Marine Geology*, **112**, 353–362.

KIDD, R.B. & DAVIES, T.A. 1978. Indian Ocean sediment distribution since Late Jurassic. *Marine Geology*, **26**, 49–70.

KIDD, R.B, BALDAUF, J.G., DAVIES, T.A., JENKINS, D.G., RAMSAY, A.T.S., SYKES, T.J.S. & WISE, S.W. 1992. An Indian Ocean framework for paleoceanographic synthesis based on DSDP and ODP results. *In:* DUNCAN, R.A., KIDD, D.K., VON RAD, U. & WEISS, J.K. (eds) *Synthesis of Results from Scientific Drilling in the Indian Ocean*. Geophyical Monograph, American Geophysical Union, **70**, 403–422.

KOLLA, V. & COUMES, F. 1985. Indus Fan, Indian Ocean. *In:* BOUMA, A.H., NORMARK, W.R. & BARNES, N.E. (eds) *Submarine Fans and Related Turbidite Systems*. Springer, New York, 129–136.

KOOP, W.J. & STONELEY, R. 1982. Subsidence history of the middle Zagros Basin. *Philosophical Transactions of the Royal Society of London, Series A*, **305**, 149–169.

KOPP, C., FRUEHN, J., FLUEH, E.R., REICHERT, C., KUKOWISKI, N., BIALAS, J. & KLAESCHEN, D. 2000. Structure of the Makran subduction zone from wide-angle and reflection profiles. *Tectonophysics*, **329**, 171–191.

KUKOWSKI, N., SCHILLHORN, T., FLUEH, E.R. & HUHN, K. 2000. Newly identified strike-slip plate boundary in the Northeastern Arabian Sea. *Geology*, **28**, 355–358.

KUKOWSKI, N., SCHILLHORN, T., HUHN, K., VON RAD, U., HUSEN, S. & FLUEH, E.R. 2001. Morphotectonics and mechanics of the central Makran accretionary wedge off Pakistan. *Marine Geology*, **173**, 1–19.

LAFONT-PETASSOU, S. & ORSZAG-SPERBER, F. 1991. Cinétique de migration du cône supérior de l'Indus depuis l'Oligocène moyen: datation par les phases tectoniques regionales et/ou les variation du niveau relatif de la mer. *Comptes Rendu de l'Académie des Sciences, Série II*, **312**, 1475–1481.

LAWRENCE, R.D. & YEATS, R.S. 1979. Geological reconnaissance of the Chaman Fault in Pakistan. In: FARAH, A. & DEJONG, K.A. (eds) *Geodynamics of Pakistan*. Geological Survey of Pakistan, Quetta, 351–357.

LEES, G.M. 1928. The physical geography of south-eastern Arabia. *Geographical Journal*, **108**, 707–712.

LEGGETT, J.K. & PLATT, J.P. 1984. Structural features of Makran forearc on Landsat imagery. In: HAQ, B.U. & MILLIMAN, J.D. (eds) *Marine Geology and Oceanography of Arabian Sea and Coastal Pakistan*. Van Nostrand Reinhold, New York, 33–44.

LEHNER, P., DOUST, H., BAKKER, G., ALLENBAC, P. & GUENEAU, J. 1983. Active margins, part 4—Makran fold belt, profile N 1804. In: BALLY, W. (ed.) *Seismic Expression of Structural Styles, a Picture and Work*. American Association of Petroleum Geologists, Studies in Geology Series, **15-3**, 3.4.2-82–3.4.2-91.

MALOD, J.A., DROZ, L., MUSTAFA KEMAL, B. & PATRIAT, P. 1997. Early spreading and continental to oceanic basement transition beneath the Indus deep-sea fan. *Marine Geology*, **141**, 221–235.

MCCALL, G.J.H. & KIDD, R.G.W. 1982. The Makran, southeastern Iran: the anatomy of a convergent margin active from Cretaceous to present. In: LEGGETT, J.K. (ed.) *Trench–Forearc Geology: Sedimentation and Tectonics of Modern and Ancient Plate Margins*. Geological Society, London, Special Publications, **10**, 387–397.

MICHAELIS, P.L. & PAUKEN, R.J. 1990. Seismic interpretation of the structure and stratigraphy of the Straits of Hormuz. In: ROBERTSON, A.H.F., SEARLE, M.P. & RIES, A.C. (eds) *The Geology and Tectonics of the Oman Region*. Geological Society, London, Special Publications, **49**, 387–395.

MILES, P.R., MUNSCHY, M. & SÉGOUFIN, J. 1998. Structure and early evolution of the Arabian Sea and the East Somali Basin. *Geophysical Journal International*, **134**, 876–888.

MINSHULL, T.A., WHITE, R.S., BARTON, P.J. & COLLIER, J.S. 1992. Deformation at plate boundaries around the Gulf of Oman. *Marine Geology*, **104**, 265–277.

MOLNAR, P. & TAPPONNIER, P. 1975. Cenozoic tectonics in Asia: effects of a continental collision. *Science*, **189**, 419–426.

MOUNT, V.S. & SUPPE, J. 1987. State of stress near the San Andreas fault: implications for wrench faulting. *Geology*, **15**, 1143–1146.

MOUNTAIN, G.S. & PRELL, W.L. 1990. A multiphase plate tectonic history of the southeast continental margin of Oman. In: ROBERTSON, A.H.F., SEARLE, M.P. & RIES, A.C. (eds) *The Geology and Tectonics of the Oman Region*. Geological Society, London, Special Publications, **49**, 725–743.

NATIONAL IRANIAN OIL COMPANY 1977. *Geological Map of Iran. Sheet 6 of south-east Iran*, 1:1 000 000 plus text.

NIAZI, M., SHIMAMURA, H. & MATSU'URA, M. 1980. Microearthquakes and crustal structure off the Makran coast of Iran. *Geophysical Research Letters*, **7**, 297–300.

PAGE, W.D., ALT, J.N., CLUFF, L.S. & PLAFKER, G. 1979. Evidence for recurrence of large-magnitude earthquakes along the Makran coast of Iran and Pakistan. *Tectonophysics*, **52**, 533–547.

PLATT, J.P., LEGGETT, J.K., YOUNG, J., RAZA, H. & ALAM, S. 1985. Large-scale underplating in the Makran accretionary prism, southwest Pakistan. *Geology*, **13**, 507–511.

POWELL, C.M. 1979. A speculative tectonic history of Pakistan and surroundings; some constraints from the Indian Ocean. In: FARAH, A. & DEJONG, K.A (eds) *Geodynamics of Pakistan*. Geological Survey of Pakistan, Quetta, 5–24.

PRELL, W.L., NIITSUMA, N., ET AL. 1989. *Proceedings of the Ocean Drilling Program, Initial Reports, 117*. Ocean Drilling Program, College Station, TX.

PRINS, M.A., POSTMA, G. & WELTJE, G. 2000. Controls on terrigenous sediment supply to the Arabian Sea during the late Quaternary: the Makran continental slope. *Marine Geology*, **169**, 351–371.

QUITTMEYER, V.N. & KAFKA, A.L. 1984. Constraints on plate motions in southern Pakistan and northern Arabian Sea from focal mechanisms of small earthquakes. *Journal of Geophysical Research*, **89**, 2444–2458.

RICATEAU, R. & RICHÉ, P.H. 1980. Geology of the Musandam peninsula (Sultanate of Oman) and its surroundings. *Journal of Petroleum Geology*, **3**, 139–152.

ROSS, D.A., UCHUPI, E. & WHITE, R.S. 1986. The geology of the Persian Gulf–Gulf of Oman: a synthesis. *Review of Geophyics*, **24**, 537–556.

SARWAR, G. & DEJONG, K.A. 1979. Arcs, oroclines, syntaxes: the curvature of mountain belts in Pakistan. In: FARAH, A. & DEJONG, K.A. (eds) *Geodynamics of Pakistan*. Geological Survey of Pakistan, Quetta, 341–349.

SEARLE, J.N., CALON, T.J. & SMEWING, J.D. 1983. Sedimentological and structural evolution of the Arabian continental margin in the Musandam Mountains and the Dibba Zone, United Arab Emirates. *Geological Society of America Bulletin*, **94**, 1381–1400.

SEIBOLD, E. & ULRICH, J. 1970. Zur Bodengestalt des nordweslichen Golfs von Oman. *'Meteor' Forschungsergebnisse Reihe C*, **3**, 1–14.

SHIPBOARD SCIENTIFIC PARTY 1989. Background and summary of drilling results—Owen Ridge. In: PRELL, W.L. & NIITSUMA, N. (eds) *Proceedings of the Ocean Drilling Program, Initial Reports, 117*. Ocean Drilling Program, College Station, TX, 35–42.

SHIRLEY, K. 1991. From Yemen to Somalia. Firms seek the rift connection. *Explorer*, **12**(1), 16–21.

SMITH, W.H.F. & SANDWELL, D.T. 1995. Marine gravity

field from declassified Geosat and ERS-1 altimetry. *EOS Transactions, American Geophysical Union*, **76**, 156.

SNEAD, R.E. 1964. Active mud volcanoes of Baluchistan, west Pakistan. *Geographical Review*, **54**, 546–560.

SNEAD, R.E. 1967. Recent morphological changes along the coast of west Pakistan. *Annual Abstracts, American Geographers*, **57**, 550–565.

SNEAD, R.E. 1969. *Physical Geography Reconnaissance: West Pakistan Coastal Zone*. University of New Mexico Publications in Geography, **1**.

SNEAD, R.E. 1970. *Physical Geography of Makran Coastal Plain of Iran: Final Report, Reconnaissance Phase*. US Office of Naval Research Geography Programs, Washington, DC.

SNEAD, R.E. & FRISHMAN, S.A. 1968. Origin of sands on the east side of the Las Bela Valley, west Pakistan. *Geological Society of America Bulletin*, **79**, 1671–1676.

TILTON, G.R., HOPSON, C.A. & WRIGHT, J.E. 1981. Uranium–lead isotopic ages of the Semail ophiolite, with implications to Tethyan ridge tectonics. *Journal of Geophysical Research*, **86**, 2763–2775.

UCHUPI, E., SWIFT, S.A. & ROSS, D.A. 2002. Morphology and late Quaternary sedimentation in the Gulf of Oman. *Marine Geophysical Researches*, in press.

ULLAH, A. 1953. Physiography and structure of S.W. Makran. *Pakistan Geographical Review*, **9**, 28–37.

UNESCO 1976. *Geological World Atlas*, Sheet 11, scale 1:10 000 000. UNESCO, Paris.

VITA-FINZI, C. 1973. Late Quaternary subsidence. *Geographical Journal*, **139**, 414–421.

VON RAD, U., BERNER, U., DELISLE, G. & 7 OTHERS 2000. Gas and fluid venting at the Makran accretionary wedge off Pakistan. *Geo-Marine Letters*, **20**, 10–19.

WEISSEL, J.K. & KARNER, G.D. 1989. Flexural uplift of rift flanks due to mechanical unloading of the lithosphere during extension. *Journal of Geophysical Research*, **94**, 13919–13950.

WHITE, R.S. 1977*a*. Recent fold development in the Gulf of Oman. *Earth and Planetary Science Letters*, **36**, 85–91.

WHITE, R.S. 1977*b*. Seismic bright spots in the Gulf of Oman. *Earth and Planetary Science Letters*, **37**, 29–37.

WHITE, R.S. 1979. Deformation of the Makran continental margin. *In:* FARAH, A. & DEJONG, K.A. (eds) *Geodynamics of Pakistan*. Geological Survey of Pakistan, Quetta, 295–304.

WHITE, R.S. 1982. Deformation of the Makran accretionary sediment prism in the Gulf of Oman (north-west Indian Ocean). *In:* LEGGETT, J.K. (ed.) *Trench–Forearc Geology: Sedimentation and Tectonics on Modern and Ancient active Plate Margins*. Geological Society, London, Special Publications, **10**, 357–372.

WHITE, R.S. 1983*a*. The Little Murray Ridge. *In:* BALLY, A.W. (ed.) *Seismic Expression of Structural Styles, a Picture and Work Atlas*. American Association of Petroleum Geologists, Studies in Geology, **15-1**, 1.3-19–1 3-23.

WHITE, R.S. 1983*b*. The Makran accretionary prism. *In:* BALLY, A.W. (ed.) *Seismic Expression of Structural Styles, a Picture and Work Atlas*. American Association of Petroleum Geologists, Studies in Geology, **15-3**, 3.4.2-178–34.2-182.

WHITE, R.S. 1984. Active and passive plate boundaries around the Gulf of Oman, north-west Indian Ocean. *Deep-Sea Research*, **31**, 731–745.

WHITE, R.S. & KLITGORD, K. 1976. Sediment deformation and plate tectonics in the Gulf of Oman. *Earth and Planetary Science Letters*, **32**, 199–209.

WHITE, R.S. & LOUDEN, K.E. 1982. The Makran continental margin: structure of a thickly sedimented convergent plate boundary. *In:* WATKINS, J.S. & DRAKE, C.L. (eds) *Studies in Continental Margin Geology*. American Association of Petroleum Geologists, Memoirs, **34**, 499–518.

WHITE, R.S. & ROSS, D.A. 1979. Tectonics of the western Gulf of Oman. *Journal of Geophysical Research*, **84**, 3479–3489.

WHITMARSH, R.B. 1979. The Owen Basin off the southeast margin of Arabia and the evolution of the Owen Fracture Zone. *Geophysical Journal of the Royal Astronomical Society*, **58**, 441–470.

WHITMARSH, R.B., WESER, O.E., ROSS, D.A., ET AL. (eds) 1974. *Initial Reports of the Deep Sea Drilling Project, 23*. US Government Printing Office, Washington, DC.

WIEDICKE, M., NEBEN, S. & SPIESS, V. 2000. Mud volcanoes at the front of the Makran accretionary complex, Pakistan. *Marine Geology*, **172**, 57–73.

WILCOX, R.E., HARDING, T.P. & SEELY, D.R. 1973. Basic wrench tectonics. *AAPG Bulletin*, **57**, 74–96.

Woodward–Noorany Consulting Engineers Inc. & Woodward–Clyde Consultants, International 1975. *Site Safety Analysis Bandar Abbas region, Iran, Phase I Site Conformation Studies*. Report prepared for Atomic Energy Organization of Iran.

ZOBACK, M.D., ZOBACK, M.L., MOUNT, V.S. & 10 OTHERS 1987. New evidence on the state of stress of the San Andreas fault system. *Science*, **238**, 1105–1111.

Paleogene magnetic isochrons and palaeo-propagators in the Arabian and Eastern Somali basins, NW Indian Ocean

A. K. CHAUBEY[1], JÉROME DYMENT[2], G. C. BHATTACHARYA[1], JEAN-YVES ROYER[2], K. SRINIVAS[1] & V. YATHEESH[1]

[1]*Geological Oceanography Division, National Institute of Oceanography, Dona Paula, Goa 403 004, India (e-mail: chaubey@darya.nio.org)*
[2]*CNRS Domaines Océaniques, Institut Universitaire Européen de la Mer, Place Copernic, 29280 Plouzané, France*

Abstract: We present a revised magnetic isochron map of the conjugate Arabian and Eastern Somali basins based on an up-to-date compilation of Indian, French, and other available sea-surface magnetic data. We have used the magnetic anomaly and the modulus of the analytical signal computed from the magnetic anomaly to identify and precisely locate the young and old edges of magnetic chrons in both basins. In addition to the major, well-defined anomalies, we have also used correlatable second-order features of the magnetic anomalies, the 'tiny wiggles', to strengthen the interpretation. The resulting isochrons and tectonic elements have been validated using the stochastic method of palaeogeographical reconstruction. The magnetic anomaly pattern in both basins depicts clear oblique offsets, characteristics of pseudofaults associated with propagating ridge segments. Our tectonic interpretation of the area revealed: (1) a complex pattern of ridge propagation between Chrons 28n (c. 63 Ma) and 25n (c. 56 Ma), with dominant eastward propagation between Chrons 26n (c. 58 Ma) and 25n; (2) numerous, systematic westward propagations between Chrons 24n (c. 53 Ma) and 20n (c. 43 Ma); (3) asymmetric crustal accretion (caused by ridge propagation and asymmetric sea-floor spreading) in the conjugate basins during the whole period; (4) a slowing of India–Somalia motion after c. 52 Ma.

The present study focuses on the Arabian and Eastern Somali basins, which are located in the NW Indian Ocean (Fig. 1). Geographically, the Arabian Basin is bounded to the west by the Owen Fracture Zone, which demarcates the transform boundary between the Indian and Arabian plates; to the south by the active Carlsberg Ridge, which separates the Indian and African plates; and to the north and east by the Laxmi Ridge and the northern portion of the Chagos–Laccadive Ridge. The Eastern Somali Basin is bounded by the Carlsberg and Central Indian ridges in the north and east, the Seychelles block and the northern Mascarene plateau to the south, and the Chain Ridge to the west. These conjugate basins were formed by sea-floor spreading on the Carlsberg Ridge since Early Tertiary time.

The evolution of these basins has probably been influenced by two major geodynamic events: the onset of Reunion hotspot activity (Deccan Trap) and the Indo-Eurasian collision. The Deccan Trap flood basalts erupted at Chron 29r (Courtillot *et al.* 1986; Vandamme *et al.* 1991; Bhattacharji *et al.* 1996) and the Deccan–Reunion hotspot has since been continuously active, forming large parts of the Chagos–Laccadives and Mascarene volcanic ridges (Duncan 1981; Morgan 1981). The Indo-Eurasian continental collision started at about Chron 24n (52 Ma; Patriat & Achache 1984) and has continued up to now. Probably because of the effect of these major geodynamic events, the tectonic fabric in the conjugate Arabian and Eastern Somali ocean basins is complex and the resulting magnetic anomalies have remained difficult to interpret. The various interpretations of magnetic anomalies in these two basins published before 1997 considered linear magnetic isochrons offset by orthogonal fracture zones (McKenzie & Sclater 1971; Whitmarsh 1974; Norton & Sclater 1979; Schlich 1982; Naini & Talwani 1983; Karasik *et al.* 1986; Mercuriev & Sochevanova 1990; Bhattacharya *et al.* 1992; Chaubey *et al.* 1995; Glebovsky *et al.* 1995). Availability of additional data in some sectors, as well as more reliable identifications of the anomalies, showed that the magnetic lineation pattern is complex as a result of the presence of a number of pseudofaults, which offset the linear magnetic

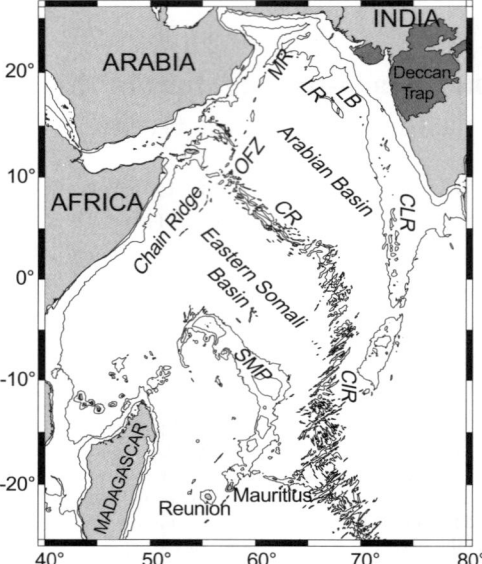

Fig. 1. Outline map of the NW Indian Ocean showing location of Arabian and Eastern Somali basins. Selected isobaths (200 and 3000 m) are based on digital data of General Bathymetric Chart of the Oceans (Intergovernmental Oceanographic Commission et al. 1997). CIR, Central Indian Ridge; CLR, Chagos–Laccadive Ridge; CR, Carlsberg Ridge; LB, Laxmi Basin; LR, Laxmi Ridge; MR, Murray Ridge; OFZ, Owen Fracture Zone; SMP, Seychelles–Mascarene Plateau.

isochrons obliquely (Miles & Roest 1993; Chaubey et al. 1998; Dyment 1998). However, although these studies were able to establish the broad pattern of propagating ridges and the fact that formation of both these basins was highly influenced by propagating spreading ridge segments during Paleogene time, details of the geometry of the propagating ridge segments and the associated pseudofaults were not mapped.

This paper presents an interpretation of the complex pattern of Paleogene magnetic anomaly lineations (63–43 Ma) in this area, based on a compilation of available sea-surface magnetic data. The resulting tectonic map delineates numerous propagating ridge segments and their associated characteristic tectonic elements (pseudofaults and transferred crust).

Data

The National Institute of Oceanography (NIO), India, and several French oceanographic research institutions have collected a large amount of sea-surface magnetic data in the Arabian and Eastern Somali basins. As part of an Indo-French collaborative research project (between NIO and Centre National de la Recherche Scientifique, Domaines Océaniques Laboratory), we have compiled all available Indian and French digital magnetic data in both basins. This dataset along with data obtained from the NGDC (National Geophysical Data Center 1998), USA, constitute the primary database for our study. This large dataset (Fig. 2) has been edited and corrected for some navigational discrepancies. The total magnetic intensity data have been reduced to residual magnetic anomalies by subtracting the International Geomagnetic Reference Field for appropriate epochs (IAGA Division V, Working Group 8, 2000). As is normally the case with deep-sea surveys, diurnal correction for the daily variation of the Earth's magnetic field was not applied. We have also digitized published magnetic contours (Karasik et al. 1986; Mercuriev & Sochevanova 1990), drawn from the dense magnetic surveys carried out by Russian scientists in the area, and used them as additional constraints. Together with the primary database, our compilation constitutes the most complete database of available magnetic and bathymetry data for the Arabian and Eastern Somali basins. In addition, existing bathy-

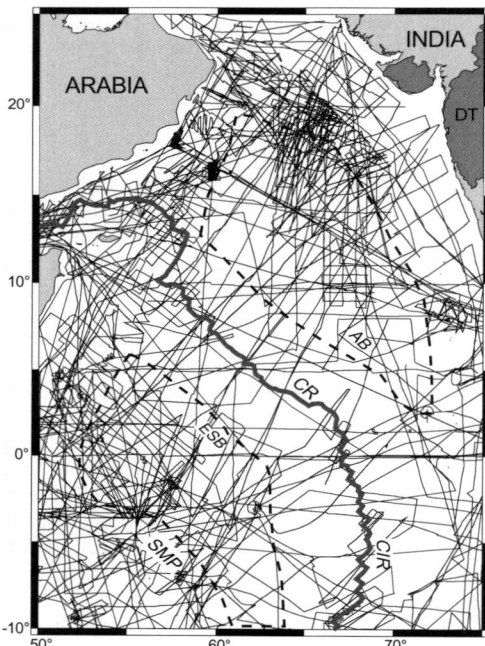

Fig. 2. Track chart of marine magnetic data coverage. Ship's tracks located within dashed blocks are used in the present study. AB, Arabian Basin; ESB, Eastern Somali Basin; DT, Deccan Trap. Other abbreviations as Figure 1.

metric contours from the General Bathymetric Chart of the Oceans (GEBCO) digital data (Intergovernmental Oceanographic Commission et al. 1997) and free-air gravity anomalies derived from satellite altimetry data (Sandwell & Smith 1997) have been extracted to constrain the structural framework of the area.

Method of analysis

Several techniques have been used to identify and precisely locate the sea-floor spreading isochrons in the study area. They include: (1) for clearly identifiable magnetic anomalies, the conventional method of inter-profile correlation and comparison with synthetic anomaly models; (2) for identification of ambiguous magnetic anomalies, the additional use of second-order, correlatable detailed features of the magnetic anomalies; (3) for precise location of the boundaries of magnetized blocks, the determination of the maxima of the modulus of the analytical signal.

We used magnetic anomaly and modulus of the analytical signal together to identify and locate selected isochrons, as we consider this more precise than the conventional method used in previous works. Such precise delineation of isochrons ensures reliable application of the stochastic method of palaeogeographical reconstruction (Chang 1988; Royer & Chang 1991; Kirkwood et al. 1999). The geomagnetic polarity time scale and the chron nomenclature proposed by Cande & Kent (1995) have been adopted throughout this paper (Table 1). The young (y) and old (o) edges of magnetic chrons refer to the end and beginning, respectively, of normal (n) polarity intervals. For example, Chron 24n.3y (52.903 Ma) corresponds to the (young) end of the third and oldest normal polarity interval of Chron 24n.

Identification of magnetic anomalies

We initially applied the conventional method of inter-profile correlation and comparison with synthetic anomaly models to identify magnetic anomalies between 27n and 20n in both basins, wherever those anomalies are clearly identifiable. Positions of selected magnetic anomaly profiles are shown in Fig. 3, whereas Fig. 4a and b shows anomaly correlation on these profiles and their correspondence to synthetics generated with a simple magnetic block model. It can be seen that some of the magnetic anomalies show good correlation. However, anomaly sequences are often complicated because they are divided into several segments by pseudofaults (Figures 3, 4a and b), which pose problems for their correlation and identification. We also observed that over a large part of both basins it is rare to find the entire normal and reverse sequences of Paleogene anomalies on any single profile, even though the profiles are long enough and oriented parallel to the spreading direction. Most profiles from the central parts of both basins display only a small part of this sequence. Such incomplete and isolated sequences of magnetic anomalies are often too short to allow reliable pattern recognition and establish correspondence to the magnetic block model. Therefore, we looked for correlatable detailed features within the magnetic anomalies, the 'tiny wiggles', which are low-amplitude short-wavelength secondary magnetic anomalies of geomagnetic origin (Cande & Kent 1992). This method was used in two earlier studies (Chaubey et al. 1998; Dyment 1998) to unambiguously identify anomalies in complex tectonic areas of both basins.

Analytical signal

The shape of marine magnetic anomalies depends upon the magnetic latitude and the orientation of the source body at the time it acquired magnetization. The oceanic crust of the Arabian and Eastern Somali basins was formed at low southern magnetic latitudes and has drifted substantially northward from its original location. The interpretation of the observed magnetic anomalies is therefore complicated by the distortion (skewness) of the anomalies, which induces significant horizontal displacements of the anomaly highs and lows with respect to the source geometry. The skewness corresponds to a phase shift applied to the reduced-to-the-pole anomaly, and depends on the inclination of the magnetization and ambient

Table 1. *Ages (Ma) of the magnetic chrons, after Cande & Kent (1995)*

Chron	Young edge	Old edge
20n	42.536	43.789
21n	46.264	47.906
22n	49.037	49.714
23n.1	50.778	50.946
23n.2	51.047	51.743
24n.1	52.364	52.663
24n.2	52.757	52.801
24n.3	52.903	53.347
25n	55.904	56.391
26n	57.554	57.911
27n	60.920	61.276
28n	62.499	63.634

The young (y) and old (o) edges of magnetic chrons refer to the end and beginning, respectively, of normal (n) polarity intervals.

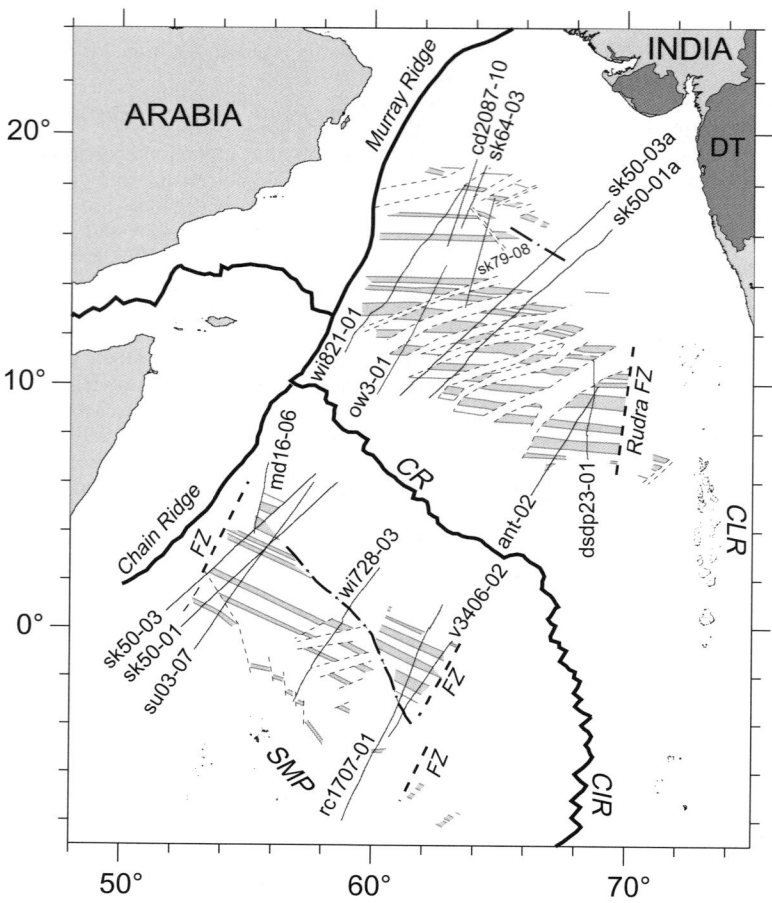

Fig. 3. Index map of selected track lines (continuous lines with label) used in Fig. 4a and b. Light shaded regions represent zones of normally magnetized crust. Thin dashed lines are pseudofaults. Thick dashed lines are fracture zones. Thick dash–dot lines are major outer pseudofaults.

geomagnetic field vectors (Schouten & McCamy 1972). To overcome the difficulty of picking polarity boundaries, we have used the modulus of the analytical signal (MAS), a zero phase signal which is easily computed from the Hilbert transform of the original signal (Nabighian 1972, 1974). MAS is derived from the square root of the sum of squares of the vertical and horizontal derivatives of the magnetic field (Roest et al. 1992). It corresponds to the envelope of all possible phase shifts of an observed anomaly. This envelope is clearly independent of the skewness of the observed profile. The envelope has maxima that are located directly above the magnetization contrasts, providing an easy way to precisely locate the polarity boundaries. These maxima of MAS correspond to the location of the polarity boundaries and their determination is independent of the Earth's magnetic field and source magnetization direction.

Initially, the locations of the polarity boundaries of the sea-floor spreading isochrons were automatically picked over the entire area. However, all detected maxima may not be associated with a real polarity boundary, either because of additional sources for secondary magnetization contrasts, such as geomagnetic field intensity variations and peculiar tectonic or volcanic features, or as a result of biases in applying MAS analysis on profiles over non-cylindrical magnetic sources. Therefore, each detected maximum was compared with the original magnetic anomaly, and any pick inconsistent with the anomaly was deleted and eventually repicked manually. Figure 5 presents an

MAGNETIC ISOCHRONS, ARABIAN AND SOMALI BASINS

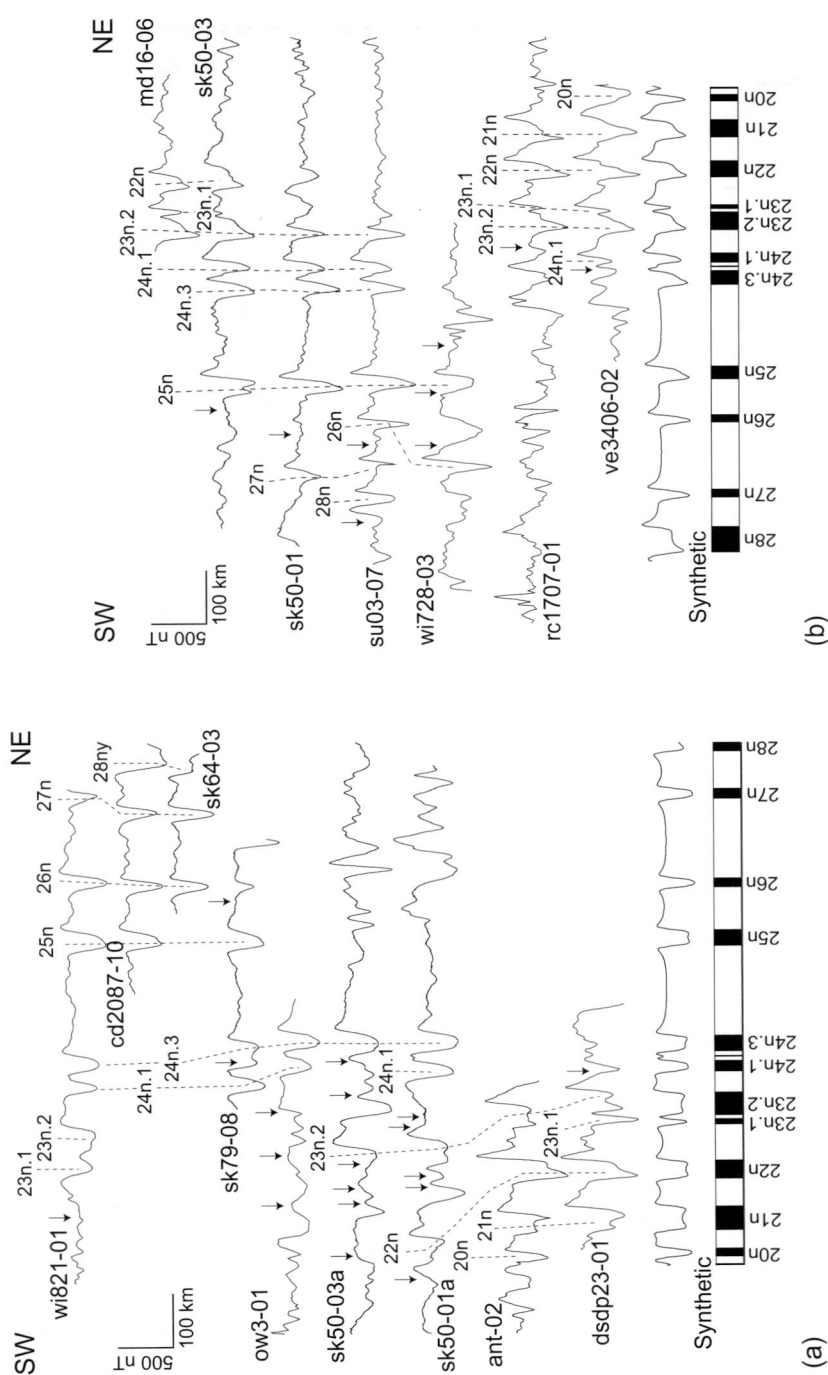

Fig. 4. (a) Selected magnetic anomaly profiles of the Arabian Basin projected at an azimuth of 0°N. Profile locations are shown in Fig. 3. Synthetic magnetic profile is computed for a ridge striking N50°W at latitude of 5°S and now observed at 12°N, 65°E. Normally magnetized blocks are indicated by black bars. Magnetized layer (susceptibility 0.01 c.g.s. unit) is flat, 2.0 km thick with its top at 5.5 km below the sea level. Half-spreading rates are as in Table 2. Magnetic anomaly identifications follow the geomagnetic polarity reversal time scale and polarity chron nomenclature proposed by Cande & Kent (1995). Thin dashed lines indicate correlations of magnetic anomalies. Arrows indicate the location where profiles intersect the interpreted pseudofaults. (b) Selected magnetic anomaly profiles of the Eastern Somali Basin projected along N30°E direction. Profile locations are shown in Fig. 3. Synthetic magnetic profile is computed for a ridge striking N50°W at latitude 5°S and now observed at 1°N, 57°E. Other details are as in (a).

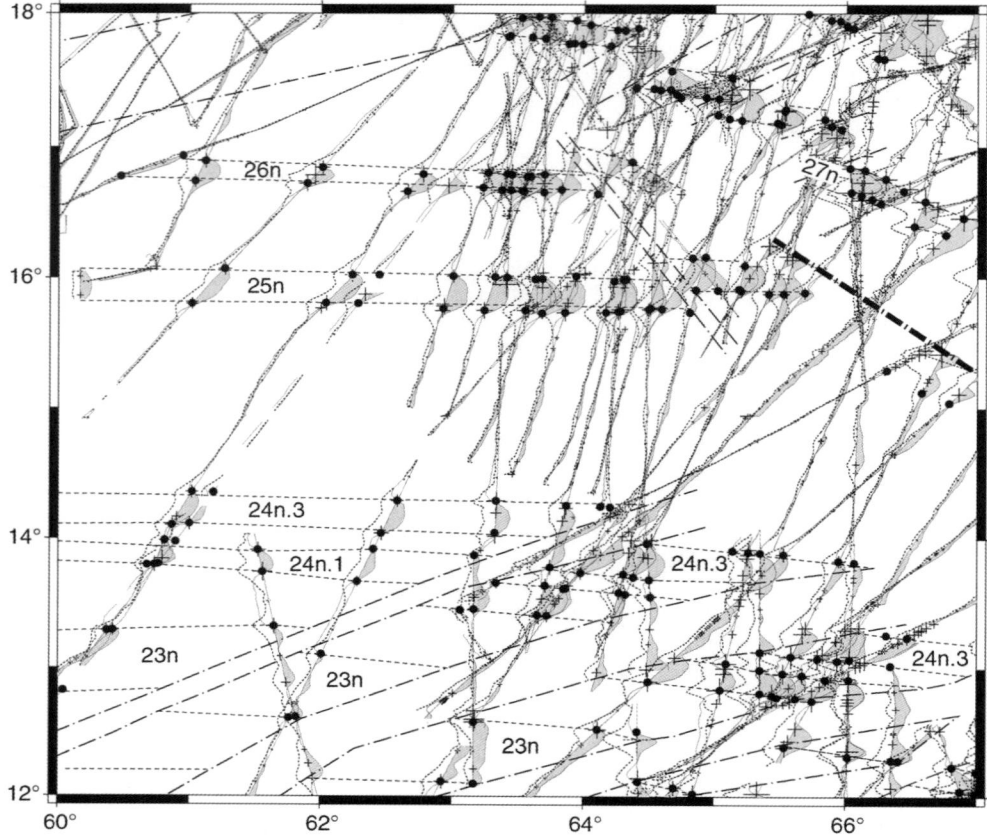

Fig. 5. Example showing joint interpretation of magnetic anomaly and modulus of the analytical signal (MAS; light dotted wiggles) computed from the magnetic anomaly to identify and precisely locate the younger and older boundaries of magnetic chrons. Negative magnetic anomalies are shaded. + (on track lines), relative amplitude of MAS; ●, digitally picked polarity boundaries. Other details are as in Fig. 7a.

example of the magnetic polarity boundaries of the sea-floor spreading isochrons determined using MAS.

Validation of interpreted isochrons

As mentioned above, the central parts of both basins contain isolated sequences of magnetic anomalies and additional normal and/or reverse events that are difficult to identify. Furthermore, several magnetic anomalies observed in the Eastern Somali Basin do not seem to have conjugates in the Arabian Basin and vice versa. This situation results from the capture of oceanic crust slivers by one basin from its conjugate during the process of spreading ridge segment propagation. To identify these anomalies and refine our interpretation in the complex area, we used plate reconstruction tools (Chang 1988; Royer & Chang 1991; Kirkwood et al. 1999) to reconstruct the magnetic lineations at various times. The rotation parameters (rotation pole and angle) computed from our consistent sets of magnetic picks 27n to 21n are given in a companion paper (Royer et al. 2002). We would like to emphasize here that the computed rotation parameters are based solely on consistent sets of conjugate magnetic picks (591 identifications) and a few available reliable fracture zones in both basins. Using the reconstruction parameters, we rotated all the magnetic data and picks of each chron, including the picks that were not used in the computation of rotation parameters because they are observed in only one of the basins. This iterative approach helped us to decide whether the interpretation of these picks is justified and consistent with the regional tectonic framework. To illustrate the use of the reconstructions, we show the magnetic anomalies (presented as wiggles projected perpendicular to the tracks), picks and interpreted lineations rotated from the Arabian Basin on to the Eastern Somali Basin at anomaly 24n.1y time (Fig. 6). This approach proved to be useful in refining our interpretations of magnetic isochrons and pseudofaults.

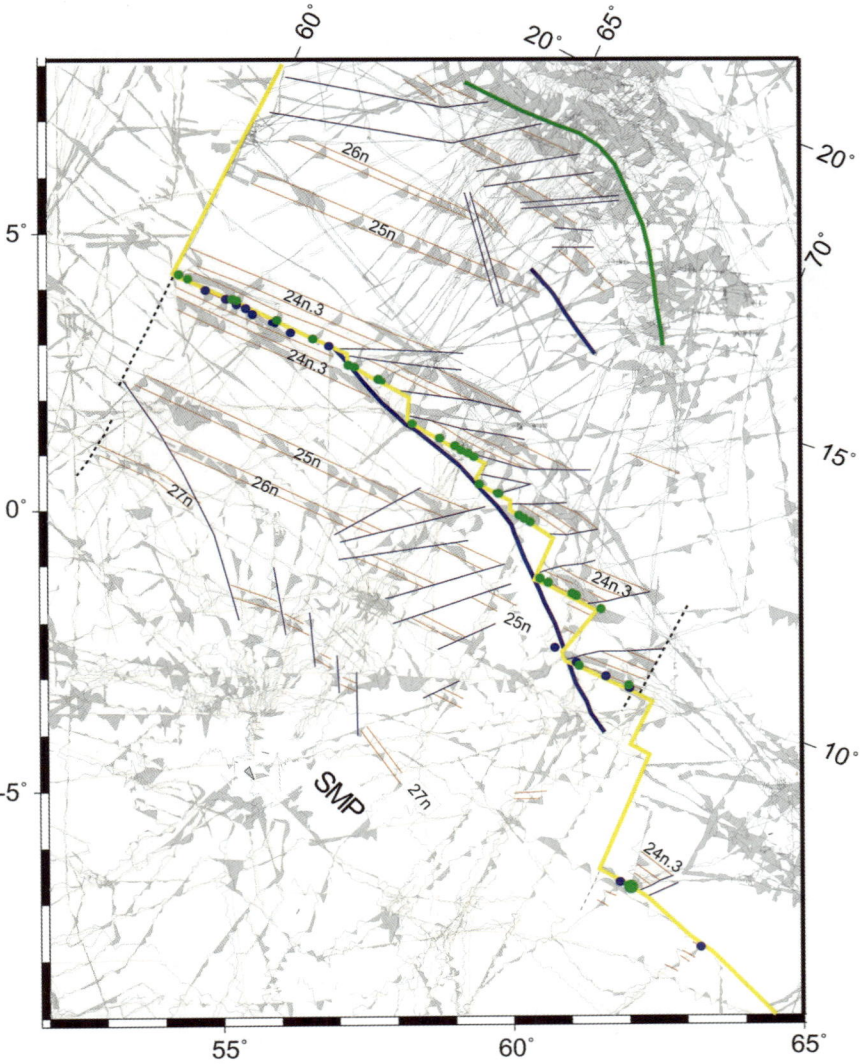

Fig. 6. Reconstructed magnetic anomaly wiggles, picks, lineations and fracture zones at Chron 24n.1y. Magnetic anomalies are plotted perpendicular to track lines and negative anomalies are shaded. An oblique Mercator projection about a pole at 65.27°N, 51.14°S is used for the rotation of Arabian Basin features relative to a fixed Somalian plate. Blue filled circles are unrotated magnetic anomaly picks (24n.1y) from the Eastern Somali Basin. Green filled circles are rotated picks from the conjugate Arabian Basin. Red lines are magnetic lineations; thin blue lines are pseudofaults; thick blue lines are major outer pseudofaults; the yellow line indicates the clip boundary between the Arabian and Eastern Somali basins (i.e. interpreted active plate boundary at Chron 24n.1y); the green line shows the axis of Laxmi Ridge gravity low.

Results and discussion

Magnetic lineation pattern

The techniques of magnetic anomaly identification and validation described above have been applied to the entire dataset for the Arabian and Eastern Somali basins. Figure 7a and b shows magnetic anomaly profiles together with magnetic anomaly picks and interpreted lineations in both basins. To avoid confusion, some of the tracks are not shown in the figure. The trend and extension of the magnetic lineations are defined by the young and old boundaries of the normally magnetized blocks. Anomalies 26n–22n are well developed and easy to correlate in the western sector of both basins.

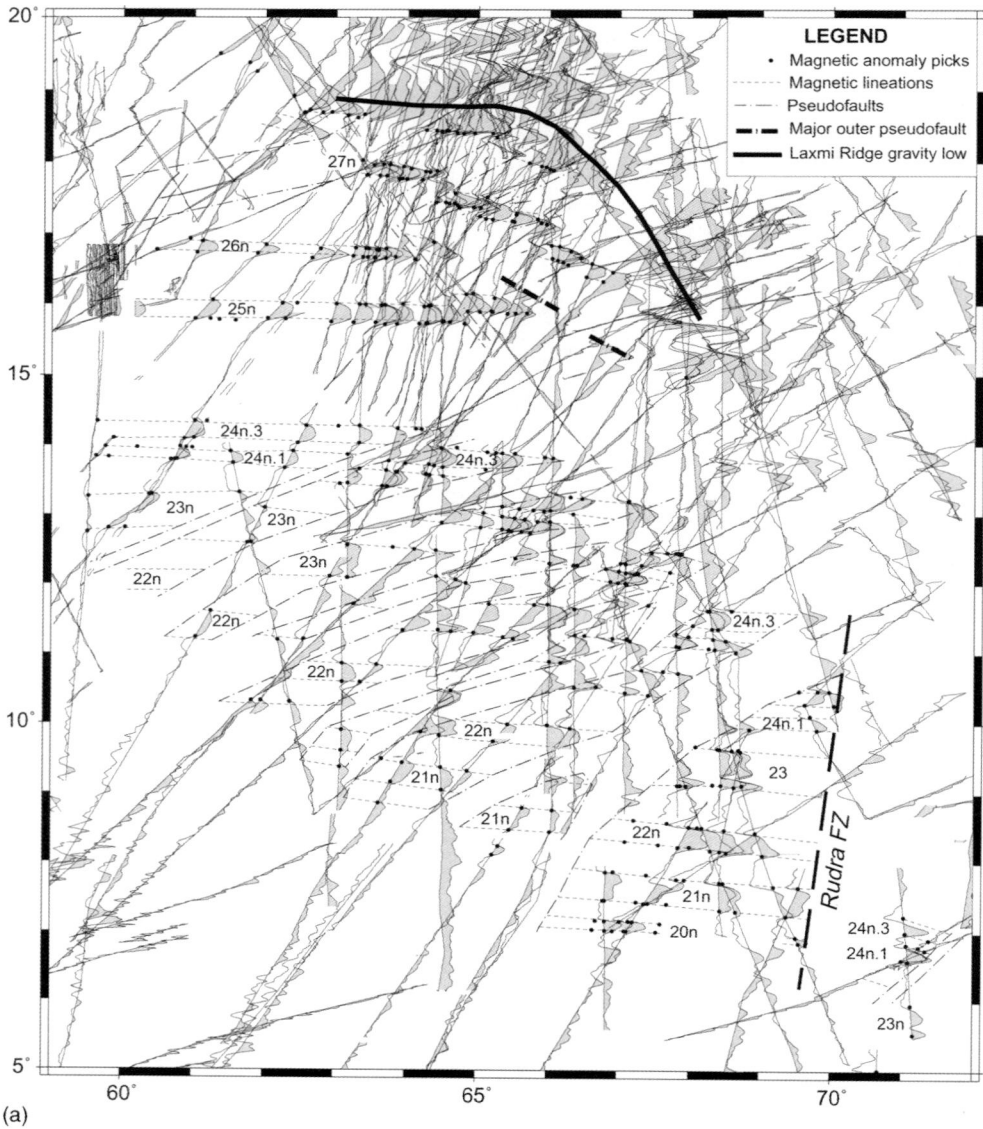

Fig. 7. Magnetic anomaly profiles along ship's tracks in the (**a**) Arabian and (**b**) Eastern Somali basins. Negative anomalies are shaded. Magnetic anomaly plot direction is N90°E and N120°E for Arabian and Eastern Somali basins, respectively. Some of the tracks are removed to avoid clustering. Fracture zones (FZ) are shown with thick dashed lines.

Similarly, anomalies 24n.3–20n are also confidently recognized in the eastern sector of both basins. Unlike the eastern and western sectors, the central sectors of both basins are characterized by complex anomaly patterns. The magnetic lineations 24n.3–20n in the Arabian Basin are short and right laterally offset (Fig. 7a), whereas corresponding conjugate anomalies in the Eastern Somali Basin (Fig. 7b) are missing. Similarly, the magnetic lineations 26n–25n in the Eastern Somali Basin are short and generally right laterally offset (Fig. 7b), whereas corresponding conjugate anomalies in the Arabian Basin are missing. This later observation is mainly deduced from Russian magnetic anomaly contours (Mercuriev & Sochevanova 1990), as our data are sparse in this area.

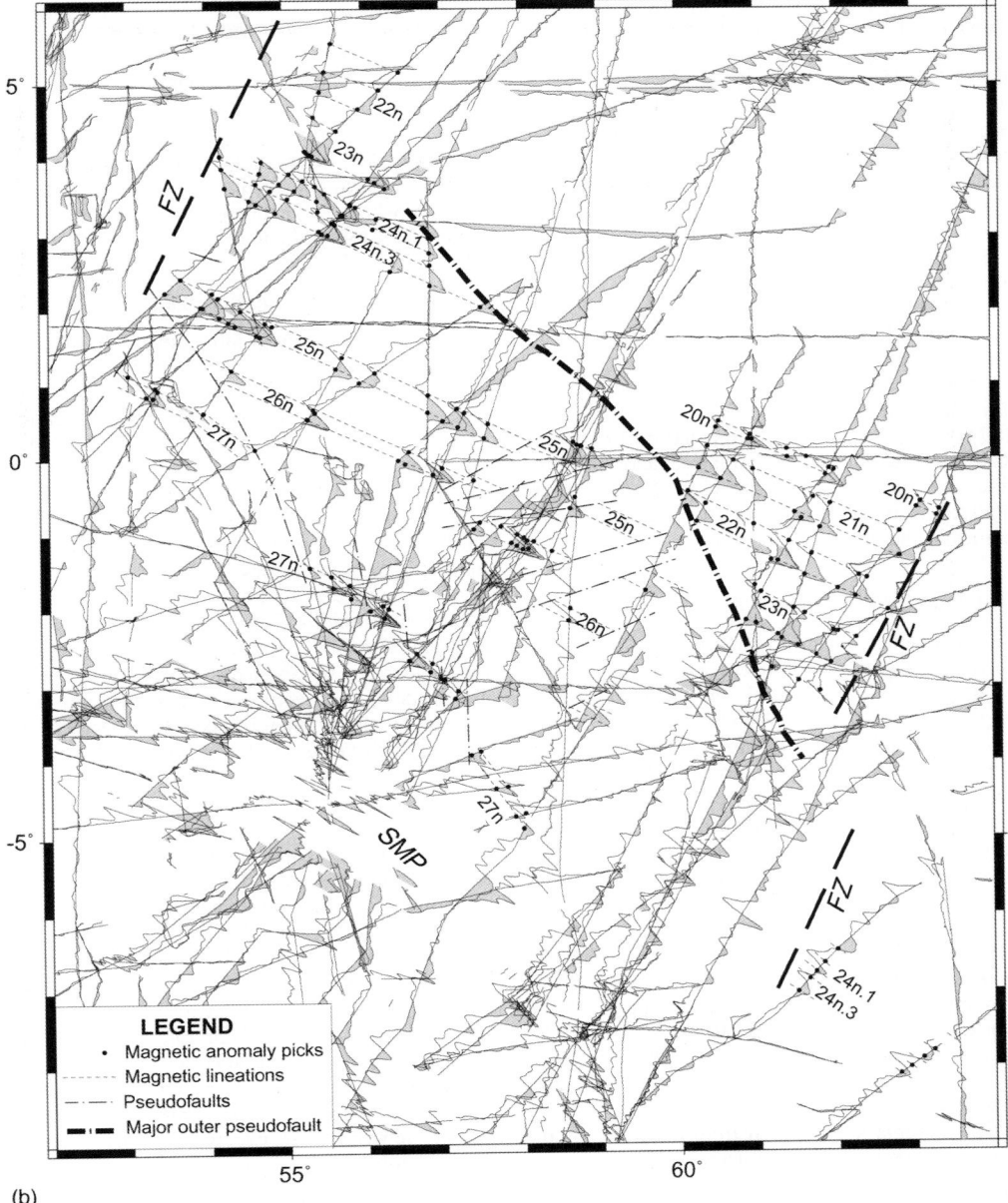

(b)

Fig. 7. (continued).

Magnetic lineation 27n and the younger edge of magnetic lineation 28n have been identified in the northern sector of the Arabian Basin near Laxmi Ridge. Magnetic lineation 27n has been identified in the southwestern sector of the Eastern Somali Basin near the Seychelles Plateau, whereas identifications of anomaly 28n still remain debatable north of the Seychelles Plateau despite the availability of numerous magnetic profiles. For instance, the high-amplitude magnetic anomaly along the Laxmi Ridge in the northwestern part has been interpreted either as anomaly 28n (Naini & Talwani 1983; Miles & Roest 1993; Chaubey et al. 1998) or as the signature of the continent–ocean boundary (Miles et al. 1998). The northeastern part of the Arabian Basin and southeastern part of the Eastern Somali Basin are characterized by a complex pattern of anomalies, which did not permit reliable identification.

In the easternmost part of both the basins, consistent sets of anomalies 24n–22n have been confidently identified and imply about 200 km right-lateral offset from the main part of the basin. This offset in the Arabian Basin is accommodated by a fracture zone (Rudra Fracture Zone), which has been inferred from seismic data (Whitmarsh 1974) and can be partly deciphered on the gravity data (dashed line at 70°E, Figs. 8 and 9). Other fracture zones depicted in Figs. 8 and 9 are inferred from linear gravity trends and discontinuities in the magnetic lineations.

Propagating ridge segments

The magnetic lineation pattern (Fig. 9) shows consistent right-lateral offsets, except at the limit of the western and central sectors (near 16.5°N, 64.5°E in the Arabian Basin and 0°, 58°E in the Eastern Somali Basin) where magnetic lineations 26n–25n show c. 20 km left lateral offset (Fig. 9). Except for the Rudra Fracture Zone, the observed offsets are clearly oblique (Figs. 8 and 9) and vary from c. 15 km to 115 km. Most of these oblique offsets coincide with similar trends on the gravity anomaly map derived from satellite altimetry data (Fig. 8). Oblique offsets of magnetic lineations are the diagnostic feature of the propagating spreading ridge segments (Hey 1977; Hey et al. 1980), similar to those defined in several parts of the world's oceans (Hey et al. 1986; Wilson 1988; Brozena & White 1990; Dyment 1993; Auzende et al. 1994; Morgan & Sandwell 1994; Wilson & Hey 1995).

The tectonic map derived from magnetic anomaly identification (Fig. 9) suggests three major stages of evolution for the Arabian and Eastern Somali basins between Chrons 28n and 20n. The first stage, between Chrons 28n and 27n, is characterized by short segments propagating westward between the Laxmi Ridge and the Seychelles Plateau. The second stage; between Chrons 26n and 25n, shows a general eastward propagation, with a long western segment and several short segments in the central sector. These short segments are observed in the Eastern Somali Basin only and do not have conjugates in the Arabian Basin. The third stage, between Chrons 24r and 20n, presents a systematic westward propagation throughout the basins, with the westernmost segment receding and the easternmost segment lengthening. The short segments in the central sector are observed in the Arabian Basin only and do not have conjugate segments in the Eastern Somali Basin. Instead, this stage is marked by a prominent tectonic feature on the gravity map (Fig. 8) which may correspond to the trace of a major outer pseudofault.

Spreading and propagation rates

Average half-spreading rates for each magnetic anomaly interval have been determined using undisturbed conjugate lineation sets of both the basins. For this purpose, magnetic lineations 28ny–27ny and 26ny–24n.1y located in the westernmost sectors (A and AA in Fig. 9) and magnetic lineations 24n.1y–20n located in the easternmost sectors (B and BB in Fig. 9) of both basins are considered. These sectors were carefully chosen because they contain conjugate lineation sets and are continuous for long distances without any major offset produced by pseudofaults or fracture zones. To estimate the spreading rate for each magnetic anomaly interval, the orthogonal distance between each magnetic lineation, defined by the anomaly picks, is measured. The measured distances between two isochrons are averaged to obtain a representative distance and the spreading rates are then estimated using the geomagnetic polarity time scale of Cande & Kent (1995). The estimated half-spreading rates between lineations 28ny and 20ny and the mean total spreading rates are presented in Table 2. The spreading rate between lineation 27ny and 26no could not be measured, because there was no undisturbed section of this segment in either Arabian or Eastern Somali basins.

Variations in half-spreading rates (Table 2) with respect to time in the conjugate Arabian and Eastern Somali basins are shown in Fig. 10. It appears that some persistent spreading asymmetry existed between these two basins. Spreading was c. 20% consistently faster in the Arabian Basin than in the Eastern Somali Basin except during the period 24n.1y–23n.2o (52.4–51.7 Ma), when Eastern Somali Basin spreading was c. 10% faster

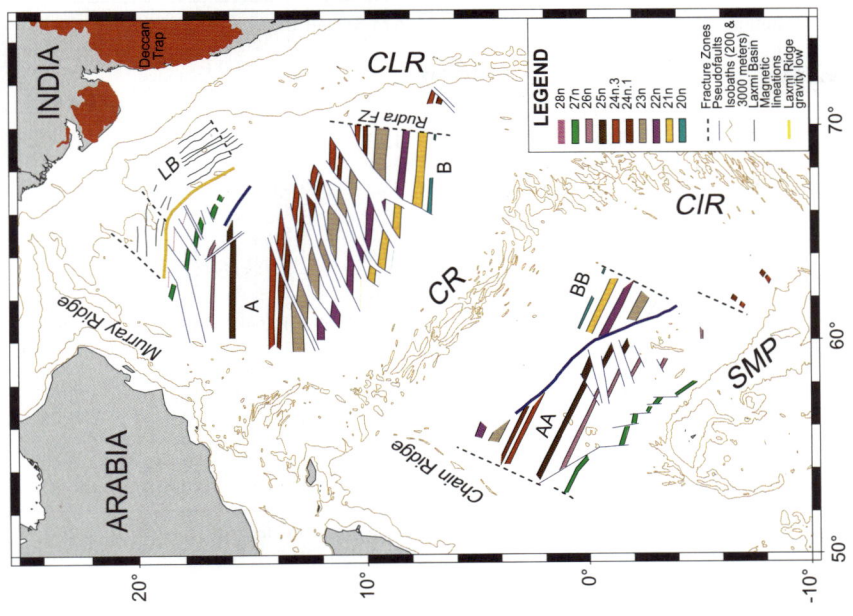

Fig. 9. Tectonic chart of the Paleogene magnetic anomaly lineations, pseudofaults, fracture zones and major tectonic features in the Arabian and Eastern Somali basins. Thick dark blue lines in the Arabian and Eastern Somali basins are major outer pseudofaults. Inferred magnetic lineations (thin black lines), in the Laxmi Basin after Bhattacharya et al. (1994) and north of the Laxmi Ridge after Malod et al. (1997). The 200 and 3000 m GEBCO isobaths are shown for reference. Other details are as in Fig. 1.

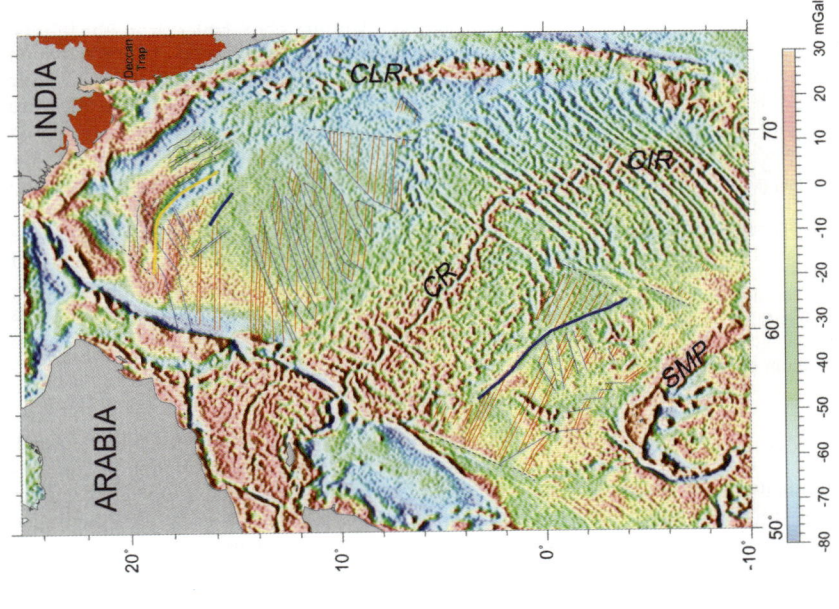

Fig. 8. Paleogene magnetic anomaly lineations (red lines), pseudofaults (thin blue lines), and fracture zones (dashed lines) overlain on satellite-derived gravity anomaly map (Sandwell & Smith 1997), which is illuminated from the north. Thick dark blue lines in the Arabian and Eastern Somali basins are major outer pseudofaults. Thick yellow line in the Arabian Basin shows axis of Laxmi Ridge gravity low. Inferred magnetic lineations (thin black lines), in the Laxmi Basin after Bhattacharya et al. (1994) and north of Laxmi Ridge after Malod et al. (1997). Other details are as in Fig. 1.

Table 2. *Half-spreading rates (HSR) for the Arabian and Eastern Somali basins, calculated from the undisturbed and well-defined conjugate magnetic lineations (A and AA, B and BB areas in Fig. 9)*

Chrons	Duration (Ma)	Arabian Basin		Eastern Somali Basin		Total	
		Distance (km)	HSR (cm a^{-1})	Distance (km)	HSR (cm a^{-1})	Distance (km)	Rate (cm a^{-1})
20ny–20no	1.253	15.8	1.26	13.3	1.06	29.1	2.32
20no–21ny	2.475	33.2	1.34	34.1	1.38	67.30	2.72
21ny–21no	1.642	45.5	2.77	33.7	2.05	79.20	4.82
21no–22ny	1.131	50.8	4.49	43.2	3.82	94.00	8.31
22ny–22no	0.677	35.0	5.17	31.5	4.65	66.50	9.82
22no–23n.1y	1.064	66.5	6.25	50.9	4.78	117.40	11.03
23n.1y–23n.2o	0.965	59.7	6.19	48.3	5.01	108.00	11.19
23n.2o–24n.1y	0.621	38.5	6.20	42.4	6.83	80.90	13.03
24n.1y–24n.3o	0.983	68.3	6.95	59.5	6.05	127.80	13.00
24n.3o–25ny	2.557	166.3	6.50	152.2	5.95	318.50	12.46
25ny–25no	0.487	30.6	6.28	25.3	5.20	55.90	11.48
25no–26ny	1.163	76.8	6.60	65.8	5.66	142.60	12.26
26ny–26no	0.357	17.5	4.90	14.9	4.17	32.40	9.08
26no–27ny	3.009	–	–	–	–	–	–
27ny–27no	0.356	19.7	5.53	15.8	4.44	35.50	9.97
27no–28ny	1.223	68.3	5.59	54.8	4.48	123.10	10.07

The young (y) and old (o) edges refer to the end and beginning, respectively, of normal (n) polarity intervals.

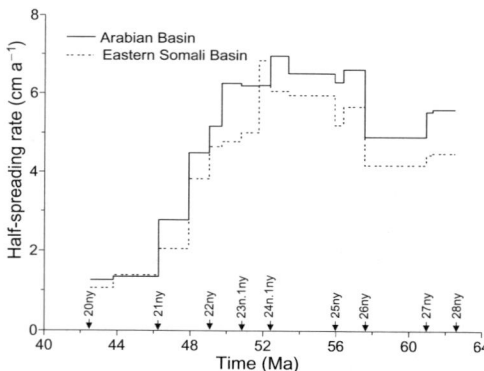

Fig. 10. Variations in half-spreading rates (Table 2) with respect to time in the conjugate Arabian and Eastern Somali basins. In general, the average spreading is *c.* 20% faster in the Arabian Basin than in the Eastern Somali Basin. It may be noted that crustal asymmetry owing to asymmetric spreading is different from the asymmetry in crustal accretion caused by propagating spreading ridge segments.

compared with its conjugate. Interestingly, a gradual decrease in spreading rates started immediately after the period 24n.1y–23n.2o and continued at least until Chron 20n in both the basins.

Variations in propagation rates have been measured in the Arabian Basin for the younger episode of westward propagation, i.e. between Chrons 24n and 20n. We have restricted the analysis to this limited set of anomalies, because the propagators defined in the Eastern Somali Basin between Chrons 25n and 26n partly rely on the digitized Russian maps and may lack the required precision. Similarly, the early episode of westward propagation between Chrons 28ny and 27n is too limited to allow reliable measurements of propagation rates. The propagation rates decrease with time from about 25 cm a^{-1} at Chron 24n to nearly zero at Chrons 21n and 20n (Fig. 11), which results in the arcuate shape of the pseudofaults in the Arabian Basin (Fig. 11). The apparent link between propagation rate and spreading rate (correlation coefficient of 0.75) suggests a possible relationship between the systematic ridge segment propagation and stress variations at plate boundaries such as those active during the initial stages of the Indo-Eurasia collision.

Conclusions

Analysis of an up-to-date compilation of marine magnetic data allowed reconfirmation of the existence of numerous spreading ridge propagation events in the Arabian and Eastern Somali basins and the preparation of a new magnetic isochron map for these basins. The dominant propagation direction was westward during Chrons 28ny–27n, eastward during Chrons 26n–25n, and westward again during Chrons 24n–20n. In addition to the

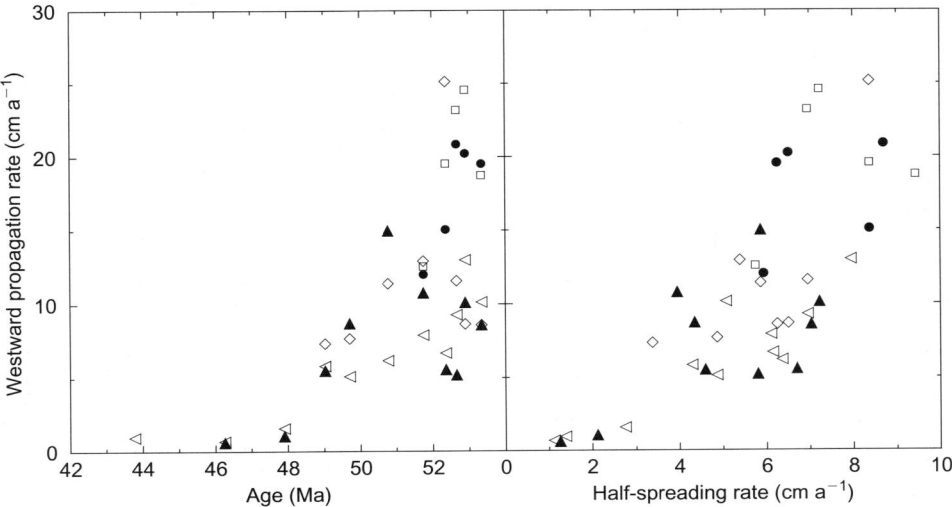

Fig. 11. Variations in westward propagation rate in the Arabian Basin between Chrons 24n and 20n v. age (left) and half-spreading rate (right). Symbols refer to the various propagating ridge segments.

transfer of crust through ridge propagation, seafloor spreading was consistently asymmetrical for 20 Ma. Measured spreading rates also show a major slow-down of the India–Somalia relative motions after Chron 24n.1y (c. 52 Ma). The magnetic anomalies and tectonic fabric of both these basins could be resolved over most of the area; however, complexities in small parts of the NE Arabian Basin and of the SE Eastern Somali Basin still remain elusive. The genesis and cause of long (>20 Ma) sustenance of these propagating ridges and their abrupt change in propagation direction are yet to be understood.

The authors are thankful to E. Desa, Director, NIO, and P. Tarits, IUEM, for support and encouragement to carry out this work. The Indo-French Centre for the Promotion of Advanced Research (IFCPAR), New Delhi, funded this work under project grant 1911-1. A.K.C. gratefully acknowledges the support provided by IFCPAR for Post-Doctoral Fellowship at CNRS UMR 6538, Institut Universitaire Européen de la Mer, Université de Bretagne Occidentale, Plouzané. J.D. acknowledges the support of the French Embassy in India at an early stage of this study. Many thanks go to D. Cardinal for his help. We thank P. Miles, P. Clift and an anonymous reviewer for their careful reviews of the manuscript and valuable comments. All figures were drafted with GMT software (Wessel & Smith 1995). This is NIO Contribution 3724.

References

AUZENDE, J.M., PELLETIER, B. & LAFOY, Y. 1994. Twin active spreading ridges in the North Fiji Basin (Southwest Pacific). *Geology,* **22,** 63–66.

BHATTACHARJI, S., CHATTERJEE, N., WAMPLER, J.M., NAYAK, P.N. & DESHMUKH, S.S. 1996. Indian intraplate and continental margin rifting, lithospheric extension, and mantle upwelling in Deccan flood basalt volcanism near the K/T boundary: evidence from mafic dike swarms. *Journal of Geology,* **104,** 379–398.

BHATTACHARYA, G. C., CHAUBEY, A. K., MURTY, G. P. S. & 5 OTHERS 1992. Marine magnetic anomalies in the north-eastern Arabian Sea. *In:* Desai, B. N. (ed.) *Oceanography of the Indian Ocean.* Oxford and IBH, New Delhi, 503–509.

BHATTACHARYA, G.C., CHAUBEY, A.K., MURTY, G.P.S., SRINIVAS, K., SARMA, K.V.L.N.S., SUBRAHMANYAM, V. & KRISHNA, K.S. 1994. Evidence for seafloor spreading in the Laxmi Basin, northeastern Arabian Sea. *Earth and Planetary Science Letters,* **125,** 211–220.

BROZENA, J.M. & WHITE, R.S. 1990. Ridge jumps and propagations in the South Atlantic Ocean. *Nature,* **348,** 149–152.

CANDE, S.C. & KENT, D.V. 1992. Ultrahigh resolution marine magnetic anomaly profiles: a record of continuous paleo-intensity variations? *Journal of Geophysical Research,* **97,** 15075–15083.

CANDE, S.C. & KENT, D.V. 1995. Revised calibration of the geomagnetic polarity timescale for the Late Cretaceous and Cenozoic. *Journal of Geophysical Research,* **100,** 6093–6095.

CHANG, T. 1988. Estimating the relative rotation of two tectonic plates from boundary crossings. *Journal of the American Statistical Association,* **83,** 1178–1183.

CHAUBEY, A.K., BHATTACHARYA, G.C., MURTY, G.P.S., SRINIVAS, K., RAMPRASAD, T. & GOPALA RAO, D. 1998. Early Tertiary seafloor spreading magnetic anomalies and paleo-propagators in the northern

Arabian Sea. *Earth and Planetary Science Letters*, **154**, 41–52.

CHAUBEY, A.K., BHATTACHARYA, G.C. & RAO, D.G. 1995. Seafloor spreading magnetic anomalies in the southeastern Arabian Sea. *Marine Geology*, **128**, 105–114.

COURTILLOT, V., BESSE, J., VANDAMME, D., MONTIGNY, R., JAEGER, J.J. & CAPPETTA, H. 1986. Deccan flood basalts at the Cretaceous/Tertiary boundary? *Earth and Planetary Science Letters*, **80**, 361–374.

DUNCAN, R.A. 1981. Hotspots in the southern oceans—an absolute frame of reference for motion of the Gondwana continents. *Tectonophysics*, **74**, 29–42.

DYMENT, J. 1993. Evolution of the Indian Ocean triple junction between 65 and 49 Ma (anomalies 28 to 21). *Journal of Geophysical Research*, **98**, 13863–13877.

DYMENT, J. 1998. Evolution of the Carlsberg Ridge between 60 and 45 Ma: ridge propagation, spreading asymmetry, and the Deccan–Reunion hotspot. *Journal of Geophysical Research*, **103**, 24067–24084.

GLEBOVSKY, V. YU., MASCHENKOV, S. P., GORODNITSKY, A.M. & 8 OTHERS 1995. Mid-oceanic ridges and deep oceanic basins: AMF structure. *In*: Gorodnitsky, A. M. (ed.) *Anomalous Magnetic Field of the World Ocean*. CRC Press, Boca Raton, FL, 67–144.

HEY, R. 1977. A new class of 'pseudofaults' and their bearing on plate tectonics: a propagating rift model. *Earth and Planetary Science Letters*, **37**, 321–325.

HEY, R., DUENNEBIER, F.K. & MORGAN, W.J. 1980. Propagating rifts on midocean ridges. *Journal of Geophysical Research*, **85**, 3647–3658.

HEY, R.N., KLEINROCK, M.C., MILLER, S.P., ATWATER, T.M. & SEARLE, R.C. 1986. Sea beam/deep-tow investigation of an active oceanic propagating rift system, Galapagos 95.5°W. *Journal of Geophysical Research*, **91**, 3369–3393.

IAGA Division V, Working Group 8 2000. International Geomagnetic Reference Field—2000. *Physics of the Earth and Planetary Interiors*, **120**, 39–42.

Intergovernmental Oceanographic Commission, International Hydrographic Organization & British Oceanographic Data Centre 1997. *General Bathymetric Chart of the Oceans (GEBCO) Digital Atlas* (CD-ROM), 1997 edn. British Oceanographic Data Centre, Birkenhead.

KARASIK, A.M., MERCURIEV, S.A., MITIN, L.I., SOCHEVANOVA, N.A. & YANOVSKIY, V.N. 1986. Main features in the history of opening of the Arabian Sea, according to data of a systematic magnetic survey. *Transactions of the USSR Academy of Science, Earth Science Section*, **286**, 933–938.

KIRKWOOD, B.H., ROYER, J.-Y., CHANG, T.C. & GORDON, R.G. 1999. Statistical tools for estimating and combining finite rotations and their uncertainties. *Geophysical Journal International*, **137**, 408–428.

MALOD, J.A., DROZ, L., KEMAL, B.M. & PATRIAT, P. 1997. Early spreading and continental to oceanic basement transition beneath the Indus deep-sea fan: northeastern Arabian Sea. *Marine Geology*, **141**, 221–235.

MCKENZIE, D. & SCLATER, J.G. 1971. The evolution of the Indian Ocean since the Late Cretaceous. *Geophysical Journal of the Royal Astronomical Society*, **25**, 437–528.

MERCURIEV, S. & SOCHEVANOVA, N. 1990. Complex patterns of the magnetic field as a consequence of an ancient tripple junction on the Carlsberg Ridge ? (in Russian). *In:* JDANOV, M. (ed.) *Electromagnetic Induction in the World Ocean*. USSR Academy of Sciences, Moscow, 48–56.

MILES, P.R. & ROEST, W.R. 1993. Earliest sea-floor spreading magnetic anomalies in the north Arabian Sea and the ocean–continent transition. *Geophysical Journal International*, **115**, 1025–1031.

MILES, P.R., MUNSCHY, M. & SÉGOUFIN, J. 1998. Structure and early evolution of the Arabian Sea and East Somali Basin. *Geophysical Journal International*, **134**, 876–888.

MORGAN, W.J. 1981. Hotspot tracks and the opening of the Atlantic and Indian Oceans. *In:* EMILIANI, C. (ed.) *The Sea, 7.* Wiley–Interscience, New York, 443–487.

MORGAN, J.P. & SANDWELL, D.T. 1994. Systematics of ridge propagation south of 30°S. *Earth and Planetary Science Letters*, **121**, 245–258.

NABIGHIAN, M.N. 1972. The analytical signal of two-dimensional magnetic bodies with polygonal cross-section: its properties and use for automated anomaly interpretation. *Geophysics*, **37**, 507–517.

NABIGHIAN, M.N. 1974. Additional comments on the analytic signal of two-dimensional magnetic bodies with polygonal cross-section. *Geophysics*, **39**, 85–92.

NAINI, B.R. & TALWANI, M. 1983. Structural framework and the evolutionary history of the continental margin of western India. *In:* WATKINS, J.S. & DRAKE, C.L. (eds) *Studies in Continental Margin Geology*. American Association of Petroleum Geology Memoir, **34**, 167–191.

NATIONAL GEOPHYSICAL DATA CENTER 1998. *GEODAS CD-ROM, Worldwide Marine Geophysical Data*, updated 4th edn. National Oceanic and Atmospheric Administration, US Department of Commerce, Washington, DC, 98-MGG-04.

NORTON, I.O. & SCLATER, J.G. 1979. A model for the evolution of the Indian Ocean and the breakup of Gondwanaland. *Journal of Geophysical Research*, **84**, 6803–6830.

PATRIAT, P. & ACHACHE, J. 1984. India–Eurasia collision chronology has implications for crustal shortening and driving mechanism of plates. *Nature*, **311**, 615–621.

ROEST, W.R., ARKANI-HAMED, J. & VERHOEF, J. 1992. The seafloor spreading rate dependence of the anomalous skewness of marine magnetic anomalies. *Geophysical Journal International*, **109**, 653–659.

ROYER, J.-Y. & CHANG, T. 1991. Evidence for relative motions between the Indian and Australian plates during the last 20 m.y. from plate tectonic reconstructions: implications for the deformation of the Indo-Australian plate. *Journal of Geophysical Research*, **96**, 11779–11802.

ROYER, J.-Y., CHAUBEY, A.K., DYMENT, J., BHATTACHARYA, G.C., SRINIVAS, K., YATHEESH, V. & RAMPRASAD, T. 2002. Paleogene plate tectonic evolution of the Arabian and Eastern Somali basins. *In:* CLIFT, P.D., KROON, D., GAEDICKE, C. & CRAIG, J. (eds) *The Geological and Climatic Evolution of the Arabian Sea Region*. Geological Society, London, Special Publications, **195**, 7–24.

SANDWELL, D.T. & SMITH, W.H.F. 1997. Marine gravity anomaly from Geosat and ERS 1 satellite altimetry. *Journal of Geophysical Research*, **102**, 10039–10054.

SCHLICH, R. 1982. The Indian Ocean: aseismic ridges, spreading centres and oceanic basins. *In:* NAIRN, A.E.M. & STEHLI, F.G. (eds) *The Ocean Basins and Margins, The Indian Ocean, 6*. Plenum, New York, 51–147.

SCHOUTEN, H. & MCCAMY, K. 1972. Filtering marine magnetic anomalies. *Journal of Geophysical Research*, **77**, 7089–7099.

VANDAMME, D., COURTILLOT, V., BESSE, J. & MONTIGNY, R. 1991. Paleomagnetism and age determination of the Deccan Traps (India): results of a Nagpur–Bombay traverse and review of earlier work. *Reviews of Geophysics*, **29**, 159–190.

WESSEL, P. & SMITH, W.H.F. 1995. New version of the Generic Mapping Tools released. *EOS Transactions, American Geophysical Union*, **76**, 329.

WHITMARSH, R.B. 1974. Some aspects of plate tectonics in the Arabian Sea. *In:* WHITMARSH, R.B., WESER, O.E. & ROSS, D.A. (eds) *Initial Reports of the Deep Sea Drilling Project, 23*. US Government Printing Office, Washington, DC, 527–535.

WILSON, D.S. 1988. Tectonic history of the Juan de Fuca ridge over the last 40 million years. *Journal of Geophysical Research*, **93**, 11863–11876.

WILSON, D.S. & HEY, R.N. 1995. History of rift propagation and magnetization intensity for the Cocos–Nazca spreading center. *Journal of Geophysical Research*, **100**, 10041–10056.

Neotectonics on the Arabian Sea coasts

CLAUDIO VITA-FINZI

Department of Mineralogy, Natural History Museum, London SW7 5BD, UK
(e-mail: cvitafinzi@hotmail.com)

Abstract: The Holocene record on the coasts of the Arabian Sea provides information on the nature and rate of deformation generated by the interaction between the Indian, Arabian and Eurasian Plates. Holocene marine terraces show that the southern Makran has been subject to the infrequent but vigorous coseismic uplift (≤ 2 m) that characterizes other subduction settings and they indicate landward rotation of the imbricate faults among which shortening is distributed. The lack of significant Holocene deformation on the SE coast of the Arabian peninsula is consistent with its position parallel to a transform, although there is evidence for large-scale buckling driven by convergence at the Strait of Hormuz in the NE. Geomorphological and tide-gauge evidence for localized uplift on the southwestern coast of India may represent compressional buckling here too in response to Himalayan collision. Bathymetric and geodetic data can help to bridge these sequences and thus enhance their value for quantifying plate rheology and dynamics, notably by linking variations in plate-margin displacement with major sea-floor strike-slip structures and by eventually confirming transitory as well as sustained compressive buckling on land.

The Arabian Sea region provides scope for examining the short-term effects of subduction and transform displacement along the margins of a portion of lithosphere. An attempt to explore the question of marginal deformation of the Arabian Plate (Vita-Finzi 2001) was hampered by a lack of seismicity and geodetic data with which to trace the modern stress field. In the Arabian Sea the inaccessibility caused by submergence is partly outweighed by a wealth of geophysical data and of instrumented earthquakes. Although the variety of terrains and tectonic regimes at issue must make any proposed generalizations very provisional, the attempt is justified by the light its outcome may shed on regional neotectonics and geophysics.

This paper is based primarily on the Holocene coastal record, which is long enough for tectonic trends to emerge but still remains within range of relatively secure ^{14}C dating during a period when sea-level fluctuations are well documented. To judge from the available literature for the region, progress in marine geophysics and in seismology has not been matched in the study of Holocene tectonics, whence the reliance on some elderly sources in this paper. The primary aim is to assess the relevant published data in the light of current ideas about regional deformation so as to highlight issues where field observations can fruitfully link geodetic measurements with structural modelling.

The shorelines in question include those along the Gulf of Aden, the SE coast of Arabia and the Gulf of Oman in the west; the Makran of Iran and Pakistan in the north; and the Indian peninsula in the east. There is little geographical agreement between plate boundaries and the present-day coasts: the sea is crossed by the transform boundary between the Arabian Plate (AR) and the Indian Plate (IN) along the Owen Fracture Zone and the Murray Ridge, and by the spreading ridge between the African Plate (AF) and Arabia along the Gulf of Aden. It covers part of the subduction front between Arabia and the Eurasian Plate (EU) along the Makran coast (Fig. 1).

In this paper the Arabian Sea is defined to the SW by the Carlsberg Ridge, which is linked to the Red Sea–East African system by the Sheba Ridge (from which it is separated by the Owen Fracture Zone), and to the SE by the Chagos–Laccadive Ridge (CLR). The spreading half-rate (anomalies 1–5) for the Carlsberg Ridge is 1.2–1.3 cm a^{-1}; the Arabian Sea anomalies are no younger than 23 (Schlich 1982) and thus have little bearing on Holocene events. Similarly, the CLR has been equated with a transform fault that was active during Indian northward movement in Cretaceous–Eocene time (McKenzie & Sclater 1971). There is growing acceptance of an embryonic division between an Indian Plate and an Australian Plate (e.g. Stein *et al.* 1990). The proposed boundary zone (Fig. 1) is relatively active seismically and there is evidence of folding in both sediments and crust.

A subdivision of the northern Arabian Plate, the Ormara microplate, has also been identified. Its

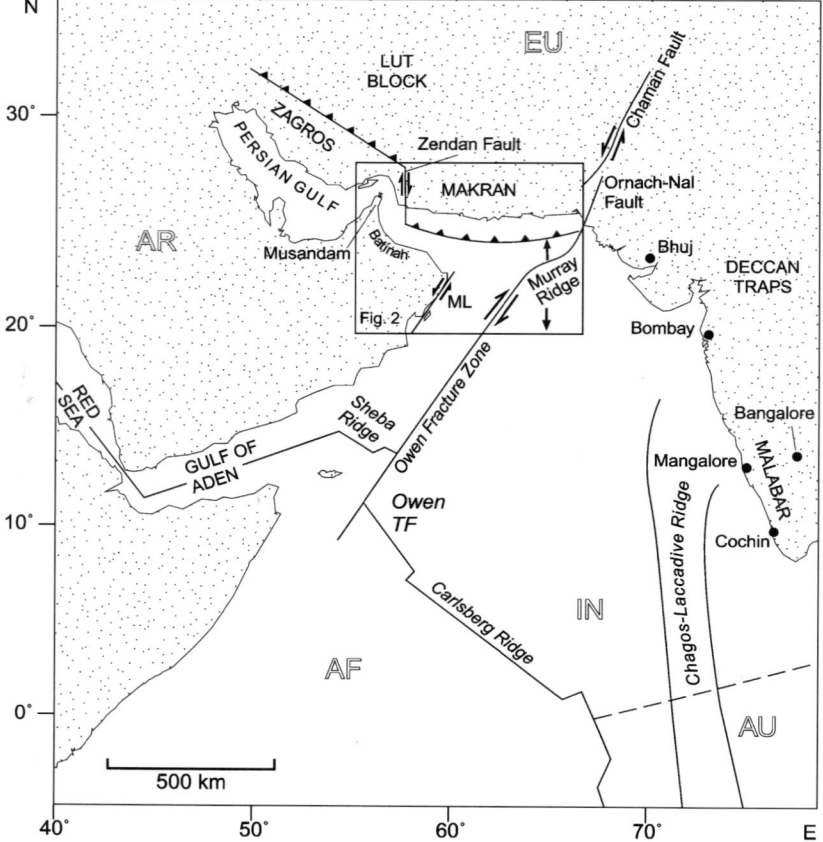

Fig. 1. Tectonic setting of region discussed in the paper. Dashed line shows proposed division between an Indian Plate (IN) and an Australian Plate (AU). EU, Eurasian Plate; AR, Arabian Plate; AF, African Plate; TF, transform fault; ML, Masirah line.

eastern boundary is the Sonne Fault (Fig. 2), which, according to Kukowski *et al.* (2000), has been active for the last *c.* 2 Ma, obliquely crosses the Makran Accretionary Wedge, offsets the Little Murray Ridge and related magnetic anomalies, and represents tearing of the Arabian Plate during its subduction. The following account of coastal Makran suggests that there may be more than one such transverse boundary in existence or in course of development.

Makran

The Makran of Iran and Pakistan is the site of low-angle subduction of the Arabian Plate beneath Eurasia. Its coast occupies part of an extensive accretionary prism (e.g. Farhoudi & Karig 1977), which measures almost 1000 km along strike. The western boundary of the Makran is complex: 'what becomes of the Zagros ranges to the east of the Zindan range?' asked Harrison (1941), p. 14). The Zendan Fault is a major structural discontinuity between the Zagros folds and the open structures to the east, and may be interpreted as a (dextral) transform that was already in operation when the Zagros folds came into existence (Shearman 1976; Stoneley 1990).

Some workers correlate the Zendan Fault, of which the Minab Fault is generally seen to be a part, with the Oman or Dibba Line in the northern part of the Oman Mountains (e.g. Falcon 1967). The tectonic significance of this link across the Strait of Hormuz varies depending on the researcher. Early accounts emphasized parallels between the Oman Mountains and the Zagros folds and thus included them in the Alpine–Himalayan fold system (e.g. Lees 1928*a*). This notion could survive alternative structural interpretations in the guise of the mid-Tertiary compressive tectonism that appears to have affected the Semail nappe

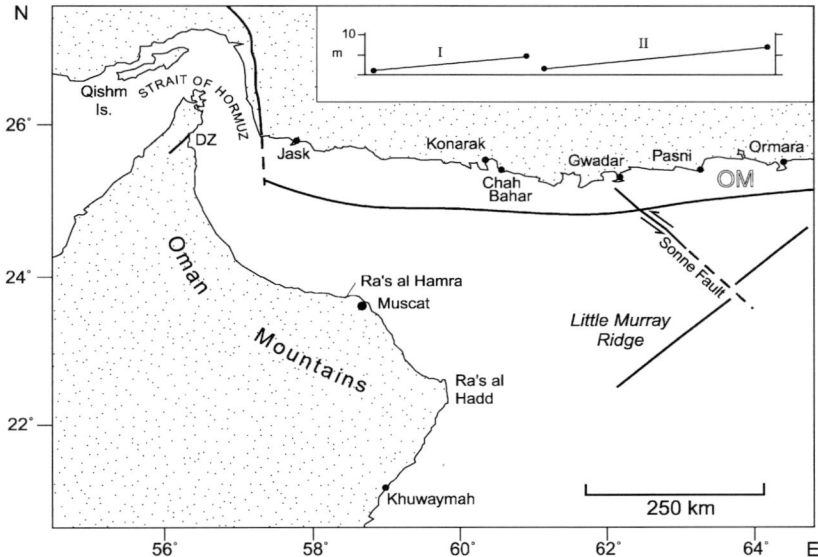

Fig. 2. Locations in Makran and Oman coast discussed in the paper; (inset) sketch sections show inferred tilt of Holocene waterlines between Jask and Konarak and between Chah Bahar and Ormara. Relevant ^{14}C ages are given in the text. OM, proposed Ormara microplate; DZ, Dibba fault zone.

emplaced onto the Arabian continental margin in Late Cretaceous time (Lippard et al. 1986). However, current practice is to present the Zendan Fault or Oman line itself as the onland boundary between continental collision to the west and oceanic subduction to the east (White & Klitgord 1976; White & Ross 1979; Lippard et al. 1986) and thus as a link between the main Zagros thrust in the north and the Makran deformation front margin in the south (Platt et al. 1988; Byrne et al. 1992; Kukowski et al. 2000).

The eastern boundary of the Makran appears less problematic. Some workers link the Ornach–Nal (sinistral) fault system (Fig. 1) north to the Chaman Fault and south to the Makran deformation front and thence to the Murray Ridge and Owen Fracture Zone (OFZ; Platt et al. 1988; Byrne et al. 1992). To reconcile dextral motion on the OFZ with sinistral motion on the Chaman system, Kukowski et al. (2000), among others, have invoked oblique extension along the Murray Ridge (see White 1982; Edwards et al. 2000). The Chaman Fault itself is complemented by a west-trending thrust whose locus has been gradually migrating south in response to sedimentation (Lawrence & Yeats 1979); in due course part of the motion was transferred to the Ornach–Nal Fault. A similar southward shift in activity was reported by Papastamatiou & Vita-Finzi (1993) south of a mountain loop that links the Sulaiman and the Kirthar ranges, where shortening by folding appears to have been usurped by strike-slip movement on the Chaman Fault.

Deformation of the Makran continental margin offshore is characterized by the development of asymmetrical frontal folds, which are replaced serially by a new fold once they attain amplitudes of c. 350 m. Within the initial 60–70 km the buckling takes place over a décollement surface c. 2.5 km beneath the sea floor, which may represent a weak layer with high pore pressures; further north the faults probably extend down to or into the basement (White 1979). Uplift occurs along imbricate faults, which steepen landwards; the deformation front has moved southwards by some 250 km since Oligocene time (Ahmed 1969; Uchupi et al. 2002). The process has continued into recent times, as shown by uplift of three Harappan ports on the coast of eastern Makran, Sutkagendor, Sotka Koh and Bala Kot, since the late third millennium (Dales 1966), and the well-documented events at Pasni in 1945 discussed below.

Marine terraces at heights of at least 800 feet (240 m) in Iran and 1500 feet (460 m) in Pakistan (Hunting Corporation 1960) point to a general eastward increase in cumulative coastal uplift (Page et al. 1979). The trend can be identified in the Holocene record. Immediately east of Jask, for example, two ^{14}C ages for a terrace 1.3 m above sea level give an average of 5730 ± 320 a bp whereas a terrace of similar age (5845 ± 350 a bp) stands at 5.5 m at Konarak; similarly, a 2 m level dated to 3670 ± 50 a bp at Chah Bahar (Reyss et al. 1998) may be contrasted with a 6 m level dated to 2710 ± 135 a bpat Ormara (Page et al. 1979) (Fig. 2, inset). Granted that such trends will be distorted by the widespread normal faulting, the data indicate an eastward upward tilt of $1/10^6$ in 6 ka, the latter one of $1/10^5$ in c. 3 ka.

Seismic and gravity data are consistent with a northward dip of $<2°$ of the subducted oceanic crust (White & Klitgord 1976; White & Louden 1982). In the synthesis of DeMets et al. (1990), the average convergence rate across the Makran is c. 40 mm a^{-1}, which would result in an average annual uplift of 0.9 mm a^{-1}, in good agreement with the rate indicated by the ^{14}C ages obtained by Page et al. (1979) at Konarak (namely, 1 mm a^{-1}). The faster rate in the east (42.0 mm a^{-1}, compared with 36.5 mm a^{-1} in the west) is in accord with the observed differential uplift even though at least some of the uplift has been coseismic, as suggested by the 2 m of uplift at Ormara that accompanied the M_w 8.1 earthquake of 28 November 1945 at Pasni. (According to Page et al. (1979), there is a possibility that initial uplift was 3.7 m and has since decayed.) The 6 m deposit at Ormara, which indicates an average uplift rate of 1.5 mm a^{-1}, can thus be equated, in the absence of significant aseismic movement, with three great earthquakes in 2700 a.

White (1982) showed that diapirism was not the driving force for deformation of the frontal sediments. In his account, folding to about 400 m takes c. 0.4 Ma and is followed by uplift along a thrust by 1.25 km, when the additional normal stress across the thrust fault of the uplifted ridge (which equals the displacement of 1.25 km of water by uplifted sediment) is 2×10^7 N m^{-2} or 200 bars. A new frontal fold is then initiated. Fault-related folding would combine the observed geometry coseismically. It would also suggest that, as in the Sumatran forearc (Vita-Finzi 1995), activity shifts among those imbricate faults that have still not been steepened by rotation to the point where they are immobilized.

The western and eastern halves of the Makran have contrasting seismic patterns, with the latter characterized by large historical events, a feature that is cited in support of segmentation of the subduction zone (Byrne et al. 1992) and, as mentioned above, an eastern (Ormara) microplate, which could, however, be an artefact of a short record. Thus a large event in 1483, which affected the Strait of Hormuz and NE Oman, may have had its epicentre in the western Makran (Ambraseys & Melville 1982). Dykstra & Birnie (1979) proposed segmentation of the subduction zone into four tracts separated by transverse breaks, which are marked by offsets in the Quaternary volcanic chain in inner Makran and which show some correlation with epicentre distribution for 1960–1975. The suggestion did not initially gain support from the offshore data or the focal mechanisms available for the Makran (Jackson & McKenzie 1984) but it may well be revived in the light of bathymetric evidence for the Sonne lineament (Kukowski et al. 2000). The neotectonic evidence reveals landward rotation of terrace units as well as widespread normal and strike-slip faulting, all of which may be the superficial products of subduction.

Regrettably, the late Quaternary offshore terrigenous record cannot distinguish between depositional pulses caused by flash floods and those triggered by earthquakes (Prins et al. 2000). The terrestrial palaeoseismological evidence is at present limited to the Chah Shirin Fault, a reverse structure on the Zendan system with 3.5 m of slip, which can be traced at the surface for 15 km. To judge from associated Neolithic artefacts and correlation with deposits dated by ^{14}C, it falls within the period 7000–1250 a bp (Vita-Finzi & Ghorashi 1978), and the poorly consolidated nature of the faulted deposits favours a coseismic interpretation. Magnitude is difficult to estimate; the failure of the nearby Khurgu earthquake (M_s 7.0) of 21 March 1977 to produce surface faulting (Berberian et al. 1977) provides some sort of minimum value, although uncertainty over the depth of the Khurgu event (a default value of 33 km is sometimes cited) hampers the comparison.

There are recent faults in the Makran proper but their ages are even more poorly constrained. On the peninsula of Konarak, for instance, normal faulting has truncated some of the older (pre-Holocene) marine terraces whose age remains uncertain. If the lowest two terraces (T1 and T2 of Page et al. (1979)) are separated by a coseismic event rather than a eustatic pulse it can at best be placed in the last 6 ka.

The historical record mentions large earthquakes on the Makran coast in ad 1483, 1765, 1851 and 1864, as well of course as the 1945 Pasni event, but no significant surface breaks are reported (Ambraseys & Melville 1982; Byrne et al. 1992). Besides the 2 m of uplift already mentioned, the 1945 earthquake, which had its epicentre off Pasni, created four new mud volcano islands 8–30 m high; it was followed by a large aftershock (M_s 7.3) in 1947. Dislocation modelling suggests that the 1945 event witnessed 6–8 m of slip on a rupture surface measuring 100 km down dip and 100–150 km along strike and amounting to rupture of, at most, one-fifth of the Makran subduction zone (Byrne et al. 1992).

Uplift on imbricate thrusts has been inferred (Ahmed 1969), and the 1945 uplift was plausibly modelled by combining 3.2–6 m of slip on a basal thrust with 3–4 m of slip on an imbricate thrust extending from it to near the surface (Byrne et al. 1992).

Several of the east Makran earthquakes for which there are adequate instrumental data (including the 1945 event) have interplate hypocentres whereas most of those in the west nucleated within (though near the top of) the downgoing slab. However, in both regions (as in other subduction settings) the shallowest part of the plate contact as well as the overlying accretionary wedge are aseismic (Byrne et al. 1992). Is subduction-related movement along imbricate faults within the wedge therefore ruled out? One possibility is aseismic slip, facilitated by wet sediments; another is coseismic slip manifested in small events. If the beach-capped headlands of the Makran (e.g. Falcon 1947) have been raised by imbricate faults, coseismic uplift can occur right up to the coast and indeed north of it, and it is noteworthy that among the events of m_b 4.6–6.0 tabulated by Byrne et al. (1992) for the Makran coast are four with dips of $>60°$. Additional steep faults may in due course be revealed by analysis of smaller earthquakes, some of them presumably the aftershocks of subduction events.

Arabia

Southeastern Arabia lies well to the rear of the Owen Fracture Zone but the straightness of the margin of eastern Oman has been ascribed to a fracture zone or transform (Fig. 1: the Masirah line; see Lippard et al. 1986), and the Yemeni coast (like that of NE Somalia) cannot be immune from the influence of the spreading Gulf of Aden.

Lees (1928a) referred to well-developed terraces at varying heights up to 1130 feet (344 m) as evidence for gentle east–west arching 'of comparatively recent date' within Oman proper. Elsewhere in eastern Arabia no such effect was visible: according to Lees, to the south recent miliolite never exceeded 150 feet (46 m) in elevation; to the north, drowned valleys indicated depression by at least 1100 feet (335 m). Miliolite is now widely considered to include predominantly aeolian material and its height is thus not a sensitive measure of uplift; moreover, Lees' dating of the arching appears to have depended on the presence of marine molluscs of recent appearance inland at elevations of up to 1230 feet (375 m) (Lees 1928b), but the shells may have been derived from middens rather than beaches.

On the other hand, upwarping with a NE–SW trend is a tenable proposition. In Musandam, the depth of a subaerial valley caliche (after allowing for the postglacial sea-level rise) indicates c. 155 m of Holocene subsidence at an average of 7.5 mm a^{-1} in the last 10 ka (Vita-Finzi 2001). The Batinah portion of the Oman coast appears to have been relatively stable during Holocene time. Marine deposits $\leqslant 1$ m above present High Water were dated by ^{14}C to c. 4000–5400 a bp at three localities (Vita-Finzi 1982; Costa & Wilkinson 1987).

Abrams & Chadwick (1994) showed that alluvial fan terraces to the rear of this coast represent climatic pulses rather than the uplift inferred by earlier workers, and suggested that subsidence of Musandam post-dates their Pliocene–early Pleistocene alluvial Unit 1. At Ra's al-Hamra, undercuts and terraces attest to shorelines up to 8 m high but the archaeological evidence suggests that a sea level close to the present was attained by 7000–6000 a bp (Durante & Tosi 1980) and consequently that the terraces are of pre-Holocene age. East of the Wahiba sands, near Khuwaymah, fossiliferous marine beds at 12–15 m with a nearshore fauna gave ^{14}C ages near the limit of the method and were interpreted as interglacial in age; shell from a beach at 1.5–2.0 m was dated by the writer to 3270 ± 600 a bp (UCL-105) (Gardner 1988).

These data are consistent with the view of Abrams & Chadwick (1994) that the southern Batinah is a 'null point' between subsidence to the north and uplift to the south in response to collision between Arabia and Eurasia, a buckling effect with a wavelength of 900 km, which is consistent with the wavelength of 500–1000 km that characterizes a crustal instability where the crustal thickness is 45 km and reasonable values are assumed for Young's modulus, Poisson's ratio, and mantle and crustal density.

In an early account of plate tectonics, Le Pichon et al. (1973) equated the Mesopotamian Plains and the Persian Gulf with a trough produced by convergence between the Arabian and Eurasian Plates. Reches & Schubert (1987) modelled the Arabian Plate as a uniform viscous plate undergoing large-scale bending. On the basis of the field evidence, the Dibba–Ra's al Khaymah line, roughly coincident with the Dibba Fault (South 1973), has been proposed as the hinge zone for subsidence of the Musandam peninsula as there is borehole and alluvial fan evidence to suggest stability of both the Persian Gulf and the Arabian Sea coasts at about this latitude (Vita-Finzi 1982). It may be that the Ruus al Jibal in Musandam represents a smaller buckle; Falcon (1973) estimated that it had undergone at least 3 km of uplift since Oligocene time. He also reported that the top of the Musandam Limestone (of Jurassic to Cretaceous age) had been found at a depth of about 4 km in a borehole on Qishm Island. The transmission of elastic energy released by seismicity from the leading to the trailing edge of the Arabian Plate (Vita-Finzi 2001) may give rise to transitory buckling, which will require continuous or repeated geodetic measurement for its detection.

India

The earliest dependable survey monuments in India date from the nineteenth-century Great Trigonometrical Survey. The South Indian Strain Measuring Experiment of 1994, which referred to a period of 125 a, detected no significant strain changes in India south of Bangalore (Paul *et al.* 1995). GPS data for 1991–1994 between Bangalore and Kathmandu indicated $<3 \pm 2$ mm a^{-1} of shortening (Paul *et al.* 2001). The period of measurement included the Latur earthquake (29 September 1993, M_w 6.2), showing that it had a negligible effect on net deformation. In other words, strain changes in peninsular India are at present 'immeasurably low' (Bilham *et al.* 1998). Whether the Bhuj earthquake of 26 January 2001 (M 7.9) will upset this pattern remains to be determined.

Buckling of the peninsular part of the Indian Plate has been inferred from hydrographic patterns, notably near latitude 13°N, where a major east–west drainage divide corresponds to a gravity high, a belt of microseismicity, crustal thinning, river migration, coastal aggradation and the tidal record (Ramasamy 1989; Subrahmanya 1996). Sukhtankar (1989) observed features characteristic of an emergent shoreline north of 18°15' and of submergence south of this latitude. Fieldwork by the writer confirmed the view of Bendick & Bilham (1999) that the geological evidence for Holocene deformation on the Malabar coast is ambiguous. For instance, modern oysters were found living at the same level as fossil oyster beds cited by previous workers in support of recent emergence. On the other hand, ^{14}C dating of a shelly deposit on Cocoanut Is. (13°28'N) gave ages of 3150 ± 800 a at 0.25 m and 2800 ± 200 a at 2.25 m above sea level (UCL-401 and -399), corresponding to uplift at c. 0.8 mm a^{-1} (Bendick & Bilham 1999). Moreover, tide-gauge records for 1950–1980 and levelling data for 1960–1980 indicate uplift in recent decades at over 3 mm a^{-1} at Mangalore (latitude 13° S) relative to Cochin to the south, consistent with buckling at a wavelength of c. 200 km (Bendick & Bilham 1999). Archaeological evidence for submergence has also been reported from Mangalore (Emery & Aubrey 1989).

Nair & Subramanya (1989) saw some of the ENE–WSW lineaments on the west coast as the extension of transform faults in oceanic crust generated by the Carlsberg Ridge. Fault movement resulting in surface folding would seem a possible source of deformation, especially as the Indian craton is considered to be too thick to buckle under the stresses resulting from collision (Bendick & Bilham 1999), although elastic deformation has been proposed in response to vertical loading by the Deccan Traps (Watts & Cox 1989; Manglik & Singh 1992). At all events, a compressional regime is indicated by *in situ* stress measurements, earthquake focal mechanisms, drainage patterns and joint orientations (Scheidegger & Padale 1982; Gowd *et al.* 1992).

Buckling of the Indian sea floor is well documented. South of the Bay of Bengal, for example, there are undulations in acoustic basement with wavelengths of about 200 km and amplitudes of as much as 3 km and striking roughly east–west (McAdoo & Sandwell 1985). Although the lineaments have been ascribed to small-scale convection (e.g. Cazenave *et al.* 1987), their association with reverse faults and with geoid anomalies is consistent with buckling in response to north–south compression (Wiens *et al.* 1986; Bull & Scrutton 1990).

The pattern of deformation can be explained by the stresses predicted by a model that incorporates slab pull and ridge push modulated in accordance with the lithosphere age and equivalent to c. 4–5 kbar (e.g. Cloetingh & Wortel 1986). Weissel *et al.* (1980) had modelled the deformation as buckling of an elastic plate 12 km thick and obtained a stress of 24 kbar. Lower stress values can be obtained by postulating a more complex rheology, notably by the inclusion of a weak subsurface layer, which will reduce the mean yield strength of the crust and favour buckling instabilities (Martinod & Molnar 1995; Gerbault 2000). The location of the buckling away from the zone of maximum predicted compressive stress in the plate can be explained by the amplification of features produced by seamount loading (Karner & Weissel 1990); a cover of weak sediment is not required to initiate buckling but it may increase the dominant wavelength and the rate at which the buckles develop (Zuber 1987).

Discussion

As sea-floor deformation has yet to be traced north beyond 8°N for reasons which, at least east of the Indian peninsula, include blanketing by sediments (Stein *et al.* 1990), its association with the Himalayan collision is still circumstantial. More generally, the stress fields for the plate as a whole predicted by (for example) Cloetingh &

Wortel (1986) cannot be tested by the available oceanic data in other than very general terms. The tentative indications of buckling in peninsular India, and of elastic energy storage and release in Arabia, likewise may echo the marine evidence but do not yet provide the wherewithal for evaluating variations in the yield strength of the lithosphere (see Gordon 2000).

The extension of gravity surveys into south Oman will make it possible to evaluate the component of lithospheric deflection caused by loading by sediments and obducted ophiolites (Ravaut et al. 1997), as well as dynamic effects such as slab rollback, so that any true folding caused by end loading can be identified and quantified. Even then, the discussion cannot progress without information on the extent of plate deformation within the sea floor, so that onshore deformation can be matched with the stress level indicated by offshore lithospheric buckling (Gerbault 2000). The analysis stands to benefit from improvements in the resolution of the vertical component of geodetic surveys by the application of techniques already routinely applied to the monitoring of the Antarctic ice cap and of the ocean surface. Similarly, some of the ^{14}C dates cited above will doubtless need revision, among other things for the apparent age of the Arabian Sea waters, which is suspected to exceed 1000 a; it seems preferable to await direct measurement of this effect, as extrapolation of modern readings to the past assumes that the local oceanic circulation has changed little.

In their discussion of the stress field in the (undivided) Indo-Australian Plate, Coblentz et al. (1998) concluded that, although the main source of the first-order intraplate stress field was ridge push, further progress in understanding deformation awaited an improved grasp of the boundary forces at the opposite plate margin. Indeed, Wiens et al. (1986) went so far as to state that plate boundary forces may lead not only to intense intraplate deformation but also to new plate boundary geometries. The current torpor of the (dextral) Owen Fracture Zone thus reflects the 'collisional resistance to further northward motion of India' (Gordon & DeMets 1989), and could be the prelude to rearrangement of plate boundaries including the incorporation of the Arabian Plate in that portion of the Indian Plate that lies north of the proposed boundary with the Australian Plate mentioned above (Stein et al. 1990; Wiens et al. 1986).

The notion that variations in boundary behaviour could lead to intraplate disruption was advanced with regard to the (sinistral) Ninetyeast Ridge by Stein & Okal (1978). The neotectonic vigour of the Zendan Fault is likewise consistent with the contrasting seismicity and deformational style of the Zagros and Makran margins of the Arabian Plate. It marks a change from shortening of continental basement in the west to subduction of oceanic crust in the east (Jackson & McKenzie 1984); and the subsurface evidence (Falcon 1976) shows that shortening across the structure is not a transitory transpressive phenomenon. Its almost aseismic character is clearly misleading.

In contrast to the Zagros, where deformation is by serial folding related to basement faulting, Arabia–Eurasia convergence in the Makran appears to be relatively uninhibited. The average rate of opening of the Red Sea is matched by shortening of the Zagros at $c.$ 20 mm a^{-1}, much of it concentrated in the coastal folds. India's measured rate of collision with southern Tibet is 20.5 ± 2 mm a^{-1} (Bilham et al. 1998), so that the slow dextral motion on the OFZ calculated by Gordon & DeMets (1989) from an Arabia–India Euler vector (a mere $c.$ 2 mm a^{-1}) turns out to be reasonable. As we have noted, deformation in the Makran is distributed across a zone 200–300 km wide and includes imbricate faults and folds offshore, but variations in convergence along strike (deMets et al. 1990) can be detected in the Holocene uplift rate, lending support to the notion that (as in the Zagros), deformation is concentrated in the frontal folds.

The fact that the Zendan fault system is dextral and the Ornach–Nal system is sinistral shows that the Makran withstands convergence more successfully than either the Zagros or the Himalayas. Its low seismicity immediately suggests that the answer lies in passive resistance, i.e. unimpeded low-angle subduction; but what seismicity there is, coupled with the marine terrace evidence, reveals that the historical seismic record is misleadingly modest, and prompts a contrary interpretation: that high-angle reverse faulting has almost run its course in the Makran (with the Lut block as backstop) whereas the Zagros can take up further shortening by fold development in the Persian Gulf; and, unsurprisingly, that, as in Indonesia (where the corresponding examples are the Baniak Islands off Sumatra and Timor), oblique convergence is more accommodating than the normal variety.

Fieldwork in India in 1998 was funded by the Royal Society and the Indian National Science Academy. I thank K. R. Subrahmanya for guidance in the field, R. Bilham, V. K. Gaur, R. G. Gordon and R. N. Singh for discussions, and P. Clift, E. Kirby and A. Robertson for their comments on the manuscript.

References

ABRAMS, M.J. & CHADWICK, O.H. 1994. Tectonic and climatic implications of alluvial fan sequences along the Batinah coast, Oman. *Journal of the Geological Society, London*, **151**, 52–58.

AHMED, S.S. 1969. Tertiary geology of part of south Makran, Baluchistan, West Pakistan. *AAPG Bulletin*, **53**, 1480–1499.

AMBRASEYS, N.N. & MELVILLE, C.P. 1982. *A History of Persian Earthquakes*. Cambridge University Press, Cambridge.

BENDICK, R. & BILHAM, R. 1999. Search for buckling of the southwest Indian coast related to Himalayan collision. *In:* MACFARLANE, A., SORKHABI, R.B. & QUADE, J. (eds) *Himalaya and Tibet: Mountain Roots to Mountain Tops*. Geological Society of America, Special Papers, **128**, 313–320.

BERBERIAN, M., PAPASTAMATIOU, D. & QORAISHI, M. 1977. *Khurgu (North Bandar Abbas, Iran) Earthquake of March 21, 1977: a Preliminary Field Report and a Seismotectonic Discussion*. Geological and Mining Survey of Iran, Report, **40**, 7–49.

BILHAM, R., BLUME, F., BENDICK, R. & GAUR, V.K. 1998. Geodetic constraints on the translation and deformation of India: implications for future great Himalayan earthquakes. *Current Science*, **74**, 213–229.

BULL, J.M. & SCRUTTON, R.A. 1990. Fault reactivation in the central Indian Ocean and the rheology of oceanic lithosphere. *Nature*, **344**, 855–858.

BYRNE, D.E., SYKES, L.R. & DAVIS, D.M. 1992. Great thrust earthquakes and aseismic slip along the plate boundary of the Makran subduction zone. *Journal of Geophysical Research*, **97**, 449–478.

CAZENAVE, A., MONNEREAU, M. & GIBERT, D. 1987. Seasat gravity undulations in the central Indian Ocean. *Physics of the Earth and Planetary Interiors*, **48**, 130–141.

CLOETINGH, S. & WORTEL, R. 1986. Stress in the Indo-Australian plate. *Tectonophysics*, **132**, 49–67.

COBLENTZ, D.D., ZHOU, S., HILLIS, R.R., RICHARDSON, R.M. & SANDIFORD, M. 1998. Topography, boundary forces, and the Indo-Australian intraplate stress field. *Journal of Geophysical Research*, **103**, 919–931.

COSTA, P.M. & WILKINSON, T.J. 1987. The hinterland of Sohar. *Journal of Oman Studies*, **9**, 1–238.

DALES, G.F. 1966. The decline of the Harappans. *Scientific American*, **241**, 92–100.

DEMETS, C., GORDON, R.G., ARGUS, D.F. & STEIN, S. 1990. Current plate motions. *Geophysical Journal International*, **101**, 425–478.

DURANTE, S. & TOSI, M. 1980. The aceramic shell middens of Ra's al-Hamra: a preliminary note. *Journal of Oman Studies*, **3**, 137–162.

DYKSTRA, J.D. & BIRNIE, R.W. 1979. Segmentation of the Quaternary subduction zone under the Baluchistan region of Iran and Pakistan. *In:* FARAH, A. & DEJONG, K.A. (eds) *Geodynamics of Pakistan*. Geological Survey of Pakistan, Quetta, 319–323.

EDWARDS, R.A., MINSHULL, T.A. & WHITE, R.S. 2000. Extension across the Indian–Arabian plate boundary: the Murray Ridge. *Geophysical Journal International*, **142**, 461–477.

EMERY, K.O. & AUBREY, D.G. 1989. Tide gauges of India. *Journal of Coastal Research*, **5**, 489–501.

FALCON, N.L. 1947. Raised beaches and terraces of the Iranian Makran. *Geographical Journal*, **109**, 149–153.

FALCON, N.L. 1967. The geology of the northeast margin of the Arabian Basement Shield. *Advancement of Science*, **24**, 31–42.

FALCON, N.L. 1973. Vertical and horizontal earth movements. *Geographical Journal*, **139**, 404–409.

FALCON, N.L. 1976. The Minab anticline. *Geographical Journal*, **142**, 409–410.

FARHOUDI, G. & KARIG, D.E. 1977. Makran of Iran and Pakistan as an active arc system. *Geology*, **5**, 664–668.

GARDNER, R.A.M. 1988. Aeolianites and marine deposits of the Wahiba Sands: character and palaeoenvironments. *Journal of Oman Studies, Special Report*, **3**, 75–94.

GERBAULT, M. 2000. At what stress level is the central Indian Ocean lithosphere buckling? *Earth and Planetary Science Letters*, **178**, 165–181.

GORDON, R.G. 2000. Diffuse oceanic plate boundaries: strain rates, vertically averaged rheology, and comparisons with narrow plate boundaries and stable plate interiors. *In:* RICHARDS, M., GORDON, R.G. & VAN DER HILST, R.D. (eds) *The History and Dynamics of Global Plate Motions*. Geophysical Monograph, American Geophysical Union, **121**, 143–159.

GORDON, R.G. & DEMETS, C. 1989. Present-day motion along the Owen fracture zone and Dalrymple trough in the Arabian Sea. *Journal of Geophysical Research*, **94**, 5560–5570.

GOWD, T.N., SRIRAMA RAO, S.V. & GAUR, V.K. 1992. Tectonic stress field in the Indian subcontinent. *Journal of Geophysical Research*, **97**, 11879–11888.

HARRISON, J.V. 1941. Coastal Makran. *Geographical Journal*, **97**, 1–17.

HUNTING CORPORATION 1960. *Reconnaissance Geology of Part of West Pakistan*. Hunting Survey Corporation, Toronto, Ontario.

JACKSON, J.A. & MCKENZIE, D. 1984. Active tectonics of the Alpine–Himalayan Belt between western Turkey and Pakistan. *Geophysical Journal of the Royal Astronomical Society*, **77**, 185–264.

KARNER, G.D. & WEISSEL, J.K. 1990. Compressional deformation of oceanic lithosphere in the central Indian Ocean: why it is where it is. *In:* COCHRAN, J.R. & STOW, D.A.V. (eds) *Proceedings of the Ocean Drilling Program, Scientific Results*, **116**. Ocean Drilling Program, College Station, TX, 279–289.

KUKOWSKI, N., SCHILLHORN, T., FLUEH, E.R. & HUHN, K. 2000. Newly identified strike-slip plate boundary in the northeastern Arabian Sea. *Geology*, **28**, 355–358.

LAWRENCE, R.D. & YEATS, R.S. 1979. Geological reconnaissance of the Chaman fault in Pakistan. *In:* FARAH, A. & DEJONG, K.A. (eds) *Geodynamics of Pakistan*. Geological Survey of Pakistan, Quetta, 351–357.

LEES, G.M. 1928a. The geology and tectonics of Oman and parts of south-eastern Arabia. *Quarterly Journal of the Geological Society, London*, **84**, 585–668.

LEES, G.M. 1928b. The physical geography of south-eastern Arabia. *Geographical Journal*, **71**, 441–470.

LE PICHON, X., FRANCHETEAU, J. & BONNIN, J. 1973. *Plate Tectonics*. Elsevier, Amsterdam.

LIPPARD, S.J., SHELTON, A.W. & GASS, I.G. 1986. *The Ophiolite of Northern Oman*. Geological Society, London, Memoirs 11.

MANGLIK, A. & SINGH, R.N. 1992. Rheological thickness and strength of the Indian continental lithosphere. *Proceedings of the Indian Academy of Sciences (Earth and Planetary Science)*, **101**, 339–345.

MARTINOD, J. & MOLNAR, P. 1995. Lithospheric folding in the Indian Ocean and the rheology of the oceanic plate. *Bulletin de la Societé Géologique de France*, **166**, 813–821.

MCADOO, D.C. & SANDWELL, D.T. 1985. Folding of oceanic lithosphere. *Journal of Geophysical Research*, **90**, 8563–8569.

MCKENZIE, D.P. & SCLATER, J.G. 1971. The evolution of the Indian Ocean since the Late Cretaceous. *Geophysical Journal of the Royal Astronomical Society*, **25**, 437–528.

NAIR, M.M. & SUBRAMANYA, K.S. 1989. Transform faults of the Carlsberg Ridge—their implication in neotectonic activity along the Kerala coast. *Geological Survey of India, Special Publications*, **24**, 327–332.

PAGE, W.D., ALT, J.N., CLUFF, L.S. & PLAFKER, G. 1979. Evidence for the recurrence of large-magnitude earthquakes along the Makran coast of Iran and Pakistan. *Tectonophysics*, **52**, 533–547.

PAPASTAMATIOU, D. & VITA-FINZI, C. 1993. Decreased tectonism of the Sui Dome. *In:* SHRODER, J.F. JR (eds) *Himalaya to the Sea*. Routledge, London, 236–250.

PAUL, J., BLUME, F., JADE, S. & 10 OTHERS 1995. Microstrain stability of peninsular India 1864–1994. *Proceedings of the Indian Academy of Sciences (Earth and Planetary Science)*, **104**, 131–146.

PAUL, J., BURGMANN, R., GAUR, V.K. & 9 OTHERS 2001. The motion and active deformation of India. *Geophysical Research Letters*, **28**, 647–651.

PLATT, J.P., LEGGETT, J.K. & ALAM, S. 1988. Slip vectors and fault mechanics in the Makran accretionary wedge, southwest Pakistan. *Journal of Geophysical Research*, **93**, 7955–7973.

PRINS, M.A., POSTMA, G. & WELTJE, G.J. 2000. Controls on terrigenous sediment supply to the Arabian Sea during the late Quaternary: the Makran continental slope. *Marine Geology*, **169**, 351–371.

RAMASAMY, S. 1989. Morpho-tectonic evolution of east and west coasts of Indian peninsula. *Geological Survey of India, Special Publications*, **24**, 333–339.

RAVAUT, P., BAYER, R., HASSANI, R., ROUSSET, D. & AL YAHYA'EY, A. 1997. Structure and evolution of the northern Oman margin: gravity and seismic constraints over the Zagros–Makran–Oman collision zone. *Tectonophysics*, **279**, 253–280.

RECHES, Z. & SCHUBERT, G. 1987. Models of post-Miocene deformation of the Arabian plate. *Tectonics*, **6**, 707–725.

REYSS, J.L., PIRAZZOLI, P.A., HAGHIPOUR, A., HATTÉ, C. & FONTUGNE, M. 1998. Quaternary marine terraces and tectonic uplift rates on the south coast of Iran. *In:* STEWART, I.S. & VITA-FINZI, C. (eds) *Coastal Tectonics*. Geological Society, London, Special Publications, **146**, 225–237.

SCHEIDEGGER, A.E. & PADALE, J.G. 1982. A geodynamic study of peninsular India. *Rock Mechanics*, **15**, 209–241.

SCHLICH, R. 1982. Aseismic ridges, spreading centers, and basins. *In:* NAIRN, A.E.M. & STEHLI, F.G. (eds) *The Ocean Basins and Margins, 6. The Indian Ocean*. Plenum, New York, 51–147.

SHEARMAN, D.J. 1976. The geological evolution of southern Iran. *Geographical Journal*, **142**, 393–410.

SOUTH, D. 1973. Pre-Quaternary geology. *Geographical Journal*, **139**, 400–403.

STEIN, S. & OKAL, E.A. 1978. Seismicity and tectonics of the Ninetyeast Ridge area: evidence for internal deformation of the Indian plate. *Journal of Geophysical Research*, **83**, 2233–2245.

STEIN, C.A., CLOETINGH, S. & WORTEL, R. 1990. Kinematics and mechanics of the Indian Ocean diffuse plate boundary zone. *In:* COCHRAN, J.R. & STOW, D.A.V. (eds) *Proceedings of the Ocean Drilling Program, Scientific Results, 116*. Ocean Drilling Program, College Station, TX, 261–277.

STONELEY, R. 1990. The Arabian continental margin in Iran during the Late Cretaceous. *In:* SEARLE, A.H.F. & RIES, A.C. (eds) *The Geology and Tectonics of the Oman Region*. Geological Society, London, Special Publications, **49**, 787–795.

SUBRAHMANYA, K.R. 1996. Active intraplate deformation in South India. *Tectonophysics*, **262**, 231–241.

SUKHTANKAR, R.K. 1989. Coastal geomorphic features in relation to neotectonics along the coastal tract of Maharashtra and Karnataka. *Geological Survey of India, Special Publication*, **24**, 319–325.

UCHUPI, E., SWIFT, S.A. & ROSS, D.A. 2002. Tectonic geomorphology of the Gulf of Oman Basin. *In:* CLIFT, P.D., GAEDICKE, C., KROON, D. & CRAIG, J. (eds) *The Tectonic and Climatic Evolution of the Arabian Sea Region*. Geological Society, London, Special Publications, **195**, 37–70.

VITA-FINZI, C. 1982. Recent coastal deformation near the Strait of Hormuz. *Proceedings of the Royal Society of London, Series A*, **282**, 441–457.

VITA-FINZI, C. 1995. Pulses of emergence in the outer-arc ridge of the Sunda Arc. *Journal of Coastal Research (special issue)*, **17**, 279–281.

VITA-FINZI, C. 2001. Neotectonics at the Arabian plate margins. *Journal of Structural Geology*, **23**, 521–530.

VITA-FINZI, C. & GHORASHI, M. 1978. A recent faulting episode in the Iranian Makran. *Tectonophysics*, **44**, 21–25.

WATTS, A.B. & COX, K.G. 1989. The Deccan Traps: an interpretation in terms of progressive lithospheric flexure in response to a migrating load. *Earth and Planetary Science Letters*, **93**, 85–97.

WEISSEL, J.K., ANDERSON, R.N. & GELLER, C.A. 1980. Deformation of the Indo-Australian plate. *Nature*, **287**, 284–291.

WHITE, R.S. 1979. Deformation of the Makran continental margin. *In:* FARAH, A. & DEJONG, K.A. (eds) *Geodynamics of Pakistan*. Geological Survey of Pakistan, Quetta, 295–304.

WHITE, R.S. 1982. Deformation of the Makran accretionary sediment prism in the Gulf of Oman (northwest Indian Ocean). *In:* LEGGETT, J.K. (ed.) *Trench–Forearc Geology.* Geological Society, London, Special Publications, **10**, 357–372.

WHITE, R.S. & KLITGORD, K.D. 1976. Sediment deformation and plate tectonics in the Gulf of Oman. *Earth and Planetary Science Letters*, **32**, 199–209.

WHITE, R.S. & LOUDEN, K.E. 1982. The Makran continental margin: structure of a thickly sedimented convergent plate boundary. *In:* WATKINS, J.S. & DRAKE, C.L. (eds) *Studies in Continental Margin Geology.* American Association of Petroleum Geologists, Memoir, **34**, 499–518.

WHITE, R.S. & ROSS, D.A. 1979. Tectonics of the western Gulf of Oman. *Journal of Geophysical Research*, **84**, 3479–3489.

WIENS, D.A., STEIN, S., DEMETS, C., GORDON, R.G. & STEIN, C. 1986. Plate tectonic models for Indian Ocean 'intraplate' deformation. *Tectonophysics*, **132**, 37–48.

ZUBER, M.T. 1987. Compression of oceanic lithosphere: an analysis of intraplate deformation in the central Indian basin. *Journal of Geophysical Research*, **92**, 4817–4825.

Constraints on India–Eurasia collision in the Arabian Sea region taken from the Indus Group, Ladakh Himalaya, India

PETER D. CLIFT[1], ANDREW CARTER[2], MICHAEL KROL[3] & ERIC KIRBY[4]

[1] *Department of Geology and Geophysics, Woods Hole Oceanographic Institution, Woods Hole, MA 02543, USA (e-mail: pclift@whoi.edu)*

[2] *Department of Geological Sciences, University College, Gower Street, London WC1E 6BT, UK*

[3] *Department of Earth Sciences and Geography, Bridgewater State College, Bridgewater, MA 02325, USA*

[4] *Institute for Crustal Studies, University of California, Santa Barbara, CA 93106, USA*

Abstract: The Indus Group is a Paleogene, syntectonic sequence from the Indus Suture Zone of the Ladakh Himalaya, India. Overlying several pre-collisional tectonic units, it constrains the timing and nature of India's collision with Eurasia in the western Himalaya. Field and petrographic data now allow Mesozoic–Paleocene deep-water sediments underlying the Indus Group to be assigned to three pre-collisional units: the Jurutze Formation (the forearc basin to the Cretaceous–Paleocene Eurasian active margin), the Khalsi Flysch (a Eurasian forearc sequence recording collapse of the Indian continental margin and ophiolite obduction), and the Lamayuru Group (the Mesozoic passive margin of India). Cobbles of neritic limestone, deep-water radiolarian chert and mafic igneous rocks, derived from the south (i.e. from India), are recognized in the upper Khalsi Flysch and the unconformably overlying fluvial sandstones of the Chogdo Formation, the base of the Indus Group. The Chogdo Formation is the first unit to overlie all three pre-collisional units and constrains the age of India–Eurasia collision to being no younger than latest Ypresian time (>49 Ma), consistent with marine magnetic data suggesting initial collision in the Arabian Sea region at *c.* 55 Ma. The cutting of equatorial Tethyan circulation north of India at that time may have been a trigger to the major changes in global palaeoceanography seen at the Paleocene–Eocene boundary. New $^{40}Ar/^{39}Ar$, apatite fission-track and illite crystallinity data from the Ladakh Batholith and Indus Group show that the batholith, representing the old active margin of Eurasia, experienced rapid Eocene cooling after collision, but was not significantly reheated when the Indus Group basin was inverted during north-directed Miocene thrusting (23–20 Ma). Subsequent erosion has preferentially removed 5–6 km (*c.* 200 °C) over much of the exposed Indus Group, but only *c.* 2 km from the Ladakh Batholith. Reworking of this material into the Indus fan may complicate efforts to interpret palaeo-erosion patterns from the deep-sea sedimentary record.

The collision of India with Eurasia, starting at least as far back as Mid-Eocene time and continuing to the present day, represents the type example of continental collision and orogenesis. The topographic growth of the Himalayas and the Tibetan Plateau since that time is often considered to have had a major impact on the global environment, intensifying the South Asian monsoon (e.g. Molnar *et al.* 1993), and leading to a worldwide, long-term cooling (Ruddiman *et al.* 1988; Raymo & Ruddiman 1992). The collision itself severed deep-water circulation in the equatorial Tethys, and may be one of the principal driving forces behind large-scale changes in oceanic circulation (i.e. the initiation of deep-water saline bottom water flow; Brass *et al.* 1982; Pak & Miller 1992;

Zachos *et al.* 1993; Bice *et al.* 2000), as well as a trigger for the Paleocene–Eocene Boundary Thermal Event (e.g. Stott *et al.* 1996). The collision has had a major effect on the tectonic development of the Arabian Sea basin, the regional climate and palaeoceanography, as well as the nature of the sedimentary record along the margins. A good understanding of this process is necessary to interpret the marine geology in the context of its collisional setting.

Unfortunately, links between the relatively well-dated oceanographic evolution and orogenic uplift are difficult to demonstrate because of the lack of understanding of how and when the early Himalaya developed. Despite suggestions of a Late Cretaceous collision between India and Eurasia, a

From: CLIFT, P.D., KROON, D., GAEDICKE, C. & CRAIG, J. (eds) 2002. *The Tectonic and Climatic Evolution of the Arabian Sea Region.* Geological Society, London, Special Publications, **195**, 97–116. 0305-8719/02/$15.00 © The Geological Society of London 2002.

general consensus puts the date at c. 50–55 Ma (Garzanti et al. 1987; Searle et al. 1987; Rowley 1996; De Sigoyer et al. 2000). The best indication of when collision started in the western Himalaya can currently be gathered through examination of the syncollisional sedimentary sequences in the Indus Suture Zone. In this paper we examine the tectonic stratigraphy and thermal evolution of the Indus Suture in Ladakh (Fig. 1), and assess how this can be used to constrain the timing of collision in the western Himalaya, as well as the later Neogene deformation of the suture zone.

The Indus Group

The Indus Group, often referred to as the Indus Molasse, is a folded and thrust sequence of carbonate and clastic formations, which rests unconformably over the Ladakh Batholith on its northern edge and on Cretaceous platform limestones (Khalsi Limestone), Indian passive margin units (Lamayuru Group) and volcanic and carbonate sediments of the Eurasian margin (Jurutze Formation) on its southern edge (Figs 2–4; Bassoullet et al. 1983; Van Haver 1984; Garzanti & Van Haver 1988; Searle et al. 1990; Sutre 1990). The position of the Indus Group within the suture between the Indian and Eurasian plates led earlier workers to propose a forearc interpretation (Mascle 1985; Garzanti & Van Haver 1988; Steck et al. 1993) followed by deposition in an early intra-montane setting after continental collision. The stratigraphy of the Indus Group is yet to be formalized, but has been divided into a number of formations by different workers (Fig. 5). In this study we follow the stratigraphic scheme of Sinclair & Jaffey (2001) in naming units of the Indus Group.

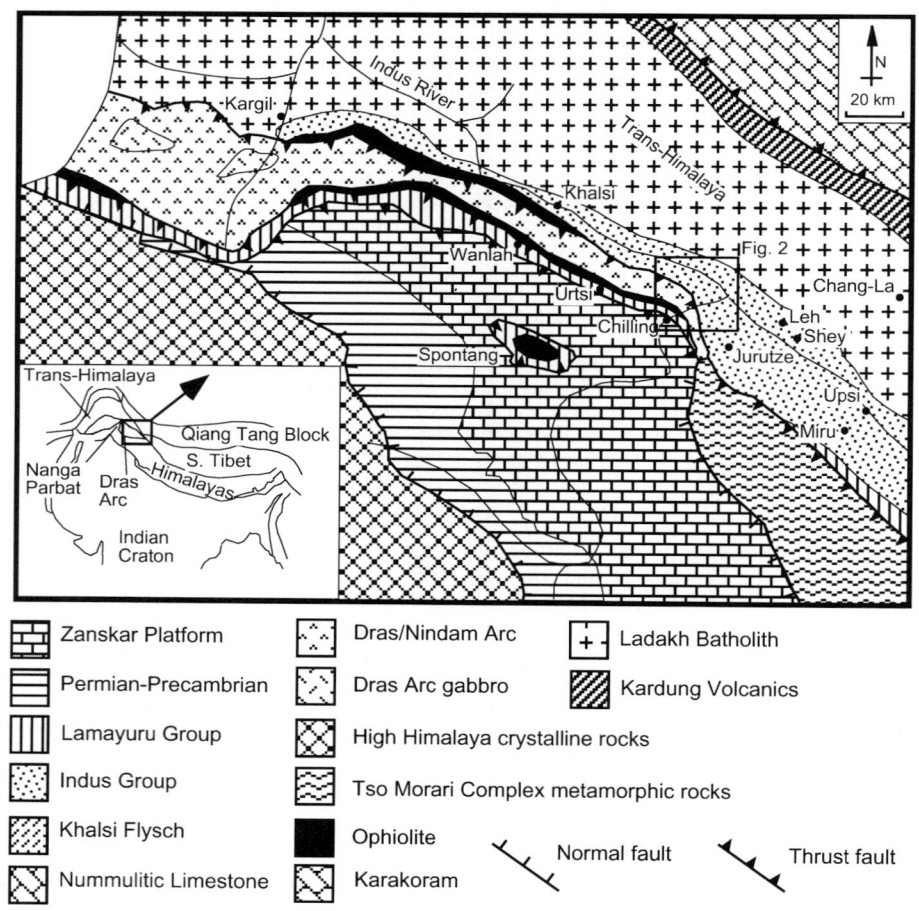

Fig. 1. Tectonic map of Ladakh. Inset shows the location of the region within the Himalaya–Tibet orogenic system.

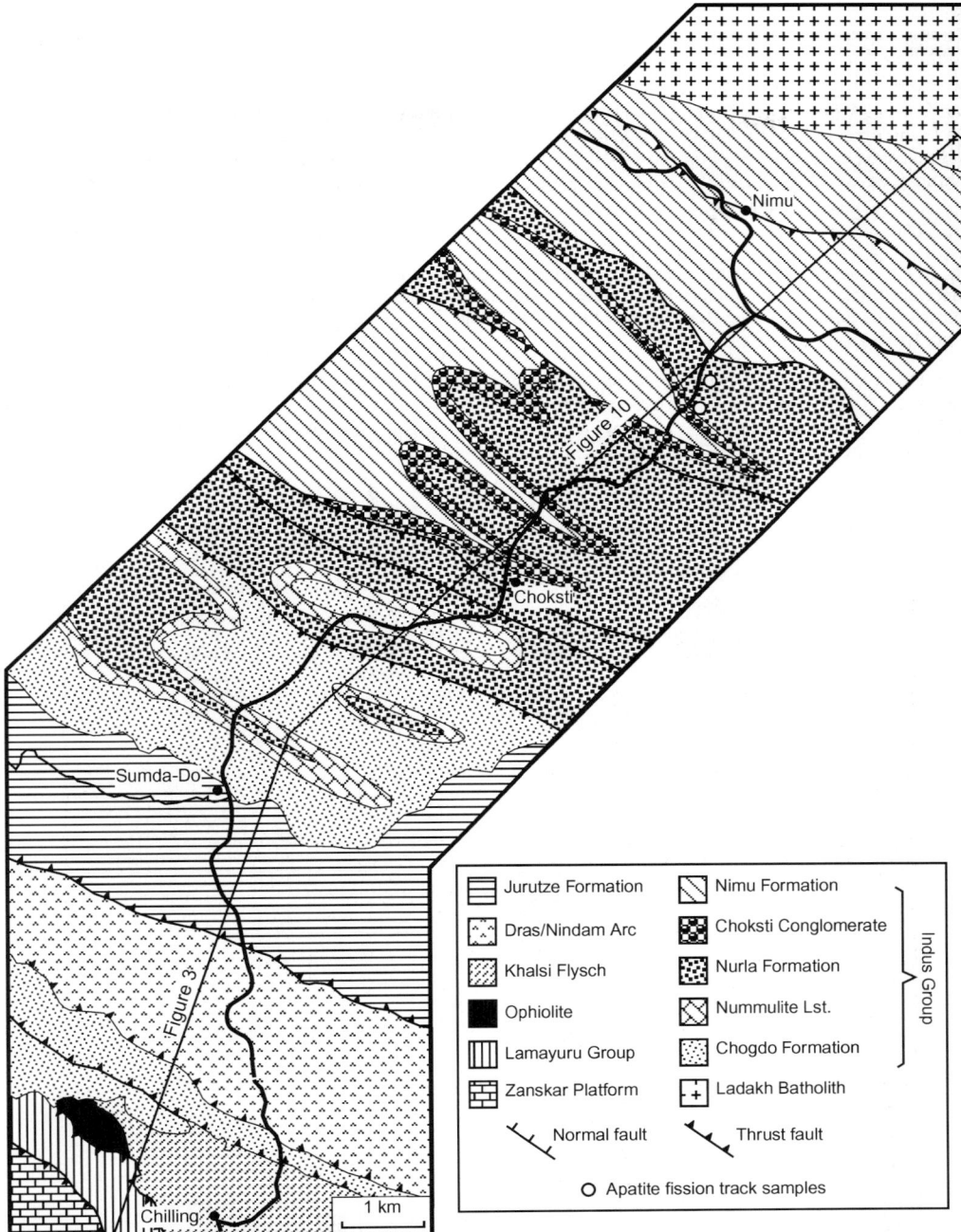

Fig. 2. Geological map of the Indus Suture along the Zanskar River in Ladakh, showing the main stratigraphic units and faults discussed in this paper, as well as principal localities.

The Ladakh Batholith, part of the Trans-Himalaya, is usually regarded as the eroded root of a pre-collisional, Eurasian, continental, magmatic arc, whose deformed forearc is exposed towards the south. The Cretaceous Nindam Formation represents the forearc basin to an oceanic volcanic island arc (Dras–Kohistan Arc) accreted to the margin of Eurasia (Cannat & Mascle 1990; Steck

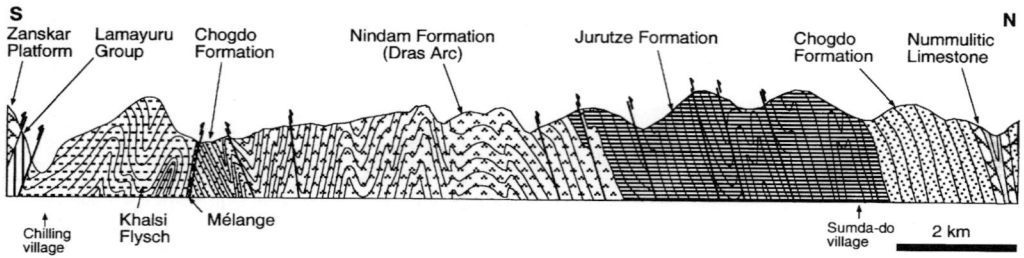

Fig. 3. Cross-section through the Indus Suture within the Zanskar Gorge, showing the Chogdo Formation lying over Lamayuru (Indian) Group turbidites near Chilling village and, together with the Nummulitic Limestone, over volcaniclastic turbidites of the Dras Arc near Sumda-Do village.

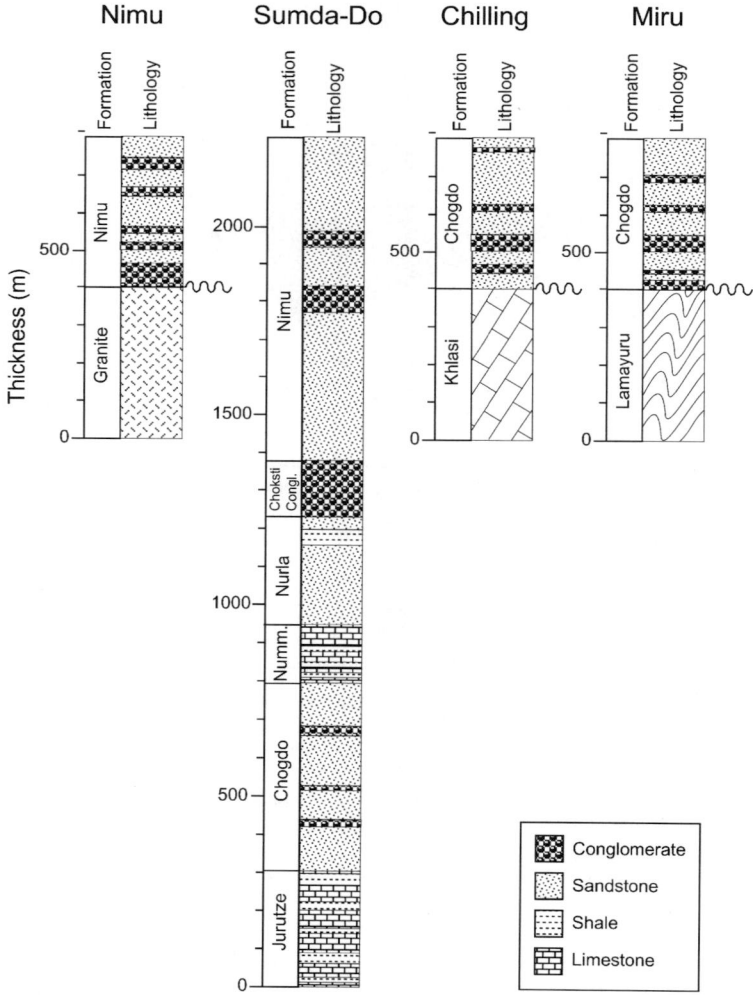

Fig. 4. Schematic stratigraphic columns showing the different contact relationships at the base of the Indus Group overlying the main tectonic units across the suture zone.

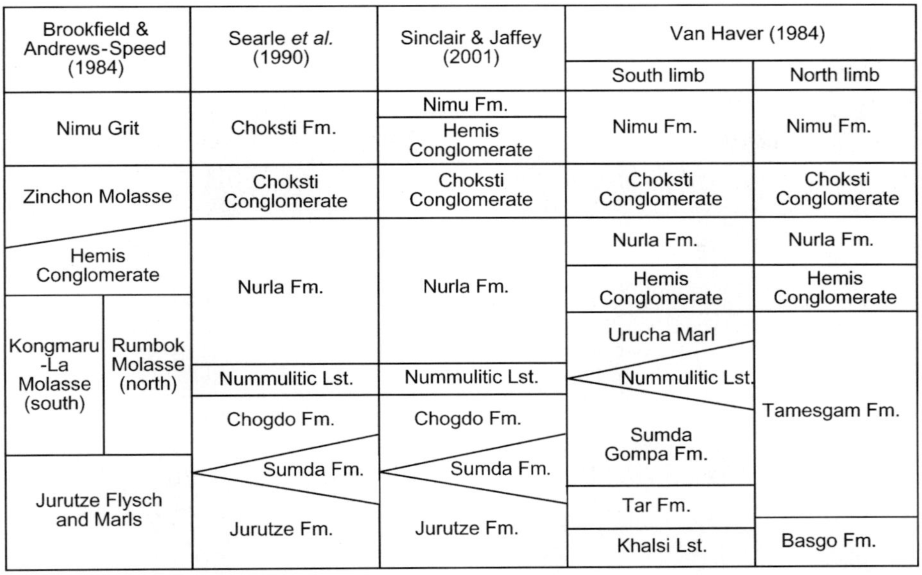

Brookfield & Andrews-Speed (1984)	Searle et al. (1990)	Sinclair & Jaffey (2001)	Van Haver (1984)	
			South limb	North limb
Nimu Grit	Choksti Fm.	Nimu Fm.	Nimu Fm.	Nimu Fm.
		Hemis Conglomerate		
Zinchon Molasse	Choksti Conglomerate	Choksti Conglomerate	Choksti Conglomerate	Choksti Conglomerate
Hemis Conglomerate	Nurla Fm.	Nurla Fm.	Nurla Fm.	Nurla Fm.
			Hemis Conglomerate	Hemis Conglomerate
Kongmaru-La Molasse (south) / Rumbok Molasse (north)	Nummulitic Lst.	Nummulitic Lst.	Urucha Marl	Tamesgam Fm.
			Nummulitic Lst.	
	Chogdo Fm.	Chogdo Fm.	Sumda Gompa Fm.	
	Sumda Fm.	Sumda Fm.		
Jurutze Flysch and Marls	Jurutze Fm.	Jurutze Fm.	Tar Fm.	
			Khalsi Lst.	Basgo Fm.

Fig. 5. Comparison of different existing stratigraphic frameworks for the Indus Group.

et al. 1993; Robertson & Degnan 1994; Clift et al. 2000a). Although it has recently been suggested that the Dras Arc was part of the Eurasian margin east of Leh (Rolland et al. 2000), the consistently intra-oceanic character of the trace element chemistry of the Nindam Formation turbidites in this area (Clift et al. 2000a), coupled with the oceanic Pb isotopic signature of much of the Ladakh Batholith (Clift et al. 2001a), makes this interpretation untenable. The Indus Group is only rarely found in direct contact with rocks of the Dras–Kohistan Arc, but is found in unconformable contact overlying Nindam Formation at Urtsi (Fig. 1; Robertson 2000).

The Jurutze Formation

Within and east of the Zanskar Gorge the Nindam Formation is depositionally overlain by the Jurutze Formation, a series of carbonate and volcaniclastic sedimentary rocks (Fig. 6d; Brookfield & Andrews-Speed 1984; Sinclair & Jaffey 2001). These rocks were called the Sumda Gompa and Tar Formations by Van Haver (1984), and have been assigned a Mid–Early Eocene age on the basis of their foraminiferal assemblage (Dainelli 1934). However, their location depositionally under the Lower Eocene Nummulitic Limestones of the Indus Group (Van Haver 1984) requires that they be of Late Cretaceous–Paleocene age, a hypothesis supported by the identification of Paleocene foraminifers in the upper part of the section (Sumda Formation) within the Zanskar Gorge (Van Haver 1984). The Jurutze Formation, like other parts of the Tar Group, was deposited within the Eurasian forearc basin before collision with India.

Compared with the deep-water shales and volcaniclastic turbidite sedimentary rocks of the underlying Nindam Formation, the Jurutze Formation is marked by shallower-water carbonate sedimentation. Volcaniclastic strata are still present and display normal grading and erosive bed bases indicating redeposition (Fig. 7). However, these are subsidiary to the dominant shales, marly limestones and limestones that make up the bulk of the section. Some limestones show a brecciation suggestive of desiccation above the low tide mark, and shallow-water shelly fauna are noted towards the top of the section (Fig. 8c). A general shallowing-upwards trend can be recognized within the formation. Some beds, up to 70 cm thick, comprise a shell hash of winnowed, mostly bivalve material. Wave ripples and parallel lamination are recognized locally, although most beds are massive. Occasional shaly intervals are stained by a rusty appearance, suggestive of input from an exposed oxidizing source, presumably the Eurasian margin. The frequency of these beds increases up-section. The Jurutze Formation is marked by a light, tan-coloured, soft-weathering exposure contrasting with the harder weathering, greyish Nindam Formation and the red of the Indus Group (Fig. 6d). At its top the Jurutze

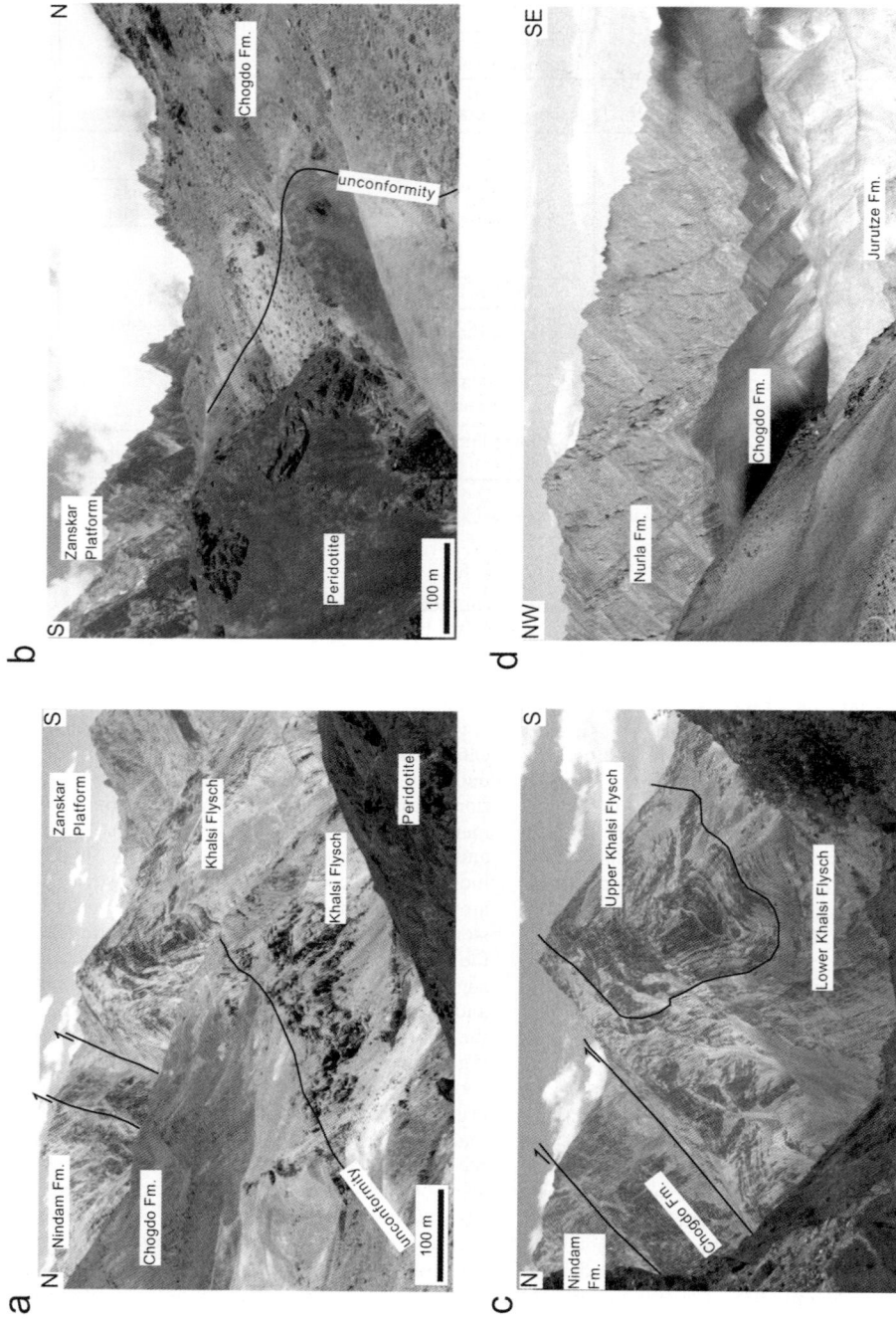

Fig. 6. Photographs of the basal relationships of the Indus Group, Chogdo Formation. (a) Chogdo Formation, WSW of Chilling. (b) Chogdo Formation unconformably overlies Khalsi Flysch, and peridotite. (c) View of the coherent, folded nature of the Khalsi Flysch sequence at Chilling. (d) View at Jurutze village towards the north, showing the light-coloured carbonate lithologies of the Jurutze Formation overlain by darker clastic units of the Chogdo Formation, overlain in turn by the prominent weathering sandstones of the Chogdo and Nurla Formations.

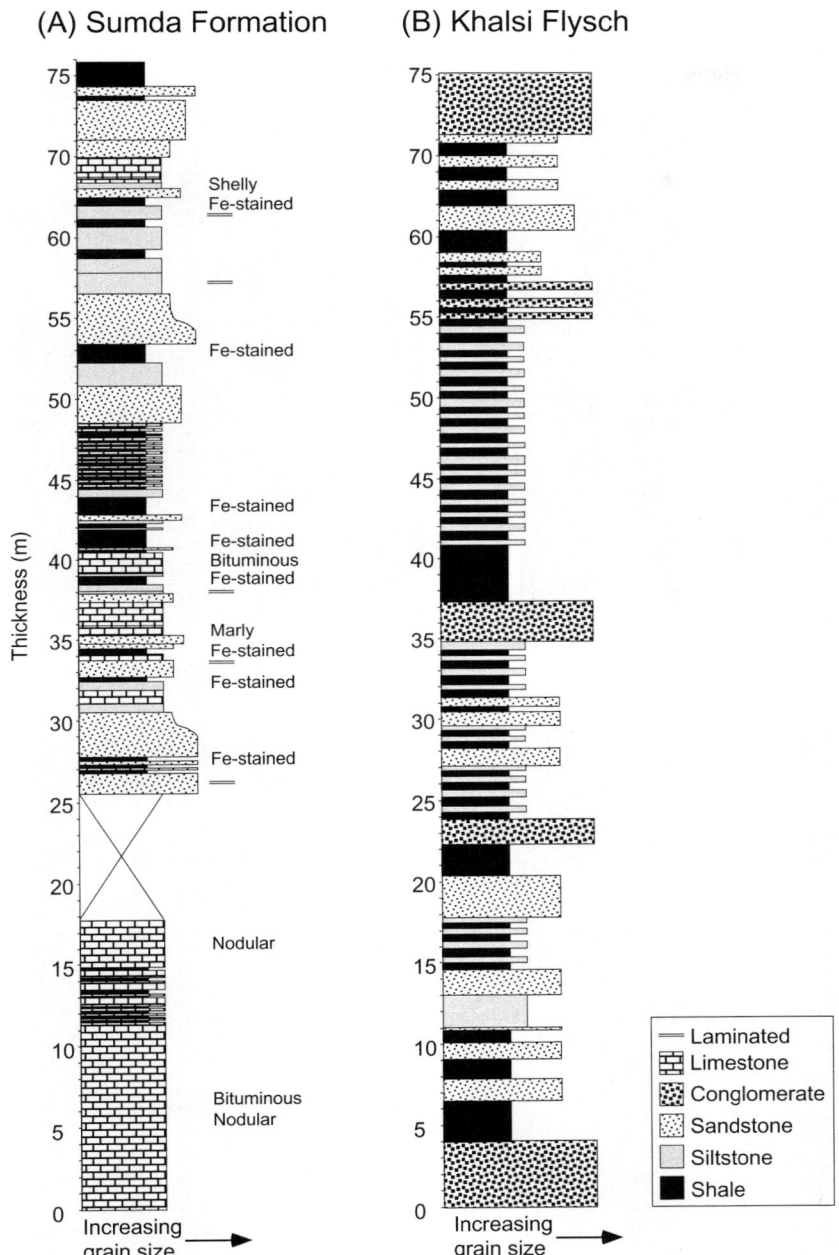

Fig. 7. Sedimentary logs of (**a**) a portion of the upper Jurutze Formation showing the shallow-water carbonate facies, contrasting with the deep-water volcaniclastic turbidites of the underlying Nindam Formation and the fluvial clastic sedimentary rocks of the younger Chogdo Formation. (**b**) Representative sedimentary log of the Khalsi Flysch WSW of Chilling.

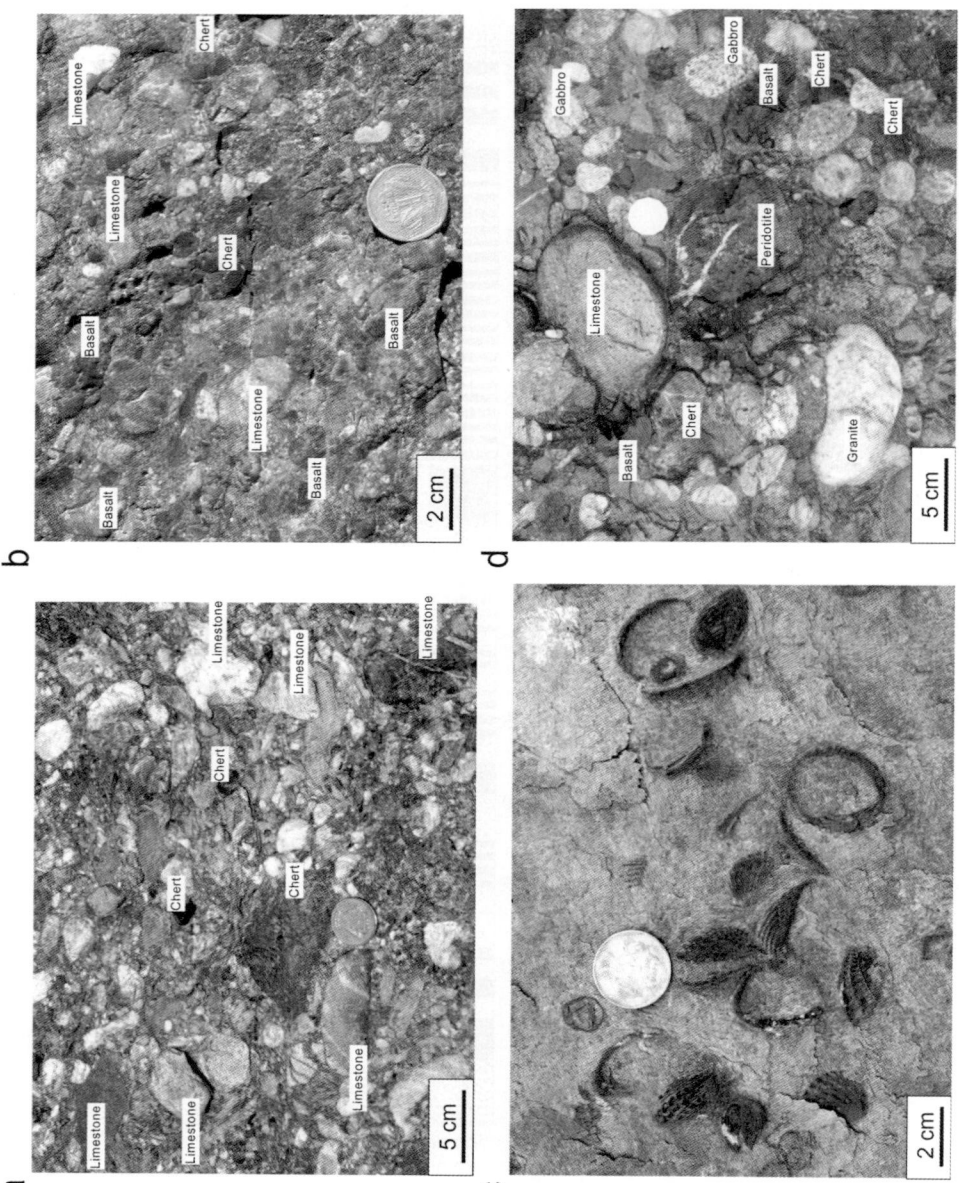

Fig. 8. Photographs of the evolving character of the Khalsi Flysch: (a) a limestone-dominated composition in the lower units; (b) increasing influx of ophiolitic clasts up-section; (c) the shallow-water carbonate facies at the top of the Jurutze Formation, south of Sumda-Do; (d) a basal Chogdo Formation conglomerate, rich in ophiolitic lithologies.

Formation passes rapidly but conformably up into the basal unit of the Indus Group, the fluvial Chogdo Formation, via an intervening limestone unit, the Sumda Formation (Fig. 7a; Garzanti & Van Haver 1988).

Southern boundary

Along its southern edge the Indus Group lies unconformably on a series of tectonic units. South of Miru (Fig. 1) red conglomerates are seen in angular erosional contact with shales of the Lamayuru Group, typically interpreted as the allochthonous Indian slope (Baud *et al.* 1984; Brookfield & Andrews-Speed 1984; Robertson & Degnan 1993), although alternatively considered as shaly intervals within a relatively coherent Zanskar (Indian) platform (Fuchs 1985, 1989). The Chogdo Formation also overlies the Lamayuru Group west of Chilling. In this region it is also seen unconformably overlying peridotite (Fig. 6b), which correlates with the southern mélange zone of Robertson (2000), exposed a short distance to the west at Urtsi (Fig. 1).

The Khalsi Flysch

Along its southern boundary the Indus Group overlies a coherent series of redeposited sedimentary rocks (Fig. 2), which we correlate here with the Khalsi Flysch of Brookfield & Andrews-Speed (1984) on the basis of the sedimentary facies and structural position. This unit was also recognized by Fuchs (1985), who called it the Chilling Formation. Its development along the length of the suture is patchy, and it does not outcrop as far west as Wanlah (Fig. 1). The Khalsi Flysch should not be confused with the conglomeratic Wanlah Formation of Cannat & Mascle (1990), which is located in a similar part of the suture along strike, but which is more similar to the Chogdo Formation discussed below. The Khalsi Flysch is a small unit, and one that has not been recognized by several studies. Garzanti & Van Haver (1988) classified it as part of the Lamayuru Group, whereas Searle *et al.* (1990) identified this thrust slice as part of the Nindam Formation, from which it differs in terms of geochemistry and sedimentary facies.

The contact between the Khalsi Flysch and the overlying red Chogdo Formation clastic sedimentary rocks is marked by both an angular discordance and a well-developed, iron-rich palaeosol up to 1 m thick (Fig. 6a). The Khalsi Flysch forms a well-defined, if folded, stratigraphy (Fig. 6c), and cannot be considered as part of the southern mélange zone of Robertson (2000). The sequence can be readily separated into a lower grey–tan weathering sequence and an upper red-coloured sequence. The colour difference can be readily related to the changing provenance moving up-section, specifically to the influx of more ophiolitic detritus, red chert and mafic rock fragments. However, interbedded siltstones and shales, intercalated with thicker sandstone beds and especially structureless, muddy, matrix-supported debris-flow conglomerates (Fig. 7b), dominate the Khalsi Flysch throughout. Texturally the conglomerates of the Khalsi Flysch are immature, being both poorly sorted and including angular clasts. Brookfield & Andrews-Speed (1984) interpreted the debris-flow deposits as being equivalent to sediment found in the upper parts of modern submarine fans.

Structurally sandwiched between the Indian slope sediments of the Lamayuru Group to the south and the Jurutze and Nindam Formations to the north, the Khalsi Flysch is here interpreted to represent a trenchward equivalent of the Jurutze Formation of Brookfield & Andrews-Speed (1984) or the Tar Group of Garzanti & Van Haver (1988). It is part of the Eurasian margin, because its location north, and therefore oceanward, of the deep-water Indian slope units makes an association with the Indian Zanskar Platform untenable. Brookfield & Andrews-Speed (1984) argued that the Khalsi Flysch depositionally overlies the Cretaceous Khalsi Limestone south of Khalsi, again consistent with it being part of the Eurasian margin before collision. We believe that the Khalsi Flysch was deposited on the same margin as the Jurutze Formation, but is distinct in terms of its facies and needs to be considered as a separate formation.

The Chogdo Formation

The Chogdo Formation is the oldest sedimentary unit to span tectonic units on either side of the Indus Suture. The sedimentary facies of this formation are very distinctive, being uniformly coarse sandy and conglomeratic, red-weathering clastic sedimentary rocks (Van Haver 1984; Searle *et al.* 1990; Sinclair & Jaffey 2001). The onlapping relations and continental alluvial facies of the Chogdo Formation show that marine conditions had been eliminated from the suture zone by the time of its deposition, implying advanced continental collision. The age of the Chogdo Formation is unknown, but it is overlain by a Nummulitic Limestone, dated to Early Eocene time by Van Haver (1984) (i.e. >49 Ma; Berggren *et al.* 1995). Consequently, the Chogdo Formation constrains India–Eurasia collision to being before, or during, Ypresian (Early Eocene) time in Ladakh. In this study we correlate two separate outcrops, north of Sumda-Do and at Chilling, as both being Chogdo Formation (Fig. 2). The correlation is based on the similar facies, colour, clast compositions and palaeo-current indicators seen in both locations.

Sedimentary provenance

The progressive collision of the Eurasian forearc with the passive margin of India is recorded by the evolving provenance of the sedimentary rocks, as well as the onlapping relationships of the Chogdo Formation. The earliest stages of the collision are recorded in the Khalsi Flysch. The colour difference between the lower and upper parts of this formation reflects a change from debris flows dominated by limestone and black cherts (Fig. 8a), to ones with a significant component of red radiolarian chert and altered, mafic volcanic clasts (Fig. 8b). The change in provenance is also reflected in the preserved palaeo-current indicators. SW-directed palaeo-currents in the lower Khalsi Flysch, and Jurutze Formation, are consistent with sediment derivation from the Eurasian margin into the forearc basin (Fig. 9). However, the upper Khalsi Flysch and Chogdo Formation show NE-directed flow, i.e. from India, a change coincident with the influx of platform sedimentary clasts, radiolarian chert and a variety of ophiolitic, mafic lithologies. The provenance change is most marked in the Chogdo Formation (Fig. 8d), which differs from the upper Khalsi Flysch in being even more ophiolitic in clast population, and also by the presence of deeper, mid- and lower-crustal rocks, most notably peridotites and gabbros, but also granites. Exhumation of deeper crustal levels is consistent with the more advanced stage of collision.

Although some of the clast lithologies, such as shallow-water limestones, are found on the Eurasian margin (Khalsi Limestone), there are no known sources there of radiolarian chert and mafic or ultramafic plutonic rocks. Their occurrence in the upper Khalsi Flysch requires that obduction of the Spontang Ophiolite onto the Zanskar (Indian) Platform had at least begun by the time of this sedimentation, because the only known source of the suite of oceanic lithologies seen in the con-

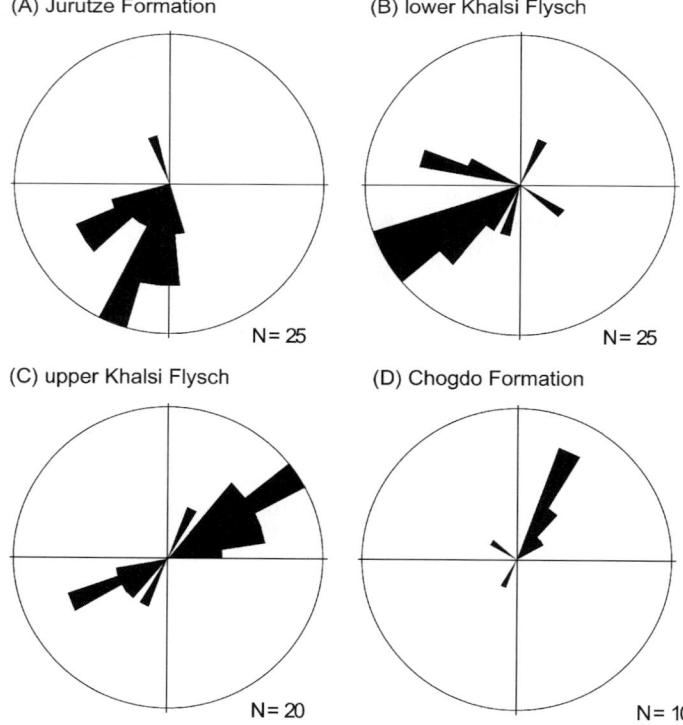

Fig. 9. Evolving palaeo-currents in the Jurutze forearc basin. Pre-collisional south- and SW-directed flow is reversed to a northeastward flow as collision occurs and detritus from the Indian margin is shed into the new Indus Group Basin.

glomerates of the Khalsi Flysch is the Spontang Ophiolite, or its eroded equivalents (Fig. 1). Although suitable lithologies are present in the southern and northern mélange zones of Robertson (2000), these cannot be source of the ophiolitic sediments in the Khalsi Flysch and Chogdo Formation because of the northward palaeo-flow and because the southern mélange zone was buried by the Chogdo Formation and was not eroded at this time. Unfortunately, the age of the Khalsi Flysch is not known, except for being pre-Chogdo Formation, i.e. of Ypresian age or older. The abundance of ophiolitic material derived from the SSW during Chogdo times requires that ophiolite obduction had been achieved before the end of Ypresian time (>49 Ma).

Regarding provenance of the rest of the Indus Group, the lack of garnet or muscovite mica grains within sandstones argues against deposition during or after the unroofing of sources abundant in these minerals within the High Himalaya, a hypothesis supported by isotopic provenance studies (Clift et al. 2001a). Radiometric dating constrains High Himalayan exposure to c. 20–23 Ma in the adjacent western Himalaya (Searle et al. 1992; Metcalfe 1993; Vance & Harris 1999; Walker et al. 1999), and suggests that the Indus Group is older than this time.

Thermal evolution

Following India–Eurasia collision in the western Himalaya, continued convergence of the plate has resulted in orogenic uplift and deformation. Deposition of the Indus Group was brought to an end by the inversion of the basin during northward thrusting of the Zanskar Himalaya during Early Miocene time (e.g. Bassoullet et al. 1980; Van Haver et al. 1984; Searle et al. 1987). Clift (2002) has pointed out that despite this uplift the Indus River, which was responsible for depositing most of the sediments in the Indus Group, continued to flow along the suture, by incising into the deforming sequences. Understanding the possible effect that this inversion had on sedimentation and erosion rates within the Indus River and Fan system requires an examination of the post-depositional thermal history of the basin and the adjacent Ladakh Batholith. To do this we collected detrital apatites from the basin and apatites from granites within the batholith for fission-track analysis. As fission tracks in apatite anneal at temperatures >125 °C, we examined authigenic illite to evaluate the degree of post-depositional heating in the basin. Finally, we analysed a suite of biotite and hornblende grains from the batholith to examine the higher-temperature portion of the thermal history.

Apatite fission-track analysis

The location of samples considered in this study is provided in Table 1, and Figs. 1 and 2. Analysis of the Ladakh Batholith provides a record of its unroofing during Cenozoic time, whereas analysis of detrital sandstones can provide a record of either the cooling history of the original source terrain, provided the section has not been thermally reset since sedimentation, or the burial and erosion of the basin itself. The application of the fission-track method for deriving denudational histories is based on the low closure temperatures for apatite and the ability to detect and quantify the level of resetting: fission tracks in apatite totally anneal at temperatures of >110–125 °C for periods of heating of 1–10 Ma (Green et al. 1989). However, between c. 60 °C and c. 110 °C the tracks are semi-stable and shorten with time (Green et al. 1989). Sample residence within this 'partial annealing zone' (PAZ), can be quantified by measuring the fission-track lengths. The track shortening process has been described mathematically (Laslett et al. 1987), which permits the deciphering of a sample's thermal history.

Samples for fission-track analysis were prepared by using standard separation and polishing techniques at University College, London. Apatite grain mounts were etched with 5 M HNO at 20 °C for 20 s and irradiated in the thermal facility (cadmium ratio for Au >400) of the Risø Reactor at Roskilde, Denmark, together with Corning glass dosimeter CN-5. Central ages were calculated by using the International Union of Geological Sciences recommended zeta calibration approach (Hurford 1990).

To derive a range of cooling-history paths compatible with the fission-track age and length data we use the method of Gallagher (1995), which is based on a form of stochastic simulation using the annealing algorithm of Laslett et al. (1987). Several thousand possible thermal models are run and compared with the analytical data to identify which aspects of the sample's thermal history are well resolved (Fig. 10). Modelled data were confined to samples derived from a single source with a single age population.

In the case of the Indus Suture, many of the samples analysed did not have sufficient numbers of tracks to allow modelling. One example that did is shown from the Ladakh Batholith (Fig. 10), exposed in Leh (Fig. 1). Cooling of this pluton through the PAZ occurred gradually between c. 30 and 10 Ma. Other samples from the Ladakh Batholith yielded central ages ranging from 27 to

Table 1. *Analytical data from the fission-track analysis of apatite crystals extracted from the Indus Group and Ladakh Batholith*

Sample no.	Location	Rock type	No. of crystals	Dosimeter pd	Dosimeter Nd	Spontaneous pd	Spontaneous Nd	Induced pd	Induced Nd	$P\chi^2$	Age dispersion RE%	Central age (Ma)	Mean track length (μm)	SD	No. of tracks
LA-97-1	Leh City	Granite	20	1.203	6667	0.733	560	5.434	4150	25	4.4	27.5 ± 1.3	13.90 ± 0.18	1.78	105
LA-97-2	Zanskar Gorge	Sandstone	15	1.203	6667	0.223	63	3.346	946	40	21.2	13.8 ± 1.9	No length data		
LA-97-41	Khalsi	Granite boulder	14	1.203	6667	0.162	72	0.9244	410	40	7.9	35.7 ± 4.7	No length data		
LA-97-83	Shey	Granite	11	1.203	6667	0.171	61	0.789	281	80	0	44.1 ± 6.3	No length data		
LA-98-47	Changa-La	Granite	18	1.241	6882	0.137	99	1.052	758	50	17.9	29.2 ± 3.3	No length data		
LA-98-48	Changa-La	Granite	20	1.241	6882	0.603	561	4.327	4024	<1	18.4	29.4 ± 1.8	14.11 ± 0.15	1.12	57
LA-98-63	Zanskar Gorge	Sandstone	9	1.241	6882	0.138	53	2.219	852	<1	55.4	13.7 ± 3.2	No length data		

Analyses by external detector method using 0.5 for the $4\pi/2\pi$ geometry correction factor. Ages calculated using dosimeter glass CN-5; analyst Carter ζCN5 = 339 ± 5 calibrated by multiple analyses of IUGS apatite and zircon age standards (see Hurford 1990). $P\chi^2$ is probability for obtaining χ^2 value for v degrees of freedom, where v = number of crystals – 1. Central age is a modal age, weighted for different precisions of individual crystals (see Galbraith 1990).

44 Ma, during which time they must have cooled through the apatite fission-track PAZ. The central ages for the samples from the Indus Group lie at c. 14 Ma, far younger than might be expected for a basin that was inverted before High Himalayan unroofing if these grains were not reset since deposition.

Illite crystallinity

Post-depositional heating of the Indus Group is important to consider when interpreting low-temperature thermochronometers, such as apatite fission-track data. A study of the thermal evolution also allows us to estimate how much erosion has occurred in the suture since inversion and how much of the Indus Fan may be derived through erosion of these units. We assess the degree of heating using the illite crystallinity technique applied to clay minerals recrystallized during diagenesis. The low grades of metamorphism of the Indus Group make the analysis of illite clay minerals a suitable technique for the investigation of the burial and unroofing history of this sequence. X-ray diffraction analysis of illite clays allows relative metamorphic temperatures to be determined, as a result of the progressive recrystallization of smectite clays to illite–smectite mixed layer clays and subsequently chlorite, illite and albite as temperature rises (e.g. Kisch 1983). A number of methods now exist that exploit this change in an attempt to quantify the degree of low-grade metamorphism (Kisch 1983, 1990), and in this study the method of Weber (1972) was employed. Weber (1972) used the sharpness of the illite peak to calculate a metamorphic grade quantified by a factor that he called 'Hb_{rel}', which decreases with increasing metamorphic grade. Once the Hb_{rel} is calculated it is possible to correlate this value with standard low-grade metamorphic zones. Pressure is unconstrained by the illite crystallinity method, but a typical continental geothermal gradient of 30–40 °C km would correspond to a depth of 5.0–6.6 km to the anchizone boundary (200 °C).

The samples were broken up by coarse jaw crushing, placed for 10 min in an ultrasonic bath, then sieved and finally centrifuged to concentrate the finest fraction material. The clays were put into an aqueous suspension and pipetted onto glass slides and dried at room temperature. Analysis was performed with Cu-Kα radiation at 40 kV, 20 mA, and a scan speed of 0.5° 2θ min. A quartz standard was also analysed. After an initial analysis the specimens were dehydrated by placing in an enclosed chamber in the presence of ethylene glycol, and were then reanalysed. This procedure was used to eliminate the effect of hydrated,

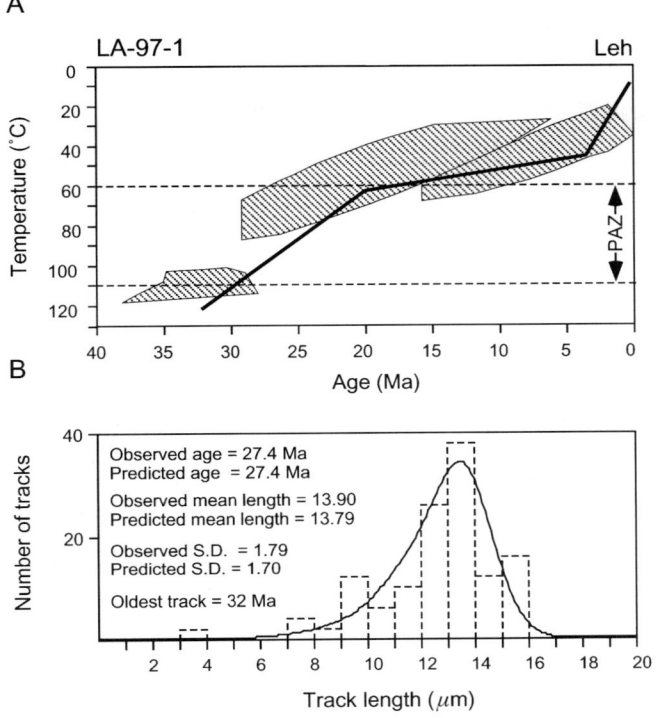

Fig. 10. (a) Best-fit thermal model for apatite fission-track data derived from a sample taken from the Ladakh Batholith at Leh. Shaded region represents best-defined (90% confidence contour) part of cooling history, with the continuous line representing the most likely cooling path. PAZ, partial annealing zone. (b) Comparison of fission-track length characteristics between model and observation followed the technique of Gallagher (1995).

expandable-layer, smectite clays, which cause broadening of the main illite peak.

The results of the illite crystallinity study are shown in Fig. 11, a transect through the Indus Group and underlying units in the Zanskar Gorge. Almost all the samples, both air-dried and glycolated, are of diagenetic grade, although values do approach the anchizone boundary at several points, at 32, 17 and 4 km from the contact with the granite. Our analysis is comparable with the anchizonal grade reported for the central part of the Indus Group exposure by Van Haver et al. (1986). However, in contrast to that study the lowest grades are seen to directly juxtapose the granite and are also found just north of Sumda-Do, at 19 km from the contact. Peak values do not occur randomly, but instead occur at the crest of increasing trends that correlate to the first order with the major identified structures in the gorge. We interpret the peak Hb_{rel} values to reflect the exhumation of deeper structural levels along thrust faults and in the cores of anticlines. Anchizone metamorphism begins at c. 200 °C (Kisch 1983; Mullis 1987), so that we estimate temperatures a little below that level for the bulk of the section. Because apatite fission tracks are totally annealed at c. 110 °C (Green et al. 1989) it is likely that the bulk of the Indus Group was reset for these low-temperature palaeo-thermometers after sedimentation.

$^{40}Ar/^{39}Ar$ geochronology

A suite of biotite and hornblende crystals from the Ladakh Batholith were examined using the $^{40}Ar/^{39}Ar$ method, to constrain the cooling of this body at higher temperatures than possible using fission tracks alone. Analyses were performed at the Massachusetts Institute of Technology, USA. Samples were irradiated at the McMaster Nuclear Reactor in Hamilton, Ontario, Canada. The technique employed was similar to that described by Coleman & Hodges (1995). Results of the analysis are shown in Fig. 12, where the spectrum of Ar gas release can be observed. The age of cooling through the Ar closure temperature is derived

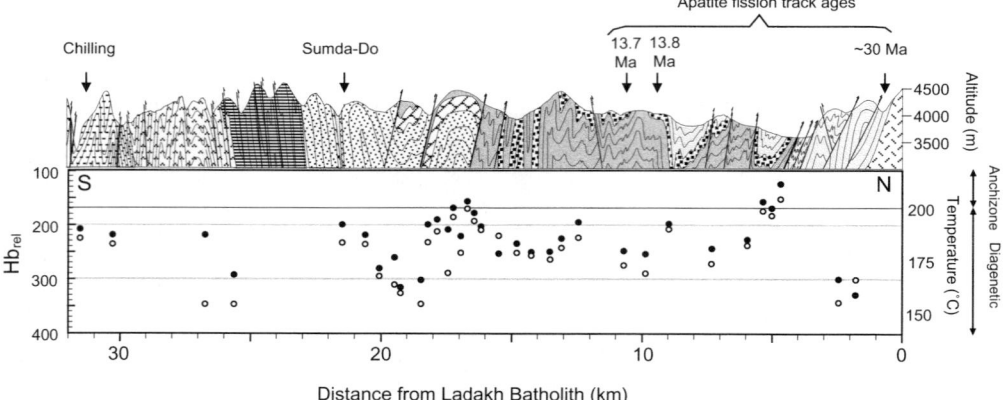

Fig. 11. Section showing the generally increasing diagenetic grade (i.e. rising post-depositional temperatures) experienced by the Indus Group moving south from the Ladakh Batholith. Illite crystallinity is determined by the method of Weber (1972). ●, samples analysed after glycolating; ○, air-dried samples.

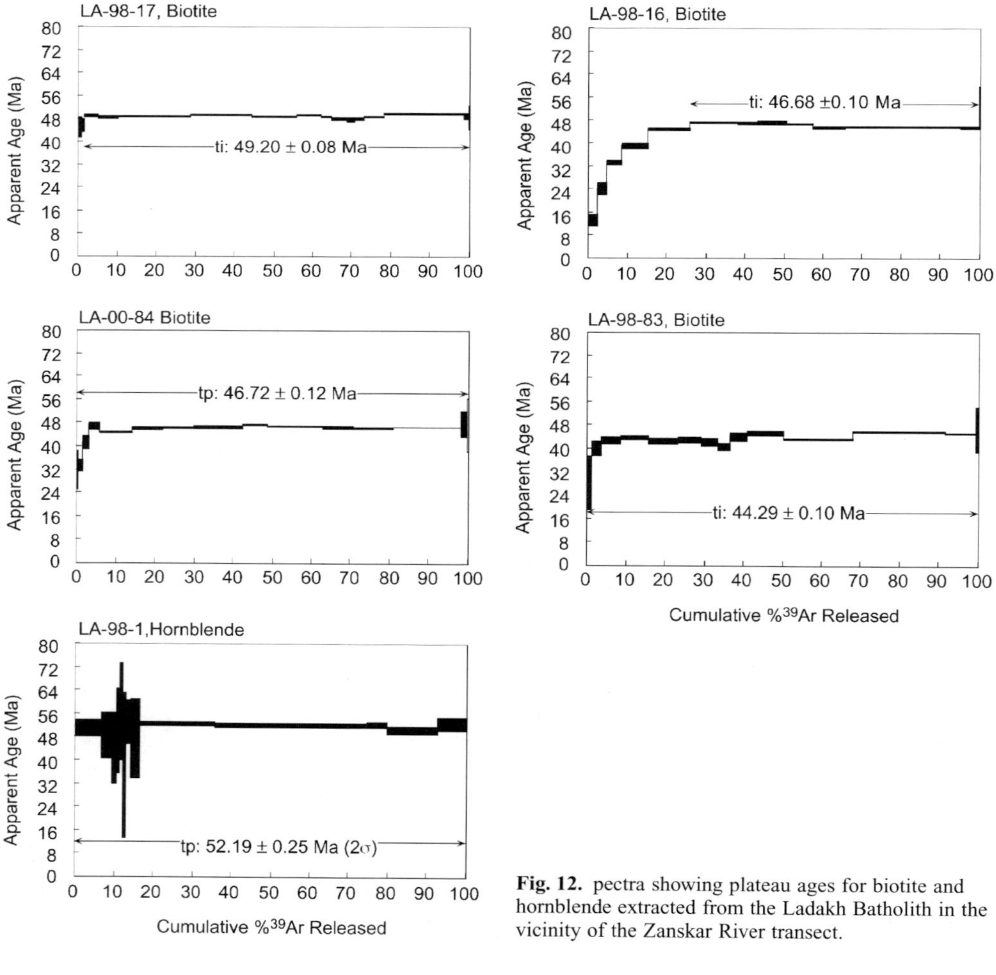

Fig. 12. pectra showing plateau ages for biotite and hornblende extracted from the Ladakh Batholith in the vicinity of the Zanskar River transect.

from the stable 'plateau' stage of gas release. A closure temperature of c. 300 °C is typically accepted for biotite (Harrison et al. 1985), whereas c. 500 °C is representative of most hornblende, despite the potential complexities introduced by exsolution and composition (Harrison & FitzGerald 1986; Leake et al. 1988). Biotite ages show a spread of 44–49 Ma, whereas the single hornblende yielded an older (52 Ma) age.

Discussion

India–Eurasia collision

The age of the India–Eurasia collision in the western Himalaya implied by the age and structural position of the Chogdo Formation is broadly consistent with other estimates of collision from the Arabian Sea region. Seismic and provenance data from the Pakistan margin and Owen Ridge show that fan sedimentation triggered by collision dates from at least Mid-Eocene time (Clift et al. 2000b, 2001b). In Pakistan and northern India synorogenic clastic sedimentation has been dated back to late Paleocene time in the foreland (DeCelles et al. 1998; Najman & Garzanti 2000), and to at least earliest Eocene time in the Katawaz Basin of western Pakistan–Afghanistan (Tapponnier et al. 1981; Qayyum et al. 1997). These dates constrain collision, at least in the western Himalaya, to latest Paleocene–Early Eocene time (56.2–49.0 Ma). Rowley (1998) argued that the lack of rapid collision-related subsidence on the Indian margin before Early Eocene time (50.8 Ma; Gaetani & Garzanti 1991) proves that collision and ophiolite obduction post-dates this time. However, because the original horizontal distance between the suture and the preserved section of the Zanskar Platform is unknown, the age of collision-related flexural subsidence there can only represent a minimum estimate of India–Eurasia collision.

Radiometric dating by De Sigoyer et al. (2000) of eclogites within the Tso Morari Complex (Fig. 1), which are of Indian plate affinity, established peak metamorphism there at 55 ± 6 Ma. Those workers suggested that the distal edge of the Indian plate was subducted during the initial contact with Eurasia, forming the eclogites that were then exhumed after collision. This interpretation was disputed by Searle (2001), who pointed out that the 55 Ma age estimate was only an average of a suite of ages, each of which needed to be considered and interpreted separately. Instead, he suggested that these rocks were formed during earlier ophiolite obduction, being partially reset during later India–Eurasia collision. The timing of India–Eurasia collision inferred from the Indus Group is consistent with the 55 Ma age of De Sigoyer et al. (2000), because here we constrain the elimination of oceanic crust within the suture to being no later than the end of Ypresian time (49 Ma). However, the sedimentary evidence from Ladakh is also compatible with the two-stage ophiolite obduction and collisional model advocated by Searle (2001). We cannot constrain the oldest possible age of collision tightly because this must simply post-date the Paleocene Sumda Formation carbonate sequences (Van Haver 1984). Our estimate is consistent with regional plate reconstructions, which have indicated the start of 'soft collision' between India and Eurasia at c. 55 Ma (Lee & Lawver 1995), marked by a rapid decrease in the convergence rate (Klootwijk et al. 1992).

It is noteworthy that this 55 Ma collision age is within error of the accepted age of the Paleocene–Eocene Boundary Thermal Event (e.g. Stott et al. 1996), as well as the initiation of deep-water saline bottom water flow. Continental collision and the resultant cutting of east–west equatorial flow may thus be possible tectonic triggers for these palaeoceanographic events, forcing large-scale reorganization of global circulation patterns. The end of equatorial flow in the Tethys north of India would have created gulfs around the collision point, where evaporation might have produced the saline waters required to initiate the deep-water bottom water flow.

Thermal evolution of suture zone

The thermal data for the Indus Group and adjacent Ladakh Batholith may be compared with existing analyses to reconstruct the regional cooling and unroofing history related to deposition within, and then inversion and destruction of the Indus Group Basin. For the purpose of this study we focus on the area close to Leh and ignore data from western Ladakh, where large-scale structural doming associated with the Nanga Parbat massif results in younger cooling ages (e.g. Sorkhabi et al. 1994). Magmatism in the Ladakh Batholith has been dated back into Cretaceous time using the Rb–Sr system (Honegger et al. 1982; Schärer et al. 1984). Recently, these ages have been confirmed by U–Pb dating of zircons, which show a spread from 70 to 50 Ma in Ladakh (Weinberg & Dunlap 2000), as do a suite of 70–80 Ma detrital micas from the Indus Group measured by K–Ar methods (Van Haver et al. 1986). The youngest magmatism in the Leh area dates from 46 ± 1 Ma, although further east, at Giak, 25–30 Ma ages are recorded (Trivedi et al. 1982). The new $^{40}Ar/^{39}Ar$ ages indicate rapid cooling of the batholith after 50 Ma (Fig. 13). The biotite ages do not substantially post-date the U–Pb zircon ages and indicate that

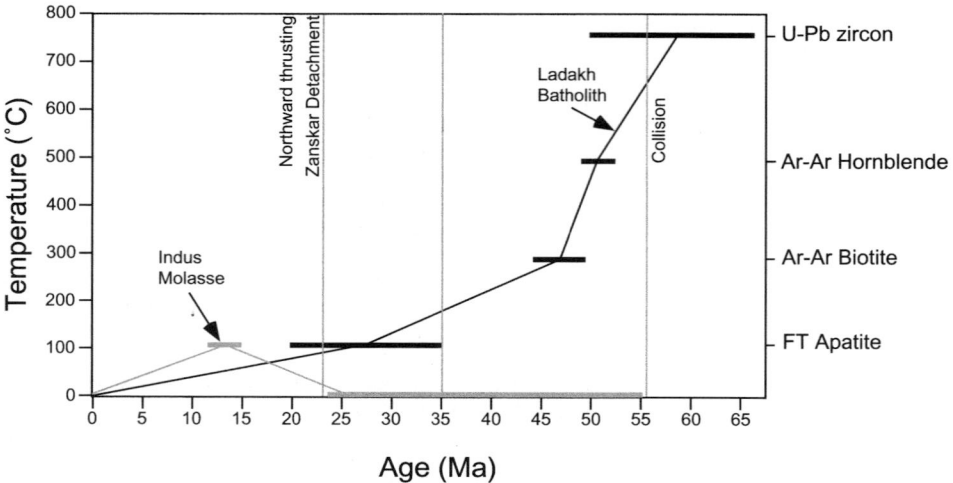

Fig. 13. Thermal evolution diagram for the Ladakh Batholith and Indus Group in the study area. (Note the increased rate of cooling of the batholith after the inferred collision close to the Paleocene–Eocene boundary.)

cooling must have been rapid at 44–49 Ma, followed by slower cooling to the present day. The fission-track data from the Ladakh Batholith are consistent with a gradual cooling during Neogene time, as modelled for the sample taken at Leh (LA-97-1; Fig. 11). The timing of the cessation of arc magmatism and subsequent rapid cooling thus immediately followed, and is consistent with, the inferred age of the India–Eurasia collision in western India and Pakistan at c. 55 Ma. The subduction of relatively buoyant thinned continental crust at the trench, together with the first compressional tectonic activity, might reasonably be expected to have generated uplift along the active margin shortly after the start of collision. As the deformation migrated south into the Himalaya with time (Le Fort 1975; Bassoullet et al. 1980; Searle 1986) the rate of uplift and unroofing in the Ladakh Batholith was reduced.

Neogene thrusting of the Indus Group northwards against the Ladakh Batholith had a major thermal impact on the strata, but is not recorded by the granites themselves or the sediments along the southern edge of the batholith. Although the illite crystallinity method is sensitive at low temperatures, and thus for shallow-level processes, it is noteworthy that the two sandstone samples from the Zanskar Gorge analysed for fission tracks have much younger ages than the adjacent granite (14 Ma compared with c. 30 Ma; Table 1). They are interpreted as having been reset during Neogene thrusting. The thrusting appears to have resulted in a preferentially greater cooling of the imbricated Indus Group Basin compared with the Ladakh Batholith. The southward increase in metamorphic grade is also consistent with the higher-grade epizonal metamorphism recorded by Garzanti & Brignoli (1989) in the Zanskar Himalaya immediately south of the studied region. We suggest that the Indus Group Basin was inverted and deformed during northward thrusting, whereas the more coherent Ladakh Batholith remained relatively unaffected. Thus although the Ladakh Batholith has suffered more cooling and erosion during Cenozoic time, during the Neogene period alone erosion here was <2 km, compared with 5–6 km in the Indus Group Basin, assuming normal geothermal gradients.

Age of the Indus Group

The fission-track data presented above allow some constraints to be placed on the enigmatic age of sedimentation of the Indus Group. The reset 14 Ma ages recorded in apatites from Nurla Formation sandstones require deposition before that time, consistent with the lack of High Himalayan minerals and the known age of unroofing of these sources (Clift et al. 2001a). However, the fact that the Nimu Formation conglomerates that overlie the Ladakh Batholith seen at Khalsi (Basgo Formation of Van Haver (1984); Fig. 1) have fission-track ages (35.7 Ma) that predate the 14 Ma thermal event recorded in the main part of the Indus Group suggests that this contact was not reset during Miocene thrusting (i.e. it remained cooler than 60 °C, c. 2 km). The fission tracks must reflect the cooling of this granite within the source before erosion. This implies that those sedimentary rocks, which lie close to the stratigraphic top of the Indus Group according to the reconstruction of Searle et al. (1990), were deposited after 35.7 Ma, the central fission-track age of the boulders, in stark contrast to the Paleocene–Upper Cretaceous age assignment that Baud et al. (1982) and Van Haver (1984) gave these rocks, which they named the Basgo and Temesgam Formations (Table 1).

The new thermal data provide no support for the suggestion by Van Haver et al. (1986) that the southern edge of the Ladakh Batholith and its cover sedimentary rocks (Nimu Formation of Sinclair & Jaffey (2001), Temesgam and Basgo Formations of Van Haver (1984)) suffered hydrothermal activity at 20–30 Ma. The low diagenetic grade seen in the illite crystallinity along the contact shows that thermal resetting of detrital minerals (micas, apatites or hornblendes) cannot have occurred, and that therefore the cooling ages recorded by Van Haver et al. (1986) correspond to cooling in the unroofing batholith before sedimentation. The earlier interpretation was based on the identification by Van Haver (1984) of Maastrichtian ostracodes in the Temesgam Formation that overlies the basal Basgo Formation. Presumably this fauna must have been reworked, as the batholith rocks on which these units were deposited were still below the PAZ at this time.

Together with the Early Eocene age of the Nummulitic Limestone at the top of the Chogdo Formation (Van Haver 1984), the new thermal data suggest at least 20 Ma of sedimentation within the Indus Group. It is likely that the basin was in semi-continuous sedimentation from the Paleocene–Eocene boundary (c. 55 Ma) until the start of Zanskar backthrusting at c. 23 Ma (Metcalfe 1993), >30 Ma.

Conclusions

The revised tectonic stratigraphy for the Indus Suture, and in particular the onlapping relationships at the base of the Indus Group, presented here reveals a Cretaceous intra-oceanic forearc sequence (Nindam Formation) overlain by an Upper Cretaceous–Paleocene continental Eurasian forearc sequence (Jurutze Formation). This formation and its distal trenchward equivalent (Khalsi Flysch) both show a rapid influx of carbonate platform and ophiolite detritus from the south (i.e. India) at the start of deposition of the fluvial Chogdo Formation. The Chogdo Formation is the oldest sediment to overlie these forearc units, ophiolitic mélanges and the Indian slope Lamayuru Group in the vicinity of Chilling in the Zanskar Gorge. The Early Eocene (Ypresian) age of the Nummulitic Limestone overlying the Chogdo Formation (Van Haver 1984) requires that Eurasia and India were actively colliding no later than the end of that period (>49 Ma), because they were separated only by a continental, fluvial basin.

The sedimentary evidence from the suture zone is consistent with geomagnetic data (Klootwijk et al. 1992; Chaubey 2002; Royer et al. 2002) in implying a rapidly slowing northward motion of India after 55 Ma caused by initial collision between the passive margin of India and the Eurasian trench. Such a collision may have triggered the global Paleocene–Eocene Boundary Thermal Event at that time, as well as the deep-water saline bottom water flow that commenced after collision, as the equatorial Tethyan gateway north of India was closed.

After India–Eurasia collision, sedimentation continued relatively unbroken in the Indus Group Basin until basin inversion at c. 23 Ma. Fission-track data constrain the end of sedimentation at 35–14 Ma, but the absence of minerals eroded from the High Himalaya suggests that basin inversion preceded unroofing of these terrains, and probably coincides with northward backthrusting of the Zanskar Platform, which imbricated the strata against a relatively inactive Ladakh Batholith. Subsequent erosion has preferentially affected the Indus Group (5–6 km) and the over-thrusting Zanskar Himalayas compared with the Ladakh Batholith (<2 km). Reworking of Indus Group sediments into the Indus Fan and foreland after basin inversion needs to be considered when the detrital history of these Neogene sequences is being interpreted in terms of evolving basement exhumation patterns, although the total volume is modest compared with that from other potential sources.

I wish to thank M. Searle, H. Sinclair, M. A. Holmes, P. Kelemen, P. Molnar, D. Norris and K. Bice for thought-provoking discussion on this work, and Woods Hole Oceanographic Institution and JOI/USSAC for support for this study. We thank L. Poppe at USGS, Woods Hole, for help with the X-ray diffraction work. The paper has benefited from constructive reviews by R. Corfield, E. Garzanti, M. Searle and an anonymous reviewer. Rockland Tour and Trek and F. Hussein in Leh provided logistical support. This is WHOI Contribution 10630.

References

BASSOULLET, J.P., COLCHEN, M., JUTEAU, T., MARCOUX, J. & MASCLE, G. 1980. L'edifice de nappes du Zanskar (Lakakh, Himalaya). (The structure of the Zanskar Nappes; Ladakh, Himalayas.). *Comptes Rendus de l'Académie des Sciences, Série D*, **290**, 389–392.

BASSOULLET, J.P., COLCHEN, M., JUTEAU, TH., MARCOUX, J., MASCLE, G. & REIBEL, G. 1983. Geological studies in the Indus Suture Zone of Ladakh (Himalayas). In: GUPTA, V.J. (ed.) Stratigraphy and Structure of Kashmir and Ladakh Himalaya. Contributions to Himalayan Geology, 2, 96–124.

BAUD, A., ARN, R., BUGNON, P. & 8 OTHERS 1982. Le contact Gondwana–Peri-Gondwana dans le Zanskar oriental (Ladakh Himalaya). (The Gondwana–peri-Gondwana contact in eastern Zanskar, Ladakh, Himalayas.). Bulletin de la Société Géologique de France, 24, 341–361.

BAUD, A., GAETANI, M., GARZANTI, E., FOIS-ERICKSON, E., NICORA, A. & TINTORI, A. 1984. Geological observations in southeastern Zanskar and adjacent Lahul area (northwestern Himalaya). Eclogae Geologicae Helvetiae, 77, 171–197.

BERGGREN, W.A., KENT, D.V., SWISHER, C.C. & AUBRY, M.P. 1995. A revised Cenozoic geochronology and chronostratigraphy. In: BERGGREN, W.A., KENT, D.V, AUBRY, M.P. & HARDENBOL, J. (eds) Geochronology, Time Scales and Global Stratigraphic Correlation. Society of Economic Paleontologists and Mineralogists, Special Publications, 54, 129–212.

BICE, K.L., SCOTESE, C.R., SEIDOV, D. & BARRON, E.J. 2000. Quantifying the role of geographic change in Cenozoic ocean heat transport using uncoupled atmosphere and ocean models. Palaeogeography, Palaeoclimatology, Palaeoecology, 161, 295–310.

BRASS, G.W., SOUTHAM, J.R. & PETERSON, W.H. 1982. Warm saline bottom waters in the ancient ocean. Nature, 296, 620–623.

BROOKFIELD, M.E. & ANDREWS-SPEED, C.P. 1984. Sedimentology, petrography and tectonic significance of the shelf, flysch and molasse clastic deposits across the Indus suture zone, Ladakh, NW India. Sedimentary Geology, 40, 249–286.

CANNAT, M. & MASCLE, G. 1990. Réunion extraordinaire de la Société Géologique de France en Himalaya du Ladakh, August 6–September 1, 1987. Bulletin de la Société Géologique de France, 6, 553–582.

CHAUBEY, A. 2002. Paleogene magnetic isochrons and palaeo-propagators in the Arabian and Eastern Somali basins, NW Indian Ocean. In: CLIFT, P.D., KROON, D., GAEDICKE, C. & CRAIG, J. (eds) The Tectonic and Climatic Evolution of the Arabian Sea Region. Geological Society, London, Special Publications, 195, 71–86.

CLIFT, P.D. 2002. A brief history of the Indus. In: CLIFT, P.D., KROON, D., GAEDICKE, C. & CRAIG, J. (eds) The Tectonic and Climatic Evolution of the Arabian Sea Region. Geological Society, London, Special Publications, 195, 237–258.

CLIFT, P.D., DEGNAN, P.J., HANNIGAN, R. & BLUSZTAJN, J. 2000a. Sedimentary and geochemical evolution of the Dras fore-arc basin, Indus Suture, Ladakh Himalaya, India. Geological Society of America Bulletin, 112, 450–466.

CLIFT, P.D., SHIMIZU, N., LAYNE, G.D., GAEDICKE, C., SCHLÜTER, H.U., CLARK, M.K. & AMJAD, S. 2000b. Fifty five million years of Tibetan evolution recorded in the Indus Fan. EOS Transactions, American Geophysical Union, 81, 277–281.

CLIFT, P.D., SHIMIZU, N., LAYNE, G. & BLUSZTAJN, J. 2001a. Tracing patterns of unroofing in the Early Himalaya through microprobe Pb isotope analysis of detrital K-feldspars in the Indus Molasse, India. Earth and Planetary Science Letters, 188, 475–491.

CLIFT, P.D., SHIMIZU, N., LAYNE, G., GAEDICKE, C., SCHLÜTER, H.U., CLARK, M. & AMJAD, S. 2001b. Development of the Indus Fan and its significance for the erosional history of the western Himalaya and Karakoram. Geological Society of America Bulletin, 113, 1039–1051.

COLEMAN, M. & HODGES, K. 1995. Evidence for Tibetan Plateau uplift before 14 Myr ago from a new minimum age for east–west extension. Nature, 374, 49–52.

DAINELLI, G. 1934. La serie dei terreni Spedizione Italiana de Filippi nell' Himalaia. Caracorum e Turchestan Cinese, 2, 1–458.

DECELLES, P.G., GEHRELS, G.E., QUADE, J. & OJHA, T.P. 1998. Eocene–early Miocene foreland basin development and the history of Himalayan thrusting, western and central Nepal. Tectonics, 17, 741–765.

DE SIGOYER, J., CHAVAGNAC, V., BLICHERT-TIFT, J. & 5 OTHERS 2000. Dating the Indian continental subduction and collisional thickening in the NW Himalaya: multichronology of the Tso Morari eclogites. Geology, 28, 487–490.

FUCHS, G. 1985. The geology of the Markha–Khurnak region in Ladakh (India). Jahrbuch der Geologischen Bundesanstalt Wien, 128, 403–437.

FUCHS, G. 1989. Arguments for the autochthony of the Tibetan Zone. Eclogae Geologicae Helvetiae, 82, 685–692.

GAETANI, M. & GARZANTI, E. 1991. Multicyclic history of the northern India continental margin (northwestern Himalaya). AAPG Bulletin, 75, 1427–1446.

GALBRAITH, R.F. 1990. The radial plot: graphical assessment of spread in ages. Nuclear Tracks, 17, 207–214.

GALLAGHER, K. 1995. Evolving temperatures histories from apatite fission-track data. Earth and Planetary Science Letters, 136, 421–435.

GARZANTI, E. & BRIGNOLI, G. 1989. Low temperature metamorphism in the Zanskar sedimentary nappes (NW Himalaya, India). Eclogae Geologicae Helvetiae, 82, 669–684.

GARZANTI, E. & VAN HAVER, T. 1988. The Indus clastics: forearc basin sedimentation in the Ladakh Himalaya (India). Sedimentary Geology, 59, 237–249.

GARZANTI, E., BAUD, A. & MASCLE, G. 1987. Sedimentary record of the northward flight of India and its collision with Eurasia (Ladakh Himalaya, India). Geodinamica Acta, 1, 297–312.

GREEN, P.F., DUDDY, I.R., LASLETT, G.M., HEGARTY, K.A., GLEADOW, A.J.W. & LOVERING, J.F. 1989. Thermal annealing of fission tracks in apatite: 4: Quantitative modeling techniques and extension to geological timescales. Chemical Geology, 79, 155–182.

HARRISON, T.M. & FITZGERALD, J.D. 1986. Exsolution in hornblende and its consequences for $^{40}Ar/^{39}Ar$ age spectra and closure temperature. Geochimica et Cosmochimica Acta, 50, 247–253.

HARRISON, T.M., DUNCAN, I. & MCDOUGALL, I. 1985. Diffusion of ^{40}Ar in biotite; temperature, pressure and compositional effects. *Geochimica et Cosmochimica Acta*, **49**, 2461–2468.

HONEGGER, K., DIETRICH, V., FRANK, W., GANSSER, A., THOENI, M. & TROMMSDORFF, V. 1982. Magmatism and metamorphism in the Ladakh Himalayas (the Indus–Tsangpo suture zone). *Earth and Planetary Science Letters*, **60**, 253–292.

HURFORD, A.J. 1990. Standardization of fission track dating calibration: recommendation by the Fission Track Working Group of the IUGS subcommission on geochronology. *Chemical Geology*, **80**, 177–178.

KISCH, H.J. 1983. Mineralogy and petrology of burial diagenesis (burial metamorphism) and incipient metamorphism in clastic rocks. *In:* LARSEN, G. & CHILINGAR, G.V. (eds) *Diagenesis and Sediments and Sedimentary Rocks*. Elsevier, Amsterdam, 289–493.

KISCH, H.J. 1990. Calibration of the anchizone: a critical comparison of illite crystallinity scales used for definition. *Journal of Metamorphic Geology*, **8**, 31–46.

KLOOTWIJK, C.T., GEE, J.S., PEIRCE, J.W., SMITH, G.M. & MCFADDEN, P.L. 1992. An early India–Asia contact; paleomagnetic constraints from Ninetyeast Ridge, ODP Leg 12. *Geology*, **20**, 395–398.

LASLETT, G.M., GREEN, P.F., DUDDY, I.R. & GLEADOW, A.J.W. 1987. Thermal annealing of fission track grains in apatite 2: A quantitative analysis. *Chemical Geology*, **65**, 1–13.

LEAKE, B.E., ELIAS, E.M. & FARROW, C.M. 1988. The relationship of argon retentivity and chemical composition of hornblende. *Geochimica et Cosmochimica Acta*, **52**, 2165.

LEE, T.-Y. & LAWVER, L.A. 1995. Cenozoic plate reconstruction of the Southeast Asia region. *Tectonophysics*, **251**, 85–138.

LE FORT, P. 1975. Himalayas; the collided range; present knowledge of the continental arc. *American Journal of Science*, **275**, 1–44.

MASCLE, G. 1985. L'Himalaya résulte-t-il du télescopage de trois chaînes. *Bulletin de la Société Géologique de France*, **1**, 289–304.

METCALFE, R.P. 1993. Pressure, temperature and time constraints on metamorphism across the Main Central Thrust zone and high Himalayan slab in the Garhwal Himalaya. *In:* TRELOAR, P.J. & SEARLE, M.P. (eds) *Himalayan Tectonics*. Geological Society, London, Special Publications, **74**, 485–509.

MOLNAR, P., ENGLAND, P. & MARTINOD, J. 1993. Mantle dynamics, the uplift of the Tibetan Plateau, and the Indian monsoon. *Reviews in Geophysics*, **31**, 357–396.

MULLIS, J. 1987. Fluid inclusion studies during very low grade metamorphism. *In:* FREY, M. (ed.) *Low Temperature Metamorphism*. Blackie, Glasgow, 162–198.

NAJMAN, Y. & GARZANTI, E. 2000. Reconstructing early Himalayan tectonic evolution and paleogeography from Tertiary foreland basin sedimentary rocks, northern India. *Geological Society of America Bulletin*, **112**, 435–449.

PAK, D.K. & MILLER, K.G. 1992. Paleocene to Eocene benthic foraminiferal isotopes and assemblages: implications for deepwater circulation. *Paleoceanography*, **7**, 405–422.

QAYYUM, M., LAWRENCE, R.D. & NIEM, A.R. 1997. Discovery of the palaeoIndus delta–fan complex. *Journal of the Geological Society, London*, **154**, 753–756.

RAYMO, M.E. & RUDDIMAN, W.F. 1992. Tectonic forcing of the late Cenozoic climate. *Nature*, **359**, 117–122.

ROBERTSON, A.H.F. 2000. Formation of mélanges in the Indus Suture Zone, Ladakh Himalaya by successive subduction-related, collisional and post-collisional processes during Late Mesozoic–Late Tertiary time. *In:* KHAN, M.A., TRELOAR, P.J., SEARLE, M.P. & JAN, M.Q. (eds) *Tectonics of the Nanga Parbat Syntaxis and the Western Himalaya*. Geological Society, London, Special Publications, **170**, 333–374.

ROBERTSON, A.H.F. & DEGNAN, P.J. 1993. Sedimentology and tectonic implications of the Lamayuru Complex; deep-water facies of the Indian passive margin, Indus suture zone, Ladakh Himalaya. *In:* TRELOAR, P.J. & SEARLE, M.P. (eds) *Himalayan Tectonics*. Geological Society, London, Special Publications, **74**, 299–321.

ROBERTSON, A.H.F. & DEGNAN, P.J. 1994. The Dras arc complex: lithofacies and reconstruction of a Late Cretaceous oceanic volcanic arc in the Indus Suture Zone, Ladakh Himalaya. *Sedimentary Geology*, **92**, 117–145.

ROLLAND, Y., PECHER, A. & PICARD, C. 2000. Middle Cretaceous back-arc formation and arc evolution along the Asian margin; the Shyok suture zone in northern Ladakh (NW Himalaya). *Tectonophysics*, **325**, 145–173.

ROYER, J.-Y., CHAUBEY, A.K., DYMENT, J., BHATTACHARYA, G.C., SRINIVAS, K., YATHEESH, V. & RAMPRASAD, T. 2002. Paleogene plate tectonic evolution of the Arabian and Eastern Somali Basins. *In:* CLIFT, P. D., KROON, D., GAEDICKE, C. & CRAIG J. (eds) *The Tectonic and Climatic Evolution of the Arabian Sea Region*, **195**, 7–24.

ROWLEY, D.B. 1996. Age of initiation of collision between India and Asia; a review of stratigraphic data. *Earth and Planetary Science Letters*, **145**, 1–13.

ROWLEY, D.B. 1998. Minimum age of initiation of collision between India and Asia north of Everest based on subsidence history of the Zhepure Mountain Section. *Journal of Geology*, **106**, 229–235.

RUDDIMAN, W.F., RAYMO, M.E., LAMB, H.H. & ANDREWS, J.T. 1988. Northern Hemisphere climate regimes during the past 3 Ma; possible tectonic connections; discussion. *In:* SHACKLETON, N.J., WEST, R.G. & BOWEN, D.Q. (eds) *The Past Three Million Years; Evolution of Climatic Variability in the North Atlantic Region*. Cambridge University Press, Cambridge, 1–20.

SCHÄRER, U., HAMET, J. & ALLÈGRE, C.J. 1984. The Transhimalaya (Gangdese) plutonism in the Ladakh region; a U–Pb and Rb–Sr study. *Earth and Planetary Science Letters*, **67**, 327–339.

SEARLE, M.P. 1986. Structural evolution and sequence

of thrusting in the High Himalayan, Tibetan-Tethys and Indus suture zones of Zanskar and Ladakh, western Himalaya. *Journal of Structural Geology*, **8**, 923-936.

SEARLE, M.P. 2001. Discussion of de Sigoyer, J., Chavagnac, V., Blichert-Toft, J., Villa, I. M., Luais, B., Guillot, S., Cosca, M. & Mascle, G., Dating the Indian continental subduction and collisional thickening in the Northwest Himalaya; multichronology of the Tso Morari eclogites. *Geology*, **29**, 192-193.

SEARLE, M.P., WINDLEY, B.F., COWARD, M.P. & 8 OTHERS 1987. The closing of Tethys and the tectonics of the Himalaya. *Geological Society of America Bulletin*, **98**, 6787-701.

SEARLE, M.P., PICKERING, K.T. & COOPER, D.J.W. 1990. Restoration and evolution of the intermontane Indus Molasse Basin, Ladakh Himalaya, India. *Tectonophysics*, **174**, 301-314.

SEARLE, M.P., WATERS, D.J., REX, D.C. & WILSON, R.N. 1992. Pressure, temperature and time constraints on Himalayan metamorphism from eastern Kashmir and western Zanskar. *Journal of the Geological Society, London*, **149**, 753-773.

SINCLAIR, H.D. & JAFFEY, N. 2001. Sedimentology of the Indus Group, Ladkah, northern India: implication of timing of initiation of the palaeo-Indus River. *Journal of the Geological Society, London*, **158**, 151-162.

SORKHABI, R., JAIN, A.K., NISHIMURA, S., ITAYA, T., LAL, N., MANICKAVASAGAM, R.M. & TAGAMI, T. 1994. New age constraints on the cooling and unroofing of the Trans-Himalayan Ladakh Batholith (Kargil area), NW India. *Proceedings of the Indian Academy of Sciences: Earth and Planetary Sciences*, **103**, 83-97.

STECK, A., SPRING, L., VANNAY, J.-C. & 5 OTHERS 1993. The tectonic evolution of the Northwestern Himalaya in eastern Ladakh and Lahaul, India. *In*: Treloar, P. J. & Searle, M. P. (eds) *Himalayan Tectonics*. Geological Society, London, Special Publications, **74**, 265-276.

STOTT, L.D., SINHA, A., THIRY, M., AUBRY, M.P. & BERGGREN, W.A. 1996. Global $\delta^{13}C$ changes across the Paleocene-Eocene boundary; criteria for terrestrial-marine correlations. *In:* KNOX, R.W.O.B., CORFIELD, R.M. & DUNAY, R.E. (eds) *Correlation of the Early Paleogene in Northwest Europe*. Geological Society, London, Special Publications, **101**, 381-399.

SUTRE, E. 1990. *Les formations de la marge nord Neotethysienne et les mélanges ophiolitiques de la zone de l'Indus en Himalaya du Ladakh*. Ph.D. thesis, University of Poitiers.

TAPPONNIER, P., MATTAUER, M., PROUST, F. & CASSAIGNEAU, C. 1981. Mesozoic ophiolites, sutures and large-scale tectonic movements in Afghanistan. *Earth and Planetary Science Letters*, **52**, 355-371.

TRIVEDI, J.R., GOPALAN, K., SHARMA, K.K., GUPTA, K.R. & CHOUBEY, V.M. 1982. Rb-Sr age of Gaik Granite, Ladakh Batholith, Northwest Himalaya. *Proceedings of the Indian Academy of Sciences: Earth and Planetary Sciences*, **91**, 65-73.

VANCE, D. & HARRIS, N. 1999. Timing of prograde metamorphism in the Zanskar Himalaya. *Geology*, **27**, 395-398.

VAN HAVER, T. 1984. *Étude stratigraphique, sedimentologique et structurale d'un bassin d'avant arc: exemple du bassin de l'Indus, Ladakh, Himalaya*. Ph.D. thesis, University of Grenoble.

VAN HAVER, T., BASSOULLET, J.P., BLONDEAU, A. & MASCLE, G. 1984. Les séries détritiques du bassin de l'Indus: nouvelles données stratigraphiques et structurales. *Rivista Italiana di Paleontologia e Stratigrafia*, **90**, 87-102.

VAN HAVER, T., BONHOMME, M.G., MASCLE, G. & APRAHAMIAN, J. 1986. Analyse K/Ar de phyllites fines des formations détritiques de 'Indus au Ladakh (Inde). Mise en evidence de l'âge Eocène supérieur du metamophisme. *Comptes Rendues de l'Académie des Sciences de Paris, Série II*, **302**, 325-330.

WALKER, J.D., MARTIN, M.W., BOWRING, S.A., SEARLE, M.P., WATERS, D.J. & HODGES, K.V. 1999. Metamorphism, melting, and extension; age constraints from the High Himalayan slab of Southeast Zanskar and Northwest Lahaul. *Journal of Geology*, **107**, 473-495.

WEBER, K. 1972. Notes on the determination of illite crystallinity. *Neues Jahrbuch fur Mineralogie, Monatshefte*, **1972**, 267-276.

WEINBERG, R.F. & DUNLAP, W.J. 2000. Growth and deformation of the Ladakh Batholith, Northwest Himalayas; implications for timing of continental collision and origin of calc-alkaline batholiths. *Journal of Geology*, **108**, 303-320.

ZACHOS, J.C., LOHMANN, K.C., WALKER, J.C.G. & WISE, S.W. 1993. Abrupt climate changes and transient climates during the Paleogene; a marine perspective. *Journal of Geology*, **101**, 191-213.

Basalt and peridotite recovered from Murray Ridge: are they of supra-subduction origin?

KLAUS-PETER BURGATH[1], ULRICH VON RAD[1], WILLEM VAN DER LINDEN[2], MARTIN BLOCK[1], ATHAR ALI KHAN[3], HANS ALBERT ROESER[1] & WOLFGANG WEISS[1]

[1] *Bundesanstalt für Geowissenschaften und Rohstoffe (BGR), PF 510153, 30631 Hannover, Germany (e-mail: k.burgath@bgr.de)*
[2] *Institute of Earth Sciences, Rijksuniversiteit Utrecht, Utrecht, Netherlands*
[3] *National Institute of Oceanography, Karachi, Pakistan*

Abstract: Petrographic and geochemical data for new basalt and peridotite samples recovered from sampling sites at the Southern Murray Ridge help to constrain models for the evolution of the Owen–Murray Ridge system, which forms the northwestern boundary of the Indian plate. Trace elements immobile during alteration (Ti, Zr, Nb, Y and rare earth elements) suggest that the altered microphyric metabasalt has affinities to magmatism of active margins (island-arc tholeiite *sensu lato*). It is distinctly different from mid-ocean ridge basalt, back-arc-basin basalt, or intra-plate Deccan Trap basalt. The sampled serpentinized harzburgite or clinopyroxene-poor lherzolite was deformed under mantle conditions and is similar to the mantle section of nearby ophiolite sequences. This association of rocks suggests that an ophiolite mélange was sampled. However, results from sampling station 462 NIOP indicate that the Murray Ridge complex also contains igneous rocks with Deccan Trap affinity. For the emplacement of the island-arc tholeiite we assume an origin in a convergent supra-subduction setting, related to the closing of a Late Cretaceous Neo-Tethyan ocean basin between the Arabian and Indian plates to the south and the Eurasian plate to the north. Since Neogene time, the Murray Ridge–Dalrymple Trough has been underlain by attenuated (?)continental crust and characterized by extensional rift tectonics.

The Murray Ridge *sensu lato* forms the central part of the dogleg-shaped northwestern boundary of the Indian plate, linking the Owen Fracture Zone in the SW to the Ornach–Nal Fault Zone in the NE (Fig. 1). Despite many geological and geophysical surveys in the northern Arabian Sea by US (e.g. Mountain & Prell 1992), British (e.g. White & Louden 1982), German and Dutch expeditions (see below), the exact nature and structure of the Murray Ridge is still not satisfactorily resolved and its geological evolution is still poorly understood.

This paper focuses on the results of the petrologic analyses of sedimentary and igneous rocks recovered from the Southern Murray Ridge during Leg D2 of the 1992–1993 *Tyro* (Netherlands Indian Ocean Programme, NIOP) and the 1998 *Sonne* cruise SO-130 of the Bundesanstalt für Geowissenschaften und Rohstoffe (BGR).

Whereas the physiography and tectonic aspects of the region based on multibeam echosounder, magnetic and seismic surveys are briefly covered, the detailed results of geophysical studies are being published elsewhere (see Clift *et al.* 2000; Edwards *et al.* 2000; Schreckenberger *et al.* 2001; Gaedicke *et al.* 2002a, 2002b). The petrological results presented here could contribute to constrain the interpretations presented thus far on the nature and evolution of the Murray Ridge complex, interpretations that are based largely on geomorphological and geophysical studies in the region.

The Murray Ridge complex

The Murray Ridge *sensu lato* (discovered by S. Seymour in 1933 during the *John Murray* cruise) extends from the continental slope south of Karachi towards the SW, where it joins the Owen Fracture Zone (Fig. 1). The 'Murray Ridge' is not a single topographic feature, but instead forms part of a number of subparallel (mostly) sediment-covered ridges, seamounts, basins and troughs. It is located between the continental

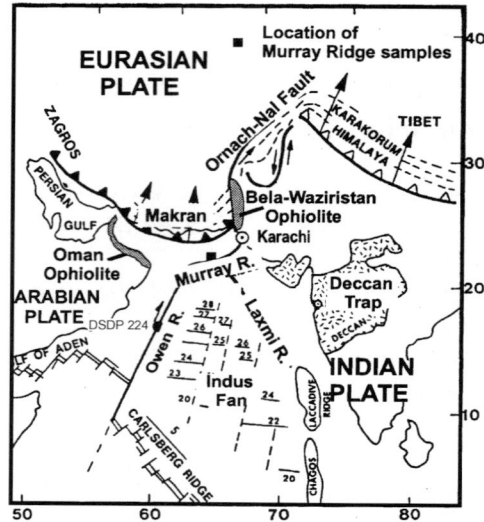

Fig. 1. Plate-tectonic sketch (modified from Coumes & Kolla 1984). Fine lines, fracture or transform faults; double lines, mid-oceanic spreading ridges; full arrowheads, subduction or collision directions; half-arrowheads, movement along fracture or transform faults; fine lines with numbers, magnetic anomalies (after Miles & Roest 1993); stippled pattern, subaerial outcrops of Deccan Trap in India.

margin of the Indian subcontinent and the Owen Ridge, an area at least 200 km across and about 400 km along strike. Hence we propose to rename this complex morphotectonic feature the 'Murray Ridge complex' (Fig. 2). From NW to SE we distinguish the following: (1) the Little Murray Ridge, apparently a fault-controlled chain of reef-capped seamounts, guyots and basement highs, that are largely covered by up to 1 km of sediment; (2) the Northern Murray Ridge, a relatively broad, flat-topped, sediment-covered ridge, that rises from the $c.$ 3000 m deep, flat-floored Oman Basin to its minimum elevation of $c.$ 2000 m; (3) the Southern Murray Ridge (Fig. 3), a narrow ridge that rises relatively steeply from a flat sea floor at 2500 m to reach a peak depth of 360 m; (4) the narrow, steep-sided (fault-controlled), >4200 m deep, sediment-filled Dalrymple Trough, between the Northern and Southern Murray Ridges; this trough and its northeastern extension are formed of a number of broad flat-floored basins ('Jinnah Trough') shoaling, from west to east, from $c.$ 4000 m to <3000 m; (5) the steep-sided Jinnah Seamount, near the northeastern end of the Dalrymple Trough, sitting right on top of the southern boundary fault of the Trough; its morphology and magnetic signature (Barker 1966) suggest that

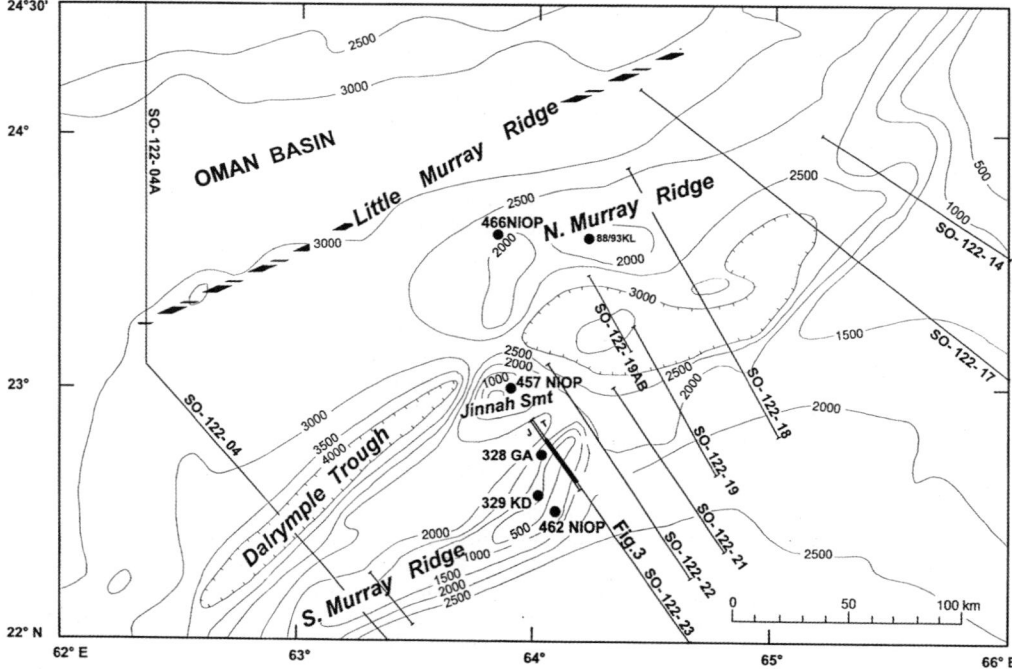

Fig. 2. Morphotectonic map of the Murray Ridge, based on data from GEBCO-CD-ROM and *Sonne* cruise SO-122. In addition, the locations of grab and dredge stations, and location of selected geophysical lines are shown. JT, Jinna Trough.

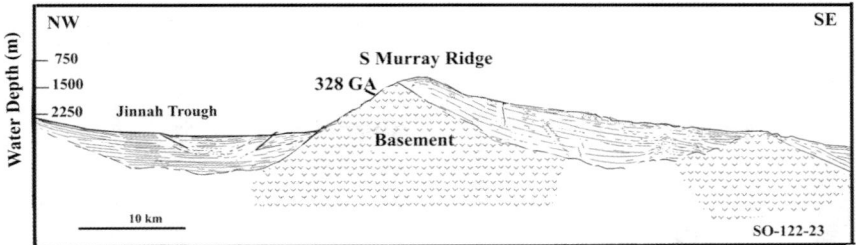

Fig. 3. Seismic profile SO-122-23 (line drawing) across the Southern Murray Ridge with location of TV-grab Site 328 GA.

this feature is a solitary volcano; (6) another (basement) ridge, on seismic line SO-122-23 (Fig. 3), which culminates at a distance of 16 km SE of the Southern Murray Ridge.

South of the Southern Murray Ridge, the northernmost identified magnetic lineation is anomaly 27 (Miles et al. 1998), just south of the northern end of the Laxmi Ridge.

Material and methods

Geophysical surveys

Magnetic traverses were run across the Murray Ridge area during the 1961–1963 H.M.S. *Dalrymple* cruise (Barker 1966). Detailed geophysical and geological data were obtained during cruises of R.V. *Charles Darwin* (1986–1988; Minshull & White 1989; Minshull et al. 1992) and the *Atlantis II* cruise (Mountain & Prell 1992). Recently, the NIOP studied the area with R.V. *Tyro* (van der Linden & van der Weijden 1994). *Sonne* cruises SO-122/1997 (Roeser & Scientific Party 1997), SO-123/1997 (Flueh et al. 1997), and, finally, SO-130 (von Rad et al. 1998) revisited the area. Geophysical surveys (SO-122, -123) included swath bathymetry, sediment echography, multi-channel seismic reflection, seismic refraction, magnetometry and gravimetry (Flueh et al. 1997; Roeser & Scientific Party 1997). Sediment echo sounding (SO-122, -123, -130) was undertaken with a Parasound system.

The position of selected SO-122 lines for the Murray Ridge area is shown in Fig. 2. The northern part of multi-channel seismic line SO-122-23 (Fig. 3) illustrates the typical appearance of the Southern Murray Ridge rising from the basement of the Indus Fan. In general, the Southern Murray Ridge is the edge of a 50 km wide basement high located to the NW of the Indus Fan. Along its whole length (300 km) the northwestern flank of the ridge is characterized by extensional tectonic structures. This northwestern flank of the ridge is characterized by basement outcrops on lines SO-122-03, -04, -21 and -23. At these flanks, Barker (1966), van der Linden et al. (1995) and BGR have sampled basement rocks.

Figure 4 is a combination of the magnetic dataset obtained during the *Sonne* cruises (1997) combined with data from Barker (1966); it shows the anomalies as wiggle traces plotted along the profile lines. The Jinnah Seamount is characterized by a positive anomaly of 600 nT and an adjoining negative anomaly of −300 nT to the north. This fits exactly the pattern expected for a seamount that to a large extent is magnetized parallel to the present field with a magnetization of about $10\,\text{A}\,\text{m}^{-1}$. In most profiles across the Southern Murray Ridge a wide negative magnetic anomaly is observed over the basement rise at the northern edge of the Indus Fan. The lower part of the southern flank of the ridge itself is marked by a positive anomaly, the upper part and the summit region by a negative anomaly. The anomaly amplitudes are about 100 nT. The peak region shows also local anomalies with amplitudes between 100 and 150 nT and wavelengths of the order of 5 km. The great difference in the wavelengths between these two types of anomalies indicates that they are due to different bodies and not just one body approaching the surface near the ridge crest. The short wavelengths of the local anomalies indicate that they are related to the outcropping basement. The observed anomalies require magnetizations of the order of $1\,\text{A}\,\text{m}^{-1}$. The wide anomalies could be modelled with a much lower magnetization of volcanic flows or of the whole basement.

Sampling

During Leg D2 of the NIOP *Tyro* cruise (van der Linden & van der Weijden 1994) a box core was lowered on a small sedimentary basin on the southwestern flank of the Jinnah Seamount (station 457 NIOP, water depth 301 m). Although no proper sediment core was recovered, the box recovered weathered, phosphorite- and carbonate-coated basalt pebbles, clasts of organic debris and two allochthonous fragments, an angular piece of

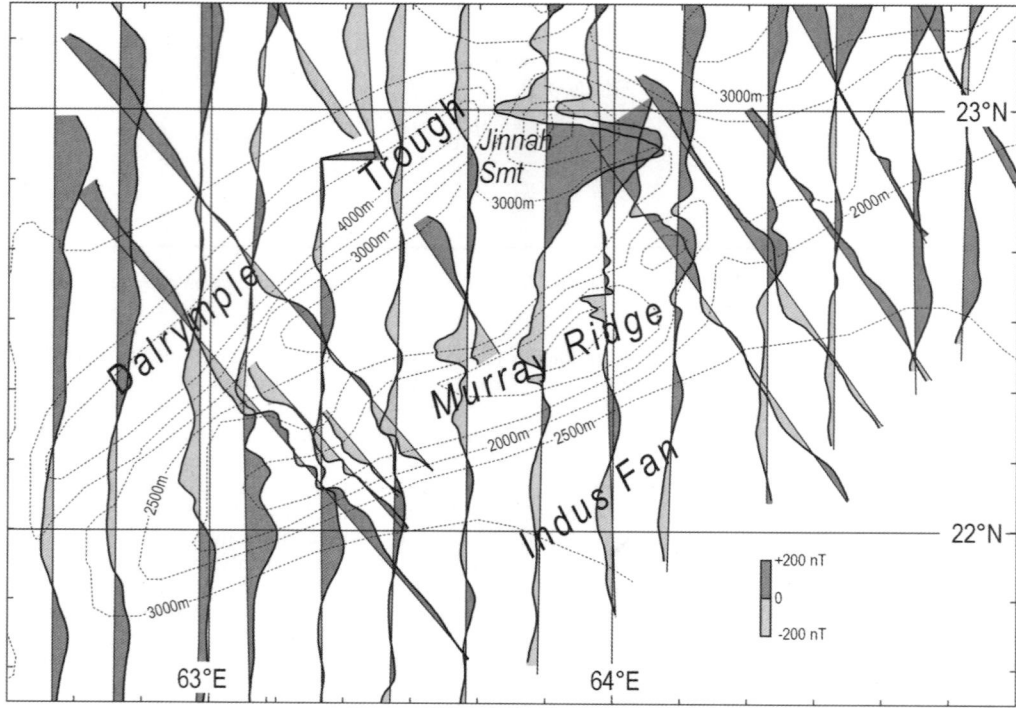

Fig. 4. Magnetic anomalies over the Southern Murray Ridge and Jinnah Seamount plotted along the survey lines. Positive anomalies are to the right or to the upper side of the lines. Dashed lines are bathymetry from the GEBCO CD-ROM.

quartz and a rounded schist pebble, both about 3 cm in size.

On *Tyro* station 462 NIOP a dredge haul, scraping a pipe dredge upslope against the top scarp of the Southern Murray Ridge from first bottom contact at −620 m depth to −510 m depth, produced a bucketful of predominantly basalt pebbles, cobbles and boulders, and one, centimetre-sized, angular piece of hardground, a volcanic breccia cemented by large-foraminiferal micrite. This is in agreement with earlier findings (dredged basalt, tuff and microdiorite fragments) obtained by H.M.S. *Dalrymple* in 1960 from a station at the northeastern end of the southern Murray Ridge (Barker 1966).

During the *Sonne* 130 cruise (von Rad *et al.* 1998) large fragments (about 70 cm across) of volcanic rocks were taken by a TV-guided grab sampler (station 328GA; 22°44.4′N, 64°2,5′E) from a water depth of 885 m on the steep NW slope of the Southern Murray Ridge, where outcrops and large boulders were observed by the TV system to be *in situ*. Subsequently (station 329 KD; 22°31.0′S, 64°59.9′E; water depth about 580 m), mainly hardground material of highly altered and bored, fossiliferous carbonate- and Mn-encrusted volcanic rock and a few fragments of peridotite and siltstone were recovered by a chain bag dredge from the uppermost slope of the Southern Murray Ridge. Good pulls were experienced close to the top of the slope, in water depths between about 600 and 500 m.

Petrographic, geochemical and palaeontological analyses

Petrological analyses of the *Tyro* and *Sonne* rock samples were carried out in Utrecht (Ickenroth & van Bergen 1994) and at BGR (K. Burgath, Hannover), respectively. The methods included petrological and geochemical investigations of igneous rock samples, obtained by a dredge and a TV-guided grab.

Six basalt samples (from two stations) were analysed at BGR with the following methods: (1) microscopy of polished thin sections under transmitted and reflected light; (2) preliminary scanning electron microscope analysis on alteration phases; (3) microprobe analysis of clinopyroxene; (4) standard X-ray fluorescence (XRF) analysis for major and minor elements; (5) inductively coupled plasma mass spectrometry (ICP-MS) analysis of trace elements and rare earth elements (REE).

Optical and geochemical analyses proved that the basaltic rocks recovered during the NIOP *Tyro* (1992) and *Sonne* 130 cruises are highly altered and are therefore not suitable for K/Ar or Ar/Ar dating. However, the fossiliferous carbonate crusts permitted an estimate of the minimum age of the volcanic rocks.

Sediment samples (siltstone, limestone) were studied by thin-section and XRF analysis (U. von Rad). Benthic and planktonic foraminifers were determined by W. Weiss (BGR) from thin sections under the microscope. From the foraminifer-rich matrix of the volcanic breccia collected on dredge station 462 NIOP, 10 thin sections were cut and analysed (Drooger 1994).

Petrography and geochemistry of igneous Murray Ridge rocks (SO-130)

Microporphyric tholeiitic metabasalt (SO-130-328 GA/1a)

Petrography. The highly altered rock contains phenocrysts (about 10 vol. %) of plagioclase An_{20-30} and mafic minerals, which are enclosed in a greenish brown groundmass of albite-rich plagioclase, magnetite and former glass. Although the mafic phenocrysts are extremely altered, we recognized two different original compositions (Fig. 5): (1) pseudomorphs of phyllosilicates after former olivine (or pyroxene ?) + Fe oxyhydroxides; (2) pseudomorphs of clear quartz after clinopyroxene (remnants of augite still present). The composition of clinopyroxene relics is presented in Table 1.

The pseudomorphs of phyllosilicates exhibit mesh texture reminiscent of olivine. Inclusions of Cr-bearing spinel, often found in mid-ocean ridge basalt (MORB) (e.g. Frenzel *et al.* 1990; Juteau & Maury 1997), are not present. This is reflected by a very low content of chromium (9 ppm) in the rock (Table 2). It is further remarkable that the investigated sample contains phenocrysts of clinopyroxene and probably olivine. This is atypical of ocean-floor basalts, which, in general, are moderately porphyric and characterized by plagioclase in combination with olivine or, more rarely, with clinopyroxene (Sun *et al.* 1979; Maury *et al.* 1982; Hekinian *et al.* 1993; Bordier 1994).

The orange–yellow filling material of these pseudomorphs shows abnormal brown to orange

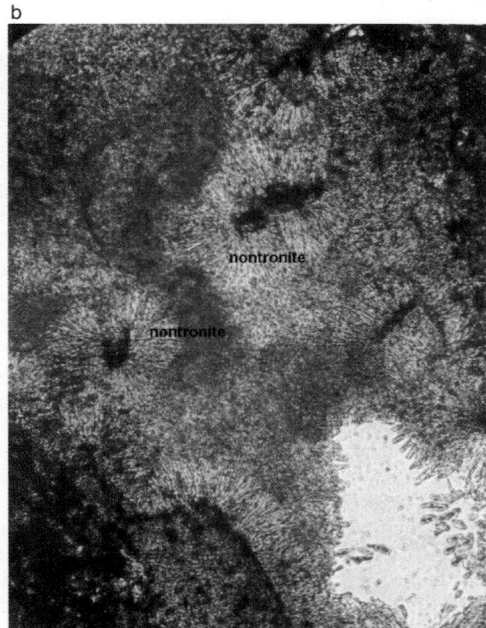

Fig. 5. Tholeiitic microporphyric basalt. (**a**) Pseudomorphs of clear quartz (qtz) after clinopyroxene (cpx; remnants still present) intergrown with pseudomorphs of mafic hydrous phases (Fe oxyhydroxides and saponite?), probably after olivine. Groundmass: plagioclase laths and dark mixture of Fe oxyhydroxides and smectite. Sample SO-130-328 GA/1. Width of field of view is 0.6 mm. (**b**) Radially oriented fibrous aggregates of nontronite in open cavity. Sample SO-130-328 GA/1. Width of field of view is 0.25 mm

Table 1. *Microprobe analyses of clinopyroxene microphenocrysts. Sample 328 GA/1a, Murray Ridge*

wt %	Number of analysis			
	1	2	3	4
SiO_2	52.50	52.48	51.29	52.28
TiO_2	0.06	0.14	0.08	0.16
Al_2O_3	1.55	1.45	1.57	1.47
FeO	7.58	8.04	7.65	8.26
MnO	0.21	0.17	0.19	0.23
MgO	16.25	15.47	16.35	15.55
CaO	21.24	21.51	21.15	21.43
NiO	0.00	0.00	0.00	0.00
Cr_2O_3	0.00	0.00	0.00	0.00
Na_2O	0.08	0.12	0.09	0.08
K_2O	0.00	0.00	0.00	0.00
Total	99.48	99.39	98.37	99.46
Cations				
Si	1.944	1.952	1.918	1.945
Ti	0.002	0.004	0.002	0.005
Cr	0.000	0.000	0.000	0.000
Al	0.067	0.064	0.069	0.064
Fe^{3+}	0.047	0.033	0.097	0.042
Fe^{2+}	0.188	0.217	0.142	0.215
Mn	0.007	0.005	0.006	0.007
Mg	0.897	0.858	0.911	0.863
Ca	0.843	0.858	0.847	0.854
Na	0.006	0.009	0.007	0.006
K	0.000	0.000	0.000	0.000

interference colours and is here correlated with Type 7 (saponite + goethite) in the classification of Laverne *et al.* (1996). The original interstitial glass is replaced by Fe oxyhydroxides (mainly goethite), a patchy mixture of the Type 7 material and very fine-grained aggregates of light brownish green nontronite. Occasionally, very fine patches of epidote are present. Acicular and radiated nontronite occupy also the rims of open or quartz-filled pores. These textures resemble Type 5 of Laverne *et al.* (1996). Other caverns are filled by pure Fe oxyhydroxide. Reflected-light microscopy revealed further that magnetite is, at least partly, altered to maghemite. Moreover, many magnetite grains are broken and cemented by Fe sulphides.

The texture and the secondary mineral assemblage suggest that the metabasalt was probably affected by at least two different alteration processes.

(1) Low-temperature alteration (oxidizing conditions): the replacement of mafic phenocrysts and the major part of the groundmass by phyllosilicates and Fe oxyhydroxides is well known from the uppermost zone of oceanic volcanic rocks down to about 400 m (T <100 °C; Alt 1995, 1999; Teagle *et al.* 1996; Alt *et al.* 1998). This zone of low-T alteration is characterized by open seawater circulation and oxidation, and the original composition of the basalt is modified by moderate chemical changes. These changes will be discussed in the section on geochemistry.

(2) Alteration at slightly higher temperature under slightly reducing conditions is indicated by the presence of Fe sulphides in the metabasalt. However, other alteration phases are not observed, e.g. zeolites and chlorite–smectite, which are characteristic for the deeper zone of alteration with restricted seawater circulation and more reducing conditions (Alt *et al.* 1996; Teagle *et al.* 1996). Albitization of plagioclase, on the other hand, is documented in thin section and is reflected by the remarkable Na_2O content of the sample (6.8 wt %; Table 2). According to Alt *et al.* (1996), albite is not formed in the uppermost (oxidation) zone of ocean-floor volcanic rocks, but appears first in the deeper part of the low-T zone under more reducing conditions (T c. 100–150 °C). This is also the zone of the first appearance of epidote (Alt 1995; Alt *et al.* 1996; Vanko & Laverne 1998).

The clinopyroxenes in the Murray metabasalt are partly replaced by quartz. It is generally recognized that clinopyroxene is not affected during low-T alteration (T <150 °C) up to greenschist facies conditions with T ≤ 350 °C. Advanced clinopyroxene decomposition starts only at the greenschist facies–amphibolite facies transition (T c. 400 °C; Gillis & Thompson 1993; Alt 1999). Additional work to obtain more precise data about these indications of higher T alteration of the Murray Ridge basalt is in progress. The expected chemical changes are discussed in the following section.

Geochemistry. Obviously, the chemical composition of the recovered metabasalt from the Murray Ridge (Table 1) has been affected by alteration and chemical changes. This must be taken into account for the assessment of the original composition and further interpretations regarding geotectonic relationships.

Seawater–rock exchange and low-T alteration under oxidizing conditions lead to the following modifications in oceanic basalt: (1) major loss of Ca, Na, Mn, Co, Ni and Cu, and uptake of Fe, K, Rb and P (the latter possibly by absorption onto Fe oxyhydroxides); (2) variable behaviour of Mg, Al and Si (Alt 1995, 1999; Teagle *et al.* 1996; Alt *et al.* 1998); (3) no noticeable changes or minor loss only for Ti, Al, Zr, Ta, Hf, Nb, Y and REE (Juteau & Maury 1997); (4) enrichment of light REE (LREE), recognized in dredged basalt samples with extensive exposure to seawater, compared with unmodified REE contents in core samples (Ludden & Thompson 1979; Staudigel *et al.* 1979; Teagle *et al.* 1996; Alt *et al.* 1998).

Table 2. Chemical composition of sample SO-130-328 GA/1 (tholeiitic metabasalt), sample SO-130-329KD/5 (serpentinized peridotite) and of dredged basalts from Site 462 NIOP in the northeastern part of the Southern Murray Ridge (NIOP data from Ickenroth & van Bergen 1994)

	328GA1	Samples from Site 462 NIOP				329KD 5
		462-1	462-2	462-4	462-5	
	Basalt	Basalt	Basalt	Basalt	Basalt	Serpentinite
wt %						
SiO_2	51.14	37.61	39.62	40.54	41.95	37.09
TiO_2	0.442	2.49	3	2.57	2.72	0.01
Al_2O_3	16.30	13.38	11.14	11.83	10.82	0.6
Fe_2O_3	12.60	12.65	10.28	11.83	10.82	8.43
MnO	0.06	0.09	0.17	0.17	0.23	0.04
MgO	1.06	13.73	10.98	11.3	4.74	35.5
CaO	7.61	6.2	14.75	13.76	8.69	0.64
Na_2O	6.81	2.02	1.89	0.95	5.09	0.35
K_2O	0.07	1.82	0.32	1.23	1.28	0.05
P_2O_5	0.454	1.28	0.43	0.47	0.96	0.454
LOI	2.93	8.15	5.66	3.78	5.36	15.1
Total	99.47	99.42	98.24	97.6	98.88	98.26
ppm						
Cr	9	449	9	521	2	2188
Ni	25	193	25	207	15	2621
Co	31	47	31	49	30	57
Sc	32.8	29	46	42	8	7
V	562	432	562	313	301	68
Rb	7	20		33	30	16
Ba	16	248	16	517	822	15
Sr	37	334	37	723	405	38
Nb	3	34.7	3	57.6	135.4	<2
Zr	27	184	27	257	395	<3
Y	16.5		20	17	33	<3
La	3.58	27.4	40.6	42.9	98.3	<20
Ce	6.28	63.2	90.1	95.5	201	<20
Nd	3.12	38.8	47.9	50.2	87.5	
Sm	0.86	5.16	7.25	7.1	10.8	
Eu	0.29	1.93	2.49	2.59	4.11	
Tb	0.29	0.72	0.81	0.77	1.03	
Dy	2.36	3.58	4.99		7.86	
Er	1.89					
Yb	2.21	1.32	1.38	1.09	2.43	

Major and trace elements in the Murray metabasalt. Considering the pseudomorphs after mafic phenocrysts, i.e. clinopyroxene (cpx) and probably olivine, and the replacement of the groundmass by nontronite, smectite, Fe oxyhydroxides and quartz, mobilization of Ca, Sr, Mg Cr (from cpx) and the metal (Mn, Cu, Ni, Zn) content in the former glass is indicated. The overall presence of Fe oxyhydroxides and the albitization of plagioclase document the uptake of Fe and Na. Uptake of K, Rb and P is typical of altered oceanic volcanic rocks (see above) and is also assumed for the sample from Murray Ridge. These chemical changes are reflected in the normalization of the investigated sample against typical fresh basalt of oceanic environments (normal MORB, evolved MORB, island-arc tholeiite (IAT); Table 2 and Fig. 6a). Mg, Sr, Mn, Ni and Co are moderately to strongly depleted, compared with all reference materials (for Cu no values were available), whereas Na, P, Rb and Fe are moderately to strongly enriched. The low content of K (0.07 wt %; Table 2) is difficult to explain and may indicate mobilization from altered glass. Ti, Zr, Nb and Y, generally recognized as immobile during low-T alteration, display a different behaviour if compared with the reference basalts. Ti is rather depleted and Zr, Nb and Y are distinctly enriched, if compared with MORB, but are close to IAT (Fig. 6a and b). A comparison of fresh and slightly altered samples

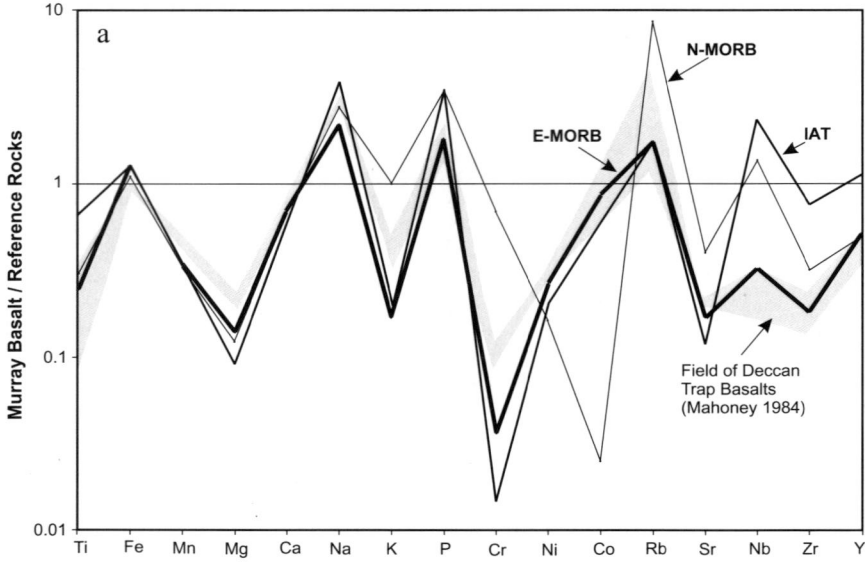

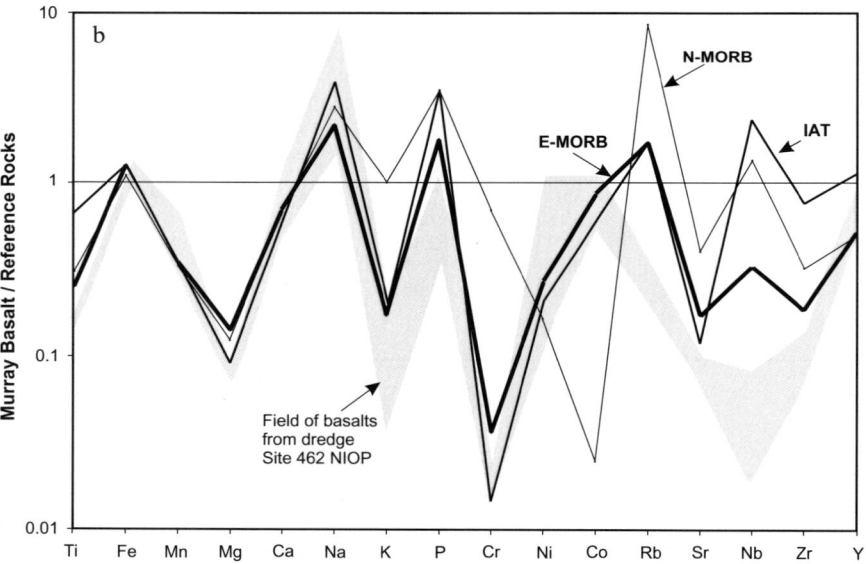

Fig. 6. (**a**) Composition of sample 328 GA/1 (tholeiitic metabasalt) normalized to N-MORB, E-MORB, IAT and Deccan Trap basalts. Reference data from Juteau & Maury (1997), pp. 30 and 35), and from Mahoney (1984). (**b**) Composition of sample 328 GA/1 (tholeiitic metabasalt) normalized to N-MORB, E-MORB, IAT and dredged basalts from Site 462 NIOP in the northeastern part of Southern Murray Ridge. Data from Ickenroth & van Bergen (1994). It should be noted that sample 328GA differs from Deccan and 462 NIOP samples (see text).

of N-MORB dredged from the same site in the SW Indian Ocean (Le Roex et al. 1983) indicates, however, that altered samples are distinctly depleted in Zr. This might indicate that the basaltic protolith of the Murray sample was originally even richer in zirconium. These results are significant for the identification of the precursor of the Murray metabasalt and will be discussed below.

Rare earth element distribution. The common use of REE contents for classification and investigation of the origin of meta-magmatic rocks is based on the assumption that REE are more immobile than other major and trace elements during weathering, alteration and metamorphic processes. Humphris (1984), however, pointed out that the mobility of REE is not uniform and is subject to numerous factors; for example, the siting of REE in primary rocks, their accommodation in secondary minerals, or the partitioning behaviour of REE between rocks and fluids. Regarding low-T alteration of oceanic basalt, a strong enrichment is reported for weathered basalt recovered from the sea floor, in contrast to rather immobile REE behaviour in samples from drill cores (Ludden & Thompson 1978; Teagle *et al.* 1996; Alt *et al.* 1998; Alt 1999). According to Juteau & Maury (1997) no mobilization of heavy rare earth elements (HREE) occurs during low-T alteration under oxidizing conditions. These results must be considered when examining the REE distribution in the metabasalt sample from Murray Ridge.

First of all, the convex-downwards pattern of the chondrite-normalized (CN) REE values of the investigated sample is noteworthy (Fig. 7). (La/Yb)$_{CN}$ and (La/Lu)$_{CN}$ are close to unity (0.9296 and 1.0921, respectively), but (La/Sm)$_{CN}$ is 2.6185. This pattern is atypical of basic volcanic rocks from marine environments with flat or convex-upwards patterns in N-type MORB and IAT or with patterns with negative slope in T-MORB, E-MORB and alkalic basalt (Saunders 1984; Johnson & Sinton 1990; Juteau & Maury 1997). It seems reasonable to assume a modification of the primary REE distribution in the Murray sample in connection with alteration and low-grade metamorphism.

The pattern is similar to the trough-shaped REE patterns of serpentinized harzburgite and dunite in ophiolite (Pallister & Knight 1981), which are generally considered as a result of fluid activity during alteration and low-grade metamorphic processes (Frey 1984). If the REE pattern of the Murray sample is examined in more detail, a slight depletion of Ce compared with La is recognized. This configuration was observed in palagonitized and smectite-bearing zeolitized basalts exposed to seawater infiltration (Menzies *et al.* 1977; Ludden & Thompson 1978). More striking is the strongly negative Eu anomaly.

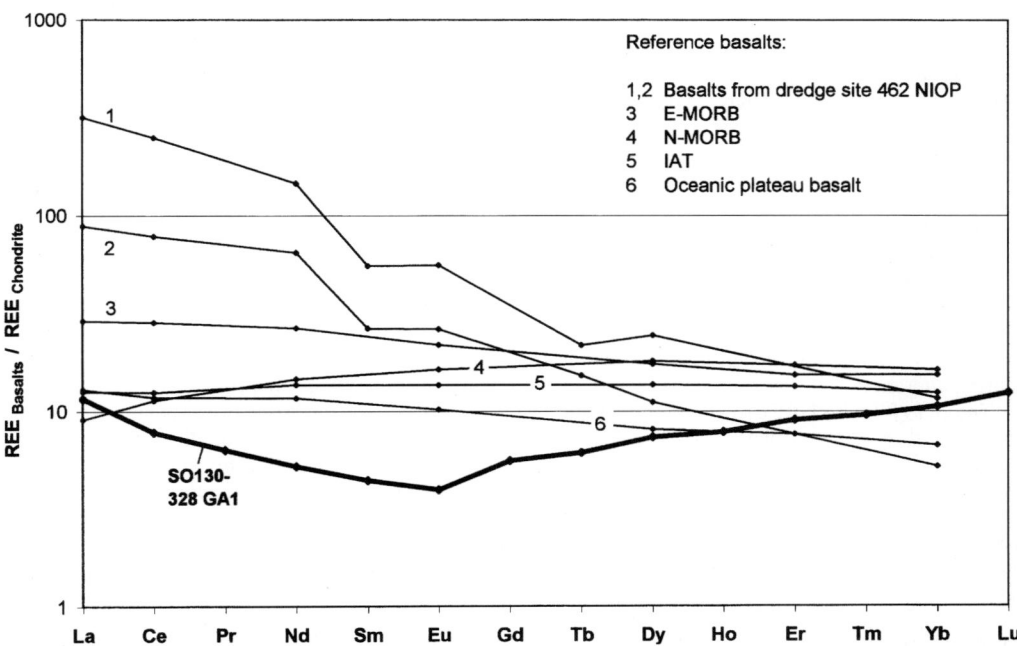

Fig. 7. Chondrite-normalized REE pattern of Murray Ridge metabasalt 328 GA and reference rocks. Data sources: 1 and 2, Ickenroth & van Bergen (1994); 3 and 4, Juteau & Maury (1997), p. 30; 5, Juteau & Maury (1997), p. 35, analysis BTH); 6, Juteau & Maury (1997), p. 33, analysis BPN).

The Murray Ridge metabasalt contains altered plagioclase and two types of replacement textures after mafic phenocrysts: pseudomorphs of quartz after clinopyroxene and pseudomorphs of green phyllosilicates, etc. after either olivine or pyroxene (no relics left). Green phyllosilicates replace also the former glass in the groundmass. Olivine has a very low mineral–melt distribution coefficient for all REE (Henderson 1984) and its alteration would not have a lasting influence on the original REE pattern of the rock. Clinopyroxene, on the other hand, is characterized by very low mineral–melt distribution coefficients for La (close to unity) and high coefficients for Ce to Lu. It is expected that its replacement strongly affects the primary REE_{CN} pattern of the host rock. The trough-shaped pattern of the Murray sample with a minimum at Eu_{CN} is explained by preferential removal of light to intermediate atomic number REE, and particularly of Eu as a result of Ca mobilization from albitized plagioclase and replacement of clinopyroxene by quartz. It is also expected that La should be affected by this mobilization, but this is not reflected by the pattern in Fig. 7. The chemical analysis of the metabasalt from Murray Ridge yielded a high content of Na_2O (6.81 wt %; Table 1), which is due to the overall albitization of plagioclase. Moreover, the strong release of Ca during the alteration is partially counteracted by the formation of some epidote. A replacement of Ca in Ca silicates and of Na in the albite by La is very likely, because Na^+ and Ca^{2+} have similar sizes to the lightest REE in their trivalent state (Henderson 1984). This may explain the slight enrichment of La_{CN} in the pattern of the metabasalt.

In summary, it is suggested that the precursor of the metabasalt from Murray Ridge suffered partial REE mobilization and removal, mainly of the intermediate atomic number REE (as a result of seawater–rock interaction, and mafic phenocryst and glass decomposition). The original content of La was probably slightly increased as a result of the formation of Na- and Ca-rich phases during alteration.

The heavy REE (HREE) part of the chondrite-normalized pattern displays a flat and smooth increase. This corresponds to the experience that alteration and low-grade metamorphism of basalts lead, in contrast to uneven $LREE_{CN}$ distributions, in most cases to flat and smooth $HREE_{CN}$ patterns, which do not strongly deviate from the distributions in the parental rocks (for examples, see Humphris 1984). If these explanations are accepted for the values of La_{CN} and the heaviest REE_{CN} (Yb, Lu) in the Murray sample, the ratios of 1.09 for $(La/Yb)_{CN}$ or 0.92 for $(La/Lu)_{CN}$ would roughly reflect the ratios in the parental rock of the metabasalt. This REE evidence would favour either early IAT or N-type MORB, respectively. The ratios of $(La/Yb)_{CN}$ are also similar in the Murray sample and in normal (not enriched) oceanic plateau basalts (e.g. 1.01 in sample BPN; Juteau & Maury 1997, p. 33). However, basalts with a pronounced negative slope of REE_{CN} (T-MORB, E-MORB, ocean-island basalt, calc-alkaline basalt) can be excluded as precursor. Continental tholeiites (including the Deccan basalts; see below), ranging from those with very low values of $LREE_{CN}$ and a positive slope with $(La/Lu)_{CN}$ of 0.5, to those with very high values of $LREE_{CN}$ and a distinct negative pattern with $(La/Lu)_{CN}$ of 7.6 are very different from the pattern obtained from the investigated Murray Ridge samples.

Comparison of the investigated Murray Ridge metabasalt with Deccan Trap basalts. Geophysical data suggest that the Southern Murray Ridge is underlain by fragments of attenuated continental crust, and several samples dredged from Site 462 NIOP (Fig. 2 and Table 2) were assigned to continental flood basalts associated with the Deccan Trap (Ickenroth & van Bergen 1994). For this reason the investigated Murray Ridge metabasalt is compared with the 462 NIOP samples and representative Deccan Trap basalts (Mahoney 1984). To minimize the influence of chemical changes connected with alteration, only elements of low mobility (Ti, Zr, Nb, Y), Cr and REE are considered. It is evident from Fig. 6a and b that the Murray metabasalt (328 GA) differs strongly from the Deccan and NIOP samples, which are more enriched in most elements.

Harzburgite or clinopyroxene-poor lherzolite, completely serpentinized (SO-130-329 KD/2)

Petrography. The coarse-grained sample is composed of large anhedral olivine (size of original grains >1 mm; grains completely serpentinized), anhedral and flattened orthopyroxene (size up to 2.5 mm across, replaced by serpentine in parallel orientation), and rather equidimensional pseudomorphs, which are filled by finer-grained serpentine (size up to 1.9 mm across; pseudomorphs after clinopyroxene?). Reddish brown spinel (translucent in thin section; size up to 1.2 mm across) with round inclusions of olivine and with corroded outlines and wormy extensions is present as an accessory phase.

The fabric of the sample displays deformation under mantle conditions and is fairly similar to the fabric of ultramafic tectonites in ophiolites or in oceanic environments (e.g. at Ocean Drilling

Program (ODP) Leg 153; Burgath et al. 1994; Ceuleneer & Cannat 1994). The deformation includes strong interfingering of former olivine and orthopyroxene, with kink banding and flattening in the latter. Recovery is indicated by small (serpentinized) olivine neoblasts with sizes up to 0.7 mm across along the edges of large deformed olivine and orthopyroxene grains. Small round grains of spinel (size <0.06 mm), yellowish translucent in thin sections, could also represent a neoblast generation.

The estimated modal ratio of the original rock is >70% olivine and <30% pyroxene (with >15% orthopyroxene). Thus the sample is assigned to refractory harzburgite or refractory clinopyroxene-poor lherzolite.

Geochemistry. The chemical composition of sample 329 KD/2 is listed in Table 2. The original nature of the olivine-rich rock and its complete serpentinization is documented by the low content of SiO_2 (37.09 wt %), the high content of MgO (35.5 wt %) and a high loss on ignition (LOI) (15.1 wt %). In the diagram of CaO v. Al_2O_3, which, from our experience (e.g. Burgath et al. 1987), is useful for further discrimination of (carbonate-free) serpentinized ultrabasic rocks, the sample plots in the field of harzburgite close to the harzburgite–lherzolite boundary. Unfortunately, the Na content, which is generally very low in harzburgite and distinctly higher in lherzolite (compare, for example, representative analyses of Coleman (1977)), cannot be used for further discrimination, because Na is probably introduced during alteration.

The sample contains Cr-bearing spinel with conserved outlines of overprint under temperature and pressure conditions of the mantle. This indicates that Cr entered spinel and therefore the sample cannot have lost a noticeable amount of its original Cr content during transformation of the primary silicates to serpentine. Moreover, there is no indication of a noticeable change of the original content of Ni in the sample during serpentinization. Ni supply, for example by lateritization, is unlikely and despite the complete alteration of olivine, the considerable content of Ni (2621 ppm) does not support a major loss of Ni. Hence we conclude that Cr and Ni can be used for further discrimination (see Discussion).

Sedimentary rocks

Carbonate crusts (SO-130-329KD/2)

Some of the basalts and ultramafic rocks are coated by carbonate crusts and the vugs are infiltrated by micritic calcite with abundant pelagic and shallow-water fossils indicating a pelagic, sublittoral palaeoenvironment (unpublished BGR data; Drooger 1994). The micritic matrix contains a planktonic foraminiferal assemblage with few benthic Foraminifera (?*Astigerina*, ?*Oridorsalis* sp.), some algal fragments and very rare other bioclasts (e.g. echinoid spines).

At the BGR the following species of planktonic foraminifers were identified: *Pulleniatina obliquiloculata, Orbulina universa, Globoquadrina altispira, Globigerinoides* spp., *Sphaeroidinellopsis* cf. *seminulina* and other species. From this fauna a Late Miocene–Early Pliocene age (N18–N19; 5.8–4.8 Ma according to Berggren et al. (1995)) was determined.

From the red foraminiferal matrix within a volcanic breccia (462 NIOP) Drooger (1994) identified two (micro)faunal elements: *Amphistegina*(dominant) and *Lepidocyclina* (frequent). Less frequent, but certainly present, are *Cycloclypeus, Operculina, Gypsina, Lenticulina* and *Globigerina*. Possibly, *Heterostegina* and Peneroplidae (*Marginopora*) are present. Also present are a few shell fragments and algal remains, as are small-benthos. Terrigenous material is missing. Drooger concluded that the fossil assemblage is an autochthonous association formed in an open-marine environment outside terrigenous influence, i.e. in the deeper part of the photic zone (between −50 and −100 m). Because *Myogypsina* sp. is lacking in the association, it is most probably of Late Tertiary age, within the range of Late Miocene (post N13/14) to Early Pliocene (N19) time.

Quartz siltstone (SO-130-329KD/4)

This evenly sized, well-sorted unfossiliferous quartz siltstone with a ferruginous–clayey matrix and trace amounts of feldspar and mica is of terrigenous origin, possibly as a distal turbidite derived from the Makran continent. The sediments must have been deposited in a basinal setting, e.g. in a proto-Dalrymple Trough, before the Murray Ridge was uplifted above the Oman Abyssal Plain.

Magnetic investigations

Two samples of dark grey basalt, which were obtained at station 328GA, were investigated magnetically. SO-130-328GA/1 is uniformly grey altered tholeiitic basalt, whereas SO-130-328GA/1b is microphyric tholeiitic basalt. For SO-130-328GA/1, susceptibility is 4×10^{-3} SI, induced magnetization is 0.16 A m^{-1} and remanent magnetization is low; for SO-130-328GA/1b, susceptibility is 20×10^{-3} SI, induced magnetization is 0.8 A m^{-1} and remanent magnetization is high (10 A m^{-1}). The other samples from SO130-

328GA belong mostly to the group with smaller susceptibilities.

According to the petrographic investigations, sample SO-130-328GA/1b contains pyrite, which has partly displaced magnetite and hematite. In view of its high magnetization, this basalt type may be responsible for the local anomalies along the crest of the Murray Ridge. The wider anomalies observed on large parts of the Murray Ridge and its surroundings can be related to the more weakly magnetized basalts. Thus, the weakness of the magnetic anomalies does not necessarily imply that basalts are absent.

The high variability of the magnetization of basalts is well known (see Telford *et al.* 1991). Titanomagnetites, which are the main carrier of the magnetization, are only an accessory constituent of basalts. Rocks that look completely alike may be very different magnetically. The amplitudes of magnetic anomalies generated by basalt complexes may vary between 1 and 1000 nT.

Apparently, highly magnetized basalts occur only in the crest region of the Murray Ridge, where their magnetic influence is observed in most of the investigated lines. It is not known whether the high magnetization is connected to the presence of the pyrites. If the crest was subaerial at the time of the maximum uplift of the ridge, subaerial conditions may have changed the properties of the basalts. All this remains speculation, in the absence of more information on the relationship between the magnetization and the mineralogical properties of the basalts.

Discussion

On the basis of the modal composition and chemical characteristics, sample 329 KD/2 dredged from the Southern Murray Ridge is classified as harzburgite or clinopyroxene-poor lherzolite. The sample displays conserved patterns of deformation under mantle conditions. This characterization is supported by the position in the plot of Ni v. Cr contents (not shown), which allows discrimination of refractory below-Moho peridotites from ultramafic supra-Moho cumulates in ophiolite complexes (Irvine & Findlay 1972). In this diagram, sample 329 KD/2 plots in the narrow field of typical 'refractory peridotites'.

From its composition, the metabasalt sample 328 GA/1 is classified as moderately altered tholeiite. Despite its postmagmatic overprint, the rock has preserved its porphyric texture with phenocrysts of clinopyroxene and probably former olivine, which is not very common in MORB. To obtain further information about the geotectonic context of the basaltic protolith, it was decided to use only geochemical discrimination diagrams, which are based on elements of low mobility (Ti, Zr, Y). The application of other common diagrams based on Cr, Ni, Sr and P is not useful, because these elements in the Murray Ridge sample were obviously affected by alteration. This is displayed by their striking 'anomalies' if they are normalized versus reference basalts from various oceanic environments (Fig. 6a and b).

It becomes apparent that the selected elements are close to the composition of IAT, but have little in common with the composition of MORB, enriched basalts and within-plate volcanic rocks, for example, rocks from the Deccan Trap, the most prominent basaltic province in the vicinity of Murray Ridge (Fig. 8a–c).

A comparison with samples of obvious Deccan Trap affinity dredged from Site 462 NIOP near the northeastern end of the Southern Murray Ridge (Ickenroth & van Bergen 1994; Fig. 1) also displays major differences, and sample 328 GA is depleted in most elements (Fig. 6b). It would require a strong release of these elements during alteration in Deccan-type basalts to achieve the actual composition of the studied Murray Ridge sample (328 GA); however, this sample does not show any indication for such an extensive loss. It is therefore unlikely that the investigated sample represents an altered and moderately metamorphosed sample of Deccan Trap-type basalt.

Further information on the origin of the Murray metabasalt is inferred from the comparison with N-MORB and various active margin basalts in an extended Coryell–Masuda plot (Fig. 9). With regard to typical N-MORB and IAT, the pattern of the Murray rock is closer to the pattern of the latter (represented by two examples from the Mariana arc; data from Woodhead (1989)), particularly with respect to elements with low mobility during alteration.

Maury *et al.* (1982) investigated volcanic rocks from ODP Leg 67 sites (Guatemala Trench) and classified most of them as LREE-depleted MORB, but samples from Hole 494-A displayed magmatic affinities of an active continental margin. Their pattern, represented by sample 33 CC (18–20 cm) in Fig. 9, is also rather similar to that of the Murray metabasalt.

The possible relationship between the southern Murray Ridge and the ophiolites in Oman and Pakistan is discussed further below. For this reason, a representative basalt from the arc-related Alley unit in the Oman ophiolite is included in Fig. 9 (sample MB 527-A; data from Beurrier *et al.* (1989)). A corresponding geochemical signature of this sample and the Murray metabasalt is obvious.

A further argument for the origin of the Murray metabasalt is provided by the composition of the

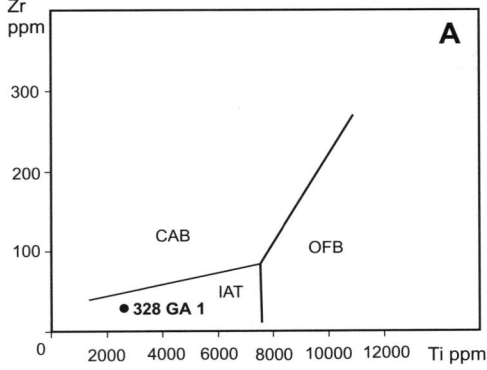
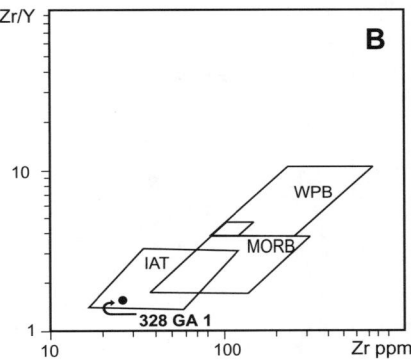
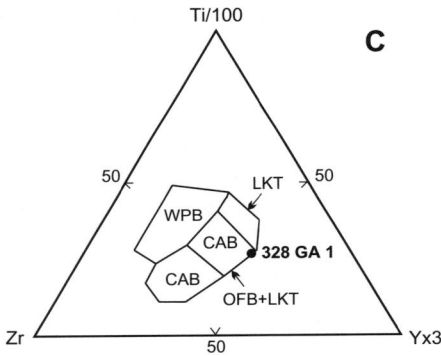

Fig. 8. Sample 328 GA (Murray Ridge metabasalt) in the discrimination diagrams of Zr v. Ti (Garcia 1978), Zr/Y v. Zr (Pearce & Norry 1979) and Ti/100–Zr–3Y (Pearce & Cann 1973). CAB, calc-alkaline basalt; IAT, island-arc tholeiite; LKT, low-K basalt; MORB, mid-ocean ridge basalt; OFB, ocean floor basalt; WPB, within-plate basalt.

relics of clinopyroxene microphenocrysts. In the discrimination diagrams of Ti (+ Cr) v. Ca and Ti v. total Al (after Leterrier et al. 1982) the composition plots in the field of pyroxenes in basalts in subduction environment (orogenic basalts or arc tholeiites, respectively).

Conclusions

Supra-subduction origin of Murray Ridge?

In conclusion, we suggest that the basalt sample 328 GA from the Murray Ridge has an affinity to IAT. Ickenroth & van Bergen (1994) collected two rather altered samples of aphyric basalt from the Murray Ridge, which (even if element mobilization is taken into account) also display IAT affinity: that is, rather high MgO and moderate SiO_2 contents (LOI corrected: MgO between 10 and 12 wt %, and SiO_2 around 50 wt %), a position inside or close to the IAT (or volcanic arc basalt) fields in all commonly used diagrams for basalt discrimination, and primordial mantle-normalized REE patterns that are shifted towards distinctly lower values than those of reference MORB. These findings support our assumption of a former island arc in the Murray Ridge area.

The investigated basalt sample from the Southern Murray Ridge has undergone strong hydrothermal alteration, but displays affinity to IAT. At first sight, the discovery of a basalt sample with IAT affinity on the Murray Ridge seems enigmatic, especially as geophysical results suggest a Neogene setting typical of rift-type extensional tectonics with an attenuated continental crust. The rifting of Dalrymple Trough may have started during Miocene times (Clift et al. 2000, 2002; Edwards et al. 2000; Gaedicke et al. 2002a,b). We postulate that at least part of the basement of the Southern Murray Ridge was formed much earlier (in Latest Cretaceous time) in an active margin (convergent or transform plate boundary) setting between the northeastward-moving Arabian plate, the northward-moving Indian plate and the Eurasian plate (Fig. 10). At this time, the small remaining oceanic basins of the Neo-Tethys were

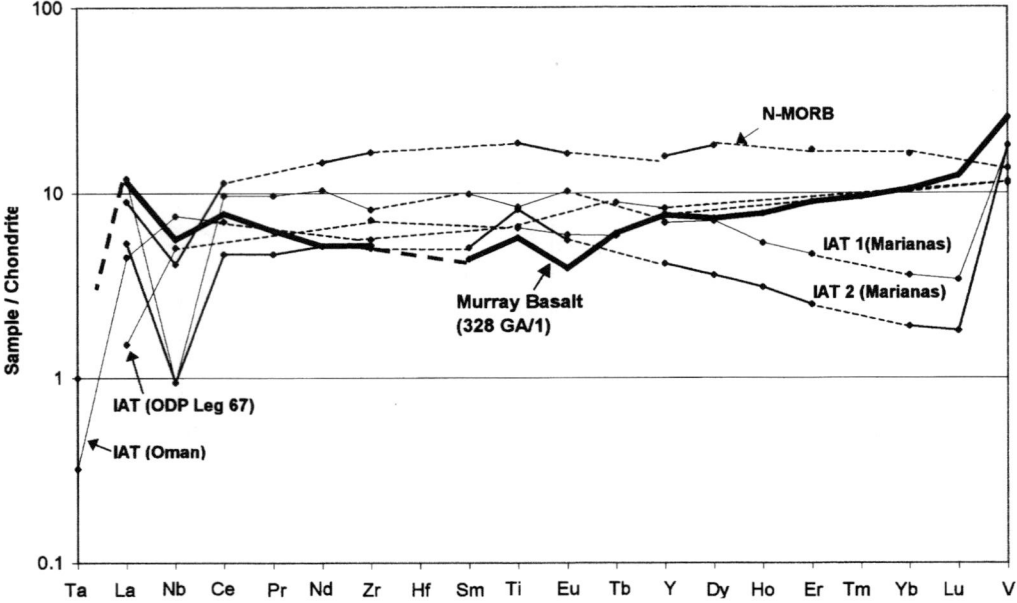

Fig. 9. Coryell–Masuda plot of the Murray metabasalt 328 GA and reference rocks from the Mid-Atlantic Ridge (typical N-MORB; data from Juteau & Maury (1997)), from the Guatemala Trench (IAT (ODP Leg 67); basic volcanic rock with 'active margin' signature; analysis 494A33CC of Maury et al. (1982)), from the Mariana arc (IAT; analyses PAF3B and PASN of Woodhead (1989)) and from the Alley unit in the Oman ophiolite (IAT; analysis MB 527-A of Beurrier et al. (1989)) In general, with respect to elements with low mobility during alteration, the geochemical pattern of the Murray metabasalt is closer to the pattern of active margin basalts.

being closed and oceanic crust (including island-arc volcanic rocks) was obducted onto or imbricated in the Arabian and Eurasian plates (Fig. 10).

The serpentinized harzburgite, dredged at the Southern Murray Ridge, is also in conflict with the interpretation of all Murray Ridge basement rocks as typical continental basalt. The harzburgite resembles ultrabasic rocks from mantle outcrops in extension zones of oceanic environments (e.g. at the Kane transform of the Mid-Atlantic Ridge, and the Hess Deep in the eastern Pacific), or from ophiolite suites, such as the Oman ophiolite, which was obducted from the Oman Basin onto the Arabian shield (e.g. Searle & Cox 1999), or the Bela–Waziristan ophiolite in the western fold belt of southern Pakistan, which was formed in a supra-subduction setting (Ahmed 1991). The Bela ophiolite was obducted during Latest Cretaceous time, when the Indian plate experienced first oblique continent–continent collision towards the NW with the Afghanistan block (Eurasian plate) along the suture zone of the Ornach–Nal Fault (Bannert 1992).

New results from seismic and magnetic surveys (Miles & Roest 1993; Malod et al. 1997; Edwards et al. 2000; Minshull et al. 2001; Schreckenberger et al. 2001; Gaedicke et al. 2002a,b) suggest that the Southern Murray Ridge and Dalrymple Trough are underlain by faulted and tilted blocks of thinned (14–20 km) continental basement, overlain by thick volcanic rock associations. In this interpretation, the Dalrymple Trough and Murray Ridge represent the northernmost extension of continental fragments of the Indian–Arabian plate forming an oblique-slip plate boundary against oceanic basement of the Oman Abyssal Plain to the north. As a result of former plate motion and ocean closure (as indicated by the obducted Semail and Bela–Waziristan ophiolites in the adjacent areas), this continental crust may have incorporated a mélange of ophiolitic fragments like the observed harzburgite and island arc-type basalt, which are suggested to have formed in a Late Cretaceous active margin setting (Fig. 10).

Today the easternmost coastal part of the Arabian Peninsula is dominated by the Oman ophiolite. This Late Cretaceous complex was initially formed in an oceanic rift environment. This evolution was simultaneous with NE dipping of Arabian continental crust including oceanic mate-

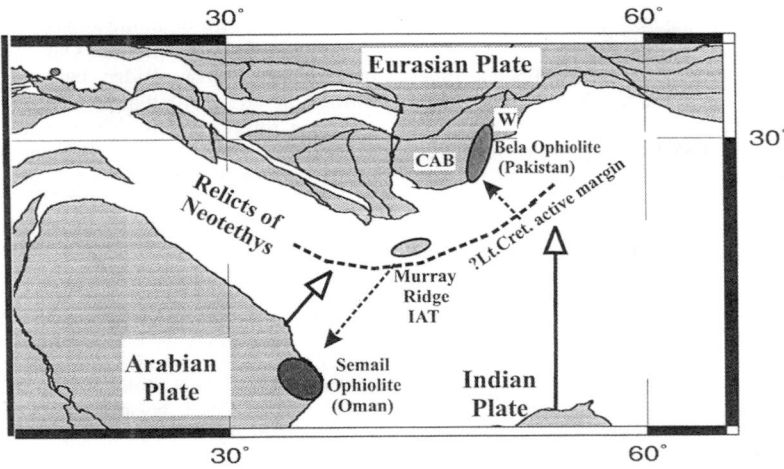

Fig. 10. Tentative sketch of envisaged plate-tectonic setting of Murray Ridge site, Semail and Bela ophiolites during latest Cretaceous times (65 Ma). CAB, Central Afghanistan Block; W, Waziristan Block. A Late Cretaceous active (convergent–transform) margin is inferred from the existence of island-arc basalts that are partly preserved as small slivers on Murray Ridge, partly obducted onto the Semail and Bela ophiolite complexes (arrows with filled arrowheads), and from the fact that the Arabian plate is moving northeastwards and the Indian plate northward (arrows with open arrowheads) closing the relict basins of the Neotethys. Plate-tectonic reconstruction from Ocean Drilling Stratigraphic Network (ODSN, GEOMAR; Hay et al. 1999).

rial, evolution of arc magmatism (Lasail arc), and obduction of this material towards the SW onto the Arabian continental margin (Fig. 10). The minimum emplacement distance of the Oman ophiolite from its site of origin in the NE is estimated to be c. 400 km (Searle & Cox 1999). The Oman ophiolite hosts a sequence of three series of volcanic rocks evolving from MORB via a transitional ridge–arc stage to immature IAT compositions. (Beurrier et al. 1989; Searle & Cox 1999). The metabasalt recovered from the Murray Ridge could be an equivalent of the latter. The dredged harzburgite from the Murray Ridge indicates advanced depletion of the mantle source and thus harmonizes also with a supra-subduction setting.

Evolution of the Murray Ridge

Tentatively, we speculate about the correlation in time and setting of parts of the Southern Murray Ridge, the younger evolution of the proto-Oman ophiolite and the formation of supra-subduction-zone parts within the proto-Bela–Warizistan ophiolite. We suggest that parts of the Southern Murray Ridge formed over a subduction zone, possibly in a transpressional domain. Slivers of the common proto-ophiolite were obducted onto the Eurasian continent (Bela ophiolite) and onto the Arabian plate (Oman ophiolite; Fig. 10), but relicts have obviously survived on the sea floor closer to the original site (Southern Murray Ridge).

For the emplacement of the tholeiitic arc-basalt and the associated harzburgite we envisage a Late Cretaceous (95–70 Ma) age, similar to that of the Oman ophiolite (Lippard et al. 1986). This agrees with data from Deep Sea Drilling Project (DSDP) Site 223 (close to the Owen Ridge; Fig. 1), where the basalts underlying Late Paleocene sediments are about >60–70 Ma old (Whitmarsh et al. 1974) and with the age of the obduction of the Bela ophiolite (Late Cretaceous to Paleocene time; Allemann 1979; Bannert 1992).

The dredged quartz siltstone from the top of Southern Murray Ridge must have been deposited in a basinal setting; it consists of siliciclastic material derived from a continental margin located to the north, before the Murray Ridge was uplifted. This correlates well with observations from the Owen Ridge, where the uplift history is characterized by the transition from turbiditic to pelagic facies in late early Miocene time (Shipboard Scientific Party 1989). The turbidites at DSDP Site 224 (Whitmarsh et al. 1974), thought to be derived from the Indus Fan, were dated to late Oligocene to early Miocene time. Apparently they were deposited on a flat abyssal plain before the Owen Ridge was uplifted as a result of compression along the fracture zone caused by changes in spreading direction associated with the continued collision of India and

Asia and the opening of the Gulf of Aden (Shipboard Scientific Party 1989).

During ?Late Miocene to Early Pliocene times the erosion of the top of the Southern Murray Ridge started. By 'footwall uplift' between the Dalrymple Trough and Southern Murray Ridge (Edwards et al. 2000), the Dalrymple Trough subsided during post-mid-Miocene times (Gaedicke et al. 2002a,b), whereas part of the basin fill (including turbidites) was uplifted along major normal and strike-slip faults. A tilted fault block encompassing basaltic basement overlain by uplifted turbidites from the Dalrymple Trough was formed and uplifted to the euphotic zone. This is evidenced by the open-marine, large benthic foraminiferal fauna (Drooger 1994) in the carbonates encrusting and cementing the volcanic and ultrabasic pebbles, and by algal fragments indicating a sublittoral environment of deposition.

Correlation of key horizons from new BGR multi-channel seismic data led Gaedicke et al. (2002a) and Clift et al. (2000, 2002) to propose the following evolution of the Oman Abyssal Plain–Indus Fan region including the Murray Ridge–Dalrymple Trough: (1) acoustic basement of ?Paleocene age is overlain by Paleogene syn-drift sediments that were tilted in Miocene time; (2) extensional block-faulting caused the uplift of the Murray Ridge during post-Early Miocene time (c. 20 Ma ago) and the subsidence of the Dalrymple Trough during post-mid-Miocene times; (3) a late Miocene–early Pliocene M-unconformity is overlain by flat-lying distal channel–levee sediments (upper Indus Fan).

From Pliocene time to the present, the Southern Murray Ridge started to subside slowly (400–1000 m) to the present water depth. Pelagic sediments including planktonic Foraminifera were deposited at sedimentation rates of about 7–10 cm ka^{-1} on top of igneous rocks. The sediments overlying the top of the ridge were partly eroded, and the basement was exposed and weathered at several places. Whereas the top of the Southern Murray Ridge looks like a submerged biogenic hardground or platform, the flanks of the ridge are draped by a southward-thickening sequence of hemipelagic sediments (up to >1 s two-way travel time (TWT) thick), possibly of post-Early Miocene age.

In summary, we realize that our interpretation is limited by the small set of samples and their extensive alteration, which prevented any radiometric dating. We hope that future sampling or deep-sea drilling will recover more and fresher material. However, our petrographic and geochemical study (using rather immobile relict components) indicates strong affinities to a supra-subduction origin of the sampled rocks, which can be compared with ophiolite suites from the nearby Oman and Bela Mountains. Our hypothesis that part of the Murray Ridge basement may have formed in an active margin setting can help to constrain future models of the geotectonic evolution of this critical tectonic element at the boundary of the Indian, Eurasian and Arabian plates.

We are very grateful to C. Gaedicke (BGR) and H.-U. Schlüter (BGR) for discussions on the nature of the Murray Ridge–Indus Fan basement, and to A. Robertson (Edinburgh, UK). R. Edwards and R. White (Cambridge, UK), as well as to the anonymous reviewers for helpful critical comments. U. Siewers and F. Melcher provided geochemical data, and S. Sturm kindly helped with the presentation of figures. The success of the SO-122 and -130 cruises would not have been possible without the dedicated, untiring commitment of all scientists, technicians and crew members of R.V. Sonne under Captains Andresen and Papenhagen. The Sonne cruises were funded by the German Federal Ministry of Education, Science, Research and Technology (Grants 03G 0122-A and 0130-A).

References

AHMED, Z. 1991. Basalt geochemistry and supra-subduction zone origin of the Bela ophiolite, Pakistan. *Acta Mineralogica Pakistanica*, **5**, 77–82.

ALLEMANN, F. 1979. Time of emplacement of Zhob Valley ophiolites and Bela ophiolites, Baluchistan (preliminary report). *In:* FARAH, A. & DE JONG, K.A. (eds) *Geodynamics of Pakistan*. Geological Survey of Pakistan, Quetta, 242–261.

ALT, J.C. 1995. Sub-seafloor processes in mid-ocean ridge hydrothermal systems. *In:* HUMPHRIS, S.E., ZIERENBERG, R.A., MULLINEAUX, L.S. & THOMSON, R.E. (eds) *Seafloor Hydrothermal Systems*. Geophysical Monograph, American Geophysical Union, **91**, 85–114.

ALT, J.C. 1999. Very low-grade hydrothermal metamorphism of basic igneous rocks. *In:* FREY, M. & ROBINSON, D. (eds) *Low-grade Metamorphism*. Blackwell Science, Oxford, 169–201.

ALT, J.C., LAVERNE, C., VANKO, D.A. & 10 OTHERS 1996. Hydrothermal alteration of a section of upper oceanic crust in the eastern equatorial Pacific: a synthesis of results from Site 504 (DSDP Legs 69, 70 and 83, and ODP Legs 111, 137, 140, and 148). *In*: Alt, J.C., Knoshita, H., Stocking, L.B. & Michael, P.J. (eds) *Proceedings of the Ocean Drilling Program, Scientific Results, 148*. College Station, TX: Ocean Drilling Program, 417–434.

ALT, J.C., TEAGLE, D.A.H., BREMER, T., SHANKS, W.C. III & HALLIDAY, A. 1998. Alteration and mineralization of an oceanic forearc and the ophiolite–ocean crust analogy. *Journal of Geophysical Research*, **103**(B6, 12), 365–380.

BANNERT, D. 1992. The structural development of the Western Fold Belt, Pakistan. *Geologisches Jahrbuch, Reihe B*, **80**, 1–60.

BARKER, P.F. 1966. A reconnaissance survey of the Murray Ridge. *Philosophical Transactions of the Royal Society of London, Series A*, **259**, 187–197.

BERGGREN, W.A., KENT, D.V., SWISHER, C.C. III & AUBRY, M.-P. 1995. A revised Cenozoic geochronology and chronostratigraphy. *In:* BERGGREN, W.A., KENT, D.V., AUBRY, M.-P. & HARDENBOL, J. (eds) *Geochronology, Time Scales and Global Stratigraphic Correlation.* Society of Economic Paleontologist and Mineralogists, Special Publications, **54**, 129–212.

BEURRIER, M., OHNENSTETTER, M., CABANIS, B., LESCUYER, J.-J., TEGYEY, M. & LE METOUR, J. 1989. Géochimie des filons doléritiques et des roches volcaniques ophiolitiques de la nappe du Semail: contraintes sur leur origine géotectonique au Crétacé supérieur. *Bulletin de la Société de Géologie de France*, **8**(2), 205–219.

BORDIER, J.L. 1994. *Le Volcanisme.* Editions BRGM, Orléans.

BURGATH, K.P., MARCHIG, V. & MUSSALLAM, K. 1994. Data report: mineralogic, structural and chemical variability of mantle sections from Holes 920B and 920D. *In:* KARSON, J., CANNAT, M., MILLER, D.J. & ELTHON, D. (eds) *Proceedings of the Ocean Drilling Program, Scientific Results, 153.* Ocean Drilling Program, College Station, TX, 505–521.

BURGATH, K.P., MOHR, M, DROZAK, J. & KLIMAINSKY, M. 1987. Origin of metamorphic peridotites and amphibolites in the Central Schwarzwald. *Terra Cognita*, 7(2-3), 168.

CEULENEER, G. & CANNAT, M. 1994. High-temperature ductile deformation of Site 920 peridotites. *In:* KARSON, J.A., CANNAT, M., MILLER, D.J. & ELTHON, D. (eds) *Proceedings of the Ocean Drilling Program, Scientific Results.* Ocean Drilling Program, College Station, TX, **153**, 5–21.

CLIFT, P., SHIMIZU, N., GAEDICKE, C., SCHLÜTER, H.-U. & CLARK, M. 2000. 55 Million years of Tibetan and Karakoram evolution recorded in the Indus Fan. *EOS Transactions, American Geophysical Union*, 81(25), 277–282.

CLIFT, P., SHIMIZU, N., LAYNE, G., GAEDICKE, C., SCHLÜTER, H.-U., CLARK, M. & AMJAD, S. 2002. Development of the Indus Fan and its significance for the erosional history of the western Himalaya and Karakoram. *Geological Society of America Bulletin*, **113**(8), 1039–1051.

COLEMAN, R.G. 1977. *Ophiolites. Minerals and Rocks,12.* Springer, Berlin.

COUMES, F. & KOLLA, V. 1984. Indus Fan: seismic structure, channel migration and sediment thickness in the upper fan. *In:* HAQ, B.U. & MILLIMAN, J.D. (eds) *Marine Geology and Oceanography of the Arabian Sea and Coastal Pakistan.* Van Nostrand Reinhold, New York, 101–110.

DROOGER, C.W. 1994. *Brok kalksteen met grote forams uit dredgestation 462 nabij de Murray Ridge in de Arabische Zee.* Preliminary Internal Report, Rijksuniversiteit Utrecht.

EDWARDS, R.A., MINSHULL, T.A. & WHITE, R.S. 2000. Extension across the Indian–Arabian plate boundary: the Murray Ridge. *Geophysical Journal International*, **142**, 461–477.

FLUEH, E.R., KUKOWSKI, N., REICHERT, C. & SCIENTIFIC PARTY 1997. *Cruise Report. SO-123 (MAMUT), Makran–Murray Traverse.* GEOMAR Report, **62**.

FRENZEL, G., MÜHE, R. & STOFFERS, P. 1990. Petrology of the volcanic rocks from the Lau Basin. *Geologisches Jahrbuch*, **D92**, 395–479.

FREY, F.A. 1984. Rare earth element abundances in upper mantle rocks. *In:* HENDERSON, P. (eds) *Developments in Geochemistry (Rare Earth Element Geochemistry).* Elsevier, Amsterdam, 153–203.

GAEDICKE, C., PREXL, A., SCHLÜTER, H.-U., MEYER, H., ROESER, H. & CLIFT, P. 2002*a*. Seismic stratigraphy and correlation of major unconformities in the northern Arabian Sea. *In:* CLIFT, P.D., KROON, D., GAEDICKE, C., & CRAIG, J. (eds) *The Tectonic and Climatic Evolution of the Arabian Sea Region.* Geological Society, London, Special Publications, **195**, 25–36.

GAEDICKE, C., SCHLÜTER, H.-U., ROESER, H., PREXL, A., SCHRECKENBERGER, B., REICHERT, C., CLIFT, P. & AMJAD, S. 2002*b*. Structure and origin of the crust below the northern Indus Fan and Murray Ridge System, northern Arabian Sea. *Tectonophysics*, in press.

GARCIA, M.O. 1978. Criteria for the identification of ancient volcanic arcs. *Earth-Science Reviews*, **14**, 147–165.

GILLIS, K.M. & THOMPSON, G. 1993. Metabasalts from the Mid-Atlantic Ridge: new insights into hydrothermal systems in slow-spreading crust. *Contributions to Mineralogy and Petrology*, **113**, 502–523.

HAY, W.W., DECONTO, R.M., WOLD, C.N. & 8 OTHERS 1999. Alternative global Cretaceous paleogeography. *In*: BARRERA, E. & JOHNSON, C.C. (eds) *Evolution of the Cretaceous Ocean–Climate System.* Geological Society of America, Special Papers, **332**, 1–47.

HEKINIAN, R., BIDEU, T., FRANCHETEAU, J., CHEMINEE, J.L., ARMIGO, R., LONSDALE, P. & BLUM, N. 1993. Petrology of the East Pacific Rise crust and upper mantle exposed in Hess Deep (Eastern Equatorial Pacific). *Journal of Geophysical Research*, **98**(135), 8069–8094.

HENDERSON, P. 1984. General geochemical properties and abundances of the rare earth elements. *In:* HENDERSON, P. (eds) *Developments in Geochemistry (Rare Earth Element Geochemistry).* Elsevier, Amsterdam, 1–32.

HUMPHRIS, S.E. 1984. The mobility of the rare earth elements in the crust. *In:* HENDERSON, P. (eds) *Rare Earth Element Geochemistry.* Elsevier, Amsterdam, 317–342.

ICKENROTH, V. & VAN BERGEN, M. 1994. *Basalts from NIOP Cruise.* Report, University of Utrecht.

IRVINE, T.N. & FINDLAY, T.C. 1972. Alpine-type peridotite with particular reference to the Bay of Islands igneous complex. *Publications of the Earth Physics Branch, Department of Energy, Mines and Resources*, **42**, 97–128.

JOHNSON, K.T.M. & SINTON, J.M. 1990. Petrology, tectonic setting and the formation of back-arc basin basalts in the North Fiji Basin. *Geologisches Jahrbuch*, **D92**, 517–545.

JUTEAU, T. & MAURY, R. 1997. *Géologie de la Croûte océanique—Pétrologie et Dynamique endogène.* Masson, Paris.

LAVERNE, C., BELAROUCHI, A. & HONNOREZ, J. 1996. Alteration mineralogy and chemistry of upper oceanic crust from Hole 896-A, Costa Rica Rift. *In:* ALT, J.C., KINOSHITA, H., STOKKING, H. & MICHAEL, P.J. (eds) *Proceedings of the Ocean Drilling Program, Scientific Results, 148*. Ocean Drilling Program, College Station, TX, 151–170.

LE ROEX, A.P., DICK, H.J.P., ERLANK, A.J., REID, A.M., FREY, F.A. & HART, S.R. 1983. Geochemistry, mineralogy and petrogenesis of lavas erupted along the Southwest Indian Ridge between the Bouvet triple junction and 11 degrees east. *Journal of Petrology*, **24**, 267–318.

LETERRIER, J., MAURY, R.C., THONON, P., GIRARD, D. & MARCHAL, M. 1982. Clinopyroxene composition as a method of identification of the magmatic affinities of paleo-volcanic series. *Earth and Planetary Science Letters*, **59**, 139–154.

LIPPARD, S.J., SHELTON, A.W. & GASS, I.G. 1986. The Ophiolite of Northern Oman. *Geological Society, London, Memoir*, 11.

LUDDEN, J.N. & THOMPSON, G. 1978. An evaluation of the behaviour of the rare earth elements during the weathering of sea floor basalts. *Earth and Planetary Science Letters*, **43**, 85–92.

MAHONEY, J.J. (1984): *Isotopic and chemical studies of the Deccan and Rajmahal traps, India: mantle sources and petrogenesis*. Ph.D. dissertation, University of California, San Diego.

MALOD, J.A., DROZ, L., MUSTAFA KEMAL, B. & PATRIAT, P. 1997. Early spreading and continental to oceanic basement transition beneath the Indus deep-sea fan. *Marine Geology*, **141**(1–4), 221–235.

MAURY, R.C., BOUGAULT, H., JORON, J.L., GIRARD, D., TRENIL, M., AZÈMA, J. & AUBOUIN, J. ET AL.1982. Volcanic rocks from Leg 67 sites: mineralogy and geochemistry. *In:* AUBOUIN, J. & HUENE, R. (eds) *Initial Reports of the Deep Sea Drilling Project*. US Government Printing Office, Washington, DC, **67**, 557–576.

MENZIES, M., BLANCHARD, D., BRANNON, J. & KOROTER, R. 1977. Rare earth geochemistry of fused ophiolitic and alpine lherzolites. *Contributions to Mineralogy and Petrology*, **64**, 53–74.

MILES, P.R. & ROEST, W.R. 1993. Earliest sea floor spreading magnetic anomalies in the north Arabian Sea and the ocean–continent transition. *Geophysical Journal*, **115**(3), 1025–1031.

MILES, P.R., MUNSCHY, M. & SEGOUFIN, J. 1998. Structure and early evolution of the Arabian Sea and East Somali Basin. *Geophysical Journal International*, **134**, 876–888.

MINSHULL, T.A. & WHITE, R.S. 1989. Sediment compaction and fluid migration in the Makran accretionary prism. *Journal of Geophysical Research*, **94**(B6), 7387–7404.

MINSHULL, T.A., WHITE, R.S., BARTON, P.J. & COLLIER, J.S. 1992. Deformation at plate boundaries around Gulf of Oman. *Marine Geology*, **104**, 265–277.

MINSHULL, T.A., EDWARDS, R.A., FLUEH, E. & KOPP, C. 2001. New constraints on the age and tectonic history of the lithosphere at Murray Ridge and Dalrymple Trough. *In:* CLIFT, P. (eds) *Geological and Climatic Evolution of the Arabian Sea Region, Abstracts, 5–6 April 2001*. Geological Society, London.

MOUNTAIN, G.S. & PRELL, W.L. 1990. Multiphase plate tectonic history of the southeast continental margin of Oman. *In:* ROBERTSON, A.H.F., SEARLE, M.P., & RIES, A.C. (ed.) *The Geology and Tectonics of the Oman Region*. Geological Society, London, Special Publications, **49**, 725–743.

PALLISTER, J.S. & KNIGHT, R.J. 1981. Rare earth element geochemistry of the Samail ophiolite near Ibra, Oman. *Journal of Geophysical Research*, **86**, 2673–2697.

PEARCE, J.A. & CANN, J.R. 1973. Tectonic setting of basic volcanic rocks determined using trace element analyses. *Earth and Planetary Science Letters*, **19**, 290–300.

PEARCE, J.A. & NORRY, M.J. 1979. Petrogenetic implications of Ti, Zr, Y, and Yb variations in volcanic rocks. *Contributions to Mineralogy and Petrology*, **69**, 33–47.

ROESER, H.A. & SCIENTIFIC PARTY 1997. *MAKRAN I: the Makran accretionary wedge off Pakistan—tectonic evolution and fluid migration (part 1)*. Cruise Report, BGR Archive No. **116 643**.

SAUNDERS, A.D. 1984. The rare earth element characteristics of igneous rocks from the ocean basins. *In:* HENDERSON, P. (eds) *Developments in Geochemistry (Rare Earth Element Geochemistry)*. Elsevier, Amsterdam, 205–236.

SCHRECKENBERGER, B., GAEDICKE, C., SCHLÜTER, H.-U. & ROESER, H.A. 2001. Submerged continental crust and volcanism in the Indus Fan and Murray Ridge area: implications from gravity and magnetic modeling. *In:* CLIFT, P. (ed.) *Geological and Climatic Evolution of the Arabian Sea Region, Abstracts, 5–6 April 2001*. Geological Society, London.

SEARLE, M. & COX, J. 1999. Tectonic setting, origin, and obduction of the Oman Ophiolite. *Geological Society of America Bulletin*, **111**, 104–122.

SHIPBOARD SCIENTIFIC PARTY 1989. Background and summary of drilling results—Owen Ridge. *In:* PRELL, W.L. & NIITSUMA, N. (eds) *Proceedings of the Ocean Drilling Program, Initial Reports*. Ocean Drilling Program, College Station, TX, **117**, 35–42.

STAUDIGEL, H., FREY, F.A. & HART, S.A. ET AL. 1979. Incompatible trace element geochemistry and $^{87}Sr/^{86}Sr$ in basalts and corresponding glasses and palagonite. *In:* DONNELLY, T. & FRANCHETEAU, J. (eds) *Initial Reports of the Deep Sea Drilling Project*. US Government Printing Office, Washington, DC, **51, 52, 53 (Part 2)**, 1137–1144.

SUN, S.S., NESBITT, R.W. & SHARASKIN, A. 1979. Geochemical characteristics of mid-ocean ridge basalts. *Earth and Planetary Science Letters*, **44**, 119–138.

TEAGLE, D.A.H., ALT, J.C., BACH, W. & ERZINGER, J. 1996. Alteration of upper ocean crust in a ridge-flank hydrothermal upflow zone: mineral, chemical, and isotopic constraints from Hole 896-A. *In:* ALT, J.C., KINOSHITA, H., STOKKING, L.B. & MICHAEL, P.J. (eds) *Proceedings of the Ocean Drilling Program, Scientific Results*. Ocean Drilling Program, College Station, TX, **148**, 119–150.

TELFORD, W.M., GELDART, L.P. & SHERIFF, R.E. 1991. *Applied Geophysics*, 2nd edition. Cambridge University Press, Cambridge.

VAN DER LINDEN, W.J.M. & VAN DER WEIJDEN, C.H.,

1994. *Geological Study of the Arabian Sea, Vol. 3.* Cruise Report, Netherlands Indian Ocean Program, **NIOP-92-93**.

VAN DER LINDEN, W.J.M., VAN BERGEN, M.J., DROOGER, C.W. & ICKENROTH, V.J.G. 1995. Rock from the Murray Ridge Complex, Northern Arabian Basin (abstract). Arabian Sea Workshop, Texel, 1995.

VANKO, D.A. & LAVERNE, C. 1998. Hydrothermal anorthization of plagioclase within the magmatic/hydrothermal transition at mid-ocean ridges: examples from deep sheeted dikes (Hole 504B, Costa Rica Rift) and a sheeted dike root zone (Oman ophiolite). *Earth and Planetary Science Letters*, **162**, 27–43.

VON RAD, U., DOOSE-ROLINSKI, H. ET AL. 1998. *SONNE Cruise SO-130—MAKRAN II, Cruise Report.* BGR, Hannover, Archive No. 117368.

WHITE, R.S. & LOUDEN, K.E. 1982. The Makran continental margin: structure of a thickly sedimented convergent plate boundary. *In:* WATKINS, J. & DRAKE, C.L. (eds) *Studies in Continental Margin Geology.* American Association of Petroleum Geologists, Memoirs, **34**, 499–518.

WHITMARSH, R.B., WESER, O.E., ROSS, D.A., ET AL. (eds) 1974: *Initial Reports of the Deep Sea Drilling Project, 23.* US Government Printing Office, Washington, DC.

WOODHEAD, J. 1989. Geochemistry of the Mariana arc (western Pacific): source composition and processes. *Chemical Geology,* **76**, 1–24.

Gas hydrates acting as cap rock to fluid discharge in the Makran accretionary prism?

G. DELISLE & U. BERNER

Bundesanstalt für Geowissenschaften und Rohstoffe, Stilleweg 2, D-30655 Hannover, Germany (e-mail: G.Delisle@bgr.de)

Abstract: We present a numerical model of the geothermal field of the Makran accretionary prism and of the slab being subducted below it. Calculated heat flow density values for the sea floor of the abyssal plain and the shelf slope are compared with *in situ* measured and bottom simulating reflector (BSR)-derived heat flow density values. The result suggests a predominance of conductive heat transport within the accretionary complex. Little evidence is found to suggest that fluid flow or frictional heat modifies the observed geothermal field to any great extent. We also studied the geothermal field associated with the decay of the potential gas hydrate layers (indicated by the presence of BSRs), as gas hydrate layers are being tectonically uplifted out of the gas hydrate stability field into shallower and warmer sea water. Theoretical considerations suggest a complete disappearance of gas hydrates at a water depth of about 750 m. The observed presence of numerous gas seeps almost exclusively at water depths of less than 800 m suggests that gas hydrate layers in the Makran accretionary prism act as a very effective cap rock to upward-directed flow of fluids containing notable amounts of dissolved gas from within the prism to the sea floor.

The Indo-Arabian Plate subducts at low angles under the Eurasian plate at a velocity of currently about 3–5 cm a^{-1} (White 1982; White & Louden 1983). The marine sediments overlying the basement are up to 9 km thick. They form a rapidly thickening sequence between the Murray Ridge (a SW–NE-trending submarine ridge system about 300 km south of the eastern sector of the Makran coast) and the Makran accretionary prism. The sediments are partially scraped off at the accretionary front and form the *c.* 100 km wide offshore part of the Makran accretionary prism. The prism continues on land as a 500 km wide fold belt, the West Pakistan fold belt (Hunting Survey Corporation Ltd 1960) of the Makran desert, which consists predominantly of Oligocene to Pleistocene compacted marine sediments. The onset of the subduction and development of the fold belt is thought to have occurred at *c.* 75 Ma, based on the emplacement age of ophiolites in the Western Ophiolite Belt of Pakistan (Gnos *et al.* 1977). The coastal ranges between Pasni and Ormara are characterized by upthrust and folded Upper Miocene sediments (Parkini Formation), overlain at the shoreline by Pleistocene to Holocene shelf deposits, which were raised by as much as 500 m above the current sea level (Page *et al.* 1979; Harms *et al.* 1983). This observation reflects the rapid and continuous deformation of the offshore deposits, which are continuously added at the southern rim of the West Pakistan fold belt.

Marine-seismic reflection and refraction surveys of the Makran accretionary prism offshore (White 1982; White & Louden 1983; Minshull & White 1989; Kopp *et al.* 2000) have revealed a sequence of uplifted and faulted ridges with interfold basins between. Their seismic reflection lines showed discontinuous traces of a bottom simulating reflector (BSR) in the accretionary prism and in the sediments of the abyssal plain. This BSR, positioned about 500–800 m beneath the sea floor, is generally believed to mark the lower boundary of the gas hydrate layer, consisting of primarily methane 'frozen' in an ice lattice (Shipley *et al.* 1979). A detailed analysis of the compressional wave velocity distribution around the BSR has led Sain *et al.* (2000) to postulate a thick free gas layer below the gas hydrate zone of the Makran accretionary prism.

Wide-angle and reflection seismic records place the depth of the décollement zone at about 8 km below sea level near the front of the accretionary prism (Fruehn *et al.* 1997; Kopp *et al.* 2000). A dip angle of 3° for the subducting plate was given by Kopp *et al.* (2000), which is somewhat steeper than the previous estimate of less than 2° by White & Louden (1983).

BSR and heat flow density

The reader is referred to the paper of Minshull & White (1989), who have discussed in detail the possibilities and potential errors of deriving heat flow values for the Makran accretionary prism from the depth of the BSR. Their data show that BSR-derived heat flow values decrease moderately from c. 40 mW m^{-2} at the accretionary front to c. 35 mW m^{-2} at a position 60 km landward (identical to the position of the disappearance of BSRs in the seismic record) of the deformation front. They attributed the only moderate decrease in heat flow in relation to the expected heat flow depression caused by tectonic thickening of the prism to an advective heat transport component related to dewatering of the compacting sediments.

This paper attempts to review the available record based on two recent surveys. New seismic lines across the Makran accretionary prism were obtained in 1997 by Roeser & Shipboard Scientific Party (1997) during the R.V. *Sonne* cruise SO-122. Several of these lines (SO122-04-A, SO122-13; SO122-13-A; Fig. 1) did reveal a BSR. Using the same approach as described by Minshull & White (1989), heat flow values were obtained at points spaced 1 km apart, wherever a BSR reflector was identified along the seismic profiles. The same procedure was applied to profile N1804 (taken from Lehner *et al.* (1983)) across the two most seaward anticlinal ridges of the prism that shows, in places, a BSR. Hutchison *et al.* (1981) had reported *in situ* measured heat flow values at 27 sites south of the accretionary front by a marine heat flow probe. Measured heat flow values from the accretionary prism were presented by Kaul *et al.* (2000). All these heat flow values (BSR-derived and from marine heat flow probes) are summarized in Fig. 2.

Trace gases in the water column

Plumes of bacterial methane had already been detected within the oxygen minimum zone in 1993 (von Rad & Scientific Shipboard Party 1993). They extended over more than 20 km at

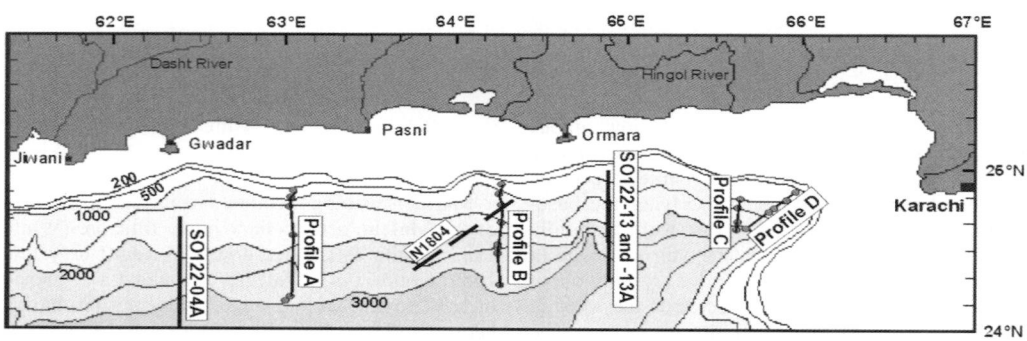

Fig. 1. Location of seismic lines (SO122-) with BSR reflections, from which heat flow density was derived. Also shown are profiles A–D, on which trace gases in the water column were measured (bathymetric data after GEBCO Digital Atlas British Oceanographic Data Centre 1994).

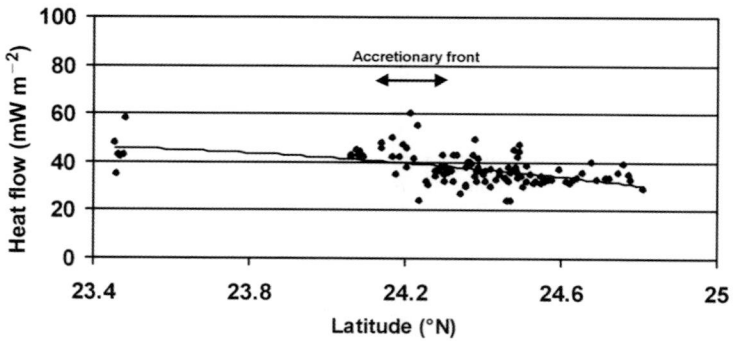

Fig. 2. Heat flow density values derived from SO122-seismic lines and literature values between latitudes 23.4° and 24°. (See text for sources of data.)

around 600–800 m water depth horizontally and southwards from the top of the accretionary prism (profile D in Fig. 1). Gas plumes are known to remain stable in the water column at least over days (DeAngelis et al. 1993). In 1998, BGR conducted a second cruise with R.V. *Sonne* (SO-130), with the objective of gaining a better understanding of the occurrence of gas hydrates, the origin of submarine gas seeps and of the activities of methane or H_2S-oxidizing bacteria and chemoautotrophic vent biota (von Rad et al. 2000). The concentration of trace gases, and in particular methane, in the water column of the Arabian Sea was remeasured in 1998 at 24 sampling stations distributed over three north–south-trending profiles across the Makran accretionary prism (Fig. 1). A horizontal gas plume at least 25 km long was observed on profile A with concentrations decreasing from 550 to 150 nl l^{-1}, which started at a water depth of 600 m on the wedge slope. The plume on profile B also extends more than 25 km laterally at around 500 m water depth into the ocean. Two plumes at 300 and 800 m water depth with lateral extents of 5 and 8 km seaward were observed on profile C (see Fig. 3a–d). At only one point were methane concentrations of about 700 nl l^{-1} found at a water depth of 2400 m. This seep was located within the 1 km wide 'Calyptogena Canyon', an erosional channel that has cut about 400 m deep into the sediments and the gas hydrate layer. Vent fauna, sustained by H_2S- and methane-rich fluids (for location, see sampling station MS324 in Fig. 3b in conjunction with Fig. 1) were found in this canyon, detected during a detailed bathymetric survey during cruise SO-130 (von Rad et al. 2000).

Numerical simulation of the geothermal state of the accretionary prism

We present three numerical models to explore key factors, which in our view influence the geothermal field of the accretionary prism. The governing equations (Carslaw & Jaeger 1959) in general are

$$\rho c \frac{\delta T}{\delta t} = \mathrm{div}(\lambda \mathrm{grad} T) + Q \quad (1)$$

and, where latent heat uptake is involved (only model 2),

$$\lambda_1 \frac{\mathrm{d}T_1}{\mathrm{d}z} - \lambda_2 \frac{\mathrm{d}T_2}{\mathrm{d}z} = L\rho_e \frac{\mathrm{d}X}{\mathrm{d}t} \quad (2)$$

where λ is thermal conductivity, ρ is density, c is heat capacity, Q is radiogenic heat production in sediments, $\lambda_1 \mathrm{d}T_1/\mathrm{d}z$ is heat flow immediately above freezing front, $\lambda_2 \mathrm{d}T_2/\mathrm{d}z$ is heat flow immediately below freezing front, L is latent heat, ρ_e is hydrate content in sediment, X is depth of phase change boundary, $\mathrm{d}X/\mathrm{d}t$ is rate of movement of phase change boundary (pore water–hydrate), T is temperature, z is depth and t is time.

The simulation of the subduction process is achieved by shifting the temperature field below the décollement zone in the direction of subduction according to the specified subduction rate. All models were calculated using an explicit finite difference scheme.

Model 1 is a large-scale model to explore the thermal effect of the subduction of the oceanic crust under the Makran accretionary prism. Subsurface temperature fields are calculated for three cases, assuming a dip angle for the subducting plate of 1°, 2° or 3°. We consider a south–north-oriented profile from the northern edge of the Murray Ridge to a position of at least (depending on assumed dip angle in model) 200 km north of the Makran coast.

Model 2, a small-scale model, considers the thermal history of a sediment column capped by a gas hydrate layer, which is uplifted in geological time by the accretionary process, from a deep-water position into shallow water near the coast. This model refers in particular to the slope segment with water depths between 750 and 1700 m, where the Makran accretionary prism shows the highest slope angle (see Fig. 1) and therefore to the area with presumably the highest uplift rate within the prism.

Model 3 investigates the disturbance of the distribution of heat flow density around ridges and backfill basins by sea bottom topography.

Model 1

The subduction of the oceanic crust under the Eurasian plate was modelled numerically (for geometry, see also Fig. 4). The subsurface temperature field in the model domain was calculated to a depth of 25.75 km. The following boundary conditions were chosen. Temperatures along the sea floor were kept constant and modelled after the measured bottom water temperature distribution as a function of water depth of the Arabian Sea (see also Fig. 6, below). Temperatures along the land surface were kept constant at 23 °C. Heat flow across the lateral boundaries and the base of the model was kept constant. Heat transport was kept constant in the direction of movement, where the subducting plate passes through the base of the model. The spacing of grid points in the vertical direction is 125 m. The horizontal resolution depends on the chosen dip angle of the oceanic plate and varies between 2385 m (3°), 3515 m (2°) and 7160 m (1°). A total of 45 360 grid points were introduced into the model. A

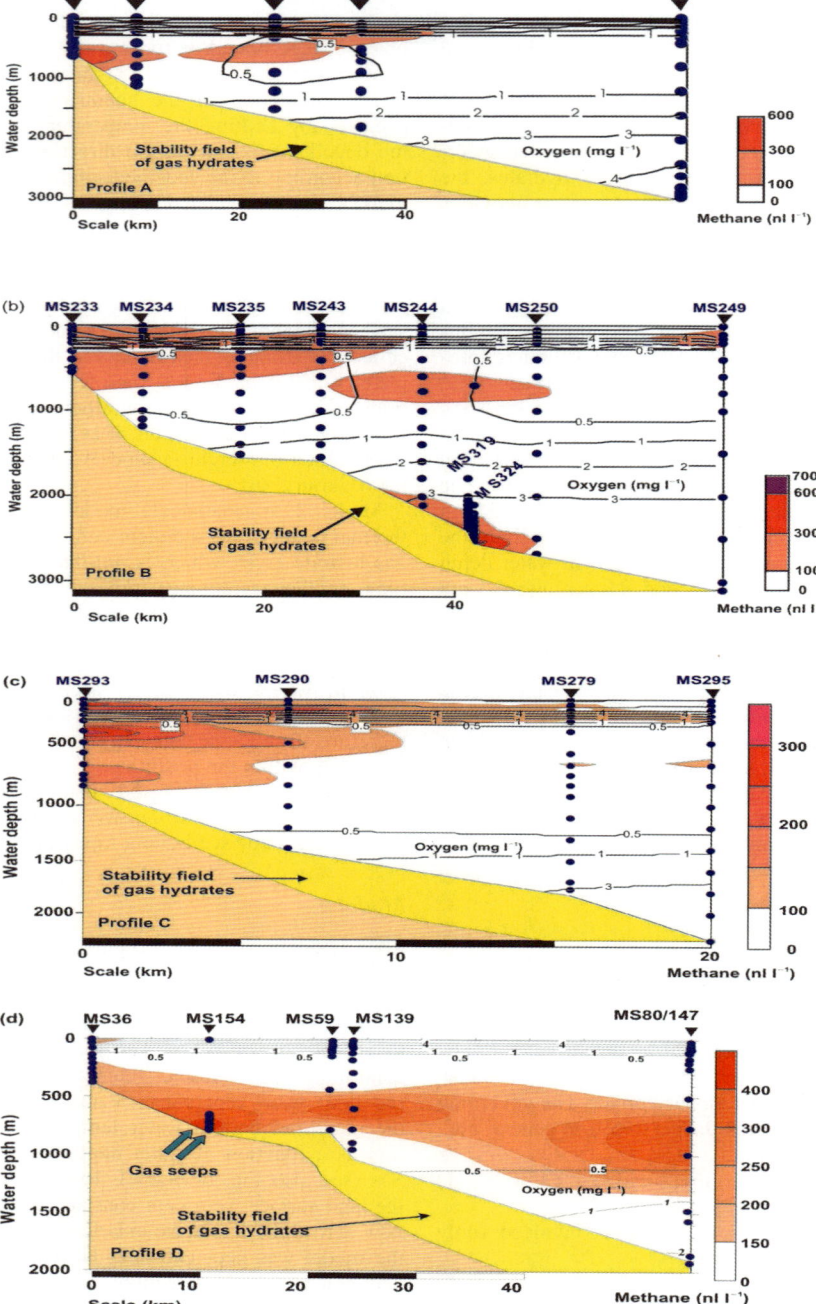

Fig. 3. (a)–(d) Fluids with elevated methane concentrations rise at water depths of less than 800 m (landward boundary of the gas hydrate stability field) and drift laterally seaward over tens of kilometres. Water sampling stations are indicated by black arrowheads (with station numbers). Blue dots show depth positions, where water samples were taken in the water column. The gas hydrate stability fields are sketched in on the basis of known BSR depths and according to the P–T stability field of gas hydrate with sea water (Englezos & Bishnoi 1988). (See Fig. 1 for location of profiles A–D.)

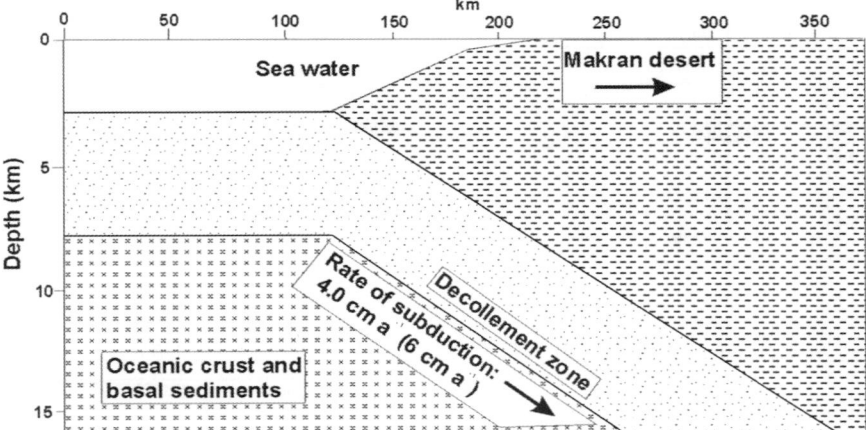

Fig. 4. Model configuration for numerical model 1, which analyses the large-scale geothermal field of the Makran accretionary prism.

value of 2.1×10^{-6} W s kg^{-1} K^{-1} was chosen as uniform value for the product of sediment density and heat capacity (ρc), and the other material parameters are listed in Table 1.

The incoming oceanic plate is exposed towards the accretionary front to high sedimentation rates. A simplified algorithm was introduced to incorporate a sedimentation rate of 1.76 km Ma^{-1} (equivalent to sedimentation of a 6 km thick sediment pile within 3.4 Ma on a plate, which travels the distance of 85 km from the Murray Ridge to the accretionary front with a velocity of 4 cm a^{-1}). The oceanic plate dips under the accretionary prism and the continent at a given angle (1°, 2° or 3°) and a velocity of 4 cm a^{-1}. The depth of the décollement zone was placed 8 km below sea level (5 km below the sea floor) seaward of the accretionary front, dipping at the same angle as the oceanic plate under the prism. The model was calculated in two versions.

(1) The accretionary prism is considered to be a static body, under which the oceanic crust subducts at a rate of 4 cm a^{-1} (version 1a).

(2) In reality, the prism grows continuously outward. A comparison of the annually accreted mass with the mass stored above the décollement zone in the submarine accretionary prism suggests a replacement of the prism within ≥5.5 Ma. This is equivalent to a seaward migration of the coastline or of the prism by ≤2 cm a^{-1}, which is equivalent to an effective rate of subduction of oceanic crust of ≤6 cm a^{-1}. Model 1 was recalculated (version 1b) assuming an effective rate of 6 cm a^{-1}.

Both versions are considered to constrain the range within which the true heat flow density distribution should be found.

The evolution of the temperature field and the heat flow distribution for all model cases was calculated for a time period of 10 Ma; a period that is significantly longer than the required time for a 'proto-thrust' to be formed in deep water, compressed, deformed and uplifted to subaerial conditions. As will be shown by the results, this time period is sufficient for the development of a quasi-stationary temperature field in the model prism.

The resulting heat flow profiles over the accretionary prism are shown in Fig. 5. Missing in these models is the thermal effect associated with the uplift of prism sediments from great water depths with cold basal waters to shallow water depths with warmer basal waters. That factor is analysed by model 2.

Model 2

The top of the Makran accretionary prism appears to be covered by a gas hydrate layer, wherever the

Table 1. *List of thermal rock parameters applied in model 1*

	Thermal conductivity (W m^{-1} K^{-1})	Radiogenic heat production (W m^{-3})
Folded prism of pre-Holocene age	1.7	0.5×10^{-6}
Marine sediments	1.7	0.25×10^{-6}
Oceanic crust	3.3	0.14×10^{-6}

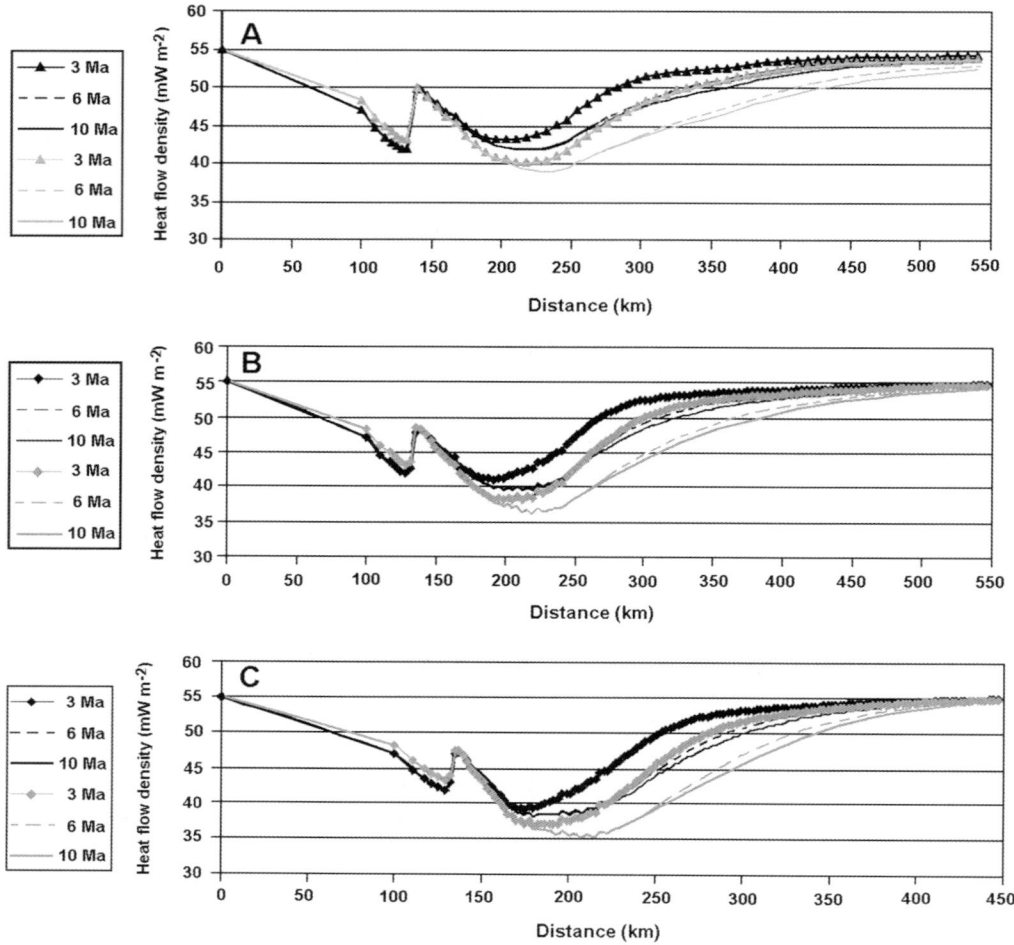

Fig. 5. Resulting heat flow density for the transect of model 1. (**a**) Dip angle of oceanic plate and décollement is 1°; (**b**) dip angle is 2°; (**c**) dip angle is 3°. Graphs in black are for subduction rate of 4 cm a^{-1}; those in grey are for 6 cm a^{-1}.

sediments are within the gas hydrate stability field. The thickness of the gas hydrate layer is 800 m near the accretionary front and decreases in thickness in the landward direction. The thinning is a reflection of the diminishing size of the gas hydrate stability field as the depth of the sea floor (and water pressure) decreases and bottom water temperature rises. One effect of the deformation of the accretionary prism and its lateral transport toward the coast is a gradual shift of gas hydrates out of the gas hydrate stability field. An approximate uplift rate can be estimated from the fact that the exposed and folded marine sediments along the shore of the Makran desert are of Pleistocene age (Harms *et al.* 1983). These sediments were raised from 3000 m water depth (abyssal plain) to sea level within 2.5 Ma, which suggests an uplift rate of *c.* 1 mm a^{-1}. However, uplift is possibly not evenly distributed across the shelf area, but is concentrated at water depths between 500 and 1500 m, where the largest slope angle has developed. Therefore, two calculations, based on uplift rates of 1 and 3 mm a^{-1}, were carried out.

Model 2 considers a 9.5 km thick sediment column, initially located at a water depth of 1700 m and capped with an initially 500 m thick gas hydrate layer (in agreement with the *P–T* conditions and the stability field of gas hydrates at this water depth). The numerical model has a grid spacing of 4 m in the column within the gas hydrate layer. Grid spacing gradually increases below the hydrate layer toward the bottom of the column. The column incorporates 280 grid points.

The temperature at the upper surface of the column is defined by the temperature–depth curve (Fig. 6) of sea water. Heat flow through the bottom of the column is 40 mW m^{-2} and constant. The gas hydrates are assumed to occupy 6% of the pore volume, linearly decreasing to 0% at the sea floor in agreement with the study of Davie & Buffett (2001) on the typical gas hydrate distribution below the sea floor. The existence of a gas high hydrate pore volume is suggested by the presence of free gas under the BSR (Sain *et al.* 2000). The column is uplifted and its top raised with time into warmer sea water. This scenario is equivalent to the situation of a land surface exposed to steady climatic warming. This model attempts to estimate the resulting reduction in heat flow density at the sea floor. The decay of the gas hydrate layer implies uptake of latent heat (54.2 kJ mol^{-1}), which has to be supplied by heat flow from deeper levels and from the bottom waters. Applying equations (1) and (2) allows the calculation of the changes in heat flow density on top of the column as a result of the uplift into successively warmer waters at rates of 1 and 3 mm a^{-1} (Fig. 6). The sediment temperature at the top of the column exceeds the boundary temperature of the methane stability field at water depths of less than 750 m. Gas hydrates can potentially exist only at depths of more than 750 m.

Model 3

This model considers the thermal field of a 4.2 km wide ridge element with a flank to the left, rising by 260 m, and falling off by 435 m on the right flank (Fig. 7), a topographic form that can be

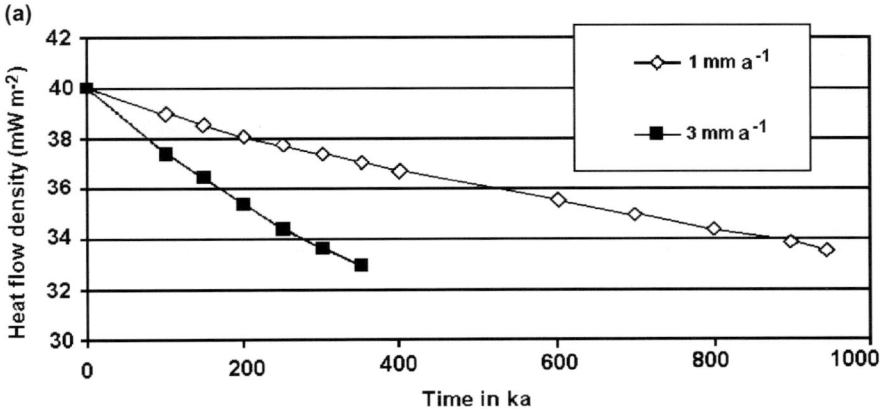

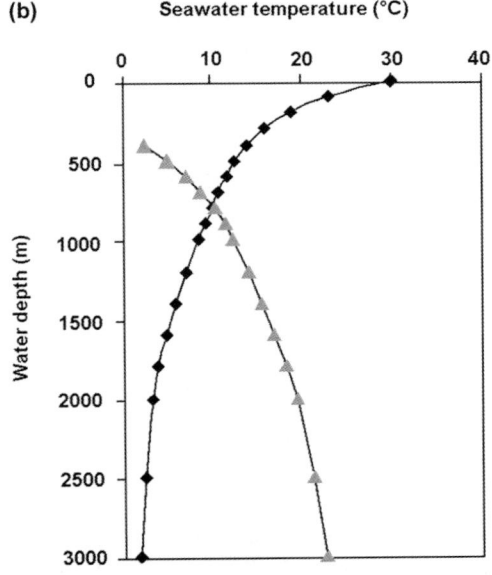

Fig. 6. (a) Thermal effect of uplift of a sediment column, capped with a gas hydrate layer, from a water depth of 1700 m into shallower and therefore warmer bottom waters. Two uplift rates (1 and 3 mm a^{-1}) are considered. (b) The measured temperature profile in the water column offshore Makran (black line) and the stability field of gas hydrates (grey line).

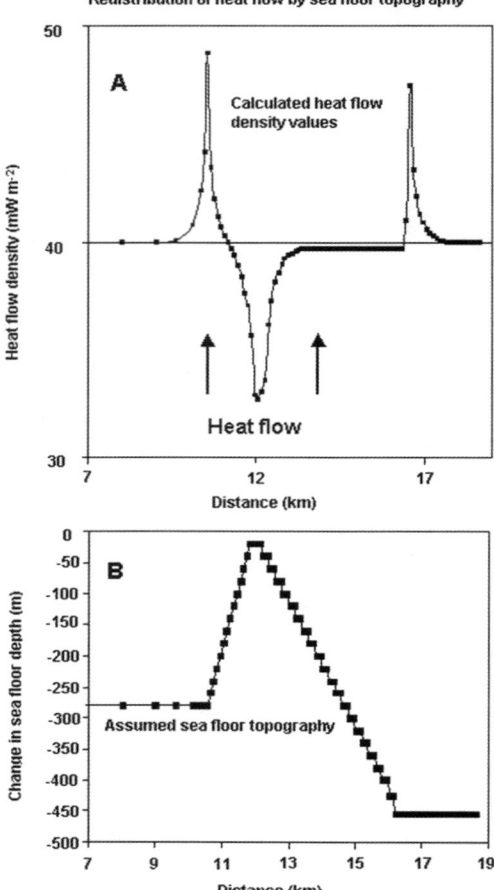

Fig. 7. Calculated effect of topographic features typical for the ridge structure of the Makran accretionary prism on the observable heat flow density distribution at the sea floor. (**a**) Heat flow diminishes on ridge top and increases at foot of ridge. (**b**) The assumed sea-floor topography.

found repeatedly on the surface of the Makran accretionary prism. The thermal conductivity of the ridge material is taken uniformly as 1.2 W m^{-1} K^{-1}. A heat flow density q of 40 mW m^{-2} rising from the base of the model was chosen. The lateral extent of the model is 28 km, and the vertical extent is 22.5 km. Vertical spacing of grid points in the area of the modelled ridge varies between 20 and 25 m, and horizontal spacing is 100 m. Spacing in the horizontal and vertical directions slowly increases beyond the model ridge area. A uniform bottom water temperature of 3 °C was assumed, which is a reasonable assumption for deep oceanic water. The variation of heat flow density in response to the topographic expression of the ridge surface is shown in Fig. 7.

Results and discussion

Model 1 suggests that a high sedimentation rate between the Murray Ridge and the accretionary front causes a gradual decline of heat flow from 55 to 43 mW m^{-2}. The small heat flow density increase at the accretionary front is caused by the sudden ending of sedimentation of cold material. The wedge itself experiences no sedimentation (equivalent to the assumption of an even balance between sedimentation and sediment slumping). In consequence, the negative geothermal anomaly contained within the subducted top sediments now under the wedge nose decays by re-equilibration of the thermal gradient. In continuation towards the coast, the heat flow flux is further reduced at a distance of 70 km inland from the accretionary front, reflecting the influence of the subduction of cold material of the oceanic crust, only to recover onshore in the Makran desert and in the interior of the continent. The landward migrating oceanic plate finds ever rising temperatures at the upper boundary (first as a result of increasingly warmer bottom waters along the rising sea floor, then on the land surface), which is one factor contributing to the heat flow density minimum observed on the slope (see, for comparison, model 2).

Comparison of this result with the available heat flow values (measured *in situ* or BSR derived) shows closest agreement with the version incorporating a dip angle of 3° (see, for comparison, Figs. 2 and 5). The measured heat flow values at the accretionary front at 24.2°N range on average between 45 and 50 mW m^{-2} and decrease within 50 km landwards to values around 35 mW m^{-2}. Little is known about heat flow values in the coastal region. Raza *et al.* (1990) reported geothermal gradients of 15–20 °C km^{-1} in the nearshore region of the Makran based on data from oil exploration boreholes (which might not have been thermally equilibrated at the time of measurement). These gradient values suggest therefore minimum heat flow values of 26–34 mW m^{-2} (assuming a typical thermal conductivity of 1.7 W m^{-1} K^{-1}).

Model 1 demonstrates three aspects, as follows.
(1) The decrease of heat flow from the far field to, on average, 45 mW m^{-2} (Fig. 5) can be well explained as a result of increasing sedimentation rates in the abyssal plain, as the accretionary front is approached.
(2) The decrease of heat flow from the accretionary front to the coast can be explained by the subduction of cold material under the prism below the décollement zone and the shift in surface

temperatures from cold to warmer bottom waters and then the hot land surface. The best fit between the measured and calculated heat flow pattern across the prism has been achieved by the model version involving a dip angle of 3° of the oceanic plate, in agreement with most recent evaluations of seismic records (Kopp *et al.* 2000 and a depth level of the décollement zone of $\geqslant 8$ km. (Nevertheless, it is worth noting that the closest fit between measured and calculated values would be achieved if the décollement zone were placed 1–1.5 km shallower). As the measured heat flow curve across the prism shows values 3–4 mW m^{-2} lower than the calculated best-fit version, we find little room for postulating the existence of positive geothermal anomalies as a potential indication for the presence of upwelling fluids across the gas hydrate layer or of the influence of frictional heating along the décollement zone. The geochemical evidence suggests the presence of upwelling fluids only for the shallow coastal water region, from where we have, however, no heat flow data.

(3) From this type of analysis it appears that the geothermal field of the Makran accretionary prism is predominantly controlled by heat conduction.

In model 2, the rise of the modelled sediment column from 1700 m water depth to less than 755 m within 945 ka lifts the capping gas hydrate layer out of the methane stability field and, in addition, raises the temperature at the water–sediment interface from 4.8 to 11 °C.

The gas hydrate in the prism will decay with no significant time delay as soon as the prism is uplifted out of the gas hydrate stability field. No significant geothermal anomaly produced by the gas hydrate decay was noted. Latent heat uptake is an insignificant process in relation to the effect of uplift of the sediments into successively warmer bottom waters, which reduces the thermal gradient below the sea floor and with it the heat flow density.

Model 2 suggests a reduction of heat flow flux by 6.4 mW m^{-2} (1 mm a^{-1} uplift) or 7 mW m^{-2} (3 mm a^{-1} uplift) landward of the accretionary front, which is the same magnitude as predicted by model 1, although on the basis of different boundary conditions. The minor influence of the uplift rate on heat flow density reduction supports the argument that the increasingly higher surface temperatures with respect to either the rising sediment column or the subducted oceanic plate shift is an important component, which is responsible for the heat flow density minimum observable about 70 km north of the accretionary front.

The fact that gas plumes occur (with the discussed exception of the Calyptogena Canyon site) at water depths of less than 800 m (i.e. strictly in areas outside potentially present gas hydrates) has led us to speculate that gas hydrate layers of this accretionary prism act as a cap rock, which prevents to a large extent the escape of fluids with dissolved gases along faults through the gas hydrate layer to the sea floor. This scenario is in excellent agreement with the above finding that the geothermal field of the prism can be well explained by a predominance of heat conduction as the principal heat transport mechanism.

Model 3 shows a distinct increase of 9 mW m^{-2} at the left foot of the ridge (7.3 mW m^{-2} at the right foot of the ridge) and a distinct decrease of heat flow density (9 mW m^{-2}) over the peak of the modelled ridge (Fig. 7). The point we wish to demonstrate by this analysis is the difficulty of identifying fluid flow from geothermal anomalies alone. The lack of elevated gas content in bottom waters above the gas hydrate layer of the Makran accretionary prism leads us to conclude that the moderate noise we see in the heat flow density data around the trend curve (Fig. 2) is primarily caused by topographic effects, which locally redistribute the heat flow.

Conclusions

The numerical modelling of the thermal field of the accretionary prism suggests the following key points:

(1) high sedimentation rates in the abyssal plain regionally reduce the heat flow density by about 10 mW m^{-2}.

(2) The geothermal field of the Makran accretionary prism is dominated by conductive heat transport. Advective or convective fluid flow and frictional heating along the décollement zone do not appear to influence significantly the temperature field of the prism.

(3) This point is further substantiated by the fact that no evidence for anomalous high heat flow in the prism was found either via direct measurement or calculated from BSR depths.

(4) The uneven sea-floor topography induces heat flow density variations of the order of less than 10 mW m^{-2}.

(5) The absence of gas plumes at water depths of >800 m (except for a local gas seep, where an erosive canyon has significantly reduced the thickness of the gas hydrate layer and faults through accretionary ridges formed conduits) implies a low permeability of the gas hydrate layer for fluids.

References

BRITISH OCEANOGRAPHIC DATA CENTRE 1994. *GEBCO (General Bathymetric Chart of the Oceans) Digital Atlas (CD-ROM)*. British Oceanographic Data

Centre, Birkenhead.

CARSLAW, H.S. & JAEGER, J.C. 1959. *Conduction of Heat in Solids.* Oxford University Press, London.

DAVIE, M.K. & BUFFETT, B.A. 2001. A numerical model for the formation of gas hydrate below the seafloor. *Journal of Geophysical Research*, **106**, 497–514.

DEANGELIS, M.A., LILLEY, M.D., OLSON, E.L. & BAROSS, J.A. 1993. Methane oxidation in deep-sea hydrothermal plumes of the Endeavour segment of the Juan de Fuca Ridge. *Deep-Sea Research I*, **40**(6), 1169–1186.

ENGLEZOS, P. & BISHNOI, P.R. 1988. Prediction of gas hydrate formation conditions in aqueous electrolyte solutions. *Journal of the American Institute Chemical Engineers*, **34**, 1718–1721.

FRUEHN, J., WHITE, R.S. & MINSHULL, T.A. 1997. Integral deformation and compaction of the Makran accretionary wedge. *Terra Nova*, **9**, 101–104.

GNOS, E., IMMENHAUSER, A. & PETERS, T. 1977. Late Cretaceous/Early Tertiary convergence between the Indian and Arabian plates recorded in ophiolites and related sediments. *Tectonophysics*, **271**, 1–19.

HARMS, J.C., CAPPEL, H.N., FRANCIS, D.C. & SHACKELFORD, T.J. 1983. Summary of the geology of the Makran coast. *In:* BALLY, A.W. (ed.) *Seismic Expression of Structural Styles.* American Association of Petroleum Geologists, Studies in Geology, **15**(3), 3.4.2-173–3.4.2-177.

HUNTING SURVEY CORPORATION LTD 1960. *Reconnaissance Geology of Part of West Pakistan.* Maracle Press, Oshawa, Ontario.

HUTCHISON, I., LOWDEN, K.E., WHITE, R.S. & VON HERZEN, R.P. 1981. Heat flow and age of the gulf of Oman. *Earth and Planetary Science Letters*, **56**, 252–262.

KAUL, N., ROSENBERGER, A. & VILLINGER, V. 2000. Comparison of measured and BSR-derived heat flow values, Makran accretionary prism, Pakistan. *Marine Geology*, **164**, 37–51.

KOPP, C., FRUEHN, J., FLUEH, E.R., REICHERT, C., KUKOWSKI, N., BIALAS, J. & KLAESCHEN, D. 2000. Structure of the Makran subduction zone from wide-angle and reflection seismic data. *Tectonophysics*, **329**(1-4), 171–191.

LEHNER, P., DOUST, H., BAKKER, G., ALLENBACH, P. & GUNEAU, J. 1983. Makran fold belt, profile N1804. *In:* BALLY, A.W. (ed.) *Seismic Expression of Structural Styles.* American Association of Petroleum Geologists, Studies in Geology, **15**(3), 3.4.2-81–3.4.2-91.

MINSHULL, T.A. & WHITE, R.S. 1989. Sediment compaction and fluid migration in the Makran accretionary prism. *Journal of Geophysical Research*, **94**(B6), 7387–7402.

PAGE, W.D., ALT, J.N., CLUFF, L.S. & PLAFKER, G. 1979. Evidence for the recurrence of larger magnitude earthquakes along the Makran coast, of Iran and Pakistan. *Tectonophysics*, **52**, 533–547.

RAZA, H.A., RIAZ, A. & ALI MANSHOOR, S. 1990. Pakistan offshore: an attractive frontier. *Pakistan Journal of Hydrocarbon Research*, **2**(2), 1–42.

ROESER, H.A. & SHIPBOARD SCIENTIFIC PARTY 1997. *The Makran Accretionary Wedge off Pakistan: Tectonic Evolution and Fluid Migration—Part 1.* BGR Open File Report, Archive No. **116 643**.

SAIN, K., MINSHULL, T.A., SINGH, S.C. & HOBBS, R.W. 2000. Evidence for a thick gas layer beneath the bottom simulating reflector in the Makran accretionary prism. *Marine Geology*, **164**, 3–12.

SHIPLEY, T.H., HOUSTON, M.H., BUFFLER, R.T., SHAUB, F.J., MCMILLEN, K.J., LADD, J.W. & WORZEL, J.L. 1979. Seismic evidence for the widespread occurrence of possible gas-hydrate horizons on continental slopes and rises. *AAPG Bulletin*, **63**, 2204–2213.

VON RAD, U. & SCIENTIFIC SHIPBOARD PARTY 1994. *PAKOMIN—Influence of the Oxygen Minimum Zone on the Sedimentation at the Upper Continental Slope off Pakistan (NE Arabian Sea), Research Cruise SO-90 with R.V. Sonne.* BGR Report, Archive No. **111987**.

VON RAD, U., BERNER, U., DELISLE, G. & 8 OTHERS 2000. Gas and fluid venting at the Makran Accretionary Wedge off Pakistan. *GeoMarine Letters*, **20**, 10–19.

WHITE, R.S. 1982. Deformation of the Makran accretionary sediment prism in the Gulf of Oman (northwest Indian Ocean). *In:* LEGGETT, J.K. (ed.) *Trench and Fore-arc Geology: Sedimentation and Tectonics on Modern and Ancient Active Plate Margin.* Blackwell Scientific, Oxford, 357–372.

WHITE, R.S. & LOUDEN, K.E. 1983. The Makran continental margin: structure of a thickly sedimented convergent plate boundary. *In:* WATKINS, J.S. & DRAKE, C.L. (eds) *Studies in Continental Margin Geology.* American Association of Petroleum Geology, Memoirs, **34**, 499–518.

A summary of the geology of the Iranian Makran

G. J. H. MCCALL

44 Robert Franklin Way, South Cerney GL7 5UD, UK (e-mail: McCall@freenetname.co.uk)

Abstract: The Iranian Makran has been entirely mapped geologically on a scale of 1:250 000, except for a narrow coastal strip, which exposes the very youngest Cenozoic sediments of the main Makran accretionary prism. The geology of the Makran is less widely known than the geology of Oman, because it has been published in detail only in reports of the Geological Survey of Iran. There is no extension of the geological formations of Oman into the Makran, the only extension of Oman ophiolitic formations into Iran being at Neyriz and Kermanshahr, hundreds of kilometres to the NW. This summary is based on field mapping, photo-interpretation being used only to connect traverse lines. The oldest rocks are metamorphic rocks, which form the basement to the Bajgan–Dur-kan microcontinental 'sliver', a narrow block that extends hundreds of kilometres from the Bitlis Massif in Turkey, through the Sanandaj–Sirjan Block of the Zagros, to north of Nikshahr in the east of the Makran. Other metamorphic rocks form the Deyader Complex near Fannuj on the southern margin of the Jaz Murian Depression. These include blueschists, and are thought to form the tip of the Tabas Microcontinental Block, largely exposed north of the depression. There is also a small microcontinental block to the east, the Birk Block, which exposes only Cretaceous platform limestones and Permian sediments. The Bajgan Metamorphic Series are overlain, with a tectonized unconformable contact, by highly deformed and disrupted platform carbonates of Early Cretaceous to Early Paleocene age (Dur-kan Complex), containing tectonic inliers of Carboniferous, Permian and, rarely, Jurassic age. Ophiolites occur in two structural positions. South of the Bajgan–Dur-kan Block, the tectonic Coloured Mélange of the Zagros continues eastwards inland of the Bashakerd Fault; this includes two layered ultramafic complexes, one with chromitites. The blocks forming the mélange include radiolarites and deep-water limestones of Jurassic to Early Paleocene age. Ophiolites developed north of the microcontinental block form three distinct igneous complexes, two layered and one with intermediate sheeted dykes. Intercalated in the volcanic rocks of these ophiolites are radiolarites and deep-water limestones ranging in age from Jurassic to Paleocene time. There are small developments of Cretaceous sediments carrying rudists in the extreme NW of the inner ophiolite tract. In the NE, ophiolites are developed in the Talkhab Mélange. All these ophiolites represent former, largely Cretaceous, tracts of deep ocean. The Cenozoic rocks form two immense accretionary prisms. The main Makran prism includes Eocene–Oligocene and Oligocene–Miocene flysch turbidite sequences, estimated as individually >10 000 m thick. Above these sequences, there is an abrupt passage up without any apparent unconformity, through reefal Burdigalian limestones, and locally a harzburgite conglomerate development, into neritic sequences with minor turbidites, extending into the Pliocene units. The Saravan accretionary prism to the east repeats tectonically three thick flysch turbidite sequences of Eocene–Oligocene age, but younger sediments are restricted here to minor Oligocene–Miocene conglomerates, unconformable on the above sequences. There is a line of Oligocene(?) granodiorite bodies within the Saravan accretionary prism. Intense folding and development of schuppen structure, dislocation and mélanging of the sediments affected the entire region in Late Miocene–Early Pliocene time. Post-tectonic uplift was followed by scattered developments of fanglomerates beneath the fault scarps. The Neogene deformation has obscured earlier deformational events. There is unconformity beneath Eocene sediments representing a mid-Paleocene disturbance. There is also evidence of a discontinuity in mid-Oligocene time. Pliocene–Pleistocene fanglomerates are unconformable on folded rocks. There are discontinuous developments of Eocene–Oligocene neritic sediments unconformably above the older rocks (ophiolites, platform limestones, metamorphic rocks), and to the north of the southern edge of the Jaz Murian Depression, the northern limit of the Makran, there is evidence of the survival here of a very shallow sea through Neogene time and the formation of small patches of reefal Oligocene–Miocene limestones, and Eocene to Pliocene shallow-water clastic sediments. A 150 km wide tract separates the coast from the trench, the total Cenozoic accretionary prism being 500 km wide. Extension from the Murray Ridge affects the extreme east of the region. The Saravan accretionary prism, it is suggested, faced a gulf, comparable with the Gulf of Oman, and this Saravan Gulf filled up and closed up by Early Oligocene time. Seismological evidence suggests that there is now active continental collision continuing along this suture.

The geology of Oman inland from the Arabian coast of the Gulf of Oman is much better known than that of the opposing Iranian Makran coast, having been widely published. The regional mapping of the geology inland from the Makran coast is, however, now almost complete except for a narrow southern coastal strip and an area near Iranshahr, but it has been published in a number of Geological Survey of Iran reports (McCall 1985a, b, c, d, e; McCall & Eftekhar-nezad 1993, 1994) in Tehran and these, although lodged in two libraries in London (at the Geological Society and the Natural History Museum) and one at Stanford, USA, are not widely accessible.

There is no extension of the geological formations of Oman into the Makran, because they are separated by an ocean basin, a relic of Neotethys. The Makran represents a separate subduction complex, active with three hiatuses from Cretaceous time to the present day. There is, however, extension of the Oman geology into Iran some hundreds of kilometres to the NW of the Makran, where the Semail Ophiolite Complex passes NW into Iran, to be represented at Neyriz and Kermanshah, a sector of continental collision (Fig. 1). The oceanic tract terminates to the NW at the Straits of Hormuz. The Persian Gulf occupies a flexural downwarp behind the underriding edge of the Arabian plate. The tract between the Gulf and the Zagros exposes folded Miocene sediments above the Arabian plate. Only the formations of the Zagros continue southwards into the Makran, to the east of the collision zone.

There is an abrupt change in the tectonic style just north of the town of Minab (Fig. 2) in the Makran and this has given rise to the concept of the 'Oman Line', a supposed significant geotectonic boundary running from that point northeastward through Iran. Regional mapping of the Makran failed to reveal any indication that such a boundary exists as a tectonic feature: at this junction the anticlines and salt domes of the hydrocarbon-rich tract west of the Zagros give way to a converse pattern of emphasized synclines and subordinate anticlines, with no salt domes at all. The structure of the Makran is also dominated by faults dipping east, north or NE, inwards away from the coast, forming a schuppen (duplex) structure. The change in tectonic style to a mirror image of the Zagros style has been attributed by the author (McCall 1985e, pp. 499 and 572) to the change from the colliding Arabian continental crust passing under the Zagros to the oceanic crust of the Indian Ocean and Oman Sea passing under the Makran. Thus the geology of the greater part of the Makran is that of a subduction zone of long history. None the less, there are complications at the eastern end bordering on Pakistan, which are believed by the author to involve another Cenozoic accretionary prism in addition to the main frontal prism of the Makran.

Geological mapping by Paragon–Contech, a joint venture set up for the purpose of mapping the Makran: logistics

The greater part of the geological mapping of the Makran was carried out on contract from 1976 to 1978 by a joint venture between an Iranian company (Paragon) and an Australian company (Contech): this will be referred to as Paragon–Contech (PC). Of the area shown with no ornament in Fig. 2, PC mapped all but the four 1:100 000 sheets bounded by dotted lines, the remainder being mapped by the Geological Survey of Iran (GSI). The cross-hatched areas in the extreme NE were mapped by Intercon–Texas Instruments (ITI). The Narreh-Now quadrangle to the north of 26°N latitude, mapped by ITI, extends the geology along strike from the Saravan quadrangle to the south and thus can be regarded as completing the Makran mapping. A small and rather complex area of outcrop to the north of Iranshahr (Fig. 2) has not, so far as is known, been yet mapped in any detail, except for isolated sheet coverage, and nor has the southern coastal strip.

The Makran is extremely rugged terrain and except in the Saravan quadrangle there are few roads, and what there are consist of little more than rough tracks. Sealed bitumen roads only traverse the Minab quadrangle diagonally from SW to NE (Minab to Kahnuj (see Fig. 4 below) and thence to Jiroft (see Fig. 30 below)) and the Nikshahr quadrangle from north to south (Iranshahr to Chahbahar on the coast; Fig. 2). Because of the inaccessibility and ruggedness of the terrain, the mapping by PC was by means of helicopter location of foot-traverses by geologists on the ground. Some mapping in monotonous terrain was carried out by means of traverses involving repeated landing of helicopter-borne teams and rapid inspection of sites (up to 70 sites were mapped in single traverses), and a few vehicle-borne road traverses were also carried out, close to the two base camps (at Kam Sefid and Sarbaz; Fig. 3). Maps were entirely compiled in draft in the field, but were then reprocessed adding photo-interpretation in Tehran to fill in detail between ground traverse lines, which were very closely spaced (about 2 km), so that the maps produced represent a reliable record of the geology on 1:250 000 scale. In addition, nine sheets in key areas were published on 1:100 000 scale with detailed side and back notes (Now-

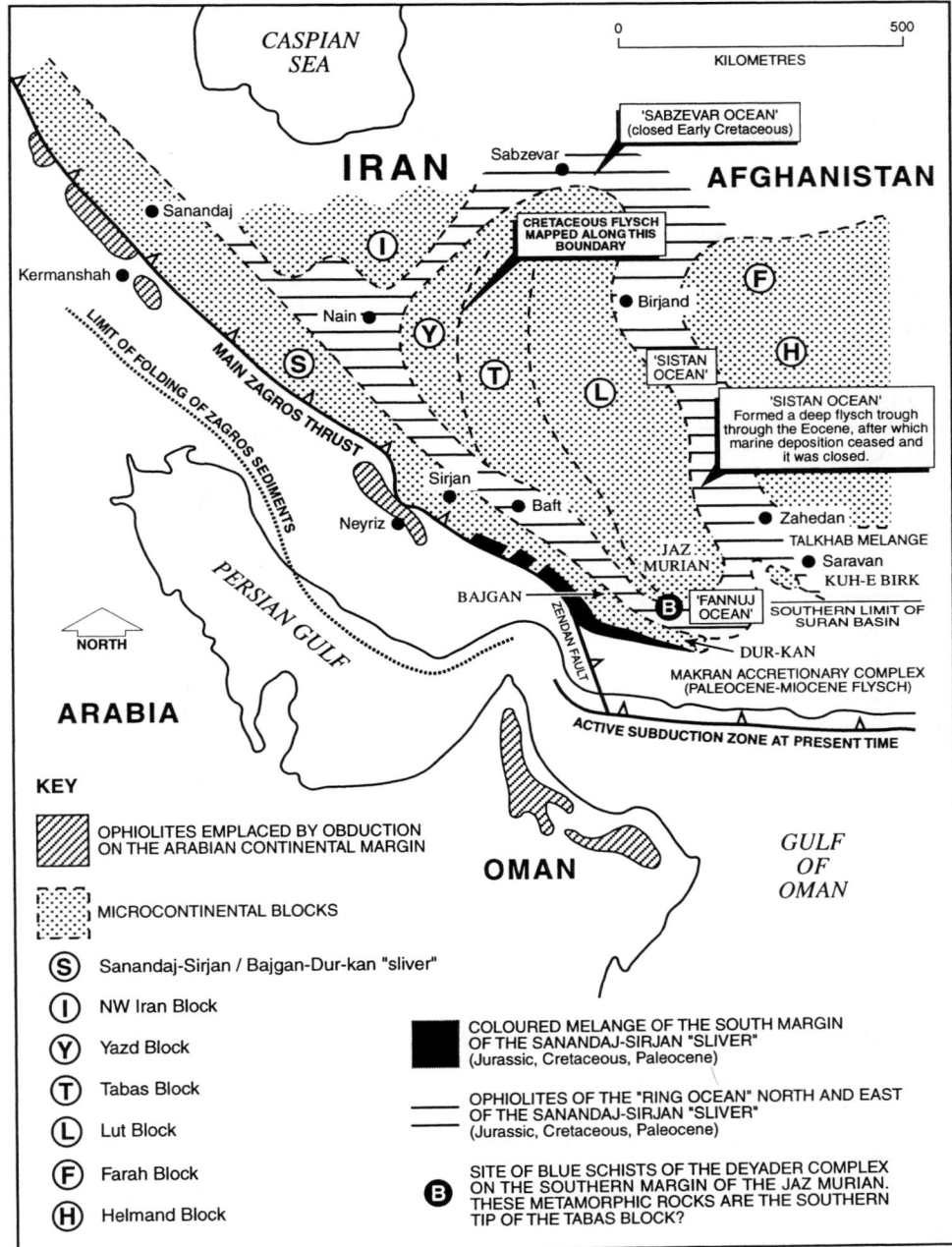

Fig. 1. Diagram showing the main geotectonic elements in southern Iran (from McCall 1997).

Dez, Kahnuj, Minab, Qal-eh Manujan, Dur-kan, Dar-Pahn, Avartin, Remeshk and Ramak). The four 1:100 000 sheets mapped by GSI were mapped on the ground without the aid of helicopters. Two of these, Fannuj and Espakeh, have been published. The mapping to complete the Saravan 1:250 000 quadrangle and of the Narreh-Now quadrangle by ITI was carried out by ground traverses. The project covered the entire Makran, except for the area of outcrop north and west of

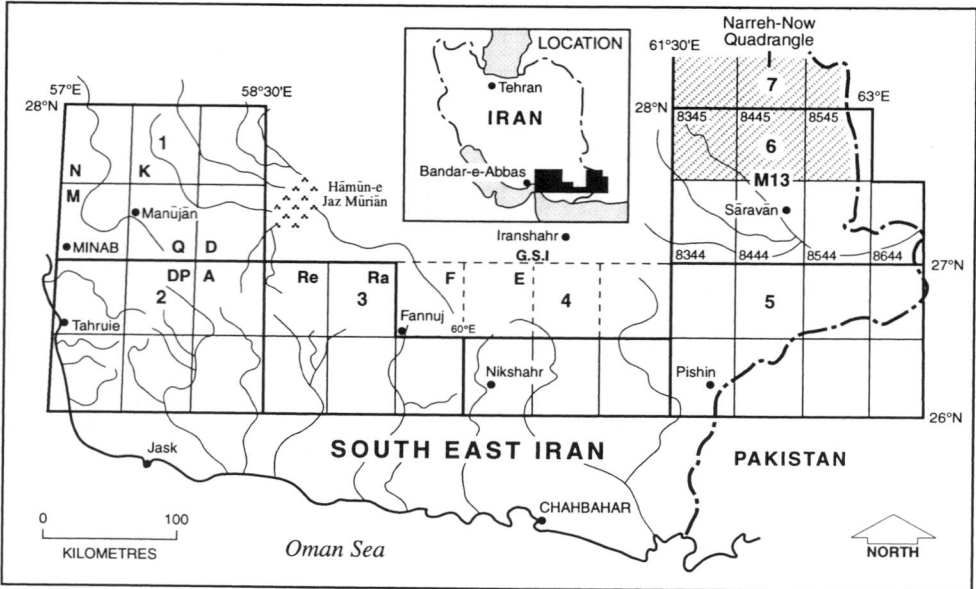

Fig. 2. The four full and two half 1:250 000 quadrangles mapped by Paragon–Contech, the four 1:100 000 sheets mapped by the GSI (dashed boundaries), and the two half-quadrangles mapped by Intercon–Texas Instruments (cross-hatched). 1, Minab quadrangle; 2, Tahruie quadrangle; 3, Fannuj quadrangle; 4, Nikshahr quadrangle; 5, Pishin quadrangle; 6, Saravan quadrangle; 7, Narreh-Now quadrangle. The 100 000 sheets: N, Now Dez; K, Kahnuj; M, Minab; Q, Q'al eh Manujan; D, Dur-kan; DP, Dar Pahn; A, Avartin; Re, Remeshk; Ra, Ramak; F, Fannuj; E, Espakeh.

Iranshahr and between the Saravan quadrangle and the Jaz Murian desert; also the immediate coastal strip in the south, which is known to expose only Miocene to Pliocene shallow-water sediments in the bedrock. Part of this coastal strip in the west near Jask (Fig. 2) had been mapped in reconnaissance fashion by Stocklin (1952) and also by the Italian hydrocarbon explorers AGIP/SIRIP (1962).

Six quadrangles (Minab, Tahruie, Fannuj, Nikhshahr, Pishin and Saravan) (Fig. 2) are covered by detailed reports (McCall, 1985a, b, c, d; McCall & Eftekhar-nezad 1993, 1994), which are accompanied by maps on 1:250 000 scale. The seventh (Narreh-Now) quadrangle is also covered by a map on the same scale (Geological Survey of Iran 1987). An overall report was prepared on the areas mapped by PC (McCall 1985e) and there is also an unpublished report on the mineral resource potential held at the GSI in Tehran. All this material, except for the last report, is lodged at the Geological Society Library, London, and the Natural History Museum Library, London.

The previous work was very limited: it was summarized by McCall (1985e, pp. 44–46). Details of the palaeontological determinations have been given by McCall (1985e), as well as in the individual specialist reports. Details of the geochemical anlayses of the igneous and sedimentary rocks, and interpretations, have also been given by McCall (1985e), pp. 437–462).

There are also follow-up publications on special aspects, based on this mapping: the broader geotectonic picture (McCall & Kidd 1982); carbonate deposition on the Tertiary accretionary prism and Miocene coral fauna (McCall et al. 1994); the logistics of the project (McCall & Simonian 1986); trace fossils (Crimes & McCall 1994); the coloured and sedimentary mélanges (McCall 1983); the Plio-Pleistocene fanglomerates (McCall 1996). A second geotectonic appraisal (McCall 1997) related the Makran mapping to the geology to the north in southern Iran.

In the following text, figures showing a number of key areas of the Makran are presented. This method is adopted because of the very large area covered in this summary; a map of the entire area would be of too small a scale to show important detail. A structural and solid geology map in two sheets on a scale of 1:500 000 is contained in the overall report (McCall 1985e). The location of the areas covered by the figures cited below is given in Fig. 3.

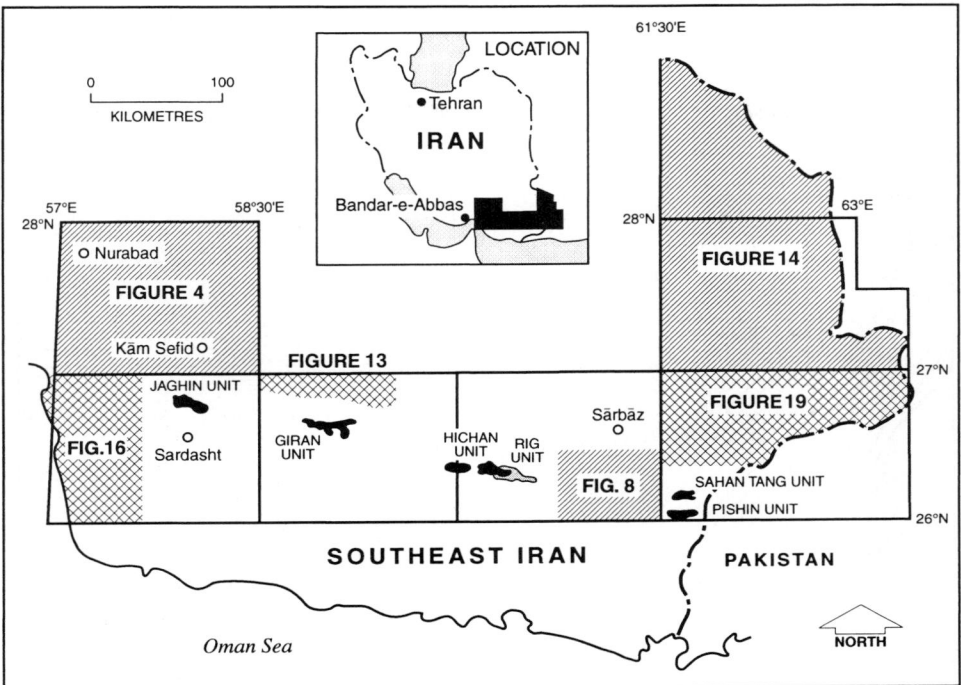

Fig. 3. Diagram showing the areas represented in Figs. 4, 13, 14, 17 and 19, the outcrop areas of some units not shown on those diagrams and the location of the two base camps, Kam Sefid and Sarbaz, also Sardasht and Nurabad.

Geology of the Makran

The geology can be divided in two parts: (1) the pre-mid-Paleocene geology; (2) the Paleogene, Neogene and Quaternary geology. Essentially, the first is restricted to the region north of the Bashakerd Fault (Fig. 4), which is a continuation swinging eastward of the Zagros Fault (the Zagros Fault diverges into this fault and the Zendan Fault (Fig. 4) running SSE east of Minab). There are exotic blocks of Mesozoic and older rocks within the Cenozoic flysch turbidites and neritic sediments to the south of the Bashakerd Fault, but continuous development of these units is restricted to the north of this fault. Likewise, there are outliers of shallow-water Eocene–Miocene sediments to the north of the Bashakerd Fault, resting unconformably on the older rocks, but continuous sequences of the flysch and neritic sediments occur only south of this major lineament.

The pre-Mid-Paleocene geology

This can be subdivided into seven essential components: (a) the Bajgan Metamorphic Complex and the overlying Dur-kan Complex; (b) the Coloured Mélange Complex; (c) the Band-e Zeyarat–Dar Anar, Ganj and Remeshk–Mokhtarabad Ophiolite Complexes; (d) the Deyader Metamorphic Complex; (e) the Kuh-e Birk Limestones; (f) minor outcrops of Permian sediments and volcanic rocks within the Saravan quadrangle (the Morghak Unit); (g) the Talkhab Mélange. These geological formations are listed with their respective age range in Table 1.

The Bajgan Complex and overlying Dur-kan Complex. The Bajgan Complex (Fig. 4) is a development of metamorphic rocks that forms a continuation of the Sanandaj–Sirjan Block, which runs northwestwards through the Zagros and connects up with the Bitlis metamorphic units in eastern Turkey, thus forming a narrow sliver-like microcontinental block several thousand kilometres in length. The development of the metamorphic rocks is wide in the Now-Dez 1:100 000 sheet in the extreme NW of the Makran, but tapers down to a narrow zone going eastwards and terminates in discontinuous lenses near Avartin at 58°20′E, being tectonically occluded. There is, however, indirect evidence from the distribution of

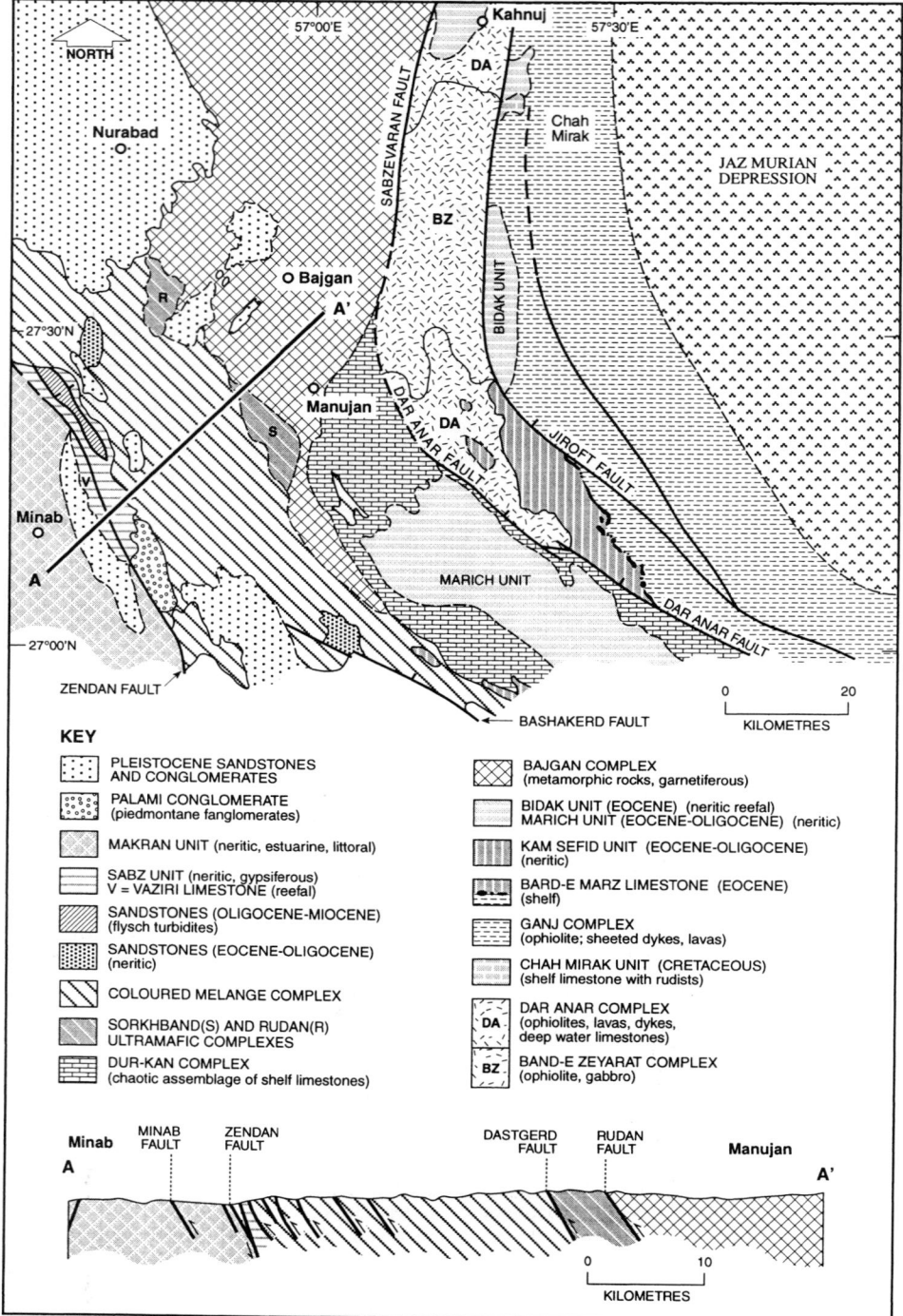

Fig. 4. Generalized diagram showing the field relationships of the Lower Paleocene and older rocks and their Cenozoic unconformable cover in part of the Minab quadrangle (scale 1:500 000; section approximately twice that scale).

Table 1. *List of the geological formations of the Iranian Makran of Early Paleocene age and older*

Formation	Age
Birk Unit	Late Cretaceous
Chah Mirak Unit	Late Cretaceous
Ganj Complex (O)	Late Cretaceous
Band-e Zeyarat Complex–Dar Anar Complex (O)	Early Cretaceous–Early Paleocene
Remeshk Complex–Mokhtarabad Complex (O)	Jurassic–Early Paleocene
Talkhab Mélange Complex (O)	Cretaceous
Coloured Mélange Complex (O)	Jurassic–Early Paleocene
Dur-kan Complex	Carboniferous, Permian, Jurassic, Cretaceous, Early Paleocene
Morghak Unit	Permian
Deyader Complex (M)	Early Palaeozoic or older
Remeshk Complex (M)	Early Palaeozoic or older

O, ophiolite; M, metamorphic.

the overlying Dur-kan Complex that this metamorphic foundation continues, at depth, at least to just beyond 60°E, within the Nikhshar quadrangle. The Bajgan Complex consists of contorted phyllites, phyllonites and schists (pelites, semipelites and psammites), calc-silicate rocks, marbles, magnesian and iron-rich rocks, abundant metavolcanic rocks, and basic and acid intrusive rocks. Serpentinites also occur within it, but it is uncertain whether they are part of the metamorphic complex or later emplacements. Metamorphism is low, greenschist facies in the south; but increases to amphibolite facies in the north (garnets are abundant), and glaucophane occurs sparingly in basic rocks and phyllonites. The complex is overlain by Jurassic sediments to the north of the Makran, and must be of either Early Palaeozoic or Precambrian age, as Carboniferous and Permian limestones are represented in the overlying Dur-kan Complex. Attempts at radiometric age dating of the extension of this complex by the K–Ar method produced only Cretaceous (Senonian) ages (two samples; see McCall 1985e, p. 81), which must have been superimposed on the much older complex. To the north of the PC project area, an early Palaeozoic age was obtained by the same method (see McCall 1985a, p. 291). There is also some evidence of a Devonian age in reported fossil finds to the north (see McCall 1985a, p. 291).

The Dur-kan Complex (Fig. 4) overlies the Bajgan Complex. Between Manujan and Avartin it forms a synclinal structure, being obscured in the centre of this by Eocene to Oligocene shallow-water sediments. It continues eastwards in a narrower outcrop belt, still patchily overlain by Eocene sediments, with its last outcrops in the Nikshahr quadrangle, within the Espakeh 1:100 000 sheet. The complex is a tectonic association of Lower Cretaceous to Paleocene shallow-water sediments (limestones dominant) and igneous rocks, with tectonic inliers or rafts of Upper Palaeozoic and Jurassic shelf sediments, mainly limestones. Permian fossils are fairly abundant in the limestones. It is essentially a tectonically mixed assemblage derived from a sequence of shelf deposits overlying the metamorphic units of the microcontinental sliver.

Lithologies of the Dur-kan Complex include massively bedded, recrystallized, white and grey or fawn coloured, commonly brecciated, clastic biomicrite and biomicrudite limestones. Interbedded with these are sequences of sandstone, shale, mudstone, polymictic conglomerate and breccia. In the eastern part of the outcrop, pillow lava, chert, hyaloclastite, lapilli tuff, volcanoclastic sediments and minor mafic and ultramafic intrusions are prominent, although such rocks are only sparse components of the complex in the west. It would require very detailed mapping to derive any sort of internal stratigraphy in this tectonically disturbed complex, and indeed it might be impossible. Echinoid fragments, solitary corals, brachiopod shell fragments and algae have been recorded, but age determinations were based on benthic foraminifera (for details, see McCall 1985e, pp. 125–126). There is an absence of Coniacian–Santonian and Turonian limestones. Jurassic sediments were found in only one locality. There were clearly gaps in the long history of shelf deposition.

The Coloured Mélange Complex (Fig. 4). The Coloured Mélange Complex crops out immediately north of the Bashakerd Fault. It is a tectonic mélange (McCall 1983), a chaotic mixture (Fig. 5) of cognate pelagic sediments (cherty and argillitic radiolarites (Fig. 6), pink micritic *Globotruncana* limestone, rare globigerinid oozes (packed biomicrites) and pillow lavas (mainly vesicular olivine basalts) (Fig. 7)). Some andesites, andesitic tuffs,

Fig. 5. The Coloured Mélange Complex exposed in a hillside, Fannuj quadrangle. The variegated hues of the rocks exposed in this hillside range from red to mauve and green, hence the term. It is a chaotic assemblage of large blocks of igneous and sedimentary rocks.

Fig. 6. Radiolarian chert within the Coloured Mélange Complex, Tahruie quadrangle.

rhyodacites, associated welded tuffs and hyalocastites are also recorded. The pillow lavas exhibit only normal sea-floor hydrothermal metamorphism, rarely more than zeolite facies. Exotic components include serpentinite, reefal limestone and other minor lithologies such as sandstone and amphibolite. The serpentinite is derived from a range of peridotite types (dunite, harzburgite, wehrlite, lherzolite and websterite); the amphibolite, derived from gabbro, fringes small serpenti-

Fig. 7. Pillow lava showing entrail-like forms, within the Coloured Mélange Complex, Tahruie quadrangle.

nite bodies. Trondhjemite is a minor component, derived from ophiolite sequences. There is no matrix in the strict sense of a continuous host to blocks (and certainly not the argillaceous matrix of the Franciscan mélange) (Hsu 1971); Sabzehei (1974) did refer to the cognate assemblage as 'matrix', but this is confusing: the formation consists of a jumble of blocks of the various lithologies. The serpentinites, which may exhibit rodingite selvages, do not form a continuous matrix, except in rare cases, but where they are abundant the mélange is increasingly chaotic. Clearly, they did exert a lubricating role, as they did together with the argillaceous matrix in the Franciscan mélange.

Ultramafic rock types also occur as two large masses within in the mélange, the largest being the Sorkhband body west of Manujan, which is 17 km long and 9 km wide and includes layered ultramafic rocks and chromitite (Fig. 8), minor pyroxenite and gabbro. The smaller Rudan body to the north of it consists of layered harzburgite, with minor serpentinized dunite, websterite and lherzolite. These bodies are both grossly and finely layered, but cumulate textures are evident only in the chromitites of the Sorkhband body, a tectonite fabric having overprinted the cumulate texture in the other layered lithologies. No cumulate textures were recognized in the Rudan body. These bodies are regarded as dismembered ophiolites.

Fig. 8. Layered chromitite, Sorkhband Ultrabasic Series within the Coloured Mélange Complex, Minab quadrangle.

Blocks in the mélange may be up to several kilometres long, the largest being of limestone and chert, and where the mélange is less chaotic and has remained largely sequential, these lithologies may form definite strata up to several kilometres long.

Rare exotic blocks include Albian *Orbitolina* limestone, phyllite, schist and amphibolite, of lithologies identical to rocks in the adjacent Durkan and Bajgan Complexes (see above). The most likely origin is tectonic spalling off at the boundary. Mapping showed that the mélange becomes increasingly chaotic from this inner boundary outwards.

Paleocene packed biomicrites give an upper age limit to the cognate mélange rocks, which are unconformably overlain by Eocene limestones of shelf facies and flysch-type shallow-water sediments. The mélange was formed before Eocene time. Campanian–Maastrichtian *Globotruncana* microfaunas dominate the pink micrites, but there are also Cenomanian, Turonian, Coniacian and Santonian microfaunas. These biomicrites occur interleaved in pillow lavas so that these must also be of late Cretaceous age. The radiolarites range from Jurassic (Pliensbachian) to Coniacian age, despite the fact that they are also intimately associated with pillow lavas and the biomicrites. This age discrepancy is unexplained. Details of palaeontological determinations have been given by McCall (1985e, pp. 101–104).

The mélange is ophiolitic in character, and the cognate sediments are of deep-water facies. The complex must represent a subduction zone active in Late Cretaceous and Early Paleocene time, and probably consists of offscraped slabs of oceanic crust. Although olistostromes may have been fortuitously caught up in the mélange, sedimentary olistostrome formation was not a significant process in producing the chaotic character of this mélange. This tectonic mélange has extensions into the Zagros and is interpreted as having been formed at the subducting NE margin of the southern arm of the Neo-Tethys ocean.

The Band-e Zeyarat–Dar-Anar, Ganj and Remeshk–Mokhtarabad Ophiolites. Three ophiolites occur within the area mapped by PC, on the inner NE side of the microcontinental sliver of the Bajgan Complex and Dur-kan Complex: (1) the Band-e Zeyarat Complex and Dar-Anar Complex (Fig. 4) are respectively the plutonic (gabbro, minor serpentinite and trondhjemite) lower part, and lava with intercalated sediment upper part of the same ophiolite; (2) the Ganj Complex is faulted against the above, and between them and the Jaz Murian desert; it is dominated by sheeted dykes and screens of lava between them; (3) the Remeshk Complex and Mokhtarabad Complex are respectively the plutonic lower part and the lava with intercalated sediment upper part of the same ophiolite.

The Band-e Zeyarat Complex and Dar-Anar Complex (Fig. 4). The Band-e Zeyarat Complex forms a whale-back feature NE of Manujan. The outcrop, bounded on the west by the Sabzevar Fault and on the east by the Jiroft Fault, covers 50 km × 15 km. Four major subdivisions were mapped in it, from the bottom up: serpentinized dunite and hornblende peridotite, forming large bodies on the eastern margin; banded leuco- and melagabbro, minor troctolite and dunite (the 'low level' gabbro) (Fig. 9); uralitized gabbro, hornblende gabbro, minor diorite and trondhjemite, accompanied by diabase and diorite dykes, locally sheeted (the 'high-level' gabbro); above these lie leucograbbro, trondhjemite, diorite and minor anorthosite. The high- and low-level gabbros form the greater part of the outcrop. Cumulate layering, with rhythmic and grading patterns, is characteristic of the low-level gabbro, but the layers are commonly lensed. The high-level gabbro does not display cumulate layering. There are areas where diabase sheeted dykes predominate; in one such area at the northern nose, in a zone up to 2 km wide, there is abundant net-veined rock (Fig. 10) and agmatite, as a result of trondhjemite and microtrondhjemite intrusion. The geochemistry of the suite is tholeiitic and K-poor, not calc-alkaline. Mohamad Ghazi (pers. comm.) has reported mid-ocean ridge basalt (MORB) composition in this complex. The complex is fringed by the Dar-Anar Complex at its northern and southern ends. Age dating by the K–Ar method on hornblende pegmatite and diorite rocks yielded Neocomian (Early Cretaceous) ages (142–140 Ma: Mahomed Ghazi, pers. comm.); by inference drawn from the Dar-Anar Complex (see below), an Early Cretaceous–Paleocene age is inferred.

The Dar-Anar Complex (Fig. 4) consists of pillowed and non-pillowed basalt to andesite flows and pillow breccias, interleaved in varying proportions with sediments: pelagic micrite and biomicrite limestone with subordinate red chert, arkose, calc-wacke, sandstone, siltstone and shale. It is intruded by numerous diabase dykes as well as dykes and sills of gabbro, plagioclase porphyry, fine-grained plagiogranite or trondhjemite, and feldspathic wehrlite. Alteration is common in the volcanic rocks, to zeolite or greenschist facies, pillow structure being obliterated. There are some old copper workings in this complex, of unknown age. The boundary with the Band-e Zeyarat Complex is faulted and that with Eocene and Eocene–Oligocene formations unconformable. The com-

Fig. 9. Layered low-level gabbro within the Band-e Zeyarat Complex, Minab quadrangle.

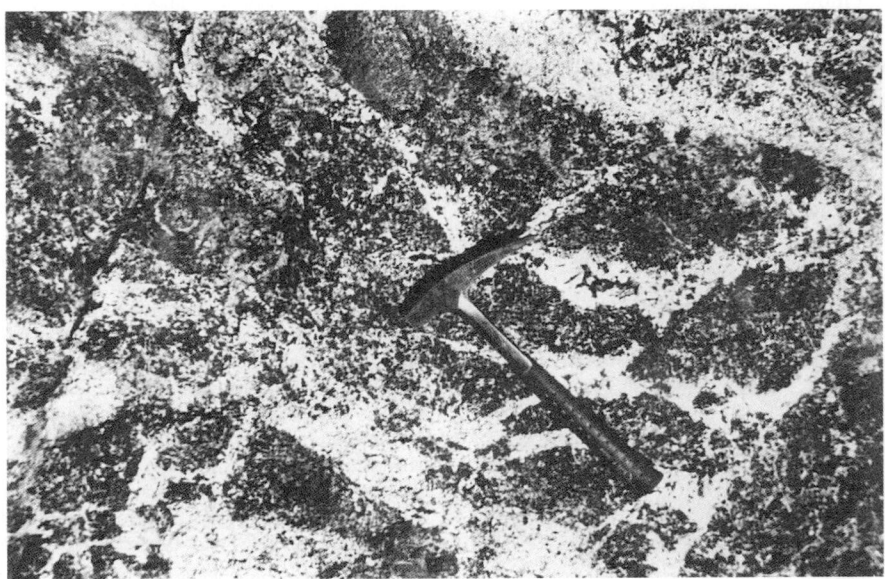

Fig. 10. Trondhjemite net veining of the high-level gabbro, Band-e Zeyarat Complex, Minab quadrangle.

plex itself displays gentle flexuring of the massive flows, whereas the limestones are folded disharmonically with the flows, into tight asymmetrical folds. Some biomicrites contain an Early Paleocene planktonic microfauna, but chert and other biomicrites have yielded three separate Cretaceous microfaunas of Aptian–Albian to Cenomanian, Aptian–Albian and Cenomanian age (for details, see McCall 1985e, pp. 106–107). The complex represents the extrusive element of a major ophiolite development, of which the Band-e Zeyarat Complex and diabase dyke developments in both complexes represent the plutonic and hypabyssal fractions. The two complexes forming this ophiolite sequence occupy an area more than 120 km long and up to 20 km wide. The thickness of the Dar-Anar Complex is about 2000 m.

The Ganj Complex (Fig. 4). This is the most extensive of the ophiolite developments of the inner Makran, occupying a broad discontinuous terrain of dissected hills 165 km long and up to 35 km wide, yet very little of the coarse-grained intrusive rocks of the magma chamber is exposed, the exposures being dominated by sheeted dykes and lava flows. This complex crops out east and NE of the Band-e Zeyarat–Dar-Anar outcrop, being separated by the major Jiroft Fault. On its other margin it is obscured by the superficial deposits of the Jaz Murian desert.

The complex includes a prolific development of sheeted, locally sinuous or lensed (Fig. 11) multiple dykes, plagioclase porphyries of andesite to rhyodacite composition; these intrude country rock that consists essentially of massive, pillowed and brecciated lava flows, the first mostly of intermediate composition whereas the pillowed and brecciated flows are mostly basalt and basaltic andesite. Pillows display glassy margins and zonation of vesicles. The sparse developments of plutonic country rock include plagiogranite (tonalite, trondhjemite, albite granite) and gabbro, mainly associated with it in agmatite. These rocks appear to be comagmatic with the dykes. The lavas are hydrothermally altered to varying degrees up to greenschist facies assemblages, but the dykes, where widely spaced, only to zeolite facies assemblages. The lavas were gently folded before dyke intrusion, but the pillows are nowhere inverted.

There are minor turbidite sediments interleaved in the country rock and these yielded a Campanian to Mastrichtian microfauna (for details of palaeontological determinations, see McCall 1985e, pp. 131–133). The presence of these sediments indicates that the lavas were extruded in a turbidite environment. Eocene and Eocene–Oligocene sediments lie unconformably on the complex. At the extreme north end of the outcrop a sequence of turbidite sandstones, mudstones and conglomerates rich in volcanic detritus lies disconformably above the Ganj Complex; these constitute the Chah Mirak Unit. In this unit are thinly bedded pelagic limestones and shallow-water limestones containing prolific rudists (*Sphaerulites boureau* Toucas) (Fig. 12), algae and foraminifera. One horizon contains numerous nereneid gastropods (*Plesioplocus karabakhensis* Pcelincev). These deposits are of Cenomanian–Santonian age (for details, see McCall 1985e, pp. 96–97). This Cretaceous sequence resembles those of the Bazman Block (Huber 1978), which obscure older continental rocks of the Central 'Lut Block' (now split into the Yazd, Tabas and Lut components; McCall 1997).

It is impossible to determine the original relationships of the Bande-e Zeyarat–Dar-Anar Complex to the Ganj Complex because of the faulting: the Ganj Complex is calc-alkaline (McCall 1985a) and the complexes are certainly separate ophiolite developments, of very different character. Age dating by the K/Ar method yielded a Senonian age for a plagioclase porphyry dyke (one determination) and Albian ages for trondhjemite country rock to the dykes (two determinations) (McCall 1985e, p. 130). Planktonic foraminifera within the turbidite country rock are of Campanian–Mastrichtian age: thus an Albian to Maastrichtian age span is assigned to this complex.

Remeshk Complex and Mohtarabad Complex (Fig. 13). The Remeshk Complex is the highly dismembered plutonic fraction of an immense ophiolite development, the volcanic and sedimentary fraction being the Mohtarabad Complex. Together they extend more than 150 km along strike and have a maximum width of outcrop of 35 km,

Fig. 11. Lenticular sheeted dykes of intermediate (andesitic) composition, Ganj Complex, Minab quadrangle. The screens between the dykes are of basaltic lavas.

Fig. 12. Rudists from the Cretaceous Chah Mirak Unit, Minab quadrangle.

within the Fannuj 1:250 000 quadrangle (Remeshk and Ramak 1:100 000 sheets). They extend eastwards into the Fannuj 1:100 000 sheet and the Espakeh 1:100 000 sheet, both mapped by the GSI, but become largely obscured there by Cenozoic sediments unconformably above them. Unlike the ophiolitic complexes above, these rocks are dissected by faults, flexed into immense open folds and are locally strongly disturbed to form mélange. There are seven subdivisions of the Remeshk Complex, from top to bottom: (7) trondhjemite, minor diorite and tonalite; (6) uralitized gabbro and hornblende gabbro (characterized by *in situ* crystallization textures); (5) gabbro (characterized by *in situ* crystallization textures); (4) leucogabbro, anorthosite (*sensu lato*) and minor troctolite (characterized by cumulate textures); (3) banded and layered felsic troctolite, troctolite, olivine gabbro and gabbro (as above); (2) serpentinized dunite, harzburgite, lherzolite, minor wehrlite, olivine gabbro and troctolite (tectonite and/or cumulate textures); (1) serpentinized ultrabasic rocks (massive and tectonized, with minor exotic blocks).

This complex is believed to represent a great mass of cooling magma, which was almost certainly subdivided into a number of magma chambers. The lowest two divisions are not distinct differentiated layers, but reflect different degrees of serpentinization. Minor rhyolitic dykes, pods and lenses occur at all levels. There is a mixed zone in divisions (5) and (6), where late siliceous phases (tonalite, trondhjemite, diorite and granite) invaded the basic rocks along cracks to produce agmatite and net-veining, although this is less widespread than in the Band-e Zeyarat Complex. The troctolite and leucotroctolite are mostly plagioclase cumulates with intercumulus olivine, but the reverse is also recorded. The anorthosites are plagioclasites, with some adcumulus plagioclase and sparse accessory minerals. The non-cumulate gabbros all display hypidiomorphic granular texture. Static hydrothermal metamorphism to greenschist facies is particularly well advanced in these rocks, and may be essentially deuteric (late stage in the cooling history) alteration. The trondhjemite is allotriomorphic, granular and usually coarse textured, and consists of albite and quartz with accessory hornblende and opaque minerals. That it is definitely the last crystallized layer is shown by field relationships.

The age of the Remeshk Complex, early Cretaceous to Paleocene time, is inferred from the overlying Mokhtarabad Complex. Not enough detailed structural work has been done to determine the age of the folding, but the fold style and field relationships suggest that the deformation relates to the complex being influenced by Neogene listric faulting and folding, evident throughout the Cenozoic accretionary prism. Geochemical analyses (McCall 1985c) carried out on this complex show it to be calc-alkaline, like the Ganj Complex, but it is believed to be a separate ophiolite, interdigitating at the boundary with the Ganj Complex to the west.

Diabase dykes intervene between the Remeshk

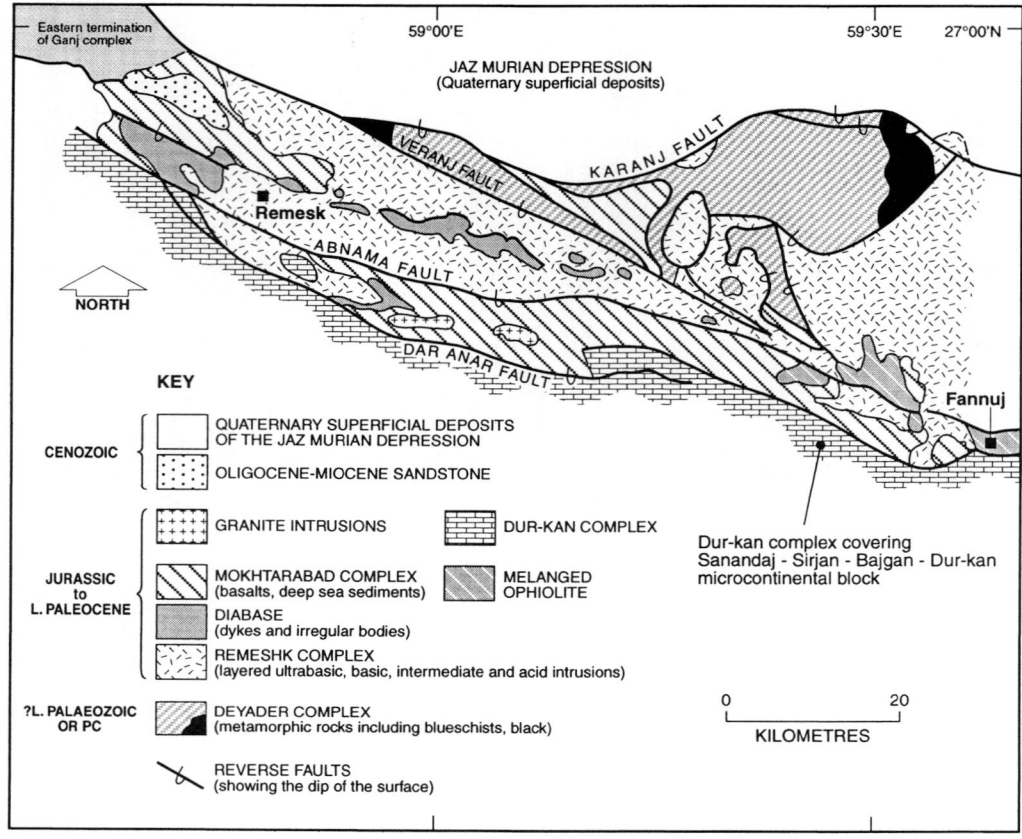

Fig. 13. Generalized diagram showing the field relationships of the Lower Paleocene and older rocks and their unconformable Cenozoic cover in part of the Fannuj quadrangle (scale 1:500 000).

Complex and the Mokhtarabad Complex, but are not as well developed individually or as abundant as in the two ophiolite sequences to the west. Locally they are sheeted with thin screens of country rocks; elsewhere they are multiple but substantial areas of country rocks intervene. The dykes pass up into the Mohtarabad Complex. There are also irregular masses of diabase. The dykes are all metadiabase with ophitic to intersertal texture, and are locally plagioclase-phyric. Hydrothermal metamorphism has produced greenschist facies assemblages. The dykes are believed to be co-magmatic with the plutonic rocks of the Remeshk Complex.

The Mokhtarabad Complex forms the upper part of the same ophiolite development as the Remeshk Complex, but generally has faulted and tectonized contacts with it. The thickness is estimated from the mapping as several thousand metres. The complex consists of massive and pillowed basaltic lava flows with subordinate grey and white recrystallized pelagic limestones, tuffaceous sandstones, red and green chert lenses, pebble and cobble conglomerates, red, green and brown volcanic arenite sandstones, purple manganiferous siltstones and grey calc-shale. The limestone beds are thin and segmented by dykes. The cherts include radiolarites, closely associated with red biomicrites. The entire sequence is traversed by diabase dykes, as well as small irregular intrusions of serpentinite, protoclastic granitoid rock, gabbro, microgabbro and diorite, granodiorite and plagiogranite.

The lavas are mostly porphyritic basalts, with developments of calcite and zeolite. Hydrothermal metamorphism is mainly of zeolite facies, but locally attains greenschist facies with developments of epidote and clinozoisite. A vesicular trachyte was also recorded, as were crystal and lithic tuffs.

Age control has been derived from biomicrite limestones, with two planktonic foraminiferal assemblages yielding Early Paleocene age and Cenomanian age. Others yielded Santonian–Maastrichtian microfaunas. One biomicrite yielded a radiolarian microfauna of Jurassic age. This evidence suggests a deep ocean from Jurassic to Early Paleocene time. Details of the palaeontological determinations have been given by McCall (1985e, p. 187).

The Mokhtarabad Complex is locally sheared, folded and dislocated, but only rarely forms mélange. It is affected by the same large-scale folds that affected the Remeshk Complex. Structurally, synclines are dominant and may be up to 20 km in length. A series of steeply dipping fault blocks, bounded by north-dipping reverse faults, dissect the complex.

The Deyader Complex (Fig. 13). Metamorphic rocks crop out within the Fannuj quadrangle, on the edge of the Jaz Murian desert, in the north of the Remeshk and NW of the Ramak 1:100 000 sheets. Several thousand metres of metamorphic rocks must be represented, everywhere in faulted relationship with other formations. These rocks are derived from volcanic rocks and sedimentary rocks (sands, silts and argillites, with minor limestones) in about equal proportions. The volcanic rocks (olivine metabasalts and meta-andesites) have mostly suffered only hydrothermal metamorphism, and locally preserve pillows. The sediments have been converted to pelites, semipelites, psammites and finely recrystallized limestones. Phyllonites, phyllites and calc-silicate rocks are represented. Pink micrites and cherts form interleaved lenses. In a small area within the Remeshk sheet and a larger area within the Ramak sheet (Fig. 13), the basaltic rocks are fully recrystallized to glaucophane schists. Within the complex, three metamorphic zones are recognized: lawsonite–pumpellyite–albite (over 80%); clinozoisite–albite (greenschist) (very minor); glaucophane schist (blueschist).

The first zone contains many relics of igneous rocks retaining their primary textures. The metamorphism was of high-pressure type characteristic of subduction zones (Jackson 1997) A Late Cretaceous foraminiferal microfauna and an Early Cretaceous radiolarian microfauna (for details, see McCall 1985e, p. 122) are preserved in the micrite and chert, respectively, but these are probably tectonic introductions and the similarity of the rocks to those of the Bajgan Complex suggests

that these are mainly Early Palaeozoic or older rocks. It is probable that they are the extremity of the Tabas block (part of what was once called the Lut Block), representing the north coast of the oceanic tract in which the ophiolites were formed (McCall 1997).

The complex displays complex structural deformation including faults, several large folds and small-scale crenulation, and appears to have suffered two stages of metamorphism. It would be an ideal subject for a detailed structural and metamorphic study.

The Birk Unit (Fig. 14). This is a shelf carbonate inlier forming a NW–SE-trending ridge containing the highest summit in the Makran at 2499 m. A large part of it lies within the Saravan quadrangle, but it extends northwestwards into the largely unmapped area north of Iranshahr. Though it consists almost entirely of folded shelf carbonate rocks (Fig. 15), these are believed to cover a small older microcontinent. The bottom of the limestone sequence is not exposed, and all contacts with younger rocks are faulted. The succession is about 800 m thick, and is dominated by shallow-water limestones, thinly to thickly bedded, buff, grey, yellow, white, brown and orange in colour. Sandy or silty calcarenites are present and the limestone may be oolitic, pisolitic, oncolitic or pelletal. There are thin beds of broken gastropod, crinoid, coral and bryozoa fragments. Some beds contain large and complete chaetetids. Algal coatings are common on clastic fragments and around rare solitary corals. There are thin intercalations of calcareous grit, granule conglomerate, shale, quartzose grit and sandstone, which may be cross-bedded and ripple cross-laminated. Micrite limestone, commonly dolomitized, occurs as thin interbeds; it is devoid of diagnostic microfossils, though it contains shell fragments. Red *Globotruncana* limestone also occurs as thin intercalations and contains abundant planktonic foraminifera: this lithology may be associated with thin shale, sandstone, porphyritic andesite, amygdaloidal dacite and fine, highly altered tuff developments. Serpentinite occurs as narrow dykes and lenses parallel to the strike.

A very poorly diagnostic microfauna in the shallow-water limestones is consistent with a Late(?) Cretaceous age. The chaetetids are in all probability of Coniacian–Santonian age, and the red micrites contain a rich Santonian (?Turonian)–Maastrichtian microfauna (for details, see McCall 1985e, pp. 92–93). The sequence is certainly of Cretaceous age, most likely of late Cretaceous age. The depositional environment was a shallow shelf with incursions of terrigenous detritus. The juxtaposition with minor deep-water sediments is

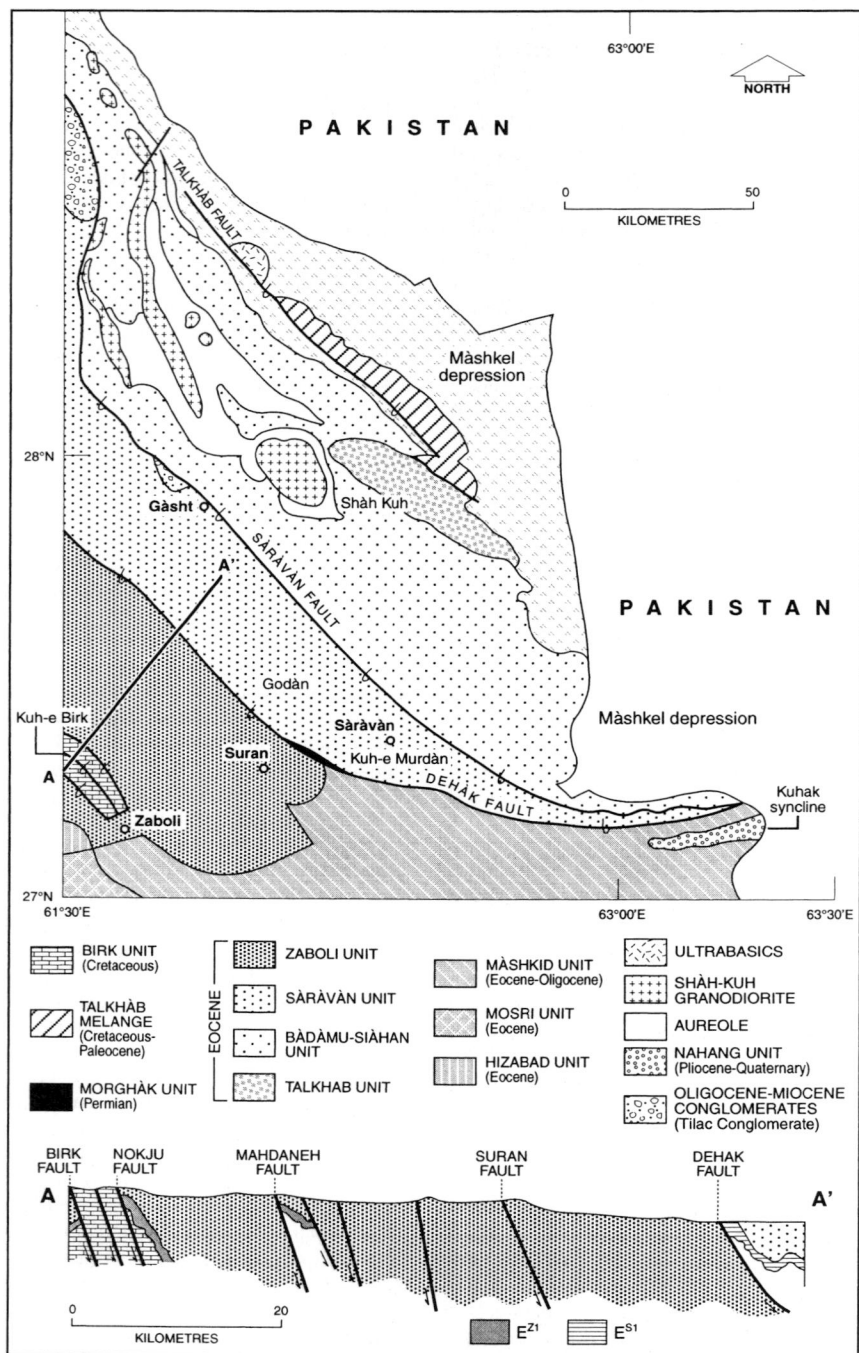

Fig. 14. Generalized diagram showing the field relationships in the Saravan and Narreh-Now quadrangles. (Scale as shown; section at about twice that scale.)

Fig. 15. Folded shallow-water limestones of Cretaceous age, Kuh-e Birk Massif, Saravan quadrangle.

unusual and not understood, but similar minor elements were found in the Dur-kan Complex (see above). The field relationships suggest a normal sequence, not tectonic intercalation.

The Birk Unit is possibly an isolated extension of the Dur-kan Complex. The large-scale structure of the pre-Cenozoic rocks of the eastern Makran involves a huge sigmoid fold with a middle limb running north from Iranshahr. Unfortunately, this middle limb has not been mapped in any detail. The Bajgan–Dur-kan Microcontinental Block (McCall 1997) could swing north and be tectonically attenuated or cut out entirely through this limb, reappearing as the Kuh-e Birk Block in the top limb of the 'S'.

Minor occurrences of Permian sediments (the Morghak Unit). Scattered occurrences of Permian sediments and volcanic rocks occur within the Godan syncline, 35 km NE of Kuh-e Birk, within the Saravan quadrangle (Fig. 14), beneath the lowest Eocene sediments of the Saravan Unit. The base of the sequence is not seen and its thickness is unknown. In equal parts, the sequence consists of calc-shale and carbonate clastic sediments, ranging from calc-siltite and calc-arenite to granule and fine pebble calc-rudite. The beds have a distinctive orange–brown coloration, which allows easy separation in the field from the Eocene sediments. The calc-arenites may contain bioclastic debris; disarticulated and broken bivalves, and more or less intact gastropods. Some of the calc-arenite is a packite of micrite and chert sand grains, together with scattered benthic foraminifera set in a brown micritic matrix. The calc-rudite tends to form large pods or boudins, associated locally with oolitic or pisolitic biomicrite. All the carbonate rocks are more or less recrystallized and many are sparry. The limestones display bioturbation. The volcanic rocks are altered fine-grained feldspar-phyric lava, with red and white infilled vesicles. The unit contains diagnostic Permian microfaunas, including large fusulinids forming small pebbles (for details, see McCall 1985e, pp. 189–190). The beds are tightly or isoclinally folded. This unit crops out not far distant from Kuh-e Birk and may represent tectonic slices of Permian shelf sediments that underlay the Birk Unit, analogous to the Permian representation within the Dur-kan Complex (see above), as suggested by McCall (1997).

The Talkhab Mélange (Fig. 14). An ophiolitic mélange crops out in the NE corner of the Saravan quadrangle. This is in the half of the quadrangle mapped by Intercon–Texas Instruments, but compiled by the author (McCall & Eftekhar-nezad 1994). It continues northwestwards into the Nar-reh-Now quadrangle also mapped by ITI, and passes southeastwards under the desert deposits of the Mashkel Depression. The mélange forms a prominent ridge and is thrust over the Eocene Talkhab Unit of shallow-water limestones and siltstones and the Eocene Badamu-Siahan Unit of flysch turbidites. The components are carbonate rocks (listvenite), serpentinite, harzburgite, lherzolite, dunite, small blocks of gabbro (some layered), metasediments, recrystallized limestone, basalt, basaltic andesite, hyaloclastite, recrystallized sandstone and red radiolarian chert; the basalts may be pillowed and are closely associated with radiolarian chert developments. It is an ophiolitic tectonic mélange, mostly metamorphosed to greenschist facies, probably by hydrothermal metamorphism. A radiolarian chert within it was dated to Cretaceous time (McCall & Eftekhar-nezad 1994). From its geotectonic position, it seems probable (McCall 1997) that this represents an inner oceanic tract, open during Mesozoic time, and is a continuation of the 'Sistan Ocean' of Sengor et al. (1988). It is in all probability equivalent to the Remeshk–Mokhtarabad Ophiolites and might well, like them, contain rocks ranging in age from Jurassic to Early Paleocene time.

Summary. The geology before Mid-Paleocene time was dominated by a cluster of microcontinental blocks or slices. These separated from Gondwana, probably during Early Triassic time (Sengor et al. 1988); the oldest deep-water sediments associated with the inner ophiolites are of Jurassic age, so the oceanic tracts that separated them until Early Paleocene time were established before Cretaceous time. These inner tracts are now represented by ophiolites. The tracts now form rather narrow outcrop belts but were once presumably tens or hundreds of kilometres wide. These ophiolites have not been dismembered to mélange in the west (Minab quadrangle), and display local dismemberment and intense folding further east (Fannuj quadrangle). The Talkhab Mélange in the NE corner of the Saravan quadrangle appears to be fully dismembered; however, it is not certain that this represents a continuation of the inner group of ophiolites.

These inner ophiolites were interpreted (McCall 1997) by means of an analogy drawn with the Mediterranean; this southern region resembles what will develop when the Mediterranean is terminated as a result of the push of Africa northwards. It was suggested that there was subduction along the southern margin of the Central Yazd–Tabas–Lut Microcontinental Block (evidence of blueschist development) and also the Helmand–Farah Microcontinental Block (evidence of ophiolitic mélange). The southern development of dismembered Coloured Mélange was at the southern margin of the Sanandaz–Sirjan and Bajgan–Dur-kan block, the northern margin of the southern branch of Neo-Tethys. At that time India was fairly close to the Arabia block (Sengor et al. 1988) and they together formed the southern margin of this tract of ocean.

An important fact is that the Coloured Mélange, Dur-kan and Birk platform limestone units, and the three inner ophiolite developments, all include sequences of ages that overlap; in other words, these geotectonic elements represent different local geological and palaeogeographical conditions coexisting between Jurassic and Early Paleocene time. After this situation there was a major tectonic event, referred to by Iranian geologists as the 'Laramide' orogeny, although such a term appears misplaced. There was a major event, however, and nothing younger in the Makran shows continuity with these older rocks, field relations being unconformable, tectonized or faulted. We know very little about this tectonic event because of the overprint of Neogene deformation.

The inner ophiolites are probably analogous developments to the Troodos of Cyprus. It would be incorrect to equate these remnant developments of narrow tracts of oceanic crust with ocean basins such as the Atlantic. The magma chamber developments represented by the ultramafic rocks, gabbros, plagioclasites and trondhjemites possibly represent fairly simple developments compared with those that must underlie the Mid-Atlantic Ridge, for although they appear to have been active for many millions of years, their ocean-floor extrusive products must have been minute in volume compared with those from the Mid-Atlantic Ridge in a comparable span of time.

The Paleogene, Neogene and Quaternary rocks

Concise description of these rocks is difficult because the rocks within the Minab, Tahruie, Fannuj, Nikshahr and Pishin quadrangles (Fig. 1) form a huge accretionary prism, parcelled up by listric faults dipping into the continent from the coast. On account of this parcelling, the stratigraphy had to be erected in the form of numerous informal units, separated by fault boundaries. Ages of these units frequently overlap, but it is not usually possible to correlate across fault boundaries. The same applies to the northern part of the Pishin quadrangle and southern part of the Saravan quadrangle. In the Saravan quadrangle and Narreh-Now quadrangle (Fig. 2), the sequences belong to what is believed to be the separate Saravan accretionary prism, and there is no flysch turbidite or neritic sedimentary cover younger than early Oligocene time, whereas in the main Makran prism such sediments extend through Miocene and into Pliocene time.

The descriptions here will be given under the

following headings, each incorporating several stratigraphic units:

(a) main Makran accretionary prism: (1) Eocene–Early Oligocene flysch sequences; (2) Oligocene–Miocene flysch sequences; (3) Miocene neritic sequences east of the Zendan fault (Fig. 16) (including reefal limestones); (4) Late Miocene–Pliocene neritic, deltaic, littoral and fluviatile sequences west of the Zendan Fault.

(b) Saravan accretionary prism: (1) Eocene–Early Oligocene flysch sequences; (2) Oligocene–Miocene conglomerates; (3) the Shah Kuh and associated granodiorite intrusions.

(c) Eocene–Oligocene neritic sediments in outliers on the Lower Paleocene, Mesozoic and older rocks.

(d) Cenozoic sediments in the Jiroft–Iranshahr tract (see Fig. 30 below).

(e) Pliocene–Pleistocene fanglomerates.

(f) Superficial deposits: the Jaz Murian Depression (see Fig. 30 below).

The superficial deposits, which form a number of Quaternary terrace deposits and sabkha along the coast, as well as the entire cover of the Jaz Murian Depression including dune sands and playa lake deposits, are not described in detail here, but descriptions have been given by McCall (1985a, b, c, d, e).

Makran accretionary prism.

Eocene and Eocene–Early Oligocene flysch sequences. These sequences include the Guredak, Ruk, Bamposht, Darban, Irafshan, Kenar, Mosri, Nargakan, Rig and Shirinzad Units. These formations are listed in Table 2 with their age ranges.

The outcrop of the Guredak Unit extends from 57°45'E (Fig. 16) to 61°E (Fig. 17). The remaining units occupy the area from that point to the Pakistan border, occupying a series of fault-bounded packages. The Ruk Unit is almost certainly the continuation of the Guredak Unit (the distinction was due to mapping logistics). The Guredak Unit can be taken as representative of these formations and will be described in some detail. Its outcrop extends from just east of the Zendan Fault within the Tahruie quadrangle to the Nikhahr quadrangle (Figs 16 and 17). The sequence thickens from about 3500 m in the west to possibly as much as 10 000 m in the east. Such estimates are apparent, derived from the mapping, but the real thickness is probably somewhat less than such estimates, although the nature of the tectonic folding suggests that the overestimation is not significant. Contacts are entirely faulted. Cyclic sandstone–shale alternations predominate, with sandstone dominant. The thickness of individual cycles commonly increases upwards. The sandstones (Fig. 18) are mainly arenites with minor wackes. There is a subordinate shaly facies. Low in the sequence shallow-water, grey allodapic nummulitic limestones were redeposited and form lenticles in the sequence, which is essentially a deep-water distal turbidite sequence. Minor interbeds include micritic limestone, tuff associated with purple and red chert, and radiolarite siltstone. Lensed pebble conglomerate intercalations include sandstone, limestone and acid to intermediate porphyry pebbles. Locally, particularly towards the east, ropy basalt flows intervene. Bioturbation includes horizontal annelid burrows and large sinusoidal tracks. In the sandy facies, load casting, prod marks, groove casts, flute casts and other sole marks occur, as do interference ripple marks, ripple cross-lamination, parallel lamination and graded bedding. Many foraminiferal microfaunas have been studied from this unit and the results have been summarized by McCall (1985e, pp. 138–142). Both the sandstone–shale sequence and intercalated limestones yielded Early–Late Eocene microfaunal ages. One microfauna from the Nikshahr quadrangle in the east could be of Paleocene–Eocene age. There was a tentative identification of one microfauna extending into Oligocene time from the Tahruie quadrangle in the west. This is a typical continental slope distal turbidite deposit, of an unstable steep-sided trough, with localized areas of quiet deposition during periods of increased stability, represented by the shaly facies and pelagic limestone. The allodapic limestones probably came from the shelf fringing the continent. The general absence of trace fossils probably indicates rapid deposition or anoxic bottom conditions. The unit is complexly folded and tectonically dislocated over wide areas. Where dislocated it contains exotic blocks, and the shales are sub-phyllitic. In the west, folding is irregular with curving periclinal synclines dominant. In the east, within the Nikshahr quadrangle, the folding is more regular and chevron styles predominate. The periclines follow the regional strike and there is a profusion of second-order folds mappable on outcrop scale, again mainly chevron folds, but also curved hinged open folds, box folds, kink folds and isoclinal folds. There is no overturning evident in sedimentary structures except on this scale. Vergence is to the south. Fold limbs tend to be cut out by reverse faults. The periclines plunge gently in either direction, and anticlines are weakly developed or cut out. Disharmonic folding is characteristic, with the finer shaly beds displaying more folding than the enclosing sandstone beds.

The Ruk Unit crops out in the east, within the Nikshahr and Pishin quadrangles, extending to Pakistan (Figs 17 and 19), and has been estimated

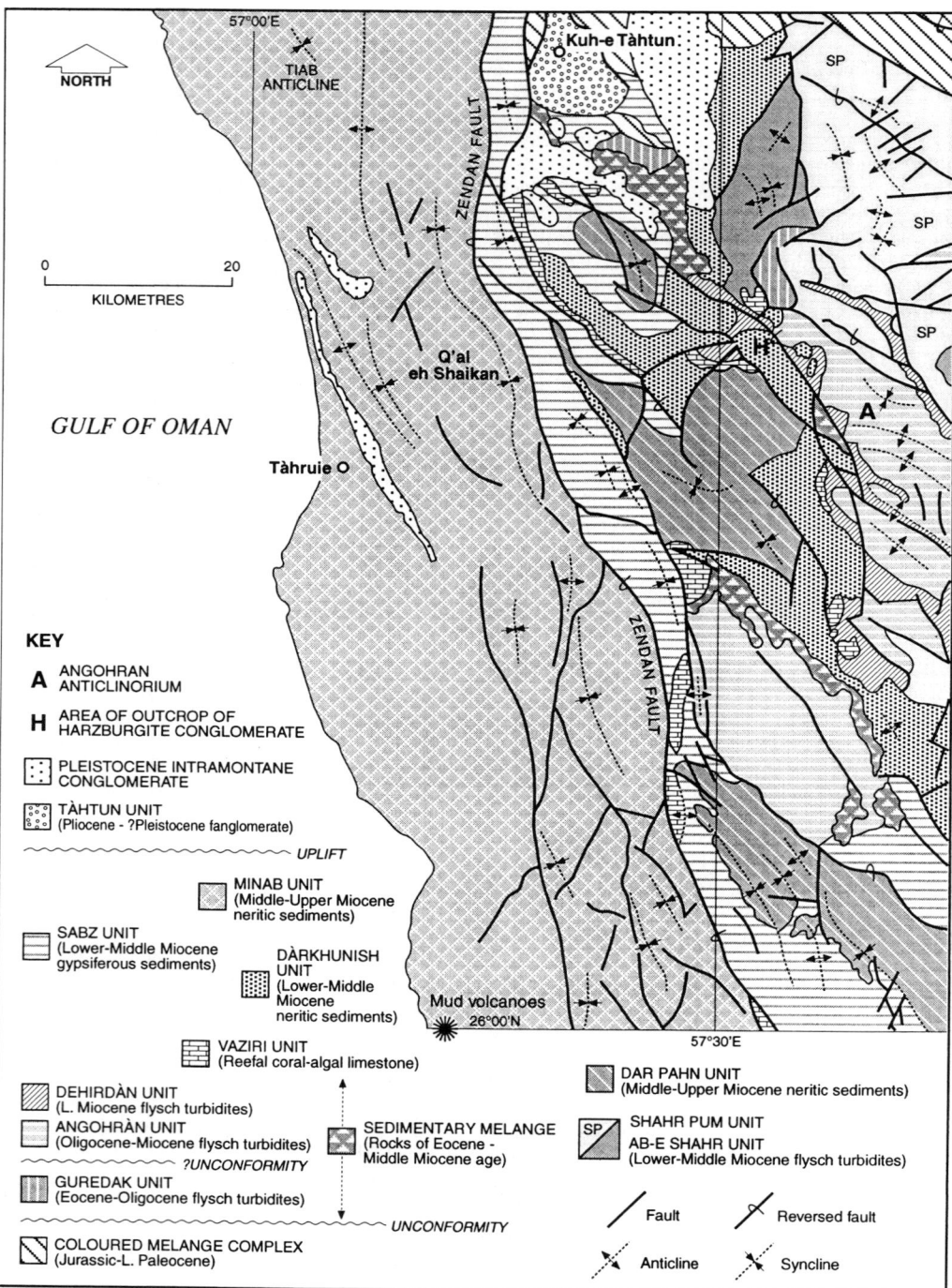

Fig. 16. Generalized diagram showing the field relationships in the western part of the Tahruie quadrangle (scale 1:500 000).

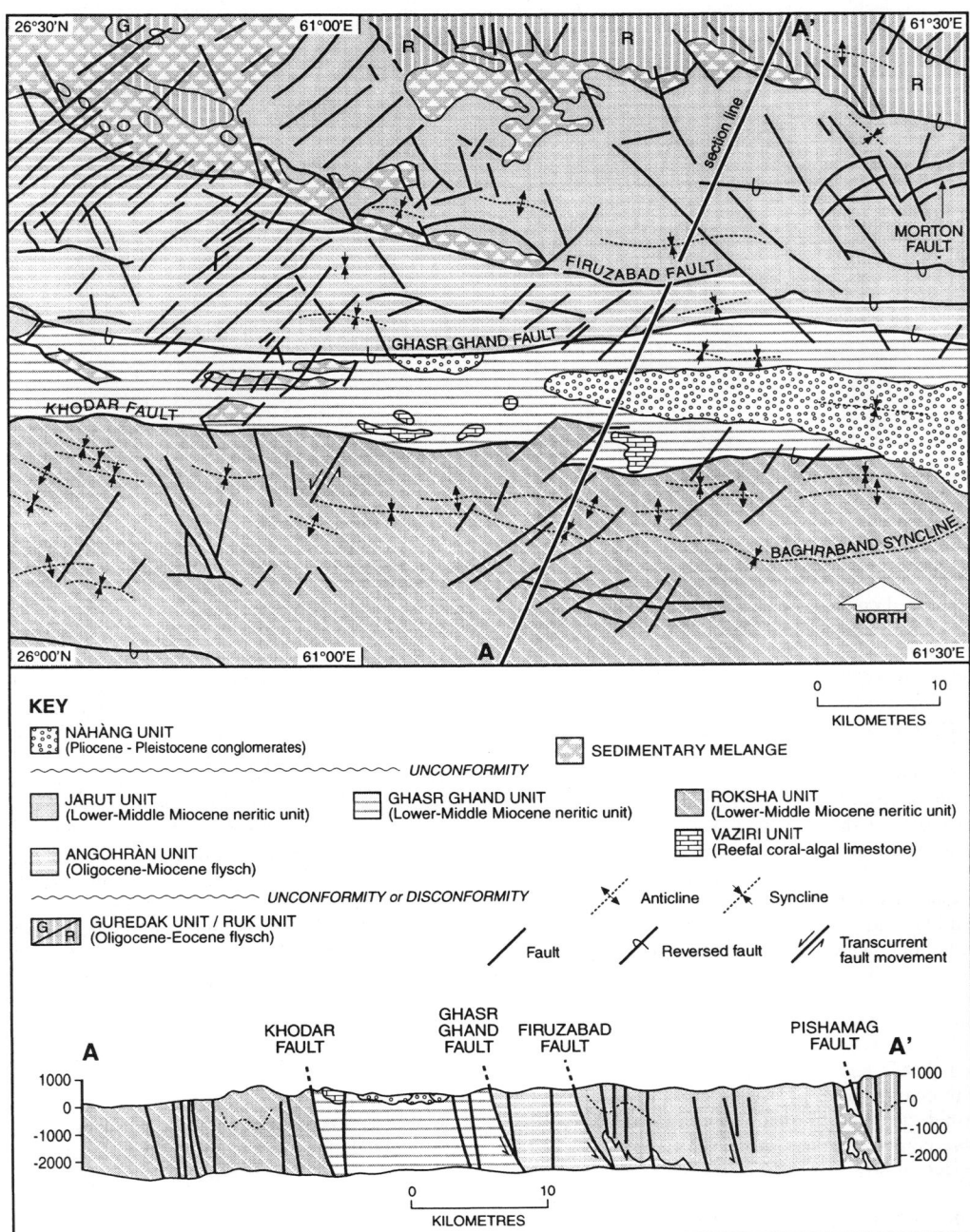

Fig. 17. Generalized diagram showing the field relationships in the SE corner of the Nikshahr quadrangle with a cross-section (scale 1:500 000). This shows the strong development of late conjugate cross-faults in the eastern part of the Iranian Makran.

Fig. 18. Eocene sandy flysch turbidites of the Eocene Guredak Unit, looking east along the outcrop belt, Tahruie quadrangle.

to be up to 14 000 m thick. It has faulted or tectonized contacts with all other units, except the Pliocene–Pleistocene Nahang Unit, which is unconformable upon it. It is dominated by thick sandstone–shale cycles; in the Nikshahar quadrangle these coarsen upwards, from 20 to 70% sandstone; further east in the Pishin quadrangle they coarsen from 10 to 90% sandstone, and individual megacycles, which contain smaller rhythms within them, may amount to thicknesses of 500 m of the sequence. Very coarse sandstone beds may have pebbly layers at their base. The sandstones display plane lamination, ripple cross-lamination, convolute lamination, minor cross-bedding and sole markings; the last include load casts, flute casts, groove casts and prod marks. Bioturbation and feeding tracks are common, including 14 deep-water types of trace fossil, the most common being *Spirorhaphe*, *Paleodictyon* and *Urohelminthoida*. A tuffaceous facies with fine calcareous and feldspathic tuff interbedded with coarse volcanic grits was recorded within the Pishin quadrangle. The micropalaeontology was documented by McCall (1985e, pp. 210–212). Besides Early to Late Eocene planktonic foraminiferal microfaunas, both Early and Late Oligocene diagnostic planktonic foraminifera were recorded. This is a distal turbidite sequence, deposited in deep water. The unit displays spectacular chevron folding within the Nikshahr quadrangle, also curved hinged, box and isoclinal folds on outcrop scale. Within the Pishin quadrangle to the east, short periclinal synclines are much in evidence. The folding is clearly integrated with north-dipping reverse faults.

Only within the Nikshar quadrangle is the coarsely conglomeratic Rig Unit mapped (Fig. 3). This has an estimated thickness of about 2000 m and is everywhere fault bounded. Crude cycles of sandstone and conglomerate dominate, with boulder size prominent. The conglomerates are polymict, well rounded and poorly sorted. Lensing and channelling of conglomerate beds is common. Scours, drag and tool marks, and convolute lamination are recorded. Siltstones within this unit yielded Mid-Eocene planktonic foraminiferal microfaunas (see McCall 1985e, p. 203). This is a deep-water unit, produced by submarine debris flows.

Of the seven separate units differentiated within the Pishin quadrangle north of the Ruk Unit outcrop, and within the southern part of the Saravan quadrangle, the Irafshan Unit, cropping out within the eastern part of the Nikshahr quadrangle and western part of the Pishin quadrangle (Fig. 19), is the oldest. It thins eastwards from an estimated 9000 m to 1000 m, and passes up concordantly into the Darban Unit. It is a flysch turbidite sequence dominated by volcanic arenites. The shales are mainly siltstones cemented with sparry calcite. Pelagic biomicrites and allodapic limestones are intercalated in the distal turbidite

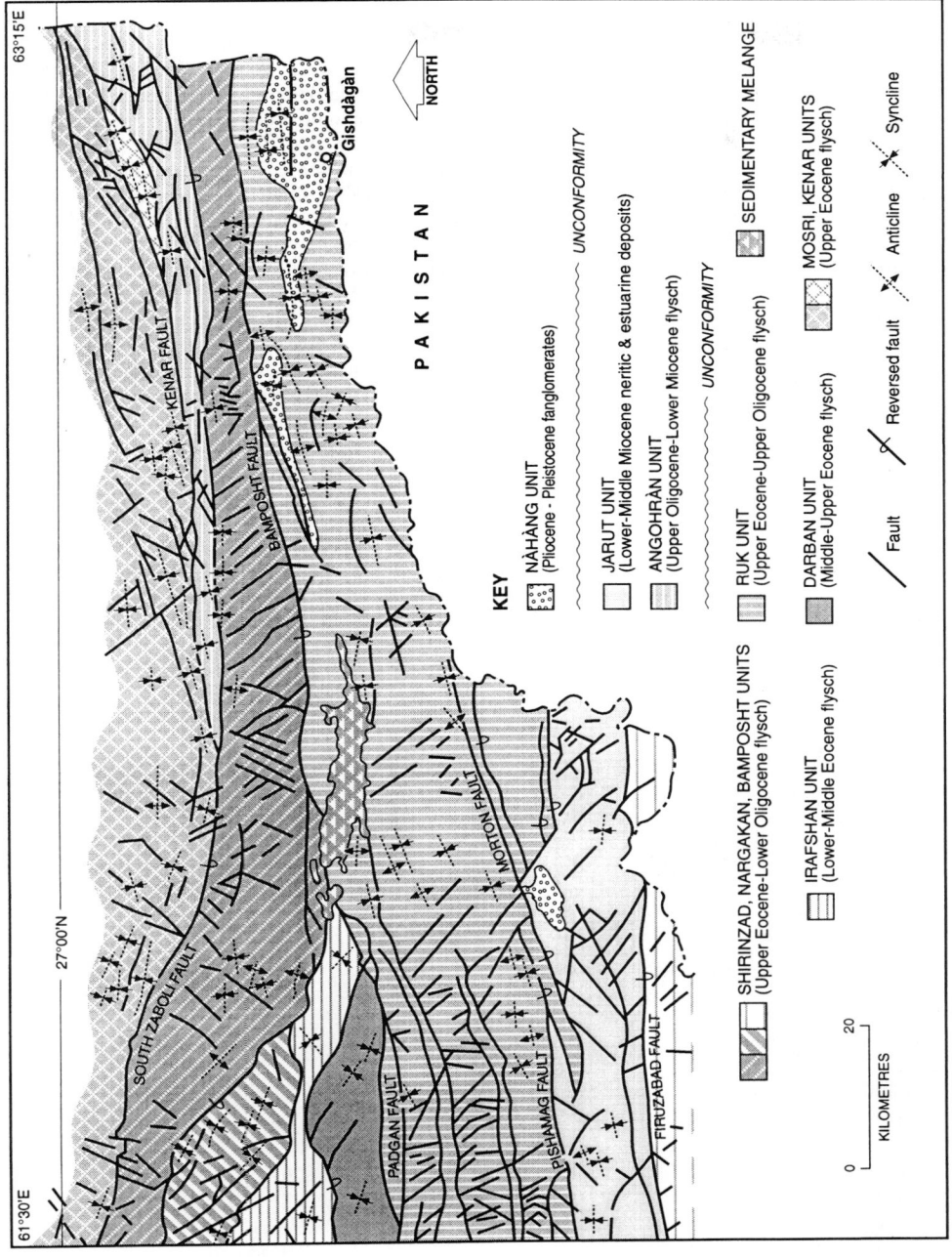

Fig. 19. Generalized diagram of the field relationships in the northern part of the Pishin quadrangle (scale 1:500 000).

sequence. Both planktonic foraminifera and displaced benthic foraminifera indicate an Early to Mid-Eocene age (see McCall 1985e, pp. 181–182). The Darban Unit above, also cropping out within the eastern part of the Nikshahr quadrangle and western part of the Pishin quadrangle (Fig. 19), has an estimated thickness of about 9000 m and is dominated by shale. It includes airfall tuff beds, pelagic biomicrites and allodapic (redeposited) shallow-water limestones. One deep-water trace fossil, *Subphyllochordia*, is recorded in this distal turbidite sequence. Both planktonic and benthic foraminifera indicate a Mid- to Late Eocene age (see McCall 1985e, p. 109). Locally, this unit passes up conformably into the Shirinzad Unit, which thins from an estimated 14 500 m in the west to about 1000 m at the Pakistan border (Fig. 19). Elsewhere, contacts of the Shirinzad Unit with other units are faulted. It is a deep-water turbidite sequence, with both sandy and shaly facies. Very large flute casts are a feature of the sandstones (Fig. 20). The trace fossils *Paleodictyon* and *Nereites* are recorded, both deep-water forms (Crimes & McCall 1994). Planktonic and benthonic foraminifera indicate a Late Eocene–Early Oligocene age for this unit (see McCall 1985e, p. 232). The Nargakan Unit, also cropping out in the west of the Pishin quadrangle (Fig. 19), is a flysch sequence, containing some pelagic limestones and red and green shales. Planktonic and benthic microfaunas indicate a Late Eocene to Early Oligocene age (McCall 1985e, p. 196). This is an extension of the Shirinzad Unit, isolated from it by faults. The Mosri Unit straddles the boundary between the Pishin and Saravan quadrangles (Figs 14 and 19) and in the former thins eastwards from 7000 to 2000 m. This unit consists of alternations of shaly and sandy flysch, of outer shelf to deeper-water facies. The shales yielded a Late Eocene planktonic microfauna and the sandstones a benthonic microfauna of the same age (McCall 1985e, pp. 191–192). The Kenar Unit, cropping out in the east of the Pishin quadrangle in a tectonically isolated area (Fig. 19), is up to 500 m thick, and consists of cyclic sandstone–shale flysch with minor interbedded pebble and cobble conglomerates. It is of a proximal to middle fan facies and Late Eocene age. The Bamposht Unit is a fault-bounded flysch unit intervening in the eastern half of the Pishin quadrangle, between the Shirinzad and Mosri Units (Fig. 19). Its outcrop belt widens towards Pakistan, being estimated to have a thickness of 4000 m at the frontier. The lower part is sandy flysch in massive beds with minor shale intercalations: and above this lies sandstone–silty shale flysch with thin pebble bed and polymict conglomerate alternations, the latter with chert, quartz, sandstone, limestone and volcanic clasts. A shaly facies lies at the top of the sequence. Planktonic foraminiferal microfaunas and displaced benthic foraminiferal microfaunas indicate a Late Eocene to Early Oligocene age (McCall 1985e, pp. 82–83). This unit also is an extension

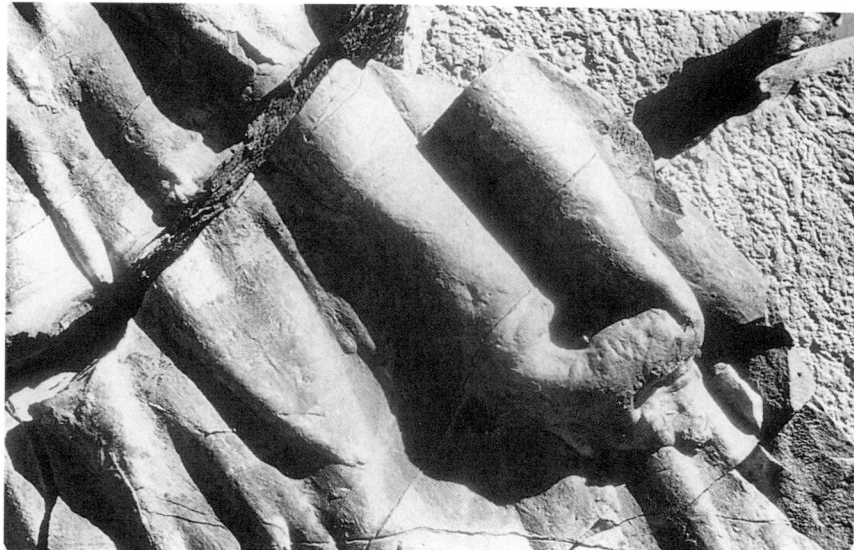

Fig. 20. Large flute casts in the base of a sandstone bed, Shirinzad Unit, Pishin quadrangle (scale shown by hammer in top right-hand corner).

of the Shirinzad Unit, isolated from it by major faults. The sequence is a shallow-water facies, consisting of proximal deposits of a prograding delta or proximal fan.

Oligocene–Miocene flysch sequences. The stratigraphic units are listed in Table 2 with their age ranges.

A tectonic event affected the Makran during Oligocene time. Regional geologists in Iran such as H. Huber (1978) have referred to a 'Mid-Oligocene Regional Tectonic Event' (McCall 1985e, p. 66). Only a single unit, the Ruk Unit, appears to straddle this event, and even this age assignment is based on a single microfaunal determination.

There are five named flysch units within the Late Oligocene–Miocene age range: the Angohran, Shahr Pum, Dehirdan, Ab-e Shahr and Hichan Units. There is also a small development of Oligocene–Miocene flysch north of Minab, which was not defined as a named unit. The Cenozoic accretionary prism tapers out here, a primary thinning, rather than a secondary, tectonically imposed taper.

The most extensive development is the Angohran Unit, which extends in a more or less continuous outcrop belt from the Tahruie quadrangle eastwards to the Pakistan frontier (Figs. 16, 17 and 19). The estimated thickness of this unit is 5000 m in the Tahruie quadrangle in the west and 7000 m in the Nikshahr quadrangle in the east. The unit has faulted contacts with other units, except with the Miocene Dehirdan Unit, which conformably overlies it. The Miocene Shahr Pum Unit is nowhere in normal concordant relationships with it. Three facies, shaly, sandy and undifferentiated, are distinguished. The shaly facies (Fig. 21) commonly underlies the sandy

Table 2. *Stratigraphic units of the Makran accretionary prism, listed with their age ranges*

Stratigraphic unit	Age range
Flysch turbidite units	
Hichan Unit	Late Early Miocene
Ab-e Shahr Unit	Early Miocene
Shahr Pum Unit	Early Miocene
Dehirdan Unit	Early Miocene
Angohran Unit	Late Oligocene–Early Miocene
Nargakan Unit	Late Eocene–Early Oligocene
Bamposht Unit	Late Eocene–Early Oligocene
Shirinzad Unit	Late Eocene–Early Oligocene
Ruk Unit	Late Eocene–Early Oligocene (?Late)
Kenar Unit	Late Eocene
Mosri Unit	Late Eocene
Rig Unit	Mid–early Late Eocene
Darban Unit	Mid–Late Eocene
Guredak Unit	Early Eocene–Early Oligocene
Reefal, neritic and piedmontane conglomerate units	
Palami Conglomerate }	Late Pliocene–Pleistocene
Tahtun Unit	
Nahang Unit	
Makran Unit	
Minab Conglomerate	Early Pliocene
Kheku Sandstone–Tiab Sandstone	Late Miocene–Early Pliocene
Gushi Marl	Late Miocene
Jaghin Unit	Mid–Late Miocene
Dar Pahn Unit	Mid–Late Miocene
Roksha Unit	Early–Mid-Miocene (?Late)
Jarut Unit	Early–Mid-Miocene
Darkhunish Shale	Early–Mid-Miocene
Band-e Chaker Unit	Early–Mid-Miocene
Ghasr Ghand Unit	Early–Mid-Miocene
Sahan Tang Unit	Early–Mid-Miocene
Sabz Unit	Early–Mid-Miocene
Pishin Unit	Early Miocene
Harzburgite Conglomerate	Early Miocene (Burdigalian)
Vaziri Unit (reefal limestone)	Early Miocene (Burdigalian)

Fig. 21. The lower shaly flysch turbidite division of the Oligocene–Miocene Angohran Unit, Tahruie quadrangle. Coral–algal limestone outcrop of the Lower Miocene (Burdigalian) Vaziri Unit is in the immediate foreground.

facies and grades up into it. This facies is dominated by micaceous shales but locally there are interbeds of well-sorted calcareous arenite. The sandy facies consists of a cyclic sequence of normally graded sandstone and shale. The sandstone beds are massive, immature, carbonate-cemented arenites, in which volcanic (dacite) fragments are abundant. There are thin beds of grit and pebble conglomerate. The subordinate shale interbeds are thin and silty, and grade upwards into calcareous siltstones. Primary structures include planar lamination, ripple marks (wavy flow, asymmetrical and oscillation types). Channelling, flute casts, groove casts, drag marks, frondescent marks, and small load casts are all evident, as well as the deep-water trace fossils *Spirorhaphe* and *Paleodiction*. Calcareous arenites and calc-wackes grade into limestones, but these are minor, as are tuffs. The undifferentiated facies consists of sandstone (up to 40%)–shale alternations, in 100–200 m thick, coarsening-upwards megacycles. B, C, D and E of the Bouma sequence are represented. Minor conglomerate has been recorded, as well as plant debris in shales and vertical burrows. These descriptions relate to the Tahruie quadrangle in the west (Fig. 16), and

in the eastern outcrops (Fig. 17) similar developments are evident, although the complete set of three facies may not be represented. In the Fannuj quadrangle, the outlines of shards are preserved in tuffs and also planktonic foraminifera, indicating air fall into sea water. In the Nikhshahr quadrangle shale clast breccias fill channels in the sandstones, which not uncommonly display a basal pebble conglomerate; the shaly facies contains orange micrite septarian concretions, and displays bounce marks, tool marks and the deep-water trace fossil *Paleodictyon carpathicum*.

Sixteen planktonic and benthic foraminiferal microfaunas and two echinoid faunas indicate a Late Oligocene to Early Miocene age for this unit (McCall 1985e, pp. 74–76). This is a submarine fan sequence, deposited rapidly at some distance from the continental shelf, downslope. The only trace fossils are those found at a depth of <1000 m. The tuff bands indicate explosive volcanism (subaerial) at no great distance away. The lack of allodapic (redeposited) limestones contrasts with the Eocene–Oligocene flysch of the Makran and indicates that the slope was steep and the shelf narrow. The microfauna is consistent with a bathyal environment, no shelf components being present except those that are reworked.

The shaly facies is contorted, widely dislocated and grades into sedimentary mélange, but does not carry exotic blocks where dislocated. The remainder is intensely folded into open to tight isoclinal folds on a mappable scale; such folds are mostly synclinal periclines. Second-order folds, on outcrop scale, include chevron folds, curved hinged open folds, box folds and isoclinal folds. Folds are vergent to the south, and both axial regions and fold limbs are cut out by reversed faults. The structure in the west includes an 8 km wide anticlinorium, the Angohran anticlinorium, plunging NW (Fig. 16). Overturned fold limbs do occur but are rare. Plunges are rarely steep. Small flexural décollements are common.

The only other flysch unit containing Oligocene microfossils is the undifferentiated unit north of Minab (Fig. 4). This occurs in a fault-bounded outcrop lens 25 km long and up to 4 km wide and a smaller lens to the south. The thickness is estimated at 3000 m. Fine to medium sandstones are interbedded with shale and mudstone. Benthonic and planktonic foraminiferal microfaunas indicate a late Oligocene to earliest Miocene age (McCall 1985e, p. 259). This is a flysch turbidite sequence, deposited rapidly in a trough of considerable depth. It is a continuation of the Angohran Unit northwards, to the east of the Zendan Fault; it was mapped before it was decided to name individual units informally and somehow escaped naming.

The Dehirdan Unit (Fig. 16) is conformable above the Angohran Unit. Above this boundary may come the Vaziri Unit (reefal limestone) or the Harzburgite Conglomerate, both of which have only patchy development; or else the Shahr Pum or Ab-e Shahr Units (flysch turbidites), or the Band-e Chaker or Darkhunish Units (neritic). Contacts with the neritic Sabz Unit are invariably faulted. The Dehirdan unit, up to 1000 m thick, crops out only in the Tahruie quadrangle, and is dominated by calcareous shale, but there are also fine- to medium-grained calcareous and micaceous sandstone interbeds. Locally there is a sandy facies developed. Graded bedding is evident but not sole marks. A benthic foraminiferal microfauna indicates an Early Miocene age (McCall 1985e, pp. 120–121). At Sardasht (Fig. 3), the unit contains six fairly closely spaced horizons of coral, algal and foraminiferal limestone, which resembles the reefal limestone (Vaziri Unit) of the neritic sequences, but the benthic foraminiferal microfauna is of Aquitanian (21.5–23.3 Ma) not Burdigalian age (16.3–21.5 Ma). These limestones are washed in from a distance and the lower clastic beds of the Dehirdan Unit are of a moderately deep-water shelf-slope facies, although, in contrast, the upper beds just below the Burdigalian reefal limestone are of fairly shallow-water facies. That deep-water environments are represented here is shown by the deep-water trace fossil *Helminthoida* immediately above the uppermost redeposited limestone bed. The environment of deposition of the Dehirdan Unit progressively shallowed up-section. Small-scale folds predominate in this thinly bedded sequence.

The second Miocene flysch unit is the Shahr Pum Unit, which crops out in the Tahruie (Fig. 16) quadrangle and extends eastwards into the Fannuj quadrangle. This unit thins eastwards from an estimated 9500 m to an estimated 4500 m. It is conformable above the Dehirdan Unit and elsewhere occurs above the Vaziri Unit (reefal limestone), where it occupies the interface between the two units. There is a facies-change boundary with the Ab-e Shahr Unit. The lower subdivision consists of megacycles of turbidites 200–300 m thick, with smaller internal rhythms. Lithologies present include sandstone, shale, minor conglomerate, siltstone, mudstone and limestone. The smaller rhythms are capped by thick sandstone beds. The middle subdivision is a sequence of fine micaceous sandstones and shales, in which megacycles are not abundant. The uppermost subdivision is dominated by fine- to coarse-grained sandstone with minor interbeds of silty calcareous shale. Megacycles are again prominent. Sedimentary structures are widely developed throughout this unit and include spectacular sole marks, among them flute casts with a peculiar superimposed frondescent pattern. Ripple marks, impact casts, groove casts and drag marks are very widespread. The trace fossils are all deep-water types: *Paleodictyon, Spirorhaphe* (Fig. 22), *Helminthoida crassa* and *Helicorhaphe*. The sandstone is mainly lithic and volcanic arenite. Foraminifera were difficult to find in this unit, but sparse planktonic

Fig. 22. Trace fossil *Spirorhaphe*, Miocene Shahr Pum Unit, Tahruie quadrangle (about actual scale).

and benthic foraminiferal microfaunas indicate a Miocene age, and the fact that the upper subdivision yielded an early Miocene benthonic microfauna suggests that the entire unit, despite its thickness of nearly 10 000 m, is of Early Miocene age (McCall 1985e, p. 229). This is a proximal turbidite sequence, deposited in fairly deep water, which received increments of coarse detritus from an uplifted foreland. The entire unit occurs in a broad anticlinorial structure and is flexed into second- and third-order folds, chevron, box, isoclinal and curved hinged in style. Vergence is to the south. The unit forms, with the Angohran and Dehirdan Units, an immensely thick sequence of turbidites, just reaching Late Oligocene age at its base and only reaching Early Miocene age at the top. It is estimated that the entire thickness of turbiditic sediments represented by these three units is of the order of 16 000 m.

The Ab-e Shahr Unit crops out in a small area in the Tahruie quadrangle (Fig. 16), It has an estimated thickness of 2000 m and has a facies change boundary with the Shahr Pum Unit, apparently concordantly underlying the upper beds of that unit. It has faulted or tectonized boundaries with the Dehirdan Unit, which stratigraphically and structurally underlies it; and is faulted against the neritic Darkhunish Unit to the west, that unit being almost certainly a lateral equivalent. Locally the Vaziri Unit (reefal limestone) intervenes between it and the Dehirdan Unit, or else the Harzburgite Conglomerate unit intervenes. This is a monotonous sequence of fine-grained turbiditic sandstones with shale intercalations. The sandstone is less calcareous and more poorly cemented than that of the Shahr Pum Unit. There are no megacycles. There are some intercalations of limestone. A calc-rudite contains a benthic foraminiferal microfauna of Early Miocene age (McCall 1985e, p. 71). As it overlies the Burdigalian Vaziri Unit, it cannot be older than Burdigalian time. This a sandy flych sequence, probably transitional between the deep-water Shahr Pum Unit and the dominantly shallow-water, neritic environment of the equivalent beds of the Darkhunish Unit to the west.

The Hichan Unit crops out only in a synclinal zone straddling the boundary between the Fannuj and Nikshahr quadrangles (Fig. 3). It is faulted against the Angohran Unit, which probably originally lay beneath it, conformably. It is a sequence of shaly flysch, estimated to be 3000 m thick. There are some thinly bedded calcareous lithic and pebbly sub-arkosic sandstone intercalations in the sequence, and also some thick sandstone strata. Sedimentary structures include ripple cross-lamination, asymmetrical ripple marks, rare cross-bedding and graded bedding, prod marks and groove casts. Casts of burrows are recorded and a single trace fossil, *Fucusopsis*, tentatively identified. Planktonic and benthic microfaunas indicate a late Early Miocene age (McCall 1985e, pp. 147–148), but the sequence may extend to mid-Miocene age. The sequence is folded into open synclines, but second-order folds on outcrop scale are rare.

Miocene neritic sequences east of the Zendan Fault. The youngest unrelieved flysch sequence was deposited in Early Miocene time, following which there was a transition to shallow-water sedimentation sequences. As these are highly variable, their correlation across faults is very difficult. Ten such sequences have been differentiated in the main accretionary prism, to the east of the major Zendan Fault (Fig. 16) and trending west-east across the accretionary prism. These units are generally exposed to the south of the Oligocene to Miocene flysch units described above, but there are cases where this is not the case: for example, the fault-bounded outcrop zones of the Jarut Unit, which intervenes between the Eocene–Oligocene Ruk Unit and the Oligocene–Miocene Angohran Unit within the Nikshahr and Pishin quadrangles (Fig. 17).

The ten units recognized are the Sabz, Bande-e Chaker, Darkhunish, Dar Pahn, Jaghin, Ghasr Ghand, Roksha, Sahan Tang, Pishin and Jarut Units. These stratigraphic units are listed in Table 2 with their age ranges.

The 6000 m thick Sabz Unit crops out in a number of fault-bounded zones in an arc from Kuh-e Zendan, north of Minab (Fig. 4), running south and eastwards through the Tahruie quadrangle (Fig. 16) and terminating in the very south of the Fannuj quadrangle. The dominant lithology in the Minab quadrangle is a pale red, green or grey calc-mudstone associated with medium-grained calc-sandstone, pale to dark brown pebble conglomerate and partly recrystallized grey to cream limestone. These beds are pale in outcrop as a result of the presence of gypsiferous material. The coarser beds display graded bedding and flute casts. The unit is commonly brecciated and dislocated. In the north of the Tahruie quadrangle there is a gypsiferous mudstone sequence together with (generally below it) an only sparsely gypsiferous turbidite sequence. In the south, extending into the Fannuj quadrangle, it is an entirely gypsiferous sequence, and the gypsiferous facies is relieved only by thin beds of brown friable sandstone. Conglomerates are rare and where present are polymict, with limestone and sandstone pebbles and cobbles. Both planktonic and benthic foraminifera are present in the beds of the Sabz Unit, and in the Tahruie quadrangle a fauna of 25 planktonic foraminifera was recorded (McCall 1985e, pp. 215-216). The age range is from Early to Mid-Miocene time, but only Mid-Miocene planktonic forms are recorded in the Fannuj quadrangle. The environment was neritic, and the abundant gypsum indicates a warm climate. The gypsum is now all secondary, but is ubiquitous and must surely have been laid down as a primary deposit. The coral-algal limestones of the Vaziri Unit occur in some places sandwiched within the Sabz Unit and these both confirm the shallow, warm sea environment and also, because they are of Burdigalian age (McCall et al. 1994), that a part of this unit is of Early Miocene age. The Sabz Unit is very incompetent and plastic, and has allowed access to exotic blocks, derived from the Coloured Mélange beneath (see McCall (1983) for the manner of emplacement of such exotic blocks). These blocks are few and scattered and the host beds are contorted more than dislocated, except in the Minab quadrangle, where there is sufficient lithological contrast to allow dislocation.

The Darkhunish Shale Unit crops out only within the Tahruie quadrangle, east of the Zendan Fault. Its thickness is estimated at 1500 m and it conformably overlies the Dehirdan Unit, and the Vaziri Unit and the Harzburgite Conglomerate where they are present. It has a facies transition in the east of the quadrangle into the upper part of the Sabz Unit. It is conformably overlain by the Dar Pahn Unit. The characteristic lithologies are thin grey shale, poorly cemented siltstone and ungraded or finely graded sandstone displaying festoon cross-bedding, flute casts and grazing track bioturbation. Planktonic and benthic foraminiferal faunas indicate an Early Miocene (Burdigalian) to Mid-Miocene (*Orbulina universa* zone) age (see McCall 1985e, pp. 111-112). The environment of deposition was neritic with intermittent sandy turbidites deposited by density currents. There was a slight deepening of the environment of deposition after the Burdigalian reefal limestone of the Vaziri Unit was deposited, but not to the extent of the time-equivalent Shahr Pum Unit. The unit is equivalent in its lower part to the Ab-e Shahr and Band-e Chaker Units. It is not so contorted as the Sabz Unit, being only rarely gypsiferous, and is flexed mainly into open folds.

The Band-e Chaker Unit unit crops out within both the Tahruie and Fannuj quadrangles, east of the Zendan Fault (Fig. 16). It thickens towards the SE from an estimated 2500 m to an estimated 5000 m, and overlies the Dehirdan and Vaziri Units conformably. The characteristic lithologies are dark shale, massive sandstone and thin pebble beds with acid-intermediate porphyry and metamorphic clasts. Shaly megacycles occur in the lower part of the sequence, which is more turbiditic than in the Darkhunish Unit. At the top are diagenetically red, yellow and green sandstone beds, sapropelic horizons, and leaves and stems of plants, and, within the Fannuj quadrangle, pebble conglomerates and shell beds, with bivalves and gastropods. Planktonic and benthic foraminiferal microfaunas are consistent with an Early (Burdigalian) to Mid-Miocene age (see McCall (1985e, pp. 84-65) for these and mollusc identifications). The environment changed throughout this sequence from fairly deep water (the trace fossil *Paleodictyon* was recorded: Fig. 23) with turbidite activity to very shallow shelf. Besides the Shahr Pum, Ab-e Shahr and Darkhunish Units, this unit is also equivalent and in many ways similar to the Roksha Unit, which crops out from the east of the Fannuj quadrangle through to the Pishin quadrangle. In the Tahruie quadrangle large open folds predominate and the unit is less tectonically active than the flysch units. Chevron and box second-order folds are less widespread, and most small folds are open. In the east, in the Fannuj quadrangle, periclines, with synclines dominant and

Fig. 23. Trace fossil *Paleodictyon imperfectum*, Miocene Band-e Chaker Unit, Tahruie quadrangle (actual scale).

anticlines weakly developed or absent, become widespread.

The Ghasr Ghand Unit (Fig. 17) occurs only in the southern part of the Nikshahr and Pishin quadrangles in the east and has an estimated thickness of 3200 m. It is faulted against all older units and the Roksha Unit. The reefal limestone of the Vaziri Unit is sandwiched within it and it has a sheared normal contact with the Sahan Tang Unit in the Pishin quadrangle, the latter unit being a facies variant. It is dominated by gypsiferous mudstones: the gypsum is secondary (diagenetic), but so abundant that this must again surely have been a primary gypsum deposit. There are thin sandstone beds and rare pebble conglomerates, and, locally, a sandy facies. The sandstones show planar lamination and ripple cross-lamination, convolute lamination and graded bedding. Foraminiferal microfaunas indicate an Early to Mid-Miocene age range (see McCall 1985e, pp. 132–133). Trace fossils include the deep-water *Paleodictyon*, but also *Paleophycus*, *Granularia* and *Fucusopsis* (the last two are deep- to shallow-water facies-crossing forms). The environment of deposition was mainly very shallow, with some turbidite involvement as well as intertidal sabkha. The unit is equivalent to the Sabz Unit to the west, and like it contains the reefal Burdigalian Vaziri Unit. It is also equivalent to the Roksha Unit. The outcrop is much obscured by white gypsiferous talus slopes. This unit is highly incompetent, and, in the Pishin quadrangle, is folded into a broad syncline, with smaller-scale tighter folding on the northern limb.

There are two facies variants of the Ghasr Ghand Unit, both cropping out in the extreme south of the Pishin quadrangle. The Sahan Tang Unit (Fig. 3), estimated at 4000 m thick, has a discordant junction with the Ghasr Ghand Unit, but probably was deposited above part of it more or less concordantly. It consists of sandstone–shale rhythms and is gypsiferous, like the Ghasr Ghand Unit. The sandstones contain volcanic fragments and small shell fragments. Trace fossils include the shallow-water *Tigillites* and the facies-crossing *Fucusopsis* and *Planolites*. Benthic foraminiferal microfaunas and rare planktonic microfaunas are consistent with a late Early to Mid-Miocene age range (see McCall 1985e, pp. 218–219) The sedimentation was by small turbidite fans prograding onto a shelf, much of it probably distal deltaic. The Pishin Unit (Fig. 3), estimated at 2500 m thick, is conformable above part of the Ghasr Ghand Unit or faulted against it. It consists of 100 m thick cycles with increasing sandstone content up-section. Festoon cross-bedding is developed in the sandstone. Planktonic and benthic foraminiferal microfaunas are consistent with an Early Miocene age. The rhythms in the lower part of the unit represent transitions from turbidite deposition to sand transport at or above wave base. The upper part of this sequence was entirely formed by turbidite deposition. Inner and outer shelf and slope environments are indicated by the microfaunas. The trace fossil *Spirorhaphe involuta*, a deep-water form, was recorded.

The Roksha Unit crops out in the SE of the Fannuj quadrangle and forms the 15 km wide Baghraband Syncline in the south of the Nikshahr quadrangle (Fig. 17), just extending eastwards into the Pishin quadrangle. It is subdivided into four divisions, the lowest 6453 m thick, the second estimated at 2000 m, the third 400 m thick and the uppermost estimated at 400 m. The lowest subdivision, which includes the reefal Vaziri Unit sandwiched within it in both the Fannuj and Nikshahr quadrangle, is faulted against the Ghasr Ghand Unit. Contacts with all other units are also faulted. In the Nikshahr quadrangle, the unit passes across the southern boundary of the mapping area and is seen to extend south into the unmapped coastal strip. The mapped subdivisions are, from the base up:

M^{r1}. Monotonous sequence of shale, siltstone and thin sandstones, all calcareous and locally gypsiferous. Grading, cross-bedding and load casts are recorded, as are frondescent marks, prod marks, burrows, and trace fossils *Ophiomorpha* and *Skolithos*.

M^{r2}. This is a sequence of coarsening-upwards cycles. Besides the lithologies mentioned above, pebbly sandstones and minor pebble conglomerate (locally diagenetically reddened) occur. These strata are calcareous but not gypsiferous. Cross-lamination, ripple marks, groove casts, flute casts and frondescent marks are recorded; also burrows and the trace fossils *Thalassanoides*, *Granularia*, *Ophiomorpha* and *Planolites*. There is sparse gastropod and bivalve debris, some entire turritellid gastropods, and also plant debris.

M^{r3}. Coarse-grained, thick beds of arenaceous sandstone predominate (95%), with minor calc-siltstone, arenaceous shale and conglomerate; graded bedding, bimodal cross-bedding, ripple cross-lamination and channelling occur. Lenses of pebbly or gravelly conglomerate, shell-wash and rip-up mud pellets occur as lag deposits in small channels. Numerous shell beds contain turritellid gastropods and bivalves, entire but transported a few hundred metres rather than in growth position. Burrows and the trace fossils *Thalassanoides*, *Tigillites*, *Ophiomorpha* and *?Arenicolites* are recorded.

M^{r4}. Poorly sorted pebble and cobble conglomerate with sandstone interbeds. Polymictic and clast supported, the conglomerate contains clasts of the older flysch lithologies, chert, basic volcanic rocks and coral from the reefal limestone of the Vaziri Unit. There is cementation by calcite and reddening of the matrix to a rusty brown. The sandstone is arenitic, and commonly graded and planar laminated, although some beds are massive and structureless.

The lower subdivision has yielded *Ditrupa* and a mixed foraminiferal microfauna of latest Early to Mid-Miocene age, also a Mid-Miocene benthonic microfauna. The subdivision includes locally the limestone of the Vaziri Unit, so must include Burdigalian time. The two lower subdivisions yielded microfaunas consistent with a latest age limit in Mid-Miocene time, but the upper two subdivisions, although yielding some foraminifera, cannot be dated on palaeontological evidence, so the age of the Roksha Unit might well extend to Late Miocene time. The lower subdivision is of open inner to outer shelf environment, the next of inner shelf and estuarine environment. The third subdivision was deposited in little more than 10 m water depth and the uppermost subdivision represents a talus cone prograding over it. This unit is equivalent to the Sabz, Ghasr Ghand and Jarut Units in age range. Broad, open folding characterizes the unit on the larger scale, but the lowermost subdivision locally displays dislocation and chevron-type folding on outcrop scale.

The Dar Pahn Unit crops out in a chain of synclinal basins within the Tahruie and Fannuj quadrangles, thickening southeastwards from an estimated 2000 m to 4000 m in one measured section within the Fannuj quadrangle. It conformably overlies the Sabz Unit, with local interdigitation, and also the Darkhunish and Band-e Chaker Units. This sequence is very diverse and only a brief summary can be given here (see McCall (1985*e*, pp. 114-116 for more details). It is characterized by thick, massive, gritty sandstones and pebble- to cobble-grade conglomerates with scattered boulders. Much of the sandstone is muddy and poorly sorted. The conglomerate is both matrix and clast supported. Calcite cement is common. There are minor shales, which are locally gypsiferous. Sandstones may be encrusted with *Ostrea*, turritellid gastropods and bivalves, and are interpreted as hard grounds. Most of the conglomerate carries shells and is marine, but locally it may be fluviatile. Burrows and the trace fossil *Ophiomorpha* are recorded. Marls, dirty micrites and mudstones have yielded mixed planktonic and benthic foraminiferal microfaunas (see McCall 1985*e*, pp. 117-118) and these indicate a Mid- to Late Miocene age range, consistent with its relationships with the three units of Early to Mid-Miocene age. Macrofauna in the Demohat Syncline in the extreme south of the Fannuj quadrangle includes *Ostrea*, gastropods and bivalves indicating brackish water conditions. This is a 'molasse' sequence, dominated by conglomerates, which are marine but related to post-orogenic uplift of the source landmass, so that one can envisage a series of shallow piedmontane, depositor basins bordered by an abrupt escarpment. Intermittent uplift and periodic flooding brought coarse detritus, and the finer argillaceous beds represent intervals of quiet deposition in shallow marine basins, possibly when the source area was planed down by erosion. Lateral shifts in drainage and depositional patterns may also have influenced the pattern of deposition. There is a general lack of features diagnostic of turbidites, although there are a few sole marks. High in the sequence there is some gypsum, now in secondary sites but representing sabkha environments. This very competent sequence experienced the effects of the orogenic deformation that all the rocks of the Makran except the Pliocene-Pleistocene fanglomerates and Quaternary superficial deposits were subjected to, but on account of its competence, the folding is almost entirely into large open synclines, separated by narrow, appressed anticlinal zones, and exposing thick, little-disturbed sedimentary sequences. The synclines form a spectacular chain, visible on Landsat imagery. These can only be tectonic structures, as is shown by the fact that the intervening anticlines correspond to the structures in older rocks to the north. The structures may, however, have been initiated while the sequence was still unlithified, because the section is thinner in the anticlinal zones, and this may be responsible for the lack of subsidence here, compared with the thick sequences of the synclines. There was renewed compression after the synclines formed and the northern limbs were steepened.

The Jaghin Unit (Fig. 3) crops out in the Tahruie quadrangle, in three subdivisions, the lower estimated at 1100 m thick, the middle at 700 m and the uppermost at 950 m. It occurs in synclinal outcrop zones, conformably above the Band-e Chaker Unit. The three subdivisions are as follows, going up-section.

M^{ja1}. Poorly sorted sandstone, lithic arenite, composed of angular rock fragments, feldspar and mica, gastropods and *Ostrea*. Matrix-supported, polymictic pebble conglomerate, with rounded to subrounded clasts of intermediate to acid porphyry and basalt. The conglomerate is lensed and channelled, and displays cross-bedding and ripple marks. Shale is minor but there are some shaly megacycles.

M^{ja2}. Buff to red sandstone with thin pebble- and cobble-grade conglomerate beds, Minor calcareous shale beds. The sandstone beds are massive and structureless, or laminated.

M^{ja3}. Shale, siltstone, sandstone and conglomerate, and shell beds. Fine- to medium-grained sandstones display cross-bedding, ripple marks and flute casts.

Mixed foraminiferal microfaunas are consistent with Mid- to Late Miocene age (McCall 1985e, pp. 153–154), which is consistent with the position over the Band-e Chaker Unit. The environments of deposition were shallow water, deltaic, paralic or estuarine. The unit was deposited in the already almost filled-up basin of the Band-e Chaker Unit, but also prograded northwards as the Shahr Pum Basin filled. The Jaghin Unit is correlated with the Dar Pahn unit. It is very competent and only flexed into large open folds.

The Jarut Unit forms an outcrop belt extending from the eastern half of the Nikshahr quadrangle though the Pishin quadrangle to the Pakistan frontier (Figs 17 and 19). The thickness is estimated at 6000 m. It has four subdivisions and a measured section through the lower subdivision covered a thickness of 2280 m. The unit is in faulted relationship with the Eocene–Oligocene flysch of the Ruk Unit to the north and the Oligocene–Miocene Angohran Unit to the south.

Both the Jaghin and Jarut Units are sandwiched by faults within the flysch, which is unusual for younger neritic sequences. The four subdivisions of the Jarut Unit are as follows, going up-section.

M^{j1}. Thin shale and subordinate sandstone cycles, with beds thickening up-section to 2 m thick sandstone beds at the top. The trace fossil *Paleodictyon* was recorded, as well as plant and leaf debris on rusty partings.

M^{j2}. Massive grey to buff calcareous sandstone beds, with individual beds very thick (1–10 m), forming the tops of very thick cycles (sandstone 60%) and forming cliffs. Cross-bedding and ripple cross-lamination are evident, as well as grading upwards from pebbly bases to fine sandstones. Such cycles form the lower part of the sequence, giving way up-section to finer and thinner shaly rhythms (sandstone 40%, in beds up to 1 m thick), siltstone and mudstone. Graded bedding, convolute lamination and carbonacoeus material are evident.

M^{j3}. Dark grey to black mudstone and siltstone with carbonaceous material. The trace fossils *Glockeria* and *Ophiomorpha* were recorded.

M^{j4}. Sandstone-dominated sequence, with red, green and purple zones discordant to the bedding. There are graded beds from pebble conglomerate to fine sandstone. Bimodal cross-bedding, channelling, flute casts, prod marks, bounce and scour casts are recorded. Desiccation cracks and rain pits indicate emergence. Purple, chocolate and greenish grey mudstone and shale make up the remainder (20%) and also display desiccation cracks.

Mixed planktonic and benthic foraminiferal microfaunas are consistent with an Early to Mid-Miocene age range (see McCall 1985e, pp. 157–158). Gastropods and bivalves in the upper subdivision are dominated by *Turritella bidentata* and *Terebralia angulata*. The lower part of the sequence is turbiditic but the remainder is shallow water, deltaic (delta front to delta top) passing up into fluviatile or mangrove swamp environments (*Terebralia* is indicative of mangrove swamps) (McCall & Eftekhar-nezad 1993). The Jarut Unit appears to have been deposited in a slowly subsiding basin. The sequence is time equivalent to the Ghasr Ghand and Roksha Units to the south, units separated from it by the Oligocene–Miocene flysch turbidite tract. The less competent lower subdivision displays small tight flexures (tight periclines in the Pishin quadrangle), but the upper more competent subdivisions are only flexed into large open folds. The unit as a whole forms a 'thrust slice' sandwiched between 'thrust slices' of older units within the regional listric fault pattern.

Two more units need to be considered here. The Vaziri Unit crops out in all four western quadrangles (Figs 4, 16 and 17), but not in the Pishin or Saravan quadrangle. Whereas it was referred to as a 'unit'; in the Paragon–Contech reports, it occurs in some cases within units and therefore in any formal stratigraphy would be a 'member'. It mostly occurs in small areas of outcrop, the largest covering 100 000 m^2, suggesting a patch reef setting. It is duplicated locally in the sequence, as in the Band-e Chaker fold structure. In one occurrence close to the Zendan Fault, there are numerous lenses of limestone within a sequence of reddened sandstones and shales measured in section as 2348 m. This unit occurs within the Sabz, Roksha and Ghasr Ghand Units, and beneath the Darkhunish and Band-e Chaker Units; it overlies the Dehirdan Unit. The limestones are reefal or reef-like and are dominated by corals (Fig. 24) and algae, together with, locally, gastropods and bivalves (Fig. 25). One shark's tooth was also found (Fig. 26). The coral limestone is commonly associated with a foraminiferal limestone carrying an extensive benthic microfauna, of Early Miocene (Burdigalian: 16.3–21.5 Ma) age (see McCall 1985e, pp. 239–244). A large collection of macro- and microfossils was taken to the Natural History Museum (London) by the author. The coral fauna was shown to comprise more than 40 genera and 90 species and to represent the largest recorded Miocene coral collection between the Mediterranean and Indonesia. It is important both intrinsically and because of the light it throws on the separation of the Mediterranean and Indian Ocean. To find such reefal limestones within an accretionary prism above a subduction zone is

Fig. 24. Coral, *Parascolymia vitiensis* (Brueggemann, 1877), Burdigalian Vaziri Unit, Minab quadrangle (×2).

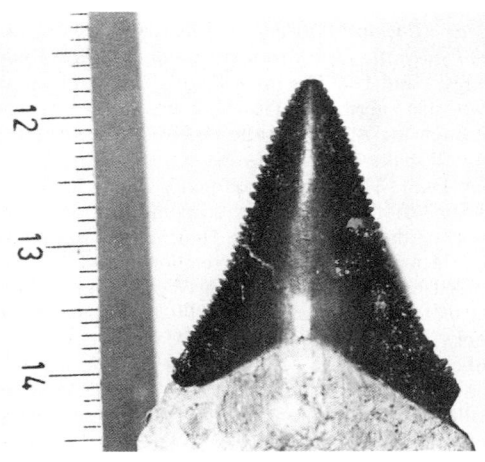

Fig. 26. Shark's tooth, Burdigalian Vaziri Unit, Minab quadrangle (scale in centimetres).

Fig. 25. Bivalve *Spondylus waylandi*, Burdigalian Vaziri Unit, Minab quadrangle (natural scale).

novel. The environment is interpreted as an intermittently developed and discontinuous carbonate shelf with patch reefs and marine grass beds, perhaps localized on a 'fore-arc' ridge: there was considerable contemporaneous erosion and reworking. The conditions were tropical to subtropical, warm shallow waters, as indicated by the still living coral species represented. These limestones have been described in considerable detail by McCall *et al.* (1994).

The Harzburgite Conglomerate crops out within the Tahruie quadrangle, within the Shahr Pum and Ab-e Shahr Units (Fig. 16). It contains limestone boulders derived from the Burdigalian Vaziri Unit and occupies much the same stratigraphic level. It is a boulder-, cobble- and pebble-grade conglomerate, associated with minor sandstone, siltstone and mudstone Cobbles of harzburgite predominate but there are a variety of ultramafic and mafic lithologies represented. Large angular blocks of the reefal limestone of the Vaziri Unit occur within the conglomerate (see illustration given by McCall 1985b, p. 138, plate 23), as well as Eocene limestone. The cobbles are sub-rounded and are water worn: they have the long axis parallel to the bedding. A Burdigalian age is consistent with the included limestone boulders and the position of the conglomerate outcrops close to outcrops of the limestone. The almost monomictic conglomerate must have a source in the Coloured Mélange Complex, possibly an isolated mass extruded up a submarine fault. The Eocene limestone probably had a similar source. This unusual occurrence suggests that movement of exotic blocks, as seen in the dislocated sedimentary units and sedimentary mélange (McCall 1983, p. 299) was continuing through Neogene time and was not entirely restricted to the terminal Late Miocene–Pliocene orogenic deformation that affected the entire Makran region.

Late Miocene–Pliocene neritic, deltaic, littoral and fluviatile deposits west of the Zendan Fault (Figs 4 and 16). The main Zagros Fault splits into two just north of the Makran and continues through the Makran in the dominantly east-trending Bashakerd Fault and the south-trending Zendan Fault. There is no continuity in the mapping of any of the formations cropping out east and west of the major Zendan Fault across the fault, nor is any lithological correlation possible. The folded rocks west of the fault (Figs. 4 and 16) are all of Late Miocene or Early Pliocene age. These rocks were described as the Makran Group by Stocklin (1952), a term retained by PC, but the term 'unit' was substituted for 'group', in line with an instruction from the GSI to erect only informal units (McCall 1985e, p. 165). Within the unit four subdivisions were recognized: (1) Gushi Marl; (2) Kheku Sandstone; (3) Tiab Sandstone; (4) Minab conglomerate. These stratigraphic sub-units are listed in Table 2 with their age ranges.

The Gushi Marl crops out in the western part of the Minab and Tahruie quadrangles (Fig. 27) and has an estimated thickness of 2250 m. It is dominated by grey, gypsiferous marl, mudstone and shale, with subordinate brown interbeds of friable sandstone (poorly sorted lithic arenite) and siltstone. There are some developments of pebble conglomerate. Discrete shell beds or a hash of broken shells are common. There is a diverse planktonic foraminiferal microfauna ranging from the *Globorotalia acostaensis* to *G. humerosa* zones, virtually the full range of Late Miocene time (McCall 1985e, pp. 165–166). Twelve molluscs are listed, including oysters, barnacles, gastropods, bivalves and echinoid fragments. Sharks' teeth were also recorded. Some beds are dominated by echinoid fragments. The environment

Table 3. *The stratigraphic units of the Saravan accretionary prism with their age ranges*

Stratigraphic unit	Age range
Piedmontane conglomerates (Shah-Kuh Granodiorite Intrusion)	Oligocene–Miocene (Oligocene)
Mashkid Unit	Late Eocene–Early Oligocene
Hizabad Unit	Mid–Late Eocene
Saravan Unit	Early–Late Eocene
Zaboli Unit	Early–Mid-Eocene
Talkhab Unit	Early–Mid-Eocene
Badamu–Siahan Unit	Early–Mid-Eocene

Fig. 27. Gypsiferous talus slopes covering outcrop of the Upper Miocene Gushi Marl, Tahruie quadrangle.

was a shallow platform sea, near-shore with a high evaporation rate (warm climate), probably including both intertidal sabkha and lagoons. There were periodic influxes of sand, as a result of uplift of the source area or climatic controls. The Gushi Marl is equivalent to the upper part of the Dar Pahn Unit east of the Zendan Fault.

The Kheku Sandstone (Fig. 28) overlies the Gushi Marl conformably, in both the Minab and Tahruie quadrangles, and has an estimated thickness of 2000 m. The lower part of the sequence consists of massive, uniform, thin grey-green sandstone and siltstone beds. They are succeeded by cycles of hard, medium- to fine-grained ferruginous sandstone, and less resistant mudstone and siltstone. Above these lies friable white sandstone with thin pebble horizons alternating with red ferruginous sandstone and pebble conglomerate lenses; and at the top comes massive pebble conglomerate. Coarsening-up cycles are characteristic of this unit. Channelling is widespread and in the Tahruie outcrops channels may be tens of metres wide: graded bedding, cross-bedding and ripple marks are also widespread. Macrofossils are abundant: gastropods, bivalves, oysters, bryozoa, ostracodes and spines. Planktonic foraminiferal microfaunas suggest that this unit extends into earliest Pliocene time (*Pulleniatina obliquiloculata*: McCall 1985e, pp. 168–169). Nine mollusc species are listed. The environment of deposition was near-shore, extremely shallow marine.

The Tiab Sandstone crops out only in a small area close to the present coast on the join between the Minab and Tahruie quadrangles, where it forms the middle subdivision of the Makran Unit within the Tiab Anticline. A total thickness of 478 m was recorded in a measured section. The base of the sequence is never seen but it is conformable beaneath the Minab Conglomerate. Red and brown calc-arenite and very fine-grained calc-sandstone of dune facies are associated with brown to light grey rubbly limestone (beach rock) and greenish grey soft fossiliferous calcareous mudstone. A rich planktonic foraminiferal microfauna suggests that the base is in the *G. humerosa* zone of Late Miocene time and that the sequence may extend into earliest Pliocene time. The unit is thus a facies variant of the Kheku Sandstone. Macrofossil debris is abundant, reworked from shallow marine and estuarine sources. A beach or offshore bar environment is indicated, with some deltaic representation.

The Minab Conglomerate crops out within both the Minab and Tahruie quadrangles. A thickness of 1355 m was recorded in a measured section. The unit overlies both Kheku and Tiab Units conformably. It consists of matrix-supported red to brown polymictic conglomerate, grading into pebble beds and sandstone intercalations. A basal sandstone–pebble conglomerate development separates this unit from the Kheku Sandstone beneath. Cross-bedding and graded bedding are widespread. Shell beds are common, including the 'Plazenta beds' of Huber (1952), which indicate a Pliocene age. As this unit overlies the Kheku Unit, it cannot be older than Pliocene time. These are

Fig. 28. Outcrops of the Upper Miocene Kheku Sandstone, Tahruie quadrangle.

mainly marine delta-top conglomerates, deposited in a period of high fluvial flow and/or strong uplift of the source. The uppermost beds on the Qal'eh Shaikan feature (Fig. 16) in the Tahruie quadrangle (Fig. 29) are fluviatile fanglomerates. The Minab Conglomerate is the youngest unit folded by the event that affects all the older rocks of the Makran region (even Pliocene rocks here have suffered strong flexure into open folds and display steep dips) and because of this it is likely to be of Pliocene age but older than the fanglomerates described by McCall (1996), which are unconformable on the folded sequence.

The sedimentary rocks of the Makran Unit are flexed into extensive open folds on a 10 km axial scale and this is consistent with general competence of the beds. The Makran Unit may have Late Miocene–Pliocene equivalent sequences in the coastal strip to the south of the area mapped by PC, but these shallow-water coastal units are very variable and it is not to be expected that they will be identifiable east of the Zendan Fault. This coastal strip was partially mapped near Jask (Fig. 2) by Stocklin (1952) and later by AGIP/SIRIP (1962), but despite the elegant maps produced by the latter, held at the GSI, there is no indication of the age of the sequences, although the geomorphology suggests that they may be entirely Late Miocene to Pliocene in age, and be equivalent in age to the Makran Unit.

Saravan accretionary prism.

Eocene flysch sequences. One of the most intractable problems of the Makran geology is explanation of the fact that the Eocene–Oligocene and Oligocene–Miocene–Pliocene sedimentary rocks of the main Makran accretionary prism continue without deflection eastwards through the Pakistan Makran, inland from the coast, whereas there is an immense development of flysch turbidite sediments sweeping down from the north, to the west of the Pakistan frontier in an arc that eventually brings them into a west–east strike behind the Eocene–Oligocene Ruk Unit outcrop and the other flysch units of like age described above. There appear to be two alternative explanations: either this represents an immense pre-Neogene dislocation (sigmoid fold structure or transcurrent fault) or it represents the development of a second accretionary prism during Paleogene time, built up in front of the Farah and Helmand Microcontinental Blocks of Afghanistan. The author favours the second alternative (McCall 1997), believing that the development here of a thick Eocene marine sequence indicates that there was an oceanic tract in existence in Eocene time, an extension of the Sistan Ocean (Sengor *et al.* 1988) (a tract analogous to the still-open Gulf of Oman to the west) and that these immense developments of flysch turbidites represent deposition in a deep basin, which terminated when the basin filled and closed up, and collided with the back of the main Makran prism. The rocks of this tract comprise a three-fold group, extremely thick units of like age ranges being separated by major faults: from east to west these are the Badamu-Siahan Unit, the Saravan Unit and the Zaboli Unit. Two other less extensive developments are probably

Fig. 29. Outcrop of the Lower Pliocene Minab Conglomerate, Q'al eh Shaikan, Tahruie quadrangle.

facies variants of the Zaboli Unit: the Mashkid Unit and the Hizabad Unit. The Talkhab Unit is a facies variant of the Badamu-Siahan Unit. The stratigraphic units of the Saravan accretionary prism are listed in Table 3 with their age ranges.

The Badamu-Siahan Unit forms a wide outcrop belt, extending from the Narreh-Now quadrangle in the north through the Saravan quadrangle and eastwards into Pakistan (Fig. 14). The thickness is estimated as a maximum of 13 000 m. No top or bottom is seen, but in the north, within the Saravan and Narreh-Now quadrangles, the unit is faulted against a shallow-water limestone and calc-siltstone sequence (Talkhab Unit) separating it from the Talkhab Mélange (McCall & Eftekhar-nezad 1994). Its western boundary is a major reverse fault, the Saravan Fault, separating it from the Saravan Unit, of equivalent age. This is a sequence of sandstones and shales, with shaly, sandy and siltstone facies differentiated on the map, and large areas undifferentiated. Coarsening-up cycles are common and may be 100–500 m thick individually. There are minor sandstone conglomerate, and, locally, conglomerate and grit bands contain volcanic fragments. Brown limestone lenses containing nummulites are common, and most of the nummulites are displaced. Algal and shell fragments may be present. The unit has suffered incipient metamorphism and surfaces of the finer lithologies glisten, but the grade is no more than sub-phyllitic in the south and lower greenschist facies in the north, where there is also contact metamorphism in the aureole surrounding the Shah-Kuh granodiorite and other similar bodies to the north of it, within the Narreh-Now quadrangle. On account of the penetrative deformation, sedimentary structures are less well preserved than in the other two units (Saravan, Zaboli), but graded bedding, flute and load casts, tool marks and ripple cross-lamination are recorded. The sandstones are sub-lithic, lithic, volcanic, quartzose and feldspathic arenites and are invariably calc-cemented. All show recrystallization. Benthic foraminifera are poorly preserved in the southern half of the Saravan quadrangle, but the indications are that this unit has an Early to Mid-Eocene age range. A number of benthic foraminiferal microfaunas from the northern half indicate that the unit is there mostly of Early Eocene age but extends into Mid-Eocene time (McCall & Eftekhar-nezad 1994, pp. 49–52). It may extend into the early Oligocene (Lattorfian) time, as suggested by two microfaunas from the north of the Saravan quadrangle. The sandy facies appears to represent turbidite fans spreading out southwards and sourced to the north; the unit as a whole represents a deep-water flysch trough environment. Persistent tight shallow anticlines and synclines, plunging in both directions, follow the arcuate strike trend, and have smaller second-order folds superimposed upon their limbs. Numerous minor faults follow the strike trend and generally dip to the NNE to north.

The Saravan Unit (Fig. 14) crops out to the west and south of the Badamu-Siahan Unit, the Saravan Fault bounding it to the east and north, and another major fault, the Dehak Fault, separating it from the Zaboli Unit. A thickness of 6850 m has been estimated for the entire unit. There are six subdivisions within this unit.

The Abedan Sub-unit is inferred to be unconformable on the Permian Morghak Unit, although the unconformable contact is not exposed. The sub-unit, up to 300 m thick, consists of granule- to boulder-grade conglomerate, with subordinate interbedded calc-arenite, calc-siltstone, marl, shale and micrite. It may pass transitionally up into the Murdan Sub-unit or, where that is absent, the Simish Sub-unit. The former, no more than 30 m thick, consists of massive white or grey nummulitic micrite beds with minor conglomerate and sandstone. Locally, red pelagic (globigerinid) siltstone overlies the micrite, and there are also developments of pelagic limestone, purple shale and siltstone. This sub-unit is locally absent, as a result of depositional and tectonic causes. The Simish Sub-unit is estimated to be up to 2000 m thick, and is a cyclic flysch turbidite sandstone and shale sequence, with minor interbeds of allodapic limestone and calcareous conglomerate. The cycles coarsen upwards, each 20–100 m thick. The tops consist entirely of sandstone beds. Graded bedding is common, but other primary structures characteristic of turbidites are rare. Red micrites and shales occur, associated with volcanic rocks (as in the Zaboli Unit). The limestone lenses are nummulitic. The Dabbah Sub-unit (c. 515 m thick) is dominated by sandstones, some conglomeratic. Both coarsening-up and rarer fining-up cycles occur, ranging from 20 to 100 m thick. Thin beds of displaced nummulite tests are common. Channelling is recorded in the coarser sandstone beds. Ripple marks, rare bottom structures and burrows are also present. The Kahan Davud Sub-unit (c. 4000 m thick) is a cyclic sandstone–shale sequence with only rare conglomerate, calcarenite and other limestone intercalations. Coarsening-up cycles are 30–100 m thick, and the entire upper part consists of sandstone. Load casts, flute casts, groove casts and tool marks are recorded, but the sandstones mostly are planar based. Graded bedding, cross-bedding and ripple marks, as well as pebble stringers, are also recorded. The Dehak Sub-unit is very poorly exposed and its thickness could not be estimated; it consists of shales with subordinate sandstone, the shales carrying minor amounts of gypsum. The sandstone displays ripple cross-lamination, graded bedding, flute casts, load casts and burrows. There are some unusually large bottom structures. Pebble beds and conglomerates are minor. This sub-unit contains some immense blocks of shallow-water limestone, apparently tectonically introduced. Cognate blocks (of the same age as the host sediments and derived from the Dabbah Sub-unit) include grey nummulitic limestone, calc-siltstone and mudstone. Older exotic blocks are of packstone and wackestone. Besides benthic foraminifera, the limestone blocks contain corals and algae, gastropods, bryozoa and echinoid spines.

Benthic foraminiferal microfaunas from the lower three sub-units (Abedan, Murdan, Simish) are of Early to Mid-Eocene age (McCall 1985e, pp. 222–226). Those from the Dabbah Sub-unit are of Mid-Eocene age, and those from the Kahan Davud Sub-unit of Mid- to Late Eocene age. Planktonic and benthic foraminiferal microfaunas from the Dehak Sub-unit are of Late Eocene age, matched by a planktonic and benthonic microfauna from the cognate limestone blocks. The older exotic limestone blocks yielded both Early and Mid-Eocene benthonic foraminiferal microfaunas.

The sedimentary evolution of the Saravan Unit shows the same progression as in the contemporaneous Zaboli Unit described below. The sequence commences with a shallow-water carbonate shelf environment (Abedan), conglomerates related to the basal unconformity and terrigenous clastic detritus being bound together by carbonate, alongside debris from patchy carbonate reefs. The reefs widened (Murdan Sub-unit) before deepening and removal of shoreline influences, introducing pelagic sediments. Inner shelf conditions followed (Simish Sub-unit), with decreased shoreline influences. Turbidite features are weakly represented, and limestones were also deposited. A proximal fan regime then developed (Dabbah Sub-unit) with a slightly more distal shelf regime to the west. This was followed by a middle fan regime, fairly deep water or shallower, a rapidly subsiding shelf regime (Kahan Davud Sub-unit), which represents something like two-thirds of the total thickness of the Saravan Unit and displays fully developed turbidite characteristics. Lastly there was an outer shelf regime (Dehak Sub-unit), but the abundant and large flute casts, up to 0.5 m long, on bottom surfaces in this sub-unit (illustrated by McCall & Eftekhar-nezad 1994) are unusual and indicate fairly high-energy turbidite flow. The entire Saravan Unit appears to have been deposited in a basin that was never very deep, and the initial carbonate shelf deposition was probably related to an extension towards the ENE of the Kuh-e Birk submarine high (Fig. 14).

The structure of the Saravan Unit is dominated by boat-shaped periclinal synclines and weakly developed or occluded anticlines. Folds are mostly upright, with vertical axial surfaces, but some are south-vergent. There are tight, second-order folds on the limbs and there is considerable dislocation. Faults, dipping to the NNE, traverse the outcrop zone.

The Zaboli Unit crops out in a broad belt parallel to that of the Saravan Unit, to the west and south of it (Fig. 14). To the east, the outcrop belt passes with a transverse join on the map into the Mashkid Unit (of Late Eocene to Early Oligocene age); this is probably a facies change. The contacts elsewhere are mostly faulted but, locally, SE of Suran, there appears to be a conformable, upwards transition. The join between the sedimentary rocks attributed to the two separate inferred accretionary prisms is confused by numerous faults. The field geologists indicated that locally there appears to be a tectonized, apparently conformable contact between the Mashkid Unit and Mosri Unit below it to the south. It is apparent that, if the separation into two separate accretionary prisms is correct, more detailed investigation on the ground is needed to understand the nature of the contact. For the purposes of this account, the Mosri Unit (see Figs. 14 and 19) is considered to be the northernmost unit belonging to the Makran accretionary prism.

The Zaboli Unit has an estimated total thickness of 6000 m. It has been divided into three lower calcareous subdivisions and two sandstone–shale subdivisions above, the latter two being undifferentiated over large areas of the outcrop. The lowest sub-unit (E^{Z1}), unconformable above or faulted against the Cretaceous Birk Unit, consists of massively bedded limestone, with reef-like patches and talus aprons, mainly biomicritic packstone and breccia. The bedded limestone contains large benthic foraminifera, reef-like coral and algal limestone fragments, gastropods, disarticulated bivalves and echinoids. Locally, a thinly bedded upper section of graded nummulitic calcarenite is developed, an allodapic limestone consisting of tests displaced from a shelf environment into a pelagic environment. Some calcareous sandstones and calc-rudites are also locally developed. The second sub-unit (E^{Z2}) consists of greenish or maroon fine-grained globigerinid biomicrite, a pelagic sedimentary rock. Subordinate marl, red mudstone and siltstone interbeds are present. Some volcanic rocks, amygdaloidal basalt and dacite flows, are interleaved. Andesite tuff and agglomerate, and volcanic arenite are also recorded. The third sub-unit (E^{Z3}) consists of green, maroon and red shale with minor siltstone and sandstone interbeds, as well as thin limestone beds and lenses. The fourth sub-unit (E^{Z4}) consists largely of weakly rhythmic shales, with local developments of sub-phyllitic chloritic shale. There are some fine-grained, ripple cross-laminated interbeds of sandstone and these display flute casts, groove casts and worm casts. The deep-water trace fossils *Spirorhaphe, Helminthoida, Urohelminthoida, ?Cosmorhaphe, ?Spirophycus* and *?Helminthopsis* are locally developed in abundance. Pebble- and cobble-grade conglomeratic sandstones form thin interbeds and display large load cast structures. The uppermost sub-unit (E^{Z5}) is a strongly cyclic sandstone–shale sequence, cycles being 30–100 m thick. Sandstone may account for up to 90% of the upper part of the cycles. The top of the cycle may consist of tens of metres of unrelieved sandstone beds. Ripple cross-lamination and graded bedding occur in the shaly lower part of the cycles, in thin sandstone beds, and sole marks on sandstone beds include flute casts, groove casts, and prod and tool marks. The shales are typically cleaved, locally sub-phyllitic, and 'pencil cleavage' occurs in tight folds. Trace fossils are again in evidence. Minor conglomerate beds are again developed, together with volcanic rocks that may form a localized sequence up to 300 m thick. They include andesite, andesite tuff and agglomerate, with minor interbeds of red pelagic shale and arenite.

The basal limestone unit yielded benthic foraminifera of Early to Mid-Eocene age. Pelagic lithologies of the lower three sub-units contain rich planktonic foraminiferal microfauna representing the greater part of Mid-Eocene time. The uppermost sub-unit has yielded both planktonic and benthic foraminifera of Mid-Eocene (Lutetian: 42.1–50.0 Ma) age (McCall 1985*e*, pp. 246–248).

In some parts of the outcrop of the Zaboli Unit it has not been differentiated into sub-units, and was mapped as E^Z.

There is a progressive change in the environment of deposition of the Zaboli Unit, from a shallow-water shelf, including both high- and low-energy depositions, and with a meagre supply of terrigenous detritus, through pelagic deposition into deep trench environments. There must have been a sudden deepening at the end of Early Eocene time. The remainder of the sequence is dominated by the flysch turbidites of the deep basin facies. Initially, the turbidite deposition was of distal facies, but a more proximal environment of increased sediment supply in prograding turbidite fans followed this. The environmental history overall was cyclic, from shallow to deep and then back towards shallow again. The Zaboli Unit, like the largely contemporaneous Saravan Unit, is characterized by initial carbonate shelf deposition, probably on the submarine Kuh-e Birk High, a microcontinental block that was more extensive than is now indicated by the narrow fault-bounded Birk mountain range (as is shown by the existence of the sub-Cretaceous Permian sequence well away from it, in the Morghak Unit).

The structure of the Zaboli Unit is complicated by numerous periclinal folds on mappable scale, and tight second-order chevron folds and isoclines developed on outcrop scale. The undifferentiated turbidite-dominated sequence, equivalent to the upper two sub-units, is widely dislocated. The unit is imbricated by numerous anastomosing reverse faults, following the strike sinuously, and dipping to the NE.

The Mashkid Unit, which straddles the boundary between the Pishin and Saravan quadrangles (Fig. 14), has an estimated thickness of 12 000 m and consists of three equal divisions, distal turbidites below, transitional facies above, and proximal turbidites at the top. Where this unit is tectonically dislocated, the three facies cannot be easily distinguished. The distal facies consists of coarsening-upwards (from 20 to 80%) cycles of sandstone and shale. Grits are also developed and there are basal nummulite death beds in the rhythms. Asymmetrical ripple marks, channelling, graded bedding and obscure sole marks also occur. Benthic foraminiferal tests display current imbrication. The transitional facies shows sandstone–shale alternations at the base of the rhythms. Bottom structures are well developed, including flute casts, groove casts and gutter casts. Ripple cross-lamination, climbing ripples and cross-bedding are evident. The cycles coarsen upwards to 80% sandstone. Turritellid gastropods and bivalves occur at the base of some sandstone beds, commonly associated with redeposited nummulites. The shallow-water facies consists of sandstone and thin conglomerate beds with minor shale and siltstone. Conglomerate beds are commonly lensed, and consist of rounded pebbles and cobbles of sandstone, quartz, chert, quartzite and volcanic rock. Clast support is most common, but matrix support is also recorded. Displaced nummulite tests and broken gastropod shells, together with *Ostrea* debris, form thin packstone beds, and nummulites may be concentrated on the foresets of cross-bedded sandstone. The trace fossil *Spirorhaphe* was recorded in the transitional facies of this unit. Planktonic and benthic foraminifera indicate a Late Eocene to Early Oligocene age for the entire unit (McCall 1985*e*, pp. 183–184) Although the broad tripartite sequence is recognized, there was much variation of the pattern throughout the extensive outcrop, and shallow-water (foredelta?) conditions may have occurred locally in the transitional facies. The Mashkid Unit contains shallow-water deposits in its top subdivision, and it may be significant that it overlies the extension to the west of the Kuh-e Birk Microcontinental Block, which formed a distinct submarine ridge trending ESE. The unit is folded into broad periclinal synclines more than 10 km in length. Dislocated zones occur between these, representing narrow anticlines with shaly beds in their cores.

There is, at the western margin of the Saravan quadrangle (Fig. 14), a small fault-bounded outcrop of a shaly deep-water flysch with volcanic tuff intercalations, named the Hizabad Unit. Silicic tuff is associated with volcanic arenite containing dacite clasts. Nummulitic packstones and allodapic limestones are recorded. Biomicritic wackestone has yielded Mid-Eocene planktonic foraminifera and benthonic foraminifera of Late Eocene age (McCall 1985e, pp. 149–150).

The Talkhab Unit, which intervenes between the Badamu-Siahan Unit and the Talkhab Mélange (Fig. 14), is dominated by orange shallow-water foraminiferal limestones and foraminiferal calcsiltstones of shallow-water, shelf facies, metamorphosed to phyllites. Orange septarian nodules are present, as well as thin sandstone beds. The unit is estimated to be just over 2000 m thick. The phyllites are metamorphosed to slates close to the Shah Kuh intrusion (Fig. 14). A number of foraminiferal microfaunas indicate an Early to Mid-Eocene age (McCall & Eftekhar-nezad 1994, pp. 58–60). The unit has a transitional facies with the Badamu-Siahan Unit, with which it is correlated, and it is in faulted contact with the Talkhab Mélange.

Oligocene–Miocene conglomerates. Sheared and folded polymict boulder conglomerates with minor sandstone, siltstone and basic lava interbeds crop out in small areas immediately SW of the Saravan Fault to the NW of Gasht, and again within the Narreh-Now quadrangle, to the north of the Saravan quadrangle (Fig. 14). Near the base of the north-dipping sequence, pebble-, cobble- and boulder-sized clasts are of volcanic rocks and sandstones, but higher in the sequence the boulders are largely of granodiorite, tonalite and quartz monzonite, derived from the nearby Shah Kuh and associated intrusions. These clasts indicate a maximum Oligocene age, based on the evidence of a Mid-Oligocene age for the Shah Kuh and associated intrusions (see below). These conglomerates are believed to be piedmontane, related to the uplift associated with these intrusions. There may have been early movement along the Saravan Fault, but these conglomerates have suffered the major end-Miocene–Pliocene tectonic deformation of the Makran, silt beds within this sequence being converted to 'pencil slates'. These are the only post-early Oligocene sedimentary rocks within the vast tract covered by the deposits of the inferred second Saravan accretionary prism.

The Shah-Kuh and associated granodiorite intrusions. There is a large intrusive body more than 30 km long straddling the boundary between the Saravan and the Narreh-Now quadrangles (Fig. 14). This is the southernmost of a chain of similar intrusions extending northwestwards to Zahedan (see Fig. 43, below). Another such intrusion in the north of the Narreh-Now quadrangle is believed to have the form of a sill, having concordant boundaries with the enclosing sediments. These bodies have some small associated outlying developments, roof pendants. The rocks forming the intrusion are mostly granodiorites, but there is marginal quartz monzonite. There are numerous late-stage aplite, porphyritic microdiorite and quartz diorite sills, dykes and irregular 'sweatouts' near the margins, within both the granodiorite and host sediments. There is a wide contact aureole around the main bodies, represented by hard slaty hornfels, with garnet spots close to the contact and chlorite spots further out, where phyllites and slates are developed. The sills and dykes have narrow contact aureoles. In some of the contact aureole rocks of the Badamu-Siahan Unit turbidite characteristics are preserved, but generally primary structures of sediments are obliterated by the contact metamorphic effects. There is slight copper staining associated with the Shah Kuh body (these bodies may be prospective for porphyry copper mineralization). Further details of these bodies have been given by McCall & Eftekhar-Nezad (1994, pp. 145–146). These intrusions apparently represent a final burst of magmatism above a subduction zone as the 'Sistan Ocean' finally closed up (McCall 1997) Radiometric dating indicates an Oligocene age for this chain (Geological Survey of Iran 1983). Granitoid intrusions of like age, apparently related to the main Makran subduction zone, occur to the north of the Jaz Murian Depression (Fig. 30) and active volcanic centres related to the present active subduction offshore are situated in Bazman, Taftan and Koh-i Soltan (Pakistan) (see Fig. 42, below). The subduction angle here appears to be very shallow (McCall 1997). The position of the Shah Kuh-Zahedan chain suggests that the subducting slab related to the Saravan accretionary prism may have been more steeply inclined.

Eocene–Oligocene outliers on the Lower Palaeozoic and older rocks. There are a number of outliers of Eocene–Oligocene sediments uncon-

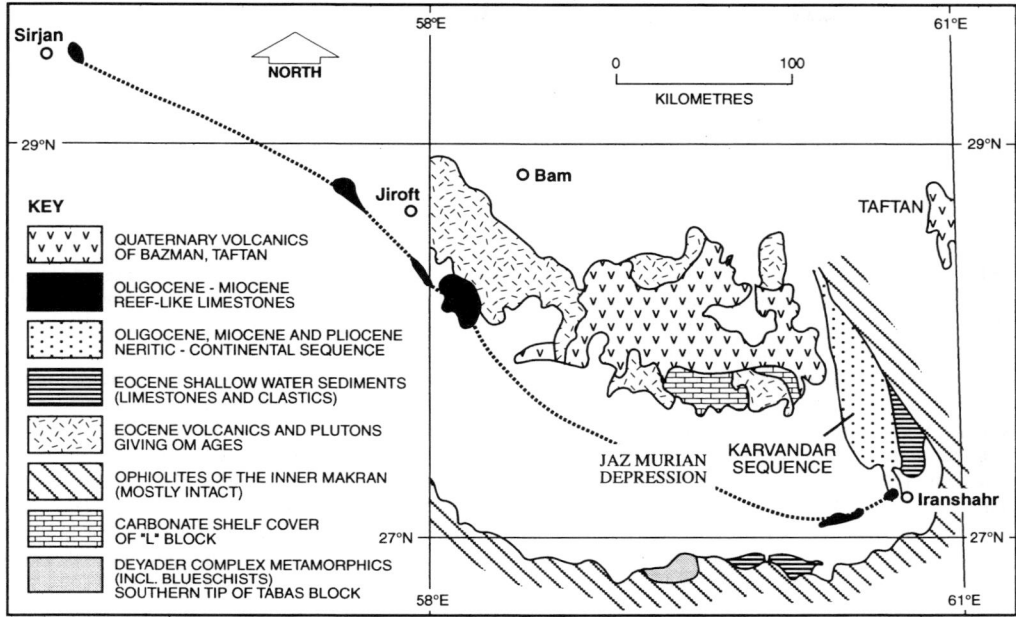

Fig. 30. Diagram illustrating the distribution of bedrock peripheral to the Jaz Murian Depression, and to the north of the Makran (from McCall 1997).

formable with Mesozoic–Lower Paleocene and older rocks. All these units are shallow-water marine deposits. They occur as structural outliers, not erosional basins superimposed on the older rocks. The main developments are the Bidak Unit, Bard-e Marz Limestone, Kam Sefid Sandstone, Marich Unit and Giran Unit. These stratigraphic units are listed in Table 4 with their age ranges.

The Bidak Unit crops out on the edge of the Jaz Murian Depression, to the east of the outcrop of the Band-e Zeyarat Complex, and curves round the nose of the complex near Kahnuj, in the north of the Minab quadrangle (Fig. 4). A minimum stratigraphic thickness of 842 m is estimated. In the eastern outcrops it overlies the Cretaceous Chah Mirak Unit unconformably and at the top interdigitates with some Eocene basalt flows. Elsewhere it rests unconformably on the Band-e Zeyarat and Dar Anar Complexes, and is inferred to rest likewise on the Ganj Complex. In the exposures close to the contact with the Chah Mirak Unit, it consists of conglomerates with volcanic clasts, associated with subordinate fossiliferous mudstone, shale, siltstone and coarse sandstone. Elsewhere, the conglomerate is subordinate and the other lithologies are well developed. There is one development of a reefal limestone, and rhyolitic tuff is also recorded. In the southern outcrops, it is reportedly intruded by numerous sills of diorite, diabase and metagabbro. Volcanic arenites and calcarenites and calc-wackes are widespread. The conglomerate clasts are mainly derived from the Ganj Complex, but also are derived from nearby rhyodacite porphyry. The tuffs are rhyolitic and include both lithic and welded tuffs. The reefal limestones contain Early and Mid-Eocene benthic foraminifera, as well as the algal species *Disticholax biserialis* and bryozoa in the upper part. The clastic sediments yield a microfauna of reworked displaced benthic foraminifera, in which *Nummulites fabianii* indicates a Late Eocene age. However, mudstones yield a planktonic foraminiferal microfauna of Early Eocene age McCall (1985e, pp. 90–91). The age

Table 4. *Stratigraphic units forming outliers on the Early Paleocene and older rocks of the Makran accretionary prism, to the north of the Coloured Mélange outcrop zone; with their age ranges*

Stratigraphic unit	Age range
Marich Unit	Mid-Eocene–Early Oligocene
Kam Sefid Sandstone	Mid-Eocene–Early Oligocene
Giran Unit	Mid–Late Eocene
Bidak Unit	Early–Late Eocene
Bard-e Marz Limestone	Early–Mid-Eocene

range of the unit is thus taken as Early to Late Eocene time. The environment of deposition was turbid, shallow water, with the reefal deposits being formed where there was restricted influx of clastic detritus. The Bidak Unit correlates with the Bard-e Marz Limestone, Kam Sefid Sandstone, Marich Unit and Giran Unit to the SE. The sequence is folded into open synclines and anticlines; shallow dips predominate, but there are locally steeply inclined beds.

The Bard-e Marz Limestone forms a conspicuous outcrop of white limestone separating the clastic sediment outcrop of the Marich Unit from the Ganj Complex. It crops out only within the Minab quadrangle (Fig. 4). The outcrop is 35 km long and averages 1 km wide, forming a ridge feature. A stratigraphic thickness of 212 m was recorded in measured section. The Bard-e Marz Limestone shows angular unconformity with the underlying Ganj Complex and subconcordant contacts with the overlying Kam Sefid Sandstone. In a typical section, there is a massive basal conglomerate with pink nodular limestone interbedded, overlain by a buff nodular limestone and then a white, massively bedded algal limestone. The unit is, however, laterally variable in lithology. The basal conglomerate contains clasts derived from the Ganj Complex and the matrix changes upwards, becoming less clastic. The limestones in the lower part of the section are dirty foraminiferal micrites, with sparse wackestones. The algal limestone is a boundstone with minor foraminifera. *Alveolina* (*Glomalveolina*) spp. in the lower part of the section and *Nummulites gizehensis* in the upper part are of Early and Mid-Eocene age, respectively (McCall 1985e, p. 88). There is an echinoid macrofauna of six species (Fig. 31), of which two are restricted to Eocene time: *Amblypygus* L. Agassiz 1840 and *Conoclypus* L. Agassiz 1839. Three bivalve species and three gastropod species were collected and *Velates* cf. *perversus* is known from the Laki and Khirthar Series of India and Pakistan, the genus becoming extinct in the Late Eocene time (McCall 1985a, p. 469). A nearshore to neritic environment is indicated. The unit has a stepped outcrop zone as a result of offsetting by minor faults and folds.

The Kam Sefid Sandstone crops out immediately west of the Bard-e Marz Limestone in a 50 km long outcrop belt (Fig. 4). It overlies the Bard-e Marz Limestone with a sheared normal contact and the Dar Anar Complex by an unconformity. At the top it is faulted out. The sequence changes progressively upwards from coarse conglomerate interbedded with immature, flaggy sandstone, minor siltstone and shale to an upper monotonous sequence of coarse sandstones, some calcareous, and siltstones. Graded bedding and ripple marks are recorded. There are a few boulder layers, with igneous and nummulitic limestone blocks up to half a metre in diameter, possibly debris-flow deposits. Pebble conglomerate and calcareous sedimentary breccias occur, as do thin lenses of nummulitic limestone. There are rare basalt sills and flows, as well as crystal and shardy tuffs. The sandstones are mainly volcanic wackes and the limestones are micrites. Both benthic and planktonic foraminifera indicate a Mid-Eocene to possibly earliest Oligocene age (McCall 1985e, pp. 160–161). Gastropods are common near the base of the sequence, apparently reworked from the Bard-e Marz Limestone beneath. An echinoid *Echinolampas*, a coral *Stylocoenia maxima* and a sea-star *Recurvaster* sp. indet. were collected from a small limestone outcrop. This is a moderately deep-water proximal turbidite sequence, of an upper slope environment, of flysch-type.

The Marich Unit is the most extensive of these units. It occurs in the Minab and Tahruie quadrangles, over a strike length of about 90 km, and has six subdivisions. Going up-section these are the Gowrt Shale, Konashamir Sandstone, Arkuran Shale, Geshmiran Sandstone, Gashulig Siltstone and Patkon Conglomerate. The total thickness is 4000–5000 m. Any of these units may be unconformable or disconformable above the Dur-kan Complex.

The Gowrt Shale, which crops out in the Minab quadrangle only, consists of contorted, thinly laminated grey and brown shales and purple-weathering, locally calcareous mudstones. There are minor sandstone, siltstone and limestone interbeds. It yielded a Mid-Eocene planktonic foraminiferal microfauna (McCall 1985e, pp. 173–174). The depositional environment was moderately deep water, shelf–slope.

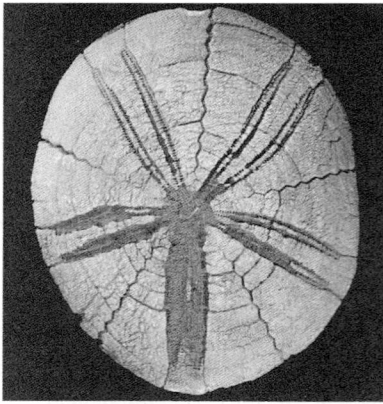

Fig. 31. Echinoid *Echinolampas sindensis*, Eocene Bard-e Marz Limestone, Minab quadrangle (actual scale).

The Konashamir Sandstone crops out in the Minab and Tahruie quadrangles, and also just extends into the Fannuj quadrangle; it is estimated to be 1500–2000 m thick. Medium to fine, well-bedded grey and brown sandstone predominates, but pebble conglomerates, calcareous shale and reefal limestone beds also occur. The limestone is foraminiferal and algal, and may be sandy and brecciated. Thin shale beds are contorted. Sandstones display graded bedding and flute casts. They are mostly volcanic arenites, although massive wacke beds also occur. In the Minab quadrangle, planktonic foraminifera from shales, limestones and siltstones, and benthic foraminifera from limestones (McCall 1985e, pp. 175–176) are consistent with a Mid-Eocene age, but from the Tahruie quadrangle there are indications from benthic foraminifera of Early Eocene sedimentation. The limestones contain algae and corals. The depositional environment of the sediments in the Minab quadrangle was a relatively shallow trough, with deposition of fine sediments being interrupted by intermittent turbidite activity with the influx of coarser detritus. Somewhat deeper-water conditions prevailed in the case of the sediments of the Tahruie quadrangle.

The Geshmiran Sandstone has an estimated thickness of 2000 m and has four sub-facies; calc-sandstone and shale; shale; sandstone; sandstone and conglomerate (Fig. 32). The sandstones are volcanic arenites and the shale is commonly silty. Graded bedding and cross-bedding are seen in the sandstone. There are some reefal limestone intercalations. Benthic and planktonic foraminifera are consistent with a Mid- to early Late Eocene age. This is a proximal turbidite sequence, deposited in an unstable basin of variable depth. Intermittent uplift of the landmass and subsidence of the basin is indicated by the cyclic influx of clastic material. The proximal fluxo-turbidites were deposited by mass-flow traction. A steep palaeoslope is inferred. The sandstones and shales are consistent in character with a mid–outer fan depositional environment. Taken together, this sequence appears to have been laid down under conditions of a narrow coastal shelf and the slope beyond it.

The Gashulig Siltstone is 1500 m thick and consists of fissile thinly bedded calc-siltstones, subordinate fine- to medium-grained calcareous sandstones and calc-shales. There are very minor beds of coarse sandstone, conglomerate, limestone and tuff. Planktonic foraminifera indicate a Late Eocene–Earliest Oligocene age (McCall 1985e, p. 179). Benthic foraminifera are also present. Limestones in this sub-unit and in the Geshmiran Sandstone beneath have yielded a rich benthic and limited planktonic foraminiferal microfauna (McCall 1985e, p. 180). The depositional environ-

Fig. 32. Conglomerate in the Geshmiran Sandstone Sub-unit of the Eocene–?Oligocene Marich Unit, Minab quadrangle.

ment was moderately deep water with redeposition of the sedimentary material from the basin margin towards the centre of the basin.

The Patkon Conglomerate is the uppermost sub-unit and is estimated at 2000 m thick. It is a localized development of polymictic pebble- to boulder-grade conglomerate with minor siltstone, shale and limestone interbeds. No microfossils are recorded. From its stratigraphic position, above the Gashulig Siltstone, it is taken to be of Late Eocene to earliest Oligocene age. The environment of deposition was shallow water, with a return to conditions more like those of the Geshmiran Sandstone.

The Giran Unit is 1320 m thick and crops out in the western half of the Fannuj quadrangle, forming a narrow synclinal outcrop zone 30 km long (Fig. 3). It has a largely faulted and tectonized unconformable contact with the Dur-kan Complex beneath. Grey-, green- and purple-weathering, immature, poorly sorted sandstones, locally epidotic, are associated with subordinate polymictic, matrix-supported conglomerate containing basalt, sandstone, limestone and shale clasts. Brown and green shale and mudstone are common, but limestone and tuff are very minor. There is a single andesite layer, either a sill or a flow. The sandstones are volcanic arenites, and have glaucophane as a minor component. The limestones are foraminiferal packstones. There is a reworked microfauna of Eocene benthic foraminifera, including nummulites (McCall 1985e, p. 134). A Mid- to Late Eocene age is inferred, because this is an extension of the very similar Marich Unit, given a separate name because of isolation of its outcrop to the east. The age of the sequence could extend to earliest Oligocene time. It is a proximal 'flysch-type' sequence, deposited in a moderately shallow basin with an elevated source of the detritus nearby. The structure is open synclinal and there are minor second-order folds.

There are a number of minor developments of Eocene to earliest Oligocene sediments that have not been named. In the Minab quadrangle, such outliers of sandstone, siltstone and pelagic limestone, with a tectonized unconformable contact with the Coloured Mélange Complex beneath, contain *Nummulites fabianii*, indicating a Late Eocene age. This is given the symbol E^s on the map. Other outliers may be as young as Oligocene time. There are also some developments of Mid- to Late Eocene flysch-type sandstone, conglomerate and limestone forming similar outliers on the Remeshk and Mokhtarabad Complexes in the eastern part of the Fannuj quadrangle (McCall 1985c, pp. 97–98). In the northern part of the Nikshahr quadrangle, there are some very small outliers of marls and limestones unconformable on the older rocks, yielding Eocene benthic foraminifera (McCall & Eftekhar-nezad 1993, pp. 78–79). All these unnamed occurrences are minor isolated extensions of the flysch-type Marich and Giran Unit sequences; they are not shown in the figures, but are represented in the relevant quadrangle 1:250 000 maps accompanying the GSI reports.

In general the Eocene (–Oligocene?) outliers are composed of very variable sequences of clastic sediments, with conglomerates being conspicuous and limestones very minor, except for the basal Bard-e Marz Limestone. Shallow-water facies are dominant, although deeper-water turbidite fan facies do occur. These sequences are contemporary with the Eocene flysch turbidite sequences, such as the Guredak and Ruk Units, but nowhere display the monotonous, unrelieved flysch turbidite character of those sequences, nor the thick internal cycles. This suggests that the older rocks beneath the unconformity formed largely high-standing areas within the Paleogene sea floor.

Cenozoic sediments of the Jiroft–Irafshan tract. To the north of the inner ophiolites of Mesozoic–Early Paleocene age (Fig. 4), the bedrock geology is largely obscured by the Jaz Murian Depression (Fig. 30). There is some evidence that this depression, now covered by desert, was initiated as far back as Eocene time (McCall 1997, p. 525). Along the southern margin of the depression are sporadic outcrops of Eocene strata. The Bidak Unit of the Minab quadrangle has already been described. In addition, there are small, low-standing developments of clastic sediments and limestones containing Eocene benthic foraminifera on the fringe of the depression within the Fannuj quadrangle and extending into the extreme north of the Nikshahr quadrangle. The greatest development of Eocene sediments is in a low north–south-trending range, immediately east of the Jaz Murian Depression, the Karvandar Sequence (Fig. 30). This has been mapped on 1:100 000 scale by the Geological Survey of Iran (1989), which indicated that these are also shallow-water clastic sedimentary rocks and limestones. Thus, within the inner arc of Mesozoic–Early Paleocene ophiolites, which turns northwards in the east to parallel the Karvandar Range and separate it from the flysch turbidite developments of the Saravan accretionary prism to the east, there are no developments of Eocene flysch turbidites, only clastic sediments and limestones deposited in shallow water. Above the Eocene sequences of the Karvandar Range are, successively, Oligocene, Miocene and Pliocene shallow-water sequences, including conglomerates. The only other developments of Oligocene–Miocene sediments are reefal limestones, which crop out to the west of Jiroft, on the western side of the Jaz Murian Depression (Fig. 30). These are a SE-trending linear extension of the reefal limestones that crop out near Sirjan at Saadabad and Chahar Gonbad (Sjerp et al. 1969; McCall et al. 1994). Similar reefal developments have been mapped by the Geological Survey of Iran (1989) to the west of Iranshahr (Fig. 30). These Oligocene–Miocene limestones are marine and contain rich fossil faunas. It is apparent that there was a residual shallow marine basin, which survived here through Neogene time, to Pliocene time, but its deposits, represented by bedrock, have been obscured by the Quaternary superficial deposits of the Jaz Murian Depression, except at its fringes. It is not known where this basin connected with the ocean, but the likelihood is that the connection was to the west.

Huber (1978) erected a 'Bazman Platform', in which Cretaceous platform sediments overlie the older rocks of the 'Lut Block' (i.e the Yazd–Tabas–Lut Block shown in Fig. 1). It is probable that the Chah Mirak Unit (Fig. 4) represents this Cretaceous cover. The Jaz Murian Depression probably largely obscures this Cretaceous cover, mantling the older rocks of the microcontinent. The Cretaceous rocks in turn underlie a thin sequence of Paleogene–Neogene clastic sediments and limestones, described above. The ophiolite belt fringing the Jaz Murian Depression to the west, south and east is apparently narrow, although it represents a once much wider extent of ocean.

Pliocene–Pleistocene fanglomerates. All the rocks so far described were affected by the intense orogenic deformations, folding and faulting, which took place at the end of Miocene time and even affected rocks of Early Pliocene age (McCall 1997) (Table 2). After this deformation there was intense uplift and locally conglomerates were deposited unconformably over the folded sequences. These are piedmontane deposits, formed against fault-scarp features. They have been described by McCall (1985e, 1996). Within the Minab quadrangle, the Palami Conglomerate forms a single development, 4500 m thick and covering an area of 18 km × 6 km, elongated parallel to the Zendan Fault. The unit consists of monotonous, well-bedded polymictic conglomerate, up to boulder grade and with sandstone lenses. From its stratigraphic position it can be no older than Late Pliocene time. To the south is another development of similar fanglomerates, the Tahtun Unit, 4935 m thick and covering an area measuring 12 km × 10 km. This forms a single fan, the front of which faces to the south, the axis of the fan parelleling the Zendan Fault. The sequence includes thin sandy and silty intercalated beds and lenses. On the southern margin, the base of the fan consists of conglomerates and sandstones with broken shell fragments, representing the passage from marine to continental deposition. A silty mudstone at the base contains death assemblages of gastropods and articulated bivalves. These basal beds are not reddened, whereas the overlying conglomerates are, like the Palami sequence. Graded bedding, ripple marks, burrows, minor scour channels and mudcracks indicating emergence characterize the basal beds. The conglomerates display coarse cross-bedding, normal and reversed grading and channelling, as well as imbrication of clasts. This is a regressive fluviatile sequence, following an estuarine base.

No further developments of fanglomerate occur to the east of the Tahtun Unit until the eastern half of the Nikshahr quadrangle, south of Sarbaz (Fig. 17), where similar conglomerates crop out in a 70 km long zone immediately south of the major Ghasr Ghand Fault and extending through the Pishin quadrangle to the Pakistan frontier. These conglomerates are grouped with three other such occurrences to the ENE, as the Nahang Unit. Of these, only in the 7 km long exposure immediately south of the Morton Fault within the Pishin quadrangle (Fig. 19) is the shape of the fan recognized, in this case with its axis parallel to the fault and prograding to the west. Two more extensive developments further east comprise the Nahang (Pishin quadrangle) and Kuhak (Saravan quadrangle) Synclines, which are 40 and 45 km long, respectively. Apparent thicknesses increase eastwards from 150 m south of Sarbaz, to 316 m in the Morton Fan, 700 m in the Nahang Syncline (near Gishdagan; Fig. 19) and 1590 m in the Kuhak Syncline. (Fig. 14). The typical lithology is poorly sorted cobble and pebble conglomerates with a friable sandstone matrix and minor silty material. In the Morton Fan the finer conglomerates are matrix supported, but the coarser conglomerates are clast supported and contain sparse boulders of limestone, although most are of sandstone (Fig. 33). In the Nahang Syncline, there is sparse calcite cement. The conglomerates are poorly sorted with respect to size of clasts. Cross-bedding, channelling, channel cross-bedding and imbrication are common. In the middle of the sequence in the Nahang Syncline, there is a development of muddy siltstone with thin layers of pebble conglomerate. A similar development occurs in the Kuhak Syncline, characterized by tabular cross-bedding with foresets 30 cm thick. At Gishdagan (Fig. 19), the ostracodes *Ilyocipris ramdohr* and *Cyprinotus* were recorded, the first species not known before Pleistocene time. However, the greater induration of the Nahang Unit, compared with some Pleistocene conglomerates in the NW corner of the Minab quadrangle, suggests that the Nahang Unit may be of Late Pliocene rather than Pleistocene age. A Late Pliocene–Pleistocene age is assigned to all these fanglomerate sequences. Although in all cases they are gently folded into synclines and traversed by faults of slight displacement, they clearly postdate the main deformation, and this is shown by the fact that the sequence of the Kuhak Sycline displays a spectacular basal unconformity over steeply dipping Eocene–Oligocene clastic sediments of the Mashkid Unit (Fig. 34). Such angular unconformity is not universal, and basal contacts are more commonly disconformable.

Superficial deposits. Superficial deposits of Pleistocene to recent age obscure the older bedrocks locally throughout the Makran. For instance, in the Nikshahr quadrangle they occupy 20% of the area. The oldest are bedded piedmontane conglomerates, extreme high-level terrace deposits which may be in part of Pliocene age. These are succeeded by unconsolidated high-level and low-level piedmont fan and valley terrace deposits, each produced by degradation of the older sur-

Fig. 33. Polymict conglomerate, Pliocene–?Pleistocene Nahang Unit, Morton Fan, Pishin quadrangle.

Fig. 34. Angular unconformity, Pliocene–?Pleistocene Nahang Unit above steeply dipping sandstone and shale of the Eocene–Oligocene Mashkid Unit, Kuhak Syncline, Saravan quadrangle.

faces. They characteristically form deflation platforms strewn with irregular rock fragments. The youngest deposits are silty outwash deposits and river channel alluvium, and these cut through wide developments of the older superficial deposits in narrow channels.

In the northwestern Makran, within the Minab quadrangle and just extending into the Tahruie quadrangle to the south (Figs 4 and 16), intramontane basin deposits of siltstone, sandstone and polymictic conglomerate deposits cover large areas, the largest of which is the Nurabad Depression (Fig. 3). These are not affected by the regional deformation, being only gently folded.

These are probably of much the same age as the high-level terrace deposits.

Within the Jaz Murian Depression (Fig. 2), there is a complete coverage of superficial deposits of Recent age: dune sands (including barchan dunes), stony deflation surfaces and saline playa lake deposits in the centre of the basin of internal drainage.

On the short section of coast in the Tahruie quadrangle (Fig. 2), there is a complex sequence of deposits, including intra-tidal mudflats (sabkha), intra- and supra-tidal ribbon sands, beach and barrier ridge deposits, island and spit deposits, as well as tidal channel, channel levee and sandy tidal flat deposits. In the extreme south of the Tahruie quadrangle, there are two mud volcanoes (McCall 1985b) (Fig. 35).

Structural overview

The regional mapping of seven quadrangles revealed a structural unity extending right through all rocks older than Early Pliocene time. This has been summarized by McCall (1985e, pp. 490–511). There is no evidence of refolding and it is apparent that the fold structures, whether they be superimposed on Neogene, Paleogene or Mesozoic rocks, were imposed in a single major tectonic episode during Late Miocene–earliest Pliocene time. The fanglomerate sequences of Late Pliocene (–Pleistocene?) age (Palami Conglomerate, and Tahtun and Nahang Units) are affected only by gentle open folds and minor faults, and are essentially post-tectonic. They were developed after and during a period of post-orogenic uplift. The uplift is still going on (as indicated by the euryhyaline foraminifera *Ammonia aotaenus* in silty sand of the Tahruie coastal plain, which shows that the upper terrace level represents an apparent higher relative sea level, explained by tectonic uplift of the landmass (McCall 1985b, p. 281)). The faulting is intimately related to the pattern of folding and is dominated by fairly steep reverse faults dipping north or NE, forming a complex 'schuppen' or 'duplex' pattern. These are interpreted as listric faults, shallowing in dip and merging in depth.

The dislocation widespread in those sediments of the accretionary prisms older than Mid-Miocene time is, again, intricately related to the main Neogene deformation. Dislocation is developed in sequences with a competence contrast because of alternating sandstone and siltstone or shale layers. Its progressive development can be followed in all its stages of development in the field. It has been more fully described by McCall (1983, pp. 295–299). The development of dislocation, marked by trains of blocks composed of the competent sedimentary layers, within a matrix of distorted less competent layers, may be accompanied by the emplacement of exotic blocks of other, mainly older lithologies, but dislocation may develop without such blocks. Deposits with these exotic blocks have been incorrectly mapped previously in

Fig. 35. Mud volcano near the shore of the Gulf of Oman, extreme south of Tahruie quadrangle (estimated diameter of the structure is 30 m).

Iran as 'wildflysch' or attributed to olistostrome formation. The process of formation is manifestly entirely tectonic. There is a tendency on the map for chains of exotic blocks to follow fault lines closely, although the blocks are not restricted to the actual fault surface. Dislocation is closely related spatially to intense folding as well as faulting (Figs. 36 and 37). Where dislocation is extreme, all stratigraphic continuity is lost and, where exotic blocks are scattered through the dislocated matrix, the rocks have been mapped as 'sedimentary mélange' (Fig. 38). This is a tectonic mélange, composed of sedimentary rocks, extremely deformed and incoherent, with many contained exotic blocks of various mainly older lithologies. The 'sedimentary mélange' has been described in more detail by McCall (1983, pp. 295–299). The exotic blocks have been emplaced from lower structural and stratigraphic levels by tectonic pressures.

The Neogene deformation has largely obscured older folding and fault structures within the rocks of Mesozoic–Early Paleocene age, but there can be no doubt from the mapping that there was significant deformation at the end of Early Paleocene time. Besides the evidence of unconformities, there are some other indications that there was faulting in this episode. For instance, the Eocene Bidak Unit trangresses in its outcrop across the boundary between the Ganj and Band-e Zeyarat ophiolite complexes, which are very different in character, and it would seem that these were brought together by pre-Eocene faulting. It is likely that there was a transcurrent fault separating them, before the later Neogene deformations took place. Again, the Oligocene–Miocene piedmontane conglomerates, in small areas of outcrop within the Saravan quadrangle, indicate that there were sizeable escarpments, fault scarps, in existence before the Neogene deformation. The evidence is mainly of older faulting; whether there was accompanying folding or not is uncertain.

Geotectonic overview

The geotectonic history and setting of the Makran have been covered in initial descriptions by Kidd & McCall (McCall & Kidd 1982; McCall 1985e, pp. 564–613; McCall 1997). McCall (1997) introduced new evidence resulting from the compilation of the Nikshahr, Saravan and Narreh-Now 1:250 000 maps and the accompanying reports on the first two quadrangles (McCall & Eftekharnezad 1993, 1994; Geological Survey of Iran 1987).

The major tectonic movements affecting the Makran are now considered.

Zagros collision

To the NW of the Makran, Arabia is pushing against and under the Zagros: the evidence sug-

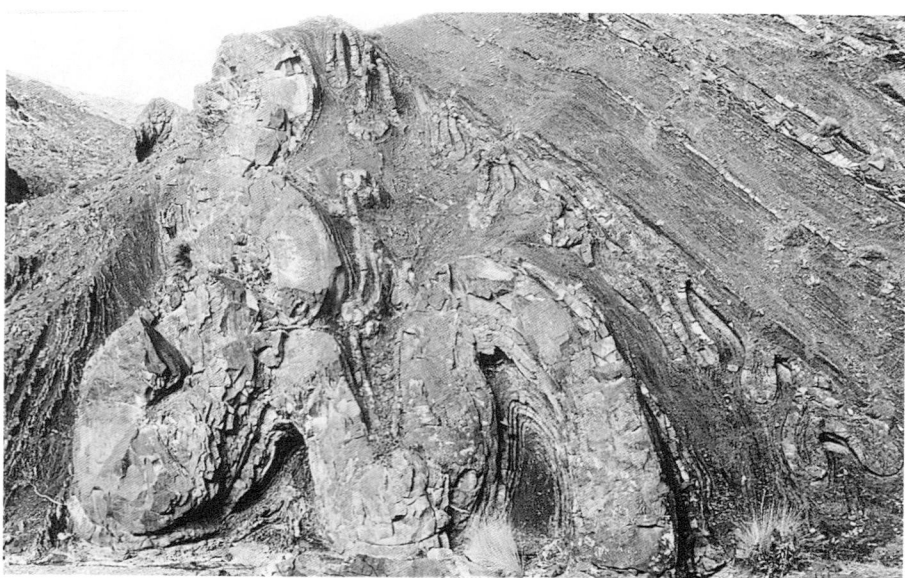

Fig. 36. Extreme contortion involving disharmonic folding of massive sandstone and shale alternations, Shirinzad Unit, Pishin quadrangle. With further deformation the sandstone suffers disruption and is preserved as separated lenses and blocks in the dislocated rock mass (field of view is c. 4 m across).

gests that the actual collison occurred after Late Miocene time (McCall & Kidd 1982). The boundary between Arabian continental crust to the NW and oceanic crust to the SE must be passing approximately under the town of Minab. This change is marked by the presence of salt domes to the NW and their absence to the SE, because the salt derives from Neoproterozoic and Cambrian saline sediments of the continental crust (Talbot 1998) (Fig. 39). The direction of movement according to Le Pichon et al. (1973) is NNE, and such a vector was also recently shown by Kukowski et al. (2000). It is indicated by the long arrows in Fig. 40. The change in tectonic style from accentuated anticlines in the Zagros sector to accentuated synclines in the Makran has been attributed to this boundary, and occurs at the position of the so-called 'Oman Line' (McCall 1985e, p. 572; Talbot 1998). The driving force behind the movement is complex: the opening of the Red Sea and spreading from the offset Sheba Ridge entering the Gulf of Aden (Fig. 40) may both contribute to this motion.

Makran front

The Coloured Mélange Zone of the Zagros continues into the Makran and traverses the Makran from west to east, inland of the Bashakerd Fault, but disappears through the Nikhshahr and Pishin

Fig. 37. A huge exotic block in a dislocated flysch turbidite matrix, Eocene Irafshan Unit, Pishin quadrangle.

Fig. 38. Sedimentary mélange, a large expanse of chaotic terrain, with a sandstone exotic block showing dark in the foreground, Fannuj quadrangle.

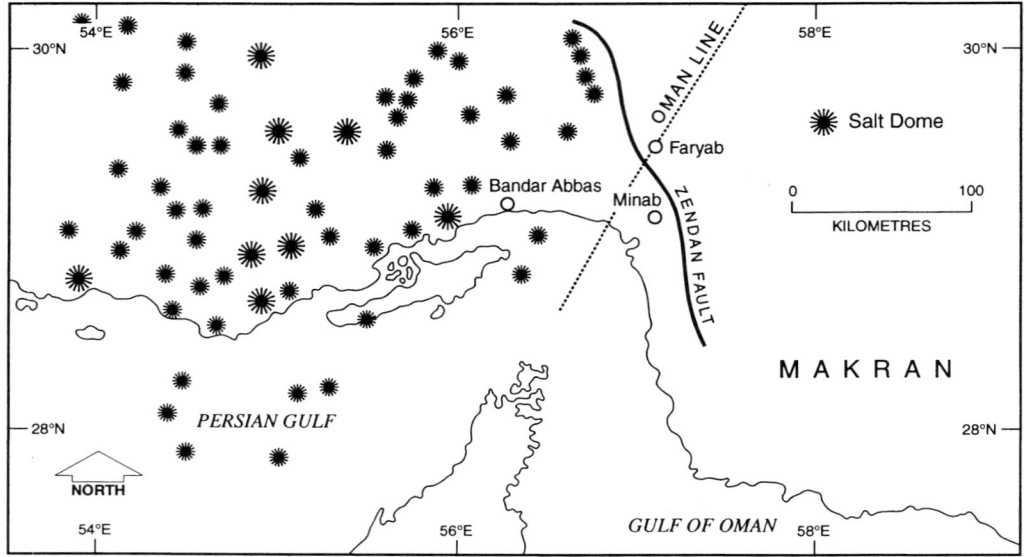

Fig. 39. Diagram showing the delineation of the underthrusting continental crust in the southern collision zone of the Zagros, picked out by salt domes and the 'Oman Line', which marks the southern boundary with subducting oceanic crust (after Shearman 1976).

quadrangles (Figs 17 and 19). This pinching out may be due to tectonic occlusion by the Eocene turbidites of the Makran accretionary prism (obscuring the older rocks) or actual termination of the ophiolite tract. Blocks and rafts of the Coloured Mélange within Cenozoic rocks suggest that the mélange continues at depth at least to the Pishin quadrangle. This mélange must record subduction that terminated in Early Paleocene time, when there was a significant tectonic event, the evidence of which is largely obscured except for the sub-Eocene unconformity. Arabia and India were closer during Mesozoic time (see Edwards *et al.* 2000) and together they formed the opposite coast to the Makran coast of this sector of Neotethys.

Subduction in Cenozoic time, when the vast accretionary prism was formed, must surely be correlated with dextral movement on the Owen Fracture Zone and spreading at the Sheba Ridge (Edwards *et al.* 2000) (Fig. 40). The present situation in the Makran is complicated by spreading from the Murray Ridge. Kukowski *et al.* (2000) have recognized, using seismological evidence, the Sonne Fault, a sinistral transcurrent displacement beneath the sea to the east of the Iranian Makran. This bounds a newly erected Ormara Microplate (Fig. 40). Seismic activity continues in this sector, reflecting the spreading from the Murray Ridge (composed of short ridge sectors offset by transform faults). The extension of the Murray Ridge (Dalrymple Trough) south of the Sonne Fault is not seismically active and has been taken by Kukowski *et al.* (2000) to be a pure transform.

Jackson *et al.* (1995) used seismological data to assess the style of deformation throughout Iran. The vectors indicated by the large arrows in Fig. 40 are consistent with their findings. However, it is noteworthy that the Iranian Makran east of the Zendan Fault has no activity, whereas to the east of Gwadar in the Pakistan Makran there is a cluster of seismicity (Fig. 41). Also, Jackson *et al.* (1995) showed the direction of movement here as northwestwards (Fig. 42). Such a vector would be compatible with the trend of the Bazman, Taftan and Koh-i Soltan active volcanic chain inland (McCall 1997) (Figs 40 and 43). Spreading off the Murray Ridge may be along this vector, and, if so, the component of spreading from the Murray Ridge does not at present affect the Iranian Makran sector to the west of the Sonne Fault (Fig. 40), any continuing subduction there of oceanic crust being related to the Sheba Ridge, far to the south.

The offshore part of the Makran accretionary prism, extending from the coast out to the deformation front, is 100–150 km wide, and characterized by a series of remarkably steep accretionary ridges of limited length (Kukowski *et al.* 2000). The Cenozoic accretionary prism is 350 km wide and the remainder of it is exposed on land,

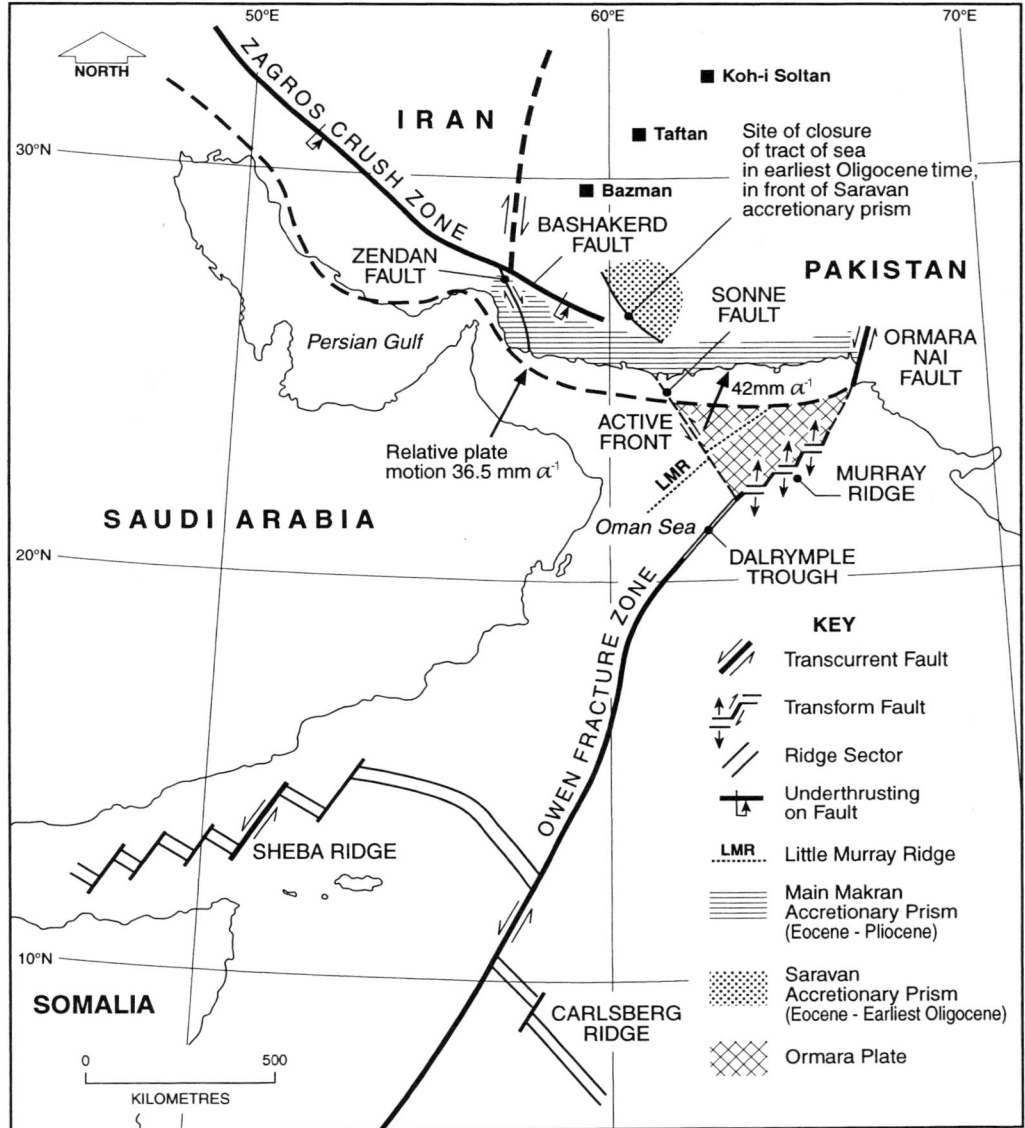

Fig. 40. Compilation of the offshore geotectonic elements in relation to the Makran.

because of uplift. The geological evidence from inshore shows that the build-up of the Cenozoic accretionary prism extended from latest Paleocene or earliest Eocene time to the present day. Kukowski et al. (2000) observed that 'a convergence zone east of Arabia and north of the (proto) Owen Fracture Zone was already active prior to the collision between India and Eurasia and probably represented the most continuous tectonic element in the Makran Region (Powell 1979)'.

Mapping of the Makran is entirely consistent with this statement, and the vector of the Cenozoic movement may well have been to the NNE, as shown in Fig. 40 for the present-day movement, but it is unlikely that past movement affecting the Makran was a steady state. Three major events punctuated the subduction on the Makran front. Besides the major disturbance producing an unconformity after early Paleocene time, the turbidite sequences of the main Makran accretionary

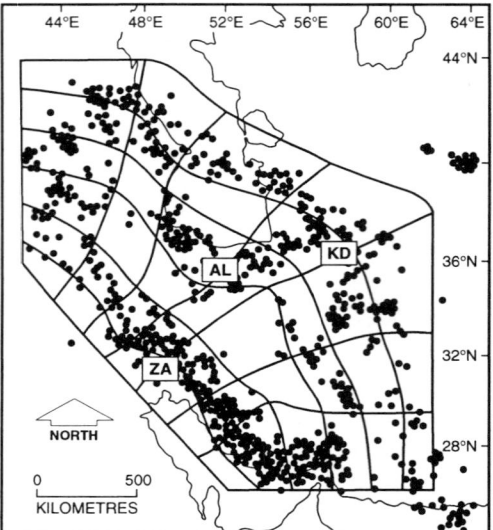

Fig. 41. Epicentres of earthquakes m_b >4.5, 1964–1990, for the Iranian region (from Jackson et al. 1995).

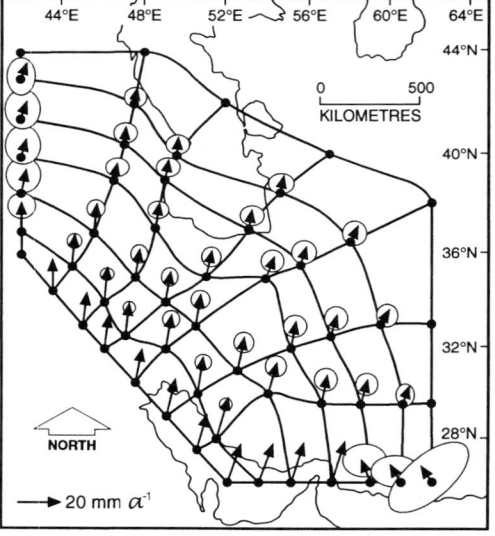

Fig. 42. Diagram showing the vectors of plate movement for the Iranian region (from Jackson et al. 1995),

prism are of Eocene–earliest Oligocene and Late Oligocene–Early Miocene age. The younger turbidites were conformably followed by thick neritic sediments, including some turbidites and early Miocene (Burdigalian) reefal limestones.

The mid-Oligocene hiatus is, according to the GSI, marked by an unconformity, and Huber (1978), working for the National Iranian Oil Company, also recognized a 'mid-Oligocene event'. In the Makran, this unconformity may be locally developed rather than universal, for there is commonly no evidence of discordance between the sequences, although tectonism may have obliterated an original discordance. Nevertheless, the hiatus certainly exists, although its cause remains obscure and this question can be resolved only by analysis of a much larger region than the Iranian Makran. It may be significant that the micropalaeontological evidence yielded only one indication of a sedimentary sequence bridging the Early to Late Oligocene divide.

The third event took place at the end of Miocene time and just affected early Pliocene neritic sediments; it was a very large-scale tectonic event that produced the schuppen (duplex) faulting, the intense folding, and the dislocation–mélanging of the accretionary prism sediments, with widespread displacement of large bodies of rock to exotic situations. There was also, immediately following, strong uplift, testified to by the huge developments of Pliocene (–Pleistocene?) fanglomerates. Gentle warping and minor faulting (renewals of movement) occurred to the present day, and affect these essentially post-tectonic sequences. The active deformation front was presumably moving progressively southwards during the prior build-up of the Cenozoic prism, but it appears likely that at this time it moved abruptly southwards.

The spreading from the Murray Ridge continues, to the east, but the inclination of the subduction surface is remarkably shallow for the active volcanic arc (Bazman, Taftan and Koh-i Soltan volcanoes), which is c. 600 km to the north of the active front (McCall 1997) (Figs 40 and 43). A similar shallow inclination is also indicated for the Cenozoic subduction beneath the Makran accretionary prism in Iran.

It will be noted that dextral transcurrent movement is assigned to the Zendan Fault, following Kukowski et al. (2000) (Fig. 30). This is in accordance with seismological evidence, but this fault has a long history and field mapping suggests that there has been considerable eastwards underthrusting on this major fracture in the past.

Kukowski et al. (2000) identified the Little Murray Ridge (Fig. 40) as a basement feature. This may well be another microcontinental sliver, analogous to the Kuh-e Birk feature, and likewise separated from Gondwana in Triassic time.

There is no extension of the geological formations of the Makran into Oman, but it is noteworthy that the ophiolitic rocks of Masirah Island have been demonstrated by Gnos & Perrin (1996)

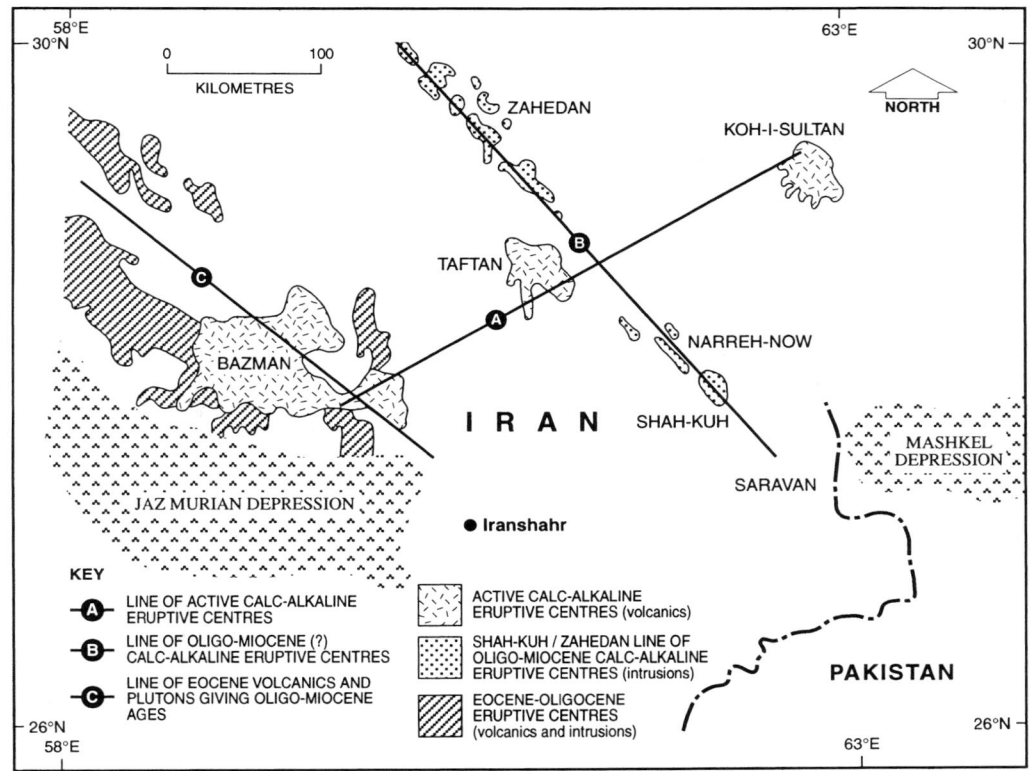

Fig. 43. Diagram showing the two chains of granodiorite intrusions of Paleogene (Eocene–Oligocene and ?Oligocene) age, respectively, north of the Jaz Murian Depression and between Shah-Kuh and Zahedan in the east: also the transverse chsin of active calc-alkaline volcanic centres (Bazman, Taftan and Koh-I Soltan).

not to be part of the Semail Ophiolite Complex of Oman, but to have been formed in Jurassic time to the west of NW India and obducted onto the Arabian continental surface 100 Ma later, during early Paleocene time (Marquer et al. 1998). This ophiolitic complex formed more or less contemporaneously with the early stages of formation of the Coloured Mélange and Remshk–Mokhtarabad Ophiolites of the Makran, but it is not possible, in the absence of any palaeomagnetic evidence, to indicate their geographical relationship at the time of formation.

The Saravan accretionary prism

The Eocene and later sediments of the Makran accretionary prism continue on an undiverted west–east trend into the Pakistan Makran. However, north of the Eocene turbidites lies an immense arc of Eocene and Eocene–early Oligocene turbidites, swinging from a NW–SE alignment to a west–east trend as they cross into Pakistan. Although the boundary with the turbidites of the Makran prism is obscure, these thick northern turbidite sequences appear to represent a second accretionary prism, composed of detritus from the Helmand–Farah Block to the NE. The age of the sequence extends just into earliest Oligocene time, and the only younger rocks here (other than Quaternary superficial deposits) are Oligocene–Miocene piedmontane conglomerates and the scattered line of Oligocene(?) granodiorite intrusions running NNW from Shah Kuh (Fig. 14) to Zahedan (Fig. 43) (McCall 1997).

There appear to be only two possible explanations: an immense sigmoid fold or fault displacement, or a second accretionary prism related to an oceanic tract comparable with the present Gulf of Oman and closed-up immediately after earliest Oligocene time, the prism colliding with the northern edge of the Makran ranges before being crumpled in the Late Miocene tectonic event. The latter solution appears to be the only acceptable on the field evidence. That there was an older Mesozoic subduction zone following this trend is evidenced by the Talkhab Mélange in the extreme north of the Saravan quadrangle (McCall & Eftekhar-nezad 1994, pp. 151–153). The oceanic tract represented by the ophiolitic Talkhab Mélange is a southeastwards extension of the 'Sistan Ocean' of Sengor et al. (1988). This sector apparently remained open into Cenozoic time. It seems reasonable to suppose that, as in the Makran prism, subduction developed here parallel to the trend of the Mesozoic Talkhab Mélange, the accretionary margin moving further southwestwards during Eocene and Early Oligocene time. Unlike the Makran prism, however, only very sparse neritic sediment deposition followed deepwater turbidite deposition and marine deposition ceased entirely in Early Oligocene time. The deformation at the end of Miocene time seems to have SW–NE compression, telescoping the prism by faulting and folding, although the driving force remains obscure. However, it is noteworthy that the age of the Neogene deformation throughout the Makran, including this second Saravan accretionary prism, is that of the Arabia–Zagros collision (McCall & Kidd 1982). This collision, collision on the Saravan Suture and possible other simultaneous collisions in the Afghanistan–Pakistan region, compressed the region into a single continental expanse extending from Arabia to India, within which compression was focused in sedimentary prisms that took the strain along faults and by crumpling.

It is noteworthy that in Fig. 41 there are two distinct zones of earthquake epicentres in southern Iran. The western zone follows the Zagros trend southeastwards and then curves eastwards along the Makran front. The eastern zone trends southeastwards, its western boundary following the line of the inferred front of the Saravan accretionary prism, sutured in Early Oligocene time.

Another indication of the reality of the Saravan accretionary prism is illustrated in Fig. 43. The active volcanic line (Bazman, Taftan and Koh-i Soltan volcanoes) is almost at right angles to the Shah Kuh–Zahedan chain of analogous calc-alkaline eruptive centres, of Oligocene(?) age. Such geometry is surely impossible unless an older Saravan accretionary prism had long ago ceased activity and been accreted against the back of the main Makran prism?

Although the alternative explanation involving large-scale transcurrent faulting or sigmoid flexure is rejected, there is a very large-scale sigmoid structure with its middle limb running northwards from Iranshahr (Fig. 2) to a sharp hinge into the NW–SE trend of the Kuh-e Birk Range (Fig. 1). This flexure (which resembles the Banda Arcs decribed by Hamilton (1977)) affects the pre-Eocene rocks and is interpreted as a pre-Eocene configuration.

Conclusion

The geology of the Iranian Makran essentially reveals a number of geotectonic zones in which the succession of the component rocks is very different. These zones are summarized in Fig. 44, reproduced here from McCall (1997). Three microcontinental blocks, two accretionary prisms largely composed of flysch turbidite sediments in very thick sequences and one former oceanic tract, exposing ophiolites, are represented in these columns, together with platform limestone and neritic clastic sediment cover. In Fig. 45, modified slightly from McCall (1997), the sequence of events is diagrammatically summarized.

The main conclusions are as follows.

(1) The Makran, together with the Zagros and all southern and central Iran, is founded on microcontinents, which separated from the edge of Gondwana in Triassic time. An analogy can be drawn with the present Mediterranean region, but here the tracts of sea have been obliterated, being represented by only Mesozoic–Early Paleocene ophiolites and the Paleogene Saravan accretionary prism. The microcontinental cluster could have parallels elsewhere, for instance the Long Range in Newfoundland.

(2) The Makran accretionary prism, of Cenozoic age, fronts the Ophiolitic Coloured Mélange development, a continuation of the Coloured Mélange of the Zagros. This tectonic mélange

Fig. 44. Generalized diagram showing the established and inferred sequences in the Makran and the region to the north of it (from McCall 1997). The columns do not represent the actual thicknesses, but simply illustrate the sequence of superposition. What emerges is that the microcontinental blocks, although submerged after separation from Gondwana in Triassic time, were nowhere covered by abyssal (flysch turbidite) sequences, but were covered by platform limestones. The ophiolite zones, representing Mesozoic deep ocean tracts, had deep ocean sediments (radiolarites, etc.) interleaved with the volcanic rocks, but were later in some areas covered by abyssal flysch turbidite sequences. Such sequences passed, in the main Makran accretionary prism, with no discordance into thick neritic clastic sediment sequences with some Miocene reef-like limestones, but in the Saravan accretionary prism there was no such development of neritic sediments (except for minor developments in the flysch turbidite sequences) and Oligocene–Miocene conglomerates followed terminal uplift much earlier than the uplift that produced the Pliocene to Pliocene fanglomerates throughout the extent of the Makran.

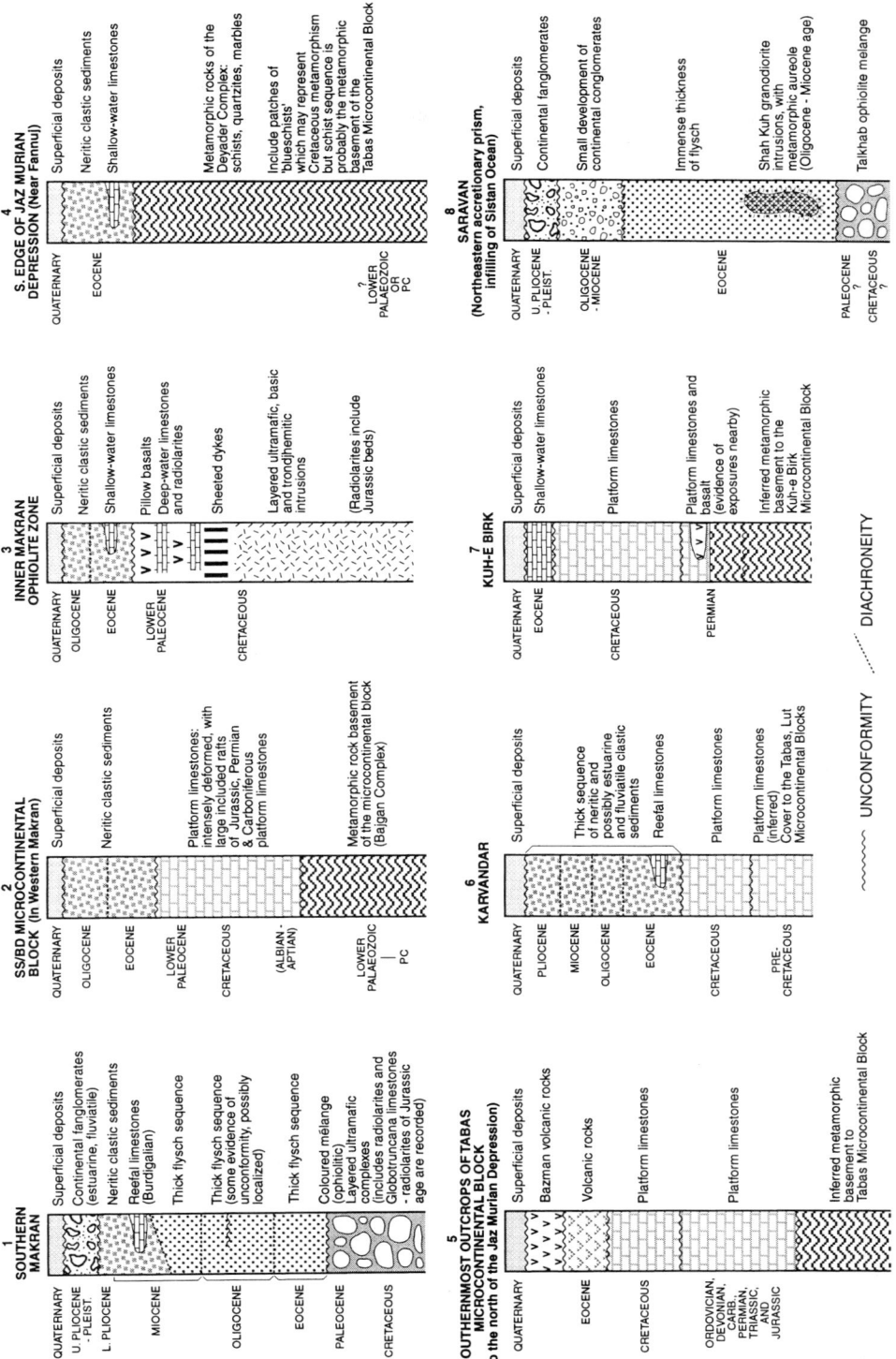

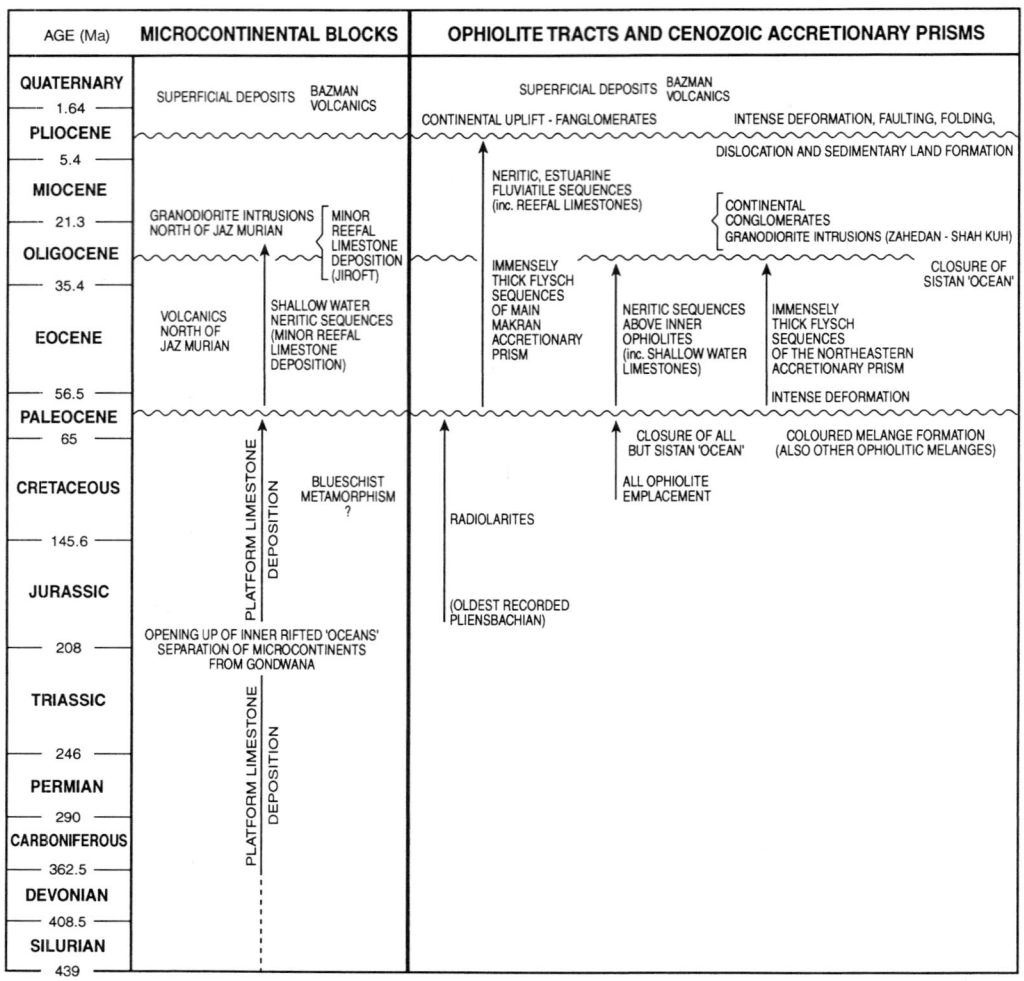

Fig. 45. Diagram illustrating the time relationships between the tectonic and other events in the Makran region (modified from McCall 1997).

represents the subduction zone fronting the narrow, several thousand kilometres long Sanandaj–Sirjan–Bajgan–Dur-kan Microcontinental Block, and occupied the north coast of the southern branch of Neotethys. After a tectonic event during Paleocene time, a very thick accretionary prism built up above a still subducting juncture; the inclination of the subduction surface was shallow, there was a tectonic event interrupting sedimentation in Mid-Oligocene time, and there was accumulation of a series of flysch turbidite fans in fairly deep water forming a remarkably thick sequence until Early Miocene time (21.5 Ma), when reefal limestones briefly appeared followed by a thick sequence of neritic clastic sediments.

(3) In the Saravan accretionary prism to the NE, after some shelf limestone deposition in Early Eocene time, deposition of flysch turbidites in fans in fairly deep water occurred, but continued only to Early Oligocene time, when this depository ceased activity. Neritic clastic sediments are minor in this accretionary prism.

(4) Throughout the Makran the main tectonic event took place in Late Miocene–Early Pliocene time, producing complex folding, listric faulting in a duplex or schuppen pattern, dislocation and formation of a sedimentary mélange, and this major event was followed by uplift and the production of scattered piedmontane fanglomerates.

(5) Calc-alkaline eruptive rocks occur in three chains: one, of Eocene–Oligocene age, inland from the Jaz Murian Depression, to the north of Jiroft; the second, of Oligocene age, intruded within the Saravan accretionary prism and extending NW from Shah Kuh to Zahedan; the third, the active chain extending from Bazman, through Taftan to Koh-i Soltan. The present-day subduction surface is clearly very shallow. The Makran accretionary prism is now 500 km wide, 150 km of it being offshore. Active mud volcanoes are scattered along the present Makran coast.

The projects in which the Iranian Makran was mapped are exceptional in that field activity by PC and ITI came to an abrupt end with the revolution in Iran at the end of 1978. Although the new government carried on the contracts, no final checking in the field was possible, nor any detailed follow-up studies. The stratigraphy developed was provisional, but, as far as the author knows, has never been formalized. The state of knowledge revealed in this summary account and the full reports and maps can only be improved now by further programmes of geological research in the field.

This account is based on the work of some 30 or so geologists who worked for PC from 1976 to 1980, and also other geologists of Intercon–Texas Instruments and the GSI, who mapped additional areas. The supervisors of the GSI played an important part in the successful outcome of this project, especially J. Eftekhar-nezad and M. Samimi-namin. R. S. White has kindly assisted the author by supplying material on which to base the final paragraphs on geotectonics, although the conclusions are the author's own. M. Lear was responsible for preparing the line diagrams from the author's drafts. The two reviewers, P. D. Clift and R. Coleman, made a wealth of invaluable suggestions for revision of the original draft.

References

AGIP/SIRIP 1962. *Unpublished Report on the Geology of the Southern Makran*. Filed at the Geological and Mineral Survey, Tehran.

CRIMES, T.P. & McCALL, G.J.H. 1994. A diverse ichnofauna from Eocene–Miocene rocks of the Makran range (S.E. Iran). *Ichnos*, **3**, 1–28.

EDWARDS, R.A., MINSHULL, T.A. & WHITE, R.S. 2000. Extension across the Indian–Arabian plate boundary; the Murray Ridge. *Geophysical Journal International*, **142**, 461–477.

GEOLOGICAL SURVEY OF IRAN 1987. *Geological Quadrangle Map of Iran No. M12, Narreh-Now, Scale 1:250 000* (compiled by J. Eftekhar-nezad).

GEOLOGICAL SURVEY OF IRAN 1989. *Karvandar 1:100 000 Map Sheet* (with side and back notes).

GNOS, E. & PERRIN, E. 1996. Formation and evolution of the Masirah Ophiolite constrained by paleomagnetic study of volcanic rocks. *Tectonophysics*, **256**, 53–64.

HAMILTON, W. 1977. Tectonics of the Indonesian Region. *US Geological Survey Professional Paper*, **1078**, 345 pp.

HSU, K.J. 1971. Franciscan mélanges as a model for eugeosynclinal sedimentation and underthrusting. *Journal of Geophysical Research*, **76**, 1162–1170.

HUBER, H. 1952. *Geology of the Western Coastal Makran Area*. Iranian Oil Company (National Iranian Oil Company) Report **GR 91B**.

HUBER, H. 1978. *Geological Map of Iran, Sheet No. 6, Southeast Iran, 1:1 000 000* (backnotes). National Iranian Oil Company, Tehran.

JACKSON, J.A. 1997. *Glossary of Geology*. American Geological Institute, Alexandria, VA.

JACKSON, J., HAINES, J. & HOLT, W. 1995. The accomodation of Arabia–Eurasia plate convergence in Iran. *Journal of Geophysical Research*, **100**, 205–215.

KUKOWSKI, N., SCHILLHORN, T., FLUSH, E.R. & HUHN, K. 2000. Newly identified strike-slip plate boundary in the northeastern Arabian Sea. *Geology*, **28**, 355–358.

LE PICHON, X., FRANCHETEAU, J. & BONNIN, J. 1973. *Plate Tectonics*. Elsevier, Amsterdam.

MARQUER, J., MERCOLLI, I. & PETERS, T. 1998. Early Cretaceous intra-oceanic rifting in the Proto-Indian Ocean recorded in the Masirah Ophiolite. Sultanate of Oman. *Tectonophysics*, **292**, 1–16.

McCALL, G.J.H. 1983. Mélanges of the Makran, southern Iran. *In:* McCALL, G.J.H. (ed.) *Ophiolitic and Related Mélanges*. Benchmark Papers in Geology, **66**, 292–299.

McCALL, G.J.H. 1985a. *Explanatory Text of the Minab Quadrangle Map, 1:250 000, No. J13*. Geological Survey of Iran, Tehran.

McCALL, G.J.H. 1985b. *Explanatory Text of the Tahruie Quadrangle Map, 1:250 000, No. J14*. Geological Survey of Iran, Tehran.

McCALL, G.J.H. 1985c. *Explanatory Text of the Fannuj Quadrangle Map, 1:250 000, No. K14*. Geological Survey of Iran, Tehran.

McCALL, G.J.H. 1985d. *Explanatory Text of the Pishin Quadrangle Map, 1:250 000, No. M14*. Geological Survey of Iran, Tehran.

McCALL, G.J.H. 1985e. *Area Report*. East Iran Project, Area No. **1**, Report no. 57.

McCALL, G.J.H. 1996. The post-tectonic fanglomerates of the Makran accretionary prism. *Geoscientist*, **6**(6), 11–13.

McCALL, G.J.H. 1997. The geotectonic history of the Makran and adjacent areas of southern Iran. *Journal of Asian Earth Sciences*, **15**, 517–531.

McCALL, G.J.H. & EFTEKHAR-NEZAD, J. 1993. *Explanatory Text of the Nikhshahr Quadrangle Map, 1:250 000, L14*. Geological Survey of Iran, Tehran.

McCALL, G.J.H. & EFTEKHAR-NEZAD, J. 1994. *Explanatory Text of the Saravan Quadrangle Map, 1:250 000, M13*. Geological Survey of Iran, Tehran.

McCALL, G.J.H. & KIDD, RG.W. 1982. The Makran, south-eastern Iran, the anatomy of a convergent plate margin, active from the Cretaceous to the Present. *In:* LEGGETT, J.K. (ed.) *Trench–Fore-arc Geology*. Geological Society, London, Special Publications, **10**, 387–397.

McCall, G.J.H. & Simonian, G.O. 1986. The Makran project—a case history. *In: Prospecting in Desert Terrain*. Institute of Mining and Metallurgy, London, 31–42.

McCall, G.J.H., Rosen, B.R. & Darrell, J.G. 1994. Carbonate deposition in accretionary prism settings: early Miocene coral-limestones and corals of the Makran Mountain Range, southern Iran. *Facies*, **31**, 141–178.

Powell, C.McA. 1979. A speculative tectonic history of Pakistan and surroundings; some constraints from the Indian Ocean. *In:* Farah, A. & DeLong, K.A. (eds) *Geodynamics of Pakistan*. Geological Survey of Pakistan, Quetta, 5–24.

Sabzehei, M. 1974. *Geology of Esfandageh*. Thesis, University of Grenoble.

Sengor, A.M.C., Altiner, D., Cin, A., Ustaomer, T. & Hsu, K.J. 1988. The origin and assembly of the Tethyside orogenioc collage at the expense of Gondwanaland. *In:* Audley-Charles, M. & Hallam, A. (eds) *Gondwanas and Tethys*. Geological Society, London, Special Publications, **37**, 119–181.

Shearman, D.J. 1976. The geological evolution of Southern Iran. *Geograpjical Journal*, **142**, 397–410.

Sjerp, T., Issakhanian, V. & Brants, A. 1969. The geological environment of the Chahar Gonbad Mine: a study in copper mineralisation. *Geological Survey of Iran Reports*, **16**(I-IV), 1–64.

Stocklin, J. 1952. *Geology of the Central Coastal Makran Area*. Iranian Oil Company (National Iranian Oil Company) Report **GR 91 C**.

Talbot, C.J. 1998. Extrusions of Hormuz salt in Iran. *In:* Blundell, D.J. & Scott, A.C. (eds) *Lyell: the Past is Key to the Present*. Geological Society, London, Special Publications, **143**, 315–334.

The geometry of structures in the Zagros cover rocks and its neotectonic implications

YOSEF SATTARZADEH[1], JOHN W. COSGROVE[2] & CLAUDIO VITA-FINZI[3]

[1]*Department of Geology, University of Tabriz, Tabriz 51664, Iran*
(e-mail: sattarzadeh@ark.tabrizu.ac.ir)
[2]*Department of Earth Sciences and Engineering, Royal School of Mines, Imperial College, London SW7 2BP, UK*
[3]*Department of Mineralogy, Natural History Museum, London SW7 5BD, UK*

Abstract: The Zagros Mountains are situated along the NE margin of the Arabian plate and are the product of complex deformation which began in Late Cretaceous time as a result of the collision between the Arabian and Central Iranian plates. During Pliocene time, deformation increased when plate convergence was accelerated by the opening of the Red Sea. This stimulated the migration of a deformation front from the collision zone towards the SW into the undisturbed Zagros basin and led to the creation of the Zagros Mountain Belt. The type and distribution of the deformation in the Zagros are controlled mainly by plate velocity, which is linked to the anticlockwise rotation of the Arabian plate around a pole in Syria, and the regional stratigraphy. The sedimentary cover and the underlying metamorphic basement decouple along an important detachment horizon, the Hormuz Salt Formation, and the uneven thickness and distribution of this salt plays a key role in determining the geometry of the deformation belt. Analysis of the distribution and geometry of the folds provides evidence for the southwestwards migration of the deformation front into the Arabian plate. The analyses are consistent with field evidence for serial folding, which indicates that each fold takes *c.* 600 ka to develop fully, and with the model of a southwestward advancing deformation front driven by the processes of serial folding and footwall collapse.

As can be seen from a tectonic map of the Middle East (Fig. 1), the Zagros Mountain Belt and extends over 1500 km along the northeastern margin of the Arabian plate from the Iraq–Turkey border to the Strait of Hormuz. The belt widens in a southeasterly direction and is bounded to the NE by the Main Zagros thrust line with an almost straight NW–SE trend, and to the SW by a sinuous mountain front. At their northwestern end the Zagros Mountains join the Taurus Mountains of southern Turkey; to the SE they are separated from the Makran Ranges by the north–south-trending Minab–Zendan Fault Zone (Fig. 1). The belt has an asymmetric topography and it rises gradually northeastward from the Mesopotamian alluvial plains at around sea level to a crestline of imbricate slabs and permanent snowfields between 3350 and 4600 m above sea level.

The range is divided into several structural zones (e.g. Oberlander 1965). The Outer (younger) Zagros, the southwestern half of the orogenic system, is a zone of intense folding produced for the most part by a late Pliocene orogeny (Falcon 1974; Alavi 1980). It is termed the Simply Folded Belt (Fig. 2) and is composed of a thick sequence of Late Precambrian to Pliocene shelf sediments without any visible angular unconformity (Fig. 3). The Inner (older) Zagros, which makes up the northeastern parts of the highland, is essentially a zone of thrust faulting. Its structures have been developed by several periods of compression beginning in Late Cretaceous time (Alavi 1980). This thrust zone is composed of a wide variety of lithologies including crushed limestones, radiolarites, ultrabasic rocks and metamorphic rocks, all of which have been intensely faulted. The thrust zone of the Inner Zagros has been subdivided longitudinally by Oberlander (1965) into a zone of horizontal overthrusts referred to simply as the Thrust Zone and a zone of high-angle thrust faults that he termed the Imbricate Zone (Fig. 2).

Although compressional deformation along the NE margin of the Arabian plate was initiated in Late Cretaceous time, the southwestward migration of the deformation front, mountain building and the creation of the Zagros Mountain Belt did not begin until Pliocene time. It is generally considered that this increase in deformation was

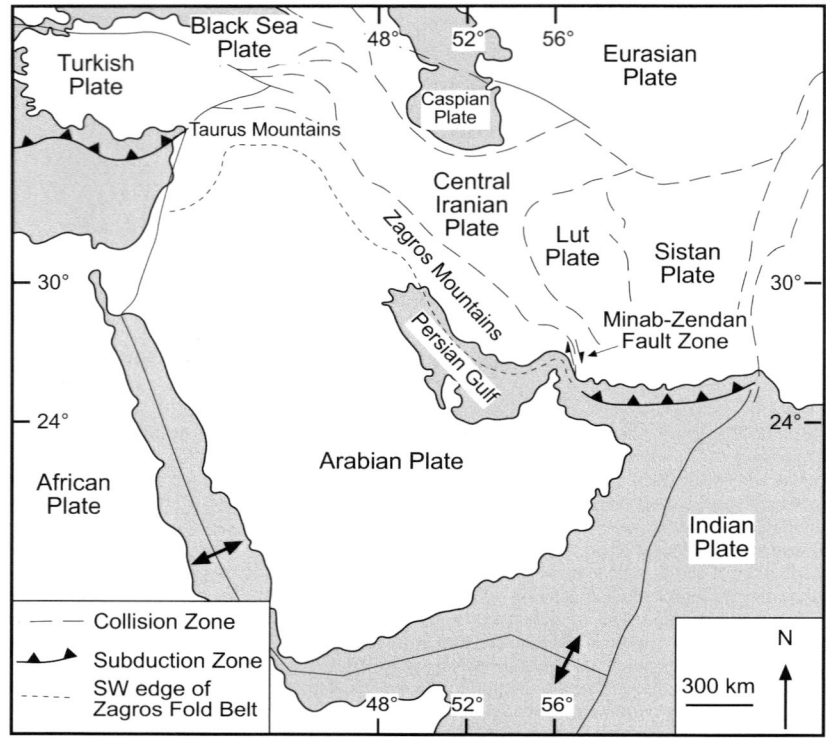

Fig. 1. Plate tectonic map of the Middle East showing the Zagros Mountains along the northeastern margin of the Arabian plate. Based on Berberian (1981) and Nowroozi (1987).

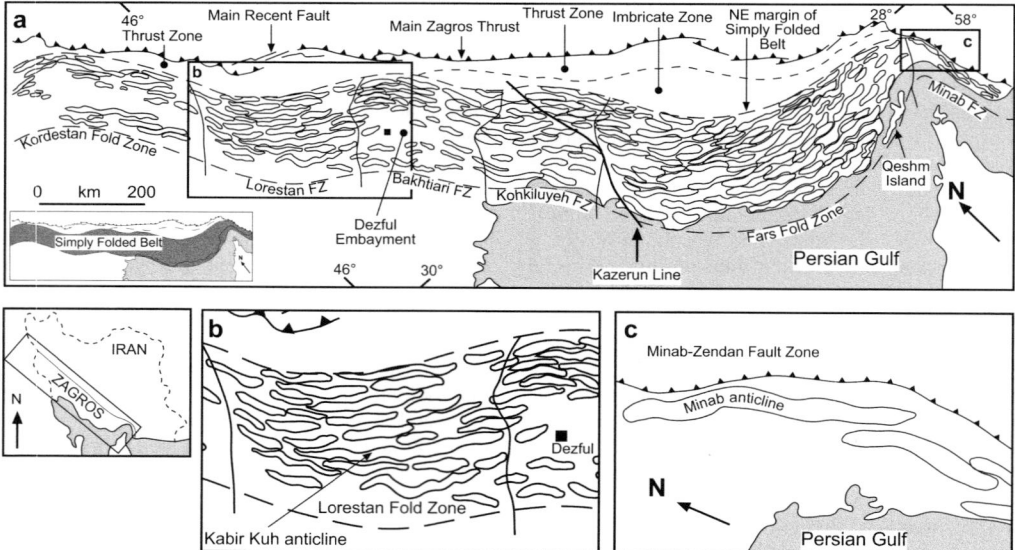

Fig. 2. (a) Distribution of folds within the 'Simply Folded Belt', which is divided into six zones. Rectangles indicate the locations of (b) and (c), which show the Kabir Kuh and Minab anticlines; these anticlines have anomalously high half wavelength/length ratios. They are not classical buckle folds but forced folds formed by reverse dip-slip and oblique reverse dip-slip movement, respectively, on underlying faults.

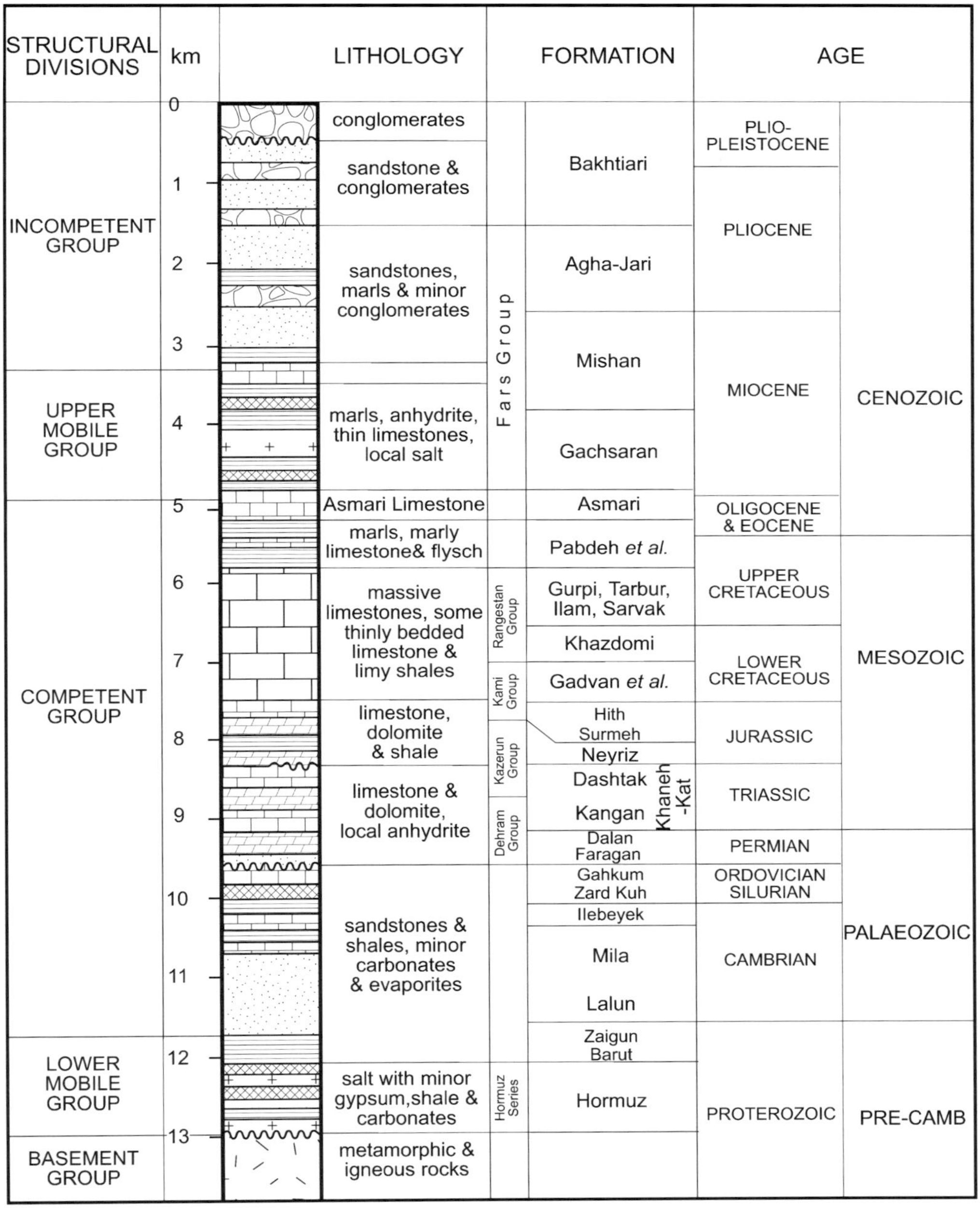

STRUCTURAL DIVISIONS	km	LITHOLOGY	FORMATION		AGE	
INCOMPETENT GROUP	0	conglomerates	Bakhtiari	Fars Group	PLIO-PLEISTOCENE	CENOZOIC
	1	sandstone & conglomerates			PLIOCENE	
	2	sandstones, marls & minor conglomerates	Agha-Jari			
	3		Mishan		MIOCENE	
UPPER MOBILE GROUP	4	marls, anhydrite, thin limestones, local salt	Gachsaran			
COMPETENT GROUP	5	Asmari Limestone	Asmari		OLIGOCENE & EOCENE	MESOZOIC
		marls, marly limestone & flysch	Pabdeh et al.		UPPER CRETACEOUS	
	6	massive limestones, some thinly bedded limestone & limy shales	Gurpi, Tarbur, Ilam, Sarvak	Rangestan Group		
	7		Khazdomi	Kami Group	LOWER CRETACEOUS	
			Gadvan et al.			
	8	limestone, dolomite & shale	Hith Surmeh	Kazerun Group	JURASSIC	
			Neyriz			
	9	limestone & dolomite, local anhydrite	Dashtak	Dehram Group	TRIASSIC	
			Kangan	Khaneh-Kat		
			Dalan Faragan		PERMIAN	
	10		Gahkum Zard Kuh		ORDOVICIAN SILURIAN	PALAEOZOIC
		sandstones & shales, minor carbonates & evaporites	Ilebeyek			
	11		Mila		CAMBRIAN	
			Lalun			
LOWER MOBILE GROUP	12		Zaigun Barut			
		salt with minor gypsum, shale & carbonates	Hormuz	Hormuz Series	PROTEROZOIC	PRE-CAMB
BASEMENT GROUP	13	metamorphic & igneous rocks				

Fig. 3. Stratigraphic column and the lithological and structural divisions of the Zagros Mountain Belt (updated from Coleman-Sadd 1978).

the result of the opening of the Red Sea, which was initiated in Miocene time (Laughton 1966) and resulted in the separation of the African and Arabian plates and an increase in the rate of motion of the Arabian plate towards the Central Iranian plate.

The stratigraphy of the Zagros region plays a critical role in controlling the deformation of the mountain belt and the evolution of all its major structures. From a mechanical point of view, the stratigraphic column can be considered to be made up of a rigid basement under a sedimentary cover including four distinct units (Fig. 3). The lowest unit is the Lower Mobile Group, which is composed predominantly of the thick Hormuz Salt Formation (about 1 km thick) and allows the complete decoupling of the basement and cover during deformation. Overlying the Lower Mobile Group are, in ascending order, the Competent Group, the salt-rich Upper Mobile Group, and the Incompetent Group. The Competent and Incompetent groups often fold harmonically. However, in places the Upper Mobile Group is very thick and salt-rich and the two groups fold disharmonically, and the deformation above and below them becomes completely decoupled (Fig. 4).

Large-scale folds, faults and salt diapirs are the dominant geological structures that have developed in the course of the structural evolution of the Zagros. The present study focuses on the spatial distribution, geometry and mechanism of formation of these structures in an attempt to improve our understanding of the processes responsible for their generation and their implications regarding the structural evolution of the belt.

Folds

The Zagros Mountain Belt of Iran is one of the best examples of a classical young orogenic belt in the world and shows a spectacular collection of folds with a large variety of fold styles. The spatial distribution, geometric characteristics and mechanisms of formation of these folds are considered in the following sections.

The Zagros folds are generally confined to what is referred to as the Simply Folded Belt (Fig. 2). Folds are also developed in the Imbricate Zone (Fig. 2) but they are commonly distorted and cut by high-angle reverse faults and were deeply dissected during a long period of erosion.

Although the belt runs approximately parallel to the Main Zagros Thrust (Fig. 2) there are variations in its trend, which give it a sinuous geometry in plan. The width of the Simply Folded Belt is not constant and generally increases from NW to SE. The northeastern margin of the Simply Folded Belt is marked clearly by the sinuous boundary between the Simply Folded Belt and the Imbricate Zone. In contrast, the southwestern margin is not well defined at the surface. Nevertheless, geological sections and seismic profiles indicate that the southwestern margin of the fold belt, like its northeastern margin, is also sinuous and runs almost parallel to but c. 50–100 km SW of the Mountain Front, the most southwesterly topographic expression of the thrusting, under the Mesopotamian Alluvial Plain and offshore under the Persian Gulf.

The population and spatial distribution of the folds vary along the Simply Folded Belt and on the basis of their morphological and structural differences the Belt can be divided into six contrasting fold zones. From NW to SE these are the Kordestan, Lorestan, Bakhtiari, Kohkiluyeh, Fars and Minab fold zones (Fig. 2).

Folds within these distinct fold zones are arranged in different ways when they are viewed in plan. However, they are generally aligned with their hinges sub-parallel to each other and are frequently offset from each other in an en echelon manner. The en echelon disposition of the folds may be either random or consistent. A random en echelon offset is a typical feature of classical buckle folding. A consistent en echelon offset of folds suggests the influence of basement strike-slip faults.

The increase in width of the fold belt when traced from NW to SE is compatible with the anticlockwise rotation of the Arabian plate. Second-order variations in the width and trend of the

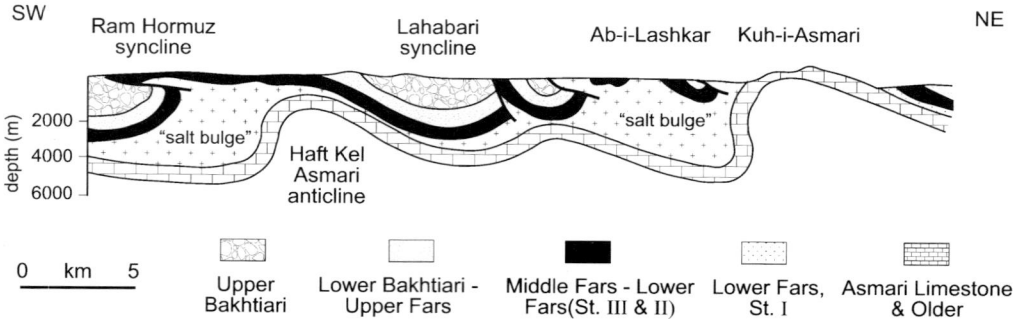

Fig. 4. NE–SW cross-section through the Masjed-e Solyman and Haft Kel areas of the Dezful Embayment (Bakhtiari Fold Zone, Fig. 2), showing the Miocene Lower Fars salt-rich evaporites (Gachsaran Formation) forming 'salt bulges' and 'flow-sheets' as a result of disharmonic folding. (Modified from O'Brien 1957.)

belt are shown below to be the result of the uneven thickness and distribution of the main detachment horizon (the Infra-Cambrian Hormuz Salt Formation) at the basement–cover interface.

The present authors consider that the anomalous bending of the Simply Folded Belt around the Musandam Peninsula and the considerably narrower width of the belt in the Minab Fold Zone east of the Strait of Hormuz result from the indentation of the Musandam Peninsula into the Zagros basin during the Pliocene orogeny (Falcon 1974; Alavi 1980). The peninsula acted as a rigid barrier to the southwestward advance of the fold front.

Fold geometry

The aspect ratio (half-wavelength/hinge length ratio) of many of the Zagros folds is about 1:5, which is typical of buckle folds (see Price & Cosgrove 1990). However, some of the Zagros folds, such as the Minab Anticline in the Minab Fold Zone and the Kabir Kuh Anticline in the Lorestan Fold Zone, are characterized by high aspect ratios of the order of 1:20 to 1:30 (Fig. 5). Sattarzadeh et al. (2000) have shown that the type of folding and its distribution are linked to the pattern of basement faulting, as well as the rheological profile of the cover sequence. En echelon arrangements of buckle folds are found above basement wrench faults. Forced folds (e.g. the Kabir Kuh Anticline, Fig. 2) occur over reverse dip-slip faults and transpressional strike-slip basement faults.

As the present study is concerned primarily with the shortening across the Zagros Mountain Belt, considerable attention has been paid to the profile geometry of the folds. To study variations in the shortening of the whole sedimentary cover across the fold belt the following measurements were made along 21 geological sections running approximately NE–SW covering the four central zones distinguished on the basis of morphological and structural characteristics (Figs. 2 and 6a): (1) the half-wavelength/amplitude ratio of the folds; (2) the elevation of the median surface of the folds; (3) the percentage shortening of the sedimentary cover.

Half-wavelength/amplitude ratio of the folds

Because the wavelength of the folds along the sections studied is rather variable the ratio of half-wavelength/amplitude was measured, as this provides a normalized value that relates directly to the stage of development of the fold and is independent of its absolute size. The Lower–Middle Cretaceous contact was chosen as a marker horizon, the position of the inflection points of each fold along the sections was marked, and the length of the folded marker horizon between two adjacent inflection points in each fold was measured as the half-wavelength. The definition of amplitude follows normal usage.

To chart the change in the half wavelength/amplitude ratio across the Mountain Belt, the Belt was subdivided into 11 strips of 20 km width parallel to its northeastern margin. The average half-wavelength/amplitude ratio of all the individual folds within each strip was calculated for each geological section. The results of the measurements are summarized in Fig. 6b and show that the half-wavelength/amplitude ratio of the folds generally increases from NE to SW. However, despite this common trend, considerable differences in the half-wavelength/amplitude ratio profiles are found in the four fold zones. The most striking is that in the sections for the Bakhtiari Fold Zone. Here the increase is extremely rapid and corresponds to very intensive and closely spaced fold development on the northeastern parts of the belt and a sudden change to more gentle and widely spaced folding in the Dezful Embayment to the SW (Fig. 2).

The anomalous distribution of folds within this fold zone is thought to be related directly to stratigraphy. The Hormuz Salt thins dramatically over a major basement strike-slip fault (the Kazerun Line), which forms the southeastern margin of the Kohkiluyeh Fold Zone. As will be seen, this reduction in the amount of salt would increase the resistance to the southwestward migration of a deformation front across the Zagros sedimentary sequence and would account for the localization of intensive folding along the NE margin of the fold belt that characterizes the higher Bakhtiari Mountains.

Elevation and inclination of the median surface of the folds

As noted above, the Zagros Mountains as a whole have an asymmetric SW-dipping topography. However, there are several major culminations and depressions in different parts of the belt both along its length and across it. To investigate the origin of this complex topography and examine the effect of folding and warping of the sedimentary cover in the creation of the present topography and its implications regarding the evolution of the Zagros deformation belt, a study of the elevation and inclination of the median surface of the folds in the four central zones of the Simply Folded Belt was carried out (Fig. 6a).

Fig. 5. Oblique aerial photograph of the Kabir Kuh Anticline. The fold is 200 km long although only 80 km can be seen on the photograph. Such long folds are typical of forced folds formed over a fault scarp, and a major blind thrust is thought to exist beneath this structure.

The marker horizon chosen to carry out this median surface analysis was again the Lower–Middle Cretaceous contact because this horizon was present along the whole of most of the section. The inflection points of every fold along each section were identified, their elevation above sea level and distance from the northeastern margin of the Simply Folded Belt were recorded, and the average elevation of the inflection points within the 20 km wide strips was calculated. The results of the measurements of the median surface are summarized in Fig. 6c, which shows that the median surface of the Zagros folds generally dips towards the SW, although not uniformly, in the four fold zones along the belt.

The most remarkable and sharp decrease occurs in the Bakhtiari Fold Zone, where the Middle Cretaceous–Upper Cretaceous marker horizon, which has an elevation of over 2 km at the NE margin of the Simply Folded Belt, drops below sea level at about 40 km from this margin. In contrast, in the neighbouring Lorestan and Kohkiluyeh Fold Zones the elevation of the median surface does not reach an elevation of 2 km and does not fall below sea level until it is about 100 km from the northeastern margin of the Simply Folded Belt.

Figure 6c also shows that the elevation of the median surface of the folds in the Fars Fold Zone is very different from that of the folds in the other zones. The marker horizon in the Fars Fold Zone is below sea level near the northeastern margin of the Simply Folded Belt. It rises above sea level at a distance of 40 km from the margin and drops below it again at 70 km from the margin. The gradient of the median surface of the folds in this zone is considerably lower than in the Bakhtiari

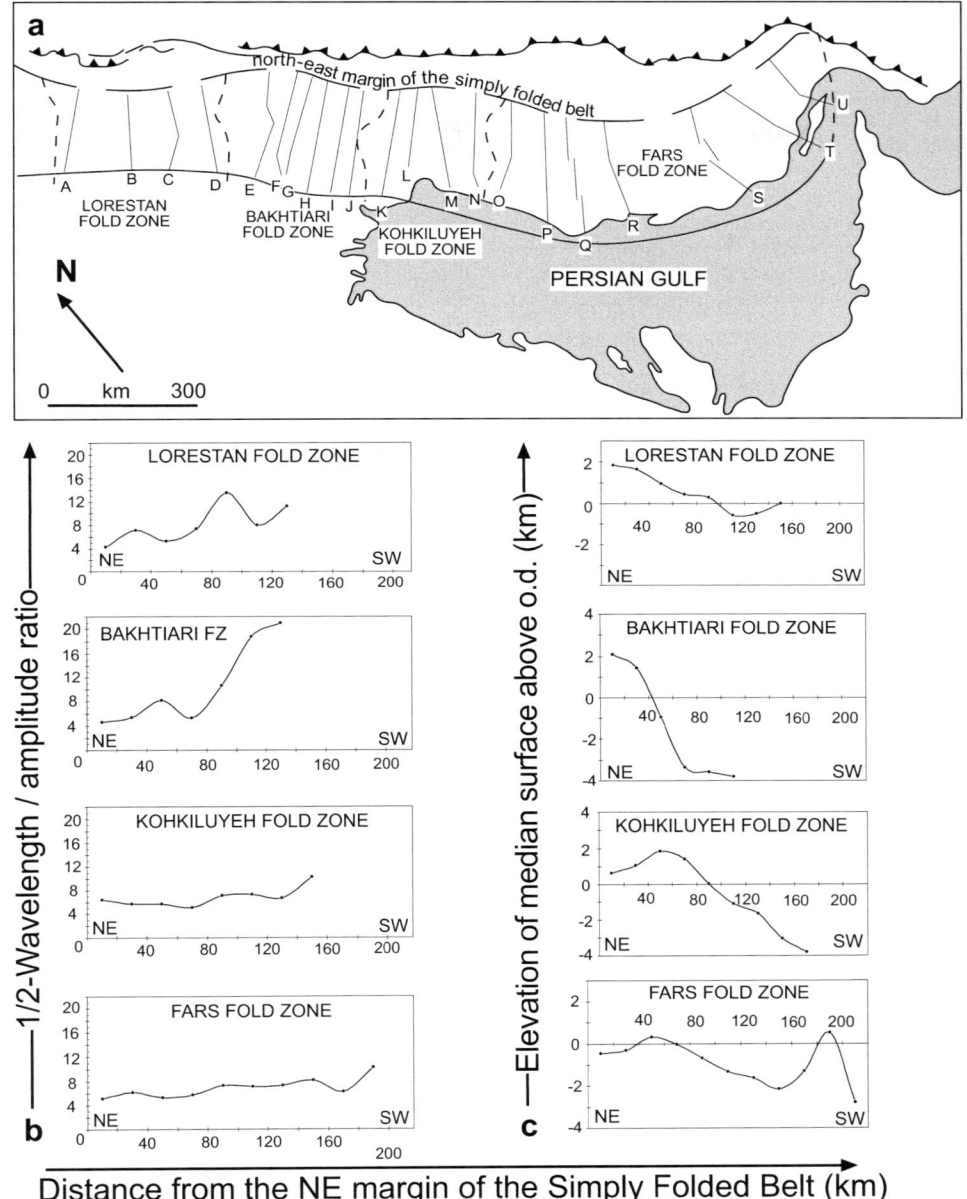

Fig. 6. (a) The location of the 21 cross-sections through the four central morphologically and structurally distinct fold zones of the Simply Folded Belt. (b) Variation of the half-wavelength/amplitude ratio of folds in the four zones. (c) Contrasting southwestwards inclination of the median surface of folds within the four adjacent fold zones. The major topographic culmination of the Bakhtiari Mountains can be seen in the Bakhtiari Fold Zone graph.

Fold Zone, but approximately the same as that in the other zones. The anomalous rise in the median surface elevation in the extreme SW of the section that can be seen in the Fars Fold Zone (Fig. 6c) is probably the result of salt diapirism and the associated emergence of Qeshm Island (Fig. 2).

Finally, as is clear from the above results, the present asymmetric and complex topography of the Zagros Mountains is wholly compatible with the geometry of the median surface of the Zagros folds, which implies that the present asymmetric and complex topography of the Zagros Mountain Belt is a direct product of both folding and warping of the whole sedimentary cover and that its major culminations and depressions such as the Bakhtiari Mountains Culmination and the Shiraz–Neyriz Depression have an internal and structural origin. As is discussed below, many of these structural variations relate directly to variations in the stratigraphy, particularly the presence or absence of salt.

Shortening of the sedimentary cover

The variation in the shortening profiles both along and across the Zagros was investigated using the 'balanced section method' (e.g. Dahlstrom 1970) on the 21 sections available. The technique is particularly appropriate where a relatively thin sequence of continental shelf sediments originally deposited on a continental-type basement has been deformed by horizontal shortening. In such regions the basement is often very competent compared with the overlying sediments and the cover is often mechanically detached from the basement along a décollement fault and involved in complex faulting and folding of a style not shared by the underlying crystalline basement. During such thin-skinned tectonic deformation (see e.g. Ramsay & Huber 1987) thrust faulting often predominates over other styles. The thrusts are usually associated with local accommodation of deformation by the development of kink folds and box folds (termed fault-bend folds and fault propagation folds by Suppe (1983, 1985)) when the layers are transported over underlying ramp–flat surfaces.

To determine the variation of shortening at different depths in the sedimentary cover, three marker horizons were selected; two from the Competent Group (Fig. 3), namely the Jurassic–Lower Cretaceous contact and the Middle–Upper Cretaceous contact, and one from the Incompetent Group, the Middle–Upper Fars (Miocene) contact (Mishan–Agha-Jari contact, Fig. 3). The graphs in Fig. 7a–d show the average shortening of the three marker horizons within each fold zone across each 20 km wide strip for all the geological sections. Figure 7e summarizes these results and represents the average variation of the shortening within the four fold zones.

Inspection of the graphs shows a decrease in shortening across the belt from 16% in the first 20 km wide strip close to the NE margin of the Simply Folded Belt to around 2% across a 20 km wide zone 200 km to the SW. This observation is compatible with the increase in half-wavelength/amplitude ratio in this direction (see Fig. 6b) and the regional decrease in elevation of the median surface of the folds to the SW (Fig. 6c).

A general decrease in shortening across the belt from NW to SE is not the only difference between the shortening profiles across the four morphologically and structurally distinct zones of the belt are not the same. The shortening across the Bakhtiari Fold Zone is concentrated close to the NE half of the zone and, as a result, the gradient of the percentage shortening graph is much steeper in this zone than in the other zones. As discussed above, the localization of folding within this zone is thought to reflect the lack or reduced thickness of the salt component of the Hormuz Salt detachment horizon in the area of the present-day Dezful Embayment, which lies to the SW of the Bakhtiari Mountains.

Inspection of the graphs shows that the shortening profiles of the two lower marker horizons are remarkably similar. However, this does not always hold for the upper marker horizon. This is particularly evident in the graphs shown in Fig. 7a and c, which relate to the Lorestan and Kohkiluyeh Fold Zones, where different shortening profiles are the result of disharmonic folding between the sediments above and below the Upper Mobile Group (Fig. 4). The folding is particularly marked where the thickness of the Upper Mobile Group is greatest. It should be noted that, although there is no apparent disharmony between the shortening profile in the three marker horizons in the Bakhtiari Fold Zone, this is only because the sections do not permit the three marker horizons to be traced across the whole of the Simply Folded Belt. However, from the geological cross-sections examined, it is evident that disharmonic folding does occur in the southwestern half of this fold zone.

When the individual shortening profiles for the four fold zones are examined, a marked periodic variation in shortening is seen to be superimposed on the general NE to SW decrease in shortening across the belt. Zones of relatively high shortening can be traced from one section to the adjacent sections along-strike and these zones are generally associated with faulting (thrusting) as well as folding. Thus it seems likely that the faults play an important role in controlling the localization of folds and therefore of shortening, and that they account for the observation that these zones have a remarkable continuity along-strike, a feature uncharacteristic of simple buckle anticlines and characteristic of fault-related (i.e. fault bend) folds. A study of the horst blocks between the adjacent thrusts shows a decrease in deformation, both of buckling and fault-bend folding, from NE to SW.

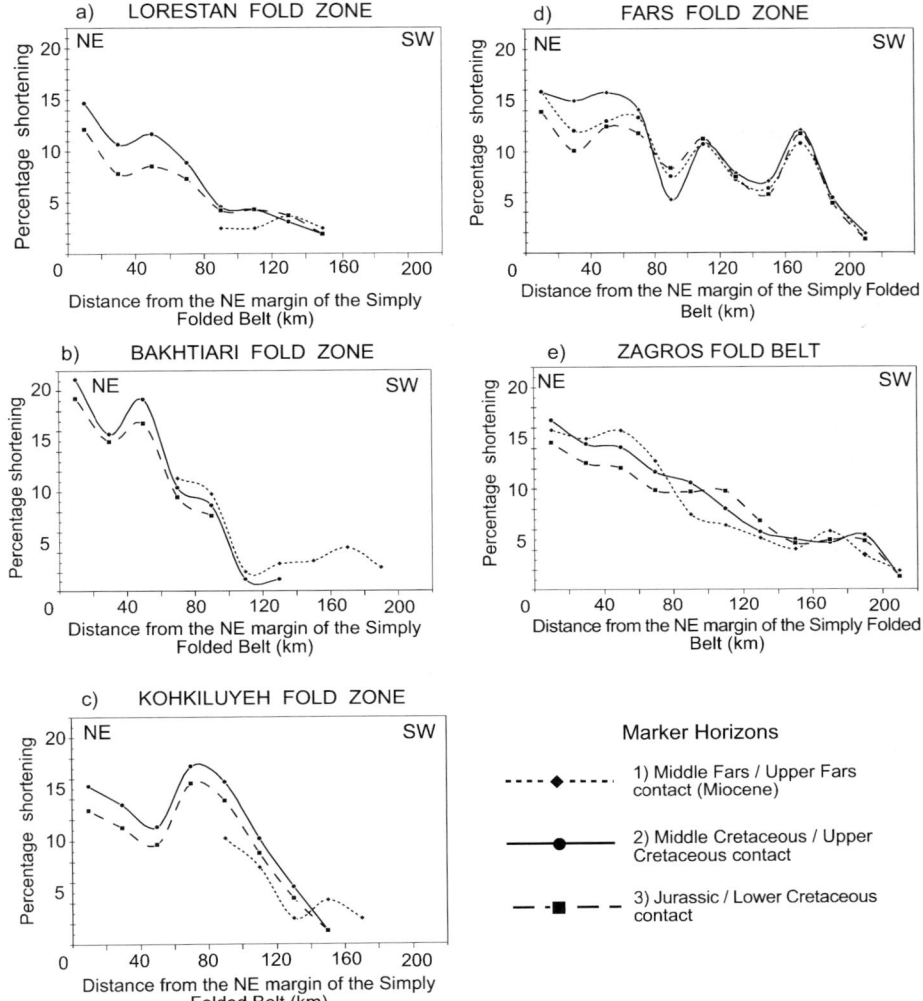

Fig. 7. Percentage of shortening of three marker horizons within the 20 km wide strips parallel to the northeastern margin of the Simply Folded Belt: (**a**) in the Lorestan Fold Zone; (**b**) in the Bakhtiari Fold Zone; (**c**) in the Kohkiluyeh Fold Zone; (**d**) in the Fars Fold Zone; (**e**) average for the central four fold zones of the Zagros Belt. The geographical location of the fold zones is shown in Fig. 6.

Interplay between folds and faults during the evolution of the Zagros

It is now clear that shortening of the cover sequence of the Zagros Mountain Belt is being achieved by both folding (ductile deformation) and thrusting (brittle deformation), and that there is often a close link between the two. In addition, it can be seen from the analyses discussed above that the amount of shortening decreases when traced across the belt from NE to SW (see Figs. 6 and 7). It is argued below that these observations are compatible with the migration of ductile and brittle deformation fronts from NE to SW by the processes of 'serial folding' and 'footwall collapse', respectively.

The process of serial folding, whereby folds develop one after the other as a deformation front propagates through a sedimentary sequence, has been fully discussed by Blay *et al.* (1977). The main conditions required for this process are all satisfied in the Zagros region. These are: (1) an undisturbed and mechanically anisotropic sedimentary sequence over a rigid basement; (2) a ductile layer of sufficient thickness and mobility to act as décollement horizon between the basement and overlying sedimentary cover; (3) a horizontal compression applied to one end of the sedimentary cover; (4) mechanical and physical conditions appropriate for classical buckling of the sedimentary cover.

Experiments designed to study serial folding show the migration of folds from a stressed margin of a model into the less deformed region by the successive initiation and amplification of adjacent folds. A related process is that of footwall collapse, which can be considered to be the brittle equivalent of serial folding a (e.g. Boyer & Elliot 1982; Mitra & Boyer 1986). As a thrust moves along a ramp cutting through a succession of beds, the overlying beds become progressively more deformed. It becomes more and more difficult for slip to occur and eventually the stress necessary to continue movement on the thrust exceeds that necessary to initiate a new thrust. The thrust is abandoned for a new thrust, which often forms in the relatively undeformed footwall. As this new thrust moves, slip becomes progressively more difficult. Eventually, the thrust locks up and the process is repeated.

A review of the tectonic and geomorphological evolution of the Zagros Mountain Belt (Sattarzadeh-Gadim 1997) shows that, except for the deep-seated high-angle reverse faults in the Imbricate and Thrust Zones, which were initiated as normal faults during the opening of the Neo-Tethys sea in Triassic time (Shearman 1976; Alavi 1980), and some major basement strike-slip faults such as the Kazarun Line (Fig. 2), almost all other Zagros faults, mainly thrust and strike-slip faults, were created during and after the collision of the Arabian and Central Iranian plates and the ensuing Pliocene orogeny.

The thrusts can be divided into two groups: those of low angle and restricted to the beds of the Incompetent Group above the Upper Mobile Group (Figs. 3 and 4), and those of relatively higher angles, which are initiated in the Hormuz Salt and cut all the overlying succession (Fig. 8). The relationship between the folds and these two groups of thrusts is different.

In the Incompetent Group, which is situated above the Upper Mobile Group, the initial structures are often the buckle folds. The formation of low-angle thrusts on the steeper flank of the folds occurs as the folds become progressively more asymmetric and overturn towards the SW. Thus if, as has been argued by Shearman (1976) and Mann & Vita-Finzi (1989), the folds develop serially, the faults would also form in a serial manner.

The second group of thrusts are larger, medium-angle dip-slip reverse faults, which generally cut the whole thickness of the sedimentary cover and link to the Hormuz Salt décollement zone. On the basis of the geometry and location of the folds associated with these thrusts, it is suggested that the folds are fault-bend folds (Suppe 1983). Thus, in contrast to the fold-thrust association in the Incompetent Group, the folds associated with the thrusts in the Competent Group postdate the faults.

The concentration of folding along the NE margin of the Bakhtiari Fold Zone as shown in Figs. 2, 6 and 7 has resulted in the Bakhtiari Mountains, the major culmination in the Zagros Range. The reason for this local concentration of folding is related to the stratigraphy of the cover rocks within the Bakhtiari Fold Zone. There is considerable evidence (e.g. the paucity of salt domes in the Dezful Embayment and the concentration of folding in the northeastern part of the Embayment) that the Hormuz Salt is either absent or considerably thinned in this zone and that the Kazerun Fault, a major strike-slip fault zone (Fig. 2), acted as a basin-bounding fault during the deposition of the salt, significantly limiting its deposition to the NW, i.e. in the region of the Bakhtiari Fold Zone.

Salt diapirism in the Bakhtiari Fold Zone (which contains the Bahktiari Mountains and the Dezful Embayment) is either absent or rare. In contrast, SE of the Kazerun Fault Zone in the Fars

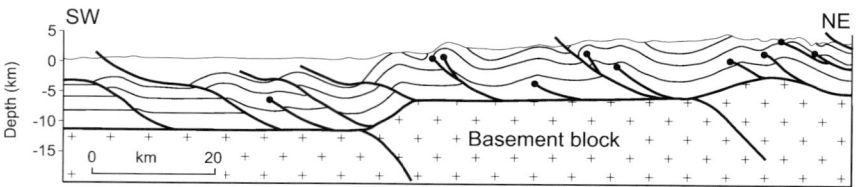

Fig. 8. Structural cross-section through the Zagros fold-thrust belt based on a NE-SW seismic profile passing through Burugen in the SE of the Bakhtiari Fold Zone (Fig. 2). It can be seen that many of the thrusts that cut the cover sediments are initiated in the Hormuz Salt at the basement-cover contact. (After Payne 1990.)

Fold Zone, which is underlain by thick Hormuz Salt, numerous approximately equant salt diapirs rise through the cover and pierce the folds in a random fashion.

The influence of salt thickness on the pattern of deformation

The presence or absence of a thick salt at the base of the cover rock succession plays a key role in controlling the rate of propagation of folds within the sediments by the process of serial folding. The mechanism of serial folding is based on the theory for the propagation of stress through an elastic plate resting on a viscous substrate (Bott & Dean 1973). As can be seen from the following equation, a change in the thickness, h_2, of the Hormuz Salt décollement horizon (the viscous substrate) will significantly influence the rate of advance of the deformation front:

$$v = \frac{Eh_1 h_2}{2\mu L}. \quad (1)$$

Equation (1) shows the relationship between the velocity of advance of the deformation front, v, the thickness of the décollement horizon h_2, its viscosity μ and the Young's modulus, thickness and length of the overlying elastic layer, E, h_1 and L, respectively.

If the salt dies out completely ($h_2 = 0$) it follows that the deformation front will become completely arrested. It is argued that the concentration of folds along the NE margin of the fold belt within this fold zone relates to the variation in wavefront velocity caused by differences in salt thickness in this and the adjacent fold zones. This has resulted in the formation of folds with anomalously high amplitudes and has resulted in the major culmination in the Zagros, namely, the Bakhtiari Mountains.

It has been shown that the ability of a salt layer to feed a growing diapir is related to the fourth power of the thickness of that layer (Price & Cosgrove 1990). Thus the dramatic boundary between the diapir-rich zone to the SE of the Kazerun Line and the virtual absence of diapirs in the Dezful Embayment to the NW is further evidence of a major thinning of the salt across this lineament. Where salt intrusions occur to the NW of the Kazerun Line they form narrow slivers rather than equant diapirs and it is clear that these salt walls are the result of salt intrusion along reverse faults, generally thrusts. In regions where the salt thickness is too low for the random intrusion of diapirs, diapirs can be initiated by faulting.

Synchronous thrust and wrench faulting

Field observations show that both thrust and strike-slip faulting is occurring in the cover rocks of the Zagros deformation belt despite the fact that the stress regimes for the formation of thrusts and wrench faults are different. Although both require a horizontal maximum principal compression, the former requires the least principal compression to be vertical and the latter the intermediate principal compression to be vertical. Thus the synchronous occurrence of thrusts and wrench faulting within the Zagros belt demands an explanation.

The vertical stress in the crust will increase with depth because of increasing overburden. Therefore, if the minimum principal stress is vertical, as the vertical stress increases with depth there will be some depth where it will equal and then exceed the intermediate stress. At this depth, thrusting will give way to wrench tectonics. In the Zagros Fold Belt the faulting of the cover sediments is dominantly by thrusting and the wrench faulting occurs mainly in the basement, which is generally decoupled from the cover by the Hormuz Salt.

However, movement along basement wrench faults will generate deformation in the cover rocks. This effect has been extensively studied both in the field and by experimental work (e.g. Oliver 1987; Richard 1990, 1991; Richard & Krantz 1991; Richard et al. 1991). A zone of strike-slip deformation is generated in the cover rocks, and the detailed geometry of this zone is determined by the nature of the basement faults, i.e. whether it is a transpressive, transtensional or pure strike-slip fault. In all these situations the strike-slip deformation is transmitted to the cover rocks. The thicker the décollement horizon between the basement and cover, the more diffuse the fault zone that develops within the cover (Richard 1991).

The wrench faults that form in the basement, where conditions are such that the intermediate principal stress is vertical or where old wrench faults can be reactivated at stress levels below those required for the initiation of thrusts, can generate wrench faulting in the cover rocks. In this way, it is possible to generate both thrusting and wrench faulting in the cover at the same time.

Implications for neotectonic deformation of the Zagros

The field data show that, as suggested by Shearman (1976), fold initiation along the NE margin of the Arabian plate is actively migrating towards the SW. This is consistent with the expectation raised by our analysis that the displacement along the thrust and deformation of the thrust block will decrease when traced from older to younger thrusts. Such a decrease can be seen in the Zagros belt, because the least deformed thrusts occur at the SW margin of the fold belt and the amount of slip and deformation of the thrust blocks increases to the NE.

In short, a deformation front is migrating southwestwards into the Arabian plate from its northeastern margin by both ductile and brittle deformation. The regional gradient in deformation discussed above is modified locally by the distribution of salt. The most impressive modification occurs in the Bakhtiari Fold Zone of the central Zagros where, in contrast to the other zones, folding is concentrated along the NE margin of the belt. In our view, this is the result of dramatic thinning or even absence of the salt in this region.

Faults, especially thrusts, are propagating in the same direction by the process of footwall collapse. Footwall collapse can be considered to be the brittle equivalent of serial folding. The fault-bend folds associated with footwall collapse will, of course, also develop sequentially from NE to SW in exactly the same way as by the process of serial buckle folding.

Seismicity within the Zagros Thrust and Imbricate zones (Fig. 2) to the NE is relatively low compared with that in the Simply Folded Belt to the SW (Nowroozi 1972; Chandra 1984; Jackson & McKenzie 1984; Ni & Barazangi 1986). This is consistent with the model of footwall collapse proposed for the deformation belt, which would drive deformation from the NE to the SW as early thrusts steepen and, together with old normal faults, become progressively more difficult to reactivate. The rate at which the deformation front migrates southwestwards cannot be derived from the seismic map, however, because there is no sharp cutoff point in activity. In a previous study (Mann & Vita-Finzi 1989) an estimate of the rate was made by accepting a Pliocene age (c. 1.8–5.3 Ma) for the start of folding, a width of 200 km for the Simply Folded Belt (Falcon 1974) and uniform advance. The resulting average is 17 mm a^{-1}, and as the average fold wavelength for the 23 sections across the Simply Folded Belt studied is 10.6 km, each fold required 600 ka to develop fully.

Support for the thesis of serial folding comes from the Holocene record. Analysis of Holocene deformation of coastal folds running north–south and east–west to the east and north, respectively, of the Musandam Peninsula at the entrance to the Gulf gave a resultant shortening of c. 20 mm a^{-1}. at 061°N over the last 6 ka (Vita-Finzi 2001).

Comparison with the regional convergence rate of 30 mm a^{-1} at 015°N (Jackson et al. 1995) suggests that, assuming much of the deformation is coseismic, it is concentrated in the frontal folds.

The Holocene data were used to compute the force required to sustain the frontal folds and thus to obtain a minimum value for the force required to initiate buckling. The result favoured a model powered by gravitational sliding of the Arabian plate (Mann & Vita-Finzi 1989; Vita-Finzi 2001). The new data presented here for southwestward inclination of the median surface show that the folds are indeed being sustained against gravity by the northeastward translation of the Arabian plate.

References

ALAVI, M. 1980. Tectonostratigraphic evolution of the Zargrosides of Iran. *Geology (Boulder)*, **8**, 144–149.

BERBERIAN, M. 1981. Active faulting and tectonics of Iran. In: GUPTA, H.K. & DELAY, F.M. (eds) *Zagros, Hindu Kush, Himalayan Geodynamic Evolution*. Geodynamics Series, American Geophysical Union, **3**, 33–69.

BLAY, P., COSGROVE, J.W. & SUMMERS, J.M. 1977. An experimental investigation of the development of structures in multilayers under the influence of gravity. *Journal of the Geological Society, London*, **133**, 329–342.

BOTT, M.H.P. & DEAN, D.S. 1973. Stress diffusion from plate boundaries. *Nature*, **243**, 339–341.

BOYER, S.E. & ELLIOT, D. 1982. Thrust systems. *AAPG Bulletin*, **66**(9), 1196–1230.

CHANDRA, U. 1984. Focal mechanism solutions for earthquakes in Iran. *Physics of the Earth and Planetary Interiors*, **34**, 9–16.

COLEMAN-SADD, S.P. 1978. Fold development in the Zagros Simply Folded Belt, SW Iran. *AAPG Bulletin*, **62**, 984–1003.

DAHLSTROM, C.D.A. 1970. Structural geology in the eastern margin of the Canadian Rocky Mountains. *Canadian Bulletin of Petroleum Geology*, **18**, 332–406.

FALCON, N.L. 1974. Southern Iran: Zagros Mountains. In: SPENCER, A.M. (eds) *Mesozoic–Cenozoic Orogenic Belts*. Scottish Academic Press, Edinburgh, 199–211.

JACKSON, J.A. & McKENZIE, D.P. 1984. Active tectonics of the Alpine Himalayan Belt between Western Turkey and Pakistan. *Geophysical Journal of the Royal Astronomical Society*, **77**, 185–264.

JACKSON, J.A., HAINES, J. & HOLT, W. 1995. The accommodation of Arabia–Eurasia plate convergence in Iran. *Journal of Geophysical Research*, **100**, 15205–15219.

LAUGHTON, A.S. 1966. The Gulf of Aden. *Philosophical Transactions of the Royal Society of London*, **A259**, 150–171.

MANN, C.D. & VITA-FINZI, C. 1989. Holocene serial folding in the Zagros. In: AUDLEY-CHARLES, M.G. & HALLAM, A. (eds) *Gondwana and Tethys*. Geological Society, London, Special Publications, **37**, 51–59.

MITRA, G. & BOYER, S.E. 1986. Energy balance and deformation mechanisms of duplexes. *Journal of Structural Geology*, **8**, 291–304.

NI, J. & BARAZANGI, M. 1986. Seismotectonics of the Zagros continental collision zone and a comparison with the Himalayas. *Journal of Geophysical Research*, **91**, 8205–8218.

NOWROOZI, A.A. 1972. Focal mechanisms of earthquakes in Persia, Turkey, West-Pakistan and Afghanistan and plate tectonics of the Middle East. *Bulletin of the Seismological Society of America*, **62**, 823–850.

NOWROOZI, A.A. 1987. Tectonics and earthquake risk of Iran. *In:* CAKMAK, A.S. (ed.) *Ground Motion and Engineering Seismology*. Elsevier, Amsterdam, 59–79.

OBERLANDER, X. 1965. *The Zagros Streams: a New Interpretation of Transverse Drainage in an Orogenic Zone. Syracuse Geographical Series 1*. Syracuse University Press, Syracuse, NY.

O'BRIEN, C.A.E. 1957. Salt diapirism in south Persia. *Geologie en Mijnbouw (n.s.)*, **19**, 357–376.

OLIVER, D. 1987. *The development of structural patterns above reactivated basement faults*. Ph.D. thesis, University of London.

PAYNE, A. 1990. *A Structural Interpretation of the Zagros Fold Belt, SW Iran, NE Iraq*. British Petroleum Company Ltd., internal report.

PRICE, N.J. & COSGROVE, J.W. 1990. *Analysis of Geological Structures*. Cambridge University Press, Cambridge.

RAMSAY, J.G. & HUBER, M.I. 1987. *Techniques of Modern Structural Geology—Volume 2, Folds & Fractures*. Academic Press, London.

RICHARD, P. 1990. *Champs de failles audessus d'un décrochement de socle: modelisation expérimentale*. Ph.D. thesis, University of Rennes.

RICHARD, P. 1991. Experiments on faulting in a two-layer cover sequence overlying a reactive basement fault with oblique slip. *Journal of Structural Geology*, **13**, 459–470.

RICHARD, P. & KRANTZ, R.W. 1991. Experiments on fault reactivation in strike-slip mode. *Tectonophysics*, **188**, 117–131.

RICHARD, P., MOCQUET, B. & COBBOLD, P.R. 1991. Experiments on simultaneous faulting and folding above a basement wrench fault. *Tectonophysics*, **188**, 133–141.

SATTARZADEH-GADIM, Y. 1997. *Active tectonics in the Zagros Mountains, Iran*. Ph.D. thesis, University of London.

SATTARZADEH, Y., COSGROVE, J.W. & VITA-FINZI, C. 2000. The interplay of faulting and folding during the evolution of the Zagros deformation belt. *In:* COSGROVE, J.W. & AMEEN, M.S. (eds) *Forced Folds and Fractures*. Geological Society, London, Special Publications, **169**, 187–196.

SHEARMAN, D.J. 1976. The geological evolution of southern Iran: the report of the Iranian Makran expedition. *Geographical Journal*, **142**, 393–413.

SUPPE, J. 1983. Geometry and kinematics of fault-bend folding. *American Journal of Science*, **283**, 648–721.

SUPPE, J. 1985. *Principles of Structural Geology*. Prentice-Hall, Englewood Cliffs, NJ.

VITA-FINZI, C. 2001. Neotectonics at the Arabian plate margins. *Journal of Structural Geology*, **23**, 521–530.

Quaternary sedimentation on the Makran margin: turbidity current–hemipelagic interaction in an active slope-apron system

DORRIK A. V. STOW[1], ALI R. TABREZ[2] & MARTEEN A. PRINS[3]

[1] Southampton Oceanography Centre, University of Southampton, Southampton SO14 3ZH, UK (e-mail: davs@soc.soton.ac.uk)
[2] National Institute of Oceanography, Karachi, Pakistan
[3] Faculty of Earth Sciences, Vreije Universiteit, 1081 HV Amsterdam, Netherlands

Abstract: The Makran slope-apron system is a stepped convergent margin across an active subduction complex. Shallow penetration piston cores have been recovered from the upper-slope region, three mid-slope basins and the abyssal plain. At most sites the upper 5–14 m of cored section is dominated by fine-grained, thin- to medium-bedded turbidites, averaging 5–10 turbidite events per metre of section. Oxygen isotope stratigraphy yields mean sedimentation rates of 50–95 cm ka^{-1} and a turbidite frequency of one event per 200–300 a. The upper-slope site has fewer turbidites and a greater proportion of hemipelagic mud. Fine-grained turbidite sequences are common, with top-cut-out and base-cut-out sequences most evident. Markov chain analysis of the transition between turbidite divisions confirms the normal T0–T8 order of sequence divisions. In some cases there is an upward gradation into a hemiturbidite facies. The range of turbidite bed thicknesses can be approximated by both power-law and log-normal distributions, typical of seismic triggering on an active margin, or of frequent river-flood sediment input. Small-scale vertical variations of turbidite bed thickness recognized by autocorrelation techniques can be interpreted as the result of bed-relief compensation effects (compensation cycles). The lateral distribution of both turbidites and hemipelagites is influenced by sediment focusing along pathways between slope basins. At a larger scale, climate, sea-level and tectonic effects have all played an important role in shaping margin sedimentation.

The continental margin off Pakistan and Iran is the offshore extension of the Makran accretionary complex, where oceanic crust of the Gulf of Oman has been subducting under the Asian continent since late Cretaceous time (Arthurton *et al.* 1982; White 1982). Progressive accretion onto the Asian plate has produced a topography of uplifted basins and intervening ridges aligned parallel to the Makran coast (White 1982, 1989). The mechanics and style of deformation has recently analysed on the basis of detailed swath-bathymetric data acquired by Kukowski *et al.* (2001). The influence of mud volcanoes throughout the accretionary complex has been well documented (e.g. Wiedicke *et al.* 2001).

Most previous studies of deep marine sedimentation in the NW Indian Ocean have concentrated on the faunal response to wind-induced upwelling, and on aeolian dust transport in relation to monsoonal climate (e.g. Prell & Streeter 1982; Prell *et al.* 1989, 1990; Clemens & Prell 1990). Recent palaeoceanographic studies in the area include those of Schulz *et al.* (1996, 1998), Reichart *et al.* (1997, 1998), 5 and Luckge *et al.* (2001). The close correspondence of climate change to style of sedimentation has been amply demonstrated by these studies. Several regional studies have also focused on clay mineral distribution in the Arabian Sea in particular (e.g. Kolla *et al.* 1976, 1981; Faugeres & Gonthier (1981); Sirocko & Lange 1991). More recently, Prins & Postma (2000) and Prins *et al.* (2000*a, b*) have attempted to distinguish between the effects of climate, sea level and tectonics on sedimentation on the Makran slope and Indus Fan.

The aim of this paper is to further document the nature of Pleistocene to Holocene sedimentation on the Makran margin, based on a series of piston five cores taken as part of the Netherlands Indian Ocean Programme (NIOP 468, 469, 470, 471 and 472) and limited 3.5 kHz data recovered from a north–south slope transect south of Gwadar. The survey area and core locations are shown in Fig. 1. Visual and X-radiograph descriptions of split cores were followed by standard textural and compositional sediment analyses, together with dating by oxygen isotope and magnetostratigraphic techniques. The full dataset has been presented by Tabrez (1995). A previous paper outlines the characteristics of hemipelagites from the Makran margin (Stow & Tabrez 1998). This paper concentrates more on details of the turbidite sedimentation. The work by Prins et al. (2000b), based on the same suite of cores (Prins 1999), focuses on the principal controls on terrigenous sediment supply.

Morphotectonic framework

The Makran continental margin forms the seaward extremity of an accretionary sediment prism that extends several hundreds of kilometres inland (White 1977; White & Louden 1982; Minshull & White 1989). This marks the zone of convergence between oceanic lithosphere of the Arabian plate and continental lithosphere of the Eurasian plate, with an estimated subduction rate of 3–5 cm a^{-1} (Quittmeyer & Kafka 1984). Because of the highly active nature of this margin, tectonic activity has undoubtedly played a key role in shaping the morphology and influencing the distribution and deposition of sediments.

Where the margin has been studied in detail by both continuous seismic reflection profiling and wide-angle seismic data, it is seen to comprise a series of ridges and intervening basins aligned parallel to the Makran coast (White & Ross 1979; Minshull & White 1989; Kopp et al. 2000; Kukowski et al. 2001). The development of this topography is exemplified in the extreme south of the margin, where gently dipping sediments beneath the abyssal plain are seen to be deformed in a frontal fold (currently at 24°10′N) to a maximum amplitude of 400 m. When crustal shortening can no longer be accommodated in this way, thrusting occurs and the frontal fold block is uplifted to over 1 km above the abyssal plain. The focus of folding moves further offshore and a proto-thrust with negligible offset marks a zone of weakness where the next thrust is likely to occur (White 1982).

A distinctive sea-floor topography is generated in this way as sediment slabs are scraped off the subducting Arabian plate. The accreted sediment is deformed into a series of imbricate thrust slices with small slope basins formed between each successive thrust slice. These basins contain an infill of post-depositional sediments derived from the nearby landmass and supplemented by slumping from the adjacent ridges, which in some cases have slopes as high as 20° (Kukowski et al. 2001). These basin sediments provide evidence for continued tectonic activity in that bedding plane dips increase with depth, and fault drag is shown by curvature of reflectors into the thrust planes.

Within such an active morphotectonic system, the development of a downslope drainage pattern is markedly affected by both the slope-parallel ridge–basin morphology and the constant changes to that morphology induced by tectonic activity. In some parts, including the study area, there are a series of incipient channels and channel-like pathways connecting basins and other topographic lows in a pattern analogous to a continental trellis drainage network. In other parts, for example further east from the study area (Kukowski et al. 2001), one or two tributary networks feed into more major canyons that have become well established (e.g. Save and Shadi Canyons). The variation in channel orientation and relief is nevertheless significantly affected by the local morphology.

Stratigraphic framework

The five piston cores recovered from the Makran margin for this study are between 5 and 14 m in length, and so penetrate only the topmost part of the stratigraphic succession. They can be correlated on the basis of distinctive lithostratigraphic characteristics, and then dated using a combination of stable oxygen isotopes and magnetostratigraphy. This latter work (by Tabrez 1995) has been more precisely calibrated by accelerating mass spectrometry (AMS) ^{14}C dates obtained by Prins (1999) and Prins & Postma (2000).

The two principal sediment facies encountered are turbidites and hemipelagites. There is a marked change in the turbidite facies, clearly recognized in four out of five cores, from a lower unit of closely spaced thin-bedded turbidites, to an upper unit of thicker-bedded but less abundant turbidites (Horizon A, Fig. 2a). This change coincides with a marked minimum in mean grain size of both the turbidite and hemipelagite facies evident in all cores (Fig. 2b). Age dating places this horizon at around 7500 a BP.

Oxygen isotope stable analyses were carried out on the planktonic foraminifer *Neogloboquadrina dutertrei* picked from some 80 closely spaced samples (every 10–20 cm in hemipelagite facies) from two long cores (469 and 472). These results are illustrated in Fig. 2b together with the standard ^{18}O isotope curve for the Arabian Sea from Sirocko et al. (1991). The marked decrease in ^{18}O values at about 5 m depth in core 469 represents the transition from the last glacial period to the Holocene interglacial (Marine Isotopic Stage MIS 2 to 1, Termination I). The distinctive two-step change in this core probably corresponds to Termination Ia and Ib, so that the lower ^{18}O values between Ia and Ib would represent the Younger Dryas event (Steens et al. 1991). The relatively

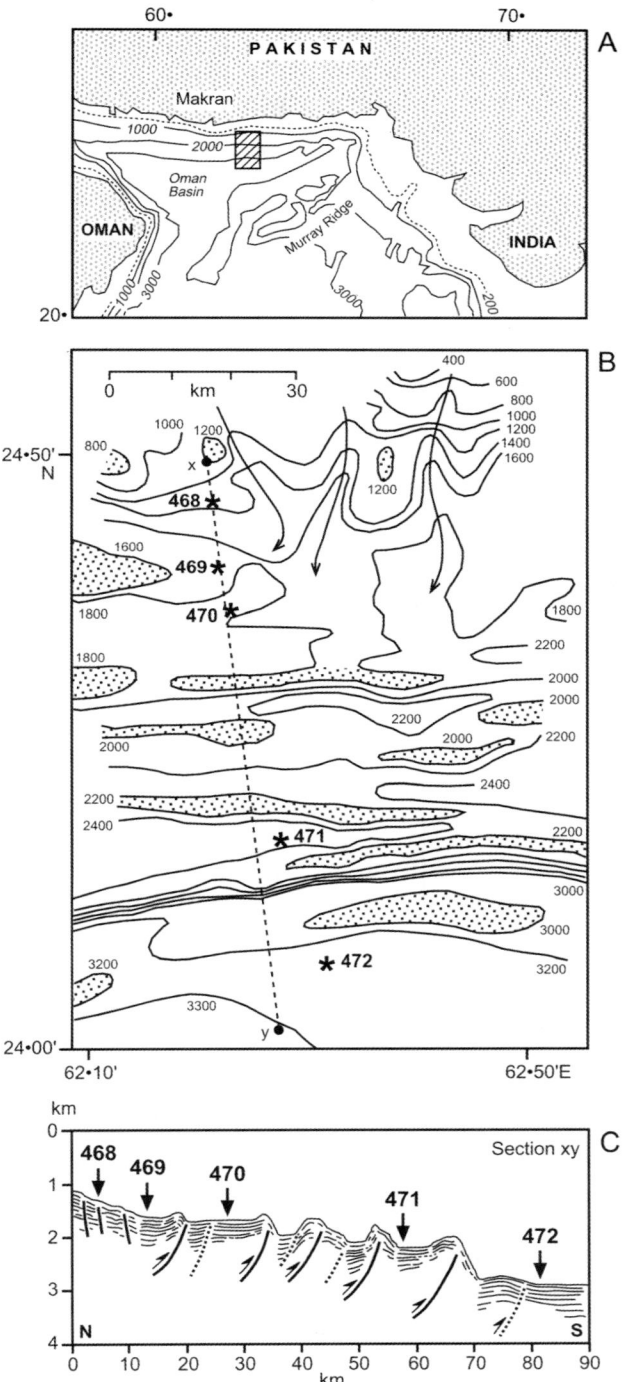

Fig. 1. (a) Location map, (b) detailed bathymetric map with core locations and (c) summary bathymetric profile of slope transect along which cores were positioned.

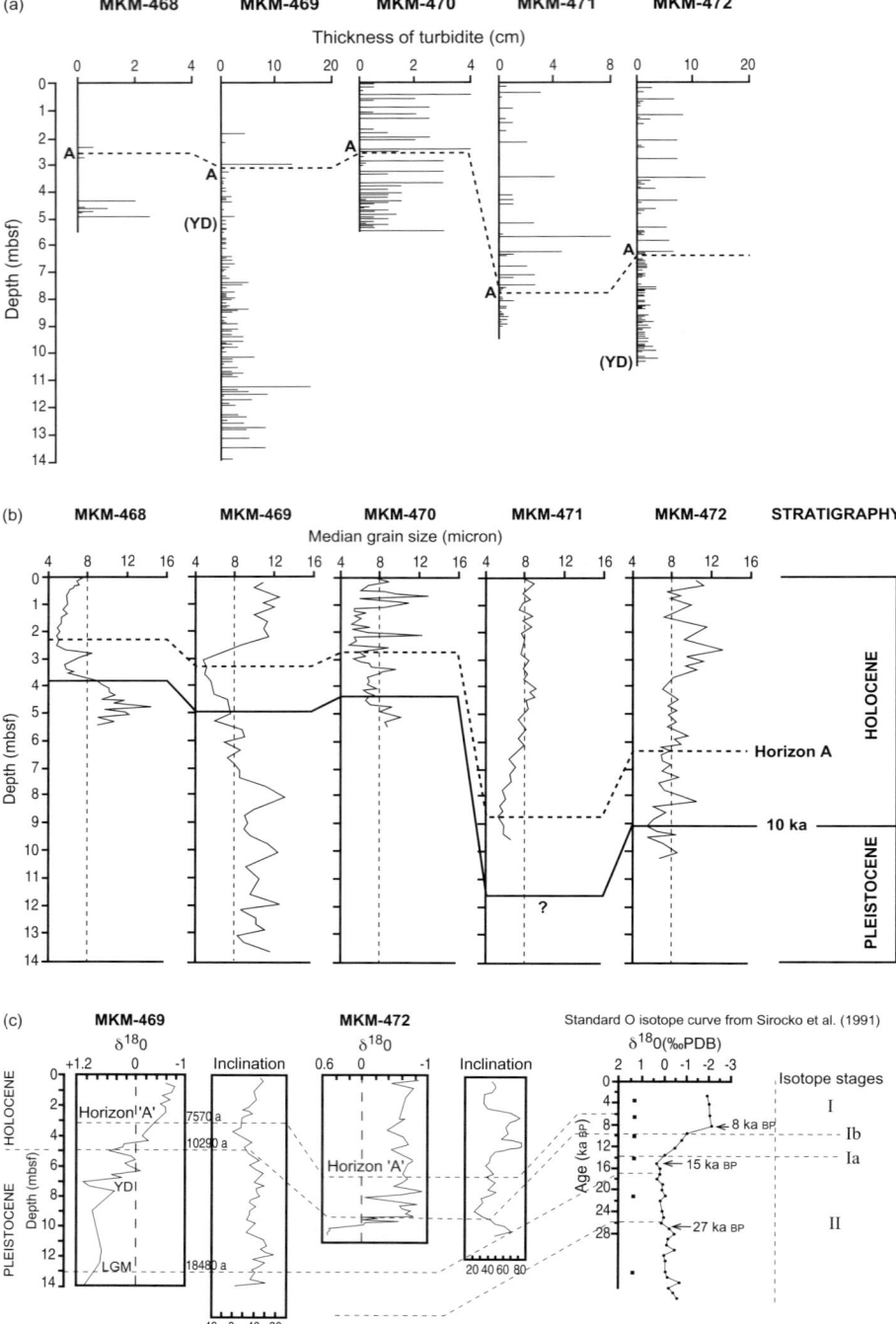

Fig. 2. Stratigraphic framework and correlation between cores. (**a**) Lithostratigraphic correlation for all five cores based on turbidite bed thickness and number. (**b**) Lithostratigraphic correlation for all five cores based on median grain size of hemipelagites. (**c**) Stratigraphic interpretation of oxygen isotope and magnetostratigraphic data from the two long cores (469 and 472). Comparison is made with the standard regional oxygen isotope curve of Sirocko et al. (1991). Figures shown in parentheses are AMS ^{14}C dates from Prins & Postma (2000).

higher ^{18}O values at about 13 m depth in core 469 may correspond to the last glacial maximum (LGM) at around 18 ka BP. By contrast, most of core 472 has ^{18}O values typical of the Holocene period, with Termination I occurring at 9–9.5 m depth.

The natural remanent magnetization (NRM) of 59 specimens from core MKM-469 and 40 specimens from MKM-472 were measured on a 2G cryogenic magnetometer. All specimens were subjected to stepwise demagnetization to a maximum field value of 40 mT. Bulk magnetic susceptibility measurements were made with a Bartington meter. For both cores, NRM values mainly vary between 10° and 45° (Fig. 2c) as would be expected for the present normal magnetozone at that latitude. A minor palaeomagnetic excursion is evident at 3.7 m in core 469 and 7.0 m in core 472. Although previously unrecognized in this area, we interpret this as a true low-latitude geomagnetic excursion (Tabrez 1995). An older possible reversal is just apparent at the base of core 469 (13.9 m depth). Using the oxygen isotope stratigraphy established above, the latest excursion is dated to occur between 6 and 8 ka BP, whereas the earlier one would be between 18 and 20 ka BP. This age is similar to the high-latitude excursions reported from Imuruk Lake (17–18 ka BP, Noltimer & Colvinaux 1976), and Baffin Bay (18 ka BP, Thouveny 1988), but probably not as old as the Mono Lake event between 24 and 9 ka BP (Denham & Cox 1971; Denham 1974; Liddicoat & Coe 1979), as we had previously thought.

Rates of sedimentation

On the basis of the stratigraphy outlined above, average sedimentation rates can be calculated for the various sites. For the Holocene period, these vary between about 0.38 m ka^{-1} (core 468) and 0.9 m ka^{-1} (core 472). The base of the Holocene sequence is probably not reached in core 471, but the estimated rate at this site for the latter part of the Holocene period is in excess of 1.4 m ka^{-1}. These values are an average for both turbidites and hemipelagites in the time period represented by each core. They are slightly higher than those found by Hutchison et al. (1981) (0.25–0.40 m ka^{-1}) for the Oman Abyssal Plain and 0.50 m ka^{-1} for the central basin of the Gulf of Oman (Stoffers & Ross 1979). For the slightly longer record recovered by core 469, the rate of sedimentation has decreased from c. 0.9 m ka^{-1} during late Pleistocene time to c. 0.5 m ka^{-1} for the Holocene period. These are very close to the rates of 0.41–1.1 m ka^{-1} reported by Wiedicke et al. (2001). Counting the number of discrete turbidites at each site yields a frequency of one event every 180–300 a.

Much higher rates of sedimentation during late Holocene time (2.3 m ka^{-1} SW of Ormara) are found further east along the upper slope within the oxygen-minimum zone, based on varve counts and ^{14}C data (von Rad et al. 1999a). In another upper slope core, directly off the Hingol River mouth, the turbidite-dominated succession accumulated at rates of 3–4 m ka^{-1}, and with an event frequency of up to 12 per century (von Rad et al. 1999a, b, and pers. comm. 2001).

Sediment facies

Facies characteristics and distribution

In the five piston cores recovered, turbidites and hemipelagites are seen to be closely interbedded, although with relatively more turbidites in the ponded basin areas and more hemipelagites on the open slope. This distribution is confirmed by study of 3.5 kHz profiles, which show transparent drape-sheet echofacies (Type A; hemipelagites) most common over the open slope and interbasinal highs, and strong parallel–multiple echofacies (Type D; turbidites) within the basins and on the abyssal plain. Basin margins and the steep flanks of interbasinal highs are characterized by irregular–chaotic echofacies (Type C; slumps and debrites). These echofacies types and interpretations follow Tabrez (1995). The relative proportions of the various facies at the core sites is given in Table 1.

The turbidites are dominantly fine grained, comprising mainly clay- and silt-grade muds, in

Table 1. Relative proportion (in percentages) of different facies at each of the core sites

Facies %	MKM-468	MKM-469	MKM-470	MKM-471	MKM-472
Turbidite	40	65	60	40	45
Hemipelagite	60	35	40	60	40
(Hemiturbidite)	(7)	(1–2)	(1–2)	(8)	(1–2)
Pelagite	0	0	0	0	15

The percentage of probable hemiturbidites is included in the overall value for hemipelagites, and also indicated separately in parentheses.

some cases with thin silts or silty sands at the base of beds. These coarse-grained intervals are typically 1–2 cm thick, and more rarely up to a maximum of 16 cm. The colour of the turbidites ranges from dark grey to blackish, becoming paler upwards and with a sharp colour change to pale green–greyish yellow in the upper part of the bed. The clay fraction of all turbidites examined is dominated by illite and chlorite, and the silt–sand fraction by quartz. Minor components include other clay minerals, feldspar, calcite and dolomite. Biogenic material is rare and organic carbon averages <0.6%. Compositional data have been presented fully by Tabrez (1995). The distribution of turbidite beds varies from core to core.

In some cases, particularly for cores 468 and 471, there are thick bioturbated units overlying the turbidites, but having a more turbiditic (terrigenous, darker colour) than hemipelagic (biogenic-rich, paler colour) nature in terms of colour and composition. These facies are probably equivalent to the hemiturbidites described by Stow & Wetzel (1990) from the distal Bengal Fan.

True hemipelagic units can be clearly distinguished from the intervening turbidites on the basis of texture, composition, sedimentary structures and colour. They comprise a foraminiferal and nannofossil biogenic fraction, and terrigenous silty clays of similar composition to the turbidites described above. They appear to be fully homogenized and mottled by bioturbation, although distinct burrow traces are not very common. The colour varies from olive grey to pale grey. The upper part of hemipelagic units is typically rich in foraminifers and nannofossils, with indistinct bioturbation. The lower contact with the underlying turbidite is usually gradational whereas the contact with the overlying turbidite is sharp. All hemipelagites are very poorly sorted, typically showing weakly developed bimodal or polymodal peaks on grain-size distribution curves. Bed thicknesses range from 5 to 50 cm.

Pelagic facies are recognized only in certain specific horizons in core 472 (forming 10–15% of the recovered core). They occur as thin to thick beds (10–40 cm), typically located towards the top of a hemipelagic unit. They have a completely gradational contact with the underlying hemipelagite and are distinguished from the latter on the basis of their paler colour and higher percentage of pelagic biogenic material (foraminifers and nannofossils). The terrigenous content is <25%, including traces of clays and fine quartz.

Bed thickness and vertical sequences

Thin-bedded, fine-grained turbidites are very common in all slope-apron, lower fan and basin plain environments, both modern and ancient, and yet many of their attributes are still poorly known. For example, small-scale cycles (<10 m) of bed thickness variation in such turbidite successions have been recognized but are little understood (Piper & Stow 1991; Hiscott et al. 1992; Forster 1995). Such variation is evident from visual inspection of Makran margin cores, in which we note both thinning-upwards and thickening-upwards sequences over a few metres of section. However, there is much discussion on the validity of sequences noted that are based solely on visual investigation of the core or outcrop and hence are liable to subjective bias introduced by preconceptions, plotting methods and so on (Hiscott 1981; Hiscott et al. 1992). Certain statistical procedures can therefore help provide a more objective assessment of sequences and their nature (Weedon 1985, 1989).

For this study we have examined both bed thickness and the sequence of bed thickness variation in Makran margin cores. Only the basal silt–sand layers of turbidites were measured, where these were in excess of 0.5 cm thick. The overlying muddy parts of the turbidite beds were ignored, partly because of the difficulty in determining the boundary between turbidite and hemipelagite, and partly because the mud thickness (2–20 cm) generally correlated closely with the sand thickness. The data were analysed using the time series autocorrelation and runs tests (Davis 1973; Forster 1995).

The bed thickness data are shown in Table 2. In general, very thin beds (<3 cm) are dominant. Cores 468, 470 and 471 show the greatest number of 0.5 cm thick beds, followed by 1 and 2 cm beds, core 472 shows equal numbers of 0.5 and 1 cm beds, whereas core 469 has equal numbers of 1 and 2 cm beds. When plotted on a cumulative frequency logarithmic plot (Fig. 3) of the number $N(t)$ of turbidite beds with a thickness $>t$ v. t, in the same way as carried out by Hiscott et al. (1992) and Forster 1995 for volcaniclastic turbidites from the Izu–Bonin forearc basin, the data approximate to a power-law distribution in which $N(t)$ is proportional to tb, where b is a positive constant, as has been proposed for turbidite bed thickness by Hsu (1983). The best-fit slope for these data over bed thicknesses from 10 to 3 cm is slightly >1.0, as found by Hiscott et al. (1992), although the decrease in slope at the thinner end of the spectrum is more marked than found previously. The data also show a relatively good fit for a log-normal distribution, and for some sites this appears to be a better fit than the power-law distribution.

The small-scale trends of bed thickness variation observed by visual inspection of the cores are

Table 2. *Data for turbidite bed thicknesses (silt–sand bases only) at each core site*

Thickness of sand–silt base (cm)	Number of beds				
	MKM-468	MKM-469	MKM-470	MKM-471	MKM-472
0.5	12	35	24	26	30
1	4	36	12	16	23
2	1	32	11	3	4
3	1	10	1	1	8
4	–	4	1	1	–
5	–	5	1	–	1
6	–	1	1	–	1
7	–	–	–	–	4
8	–	3	–	–	–
12	–	–	–	–	–
13	–	1	–	–	1
16	–	1	–	–	–
Total	18	128	61	47	72

not evident in any of the statistical tests applied (e.g. runs test), which is similar to the result found by Forster (1995). However, results of the autocorrelation study do show some cyclicity, with the most significant cycles occurring at three-bed and 10-bed intervals for core 469, at five-bed intervals for cores 470 and 471, and at six-bed intervals for core 472. The three- and six-bed cycles are significant at the 95% confidence level; the others are close to the 95% level. Other lags are not significant. Selected autocorrelation correlograms are shown in Fig. 4.

Several significant observations and interpretations can be made as follows:

(1) Thin-bedded, fine-grained turbidites are typical of the Makran slope, and are commonly arranged in small-scale (<10 m) vertical sequences of thinning or thickening upwards. These are probably caused by a mechanism analogous to that for lobe progradation and migration on submarine fans. The lack of statistical validation of these sequences is partly due to the unsatisfactory nature of the tests themselves (Forster 1995).

(2) The power-law distribution of bed thickness observed implies a large number of thin beds and small number of thick beds, as might be expected either from frequent seismic triggering of turbidity current events on an active margin (Hiscott *et al.* 1992), or from frequent river flood events. Either mechanism coupled with relatively short transport pathways to small basin plains will favour the deposition of fine- and medium-bedded turbidites, and hence a log-normal distribution of bed thickness.

(3) The very small-scale cycles noted by autocorrelation (3–6 beds) are probably best interpreted as compensation cycles (Mutti & Sonnino 1981; Forster 1995), in which the depositional relief of each turbidite deposited subtly affects the flow pathway and depositional locus of the subsequent event.

Sedimentary structures and sequence

Fine-grained turbidites show a distinctive suite of structures not seen in the classical (medium-grained) turbidites characterized by the Bouma sequence (Piper & Stow 1991; Stow *et al.* 1996). Piper (1978) first refined the scheme of Bouma (1962) to cover sedimentary features noted in turbidite muds, by dividing the E division into E1, E2 and E3. Stow (1977) further recognized nine structural divisions in fine-grained silt–mud turbidites, which he termed T0–T8 (Stow & Shanmugam 1980) (Fig. 5). The complete set of structures T0–T8 is rarely present in a single bed although the order of the sequence is believed to be correct (e.g. van Weering & van Iperen 1984; Lash 1988). However, the study by Porebski *et al.* (1991) is the first and, to our knowledge, the only one to have applied Markov chain analysis to transitions between the T0–T8 divisions to statistically verify their order of occurrence.

The Makran margin turbidites have been described in great detail, both visually and by X-radiographs, and so provide a good opportunity to further examine the Stow sequence. About 200 fine-grained turbidites have been examined in detail and their structural divisions carefully logged. All the T0–T8 divisions have been recognized, though very rarely as a complete sequence. The order of these structural divisions was studied using Markov chain analysis techniques (see Lindholm (1987) and Tucker (1988) for the full method). The raw data consist of an observed number of upward transitions, which are plotted in

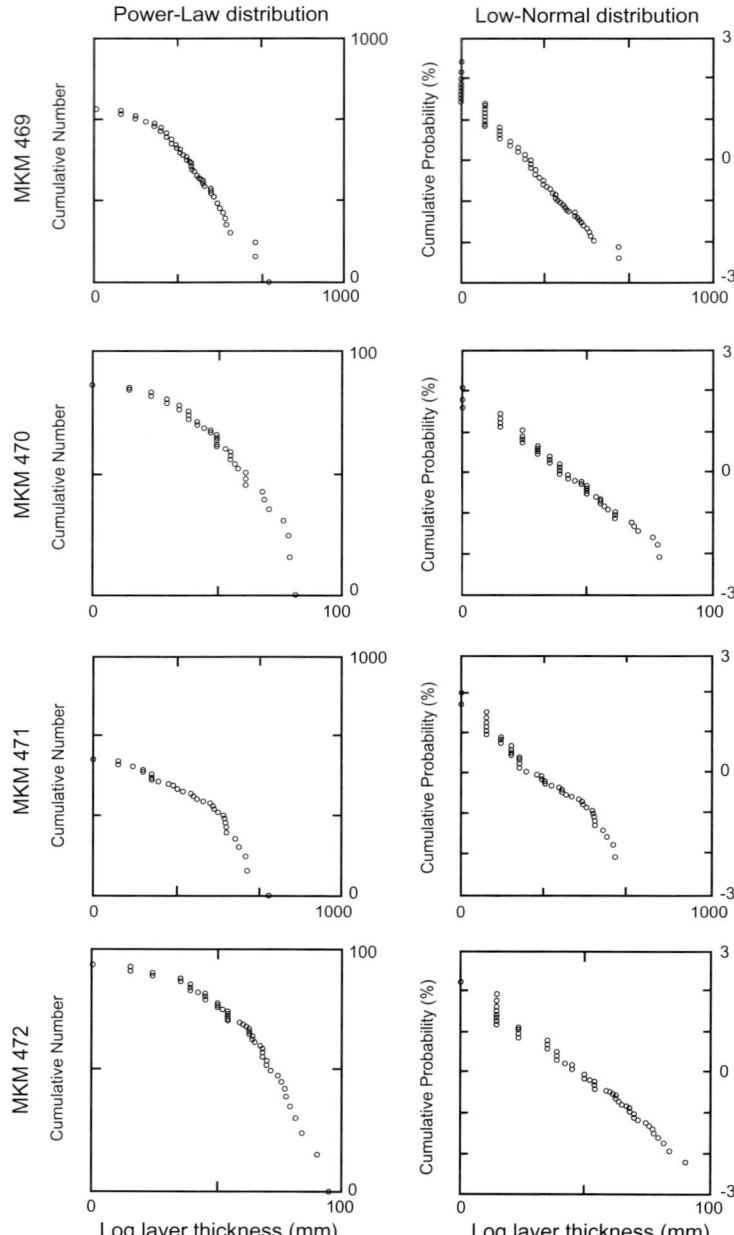

Fig. 3. Cumulative frequency plots for turbidite bed thickness for four of the five sites. Data are plotted to show best fits for both power-law and log-normal distributions. It should be noted that the scales for the different plots are necessarily varied to fit into the same figure.

matrix form. This tally matrix is converted to transition probabilities by dividing the number of transitions in each cell by the total number of transitions for the row containing the cell (row total) in the tally matrix. The difference matrix then derived shows the observed minus the calculated random probabilities with a range of values from -1.0 to $+1.0$. Positive values indicate transi-

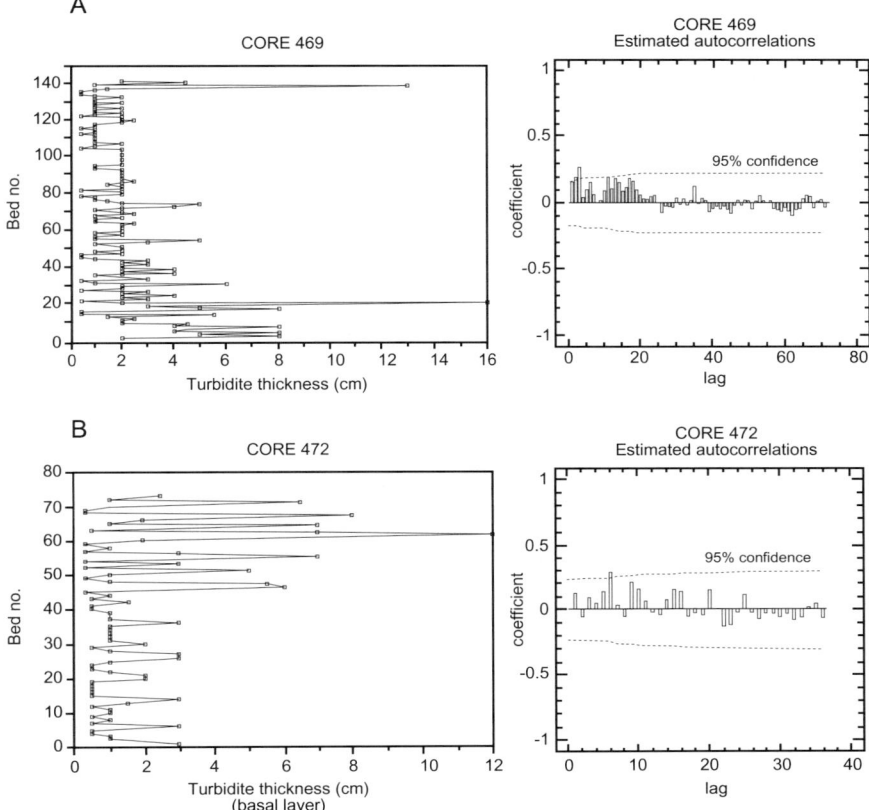

Fig. 4. Turbidite bed thickness data from selected sites. (**a**) Turbidite bed thickness for core 469 together with autocorrelation correlogram for these data. Dashed lines show the 95% confidence limit. (**b**) Turbidite bed thickness for core 472, together with autocorrelation correlogram for these data. Dashed lines show the 95% confidence limit.

tions more frequent than random, significantly so where the value is $>+0.1$, whereas negative values are taken as indicative of random transitions.

Results of this Markov chain analysis for all cores combined are given in Table 3 and illustrated in a facies relationship diagram (Fig. 6), which summarizes the principal transitions between turbidite divisions, and indicates both their frequency and direction. Several very interesting features are apparent:

(1) The most common transitions are those to the upper right of the diagonal, and the single most common transition from each division is to that immediately overlying it in the Stow (1977) sequence. T0–T2 is the only exception to this standard order, suggesting that the T1 division is either not very common or not so readily recognized.

(2) The transition from any of the divisions to T0 is also relatively common. These are taken as representing top-cut-out sequences capped by a new T0 basal division. Minor numbers of 'reverse' transitions to T3 and T4 can also be taken as top-cut-out sequences.

(3) Hemipelagites (HP) can show transitions back to any of the turbidite divisions, although T0 and T3 are the most common basal divisions observed (126 and 98 occurrences, respectively). The transition from any of the divisions to HP is also common.

(4) In most cases, the transitions that are common numerically also have a relatively high positive difference value, which is often highly significant ($>+0.2$). It is these difference values, rather than the actual numbers of transitions, that are used to construct the facies relationship diagram (Fig. 6). This demonstrates clearly that the

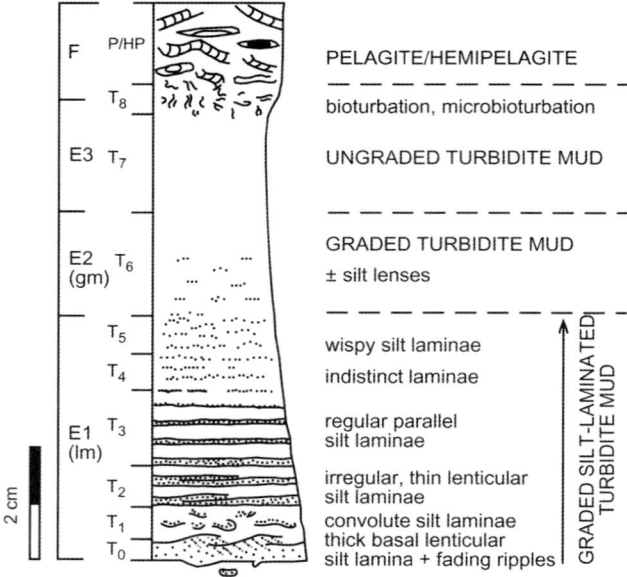

Fig. 5. Standard sequence of structures in fine-grained turbidites (after Stow 1977; Stow & Shanmugam 1980). The T0–T8 divisions are compared with the E1–3 divisions of Piper (1978) and the lm and gm divisions of Einsele (1991).

dominant transitions are all in the order predicted by the Stow sequence.

When the data for individual cores are considered separately (see Tabrez 1995), and also from visual inspection, it is clear that important differences exist between sites. The most common basal divisions vary from one site to another, and different numbers of base-cut-out, mid-cut-out and top-cut-out sequences are present. To characterize these differences, a Proximity Index for fine-grained turbidites (PI_{fgs}) has been devised, analogous to that for medium-grained turbidites that show Bouma sequences (Walker 1984). This index is given by the formula

$$PI_{fgs} = (T_{0123}/T) \times 100\%$$

where T_{0123} is the number of beds beginning with T0, T1, T2 or T3 divisions and T is the total number of beds. When applied to the Makran margin sites, this gives: core 468, 16%; core 469, 42%; core 470, 41%; core 471, 41%; core 472, 52%.

Although it is the most upslope, the location of core 468 is apparently less 'proximal', presumably because it is bypassed by turbidity currents. Conversely, the most distal site, for core 472 on the abyssal plain, has the highest Proximity Index. This is either because the larger turbidity currents deposit their load only when they reach the basin plain and/or because there is a separate input of turbidity currents deflected from the slope region further east, which is cut by more major submarine channels (Kukowski et al. 2001).

Sediment grain size

For this study grain size has been measured using a Malvern 2600 Particle Analyzer with a 100 mm focal length. This configuration provided measurements in 32 discrete size intervals between 0.5 and 188 μm (equivalent volume diameter). Some 300 samples were taken from hemipelagic facies at all sites (5–20 cm intervals), and 65 samples from all distinctive turbidites in the two longer, turbidite-dominated cores 469 and 472. Results have been computer plotted and moment statistics determined including median, mean, standard deviation and skewness.

The median grain size of all hemipelagite samples analysed varies between about 5 and 15 μm, and most of these samples lie between the tighter range of 5–10 μm. The mean size values tend to be slightly higher, typically 7–16 μm, and ranging to 27 μm. Sorting is generally very poor. Most samples show slight to moderate positive skewness. Three shapes of grain-size distribution curves can be recognized: very fine unimodal, fine bimodal and coarser polymodal (Fig. 7).

The vertical variation of grain size for hemipelagites analysed throughout the cored intervals

QUATERNARY SEDIMENTATION ON MAKRAN MARGIN 229

Table 3. *Markovian tally matrix for 651 observed transitions between turbidite divisions (T0–T8) and hemipelagite units (HP)*

	T0	T1	T2	T3	T4	T5	T6	T7	T8	HP	Total
T0		15 0.12 0.08	19 0.15 0.09	17 0.13 0.08	2 0.02 −0.03	5 0.04 0	18 0.14 0.08	1 0.01 −0.04	6 0.05 0.02	43 0.34 0.29	126
T1	2 0.06 0.02		7 0.35 0.29	4 0.20 0.09	1 0.05 0.01	1 0.05 0.01	2 0.10 0.02		1 0.05 0.01	4 0.20 0.09	20
T2				17 0.50 0.34	4 0.12 0.06	1 0.03 −0.01	4 0.12 0.06	1 0.03 −0.01		5 0.15 0.10	34
T3	6 0.06 0.04	1 0.01 −0.03	3 0.03 −0.03		46 0.46 0.36	2 0.02 −0.02	7 0.07 0.05	2 0.02 −0.03	3 0.03 −0.04	28 0.26 0.21	98
T4	8 0.14 0.10		1 0.02 −0.08	11 0.19 0.15		11 0.19 0.15	8 0.14 0.10		5 0.09 0.02	13 0.23 0.20	57
T5	4 0.17 0.08						15 0.65 0.56			4 0.17 −0.08	23
T6	14 0.19 0.12		2 0.03 −0.01	7 0.09 0.07	2 0.03 −0.01	2 0.03 −0.01		21 0.28 0.23	12 0.16 0.10	14 0.19 0.12	74
T7	5 0.16 0.08	1 0.03 −0.01		2 0.06 0.01					14 0.45 0.38	9 0.29 0.14	31
T8	14 0.27 0.18			4 0.08 0.06			1 0.02 −0.07			33 0.63 0.38	52
HP	69 0.51 0.38	2 0.01 −0.02	8 0.06 0.06	36 0.26 0.16	3 0.020 .02	3 0.02 0.02	7 0.05 0.04	2 0.01 −0.02	6 0.04 0.03		136
	122	19	40	98	58	25	62	27	47	153	651/651

It should be noted that the top row of values for each division shows the actual number of observed transitions. The middle row is the observed probability. The bottom row shows the difference between the observed and random probabilities and is taken as a measure of the statistical significance of a particular transition. Positive values suggest ordered and negative values random transitions; a value of > +0.1 is taken as a significant ordered transition.

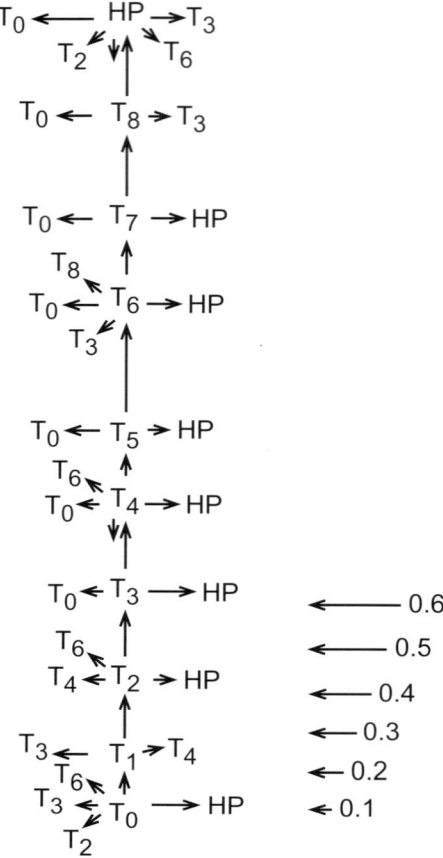

Fig. 6. Facies relationship flow diagram based on the Markovian difference matrix for 651 observed vertical transitions between turbidite divisions (T0–T8) and hemipelagite units (HP). The relative lengths of the arrows indicate the relative probabilities of the various transitions, rather than the actual numbers of transitions. The dominance of the standard sequence, and the presence of partial sequences showing the standard direction of transitions, should be noted.

at each site is extremely interesting. First, there is a zone of minimum grain size at between 2 and 3 m depth in cores 468, 469 and 470, and between 7 and 10 m in cores 471 and 472. This has been used in our lithostratigraphic correlation between cores (see above, Fig. 2a). Second, there is a small-scale oscillation or cyclicity of median grain size evident over a vertical distant of 0.2–1.0 m. This oscillation is particularly marked, with variation in median grain size of 4–6 μm, only at certain horizons in each of the cores.

Grain-size characteristics for the turbidites analysed from cores 469 and 472 are very different from those of the hemipelagites described above. Where an individual turbidite has been analysed in detail (i.e. three or more samples), there is evidence of a clear gradation in properties from base to top. One relatively thick (30 cm) silt-laminated turbidite from core 472 is illustrated in Fig. 8. The nine separate analyses show a change from base to top in median grain size (from 34.8 to 6.3 μm), mean grain size (from 37.1 to 8.7 μm), sorting (from 7.8 to 26.5 μm), and skewness (from unity to 2.5). The shape of the grain-size distribution curves change from a distinctly peaked, well-sorted curve with a marked fine tail (positive skewness) at the base, to a broad somewhat irregular curve at the top. Similar characteristics are observed throughout, although the basal silt–sand layers are slightly coarser in core 469 (mean 20–70 μm, median 12–56 μm), than in core 472 (mean 12–64 μm, median 9–57 μm). The shapes of the grain-size distribution curves range from sharply peaked, well-sorted with a small fine tail for the coarser layers, to broadly domed, irregular, poorly sorted and with a very large fine tail for the finer layers.

In summary, we recognize and interpret the following features for Makran margin hemipelagites:

(1) The grain-size characteristics are similar to those of mixed biogenic–clastic hemipelagites reported in previous work, and are taken as typical for a terrigenous fluvial–aeolian source (Stow & Tabrez 1998). Fluvial input is probably dominant in this case (von Rad, pers. comm. 2001).

(2) The unimodal distribution reflects an input of nannofossils and fine clays only, the latter probably from fluvial suspension; the bimodal distribution reflects an input of nannofossils together with mixed terrigenous material of aeolian and fluvial derivation; and the polymodal distribution is caused by a heterogeneous input of nannofossils and foraminifers, and terrigenous mud, silt and fine sand of mixed source, particularly with a greater influence of direct fluvial input at lowered sea level.

(3) The vertical variation of median grain sizes over the length of the cores can be interpreted in terms of climatic and sea-level controls: the coarser grain size at depth reflects sea-level lowstand during the last glacial period and a multi-sourced input; the mid-core grain-size low reflects sea-level rise and a fining of the fluvial input from early to mid-Holocene time; and the subsequent grain-size increase may then reflect increased aeolian input as a result of onshore aridification.

(4) The vertical oscillation of grain size over 0.2–1.0 m has a time scale of about 500–1000 a and probably reflects slight climatic variation of this order, which in turn influences aeolian–fluvial input and/or biogenic productivity.

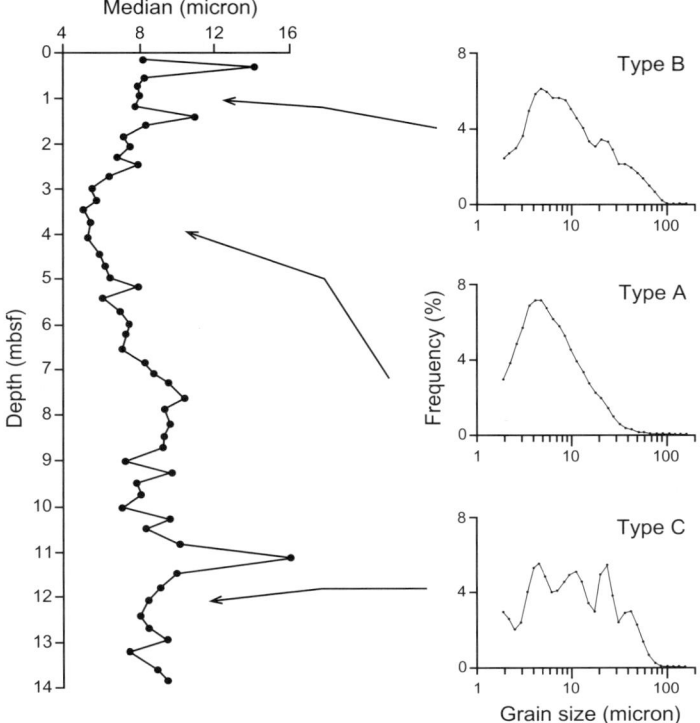

Fig. 7. Typical grain-size distribution curves for selected hemipelagite facies from core 469. Type A, very fine, unimodal; Type B, fine, bimodal; Type C, mixed, poorly sorted, polymodal. The median size plot is based on carbonate-free samples, whereas the grain-size distribution curves include the carbonate fraction.

(5) The lack of any clear-cut onshore–offshore trend in grain size is in part due to the morphological complexity of margin, and hemipelagic dispersion that generally follows the turbidite pathways, and in part due to the multi-sourced composition.

For the fine-grained turbidites, the following points can be made:

(1) The grain-size characteristics of the Makran margin turbidites are very typical of those of fine-grained turbidites elsewhere (e.g. Piper & Stow 1991; Stow *et al.* 1996) and very different from those of the hemipelagites. The upward change through a single graded, laminated unit is also typical.

(2) There is a relatively clear offshore decrease in grain size from core 469 to 472, reflecting the downslope distribution pathway, but apparently not coincident with the inverse nature of the proximality index between these sites, as noted above.

(3) There is no clear relationship between grain size and bed thickness for the measured turbidites.

Discussion

Turbidite statistics

Makran margin sedimentation is characterized by a relative abundance of fine-grained turbidites, approximately one event per 180–300 a. Careful study therefore allows a more rigorous or semi-quantitative analysis of turbidite features than is normally possible.

Statistical analysis of bed thickness data shows best-fit lines with either an imperfect power-law distribution or a log-normal distribution. Whereas Hiscott *et al.* (1992) argued that a power-law distribution should be expected for turbidite successions because of the likely scale-invariance of the (seismic) triggering mechanism, Forster (1995) demonstrated that both log-normal and power-law distributions are present in different turbidite successions and concluded that the former are the result of various external effects on the transport and deposition by turbidity flows subsequent to their initiation. In the case of the Makran turbidites, it may be that the local morphology of

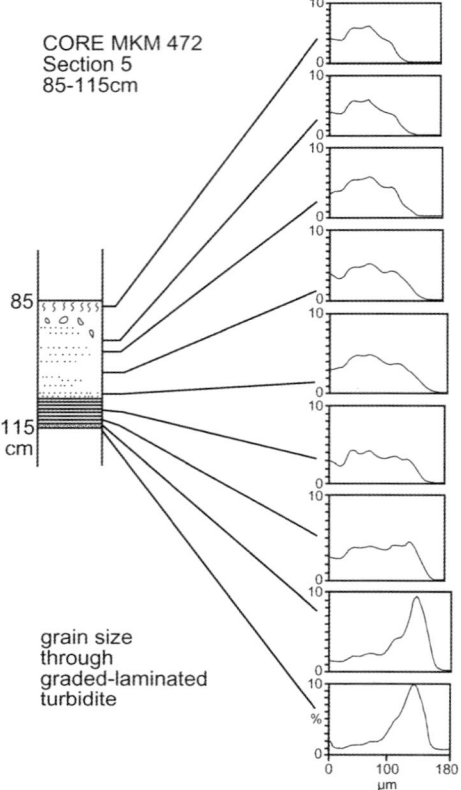

Fig. 8. Series of samples taken through a silt-laminated mud turbidite of core 472, showing typical grain-size distribution curves (whole fraction) from base to top of the bed.

than repeatedly thickening upwards as originally proposed by Mutti & Sonnino (1981).

Markov chain analysis has allowed an objective assessment of the sequence of structures found in fine-grained turbidites, thereby testing the validity of the standard model proposed by Stow (1977) and Stow & Shanmugam (1980). On the basis of data from 651 transitions between divisions, the order of the Stow sequence (T0–T8) is found to be fully valid for over 95% of the transitions. This supports an earlier testing of the sequence by Porebski et al. (1991) in turbidites from the Zaire Fan.

Furthermore, partial sequences of all turbidites are the norm, with mid-cut-out sequences being most common and base-cut-out sequences second in importance. The latter might be expected of relatively 'distal' turbidites across a slope showing a distinct lack of channelization. The absence of the middle divisions may reflect relatively rapidly deposition (at least of the upper divisions) from a partially ponded turbidity current on entering the slope basins. However, the presence of well-defined silt laminae (T0 and T3–T4 divisions especially) implies relatively mature turbidity currents and sufficiently slow deposition (at least of the basal divisions) to allow depositional sorting through the boundary layer (Stow & Bowen 1980). A Proximity Index for fine-grained turbidites (see Walker (1967, 1984) for medium-grained turbidites) has been found useful for characterizing suites of turbidites from different sites.

In a very few of the turbidites studied, non-standard sequences or repetition of certain divisions were noted. These could be explained by either (1) reflected and reversed flows in ponded basins, or (2) a multiple flow mechanism.

partly interconnected slope basins is sufficiently uniform that it favours the relatively narrow range of turbidite thicknesses shown by the log-normal distribution. Alternatively, the frequency of seismic or river flood trigger events coupled with a steady input of sediment to the shelf–upper-slope source area, might favour a particular range of flow dimensions and hence turbidite thicknesses.

Bed thickness variation observed using an autocorrelation statistical technique reveals some small-scale cyclicity (3–6 beds), which can most readily be interpreted as the result of compensation cycles forming in the small basin settings (Mutti & Sonnino 1981). The main mechanism for their development is lateral shifts in the site of turbidity current deposition as a result of the microtopography created by previous deposits. We would also agree with the findings of Forster (1995), from other turbidite series, that the variation is cyclic and typically symmetrical, rather

Sediment source, distribution and controls

There are several potential sources of terrigenous material for turbidite and hemipelagite sedimentation on the Makran margin. These include: (1) local riverborne material from the Oman–Makran hinterland (e.g. Hub, Hingol and other rivers); (2) airborne material from arid regions of Pakistan and NW India; (3) airborne material from the Arabian–African deserts; (4) direct or indirect supply from the Indus River system.

In general, the results of clay and silt mineralogical analysis (see also Tabrez 1995) show a relatively uniform sediment composition of materials derived mainly from the north (i.e. (1) and (2) above). There is little evidence for a marked palygorskite–smectite clay mineral assemblage typical of the Arabian–African desert source, and also little possibility for much Indus-derived material to find its way onto the Makran slope. The

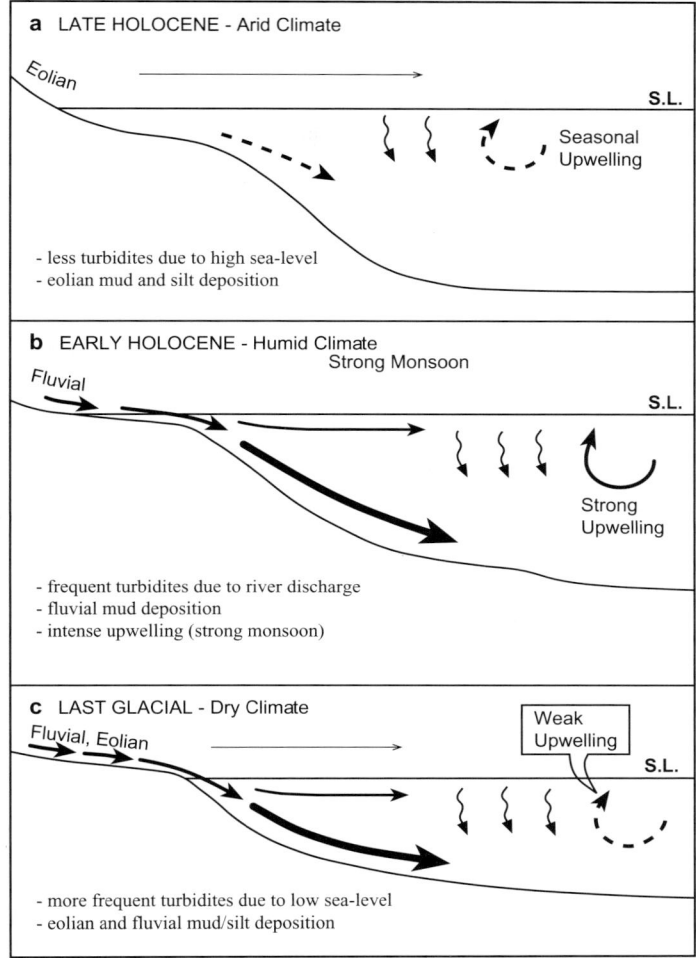

Fig. 9. Principal controls and sediment supply routes for the Makran margin during the last glacial, and early and late Holocene periods. This figure shows a schematic summary of controls; for a more detailed overview, see Prins *et al.* (2000*a, b*).

abyssal plain site for core 472 could, physiographically, have a more direct connection with an Indus source, but there is little mineralogical support for this.

Although distinct channels are incised into the steeper, upper part of the slope, turbidity current pathways rather than a true channel system can be observed across much of the rest of the study area (Fig. 1). Further east, however, channel development and incision has continued across the whole slope (Kukowski *et al.* 2001). Turbidity currents generated on the outer shelf to upper slope may initially be channelized, but then apparently follow a somewhat irregular and sinuous pathway that cuts through gaps in the fold ridges and flows either to the east or west through the semi-enclosed slope basins. Hemipelagic dispersion is believed to be affected and partly focused by the same distribution pathways. In some cases, hemiturbiditic sedimentation (Stow & Wetzel 1990) is recognized, perhaps resulting from the interaction of turbidity currents with the basin–ridge topography.

Tectonic activity, fluvial input, sea-level fluctuation and climate variation have all influenced sedimentation on the Makran margin during the time period represented in this study (Fig. 9), although their relative significance has varied and is not always easy to ascertain.

(1) During the last glacial period, sea level was lower and the climate relatively arid. Turbidite

input was frequent, and hemipelagic sedimentation was multi-sourced, with less direct fluvial discharge and more aeolian dispersion of terrigenous material, as well as a biogenic pelagic contribution. A similar interpretation for the region south of the Indus delta was presented by von Rad et al. (1999a).

(2) During deglaciation, sea level rose and the climate became more humid. Turbidite input remained frequent, although with a trend towards thinner beds evident in some of the cores, perhaps related to a retreat of the fluvial input points and fining of source material. Hemipelagic sedimentation was dominated by very fine-grained, distal, fluvial dispersions, perhaps partly as hyperpycnal discharge. A stronger monsoonal upwelling system developed, leading to primary biogenic input of mainly nannofossils.

(3) The post-glacial (late Holocene) period is one of sea-level highstand, and a climate that has resulted in generally increased aridity in the drainage basin. Von Rad et al. (1999a) carried out careful study of a piston core recovered from the upper slope off Karachi. They recognized cyclicity in varved sediments at decadal to centennial freqencies, which they related to climatic variability over the past 5 ka. Periods of higher precipitation and river discharge lead to thicker varves as well as more frequent turbidites. These alternate with more arid conditions, thinner varves and less frequent turbidites. With turbidite sedimentation much less frequent in our study area, the same degree of resolution is not possible. Hemipelagites are characterized by very fine-grained fluvial and aeolian input and pelagic supply of nannofossils under a more seasonal upwelling regime.

(4) The precise triggering mechanism for turbidity currents is always problematic. In the case of the Makran margin, turbidite thickness data from our study area indicate a relatively large number of events triggering frequent but small turbidity currents. This led to deposition of mainly thin- to medium-bedded turbidtes (<1 cm to 16 cm sandy divisions) with a frequency of one event every 200–300 a. However, we cannot distinguish between seismic activity or major river flood events as the principal trigger.

Much of the work for this paper was carried out by A.R.T. during tenure of a Pakistan Government Scholarship at the University of Southampton, and by M.A.P. as part of his Ph.D. research at Utrecht University, under the supervision of G. Postma. We also express our thanks to the National Institute of Oceanography, Pakistan, for encouraging the secondment of A.R.T. to undertake this project. The cores were collected as part of a joint Pakistan–Netherlands Indian Ocean Programme (NIOP) under the direction of W. J. M. van der Linden and C. H. van der Weijden. The Chief Scientist, officers and crew on board R.V. *Tyro* are thanked for their part in the success of this programme; as also are G. van der Linden, J. Stel and G. Postma at the University of Utrecht. We thank these individuals, in particular, for useful discussions and data exchange. We further acknowledge the excellent secretarial and technical support at our respective institutes, and a full and helpful review by U. von Rad.

References

ARTHURTON, R.S., FARAH, A. & AHMED, W. 1982. The Late Cretaceous–Cenozoic history of western Baluchistan, Pakistan—the northern margin of the Makran subduction complex. *In:* LEGGETT, J.K. (ed.) *Trench–Forearc Geology: Sedimention and Tectonics on Modern and Ancient Active Plate Margins.* Geological Society, London, Special Publications, **10**, 373–385.

BOUMA, A.H. 1962. *Sedimentology of some Flysch Deposits: a Graphic Approach to Facies Interpretation.* Elsevier, Amsterdam.

CLEMENS, S.C. & PRELL, W.L. 1990. Late Pleistocene variability of Arabian Sea summer monsoon winds and continental aridity: eolian records from the lithogenic component of deep-sea sediments. *Paleoceanography*, **5**, 109–145.

DAVIS, J.C. 1973. *Statistics and Data Analysis in Geology.* Wiley, New York, 232–235.

DENHAM, C.R. 1974. Counter-clockwise motion of paleomagnetic directions 24,000 years ago at Mono Lakes, California. *Journal of Geomagnetism and Geotectonics*, **26**, 487–498.

DENHAM, C.R. & COX, A. 1971. Evidence that the Laschamp polarity event did not occur 13,300–30,400 years ago. *Earth and Planetary Science Letters*, **13**, 181–190.

EINSELE, G. 1991. Submarine mass flow deposits and turbidites. *In:* EINSELE, G., RICKEN, W. & SEILACHER, A. (ed.) *Cycles and events in stratigraphy.* Springer-Verlag, Berlin, 313–339.

FAUGERES, J.C. & GONTHIER, E. 1981. Les argiles de sédiments marins du Quaternaire récent dans le Golfe d'Aden et la Mer d'Oman. *Oceanologica Acta*, **4**, 4.

FORSTER, C. M., 1995. *The analysis of bed thickness trends in turbidite successions—a quantitative approach for the prediction of vertical sequences.* PhD thesis, University of Southampton.

HISCOTT, R.N. 1981. Deep-sea fan deposits in Macigno Formation (mid to upper Oligocene) Oligocene) of the Gordana valley, northern Apennines, Italy—discussion. *Journal of Sedimentary Petrology*, **51**, 1015–1021.

HISCOTT, R.N., COLELLA, A., PEZARD, P.A., LOVELL, M.A. & MALINVERNO, A. 1992. Sedimentology of deep-water volcaniclastics, Oligocene Izu–Bonin forearc Basin, based on Formation MicroScanner images. *In:* TAYLOR, B. & FUJIOKA, K. (eds) *Proceedings of the Ocean Drilling Program, Scientific Results,* **126**. Ocean Drilling Program, College Station, TX, 75–94.

Hsu, K.J. 1983. Actualistic catatrophism: address of the retiring President of the International Association of Sedimentologists. *Sedimentology*, 3–9.

HUTCHISON, I., LOUDEN, K.E. & WHITE, R.S. 1981. Heat flow and age of the Gulf of Oman. *Earth and Planetary Science Letters*, **56**, 256–262.

KOLLA, V., HENDERSON, L. & BISCAYE, P.E. 1976. Clay mineralogy and sedimentation in the western Indian Ocean. *Deep-Sea Research*, **23**, 949–961.

KOLLA, V., KOSTECKI, J.A., ROBINSON, F., BISCAYE, P.E. & RAY, P.K. 1981. Distribution and origin of clay-minerals and quartz in surface sediments of the Arabian Sea. *Journal of Sedimentary Petrology*, **51**, 563–569.

KOPP, C., FRUEHN, J., FLUEH, E.R., REICHERT, C., KUKOWSKI, N., BIALAS, J. & KELISCHEN, D. 2000. Structure of the Makran subduction zone from wide-angle and reflection seismic data. *Tectonophysics*, **329**, 171–191.

KUKOWSKI, N., SCHILLHORN, T., HUHN, K., VON RAD, U., HUSEN, S. & FLUEH, E.R. 2001. Morphotectonics and mechanics of the central Makran accretionary wedge off Pakistan. *Marine Geology*, **173**, 1–19.

LASH, G.G. 1988. Sedimentology and evolution of the Martinsburg Formation (Upper Ordovician) fine-grained turbidite depositional system, central Appalachians. *Sedimentology*, **35**, 429–447.

LIDDICOAT, J.C. & COE, R.S. 1979. Mono Lake geomagnetic excursion. *Journal of Geophysical Research*, **84**, 261–271.

LINDHOLM, R.C. 1987. *A Practical Approach to Sedimentology*. Allen & Unwin, London.

LUCKGE, A., DOOSE-ROLINSKI, H., KHAN, A.A., SCHULZ, H. & VON RAD, U. 2001. Monsoonal variation in the NE Arabian Sea during the past 5000 years: geochemical evidence from laminated sediments. *Palaeogeography, Palaeoclimatology, Palaeoecology*, **167**, 273–289.

MINSHULL, T.A. & WHITE, R.S. 1989. Sediment compaction and fluid migration in the Makran accretionary prism. *Journal of Geophysical Research*, **94**(B6), 7387–7402.

MUTTI, E. & SONNINO, M. 1981. Compensation cycles: a diagnostic feature of sandstone lobes. *IAS 2nd European Meeting, Abstract Volume*, 120–123.

NOLTIMER, R.M. & COLVINAUX, P.A. 1976. Geomagnetic excursion from Imuruk, Alaska. *Nature*, **259**, 197–200.

PIPER, D.J.W. 1978. Turbidite muds and silts on deep-sea fans and abyssal plains. *In:* STANLEY, D.J. & KELLING, G. (eds) *Sedimentation in Submarine Canyons, Fans, and Trenches*. Dowden, Hutchinson and Ross, Stroudsburg, PA, 163–176.

PIPER, D.J.W. & STOW, D.A.V. 1991. Fine-grained turbidites. *In:* EINSELE, G. & SEILACHER, A. (eds) *Cycles and Events in Stratigraphy*, 2nd. Springer, Berlin, 360–376.

POREBSKI, S.J., MEISCHNER, D. & CORLICH, K. 1991. Quaternary turbidites from South Shetland Trench, West Antarctica: recognition and implications for turbidite facies modelling. *Sedimentology*, **38**, 691–715.

PRELL, W.L. & STREETER, H.F. 1982. Temporal and spatial patterns of monsoonal upwelling along Arabia: a modern analogue for the interpretation of Quaternary SST anomalies. *Journal of Marine Research*, **40**, 143–155.

PRELL, W.L., NIITSUMA, N., ET AL. (eds) 1989. *Proceedings of the Ocean Drilling Program, Initial Reports, 117*. Ocean Drilling Program, College Station, TX.

PRELL, W.L. ET AL.1990. Neogene tectonics and sedimentation of the SE Oman continental margin: results from ODP Leg 117. *In:* ROBERTSON, A.H.F., SEARLE, M.P. & RIES, A.C. (eds) *The Geology and Tectonics of the Oman Region*. Geological Society, London, Special Publications, **49**, 745–758.

PRINS, M.A., 1999. *Pelagic, hemipelagic and turbidite sedimentation in the Arabian Sea during the late Quaternary*. PhD thesis, Utrecht University.

PRINS, M.A. & POSTMA, G. 2000. Effects of climate, sea level and tectonics unravelled for last deglaciation turbidite records of the Arabian Sea. *Geology*, **28**, 375–378.

PRINS, M.A., POSTMA, G., CLEVERINGA, J., CRAMP, A. & KENYON, N.H. 2000a. Control on terrigenous sediment supply to the Arabian Sea during the late Quaternary: the Indus Fan. *Marine Geology*, **169**, 327–349.

PRINS, M.A., POSTMA, G. & WELTJE, G.J. 2000b. Control on terrigenous sediment supply to the Arabian Sea during the late Quaternary: the Makran Continental Slope. *Marine Geology*, **169**, 351–371.

QUITTMEYER, R.C. & KAFKA, A.L. 1984. Constraints on plate motions in southern Pakistan and the northern Arabian Sea from the focal mechanisms of small earthquakes. *Journal of Geophysical Research*, **89**, 2444–2458.

REICHART, G.J. ET AL. 1997. A 225 ky record of dust supply, paleoproductivity and the oxygen minimum zone from the Murray Ridge (northern Arabian Sea). *Palaeogeography, Palaeoclimatology, Palaeoecology*, **134**, 149–169.

REICHART, G.J., LOURENS, L.J. & ZACHARIASSE, W.J. 1998. Temporal variability in the northern Arabian Sea oxygen minimum zone (OMZ) during the last 225,000 years. *Paleoceanography*, **13**, 607–621.

SCHULZ, H., VON RAD, U. & VON STACKELBERG, U. 1996. Laminated sediments from the oxygen minimum zone of the NE Arabian Sea. *In:* KEMP, A.E.S. (eds) *Paleoclimatology and Paleoceanography from Laminated Sediments*. Geological Society, London, Special Publications, **116**, 185–207.

SCHULZ, H., VON RAD, U. & ERLENKEUSER, H. 1998. Correlation between Arabian Sea and Greenland climate oscillations of the past 110,000 years. *Nature*, **393**, 54–57.

SIROCKO, F. & LANGE, H. 1991. Clay-mineral accumulation rates in the Arabian Sea during the late Quaternary. *Marine Geology*, **97**, 105–119.

SIROCKO, F., SARNTHEIN, M., ERIENKEUSER, H., LANGE, H., ARNOLD, M. & DUPLESSY, J.C. 1993. Century-scale events in monsoonal climate over the past 24,000 years. *Nature*, **364**, 322–324.

STEENS, T.N.F., KROON, D., TEN KATE, W.G. & SPRENGER, A. 1991. Late Pleistocene periodicities of oxygen isotope ratios, calcium carbonate contents and magnetic susceptibilities of western Arabian Sea margin, Hole 728A. *In:* PRELL, W.L. & NIITSUMA, N. (eds) *Proceedings of the Ocean Drilling Program, Scientific Results, 117.* Ocean Drilling Program, College Station, TX, 309–320.

STOFFERS, P. & ROSS, D.A. 1979. Late Pleistocene and Holocene sedimentation in the Persian Gulf–Gulf of Oman. *Sedimentary Geology*, **23**, 181–208.

STOW, D.A.V., 1977. *Late Quaternary stratigraphy and sedimentation on the Nova Scotian outer continental margin.* PhD thesis, Dalhousie University, Canada.

STOW, D.A.V. & BOWEN, A.J. 1980. Physical model for the transport and sorting of fine-grained sediment in turbidity currents. *Sedimentology*, **27**, 31–46.

STOW, D.A.V. & SHANMUGAM, G. 1980. Sequence of structures in ancient and modern fine-grained turbidites. *Sedimentary Geology*, **25**, 23–42.

STOW, D.A.V. & TABREZ, A. 1998. Hemipelagites: facies, processes and models. Geological Society, London, Special Publications, **129**, 317–338.

STOW, D.A.V. & WETZEL, A. 1990. Hemiturbidite: a new type of deep water sediment. *In: Proceedings of the Ocean Drilling Program, Scientific Results, 116.* Ocean Drilling Program, College Station, TX, 25–34.

STOW, D.A.V., READING, H.G. & COLLINSON, J. 1996. Deep seas. *In:* READING, H.G. (eds) *Sedimentary Environments and Facies, 3rd.* Blackwell Scientific, Oxford, 380–442.

TABREZ, A.R. 1995. *Slope sedimentation around the NW Indian Ocean.* PhD thesis, University of Southampton.

THOUVENY, N. 1988. High resolution paleomagnetic study of late Pleistocene sediments from Baffin Bay: first results. *Canadian Journal of Earth Sciences*, **25**, 833–843.

TUCKER, M.E. 1988. *Techniques in Sedimentology.* Blackwell Scientific, Oxford.

VAN WEERING, T.C.E. & VAN IPEREN, J. 1984. Fine-grained sediments of the Zaire deep-sea fan, southern Atlantic Ocean. *In:* STOW, D.A.V. & PIPER, D.J.W. (eds) *Fine-Grained Sediments: Deep-Water Processes and Facies.* Geological Society, London, Special Publications, **15**, 95–114.

VON RAD, U., SCHAAF, M., MICHELS, K.H., SCHULZ, H., BERGER, W.H. & SIROCKO, F. 1999a. A 5000-year record of climatic change in varved sediments from the oxygen minimum zone off Pakistan, NE Arabian Sea. *Quaternary Research*, **51**, 39–53.

VON RAD, U., SCHULZ, H., RIECH, V., DEN DULKE, M., BERNER, U. & SIROCKO, F. 1999b. Repeated monsoon-controlled breakdown of oxygen minimum conditions during the past 30,000 years documented in laminated sediments off Pakistan, NE Arabian Sea. *Palaeogeography, Palaeoclimatology, Palaeoecology*, **152**, 129–161.

WALKER, R.G. 1967. Turbidite sedimentary structures and their relationship to proximal and distal depositional environments. *Journal of Sedimentary Petrology*, **37**, 25–43.

WALKER, R.G. 1984. *Facies Models, Geoscience Canada Reprint Series I, 2nd.* Geological Society of Canada, Waterloo, Ont.

WEEDON, G.P. 1985. Hemipelagic shelf sedimentation and climatic cycles: the basal Jurrassic (Blue Lias) of South Britain. *Earth and Planetary Science Letters*, **76**, 321–375.

WEEDON, G.P. 1989. The detection and illustration of regular sedimentary cycles using Walsh power spectra and filtering, with examples from the Lias of Switzerland. *Journal of the Geological Society, London*, **146**, 133–144.

WHITE, R.S. 1982. Deformation of the Makran accretionary sediment prism in the Gulf of Oman (northwest Indian Ocean). *In:* LEGGETT, J.K. (ed.) *Trench–Forearc Geology: Sedimentation and Tectonics on Modern and Ancient Active Plate Margins.* Geological Society, London, Special Publications, **10**, 357–372.

WHITE, R.S. & LOUDEN, K.F. 1982. The Makran continental margin: structure of a thickly sedimented convergent plate boundary. *AAPG Bulletin*, **34**, 499–518.

WHITE, R.S. & ROSS, D.A. 1979. Tectonics of the western Gulf of Oman. *Journal of Geophysical Research*, **84**, 3479–3489.

WIEDICKE, M., NEBEN, S. & SPIESS, V. 2001. Mud volcanoes at the front of the Makran accretionary complex, Pakistan. *Marine Geology*, **172**, 57–73.

A brief history of the Indus River

PETER D. CLIFT

*Department of Geology and Geophysics, Woods Hole Oceanographic Institution, Woods Hole,
MA 02543, USA (e-mail: pclift@whoi.edu)*

Abstract: The Indus River system is one of the largest rivers on the Asian continent, but unlike the Ganges–Brahmaputra system, the drainage of the Indus is dominated by the western Tibetan Plateau, Karakoram and tectonic units of the Indus Suture Zone, rather than the High Himalaya. The location of the river system relative to the Indus Suture Zone explains the deep exhumation north of that line in the Karakoram, compared with the modest erosion seen further east in Tibet. The modern Indus cuts Paleogene fluvial sedimentary rocks of the Indus Group located along the Indus Suture Zone in Ladakh, northern India. After the final marine incursion within the Indus Group in the early Eocene (<54.6 Ma), palaeo-current indicators changed from a north–south flow to an axial, westward pattern, synchronous with a marked change in sediment provenance involving erosion of South Tibet. The Indus probably was initiated by early Tibetan uplift following the India–Asia collision. The river has remained stationary in the suture since Early Eocene time, cutting down through its earlier deposits as they were deformed by northward folding and thrusting associated with the Zanskar backthrust at $c.$ 20 Ma. The Indus appears to have been located close to its present position within the foreland basin since at least Mid-Miocene time ($c.$ 18 Ma), and to have migrated only $c.$ 100 km east since Early Eocene time. In the Arabian Sea Paleogene fan sedimentation was significant since at least Mid-Eocene time ($c.$ 45 Ma). Sediment flux to the mid fan and shelf increased during Mid-Miocene time (after 16 Ma) and can be correlated with uplift of the Murray Ridge preventing sediment flow into the Gulf of Oman, tectonic uplift and erosion in the Karakoram and western Lhasa Block, and an enhanced monsoon triggered by that same uplift. Sedimentation rates fell during Late Miocene to Recent time. The Indus represents 18% of the total Neogene sediment in the basins that surround Asia, much more than all the basins of Indochina and East Asia combined ($c.$ 11%). Unlike the rivers of East Asia, which have strongly interacted as a result of eastward propagating deformation in that area, the Indus has remained uninterrupted and represents the oldest known river in the Himalayan region.

The Indus River is the principal river of the western Himalaya. It drains an area of $c.$ 1×10^6 km^2, with peak discharge during the summer months as a result of seasonal glacial melting and the increased runoff generated by the summer monsoon (Milliman *et al.* 1984). Before damming, discharge was 450×10^6 tons a^{-1}, comparable with the Mississippi (Listzin 1972). The sediment transported by the river is deposited in the Arabian Sea as the Indus Fan, one of the largest deep-sea fans in the world, totalling $c.$ 5×10^6 km^3 (Naini & Kolla 1982). Sediment in the Indus system is preferentially eroded from the western Tibetan Plateau and Karakoram (Clift *et al.* 2000, 2001*b*), in contrast to the Ganges–Brahmaputra system, which is dominated by the High Himalaya (France-Lanord *et al.* 1993). Consequently, the Indus system represents an important potential source of information on the uplift and erosion history of the western Himalaya, and crucially Tibet, whose growth has been linked to the intensification of the SW monsoon (Prell & Kutzbach 1992; Molnar *et al.* 1993).

The Indus River rises in western Tibet near Mount Kailas, and follows the NW–SE trend of the Karakoram Fault, before cutting westwards into the Indus Suture Zone (Schroder & Bishop 1999; Searle & Owen 1999). The Indus continues to follow the strike of the suture before cutting orthogonally through the Himalaya in NW Pakistan and running south to the Arabian Sea. The other tributaries to the Indus, such as the Chenab, Jellum and Sutlej (Fig. 1) do drain the crystalline High Himalaya, but do so in an area where its topography is much reduced, compared with the central and eastern Himalaya (Fig. 2). In contrast, the Bengal Fan's main feeder rivers, the Ganges and Brahmaputra, follow the High Himalaya along strike for much of the length of the orogen. In practice, this means that the Bengal Fan is swamped by the large volume of material derived from the rapidly unroofing High Himalaya (France-Lanord *et al.* 1993), with minor input from the Indian Shield in its distal region (Crowley *et al.* 1998). In contrast, the Indus Fan is

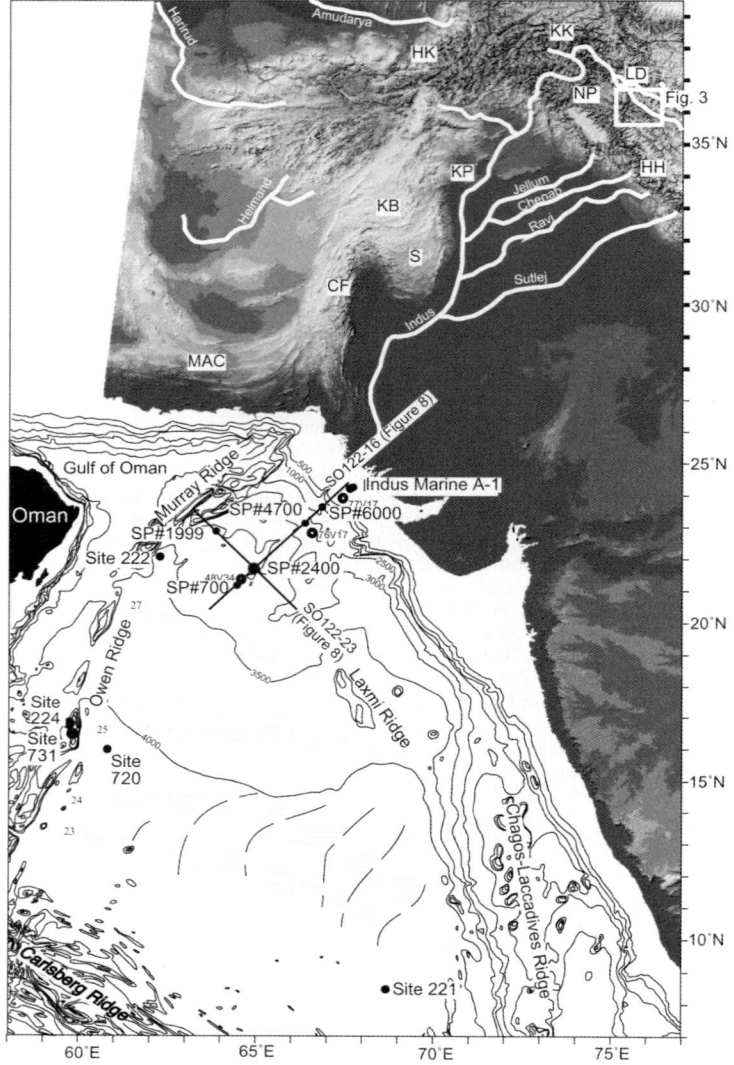

Fig. 1. Digital topographic map of the Indus drainage basin and GEBCO bathymetry for the offshore region; depths in metres. HH, High Himalaya; KK, Karakoram; LD, Ladakh; NP, Nanga Parbat; HK, Hindu Kush; CF, Chaman Fault; KB, Katawaz Basin; S, Sulaiman Ranges; KP, Kohat Plateau; MAC, Makran Accretionary Complex. Sea-floor magnetic anomalies from Miles *et al.* (1998). ○, Lamont–Doherty Earth Observatory sonobuoy locations.

dominated by tectonic units adjacent to the suture zone, including western Tibet (Clift *et al.* 2001*b*). This allows their erosional signal to be more readily isolated in the Indus Fan compared with the Bengal Fan.

In this paper I review the evidence for evolution of the Indus River at various stages of its route to the Arabian Sea, and attempt to understand the significance of the Indus to the wider issue of evolving drainage in Asia during the Cenozoic period.

The Indus River in the Indus Suture Zone

The collision of India and Asia is recorded in the sedimentary sequences now exposed in the Indus Suture Zone, and is best documented in the Ladakh Himalaya of northern India (Garzanti *et*

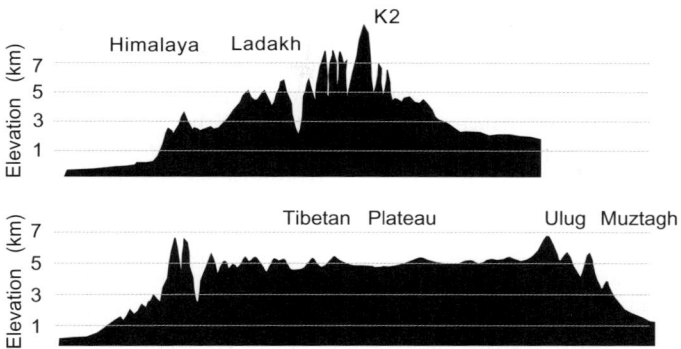

Fig. 2. Topographic cross-sections through the Himalaya showing the reduced height of the High Himalaya compared with other ranges in the western Himalaya and compared with the central and eastern Himalaya (modified after Searle 1991).

al. 1987). Here the sediments can be divided into pre-collisional forearc deposits, the Jurutze and Nindam Formations, and the syn-collisional Indus Group (Brookfield & Andrews-Speed 1984; Garzanti & van Haver 1988; Searle et al. 1990; Clift et al. 2002a). Mapping of the Indus Group shows that the Cretaceous clastic–carbonate Jurutze Formation, interpreted as the forearc basin to the active margin of Asia before collision (Garzanti & van Haver 1988; Clift et al. 2002a), passes conformably into the first synorogenic sediments, a series of red clastic sedimentary rocks, called the Chogdo Formation (Searle et al. 1990; Fig. 3). In practice, this change represents the onset of India–Asia collision. This age is only loosely constrained by the age of the Nummulitic Limestone that overlies the Chogdo Formation, dated to Lower Eocene (van Haver 1984), and

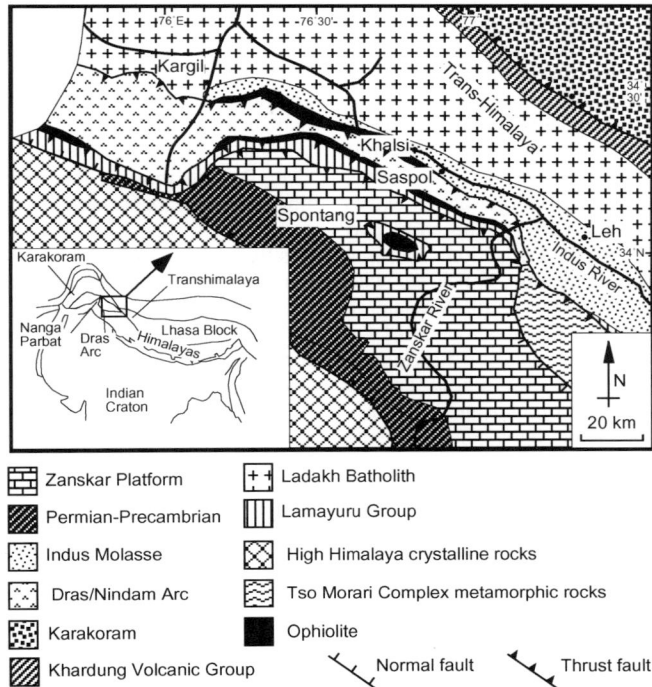

Fig. 3. Regional geological map of the Indus Suture Zone in Ladakh, showing the location of the Indus Group and modern Indus River.

most recently to late Ypresian–early Lutetian age (52–49 Ma; O. Green, pers. comm.). Because the clastic units of the Indus Group that overlie the Nummulitic Limestone are fluvial, they provide important information on the drainage system operating in the Indus Suture during Paleogene time.

Palaeo-current data

Recently, Sinclair & Jaffey (2001) have argued, using new palaeo-current data from the Indus Group exposed in the Zanskar Gorge (Fig. 3), that drainage was basically north–south throughout the group, and that therefore the Indus Molasse Basin was internally drained. Those workers inferred that the Indus River postdates the cessation of sedimentation of the Indus Molasse. However, the detailed local palaeo-current data of Sinclair & Jaffey (2001) contradict the analysis of Searle *et al.* (1990) and Clift *et al.* (2000, 2001*a*), who reported a dominant east to west flow, similar to the modern Indus River in Ladakh. On the basis of such data, Searle & Owen (1999) have suggested that the Indus Group, including the basal Chogdo Formation, represents the deposits of an ancient Indus River flowing through the suture as long ago as mid-Eocene time. Searle & Owen (1999) noted that an Eocene age for the Indus was consistent with the cross-cutting relationship of the river to the High Himalaya, indicating that the drainage pattern predated the uplift of this range in Early Miocene time.

Figure 4 shows the spread of palaeo-current indicators for the Indus Group recorded by this author at various localities within the suture zone, including the critical Zanskar Gorge section examined by Sinclair & Jaffey (2001). The basal part of the section, the Chogdo Formation, shows a south to north transport direction, consistent with the internal drainage hypothesis of Sinclair & Jaffey (2001). However, in the upper parts of the group, from where all the data of Sinclair & Jaffey (2001) were derived, two modes can be seen. A WSW–ENE group of indicators (primarily cross-beds, with minor flutes and elongate pebbles) dominates and suggests an axial, suture-parallel drainage pattern after the final marine incursion.

None the less, a second set of current indicators continue to show north–south flow, consistent with the work of Sinclair & Jaffey (2001). These data are especially common towards the north of the Indus Group outcrop. The structural restoration of the Indus Group Basin conducted by Searle *et al.* (1990) demonstrated that because of the north-vergent deformation of the Molasse under the Early Miocene Zanskar Thrust, the most northerly exposures represent the youngest part of the Indus Molasse stratigraphy. These units also restore to a location adjacent to the Ladakh Batholith. Thus, the WSW–ENE palaeo-current indicators can be considered to represent an axial proto-Indus River, whereas the north–south flow represents the erosional outwash of the Ladakh Batholith into the basin. The situation can be directly compared with the modern Indus, where east–west flow in the river contrasts with the north–south flow on the alluvial fans draining the Ladakh Range.

Provenance indicators

Isotopic provenance data from the Indus Group support the basin flow reconstruction outlined above. Nd isotopic analysis of the finest grain size performed by Clift *et al.* (2001*a*) shows isotopically positive (ϵ_{Nd} of +1 to −2) values in the lower Chogdo Formation, consistent with a source dominated by the local Transhimalayan Arc (Ladakh Batholith) and the Spontang Ophiolite, exposed immediately towards the south. Going up-section, ϵ_{Nd} shows a general shift to lower values, indicative of mixing with a new, more continental source (i.e. with an older model age, reflecting earlier extraction from the mantle). The Pb isotopic values of detrital K-feldspars and the palaeo-current constraints in the same rocks rule out the Indian Plate as a possible source, and suggest the Lhasa Block as being the most likely additional component. Clift *et al.* (2001a) inferred that the increasing material flux from the Lhasa Block into the Indus Basin requires a regional-scale axial river flowing through the suture shortly after the final marine transgression, i.e. a proto-Indus River. Interestingly, there are a small number of high positive ϵ_{Nd} values in the uppermost Choksti (or Nimmu) Formation that are out of sequence with the general stratigraphic trend. These correspond to intervals of north–south flow and appear to represent alluvial fans deriving material directly from the isotopically positive Ladakh Batholith. It is possible that the dominant north–south flow recorded by Sinclair & Jaffey (2001) is an artefact of recording a larger number of these fan-type units, as opposed to the ENE–WSW axial river units.

Isotopic changes after the final marine incursion, heralding the onset of a proto-Indus River within the suture zone, are matched by changes in the clast compositions of conglomerates seen at outcrop. During the initial passage from forearc basin to internally drained intermontane basin (Jurutze to Chogdo Formation), the dominant clastic component changes from being quartzose or volcaniclastic to quartzose and ophiolitic, with

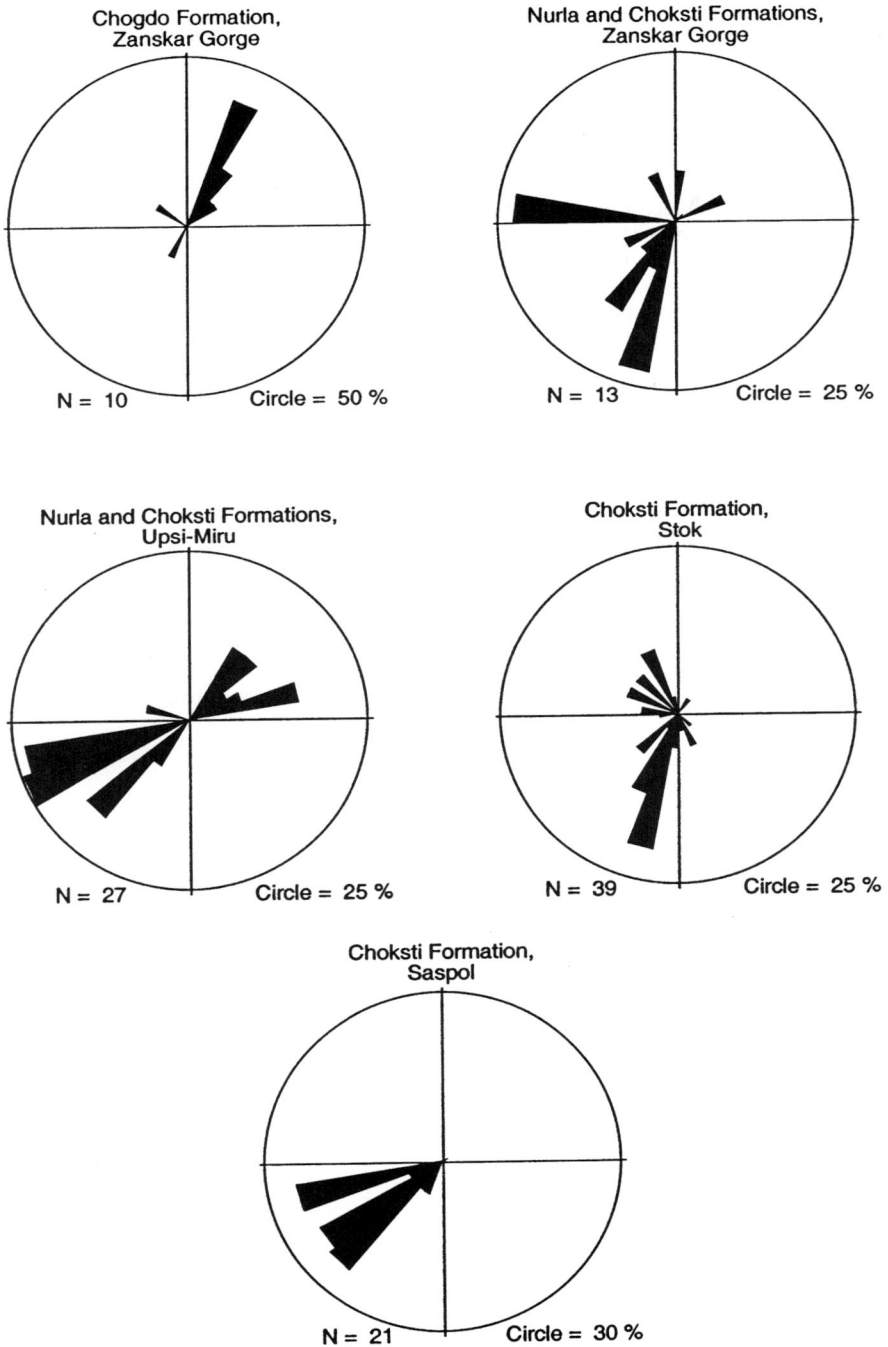

Fig. 4. Evolving palaeo-current indicators within the Indus Group, showing the change from a north–south internally drained basin to an east–west axial river system after the final marine incursion during latest Paleocene time.

clasts of red radiolarian chert, altered volcanic rocks and serpentinized ultramafic rocks (Garzanti & van Haver 1988; Clift et al. 2002a). These clasts are interpreted to have been derived from the Spontang Ophiolite that structurally overlies the Zanskar Platform of the Indian margin, now exposed just to the south of the Molasse (Fig. 3; Reuber 1986; Searle et al. 1997a). However, the quartzo-feldspathic component of Chogdo Formation sediment cannot be sourced from the Zanskar Platform. Pb isotopic constraints on these feldspars require a relatively juvenile source that is not the Indian Plate (Clift et al. 2001a), explicable by simple erosion of the Transhimalayan Ladakh Batholith.

Clay mineralogy

Sediment provenance after the onset of an ENE–WSW river shows a strong granitic component, presumably partly derived from the Transhimalayan Ladakh Batholith (Fig. 3). The upward shift to isotopically more negative ϵ_{Nd} values (Fig. 5) is not readily matched with large changes in the composition of conglomerate clasts seen at outcrop, but does parallel a change in the clay mineralogy. X-ray diffraction (XRD) analysis of the >0.2 μm fraction of the sediment allows a simple quantitative analysis of the bulk mineralogy to be performed using the method of Biscaye (1965). Slides were produced by settling of clay through water onto glass slides to present the c-axis orientation to the X-ray beam. Analysis was performed through a range of $0-30° 2\theta$, first using simple air-dried samples, then with those exposed to ethylene glycol to dehydrate the sample and permit analysis of smectite that interferes with the illite peak when hydrated. The results of the analysis are shown in Fig. 6. As might be expected for a sequence whose metamorphic grade approached anchizone (c. 200 °C; Kisch 1990; Clift et al. 2002a), much of the smectite has been converted to illite, which is the dominant mineral. None the less, a clear up-section evolution is seen towards greater contributions from chlorite, especially in the uppermost Choksti (Nimmu) Formation. Chlorite is typically interpreted as the result of mechanical weathering of thermally mature, metamorphic terranes, for example by glaciation, as opposed to either erosion of shallow-buried sedimentary rocks or chemical weathering in a wet, slowly eroding setting, which promotes the formation of kaolinite and smectite (e.g. Chamley 1989; Lauer-Leredde et al. 1998). The implication is that the proto-Indus River was transporting sediment from increasingly vigorously eroding source areas, probably reflecting the progressive uplift of the source regions (mostly the Lhasa Block and Transhimalaya) after collision.

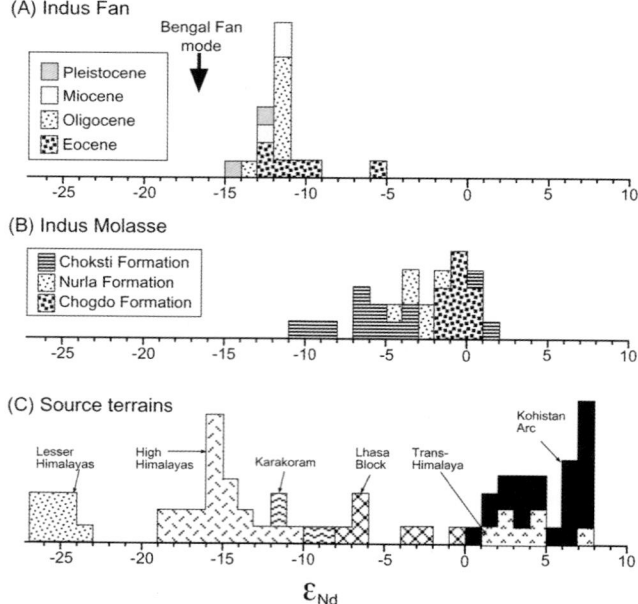

Fig. 5. Evolving Nd isotopic composition of the Indus Group (data from Clift et al. 2001a), showing influx of material from a new isotopically negative source after collision.

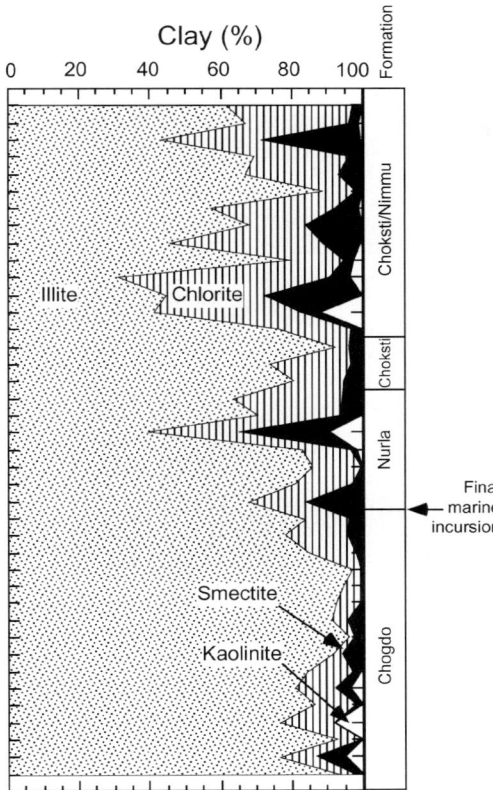

Fig. 6. showing the up-section variations in clay mineralogy within the Indus Group determined by XRD work. The base of the section is dominated by illite, whereas chlorite become more important up-section, reflecting increased mechanical weathering.

Structural inversion

The fact that the modern Indus River is now cutting its own Paleogene deposits (Fig. 7) provides powerful evidence that the location of the river has not changed substantially since its initiation in Early Eocene time. This is despite the fact that the Indus Group Basin has been tectonically shortened and inverted since that time. This event apparently occurred just before the unroofing of the High Himalaya (Hodges et al. 1992; Edwards & Harrison 1997), as these ranges did not contribute their distinctive muscovites and garnets to the sedimentary rocks of the Indus Group (Garzanti & van Haver 1988). Imbrication of the sequence, as a result of the northward thrusting along the Zanskar Thrust (Searle et al. 1988; Steck et al. 1993), caused uplift, which erosion by the Indus River was able to match so as to maintain its course. The degree of erosion has been estimated by Clift et al. (2002a) as c. 2–4 km since Early Miocene time, based on apatite fission-track data that limit cooling to 60–110 °C close to the Ladakh Batholith, and increasing towards the south. This figure assumes an elevated continental geothermal gradient of 30 °C km. Although the degree of cooling and unroofing in the Indus Suture is rather less than that measured in the adjacent sections of the High Himalaya (see Searle et al. 1992; Metcalfe 1993), significant erosion has occurred.

In summary, data from the Indus Group point to a regional proto-Indus river flowing from the western Lhasa Block through the suture shortly after India–Asia collision, probably in Early Eocene time, remaining in the same location despite major basin inversion during Early Miocene time.

The Indus River in the foreland basin

After running through the Indus Suture Zone the river cuts southward through the Himalaya in the region of the Nanga Parbat–Harramosh Massif. Although the uplift of the present massif is very recent (in Plio-Pleistocene time, e.g. Zeitler et al. 1993), it has been suggested that the western syntaxis of the Himalaya represents an ancient continental promontory to the Indian Plate that collided with the active margin of Asia, dividing the volcanic arc developed there (Searle 1991). The Nanga Parbat–Harramosh Massif is thus a new development in an area of long-lived heterogeneity in the orogen. The syntaxis was exploited by the Indus River in its attempts to break out of the suture zone and reach the ocean (Schroder & Bishop 1999). Downstream from Nanga Parbat the Paleogene Indus delta has been recognized in the Katawaz Basin (Qayyum et al. 1997) of Pakistan. In contrast, the Pakistan foreland basin is dominated by Paleocene–Eocene carbonate successions as shown by recent redating of the clastic Murree and Kamlial Formations to Oligocene time (Najman et al. 2001); these formations had previously been thought of as being of Eocene age (Bossart & Ottiger, 1989). The onset of clastic deltaic sedimentation close to the Paleocene–Eocene boundary following pelagic carbonate accumulation in this region (Cassigneau 1979) agrees well with the inferred age of Indus River initiation in the suture zone. Recent provenance work (Qayyum et al. 2001) on the Katawaz sedimentary rocks demonstrates erosion of a progressively more metamorphosed source terrane after delta initiation, consistent with the trend to increasing chlorite flux in the Indus Group.

The Katawaz Basin lies c. 100 km NW of the modern river course, which has been displaced eastwards by the thrust deformation that first

Fig. 7. Photograph of the Indus Group at Saspol in Ladakh, showing the modern Indus River cutting the older fluvial sequences, deformed in place during Early Miocene time.

inverted the Katawaz Basin, and then emplaced the Sulaiman Ranges towards the SE (Fig. 1). Growth of the Sulaiman Ranges must have pushed the course of the Indus south by 200–300 km in this region. However, the dominantly strike-slip character of the plate boundary in western Pakistan (Lawrence et al. 1992; Bernard et al. 2000), marked by the development of the Chaman Fault Zone (Fig. 1), means that the degree of migration of the thrust front has been small compared with the situation in the Indo-Gangetic Plain, simply reflecting the differing orientations of the basin relative to the India–Asia convergence direction. The Paleogene Indus Delta consequently lies close to the modern course of the river in eastern Pakistan (Fig. 1).

Migration of the Indus River during Neogene time is also seen to be limited in foreland basin strata. Distinctive blue–green amphiboles derived from the Kohistan Arc and eroded by the Indus, are found in Miocene Siwalik Group sediments, but only in the Kohat Plateau of western Pakistan, close to the modern Indus (Fig. 1; Cerveny et al. 1989). Further east the distinctive amphibole is missing because erosion in those regions is from the High Himalaya and Lesser Himalaya alone. These data imply that the Indus has flowed from the mountain front in approximately the same location since 18 Ma. Similarly, exotic conglomerate boulders, derived from crystalline terranes unique to the Indus drainage basin, are found over a zone of 25 km width within Siwalik Group deposits dated at 1–9 Ma (Abbasi & Friend 2000). These outcrops straddle the modern Indus and imply no significant migration since that time. Friend et al. (1999) have noted that in the Miocene–Recent Siwalik sedimentary rocks a clear change in palaeo-current directions is seen between those exposures east and west of the modern river. Those rocks exposed to the east tend to flow to the SE; however, the strata to the west show palaeo-flow in the north to south orientation, similar to the modern Indus, suggesting that these units represent the former locations of the river as it was pushed gradually eastwards by the growth of the Sulaiman fold and thrust belt. Such a reconstruction is consistent with the continuous Indus Fan sedimentation during Neogene time (see below), but is at odds with the suggestion of Raynolds (1982) and Burbank et al. (1996) that the Indus was temporarily captured by the Ganges during late Miocene time and flowed eastward through the Salt Ranges and Potwar Plateau.

The Indus Fan

The age of the Indus Fan and the history of mass flux into the Arabian Sea has been a controversial topic. A popular view holds that the Indus and Bengal Fans are both Miocene features, whose development was triggered by the rapid unroofing and erosion of the High Himalayas during Early Miocene time (c. 23–20 Ma; Hodges et al. 1992; Searle et al. 1992; Harrison et al. 1995). This view was based on reconstructions of the mass flux onto the Indus Fan that show an acceleration in sedimentation at the same time (e.g. Davies et al. 1995). However, such mass flux calculations are based on rather sparse and potentially unrepresentative scientific drill sites and their results are strongly debated. Rea (1992) used the same dataset to instead propose increased mass flux at 9–11 Ma, whereas Métivier et al. (1999) have proposed modest accumulation rates in the Indus until a major increase during Pleistocene time (c. 1.8 Ma).

The principal problem behind the lack of consensus is the inability to clearly date the thickest part of the Indus Fan, where the bulk of the sediment is stored. Work on the foreland sequences is biased because of the cycles of preferential sedimentation and erosion of these basins are linked to variations in eustatic sea level. In practice, they are always full and cannot be considered as good measures of mass flux through the Indus system, as accumulation will be linked to sea-level variations and to the changing width and depth of the foreland trough. Offshore drill sites are typically flawed either by penetrating the distal toe of the fan (e.g. Deep Sea Drilling Project (DSDP) Site 221; Fig. 1) or by sampling on bathymetric highs, such as the Murray or Owen Ridge (Ocean Drilling Program (ODP) Site 731). In the former case, switching of depositional lobes and the steady progradation of the fan towards the south will necessarily mean that this site cannot date the onset of fan sedimentation or record the regional rate of mass accumulation. In the latter case, the drilled sections can be used as guides only up to the time of ridge uplift. None the less, several important features of Indus Fan accumulation history can be gathered from these sources. Although it has been suggested that the palaeo-Indus Fan flowed west into the Makran region (Critelli et al. 1990) before the uplift of the Murray Ridge in Early Miocene time (Mountain & Prell 1990), it is now clear that significant volumes of Paleogene sediment are present in the Arabian Sea (Fig. 8; Clift et al. 2000, 2001b).

Drilling at DSDP Site 221 on the distal fan toe showed that even here sedimentation began during Late Oligocene time, and might be expected to be older further north. Although large thicknesses of Paleogene turbidites are found in the Makran Accretionary Complex (McCall & Kidd 1982; McCall 2002), significant volumes of mostly muddy Eocene–Oligocene turbidites have been drilled on the Owen Ridge at ODP Site 731 and DSDP Site 224 (Fig. 1). Pb isotopic provenance work on the detrital feldspars in these sediments has now demonstrated that these are not of Indian Plate provenance and must be derived instead from within or north of the Indus Suture (Clift et al. 2000, 2001b). Such a conclusion requires an Indus River flowing into the Arabian Sea by mid-Eocene time (c. 45 Ma; the age of the oldest sand analysed). It is none the less important to recognize that because the bulk ϵ_{Nd} values of the middle Eocene sediments from ODP Site 731 and DSDP Site 224 are much more negative than their equivalents in the Indus Molasse (Fig. 5) there must be mixing with an isotopically negative source after leaving the suture zone. This is inferred to be the Indian Plate, probably mostly from a proto-Himalayan mountain chain.

Seismic constraints

New seismic data from the proximal Indus Fan collected by Roeser et al. (1997) have allowed the thicknesses of Paleogene sediment in the Murray Ridge to be estimated. Clift et al. (2001b) used long distance correlations with dated horizons within petroleum well site Indus Marine A-1 (Fig. 1) to demonstrate that a large fraction of the sediment under the proximal fan is of Paleogene age (Fig. 8). Paleogene sediment on the tilted Murray Ridge are >2 km thick after a time–depth conversion has been made, although thicknesses change rapidly laterally because of the significant basement topography in this region before ridge uplift (Gaedicke et al. 2002). Clearly, large thickness of strata predate Murray Ridge tilting, generally considered to be of late Early Miocene age by comparison with the known age of uplift on the Owen Ridge (Mountain & Prell 1990). Material older than the age of Murray Ridge tilting is presumed to be mostly of Paleogene age because the age of rifting of the Pakistan Arabian Sea margin dates from c. 65 Ma. The oldest identified marine magnetic anomaly is Chron 27 (Miles et al. 1998), i.e. 60.9 Ma (Berggren et al. 1995), but presumably sedimentation would have started over acoustic basement during the rifting period, before the onset of sea-floor spreading.

Further evidence for major regional Paleogene fan sedimentation can be seen in the Pakistan Shelf region. Seismic survey shows c. 10 km of sediment under the shelf in this area (Clift et al. 2001b), after making a time–depth conversion of the reflection record using the stacking velocities derived during the processing of the multichannel data, cross-checked by existing sonobuoy data for the area (Fig. 1). Drilling at Indus Marine A-1 penetrated 2725 m below sea floor and reached deep into the Middle Miocene sequence, implying a great thickness of sediment older than that level. Exactly how much of the Paleogene sequence predates the Indus River and represents normal passive margin sedimentation is not known. How-

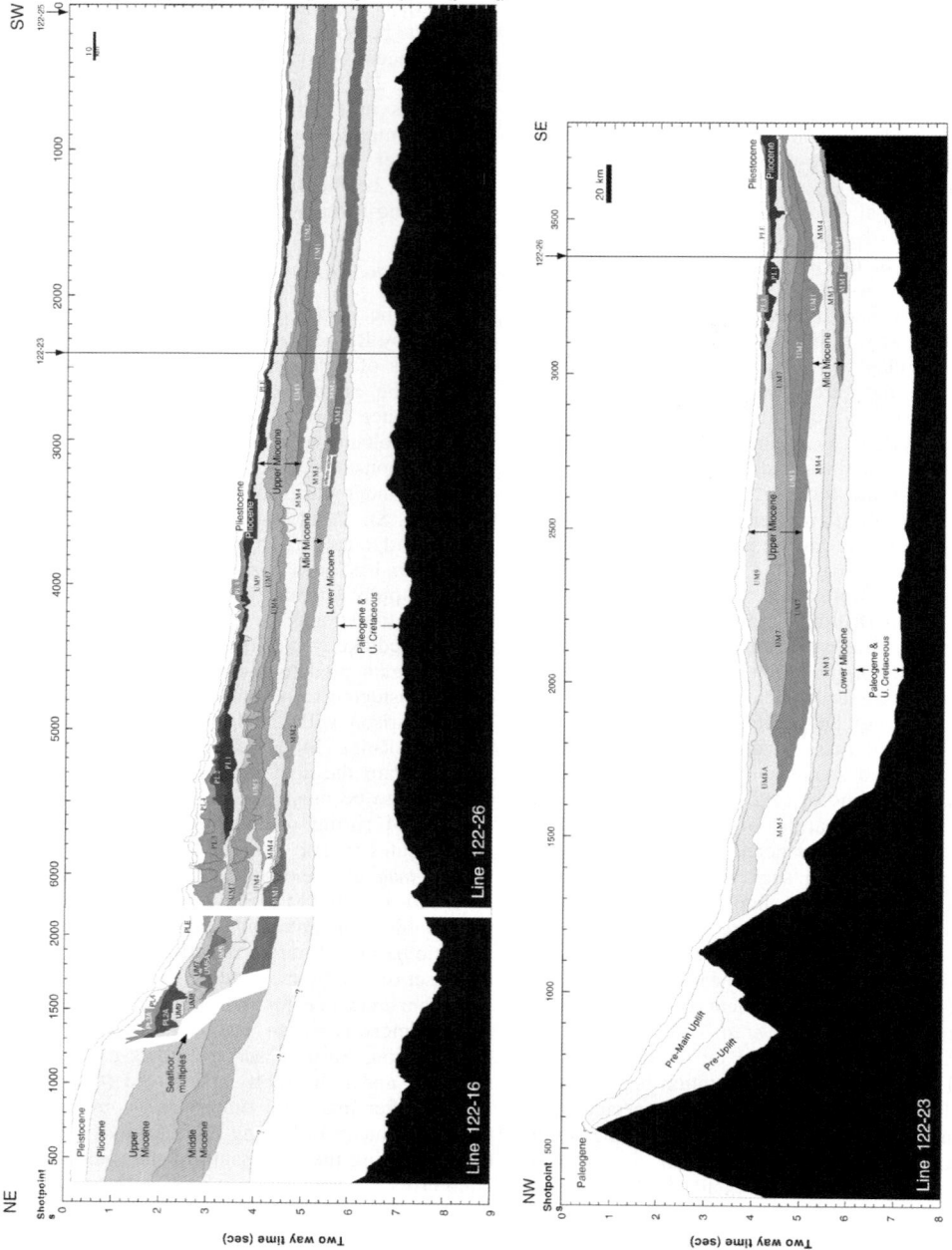

Fig. 8. Interpreted multichannel seismic profiles perpendicular to the Pakistan coast and Murray Ridge, showing the overall proximal structure of the Indus Fan (from Clift et al. 2001b, reproduced with permission of Geological Society of America).

ever, the modest time period between rifting and earliest India–Asia collision at c. 55 Ma does suggest that much of the section is Indus related.

Accumulation rates in the proximal delta area can be assessed by subsidence analysis of the drilled and seismically imaged stratigraphy, using the biostratigraphic dates of Shuaib (1982) for Indus Marine A-1 and the backstripping technique of Sclater & Christie (1980). In this approach the section is progressively unloaded and underlying units are decompacted to assess what the thickness of each section would be at the time of its deposition. The accumulation rate at a series of points across the proximal upper fan can therefore be constrained without errors related to burial interfering. Figure 9 shows the results of the decompaction exercise for Indus Marine A-1, DSDP Sites 221 and 222, and for a series of 'pseudo-wells' derived from the seismic data on the mid-fan (Clift & Gaedicke 2002). These 'pseudo-wells' are considered to assess relative sedimentation rates downslope and across the margin so as to overcome the problem of lobe switching and progradation. Dating at the 'pseudo-wells' is provided by long-distance correlation from the biostratigraphy at Indus Marine A-1 and by correlation of a dated horizon from the Murray Ridge that corresponds to the end of Early Miocene tilting of that structure (Clift et al. 2001b).

Sedimentation rates and the South Asian monsoon

The decompaction exercise reveals an acceleration of sedimentation during Neogene time matched by the first appearance of channel–levee complexes during Mid-Miocene time, consistent with a major influx of sediment at that time (Clift & Gaedicke 2002). The influx occurred after or close to c. 16 Ma, the start of Mid-Miocene time, somewhat after the 21 Ma pulse of Davies et al. (1995), and before the event recorded by Rea (1992) at 9–11 Ma. The apparent accumulation rate in the proximal fan is precisely opposite to the model of Métivier et al. (1999), suggesting low accumulation rates until Pleistocene time. The age of the sediment flux does not correlate with the accepted age of monsoon strengthening in the Arabian Sea at 8.5 Ma (Kroon et al. 1991; Prell et al. 1992), although the slackening rates in Late Miocene to Recent time are consistent with the sediment budget of Burbank et al. (1993) in which monsoonal strengthening is associated with reduced erosion in the Indian foreland basin and the Bengal Fan. The decrease in sedimentation rates inferred for the 'pseudo-wells' parallels a similar trend seen at DSDP Site 222, and suggests that the imposed age model is robust. It is, however, noteworthy that the sedimentation rates at DSDP Site 221 are fairly uniform, suggesting that such distal sites are not good records of the temporal variations in sedimentation rate, and that analysis of the thick, proximal fan is required to address mass flux issues.

Rather than assuming that the Indus does not show an erosional response to an 8.5 Ma monsoon initiation the sedimentation rate data may suggest a major misdating of this event. If the monsoon began at c. 16 Ma, then the increased sedimentation would match increased precipitation. In this scenario the various oceanographic and climatic changes noted at 8.5 Ma would reflect a secondary, later change in regional climate. Support for this view is given by analysis of the clay mineral

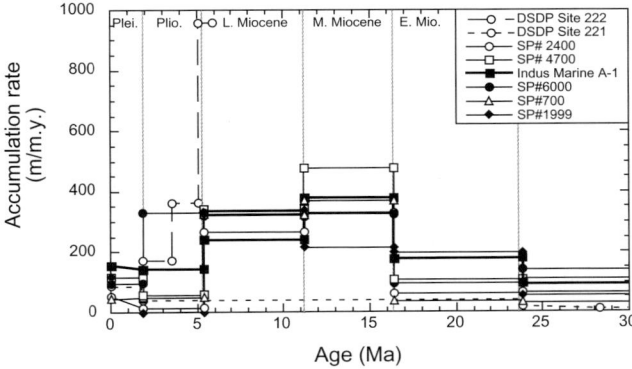

Fig. 9. Calculated mass accumulation rates at a variety of wells and 'pseudo-wells' on the Indus Fan derived from backstripping the stratigraphy using the methods of Sclater & Christie (1980) (from Clift & Gaedicke 2002, reproduced with permission of Geological Society of America).

evolution in the South China Sea, where no change in mineralogy, and by implication terrestrial weathering conditions, is noted at 8.5 Ma, but a major change from smectite-dominated to illite-dominated sedimentation is seen at 16 Ma, which was also a time of peak sedimentation rates (Clift et al. 2002b). The correlation of high sedimentation rates in the Indus Fan with an early monsoon is consistent with observations from the modern Himalaya, where erosion is faster where the monsoon is stronger (Galy & France-Lanord 2001). In addition, during the recent geological past the rates of sedimentation in the Ganges delta have been fastest during the strongest monsoonal periods (Goodbred & Kuehl 2000).

Some of the increased sedimentation starting in Mid-Miocene time can be explained by the diversion of the Indus Fan away from the west following uplift of the Murray Ridge, thus concentrating accumulation on the abyssal sea floor to the east. Major deformation and granite genesis is known from the Hindu Kush and Karakoram at 20–25 Ma (e.g. Hildebrand et al. 1998; Schärer et al. 1990), followed by a period of rapid cooling from 17 to 5 Ma (Searle et al. 1998), potentially providing a major new source of sediment during Mid-Miocene time. In addition, thermochronology work along the Karakoram Fault indicates >20 km of exhumation between 18 and 11 Ma (Searle et al. 1998). In the western High Himalaya rapid cooling and uplift is dated at c. 20–23 Ma in Pakistan (Treloar et al. 1989), as well as in Indian Zanskar (Walker et al. 1999) and Garhwal Himalaya (Metcalfe 1993), somewhat before the sediment flux. This latter mismatch may in part reflect the lesser influence of the High Himalaya on Indus sedimentation, and also the mostly tectonic mechanism of unroofing considered for these terranes.

In western Tibet new radiometric ages for dykes from the Lhasa Block at 18–19 Ma indicate that at least this part of the plateau had begun to extend, and thus may have reached its maximum height at that time (Williams et al. 2001). Such uplift would provide a rejuvenated source for sediment along the western margin of the plateau. What is noteworthy is that because the height of the Tibetan Plateau appears to be a major forcing parameter governing the strength of the monsoon (Prell & Kutzbach 1992; Molnar et al. 1993) the new age of Lhasa Block extension raises the possibility of an earlier initiation of the monsoon than 8.5 Ma. If the monsoon did strengthen markedly in Early–Mid-Miocene time then the increased sedimentation rates seen over that period may reflect the increased precipitation associated with that climate system, as much as they do tectonic rejuvenation of source areas. If that were true, then the erosion and sedimentation pattern shows a positive response to monsoon initiation, as might be expected from the behaviour of the modern and recent system, and not the negative response inferred by Burbank et al. (1993). Extension and collapse of the plateau since that time might weaken the monsoon, and reduce precipitation and erosion in the fashion shown.

In summary, it seems that there has been rapid Himalayan-derived sedimentation on the Indus Fan since at least Mid-Eocene time, c. 45 Ma, and probably earlier. Thick sequences of Paleogene clastic sediment are known from the Indus Shelf (c. 6 km) and Murray Ridge (c. 2 km) regions. Together with their equivalents in the Makran Accretionary Complex they represent the deep-water fan equivalents to the Katawaz Delta and confirm the presence of a proto-Indus system in place shortly after India–Asia collision. Sedimentation reached a peak in Mid-Miocene time, after 16 Ma, broadly correlating with strong tectonic activity in the Karakoram, southern Tibet and Hindu Kush at that time, and possibly with an enhanced monsoon, triggered by plateau uplift in late Early Miocene time.

Comparison with the Ganges–Brahmaputra system

As noted above, the drainage basin of the Indus is different from that of the Ganges–Brahmaputra, a fact manifest by the provenance of the respective fan sediments. Nd isotopes, in particular, show generally more negative (more Indian) ϵ_{Nd} values in the Bengal turbidites compared with the Indus (Fig. 5; Bouquillon et al. 1990; France-Lanord et al. 1993; Clift et al. 2001b). Indeed, the very location of the Indus River system draining blocks within and north of the suture may explain why these are now deeply exhumed whereas their equivalents further east in Tibet show only moderate erosion. Mass balances between the isotopic character and volume of the fan sediment and the eroded volumes (estimated from the area of a given source and the degree of cooling) onshore allow the relative contribution of each source terrain to be assessed for each fan and then compared. An appropriate correction is made for sediment compaction to convert sediment volume into rock volume. Clift et al. (2001b) performed this exercise for the Neogene Indus Fan (Fig. 10a), and suggested that only c. 40% of this fan was derived from the erosion of Indian Plate sources (i.e. the High and Lesser Himalaya). Understanding the difference between the Indus and Bengal Fans is crucial to any greater understanding of Himalaya–Tibetan erosion. The same mass balan-

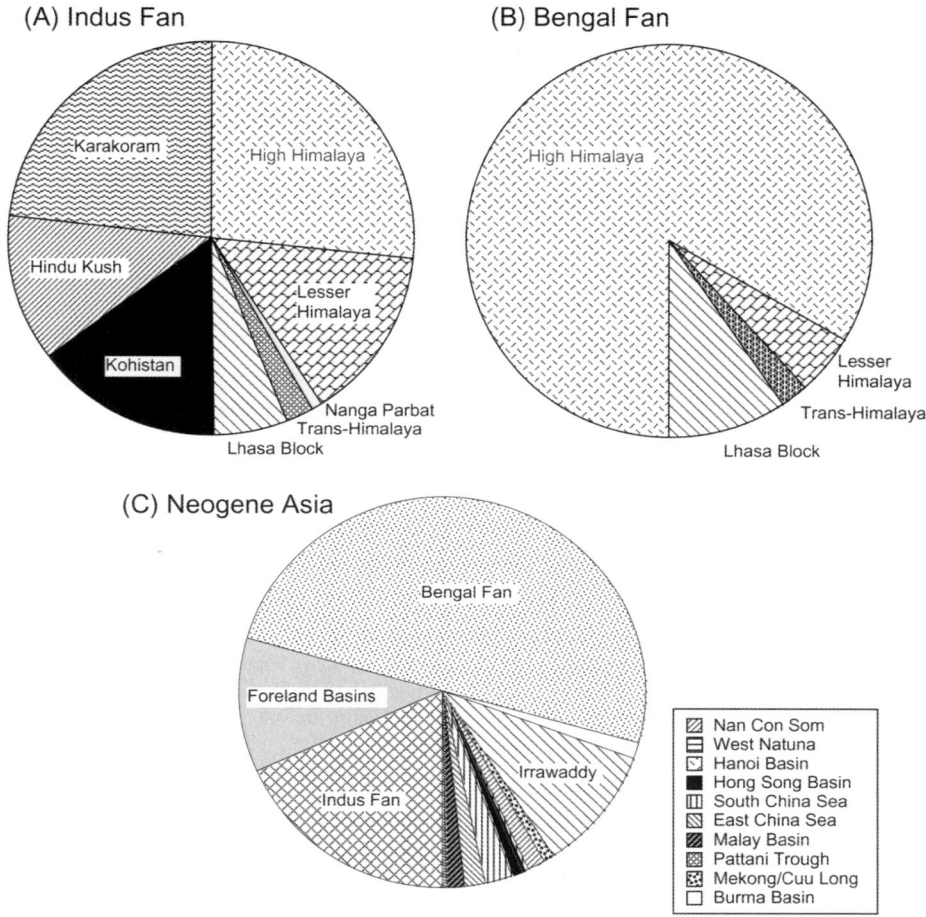

Fig. 10. Proportions of different major source terranes and their contribution during Neogene time to (**a**) the Indus Fan, and (**b**) the Bengal Fan. (**c**) The relative size of major Neogene sediment accumulations in the South and East Asia region.

cing approach can be attempted for the Bengal Fan to highlight the differences between the two systems. I consider only Neogene time in this study because the record of Paleogene cooling in the source areas is much less well constrained than that for the Neogene period.

Estimating sediment volumes

In estimating the amount of material eroded into the Bengal Fan the orogen is divided into a series of units whose cooling and geological histories are known to be different. We consider erosion of the Lesser and High Himalaya, Transhimalayan Batholith and the Lhasa Block for the Bengal Fan system. When estimating the amount of sediment in the basins the volumes of the Indian foreland basin, the Irrawaddy Fan and the associated Burma Basin (Bender *et al.* 1983) also need to be accounted for, because these are all principally filled by sediments eroded from the same regions as the Bengal Fan. The volume of the Bengal Fan is taken from the work of Curray & Moore (1971) and Le Pichon *et al.* (1992). The fan is split into a Neogene and Paleogene section in a 7:3 ratio following the dating of the Indus Fan (Clift *et al.* 2001b), as no deep age information is available for the Bengal Fan itself. The Andaman Sea is largely a Miocene Basin (Guzman-Speziale & Ni 1993); however, seismic and drilling data suggest that the adjacent Burma Basin is c. 55% of Paleogene age (Pivnik *et al.* 1998), implying a significant Paleogene thickness in the associated offshore Irrawaddy Basin. As the Burma Basin is largely inverted the proportion of Neogene material offshore is presumed to be higher and is here estimated at 45% of the total stratigraphy. The volume of the Irrawaddy is estimated from the compilation map of Letouzey & Sage (1988). I follow the division of Clift et al. (2001b) and approximate the volume of sediment in the Ganges foreland as being the eastern two-thirds of the entire Himalayan foreland between the Nanga Parbat and Namche Barwe syntaxes. The basin depths and widths are derived from Burbank et al. (1996). The volume of the Paleogene material in the foreland is minor compared with the Neogene sequence and is mostly present as small accreted thrust slices, now incorporated into the frontal ranges of the Lesser Himalaya (Najman et al. 1993).

Estimating eroded volumes

The Transhimalaya experienced much of their erosion during Paleogene time. The cooling of the range is best known in Ladakh (Fig. 1), although the range is well exposed throughout the Bengal drainage basin. Zircon crystallization ages for this granite are c. 65 Ma (Weinberg & Dunlap 2000), biotite and hornblende Ar/Ar ages (cooling below 300 °C and 500 °C) are c. 40–45 Ma (Reynolds et al. 1983). This implies rapid cooling of the batholith during Paleogene time, with limited cooling during Neogene time. The adjacent areas of the Lhasa Block are generally considered to be only lightly eroded, although the modern eastern Tibetan Plateau is clearly eroded along its edge by deep river valleys. Neogene erosion of the Lhasa Block is limited to an average of not more than 1 km for the purpose of this exercise.

The southernmost source terranes of the orogen are the High and Lesser Himalaya. Fast erosion of the Namche Barwe syntaxis region is not considered because even at the super-fast rates known from its western equivalent at Nanga Parbat (e.g. Zeitler et al. 1993), the small area of the massif makes it volumetrically insignificant. The cooling history of the High Himalaya is well known, but much of the measured cooling may be tectonically driven rather than erosional, i.e. due to the unroofing effects of the South Tibetan–Zanskar Detachment. The cooling history of this block is well known from the Garhwal and Nepalese High Himalaya (Hodges & Silverberg 1988; Guillot et al. 1994; Hodges et al. 1996; Searle 1996; Searle et al. 1997b). Unroofing of the Lesser Himalaya is more difficult to constrain as these are unmetamorphosed or low-grade rocks, suggesting that erosion has been limited. However, models of mass flux in orogenic belts (e.g. Willett et al. 1993; Royden 1996) suggest that erosion can result in large fluxes of material from the subducting or colliding plate (in this case India), without erosion unroofing any deeper structural level at any one place through time because of lateral advection. If this type of model is applied to the High and Lesser Himalaya it follows that the cooling measured by radiometric means would represent minima of the actual rock volumes eroded from each of these belts. In practice, this means that although the thickness and depth of burial of the Lesser Himalayan slab is less than that of the High Himalayan, the orogen-perpendicular length of that slab eroded from the mountain front is the same.

Because the depth of erosion of those terranes whose unroofing is tectonically driven is basically unconstrained, the volume of sediment eroded is estimated so as to achieve a mass balance, given the known sediment volumes in the sinks and the volumes eroded from other sources. This approach means that errors in estimating the exact area of the High and Lesser Himalaya drained into the Bengal Fan do not translate into errors in the rock volume eroded from these terranes. Instead, errors will result in inaccurate depths of erosion. For the purpose of this study I focus on the Neogene period because of the poor knowledge of Paleogene erosion before unroofing of the High Himalaya.

This mass balancing approach assumes that the Bengal drainage basin has covered a similar area to that seen today. The Nd isotopic signature of the Bengal Fan turbidites is basically constant in the drilled record back to c. 18 Ma (France-Lanord et al. 1993), consistent with this simplification. Although the drainage system must have evolved during that time, no detailed reconstruction is currently available. In the foreland there is evidence of a changing drainage pattern, involving expansion of the Ganges towards the west during Neogene time (Burbank et al. 1996). Consequently, the estimated depths of erosion for the Lesser and High Himalaya, which feed material into the foreland towards the east, are likely to be minima. This is because the volume of material required for the mass balance would in fact have been derived from a shorter expanse of the orogen than assumed here.

The accuracy of our approach in estimating eroded volumes can be cross-checked by the Nd isotope characteristics of the sediments measured from the Bengal Fan. The average ϵ_{Nd} can then be compared with the composition predicted from the known average compositions of the sources, and their relative contributions.

The results of the mass balance are shown in Tables 1 and 2, as well as graphically in Fig. 10b, the contrast to the Indus being striking in terms of

Table 1. *Calculated depths of erosion of the major source terranes within the drainage basin of the Bengal Fan, derived by balancing of rock volumes and matching Nd isotopic data*

Source terrane	Refs.	Length (km)	Width (km)	ε_{Nd}	Depth of erosion (km)	Volume eroded (km^3)	Proportion of total (%)
High Himalaya	a–e	2000	150	−16	15.7	5250000	55
Lesser Himalaya	a–e	2000	75	−25	15.7	2625000	28
Transhimalaya	g, f	2000	30	3	8.0	480000	6
Lhasa Block		2000	500	−8	1.0	1000000	12
Total						9355000	100

Predicted ε_{Nd} is −16.5; actual ε_{Nd} is −16 to −17. a, Hodges & Silverberg (1988); b, Guillot et al. (1994); c, Hodges et al. (1996); d, Searle (1996); e, Searle (1996); f, Reynolds et al. (1983); g, Weinberg & Dunlap (2000).

Table 2. *Sediment volumes for the various sediment sinks within the Ganges–Brahmaputra system compared with the Indus Fan*

	Length (km)	Width (km)	Total volume (km^3)	Compaction factor	Paleogene		Neogene	
					Rock volume (km^3)	Proportion of total (%)	Rock volume (km^3)	Proportion of total (%)
Bengal Fan	3000	1000	12500000	1.217	3081348	73	7189811	84
Foreland basin	2000	150	700000	1.077	194986	5	454968	5
Indo-Burma Ranges	700	200	140000	1.077	129991	3	0	0
Burma Basin	640	80	256000	1.077	130734	3	106964	1
Irrawaddy Fan	1120	320	1792000	1.217	662613	16	809860	9
Total			15388000		4199671	100	8561603	100
Indus Fan	1500	960	5000000	1.217	1437962		2670501	

erosion from north and south of the Indus Suture. Of the total eroded sediment in the Bengal system, 84% is predicted to come from the Indian Plate, split 56% and 28% between the High and Lesser Himalaya, respectively. Only 5% is derived from the Transhimalaya and 11% from the Lhasa Block. The contrast with the Indus is noteworthy because in this fan c. 58% of the Neogene sediment is sourced from north of and within the Indus Suture Zone (Clift et al. 2001b). The two fans can thus be seen to provide very different erosional records of the Himalaya–Tibet Orogen.

Regional context of the Indus system

A simple volume calculation allows the relative contribution of the Indus to the Neogene erosion of Asia to be assessed. Apart from the Bengal and Irrawaddy Fans, sediment from the collision zone is found in a series of basins spread throughout SE Asia, stretching from the Malay Peninsula to the East China Sea (Fig. 11). These basins and their approximate dimensions are shown in Table 3. Sediment thicknesses in these basins can be significant, most notably in the Mekong and Nam Con Son Basins fed by the Mekong River, the Hong Song–Yingehai Basin, which is fed by the Red River, and the Malay Basin filled by the Salween River. In each of these areas sediment thicknesses approach and even exceed 10 km. None the less, the total Neogene sediment in these basins can be seen to be minor (combined 11% of the total Neogene material), compared with the Bengal (47%) and Indus (18%) Fans, as well as the Indian foreland basins (10%). However, like the Indus Fan, their provenance differs from that of the Bengal Fan, as they are mostly eroding the Tibetan Plateau and the adjacent North and South China Blocks. Consequently, the sediment record of these basins may provide important information on the erosion of the total orogen in places where this record is not swamped by Himalayan erosional material as is the case in the Indian Ocean fans.

Discussion

The Indus River and Fan are revealed as major components within the system of sedimentary basins that surround South and East Asia. Volumetrically the Indus is significant, if only a little more than one-third the size of the Bengal Fan.

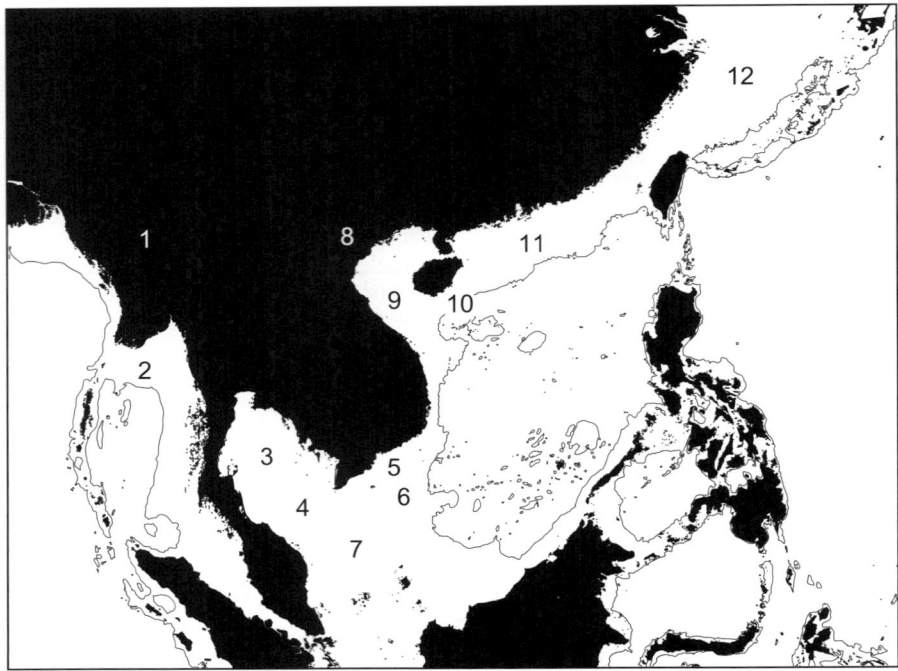

Fig. 11. Location of basins compared with the Indus Fan in Fig. 9. 1, Burma Basin; 2, Irrawaddy Fan; 3, Pattani Trough; 4, Malay Basin; 5, Mekong Basin; 6, Nam Con Son Basin; 7, West Natuna Basin; 8, Hanoi Basin; 9, Hong Song Basin; 10, SE Hainan Basin; 11, Pearl River Mouth Basin; 12, East China Sea.

Table 3. *Sediment volumes in the basins of SE Asia for comparison with the large fan systems of the Indian Ocean; although individually significant, their contribution to total Asian erosion is minor*

Basin	Length (km)	Width (km)	Maximum sediment thickness (km)	Paleogene/ Neogene volume	Compaction factor	% of SE Asia Neogene volume	Key reference
Mekong	300	120	9.0	0.4	1.217	6.5	Huchon *et al.* (1998)
Nam Con Son	250	200	11.0	0.2	1.217	14.0	Matthews *et al.* (1997)
West Natuna	200	100	4.5	0.3	1.217	1.6	Tjia & Liew (1996)
Hanoi Basin	100	35	5.0	0.2	1.077	0.4	Rangin *et al.* (1995)
Hong Song Basin	500	200	9.0	0.2	1.217	22.0	Rangin *et al.* (1995), Roques *et al.* (1997)
SE Hainan	300	200	8.0	0.2	1.217	12.0	Liu (1992)
South China Sea	800	200	5.0	0.2	1.217	16.0	Liu (1992), Clift *et al.* (2002c)
East China Sea	700	240	5.0	0.4	1.217	12.6	Liu (1992)
Malay Basin	400	120	12.0	0.35	1.217	12.5	Letouzey & Sage (1988)
Pattani Trough	360	80	4.5	0.35	1.217	2.6	Letouzey & Sage (1988)

What is most important about the Indus is that it is the only river system to provide erosional data on the mountains of the Karakoram, Hindu Kush and the western Lhasa Block. The fact that this region is also strongly associated with an intense and well-documented monsoon system makes this the place to study long-term interactions between orogenic growth and climate. The low-resolution record of sediment flux into the Arabian Sea currently allows a reduction in sedimentation rate during Late Miocene time to be linked to monsoonal strengthening. Tectonically driven uplift in the Karakoram and Lhasa Block during Mid-Miocene time correlates with a major influx of sediment into the Arabian Sea at that time. This event may reflect the initial strengthening of the monsoon before the major climatic changes already documented at 8.5 Ma (Kroon et al. 1991; Prell et al. 1992).

Although the rivers of China and Indochina do provide erosional records of Tibet, and are probably less contaminated by other sources than the Indus is by the Himalaya, these can record erosion of only the eastern plateau, which, given the size of this structure, cannot be assumed to have behaved as a single coherent block. More importantly, the rivers of East Asia are known to interact with each other in response to the evolving tectonics of that region, primarily in the form of stream capture of one river's drainage basin by another. This means that changes in the provenance of sand need not necessarily reflect changing erosion patterns, but instead stream capture. For example, Tregear (1965) suggested that the upper Yangtze formerly flowed down the course of the modern Mekong, and the middle Yangtze was routed along the modern Red River. Seeber & Gornitz (1983) proposed rerouting of the Tsangpo from an original path along the modern Luhit and Parlung Rivers, followed by capture by the Brahmaputra. Brookfield (1998) further developed the idea of the Brahmaputra capturing a succession of rivers, namely the Tsangpo, Parlung, Luhit and Irrawaddy as northeastward propagation of deformation around the Namche Barwe syntaxis continued.

Most recently, Royden et al. (2000) proposed that tectonic uplift of eastern Tibet did not cause a rerouting of rivers, because they usually seem able to down-cut at least as fast as the uplift occurs. However, the eastward growth of Tibet and especially the NE migration of the Namche Barwe syntaxis has had the effect of compressing the drainage basins of the original rivers of the area, forcing them to interact. The fact that the Indus does not have any other major rivers to interact with means that the drainage situation in the western Himalaya is much simpler than in the east. The only rivers that do not funnel material into the Indus Fan are those that drain the ranges to the north and west, i.e. the Amudarya and Helmand Rivers, respectively (Fig. 1). These are relatively minor in volume of runoff compared with the Indus, and drain into enclosed continental basins, the Aral Sea and Sistan Depression. As they lie in poorly known regions, their long-term history is obscure.

The lack of migration of the Indus River differs from several other major rivers in that whereas flow of the Nile (Said 1981), Colorado (Elston & Young 1991) and Amazon (Hoorn et al. 1995) have not been halted by regional uplift in their hinterlands, their courses have been significantly displaced as a result, so that in each case the present course is very different from that seen in Late Miocene or earlier times. The unusual situation of the Indus pinned in the suture, and running parallel alongside a dominantly strike-slip plate boundary within its foreland, means that the Indus is peculiar in being located in an active tectonic region, but without experiencing significant lateral migration. The initiation of the Indus during earliest Eocene time makes it one of the oldest documented rivers globally.

Conclusions

The Indus River appears to have formed shortly after India–Asia collision. Its drainage, then and now, is dominated by western Tibet, the Indus Suture Zone and Karakoram, with a significant minority influx from the Indian Plate. Erosion from rapidly uplifting massifs, most notably the Nanga Parbat–Harramosh Massif, does not represent a major source of fan sediment. The Indus does not seem to have migrated to any significant degree despite strong deformation and uplift along its course. Start of Indus flow in the suture immediately postdates collision, suggesting a link between collision, uplift of the Lhasa Block and river initiation. Changes in provenance and clay mineralogy indicate that the Indus was draining an uplifting, mechanically eroding Lhasa Block shortly after collision.

Unlike the rivers of east Asia, the Indus has not been affected by stream capture as a result of ongoing deformation. Instead, trapped in the suture zone soon after collision, it has remained in place there despite major tectonic uplift. In the foreland basin, moderate east-vergent thrusting in the Sulaiman Ranges has displaced the Indus towards the east, but by only c. 100 km since Early Eocene time, reflecting the dominant strike-slip mode of this plate boundary. No major migration appears to have occurred in the foreland since at least 18 Ma. The Eocene delta of the Indus has migrated south with time from its original position in the Katawaz Basin (Qayyum et al. 1997) to the modern position, as compressional tectonics eliminated the marine basin there, and because of seaward delta progradation.

Himalaya-derived clastic sediment first reached the Arabian Sea via a proto-Indus River no later than Mid-Eocene time (c. 45 Ma; Clift et al. 2001b), thus allowing a major fan to accumulate in the region during Paleogene time. In addition, significant sediment volumes were also transported towards the west, where they became part of the Makran Accretionary Complex (Critelli et al. 1990), before uplift of the Murray Ridge in Early Miocene time (c. 20 Ma; Mountain & Prell 1990). Increased sediment accumulation on the mid-fan during Mid-Miocene time reflects Murray Ridge uplift and strong erosion in the Karakoram and the western Lhasa Block, driven by tectonic uplift and exhumation, as well as an enhanced monsoon triggered by Tibetan uplift during late Early Miocene time (c. 18 Ma). The widely accepted age for onset of the South Asian monsoon at 8.5 Ma occurs during a phase of reduced sedimentation. Detailed analysis of the clastic record preserved in the fan offers the best hope of testing climate–solid Earth coupling models in this global type area.

I wish to thank P. Hildebrand, M. Searle, M. Clark and H. Sinclair for thought-provoking discussion on this work, and Joint Oceanographic Institutions (JOI) for support for this study. The work was improved thanks to reviews from M. Khan, T. Daley and P. Friend. L. Poppe at the US Geological Survey in Woods Hole and Jae Il Lee are thanked for assistance with the XRD analysis. M. Clark at MIT produced the digital topography of Asia in Fig. 1. Wintershall AG is thanked for releasing data from Indus Marine A-1. Rockland Tour and Trek, and F. Hussein in Leh provided excellent logistical support for work in Ladakh, India. This is Woods Hole Oceanographic Institution Contribution 10631.

References

ABBASI, I.A. & FRIEND, P.F. 2000. Exotic conglomerates of the Neogene Siwalik succession and their implications for the tectonic and topographic evolution of the western Himalaya. *In:* KHAN, M.A., TRELOAR, P.J., SEARLE, M.P. & JAN, M.Q. (eds) *Tectonics of the Nanga Parbat Syntaxis and the Western Himalaya*. Geological Society, London, Special Publications, **170**, 455–466.

BENDER, F., BANNERT, D., BRINCKMANN, J., GRAMANN, F. & HELMCKE, D. 1983. *Geology of Burma*. Beitraege zur Regionalen Geologie der Erde, **16**.

BERGGREN, W.A., KENT, D.V., SWISHER, C.C. & AUBRY, M.P. 1995. A revised Cenozoic geochronology and chronostratigraphy. *In:* BERGGREN, W.A., KENT, D.V. & HARDENBOL, J. (eds) *Geochronology, Time Scales and Global Stratigraphic Correlation*. Society of Economic Paleontologists and Mineralogists, Special Publication, **54**, 129–212.

BERNARD, M., SHEN-TU, B., HOLT, W.E. & DAVIS, D.M. 2000. Kinematics of active deformation in the Sulaiman Lobe and Range, Pakistan. *Journal of Geophysical Research*, **105**, 13253–13279.

BISCAYE, P.E. 1965. Mineralogy and sedimentation of Recent deep-sea clay in the Atlantic Ocean and adjacent seas and oceans. *Geological Society of America Bulletin*, **76**, 803–831.

BOSSART, P. & OTTIGER, R. 1989. Rocks of the Murree Formation of Northern Pakistan: indicators of a descending foreland basin of late Paleocene to middle Eocene age. *Eclogae Geologicae Helveticae*, **82**, 133–165.

BOUQUILLON, A., FRANCE-LANORD, C., MICHARD, A. & TIERCELIN, J. ET AL. 1990. Sedimentology and isotopic chemistry of the Bengal Fan sediments: the denudation of the Himalaya. *In:* COCHRAN, J.R., STOW, D.A.V. ET AL. (eds) *Proceedings of the Ocean Drilling Program, Scientific Results, 116*. Ocean Drilling Program, College Station, TX, 43–58.

BROOKFIELD, M.E. 1998. The evolution of the great river systems of southern Asia during the Cenozoic India–Asia collision; rivers draining southwards. *Geomorphology*, **22**, 285–312.

BROOKFIELD, M.E. & ANDREWS-SPEED, C.P. 1984. Sedimentology, petrography and tectonic significance of the shelf, flysch and molasse clastic deposits across the Indus suture zone, Ladakh, NW India. *Sedimentary Geology*, **40**, 249–286.

BURBANK, D.W., BECK, R.A. & MULDER, T. 1996. The Himalayan foreland basin. *In:* YIN, A. & HARRISON, T.M. (eds) *The Tectonic Evolution of Asia*. Cambridge University Press, New York, 149–188.

BURBANK, D.W., DERRY, L.A. & FRANCE-LANORD, C. 1993. Reduced Himalayan sediment production 8 Myr ago despite an intensified monsoon. *Nature*, **364**, 48–50.

CASSIGNEAU, C. 1979. *Contribution à l'étude des sutures Inde–Eurasie: la zone de suture de Khost (SE Afghanistan)*. Thèse spécialité, Université de Montpellier.

CERVENY, P.F., JOHNSON, N.M., TAHIRKHELI, R.A.K. & BONIS, N.R. 1989. Tectonic and geomorphic implications of Siwalik Group heavy minerals, Potwar Plateau, Pakistan. *In:* MALINCONICO, L.L. & LILLIE, R.J. (eds) *Tectonics of the Western Himalaya*. Geological Society of America, Special Paper, **232**, 129–136.

CHAMLEY, H. 1989. *Clay Sedimentology*. Springer, Berlin.

CLIFT, P.D. & GAEDICKE, C. 2002. Accelerated mass flux to the Arabian Sea during the Middle–Late Miocene. *Geology*, **30**, 207–210.

CLIFT, P.D., SHIMIZU, N., LAYNE, G.D., GAEDICKE, C., SCHLÜTER, H.U., CLARK, M.K. & AMJAD, S. 2000. Fifty five million years of Tibetan evolution recorded in the Indus Fan. *EOS Transactions, American Geophysical Union*, **81**, 277–281.

CLIFT, P.D., SHIMIZU, N., LAYNE, G. & BLUSZTAJN, J. 2001a. Tracing patterns of unroofing in the Early Himalaya through microprobe Pb isotope analysis of detrital K-feldspars in the Indus Molasse, India. *Earth and Planetary Science Letters*, **188**, 475–491.

CLIFT, P.D., SHIMIZU, N., LAYNE, G., GAEDICKE, C., SCHLÜTER, H.U., CLARK, M. & AMJAD, S. 2001b. Development of the Indus Fan and its significance for the erosional history of the western Himalaya and Karakoram. *Geological Society of America Bulletin*, **113**, 1039–1051.

CLIFT, P.D., CARTER, A., KROL, M. & KIRBY, E. 2002a. Constraints on India–Eurasia collision in the Arabian Sea region taken from the Indus GroupLadakh Himalaya, India. *In:* CLIFT, P.D., KROON, D., GAEDICKE, C. & CRAIG, J. (eds) *The Tectonic and Climatic Evolution of the Arabian Sea Region*. Geological Society, London, Special Publications, **195**, 97–116.

CLIFT, P.D., LEE, J.I., CLARK, M.K. & BLUSZTAJN, J. 2002*b*. Erosional response of South China to arc rifting and monsoonal strengthening recorded in the South China Sea. *Marine Geology*, **184**, 207–226.

CLIFT, P.D., LIN, J. & ODP LEG 184 SCIENTIFIC PARTY. 2002*c*. Patterns of extension and magmatism along the continent–ocean boundary, South China Margin. *In:* Wilson, R. C. L., Whitmarsh, R. B., Taylor, B. & Froitzheim, N. (eds) *Non-volcanic Rifting of Continental Margins: a Comparison of Evidence from Land and Sea*. Geological Society, London, Special Publications, **187**, 489-510.

CRITELLI, S., DE ROSA, R. & PLATT, J.P. 1990. Sandstone detrital modes in the Makran accretionary wedge, Southwest Pakistan; implications for tectonic setting and long-distance turbidite transportation. *Sedimentary Geology*, **68**, 241–260.

CROWLEY, S.F., STOW, D.A.V. & CROUDACE, I.W. 1998. Mineralogy and geochemistry of Bay of Bengal deep-sea fan sediments, ODP Leg 116; evidence for an Indian subcontinent contribution to distal fan sedimentation. *In:* CRAMP, A., MACLEOD, C.J., LEE, S.V. & JONES, E.J.W. (eds) *Geological Evolution of Ocean Basins; Results from the Ocean Drilling Program*. Geological Society, London, Special Publications, **131**, 151–176.

CURRAY, J.R. & MOORE, D.G. 1971. Growth of the Bengal deep-sea fan and denudation in the Himalayas. *Geological Society of America Bulletin*, **82**, 563–572.

DAVIES, T.A., KIDD, R.B. & RAMSEY, A.T.S. 1995. A time slice approach to the history of Cenozoic sedimentation in the Indian Ocean. *Sedimentary Geology*, **96**, 157–197.

EDWARDS, M.A. & HARRISON, T.M. 1997. When did the roof collapse? Late Miocene north–south extension in the high Himalaya revealed by Th–Pb monazite dating of the Khula Kangri granite. *Geology*, **25**, 543–546.

ELSTON, D.P. & YOUNG, R.A. 1991. Cretaceous–Eocene (Laramide) landscape development and Oligocene–Pliocene drainage reorganization of transition zone and Colorado Plateau, Arizona. *Journal of Geophysical Research*, **96**, 12398–12406.

FRANCE-LANORD, C., DERRY, L. & MICHARD, A. 1993. Evolution of the Himalaya since Miocene time: isotopic and sedimentological evidence from the Bengal Fan. *In:* TRELOAR, P.J. & SEARLE, M.P. (eds) *Himalayan Tectonics*. Geological Society, London, Special Publications, **74**, 603–622.

FRIEND, P.F., RAZA, S.M., SAMAD-BAIG, M.A. & KHAN, I.A. 1999. Geological evidence of the ancestral Indus from the Himalayan foothills. *In:* MEADOWS, A. & MEADOWS, P.S. (eds) *The Indus River; Biodiversity, Resources, Humankind*. Oxford University Press, Oxford, 103–113.

GAEDICKE, C., PREXL, A., SCHLÜTER, H.-U., MEYER, H., ROESER, H. & CLIFT, P. 2002. Seismic stratigraphy and correlation of major unconformities in the northern Arabian Sea. *In:* CLIFT, P.D., KROON, D., GAEDICKE, C. & CRAIG, J. (eds) *The Tectonic and Climatic Evolution of the Arabian Sea Region*. Geological Society, London, Special Publications, **195**, 25–36.

GALY, A. & FRANCE-LANORD, C. 2001. Higher erosion rates in the Himalaya; geochemical constraints on riverine fluxes. *Geology*, **29**, 23–26.

GARZANTI, E. & VAN HAVER, T. 1988. The Indus clastics: forearc basin sedimentation in the Ladakh Himalaya (India). *Sedimentary Geology*, **59**, 237–249.

GARZANTI, E., BAUD, A. & MASCLE, G. 1987. Sedimentary record of the northward flight of India and its collision with Eurasia (Ladakh Himalaya, India). *Geodinamica Acta*, **1**, 297–312.

GOODBRED, S. & KUEHL, S.A. 2000. Enormous Ganges–Brahmaputra sediment discharge during strengthened early Holocene monsoon. *Geology*, **28**, 1083–1086.

GUILLOT, S., HODGES, K., LE FORT, P. & PECHER, A. 1994. New constraints on the age of the Manaslu leucogranite: evidence for episodic tectonic denudation in the central Himalayas. *Geology*, **22**, 559–562.

GUZMAN-SPEZIALE, M. & NI, J.F. 1993. The opening of the Andaman Sea; where is the short-term displacement being taken up?. *Geophysical Research Letters*, **20**, 2949–2952.

HARRISON, T.M., MCKEEGAN, K.D. & LE FORT, P. 1995. Detection of inherited monazite in the Manaslu leucogranite by ^{208}Pb/^{232}Th ion microprobe dating: crystallization age and tectonic implications. *Earth and Planetary Science Letters*, **133**, 271–282.

HILDEBRAND, P.R., NOBLE, S.R., SEARLE, M.P., PARRISH, R.R. & SHAKIRULLAH, 1998. Tectonic significance of 24 Ma crustal melting in the eastern Hindu Kush, Pakistan. *Geology*, **26**, 871–874.

HODGES, K.V. & SILVERBERG, D.S. 1988. Thermal evolution of the Greater Himalaya, Garhwal, India. *Tectonics*, **7**, 583–600.

HODGES, K.V., PARRISH, R.R., HOUSH, T.B., LUX, D.R., BURCHFIELD, B.C., ROYDEN, L.H. & CHEN, Z. 1992. Simultaneous Miocene extension and shortening in the Himalayan Orogen. *Science*, **258**, 1466–1470.

HODGES, K.V., PARRISH, R.R. & SEARLE, M.P. 1996. Tectonic evolution of the central Annapurna Range, Nepalese Himalayas. *Tectonics*, **15**, 1264–1291.

HOORN, C., GUERRERO, J., SARMIENTO, G.A. & LORENTE, M.A. 1995. Andean tectonics as a cause for changing drainage patterns in Miocene northern South America. *Geology*, **23**, 237–240.

HUCHON, P., NGUYEN, T.N.H. & CHAMOT-ROOKE, N. 1998. Finite extension across the South Vietnam basins from 3D gravimetric modelling; relation to South China Sea kinematics. *Marine and Petroleum Geology*, **15**, 619–634.

KISCH, H.J. 1990. Calibration of the anchizone: a critical comparison of illite crystallinity scales used for definition. *Journal of Metamorphic Geology*, **8**, 31–46.

KROON, D., STEENS, T. & TROELSTRA, S.R. 1991. Onset of monsoonal related upwelling in the western Arabian Sea as revealed by planktonic foraminifers. In: PRELL, W.L. & NIITSUMA, N. (eds) Proceedings of the Ocean Drilling Program, Scientific Results, 117. Ocean Drilling Program, College Station, TX, 257–263.

LAUER-LEREDDE, C., PEZARD, P.A., ROBERT, C. & DEKEYSER, I. 1998. Mineralogical association and physical properties of sediments with palaeoclimatic implications (ODP Site 798B, Japan Sea); a comparative study from core and downhole measurements. Marine Geology, 150, 73–98.

LAWRENCE, R.D., HASAN KHAN, S. & NAKATA, T. 1992. Chaman Fault, Pakistan–Afghanistan. Annales Tectonicae, 6, 196–223.

LE PICHON, X., FOURNIER, M. & JOLIVET, L. 1992. Kinematics, topography, shortening, and extrusion in the India–Eurasia collision. Tectonics, 11, 1085–1098.

LETOUZEY, J. & SAGE, L. 1988. Geological and Structural Map of Eastern Asia, 1:2,500,000. American Association of Petroleum Geologists, Tulsa, OK.

LISTZIN, A.P. 1972. Sedimentation in the World Ocean. SEPM Special Publication, 17.

LIU, G. 1992. Map Series of the Geology and Geophysics of China Seas and Adjacent Regions. Geological Publishing House, Ministry of Geology and Natural Resources, Beijing.

MATTHEWS, S.J., FRASER, A.J., LOWE, S., TODD, S.P. & PEEL, F.J. 1997. Structure, stratigraphy and petroleum geology of the SE Nam Con Son Basin, offshore Vietnam. In: FRASER, A.J., MATTHEWS, S.J. & MURPHY, R.W. (eds) Petroleum Geology of Southeast Asia. Geological Society, London, Special Publications, 126, 89–106.

MCCALL, G.J.H. 2002. A summary of the geology of the Iranian Makran. In: CLIFT, P.D., KROON, D., GAEDICKE, C. & CRAIG, J. (eds) The Tectonic and Climatic Evolution of the Arabian Sea Region. Geological Society, London, Special Publications, 195, 147–204.

MCCALL, G.J.H. & KIDD, R.G.W. 1982. The Makran, southeastern Iran; the anatomy of a convergent plate margin active from Cretaceous to present. In: LEGGETT, J. (ed.) Trench–Forearc Geology; Sedimentation and Tectonics on Modern and Ancient Active Plate Margins. Geological Society, London, Special Publications, 10, 387–397.

METCALFE, R.P. 1993. Pressure, temperature and time constraints on metamorphism across the Main Central Thrust zone and high Himalayan slab in the Garhwal Himalaya. In: TRELOAR, P.J. & SEARLE, M.P. (eds) Himalayan Tectonics. Geological Society, London, Special Publications, 74, 485–509.

MÉTIVIER, F., GAUDEMER, Y., TAPPONNIER, P. & KLEIN, M. 1999. Mass accumulation rates in Asia during the Cenozoic. Geophysical Journal International, 137, 280–318.

MILES, P.R., MUNSCHY, M. & SÉGOUFIN, J. 1998. Structure and early evolution of the Arabian Sea and east Somali Basin. Geophysical Journal International, 134, 876–888.

MILLIMAN, J.D., QURAISHEE, G.S. & BEG, M.A.A. 1984. Sediment discharge from the Indus river to the ocean: past, present and future. In: HAQ, B.U. & MILLIMAN, J.D. (eds) Marine Geology and Oceanography of the Arabian Sea and Coastal Pakistan. Van Nostrand Reinhold, New York, 65–70.

MOLNAR, P., ENGLAND, P. & MARTINOD, J. 1993. Mantle dynamics, the uplift of the Tibetan Plateau, and the Indian monsoon. Reviews in Geophysics, 31, 357–396.

MOUNTAIN, G.S. & PRELL, W.L. 1990. A multiphase plate tectonic history of the southeast continental margin of Oman. In: ROBERTSON, A.H.F., SEARLE, M.P. & RIES, A.C. (eds) The Geology and Tectonics of the Oman Region. Geological Society, London, Special Publications, 49, 725–743.

NAINI, B.R. & KOLLA, V. 1982. Acoustic character and thickness of sediments of the Indus Fan and the continental margin of western India. Marine Geology, 47, 181–185.

NAJMAN, Y., CLIFT, P.D., JOHNSON, M.R.W. & ROBERTSON, A.H.F. 1993. Early stages of foreland basin evolution in the Lesser Himalaya, N. India. In: TRELOAR, P.J. & SEARLE, M.P. (eds) Himalayan Tectonics. Geological Society, London, Special Publications, 74, 541–558.

NAJMAN, Y., PRINGLE, M., GODIN, L. & OLIVER, G. 2001. Dating of the oldest continental sediments from the Himalayan foreland basin. Nature, 410, 194–197.

PIVNIK, D.A., NAHM, J., TUCKER, R.S., SMITH, G.O., NYEIN, K., NYUNT, M. & MAUNG, P.H. 1998. Polyphase deformation in a fore-arc/back-arc basin, Salin Subbasin, Myanmar (Burma). AAPG Bulletin, 82, 1837–1856.

PRELL, W.L. & KUTZBACH, J.E. 1992. Sensitivity of the Indian monsoon to forcing parameters and implications for its evolution. Nature, 360, 647–651.

PRELL, W.L., MURRAY, D.W., CLEMENS, S.C. & ANDERSON, D.M. 1992. Evolution and variability of the Indian Ocean Summer Monsoon: evidence from the western Arabian Sea drilling program. In: DUNCAN, R.A., KIDD, D.K., VON RAD, U. & WEISSEL, J.K. (eds) Synthesis of Results from Scientific Drilling in the Indian Ocean. Geophysical Monograph, American Geophysical Union, 70, 447–469.

QAYYUM, M., LAWRENCE, R.D. & NIEM, A.R. 1997. Discovery of the palaeoIndus delta-fan complex. Journal of the Geological Society, London, 154, 753–756.

QAYYUM, M., NIEM, A.R. & LAWRENCE, R.D. 2001. Detrital modes and provenance of the Paleogene Khojak Formation in Pakistan; implications for early Himalayan Orogeny and unroofing. Geological Society of America Bulletin, 113, 320–332.

RANGIN, C., KLEIN, M., ROQUES, D., LE PICHON, X. & TRONG, L.V. 1995. The Red River fault system in the Tonkin Gulf, Vietnam. Tectonophysics, 243, 209–222.

RAYNOLDS, R.G.H. 1982. Did the ancestral Indus flow into the Ganges drainage?. Geological Bulletin of the University of Peshawar, 14, 141–150.

REA, D.K. 1992. Delivery of Himalayan sediment to northern Indian Ocean and its relation to global climate, sea level, uplift and seawater strontium. *In:* DUNCAN, R.A., REA, D.K., KIDD, R.B., VON RAD, U. & WEISSEL, J.K. (eds) *Synthesis of Results from Scientific Drilling in the Indian Ocean.* Geophysical Monograph, American Geophysical Union, **70**, 387–402.

REUBER, I. 1986. Geometry of accretion and oceanic thrusting of the Spontang Ophiolite, Ladakh Himalaya. *Nature*, **321**, 592–596.

REYNOLDS, P.H., BROOKFIELD, M.E. & McNUTT, R.H. 1983. The age and nature of Mesozoic–Tertiary magmatism across the Indus suture zone in Kashmir and Ladakh (N.W. India and Pakistan. *Geologisches Rundschau*, **72**, 981–1003.

ROESER, H.A. & SONNE 122 SCIENTIFIC PARTY 1997. *MAKRAN I: the Makran Accretionary Wedge off Pakistan: Tectonic Evolution and Fluid Migration—Part I. Cruise Report SONNE cruise SO 122 (7 August–6 September 1997.* Report of Bundesanstalt für Geowissenschaften und Rohstoffe, **116643**.

ROQUES, D., MATTHEWS, S.J. & RANGIN, C. 1997. Constraints on strike-slip motion from seismic and gravity data along the Vietnam margin offshore Da Nang; implications for hydrocarbon prospectivity and opening of the East Vietnam Sea. *In:* FRASER, A.J., MATTHEWS, S.J. & MURPHY, R.W. (eds) *Petroleum Geology of Southeast Asia.* Geological Society, London, Special Publications, **126**, 341–353.

ROYDEN, L. 1996. Coupling and decoupling of crust and mantle in convergent orogens; implications for strain partitioning in the crust. *Journal of Geophysical Research*, **101**, 17679–17705.

ROYDEN, L.H., CLARK, M.K., WHIPPLE, K.X. & BURCHFIEL, B. 2000. River incision and capture related to tectonics of the Eastern Himalayan Syntaxis (abstract). *EOS Transactions, American Geophysical Union*, **81**, S413.

SAID, R. 1981. *The Geological Evolution of the River Nile.* Springer, Heidelberg.

SCHÄRER, U., COPELAND, P., HARRISON, T.M. & SEARLE, M.P. 1990. Age, cooling history and origin of post-collisional leucogranites in the Karakoram batholith: a multi-system isotope study. *Journal of Geology*, **98**, 233–251.

SCLATER, J.G. & CHRISTIE, P.A.F. 1980. Continental stretching; an explanation of the post-Mid-Cretaceous subsidence of the central North Sea basin. *Journal of Geophysical Research*, **85**, 3711–3739.

SEARLE, M.P. 1991. *Geology and Tectonics of the Karakoram Mountains.* Wiley, New York.

SEARLE, M.P. 1996. Cooling history, erosion, exhumation and kinematics of the Himalaya–Karakorum–Tibet orogenic belt. *In:* YIN, A. & HARRISON, T.M. (eds) *The Tectonic Evolution of Asia.* Cambridge University Press, Cambridge, 110–137.

SEARLE, M.P. & OWEN, L.A. 1999. The evolution of the Indus River in relation to topographic uplift, climate and geology of western Tibet, the Trans-Himalaya and High-Himalayan Range. *In:* MEADOWS, A. & MEADOWS, P.S. (eds) *The Indus River, Biodiversity, Resources, Humankind.* Oxford University Press, Oxford, 210–230.

SEARLE, M.P., COOPER, D.J.W. & REX, A.J. 1988. Collision tectonics of the Ladakh–Zanskar Himalaya. *Philosophical Transactions of the Royal Society, London, Part A*, **326**, 117–150.

SEARLE, M.P., PICKERING, K.T. & COOPER, D.J.W. 1990. Restoration and evolution of the intermontane Indus Molasse Basin, Ladakh Himalaya, India. *Tectonophysics*, **174**, 301–314.

SEARLE, M.P., WATERS, D.J., REX, D.C. & WILSON, R.N. 1992. Pressure, temperature and time constraints on Himalayan metamorphism from eastern Kashmir and western Zanskar. *Journal of the Geological Society, London*, **149**, 753–773.

SEARLE, M.P., CORFIELD, R.I., STEPHENSON, B. & McCARRON, J. 1997*a*. Structure of the North Indian continental margin in the Ladakh–Zanskar Himalayas: implications for the timing of obduction of the Spontang Ophiolite, India–Asia collision and deformation events in the Himalaya. *Geological Magazine*, **134**, 297–316.

SEARLE, M.P., PARRISH, R.R., HODGES, K.V., HURFORD, A., AYERS, M.W. & WHITEHOUSE, M.J. 1997*b*. Shisha Pangma leucogranite, South Tibetan Himalaya: field relations, geochemistry, age, origin, and emplacement. *Journal of Geology*, **105**, 295–317.

SEARLE, M.P., WEINBERG, R.F. & DUNLAP, W.J. ET AL. 1998. Transpression tectonics along the Karakoram fault zone, northern Ladakh; constraints on Tibetan extrusion. *In:* HOLDSWORTH, R.E. (ed.) *Continental Transpressional and Transtensional Tectonics.* Geological Society, London, Special Publications, **135**, 307–326.

SEEBER, L. & GORNITZ, V. 1983. River profiles along the Himalayan Arc as indicators of active tectonics. *Tectonophysics*, **92**, 335–367.

SCHRODER, J.F. & BISHOP, M.P. 1999. Tindus to the sea: evolution of the system and Himalayan geomorphology. *In:* MEADOWS, A. & MEADOWS, P.S. (eds) *The Indus River, Biodiversity, Resources, Humankind.* Oxford University Press, Oxford, 231–248.

SHUAIB, S.M. 1982. Geology and hydrocarbon potential of offshore Indus Basin, Ladakh Himalaya, India. *AAPG Bulletin*, **66**, 940–946.

SINCLAIR, H.D. & JAFFEY, N. 2001. Sedimentology of the Indus Group, Ladakh, northern India: implication timing of initiation of the palaeo-Indus River. *Journal of the Geological Society, London*, **158**, 151–162.

STECK, A., SPRING, L., VANNAY, J.-C. & 5 OTHERS 1993. The tectonic evolution of the Northwestern Himalaya in eastern Ladakh and Lahaul, India. *In:* Treloar, P. J. & Searle, M. P. (eds) *Himalayan Tectonics.* Geological Society, London, Special Publications, **74**, 265–276.

TJIA, H.D. & LIEW, K.K. 1996. Changes in tectonic stress field in northern Sunda Shelf basins. *In:* HALL, R. & BLUNDELL, D.J. (eds) *Tectonic Evolution of Southeast Asia.* Geological Society, London, Special Publications, **106**, 291–306.

TREGEAR, T. 1965. *A Geography of China.* University of London Press, London.

TRELOAR, P.J., REX, D.C., GUISE, P.G. & 6 OTHERS 1989. K–Ar and Ar–Ar geochronology of the Himalayan collision in NW Pakistan: constraints on the timing of suturing, deformation, metamorphism and uplift. *Tectonics*, **8**, 881–909.

VAN HAVER, T. 1984. *Étude stratigraphique, sédimentologique et structurale d'un bassin d'avant arc: exemple du bassin de l'Indus, Ladakh, Himalaya.* PhD thesis, University of Grenoble.

WALKER, J.D., MARTIN, M.W., BOWRING, S.A., SEARLE, M.P., WATERS, D.J. & HODGES, K.V. 1999. Metamorphism, melting, and extension; age constraints from the High Himalayan slab of Southeast Zanskar and Northwest Lahaul. *Journal of Geology*, **107**, 473–495.

WEINBERG, R.F. & DUNLAP, W.J. 2000. Growth and deformation of the Ladakh Batholith, Northwest Himalayas; implications for timing of continental collision and origin of calc-alkaline batholiths. *Journal of Geology*, **108**, 303–320.

WILLETT, S., BEAUMONT, C. & FULLSACK, P. 1993. Mechanical model for the tectonics of doubly vergent compressional orogens. *Geology*, **21**, 371–374.

WILLIAMS, H., TURNER, S., KELLEY, S. & HARRIS, N. 2001. Age and compositin of dikes in southern Tibet: new constraints on the timing of east–west extension and its relations to postcollisional volcanism. *Geology*, **29**, 339–342.

ZEITLER, P.K., CHAMBERLAIN, C.P. & SMITH, H.A. 1993. Synchronous anatexis, metamorphism and rapid denudation at Nanga Parbat (Pakistan Himalaya). *Geology*, **21**, 347–350.

Seismic stratigraphy of the offshore Indus Basin

TIM DALEY[1] & ZAHEER ALAM[2]

[1] *Lasmo (ENI) Venezuela BV, Piso 11, Torre Este, Centro Lido, El Rosal, Avenida Francisco de Miranda, El Rosal, Caracas, Venezuela*

[2] *Government of Pakistan, Director General, Petroleum Concession, Department of Petroleum and Energy Resources, Ministry of Petroleum and Natural Resources, 1019-A, 19-A, Pak Plaza, Fazl-e-Haq Road, Blue Area, Islamabad, Pakistan*

Abstract: In 1997 Lasmo Oil Pakistan Ltd (Lasmo) gained a significant position in the offshore Indus Basin with the award of the Indus A and B Blocks. The main hydrocarbon play comprises Miocene shelf–delta sands interbedded with intraformational shale seals and sourced by gas-prone offshore equivalents. Approximately 12 000 km of seismic data have been interpreted in the detailed evaluation of these blocks. However, only four wells have tested the preferred play type and no core or rock data were available to provide further insights into facies or age dating. Log data from two key wells in the offshore Indus area record the initial infilling of the basin by shale-dominated basinal or outer shelf sediments, followed by stacked thin-bedded sandstone–shale sequences of a shelf–delta nature. A zone of progradational sequences marks the transition between the two, but no other workable stratigraphic divisions were apparent. Regional seismic correlation established the diachronous nature of the prograding shelf package and this was matched by distinct bands of seismic progrades. A series of simple palaeogeographies of the prograding shelf margin were developed showing initial sediment input from the north and rapid progradation towards the south and west. The Oligo-Miocene basin fill of the offshore Indus Basin appears to be a 'one-step' fill process of a significant depocentre created between the Karachi Platform and the Murray Ridge. Canyons are a very distinct feature on seismic profiles and two main phases of development are apparent. The earlier phase is interpreted to be of Early Miocene age. Downcutting at this time rarely exceeded 400 m. The second phase of canyon development occurred during Plio-Pleistocene time, and these younger canyons often dominate the shallow section, with multiple phases of downcutting sometimes exceeding 1000 m. Where drilled, canyons of both ages have been found to be shale prone. These drilled canyons are interpreted to be on the palaeo-slope where erosion and sediment by-pass occurred during the active phase, and were subsequently filled by fine-grained deposits after abandonment. The two phases of canyoning are considered to relate to phases of tectonic activity in the collision zone between the western margin of the Indo-Pakistan plate and the Eurasian plate.

Lasmo Oil Pakistan Ltd ('Lasmo') signed exploration agreements with DGPC of Pakistan on 21 September 1997 for two substantial offshore blocks, Indus A and B, totalling 14 704 km². The initial exploration term was for 3 years, but at each anniversary, the blocks could be relinquished entirely, or renewed upon committing to a pre-agreed minimum work programme. The first renewal required 1500 km seismic acquisition per block and the second required one exploratory well to be drilled in each block. On the basis of the prospectivity perceived after Year 1, Lasmo elected to renew the A Block but relinquished the B Block. Offshore Indus F Block was a new block successfully applied for, and awarded on 1 December 1998. It covered 67% of the original B Block and carried a first year commitment of 850 km of seismic acquisition. The outlines of the Indus A and F blocks are shown in figures as reference to the main study area.

A seismic acquisition campaign collected 3659 km of 2D data across the A and F blocks in November and December 1998, infilling and extending existing coverage across both blocks. During 1999, Lasmo interpreted and incorporated all the new data but did not elect to renew either licence on their anniversary dates.

The Indus A and F blocks lie in the nearshore zone, south and east of Karachi and offshore from the present-day Indus delta. The regional water depth map (Fig. 1) shows that the blocks are located within the broad (60–80 km wide) shelf. Water depths increase gradually to 200 m and thereafter, slopes increase to 1000 m or more.

Figure 1 shows clearly the present Indus canyon cutting into the shelf east of F Block and also the NE-trending sea-bottom topography of the Murray Ridge, to the NW of the A Block.

Despite the relatively shallow waters, the offshore Indus province remains rather unexplored with respect to hydrocarbons. Before Lasmo's involvement, there had been extensive 2D seismic coverage across the region, amounting to in excess of 28 000 km. However, of this, 15 000 km pre-dated 1972 and the most recent survey was from 1988. Exploration drilling offshore is very sparse, totalling only five wells in the nearshore area and another five further offshore. Encouragingly, Pakcan-1 tested gas in 1986 from a Miocene clastic reservoir at rates of up to 8.3 million cubic feet of gas per day.

During the course of this evaluation, 5519 km of seismic data were reprocessed and a further 2567 km purchased as processed digital tapes. These data, when combined with the new acquisition of 1998, resulted in a total of 11 745 km of seismic data being available for a workstation-based interpretation. Across much of the high-graded area, there is now a grid of data spaced at 1–11/2 km by 2–3 km. Figure 2 shows the distribution of the available well and digital seismic data.

Tectonic elements

The main tectonics of the region are depicted in Fig. 3, adapted from Kolla & Coumes (1990), with Neogene tectonic elements emphasized. The

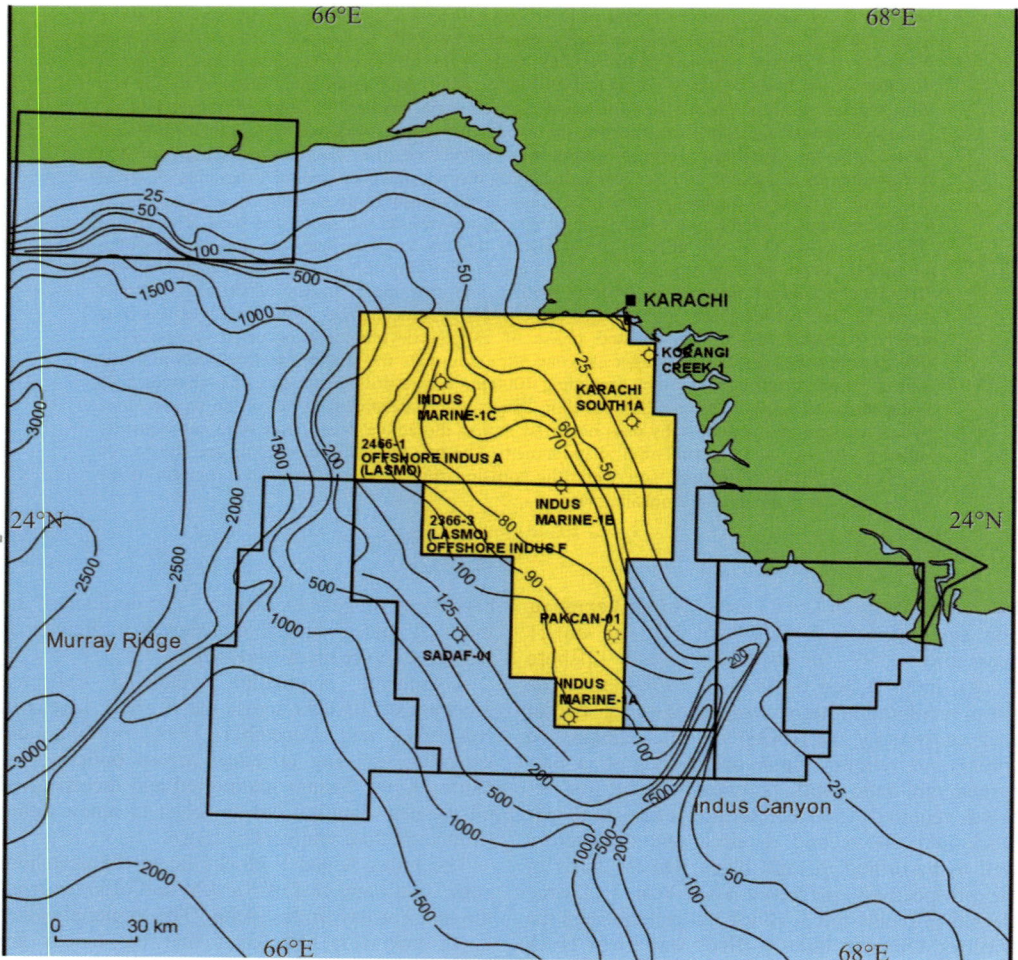

Fig. 1. Regional water depth (contours in metres). Line outlines indicate licensed area at the time of the study, and the Lasmo A and F blocks are indicated. Exploration wells within or close to the Lasmo blocks are labelled.

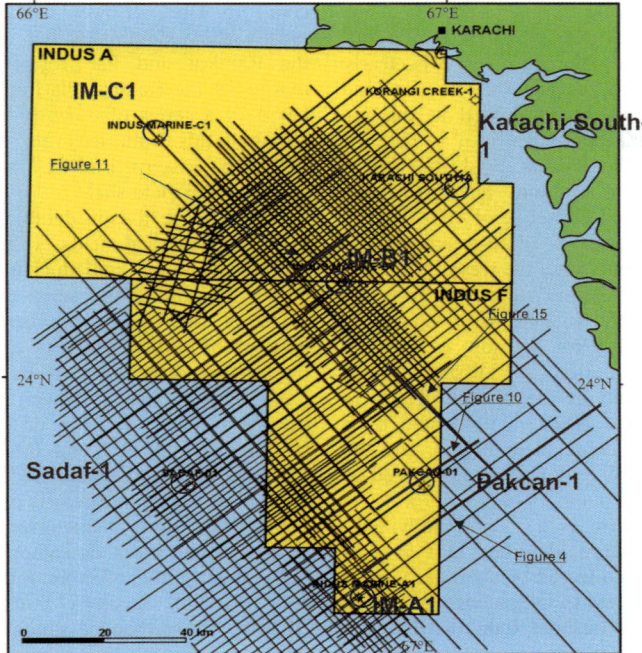

Fig. 2. Digital seismic and well database. Locations of seismic figures are indicated.

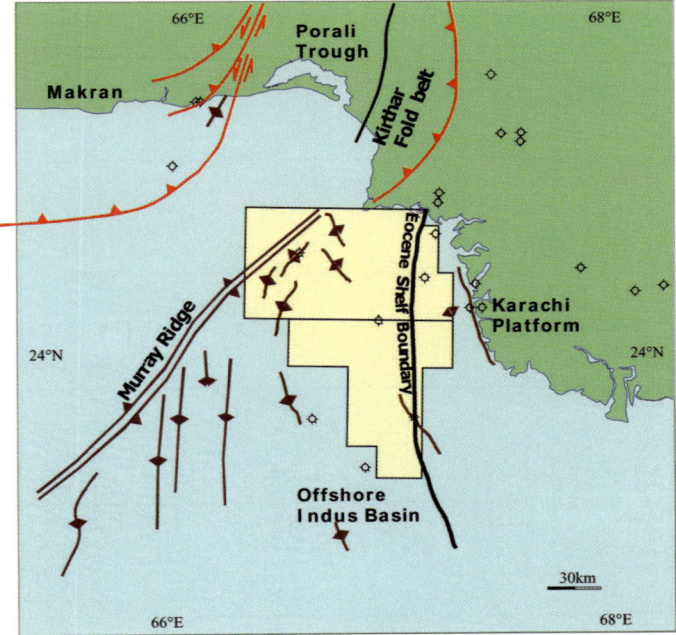

Fig. 3. Miocene tectonic elements. Main Murray Ridge Anticlinorium and other anticline axes are labelled. Eastern limits of Makran and Kirthar Fold Belt provinces are indicated by thrust fault with triangles symbol. Lasmo Indus A and F blocks are highlighted.

Indus A and F blocks lie mostly to the SE of the NE-trending Murray Ridge, which is poorly imaged by sparse seismic data. However, it is clearly a major structural high because the Miocene (and younger) beds are tilted and eroded at the sea floor on its flanks. Geographically, it is the offshore extension of the Kirthar Fold Belt but a detailed study of the tectonic history of the Murray Ridge is beyond the scope of this study.

Figure 3 shows the north–south-trending shelf boundary at mid-Eocene time, the western limit of the Karachi Platform. Between the Murray Ridge and the Karachi Platform there is a significant depocentre containing 6 km or more of Tertiary sediments and here called the offshore Indus Basin. The relationship of the Karachi Platform and the offshore Indus Basin is illustrated by a regional seismic line depicted in Fig. 4, and the stratigraphy of the two regions is summarized in Fig. 5.

The Karachi Platform is a monoclinally SW-dipping unit. The portion identified in Fig. 4 as 'Paleogene shelf' is dominated by shelf carbonates ranging from mid-Eocene to Oligocene age, which are equivalents of the onshore Laki, Kirthar and Nari formations. Deeper penetrations from some of the nearshore drilling have encountered older formations, ranging down through Ranikot (Paleocene) to Goru (Lower Cretaceous) formations, mostly shale dominated but with some sands in the Ranikot and Pab formations. The western margin of the Karachi Platform is abruptly terminated by a series of down-to-basin normal faults, although this limit may also coincide in places with the western limit of progradation of the Tertiary carbonate shelf.

To the west, a considerable thickness of post-Lower Eocene sediments is observed, much of it seismically rather indistinct and appearing to be passively infilling the steep eastern margin. Because of their great thickness, these strata are untested by drilling and they are therefore inferred by seismic correlation to be of Oligocene and Early Miocene age. However, the uppermost parts of this basinal fill were drilled by the deepest sections of Pakcan-1 and IM-B1, both of which encountered shaley sections. The apparently 'passive' fill of the depocentre is capped by a prograding seismic unit characterized by seismic-scale sigmoidal reflectors downlapping into the basin. Subsequently, the offshore Indus Basin and Karachi Platform received thick deposits (1000–2000 m) of interbedded sands and shales of Mid- to Late Miocene age that are interpreted as shallow-water shelf or deltaic deposits. Overlying this is a widespread muddy shelf unit of Pliocene

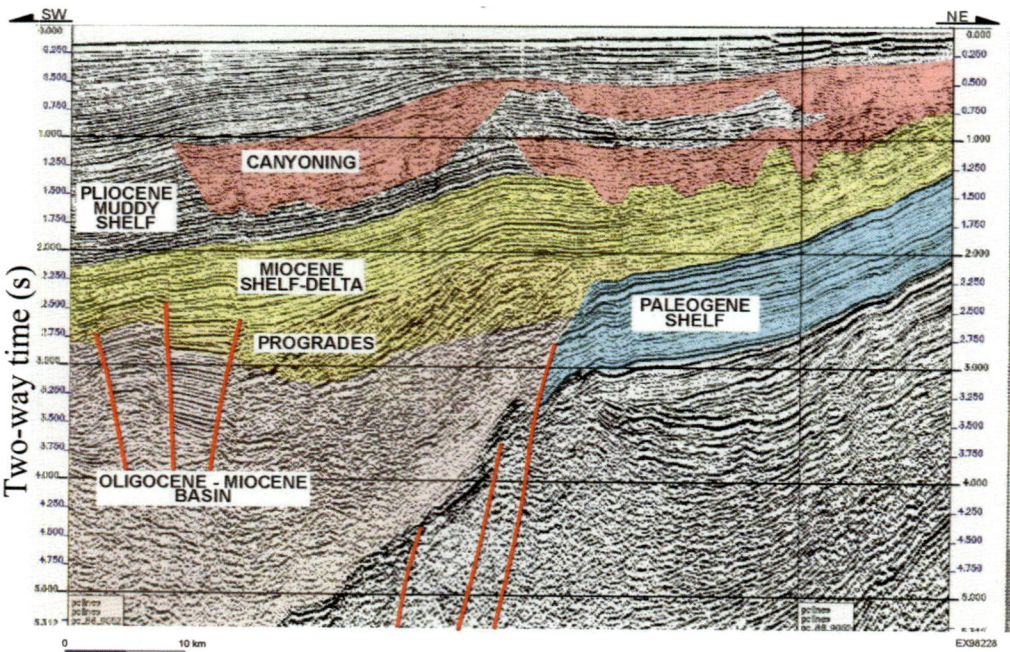

Fig. 4. Regional seismic line.

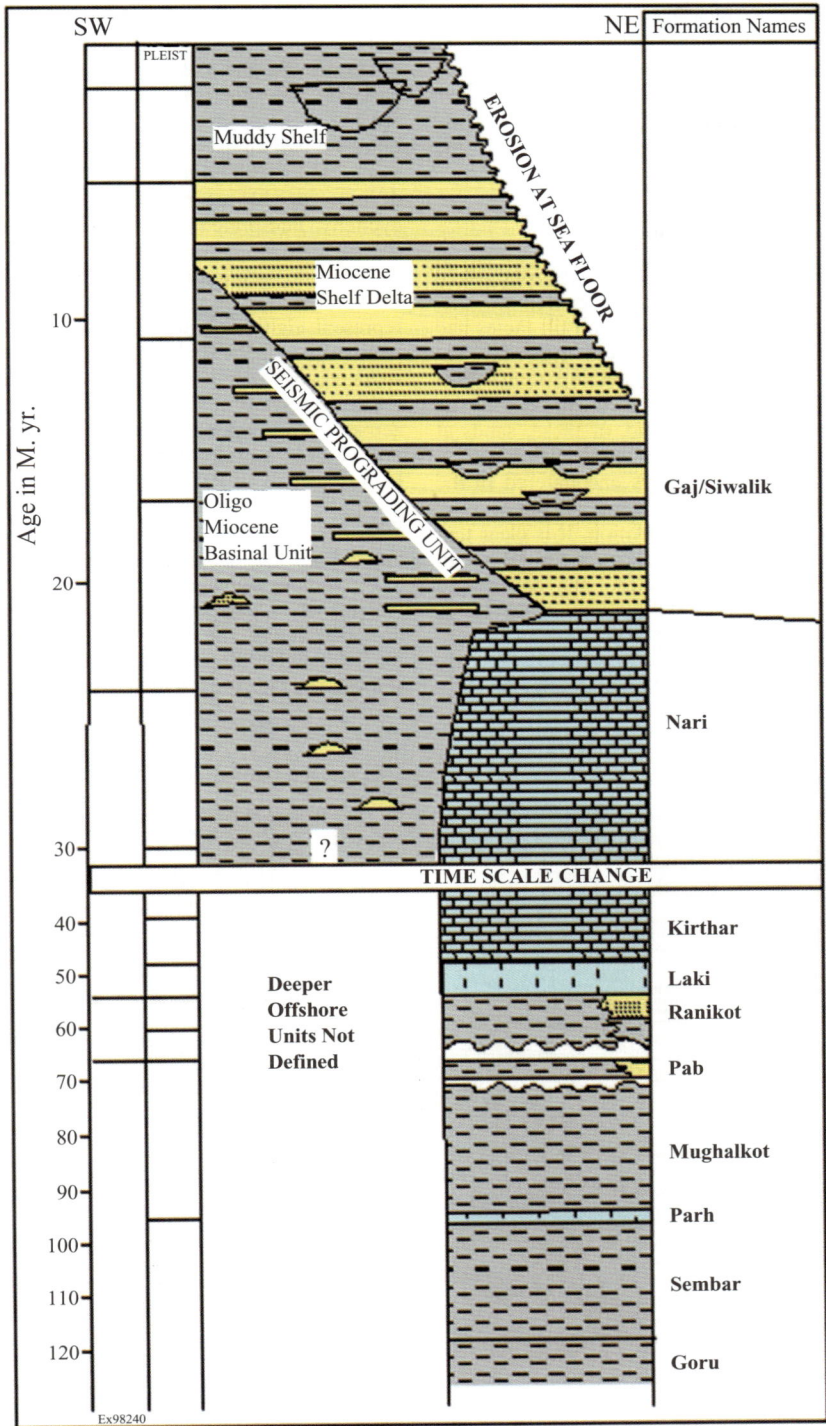

Fig. 5. Offshore Indus stratigraphic column.

and younger age, which has been successively cut by extensive canyon phases of the proto-Indus.

Unfortunately, biostratigraphic ages from well reports are of only broad help at best and no samples or core were available during this study to confirm ages of the formations, or to aid the interpretation of their facies.

Seismic data and stratigraphy

Regional seismic correlations quickly established the relationships of the sections drilled to date, and it was recognized where canyon fill facies had been penetrated, helping to put reported age determinations into context (Fig. 6). Only three wells have penetrated the targeted Miocene clastic rock reservoir-seal system and, of these, sands at Sadaf-1 are thinner and more distal in nature. Several of the shale packages encountered in the existing wells are canyon fill deposits, and at some levels, this impedes log correlation. It is clear that the sandy shelf facies are diachronous, although the overlying Pliocene muddy unit appears to represent a fairly abrupt and possibly synchronous event across the region. It is also worth emphasizing that this correlation shows that IM-1A only just penetrated the Upper Miocene sediments.

Examining well log data across the sandy Miocene shelf strata from Pakcan-1 and IM-B1 did not reveal any clear stratigraphic divisions and certainly few distinctive seismic packages were notable. The log of IM-B1 (Fig. 7) illustrates the rather monotonous stacking of interbedded sandstones and shales from 600 to 2200 m depth. The deepest logged section does, however, reveal a cyclicity, interpreted as coarsening-upwards parasequence sets. Several distinctive low-velocity shales in this section tie to high-amplitude seismic events (Fig. 8). These 'slow' shales are thought to represent higher organic content argillites, perhaps relatively condensed and deposited during a time spanned by a maximum flooding event. The resulting seismic markers are readily correlatable across a semi-regional extent.

Well log data from close to the base of the logged section in Pakcan-1 show a similar pattern of log cyclicity to that at IM-B1 (Fig. 9). A sequence stratigraphic scheme has been tentatively applied during seismic interpretation. However, the complete sequence of 250 m corresponds to less than two seismic wavelets. This log package does correlate in the Pakcan-1 area with the top of the obliquely dipping, seismic-scale progrades shown in Fig. 10, and is interpreted as the

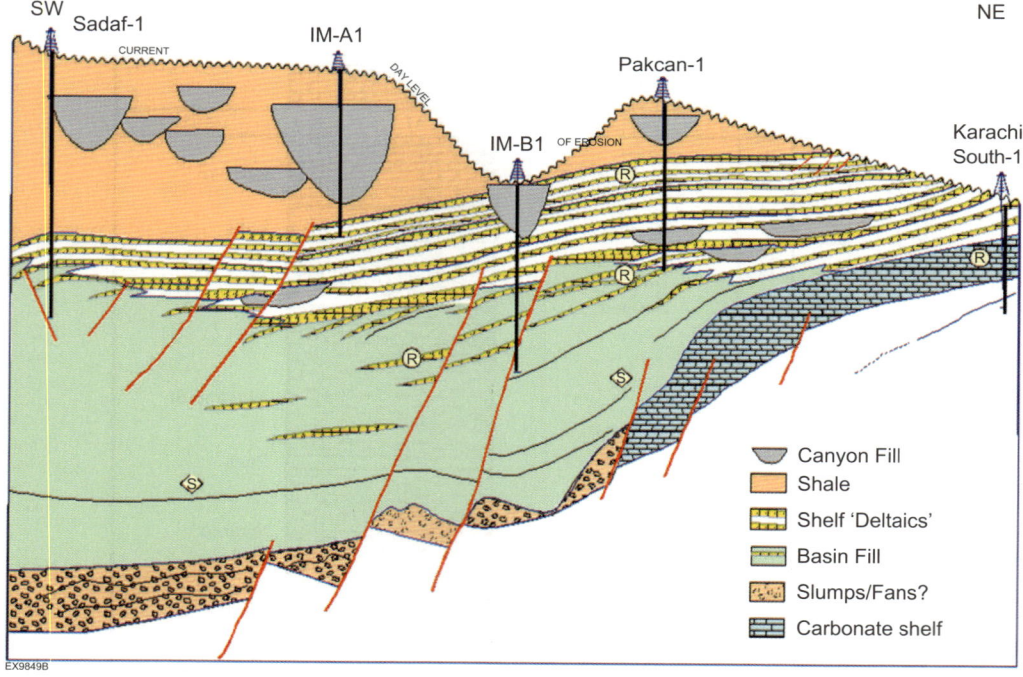

Fig. 6. Schematic correlation of Oligo-Miocene lithostratigraphy. Vertical section *c.* 6 km, section length *c.* 100 km. R in circle indicates possible hydrocarbon reservoirs, S in diamonds indicates possible hydrocarbon source.

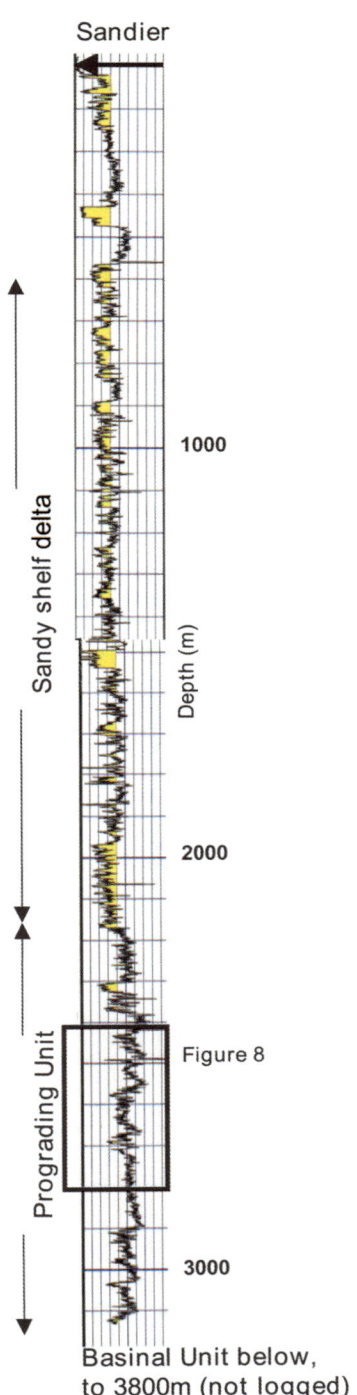

Fig. 7. Gamma well log from IM-B1, indicating the main sandier v. shale sections.

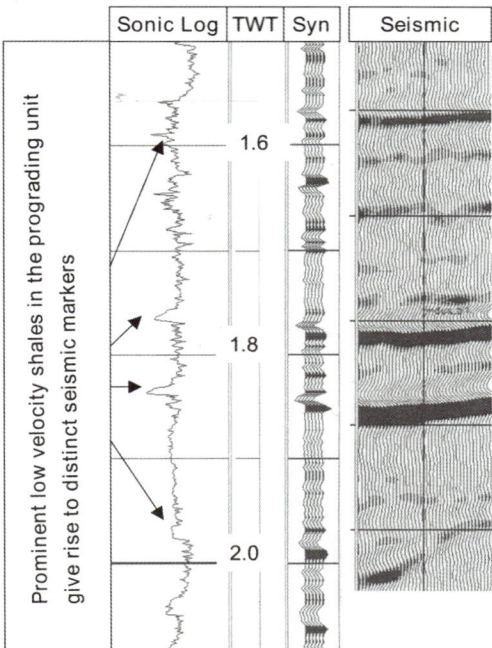

Fig. 8. Part of well-to-seismic tie for IM-B1. 'Syn', synthetic seismogram.

transition between dominantly basinal facies below and stacked shelf facies above: the result of overall basin fill and subsequent progradation of the shelf.

The seismic line shown in Fig. 11, a dip line along strike to IM-B1, highlights the compelling correlations that the higher-amplitude seismic events allow across faults. However, further east, their character fades, perhaps reflecting the proximal limit of these maximum flooding surfaces as distinctive litho-facies. Mapping at an intra-Lower Miocene marker (Fig. 12) shows the monoclinal dip down to the SW, ranging from two-way time of 1 s (c. 1500 m) to 3 s (c. 4500 m). The normal, down-to-the-basin faults, seen as fairly planar in seismic cross-sections, are part of a 60 km long, gently curved, fault system, with antithetic faults. Sedimentary growth across these faults is mostly contained within shale-prone 'pre-shelf' sediments, and their throws decline upwards to zero into the sandy clastic rocks above.

Refining regional seismic correlations and delimiting the seismic prograding units leads to a simple, relative age, stratigraphy and facies scheme (Fig. 13). Progradation of the shelf was mapped (Fig. 14), which appears to originate in the north, and at first has a strong southerly component. Later, a more southwestward progra-

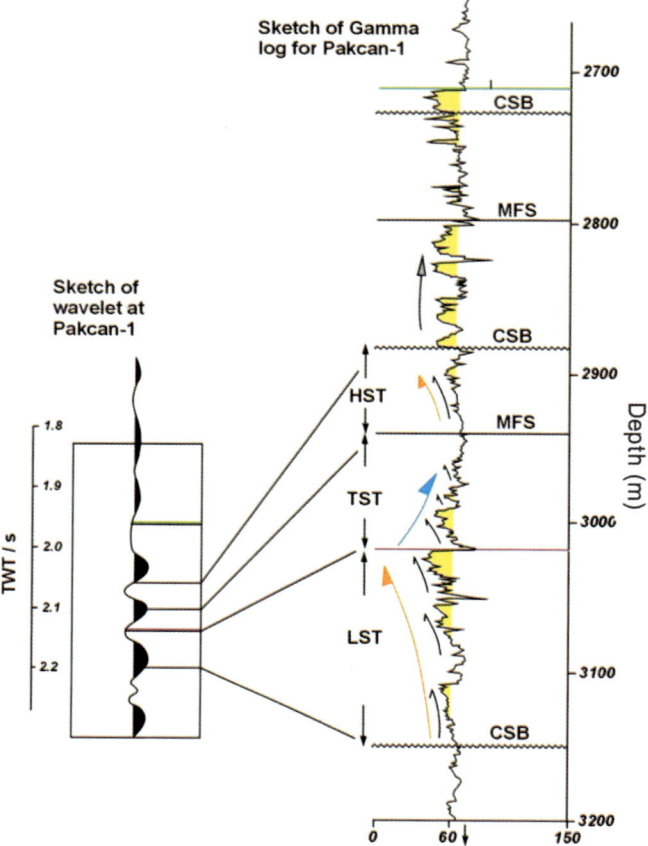

Fig. 9. Part of well log from Pakcan-1 across 'Prograding Unit'. CSB, candidate sequence boundary; HST, highstand systems tract; TST, transgressive systems tract; LST, lowstand systems tract; MFS, maximum flooding surface; TWT, two-way travel time.

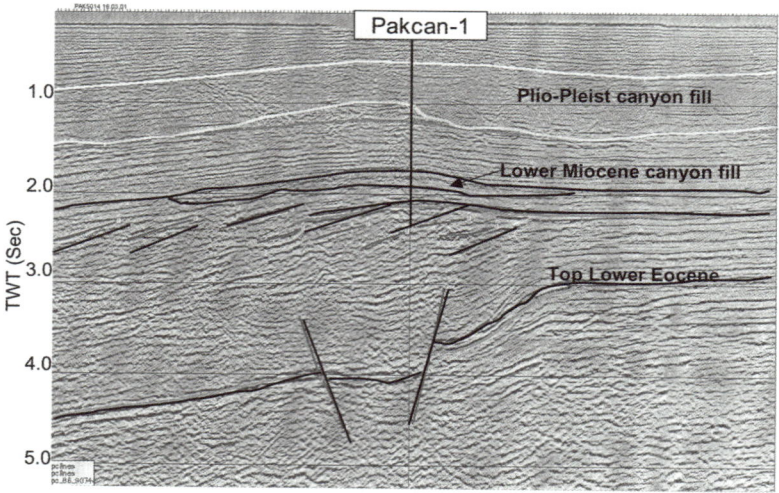

Fig. 10. Seismic scale progrades illustrated in seismic data through Pakcan-1.

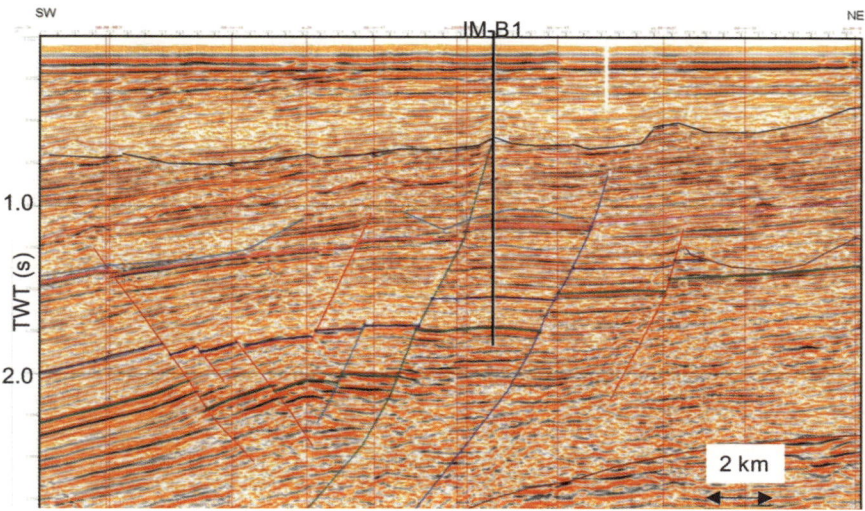

Fig. 11. Seismic dip line, along strike to IM-B1, showing down-to-basin normal faults (SW dipping) and anthithetics (NE dipping).

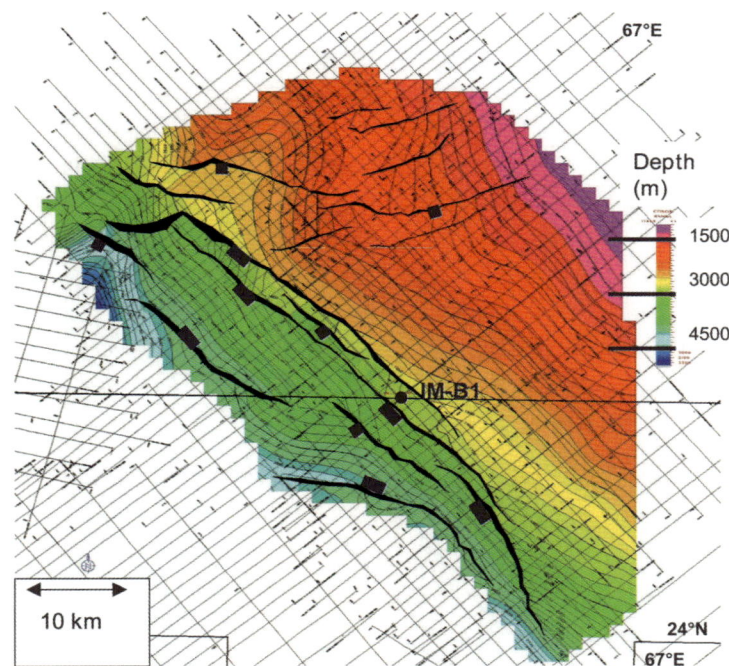

Fig. 12. Intra-Lower Miocene Marker two-way travel time map, in centre of project area with IM-B1 well shown for reference.

dation direction resulted as increasing sediment contributions arrived from the Karachi Platform. Sadaf-1 well appears to be close to the western limit that the coarser Miocene clastic deposits reached, before their termination and blanketing by muddy shelf deposits of Pliocene times.

This work therefore implies that the Oligo-Miocene sedimentary fill of the offshore Indus

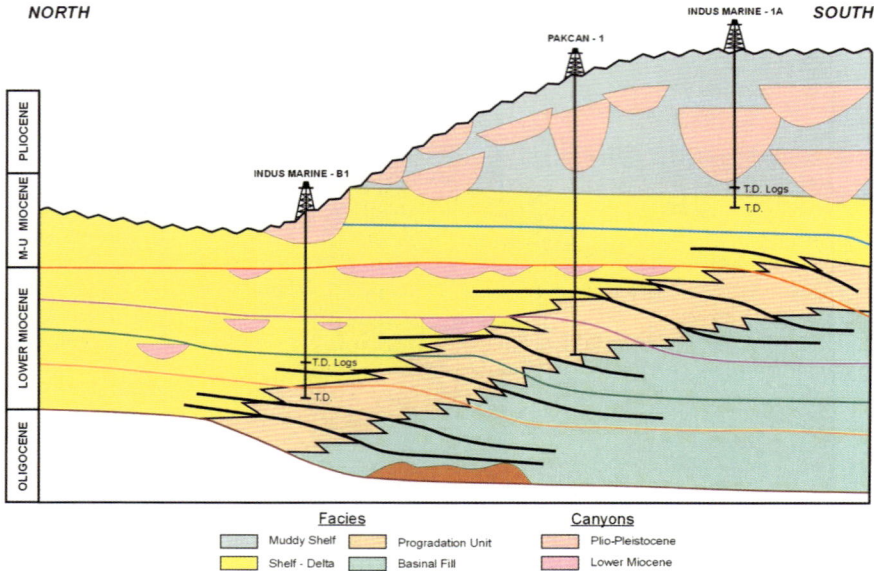

Fig. 13. Miocene stratigraphy and facies scheme.

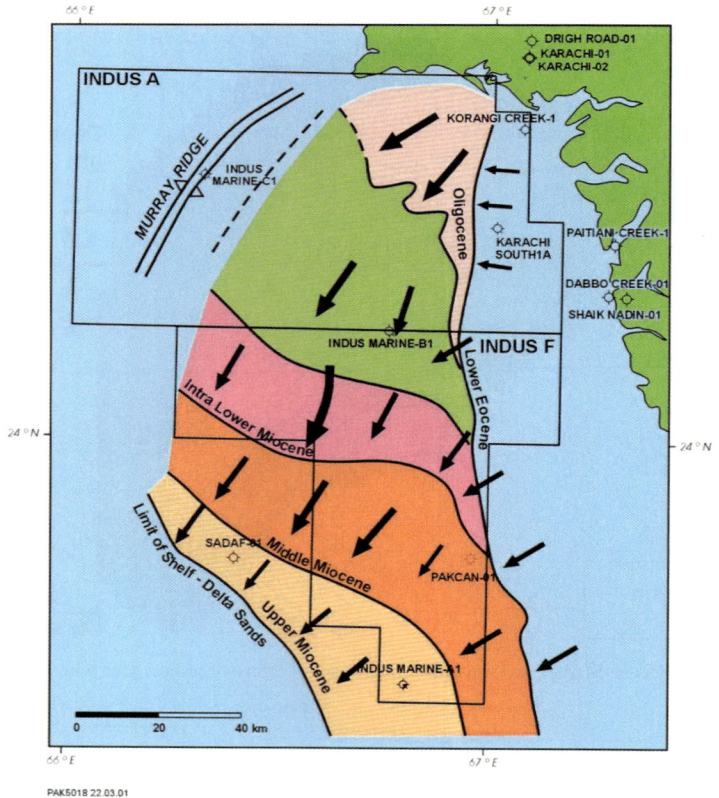

Fig. 14. Map of progradation of Oligo-Miocene shelf. Outlines of Indus A and F blocks are shown. Arrows suggest direction of sedimentation through time.

Basin was a 'one-step fill', from basinal to shelfal, with sedimentation originating from the north.

Canyons

Canyon cut and fill facies dominate the shallow section across much of the study area and have been the subject of numerous references in the region. This study area covers the 'degradation zone' as defined by McHargue & Webb (1986), where the upper erosional canyon complexes are filled by pro-delta deposits. Seismic illustrations in Figs 4, 10 and 11 show the form and dimensions of the canyons, and their relative positions in the sequence have been indicated schematically in Figs. 6 and 13. The canyon cuts and fills observed divide into two main age categories, identified in Fig. 15, whose distributions are mapped at four time intervals in Fig. 16a-d.

An older group, of Early Miocene age, is generally limited to a maximum downcutting of 400 m. These commence as relatively isolated and individually mappable units cutting 70–80 km back into the shelf (Fig. 16a) and extending out to the shelf–slope break of the time. At a time believed to correspond to the end of Early Miocene time, canyons became widespread across much of the shelf of the study area (Fig. 16b) and many have laterally coalesced, or eroded into previous canyon fills. Thereafter, canyon activity wanes, although three examples have been preserved of canyons that appear to have been active during Mid- or Late Miocene time (Fig. 16c).

The younger canyon phase cuts through the Pliocene muddy shelf, and in many cases, down into the coarser Middle and Upper Miocene clastic rocks below. It is widespread (but excluded from A Block, Fig. 16d), multiphase (at least eight levels can be demonstrated), and downcutting frequently exceeds 1000 m. Preserved individual canyon fills are rare, such was the intensity of repeated cut and fill events, although Kolla & Coumes (1987) separated three canyon complexes cutting the shelf west of the current Indus canyon. They indicated an easterly migration of canyon fill events, although Kolla & Coumes recognized 'jumps' in canyon location, and dated them to Late Miocene to Late Pliocene time.

Exploratory drilling to date has penetrated Plio-Pleistocene canyon fill facies on three occasions but an Early Miocene canyon fill facies has been encountered only once. The lithologies logged were shale dominated with occasional silty levels. This suggests that the position in which these canyons were drilled lay in the slope zone of erosion and sediment by-pass at their time of formation. The subsequent fill would have occurred after abandonment, by passive basinal or outer shelf fine-grained sediments.

It is thought that the magnitude and intensity of canyon events should reflect the scale and proximity of onshore tectonics, as well as sedimentation rates and relative sea level, and therefore could match the timing of uplift and inversion events. Early Miocene time, in the onshore Kirthar Fold Belt province, is considered a time of active inversion and uplift (Smewing et al. 2002), perhaps correlating with the older canyons phase. However, Plio-Pleistocene time is the main phase of uplift in western Pakistan, coinciding with the timing of the considerable offshore canyon development. This event is related to final docking

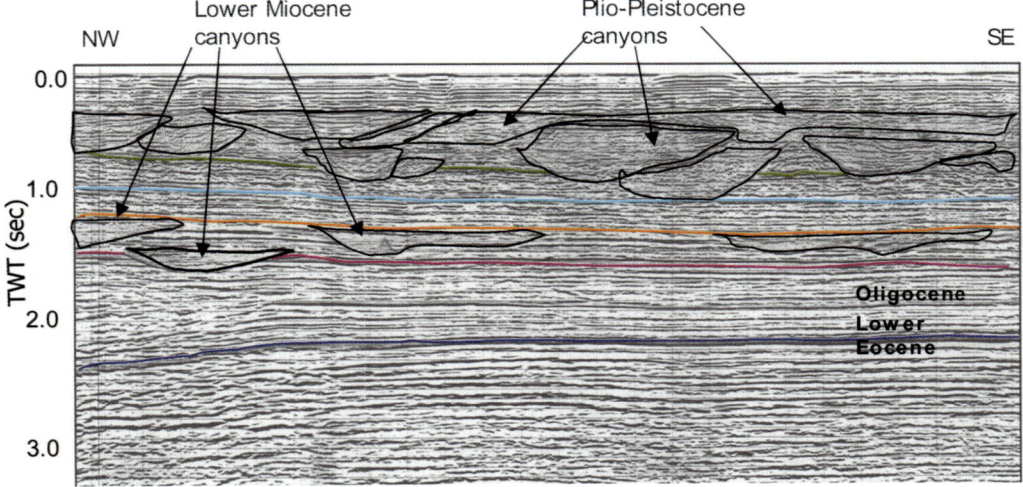

Fig. 15. Seismic strike line illustrating canyon cut and fill facies.

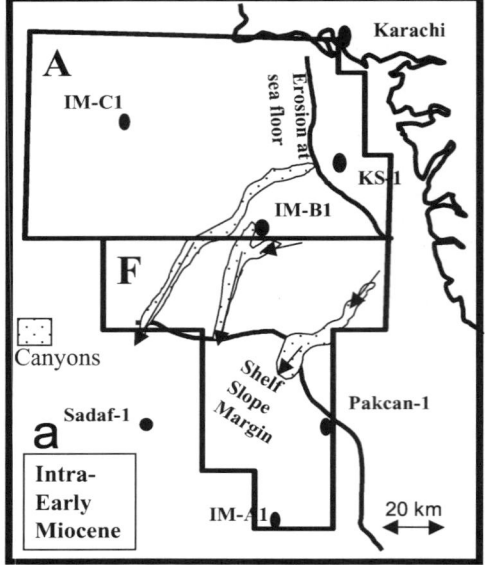
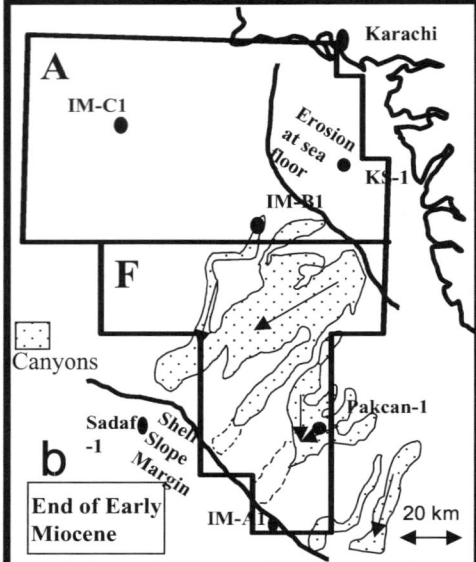
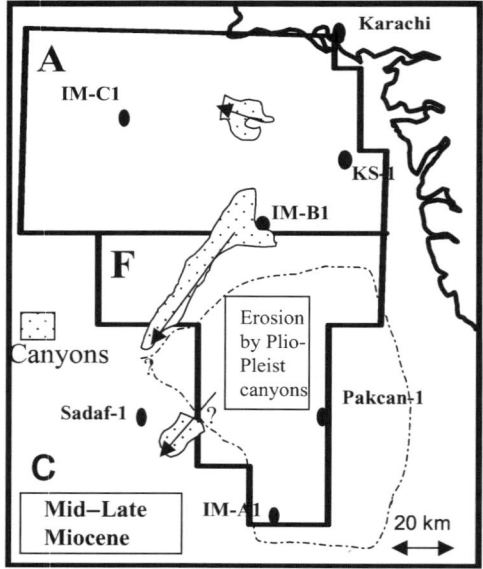
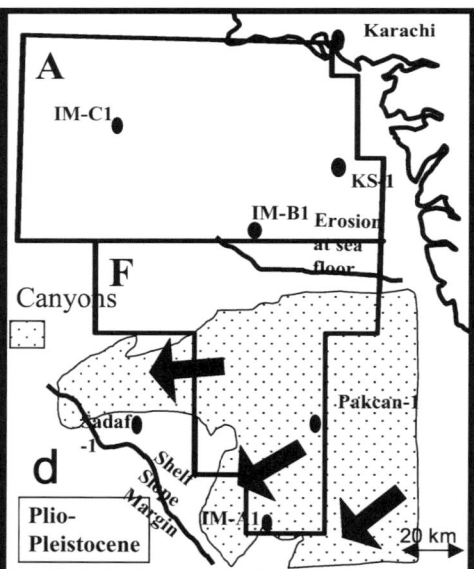

Fig. 16. Distribution of canyon incisions at four time intervals: (**a**) Intra-Early Miocene time; (**b**) end of Early Miocene time; (**c**) Mid–Late Miocene time; (**d**) Plio-Pleistocene time.

of the Kirthar margin with the Afghan plate (Smewing *et al.* 2002). Reliable age dating of canyon-fill sediments and onshore tectonic events is required to expand further the theme of correlating phases of canyon development with tectonic events.

Conclusions

The sedimentological history of the Oligo-Miocene offshore Indus Basin appears to record a broadly 'one-step' basin fill. The basinal section is essentially undrilled, up to 6 km thick, but interpreted to be mostly a passive fine-grained fill facies and mostly of Oligocene to Early Miocene age. Coarser clastic deposits may have reached the basin depocentre via Early Miocene, and possibly Oligocene, canyon systems.

The overlying shelf section comprises stacked sands and shales of Mid- and Late Miocene age, up to 2000 m thick. The transition between basin fill and shelf facies is marked by cyclic parasequence sets in well logs and oblique basinward progrades in seismic data. The overall dominance of monotonous stacking of apparently similar log facies (transition zone excepted) suggests that subsidence and sediment input rates are dominant over any eustatic imprint.

The offshore Indus Basin appears to have formed during Oligocene time, as large faults at the eastern margin with the Karachi Platform do not extend beyond Oligocene reflectors. A substantial fault zone with the northwestern boundary to the Murray Ridge also appears likely, as there is apparently only a much thinned Oligocene section on the current high, whereas a thickness varying from 2 to 4 km can be correlated into the adjacent part of the basin from nearshore well ties. According to these correlations, the majority of the basin fill is of Oligocene and Early Miocene age and does not confirm within this study area the possible presence of Indus Fan sediments dating back to mid-Eocene time (Clift 2002). The basin initially filled from the north in Oligocene time, sediments rapidly prograding southwards. By Mid- and Late Miocene time, progradation was along a wider front and adopted a south-westerly direction with an increasing sediment contribution from the Karachi Platform.

Canyon cut and fill facies form a distinctive feature in the seismic data and separate into Early Miocene and Plio-Pleistocene phases, with maximum downcutting of 400 m and 1000+ m, respectively. The former could be coeval with inversion and uplift in the Kirthar Fold Belt and the latter related to the final collision of the western Pakistan margin with the Afghan plate.

We wish to acknowledge the contributions to the project by our coworkers, principally N. Ahmed, S. Alam, S. Beswetherick and A. Qadri. We also thank DGPC of Pakistan for its support throughout our studies. J. Craig, A. Sharp and P. Clift are thanked for critically reviewing this paper.

References

CLIFT, P.D. 2002. A brief history of the Indus River. In: CLIFT, P.D., KROON, D., GAEDICKE, C. & CRAIG, J. (eds) *The Tectonic and Climatic Evolution of the Arabian Sea Region*. Geological Society, London, Special Publications, **195**, 237–258.

KOLLA, V. & COUMES, F. 1987. Morphology, internal structure, seismic stratigraphy, and sedimentation of the Indus Fan. *AAPG Bulletin*, **71**, 650–677.

KOLLA, V. & COUMES, F. 1990. Extension of structural and tectonic trends from the Indian subcontinent into the eastern Arabian Sea. *Marine and Petroleum Geology*, **7**, 188–196.

McHARGUE, T.R. & WEBB, J.E. 1986. Internal geometry, seismic facies, and petroleum potential of canyons and inner fan channels of the Indus Submarine Fan. *AAPG Bulletin*, **70**, 161–180.

SMEWING, J.D., WARBURTON, J., DALEY, T., COPESTAKE, P. & UL-HAQ, N. 2002. Sequence stratigraphy of the southern Kirthar Fold Belt and Middle Indus Basin, Pakistan. In: CLIFT, P.D., KROON, D., GAEDICKE, C. & CRAIG, J. (eds) *The Tectonoic and Climatic Evolution of the Arabian Sea Region*. Geological Society, London, Special Publications, **195**, 273–300.

Sequence stratigraphy of the southern Kirthar Fold Belt and Middle Indus Basin, Pakistan

JOHN D. SMEWING[1], JOHN WARBURTON[2,5], TIM DALEY[3], PHILIP COPESTAKE[4] & NAZIR UL-HAQ[2]

[1] *Earth Resources Ltd, Innovation Centre, Singleton Park, Swansea SA2 8PP UK*
(e-mail: erl@clara.co.uk)
[2] *LASMO Oil Pakistan Ltd, 5th Floor, The Forum, Khayaban-e-Jami, Clifton, Karachi-75600, Pakistan*
[3] *LASMO Venezuela BV, Avenue Francisco de Miranda, Centre Lido, Torre E, Piso 11, El Rosal, Caracas 1060, Venezuela*
[4] *IHS Energy Group, Enterprise House, Cirencester Road, Ilsom, Tetbury GL8 8RX, UK*
[5] *Present address: ROC Oil Company Ltd, 100 William Street, Sydney, NSW 2011, Australia*

Abstract: The southern Kirthar Fold Belt (KFB) and the contiguous Middle Indus Basin (MIB) constitute a major oil and gas province on the southern Pakistan foreland. In the Middle Indus Basin, gas is reservoired in Early Cretaceous marginal marine sandstones sealed by Paleocene shales. Reservoir quality of the Early Cretaceous sediments deteriorates towards the fold belt, but recent discoveries of gas in Upper Cretaceous sandstones sealed by Paleocene shales have highlighted its potential. To understand better the petroleum systems of this region and provide a potential correlation scheme for comparison with other areas of the country, a sequence stratigraphic interpretation was carried out on the Jurassic to Recent sediments of the KFB and MIB. On the basis of outcrop and well data, 23 depositional sequences have been identified: five in the Jurassic units, 10 in the Cretaceous units and eight in the Tertiary units. Sequence boundaries have been defined according to the Exxon method of identifying unconformities and their correlative conformities. However, equal importance has been given to identifying the potentially more chronostratigraphically significant maximum flooding surfaces between these sequence boundaries so as to define accurately the component systems tracts of each sequence. The depositional systems are described in terms of their relationship to the existing lithostratigraphic framework and interpreted in terms of sedimentary responses to external (eustatic) or local (tectonic) events. Notwithstanding the presence of a eustatic signature on some sequences, the majority appear to be tectonically driven and can be related to plate margin events affecting the NW margin of the Indo-Pakistan Plate during its rift–drift–collision history.

In 1994, LASMO Oil Pakistan Ltd (LOPL) and its partners acquired the Kirthar exploration licence in the eastern part of the southern Kirthar Fold Belt (KFB) in southern Pakistan (Fig. 1). In 1997, the group extended its fold belt exploration licences by acquiring the Kirthar West and Pab Blocks to the west. Together these three exploration licences cover an area of almost 13 000 km² from the external to internal parts of the fold belt. Since 1987, LOPL has also held interests in the Middle Indus Basin (MIB) to the east.

Although the lithostratigraphy of the southern KFB has been rationalized in a number of publications (e.g. Williams 1959; Hunting Survey Corporation 1960; Shah 1977), the same is not the case for the MIB. A disparate lithostratigraphy has emerged for this basin, based upon the use of a plethora of poorly defined terms by the various operating companies working there. As a result, the KFB and MIB were generally seen in isolation when in fact they constitute part of a contiguous depositional system (Fig. 1). For LASMO and partners, with petroleum interests in both areas, the situation was clearly unsatisfactory. One of the primary aims of this study therefore was to develop a scheme that linked the stratigraphy of the KFB and the MIB. Furthermore, it was felt that LOPL and partners' exploration interests would be best served by developing a stratigraphic scheme based on depositional sequences rather than conventional lithostratigraphy. The benefits in adopting this approach were seen as follows: (1) identifying inconsistencies of terminology in the existing lithostratigraphy-based scheme; (2) pro-

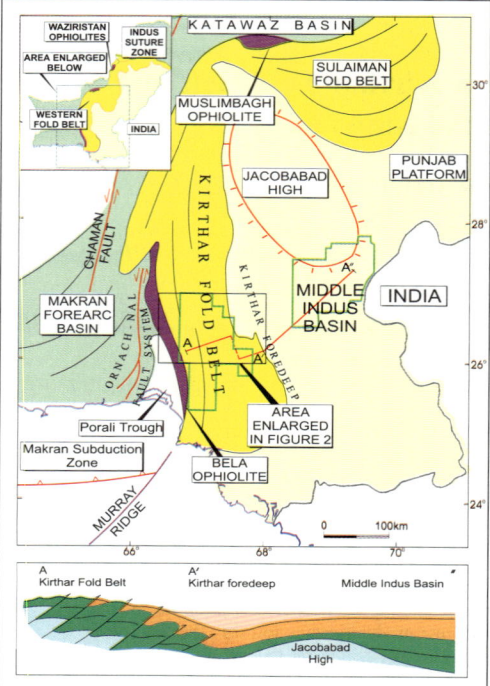

Fig. 1. Location map of Kirthar Fold Belt and Middle Indus Basin with schematic cross-section. LOPL and partners' fold belt and Middle Indus exploration licences are shown in green.

viding a greater understanding of the factors controlling relative sea-level change (tectonics, eustacy, sediment supply); (3) determining the role of these factors in controlling the distribution of source, reservoir and seal facies; (4) providing better constraints on stratigraphic traps; (5) exploiting the benefits of high-resolution biostratigraphy to produce a sound chronostratigraphic framework upon which to base the sequence stratigraphic interpretation; (6) providing a robust scheme for correlation with other parts of the country and other continental plates (compare the recent sequence stratigraphic study by Sharland et al. (2001) for the Arabian Plate).

On the basis of outcrop and well data, we present an interpretation of the stratigraphy of the southern KFB and MIB in terms of depositional sequences (Vail et al. 1977). We describe the key surfaces that define these depositional sequences and the systems tracts within them, and we interpret their development within the geological history of the KFB and MIB.

The database for this interpretation is based upon the following: (1) field analysis of 51 outcrop stratigraphic sections measured through the sediments of the southern KFB; (2) thin-section and biostratigraphic analysis of samples collected from these outcrop sections; (3) regional 1:50 000 scale geological mapping in and around LOPL's KFB exploration licences; (4) wireline log and biostratigraphic data on nine KFB and 20 MIB wells; (5) the results of an earlier study (Copestake et al. 1996, 1997) on the sequence stratigraphy of the Lower Goru and Sembar Formations of the Lower and Middle Indus Basins. Throughout this study, the time scale of Gradstein & Ogg (1996) has been used.

Tectonic setting

The KFB forms the southern part of the Western Fold Belt of Pakistan (Bannert et al. 1992) located adjacent to the present-day strike-slip western margin of the Indo-Pakistan Plate represented by the Ornach–Nal Fault System and the Chaman Fault (Fig. 1). The southern KFB consists of Jurassic to Miocene sediments exposed in a series of doubly plunging north–south to NNW–SSE anticlines and elongate anticlinal ridges, separated by broad synclines. The folds die out eastwards beneath the Kirthar Foredeep, a deep trough infilled with up to 5 km of Plio-Pleistocene molasse (cross-section in Fig. 1). The eastern margin of the foredeep overlies the Jacobabad High, a broad basement uplift underlying the western MIB. The Jacobabad High was in evidence from at least as early as Late Jurassic time; Lower Cretaceous sediments onlap its eastern side. For much of Early Cretaceous time the Jacobabad High was emergent, exposing Jurassic carbonates at the surface and separating basins to the east and west.

The KFB and MIB are located on the NW margin of the Indo-Pakistan Plate. The rift–drift–collision history of this plate is well documented (Patriat & Achache 1984; Besse & Courtillot 1988; Searle et al. 1997). The oldest sediments exposed in the KFB or penetrated by wells in the MIB are limestones with minor clastic deposits of Early to Mid-Jurassic age. These were deposited during a lengthy period of passive subsidence before separation of the Indo-Pakistan Plate from the Afro-Arabian margin in Oxfordian time. During the pre-rift, rift and drift phases of this margin up to 9 km of sediment were deposited in the KFB and MIB.

The drift phase was terminated by initial collision of the northern margin of the plate with Eurasia at 55 Ma (Searle et al. 1997), around the Paleocene–Eocene boundary. The first northerly-derived Himalayan-sourced clastic rocks were deposited in late Mid-Eocene time (40 Ma), and up to 2.5 km of late Paleogene flysch (Kohan Jhal Fm) were then deposited in a narrowing seaway between the Indo-Pakistan Plate and the Afghan Plate (Treloar & Izatt 1993). Initial uplift of the mountain belt dates from Miocene time, when the depocentre of the Himalayan-sourced clastic deposits switched from a position west of the mountain front to the developing Kirthar Foredeep to the east. The main phase of deformation and uplift, however, is of Plio-Pleistocene age and is related to the final collision of the Kirthar margin with the Afghan Plate along the proto-Chaman Fault (Treloar & Izatt 1993). This event marks the first appearance of significant westerly, fold-belt-derived clastic deposits in the Kirthar Foredeep, replacing the hitherto northerly-sourced clastic deposits.

Palaeomagnetic data for igneous units in the KFB (Bela Ophiolite, Cretaceous sills) indicate that they have been rotated anticlockwise by up to 70° since emplacement or intrusion (Hailwood & Ding 1998). A similar amount of clockwise rotation has been found in the western Sulaiman Fold Belt (Klootwijk et al. 1981). The significance of this observation for the KFB is that the present NNW–SSE strike restores to a NE–SW orientation, which is the depositional strike of the pre-collisional margin.

Although the age of rotations is not constrained by the palaeomagnetic data, their magnitude and regional extent would indicate that the most likely timing and mechanism is the Plio-Pleistocene collision of the NW margin of the Indo-Pakistan Plate with the Afghan Block. The senses of rotation would suggest that during collision the crustal block underlying the Jacobabad High became detached from the crustal blocks underlying the Kirthar and Sulaiman Ranges along two NW–SE strike-slip faults. Rotation of the sedimentary cover of the Kirthar and Sulaiman crustal blocks adjacent to these faults was achieved by the retardation of the northwestward movement of these two blocks relative to the Jacobabad crustal block.

The orthogonal orientation of the two proposed strike-slip faults with respect to the restored NE–SW orientation of the Kirthar margin and the evidence that the Jacobabad High originated in Late Jurassic time would suggest that the faults are continental transforms originating during separation of the Indo-Pakistan Plate from Afro-Arabia.

Inversion of earlier extension faults played a role in the uplift of the mountain belt and was also instrumental in structural growth in the MIB. Faults occur in the KFB with contractional structures in their footwalls and hanging walls yet still showing net extension. Stratigraphic evidence suggests that inversion may have commenced as early as Late Paleocene time; that is, from the initial stages of collision of the northern margin of the Indo-Pakistan Plate with Eurasia but significantly earlier than collision of the Kirthar margin with the Afghan Block. The evidence for this comes from the pattern of erosional surfaces as old as Late Paleocene time downcutting progressively towards faults in their hanging walls, indicating structural growth by inversion.

The southern KFB is bound to the west by the Bela Ophiolite (Figs. 1 and 2). This ophiolite was formed in Maastrichtian time at around 70 Ma and obducted onto the continental margin shortly thereafter, with final emplacement taking place in Early Eocene time (c. 50 Ma) (Gnos et al. 1998); that is, spanning the time interval from the terminal phase of northward drift to initial collision with Eurasia.

Within the KFB itself, the western margin of the Kirthar Range marks the line of the shelf-slope break from Maastrichtian time until emergence of the mountain range in Neogene time (Figs. 2 and 3). Abrupt changes in lithofacies take place across this line, particularly evident in Eocene–Oligocene sediments, which form the bulk of the outcrop on either side of the line. This is recognized by referring to the area to the west of the line as the Pab Province and the area to the east as the Kirthar Province, each with separate lithostratigraphic nomenclatures (key to Fig. 2; Fig. 3).

To the west of the KFB is the Porali Trough, an alluvial-covered embayment between the Kirthar and Makran Ranges. On its eastern side, adjacent to the KFB, gravity data indicate that the Bela Ophiolite continues into the subsurface below the alluvial cover and extends southwards into the Arabian Sea (Nayyer & Mallick 1994).

Sequence stratigraphy

Methods

Sequence boundaries in this study were defined according to the Exxon technique (Vail et al. 1977) of utilizing unconformities (or correlative conformities) rather than the Galloway (1989) approach of utilizing maximum flooding surfaces. A number of recent studies have followed the Galloway method (e.g. Copestake et al. 1996, 1997; Sharland et al. 2001), in recognition of the fact that of the two surfaces, the maximum flooding surface is likely to be the more significant chronostratigraphically. However, the danger in this approach is that the importance of the unconformities or correlative conformities between the sequence-defining maximum flooding

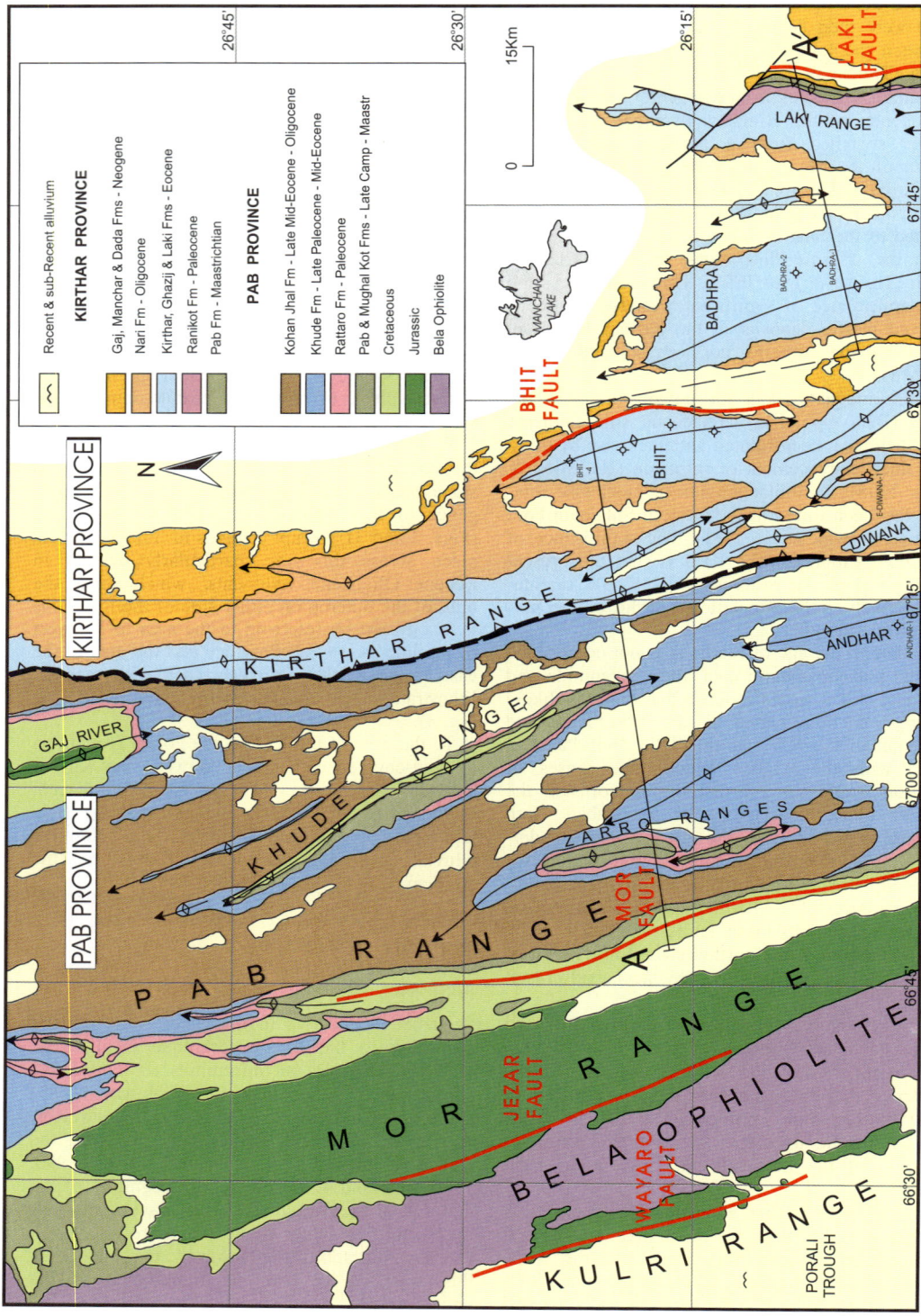

Fig. 2. Geological map of part of southern Kirthar Fold Belt based on 1:50 000 scale regional mapping. A–A' indicates line of section in Fig. 3.

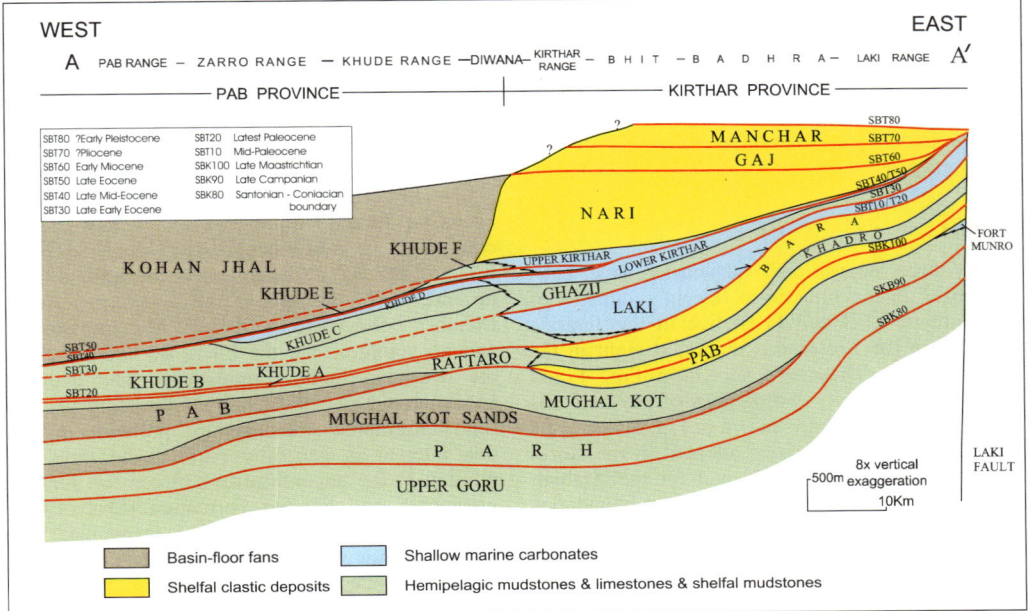

Fig. 3. Pre-Pliocene restoration of southern Kirthar margin with superimposed sequence boundaries interpreted in this study. Line of section is shown in Fig. 2. It should be noted that pre-Parh units have been omitted because of poor well or outcrop control along line of section.

surfaces is downplayed. Consequently, the constituent systems tracts of the depositional sequences become less precisely defined. For this reason, in the present study, emphasis was placed on the accurate definition of both of these key surfaces using the following criteria: (1) sequence boundaries (and correlative conformities): unconformities; junction of progradationally stacked and retrogradationally (or aggradationally) stacked parasequences; the base of basin-floor siliciclastic fans; (2) maximum flooding surfaces: junction of retrogradationally (or aggradationally) stacked and progradationally stacked parasequences; maximum marine onlap; maximum advance of slope and basin bio- and lithofacies onto platform; condensed horizons with development of authigenic minerals such as phosphate and glauconite and commonly showing enrichment in organic matter, characteristics that combine to produce maxima on wireline gamma-ray logs.

The stratigraphic framework for the interpretation herein is based on the identification and correlation of these key surfaces. Initially, this exercise was confined to the KFB, where the results from analysis of the outcrop data were calibrated against wireline log data and detailed biostratigraphic results from three of LOPL's recent exploration wells, Andhar-1, Badhra-2 and Bhit-3 (Fig. 2).

From this exercise a robust sequence stratigraphic interpretation emerged for the southern KFB, although sequence definition in the Lower Cretaceous units was hampered by the dominantly hemipelagic facies of this time interval and their consequent insensitivity to relative sea-level change. In this regard, the study of Copestake *et al.* (1996, 1997) on the Lower Goru and Sembar Formations of the MIB proved to be very beneficial to the overall scheme. In the MIB the Lower Goru and Sembar are developed in marginal marine facies with readily definable depositional sequences. Consequently, the Lower Cretaceous sequences of the overall scheme were primarily defined on the MIB data of Copestake *et al.* (1996, 1997).

One problem, however, that had to be overcome to incorporate this scheme into the present study was that the sequences of Copestake *et al.* (1996, 1997) were defined according to the Galloway (1989) method of using maximum flooding surfaces as sequence boundaries rather than the Exxon approach (Vail *et al.* 1977) of using unconformities or their correlative conformities as adopted in this study. Accordingly, Exxon-type sequence boundaries were interpreted from maximum regression surfaces in the previous scheme to rationalize this interpretation with the current scheme.

Sufficient data were available in the 20 selected MIB wells to allow the key surfaces identified in the Tertiary and Upper Cretaceous sediments of the KFB to be extended eastwards with confidence. Unfortunately, this is not the case in the Jurassic part of the section, as a result of insufficient well penetrations of the deeper sediments.

Results

Twenty-three depositional sequences have been recognized in the Jurassic to Recent sediments of the southern KFB and MIB: five in the Jurassic units, 10 in the Cretaceous units and eight in the Tertiary units. These are shown in Fig. 4 and described in Table 1. Photographs of key outcrops in the southern KFB are shown in Fig. 5.

In common with practice adopted in the North Sea (Partington et al. 1993) and for the Arabian Plate (Sharland et al. 2001), sequence boundaries are assigned the key letters J (Jurassic), K (Cretaceous) and T (Tertiary). They are numbered in decimal units, (J10, J20, J30, etc.) to allow scope for modification by inserting further key surfaces between those identified herein. By adopting this scheme, however, no correspondence is implied between similarly labelled sequences in the three areas (i.e. Sequence J10 herein does not correlate with Sequence J10 of Partington et al. (1993) or Sharland et al. (2001)).

Depositional sequences and the maximum flooding surfaces within them take the same identifier as the basal sequence boundary. For the two depositional sequences that overlap the J–K and K–T stage boundaries, this results in the MFS, and, for that matter, the bulk of the depositional sequence itself, falling in the younger period despite being named from the older period.

The use of the identifier 'SB' does not distinguish between an unconformable sequence boundary and its correlative conformity.

No attempt has been made to 'order' the sequences in the sense of Vail et al. (1977). Neither has the stratigraphic interval under consideration been subdivided into tectonic megasequences (Hubbard 1988) by major unconformities, although clearly some key surfaces, e.g. SBJ50 and SBK100 are candidates. Although the Haq et al. (1988) eustatic curve is shown against the stratigraphic column in Fig. 4, this is for reference only and no detailed comparisons between the key surfaces in this study and the curve are drawn.

The lithostratigraphic terminology used in this study is that which is in current use in the KFB and MIB and follows Williams (1959), Hunting Survey Corporation (1960) and Shah (1977) with a number of modifications listed (1)– (6) below, based on recent work by the authors. These modifications involve the introduction of new and amended lithostratigraphic terms. As these are yet to be accepted by the Stratigraphic Commission of Pakistan they should be considered as informal proposals at this stage. Full descriptions of the type sections for these proposed new and amended terms are given in the Appendix.

(1) Thin red beds and fine turbiditic sandstones occur on Jurassic limestones and below Cretaceous Sembar mudstones in the Mor and Kulri Ranges. The age of these beds is constrained to be no older than the age of the underlying Anjira Member, believed to be of latest Mid-Jurassic age and no younger than the Early Berriasian age of the overlying Sembar mudstones. The deduced Late Jurassic age of these beds therefore poses a lithostratigraphic nomenclature problem in that this interval is generally part of the time gap missing at the unconformity between the Jurassic limestones and Sembar Formation (Shah 1977). Copestake et al. (1996, 1997) recognized thin Upper Jurassic sediments on the Chiltan platform in parts of the Lower and Middle Indus Basins. We follow those workers by referring the sediments of the Mor and Kulri Ranges to the Sembar Formation and thereby extending the age of the base of the Sembar Formation to Oxfordian time.

(2) In common with practice in the MIB, the Goru Formation of the KFB is split into upper and lower members. In the KFB, a dark grey organic-rich shale of Late Cenomanian–Early Turonian age occurs at the base of the upper member.

(3) Most of the sandstones in the Pab Range are reassigned from the Pab Formation to the Mughal Kot Formation on the basis that they range to an earliest age of Late Campanian time, rest with abrupt but conformable contact upon the Parh Formation and are separated by Mughal Kot mudstones from overlying Upper Maastrichtian sandstones, age equivalent to the Pab Formation of the Kirthar Province.

(4) The Kirthar Formation is split into upper and lower members based on the presence of an unconformity or suite of unconformities representing a time gap of Late Eocene to earliest Oligocene time. The lower member is further split into a Lutetian Lower Kirthar 1 Unit and a Bartonian Lower Kirthar 2 Unit, based on the presence of a major flooding surface at the Bartonian–Lutetian boundary and the development of pelagic limestones upon this flooding surface (at the base of Lower Kirthar 2) that are correlatable with the Khude E Member of the western KFB (see below). Lower Kirthar 1 and 2 Units are progressively cut out eastwards across the eastern part of the KFB as the intra-Kirthar unconformity (or suite of unconformities) downcuts from west to east. As a result, in the Laki Range, the Lower Kirthar Member is

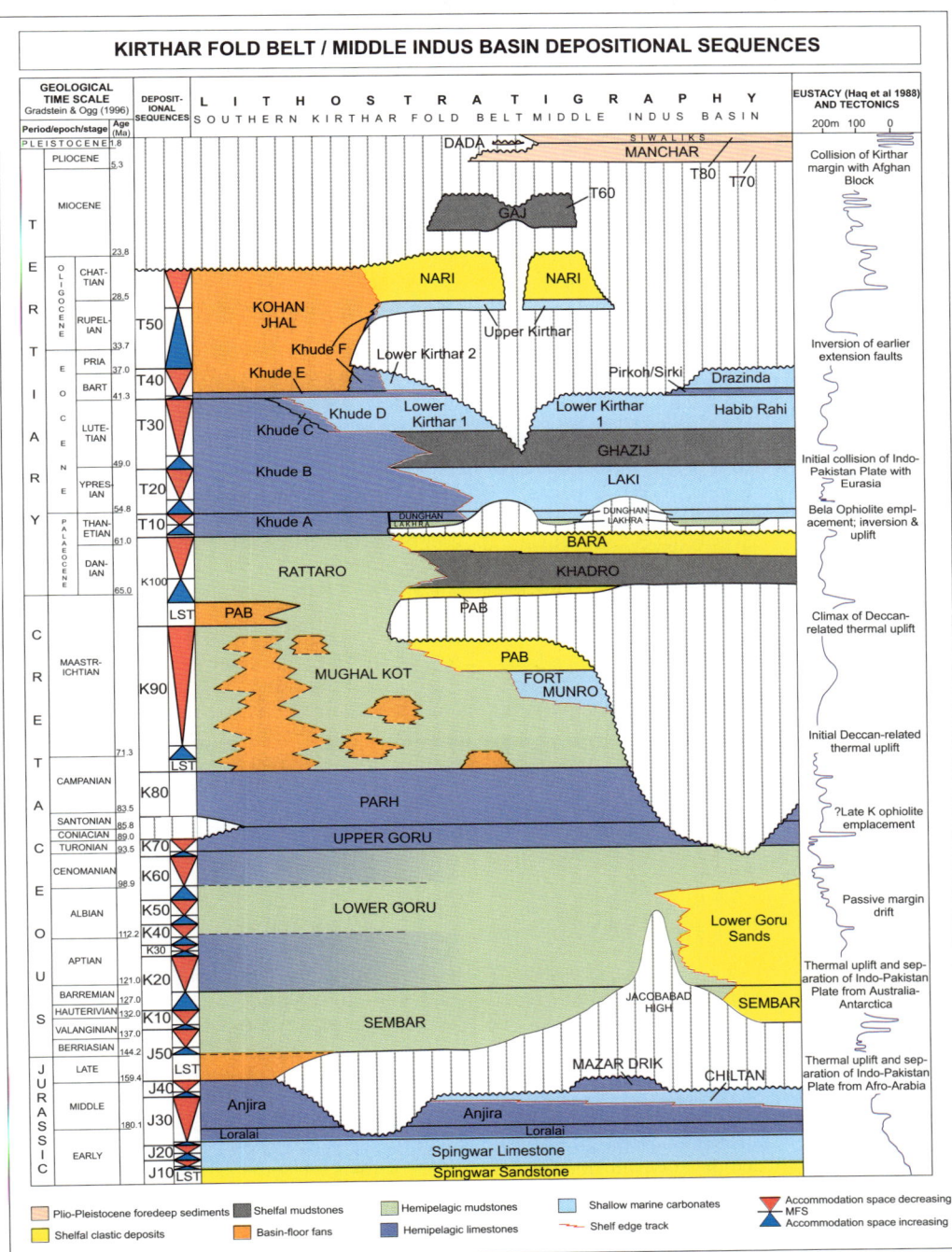

Fig. 4. Depositional sequences of southern Kirthar Fold Belt and Middle Indus Basin. The eustatic curve shown on the right is rescaled from Haq *et al.* (1988) and is for reference only.

Table 1. *Depositional sequences of southern Kirthar Fold Belt and Middle Indus Basin*

Sequence	Age	Lithostratigraphic equivalence	Regional control and thickness	SB	LST	TST	MFS	HST
T80	?Early Pleistocene	Dada Fm–Siwaliks	Limited in outcrop to eastern KFB (>400 m, top not seen)	Regional unconformity or set of unconformities with angular discordance	Not resolved	Not resolved	Not resolved	Not resolved
T70	?Pliocene	Manchar Fm–Siwaliks	Limited in outcrop to eastern KFB (up to 2500 m) but thickens eastwards into foredeep	Regional unconformity that downcuts to Sequence T30 in hanging wall of Laki Fault	Not resolved	Not resolved	Not resolved	Not resolved
T60	Miocene	Gaj	Limited in outcrop to eastern KFB (up to 1800 m); removed by erosion at SBT70 (base Siwaliks) in MIB	Regional unconformity particularly clear in Laki Range where it downcuts from Nari on western flank to Ghazij on eastern flank	Not developed	Upward increasing mudstone content in basal Gaj cycles on east flank of Bhit marks initiation of a TST above SBT60	Upper part of T60 removed by erosion at SBT70	Not resolved
T50	Late Eocene–Oligocene	Kohan Jhal Fm and Khude F Mbr in Pab Province; Nari and Upper Kirthar Fms in Kirthar Province	Widely exposed in KFB (up to 1500 m in Kirthar Province and 2000 m in Pab Province); removed by erosion at SBT70 (base Siwaliks) in MIB	Major unconformity in eastern KFB which downcuts progressively eastwards from Lower Kirthar 2 in Kirthar Range through Lower Kirthar 1 in Bhit and Badhra to top of Ghazij in Laki Range; in Pab Province, the correlatable conformity lies within Kohan Jhal–Khude F	Not developed	Basal Upper Kirthar in Kirthar Province; not recognizable in Pab Province	Of Mid-Oligocene age in lower part of Upper Kirthar in Kirthar Province; indeterminate in Pab Province	Upper part of Upper Kirthar in Kirthar Province; indeterminate in Pab Province
T40	Late Mid-Eocene (Bartonian)–Late Eocene	Khude E and F Mbrs, and Kohan Jhal Fm in Pab Province; Lower Kirthar 2 in Kirthar Province and Drazinda, Pir Koh and Sirki Mbrs of Kirthar Fm in MIB	Widely exposed in KFB (0–200 m in Kirthar Province, up to 500 m in Pab Province) and present throughout MIB although top Drazinda truncated by erosion at SBT70 (base Siwaliks) in MIB	Coincides with a major flooding surface, which led to pelagic limestones of Khude E advancing onto platform at least as far east as Kirthar Range; equivalent to base Pirkoh–Sirki in MIB	Not developed	2–3 m of rapidly deepening-upward limestones at base of Khude E Mbr and equivalent horizon at base of Lower Kirthar 2	Within planktonic foram calcisiltites and lime mudstones 2–3 m above base of Khude E or equivalent horizon at base of Lower Kirthar 2	Pelagic limestones pass upwards into shallow marine limestones of Lower Kirthar 2, which pass basinwards of into Khude F slope apron downlapping onto Khude E
T30	Latest Early Eocene–Early Mid-Eocene (Lutetian)	Khude B, C and D Mbrs in Pab Province; Ghazij and Lower Kirthar 1 Fms in Kirthar Province; Ghazij Fm and Habib Rahi Mbr of Kirthar Fm in MIB	Widely exposed in KFB (200–300 m in Kirthar Province, 400–600 m in Pab Province) and present throughout MIB (c. 500 m)	Conformable boundary throughout KFB and by inference in MIB; equivalent to Laki–Ghazij contact in Kirthar Province; within Khude B in Pab Province	Not developed	Retrogradationally stacked mudstone–limestone parasequences in lower part of Ghazij Fm	Junction of retrogradationally and progradationally stacked parasequences in lower Ghazij	Regressive mudstone-dominated parasequences passing up into limestone-dominated parasequences which contain the Ghazij–Lower Kirthar contact; HST culminates in progradation of Khude D platform out into Pab Province
T20	Latest Paleocene–Early Eocene	Khude B Mbr in Pab Province, Laki Fm in Kirthar Province	Exposed in Laki Range and west of Kirthar Range (200–600 m) and developed throughout MIB (c. 400 m)	Top Khude A–top Dunghan	Not resolved	Not resolved	Not resolved	Not resolved
T10	Late Paleocene	Khude A in the basin; Lakhra and Dunghan on the platform	Khude A is widely exposed in Pab Province (40–70 m); Lakhra and Dunghan exposed in KFB north and south of study area and widely developed in MIB (up to 500 m)	Sharp conformable contact between Khude A and Rattaro in Pab Province passing eastwards into short duration unconformity in Kirthar Province and MIB, except over Laki and Jacobabad Highs, where flooding was delayed until Eocene time	Red mudstones at base of Khude A from reworking of palaeosols on exposed platform	Lower part of Lakhra claystones	Within Lakhra claystones and green–grey shales in Khude A	Upper part of Lakhra claystones and whole of Dunghan limestones

Sequence	Age	Formation	Distribution	Lower boundary	Lowstand	Transgressive systems tract	Maximum flooding surface	Highstand
K100	Late Maastrichtian–Mid-Paleocene	Pab, Mughal Kot and Rattaro Fms in the basin; upper part of the Pab and Khadro and Bara Fms on platform to east	Widely exposed on the eastern flank of the Pab Range, in the Khude and Zarro Ranges (up to 800 m) and in the Laki Range and Gaj River area (up to 1000 m); also developed throughout MIB	Unconformity above lower Pab in Laki Range, passing westwards into a correlatable conformity above Mughal Kot in Pab Province	Up to 400 m of gravelly Pab sandstone turbidites in central Pab Range and Zarro and Khude Ranges	Retrogradational deltaic Pab sands in Kirthar Province terminated at a major transgressive surface at K–T boundary within TST and overlain by deepening-upward Khadro shales	*Cordita beaumonti* horizon in Khadro Fm of Laki Range	Progradation of Bara deltas out over Khadro shales
K90	Late Campanian–Late Maastrichtian	Pab, Mughal Kot and Fort Munro Fms	Widely exposed in Pab, Zarro and Khude Ranges in Pab Province (up to 600 m) and in Laki Range in Kirthar Province (>200 m, base not seen); cut out by erosion at SBK100 (base Khadro unconformity) in MIB	Sharp conformable contact between Parh Fm below and either turbiditic sand or mudstone facies of Mughal Kot Fm above	Falling stage basin floor fans downlapped by rising stage mudstones	Continuation of mud-dominated sediments	Within mud-dominated sediments	Progradation of Fort Munro carbonate platform and lower Pab fluvio-deltaic deposits in Laki Range; shallowing-upward mudstones at top of Mughal Kot Fm in Pab Province
K80	Santonian–Late Campanian	Parh Fm	Crops out in Gaj River, western flank of Pab Range and cores of Khude and Zarro Ranges (50–400 m); generally removed by erosion below SBK100 (base Khadro unconformity) in the MIB	Unconformity or submarine onlap surface in western KFB; passes eastward into correlative conformity	Indeterminate: a constant vertical profile of clean pelagic lime mudstones is maintained throughout this sequence			
K70	Late Cenomanian–Coniacian	Upper Goru Fm and topmost Lower Goru Fm	Exposed in western foothills of southern Pab Range and in Gaj River (50–80 m); widespread in subsurface of MIB although partly removed by erosion at SBK100 (base Khadro unconformity)	A sharp conformable contact within uppermost Lower Goru between thickening- and cleaning-upward limestone-dominated cycles at top of K60 and dark grey shales with thin limestones at base of K70	Not developed	Only 0.5 m thick in Gaj River: rapidly shaling-upward thin limestones and shales	Within a 1 m thick black shale above the 0.5 m thick TST	Rapidly thickening-upward limestone–mudstone cycles in which limestone content increases progressively upwards
K60	Late Mid-Albian–Late Cenomanian	Lower Goru Fm (part)	Present in MIB subsurface (0–400 m); sediments of this age are present in KFB but sequence boundaries are not evident	Maximum basinward progradation of sand package at top of Sequence K50	Not developed	Poorly defined	Flooding surface at Albian–Cenomanian boundary	Poorly defined; marls in MIB
K50	Early Albian–Late Mid-Albian	Lower Goru Fm (part)	Present in MIB subsurface (0–200 m); sediments of this age are present in KFB but sequence boundaries are not evident	Poorly defined in MIB; extrapolated from LIB	Not developed	Poorly defined in MIB; extrapolated from LIB	Major flooding surface; all previously emergent areas between KFB and MIB flooded at this time (see Fig. 4)	Progradational shelf sands in MIB
K40	Late Aptian–Early Albian	Lower Goru Fm (part)	Present in MIB subsurface (0–200 m); sediments of this age are present in KFB but sequence boundaries are not evident	Maximum basinward progradation of Lower Goru sands at top of Sequence K30	Not developed	Poorly defined in MIB; extrapolated from LIB	Major flooding surface in MIB at Aptian–Albian boundary	Poorly defined in MIB; extrapolated from LIB
K30	Late Early to Early Late Aptian	Lower Goru Fm (part)	Present in MIB subsurface (0–300 m); sediments of this age are present in KFB but sequence boundaries are not evident	Maximum basinward progradation of Lower Goru sands at top of Sequence K20	Not developed	Retrogradational sand package in MIB	Major flooding surface in MIB	Progradational sand package in MIB

Table 1. (continued)

Sequence	Age	Lithostratigraphic equivalence	Regional control and thickness	SB	LST	TST	MFS	HST
K20	Late Hauterivian–Late Early Aptian	Top of Sembar Fm, base of Lower Goru Fm; Sembar–Goru contact drawn at MFSK20	Present in MIB sub-surface (0–500 m); sediments of this age are present in KFB but sequence boundaries are not evident	Maximum basinward progradation of Sembar sands at top of Sequence K10	Not developed	Retrogradational sand package in MIB	Maximum advance of basinal mudstones onto platform in mid Barremian (equivalent to top Sembar)	Maximum westward advance of offlapping fluvio-deltaic wedges over shelf sediments in MIB
K10	Early Valanginian–Late Hauterivian	Sembar Fm (part)	Present in MIB sub-surface (0–400 m); sediments of this age are present in KFB but sequence boundaries are not evident	Maximum basinward progradation of Sembar sands in MIB at top of Sequence J50	Not developed	Retrogradational sand package in MIB	Maximum advance of basinal mudstones onto platform in early Late Valanginian	Progradation of fluvio-deltaic sandstones westwards, over shelf sediments in MIB
J50	Oxfordian–Early Valanginian	Lower part of Sembar Fm	Top not defined in KFB; absent in MIB	Main break-up unconformity represented on platform by exposure surface at top of Chiltan platform, or the Mazar Drik where preserved; in the Gaj River, it is an unconformity between the Loralai and Sembar; in the basin to the west, the unconformity passes into a correlatable conformity at the top of the Anjira below the red beds at the base of the Sembar	Up to 110 m of fine sandstone turbidites overlying red beds at the base of the Sembar east and west of Mor Range	Thin, rapidly deepening-upward limestones and mudstones at base of Sembar on platform	Glauconite–marcasite horizon in Gaj River, 3 m above base of Sembar	Silty calcareous radiolarian mudstones in metre-scale cleaning-upward cycles
J40	Late Bathonian–Early Callovian	Mazar Drik Fm	Southern Sulaiman Range–northern MIB (0–30 m)	Drowning unconformity at top of Chiltan highstand carbonates	Not resolved	Not resolved	Not resolved	Not resolved
J30	Toarcian–Bathonian	Chiltan Fm and Anjira, Loralai and uppermost Spingwar Limestone Mbr of Shirinab Fm	Exposed in Gaj River (140 m), Mor and Kulri Ranges (up to 500 m); undifferentiated in MIB	3 m below top of Spingwar Limestone Mbr in Gaj River; turn-round point between thickening-upward and thinning-upward limestones	Not developed	3 m of thinning-upward limestones at top of Spingwar Limestone Mbr in Gaj River	Probably coincides with a horizon of marcasite nodules in dark limestones and shales at base of Loralai in Gaj River	Loralai and Anjira; Chiltan Fm presumed to prograde over Anjira at top of HST
J20	Early Jurassic	Middle and upper parts (but not uppermost) of Spingwar Limestone Mbr of Shirinab Fm	Exposed in Gaj River, Mor and Kulri Ranges (2000 m); unproven in MIB	A prominent iron-rich tan–brown omission surface at the top of the limestone-dominated cycles of the lower Spingwar Limestone in Kharrari Nai, Mor Range	Single bed of lithoclastic packstone draping SBJ20	Mid-ramp shale–limestone cycles of middle part of Spingwar Limestone Mbr	Within shale–limestone cycles of middle part of Spingwar Limestone Mbr	Shale–limestone cycles of middle part of Spingwar Limestone Mbr becoming more limestone-rich in upper part
J10	Early Jurassic	Spingwar Sandstone Mbr and lower part of Spingwar Limestone Mbr of Shirinab Fm	Exposed in Mor and Kulri Ranges, (>400 m, base not seen); unproven in MIB	Not exposed	>200 m of fluvio-deltaic sandstones (Spingwar Sandstone Mbr)	Lower part of mudstones at base of Spingwar Limestone Mbr	Within mudstones at base of Spingwar Limestone Mbr	Mudstones passing up into 4–5 m thick limestone cycles with subordinate shale in lower part of Spingwar Limestone Mbr

SB, sequence boundary or correlative conformity; LST, lowstand systems tract; TST, transgressive systems tract; MFS, maximum flooding surface; HST, highstand systems tract.

Fig. 5. Key outcrops of the southern Kirthar Fold Belt. (a) Top of carbonate debris flow in Anjira Member of Shirinab Formation in Wayaro Dhora at 26°00.5′N, 66°34.5′E. Individual matrix-supported flow units between 1 and 3 m thick can be picked out with grey subrounded platform limestone blocks in a yellowish sheared mudstone matrix. Overlain by grey bedded pelagic limestones. (b) SBJ50 (arrowed) in Tungin Jhal in the core of the Gaj Anticline at 26°55.9′N, 67°01.7′E. The unconformity is between Berriasian Sembar and Early Jurassic Loralai units and is marked by a few centimetres of erosional relief abruptly overlain by a thin transgressive sequence of dark ferruginous mudstone passing up into a tan–brown stained rapidly fining-upward packstone, 25 cm thick. The remainder of the Sembar sequence comprises recessive mudstones, poorly exposed in the hills behind. (c) SBK90 (arrowed) on west face of central Pab Range, showing 400 m of grey Parh pelagic lime mudstones abruptly but conformably overlain by brown-weathering gravelly sandstone turbidites of the Mughal Kot basin-floor fan. Pelagic limestones are interbedded with the turbidites and confirm their Campanian age. (d) Part of progradational HSTK90 in Bara Nala in the Laki Range at 26°07.0′N, 67°53.6′E showing white, shallow marine Fort Munro limestones at the base overlain by a transitional yellow–brown weathering sandy limestone facies and finally a thickening- and coarsening-upward succession of marginal marine to fluvial channel Pab sandstones. (e) Red mudstones of LSTT10 at the base of the Khude A Member abruptly but conformably overlying brownish grey Rattaro mudstones in Khatto Dhora at 25°44.9′N, 67°10.5′E. The red mudstones are believed to be a basinally redeposited palaeosol derived from exposed shelfal areas to the east. (Note pale yellow pelagic limestones (Dunghan equivalent) at the top of Khude A below the gravel terrace.) Height of cliff 100 m.

cut out completely and the Upper Kirthar Member rests directly on the Ghazij Formation.

(5) The term 'Khude Formation', originally defined by Hunting Survey Corporation (1960) is retained for the slope to basinal facies carbonate-dominated successions of Late Paleocene to Eocene age of the Pab Province. However, it has been found practical to split the formation into six members, A–F, and distinguished as follows:

Khude A: red calcareous mudstones at base passing up into greyish green shales and yellowish grey planktonic foram lime mudstones and graded foram grainstone turbidites; 40–70 m thick.

Khude B: calcareous and non-calcareous mudstones, planktonic foram lime mudstones, foram grainstone turbidites, massive, erosive-based limestone conglomerates and minor sandstone; up to 540 m thick.

Khude C: grey, weathering greenish grey calcareous mudstones with an increasing content of muddy limestone beds in the upper part; 100–300 m thick.

Khude D: limestones, argillaceous limestones and mudstones–marls arranged in metre-scale cycles, which overall coarsen, thicken and clean upwards, culminating in a massive, cliff-forming coarse biopackstone–wackestone with large algal oncoids and algal-encrusted foraminifera; up to 200 m thick.

Khude E: a highly distinctive and mappable horizon of grey, weathering light lemon yellow-grey, thin- to medium-bedded planktonic foram lime mudstones; 10–40 m thick.

Khude F: yellowish brown medium-bedded argillaceous lime mudstones passing up into grey peloidal biopackstones and pinkish calcareous shales. Up to 170 m thick.

(6) A new term, the Kohan Jhal Formation, is introduced to describe turbiditic sandstones with interbedded mudstones and limestone conglomerates of late Mid-Eocene to Oligocene age in the Pab Province previously included in the Nari Formation (Hunting Survey Corporation 1960).

The track of the shelf-slope break is shown in Fig. 4. With the exception of the fluvial Spingwar Sandstone and the Chiltan limestones in Mid-Jurassic time, the first appearance of shelfal sediments in the KFB was not until Late Maastrichtian time, when the Fort Munro shelf developed in the east. This shelf was overridden by Pab fluvio-deltaic deposits, which prograded as far west as the present-day Kirthar Range, establishing a shelf-slope break on its western margin. The shelf-slope break remained here until Neogene emergence of the mountain belt, apart from a short basinward excursion in Mid-Eocene time (Khude D) followed by a rapid backstep in response to a major relative sea-level rise in late Mid-Eocene time (Khude E).

Basinward of the shelf-slope break, sediments of the KFB comprise hemipelagic limestones and mudstones punctuated by siliciclastic basin-floor fans. Four major basin-floor fan systems can be recognized, one each in the Sembar, Mughal Kot, Pab and Kohan Jhal Formations. The Sembar and Pab fans were deposited during lowstands on the adjacent shelf. The Mughal Kot and Kohan Jhal fans, however, were deposited mainly during transgressive and highstand systems tracts. For these two systems the adjacent shelf was not the source of the clastic deposits. Instead, they were both derived from considerable distances along depositional strike, from the uprising Himalayas for the Kohan Jhal fan and from uplifted regions of the Indian Shield to the south for the Mughal Kot fan. The decoupling of these axially flowing turbidites from the transgressive–regressive behaviour of the shelf shows that basinal and shelfal processes were effectively independent at these times. This is further emphasized by the presence of penecontemporaneous shallow marine carbonates on the adjacent shelf during deposition of the Kohan Jhal fan. The Kirthar margin cannot simply be modelled in terms of 2D dip profiles. With major packages of sediment entering the system from along strike, the third dimension assumes importance, and changes in relative sea-level have to be considered with this in mind.

Discussion

The 23 depositional systems and their component systems tracts are sedimentary responses to external (eustatic) or local (tectonic) factors affecting the NW margin of the Indo-Pakistan Plate during its rift–drift–collision history. A tectonic signature can be recognized on more than half of the key surfaces defining these depositional sequences that can be resolved from the data (Table 2). It would be wrong, however, to assume the remainder are eustatic by default; some may be related to tectonic events occurring elsewhere on the plate with only the resultant change in relative sea level being manifest in the KFB and MIB. Nevertheless, a purely eustatic signature is suggested for some surfaces where correlation with other continental plates seems likely. One such example is the Turonian–Cenomanian maximum flooding surface of depositional sequence K70. This would appear to correlate with MFSK140 of Sharland *et al.* (2001) on the Arabian Plate. Indeed, on both plates, similar organic-rich lithofacies are developed on the maximum flooding surface, reinforcing the correlation.

However, to return to the tectonic theme, it is perhaps not surprising in view of the location of the MIB and KFB on the NW margin of the Indo-

Table 2. *Summary of interpreted ages, dates and tectonic signatures on key surfaces in the southern Kirthar Fold Belt and Middle Indus Basin*

SB	MFS	Biostratigraphic age	Age (Ma)	Tectonic signatures
	T80	Not resolved	–	–
T80		?Early Pleistocene	?1	Inversion and uplift
	T70	Not resolved	–	–
T70		?Pliocene	?5	Pre-molasse uplift
	T60	Not resolved	–	–
T60		Early Miocene	24	Inversion and uplift
	T50	Mid-Oligocene	30	Post-inversion subsidence
T50		Late Eocene	37	Inversion and uplift
	T40	Late Mid-Eocene	41	–
T40		Late Mid-Eocene	41	–
	T30	Early Mid-Eocene	48	–
T30		Late Early Eocene	50	–
	T20	Early Early Eocene	53	Post-inversion subsidence
T20		Latest Paleocene	55	Pause in subsidence
	T10	Late Paleocene	57	Bela Ophiolite: decay of peripheral bulge
T10		Mid-Paleocene	60	Bela Ophiolite: inversion and uplift
	K100	Early Paleocene	64	Thermal decay
K100		Late Maastrichtian	66	Deccan-related uplift and erosion
	K90	Early Maastrichtian	70	–
K90		Late Campanian	75	Early Deccan uplift
	K80	Not resolved	–	–
K80		Santonian–Coniacian boundary	86	?Late K ophiolite emplacement
	K70	Turonian–Cenomanian boundary	93	–
K70		Late Cenomanian	94	–
	K60	Albian–Cenomanian boundary	99	–
K60		Late Mid-Albian	103	–
	K50	Early Albian	107	Syndepositional faulting
K50		Early Albian	109	–
	K40	Albian–Aptian boundary	112	Syndepositional faulting
K40		Early Late Aptian	113	–
	K30	Early Late Aptian	114	Syndepositional faulting
K30		Late Early Aptian	115	–
	K20	Mid-Barremian	124	Syndepositional faulting
K20		Late Hauterivian	128	–
	K10	Late Valanginian	134	Syndepositional faulting
K10		Early Valanginian	135	–
	J50	Early Berriasian	142	Plate margin subsidence
J50		Early Oxfordian	159	Break-up unconformity
	J40	Late Bathonian	165	–
J40		Late Bathonian	166	–
	J30	Early Jurassic	?190	Southward-propagating Tethyan rift
J30		Early Jurassic	?191	–
	J20	Early Jurassic	?195	–
J20		Early Jurassic	?199	–
	J10	Early Jurassic	?202	Southward-propagating Tethyan rift
J10		Not exposed or penetrated	–	–

Based on time scale of Gradstein & Ogg (1996). SB, sequence boundary–correlative conformity; MFS, maximum flooding surface.

Pakistan Plate that plate margin tectonic events have played such a major role in sculpting the sequence architecture of these basins. In the following discussion we interpret sequence development in these basins in the context of the rift–drift–collision history of this margin. Key tectonic events are summarized against the stratigraphic column in Fig. 4.

Rift (Early to Late Jurassic time)

The pre-rift Early and Mid-Jurassic sediments of the KFB show a progressive drowning from fluvial Spingwar sandstones, through inner shelf carbonate-dominated then mid–outer shelf mud-dominated Spingwar limestone cycles to the pelagic limestones of the Loralai and Anjira Members.

This trend is reversed only in the highstand at the top of depositional sequence J30, when the Chiltan carbonate platform prograded rapidly westwards over the Anjira Member. This was probably due to a reduction in the rate of generation of accommodation space in response to thermal uplift preceding continental separation.

The progressive deepening is related to the early stages of rifting of India–Madagascar from Afro-Arabia and the opening of the Somali and Proto-Owen Basins. The oldest ocean crust related to this spreading event in the Somali Basin is of Oxfordian age (Rabinowitz et al. 1983), and of Tithonian age in the Proto-Owen Basin (Loosveld et al. 1996). However, the effects are seen on the margins of these basins much earlier. In Madagascar, continental Karroo sediments of Early Jurassic age are terminated by a regional Early Toarcian marine transgression (Besairie 1972). In the KFB, this flooding event is recognized at the base of the Loralai ammonitic limestones and shales. However, in the KFB these ammonitic beds overlie marine limestones, indicating that an earlier transgression had already introduced marine conditions to the KFB. Marine sediments as old as Permian time occur in the Quetta area (Williams 1959). The earlier transgression occurred during Permo-Triassic rifting of the Afghan Block from the NW margin of the Indo-Pakistan Plate and the opening of the Neo-Tethys Ocean (Sharland et al. 2001).

The progressive drowning of the Jurassic sediments was accomplished by passive subsidence until Bajocian time, when extension commenced on NE–SW-oriented normal faults, downthrowing to the NW.

The evidence for Mid- to Late Jurassic extension is best preserved in the Kulri and Mor Ranges in the hanging walls of the Wayaro and Mor Faults (Fig. 2). In the Kulri Range, debris flows occur within a shaly section of 15 m thickness within the pelagic limestones of the Anjira Member and comprise individual flow units 1–3 m thick carrying dark grey shelly limestone blocks of platformal character up to 1 m across in a yellow calcareous mudstone matrix (Fig. 5a).

In addition to the debris flows, huge tabular slide blocks of interbedded limestone and shale, up to 2 km across, occur near the top of the Anjira Member, lying concordant with the surrounding bedding.

Although the constituents of the debris flows and slide blocks show a superficial resemblance to the Spingwar Limestone Member, the presence of shells floating in the matrix of the debris flows identical to those within the blocks themselves indicates that they were sourced by semi-lithified carbonates and that they are therefore unlikely to be derived from the older, deeper and presumably lithified Spingwar Limestone of pre-Toarcian age. This suggests that they were sourced from a penecontemporaneous carbonate platform up-dip to the SE.

This carbonate platform would have been the Bajocian–Bathonian Chiltan Formation, which occurs in the MIB and crops out around Quetta. It does not occur in the southern KFB, although its former presence in the Gaj River area cannot be ruled out, as here all post-Toarcian sediments have been removed at the unconformable SBJ50 below Neocomian Sembar (Fig. 5b). The Mor Fault may have marked the basinward limit of Chiltan progradation. The break-up of the platform at the Mor Fault seems the most likely source of the debris flows and slide blocks in the Kulri and Mor Ranges (Fig. 6a).

Three further examples of NE–SW-oriented normal faults, downthrowing to the NW, have been recognized in the KFB: the Jezar, Bhit and Laki Faults (Figs. 2 and 6a). The evidence for their extensional origins is based on Tertiary inversion structures associated with them. The Jezar Fault is still in net extension despite the presence of contractional structures in both its footwall and hanging wall. Depositional cycles in the Nari Formation in the hanging wall of the Bhit Fault thin along the fault towards a central portion where subsidence appears to have been lower than the regional norm. This is interpreted to be due to maximum inversion of that part of the fault, which had the highest amount of extensional displacement. For the Laki Fault, a series of erosional surfaces dating from Late Paleocene time downcut progressively through the hanging wall towards the fault, indicative of inversion (Fig. 3).

None of this evidence, however, dates the age of the original extension on these faults. For the Bhit and Laki Faults and any other potential candidates in the eastern KFB, the lack of pre-Late Cretaceous exposures precludes an analysis of their early history from outcrop. Clearly, the Jurassic section of the Kulri and Mor Ranges shows evidence of synsedimentary extensional tectonics and there is no reason to suspect that this extension did not continue towards the east. The Bhit and Laki Faults appear to represent the best candidates for the foci of this extension.

Drift phase 1, Late Jurassic to Santonian time

Progradation of the Chiltan platform was driven by regional thermal uplift preceding continental separation. Continuing thermal uplift led to regional exposure of the Chiltan platform at the end of Mid-Jurassic time at SBJ50 (Fig. 6b). This heralded a prolonged lowstand that was to last until the end of the Jurassic period. During this time,

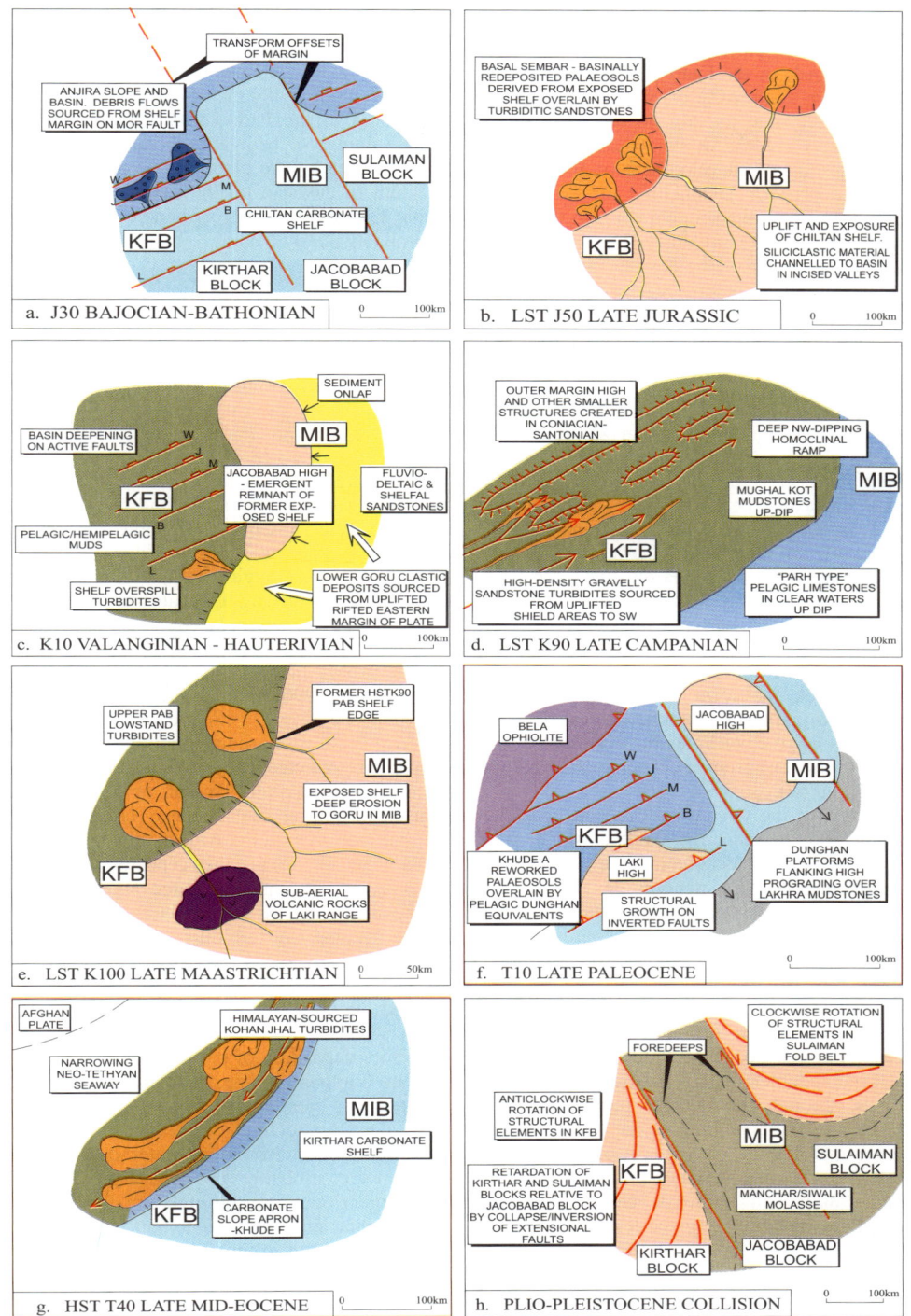

Fig. 6. Jurassic to Recent evolution of the KFB and MIB portrayed in the context of the geological history of the NW margin of the Indo-Pakistan Plate. Inferred Jurassic rift faults in KFB: L, Laki Fault; B, Bhit Fault; M, Mor Fault; J, Jezar Fault; W, Wayaro Fault.

palaeosols developed on the exposed Jurassic platform. These were partly redeposited as the basal Sembar red beds of the Kulri and Mor Ranges on the correlative conformity of SBJ50. The palaeosols on the platform contain no sand, but in the basin, sand is present in the basal Sembar red beds, suggesting the development of incised valleys on the platform channelling the first clastic deposits from the Indian Shield out into the basin. These clastic deposits are believed to be related to initial thermal uplift of the eastern margin of the Indo-Pakistan Plate before separation of India-Madagascar from Australia-Antarctica in Late Valanginian time (Besse & Courtillot 1988).

Red bed deposition in the basin terminated fairly abruptly and up to 110 m of fine turbiditic sandstones were deposited in the basin during the J50 lowstand. Turbidite deposition was drastically reduced at the end of the Jurassic period when the exposed platform to the east was partly flooded at MFSJ50. Clastic deposits were subsequently trapped in fluvio-deltaic and marginal marine environments in the eastern Lower and Middle Indus Basins. What turbidites did reach the basin were limited to discrete flows of siltstone or very fine sandstone in background mudstones. Part of the exposed shelf over the Jacobabad High remained emergent following the flooding event. This area took the form of a broad emergent median high exposing Jurassic limestones and separating a narrow north-south, west-facing clastic shelf to the east from a deeper basinal area to the west (Fig. 6c).

The basin to the west was deepened by movement on the Jurassic rift faults, which were evidently active in Neocomian time as indicated by sediment thickness patterns in the Sembar Formation. In the hanging wall of the Mor Fault the Sembar Formation thickens eastwards towards the fault from 300 m to >2000 m over a horizontal distance of 10-12 km.

The volume of clastic deposits reaching the KFB and MIB from the Indian Shield clearly increased through Sequence J50 from the disseminated sand recorded in the basinally redeposited red beds, through the Sembar turbidites to the thick fluvio-deltaic sediments in the highstand systems tract (HST) on the clastic shelf. This pattern of increasing volumes of clastic deposits continued until Aptian time, when a combination of long-term eustatic rising sea level and denudation of source areas led to a retreat of the clastic deposits to the east.

At the time of maximum advance of the Lower Goru deltas in Aptian time, mixed clay- and carbonate-dominated bathyal mudstones were being deposited in the KFB. As the Lower Goru deltas retreated in Albian time, increasingly carbonate-rich mud was deposited in the Upper Goru in the expanding basin. This process culminated in Late Santonian-Campanian time with the deposition of up to 400 m of clean bathyal lime mudstones in the Parh Formation. Despite these huge volumes of bathyal carbonates, no contemporaneous carbonate platforms occurred anywhere in southern Pakistan for the Upper Goru and Parh units. The origin of this carbonate is unknown; it is probably purely pelagic.

Drift phase 2, Santonian to Maastrichtian time

The first tectonic event to affect the Indo-Pakistan Plate during its passive northward drift occurred in Late Coniacian-Early Santonian time, when structures with up to 400 m of bathymetric relief were created on the floor of the basin. These structures were formed at the end of Goru deposition and were subsequently onlapped by the Parh limestones. The structures are considered to be compressional although there is no direct evidence for this. The driving mechanism may well be related to ophiolite emplacement, which is known to have been taking place at this time elsewhere on the margin of the plate, in Waziristan (Fig. 1) (Beck et al. 1996) and also possibly in the Indus Suture Zone (Fig. 1) (Searle et al. 1997). Any putative Late Cretaceous ophiolite in the KFB would be predicted either to underlie the Recent sands of the Porali Trough or to have been consumed during the Pliocene collision of the Kirthar margin with the Afghan Block.

Parh pelagic limestone deposition across the KFB and MIB was terminated abruptly in Late Campanian time by the emplacement of the first of the Mughal Kot turbiditic sandstones (Fig. 5c). High-density gravelly sandstone turbidites were fed into the basin from the SW, sourced from beyond the present-day shoreline of Pakistan (Fig. 6d).

Before deposition of the first Mughal Kot sands in Late Campanian time, the entire KFB and MIB comprised a gently NW-dipping (restored direction) deep homoclinal ramp covered by pelagic Parh limestones. Minimum water depths on this ramp were probably no less than 200 m. A submergent outer margin high, created during the Late Coniacian-Early Santonian tectonic event, may have existed in the western KFB to constrain off-margin transport of the Mughal Kot sands and explain their maximum accumulation (up to 500 m) along the line of the present-day Pab Range. Certainly the dispersal patterns of the Mughal Kot sandstones elsewhere in the KFB record the continued influence of other Late

Coniacian–Early Santonian bathymetric highs on the floor of the basin.

The mineralogy of the Mughal Kot and Pab sandstones indicates derivation from the Indian Shield (Waheed & Wells 1990). It is proposed that uplift of the Indian Shield was driven by passage of the Indo-Pakistan Plate over the Réunion hotspot, which ultimately provided the source of the K–T boundary Deccan Trap volcanic rocks (Courtillot et al. 1986; Duncan & Pyle 1988; Melluso et al. 1995).

With a provenance of the Mughal Kot sands from the SW there is no indication that the cratonic areas bordering the MIB to the east were uplifted at this time; these areas were probably still covered by the Parh ramp. Instead, it would appear that uplift commenced on that portion of the Indo-Pakistan Plate along strike to the SW of the KFB; that is, the part that is now buried below the present-day Indus Fan.

The first effects of uplift of the adjacent Indian craton were seen in Late Maastrichtian time, when, on the up-dip parts of the ramp in the eastern KFB and MIB, the Parh passes up into the shallow marine limestones of the Fort Munro platform. This platform did not prograde significantly westwards, its basinward limit lying to the east of Badhra on what is believed to have been a major basin-bounding fault in Late Maastrichtian time.

Soon after the establishment of this carbonate platform, carbonate production on it was terminated by the first arrival of pro-delta muds from areas of the Indian craton being uplifted to the east. The Badhra declivity was back-filled with this fine material and a platform was established that allowed the Pab deltas to prograde as far west as the present-day Kirthar Range. In Bara Nala in the Laki Range this interval is represented by 70 m of progradational, coarsening-upward sandstones, representing mouth bar, delta plain and fluvio-deltaic channel deposits (Fig. 5d).

Continued uplift of the Indian Shield in Late Maastrichtian time was responsible for erosion of the very sediments produced by the uplift, i.e. the Fort Munro and Pab deposits, over much of the MIB. This is the major base-Khadro SBK100 unconformity recognizable throughout the Middle and Lower Indus Basins. The Khadro Formation oversteps successively deeper pre-Tertiary sediments from west to east below this erosional surface, culminating in Lower Goru subcropping below the unconformity in parts of the eastern MIB (Fig. 4).

SBK100 is well exposed in Bara Nala in the Laki Range. It is defined by a palaeosol above a basaltic lava flow upon the fluvial channel sandstones at the top of the K90 highstand and a coincident change from a progradational to a retrogradational stacking pattern. To the west, however, it has not been recognized and has been interpreted as a correlative conformity.

The erosion products from SBK100 can be recognized as a basin-floor lowstand fan upon this correlative conformity in the Pab Range (Fig. 6e). These are the youngest Cretaceous sands in the basin (of latest Maastrichtian age) and complete the assemblage of Mughal Kot and Pab turbidites.

Petrographically, these Pab lowstand sands in the basin can be distinguished from the Mughal Kot sands by the presence of common volcanic clasts; volcanic clasts do occur in the upper Mughal Kot sands but not in the abundances recorded in the Pab deposits. These are interpreted to have been derived from the volcanic rocks on the exposed SBK100 surface to the east, as seen at Bara Nala. Shallow marine bioclasts are also found and these could only have been incorporated into the Pab flows after the growth (and demise) of the Fort Munro platform in Late Maastrichtian time.

Drift phase 3, Paleocene–Eocene time

The Bela Ophiolite was constructed in Late Maastrichtian time oceanward of the foreland basin in which the Pab sandstones were accumulating. The ophiolite consists of upper and lower units (Gnos et al. 1998). The upper unit shows an oceanic crust and upper-mantle structure and has an enriched mid-ocean ridge basalt chemistry (Ahsan et al. 1988). $^{39}Ar/^{40}Ar$ dates for amphibole separates from ophiolite gabbros give formation ages of 68–70 Ma (Gnos et al. 1998). The lower unit is undated and is dominated by a volcano-sedimentary unit intruded by gabbro sills. It shows a geochemical affinity to the Deccan Trap volcanic rocks (Smewing, unpubl. data), with which it is considered to be coeval. It is thought to have formed in a narrow rift basin above thinned continental crust on the NW margin of the Indo-Pakistan Plate rather than in an ocean basin as indicated for the upper unit by its structure and geochemistry.

The two units of the Bela Ophiolite were juxtaposed in latest Maastrichtian time and obducted onto the continental margin, with final emplacement on the western side of the present-day KFB taking place in Early Eocene time (c. 50 Ma) (Gnos et al. 1998).

This time interval in the KFB and MIB is represented by flooding of the SBK100 erosional surface followed by progradation of the Bara shelfal clastic deposits as far west as the present-day western margin of the Kirthar Range. In Late Paleocene time the shelf was briefly exposed

(SBT10) before rapid flooding and deposition of the glauconitic claystones of the Upper Paleocene Lakhra Formation. Parts of the shelf, however, remained emergent for a longer period and were not flooded until Early Eocene time. One such example is in the Laki Range, where a Late Paleocene erosion surface is overlain by the Lower Eocene Laki Formation. This relationship extends over a broad area west of the Laki Range as far as Bhit and is attributed to Late Paleocene inversion of the Laki Fault, creating a structural high in its hanging wall (Fig. 6f).

West of the shelf-slope break, red calcareous mudstones at the base of the Late Paleocene Khude A Member rest conformably upon grey mudstones of the Rattaro Formation (Fig. 5e). These are interpreted to be the basinally redeposited products of palaeosols derived from the exposed shelf to the east.

The Jacobabad High also re-emerged at this time, possibly as a large pop-up between its two bounding faults. These are the first of the Tertiary inversion events; compression of the continental margin at this time is attributed to emplacement of the Bela Ophiolite. Inversion is superimposed upon a fall in relative sea level, briefly exposing the shelf, followed by a rapid rise. This may be related to the foreland-directed migration of a peripheral bulge associated with ophiolite emplacement through the continental margin.

As the rate of sea-level rise slowed in Late Paleocene time, Dunghan carbonate platforms became established on the flanks of the highs (Fig. 6f). The platforms prograded basinwards during the HST of Sequence T10. Flooding of all highs was subsequently achieved during the ensuing T20 transgressive systems tract (TST) and a transgressive–regressive couplet of Laki limestones was deposited, becoming generally more shaly westwards. West of the Eocene shelf-slope break, the basinal equivalent of the Laki Formation, the Khude B Member, was deposited.

Collision

Although estimates of the timing of initial collision between the Indo-Pakistan Plate and Eurasia vary from c. 65 to c. 43 Ma (Beck et al. 1996), most workers would probably agree that an Early Eocene age between 55 and 50 Ma is most likely (Besse & Courtillot 1988; Treloar & Izatt 1993; Searle et al. 1997).

This time interval in the KFB is represented by the advance and retreat of a carbonate-dominated margin into a foreland basin to the NW. The first northerly-derived Himalayan-sourced clastic sediments were deposited in this basin in late Mid-Eocene time at c. 40 Ma as the turbiditic sandstones of the Kohan Jhal Formation (Fig. 6g). This age is comparable with the first appearance of Himalayan-sourced clastic deposits in the Katawaz Basin to the north (Qayyum et al. 1996) and in the offshore Indus Fan to the south (Clift et al. 2002). By Early Oligocene time, Himalayan-sourced clastic deposits had spilt onto the adjacent shelf as the mixed carbonate–clastic cycles of the Nari Formation.

Treloar & Izatt (1993) have shown that continent–continent collision in western Pakistan did not take place until Pliocene time, when the NW margin of the Indo-Pakistan Plate collided with the Afghan Block (Fig. 6h). This block had been displaced southwards and rotated anti-clockwise towards the western Pakistan margin in response to indentation of the northern margin of the plate into Eurasia. Indeed, the sedimentary record of the KFB supports the late timing of this collision; it is only in Plio-Pleistocene time that significant volumes of westerly-derived clastic sediments from the KFB are seen in the Kirthar Foredeep.

Uplift of the mountain belt at this time was accompanied by inversion of earlier rift faults, evidence for which can be seen in the Mor Range. Here, despite the presence of contractional structures in the footwalls and hanging walls of these faults, they are still in net extension.

The KFB plunges southwestwards into the Arabian Sea, SE of the Murray Ridge (Fig. 1). Recent dredging and seismic work in the northern Arabian Sea have revealed valuable new data on the crustal structure of offshore Pakistan (Daley et al. 2001; Edwards et al. 2001; Schlüter et al. 2001). Edwards et al. (2001) have shown that the NW flank of the Murray Ridge marks the boundary between oceanic crust flooring the Gulf of Oman and thinned continental crust of the NW margin of the Indo-Pakistan Plate to the SE. Schlüter et al. (2001) have recovered ophiolitic serpentinite from the NE part of the Murray Ridge.

Nayyer & Mallick (1994) have established from gravity profiling that the Bela Ophiolite continues south of its outcrop belt beneath the alluvial cover of the western side of the Porali Trough into the Arabian Sea. The discovery of ophiolitic serpentinite on the Murray Ridge suggests that it could represent the offshore continuation of the Bela Ophiolite, at least in its northeastern part.

The evidence for a series of subparallel rift faults in the KFB suggests that its basement would have been similar to the thinned continental crust interpreted to underlie the Arabian Sea sector SE of the Murray Ridge (Edwards et al. 2001). It is tempting therefore to speculate that this region of offshore Pakistan differs from the KFB only by the lack of inversion of these early rift faults because this part of the Indo-Pakistan Plate lay south of the collision zone with the Afghan Block in Pliocene time. If this is so, then correlation of fold-belt sequence stratigraphy with the offshore stratigraphy becomes a distinct possibility.

Conclusions

Analysis of the depositional sequences that make up the Jurassic to Recent stratigraphy of the KFB and MIB has shown that many can be attributed to tectonic events taking place on the NW margin of the Indo-Pakistan Plate during its rift–drift–collision history (Table 2). An understanding of the sedimentary responses to these tectonic events provides a valuable tool for stratigraphic correlation with other areas in Pakistan.

Events that appear to be significant and hence offer correlation potential are the following: (1) regional thermal uplift preceding the separation of the Indo-Pakistan Plate from Afro-Arabia in Late Jurassic time; (2) thermal uplift of the eastern margin of the Indian Shield preceding separation of the Indo-Pakistan Plate from Australia–Antarctica in Late Valanginian time; (3) structural growth at the Goru–Parh boundary in Late Coniacian–Early Santonian time possibly related to the emplacement of Late Cretaceous ophiolites onto the NW margin of the Indo-Pakistan Plate; (4) thermal uplift associated with the passage of the Indo-Pakistan Plate over the Réunion hotspot in Campanian–Maastrichtian time; (5) Late Paleocene inversion of extensional faults superimposed upon regional flexural uplift related to the emplacement of the Bela Ophiolite; (6) initial collision of the northern margin of the Indo-Pakistan Plate with Eurasia in Early Eocene time and uplift of the Himalayas; (7) further inversion of extensional faults in Late Eocene–Early Oligocene time; (8) collision of the Kirthar margin with the Afghan Block in Pliocene time.

This study arose from a desire to move beyond the limitations of a lithostratigraphy-based nomenclature scheme in southern Pakistan. In keeping us focused upon this aim and for numerous stimulating discussions, we thank our many colleagues in LASMO's London and Karachi offices, including M. Ali, S. Beswetherick, A. Brown, J. Craig, J. Fowler, R. Graham, C. Hamblett, R. Hedley, A. Palekar and J. Smart. We thank J. Craig and P. Clift for convening the Arabian Sea Meeting and for inviting us to submit this paper to the proceedings volume. We benefited from a perceptive and thorough review of this paper by P. Sharland. OGDC and DGPC are thanked for technical discussions and assistance in the preparation of this paper. Finally we thank LASMO Oil Pakistan Ltd in Karachi and Eni-LASMO in London for granting permission to publish this paper.

Appendix

Lithostratigraphic unit: Lower Kirthar 1 Unit
Brief description: thick-bedded foraminiferal biopackstones
Distribution: widespread in eastern KFB
Age: Lutetian (Early Mid-Eocene)
Type section: Central Kirthar Range; 26°16.6′N, 67°19.7′E

Interval	Description	Thickness (m) Interval	Cumulative
1	Overlain conformably by Lower Kirthar 2 Unit (see Lower Kirthar 2 Unit type section below) Very thick-bedded, well-sorted, medium to coarse buff grey foraminiferal biopackstones arranged in 2–3 m thick cleaning-upward units; *Alveolina* cf. *elliptica*, *Nummulites* cf. *discorbinus*, *N. aff. burdigalensis*, *Linderina* cf. *rajasthanensis*, *Lockhartia hunti*, *Rotalia* sp., *Orbitolites* sp. Base not exposed	47	47

Lithostratigraphic unit: Lower Kirthar 2 Unit
Brief description: thick-bedded foraminiferal biopackstones with planktonic foram calcisiltites and fine calcarenites at base
Distribution: Kirthar Range; cut out by unconformity at base of Upper Kirthar Member to east
Age: Bartonian (Late Mid-Eocene)
Type section: Central Kirthar Range; 26°16.6'N, 67°19.7'E

Interval	Description	Thickness (m)	
		Interval	Cumulative
	Conformably overlain by Upper Kirthar Member		
5	Thick-bedded grey fine calcarenites at base with small benthic foraminifera, coarsening and thickening upwards into coarse light grey biopackstones with larger forams and calcareous algae; occasional coral-rich horizons; rare planktonic foraminifera; common mudstone–wackestone intraclasts; *Nummulites* cf. *discorbinus*, *N.* cf. *gizehensis*, *Operculina* sp., *Amphistegina* sp., *Fabiania* sp., *Discocyclina* sp.	75	118
4	Brownish grey, weathering light brown, slightly ferruginous argillaceous limestones	12	43
3	Medium- to thick-bedded grey planktonic foram calcarenites parted by argillaceous limestones	14	31
2	Medium- to thin-bedded grey calcisiltites and fine calcarenites with glauconite and planktonic foraminifera inlcuding *Acarinina* spp., *Hantkenina* spp., *Turborotalia* spp., *Globigerina* spp., *Orbulinoides beckmanni*	12	17
1	Covered	5	5
	Conformably underlain by Lower Kirthar 1 Unit (see Lower Kirthar 1 Unit type section above)		

Lithostratigraphic unit: Upper Kirthar Member
Brief description: thick-bedded foraminiferal biopackstones
Distribution: widespread in eastern KFB
Age: Early to Mid-Oligocene
Type section: Garel Nala, Kirthar Range; 26°13.8'N, 67°20.2'E

Interval	Description	Thickness (m)	
		Interval	Cumulative
	Overlain conformably by Nari Formation		
6	Cream, weathering grey, meganodular foraminiferal biograinstones arranged in 1–2 m thick cleaning-upward cycles with yellowish marly nodular limestone bases; *Nummulites* spp. and *Lepidocyclina* (*Eulepidina*) *dilatata*	52	187
5	Massive, creamy meganodular foraminiferal biopackstones	25	135
4	Pinkish weathering, thick-bedded coarse foraminiferal biopackstones with *Nummulites* spp. and *Lepidocyclina* (*Eulepidina*) *dilatata*	7	110
3	Cream, weathering grey, massive meganodular foraminiferal biopackstones with *Nummulites* spp. and *Lepidocyclina* (*Eulepidina*) *dilatata*	20	103
2	Thin- to very thin-bedded fine platy calcarenites with disseminated quartz grains. *Nummulites* spp. and *Lepidocyclina* (*Eulepidina*) *dilatata*	1	83
1	Cream, weathering grey, massive, meganodular foraminiferal biopackstones with corals at base; miliolids, *Nummulites* spp., *Lepidocyclina* (*Eulepidina*) *dilatata*, *Austrotrillina* sp. Base not exposed	82	82

Lithostratigraphic unit: Khude A Member
Brief description: red mudstones at base passing up into planktonic foram lime mudstones
Distribution: widespread in western KFB
Age: Late Paleocene
Type section: Pahvi Nala, Khude Range; 26°25.0'N, 67°03.0'E

		Thickness (m)	
Interval	Description	Interval	Cumulative
2	Overlain conformably by Khude B Member Reddish calcareous mudstones and planktonic foram lime mudstones with thick, sharp-based, yellowish grey foraminiferal grainstone interbeds; soft sediment deformation and flute casts; *Globigerina* spp., *Morozovella velascoensis*, *M. rex*, *M. aequa* and *M. formosa* in mudstones; *Discocyclina* spp., *Operculina* sp., *Distichoplax biserialis* and *Assilina* cf. *laxispira* in grainstones	20	40
1	Poorly exposed reddish calcareous mudstones Conformably overlies Rattaro Formation	20	20

Lithostratigraphic unit: Khude B Member
Brief description: mudstones, calciturbidites and limestone conglomerates
Distribution: widespread in western KFB
Age: Ypresian–Lutetian (Early to Early Mid-Eocene)
Type section: north of Ari Pir; 26°14.3'N, 67°11.5'E

		Thickness (m)	
Interval	Description	Interval	Cumulative
11	Overlain conformably by Khude C Member (see Khude C Member type section below) Poorly sorted, monomict, matrix-supported grey conglomerate passing up into grey medium-bedded very fine peloidal biopackstone; clasts in conglomerate are angular to subrounded very fine peloidal biopackstones up to 0.25 m diameter in a matrix of the same composition; pyritized top surface to unit with horizontal burrows	3	234
10	Grainstone below abruptly truncated by argillaceous limestone passing up into yellowish grey weathering thin–medium beds of very fine peloidal biopackstone	11	231
9	Laterally persistent, highly compacted, polymict, clast-supported, poorly sorted conglomerate with clasts up to 0.25 m diameter comprising medium packstones with alveolinids, fine peloidal biopackstones with discocyclinids and sparse *Nummulites* and abundant white algal clasts; *Assilina* and large *Nummulites* largely confined to matrix; conglomerate passes up into coarse grainstone with small pebbles and algal clasts, discocyclinids and small *Nummulites*	18	220
8	Poorly exposed light yellow–green mudstones	34	202
7	Thick-bedded meganodular very fine cherty peloidal biopackstones with some coarser grainstones with reworked *Nummulites* and silicified convex-up whole oysters; these limestone beds are parted by mudstones and argillaceous very fine peloidal biopackstones; *Nummulites beaumonti*, *N.* cf. *perforatus*, *Discocyclina* spp.	39	168

Interval	Description	Thickness (m) Interval	Cumulative
6	Covered	83	129
5	Yellowish argillaceous weathered very fine peloidal biopackstone bed with abundant horizontal burrows	1	46
4	Covered	14	45
3	Yellowish grey thin-bedded argillaceous very fine peloidal biopackstones capped by medium-bedded grey–brown grainstones with horizontal burrowing traces	7	31
2	Grey, medium- to thick-bedded very fine peloidal biopackstones parted by mudstones and argillaceous limestones	9	24
1	Massive grey limestones, mostly comprising very fine peloidal biopackstones with poorly sorted, poorly aligned floating larger benthic foraminifera; also foraminiferal grainstones with oncoid pebbles and foraminiferal grainstone lithoclasts; *Nummulites* cf. *cuvillieri*, *Alveolina* spp., *Assilina* sp., *Operculina* sp., *Gypsina* sp., *Fabiania* spp., *Eorupertia* sp. Base cut out by fault	15	15

Lithostratigraphic unit: Khude C Member
Brief description: calcareous mudstones with increasing limestone content upwards
Distribution: widespread in western KFB
Age: Lutetian (Early Mid-Eocene)
Type section: north of Ari Pir; 26°14.3′N, 67°12.0′E

Interval	Description	Thickness (m) Interval	Cumulative
	Overlain conformably by Khude D Member (see Khude D Member type section below)		
5	Poorly exposed green weathering mudstones	22	308
4	Yellowish weathered medium-bedded argillaceous very fine peloidal biopackstones with floating, poorly sorted and poorly aligned large foraminifera including abundant discocyclinids and common *Nummulites*; some clean grey tops	12	286
3	Covered	23	274
2	Light brown very fine argillaceous limestones and orange–red weathered partly nodular argillaceous lime mudstones; planktonic foraminifera	50	251
1	Poorly exposed grey–brown calcareous mudstones Conformably overlies Khude B Member (see Khude B Member type section above)	201	201

Lithostratigraphic unit: Khude D Member
Brief description: mainly limestones at base passing up into clean, coarse foraminiferal biopackstones
Distribution: widespread in western KFB
Age: Lutetian (Early Mid-Eocene)
Type section: north of Ari Pir; 26°14.3′N, 67°12.3′E

Interval	Description	Thickness (m) Interval	Cumulative
	Overlain conformably by Khude E Member		
17	Coarsening-upward light grey grainstones with large alveolinids, *Nummulites* and algal oncoids at top	23	178
16	Rubbly grey biograinstones with sparse echinoid spines, *Nummulites* and discocyclinids	12	155

Interval	Description	Interval	Cumulative
15	Medium- to thick-bedded medium to coarse grainstones	9	143
14	Medium- to thick-bedded alternately clean and argillaceous grey grainstones	3	134
13	Thick light brown medium to coarse cross-bedded grainstones	2	131
12	Grey thick-bedded grainstone	1	129
11	Medium-bedded fine light brown packstones at base becoming thick bedded medium grained upwards with wavy current laminations	25	128
10	Grey thick-bedded packstone with common *Nummulites*; argillaceous base in sharp contact with underlying unit	1	103
9	Sharp-based massive, becoming medium-bedded, light brown current-bedded fine grainstone with *Nummulites*, alveolinids and other foraminifera at base of unit	20	102
8	Grey and yellowish grey medium- to thick-bedded fine biopackstones with *Nummulites* and *Assilina*; argillaceous at top	9	82
7	Grey and yellow–grey argillaceous nodular fine packstones with *Nummulites*, *Assilina* and alveolinids	20	73
6	Grey rubbly fine packstones with *Nummulites*, *Assilina* and alveolinids	9	53
5	Poorly exposed marls and nodular argillaceous limestones with *Nummulites* and *Assilina*	26	44
4	Rubbly grey very fine peloidal biopackstone with abundant *Nummulites*; also *Assilina*	2	18
3	Covered	5	16
2	Cleaning-upward cycle with mudstone base, *Nummulites*–discocyclinid argillaceous lime mudstone centre and grey clean medium packstone top with abundant and diverse foraminifera including alveolinids and *Assilina*	5	11
1	Coarsening-upward medium- to thick-bedded grey very fine peloidal biopackstones with sparse foraminifera at base and abundant various foraminifera at top including *Assilina spira*, *Nummulites* cf. *discorbinus*, miliolids, *Orbitolites* sp., *Lockhartia* cf. *hunti*, *Alveolina elliptica* and *Asterigerina* sp.; also coralline red algae and gastropods Conformably overlies Khude C Member (see Khude C Member type section above)	6	6

Lithostratigraphic unit: Khude E Member
Brief description: planktonic foram lime mudstones, calcisiltites and fine calcarenites
Distribution: distinctive marker horizon in western KFB
Age: Bartonian (Late Mid-Eocene)
Type section: Kirri Kumbi, west of Gorag Ridge; 26°42.9°N, 67°05.3°E

		Thickness (m)	
Interval	Description	Interval	Cumulative
5	Overlain conformably by Kohan Jhal Formation Argillaceous lime mudstones capped by 10 cm thick tan–brown planktonic foram calcarenite with *Acarinina* spp., *A. bullbrooki*, *Hantkenina* spp., *Turborotalia* spp., *T. cerroazulensis* s.l., *Globigerinatheka* spp., *Globigerina* spp., *G. eocaena* and *Orbulinoides beckmanni*	3	28
4	Light yellow–grey thin-bedded marly lime mudstones	18	25
3	Grey graded grainstone; coarse fluted base with benthic forams (*Discocyclina* spp., *D.* cf. *assamica*, *D.* cf. *augustae*, *D.* cf. *dispansa*, *Asterocyclina* sp., *Nummulites* spp., *N.* cf. *striatus/pengaroensis*, *N.* cf. *gizehensis/lyelli*, *Heterostegina* cf. *nuda*, *Amphistegina* sp.) and mudstone intraclasts grading up into well-sorted planktonic foram calcarenite with *Discocyclina* spp., *Nummulites* spp., *Acarinina* spp., *A. bullbrooki*, *Hantkenina* spp., *H. alabamensis*, *Turborotalia* spp., *Globigerinatheka* spp., *G. mexicana*, *Globigerina* spp. and *Orbulinoides beckmanni*	1	7
2	Light yellow–grey thin-bedded marly lime mudstones	3	6
1	Thick-bedded yellowish grey lime mudstones Conformably overlies Khude B Member	3	3

Lithostratigraphic unit: Khude F Member
Brief description: argillaceous lime mudstones, peloidal biopackstones and calcareous shales
Distribution: limited to a narrow north–south strip west of the Gorag Ridge in the central KFB
Age: Late Mid-Eocene to Mid-Oligocene
Type section: Musafiri; 26°21.7′N, 67°13.0′E

Interval	Description	Thickness (m) Interval	Cumulative
	Overlain conformably by Kohan Jhal Formation (see Kohan Jhal Formation type section below)		
20	Lemon yellow marls with a few thin argillaceous limestone cycle tops at base, cleaning upwards into thin-bedded argillaceous limestone capped by resistant brownish grey very fine biopackstone with planktonic foraminifera, small *Nummulites* and *Lepidocyclina* (*Eulepidina*) *dilatata*	9	167
19	Grey medium- to thick-bedded very fine peloidal biopackstones parted by calcareous shales, grey blocky argillaceous limestones and recessive mudstone horizons up to 1 m thick	18	158
18	Poorly exposed reddish and greenish slightly calcareous mudstones with thin bed of very fine yellowish argillaceous peloidal biopackstone at base	36	140
17	Grey medium- to thick-bedded very fine peloidal biopackstones with sparse planktonic foraminifera, vertical and horizontal tube-like burrows and pelletal burrow infills; parted by calcareous shales	13	104
16	Covered	5	91
15	Medium- to thick-bedded grey very fine packstones with pelletal burrow infills, parted by pinkish calcareous shales; common vertical and horizontal tube-like burrows	11	86
14	Covered	2	75
13	Cleaning-upward interval comprising yellowish brown blocky thick-bedded argillaceous very fine peloidal biopackstones at base becoming medium to thick bedded and less argillaceous upwards and tending to be more parted by calcareous shales	15	73
12	Covered	1	58
11	Brownish mudstone	1	57
10	Yellowish brown argillaceous very fine peloidal biopackstone	3	56
9	Covered	9	53
8	Medium- to thick-bedded grey very fine peloidal biopackstones with sparse planktonic foraminifera parted by pinkish grey calcareous shales, cleaning up into grey very fine planktonic foram peloidal biopackstones at top with *Acarinina bullbrooki*, *Morozovella* spp. and *Globigerina* spp.	7	44
7	Covered	3	37
6	Grey, medium- to thick-bedded very fine planktonic foram peloidal biopackstones parted by pinkish grey shales; vertical and horizontal tube-like burrows and *Zoophycos*, latter particularly common at top of interval	11	34
5	Covered	1	23
4	Yellowish brown bed of argillaceous lime mudstone	1	22
3	Underlying unit passes gradationally upward with decreasing limestone content and increasing mudstone content into poorly exposed yellowish and reddish mudstones with a few thin yellowish brown argillaceous limestone beds at base	7	21
2	Yellow–brown thick-bedded argillaceous lime mudstones with reddish mudstones at base; *Zoophycos* trace fossils in top bed	8	14
1	Poorly exposed argillaceous lime mudstones and pinkish mudstones Conformably overlies Khude E Member	6	6

Lithostratigraphic unit: Kohan Jhal Formation
Brief description: turbiditic sandstones with interbedded mudstones and limestone conglomerates
Distribution: widespread in eastern KFB
Age: Late Mid-Eocene to Oligocene
Type section: Musafiri; 26°21.8'N, 67°13.4'E

Interval	Description	Thickness (m) Interval	Cumulative
	Top covered by Recent alluvium		
16	Greenish brown silty mudstone with a few thin, very fine sandstone beds at top	217	358
15	Erosive-based yellow–brown limestone conglomerate arranged in layers of clast-supported, poorly sorted conglomerate with angular clasts up to 20 cm diameter of foraminiferal limestone and whole corals with larger benthonic foraminifera in matrix and better sorted, flow aligned, highly compacted sandy foram–lithic grainstone with coralline red algae, *Gypsina* sp., *Lepidocyclina* (*Eulepidina*) *dilatata*, *Nummulites vascus*, *Heterostegina* spp., *Borelis* cf. *pygmaeus* and *Austrotrillina* sp.	2	141
14	Greenish brown silty mudstone	3	139
13	Greenish mudstone with thin beds of rubbly argillaceous very fine greenish brown sandstone, which tend to become better defined, thicker bedded and more numerous upwards	16	136
12	Covered	13	120
11	Poorly exposed greenish mudstones with laminations and very thin beds of very fine sandstone turbidites	22	107
10	Very thin-, thin- and medium-bedded greenish brown very fine T_{cde} turbidites with interlaminated mudstones; thin ferruginous coatings on upper surfaces of some sandstones; many of the sandstones are amalgamated on a centimetre scale, individual flows being marked by ferruginous horizons; top of unit shows decreasing abundance and thickness of sandstones into mudstones above	8	85
9	Reddish and greenish mudstones	18	77
8	Covered	17	59
7	Interlaminated and thinly to very thinly interbedded green mudstones and ripple cross-laminated very fine greenish brown sandstones	4	42
6	Pinkish brown and greenish iron-stained non-calcareous mudstones	9	38
5	Grey–brown blocky argillaceous very fine biopackstones	3	29
4	Green mudstones	13	26
3	Grey medium- to thick-bedded very fine biopackstones	2	13
2	Light yellowish weathered argillaceous limestones	1	11
1	Covered	10	10
	Conformably overlies Khude F Member (see Khude F type section above)		

References

AHSAN, S.N., AKHTAR, T. & KHAN, Z.A. 1988. *Petrology of Bela-Khuzdar Ophiolites, Baluchistan*. Pakistan Geological Survey Information Release, **307**.

BANNERT, D., CHEEMA, A., AHMED, A. & SCHÄFFER, U. 1992. The structural development of the Western Fold Belt, Pakistan. *Geologisches Jahrbuch*, **B80**, 3–60.

BECK, R.A., BURBANK, D.W., SERCOMBE, W.J., KHAN, A.M. & LAWRENCE, R.D. 1996. Late Cretaceous ophiolite obduction and Palaeocene India–Asia collision in the westernmost Himalaya. *Geodinamica Acta*, **9**, 114–144.

BESAIRIE, H. 1972. *Géologie de Madagascar: Les Terrains sédimentaires*. Annales Géologiques Madagascar, **35**.

BESSE, J. & COURTILLOT, V. 1988. Palaeogeographic maps of the continents bordering the Indian Ocean since the Early Jurassic. *Journal of Geophysical Research*, **92**, 11791–11808.

CLIFT, P.D., SHIMIZU, N., LAYNE, G.D. & 5 OTHERS 2002. Development of the Indus Fan and its significance for the erosional history of the western Himalaya and Karakoram. *Geological Society of*

America Bulletin, **113**, 1039-1051.
COPESTAKE, P., COOPER, B.A., SLATFORD, M., VANSTONE, S., MAQSOOD, T. & ASHRAF, M. 1996. Sequence stratigraphy of the Upper Jurassic-mid Cretaceous of the Indus Basin, Pakistan and the Rajasthan area of India. *In:* CAUGHEY, C.A., CARTER, D.C., CLURE, J., GRESKO, M.J., LOWRY, P., PARK, R.K. & WONDERS, A. (eds) *Proceedings of the International Symposium on Sequence Stratigraphy in SE Asia, May 1995*. Indonesian Petroleum Association, Jakarta, 477-480.
COPESTAKE, P., SLATFORD, M. & VANSTONE, S. 1997. Sequence stratigraphy to understand the prospectivity of Lower Cretaceous to Upper Jurassic of Jaisalmer Basin Rajasthan, and the Indus Basins, Pakistan. *Proceedings, Second International Petroleum Conference and Exhibition. PETROTECH-97*, 137-146.
COURTILLOT, V., BESSE, J., VANDAMME, D., MONTIGNY, R., JAEGER, J.J. & CAPETTA, H. 1986. Deccan flood basalts at the Cretaceous/Tertiary boundary. *Earth and Planetary Science Letters*, **80**, 361-374.
DALEY, T., SIMONS, N. & BESWETHERICK, S. 2001. Seismic stratigraphy of the offshore Indus Basin (abstract). *Meeting on Geological and Climatic Evolution of the Arabian Sea Region, 5-6 April 2001*. Geological Society, London.
DUNCAN, R.A. & PYLE, D.G. 1988. Rapid eruption of the Deccan flood basalts at the Cretaceous/Tertiary boundary. *Nature*, **333**, 841-846.
EDWARDS, R. A., MINSHULL, T. A., KOPP, C. & FLUEH, E. R. 2001. Crustal structure of the Dalrymple Trough and Murray Ridge—implications for the location of the ocean-continent transition west of India (abstract). *Meeting on Geological and Climatic Evolution of the Arabian Sea Region, 5-6 April 2001*. Geological Society, London.
GALLOWAY, W.E. 1989. Genetic stratigraphic sequences in basin analysis I: architecture and genesis of flooding surface bounded depositional units. *AAPG Bulletin*, **73**, 125-142.
GNOS, E., KHAN, M., MAHMOOD, K., KHAN, A.S., SHAFIQUE, N.A. & VILLA, I.M. 1998. Bela oceanic lithosphere assemblage and its relation to the Réunion hotspot. *Terra Nova*, **10**, 90-95.
GRADSTEIN, F.M. & OGG, J. 1996. A Phanerozoic time scale. *Episodes*, **19**, 3-5.
HAILWOOD, E. & DING, F. 1998. *A Supplementary Palaeomagnetic Study of the Bela Ophiolite, Southwestern Pakistan*. Core Magnetics Report, **CM9808**.
HAQ, B.U., HARDENBOL, I. & VAIL, P.R. 1988. Mesozoic and Cenozoic chrono-stratigraphy and cycles of sea-level change. *In:* WILGUS, C.K., HASTINGS, B.S., KENDALL, C.G.ST.C., POSAMENTIER, H., ROSS, C.A. & VAN WAGONER, J.C. (eds) *Sea Level Changes, an Integrated Approach*. Society of Economic Paleontologists and Mineralogists, Special Publication, **42**, 71-108.
HUBBARD, R.J. 1988. Age and significance of sequence boundaries on Jurassic and Early Cretaceous rifted continental margins. *AAPG Bulletin*, **72**, 49-72.
HUNTING SURVEY CORPORATION 1960. *Reconnaissance Geology of Part of West Pakistan*. Maracle Press, Toronto, Ontario.

KLOOTWIJK, C.T., NAZIRULLAH, R., DE JONG, K.A. & AHMED, H. 1981. A palaeomagnetic reconnaissance of northeastern Baluchistan, Pakistan. *Journal of Geophysical Research*, **86**(B1), 289-306.
LOOSVELD, R.J.H., BELL, A. & TERKEN, J.J.M. 1996. The tectonic evolution of interior Oman. *GeoArabia*, **1**, 28-51.
MELLUSO, L., BECCALUVA, L., BROTZU, P., GREGNANIN, A., GUPTA, A.K., MORBIDELLI, L. & TRAVERSA, G. 1995. Constraints on the mantle sources of the Deccan Traps from the petrology and geochemistry of the basalts of Gujarat State (western India). *Journal of Petrology*, **36**, 1393-1432.
NAYYER, Z.A. & MALLICK, K.A. 1994. Sub-surface continuation of the ophiolites in the Bela plain of Baluchistan, Pakistan. *Ofioliti*, **19**, 269-278.
PARTINGTON, M.A., MITCHENER, B.C., MILTON, N.J. & FRASER, A.J. 1993. Genetic sequence stratigraphy for the North Sea Late Jurassic and Early Cretaceous: distribution and prediction of Kimmeridgian-Late Ryazanian reservoirs in the North Sea and adjacent areas. *In:* PARKER, J.R. (ed.) *Petroleum Geology of Northwest Europe, Proceedings of the 4th Conference*. Geological Society, London, 347-370.
PATRIAT, P. & ACHACHE, J. 1984. India-Eurasia collision chronology and its implications for crustal shortening and driving mechanisms of plates. *Nature*, **311**, 615-621.
QAYYUM, M., NIEM, A.R. & LAWRENCE, R.D. 1996. Newly discovered Palaeogene deltaic sequence in Katawaz basin, Pakistan, and its tectonic implications. *Geology*, **24**, 835-838.
RABINOWITZ, P.D., COFFIN, M.D. & FALVEY, D. 1983. The separation of Africa and Madagascar. *Science*, **220**, 67-69.
SCHLÜTER, H. U., GAEDICKE, CH., ROESER, H. A., PREXL, A., REICHERT, CH. & MEYER, H. 2001. The easternmost Murray Ridge-Makran accretionary wedge: evidence for a continent-continent collision? (abstract) *Meeting on Geological and Climatic Evolution of the Arabian Sea Region, 5-6 April 2001*. Geological Society, London.
SEARLE, M.P., CORFIELD, R.I., STEPHENSON, B. & MCCARRON, J. 1997. Structure of the north Indian continental margin in the Ladakh-Zanskar Himalayas: implications for the timing and obduction of the Spontang ophiolite, India-Asia collision and deformation events in the Himalaya. *Geological Magazine*, **134**, 297-316.
SHAH, S. M. I. (ed.) 1977. *Stratigraphy of Pakistan*. Geological Survey of Pakistan, Memoir, **12**.
SHARLAND, P.R., ARCHER, R., CASEY, D.M. & 5 OTHERS 2001. *Arabian Plate Sequence Stratigraphy*. GeoArabia Special Publication, **2**.
TRELOAR, P.J. & IZATT, C.N. 1993. Tectonics of the Himalayan collision between the Indian plate and the Afghan block: a synthesis. *In:* TRELOAR, P.J. & SEARLE, M.P. (eds) *Himalayan Tectonics*. Geological Society, London, Special Publications, **74**, 69-87.
VAIL, P. R., MITCHUM, R. M., TODD, R. G. & 5 OTHERS 1977. Seismic stratigraphy and global changes of sea-level. *In:* PAYTON, C. E. (ed.) *Seismic Stratigra-*

phy—Applications to Hydrocarbon Exploration. American Association of Petroleum Geologists, Memoir, **26**, 49–212.

WAHEED, A. & WELLS, N.A. 1990. Changes in palaeocurrents during the development of an obliquely convergent plate boundary (Sulaiman fold-belt, SW Himalayas, west–central Pakistan). *Sedimentary Geology*, **67**, 237–261.

WILLIAMS, M. D. 1959. Stratigraphy of the Lower Indus Basin, West Pakistan. *Proceedings of the Fifth World Petroleum Congress, New York.* Section I, Paper 19, 377–391.

Quaternary climatic changes over Southern Arabia and the Thar Desert, India

K. W. GLENNIE[1], A. K. SINGHVI[2], N. LANCASTER[3] & J. T. TELLER[4]

[1]*Department of Geology and Petroleum Geology, University of Aberdeen, Aberdeen AB9 2UE, UK (e-mail: glennie_ken@hotmail.com)*
[2]*Earth Science Division, Physical Research Laboratory, Ahmedabad, 380 009, India*
[3]*Desert Research Institute, Reno, NV 89512, USA*
[4]*Department of Geological Sciences, University of Manitoba, Winnipeg, Man. R3T 2N2, Canada*

Abstract: The distribution of sand dunes over the southern half of Arabia conforms to the influence of two wind systems: the northern Shamal, which is a strong wind that blows to the SSE down the Persian (Arabian) Gulf and then swings to the SW across the hyperarid Rub al Khali towards North Yemen; and the strong winds of the SW Monsoon system, which were responsible for forming linear dunes that trend north–south in the Wahiba Sands of eastern Oman and SW–NE in the Thar Desert (NW India). In the Thar Desert, the SW Monsoon alternates with the weaker NE Monsoon. The dating of exposures of older dune systems by isotopic, radiometric and optically stimulated luminescence (OSL) analyses has shown that the Shamal was active throughout the latter part of the Quaternary period, and probably as long ago as Mid-Miocene time (*c.* 15 Ma). At times of glacial maxima, when global sea level was some 100–120 m or more lower than now, siliciclastic and carbonate grains were deflated from the exposed surface of the Persian Gulf and transported into the NE Rub al Khali within the United Arab Emirates. It is suspected that occasionally the Shamal also transported some quartz sands from the NW onto the exposed narrow continental shelf of SE Arabia, with silt-size particles being carried into the Arabian Sea. The SW Monsoon, on the other hand, was re-established over the coast of SE Arabia several millennia after the last glacial maximum and was fully established near the coast of SE Arabia during the early Holocene interglacial after the atmospheric high-pressure system associated with the glacial period had become weaker. Early during the Holocene interglacial periods when the SW Monsoon dominated, a combination of quartz and carbonate sands was deflated from the exposed continental shelf and transported to the north into the Wahiba Sands. Aeolian activity in the Thar Desert also peaked during this period of transition from full glacial to interglacial conditions. The dune systems of SE Arabia overlie the distal edges of older alluvial fans that in Oman date back at least 350 ka. The sediments of some of these fluvial sequences in Oman reached the Arabian Sea via Wadi Batha, only to be removed by along-shore currents driven by the SW Monsoon. In the Thar Desert, the supply of aeolian sediment is mostly from fluvial sources. Marine sediments from the Arabian Sea between Arabia and Thar record the contrasting effects of the Shamal and the SW Monsoon: the former mostly as a source of wind-blown dust from Arabia and the latter by causing upwelling of nutrient-rich waters leading to organic blooms.

The expansion and contraction of deserts and their relationship to changes in climate including the albedo and atmospheric circulation patterns, have long been investigated (e.g. Tricart *et al.* 1957; Fairbridge 1962; Sarnthein 1978; McClure 1978, 1984; Dhir *et al.* 1992; Sanlaville 1992a, 1992b; Yan & Petit-Maire 1994). Conventionally, the arid expansion episodes have been associated with high-latitude glacial epochs and the contraction episodes with more humid phases. A synchronous response of different desert systems to climate change has been implicitly assumed. However, more recent luminescence dating studies indicate that aeolian activity occurs over more specific time windows. The window of opportunity for aeolian activity depends on a variety of factors that regulate the sediment production, supply, transport and preservation (Kocurek & Lancaster 1999). Consequently, both local and global factors influence this activity, implying that the event chronology in different deserts under apparently identical global climatic regimes need not be in phase, *sensu stricto*.

The desert in SE Arabia is influenced by two wind systems, the Shamal (Arabic for north) and the SW Monsoon, and in the Thar Desert (NW India) by the SW Monsoon and, marginally, by the NE winter monsoon. The contrasting wind regimes consequently provide a good opportunity to examine some of the key issues in desert paleoclimatology. These include the effect of what may have been asynchronous fluctuations in the two wind systems, in their style of aeolian activity, the limitations imposed by sea-level changes, changes in sediment supply and changes in the preservation potential of aeolian sands in these regions. Both the Arabian and Thar deserts possess excellent sedimentary records that indicate a wide range of depositional environments including dunes made of siliciclastic sand, sands rich in carbonate grains, sabkhas, fluvial gravels and sands, together with pedogenic and groundwater carbonates.

In this study we collate and present a synthesis of the published literature and our own research on the interpretation and chronometry of the desert landforms in SE Arabia and NW India, and the effects that associated wind systems may have had on sedimentation over the intervening Arabian Sea.

Arabian Desert

Introduction

The present distribution of dune sands over the southern half of Arabia conforms to the influence of two wind systems: (1) the major Shamal system, which is controlled by winds that blow to the SSE down the Arabian Gulf and then swing to the SW across the Rub al Khali towards North Yemen (Fig. 1a); (2) the SW Monsoon system, which is responsible for forming the much smaller north–south-trending Wahiba Sands, north of the Arabian Sea coast in SE Oman (Glennie 1987, 1998).

Isotopic and radiometric analyses based on $^{87}Sr/^{86}Sr$, U–Th and ^{14}C and optically stimulated luminescence (OSL) dating of aeolian sands in the Emirates and Oman (Fig. 2; Sanlaville 1992a,b; Glennie & Singhvi 2002), together with associated relative ages (older or younger than), have allowed the reconstruction of an event stratigraphy (Table 1).

In later Quaternary time, the repeated transportation and deposition of aeolian sands by the Shamal and SW Monsoon wind systems were affected by the growth and decay of the Northern Hemisphere ice sheets, and their effects on global sea levels and atmospheric circulation patterns (Glennie 1987, 1998; Nanson et al. 1992; Schulz et al. 1998). On a global scale, Brooke (2001) has listed the deposition of aeolianites during inter-glacial and interstadial periods for perhaps the past million years. Many coastal aeolianites have ages that straddle the Holocene or earlier interglacial highstands, thereby suggesting exposure and deflation of the adjacent continental shelf before and after the highest sea levels were reached or, on a limited scale, aeolian reworking of beach sands during the highest stands (e.g. stage 5e) and their transport inland. Along the southern Persian Gulf coast, for example, beaches and a few associated aeolianites reflect the mid-Holocene highstand in that area as indicated by dates of 4–6 ka (Table 1; see also Evans et al. 1969; Evans 1995). Over Arabia, changes in humidity and wind strength are expressed as alternations of fluvio-lacustrine and aeolian sedimentary sequences, some on a sub-Milankovitch time scale. McClure (1976) has described various lake beds that existed in Saudi Arabia during the last glaciation.

Such climatic changes are also reflected in the sedimentary record of wind-blown dust in the adjacent Arabian Sea, and a more complete record exists there than on the continent because deflation and fluvial erosion have resulted in 'missing intervals'. Marine deposits will be considered after the land areas of Arabia and Thar have been discussed.

Most Quaternary sediments of Arabia fall naturally into two categories: (1) older fluvially dominated sequences; (2) relatively younger wind-transported ones. Both types of sediment have sufficient age control to show that over the past few hundred thousand years there have been several alternations of humid and arid climatic conditions. Indeed, aeolian activity occurred as early as Mid-Miocene time (c. 15 Ma) in western Abu Dhabi (Bristow 1999; Peebles 1999), and was followed in later Miocene time by a period of higher rainfall when the 'Baynunah River', with an associated vertebrate fauna, flowed eastward through western Abu Dhabi (Friend 1999) as a probable tributary to a proto-Tigris–Euphrates River. This latter river probably debouched into the Gulf of Oman between the newly emerging Zagros Mountains and the Musandam Peninsula of northern Oman (see, e.g. Ziegler 2001).

Fluvial sequences: more humid periods

The Oman Mountains are flanked by a series of large alluvial fans that are clearly seen on Landsat imagery. Their visible extent is narrow in the northern Emirates (although extending below the waters of the Arabian Gulf, thereby indicating a lower relative sea level at that time), widening to

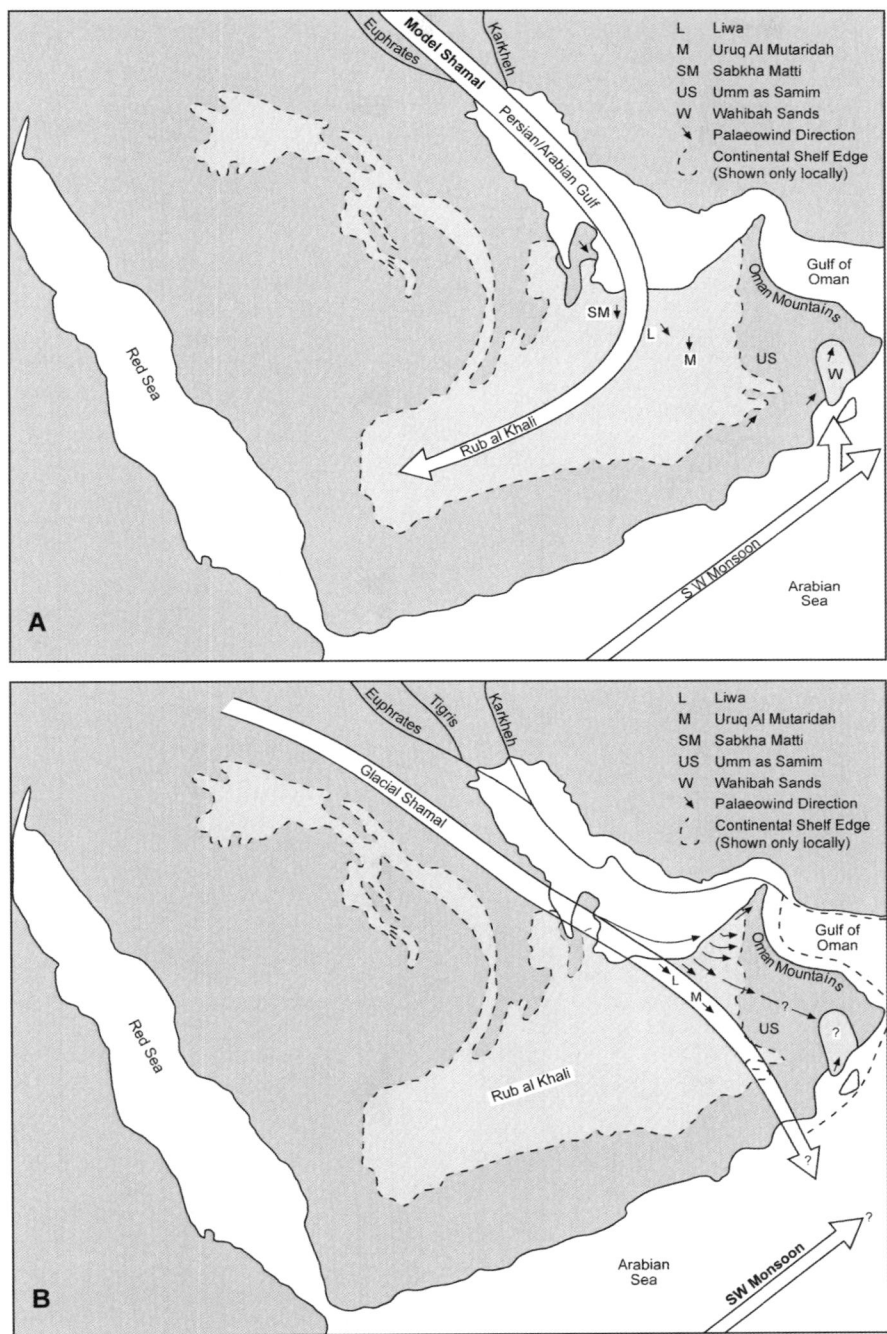

Fig. 1. The contrasting Shamal and SW Monsoon wind systems in southern Arabia. (**a**) Holocene systems, deduced in part from the axial orientations of 'modern' sand dunes. (**b**) Wind systems based on the bedding attitudes of exhumed dune sand, and the path possibly followed by the Shamal during the last and perhaps some earlier Glacial maxima. It should be noted that the alluvial fans SW of the Oman Mountains were probably subjected to considerable deflation at these times, whereas the SW Monsoon was effective only in reworking earlier-formed linear dunes into transverse forms or by the development of soil horizons over the area of the Wahiba Sands in SE Oman.

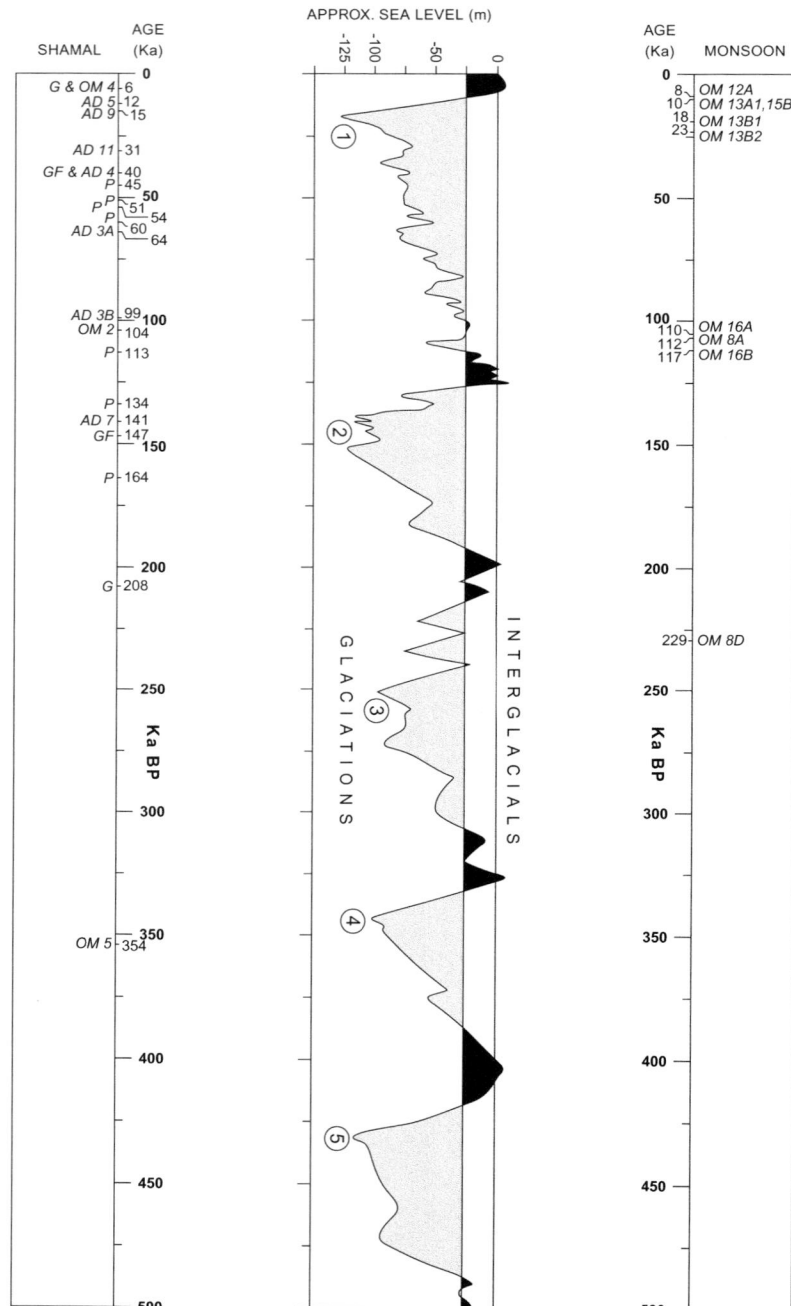

Fig. 2. OSL dates on aeolian and fluvial sands from SE Arabia seen in relation to the time scale of the last five glaciations and interglacials (Boulton 1993). Most ages are from Juyal et al. (1998) with additions from Goodall (1995) and Pugh (1997). Sample number prefixes: AD, Abu Dhabi; OM, Oman; G, Goodall; GF, Goodall fluvial; P, Pugh.

Table 1. *Age and lithological data summarized from a variety of sources; of necessity, most of these ages and thicknesses are very generalized Arabian climatic indicators and ages*

Age	Geological unit	Lithology	Thickness (m)	Age (ka unless otherwise indicated) OSL	^{14}C	Location	Reference
Emirates							
Holocene	Raised beach	Ooids, miliolids, coral	1–2		c. 4.2	Sabkha Matti coast	New
Holocene	Raised beach	Ooids, miliolids, coral	1–2		c. 5.3	Sabkha Matti coast	New
Holocene	Raised beach	Ooids			c. 5.5	60 km E of J. Dhanna	New
Holocene	Raised beach	Clam			c. 5.6	60 km E of J. Dhanna	New
Holocene	Raised beach	Minor siliciclastic sand			5.3	Sabkha Matti	Hadley *et al.* 1998
Holocene	Coastal sabkha	Laminated anhydritic quartz	2–3	c. 5		W Abu Dhabi coast	Sanlaville. 1992*a,b*
Holocene	Recent dunes	Quartz–carbonate sand	?		0–5	W Dhaid, E Emirates	Goudie *et al.* 2000
Holo.– Pleist.	NE Emirates dunes	Fine– medium grained	40	9–13		Ras al Khaimah	New
Pleistocene	Palaeodunes	Quartz and carbonate sand			21	48 km SSE of J. Dhanna	New
Pleistocene	Zeugen	Quartz and carbonate sand	4	c. 43		Al Abjan, Abu Dhabi	New
Pleistocene	NE Rub al Khali dunes	Quartz and carbonate sand	1–170	208		Sabkha Matti	Goodall 1995
Late Pleistocene	Interdune areas	Quartz and carbonate sand	18+	1–70+		W Abu Dhabi	Hadley *et al.* 1998
Late Pleistocene	Alluvial fan	Sands and conglomerates	1–50+	31		SW of Al Ain	Juyal *et al.* 1998
Mid-Pleistocene	Palaeo inland sabkha	Quartz sand	c. 1	<40–140		Liwa	Juyal *et al.* 1998
Late Miocene	Baynunah Formation	Sandstone with vertebrates	120	8–5 Ma		40 km E of J. Dhanna	Friend 1999
Mid-Miocene	Shuweihat Formation	Dune sands with dikaka	11	c. 15 Ma		20 km SW of J. Dhanna	Bristow 1999
Oman							
Holo.– Pleist.	Wadi Batha area	Aeolian activity	1–30	0.6–26		Glennie & Singhvi, 2002 N of W. Batha	Glennie & Singhvi 2002
		Fluvial activity		6.25–16		Glennie & Singhvi, 2002 N of W. Batha	Glennie & Singhvi 2002
Holocene	Wahiba Sands	Unconsolidated dune sands	1–3	>6		W Umm as Samim	New
Holocene	Fluvial	Sands and gravels	10+	6–16		N of Wahiba Sands	Juyal *et al.* 1998
	Aeolian	Quartz sands	20+	8–25		N of Wahiba Sands	New
Pleistocene	Aeolian	Quartz sands	1–150	1–117+		N Wahiba	Juyal *et al.* 1998
Pleistocene	Wahiba Sands	Aeolianite	1–>15	18–229		S Wahiba	Juyal *et al.* 1998
Pleistocene	Alluvial fan	Sand lens in gravel	0.2	104 and 354		SW Oman	Juyal *et al.* 1998
Pleistocene	Pre-Wahiba alluvium	Sands, gravels and (aeolianites)	75–300	>117		N Wahiba	Juyal *et al.* 1998
Pleistocene	Palaeodune	quartz and minor carbonate sand	20+	<117+		N Wahiba	New
Pleistocene	Palaeodune	quartz and carbonate sand	10	130		Coast W of Muscat	New
Pleistocene	Palaeodunes and soil	quartz and carbonate sand	10	112 and 229		SW Wahiba	Juyal *et al.* 1998

over 200 km in the south immediately west of the Wahiba Sands (Fig. 3); NE of the mountains the fans extend only some 30 km before reaching the Batinah coastline. NE of Ras al Khaimah (Fig. 3), the relative ages of the fans and the fluvial terraces that cut them can be deduced from the degree of fracturing of the larger clasts and the development of a pavement of interlocking smaller clasts (Al-Farraj & Harvey 2000). Along the southern edge of the mountains the fluvial sands and gravels overlie the Permian to Cretaceous rocks of the Hawasina Series, whereas their distal counterparts can be underlain by lacustrine and shallow-marine Miocene strata (Le Méteur et al. 1995). Some of the fans are old enough for fan-head entrenchment to be clearly visible on Landsat images, and elsewhere recently active fluvial channels cut into older alluvium (Glennie 1998) and extend to the eastern limit of the Rub al Khali and the large salt polygon-covered Umm as Samim (Mother of Poisons).

West of the central and northern Wahiba Sands are two fan-like areas that on Landsat imagery are distinctly lighter in colour. They comprise at least 14 vertically stacked but partly exhumed sequences of sands and gravels that are rich in ophiolite components derived from the Oman Mountains, some now diagenetically altered to dolomite, which Maizels (1988) termed barzamanite. Landsat-assisted mapping (Maizels 1988; Maizels & McBean 1990) showed an upward change from meandering channels to straight channels, each of which is picked out by a black coating of desert varnish. Maizels (1988) inferred that the difference in channel type indicates a change from a more humid climate to one in which flash floods, such as experienced in today's desert climate, was more common. Maizels (1988) suggested that the state of diagenetic alteration to barzamanite could indicate deposition as early as late Pliocene time. The present exhumed state of these channels is probably the outcome of a long period of differential deflation by winds associated with the Shamal, the SW Monsoon, or alternations of both.

The extent of these huge alluvial fans indicates a climate that was much more humid than at present, but there are only limited data on their

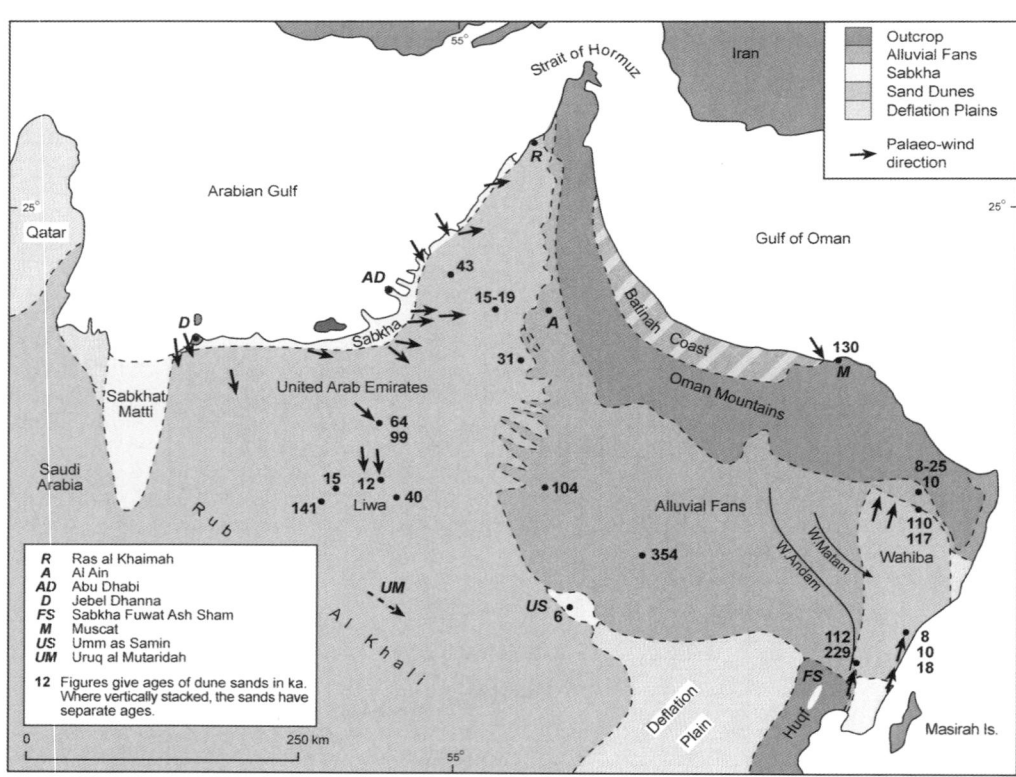

Fig. 3. Outline map of SE Arabia showing the OSL ages and, where known, the wind directions deduced from exhumed dune sand.

age. There is an OSL date of 354 ka (Table 1, Figs. 2 and 3) on a lens of aeolian sand interbedded with fluvial deposits in a mid-fan location; the base of the fan is not visible in the sample area, implying that the humid interval of fan deposition was of more extended duration. Younger sands in a fan in northern Oman (Figs. 2 and 3) date to 104 ka (Juyal et al. 1998), whereas another sample of fluvial sand in Sabkhat Matti was dated to 147 ka (Fig. 3). The latter sample was derived from deposits of a stream that presumably had crossed the Rub al Khali from the west or SW (fig. 3.11 of Goodall 1995).

Sanlaville (1992a, 1992b) has provided radiocarbon ages of 35–42 ka on carbonate grains in dune sands from a boring on the coastal plain of the NE Emirates; therefore underlying conglomerates are older than this. The upper dune sands were transgressed by marine deposits at c. 6.6 ka (maximum Holocene sea level), and in turn are overlain by sand dunes at the surface. A luminescence age of 31 ka on thin aeolian sands within another set of fluvial gravels SW of Al Ain (Juyal et al. 1998) indicates that, despite the evidence of a higher rainfall in the catchment area, windblown sands were still being deposited at this time, although their preservation potential was low.

High-resolution U–Th ages on cave speleothems from the Hoti caves in Oman also indicate more humid phases between 9.6 and 6.2 ka, and another around 125 ka, at the base of Oxygen Isotope Stage 5, and a transition from wet to dry conditions occurred at 117 and 6.2 ka (Burns et al. 1998). It was suggested that these humid episodes were linked closely to glacial–interglacial transitions. These data support the association of humid events on the Arabian Desert with interglacials, one during early Holocene time and another in Isotope Stage 5 (see Fig. 2).

In this context, Sanlaville (1992b) reported two glacial-age radiocarbon dates of 20 and 27 ka on terrestrial shells from Ras al Khaimah (NE Emirates), whereas land gastropods in sands and gravels of Wadi Dhaid, NE Emirates, give a younger ^{14}C depositional age of c. 9.4 ka (Sanlaville 1992b), consistent with the start of an early Holocene humid phase.

There is evidence for much more widespread pluvial activity in the region. McClure (1984) recorded ^{14}C ages on shells from interdune lakes of the Rub al Khali that range from around 5 to 38 ka bp. He showed a cluster of lake-bed marl and shell ages between 10 and 5 ka that peak in the range 7–8 ka bp, with far fewer dates spread fairly evenly between 20 and 32 ka, and suggested that both periods of lacustrine deposition resulted from monsoonal rainfall, which ended at c. 5 ka.

The above glacial-age dates reported by McClure (1984) and Sanlaville (1992b) imply at least scattered rainfall over Arabia at that time.

Using radiocarbon dating in Yemen, Lézine et al. (1998) considered that Holocene interdune lake development from 7.8 to 7.2 ka illustrates the influence of the SW Monsoon on local climate. An early Holocene increase in humidity similar to that which occurred in the Rub al Khali is recorded in the northern Huqf, Oman (Fig. 3), where the site of a former lake was almost certainly dependent on the SW Monsoon for its supply of water. Nearby human artefacts indicate earlier habitation, and just to the north other artefacts have been assigned nominal dates of 8 to 5 or 4 ka bp (Edens, 1988). Increasing aridity since c. 5 ka has resulted in the lake being abandoned as a site of habitation as it became today's Sabkha Fuwat ash Sham (Sabkha of Ill Omen), which now has some NNE-trending linear dunes crossing its surface (fig. 1 of Glennie et al. 1998). Indeed, not only is the lake now a sabkha, but farther south in the Huqf, the limited flow of water in Wadi Jurf is mostly just below the salt-encrusted surface, a testament to current aridity in the area. Also, on the Oman–Saudi Arabia border, the extensive salt-covered Umm as Samim is flanked along its western margin by dune sands dated to 6 ka (Fig. 3; Juyal et al. 1998), which again may indicate the start of late Holocene aridity. The earlier lacustrine periods in these sabkhas may have resulted from increased monsoonal rainfall at the start of Holocene time or from a higher water table related to the postglacial rise in global sea level.

The interplay between arid and humid periods is well illustrated in the Wahiba Sands area of east Oman, although here the SW Monsoon is the main controlling factor rather than the Shamal as seen in the Emirates. At the bottom of a well at Hawiyah, at the northern end of the Wahiba Sands, unconsolidated dune sand, dated to 117 ka, overlies undated wadi gravels, whereas within Wadi Batha just to the north, near-surface exposures of fluvial gravels are dated to 10 ka (Fig. 4). Some 10–15 km to the east, the base of a water well of 300 m depth was still in alluvium at a depth of some 60 m below present sea level, whereas in the central Wahiba Sands, undated alluvium is present 200 m below an interdune area (Jones et al. 1988) the surfaces of which are at an elevation of some 230–240 m.

In the west–central Wahiba area and clearly visible on Landsat imagery, gravels of the SE-flowing Wadi Mantam (Fig. 3) become buried beneath fairly small, north–south-trending sand dunes of possible late Holocene age. To the SE near the coast and on trend with Wadi Mantam,

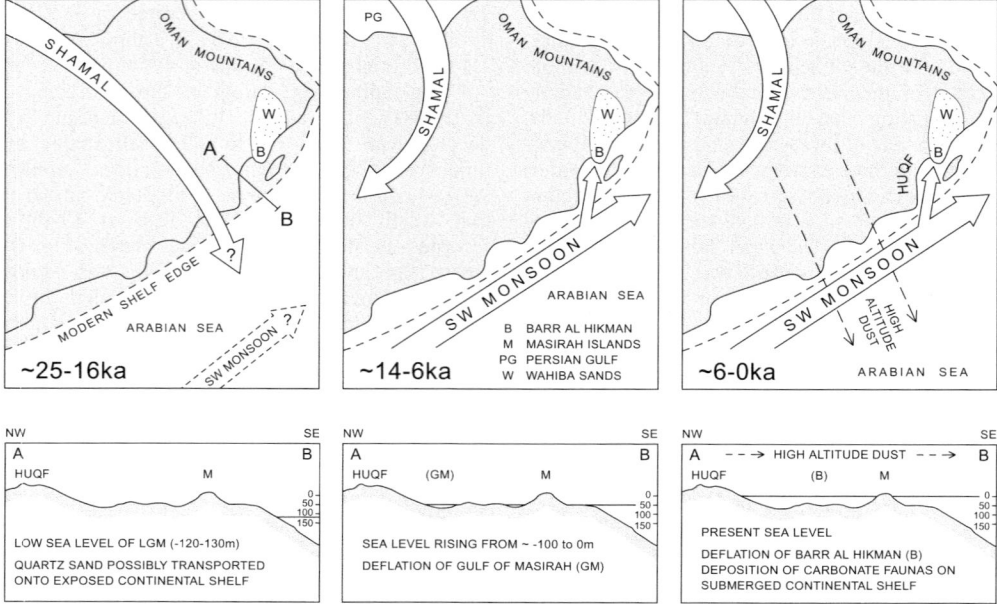

Fig. 4. Schematic depiction of the brief post-glacial time span when deflation of carbonate grains from the continental shelf is possible. The process is much more effective if applied in reverse; then, a long and irregular post-interglacial fall in sea level (e.g. Oxygen Isotope Stages 5, 4 and 3; see Fig. 2) progressively exposed the previously deposited carbonate-rich organic matter to deflation, in this example by the SW Monsoon when effective over the Arabian coastline.

the top and base of an exhumed fluvio-lacustrine and aeolian sequence have been dated to 8 and 10 ka, indicating a wetter early Holocene period. They overlie well-cemented aeolianites dated to 18 ka (Juyal *et al.* 1998) and are about to be overwhelmed from the south by a migrating dune. Beneath a thin dune cover at the southern end of the Wahiba Sands is an undated conglomerate bearing ophiolite pebbles derived from the Oman Mountains some 200 km to the north.

The variety of dates and age relationships between aeolian and fluvial or lacustrine sediments indicate that wetter intervals existed at least in the catchment areas of eastern Arabia before, or between, separate phases of aeolian deposition of the Wahiba Sands. Throughout Arabia, there appears to have been at least a brief wet phase during the last interglacial (Oxygen Isotope Stage 5e). Aeolian activity seems to have been preceded by a wetter climate as far back as at least earlier Pleistocene time, and yet drier intervals permitted the presence of both thin and more extensive aeolian sands within conglomerates wherever an active wind could deflate an unconsolidated source of sand.

Aeolian sequences: drier periods

The distribution of linear and transverse dunes of the Rub al Khali, Nafud and Dahna in southern Arabia, coupled with the attitude of bedding foresets, indicate that they were formed by a northwesterly wind known as the Shamal (Arabic for north) that veered to the south and then SW towards the mountains of North Yemen. Because of a low global sea level during the last glaciation (Oxygen Isotope Stage 3) much of the Persian (Arabian) Gulf was dry. This allowed deflation of quartz and carbonate sand from several sources: (1) the exposed floor of the Gulf (this material was transported SE to the Emirates; Sarnthein (1972), for instance, described aeolian sand dunes beneath the northern Persian Gulf, and Al-Hinai *et al.* (1987) have mapped possible sand dunes beneath the waters of the Gulf of Salwa between Bahrein Island and the Qatar peninsula); (2) quartz sand from the rising western ridges of the Zagros Mountains (calcareous sandstones of the Agha Jari and Bakhtiari formations; Alsharhan & Nairn 1997; Ziegler 2001); (3) the extended Tigris–Euphrates River; (4) carbonate grains that had been deposited under marine conditions during the previous interglacial (Teller et al. 2000). With respect to the present interglacial, Evans (1995) reviewed the productive status of the Arabian (Persian) Gulf as a carbonate–evaporite factory, and Kirkham (1998a) and Walkden & Williams (1998) described some of the complexities of marine carbonate deposition in the southern Persian Gulf that resulted from several glacially induced changes in sea level over a very shallow basin. Although carbonate grains can be detected up to 80 or 100 km south of the Emirates' present coastline, only the quartz fraction survived chemical dissolution and the abrasion involved in long-distance transport (possibly 200 km or more) and could move far into the Rub al Khali basin in Saudi Arabia.

In the context of glacially induced fluctuations in sea level, it is worth noting that just west of Muscat, aeolianites, transported landward to the SSE and dated to 130 Ma (Table 1), extend below present sea level; partial dissolution of aragonitic clasts must have occurred during later (interglacial?) wetter phases (Glennie & Gökdag 1998).

Between about 12 and 6 ka, the sea transgressed a distance of over 1000 km across the floor of the Persian Gulf and, in so doing, cut off the supply of south-moving sands to the Emirates coast (Lambeck 1998; Teller et al. 2000). A continuing Shamal wind resulted in deflation of the coastal dunes down to the water table (thereby sourcing other dunes downwind), which, in a climate that was still arid, then prograded into the classical coastal sabkha (Evans et al. 1964) in which the typical sequence of intertidal and supratidal algal mats and gypsum–anhydrite evaporites rarely exceeds 1 m (Kendall et al. 1998). This thin sabkha sequence covers a coastal area up to c. 20 km wide (United Arab Emirates University 1993), and its origin can be dated to c. 5 ka (Table 1; Hadley et al. 1998; Teller et al. 2000), which was shortly after the time of maximum Holocene transgression (Evans et al. 1969; Felber et al. 1978; Lambeck 1996).

At a few localities, unconsolidated dune sand at the landward margin of the coastal sabkha is underlain by well-cemented but generally undated aeolianite. Farther south, the degree of cementation decreases as the carbonate content of the sands declines. These less well-cemented sandstones have ages of between c. 22 and 141 ka (Fig. 3; Hadley et al. 1998; Juyal et al. 1998; Teller et al. 2000), with field relationships that indicate several intervals of lake or sabkha development. Dunes in the northern Emirates have been dated at 9–20 ka by Dalongeville et al. (1993; see Kirkham 1998b).

The dunes of the eastern Rub al Khali are partly stabilized, probably because of higher rainfall during early Holocene time. As already mentioned, they overlie fluvio-lacustrine sequences dated between c. 30 and 20 ka (Sanlaville 1992a, 1992b). Together with limited age dating (15–12 ka; Juyal et al. (1998) and previously unpublished age of 19–15 ka west of Al Ain (Fig. 3)), this indicates that they probably acquired their present linear morphology during or shortly after the last glacial maximum (c. 16–20 ka). However, one must beware of such generalizations, as rates of deposition can vary considerably. For example, Goudie et al. (2000) described a dune south of Ras al Khaimah town that between c. 11 and 10 ka accumulated at the rate of c. 3.3 m ka^{-1} whereas, within the past 1000 a, another dune further inland added 20 m to its height within 270 a.

Clearly seen on aerial photographs and on Landsat imagery, the axial trends of linear dunes within the eastern Rub al Khali of the UAE and NW Oman vary from north–south in the west to west–east and even WSW–ESE in the NE (see fig. 3 of Glennie 1998). The attitudes of dune bedding exposed in outcrop (aeolianites), in roadside borrow pits and in deflation hollows indicate a similar pattern of former winds, which in the west blew from north to south but in the east blew towards the SE, east and ENE (Fig. 3; fig. 6 of Glennie 1998). OSL dates on some of these exposed dune sandstones range between 141 and 12 ka in the Liwa area, are around 43 ka for a zeugen of aeolianite ENE of Abu Dhabi, and between 15 and 19 ka for essentially uncemented dune sand near Sweihan, between Abu Dhabi and Al Ain (Fig. 3). Some of the NE-trending linear dunes of the NE UAE show clear reworking of their crestal areas, with sand streaks crossing the interdune areas towards the east (fig. 7 of Glennie 1999). Many of the ages of dune deposition given above coincide with or are close to one of the last three time-spans of major high-latitude glacial expansion when global sea level was as much as 120–130 m lower than today (Fig. 2). The associated palaeowind directions indicate that the megadunes of the UAE were sourced from beneath the present Arabian (Persian) Gulf (Fig. 3).

Within the eastern Emirates and northern Oman the NW–SE alignment of linear dunes (see Landsat image fig. 3 and interpretive map fig. 6 of Glennie 1998), and local exposures of cross-bedding indicate sediment transport to the SE. From this evidence, it is believed that, during times of maximum high-latitude glaciation, the area of atmospheric high pressure over Arabia, and its associated clockwise wind pattern, was pushed much farther south than its present location, so that some quartz sands were possibly blown southward onto the exposed continental shelf of SE

Oman (Fig. 2b). In general, it is known that monsoonal circulation was enhanced and shifted southward during glaciations, leading to more precipitation in parts of the tropics. At that time, the SW Monsoon belt probably either shifted hundreds of kilometres to the south or was not active over the Arabian Sea shelf along the coast of Oman (e.g. Anderson & Prell 1991; Sirocko et al. 1996).

During the last deglaciation, the arid Shamal wind system weakened and its southern limit shrank northward (Sirocko et al. 1996). This allowed the axis of the more humid SW Monsoon to shift to the north and blow across Arabia's SE margin. Because the Wahiba Sands have little relief, they did not induce orographic rainfall such as probably occurred farther north in the Oman Mountains; instead, the monsoon winds deflated quartz and carbonate grains from the narrow continental shelf and deposited them in the south-to-north Wahiba dune system (Gardner 1988). We have already seen that the northern end of these dunes, which reach heights of 60–100 m or more in the north, seem to have been truncated around 10 ka bp by floodwaters associated with Wadi Batha.

With a rise in post-glacial sea level, the source of siliciclastic and carbonate sand at the up-wind (southern) end of the Wahiba dune system was cut off (Fig. 4). As in the Emirates, the dunes were subsequently deflated down to the level of the recently raised water table, and the exposed Mio-Pliocene strata became flanked by newly formed sabkhas. Continuing deflation by the SW Monsoon has caused the southern end of the Wahiba Sands (Barr al Hikman) to migrate northward, the deflated sand continuing to supply other dunes farther north.

Mineralogical studies along a south–north transect show that the Wahiba Sands comprise about 50% quartz and 30–45% carbonate grains of mostly marine origin, with a 20% maximum for grains derived from the Oman Mountains ophiolite reached only in the NE Sands (Allison 1988). The fluvial sediments originating from the Oman Mountains are poor in quartz; they are dominated by dark basic components from the Semail ophiolites and Mesozoic carbonates from turbidites of the Hawasina Series, with minor amounts of associated red radiolarian chert making a limited quartz contribution. Dune orientation and bedding attitudes indicate a fairly consistent palaeowind direction towards the north or NNE, so the source of the Wahiba aeolian sands must have been on the narrow continental shelf to the west and SW of Masirah Island (Fig. 3). A small fluvial contribution may have come from the Upper Proterozoic and Palaeozoic outcrops of the Huqf area (Fig. 3), but any direct aeolian transport from that area is likely to have been mostly to the west of the Wahiba. An alternative source may have been from the NW, from the Rub al Khali.

A cliff of aeolianite flanking Wadi Andam in the SW Wahiba gives two ages, 112 and 229 ka, the sample horizons being separated by an undated soil horizon. A similar but undated sequence occurs on the coast to the east. There, the surface of the aeolianite comprises relatively small-scale north-directed transverse bedding (C. Hern, pers. com.), which possibly represents low-velocity late-interglacial aeolian reworking of a linear dune system.

Apart from the 18 ka aeolianite (Fig. 3) beneath the lower Holocene fluvio-lacustrine sequence already mentioned, we have no other dated dune sands between there and Hawiyah in the north, where unconsolidated dune sands 50 cm apart at the bottom of a well give ages of 110 and 117 ka (Fig. 5). At this locality, there is no sign of cemented aeolianites being present. However, Jones et al. (1988) logged the presence of aeolianite at various depths up to 176 m below the surface and spread over mostly the eastern Wahiba area, together with other occurrences along its west–central margin and to the west of Wadi Andam. There are no lithological descriptions of these aeolianites, so it is not known if they include the abraded tests of delicate Foraminifera.

The Thar Desert of India

The presence of numerous dune-forms within the relatively small Thar Desert of India (Fig. 6) makes these dunes an ideal laboratory to understand factors determining dune morphology and the role of local versus global factors (see Dhir et al. 1992). The presence of fossil dune sands in regions where modern precipitation is several times higher than will allow dune activity (175–200 mm a^{-1}; Goudie et al. 1973) implies large-amplitude changes in the dune-accumulation rate of the region. It is now generally accepted that past and present fluvial and aeolian processes in the region are largely controlled by winds and precipitation associated with the SW Monsoon (Dhir et al. 1992; Thomas et al. 1999). The near absence of regionally extant fluvial or aeolian deposition during the last glacial epoch reflects this. Despite aridity and an abundant sand supply, dune building did not occur in Thar during glacial epochs because of a weakened SW Monsoon (Thomas et al. 1999). Luminescence chronometry of closely spaced samples suggests that the window of opportunity for aeolian aggradation occurred only for a limited duration that peaked at times of transition from glacial to interglacial epochs, and during the time the monsoon was re-

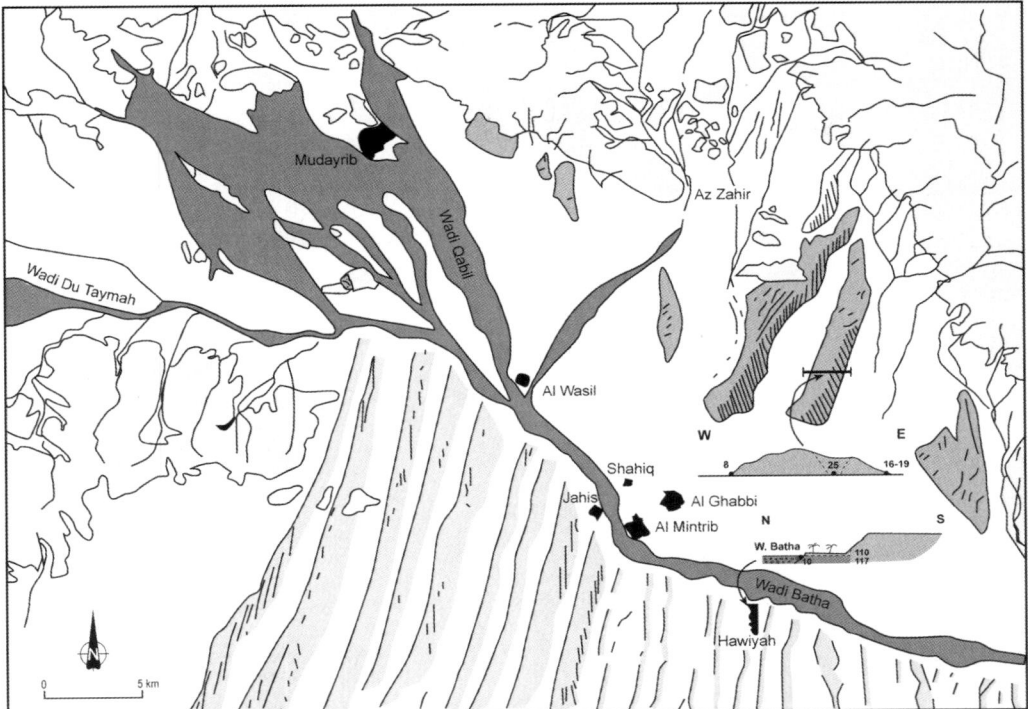

Fig. 5. The area of tributaries feeding Wadi Batha between the northern Wahiba Sands and eastern Oman Mountains. (Note the relationship between dated dune sands and wadi gravels at Hawiyah, and the isolated megadunes on the northern margin of the wadi system.)

establishing itself in the region (Fig. 6; Chawla et al. 1992). The chronometry of aeolian deposition also indicates a spatial variability within the Thar Desert, suggesting changes in the monsoon gradients during Holocene time.

Surface sands at the southern extremity of the Thar Desert stopped accumulating at some time during the early Holocene period, whereas dune accretion over central Thar continued until recently (i.e. 700–800 a ago; Juyal et al. 2002). This implies a progressive northward shift in dune activity.

The sources of dune sediment in the Thar Desert include the exposed fluvial channels that flank its western and SE margins. Other potential source areas, such as the continental shelf and Ranns of Kutch, were cut off by marine flooding during the post-glacial rise in sea level. In this context, Rao et al. (1989) have identified over 80 species of Foraminifera within the Thar dune sands, some as much as 800 km from the present coast. These could have been derived from the Arabian Sea coast, the Ranns of Kutch or saline lakes within Thar during early Holocene time. Kutch and Saurastra no longer have active inland dunes, but the presence of abraded foraminiferal tests and oolitic aeolianites (miliolite; Biswas 1971), which range in age from c. 21 to 210 ka (Patel & Bhatt 1995) imply exposure of the adjacent continental shelf. This association, coupled with dune foresets, indicates that the deflated carbonate-rich sands were transported to the NE under the influence of the SW Monsoon. Many inland sequences of miliolite are separated by soil horizons that point to repeated increases in humidity (Patel & Bhatt 1995).

Isotopic studies on pedogenic carbonates within the aeolian sands of the Thar Desert also indicate a more humid phase (with normal monsoon-like conditions) during Oxygen Isotope Stage 3 (Andrews et al. 1998). This is also seen as a red soil horizon in fluvial deposits. During Holocene time, a change in aridity is recorded by lakes that are now saline but had fresh water during early–mid-Holocene time followed by a period of desiccation (Thomas et al. 1999; Kar et al. 2000). The lakes and dunes appear to present a 1.5 ka cyclicity with a phase lag of a few centuries, which compares reasonably well with the climatic oscillations recognized by Sirocko et al. (1996) in Greenland ice and Atlantic and Arabian Sea marine sediments. Similar events in the Arctic and Arabian Sea areas were correlated by Schulz et al.

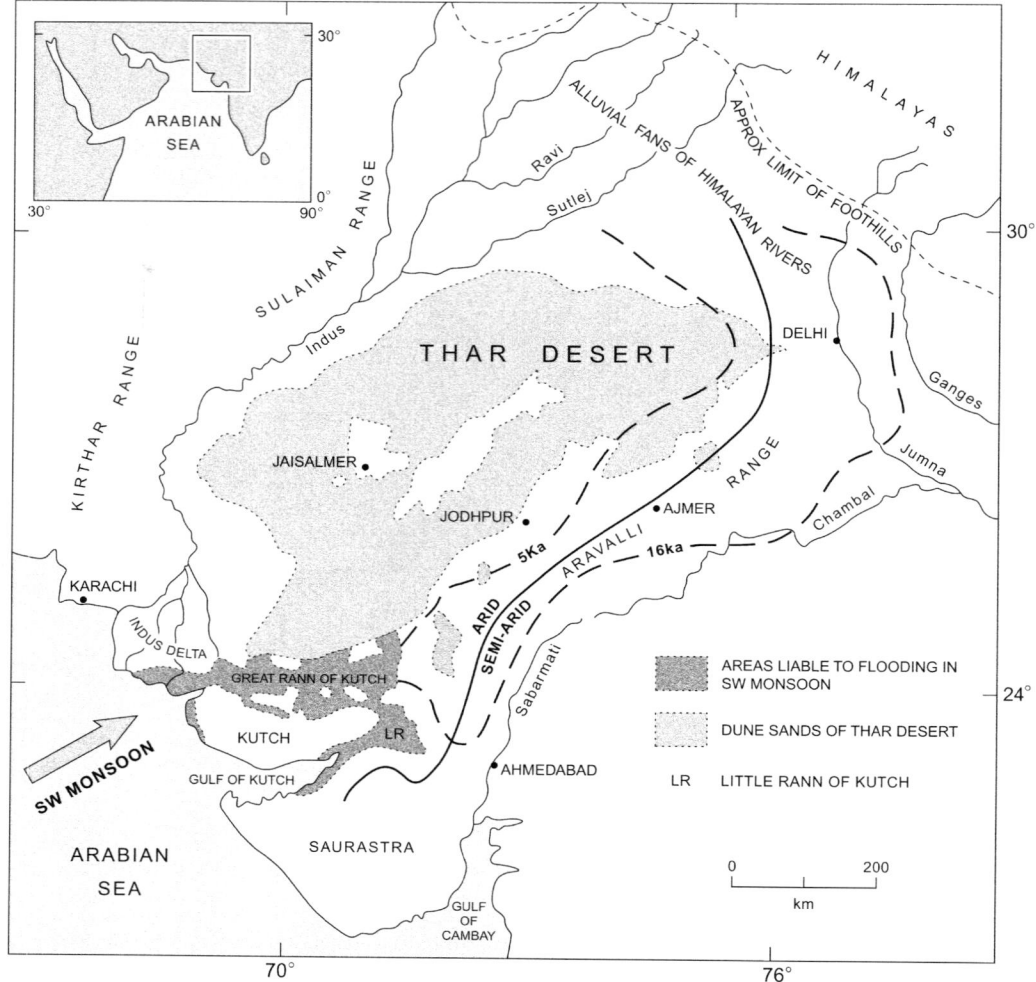

Fig. 6. Outline map of the Thar Desert and Ranns of Kutch relative to major river systems of the area and the track of the SW Monsoon. (Note the shift in the arid–semi-arid boundary between Late Glacial time (16 ka) and the Mid-Holocene time (5 ka).)

(1998); they recognized in Arabian Sea sequences the effects of Greenland interstadials and glacial Heinrich events, as well as changes in the total organic content of the sea floor related directly to monsoon activity, thereby emphasizing the glacial control on climate and sediment deposition within and adjacent to the Arabian Sea.

Supporting data from the Arabian Sea

Overall, the published data on the Quaternary marine sequences of the NW Arabian Sea give general support to the varying influence of the Shamal and SW Monsoon wind systems (e.g. Anderson & Prell 1992, 1993; Lückge *et al.* 2001), and also from climatic changes recorded especially in North Africa and SE Asia (e.g. Petit-Maire 1994; Yan & Petit-Maire, 1994; deMenocal *et al.* 2000). The chronometry of desert sequences indicates significant differences between records from SE Oman and the Thar Desert of India as compared with the more northern Arabian records, and together these permit reconstruction of past changes in circulation patterns in the region. The present study also indicates that the roles of both global and local factors in shaping desert landscapes are important, and that both landforms and sedimentary records of deserts must be studied and dated if we are to understand the complex relationships and spatial variability of climatic change.

There seems to be a close correlation in the timing of terrestrial Monsoon events over SE Arabia and the Thar Desert of India, and this can be matched fairly closely to the interpreted history of the Quaternary marine record in the Arabian Sea, which lies between the two areas. The key sediments comprise wind-blown dust-size particles and pollen, originating mostly in Arabia with minor amounts from the Horn of Africa and from the NE. Clays derived from Asian rivers such as the Indus (Sirocko & Lange 1991) are also included, as are organisms related to planktonic blooms associated with monsoonal wind-driven upwelling of nutrient-rich waters (e.g. Kroon et al. 1991; Anderson & Prell 1992, 1993). Anderson & Prell (1991) suggested that adjacent to the Arabian landmass, monsoon-induced upwelling occurred during each interglacial back to 300 ka bp but was absent or diminished during glacial times. Farther south, over the Owen Ridge, the sediments are not nearly so variable as at nearshore sites and show little glacial–interglacial (100 ka) variance (Anderson & Prell 1992).

In numerous studies, ocean-floor samples from the Arabian Sea were found to contain dust of differing mineralogy that varied in amount between late glacial time and the present (e.g. Kolla & Biscaye 1977; Sirocko & Lange 1991; Pease et al. 1998). Among the more unusual wind-blown clays found in cores and discussed by Sirocko & Lange (1991) is palygorskite. Although found only in small quantities, palygorskite makes a clear connection with Arabia and perhaps the Horn of Africa, where it is typical of sabkhas, desert soils and wadis with fluctuating water tables. Pollen, on the other hand, was derived from areas with sufficient monsoonal rain; for example, Somalia or the highlands of Yemen or eastern Oman (Van Campo 1991). Sirocko & Sarnthein (1989) presented maps of the modern monthly wind pattern over the NW Arabian Sea. These show that between May and August a considerable high-altitude (1–4.5 km) offshore wind crosses the low-level winds of the SW Monsoon almost orthogonally. This wind originates in the north as the Shamal (e.g. fig 5.3 of Pye 1987).

In a study of aerosols over the Arabian Sea, Pease et al. (1998) found that the highest dust levels occurred off the coast of SE Oman and over the Gulf of Oman. They inferred that most of the grains of dust, some of which are frosted, were transported into the Arabian Sea by dry northern winds, which were known to be associated with periods of increased offshore haze. A series of nine maps give average wind directions and relative strengths during a series of ship cruises spread over 1995, and indicate the varying influence at sea level of the SW and NE monsoons and winds connected with the Shamal. The influence of the NE Monsoon is often neglected.

Clemens & Prell (1990) showed that there is a close correlation between the dust content of both Arabian Sea cores and Vostok ice with glacial maxima. Petit-Maire (1994) and Yan & Petit-Maire (1994) traced a similar history of glacial–interglacial events from North Africa via Arabia and India to NE China; they also discussed the climatic importance of the SW Monsoon and, on a more global scale, its time equivalents elsewhere over the past 130 ka. Clemens et al. (1996) linked monsoonal fluctuations over the Arabian Sea to the development of Pleistocene ice sheets. The alternating effects of glacial and interglacial events are refined with the recognition in Arabian Sea sequences of short-term episodes of both millennial and shorter durations (Heusser & Sirocko 1997; Schulz et al. 1998; compare Chapman & Shackleton 1998).

Conclusions

The continental sedimentary sequences of SE Arabia indicate that the area has experienced alternations from hyperaridity to relatively humid conditions during the later part of Quaternary time. These alternations were the result of variations in the direction and strength of the Shamal and the SW Monsoon wind systems. During glaciation, the Shamal was probably at its strongest, contributing to increased aridity and mobilization of siliciclastic sands by these northwesterly winds. Dune sands were derived from older dunes and from the floor of the exposed Persian Gulf and continental shelf along the Arabian Sea, and included detrital carbonate composed of ooids and shells. Dust was carried out into the Arabian Sea. At these times the SW Monsoon had little influence on land in Arabia or in the Thar Desert of NW India.

During interglacials (and perhaps also some interstadials) the Shamal weakened and its influence retreated northward, thus allowing the SW Monsoon to cross the SE margin of Arabia as well as the Thar Desert. Monsoonal wind activity over these areas seems to have been strongest at times of transition between glacial and interglacial epochs, when siliciclastic and carbonate sands were deflated from the partly exposed continental shelf and transported down-wind to the dunes of Wahiba and Thar.

A post-glacial rise in sea level cut off the supply of sand and the upwind dunes were deflated down to the water table. The period of higher rainfall recorded around the start of Holocene time probably is reflective of similar events at the start of other interglacials and may have resulted in accumulation of the extensive alluvial fans west of the Oman Mountains during Pleistocene time. Associated flooding is thought to have cut through the northern end of the Wahiba Sands, leaving isolated remnants to the north. Coastal evidence of such floods was probably soon removed by along-shore currents driven by the SW Monsoon.

This history of events seems to have been matched in the sedimentary column of the Arabian Sea by Shamal-driven dust from Arabia and by monsoon-driven upwelling of nutrient-rich water leading to blooms of plankton and to sea-bottom clays rich in organic matter.

The Abu Dhabi Company for Onshore Oil Exploration (ADCO) is thanked for providing the authors with logistical support in the field in 1998, and Petroleum Development Oman (PDO) for similar support for K.W.G. on earlier occasions. J.T. acknowledges a University of Manitoba research development grant for radiocarbon dating of several samples. C. Hern is thanked for collecting samples for OSL dating from the isolated dunes north of the Wahiba Sands (Fig. 5) and for checking the penultimate draft, and the reviewers D. Benn and F. Sirocko for suggesting improvements. Shell UK Ltd and B. Fulton of Aberdeen University are thanked for draughting the figures.

References

AL-FARRAJ, A. & HARVEY, A.M. 2000. Desert pavement characteristics on wadi terrace and alluvial fan surfaces. *Geomorphology*, **35**(3-4), 279–297.

AL-HINAI, K.G., MOORE, J.M. & BUSH, P.R. 1987. Landsat image enhancement study of possible submerged sand-dunes in the Arabian Gulf. *International Journal of Remote Sensing*, **8**, 251–258.

ALLISON, R. 1988. Sediment types and sources in the Wahiba Sands. *Journal of Oman Studies, Special Report*, **3**, 161–168.

ALSHARHAN, A.S. & NAIRN, A.E.M. 1997. *Sedimentary Basins and Petroleum Geology of the Middle East.* Elsevier, Amsterdam.

ANDERSON, D.M. & PRELL, W.L. ET AL. 1991. Coastal upwelling gradient during the late Pleistocene. *In:* PRELL, W.L. & NIITSUMA, N. (eds) *Proceedings of the Ocean Drilling Program, Scientific Results, 117.* Ocean Drilling Program, College Station, TX, 265–276.

ANDERSON, D.M. & PRELL, W.L. 1992. The structure of the southwest monsoon winds over the Arabian Sea during the Late Quaternary: observations, simulations and marine geologic evidence. *Journal of Geophysical Research*, **97**(C10), 15481–15487.

ANDERSON, D.M. & PRELL, W.L. 1993. A 300 kyr record of upwelling off Oman during the Late Quaternary: evidence of the Asian Southwest Monsoon. *Paleoceanography*, **8**(2), 193–208.

ANDREWS, J.E., SINGHVI, A.K., KAILATH, A.J., KUHN, R., DENNIS, P.F., TANDON, S.K. & DHIR, R.P. 1998. Do stable isotope data from calcrete record Late Pleistocene climatic variation in the Thar Desert of India? *Quaternary Research*, **50**, 240–251.

BISWAS, S.K. 1971. The Miliolite rocks of Kutch and Kathiawar (western India). *Sedimentary Geology*, **5**, 147–164.

BOULTON, G.S. 1993. Ice ages and climate. *In:* DUFF, P.McL.D. (eds) *Holmes' Principles of Physical Geology*, 4th. Chapman & Hall, London, 439–469.

BRISTOW, C.S. 1999. Aeolian and sabkha sediments in the Miocene Shuwaihat Formation, Emirate of Abu Dhabi, United Arab Emirates. *In:* WHYBROW, P.J. & HILL, A. (eds) *Fossil Vertebrates of Arabia.* Yale University Press, New Haven, CT, 50–60.

BROOKE, B. 2001. The distribution of carbonate eolianite. *Earth-Science Reviews*, **55**, 135–164.

BURNS, S.J., MATTER, A., FRANK, N. & MANGANI, A. 1998. Speleothem based paleoclimatic record from Northern Oman. *Geology*, **26**(6), 499–502.

CHAPMAN, M.R. & SHACKLETON, N.J. 1998. Millennial-scale fluctuations in North Atlantic heat flux during the last 150,000 years. *Earth and Planetary Science Letters*, **159**, 57–70.

CHAWLA, S., DHIR, R.P. & SINGHVI, A.K. 1992. Thermoluminescence chronology of sand profiles in the Thar desert and their implications. *Quaternary Science Reviews*, **11**, 25–32.

CLEMENS, S. & PRELL, W. 1990. Late Pleistocene variability of Arabian sea summer monsoon winds and continental aridity: eolian records from the lithogenic component of deep-sea sediments. *Paleoceanography*, **5**, 109–145.

CLEMENS, S.C., MURRAY, D.W. & PRELL, W.L. 1996. Nonstationary phase of the Plio-Pleistocene Asian monsoon. *Science*, **274**, 943–947.

DALONGEVILLE, R., BERNIER, P., DUPUIS, B. & DE MEDWICKI, V. 1993. Les variations Récentes de la Ligne di Rivage dans le Golfe Persique: l'ememple de la lagune d'Umm Al-Qowayn (Emirates Arabes Unis). *Bulletin de l'institut de Geologie du Bassin Aquitaine*, **53**, 179–192.

DEMENOCAL, P., ORTIZ, J., GUILDERSON, T., ADKINS, J., SARNTHEIN, M., BAKER, L. & YARUSINSKY, M. 2000. Abrupt onset and termination of the African Humid Period: rapid climate responses to gradual insolation forcing. *Quaternary Science Reviews*, **19**, 347–361.

DHIR, R.P., KAR, A., WADHAWAN, S.K., RAJAGURU, S.N., MISRA, V.N., SINGHVI, A.K. & SHARMA, S.B. 1992. *Thar Desert in Rajasthan: Land, Man and Environment.* Geological Society of India, Bangalore.

EDENS, C. 1988. Archaeology of the Sands and adjacent portions of the Sharqiyah. *Journal of Oman Studies, Special Report*, **3**, 113–130.

EVANS, G. 1995. The Arabian Gulf: a modern carbonate–evaporite factory: a review. *Cuadernaos de Geologia Iberica*, **19**, 61–96.

EVANS, G., KENDALL, C.G.ST.C. & SKIPWITH, R. 1964. Origin of the coastal flats, the sabkha, of the Trucial Coast, Persian Gulf. *Nature*, **202**, 759–761.

EVANS, G., SCHMIDT, V., BUSH, P. & NELSON, H. 1969. Stratigraphic and geologic history of the Sabkha, Abu Dhabi, Persian Gulf. *Sedimentology*, **12**, 145–159.

FAIRBRIDGE, R.W. 1962. World Sea-Level and Climatic Changes. *Quaternaria*, **6**, 111–134.

FELBER, H., HÖTZL, H., MAURIN, V., MOSER, H., RAUERT, W. & ZÖTL, J.G. 1978. Sea level fluctuations during the Quaternary Period. *In:* AL-SAYARI, S.A. & ZÖTL, J.G. (eds) *Quaternary Period in Saudi Arabia*. Springer, Vienna, **1**, 50–57.

FRIEND, P.F. 1999. Rivers of the Lower Baynunah Formation, Emirate of Abu Dhabi, United Arab Emirates. *In:* WHYBROW, P.J. & HILL, A. (eds) *Fossil Vertebrates of Arabia*. Yale University Press, New Haven, CT, 39–49.

GARDNER, R.A.M. 1988. Aeolianites and marine deposits of the Wahiba Sands: character and palaeoenvironments. *Journal of Oman Studies, Special Report*, **3**, 75–94.

GLENNIE, K.W. 1987. Desert sedimentary environments, present and past—a summary. *Sedimentary Geology*, **50**, 135–165.

GLENNIE, K.W. 1998. The desert of southeast Arabia: a product of quaternary climatic change. *In:* ALSHARHAN, A.S., GLENNIE, K.W., WHITTLE, G.L. & KENDALL, C.G.ST.C. (eds) *Quaternary Deserts and Climatic Change*. Balkema, Rotterdam, 279–291.

GLENNIE, K.W. 1999. Dunes as indicators of climatic change. *In:* SINGHVI, A.K. & DERBYSHIRE, E. (eds) *Paleoenvironmental Reconstruction in Arid Lands*. Oxford & IBH, New Delhi, 153–174.

GLENNIE, K.W. & GÖKDAG, H. 1998. Cemented Quaternary dune sands, Ras Al Hamra housing area, Muscat, Sultanate of Oman. *In:* ALSHARHAN, A.S., GLENNIE, K.W., WHITTLE, G.L. & KENDALL, C.G.ST.C. (eds) *Quaternary Deserts and Climatic Change*. Balkema, Rotterdam, 109–116.

GLENNIE, K.W. & SINGHVI, A.K. 2002. Event stratigraphy, palaeoenvironment & chronology of SE Arabian Deserts. *Quaternary Science Reviews (Special Issue)*, **21**, 853–869.

GLENNIE, K.W., AL-BELUSHI, J. & AL-MASKERY, S. 1998. The inland sabkhas of the Huqf, Oman: a product of past extremes of humidity and aridity. *In:* ALSHARHAN, A.S., GLENNIE, K.W., WHITTLE, G.L. & KENDALL, C.G.ST.C. (eds) *Quaternary Deserts and Climatic Change*. Balkema, Rotterdam, 279–291.

GOODALL, T.M. 1995. *The geology and geomorphology of the Sabkhat Matti region (United Arab Emirates): a modern analogue for ancient desert sediments of north-west Europe*. PhD thesis, Univeristy of Aberdeen.

GOUDIE, A.S., ALCHIN, B. & HEGDE, T.M. 1973. The former extensions of the Great Indian Sand Desert. *Geographical Journal*, **139**, 243–257.

GOUDIE, A.S., COLLS, A., STOKES, S., PARKER, A., WHITE, K. & AL-FARRAJ, A. 2000. Latest Pleistocene and Holocene dune construction at the northeastern edge of the Rub Al Khali, United Arab Emirates. *Sedimentology*, **47**, 1011–1021.

HADLEY, D.G., BROUWERS, E.M. & BOWN, T.M. 1998. Quaternary paleodunes, Arabian Gulf Coast, Abu Dhabi Emirate: age and paleoenvironmental evolution. *In:* ALSHARHAN, A.S., GLENNIE, K.W., WHITTLE, G.L. & KENDALL, C.G.ST.C. (eds) *Quaternary Deserts and Climatic Change*. Balkema, Rotterdam. 123–139.

HEUSSER, L.E. & SIROCKO, F. 1977. Millennial pulsing of environmental change in southern California from the past 24 ky: a record of Indo-Pacific ENSO events *Geology*, **25**, 243–246.

JONES, J.R., WEIER, H. & CONSIDINE, P. 1988. The subsurface geology and hydrogeology of the Wahiba Sands. *Journal of Oman Studies, Special Report*, **3**, 75–94.

JUYAL, N., KAR, A., RAJAGURU, S.N. & SINGHVI, A.K., 2002. Luminescence chronology of aeolian accretion in the southern margin of Thar Desert, India. *Quaternary International* (submitted).

JUYAL, N., SINGHVI, A.K. & GLENNIE, K.W. 1998. Chronology and paleoenvironmental significance of Quaternary desert sediment in doutheastern Arabia. *In:* ALSHARHAN, A.S., GLENNIE, K.W., WHITTLE, G.L. & KENDALL, C.G.ST.C. (eds) *Quaternary Deserts and Climatic Change*. Balkema, Rotterdam, 315–325.

KAR, A., SINGHVI, A.K., RAJAGURU, S.N., JUYAL, N., THOMAS, J.V., BANERJEE, D. & DHIR, R.P. 2000. Reconstruction of the Late Quaternary environments of the lower Luni plains, Thar Desert, India. *Journal of Quaternary Science*, **16**, 61–68.

KENDALL, C.G.ST.C., ALSHARHAN, A.S. & WHITTLE, G.L. 1998. The flood recharge sabkha model supported by recent inversions of anhydrite to gypsum in the UAE sabkha. *In:* ALSHARHAN, A.S., GLENNIE, K.W., WHITTLE, G.L. & KENDALL, C.G.ST.C. (eds) *Quaternary Deserts and Climatic Change*. Balkema, Rotterdam, 29–42.

KIRKHAM, A. 1998*a*. A Quaternary proximal foreland ramp and its continental fringe, Arabian Gulf, UAE. *In:* WRIGHT, V.P., & BURCHETTE, T.P. (eds) *Carbonate Ramps*. Geological Society, London, Special Publications, **149**, 15–41.

KIRKHAM, A. 1998*b*. Pleistocene Carbonate Seif dunes and their role in the development of complex past and Present coastlines of the U.A.E. *GeoArabia*, **3**, 19–32.

KOCUREK, G. & LANCASTER, N. 1999. Aeolian system sediment state: theory and Mojave Desert dune field example. *Sedimentology*, **46**, 505–515.

KOLLA, V. & BISCAYE, P.E. 1977. Distribution and origin of quartz in the sediments of the Indian Ocean. *Journal of Sedimentary Petrology*, **47**, 642–649.

KROON, D., STEENS, T. & TROELSTRA, S.R. ET AL. 1991. Onset of monsoonal related upwelling in the western Arabian Sea as revealed by planktonic foraminifers. *In:* PRELL, W.L. & NIITSUMA, N. (eds) *Proceedings of the Ocean Drilling Program, Scientific Results*, *117*. Ocean Drilling Program, College Station, TX, 257–264.

LAMBECK, K. 1996. Shoreline reconstructions for the Persian Gulf since the last glacial maximum. *Earth and Planetary Science Letters*, **142**, 43–57.

LAMBECK, K. 1998. Shoreline reconstructions for the Persian Gulf since the last glacial maximum. *Earth and Planetary Science Letters*, **142**, 43–57.

LE METEUR, J., MICHEL, J.C., BECHENNEC, F., PLATEL, J.P. & ROGER, J. 1995. *Geology and Mineral wealth of the Sultanate of Oman*, Sultanatge Ministry of Petroleum and Minerals, Directorate General of Minerals.

LÉZINE, A.M., SALIEGE, J.-F., ROBERT, C., WERTZ, F. &

INIZAN, M.-L. 1998. Holocene lakes from Ramat as-Sab'atayan (Yemen) illustrate the impact of monsoon activity in Southern Arabia. *Quaternary Research*, **50**, 290–299.

LÜCKGE, A., DOOSE-ROLINSKI, H., KHAN, A.A., SCHULZ, H. & VON RAD, U. 2001. Monsoonal variability in the northeastern Arabian Sea during the last 5000 years: geochemical evidence from laminated sediments. *Palaeogeography, Palaeoclimatology, Palaeoecology*, **167**, 273–286.

MAIZELS, J. 1988. Palaeochannels: Plio-Pleistocene raised channel systems of the western Sharqiyah. *Journal of Oman Studies, Special Report*, **3**, 95–112.

MAIZELS, J. & MCBEAN, C. 1990. Cenozoic alluvial fan system of interior Oman: palaeoenvironmental reconstruction based on discrimination of palaeochannels using remotely sensed data. *In:* ROBERTSON, A.H.F., SEARLE, M.P. & RIES, A.C. (eds) *The Geology and Tectonics of the Oman Region*. Geological Society, London, Special Publications, **49**, 565–582.

MCCLURE, H.A. 1976. Radiocarbon chronology of late Quaternary lakes in the Arabian Desert. *Nature*, **263**, 755–756.

MCCLURE, H.A. 1978. The Rub al-Khali. *In:* AL-SAYARI, S.A. & ZOTL, J.G. (eds) *Quaternary Period in Saudi Arabia*. Springer, Vienna, **1**, 252–263.

MCCLURE, H.A., 1984. *Late Quaternary palaeoenvironments of the Rub' al Khali*. PhD thesis, University of London.

NANSON, G.C., PRICE, D.M. & SHORT, S.A. 1992. Wetting and drying of Australia over the past 300 ka. *Geology*, **20**, 791–794.

PATEL, M.P. & BHATT, N. 1995. Evidence of palaeoclimatic fluctuations in miliolite rocks of Saurashtra, western India. *Journal of the Geological Society of India*, **45**, 191–200.

PEASE, P.P., TCHAKERIAN, V.P. & TINDALE, N.W. 1998. Aerosols over the Arabian Sea: geochemistry and source areas ofor aeolian desert dust. *Journal of Arid Environments*, **39**, 477–496.

PEEBLES, R.G. 1999. Stable isotope analyses and dating of the Miocene of the Emirate of Abu Dhabi, United Arab Emirates. *In:* WHYBROW, P.J. & HILL, A. (eds) *Fossil Vertebrates of Arabia*. Yale University Press, New Haven, CT, 88–105.

PETIT-MAIRE, N. 1994. Natural variability of the Asian, Indian and African monsoons over the last 130 ka. *In:* DESBOIS, M. & DÉSALMAND, F. (eds) *Global Precipitation and Climatic Change*. NATO ASI Series, **126**, 5–26.

PUGH, J.M., 1997. *The Quaternary desert sediments of the Al Liwa area, Abu Dhabi*. PhD thesis, University of Aberdeen.

PYE, K. 1987. *Aeolian Dust and Dust Deposits*. Academic Press, London.

RAO, K.K., WASSON, R.J. & KUTTY, M.K. 1989. Foraminifera from Late Quaternary dune sands of the Thar Desert, India. *Society of Economic Paleontologists and Mineralogists. Research Reports*, 168–180.

SANLAVILLE, P. 1992a. Changements climatiques dans la péninsule arabique durant le Pléistocène supérieure et l'Holocène. *Paléorient*, **18**, 5–26.

SANLAVILLE, P., 1992b. Paleoenvironment in the Arabian peninsula during the upper Pleistocene and the Holocene (preprint). International Symposium on Evolution of Deserts (IGCP-252), Ahmedabad.

SARNTHEIN, M. 1972. Sediments and history of the postglacial transgression in the Persian gulf and Northwest Gulf of Oman. *Marine Geology*, **12**, 245–266.

SARNTHEIN, M. 1978. Sand deserts during glacial maximum and climatic optimum. *Nature*, **272**, 43–46.

SCHULZ, H., VON RAD, U. & ERLENKEUSER, H. 1998. Correlation between Arabian Sea and Greenland climate oscillations of the past 110,000 years. *Nature*, **393**, 54–57.

SIROCKO, F. & LANGE, H. 1991. Clay-mineral accumulation rates in the Arabian Sea during the late Quaternary. *Marine Geology*, **97**, 105–119.

SIROCKO, F. & SARNTHEIN, M. 1989. Wind-borne deposits in the northwestern Indian Ocean: record of Holocene sediments versus modern satellite data. *In:* LEINEN, M. & SARNTHEIN, M. (eds) *Paleoclimatology and Paleometeorology: Modern and Past Pattern of Global Atmospheric Transport*. Kluwer Academic, Dordrecht, 401–433.

SIROCKO, F., GARBE-SCHÖNBERG, D., MCINTYRE, A. & MOLFINO, B. 1996. Teleconnections between the subtropical monsoons and high-latitude climates during the last deglaciation. *Science*, **272**, 526–529.

TELLER, J.T., GLENNIE, K.W., LANCASTER, N. & SINGHVI, A.K. 2000. Calcareous dunes of the United Arab Emirates and Noah's Flood: the postglacial reflooding of the Persian (Arabian) Gulf. *Quaternary International*, **68–71**, 297–308.

THOMAS, J.V., KAR, A., KAILATH, A.J., JUYAL, N., RAJAGURU, S.N. & SINGHVI, A.K. 1999. Late Pleistocene–Holocene history of aeolian accumulation in the Thar Desert, India. *Zeitschrift für Geomorphologie, N.E. Supplementband*, **116**, 181–194.

TRICART, J., MICHEL, P. & VOGT, J. 1957. Oscillations climatiques quaternaries en Afrique Occidental. Abstract, V Congrès International INQUA (Madrid–Barcelona), 187–188.

UNITED ARAB EMIRATES UNIVERSITY 1993. *National Atlas of the United Arab Emirates*. UAE University, Al Ain.

VAN CAMPO, E. 1991. Pollen transport into Arabian Sea sediments. *In:* PRELL, W.L. & NIITSUMA, N. et al. (eds) *Proceedings of the Ocean Drilling Program Scientific Results* **147**, 277–287.

WALKDEN, G. & WILLIAMS, A. 1998. Carbonate ramps and the Pleistocene—Recent depositional systems of the Arabian Gulf. *In:* WRIGHT, V.P. & BURCHETTE, T.P. (eds) *Carbonate Ramps*. Geological Society, London, Special Publications, **149**, 43–53.

YAN, Z. & PETIT-MAIRE, N. 1994. The last 140 ka in the Afro-Asian arid/semi-arid transitional zone. *Palaeogeography, Palaeoclimatology, Palaeoecology*, **110**, 217–233.

ZIEGLER, M.A. 2001. Late Permian to Holocene paleofacies evolution of the Arabian Plate and its hydrocarbon occurrences. *GeoArabia*, **6**(3), 445–504.

Calcareous cyst-producing dinoflagellates: ecology and aspects of cyst preservation in a highly productive oceanic region

INES WENDLER, KARIN A. F. ZONNEVELD & HELMUT WILLEMS

Universität Bremen, FB 5, Postfach 330 440, D-28334 Bremen, Germany
(e-mail: flatter@uni-bremen.de)

Abstract: Absolute and relative abundances of calcareous dinoflagellate cyst species in surface sediment samples from the Arabian Sea are compared with environmental parameters of the upper 100 m of the water column to gain information on their largely unknown autecology. Ten species or morphotypes were encountered of which four occurred only as accessories. On the basis of the distribution patterns of the six more abundant species or morphotypes, the studied area is subdivided into three provinces, demonstrating a clear relationship to monsoon-controlled upper-ocean conditions. The two dominant species, *Thoracosphaera heimii* and *Orthopithonella granifera*, show opposite trends in distribution of both their absolute and relative abundances. In the NE Arabian Sea, low absolute and relative abundances of *T. heimii* are mainly attributed to enhanced dissolution of the small tests in this region, whereas elevated concentrations of *O. granifera* seem to be related to higher water temperatures and the influence of the Indus River. *Sphaerodinella albatrosiana* and *Calciodinellum operosum* are most abundant in the open ocean, associated with lower nutrient levels, relatively high temperatures and low seasonality. Spiny cysts (mainly represented by *Scrippsiella trochoidea*), in contrast, exhibit a more shelf-ward distribution and are most abundant in regions that are influenced by coastal upwelling, characterized by eutrophic and rather unstable conditions with seasonally lower temperatures and a shallow thermocline. A generally negative correlation of calcareous dinoflagellate cysts with primary productivity or high nutrient concentrations, as proposed by other workers, cannot be confirmed. Cyst accumulation rates off Somalia show that strong turbulence and high current speeds are unfavourable for calcareous dinoflagellates, suggesting that these organisms are more successful under rather stratified conditions.

Dinoflagellates represent one of the major phytoplankton groups in the oceans. Some species produce a fossilizable calcareous stage as part of their life cycle and are hereafter referred to as calcareous dinoflagellates. These dinoflagellates are phototrophic and thus inhabit the photic zone. First studies on calcareous dinoflagellate cysts in sediment cores from the Atlantic Ocean revealed distinct temporal changes in absolute and relative cyst abundance, and the comparison of these data with other proxies provided valuable information on the (palaeo-) ecological significance of this organism group (Höll *et al.* 1998, 1999; Höll & Kemle-von Mücke 2000; Esper *et al.* 2000; Vink *et al.* 2002a, b). The application of calcareous dinoflagellate cysts for the reconstruction of environmental changes requires autecological knowledge of the individual species. However, such information, particularly from highly productive areas, is still sparse. To improve the use of calcareous dinoflagellate cysts as palaeo-environmental proxy we studied surface sediment samples from several parts of the Arabian Sea (Fig. 1). This highly productive oceanic region is characterized by strong seasonality in atmospheric and oceanic conditions and provides a wide spectrum of environmental settings, which can be compared with cyst distribution patterns. Whereas this study focuses on the ecology of calcareous dinoflagellates, in an accompanying paper the preservation of cysts is discussed in greater detail (Wendler *et al.* 2002).

Climatic and oceanographic setting

The climatic and oceanographic processes in the Arabian Sea are largely determined by strong monsoon winds, which reverse semi-annually as a result of the shifting position of the Inter-Tropical Convergence Zone (ITCZ). During the summer, a pressure gradient between Central Asia and the southern Indian Ocean results in a strong, topographically steered southwesterly wind (SW monsoon), which from May to September forms a strong low-level jet stream (also called the Findlater Jet; Findlater 1971) and extends across the Arabian Sea parallel to the coast of the Arabian Peninsula. The ocean reacts with the formation of

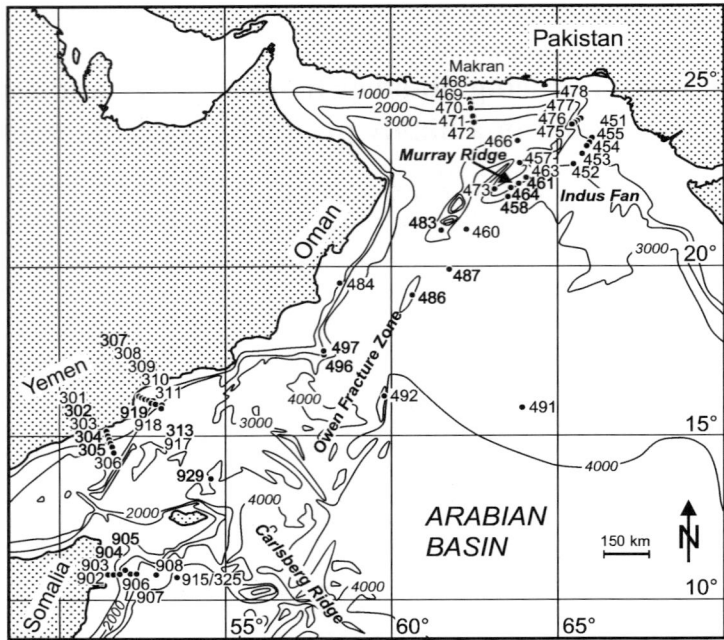

Fig. 1. Studied area with sample locations, bathymetry (m).

approximately clockwise surface currents (Wyrtki 1971; Shetye et al. 1994; Shi et al. 1999) including the strong Somali Boundary Current. A prominent feature of this current is the development of clockwise-rotating eddies (Bruce 1979; Schott 1983) that move north to NE at the end of the SW monsoon (Schott 1983; Fisher et al. 1996). Recent measurements based on acoustic Doppler current profiler and near-surface drifter tracks reveal an eastward transport of upper water masses south of 15°N and a strong dominance of upper-ocean currents in the northern Arabian Sea by large eddies (Elliot & Savidge 1990; Molinari et al. 1990; Flagg & Kim 1998). Ekman dynamics lead to coastal upwelling along the Somali and Arabian coasts, introducing cold, nutrient-rich water into the photic zone (e.g. Sastry & D'Souza 1972; Currie et al. 1973; Bruce 1974; Prell & Curry 1981; Prell & Streeter 1982) which raises primary productivity (e.g. Quraishee 1988; Brock et al. 1992; Smith et al. 1998). Coastal filaments, which are colder and fresher than their surroundings, carry nutrient-rich, highly productive waters into the central basin (Keen et al. 1997; Arnone et al. 1998; Manghnani et al. 1998; Lendt et al. 1999; Lee et al. 2000). Inshore of the wind-stress maximum (NW of the Findlater Jet axis) open oceanic upwelling occurs (e.g. Smith & Bottero 1977; Brock et al. 1992). Both offshore advection of coastally upwelled water and upward Ekman pumping counter the wind-driven entrainment and maintain upper-ocean stratification in this region (Lee et al. 2000). Deepening of the mixed layer in the central Arabian Sea (SE of the Findlater Jet axis) is attributed to convergence in the Ekman layer caused by negative wind-stress curl (Rao et al. 1989; Bauer et al. 1991), and to wind-driven entrainment (Lee et al. 2000). The SW monsoon drives strong evaporative salinity enhancement in the central basin, whereas the upwelling waters near the coasts are relatively fresh, resulting in a salinity gradient in the upper water layers. The mixed-layer waters cool and freshen during the autumn intermonsoon. This period is characterized by variable to northward surface currents, shoaling of pycnoclines and very shallow mixed layers (e.g. Dickey et al. 1998; Lee et al. 2000).

During winter, the dry and cold NE monsoon prevails, with generally lower wind stress magnitudes (2 dyn cm^{-2}) as compared with the SW monsoon (6 dyn cm^{-2}, Shetye et al. 1994). Also, the mean basin-wide flow of surface currents reverses during the NE monsoon to approximately anti-clockwise (Wyrtki 1971; Shetye et al. 1994; Shi et al. 1999). The NE monsoon leads to cooling of the surface waters, especially in the NE of the basin, which results in deep vertical mixing and dramatic deepening of the mixed layer with distance offshore (Bauer et al. 1991; Lee et al. 2000). Stratification is stronger at the base of the shallower mixed layers near the coast than beneath deep mid-basin mixed layers (Lee et al. 2000). Surface-water salinity is enhanced in the central basin and in the northern Arabian Sea, as a result of evaporation driven by the NE monsoon and advection of high-salinity Gulf of Oman water, respectively (Lee et al. 2000; Wiggert et al. 2000). The cool and salty surface water drives convective overturning, which causes repletion of the upper layers with nutrients and stimulates primary production especially in the northeastern part of the Arabian Sea (Banse & McClain 1986; Madhupratap et al. 1996; Dickey et al. 1998; Smith et al. 1998; Weller et al. 1998). During the spring intermonsoon, weakened wind forcing and strong surface heating lead to warming and re-stratification of the upper water layers and to shoaling of the mixed layer from depths as great as 120 m in February to c. 20 m in April (Gardner et al. 1999; Lee et al. 2000). Small mixed-layer variations limit the mixing of nutrients into the surface layer, thus maintaining oligotrophic conditions and low primary production with a subsurface chlorophyll maximum during this period (Gardner et al. 1999).

Except for the spring intermonsoon, the surface waters in the Arabian Sea are very fertile, especially in the northeastern part of the basin and off the Somali and Arabian coasts (Fig. 2), which makes the Arabian Sea one of the world's most productive oceanic provinces (Ryther et al. 1966). Apart from upwelling processes and convective overturning, further sources of nutrients are the Indus River discharge and deposition of aerosols. Recycling of large amounts of organic matter in combination with reduced mid-water aeration create a permanent and intense oxygen minimum zone (OMZ), which is a characteristic feature of the Arabian Sea. The oxygen-deficient zone impinges on the continental slopes of the surrounding landmasses at water depths ranging from 200 to 1200 m (e.g. von Rad et al. 1995), whereby the eastern Arabian Sea exhibits lower oxygen concentrations than the western region at the same latitude (Slater & Kroopnick 1984; Paropkari et al. 1992). In the cruise report of the Netherlands Indian Ocean Programme (NIOP) it was concluded 'that the contrast between the two monsoonal periods is very marked in the upper 100 m of the water column, and that there are hardly

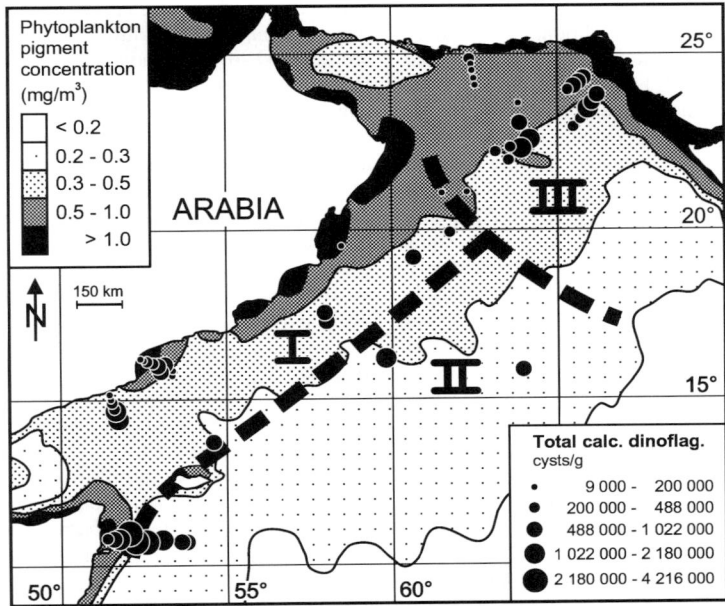

Fig. 2. Map showing the total abundance of calcareous dinoflagellate cysts in relation to phytoplankton pigment concentrations (mg m^{-3}; after composite satellite images over 8 years from NASA–GSFC) which reflect nutrient supply in surface waters. Black dashed line separates three provinces (I–III) based on distribution patterns of calcareous dinoflagellate cysts, and divides the Arabian Sea into a northeastern (III) and a southwestern part (I and II) as used in this paper.

differences below 150 m depth' (van Hinte et al. 1995).

Material and methods

Fifty-five surface sediment samples from the Arabian Sea (Fig. 1) were analysed for their content of calcareous dinoflagellate cysts. The samples represent the upper centimetre of box cores that were recovered during the NIOP in 1992–1993 (van Hinte et al. 1995). (For details on positions and water depths, see Appendix B.) Approximately 0.5 g of the dried sediment was weighted and disintegrated in water (containing a few drops of ammonia to prevent calcite dissolution) by ultrasound treatment of <1 min. The sediment was subsequently sieved over 63 μm and 20 μm stainless steel sieves to concentrate the larger cysts. The <20 μm and 20–63 μm fractions were concentrated to 100 ml and 15 ml of water, respectively. A split (50 μl or 100 μl) of the homogenized material of the two fractions was separately placed on a cover slip, dried in an oven or on a heating plate, and finally fixed with Spurr's resin. More detailed information on the preparation method has been given by Vink et al. (2000).

The cysts were counted under a light microscope using polarized light (Janofske 1996). At least one slide per fraction and sample was scanned. If there were fewer than 200 specimens in any slide for a fraction, additional slides were analysed. We use the same taxonomy as was used by Wendler et al. (2002). It should, however, be mentioned that a new taxonomic concept is currently in preparation (Meier et al., pers. comm.). The species discussed in the present paper are illustrated in Fig. 15, below, and a list of their new generic attribution is given in Appendix A. The spiny cysts, in the studied sediments, show a large morphological variation in shape and size as well as the shape and number of calcite crystals and spines. Most spiny cysts in the studied material appear to belong to *Scrippsiella trochoidea* but the group may also contain *Scrippsiella regalis* (and possibly other undescribed spiny cysts). A clear separation of the various species of spiny cysts under the light microscope was often not possible because of organic matter between the spines hiding the characteristic shape of the calcite crystals.

Absolute abundance (in cysts g^{-1} of dry sediment, Appendix B) and cyst accumulation rates (cyst AR, in cysts cm^{-2} ka^{-1}) were calculated as described by Wendler et al. (2002). Furthermore, the relative abundance of each species or morphotype was calculated. Two species, *Thoracosphaera heimii* and *Orthopithonella granifera*, clearly dominate the association, and their relative abundance is given in per cent of the whole association. For the less abundant species or morphotypes, the relative abundance is based on the association excluding the two dominant species. The geographical distributions of absolute and relative abundance of each species are illustrated in Figs 3–9. The chosen limits for dot sizes in Figs 4–9 are based on natural breaks. The distribution patterns of the individual species or morphotypes have been visually compared with physical parameters in the upper 100 m of the water column in five standard water depths: 0 m (mean of values from 0 to 5 m depth); 25 m (mean of values from 20 to 30 m depth); 50 m (mean of values from 45 to 55 m depth); 75 m (mean of values from 70 to 80 m depth); 100 m (mean of values from 95 to 105 m depth); during four periods: NE monsoon (December–February); spring intermonsoon (March–May); SW monsoon (June–September); autumn intermonsoon (October–November).

Mean temperature and salinity values for the last 92 years of 1° latitude and longitude square blocks were obtained from the National Oceanographic Data Center, Washington, DC. Density and Brunt–Väisälä frequency (as a measure of stratification) were calculated as given by Vink et al. (2000). In the comparison of cyst distributions with environmental parameters we pay special attention to conditions in water depths between 50 and 100 m, as field and laboratory studies indicate that *T. heimii* and possibly also the other species are adapted to low-irradiance conditions and preferentially inhabit the lower part of the photic zone (Karwath 2000). Statistical analyses have been carried out but did not provide significant additional information. In contrast, the influence of regionally varying and species-selective calcite dissolution in the studied area (see Discussion) bias the ecological signal in the statistical analyses. This effect cannot be eliminated from statistics as it cannot be quantified, making statistical interpretations potentially hazardous.

Results

All of the investigated samples contained calcareous dinoflagellate cysts. Of the 10 species recovered only six occurred in significant concentrations: *Thoracosphaera heimii*, *Orthopithonella granifera*, *Sphaerodinella albatrosiana*, *Sphaerodinella tuberosa* var. 2, a group of spiny cysts (mainly *Scrippsiella trochoidea*) and *Calciodinellum operosum*. These species were found throughout the Arabian Sea in varying concentrations. The four rare species (Appendix C) do not have distinct distribution patterns, except for *S. tuberosa* var. 1, which was found only near the Gulf of Aden.

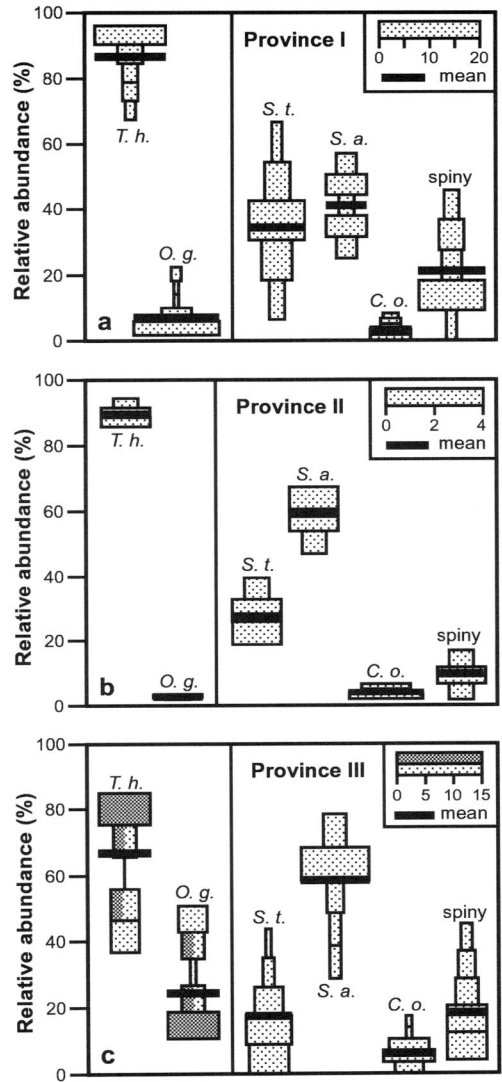

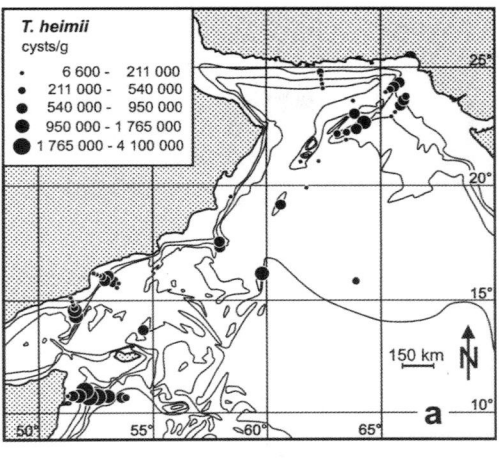

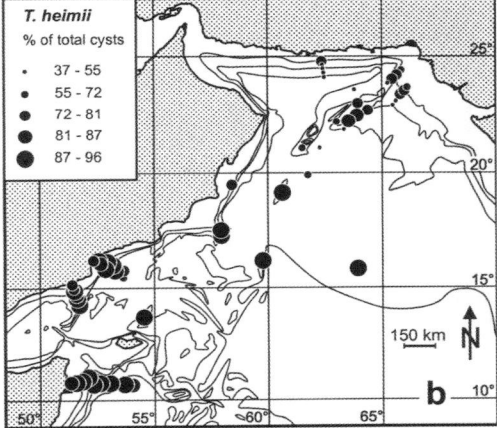

Fig. 4. Surface sediment distribution map of *Thoracosphaera heimii*: (**a**) absolute abundance; (**b**) relative abundance (per cent of whole association).

Fig. 3. Comparison of relative abundances within the three cyst provinces (see Fig. 2). *T. heimii* and *O. granifera* are given in per cent of the whole association, the other species in per cent of the association without *T. heimii* and *O. granifera*. Similar to a histogram, five (**a**, **c**) or three (**b**) clusters were separated, whereby the width of the bars indicates the number of samples contained in the cluster (scale in upper right corner). Dark grey in (**c**) indicates samples within the OMZ for *T. heimii* and *O. granifera*. Species abbreviations: *T. h.*, *Thoracosphaera heimii*; *O. g.*, *Orthopithonella granifera*; *S. t.*, *Sphaerodinella tuberosa* var. 2; *S. a.*, *Sphaerodinella albatrosiana*; *C. o.*, *Calciodinellum operosum*; spiny, spiny cysts.

Highest total cyst concentrations of up to 4 million cysts g^{-1} were seen off Somalia and on the Murray Ridge, in water depths between 900 and 2000 m. Low cyst concentrations were observed in samples from the Makran margin, in the shallowest samples off Yemen, Somalia and Oman, and in samples below 1500 m in the NE Arabian Sea (Figs 2 and 4–9). In most samples, the association is clearly dominated by *T. heimii* (Fig. 4b). The second most abundant species is *O. granifera* with maximal 51% (Fig. 5b). Together, these two species form 76–98% of the association.

Distribution of individual species

T. heimii. High absolute and relative abundance of *T. heimii* was found mainly in the SW of the

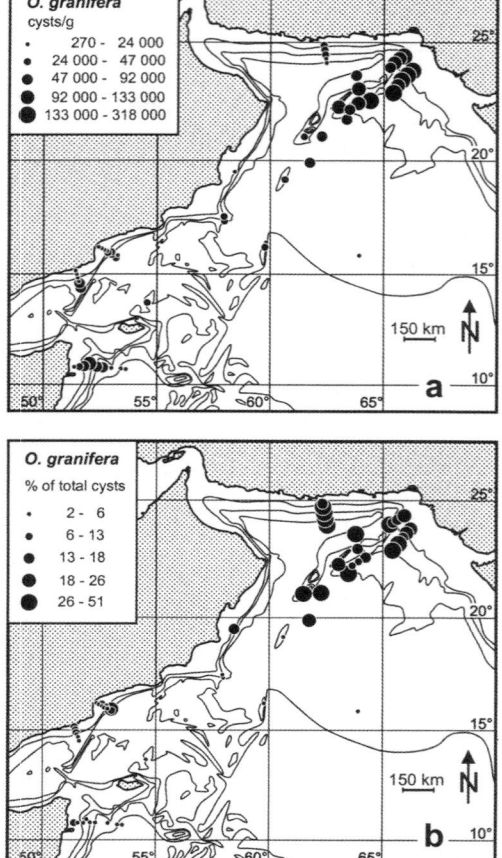

Fig. 5. Surface sediment distribution map of *Orthopithonella granifera*: (**a**) absolute abundance; (**b**) relative abundance (per cent of whole association).

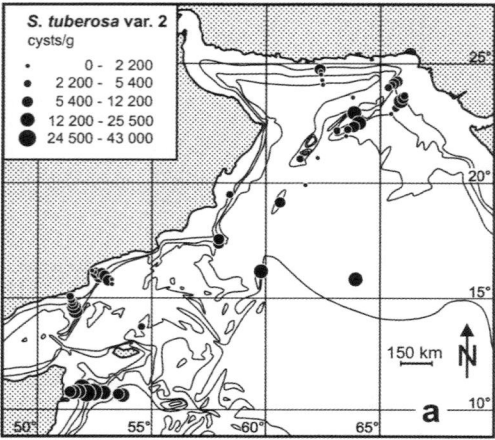

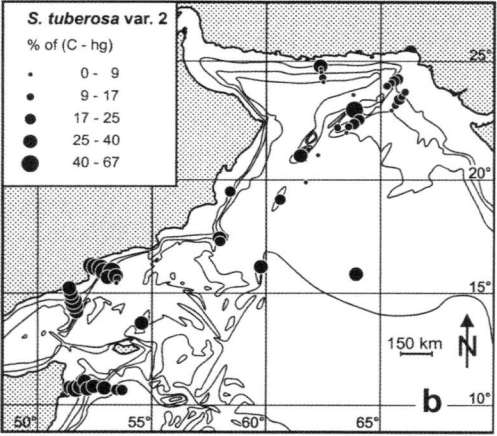

Fig. 6. Surface sediment distribution map of *Sphaerodinella tuberosa* var. 2: (**a**) absolute abundance; (**b**) relative abundance (per cent of association excluding *T. heimii* and *O. granifera*).

studied area (Fig. 4). In the NE of the basin, some elevated values occur on the Murray Ridge and on the Indus Fan, in water depths shallower than 1300 m. Fragmentation of this species is high (up to 22% of total specimens) in those samples with low absolute and relative abundance, namely in the NE Arabian Sea except for samples from water depths shallower than 1300 m (Fig. 10). The percentage of fragments is low (mainly 1–3%, maximal 7%) in the SW of the area. The ratio of other species to *T. heimii* in the SW decreases from the shallow samples towards water depths of 1000 m and remains low down to 3000 m, with a slight increase again below that depth (Fig. 11a). In the NE, this ratio is generally higher than in the SW and shows a drastic increase below 1500 m depth (Fig. 11b).

O. granifera. In contrast to *T. heimii*, *O. granifera* has high absolute and relative abundances only in the NE Arabian Sea, with the exception of the Makran margin, where high relative but low absolute abundances are found (Fig. 5).

S. tuberosa var. 2. The distribution pattern of *S. tuberosa* var. 2 is comparable with that of *T. heimii*, with generally higher concentrations in the SW of the area (Fig. 6). It also shows elevated values in the samples from water depths shallower than 1300 m on the Murray Ridge and on the Indus Fan. High relative abundances are seen especially off Yemen and Somalia.

S. albatrosiana. *S. albatrosiana* is generally more abundant in the open ocean and in the NE of the

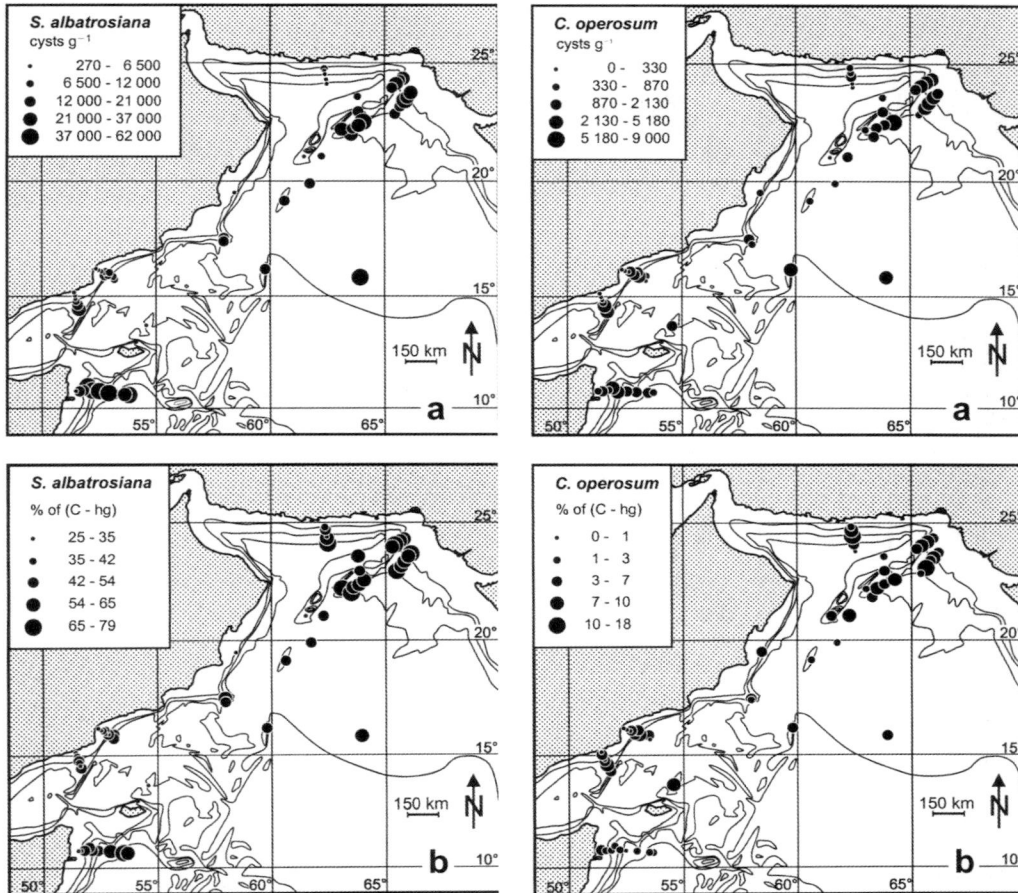

Fig. 7. Surface sediment distribution map of *Sphaerodinella albatrosiana*: (**a**) absolute abundance; (**b**) relative abundance (per cent of association excluding *T. heimii* and *O. granifera*).

Fig. 8. Surface sediment distribution map of *Calciodinellum operosum*: (**a**) absolute abundance; (**b**) relative abundance (per cent of association excluding *T. heimii* and *O. granifera*).

Arabian Sea, whereas low concentrations and relative abundances can be observed close to the Somali and Arabian coasts (Fig. 7).

C. operosum. The distribution pattern of *C. operosum* is similar to that of *S. albatrosiana*, with higher absolute and relative abundances in the open ocean and in the NE Arabian Sea (Fig. 8).

Spiny cysts. In contrast to *S. albatrosiana*, the spiny cysts are less abundant in samples from the open ocean (Fig. 9). Their absolute and relative abundance is especially high offshore Oman (Owen Ridge area) and off Somalia and Yemen.

Cyst accumulation rate (AR)

The correction of cyst concentrations for sedimentation rates along the Indus Fan and Somali profiles results in a relative increase of values near the coast, a decrease at deeper stations and a slight shift of maximal concentrations towards shallower water depths, whereby the general shape of the curve does not change significantly (Fig. 12). Cyst AR values in the profile off Somalia are higher than in the Indus Fan profile (Fig. 13). In the Somali transect, most species have maximum AR values between 1000 and 2000 m water depth (Fig. 13a). Exceptions are *S. tuberosa* var. 2, with a maximum that lies closer to the coast at about

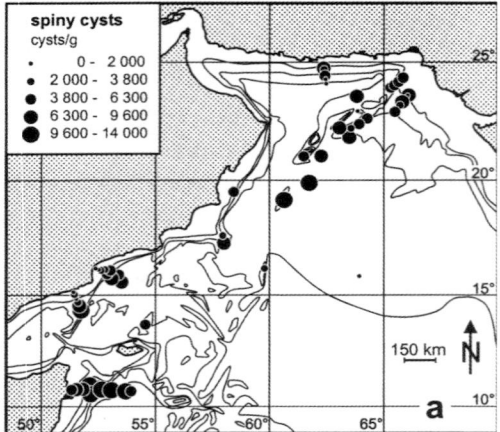

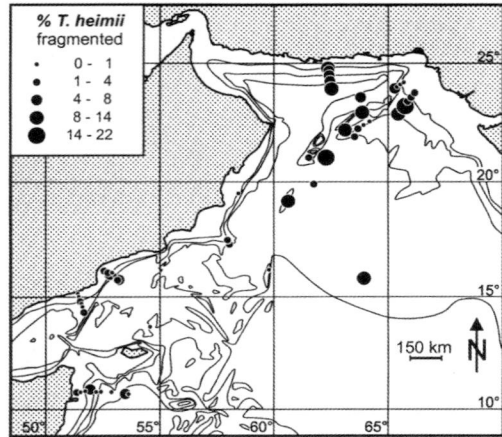

Fig. 10. Surface sediment distribution map showing fragmentation of *Thoracosphaera heimii*.

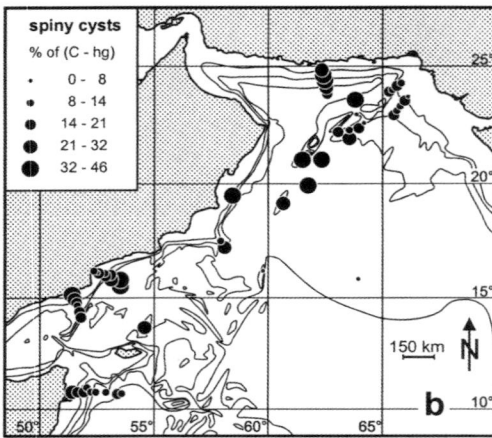

Fig. 9. Surface sediment distribution map of spiny cysts (mainly *Scrippsiella trochoidea*): (**a**) absolute abundance; (**b**) relative abundance (per cent of association excluding *T. heimii* and *O. granifera*).

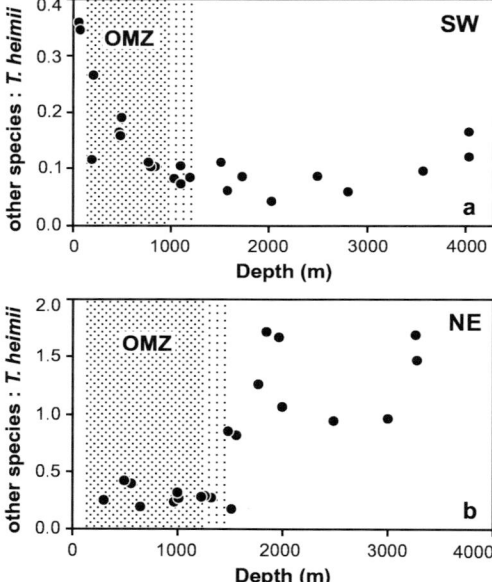

Fig. 11. Ratio of other species to *Thoracosphaera heimii* v. water depth in the SW (**a**) and NE (**b**) of the studied region (see Fig. 2 for separation of the two parts). Grey area marks the depth interval where the OMZ intersects the slope. (Note the marked increase of values at the lower boundary of the OMZ in the NE and the generally much lower values in the SW.)

800 m depth, and the spiny cysts, which decrease continuously with distance from the coast. All species show strongly decreased AR values below 3000 m depth, whereby the AR curve of *T. heimii* exhibits the steepest slope. In the Indus Fan profile the cyst AR values are high in the upper samples and drop significantly between 1250 and 1500 m depth, at the lower boundary of the OMZ.

Cyst provinces

Combining the distribution patterns of absolute and relative abundances of all species, three provinces in the studied area were selected by eye (Fig. 2):

province I: NW Arabian Sea with upper Somali continental slope, shelf areas of Arabia and adjacent deeper parts NW of the Owen Fracture Zone;

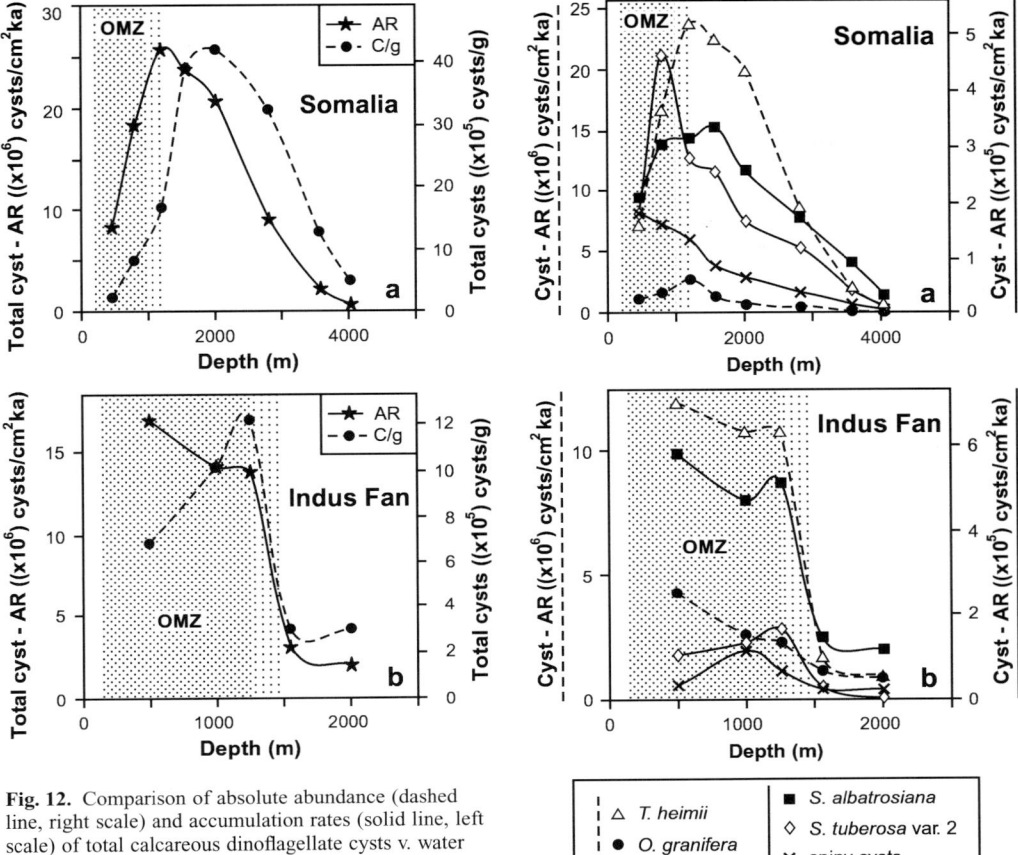

Fig. 12. Comparison of absolute abundance (dashed line, right scale) and accumulation rates (solid line, left scale) of total calcareous dinoflagellate cysts v. water depth in the profile off Somalia (**a**) and the Indus Fan profile (**b**).

Fig. 13. Cyst accumulation rates (AR) of the different species v. water depth in the profile off Somalia (**a**) and in the Indus Fan profile (**b**). Grey area marks the depth interval where the OMZ intersects the slope. Left scale applies for the two dominant species *T. heimii* and *O. granifera*. The marked drop in AR values at the lower boundary of the OMZ in the Indus Fan profile should be noted. In the Somali section AR values of most species are lower near the coast at sites of active coastal upwelling.

province II: open ocean with the central Arabian Basin, northern Somali Basin and adjacent lower slope;

province III: NE Arabian Sea including the Makran margin, Murray Ridge and the upper Indus Fan.

The characterizing species are listed in Table 1. A comparison of their relative abundance within the provinces is given in Fig. 3.

Discussion

In the Arabian Sea, the same species of calcareous dinoflagellate cysts were observed as are reported from the tropical Atlantic Ocean and the Caribbean Sea (Höll *et al.* 1998, 1999; Vink *et al.* 2000; Zonneveld *et al.* 2000). The Arabian Sea differs in the low abundance of *S. tuberosa* var. 1 and *S. tuberosa* var. 2 and the high abundance of *O. granifera* (Höll *et al.* 1999; Zonneveld *et al.* 2000; Vink *et al.* 2002*b*). The mean AR values of *T. heimii* and *S. albatrosiana* are slightly higher than Höll *et al.* (1999) reported from the eastern and western equatorial Atlantic but considerably lower than in the Caribbean Sea (Vink *et al.* 2002*b*). To understand to what extent the variety in absolute and relative abundance within the Arabian Sea and the differences from other oceanic regions are ecologically controlled, it is necessary to assess the impact of factors that modify

Table 1. *Provinces based on the distribution patterns of absolute and relative cyst abundances and their characteristic species*

Characteristic species	I	II	III
T. heimii	×	×	
S. tuberosa var. 2	×	×	
Spiny cysts	×		×
C. operosum		×	×
S. albatrosiana		×	×
O. granifera			×

the primary signals, such as transport, dilution and calcite dissolution.

Transport and dilution

Calcareous dinoflagellate cysts may be laterally transported in the water column or redistributed after settling. Lateral transport of the motile cells and cysts in the water column might occur via eddies and filaments; for example, in the 'great whirl' off Somalia (van Weering *et al.* 1997) and in offshore transporting eddies and topographically induced coastal squirts at the Oman shelf (e.g. Brock *et al.* 1992; Arnone *et al.* 1998; Latasa & Bidigare 1998; Manghnani *et al.* 1998; Lendt *et al.* 1999; Rixen *et al.* 2000). However, the very high primary production in the Arabian Sea (induced by seasonal upwelling) favours aggregation of smaller particles, which leads to fast settling of the sediment and reduces horizontal transport in the water column. Furthermore, strong diel variations in mixed-layer depths is reported for the NE Arabian Sea during the NE monsoon (Gardner *et al.* 1999), whereby restratification supports settlement of particles in deeper waters, which are unaffected by mixing in surface layers (Gardner *et al.* 1995). Indeed, fast settling of particles especially during the SW monsoon has been documented from sediment trap studies off Oman (Honjo *et al.* 1999) and in the Somali region. On the basis of a sediment trap study of sites 905 and 915, Zonneveld & Brummer (2000) found no evidence for lateral relocation of organic-walled dinoflagellate cysts during transport to the sea floor. Horizontal transport of small particles in the water column might therefore be of minor importance there.

Transport can also result from turbidites, bottom-water currents and bioturbation. The last may play a role above and below the OMZ but is very reduced or lacking within it, as is evident from laminated sediments that are common within the OMZ (e.g. Schulz *et al.* 1996; van der Weijden *et al.* 1999; von Rad *et al.* 1999; Smith *et al.* 2000).

Meadows *et al.* (2000) stated that microbiological rather than macrobenthic activity is the driving force in the processes which lead to the typical geochemical characteristics of the NE Arabian Sea sediments.

Gundersen *et al.* (1998) observed a deep particle maximum below the mixed layer in the northern Arabian Sea during both the SW and NE monsoon, which they interpreted as resulting from advection of resuspended sediment from the continental margin. Local winnowing by bottom-water currents is known from some stations at the Oman margin, on the Owen Ridge and on the Murray Ridge (Prins *et al.* 1994). The winnowed sediments are enriched in foraminifers and depleted in the fine fraction. Three samples of the studied material (457, 461, 484) showed these characteristics and have to be interpreted with care. At station 457, the cyst association differs from the surrounding samples (higher percentage of *S. tuberosa* var. 2 and lower *S. albatrosiana* values), which could indicate that recent material has been eroded from the continental margin.

Off Oman and Yemen, irregular sea-bed topography and frequently disturbed surface sediments, especially between 1000 and 1500 m water depth, were described (van Weering *et al.* 1997). Heier-Nielsen *et al.* (1995) reported frequent reworking of the inner shelf surface sediments off Yemen by slumping, bioturbation and mechanical mixing as a result of wave action during the SW monsoon. They assumed turbiditic flow processes to play an important role in transporting sediment from the upwelling zone off Yemen to the adjacent basin, and regarded deposition of older, reworked organic matter as being the reason for the large discrepancies in ^{14}C ages derived from organic matter and foraminifera. Episodic down-slope movement of sediment is also characteristic for the Makran margin, which is an active continental margin with high sedimentation rates (e.g. Prins *et al.* 1994). Accordingly, no reliable sedimentation rates are available for the Makran and the two Yemen profiles, and cyst AR values could not be calculated for these samples. Interpolation between the dated samples on the Murray Ridge and Owen Ridge or application of regional average sedimentation rates (e.g. given by Sirocko *et al.* 1991) would cause uncertainties that are larger than the variability in the dataset of cyst concentrations. Because of the down-slope transport at the Makran and Yemen margins, any distribution trends within the profiles have to be considered with care. Nevertheless, these samples give information on the general cyst association in these regions. It should be noted, however, that low cyst contents especially in the shallow samples of these three profiles are at least partly caused by dilution as a result of high near-coast sedimentation rates, whereby terrigenous material plays an important role (Kolla et al. 1981; Sirocko & Lange 1991; Sirocko et al. 1991).

Reliable sedimentation rates were available for the Indus Fan profile, for some samples on the Murray Ridge and for the Somali transect. Although there are hints for some across-slope transport of resuspended sediment at the Somali slope (Brummer 1995) most of the material is thought to be autochthonous, as the sedimentation rates in the profile decrease continuously with distance from the active upwelling zone, as would be expected. This assumption is also strengthened by studies with a long-term deployed tripod lander at 1500 m depth in the transect, which measured low current speeds in the boundary layer, implying that post-depositional removal of fine-grained sediment is not likely to occur (van der Land & Stel 1995). A further argument comes from continuously decreasing AR values of spiny cysts with depth along this profile (Fig. 13), which is thought to reflect their original distribution in coastal waters (see discussion below). Therefore, although some minor offshore transport may occur, the general trends in primary cyst production seem to be preserved in the Somali transect.

Although some small-scale transport may change local cyst distribution patterns there is no indication for large-scale transport in the Arabian Sea. This supposition is supported by the results of Zonneveld (1997), who studied organic-walled dinoflagellate cysts in the same samples and found no relation of the variance in the association to the ocean current system in the Arabian Sea.

Calcite preservation

Carbonate dissolution in the northern Somali Basin starts strongly at 3500 m depth and below, but the calcite compensation depth (CCD) is not reached with the deepest station of the studied transect (Troelstra et al. 1995). For the NE Arabian Sea, Millero et al. (1998) reported undersaturation with respect to calcite below 3400 m. Two stations from the Somali transect (908, 915), the deepest station at the Makran margin (472) and four stations on the abyssal plain of the Arabian Basin (458, 460, 487 and 491) were retrieved from water depths greater than 3000 m and could be affected by calcite dissolution as a result of deep-water undersaturation. The two deepest samples from Somalia indeed exhibit very low cyst AR values compared with the shallower stations (Fig. 13). Also, the four deep samples from the NE Arabian Sea show very low absolute abundances (e.g. Fig. 4a), especially for *T. heimii*, which is regarded as the most dissolution sensitive of the studied species (Wendler et al. 2002). In the sample from the central Arabian Basin (491), however, relatively high cyst concentrations of *S. albatrosiana* and *S. tuberosa* var. 2 and intermediate abundance of *T. heimii* were found, although water depth at this station is almost 3800 m. This could be caused by increased cyst production, better calcite preservation and/or lower sedimentation rates in the central than in the NE Arabian Sea.

For the NE Arabian Sea (province III) it was shown that the preservation of calcareous dinoflagellate cysts is enhanced within the OMZ, most probably as a result of reduced rates of organic matter decay in this zone of very low-oxic bottom water (Wendler et al. 2002). Variations in absolute and relative cyst abundances within province III are therefore mainly caused by differences in early diagenetic calcite dissolution within and below the OMZ. These secondary processes are assumed to play only a minor role in the SW of the studied area (provinces I and II), because a relation of cyst abundances to the OMZ was not notable there (Wendler et al. 2002), possibly related to fast sedimentation because of particle aggregation and reduced thickness and intensity of the OMZ in this area. We assume that the variations in cyst abundances in provinces I and II largely reflect differences in primary cyst production, with the exception of the deepest stations where calcite dissolution as a result of deep-water undersaturation has to be taken into account.

Ecology

Considering the large spatial and temporal variation of upper-ocean conditions in the Arabian Sea, the basin-wide presence of the six species discussed indicates that these species are tolerant of a relatively wide range of ecological conditions. The prevalence of *T. heimii* and *O. granifera* in the calcareous dinoflagellate associations can be explained by the dominance of the shelled stage during the life cycles of both species, and especially for *T. heimii* by its ability to produce large numbers of calcareous spheres (representing a vegetative–coccoid life-stage) in a relatively short period of time (Tangen et al. 1982; Inouye & Pienaar 1983; Karwath 2000; Karwath et al. 2000*a*).

If comparing the three provinces based on the cyst abundances (Fig. 2) with the distribution of monsoon-controlled upper-ocean conditions (see the section 'Climatic and oceanographic setting' and Table 2) it is most striking that the dividing line between provinces I and II coincides with the mean position of the Findlater Jet axis. Sites of province I are strongly influenced by coastal

Table 2. *Mean salinities, temperatures and temperature variations within cyst provinces per season and water depth*

Province	Mean salinity at 0–100 m (‰)				Mean annual salinity (‰)					
	NE monsoon	Spring intermonsoon	SW monsoon	Autumn intermonsoon	0 m	25 m	50 m	75 m	100 m	0–100 m
I	35.8	36.0	35.7	35.7	35.9	35.9	35.8	35.7	35.7	35.8
II	36.0	36.0	35.8	35.8	36.0	35.9	35.9	35.9	35.8	35.9
III	36.5	36.4	36.3	36.2	36.4	36.4	36.3	36.2	36.2	36.3
	Mean temperature at 0–100 m (°C)				Mean annual temperature (°C)					
	NE monsoon	Spring intermonsoon	SW monsoon	Autumn intermonsoon	0 m	25 m	50 m	75 m	100 m	0–100 m
I	24.0	25.4	21.4	22.4	26.6	25.1	23.2	21.4	20.2	23.3
II	24.2	25.8	24.2	24.1	26.6	26.0	24.8	23.4	22.1	24.6
III	23.1	23.9	24.1	24.6	26.2	25.3	23.6	22.7	21.8	23.9
					Mean annual temperature variation (°C)					
					0 m	25 m	50 m	75 m	100 m	0–100 m
I					3.2	4.4	5.0	5.3	4.5	4.5
II					3.2	3.6	3.5	3.6	3.4	3.5
III					4.9	3.4	1.7	5.2	2.2	3.5

upwelling during the SW monsoon accompanied by low water temperatures (down to 15°C), high nutrient concentrations, a shallow thermocline (about 10–40 m), reduced salinity (down to 35.2‰) and comparably high yearly temperature variations at 50–100 m water depth (up to 8°C). Sites of province II are characterized by open oceanic conditions with lower nutrient concentrations, higher water temperatures (20–29°C) and salinity (mainly >36‰), and a deep thermocline during most of the year (up to 120 m). Province III is affected by the influence of the Indus River discharge and by the NE monsoon deep winter mixing, accompanied by high nutrient concentrations, relatively high temperatures to 100 m depth (>20°C) and small seasonal temperature variations at 50–100 m depth (often <2°C). Yearly mean temperatures at 50–100 m depth as well as the seasonal minimum and maximum temperatures are higher in province III (20–29°C) than in the other two provinces (15–26°C). Yearly mean values of Brunt–Väisälä frequency between 0 and 100 m water depth are higher in the SW (representing more stratified conditions) than in the NE of the area and reflect the conditions during the NE monsoon and the two intermonsoon periods.

T. heimii and *S. tuberosa* var. 2. Both species are characteristic for provinces I and II whereas they are less abundant in province III. It has been shown that these two species are more dissolution sensitive than the other species discussed here (Wendler *et al.* 2002), and the observed lower abundance in the NE Arabian Sea (except for samples from within the OMZ) may not reflect ecological conditions but could be the result of increased calcite dissolution under oxic bottom-water conditions in this region. The increased ratio of other cysts to *T. heimii* below the OMZ in the NE and below 3000 m in the SW (Fig. 11) indicates that the tests of *T. heimii* (which are smaller than the cysts of the other species) are preferentially dissolved. To separate the primary signal from secondary alteration, samples from the SW are compared with samples from within the OMZ in the NE (Fig. 14), because these samples are assumed to be largely unaffected by early diagenetic calcite dissolution, which is also expressed in their low percentage of fragments of *T. heimii* (Fig. 10). Maximal AR values of *T. heimii* are higher off Somalia than within the OMZ on the Indus Fan and the Murray Ridge. This would mean that the reduced abundance of *T. heimii* (and possibly also of *S. tuberosa* var. 2) in the NE Arabian Sea reflects both increased dissolution and lower production than in the SW.

The slightly lower surface-water salinity in the SW Arabian Sea is not likely to have influenced the primary cyst production, as culturing experiments indicate that these organisms are rather tolerant to salinity: reproduction and cyst production of some species still continued under values as high as 50‰ (Höll & Karwath, pers. comm.).

Seasonal water temperatures during the NE monsoon and spring intermonsoon are related positively to the distribution of the two species, and temperatures during the SW monsoon and autumn intermonsoon are related negatively to

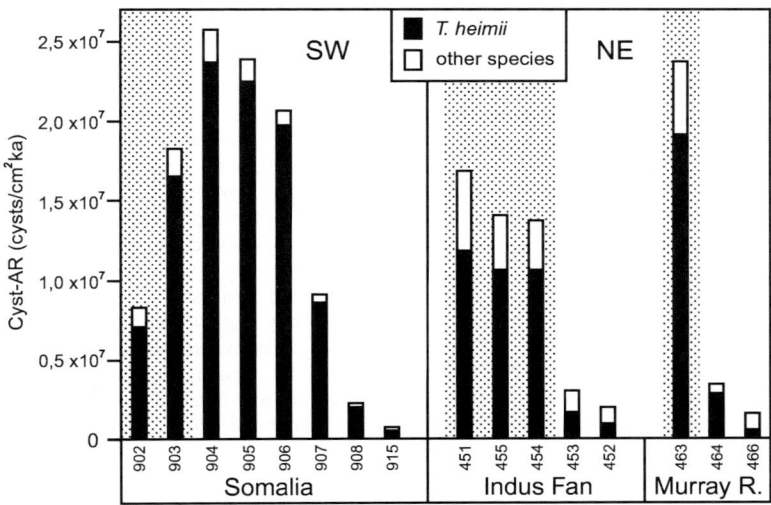

Fig. 14. Comparison of accumulation rates of *Thoracosphaera heimii* and other species in three regions in the Arabian Sea: in the SW a transect off Somalia, in the NE a profile on the Indus Fan and samples from the Murray Ridge. Grey background marks samples from within the OMZ.

this. To date, it is not known whether there is any seasonality in the production of the cysts and if so, at which time most cysts are formed. However, although the temperature gradients reverse twice each year, the absolute temperatures (maximum and minimum values) as well as the yearly mean temperatures, especially in the lower photic zone, are lower in the SW. This would imply a larger tolerance of *T. heimii* and *S. tuberosa* var. 2 to lower temperatures, which is in accordance with the results of other studies from the Atlantic Ocean (Karwath et al. 2000b; Vink et al. 2000; Zonneveld et al. 2000). In culturing experiments under controlled laboratory conditions it was shown that *T. heimii* developed less at high temperatures, with the final yield about five times higher at 16°C than at 27°C (Karwath et al. 2000a). However, these experiments show that *T. heimii* grows in a wide temperature range (14–27°C), which indicates large temperature tolerance. This can be expected for a species typical for province I, where high yearly temperature variations are caused by seasonal upwelling.

Stratification of the upper water column has been proposed to influence the distribution of calcareous dinoflagellates (Höll et al. 1998, 1999; Vink et al. 2000, 2002a). Well-stratified water can hamper the vertical migration of motile cells or could act positively as a barrier for the calcareous cysts, preventing them from sinking at depths where return to the photic zone is impossible. For several dinoflagellate species producing red tides, it is known that blooms occur in periods of calm weather and warm, stratified waters (e.g. Allen 1946; Marasovic 1989). In the Arabian Sea, the yearly mean stratification is stronger in the SW of the basin, which means that this variable is positively related to the distributions of *T. heimii* and *S. tuberosa* var. 2. This relation suggests preference of both species for stratified conditions, which is consistent with the earlier findings. However, because of the strong seasonality especially of water temperatures in the Arabian Sea, stratification is generally very variable (up to 100 m seasonal variation; see the section 'Climatic and oceanographic setting'), and its role as a controlling factor for the observed cyst distributions in the studied area is therefore not clear.

Within the SW Arabian Sea, where calcite dissolution at water depths above 3000 m seems to be negligible (Wendler et al. 2002), the relative abundances of *T. heimii* and *S. tuberosa* var. 2 exhibit no large variations. Their absolute abundance, however, is very low near the coasts, in zones of active coastal upwelling. The same pattern can be seen in the AR values off Somalia (Fig. 13a); this indicates not only enhanced dilution by other particles near the coasts but also reduced cyst production. The production rates could be lower under the turbulent conditions of coastal upwelling because too much turbulence can preclude the build-up of a standing stock of phototrophic organisms. Veldhuis et al. (1997) observed high primary production during the upwelling season off Somalia only in more mature water, whereas productivity in freshly upwelled water was relatively low despite high nutrient concentrations. Those workers considered high surface current speeds and deep vertical mixing to be the reason for this phenomenon. Various laboratory studies have emphasized the sensitivity of dinoflagellates to water motion (Thomas & Gibson 1990; Thomas & Gibson 1992, and references therein). Cyst production of various species of calcareous dinoflagellates was considerably higher under non-agitated conditions (Höll, pers. comm.), which is in accordance with their adaptation to low irradiance conditions, as with increasing water depth both light and turbulence decrease.

On the basis of Atlantic Ocean studies it has been suggested that calcareous dinoflagellates might be adapted to oligotrophic conditions (Höll et al. 1998, 1999; Esper et al. 2000; Vink et al. 2000). However, in these regions it is difficult to separate the effects of oligotrophy and stratification. In the Arabian Sea, *T. heimii* and *S. tuberosa* var. 2 are characteristic for both the relatively oligotrophic province II as well as the nutrient-rich province I. From this it can be inferred that these species are tolerant to different nutrient levels, and that their distribution is not primarily controlled by differences in nutrient supply. This supposition is supported by results of culturing experiments with *T. heimii*, which did not show differences in growth rate and final yield under different nutrient levels at constant temperature (Karwath et al. 2000a). A similar conclusion was drawn by Zonneveld et al. (2000), who compared surface sediments in the Atlantic Ocean and stated that, after correction for sedimentation rates, the differences in concentrations of *T. heimii* between the oligotrophic open ocean and the eutrophic Benguela area would largely be compensated.

The extremely low abundance of *S. tuberosa* var. 1 in the Arabian Sea might be explained by the generally high trophy of this oceanic region, because, as is evident from surface sediments of the Atlantic Ocean, this species appears to be adapted to low-nutrient conditions (Vink et al. 2000; Zonneveld et al. 2000; Vink pers. comm.).

O. granifera. The absolute and relative abundance of *O. granifera* in the Arabian Sea is considerably higher than that known from the Atlantic Ocean so far (Höll et al. 1998, 1999; Esper et al. 2000;

Vink et al. 2000; Zonneveld et al. 2000). Concentrations are especially high in the NE Arabian Sea, where the species forms up to 51% of the association (Fig. 5b). The higher absolute and relative abundance in the NE cannot be explained by early diagenetic calcite dissolution, as it is stronger in the NE. The fact that the cysts of *O. granifera* appear to be least sensitive to dissolution compared with the other species (Wendler et al. 2002) could be the reason for an increase of its relative abundance in the NE but would not explain the higher absolute abundances and AR values there (Fig. 5a). Accordingly, it is likely that variations in primary cyst production are reflected.

A possible relation of higher concentrations of *O. granifera* to relatively low salinity as observed by Vink et al. (2000) cannot be seen in the Arabian Sea using the available salinity dataset, which shows higher salinity (up to 36.8‰) in the NE, although this area is influenced by the Indus River discharge. Obviously during the last decades, when most of the salinity measurements were made, the freshwater input was compensated by high evaporation rates. However, the mean annual water and suspended sediment discharge load was much higher before 1950, when damming and channelling of the Indus River and utilization for agriculture were lower (Milliman et al. 1984), and lower salinity values have to be assumed. Sedimentation rates near the Indus outlet indicate that the studied surface samples contain 43–213 years of deposition, so a large part of the signal reflects oceanographic conditions before 1950. Therefore, it is likely that also in the Arabian Sea *O. granifera* is related to lower salinity. On the other hand, Schulz et al. (1996) found no freshwater signal in the distribution of stable isotopes in surface sediments from the Indus Fan. However, elevated *O. granifera* concentrations were also observed in the Atlantic Ocean in regions that are influenced by river outflow such as the Amazon or the Congo River, where surface-water salinities of 29.0–34.5‰ were measured (Vink et al. 2000; Vink, pers. comm.). It seems very likely that this species is adapted to conditions that are related to fluvial input, be it a lowered salinity and/or some other abiotic (e.g. specific nutrients) or biotic factors (competition or symbiosis).

Most areas that are characterized by river discharge are rather stratified as a result of the lower density of fresh water. In the NE Arabian Sea, however, the upper water masses are less stratified than in the SW of the basin most of the year, which is mainly caused by convective turnover. This results in a negative correlation of yearly mean stratification and the distribution of *O. granifera* in the Arabian Sea. Presumably, the species is tolerant to various levels of stratification.

Beside the fluvial influence, generally high temperatures and low seasonality in the NE Arabian Sea may be favourable for *O. granifera*. This is supported by culturing experiments, which have shown that the species grows better under relatively high temperatures (Höll, pers. comm.). In the oceans, *O. granifera* is so far reported exclusively from (sub)tropical regions, whereas the other species were also found in warm or temperate regions (Zonneveld et al. 1999). A positive correlation of *O. granifera* with temperature has also been reported by Vink et al. (2000). However, in the SW Arabian Sea, where cold deep water wells up near the coasts during summer and temperatures at depths of 50–100 m are about 15–20°C, concentrations of the species are still higher than in most studied regions of the Atlantic Ocean. This shows that temperature cannot be the only controlling factor.

One of the main characteristics of the Arabian Sea is the extremely high primary production caused by high nutrient concentrations. Lowest concentrations of *O. granifera* were observed in province II, which is the most oligotrophic part of the studied area (Figs. 2 and 5). From this we glean that *O. granifera* is adapted to rather high nutrient concentrations.

S. albatrosiana and *C. operosum*. Both species show lower absolute and relative abundance in province I. This distribution pattern cannot be explained by calcite dissolution because, as mentioned above, more dissolution would be expected in the deep samples from the open ocean and in province III. Low concentrations in samples close to the Somali and Arabian coasts give a negative correlation with characteristics that are typical for seasonal coastal upwelling, such as large seasonality, strong turbulence, high nutrient concentrations and low temperatures. Within the lower photic zone in the NE Arabian Sea and especially in the open ocean, seasonality is much smaller and yearly mean temperatures are relatively high as a result of downwelling-induced deepening of the mixed layer, causing a relatively uniform vertical temperature distribution of about 20–24°C down to 100 m depth.

Although *S. albatrosiana* might be less successful in upwelling areas with very high nutrient concentrations, the species seems to be adapted to a relatively wide range of nutrient levels, as high cyst concentrations have been found in the eutrophic NE Arabian Sea as well as in the oligotrophic open ocean. A mainly open oceanic distribution of *S. albatrosiana* and a negative relation to nutrient concentrations have been reported by Vink et al. (2000) for the western equatorial Atlantic Ocean. Furthermore, Zonneveld et al. (2000) described a trend to higher concentrations of *S. albatrosiana* from onshore to offshore areas in the Benguela region and observed a positive relation to water temperatures by comparing samples from different regions in the equatorial and South Atlantic Ocean. These observations confirm the interpretation of *S. albatrosiana* as being typical for open oceanic, rather oligotrophic environments with low seasonality and relatively high temperatures. In view of the usually very low cyst concentrations of *C. operosum* it appears uncertain whether similar conclusions can be drawn for this species.

Spiny cysts. The low abundance of the spiny cysts (which in the studied material mainly belong to *Scrippsiella trochoidea*) in samples from the open ocean of the Arabian Sea is in general agreement with observations from other studies, where *S. trochoidea* is reported only from neritic environments (Janofske 2000). However, in the studied area there is no restriction of the spiny cysts to coastal waters: relatively high concentrations were also found offshore Oman and in the westernmost samples from the Murray Ridge (Fig. 9). This could indicate that *S. trochoidea* is not restricted to coastal environments. On the other hand, as noted above, some basin-ward transport of cysts in the water column can be expected via eddies and filaments off Oman, which move east to NE. Furthermore, there might be some contribution of *S. regalis* to the group of spiny cysts. This species is described from the open ocean (Vink et al. 2000; Janofske 2000, and references therein), so the distribution pattern of all spiny cysts could represent a combined oceanic and coastal signal.

The group of spiny cysts is the only morphotype that is abundant in areas of active upwelling and shows continuously decreasing AR values with depth (Fig. 13a). This suggests that the dominant species *S. trochoidea* is adapted to eutrophic and rather unstable environments (large seasonality) with seasonally lower temperatures and a shallow thermocline. The interpretation is supported by a study of surface sediments of the Benguela upwelling region, where high abundance of spiny cysts is related to high nutrient concentrations and strong seasonality (Zonneveld et al. 2000). In surface sediments from different parts of the world's oceans, *S. trochoidea* is mainly reported from temperate regions (Zonneveld et al. 1999).

Spiny cysts found in the Arabian Sea show a large morphological variety, and fragile cysts (consisting of calcite crystals loosely attached to an organic layer) are particularly well preserved within the OMZ. More work on the taxonomy of this type of cysts is necessary to clearly separate species and to gain information on their ecology.

Conclusions

The distribution of calcareous dinoflagellate cysts in surface sediments of the Arabian Sea is controlled by a combination of ecology and early diagenetic calcite dissolution. Sediments in the SW of the studied area largely reflect ecologically controlled variations in dinoflagellate cyst production, in contrast to the NE where cyst accumulation rates (AR) are strongly related to bottom-water oxygen concentrations and are thought to be determined by differences in calcite dissolution within and below the OMZ. Not all of the basin-wide trends within cyst distribution patterns can be explained by early diagenetic processes and are interpreted to result from different ecological conditions within the (lower) photic zone.

The two dominant species of calcareous dinoflagellates in surface sediments of the Arabian Sea are *T. heimii* and *O. granifera*, which show distribution patterns opposite to each other. Lower AR values and relative abundance of *T. heimii* in the NE may mainly result from increased dissolution, whereas high absolute and relative abundance of *O. granifera* in this region is related to higher water temperatures, low seasonality and the influence of the Indus River. *S. tuberosa* var. 2, which has a similar distribution to *T. heimii*, is negatively related to temperature and appears to be tolerant to a wide range of nutrient concentrations. However, the distribution of *S. tuberosa* var. 2 may also be affected by enhanced dissolution in the NE. Higher abundance of *S. albatrosiana* in the open ocean and NE of the basin can be related to higher temperatures and a deep thermocline. The species seems to be less successful under upwelling conditions and is probably adapted to stable environments in the open ocean and to intermediate to low nutrient concentrations. Spiny cysts in the studied material mainly belong to *S. trochoidea*, which is known from neritic environments and appears to be adapted to eutrophic and probably cool, rather unstable conditions. The extremely low abundance of *S. tuberosa* var. 1 in the Arabian Sea is attributed to the species' preference of oligotrophic environments. However, a general relation of calcareous dinoflagellates to oligotrophic conditions, as was proposed earlier (e.g. Höll et al. 1998, 1999), cannot be confirmed. It should be carefully evaluated to what extent the observed negative correlation of calcareous dinoflagellate cysts with content of organic carbon in sediment cores might be caused by enhanced calcite dissolution, which is driven by metabolic

CO_2 during times of high primary production and increased organic matter decay. Low cyst concentrations and AR values in zones of active coastal upwelling off Somalia and Yemen indicate that strong turbulence and high current speeds are unfavourable for calcareous dinoflagellates. This is encouraging for the belief that these organisms are more successful under rather stratified conditions.

We highly appreciate the technical help of G. Graser. The research was funded by the Deutsche Forschungsgemeinschaft through the Graduierten-Kolleg 'Stoff-Flüsse in marinen Geosystemen'.

Scrippsiella trochoidea (Stein 1883) Loeblich III 1965
Scrippsiella regalis (Gaarder 1954) Janofske 2000
Melodomuncula berlinensis Versteegh 1993
Calciperidinium asymmetricum Versteegh 1993
Calcigonellum infula (Deflandre 1947) Montresor 1999

Appendix A: Taxonomic information

The calcareous dinoflagellate cyst species or morphotypes cited in this paper are listed below and illustrated in Fig. 15. Their taxonomy has been given by Keupp (1987), Keupp & Versteegh (1989), Keupp & Kohring (1993), Versteegh (1993), Hildebrand-Habel et al. (1999) and Janofske (2000). Most specimens of spiny cysts fall within the '*Scrippsiella trochoidea*-complex' described by D'Onofrio et al. (1999) and are comparable with *Rhabdothorax* sp. 1 of Vink et al. (2000). Only a few specimens were identified as *Scrippsiella regalis*. Designation of morphotypes of *S. tuberosa* is the same as that by Vink et al. (2000): *S. tuberosa* var. 1 is composed of relatively large, block-like individual crystals that do not interfinger with each other (fig. 2 in plate 1 of Vink et al. 2000), whereas *S. tuberosa* var. 2 consists of smaller, roughly triangular, interfingering crystals (Fig. 15c). It should be noted that a new taxonomic concept for the calcareous cyst-producing species is currently in preparation (Meier et al., pers. comm.; Karwath 2000). The new generic attribution of some of the species according to this concept is given in square brackets.

Thoracosphaera heimii (Lohman 1920) Kamptner 1944
Sphaerodinella tuberosa (Kamptner 1963) Hildebrand-Habel et al. 1999
—*S. tuberosa* var. 1
   [*Pernambugia tuberosa*]
—*S. tuberosa* var. 2
   [*Calciodinellum* sp.]
Orthopithonella granifera (Fütterer 1977) Keupp & Kohring 1993
   [*Leonella granifera*]
Sphaerodinella albatrosiana (Kamptner 1963) Keupp & Versteegh 1989
   [*Calciodinellum albatrosianum*]
Calciodinellum operosum (Deflandre 1947) emend. Montresor et al. 1997

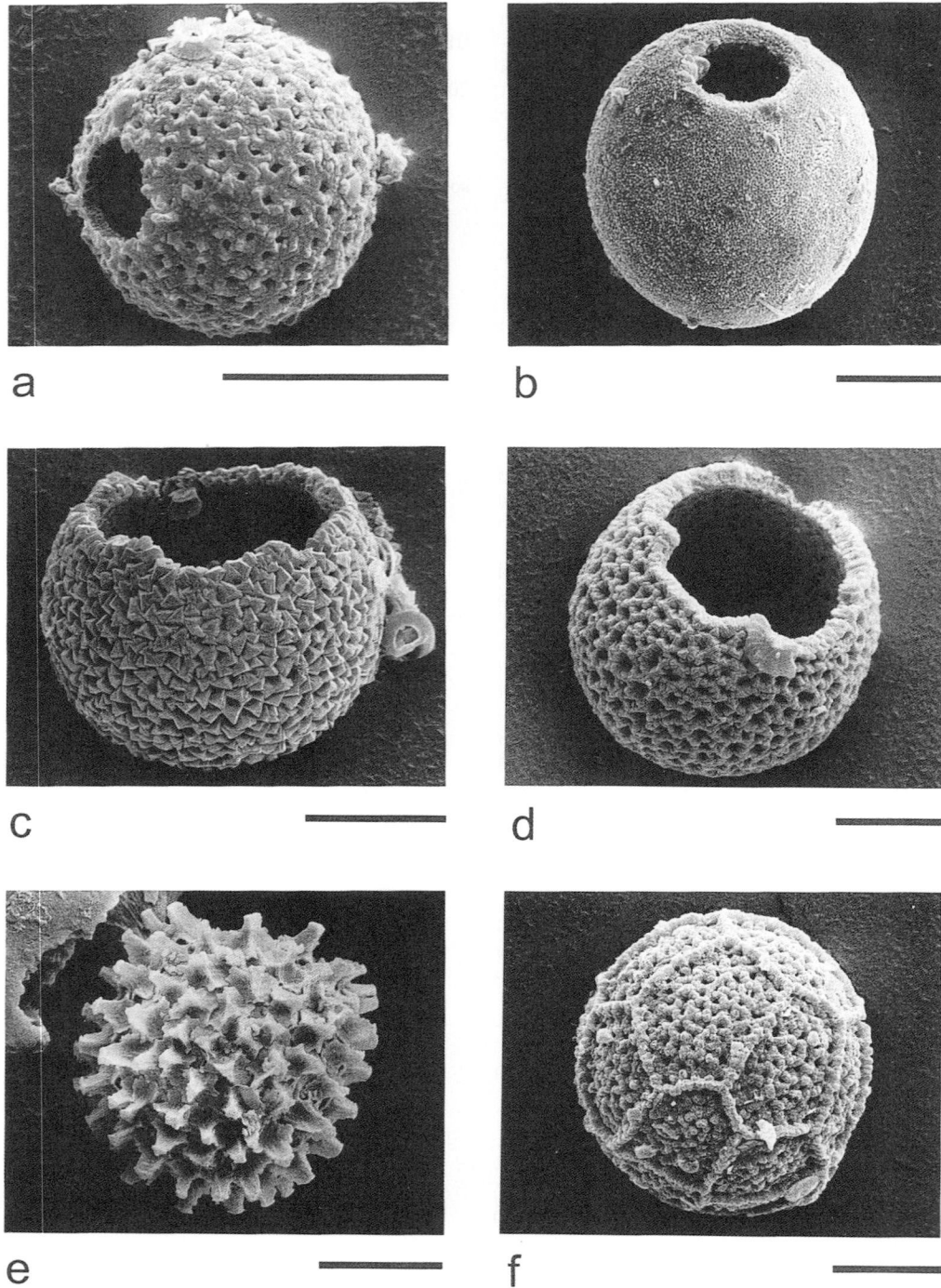

Fig. 15. Scanning electron micrographs of the calcareous dinoflagellate cyst species; sample 457; scale bars represent 10 μm. (**a**) *Thoracosphaera heimii*; (**b**) *Orthopithonella granifera*; (**c**) *Sphaerodinella tuberosa* var. 2; (**d**) *Sphaerodinella albatrosiana*; (**e**) *Scrippsiella trochoidea*; (**f**) *Calciodinellum operosum*.

Appendix B: Position and water depth of surface samples and absolute cyst abundances

Sample no.	Depth (m)	Latitude (N)	Longitude (E)	Absolute abundance (cysts g^{-1} of dry sediment)					
				T. heimii	*O. gran.*	*S. alba.*	*S. tub. 2*	*C. oper.*	spiny
301	74	15.08	51.25	38339	5021	2738	2738	0	2738
302	208	15.00	51.27	48255	5709	2585	2262	323	1939
303	474	14.51	51.29	173124	16909	5039	4031	0	1680
304	770	14.47	51.31	311999	14836	8122	8772	650	2274
305	1098	14.43	51.34	1234126	77242	17019	21602	3928	9164
306	1504	14.30	51.37	1246473	75246	31424	17956	3206	9619
307	50	16.11	52.23	6684	267	535	1337	0	267
308	196	16.08	52.30	13847	804	268	536	0	0
309	487	16.05	52.37	193671	22679	5380	6329	0	2532
310	810	16.04	52.42	356822	16158	8752	9425	1346	2020
311	1087	16.02	52.46	810395	30778	9582	12195	871	5226
313	2215	15.53	53.01	93159	29703	7201	1350	0	7201
325	4035	10.41	53.31	564115	17147	45477	16402	1491	12674
451	495	23.41	66.02	485103	173955	23787	4325	1442	1442
452	2001	22.56	65.28	146951	133260	17707	548	548	3834
453	1555	23.14	65.44	168002	114049	14423	3205	3205	2404
454	1254	23.27	65.52	945864	204359	45700	14764	3515	6328
455	998	23.33	65.57	769420	186190	34032	9618	5179	8138
457	301	22.58	63.51	600388	112850	20216	18139	1662	1662
458	3000	22.00	63.30	113176	74588	23294	1412	2118	7765
460	3262	21.43	62.55	46623	59814	10234	682	1365	6823
461	643	22.23	63.50	855004	128605	23404	7801	2128	5831
463	970	22.33	64.03	1764779	317437	60358	20864	8942	5961
464	1511	22.15	63.35	378735	47423	12343	2599	1949	2599
466	1960	23.36	63.48	57218	77465	11092	0	528	6514

468	1318	24.46	62.21	251785	46504	8969	7308	332	6643
469	1768	24.40	62.22	27477	27579	3371	613	613	2451
470	1840	24.36	62.22	26358	34288	3744	441	1909	4993
471	2482	24.18	62.27	32750	23172	4634	927	309	1854
472	3274	24.07	62.29	15816	14324	6416	671	0	1791
473	1877	22.13	63.06	245218	92007	23555	3624	604	6644
475	1472	24.05	65.27	133988	90390	13696	3424	1370	4793
476	1226	24.06	65.28	490909	98409	27955	4773	3409	4773
477	1000	24.08	65.31	575567	116709	19521	8186	2519	5038
478	556	24.13	65.40	605106	191277	27447	10851	2553	5106
483	2734	21.02	61.29	107338	42497	3943	3943	657	5257
484	527	19.30	58.26	78451	16386	3021	2472	549	3845
486	2070	19.09	60.37	737472	28825	17738	7539	665	10421
487	3566	19.54	61.43	184448	61081	14467	1808	603	10850
491	3797	15.50	63.55	539771	20949	37316	16694	3928	982
492	1917	16.11	59.46	1451045	39604	20902	17602	2750	3300
496	1900	17.26	57.57	683486	36927	17890	8257	688	8257
497	1890	17.28	57.57	646991	28472	13889	6944	1389	2083
902	459	10.46	51.34	210943	17409	6104	5426	678	5426
903	789	10.46	51.39	743739	34974	13601	20725	1295	7124
904	1194	10.47	51.46	1549934	83297	20548	18559	663	8617
905	1567	10.54	51.56	3693827	108473	55236	42239	2999	13997
906	2020	10.48	52.07	4044065	69415	51883	33404	2843	13504
907	2807	10.48	52.14	3078704	61728	61343	41281	1157	12731
908	3572	10.46	52.54	1158358	23636	50924	24495	1934	10314
915	4035	10.41	53.31	435524	16783	24336	6713	839	4196
917	2225	15.54	53.01	133435	20244	6530	1959	653	5224
918	1716	15.58	52.50	672260	26803	9806	11767	2615	7191
919	1030	16.00	52.44	758493	34556	12231	8736	1165	4659
929	2484	13.42	53.14	729687	46452	5447	4766	1362	5447

Appendix C: *Rare species: distribution (sample number) and total number of occurrences*

	Sphaerodinella tuberosa var. 1	*Melodomuncula berlinensis*	*Calcigonellum infula*	*Calciperidinium asymmetricum*
Somalia	903, 906, 907	902	915	–
Yemen	305, 306	–	313, 918, 919	307
Pakistan	–	476, 477, 455	452	–
Murray Ridge	–	457, 463	463	457
Total number of occurrences	7	7	8	2

References

ALLEN, W.E. 1946. Red water in La Jolla Bay in 1945. *Transactions of the American Microscopical Society*, **65**, 149–153.

ARNONE, R.A., GOULD, R.W., KINDLE, J., MARTINOLICH, P., BRINK, K. & LEE, C. 1998. Remote sensing of coastal upwelling and filaments off the coast of Oman. *Oceanography*, **11**, 33.

BANSE, K. & MCCLAIN, C.R. 1986. Winter blooms of phytoplankton in the Arabian Sea as observed by the coastal zone color scanner. *Marine Ecology Progress Series*, **34**, 201–211.

BAUER, S., HITCHKOCK, G.L. & OLSON, D.B. 1991. Influence of monsoonally-forced Ekman dynamics upon surface layer depth and plankton biomass distribution in the Arabian Sea. *Deep-Sea Research*, **38**, 531–553.

BROCK, J.C., MCCLAIN, C.R., ANDERSON, D.M., PRELL, W.L. & HAY, W.W. 1992. Southwest monsoon circulation and environments of recent planktonic foraminifera in the northwestern Arabian Sea. *Paleoceanography*, **7**, 799–813.

BRUCE, J.G. 1974. Some details of upwelling of the Somali and Arabian coasts. *Journal of Marine Research*, **32**, 419–423.

BRUCE, J.G. 1979. Eddies off the Somali Coast during the southwest monsoon. *Journal of Geophysical Research*, **84**(C12), 7742–7748.

BRUMMER, G.J.A. 1995. Sediment traps and particle dynamics. *In:* VAN HINTE, J.E., VAN WEERING, T.C.E. & TROELSTRA, S.R. (eds) *Tracing a Seasonal Upwelling. Report on Two Cruises of R.V.* Tyro *to the NW Indian Ocean in 1992 and 1993, Cruise Reports*. National Museum of Natural History, Leiden, **4**, 55–61.

CURRIE, R.I., FISHER, A.E. & HARGREAVES, P.M. 1973. Arabian Sea upwelling. *In:* ZEITZSCHEL, B. & GERLACH, S.A. (eds) *Biology of the Indian Ocean*. Springer, Berlin, 37–52.

DICKEY, T., MARRA, J., SIGURDSON, D.E. & 5 OTHERS 1998. Seasonal variability of bio-optical and physical properties in the Arabian Sea: October 1994–October 1995. *Deep-Sea Research II*, **45**, 2001–2025.

D'ONOFRIO, G., MARINO, D., BIANCO, L., BUSICO, E. & MONTRESOR, M. 1999. Toward an assessment on the taxonomy of dinoflagellates that produce calcareous cysts (Calciodinelloideae, Dinophyceae): a morphological and molecular approach. *Journal of Phycology*, **35**, 1063–1078.

ELLIOT, A.J. & SAVIDGE, G. 1990. Some features of upwelling of Oman. *Journal of Marine Research*, **48**, 319–333.

ESPER, O., ZONNEVELD, K.A.F., HÖLL, C. & 6 OTHERS 2000. Reconstruction of palaeoceanographic conditions in the South Atlantic Ocean at the last two Terminations based on calcareous dinoflagellate cysts. *International Journal of Earth Sciences*, **88**, 680–693.

FINDLATER, J. 1971. *Mean Monthly Airflow at Low Levels over the Western Indian Ocean*. Geophysical Memorial, **115**.

FISHER, J., SCHOTT, F. & STRAMMA, L. 1996. Currents and transport of the Great Whirl–Socotra Gyre system during the summer Monsoon, August 1993. *Journal of Geophysical Research*, **101**, 3573–3587.

FLAGG, C.N. & KIM, H.-S. 1998. Upper ocean currents in the northern Arabian Sea from shipboard ADCP measurements collected during the 1994–1996 U.S. JGOFS and ONR programs. *Deep-Sea Research II*, **45**, 1917–1959.

GARDNER, W.D., CHUNG, S.P., RICHARDSON, M.J. & WALSH, I.D. 1995. The oceanic mixed-layer pump. *Deep-Sea Research II*, **42**, 757–775.

GARDNER, W.D., GUNDERSEN, J.S., RICHARDSON, M.J. & WALSH, I.D. 1999. The role of seasonal and diel changes mixed-layer depth on carbon and chlorophyll distributions in the Arabian Sea. *Deep-Sea Research II*, **46**, 1833–1858.

GUNDERSEN, J.S., GARDNER, W.D., RICHARDSON, M.J. & WALSH, I.D. 1998. Effects of monsoon on the seasonal and spatial distributions of POC and chlorophyll in the Arabian Sea. *Deep-Sea Research II*, **45**, 2103–2132.

HEIER-NIELSEN, S., KUIJPERS, A. & TROELS, L. 1995. Holocene sediment deposition and organic matter burial in the upwelling zone off Yemen, Northwest Indian Ocean. *In:* VAN HINTE, J.E., VAN WEERING, T.C.E. & TROELSTRA, S.R. (eds) *Tracing a Seasonal Upwelling, Report on Two Cruises of R.V.* Tyro *to the NW Indian Ocean in 1992 and 1993, Cruise Reports*. National Museum of Natural History, Leiden, **4**, 111–119.

HILDEBRAND-HABEL, T., WILLEMS, H. & VERSTEEGH, G.J.M. 1999. Variations in calcareous dinoflagellate associations from the Maastrichtian to Middle Eocene of the western South Atlantic Ocean (São Paulo Plateau, DSDP Leg 39, Site 356. *Review of Palaeobotany and Palynology*, **106**, 57–87.

HÖLL, C. & KEMLE-VON MÜCKE, S. 2000. Late Quaternary upwelling variations in the Eastern Equatorial

Atlantic Ocean as inferred from dinoflagellate cysts, planktonic foraminifera, and organic carbon content. *Quaternary Research*, **54**, 58–67.

HÖLL, C., ZONNEVELD, K.A.F. & WILLEMS, H. 1998. On the ecology of calcareous dinoflagellates: the Quaternary Eastern Equatorial Atlantic. *Marine Micropaleontology*, **33**, 1–25.

HÖLL, C., KARWATH, B., RÜHLEMANN, C., ZONNEVELD, K.A.F. & WILLEMS, H. 1999. Palaeoenvironmental information gained from calcareous dinoflagellates: the late Quaternary eastern and western tropical Atlantic Ocean in comparison. *Palaeogeography, Palaeoclimatology, Palaeoecology*, **146**, 147–164.

HONJO, S., DYMOND, J., PRELL, W. & ITTEKKOT, V. 1999. Monsoon-controlled export fluxes to the interior of the Arabian Sea. *Deep-Sea Research II*, **46**, 1859–1902.

INOUYE, I. & PIENAAR, R.N. 1983. Observations on the life cycle and microanatomy of *Thoracosphaera heimii* (Dinophyceae) with special reference to its systematic position. *South African Journal of Botany*, **2**, 63–75.

JANOFSKE, D. 1996. Ultrastructure types in Recent calcispheres. *Bulletin de l'Institut Océanographique, Monaco, No. spécial*, **14**(4), 295–303.

JANOFSKE, D. 2000. *Scrippsiella trochoidea* and *Scrippsiella regalis* nov. comb. (Peridiniales, Dinophyceae): a comparison. *Journal of Phycology*, **35**, 1–12.

KARWATH, B. 2000. *Ecological studies on living and fossil calcareous dinoflagellates of the Equatorial and Tropical Atlantic Ocean*. Berichte, Fachbereich Geowissenschaften, Universität Bremen, **152**.

KARWATH, B., JANOFSKE, D., TIETJEN, F. & WILLEMS, H. 2000a. Temperature effects on growth and cell size in the marine calcareous dinoflagellate *Thoracosphaera heimii*. *Marine Micropaleontology*, **39**, 43–51.

KARWATH, B., JANOFSKE, D. & WILLEMS, H. 2000b. Spatial distribution of the calcareous dinoflagellate *Thoracosphaera heimii* in the upper water column of the tropical and equatorial Atlantic. *International Journal of Earth Sciences*, **88**, 668–679.

KEEN, T.R., KINDLE, J.C. & YOUNG, D.K. 1997. The interaction of southwest monsoon upwelling, advection and primary productivity in the Northwest Arabian Sea. *Journal of Marine Systems*, **13**, 61–82.

KEUPP, H. 1987. Die kalkigen Dinoflagellatenzysten des Mittelalb bis Untercenoman von Escalles/Boulonnais (N-Frankreich). *Facies*, **16**, 37–88.

KEUPP, H. & KOHRING, R. 1993. Kalkige Dinoflagellatenzysten aus dem Obermiozän von El Medhi (Algerien). *Berliner Geowissenschaftliche Abhandlungen (E)*, **9**, 25–43.

KEUPP, H. & VERSTEEGH, G.J.M. 1989. Ein neues systematisches Konzept für kalkige Dinoflagellaten-Zysten der Subfamilie Orthopithonelloideae Keupp 1987. *Berliner Geowissenschaftliche Abhandlungen (A)*, **106**, 207–219.

KOLLA, V., KOSTECKI, J.A., ROBINSON, F., BISCAYE, P.E. & RAY, P.K. 1981. Distributions and origins of clay minerals and quartz in surface sediments of the Arabian Sea. *Journal of Sedimentology and Petrology*, **51**, 563–569.

LATASA, M. & BIDIGARE, R.R. 1998. A comparison of phytoplankton populations of the Arabian Sea during the Spring Intermonsoon and Southwest Monsoon of 1995 as described by HPLC-analysed pigments. *Deep-Sea Research II*, **45**, 2133–2170.

LEE, C.M., JONES, B.H., BRINK, K.H. & FISCHER, A.S. 2000. The upper-ocean response to monsoonal forcing in the Arabian Sea: seasonal and spatial variability. *Deep-Sea Research II*, **47**, 1177–1226.

LENDT, R., HUPE, A., ITTEKKOT, V. & 5 OTHERS 1999. Greenhouse gases in cold water filaments in the Arabian Sea during the Southwest Monsoon. *Naturwissenschaften*, **86**, 489–491.

MADHUPRATAP, M., PRASANNA KUMAR, S., BHATTATHIRI, P.M.A., DILEEP KUMAR, M., RAGHUKUMA, S., NAIR, K.K.C. & RAMAIAH, N. 1996. Mechanisms of the biological response to winter cooling in the northeastern Arabian Sea. *Nature*, **384**, 549–552.

MANGHNANI, V., MORRISON, J.M., HOPKINS, T.S. & BÖHM, E. 1998. Advection of upwelled waters in the form of plumes off Oman during the Southwest Monsoon. *Deep-Sea Research II*, **45**, 2027–2052.

MARASOVIC, I. 1989. Encystment and excystment of *Gonyaulax polyedra* during a red tide. *Estuarine, Coastal and Shelf Science*, **28**, 35–41.

MEADOWS, A., MEADOWS, P.S., WEST, F.J.C. & MURRAY, J.M.H. 2000. Bioturbation, geochemistry and geotechnics of the sediments affected by the oxygen minimum zone on the Oman continental slope and abyssal plain, Arabian Sea. *Deep-Sea Research II*, **47**, 259–280.

MILLERO, F.J., DEGLER, E.A., O'SULLIVAN, D.W., GOYET, C. & EISCHEID, G. 1998. The carbon dioxide system in the Arabian Sea. *Deep-Sea Research II*, **45**, 2225–2252.

MILLIMAN, J.D., QURAISHEE, G.S. & BEG, M.A.A. 1984. Sediment discharge from the Indus river to the ocean: past, present and future. *In:* HAQ, B.U. & MILLIMAN, J.D. (eds) *Marine Geology and Oceanography of the Arabian Sea and Coastal Pakistan*. Van Nostrand Reinhold, New York, 65–70.

MOLINARI, R.L., OLSON, D. & REVERDIN, G. 1990. Surface current distributions in the tropical Indian Ocean derived from compilations of surface buoy trajectories. *Journal of Geophysical Research*, **95**, 7217–7238.

PAROPKARI, A.L., BABU, C.P. & MASCARENHAS, A. 1992. A critical evaluation of depositional parameters controlling the variability of organic carbon in Arabian Sea sediments. *Marine Geology*, **107**, 213–226.

PRELL, W.L. & CURRY, W.B. 1981. Faunal and isotopic indices of monsoonal upwelling: western Arabian Sea. *Oceanologica Acta*, **4**, 91–98.

PRELL, W.L. & STREETER, H.F. 1982. Temporal and spatial patterns of monsoonal upwelling along Arabia: a modern analogue for the interpretation of Quaternary SST anomalies. *Journal of Marine Research*, **40**, 143–155.

PRINS, M.A., REICHART, G.J., VISSER, H.J., POSTMA, G., ZACHARIASSE, W.J., VAN DER LINDEN, W.J.M. & CRAMP, A. 1994. Sediments recovered during NIOP cruises D1–D3. *In:* VAN DER LINDEN, W.J.M. & VAN DER WEIJDEN, C.H. (eds) *Geological Study of the Arabian Sea. Report on Three Cruises of R.V. Tyro to the NW Indian Ocean in 1992, Cruise Reports.* National Museum of Natural History, Leiden, **4**, 87–96.

QURAISHEE, G.S. 1988. Arabian Sea cooling and productivity. *In:* THOMPSON, M.F. & TIRMIZI, N.M. (eds) *Marine Science of the Arabian Sea.* American Institute of Biological Science, Washington, DC, 59–66.

RAO, R.R., MOLINARI, R. & FESTA, J. 1989. Evolution of the climatological near-surface thermal structure of the tropical Indian Ocean. *Journal of Geophysical Research*, **94**, 10801–10815.

RIXEN, T., HAAKE, B. & ITTEKKOT, V. 2000. Sedimentation in the western Arabian Sea the role of coastal and open-ocean upwelling. *Deep-Sea Research II*, **47**, 2155–2178.

RYTHER, J.H., HELL, J.R., PEASE, A.K., BAKUN, A. & JONES, M.M. 1966. Primary production in relation to the chemistry and hydrography of the western Indian Ocean. *Limnology and Oceanography*, **11**, 371–380.

SASTRY, J.S. & D'SOUZA, R.S. 1972. Upwelling and upward mixing in the Arabian Sea. *Indian Journal of Marine Science*, **1**, 17–27.

SCHOTT, F. 1983. Monsoon response of the Somali Current and associated upwelling. *Progress in Oceanography*, **12**, 357–381.

SCHULZ, H., VON RAD, U. & VON STACKELBERG, U. 1996. Laminated sediments from the oxygen-minimum zone of the northeastern Arabian Sea. *In:* KEMP, A.E.S. (eds) *Palaeoclimatology and Palaeoceanography from Laminated Sediments.* Geological Society, London, Special Publications, **116**, 185–207.

SHETYE, S.R., GOUVEIA, A.D. & SHENOI, S.S.C. 1994. Circulation and water masses of the Arabian Sea. *In:* LAL, D. (eds) *Biogeochemistry of the Arabian Sea.* Proceedings, Indian Academy of Sciences (Earth and Planetary Science), **103(2)**, 9–25.

SHI, W., MORRISON, J.M., BÖHM, E. & MANGHNANI, V. 1999. Remotely sensed features in the US JGOFS Arabian Sea Process Study. *Deep-Sea Research II*, **46**, 1551–1575.

SIROCKO, F. & LANGE, H. 1991. Clay mineral accumulation rates in the Arabian Sea during the late Quaternary. *Marine Geology*, **97**, 105–119.

SIROCKO, F., SARNTHEIN, M., LANGE, H. & ERLENKEUSER, H. 1991. Atmospheric summer circulation and coastal upwelling in the Arabian Sea during the Holocene and last Glaciation. *Quaternary Research*, **36**, 72–93.

SLATER, R.D. & KROOPNICK, P. 1984. Controls of dissolved oxygen distribution and organic carbon deposition in the Arabian Sea. *In:* HAQ, B.U. & MILLIMAN, J.D. (eds) *Marine Geology and Oceanography of Arabian Sea and Coastal Pakistan.* Van Nostrand Reinhold, New York, 305–313.

SMITH, C.R., LEVIN, L.A., HOOVER, D.J., MCMURTRY, G. & GAGE, J.D. 2000. Variations in bioturbation across the oxygen minimum zone in the northwest Arabian Sea. *Deep-Sea Research II*, **47**, 227–257.

SMITH, R.L. & BOTTERO, J.S. 1977. On upwelling in the Arabian Sea. *Deep-Sea Research Supplement*, 291–304.

SMITH, S.L., CODISPOTI, L.A., MORRISON, J.M. & BARBER, R.T. 1998. The 1994–1996 Arabian Sea Expedition: an integrated, interdisciplinary investigation of the response of the northwestern Indian Ocean to monsoonal forcing. *Deep-Sea Research II*, **45**, 1905–1915.

TANGEN, K., BRAND, L.E., BLACKWELDER, P.L. & GUILLARD, R.R. 1982. *Thoracosphaera heimii* (Lohmann) Kamptner is a dinophyte: observations on its morphology and life cycle. *Marine Micropaleontology*, **7**, 193–212.

THOMAS, W.H. & GIBSON, C.H. 1990. Effects of small-scale turbulence on microalgae. *Journal of Applied Phycology*, **2**, 71–77.

THOMAS, W.H. & GIBSON, C.H. 1992. Effects of quantified small-scale turbulence on the dinoflagellate, *Gymnodinium sanguineum* (*splendidens*): contrasts with *Gonyaulax* (*Lingulodinium*) *polyedra*, and fishery implication. *Deep-Sea Research*, **39**, 1429–1437.

TROELSTRA, S.R., GANSSEN, G.M., VAN WEERING, T.C.E., KUYPERS, T., KARS, S. & OKKELS, E. 1995. Sedimentology. *In:* VAN HINTE, J.E., VAN WEERING, T.C.E. & TROELSTRA, S.R. (eds) *Tracing a Seasonal Upwelling. Report on Two Cruises of R.V. Tyro to the NW Indian Ocean in 1992 and 1993, Cruise Reports.* National Museum of Natural History, Leiden, **4**, 103–110.

VAN DER LAND, J. & STEL, J.H. 1995. Summary and acknowledgements. *In:* VAN HINTE, J.E., VAN WEERING, T.C.E. & TROELSTRA, S.R. (eds) *Tracing a Seasonal Upwelling. Report on Two Cruises of R.V. Tyro to the NW Indian Ocean in 1992 and 1993, Cruise Reports.* National Museum of Natural History, Leiden, **4**, 9–10.

VAN DER WEIJDEN, C.H., REICHART, G.J. & VISSER, H.J. 1999. Enhanced preservation of organic matter in sediments deposited within the oxygen minimum zone in the northeastern Arabian Sea. *Deep-Sea Research I*, **46**, 807–830.

VAN HINTE, J.E., VAN WEERING, T.C.E. & TROELSTRA, S.R. (eds) 1995. *Tracing a Seasonal Upwelling. Report on Two Cruises of R.V. Tyro to the NW Indian Ocean in 1992 and 1993, Cruise Reports,* Vol. 4. National Museum of Natural History, Leiden.

VAN WEERING, T.C.E., HELDER, W. & SCHALK, P. 1997. The Netherlands Indian Ocean Expedition 1992–1993, first results and an introduction. *Deep-Sea Research II*, **44**, 1177–1193.

VELDHUIS, M.J.W., KRAAY, G.W., VAN BLEIJSWIJK, J.D.L. & BAARS, M.A. 1997. Seasonal and spatial variability in phytoplankton biomass, productivity and growth in the northwestern Indian Ocean: the southwest and northeast monsoon, 1992–1993. *Deep-Sea Research I*, **44**, 425–449.

VERSTEEGH, G.J.M. 1993. New Pliocene and Pleistocene calcareous dinoflagellate cysts from southern Italy and Crete. *Review of Palaeobotany and Palynology*, **78**, 353–380.

VINK, A., ZONNEVELD, K.A.F. & WILLEMS, H. 2000. Distributions of calcareous dinoflagellate cysts in surface sediments of the western equatorial Atlantic Ocean, and their potential use in paleoceanography. *Marine Micropaleontology*, **38**, 149–180.

VINK, A., BRUNE, A., ZONNEVELD, K.A.F., HÖLL, C. & WILLEMS, H. 2002a. On the response of calcareous dinoflagellates to oligotrophy and stratification of the upper water column in the Equatorial Atlantic Ocean. *Palaeogeography, Palaeoclimatology, Palaeoecology*, **178**, 53–66.

VINK, A., RÜHLEMANN, C., ZONNEVELD, K.A.F., MULITZA, S., HÜLS, M. & WILLEMS, H. 2002b. Shifts in the position of the North Equatorial Current and rapid productivity changes in the western Tropical Atlantic during the last glacial. *Paleoceanography*, **16**(5), 479–490.

VON RAD, U. & SCHULZ, H. von Rad, U., Schulz, H. & Sonne-90 Scientific Party 1995. Sampling the oxygen minimum zone off Pakistan:glacial–interglacial variations of anoxia and productivity (preliminary results). *Marine Geology*, **124**, 7–19.

VON RAD, U., SCHULZ, H., RIECH, V., DEN DULK, M., BERNER, U. & SIRICKO, F. 1999. Multiple monsoon-controlled breakdown of the oxygen minimum conditions during the last 30,000 years documented in laminated sediments off Pakistan. *Palaeogeography, Palaeoclimatology, Palaeoceanography*, **152**, 129–161.

WELLER, R.A., BAUMGARTNER, M.F., JOSEY, S.A., FISCHER, A.S. & KINDLE, J.C. 1998. Atmospheric forcing in the Arabian Sea during 1994–1995: observations and comparison with climatology and models. *Deep-Sea Research II*, **45**, 1961–1999.

WENDLER, I., ZONNEVELD, K.A.F. & WILLEMS, H. 2002. Oxygen availability effects on early diagenetic calcite dissolution in the Arabian Sea as inferred from calcareous dinoflagellate cysts. *Global and Planetary Change*, in press.

WIGGERT, J.D., JONES, B.H., DICKEY, T.D., BRINK, K.H., WELLER, R.A., MARRA, J. & CODISPOTI, L.A. 2000. The Northeast Monsoon's impact on mixing, phytoplankton biomass and nutrient cycling in the Arabian Sea. *Deep-Sea Research, II*, **47**, 1353–1385.

WYRTKI, K. 1971. *Oceanographic Atlas of the International Indian Ocean Expedition*. NSF-IDOE-1, Washington, DC.

ZONNEVELD, K.A.F. 1997. Dinoflagellate cyst distribution in surface sediments of the Arabian Sea (Northwestern Indian Ocean) in relation to temperature and salinity gradients in the upper water column. *Deep-Sea Research II*, **44**, 1411–1444.

ZONNEVELD, K.A.F. & BRUMMER, G.A. 2000. Ecological significance, transport and preservation of organic walled dinoflagellate cysts in the Somali Basin, NW Arabian Sea. *Deep-Sea Research II*, **9**, 2229–2256.

ZONNEVELD, K.A.F., HÖLL, C., JANOFSKE, D., KARWATH, B., KERNTOPF, B., RÜHLEMANN, C. & WILLEMS, H. 1999. Calcareous dinoflagellate cysts as palaeo-environmental tools. *In:* FISCHER, G. & WEFER, G. (eds) *Use of Proxies in Paleoceanography: Examples from the South Atlantic*. Springer, Berlin, 145–164.

ZONNEVELD, K.A.F., BRUNE, A. & WILLEMS, H. 2000. Spatial distribution of calcareous dinoflagellate cysts in surface sediments of the Atlantic Ocean between 13°N and 36°S. *Review of Palaeobotany and Palynology*, **111**, 197–223.

Centennial–millennial-scale monsoon variations off Somalia over the last 35 ka

SIMON J. A. JUNG[1], EKATARINA IVANOVA[1], GERT JAN REICHART[2], GARETH R. DAVIES[1], GERALD GANSSEN[1], DICK KROON[1] & JAN E. VAN HINTE[1]

[1]*Department of Isotope Geology, Institute of Earth Sciences, Free University Amsterdam, de Boelelaan 1085, 1081 HV Amsterdam, Netherlands (e-mail: Jung@geo.vu.nl)*
[2]*University of Utrecht, Faculty of Earth Sciences, Budapestlaan 4, 3584 CD Utrecht, Netherlands*

Abstract: We present a multi-proxy study of sediment Core 905 from the Arabian Sea offshore Somalia to assess the validity of a number of proxies for productivity, temperature and wind strength, to reconstruct the monsoon history in the western Arabian Sea. The present-day seasonal variation in productivity in the modern Arabian Sea off Somalia reflects the change from the high-productivity SW monsoon to the low-productivity NE monsoon seasons. Annual productivity is therefore largely controlled by SW monsoon driven upwelling. The geochemical records of Core 905 document millennial-scale variations, for example, in Ba/Al and C_{org} content. The Younger Dryas and the time equivalent period to Heinrich event 1 show low annual productivity whereas the early Holocene and Bølling–Allerød periods are characterized by high productivity. The upwelling–productivity peaked during Early Holocene time and was followed by a decrease toward the modern values. The total flux of planktic foraminifera and the concentration of the planktic foraminifera *G. bulloides* are not always controlled by the total productivity. Variations in calcite dissolution, the advection of expatriate fauna or a seasonal decoupling of primary and secondary production appear to hamper straightforward interpretations of those foraminifera records. We conclude that at significantly changed climatic boundary conditions compared with the present day, bulk-sediment-related proxies of productivity more consistently record the local upwelling history than foraminifer-based productivity proxies.

To improve our understanding of the Earth's climate we require a detailed understanding of the processes that produce short-term climate change. Climatically sensitive regions of the world are the most logical regions to obtain raw data to infer the nature of these processes. The present-day Asian monsoon system is an example where the variability in large-scale atmospheric pressure gradients is particularly delicately balanced and results in a seasonal change in prevailing wind direction. Interaction of these wind systems with the surface ocean in the Arabian Sea induces high productivity during the SW monsoon and low productivity during the NE monsoon (Curry *et al.* 1992; Conan *et al.* 1995). These variations in bioproduction are recorded in foraminiferal assemblages found in deep-sea sediments off Somalia (Ivanova 2000).

Recently, a number of studies on deep-sea sediments from the Arabian Sea have provided basic insights into the mainly orbitally induced monsoonal variability (Clemens *et al.* 1991; Anderson & Prell 1993; Reichart *et al.* 1998). Stimulated by the discovery of short-term climate fluctuations in Greenland ice cores, climate instabilities on similar time scales have also been documented in the Arabian Sea (Sirocko *et al.* 1993; Naidu 1995; Reichart *et al.* 1998; Schulz *et al.* 1998).

A general problem related to palaeoclimate reconstruction is that the proxies used are mainly validated in studies on glacial–interglacial time scales. The concentration of the planktic foraminifera *G. bulloides*, for example, has been established as a productivity proxy in the Arabian Sea (Kroon 1991). Similarly, organic carbon (C_{org}) concentrations or fluxes have been used to assess productivity changes (Mueller & Suess 1979). The conventional interpretation of a proxy relies on a simple but inextricably empirical way to validate it, by finding a parameter that varies strongly with that proxy (e.g. temperature or productivity) in the modern ocean. The effects of changes in the boundary conditions (e.g. atmospheric or oceanic circulation changes), however, could move parameters beyond the proxy validation database. Interpretations in such a scenario would be invalid. An approach using several proxies for the same parameter may improve our understanding of each of the proxies. Such multi-proxy studies at a high temporal resolution are rare. Here we present a multi-proxy study of sediment Core 905 from the Arabian Sea offshore Somalia (Fig. 1). On the basis of a number of centennial- to millennial-scale proxy records we assess their individual limitations and the general implications. Our main objective is to test whether proxies that have been used on glacial–interglacial time scales are applicable on a century–millennial time scale. We also discuss which proxies appear to give consistent results in terms of bioproductivity and temperature in the ocean and those that do not and hence are interpreted to reflect changes in general climatic boundary conditions or have been affected by additional processes.

Modern annual monsoon dynamics

The pacemaker of the monsoon circulation in the Arabian Sea is the seasonally varying insolation distribution, which affects the subtropical high-pressure cell over the southern Indian Ocean and the Intertropical Convergence Zone (ITCZ). The position of the ITCZ and the strength of the subtropical high-pressure cell change throughout the year and control the winds in the Arabian Sea. In the southern Indian Ocean this system induces the SE trade winds that partly feed into the SW monsoon, which dominates the Arabian Sea during summer (Fig. 1). During winter the wind direction reverses and the NE monsoon prevails. As a result of the wind–ocean interaction the SW monsoon triggers coastal upwelling off NE Africa and Arabia. During the NE monsoon no upwelling occurs off Somalia and the bioproductivity is low (Conan & Brummer 2000).

Methods

Sediment Core 905 was retrieved from 1600 m depth off Somalia (Table 1, Fig. 1). The age model is based on 12 acceleration mass spectrometry (AMS) ^{14}C dates from strategic depth intervals covering the last 35 ka BP (Fig. 2; Ivanova 2000; Jung et al. 2002). The sediment accumulation rate in Core 905 varies between roughly 13 cm ka^{-1} during the glacial period and up to 36 cm ka^{-1} during Holocene time. The dry bulk density (DBD, g cm^{-3}) was determined by weighing a fixed volume of wet sediment after drying at 50 °C. Subsequently, the samples were soaked for 1 h in water containing 2% hypochloride and washed over 75, 125, 150 and 250 µm mesh sieves. The dried samples were split with an Otto microsplitter until they contained 200–300 specimens of planktic foraminifera, which were identified and counted in the size fractions 125–150, 150–250 and >250 µm. Subsequently, 14–17 specimens of G. bulloides, 8–10 specimens of G. ruber (both from size fraction 250–355 µm) and 2–3 specimens of N. dutertrei (size fraction 355–420 µm) were analysed for oxygen and carbon isotope ratios with a MAT 251 mass spectrometer at the Vrije Universiteit Amsterdam. The internal reproducibility for δ^{18}O measurements of standards is ±0.05‰, resulting in an analytical detection limit for δ^{18}O variations of 0.1‰.

The concentration of organic carbon of the carbonate-free samples was measured on a CNS analyser (Fisons NA 1500) at the University of Utrecht. The analytical precision and accuracy were determined by replicate analyses and by comparison with international (BCR-71) and in-house F-TURB and MM-91 standards. The repro-

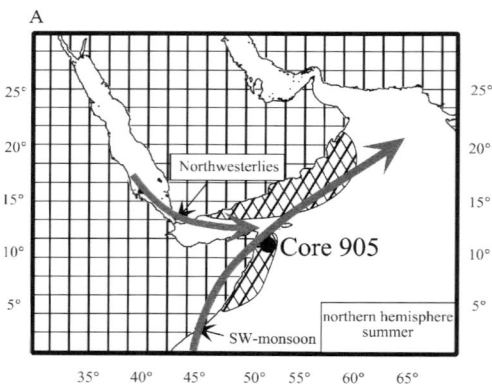
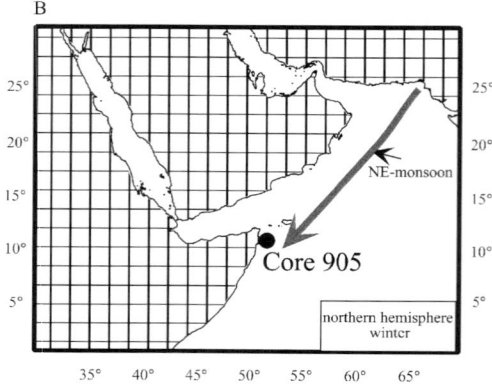

Fig. 1. Location of Core 905. Hatched areas in (**a**) indicate summer upwelling areas. Arrows indicate prevailing wind systems.

Table 1. Summary of the main results of sediment Core 905

Proxy type	Measurement	Time period		
		Holocene	Termination I	O-isotope stage 2-Termination I
Temperature related	$\delta^{18}O$ records (‰)	generally lowest values; 0.5‰ variation in *G. ruber* (most pronounced at 7–8.5 ka BP); 0.3–0.5‰ general increase in *G. bulloides* around 5–4 ka BP; *N. dutertrei* relative stable, minor maximum at 9.5 and 4 ka BP	all records sharp decrease at 16–14.8 ka BP (by 0.6–0.8‰); Bø.– Alø. low values; all records show YD maximum (most pronounced in *N. dutertrei*); Ib decrease by 0.5–0.8‰	generally low values; decrease toward Ia; 0.5‰ variation in *N. dutertrei*; mean large difference between *N. dutertrei* and *G. bulloides*, small difference between *G. bulloides* and *G. ruber*
Bio-productivity	Ba/Al	10–5 ka BP maximum values; local maximum spike at 7–6 ka BP; 5–2.5 ka BP decrease to modern values	onset Ia lowest values; maximum during Bø.– Alø. decrease onset YD; rise along with Ib	mean similar to modern; minima at c. 28–29 and 25–23 ka BP; decrease towards Ia
	C_{org} (g m^{-2} a^{-1})	slow further rise to >3 at c. 6 ka BP; subsequent sharp drop to modern type values around 2	rise to modern type values at 16–11.8 ka BP; subsequent 11.8 ka BP sharp rise to 2.8	generally low values around 1.5; subsequent to 19 ka BP decrease toward Ia
	C_{org} (%)	10–2 ka BP values of 1.4–1.7; local maxima at 6–5 and 3 ka BP; sharp rise to values around 2.2 at 2–1.8 ka BP	lowest values <1 at onset Ia; rise to 1.2 along Ia; moderately high values during Bø.– Alø.; local minimum during YD; slow rise during Ib	values of 1–1.6; local minima at 27.5–27, 25.5–22.8 and toward 16 ka BP (termination Ia)
	G. bulloides (%)	generally high values; early Holocene (to 7.5 ka BP) slightly higher than late Holocene values	highly variable; minima at 15.8 and 14 ka BP; YD maximum spike; sharp rise to modern type values at late Ib	generally decrease until 18.5 ka BP with variations superimposed; maximum spike at 16.8 ka BP; decrease toward Ia

(continued overleaf)

Table 1. (continued)

Proxy type	Measurement	Holocene	Termination I	O-isotope stage 2-Termination I
	Total flux planktic forams	medium values until 8.5 ka BP; subsequently sharp rise; highest values (with variation) between 8.5 and 3 ka BP; 3–0 ka BP decrease to medium values	onset Ia to YD negative oscillation; YD high values, with sharp drop at end; increase to medium values during first part of Ib	generally low values with variation; maxima at 26–23, 22–19 and 16.8 ka BP
Atmospheric	Ti/Al	up to roughly 6 ka BP minor variations around 560; 6–0 ka BP rise to values up to 680	continuous decrease toward YD; sharp rise at Ib to values at 560	values between 470 and 570; broad maximum at 24–18 ka BP; low values early stage 2; decrease after 18 ka BP

Dimensions of numbers given refer to measurement column (except age estimates). Termination I is abbreviated to I (a and b), and Bolling–Allerod to Bø.– Alø.

ducibility was better than 3%. We calculated accumulation rates of organic carbon (C_{org}; g cm^{-2} ka^{-1}) based on the C_{org} content of the sediments and the mass accumulation rate.

Elemental compositions were measured by inductively coupled plasma–atomic emission spectrometry (ICP-AES, Perkin Elmer Optima 3000) after subsamples were freeze-dried, thoroughly ground in an agate mortar and subjected to $HClO_4$, HNO_3 and HF digestion. Comparison with an international (SO-1) and in-house standard (MM-91) and the analysis of duplicate samples demonstrate that the accuracy is better than 5% for all elements discussed. To distinguish changes that are not caused by dilution of $CaCO_3$ or a variable input of terrestrial detritus, elements are presented normalized to Al (Shimmield & Mowbray 1991).

Proxies used in this study

The $\delta^{18}O$ values in foraminifera depend on the oxygen isotope composition and the temperature of the ambient water during calcification (e.g. Erez & Luz 1983). At Site 905, *G. bulloides* almost exclusively occurs during the SW monsoon (upwelling season) as an opportunistic species able to quickly adapt to the enhanced food supply during the upwelling season (Kroon *et al.* 1991; Prell *et al.* 1996). Sediment trap results show that about 95% of all *G. bulloides* form during SW monsoon (Conan *et al.* 1995). In contrast, *G. ruber* occurs at roughly constant frequencies throughout the year. Net samples from that region show that both species calcify at similar depth (Peeters 2000), a finding that is supported by the observation of almost equal $\delta^{18}O$ values from both species in sediment trap samples from the upwelling season at Site 905 (S. Conan, pers. com.). The different annual reproduction frequency of *G. ruber* and *G. bulloides* in combination with indistinguishable $\delta^{18}O$ values in sediment traps during the SW monsoon show that any down-core variation in the difference in $\delta^{18}O$ records between *G. bulloides* and *G. ruber* must result from productivity–temperature changes during the NE monsoon season.

We will also use variations of the concentration of *G. bulloides* as a proxy for palaeoproductivity in accordance with earlier studies (e.g. Prell *et al.* 1980; Kroon *et al.* 1991*a*; Anderson & Prell 1993; Reichart *et al.* 1997). A second proxy for productivity is the total flux of planktic foraminifera (e.g. Anderson & Prell 1993). Fluxes of most species increase during the upwelling season (Curry *et al.* 1992; Ortiz & Mix 1992; Conan *et al.* 1995). An increased flux of planktic foraminifera, however, may also occur during winter as a result of deep

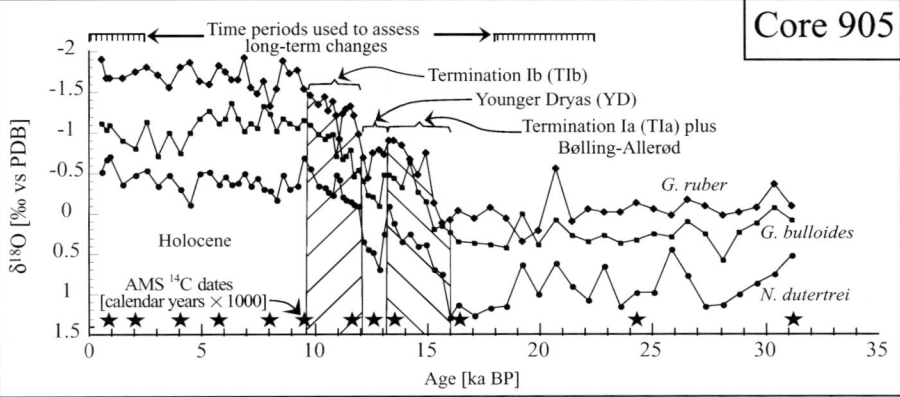

Fig. 2. $\delta^{18}O$ records of the planktic foraminifera species *G. ruber*, *G. bulloides* and *N. dutertrei*. Stars indicate AMS ^{14}C dates (Ivanova 2000; Jung et al. 2002).

convective mixing caused by the NE monsoon (Curry et al. 1992; Conan et al. 1995).

The increase in productivity during the SW monsoon also causes an increase in the flux of organic matter to the sea floor (Sarnthein et al. 1988). Reichart et al. (1997) and Den Dulk et al. (1998) found a correlation between the organic carbon content (C_{org}) and the percentage of *G. bulloides* in sediments from the northern Arabian Sea reflecting productivity variations at the sea surface. The amplitude of the variation in the C_{org} depends on, amongst other factors, the oxygenation of the ambient seawater and sediment accumulation rates, which affect the preservation of the organic matter and its dilution by other sedimentary components. The calculated C_{org} accumulation rates can be roughly corrected for variable sedimentation rates and hence serve as a productivity proxy (for summary, see Sarnthein et al. 1988).

The Ba/Al ratio of bulk sediments is used as a proxy for productivity. The assumption is that barite precipitation occurs in decaying particulate organic matter while it sinks to the sea floor (e.g. Dehairs et al. 1980; Bishop 1988; Dymond & Collier 1994). Enhanced Ba/Al ratios below high-productivity areas support this assumption (e.g. Payton et al. 1996; Nürnberg et al. 1997). Generally, because pore waters are saturated with respect to barite there is a lower diagenetic effect on barite than, for example, on C_{org} (e.g. Dymond & Collier 1996). Recent evidence suggests that barite preservation is also influenced by other factors such as the degree of barite saturation in bottom waters and organic matter degradation (e.g. Schenau et al. 2001). In extremely productive areas such as off Somalia, however, the barite flux appears to be overwhelmingly controlled by surface ocean productivity. Hence, the Ba/Al record of Core 905 may be used as a proxy for productivity variations in the surface ocean. This conclusion follows earlier studies that have successfully used Ba/Al variations to assess the productivity history of the western and northern Arabian Sea (Shimmield & Mowbray 1991; Reichart et al. 1997).

To assess changes in the past wind intensity we use the ratio of titanium to aluminium (Ti/Al) in sediments. This ratio has been used to deduce the monsoon strength in the Arabian Sea (Kroon et al. 1991a; Weedon & Shimmield 1991; Reichart 1997). Titanium is primarily concentrated in the coarser sediment fraction, particularly in heavy mineral assemblages containing ilmenite, rutile, titanomagnetite and augite (Schmitz 1987). Because the distribution of heavy minerals in deep-sea sediments is affected by changes in the grain size of wind-blown dust, it has been argued that variations of the Ti/Al ratio dominantly indicate variations in wind strength. Changes in weathering in the dust source regions may also result in variations of the Ti/Al ratio by controlling the probability of potential dust particles becoming entrained in the dust-carrying wind. In particular, variable weathering of the Ti-rich volcanic rocks of the East African Rift system may affect the Ti concentration carried to the Arabian Sea by the SW monsoon. Recently, land-use changes by early human civilizations in NE Africa may also have affected the dust export. For example, the burning of woodland would reduce soil stability and enhance dust uptake.

Patterns of monsoonal change

In this section we assess both the interglacial–glacial and the short-term variations in productivity, temperature and atmospheric circulation based on a combination of proxies. The results of the various records are summarized in Table 1 and Figs. 2 and 3. The oxygen isotope records of three planktic foraminifera species clearly indicate the change from the high $\delta^{18}O$ values during the last glacial period to low Holocene values (Fig. 2). These records also show short-term variability on a centennial to millennial time scale that is most pronounced during Termination I and the Holocene period. The pronounced oscillation in $\delta^{18}O$ between 16 ka BP and midway Termination I is a distinct expression of the well-known succession of the Bølling–Allerød warm period and the Younger Dryas (YD) cold spell.

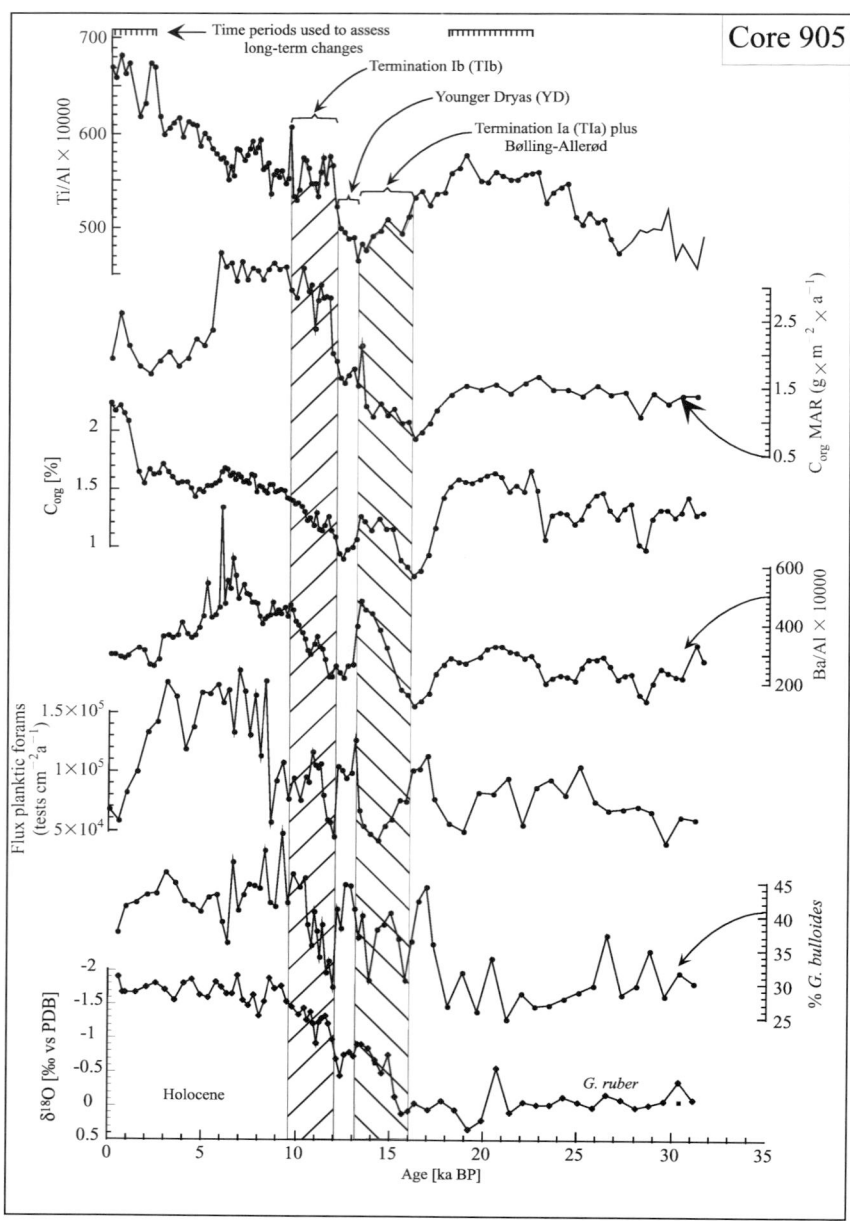

Fig. 3. Plot of the Ti/Al, C_{org} (%), C_{org} (flux), Ba/Al, per cent *G. bulloides* and planktic foraminifera (total flux) records, and $\delta^{18}O$ record of *G. ruber* of Core 905.

Productivity variations in the western Arabian Sea

A large number of studies have assessed the long-term monsoon variations in the Arabian Sea. Widespread agreement exists that the glacial productivity in the western Arabian Sea was lower as a result of a less intense SW monsoon (e.g. Clemens et al. 1991; Anderson & Prell 1993; Reichart et al. 1998; Ivanova 2000; Sirocko et al. 2000). Within our dataset the occurrence of generally low glacial concentrations of G. bulloides agrees with this conclusion. A more detailed application of other proxies, however, does not fully support such a simple interpretation. There are conflicts in the proxy data on glacial–interglacial and, to an even greater extent, on a century to millennial time scale.

A first 'proxy conflict' occurs in comparing concentration of G. bulloides with the Ba/Al record, total flux of planktic foraminifera and the accumulation rate of C_{org}. The similarity of values in these three records between the glacial period around 18–22.5 ka BP and the core top (0–2.5 ka BP) suggests that maximum productivity during the glacial period was similar to that in the modern Arabian Sea, an observation in striking contrast to previous studies (e.g. Clemens et al. 1991; Anderson & Prell 1993; Reichart et al. 1998).

A closer look at the various records on millennial to centennial time scales highlights the problems in obtaining a consistent view of the productivity history off Somalia. A pronounced mismatch occurs between conventional interpretations of the various proxies around 15–18 ka BP. During this time period the concentration of G. bulloides and the total flux of planktic foraminifera reach modern and higher values, indicating a productivity spike before the onset of Termination Ia, whereas the Ba/Al ratio, and the C_{org} concentrations and fluxes suggest minimum productivity (Fig. 3). On the basis of the current understanding of the individual proxies a consistent view of the productivity history is difficult to establish. The marked change in all proxies occurring between 15 and 18 ka BP, however, clearly documents a major reorganization of the oceanic boundary conditions.

The inconsistency between the bulk-sediment-related and foraminiferal productivity records persists throughout Termination I. The Ba/Al, C_{org} flux and C_{org} concentrations indicate a rise in productivity at the Termination Ia to Bølling–Allerød transition. The onset of the YD is marked by a drop to low values in the Ba/Al and C_{org} percentage records. C_{org} flux values are low and constant at that time. Subsequently, these three records indicate a rising productivity towards high Early Holocene values. The foraminifera-derived productivity history is strikingly different. The total flux of foraminifera and the concentration of G. bulloides suggest a decrease in productivity during the Termination Ia to Bølling–Allerød transition. The YD is marked by maximum values, whereas Termination Ib (subsequent to a sharp drop at the end of the YD) is marked by rising values. These observations show that the proxies are not consistent on century–millennial time scales.

The Holocene productivity variation indicated by the total flux of planktic foraminifera and the concentration of G. bulloides resembles the history deduced from bulk-sediment-based proxies. The slightly higher concentrations of G. bulloides during Early Holocene time point to an enhanced bioproduction during this period, a finding that, in general, is supported by the high total flux rates of planktic foraminifera and also by maxima in the Ba/Al and the C_{org} flux records.

Atmospheric variations

The wind strength variations resemble the geochemical bulk-sediment-based productivity record. If we take the variations in Ti/Al ratio at face value the average wind strength during the last glacial period was lower than during Holocene time, although around 18–22.5 ka BP values are as high as those found during Early Holocene time. During the time equivalent to Heinrich event 1 the Ti/Al record reaches minimum values. Subsequent to the YD, roughly constant and moderately high values prevail to 5–6 ka BP, followed by rising Ti/Al values. Highest values are recorded over the last 3–4 ka BP, interestingly a period when the Ba/Al, C_{org} flux, concentration of G. bulloides and accumulation rate of planktic foraminifera suggest decreasing productivity and by implication wind strength.

Synopsis

Long-term productivity variations

To start our assessment of the consistency of the various proxies we use down-core oxygen isotope records of coexisting planktic foraminifera species living at different times of the year to assess the change in water column structure and temperature (Kroon & Ganssen 1989). A comparison of the change in $\delta^{18}O$ in G. bulloides and G. ruber between the last glacial period around 18–22.5 ka BP and the core-top (0–2.5 ka BP) is presented in Fig. 2. This glacial time period reflects a period of relative stability as suggested by the bulk-sediment-related proxies for productivity. Comparison of productivity proxies from the last glacial period around 18–22.5 ka BP with the modern (0–2.5 ka BP) values therefore allows assessment of the long-term climate changes. The changes in *G. bulloides* and *G. ruber* are c. 1.3 and c. 1.75‰, respectively (Fig. 2); hence, both records are largely controlled by the 1.2–1.3‰ variation induced by the change in global ice volume change (Labeyrie et al. 1987; Fairbanks 1989). The change in $\delta^{18}O$ of 1.3‰ in *G. bulloides* almost exclusively reflects the change during the SW monsoon, as this species overwhelmingly occurs during the upwelling season. In contrast, the change in $\delta^{18}O$ found in *G. ruber* reflects an annual signal, as *G. ruber* reproduces at a constant rate throughout the year. Assuming that the modern annual frequency distribution of *G. ruber* applies (see Conan & Brummer 2000), 50% of the population of *G. ruber* should be affected by changes during the upwelling season, and 50% by the non-upwelling monsoon season. The larger $\delta^{18}O$ change in *G. ruber* of roughly 0.4–0.5‰ compared with that of *G. bulloides* has important implications. First, around 18–22.5 ka BP a significant proportion of *G. ruber* must have formed during the non-upwelling season to change $\delta^{18}O$ more than during the upwelling season alone. If we assume that the identical $\delta^{18}O$ values of *G. ruber* and *G. bulloides* during the modern SW monsoon also applied during the last glacial period, we conclude that the glacial NE monsoon season must have been cooler relative to the SW monsoon than it is today. If we further assume that 50% of the *G. ruber* reproduced during the NE monsoon, a decrease in temperature of roughly 3–4 °C relative to the SW monsoon season is needed to explain the $\delta^{18}O$ data.

Given the inferred extra cooling around 18–22.5 ka BP, the NE monsoon winds may have facilitated a deeper mixing, enhanced advection of nutrient-rich subsurface water and possibly elevated bioproductivity during the winter season (Reichart et al. 1998). If true, this scenario would imply that the relatively high productivity levels around 18–22.5 ka BP, as suggested by the bulk-sediment proxies, record significant NE monsoon related productivity, in contrast to the present day, when productivity is largely controlled by the SW monsoon. The modern-type flux rates of planktic foraminifera around 18–22.5 ka BP appear to support this conclusion.

The productivity history suggested by the bulk-sediment proxies also contrasts with abundances of *G. bulloides* around 18–22.5 ka BP, which are significantly lower than present day values. In the modern ocean *G. bulloides* quickly adapts to enhanced nutrient supply by enhanced reproduction rates. If we assume that the glacial productivity reached roughly present-day levels as suggested by the Ba/Al and C_{org} records, we would expect that the glacial concentration of *G. bulloides* would roughly reach modern-day values. This expectation contrasts with the low glacial concentrations of *G. bulloides* (Fig. 3). Both the flux rates of planktic foraminifera and the concentration of *G. bulloides* may have been affected by variations in the calcite dissolution rate. Figure 4 displays the flux of planktic foraminifera along with the change in the percentage of fragments of planktic foraminifera. The number of fragments of planktic foraminifera has been frequently used as a qualitative measure of calcite dissolution (Berger & Killingley 1977). The low concentration of fragments around 18–22 ka BP suggests that the effect of dissolution on planktic foraminifera was relatively small during this period compared with Late Holocene time. If true, we may speculate that the comparably low flux of total foraminifera during Late Holocene time may have resulted from enhanced calcite dissolution as suggested by the higher concentrations of fragments of planktic foraminifera. Consequently, the Late Holocene flux rates of planktic foraminifera were probably higher. This conclusion reinforces the difference in productivity histories inferred from the bulk-sediment-related proxies and the flux of planktic foraminifera for Late Holocene time and the last glacial period between 18 and 22.5 ka BP. Similarly, we conclude that the concentration of *G. bulloides* was only little altered during the latter period as a result of relatively low calcite dissolution. *G. bulloides* is a rather fragile species, so calcite dissolution has a more marked effect than on other more robust species (Berger & Killingley 1977). We can therefore assume that calcite dissolution affects *G. bulloides* faster than the other species, resulting in a lowering of the concentration of *G. bulloides*. Hence, the high concentration of fragments of planktic foraminifera suggests that the Late Holocene drop in the concentration of *G. bulloides* may be an artefact of enhanced calcite dissolution, strengthening the mismatch between bulk-sediment-related and foraminifera-related proxies. Around 18–22.5 ka BP we therefore observe a decoupling of primary and secondary production variations (at least for *G. bulloides*). Possible reasons for the decoupling may be a change in the composition of the food chain or an effect of the generally different ocean–atmospheric circulation during glacial conditions. In particular, the latter alternative follows suggestions by Kroon (1991) and Vergnaud-Grazzini et al. (1995). It is possible that the planktic foraminifera were sensitive to major changes in Indian Ocean circulation rather than

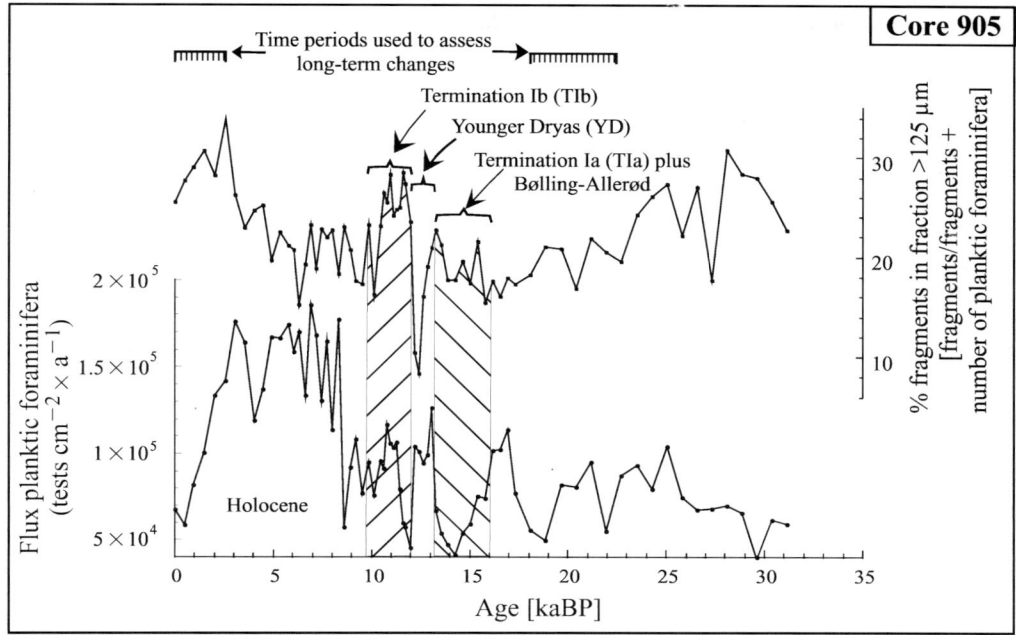

Fig. 4. Comparison of the flux record of planktic foraminifera and the fragmentation record of Core 905.

just variations in upwelling alone. It is therefore possible that some of the peaks in the concentration of *G. bulloides* represent expatriate transitional faunas.

In summary, we conclude that the changes in surface water properties between 18–22.5 ka BP and the core top (0–2.5 ka BP) appear to provide an explanation of why bulk-sediment-related proxies record comparable productivity for both time periods. The concentration of *G. bulloides* and the total flux of planktic foraminifera in contrast are inconsistent with the productivity as suggested by the bulk-sediment-related proxies. Hence, under significantly different climatic boundary conditions compared with those of the present day, bulk-sediment-related proxies appear to record the productivity history more reliably than foraminifera proxies. When boundary conditions differ only slightly, however, bulk-sediment-related and foraminifera-based proxies should provide a consistent picture of past productivity changes. In particular, the small difference between the productivity histories for Early Holocene time, as suggested by the various proxies, supports this view. All productivity records indicate higher values during for this period (compare, e.g. Prell *et al.* 1984; Anderson & Prell 1993; Sirocko *et al.* 1993). The flux of C_{org}, the Ba/Al ratio, the total flux of planktic foraminifera and the concentration of *G. bulloides* also suggest a lowered productivity during the Late Holocene period, although the timing of this reduction differs between the various records.

Short-term productivity history

The bulk-sediment proxies for productivity appear to imply that the productivity oscillated in a period that began around 18 ka BP and lasted into the YD. Before Termination Ia a lowering in production levels occurred, followed by a rise during the Bølling–Allerød warm interval and then lower values during the YD (Fig. 3). Problems arise, however, if we compare these data with the productivity history as suggested by the flux of planktic foraminifera and the concentration of *G. bulloides*. During the Termination Ia to Bølling–Allerød oscillation the bulk-sediment proxies suggest a higher productivity whereas the foraminifera-flux proxies imply the opposite. We speculate that a scenario similar to that outlined in the previous section to explain the comparison between the core-top and glacial situation may be also applicable for the Termination Ia to Bølling–Allerød period. Again, this scenario implies that the NE monsoon driven productivity primarily affected the primary production. The secondary producers, in particular the planktic foraminifera, did not respond to a primary production spike in this area. There is a large change in $\delta^{18}O$ records during the Termination Ia to Bølling–Allerød period that is largely controlled by global ice-volume change. Consequently, we cannot evaluate if there was indeed extra cooling during the NE monsoon as we proposed for the period around 18–22.5 ka BP.

A second mismatch in proxy data occurs during the YD, when the bulk-sediment proxies suggest low production, whereas the foraminifera-based proxies show peak values. Assuming that the general boundary conditions during the YD more closely resemble glacial conditions than the modern situation, we conclude that the bulk-sediment-related proxies more reliably depict the local productivity history. Consequently, we further conclude that the productivity off Somalia was lower during the YD. This raises the question of the mechanism that could cause the high fluxes of planktic foraminifera and the high concentrations of *G. bulloides*, without a primary production peak. This type of mismatch has also been found by Vergnaud-Grazzini et al. (1995) in the tropical Indian Ocean. Those workers also suggested that a decoupling between the SW monsoon and the foraminifera production might occur as a result of a general change in boundary conditions. We speculate that the 'foraminifera spikes' were the result of an advection of foraminifera, probably from the south.

Another explanation for the mismatch between bulk-sediment- and foraminifera-deduced productivity histories for the entire period before Termination Ia and into the YD could be variations in calcite dissolution. If we take the productivity history as suggested by the bulk-sediment-related proxies at face value we conclude that the varying flux of C_{org} to the sea floor and its subsequent oxidation may have modulated the calcite dissolution. The variation in the fragments of planktic foraminifera, for example, shows low values (little dissolution) around 16–17 ka BP, where there is a maximum spike in *G. bulloides* concentration and the total flux of planktic foraminifera. The productivity minimum during this time period suggested by the bulk-sediment-related proxies implies that the maxima in the foraminifera records may result from low calcite dissolution rate. Interestingly, this time period roughly coincides with the well-known Heinrich event 1 from the North Atlantic. Subsequently, higher concentrations of fragments of planktic foraminifera during Termination Ia and lower numbers during the YD suggest an alternation of high and low calcite dissolution rates. These data again vary in antiphase with the Ba/Al record and the C_{org} concentration. In that sense the minimum–maximum oscillation of the flux of planktic foraminifera and the concentration of *G. bulloides* for this period may at least in part reflect variations in the calcite dissolution, probably as a combination of variation in degradation of organic matter and a change in water mass properties.

In contrast to the mismatch in the proxy records around Termination Ia and the YD, the various proxy records generally consistently suggest rising productivity during Termination Ib. Minor deviations between the various records probably reflect combinations of the factors outlined above that affect those records. By their nature, these small differences are, however, difficult to assess.

Atmospheric variations and productivity

Overall, the wind strength history suggested by the Ti/Al record resembles the productivity record implied by the bulk-sediment proxies. In the modern Arabian Sea the dust input largely occurs during the SW monsoon season. The main source for the dust is the Arabian Peninsula (for a summary, see Sirocko et al. 2000). As a result of the close proximity to the African coast, however, Core 905 mainly receives dust from the African continent (see Sirocko et al. 2000). Hence, the Ti/Al record should mainly depict variations of the SW monsoon, although the ratio itself is not provenance specific. Consequently, any interpretation of the Ti/Al data is limited by possible variations of grain size, change of the source region or varying weathering rates. If we interpret the Ti/Al ratio in a conventional way as a proxy for wind strength, the low glacial Ti/Al values suggest that the SW monsoon was generally weaker during the last glacial period than over the period 3–4 ka BP. A weaker SW monsoon might in turn have resulted in a reduced upwelling and hence lower productivity at that time. This implication is not in conflict with the enhanced productivity deduced for the glacial NE monsoon around 18–22.5 ka BP because in the modern Arabian Sea the imprint of dust supply to Core 905 off Somalia via the NE monsoon is negligible. Hence, an enhanced productivity during the NE monsoon would not be recorded in the Ti/Al record.

The low values in the Ti/Al record during the YD support the low productivity as indicated by the bulk-sediment proxies. They would, however, conflict with the conclusion on the possible advection of foraminifera from the south. The reason for this conflict is that the high concentrations of *G. bulloides* and the flux rates of foraminifera cannot be reconciled with a weak SW monsoon, which would be responsible for the northward advection of the foraminifera. The opposite scenario, where the advection of foraminifera occurred during the NE monsoon, could explain these findings. It would, however, imply a drastic change in surface and subsurface currents during the NE monsoon. In contrast to today, when a north to northeastward subsurface flow prevails during the NE monsoon, the outlined scenario would probably imply a reversed subsurface flow during that season.

A further obvious discrepancy with the productivity records occurs over the last 5 ka. The peak Ti/Al ratios suggest highest wind speeds in Late Holocene time, which correlates with a drop in productivity. This striking conflict with the modern SW monsoon-induced fertility offshore Somalia, its Holocene history and the change in dust supply cannot be resolved based on the Ti/Al ratio alone. Terrestrial sequences suggest that around Mid-Holocene time, North Africa witnessed a dramatic change from a wet Early to a dry Late Holocene period (Ritchie et al. 1985). Accordingly, the enhanced Ti/Al maximum during Late Holocene time probably reflects more effective dust entrainment rather than a change in wind speed. Given the uncertainties of the bulk-sediment Ti/Al ratio proxy, additional, provenance-specific, data are required to assess variations in wind provenance intensity.

Concluding remarks

On the basis of a detailed comparison of a number of productivity proxies from high-resolution Core 905 off Somalia, a number of conclusions can be drawn, as follows.

(1) The bulk-sediment-related proxies appear to most reliably record the local productivity. Although the foraminifera-based productivity proxies mirror the long-term changes on glacial–interglacial time scales, they conflict with the bulk-sediment proxies on centennial to millennial time scales. The most likely explanation is a combination of the effect of variations in calcite dissolution and insufficient understanding of the ecology of a foraminifera species, in particular at the generally changed glacial boundary conditions in the Arabian Sea.

(2) The seasonal balance in productivity off Somalia shifted during the glacial period from the modern SW monsoon controlled system towards a system with a significant productivity input during the NE monsoon. The latter season may have been c. 4 °C colder relative to the modern temperature difference between the SW and NE monsoon season. The annual glacial productivity values compare with the modern core-top values at Site 905. During the YD the productivity dropped. The upwelling–productivity peaked during Early Holocene time, followed by a decrease toward modern values.

We thank two anonymous reviewers for helping to improve this paper. This study is part of the EU-TMR network ERBFMRXCT960046 'The marine record of continental tectonics and erosion' and is NSG Publication 20020402.

References

ANDERSON, D.M. & PRELL, W.L. 1993. A 300 kyr record of upwelling off Oman during the late Quaternary; evidence of the Asian southwest monsoon. *Paleoceanography*, **8**(2), 193–208.

BERGER, W.H. & KILLINGLEY, J.S. 1977. Glacial–Holocene transition in deep-sea carbonates: selective dissolution and the stable isotope signal. *Science*, **197**, 563–566.

BISHOP, J.K.B. 1988. The barite–opal–organic carbon association in oceanic particulate matter. *Nature*, **332**, 341–343.

CLEMENS, S., PRELL, W., MURRAY, D., SHIMMIELD, G. & WEEDON, G. 1991. Forcing mechanisms of the Indian Ocean monsoon. *Nature*, **353**, 720–725.

CONAN, S., IVANOVA, K., PEETERS, F., GANSSEN, G., VAN HINTE, J.E., TROELSTRA, S.R. & VAN WEERING, T. 1995. Tracing a monsoonal upwelling in the Indian Ocean. *Terra Nostra*, **1**, 57.

CONAN, S. & BRUMMER, G.-J.A. 2000. Fluxes of planktic foraminifera in response to monsoonal upwelling on the Somalia Basin margin. *Deep-Sea Research II*, **47**, 2207–2227.

CURRY, W.B., OSTERMANN, D.R., GUPTHA, M.V.S. & ITTEKOT, V. 1992. Foraminiferal production and monsoonal upwelling in the Arabian Sea: evidence from sediment traps. *In:* SUMMERHAYS, C.P.P., PRELL, W.L. & EMEIS, K.C. (eds) *Upwelling Systems: Evolution Since the Early Miocene*. Geological Society, London, Special Publications, **64**, 93–106.

DEHAIRS, F., CHESSELET, R. & JEDWAB, J. 1980. Discrete suspended particles of barite and the barium cycle in the open ocean. *Earth and Planetary Science Letters*, **49**, 528–550.

DEN DULK, M., REICHART, G.J., MEMON, G.M., ROELOFS, E.M.P., ZACHARIASSE, W.J. & VAN DER ZWAAN, G.J. 1998. Benthic foraminiferal response to variations in surface water productivity and oxygenation in the northern Arabian Sea. *Marine Micropaleontology*, **35**(1-2), 43–66.

DYMOND, J. & COLLIER, R. 1994. Particulate barium fluxes; relationships to biological processes and ocean productivity. *Geological Society of America, Abstracts with Programs*, **26**(7), 20.

DYMOND, J. & COLLIER, R. 1996. Particulate barium fluxes and their relationships to biological productivity. *Deep-Sea Research II*, **43**, 1283–1308.

EREZ, J. & LUZ, B. 1983. Experimental paleotemperature equation for planktonic foraminifera. *Geochimica et Cosmochimica Acta*, **47**, 1025–1031.

FAIRBANKS, R.G. 1989. A 17,000-year glacio-eustatic sea level record: influence of glacial melting rates on the Younger Dryas event and deep-ocean circulation. *Nature*, **342**, 637–642.

IVANOVA, E. 2000. *Late Quaternary monsoon history and paleoproductivity of the western Arabian Sea*. PhD thesis, Free University, Amsterdam.

JUNG, S.J.A., GANSSEN, G. & DAVIES, G.R. 2002. Multidecadal variations in the Early Holocene outflow of Red Sea Water into the Arabian Sea. *Paleoceanography*, **16**, 658–668.

KROON, D. & GANSSEN, G. 1989. Northern Indian Ocean upwelling cells and the stable isotope composition of living planktonic foraminifers. *Deep-Sea Research*, **36**(8), 1219–1236.

KROON, D. 1991. Distribution of extant planktic foraminiferal assemblages in Red Sea and northern Indian Ocean surface waters. *Revista Espanola de Micropaleontologia*, **23**(1), 37–74.

KROON, D., BEETS, K., MOWBRAY, S., SHIMMIELD, G. & STEENS, T. 1991*a*. Changes in northern Indian Ocean monsoonal wind activity during the last 500 ka. *Memorie della Societa Geologica Italiana*, **44**, 189–207.

KROON, D., STEENS, T.N.F. & TROELSTRA, S.R. ET AL. 1991*b*. Onset of monsoonal related upwelling in the western Arabian Sea as revealed by planktonic foraminifers. *In:* PRELL, W.L. & NIITSUMA, N. (eds) *Proceedings of the Ocean Drilling Program, Scientific Results, 117*. Ocean Drilling Program, College Station, TX, 257–263.

LABEYRIE, L.D., DUPLESSY, J.-C. & BLANC, P.L. 1987. Variations in mode of formation and temperature of oceanic deep waters over the past 125,000 years. *Nature*, **327**, 477–483.

MUELLER, P.J. & SUESS, E. 1979. Productivity, sedimentation rate, and sedimentary organic matter in the oceans; I, Organic carbon preservation. *Deep-Sea Research, Part A: Oceanographic Research Papers*, **26**(12), 1347–1362.

NAIDU, P.D. 1995. A 2,200 years periodicity in the Asian monsoon system. *Geophysical Research Letters*, **22**(17), 2361–2364.

NÜRNBERG, C.C., BOHRMANN, G. & SCHLÜTER, M. 1997. Barium accumulation in the Atlantic sector of the southern ocean: results from 190,000-year records. *Paleoceanography*, **12**, 594–603.

ORTIZ, J.D. & MIX, A.C. 1992. The spatial distribution and seasonal succession of planktonic foraminifera in the California Current off Oregon, September 1987–September 1988. *In:* SUMMERHAYS, C.P., PRELL, W.L. & EMEIS, K.C. (eds) *Upwelling Systems: Evolution Since the Early Miocene*. Geological Society, London, Special Publications, **64**, 197–213.

PAYTON, A., KASTNER, M. & CHAVEZ, F.P. 1996. Glacial to interglacial fluctuations in productivity in the Equatorial Pacific as indicated by marine barite. *Science*, **274**(5291), 1355–1357.

PEETERS, F. 2000. *The distribution and stable isotope composition of living planktic foraminifera in relation to seasonal changes in the Arabian Sea*. PhD thesis, Free University Amsterdam.

PRELL, W.L., HUTSON, W.H., WILLIAMS, D.F., BE, A.W.H., GEITZENAUER, K. & MOLFINO, B. 1980. Surface circulation of the Indian Ocean during the last glacial maximum, approximately 18,000 yr B.P. *Quaternary Research*, **14**(3), 309–336.

PRELL, W. L. & SHIPBOARD PARTY OF ODP LEG 117, 1984. Monsoonal climate of the Arabian Sea during the late Quaternary: a response to changing solar radiation. *In:* Berger, A. (ed.), *Milankovitch and Climate, Part 1*. D. Reidel, Dordrecht, 349-366.

PRELL, W.L., MURRAY, D.W. & CURRY, W.B. 1996. Sediments off the western Arabian Sea; JGOFS sediment trap locations and the Oman Margin OMZ. *EOS Transactions, American Geophysical Union*, **77**(3), 5.

REICHART, G.J., DEN DULK, M., VISSER, H.J., VAN DER WEIJDEN, C.H. & ZACHARIASSE, W.J. 1997. A 225 kyr record of dust supply, paleoproductivity and the oxygen minimum zone from the Murray Ridge (northern Arabian Sea). *Palaeogeography, Palaeoclimatology, Palaeoecology*, **134**(1-4), 149–169.

REICHART, G.J., LOURENS, L.J. & ZACHARIASSE, W.J. 1998. Temporal variability in the northern Arabian Sea oxygen minimum zone (OMZ) during the last 225,000 years. *Paleoceanography*, **13**(6), 607–621.

REICHART, J. 1997. Late Quaternary variability of the Arabian Sea monsoon and oxygen minimum zone. *Geologica Ultraiectina*, , 154.

RITCHIE, J.C., EYLES, C.H. & HAYNES, C.V. 1985. Sediment and pollen evidence for an early to mid-Holocene humid period in the eastern Sahara. *Nature*, **314**, 352–355.

SARNTHEIN, M., WINN, K., DUPLESSY, J.-C. & FONTUGNE, M.R. 1988. Global variations of surface ocean productivity in low and mid latitudes: influence on CO_2. *Paleoceanography*, **3**, 361–399.

SCHENAU, S.J., PRINS, M.A., DE LANGE, G.J. & MONNIN, C. 2001. Barium accumulation in the Arabian Sea: controls on barite preservation in marine sediments. *Geochimica et Cosmochimica Acta*, **65**(10), 1545–1556.

SCHMITZ, B. 1987. The TiO_2/Al_2O_3 ratio in the Cenozoic Bengal abyssal fan sediments and its use as a paleostream energy indicator. *Marine Geology*, **76**(3-4), 195–206.

SCHULZ, H., VON RAD, U. & ERLENKEUSER, H. 1998. Correlation between Arabian Sea and Greenland climate oscillation of the past 110,000 years. *Nature*, **393**, 54–57.

SHIMMIELD, G.B. & MOWBRAY, S.R. ET AL. 1991. The inorganic geochemical record of the Northwest Arabian Sea; a history of productivity variation over the last 400 k.y. from sites 722 and 724. *In:* PRELL, W.L. & NIITSUMA, N. (eds) *Proceedings of the Ocean Drilling Program, Scientific Results, 117*. Ocean Drilling Program, College Station, TX, 409–429.

SIROCKO, F., SARNTHEIN, M. & ERLENKEUSER, H. 1993. Century-scale events in monsoonal climate over the past 24,000 years. *Nature*, **364**, 322–324.

SIROCKO, F., GARBE-SCHÖNBERG, D. & DEVEY, C. 2000. Processes controlling trace element geochemistry of Arabian Sea sediments during the last 25,000 years. *Global and Planetary Change*, **26**, 218–303.

VERGNAUD-GRAZZINI, C., VÉNEC-PEYRÉ, M.T., CAULET, J.P. & LERASLE, N. 1995. Fertility tracers and monsoon forcing at an equatorial site of the Somali basin (Northwest Indian ocean). *Marine Micropaleontology*, **26**, 137–152.

WEEDON, G.P. & SHIMMIELD, G.B. ET AL. 1991. Late Pleistocene upwelling and productivity variations in the Northwest Indian Ocean deduced from spectral analyses of geochemical data from sites 722 and 724. *In:* PRELL, W.L. & NIITSUMA, N. (eds) *Proceedings of the Ocean Drilling Program, Scientific Results, 117*. Ocean Drilling Program, College Station, TX, 431–443.

Monsoon-driven export fluxes and early diagenesis of particulate nitrogen and its $\delta^{15}N$ across the Somalia margin

G. J. A. BRUMMER, H. T. KLOOSTERHUIS & W. HELDER

Department of Marine Chemistry and Geology, Royal Netherlands Institute for Sea Research (NIOZ), PO Box 59, 1790 AB Den Burg, Texel, The Netherlands (e-mail: brummer@nioz.nl)

Abstract: Settling nitrogen fluxes intercepted by sediment traps on the mid-slope and in the deep basin off Somalia show a consistent annual range of 3.4 ± 0.2‰ in their stable isotope composition. Seasonal minima in $\delta^{15}N$ of 3.7‰ are associated with the moderate N fluxes derived from coastally upwelled water, which is rapidly carried offshore along eddy margins passing over the mooring sites during the SW monsoon (June–September). Coastal upwelling, offshore transport and deep wind mixing cease at the end of the SW monsoon, leading to enhanced utilization of the up to 20 µM of NO_3^- in the photic layer, maxima in the N export flux, and an increasing $\delta^{15}N$ by Rayleigh distillation. Yet as stratification develops, nutrient exhaustion follows and export production collapses as the $\delta^{15}N$ increases to over 7‰. Cyanobacterial N_2 fixation probably diminishes the $\delta^{15}N$ by 0.4–1.6‰ during the autumn intermonsoon (November–December) when settling N fluxes are lowest. Nutrient utilization remains high during the NE monsoon (January–March), when nutrient entrainment by deep wind mixing results in enhanced N export with maxima in $\delta^{15}N$ of up to 7.4‰. Annual N fluxes have virtually the same $\delta^{15}N$ of 6.0‰ in all traps despite considerable differences in both N flux and $\delta^{15}N$ between the traps during the year and at different depths. In comparison with the annual $\delta^{15}N$ of 6.0‰ arriving on the sea floor, core-top sediments are enriched by +0.6‰ on the upper slope (at 487 m) increasing to +2.9‰ in the deep basin (at 4040 m), whereas the N sediment burial efficiency declines from about 17% to 3%. It appears that the extent of oxic decomposition at the sediment–water interface is the most likely cause of such isotope enrichment. Similar positive gradients in $\delta^{15}N$ with bottom depth have been reported from other continental margin transects and are generally attributed to increased nutrient utilization in the photic ocean with distance offshore. As for Somalia, nitrogen isotope fractionation as a result of oxic decomposition on the bottom rather than nutrient utilization at the ocean surface may account for the observed increase of sedimentary $\delta^{15}N$ down continental margins in general.

Assimilation of nitrate by phytoplankton is accompanied by preferential uptake of $^{14}NO_3$ relative to the slightly less reactive $^{15}NO_3^-$. Upon consumption of the nitrate pool, the remaining nitrate will become progressively more enriched in ^{15}N and its nitrogen isotope ratio, $\delta^{15}N$, increases. Subsequently, the remaining enriched nitrate is also incorporated by the phytoplankton and eventually passed on to the export flux of particulate nitrogen (e.g. Wada 1980; Owens 1987). As a result, temporal and spatial changes in the balance between NO_3^- supply and consumption in the upper ocean will lead to corresponding changes in the $\delta^{15}N$ of particulate N settling out of the productive zone towards the sea floor and buried in the sediment (Altabet & Francois 1994a; Altabet 2001). Temporal changes in nutrient utilization by plankton ecosystems may be rapid and intermittent when storm induced (Montoya et al. 1991), or slow and periodic by monsoonal upwelling (Schaefer & Ittekkot 1995, 1996) and spring bloom restratification (Altabet 1996). Spatial changes resulting from these processes leave a distinct imprint as a positive gradient in the $\delta^{15}N$ of sedimentary nitrogen with distance from the equatorial upwelling zone (Altabet & Francois 1994a) as well as with distance offshore (Holmes et al. 1996). Downcore $\delta^{15}N$ analyses of bulk sediment, diatoms and foraminifers show a Glacial minimum in nutrient utilization in the equatorial Pacific (Farrell et al. 1995), Southern Ocean (Shemesh et al. 1993) and equatorial Atlantic (Altabet & Curry 1989) consistent with lower to slightly increased biological productivity compared with the modern ocean.

Several processes other than that of new production–Rayleigh distillation may contribute to the $\delta^{15}N$ signal. Cyanobacterial N_2 fixation (Wada & Hattori 1990; Capone et al. 1997) generates particulate organic nitrogen (PON) with a $\delta^{15}N$

value near that of atmospheric N_2 in equilibrium with sea water (0.6‰; Emerson et al. 1991). Thus, N_2 fixation will lower the bulk $\delta^{15}N$ value proportionally to its contribution to the N flux (see Brandes et al. 1998; Kerhervé et al. 2001). Also, food web effects may contribute to temporal variations in the $\delta^{15}N$ value of the export flux. Increases of about 3‰ in $\delta^{15}N$ per trophic level have been reported in the literature (e.g. Fry 1988; Montoya 1994). Bacterial denitrification is associated with a high fractionation factor of around 27‰ with respect to the source nitrate (Brandes et al. 1998). Denitrification will lead eventually to substantial ^{15}N enrichment of PON once the remaining nitrate is mixed back into the photic zone and assimilated again (Cline & Kaplan 1975; Altabet et al. 1995; Brandes et al. 1998). Mass balance considerations require that the $\delta^{15}N$ of the PON formed by primary production should equal that of the exported flux. Indeed, recent studies confirm that tight links exist between the $\delta^{15}N$ of the dissolved nitrate and that of the produced PON at depth (Altabet & Francois 2001; Voss et al. 2001).

Degradation of the settling PON does not appear to be accompanied by any major change in its $\delta^{15}N$, until after its deposition on the sea floor. Altabet & Francois (1994a, b) showed that the PON contained in deep ocean sediments across the equatorial Pacific is enriched by about 4‰ with respect to the PON in the water column. Consequently, the downcore record of $\delta^{15}N$ should mirror nitrogen cycling of the past on vastly longer time scales; for example, the changing inputs from denitrification in the Glacial Arabian Sea (Altabet et al. 1995, 2002). On continental margins, surficial sediments generally show large gradients with bottom depth in $\delta^{15}N$ as a result of differences in nutrient supply and utilization in the upper water column (e.g. Holmes et al. 1998, 1999). In contrast to open ocean sediments, diagenetic enrichment in $\delta^{15}N$ with respect to settling PON appears much smaller, particularly where bottom waters are anoxic (Altabet et al. 1999), owing to the higher export fluxes and sediment burial rates, and oxygen-limited decay of the PON after deposition. On the other hand, downslope transport may deliver large amounts of aged sedimentary PON and/or inorganic NH_4^+ adsorbed to clay minerals, which commonly have a different $\delta^{15}N$ than is deposited from purely pelagic sources on the margin (Freudenthal et al. 2001a, 2001b; Schubert & Calvert 2001).

Our objective is to trace nitrogen fluxes and their isotopic composition from surface ocean production through deposition to sediment burial off Somalia (Fig. 1), as part of the Netherlands Indian Ocean Programme (NIOP; van Weering et

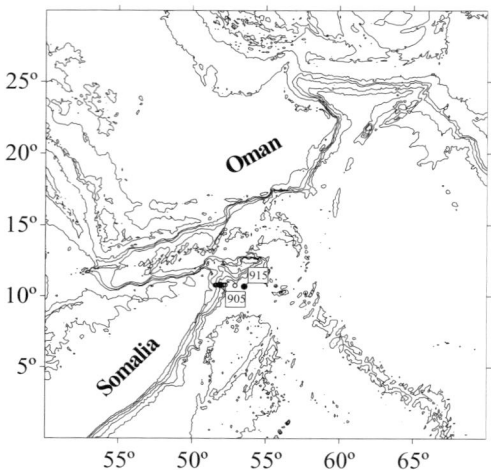

Fig. 1. Physiography of study area with location of moorings (905 and 915) and bottom stations along the Somalia transect (Fig. 2).

al. 1997). Time-series moored sediment traps were used to trace the supply and utilization of N nutrients by the $\delta^{15}N$ of settling particulate nitrogen during the monsoonal upwelling cycle. Integrated over time, the sediment traps provide an estimate for the annual flux of particles deposited on the mid-slope and deep basin floor, respectively (Koning et al. 1997; Conan & Brummer 2000). To assess the preservation of this material across the sediment–water interface, core tops were analysed from a depth transect of box-core samples across the mooring sites (Fig. 2; Helder et al. 1995; Koning et al. 1997).

On the Somalia ocean margin, coastal upwelling introduces nitrate in concentrations of over 20 μM in the surface during the SW monsoon (June–September), which is advected offshore along large ocean eddies in response to atmospheric forcing (Schott et al. 1990; Baars et al. 1994; Fischer et al. 1996). As a result, primary productivity in the Somalia upwelling area is among the highest in the world (Veldhuis et al. 1997), leading to high export fluxes of particulate matter settling to the sea floor (Brummer 1995; Koning et al. 1997; Van Weering et al. 1997) and high rates of sediment accumulation (Ivanova 1999). Typically, the water column is strongly depleted in oxygen between about 150 and 1000 m depth, and at the surface by upwelling during the SW monsoon, but neither anoxia nor denitrification occur, contrary to the Arabian Sea proper (De Wilde & Helder 1997). Although the preservation of organic matter is enhanced in the oxygen-poor water, the sediment richest in organic matter is found well below the oxygen minimum

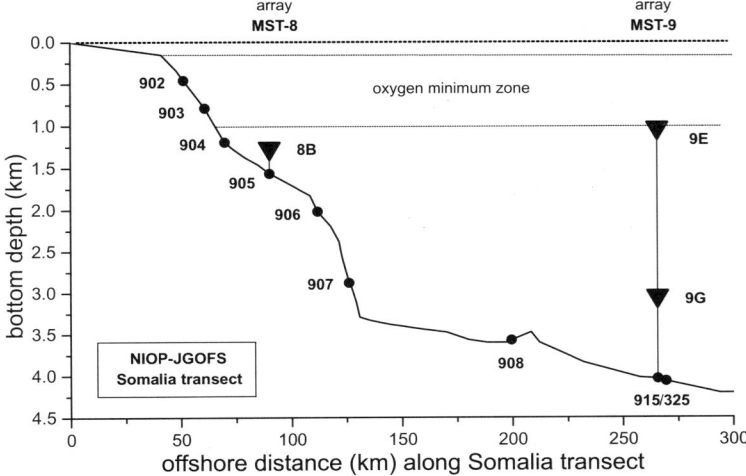

Fig. 2. Location of sediment trap moorings and coring stations as well as the extent of the oxygen minimum zone along the Somalia transect.

zone, i.e. around 1550 m depth on the mid-slope (Ivanova 1999), where one of the sediment trap arrays was moored. At that site continuous sedimentation has prevailed for at least the past 100 ka (Ivanova 1999; Koning et al. 2001).

Materials and methods

Particle fluxes were intercepted by conical sediment traps with a large, baffled aperture, moored at two sites on the Somalia transect (Figs. 1 and 2; Table 1). Samples were collected simultaneously, at 1–2 week intervals from 7 June 1992 to 14 February 1993 (Table 2). To minimize decay of the intercepted particulate material, we added a biocide ($HgCl_2$) and a pH buffer ($Na_2B_4O_7$) to sea water collected from the actual site and depth, which was used to fill the collecting cups before deployment (see Brummer (1995) for details). After recovery, samples were stored at 4°C in the dark until further processing in the laboratory. As part of the processing protocol, residues were gently dispersed over a 1 mm mesh sieve and handpicked for large 'swimmers' that entered the trap alive, as well as 'flux rarefacts' that do belong to the settling flux, commonly pteropod shells but also entire swimming crabs in two samples. Flux rarefacts generally accounted for less than 1% of the total mass flux and on one occasion for less than 3%. For organic matter analysis, the residue

Table 1. *Location and depth of sediment traps and box-coring stations off Somalia*

Station code	Latitude (°N)	Longitude (°E)	Bottom depth (m)	Trap depth (m)	Weight % N	Bulk $\delta^{15}N$ (‰)
902	10°46.72	51°34.64	459		0.18	6.4
903	10°46.97	51°39.48	789		0.29	6.5
904	10°47.27	51°46.23	1194		0.28	6.7
MST-8B	10°45.44	51°56.66	1533	1265	0.66	5.8
905	10°54.94	51°56.65	1567		0.36	6.8
906	10°48.70	52°07.76	2020		0.24	7.0
907	10°48.24	52°14.96	2807		0.20	7.4
908	10°46.66	52°54.88	3572		0.14	8.6
915	10°41.37	53°31.41	4035		0.13	9.0
MST-9E	10°43.07	53°34.42	4047	1032	0.92	6.1
MST-9G	10°43.07	53°34.42	4047	3047	0.57	5.9
325	10°41.0	53°33.0	4065		0.14	8.6

For the core tops, the bulk weight per cent of nitrogen and $\delta^{15}N$ are given; for the sediment traps the values given are flux integrated over the entire deployment period.

Table 2. Particulate nitrogen flux, weight per cent and $\delta^{15}N$ intercepted by each sediment trap during the sampled period

Date start collecting interval	MST-8 sample series	Duration (days)	N flux (mg m^{-2} day^{-1})	N content (wt %)	δ^{15}N (‰)	MST-9 sample series	Duration (days)	N flux (mg m^{-2} day^{-1})	N content (wt %)	δ^{15}N (‰)	MST-9 sample series	Duration (days)	N flux (mg m^{-2} day^{-1})	N content (wt %)	δ^{15}N (‰)
7 Jun. 92	B-2	14	5.72	0.95	7.2	E-2	14	0.49	1.02	6.9	G-2	14	1.32	0.75	6.7
21 Jun. 92	B-3	14	2.87	0.87	5.4	E-3	14	1.20	1.38	3.9	G-3	14	3.18	0.88	5.1
5 Jul. 92	B-4	14	7.75	0.60	5.1	E-4	14	1.53	0.77	5.2	G-4	14	3.39	0.83	4.1
19 Jul. 92	B-5	14	4.74	0.56	5.3	E-5	7	0.56	0.42	6.1	G-5	7	0.79	0.43	6.2
26 Jul. 92						E-6	7	0.66	0.82	4.7	G-6	7	0.63	0.57	5.2
2 Aug. 92	B-6	14	5.23	0.55	4.4	E-7	7	1.06	0.85	4.9	G-7	7	1.20	0.61	4.3
9 Aug. 92						E-8	7	0.76	0.58	3.7	G-8	7	1.72	0.59	4.3
16 Aug. 92	B-7	14	2.75	0.54	3.9	E-9	7	0.49	0.45	4.7	G-9	7	1.68	0.53	4.8
23 Aug. 92						E-10	7	1.17	0.60	4.4	G-10	7	1.46	0.46	4.9
30 Aug. 92	B-8	7	4.09	0.55	4.3	E-11	7	1.54	0.64	4.9	G-11	7	1.53	0.50	4.9
6 Sep. 92	B-9	7	4.31	0.55	4.3	E-12	7	1.36	0.53	5.6	G-12	7	1.60	0.46	6.0
13 Sep. 92	B-10	14	3.21	0.63	4.4	E-13	14	1.79	0.60	6.0	G-13	14	2.62	0.50	5.7
27 Sep. 92	B-11	14	11.60	0.36	5.6	E-14	14	1.27	0.50	6.4	G-14	14	1.55	0.42	5.8
11 Oct. 92	B-12	14	17.18	0.40	6.3	E-15	14	1.26	0.75	6.5	G-15	14	1.12	0.44	5.8
25 Oct. 92	B-13	14	2.01	0.67	6.4	E-16	14	1.00	0.88	6.3	G-16	14	1.56	0.55	6.3
8 Nov. 92	B-14	14	1.14	0.70	6.6	E-17	14	0.63	1.11	7.4	G-17	14	1.19	0.50	7.3
22 Nov. 92	B-15	14	1.01	0.72	6.2	E-18	14	0.51	1.03	6.7	G-18	14	0.36	0.53	5.8
6 Dec. 92	B-16	14	1.21	0.77	6.1	E-19	14	0.30	1.34	7.0	G-19	14	0.54	0.53	6.7
20 Dec. 92	B-17	14	1.08	0.68	6.2	E-20	14	0.33	1.28	7.2	G-20	14	1.16	0.60	6.4
3 Jan. 93	B-18	14	2.03	0.85	6.9	E-21	14	1.16	1.20	6.7	G-21	14	0.72	0.55	7.0
17 Jan. 93	B-19	14	1.08	0.73	6.8	E-22	14	0.34	1.23	7.2	G-22	14	0.63	0.46	7.0
31 Jan. 93	B-20	14	2.19	0.79	6.8	E-23	14	0.51	0.93	7.1	G-23	14	0.81	0.56	7.1
14 Feb. 93	end	252	4.28	0.66	5.8	end	252	0.90	0.92	6.1	end	252	1.41	0.57	5.9

was subdivided into multiple aliquots using a Folsom splitter to an aliquot equivalent to about 250 mg in dry weight and picked thoroughly for small 'swimmers'. Sieving, swimmer picking and splitting were carried out using the original supernate solution only. The residue was filtered onto acid-cleaned, pre-weighed 0.45 μm cellulose acetate filters, briefly rinsed with cold Milli-Q to remove the salts, dried at 55°C and back-weighed to determine the total dry weight. The dry filter cake was peeled from the filter and homogenized in an agate mortar before mass spectrometric analysis. Additionally, a swimming crab and several larger and smaller 'swimmers' were analysed separately.

Sediment samples were collected at nine stations along a depth transect (Figs. 1 and 2; Table 1) using an NIOZ-designed cylindrical box corer (50 cm diameter), equipped with a hydraulically dampened closing lid to retain the original sediment–water interface and overlying bottom water. Sub-cores of 6 cm diameter were cut into 0.5 cm to 2.0 cm slices, squeezed to retrieve the pore-water and stored frozen (for details, see Helder et al. 1995). Several box-cores were taken at each site. Their core-top recovery was checked by visual inspection and confirmed by comparison of measured oxygen and nutrient concentrations at 5 m above the bottom, the overlying water in the box-core, the squeezed pore-water and across the sediment–water interface (in situ benthic lander incubations, deck incubations). In the laboratory, the squeezed sediment was dried and homogenized in an agate mortar mill for 5 min at maximum speed. Bulk sediment accumulation rates, calculated from dry bulk-density measurements and ^{14}C accelerated mass spectrometry (AMS) dating of foraminifer shells at 3–5 depths in the box-cores (Ivanova 1999; Koning et al. 2001), are available for four stations down the Somali ocean margin, i.e. 902, 905 (at the mooring site of MST-8), 907 and 915 (at the mooring site of MST-9). Nitrogen burial fluxes were then calculated by multiplying the sediment accumulation rate by the weight per cent of nitrogen in the bulk sample (Table 3).

Sediment trap and core samples as well as external reference standards were packed into tin cups and loaded into the 50-place autosampler disc of the Carlo–Erba NA-1500 series II CN-analyser interfaced with a VG-Optima triple-collector mass spectrometer operated under continuous flow (Owens & Rees 1989). Before δ^{15}N determination, samples were analysed for total N content in a stand-alone Carlo–Erba NA-1500 system to standardize measurements of both samples and external reference material to 50 μg N for mass spectrometry. Replicate analyses of empty tin cups consistently produced 0.1 ± 0.1 μg N blanks or smaller. Samples were at least analysed in duplicate for δ^{15}N, which was measured against pure tank N_2 as the internal reference gas, corrected for any residual mass dependence using the accompanying external standard, and calibrated against various International Atomic Energy Agency (IAEA) reference and laboratory δ^{15}N standards included in each run. Values are reported in the conventional delta notation against the IAEA-N1 standard of δ^{15}N of 0.43 ± 0.07‰ with respect to atmospheric N_2 (Böhlke & Coplen 1995). Analytical precision on equivalents of 50 μg N of this standard, IAEA-N2, -N3 and several laboratory standards was better than 0.09‰ during sample runs. For the box-core and sediment trap samples, a precision was attained of ±0.12‰ based on the averaged deviation between duplicate analysis of all 86 (sub)samples.

As the sediment traps are deployed at 1000–3000 m water depth, horizontal displacement by currents during particle settling may render it difficult to link the week-integrated N flux to upper ocean conditions at a specific station. During the SW monsoon, surface water current

Table 3. *Core-top sediment properties at key stations on the Somalia ocean margin, including mooring stations 905 on the mid-slope and 915 in the deep basin*

	Coring (trap) station			
	902	905 (MST-8)	907	915 (MST-9)
Bottom depth (m)	459	1567	2807	4035
δ^{15}N (‰)	6.4	6.8	7.4	9.0
Dry weight % N	0.18	0.37	0.20	0.13
Dry bulk density (g cm^{-3})	0.47	0.45	0.43	0.40
Burial rate (cm ka^{-1})	40	18	7	4
N burial flux (g m^{-2} a^{-1})	0.36	0.30	0.07	0.025
Burial efficiency (%)		19		3.5

velocities are over 200 cm s^{-1} and amount to 3–15 cm s^{-1} at 1235 m depth at the mid-slope mooring site and down to about <1–12 cm s^{-1} at 3000 m in the deep Somali Basin (Brummer 1995). Because bulk particle settling velocities are of the order of 150 m day^{-1} or 0.2 cm s^{-1} (e.g. Knappertsbusch & Brummer 1995) it takes about 1–2 weeks before the export flux arrives in the traps, thus leaving a considerable time for horizontal displacement by surface and deeper (counter) currents. Consequently, the sediment trap records of exported fluxes will not match exactly with the place and/or the time of production (Von Gyldenfeldt et al. 2000; Waniek et al. 2000). Particularly during the SW monsoon, small temporal variations in the position of the 'Great Whirl' (Fig. 3) will lead to large changes in the export flux and its isotopic composition as received by the sediment traps at each mooring site. In spite of such difficulties, a good coherence is observed between major hydrographic features and the spatio-temporal trends in the exported flux as outlined below.

Regional hydrography and pelagic system

In the course of May, southwesterly winds rapidly gain strength along Somalia and persist at high velocities during the entire SW monsoon until late September. In response, coastal upwelling is generated and mesoscale eddies become apparent at the ocean surface, migrating northward along with the northern branch of the Somalia Current (see Baars et al. 1994; Fischer et al. 1996; Schott et al. 1990; Koning et al. 2001). These eddies are anticyclonic and carry coastally upwelled water offshore, producing conspicuous 'cold wedges' projecting from the continental margin (Fig. 3). Off Ra's Hafun such a wedge remains almost continuously present during the SW monsoon, a feature generally shown off prominent capes in the Arabian Sea. Further offshore, the wedge merges into a filament or jet curving along an eddy margin and advecting high concentrations of upwelled nutrients in cold water over both mooring sites at velocities exceeding 2 m s^{-1}. These frontal zones between eddies are marked by sharp gradients in temperature so that their spatial and temporal development could be traced relative to the sample sites, using the thermal images kindly provided by O. Brown (Rosenstiel School of Marine and Atmospheric Sciences) that cover the period from 17 May to 28 September 1992 (Figs. 3 and 4; see Koning et al. 2001, fig. 1). Shipboard measurements carried out on three cruises during June–August 1992 were used for ground-truthing (Fig. 4) but also for analysis of the associated N-nutrient concentrations in the surface water and with depth (Baars et al. 1994; Van Hinte et al. 1995; Veldhuis et al. 1997).

Southwest monsoon

As in every year, the sample area in 1992 was affected by the interplay of at least three major eddies (Fig. 3), including the Great Whirl (Baars et al. 1994; Fischer et al. 1996; Schott & Fischer 2000) and several minor eddies in the course of the SW monsoon (Koning et al. 2001). At the end of May 1992, the onset of the SW monsoon was marked by thermal doming along coastal Somalia and offshore transport from cape Ra's Hafun in the frontal zone between the Northern Gyre and the Great Whirl (Fig. 3a and b). Shipboard measurements at the mid-slope site on 2 June 1992 show a core of advected coastal waters in the frontal zone with surface temperatures lower by about 3–26°C, an enhanced NO$_3^-$ concentration of up to 3 µM, and a high integrated primary production of 1.4 g C m^{-2} day^{-1} (Baars et al. 1994; Veldhuis et al. 1997). On 4 June (Fig. 3b), the core of the frontal zone advected slightly colder water to the mid-slope site, depleted in NO$_3^-$ (≤0.1 µM) by the highest integrated primary productivity measured during the NIOP programme (2.8 g C m^{-2} day^{-1}). Meanwhile, the deep basin site remained well within the oligotrophic, nutrient-exhausted and warm water of the Great Whirl (Fig. 3). As coastal upwelling and eddy circulation intensified, much colder and more nutrient-rich water is advected much further into the basin along the migrating Great Whirl front (Fig. 3c). From mid-June to early July, the Great Whirl front passed over the open basin site (Fig. 3b–e), resulting in a pronounced minimum in surface temperature compared with the warm waters inside the eddies (Fig. 4b). Meanwhile, the mid-slope site was affected by migration of the core frontal zone to just south of the mooring (Fig. 3c). In late June, conditions at the deep basin site briefly returned to warm water north of the Great Whirl, whereas the Great Whirl front now extended over the mid-slope mooring (Fig. 3d and e).

In early July, the 'Southern Gyre' approached from the south and nearly merged with the Great Whirl (Fig. 3e–g), causing massive advection of coastally upwelled water off Ra's Hafun across both mooring sites by late July. Temperatures dropped as low as 17°C and NO$_3^-$ concentrations increased to 20 µM at the surface; yet primary productivity only attained 0.7–0.9 g C m^{-2} day^{-1}, because of vigorous vertical mixing to well below the photic zone (Baars et al. 1994; Veldhuis et al. 1997). From late July to mid-August, surface water circulation changed rapidly as small-scale eddies and associated fronts budded off, warm

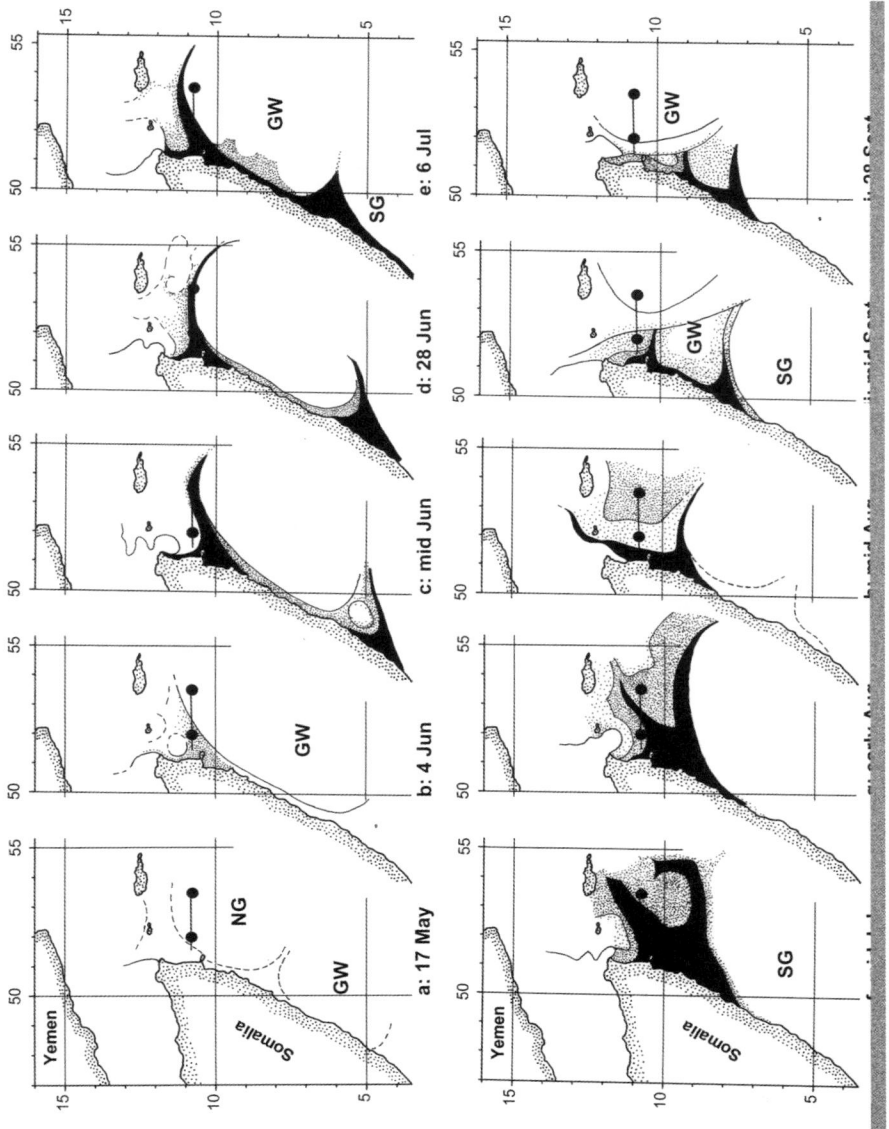

Fig. 3. Sea surface temperature during the SW monsoon of 1992 showing the development of coastal upwelling and offshore transport along eddy margins with respect to the mooring sites and box-core transect. Note the migration of the Great Whirl frontal zone across the deep basin site in June and the massive advection of upwelled water at the end of July across both sites. Schematized after IR remote sensing images and shipboard measurements (see also Fig. 4; for actual images see Koning *et al.* 2001). Black areas represent freshly upwelled water (T <22°C, [NO_3^-] >10 µM); stippled areas represent aged/shallow upwelled water and mixed water (T 22–25°C, [NO_3^-] 0.8–10 µM) and non-upwelled water (T >25°C, [NO_3^-] <0.8 µM). NG, Northern Gyre. GW, Great Whirl. SG, Southern Gyre.

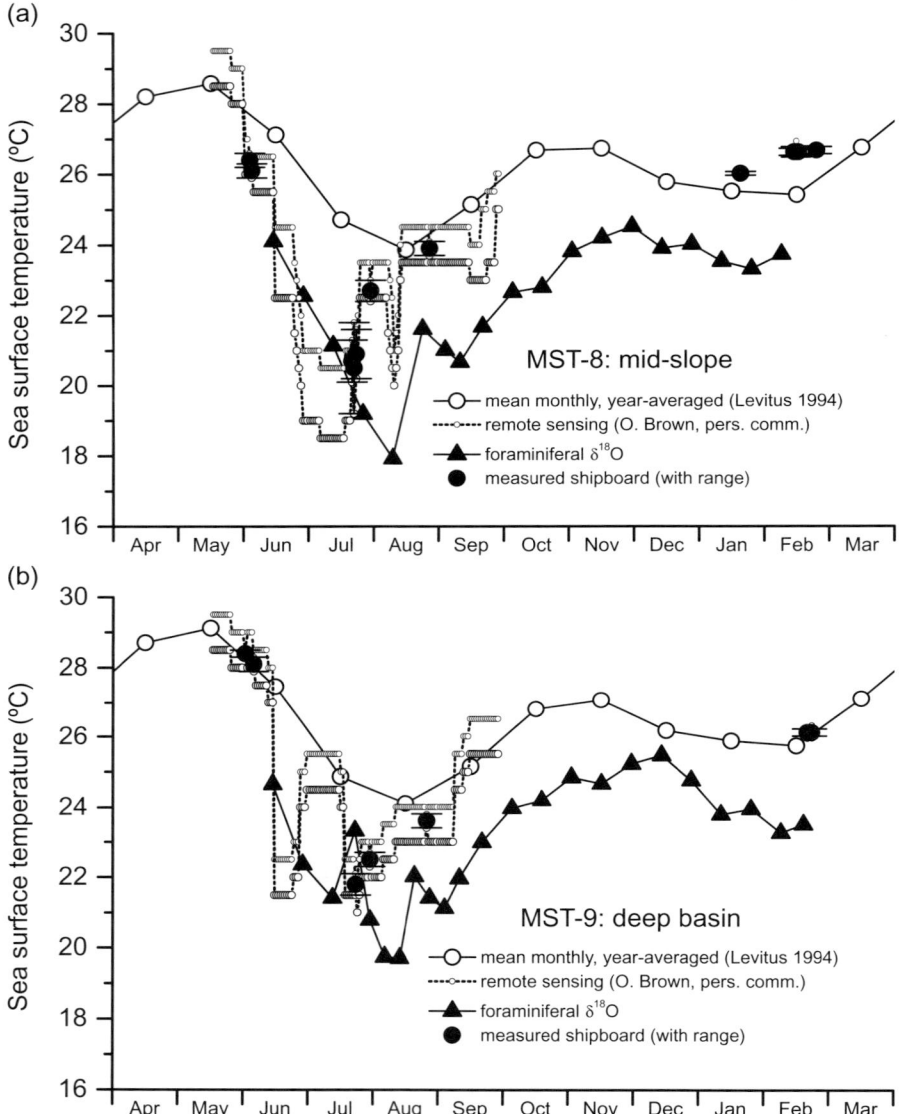

Fig. 4. Sea surface temperature at mooring sites MST-8 (**a**) and MST-9 (**b**) based on remote sensing, shipboard measurements and foraminiferal $\delta^{18}O$ (Conan *et al.*, unpubl. data) for the actual period of sampling as well as the average monthly values given by Levitus (1994).

Gulf of Aden water approached the mid-slope site and cold frontal waters from the Southern Gyre were advected to the deep basin site. As the Southern Gyre retreated southward by mid-August (Fig. 3h), it left a deep mixed layer in the Great Whirl with a low temperature of about 24°C and rich in nitrate (6 μM) at both sites. The mid-slope site remained adjacent to the coastal upwelling front of the Great Whirl. By early September surface waters show a rapid warming to about 26°C, first over the deep basin and then at the mid-slope site (Fig. 3h–j). Meanwhile, coastal upwelling diminished and the associated front migrated shoreward, away from the mid-slope site (Fig. 3j). Winds abated at the end of September, signalling the end of the SW monsoon, and did not increase in the source area until mid-December when the NE monsoon started.

Autumn intermonsoon and NE monsoon

During the autumn intermonsoon, surface mixed layer temperatures rose and peaked in November (Fig. 4), as a result of summer heating as winds dropped to very low speeds. Nitrate concentrations in the photic zone became exhausted, limiting primary productivity towards a regenerative regime like the 'Typical Tropical Structure' in the sense of Herbland & Voituriez (1979). Such conditions probably persisted while winter cooling lowered sea surface temperatures. From mid-December, deep wind mixing caused by strong NE monsoon winds increased nitrate concentrations to 0.8 μM and primary productivity to 0.7 ± 0.2 g C m^{-2} day^{-1} in mid-January (Baars et al. 1994; Veldhuis et al. 1997).

Results

All traps show a consistent 3.4 ± 0.2‰ change in the stable nitrogen isotope composition in the course of the year, yet the particulate N fluxes at the mid-slope site are much higher than collected offshore by either of the traps in the deep basin (Fig. 5; Table 2). Minima in δ^{15}N down to 3.7‰ and high N fluxes are found during the SW monsoon, following the offshore advection of coastally upwelled water with high nitrate concentrations in the eddy frontal zones. Conversely, high

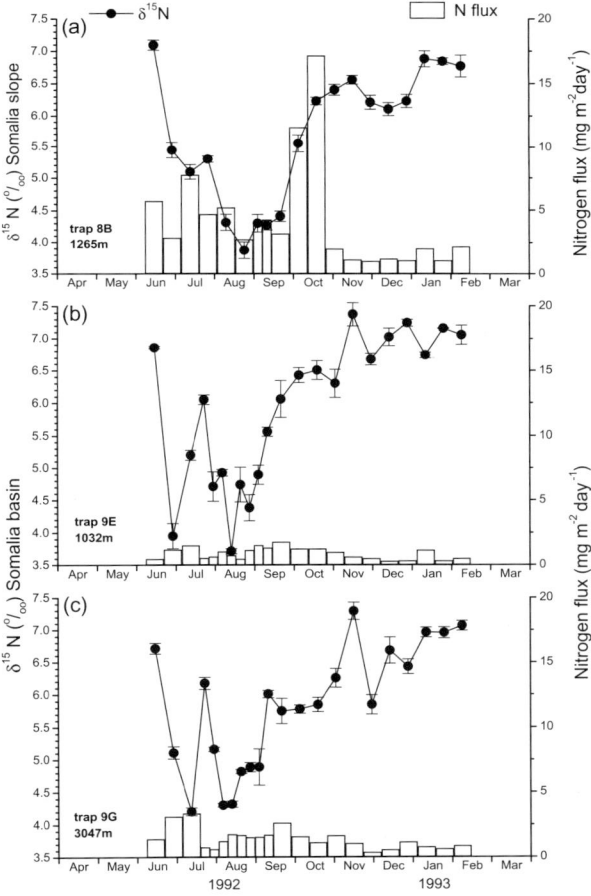

Fig. 5. Settling N flux and δ^{15}N intercepted by a sediment trap on the mid-slope (**a**) and two traps in the deep Somali Basin (**b, c**). Minima as low as 3.7‰ are associated with coastally upwelled water with N-nutrient concentrations of up to 20 μM transported over the mooring sites during the SW monsoon (June–September). Full utilization is achieved during the non-upwelling autumn intermonsoon and the NE monsoon (January–March) with low to moderate N fluxes and δ^{15}N values of up to 7.4‰. Cyanobacterial fixation of N$_2$ may have contributed to lowering the δ^{15}N during the SW–NE intermonsoon (November–December) when settling fluxes are lowest.

δ^{15}N values of up to 7.4‰ and low N fluxes are associated with both the intermonsoon and NE monsoon, when nitrate concentrations in the deep surface mixed layer were low. Superimposed on this general pattern are substantial deviations on smaller time scales that show a remarkable correspondence in δ^{15}N between the traps at 1032 and 3047 m depth on the same mooring in the deep basin (Fig. 5; Table 2). Minima in the δ^{15}N during late June–early July and August as well as maxima in mid-June, late July and early November occur in both traps at about the same time and some can be recognized in the mid-slope record as well. Nitrogen fluxes show a similar but inverse pattern, and larger differences in absolute value, compared with the δ^{15}N in both traps in the deep basin (Fig. 5; Table 2). Apparently, current displacement during gravitational settling did not lead to sampling export fluxes from different sources at the same time 2000 m apart, the equivalent of about 2 weeks of settling time. Although the temporal differences in N fluxes may be due to this effect, the temporal coherence between the isotope records argues that the sediment traps, on a 2-week time scale, did sample from the same coeval particle sources and can be compared with the record of the trap on the mid-slope at 1250 m depth to assess spatial variation.

Southwest monsoon

On the slope, the N flux intercepted by the trap in mid-June was high, as was its δ^{15}N of 7.1‰ (Fig. 5a), following the high primary productivity and low nitrate concentrations in early June. Meanwhile, since the onset of the SW monsoon, the oligotrophic and nitrate-depleted waters of the Great Whirl generated low export fluxes with a high δ^{15}N of 6.8‰ in the open Somali Basin (Fig. 5b and c). By the end of June, a maximum in the N flux associated with a pronounced minimum in the δ^{15}N of about 4.1‰ was observed, first in the shallow trap (1000 m) and then in the deep trap (3000 m), shortly after the frontal zone bordering the Great Whirl had passed over the Somali Basin site (Fig. 3). By mid-July, the N fluxes dropped whereas the δ^{15}N increased to 6.1‰ in both traps after the oligotrophic waters of the Great Whirl returned to the basin site (Fig. 5b and c). The same pattern was observed at the slope site, although seemingly less pronounced, as the δ^{15}N remained at about 5.3‰ at the end of July (Fig. 5a). However, integration of the 1 week intervals sampled in the deep basin to the matching 2 week intervals on the slope yielded almost the same δ^{15}N values for both sites.

From the end of July to mid-August, δ^{15}N values dropped to a minimum of 3.9 ± 0.2‰ in all traps (Fig. 5), following the massive advection of coastally upwelled water by the Great Whirl and the approaching Southern Gyre (Fig. 3). However, N export fluxes remained at moderate levels at both mooring sites and were lower than observed during the previous passage of a Great Whirl front (Figs. 3 and 5). By September, both the δ^{15}N and the N export flux increased, first in the deep basin and then on the slope, following the collapse of upwelling and eddy advection as the SW monsoon came to an end (Fig. 3). Maxima in the N flux were observed at the end of September and in October, on the slope and in the basin, respectively, and have similar δ^{15}N values of about 6.5‰.

Autumn intermonsoon and NE monsoon

During the autumn intermonsoon, the high N fluxes diminished drastically whereas the δ^{15}N values increased to a maximum of 7.4‰ in November (Fig. 5), when a regenerative nutrient regime developed. In December, at the height of the autumn intermonsoon, the exported N flux had dropped to a minimum, accompanied by a lower δ^{15}N. Following the onset of the NE monsoon, N fluxes and their δ^{15}N values increased as a deep wind-mixed layer developed.

Core-top sediments

The topmost sediment accumulating on the Somali margin shows a conspicuous positive gradient in δ^{15}N with increasing bottom depth and distance offshore (Fig. 6; Tables 2 and 3). From the shallowest core station (487 m) down to the mooring station on the mid-slope (1567 m) sedimentary δ^{15}N increases from 6.4‰ to 6.8‰, attaining 7.4‰ near the base of the slope (2200 m) and values up to 9.0‰ beneath the mooring in the deep basin (4040 m). Bulk dry weight per cent N in the surficial sediment shows a pronounced maximum on the mid-slope of 0.37% (1567 m) beneath the mooring, declining to 0.18% at the shallowest station (487 m) and to 0.13% in the deep basin (4040 m) below the other mooring (Table 3). However, the sedimentation rate at the shallowest station (about 40 cm ka^{-1}) is much higher than on the mid-slope (18 cm ka^{-1} at the mooring station). Consequently, the sediment burial flux of nitrogen is highest at the shallowest station, decreasing downslope from 0.36 g m^{-2} a^{-1} to 0.025 g m^{-2} a^{-1} in the deep basin (Table 3). Core tops collected during the SW monsoon of 1992 and the NE monsoon of 1993 are very similar in composition (stations 325 and 915, respectively), not showing a significant monsoonal and/or spatial difference.

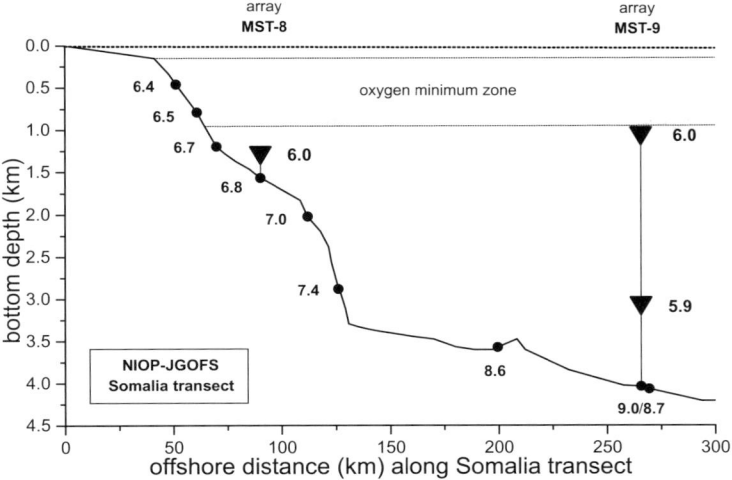

Fig. 6. Year-integrated $\delta^{15}N$ of settling N in the sediment traps and $\delta^{15}N$ of core-top sediments (upper 0.25–0.5 cm) along the Somalia transect. Settling fluxes have virtually the same annual $\delta^{15}N$ of 6.0‰ despite temporal and spatial differences between the traps in both N flux and $\delta^{15}N$. Core-top sediments are enriched in positive gradient with depth along the transect by increased sedimentary diagenesis downslope rather than by increased nutrient utilization by the plankton offshore.

At the mid-slope mooring site the downcore $\delta^{15}N$ is about the same as in the core top. Values for $\delta^{15}N$ in successive 5 mm depth intervals are 6.7, 6.7, 6.7 and 6.6‰, whereas the weight percentage of N decreases from 0.37 to 0.34%.

Discussion

Lateral transport and annual export

Lateral advection of particulate matter in addition to vertical settling may contribute considerably to the mass fluxes as intercepted by sediment traps, particularly on continental margins and close to the bottom (e.g. Freudenthal et al. 2001a). In October 1992, the sediment trap at 265 m above the mid-slope collected extremely high mass fluxes that were largely composed of silt-sized detrital carbonate from bottom sediments resuspended from the outer shelf and upper slope (Van Weering et al. 1997; Conan & Brummer 2000). Although the N fluxes sampled during both 2 week intervals are the highest found in the time-series (Fig. 5), the weight per cent N of the total mass is the lowest observed (Table 2). This suggests that the sediment-derived material is much poorer in N than the pelagic flux at that time, consistent with the much lower weight per cent N of the core top sediments (Table 2). Consequently, dilution with detrital carbonate will have caused the apparent decrease in bulk weight per cent N whereas the N flux increases, as Koning et al. (1997) showed for the biogenic silica in the same material. Interpolation between the $\delta^{15}N$ of the adjoining late September and late October intervals yields values that are only 0.2‰ higher than the bulk values measured in October between these times (Table 2). Consequently, the bottom sediment derived N would have a $\delta^{15}N$ only slightly higher than the inferred pelagic input. For the N flux a similar interpolation procedure yields values for the apparent pelagic flux, which can be used in the isotope mass balance:

$$\delta^{15}N_{bulk} \times JN_{bulk} = (\delta^{15}N_{pel} \times JN_{pel}) +$$
$$(\delta^{15}N_{sed} \times JN_{sed})$$
$$JN = JN_{pel} + JN_{sed}$$

where $\delta^{15}N$ is the nitrogen isotope ratio, JN is the export flux of nitrogen, and the subscripts pel and sed refer to the pelagic and bottom sediment source of the bulk nitrogen, respectively. Solving for $\delta^{15}N_{sed}$ gives a value of 6.2‰ associated with 55% of the bulk N flux, whereas the pelagic input would have a $\delta^{15}N$ of 5.5‰ contributing 45% of the bulk N flux. Core-top sediments range from 6.8‰ at the mid-slope mooring station to 6.4‰ at the shallowest station (487 m) and thus are close to the estimated $\delta^{15}N$ of resuspended sediment intercepted at 265 m above the sea floor.

Other normative calculations (e.g. using the Ca, biogenic silica and lithogenic matter content to estimate the detrital or pelagic fraction of bulk nitrogen) give similar results that are also consistent with the pelagic gradient at the end of the SW monsoon in the shallow trap offshore (Fig. 5).

At other times during the deployment period there is no evidence for any significant sediment-derived input in the traps, although the 'background' levels of various tracers are higher on the slope than found at either depth in the basin (Van Weering et al. 1997; Conan & Brummer 2000; Zonneveld & Brummer 2000; Koning et al. 2001). Enhanced but small contributions are associated with the passage of frontal zones during the SW monsoon and are relatively more important on the slope than in the deep basin traps. These were probably advected directly from the Somali shore along the eddy margins, as suggested by benthic marine diatoms in all traps (Koning et al. 2001). Consequently, the N fluxes and the $\delta^{15}N$ values derived from the traps are considered a reliable reflection of the particulate N-export production during the deployment period, but with less certainty for the SW monsoon because of the strong sediment-derived input.

To estimate the annual N flux over a full monsoonal cycle, we assume that the N export during the missing intermonsoon in spring is as low as was found for the autumn intermonsoon in our traps and has the same high $\delta^{15}N$ value. However, the effect on the year-integrated $\delta^{15}N$ is rather small, despite the length of the missing period from our trap records, because N fluxes are much higher during the longer period we did sample. In the open Arabian Sea, the spring intermonsoon is the least productive of the year, generating even lower N export fluxes than in autumn, yet with a similar $\delta^{15}N$ (Schaefer & Ittekkot 1993, 1995). Year-integrated $\delta^{15}N$ values were corrected for the sediment-derived contribution as outlined above, although the difference is very small. The annual $\delta^{15}N$ values of 5.9–6.0‰ are surprisingly similar for all traps despite the offshore decrease in the annual N flux (Fig. 6).

It appears that N fluxes with a low $\delta^{15}N$ from freshly upwelled waters during the height of the SW monsoon are balanced by the N export with a high $\delta^{15}N$ during the intermonsoons. In both cases, the N fluxes are relatively low with respect to the adjoining periods, as productivity was suppressed by vigorous vertical mixing to far below the euphotic zone, and limited by nutrient exhaustion, respectively (Baars et al. 1994; Van Weering et al. 1997; Veldhuis et al. 1997). High N fluxes are usually associated with steep gradients in the $\delta^{15}N$ following the passage of frontal zones between migrating eddies. Differences in nutrient utilization between the trap sites are expected to be small, as both are oriented parallel to the frontal zones and are outside the coastal upwelling zone proper.

Thus, the along-margin movement of the frontal zones has a comparable impact at both trap sites. Although interannual differences in the $\delta^{15}N$ and N flux will certainly occur, the apparent coupling between the two in the monsoonal mass balance should yield about the same $\delta^{15}N$ being deposited on an annual basis and integrate seasonally transient imbalances in trophic structure, particularly at the end of the monsoon.

Controls on the $\delta^{15}N$ of the export production

Although no measurements of the $\delta^{15}N$ of NO_3^- in the Somali Basin are available, the export flux can be used to constrain the $\delta^{15}N$ of the source nitrate, assuming that the lowest $\delta^{15}N$ observed represents that of the unutilized NO_3^- given Rayleigh fractionation kinetics. Taking an uptake fractionation factor (ϵ) of 5.5‰ (Owens 1987; Montoya 1994; Sigman et al. 1997; Altabet 2001) adds up to a $\delta^{15}N$ of 9‰ for the source NO_3^- in freshly upwelled water with NO_3^- concentrations of 15–20 μM, similar to the 6–10‰ found in the Oman upwelling region (Brandes et al. 1998). The seasonal difference of 3.4‰ in the particle export flux constrains the Rayleigh distillation of NO_3^- in freshly upwelled water to the nanomolar concentrations during non-upwelling, implying that the $\delta^{15}NO_3$ would have increased to 12.4‰. However, nitrate limitation during the non-upwelling periods would lead to indiscriminate use of the little NO_3^- available (e.g. Altabet 1996) thus reducing isotopic fractionation. In addition, other N nutrients assimilated with a different fractionation would become important for PON export production (Ostrom et al. 1997), such as ammonia (Montoya 1994). Also, a gradual shift from instantaneous product towards the accumulated product with respect to the source nitrate (see Altabet 1996) would be expected by Rayleigh fractionation kinetics as the utilized fraction of the initial nitrate decreased by two orders of magnitude. Consequently, the export fluxes of PON during the autumn intermonsoon will have a less extreme $\delta^{15}N$ than predicted by the utilization, Rayleigh fractionation and isotopic assimilation of nitrate alone. Although the new production–Rayleigh distillation model is consistent with the data, there are other processes drawing on different source N that may contribute as well, as discussed below.

For the central Arabian Sea, Brandes et al. 1998 reported that as much as 30% of the nitrogen may have come from N_2 fixation in the mass balance given the concentration and isotope profiles of NO_3^- and N_2. Fixation of N_2, for example, by cyanobacteria (Wada & Hattori 1990;

Capone et al. 1997), generates PON with a $\delta^{15}N$ near that of atmospheric N_2 in equilibrium with sea water (0.6‰; Emerson et al. 1991), which will lower the bulk $\delta^{15}N$ proportionally to the size of its contribution to the N flux (see Brandes et al. 1998). As N_2 fixation is energetically costly relative to, for instance, NO_3^-, any sizeable contribution is expected only during the intermonsoon periods when other N nutrients are depleted and N export fluxes are lowest. Such conditions probably prevailed in December, when N_2 fixation may have depressed the signal by as much as 1.6‰ with respect to the values observed both earlier and later during the autumn intermonsoon (Fig. 5). In the time-series record, however, the bulk $\delta^{15}N$ is much higher than during upwelling, indicating that the yearly export of PON derived from N_2 fixation is insignificant.

Food web effects may also have contributed to the temporal variation of $\delta^{15}N$ in the N export flux off Somalia, considering the about 3‰ increase in $\delta^{15}N$ per trophic level reported in the literature (e.g. Fry 1988; Montoya 1994). Indeed, copepod 'swimmers' as well as the carapax of the oceanic crab *Charybdis smithii* (Van Couwelaar et al. 1997) found in the traps have a $\delta^{15}N$ about 2‰ higher than that of the bulk N from the same samples. Consequently, a higher proportional contribution of secondary producers, not recognizable as such, to the N export flux during the non-upwelling periods may add to the seasonal changes observed in the bulk $\delta^{15}N$, for example, during the NE monsoon. However, although the biomass may have a high $\delta^{15}N$, the egested PON as well as the released metabolic ammonia generally have a $\delta^{15}N$ equal to or depleted with respect to the food source (e.g. Montoya 1994; Altabet 1996). Trophic isotope enrichment in the biomass may thus suppress the $\delta^{15}N$ in the metabolic export of particulate N, as is also suggested by the difference in $\delta^{15}N$ between the swimmer copepods and crab on one hand and the bulk PN on the other. Trophic effects may also appear as the result of transient imbalances in primary and secondary–tertiary production, particularly at the rapid monsoon changeovers from oligotrophic to eutrophic conditions and vice versa. Such a situation may have occurred at the onset of the SW monsoon on the mid-slope, when the N content in the initial June interval was higher than observed at any time during the time series. Also, the N flux was much higher than expected given its high $\delta^{15}N$ and that of the neighbouring traps in the deep basin (Fig. 5), possibly because of mass mortality or mass activity of previously dormant zooplankton.

Bacterial denitrification is associated with a high fractionation of around $\epsilon = 27$‰ with respect to the source nitrate (Brandes et al. 1998) and will lead eventually to substantial ^{15}N enrichment of the particulate N after the remaining nitrate is assimilated again (Cline & Kaplan 1975; Altabet et al. 1995; Brandes et al. 1998). As in other ocean basins where anoxia occurs in the water column, denitrification in the NW Indian Ocean contributes substantially to raising the $\delta^{15}N$ of the nitrate as dissolved oxygen drops to zero below the surface mixed layer to a depth of about 1 km (e.g. Altabet et al. 1995; Brandes et al. 1998; Naqvi et al. 1998). However, De Wilde & Helder (1997) demonstrated that denitrification does not occur off Somalia because the dissolved oxygen concentrations are still too high, and that nitrification prevails in the water column during both the SW and NE monsoon. Nitrification is also associated with a high fractionation factor with a reported ϵ range from 19 to 29‰ with respect to the source ammonia (see Montoya 1994), and could also lead to ^{15}N enrichment of the particulate N after assimilation of the formed nitrate. Consequently, nitrification rather than denitrification may be responsible for the enhanced $\delta^{15}N$ values of 9‰ we estimated for the upwelled nitrate utilized for primary and export production off Somalia (see above).

Without measurement of the concentration, flux and isotope composition of each N compound involved in the N cycle, it is difficult to estimate the specific contribution by particular processes that generate the N export flux as intercepted by the sediment traps. Indeed, in the open Arabian Sea where extensive denitrification occurs in the fully depleted oxygen minima, the $\delta^{15}N$ of the annual N export amounts to 7.2‰ (Schaefer & Ittekkot 1995), i.e. higher by 1.2‰ than off Somalia where only nitrification occurs. This excess caused by denitrification may even be larger as Arabian Sea water is advected to our mooring area, particularly during the NE monsoon (Fischer et al. 1996; Schott et al. 1990). Consequently, the NO_3^- remaining after denitrification could have contributed to the high $\delta^{15}N$ values during the non-upwelling periods, enhancing the effect of Rayleigh distillation.

Diagenesis and burial

Irrespective of differences in the actual source values of the N nutrients, the annual export as intercepted by the traps and delivered to the sediment has virtually the same $\delta^{15}N$ value of 5.9 ± 0.1‰. However, all core-top sediments show an increase in $\delta^{15}N$ with respect to the traps and a pronounced positive gradient with bottom depth. From the shallowest core station to the deepest, the $\delta^{15}N$ enrichment increases from +0.5‰ to +2.9‰ with respect to the annual N flux (Fig. 6).

Upward admixing of old PON into the surficial sediment should not be significant given the high modern sedimentation rates and shallow depth of bioturbation (see Koning et al. 1997, 2001; Ivanova 1999). Even in the deep basin, at least twice to six times the modern bioturbation depth of about 5 cm is required to bring up old sediments with a different $\delta^{15}N$ compared with that accumulating nowadays. Isotopic analyses currently being carried out on a piston core taken from the mid-slope site (Ivanova 1999) show very little change in downcore $\delta^{15}N$ over the last c. 7000 a (Ganeshram, pers. comm.). For the Oman margin, Altabet et al. (1995, 1999, 2002) showed that $\delta^{15}N$ values have not changed during the past c. 2500 a, and deeper downcore decreased to a minimum around 8000 a ago. Consequently, if any input of old PON by bioturbation occurred, even a large admixture would be unlikely to cause a significant departure in the $\delta^{15}N$ of the core tops taken on the Somali transect, as it has about the same isotopic composition.

Given that the $\delta^{15}N$ of the total nitrogen was measured, inorganic N may also have contributed to the bulk values found. Ammonia sorbed to mineral surfaces resulted in lowered $\delta^{15}N$ values for bulk sediments (Freudenthal et al. 2001a; Schubert & Calvert 2001). On the Somali margin both the export flux and the accumulating sediments are rather poor in terrigenous matter (Van Weering et al. 1997; Ivanova 1999), including the clay minerals (Koning et al. 1997, 2001) that contain the NH_4^+ in their interlayers. In addition, they are rich in organic matter, which would overwhelm the inorganic contribution (Altabet et al. 1999), except maybe in the deep basin where the least PON is deposited and preserved, and clay minerals are selectively enriched by carbonate dissolution (Ivanova 1999). Sediment pore-water profiles show only low NH_4^+ concentrations relative to the bottom water and the calculated efflux of NH_4^+ from the sediment is low as well (Helder et al. 1995). Consequently, the contribution by NH_4^+ should be considered as far too small to cause the observed depth gradient, particularly because NH_4^+ tends to be isotopically lighter than its source PON, thus suppressing the gradient towards the deep basin. As the clay mineral content remains low too, little adsorbed NH_4^+ from terrigenous sources is contributed that could change the bulk $\delta^{15}N$ of sediment appreciably. Apparently, the $\delta^{15}N$ of the arriving N flux is altered at or very close to the sediment–water interface, as evidenced by the contrast between the trap samples and the core top as well as downcore sediments.

There are now effectively two pathways for generating the positive gradient in $\delta^{15}N$ with bottom depth and distance offshore: (1) Rayleigh distillation during remineralization of bulk $\delta^{15}N$; (2) selective removal of compounds with low $\delta^{15}N$. Both pathways imply a decreasing preservation downslope, which is consistent with the lower burial efficiencies of N with depth, the increasing bottom water oxygen concentrations and the deepening of the oxygen penetration into the sediment along the transect (Helder et al. 1995). Consequently, diagenetic enrichment should be caused by oxic remineralization as, even at the site with the lowest bottom water concentration, the oxygen penetration depth in the sediment still extended to 2 mm (Helder et al. 1995). Indeed, incubation experiments showed that oxic remineralization left the residual PON enriched with respect to the original substrate (Wada 1980; Holmes et al. 1999) and field studies confirm this (e.g. De Lange et al. 1994; Altabet et al. 1995; Freudenthal et al. 2001a, b). Altabet & Francois (1994a) inferred a diagenetic enrichment of about 4‰ based on the difference in $\delta^{15}N$ between core-top sediments and the annual N fluxes intercepted by deep-moored sediment traps in the open ocean (see also Fig. 7). In the open ocean, oxygen consumption in the sediment is limited by the low organic matter deposition fluxes and, as a consequence, oxygen usually penetrates deeply into sediment.

The positive gradient in sedimentary $\delta^{15}N$ with depth and distance offshore is not restricted to Somalia but appears as a typical feature of ocean margin sediments and has been reported from off Oman (Altabet et al. 1995), off Zaire, Angola and Namibia in the SE Atlantic (Holmes et al. 1997, 1998), and off Ireland in the NE Atlantic (Fig. 8). The differences along the transects are generally within both the range expected from Rayleigh distillation of NO_3^- in the euphotic zone and that found for oxic decomposition along the Somali transect. Holmes et al. (1998) pointed out that the $\delta^{15}N$ increases with diminishing C_{org} content of the sediment off Namibia as it does for Somalia in terms of N burial fluxes and efficiencies. Fractionation by enhanced diagenesis may explain why some transects (e.g. off Namibia) show strong positive gradients with bottom depth but no perceivable one in nutrient utilization in the overlying water offshore, which is generally invoked as a causal mechanism for changing the $\delta^{15}N$ of sedimentary N. As for Somalia, diagenetic fractionation at the sediment–water interface downslope would enhance the positive gradient in $\delta^{15}N$ generated by nutrient limitation in the euphotic zone offshore (Altabet et al. 1995; Holmes et al. 1998).

For deep ocean sediments far offshore, diagenetic fractionation of the arriving N export flux

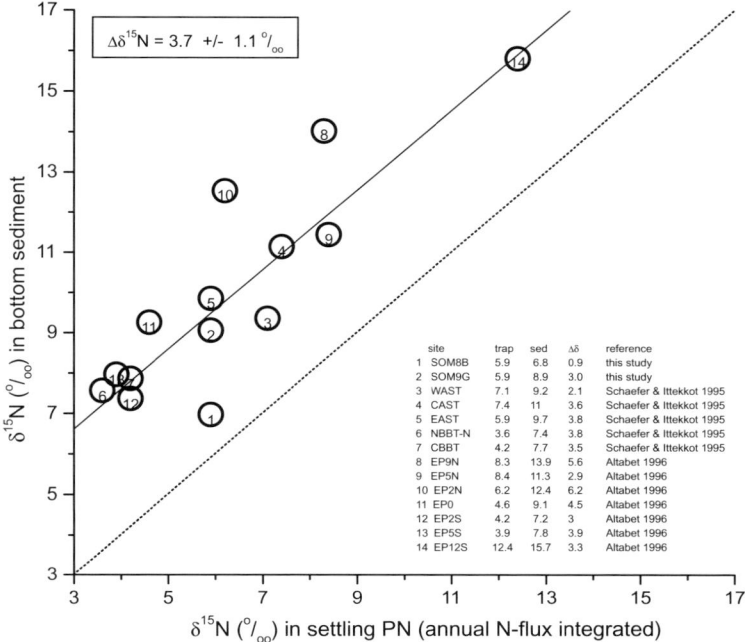

Fig. 7. Diagenetic fractionation of settling N during burial amounts to 3.7 ± 1.1‰ judging from a compilation of sediment trap data with associated sediment cores, reported mainly from open-ocean sites, i.e. the Somali Basin (1 and 2; this study), the open Arabian Sea (3–5; Schaefer & Ittekkot 1995), the Bay of Bengal (6 and 7; Schaefer & Ittekkot 1995) and the equatorial Pacific (8–14; Altabet 1996).

seems to level off at a fairly constant 3.7‰ over a 9‰ range (Fig. 7), as indicated earlier by Altabet & Francois (1994b), Altabet (1996) and Francois et al. (1997). Apparently, extended oxic remineralization may diminish the burial of N further but without affecting the $\delta^{15}N$ substantially, whether by the Rayleigh distillation of sedimentary N or the selective remineralization of specific N compounds. The consistent differences observed between transects are generally larger than along each transect (Fig. 8), and are too large to be consistent in terms of either nutrient utilization or oxic decomposition. Probably, regional differences in the $\delta^{15}N$ of the substrate N nutrients in the euphotic zone are involved (Francois et al. 1997), which are preserved in the sedimentary N despite differential diagenesis.

Sedimentary $\delta^{15}N$ has been identified as the only tracer available to assess the degree of surface ocean NO_3^- utilization and denitrification from the fossil record (Francois et al. 1992; Altabet et al. 1995). It appears that differential oxic diagenesis should also be considered as a contributor to the $\delta^{15}N$ record downcore, particularly as it leads to overestimation when taken as a measure for nutrient utilization across ocean margins in the past. On the other hand, diagenetic fractionation seems to be inversely proportional to the N burial efficiency on ocean margins, which in turn varies with the magnitude of the exported N flux and thus relates to the utilization of nutrients as well.

Conclusions

Combined sediment trap and box-core records of $\delta^{15}N$ and nitrogen fluxes off Somalia show the following features.

(1) Minima in the $\delta^{15}N$ of settling N are derived from poorly utilized nutrient-rich waters originating from coastal upwelling and advected to the sites along eddy margins during the SW monsoon. Maxima are associated with nitrate depletion in stratified waters during the autumn intermonsoon.

(2) The annual $\delta^{15}N$ values differ little between the traps, despite substantial temporal and spatial differences in the $\delta^{15}N$ and the N fluxes over the monsoonal cycle.

(3) The annual $\delta^{15}N$ values off Somalia are 1.2‰ lower than reported earlier from the open Arabian Sea, suggesting a proportionally smaller contribution of denitrification.

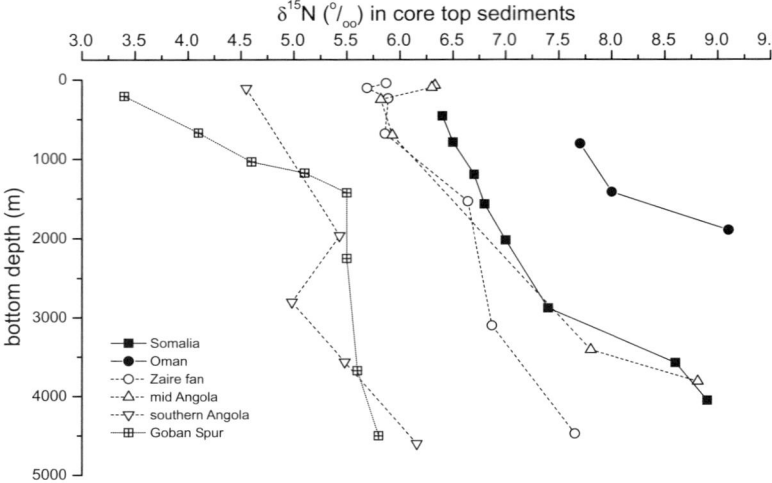

Fig. 8. Positive gradients in $\delta^{15}N$ with depth seem typical for core-top sediments on ocean margins and are reported from off Oman in the NW Indian Ocean (Altabet et al. 1995), off Zaire and Angola in the SE Atlantic (Holmes et al. 1997) and off Ireland in the NE Atlantic (Brummer et al., unpubl. data). As for Somalia, diagenetic fractionation rather than nutrient utilization can account for much of the increased $\delta^{15}N$ downslope. The consistent offset between the transects may be due to a variety of causes, e.g. areal differences in the $\delta^{15}N$ of the substrate N nutrients and their specific burial efficiencies.

(4) Oxic sedimentary decomposition accounts for the positive downslope gradient in $\delta^{15}N$ off Somalia, rather than differential nutrient utilization. Other ocean margins all show similar positive gradients. In the deep open ocean such diagenesis seems to be fairly constant at +3.7‰ relative to that of the arriving N flux.

We thank the captain and crew of the R.V. *Tyro* and the NIOZ technicians for deploying the moorings. S. Conan and S. Broerse of the Free University (VU) Amsterdam are acknowledged for their help in processing the samples, as are the NIOZ-SOZ technicians for the shipboard nutrient analyses. S. Conan and G. Ganssen (VU) provided the foraminiferal oxygen isotope data for calculating sea surface temperatures from the traps. J. Kindle (Naval Research Laboratory Stennis Space Center), O. Brown (RSMAS) and M. Baars (NIOZ) allowed us to access remote sensing data on real-time wind speeds and sea surface temperatures in the deployment area. Radiocarbon dates were made available by E. Ivanova and S. R. Troelstra (VU), which allowed calculation of sediment burial fluxes. We thank M. J. Higginson, G. J. Reichart, M. Voss, C. van der Zee and an anonymous reviewer for reviewing earlier versions of the manuscript. The Netherlands Indian Ocean Programme was funded and co-ordinated by the former Netherlands Marine Research Foundation (SOZ) of the Netherlands Organization for Scientific Research (NWO). This is Publication 3675 of the Royal Netherlands Institute for Sea Research (NIOZ).

References

ALTABET, M. 1996. Nitrogen and carbon isotopic tracers of the source and transformation of particles in the deep sea. *In:* ITTEKOT, V., SCHAEFER, P., HONJO, S. & DEPETRIS, P.J. (eds) *Particle Flux in the Ocean, SCOPE Report, 57*. Wiley, Chichester, 155–184.

ALTABET, M. 2001. Nitrogen isotopic evidence for micronutrient control of fractional NO_3^- utilization in the equatorial Pacific. *Limnology and Oceanography*, **46**, 368–380.

ALTABET, M. & CURRY, W.B. 1989. Testing models of past ocean chemistry using foraminifera $^{15}N/^{14}N$. *Global Biogeochemical Cycles*, **3**, 107–119.

ALTABET, M. & FRANCOIS, R. 1994a. Sedimentary nitrogen isotopic ratio as a recorder for surface ocean nitrate utilization. *Global Biogeochemical Cycles*, **8**, 103–116.

ALTABET, M. & FRANCOIS, R. 1994b. The use of nitrogen isotopic ratio for reconstruction of past changes in surface ocean nutrient utilization. *In:* ZAHN, R., PEDERSEN, T.F., KAMINSKI, M.A. & LABEYRIE, L. (eds) *Carbon Cycling in the Glacial Ocean: Constraints on the Ocean's Role in Global Change*. NATO ASI Series, **117**, 281–306.

ALTABET, M. & FRANCOIS, R. 2001. Nitrogen isotope biogeochemistry of the Antarctic Polar Frontal Zone at 170°W. *Deep-Sea Research II*, **48**, 4247–4273.

ALTABET, M.A., FRANCOIS, R., MURRAY, D.W. & PRELL, W.L. 1995. Climate-related variations in denitrification in the Arabian Sea from sediment $^{15}N/^{14}N$ ratios. *Nature*, **373**, 506–509.

ALTABET, M.A., PISKALN, C., THUNELL, R., PRIDE, C., SIGMAN, D., CHAVEZ, F. & FRANCOIS, R. 1999. The nitrogen isotope biogeochemistry of sinking particles from the margin of the Eastern North Pacific. *Deep-Sea Research I*, **46**, 655–679.

ALTABET, M.A., HIGGINSON, M. & MURRAY, D.W. 2002. The effect of millennial-scale changes in Arabian Sea denitrification on atmospheric CO_2. *Nature*, **415**, 159–162.

BAARS, M. A., BAKKER, K. M. J., DE BRUIN, T. F. & 9 OTHERS 1994. Seasonal fluctuations in plankton biomass and productivity in the ecosystems of the Somali Current, Gulf of Aden and southern Red Sea. *In*: Baars, M. A. (ed), *Monsoons and Pelagic Systems. Cruise Reports Netherlands Indian Ocean Programme, 1*. National Museum of Natural History, Leiden, 13–34.

BÖHLKE, J. K. & COPLEN, T. B. 1995. Interlaboratory comparison of reference materials for nitrogen-isotope-measurements. In: *Reference and Intercomparison Materials for Stable Isotopes of Light Elements. IAEA-TECDOC, 825*. International Atomic Energy Agency, Vienna, 51–66.

BRANDES, J.A., DEVOL, A.H., YOSHINARI, T., JAYAKUMAR, D.A. & NAQVI, S.W.A. 1998. Isotopic composition of nitrate in the central Arabian Sea and eastern tropical North Pacific: a tracer for mixing and nitrogen cycles. *Limnology and Oceanography*, **43**, 1680–1689.

BRUMMER, G.-J.A. 1995. Sediment traps and particle dynamics. *In*: VAN HINTE, J.E., VAN WEERING, T.C.E. & TROELSTRA, S.R. (eds) *Tracing a Seasonal Upwelling. Cruise Reports Netherlands Indian Ocean Programme*. National Museum of Natural History, Leiden, **4**, 55–61.

CAPONE, D.G., ZEHR, J.P., PAERL, H.W., BERGMAN, B. & CARPENTER, E.J. 1997. Trichodesmium, a globally significant marine cyanobacterium. *Science*, **276**, 1221–1229.

CLINE, J.D. & KAPLAN, I.R. 1975. Isotopic fractionation of dissolved nitrate during denitrification in the eastern tropical North Pacific Ocean. *Marine Chemistry*, **3**, 271–299.

CONAN, S.M.H. & BRUMMER, G.-J.A. 2000. Fluxes of planktonic foraminifera in response to monsoonal upwelling on the Somali Basin margin. *Deep-Sea Research II*, **47**, 2207–2228.

DE LANGE, G. J., VAN OS, B., PRUYSERS, P. A. & 6 OTHERS 1994. Possible early diagenetic alteration of palaeo proxies. *In*: Zahn, R., Pedersen, T. F., Kaminski, M. A. & Labeyrie, L. (eds) *Carbon Cycling in the Glacial Ocean: Constraints on the Ocean's Role in Global Change*. NATO ASI Series, **117**, 225–258.

DE WILDE, H.P.J. & HELDER, W. 1997. Nitrous oxide in the Somali basin: the role of upwelling. *Deep-Sea Research*, **44**, 1319–1340.

EMERSON, S.P., QUAY, P., STUMP, C., WILBUR, S. & KNOX, M. 1991. O_2, Ar, N_2 and ^{222}Rn in surface waters of the subarctic ocean: net biological O_2 production. *Global Biogeochemical Cycles*, **5**, 49–69.

FARRELL, J.W., PEDERSEN, T.F., CALVERT, S.E. & NIELSEN, B. 1995. Glacial–interglacial changes in nutrient utilization in the equatorial Pacific Ocean. *Nature*, **377**, 514–517.

FISCHER, J., SCHOTT, F. & STRAMMA, L. 1996. Currents and transports of the Great Whirl–Socotra Gyre system during the summer monsoon, August 1993. *Journal of Geophysical Research*, **101**, 3573–3587.

FRANCOIS, R., ALTABET, M.A. & BURKLE, L.H. 1992. Glacial to interglacial changes in surface nitrate utilisation in the Indian sector of the Southern Ocean as recorded by sediment $\delta^{15}N$. *Paleoceanography*, **7**, 589–606.

FRANCOIS, R., ALTABET, M.A., YU, E.-F. & 6 OTHERS 1997. Contribution of Southern Ocean surface-water stratification to low atmospheric CO_2 concentrations during the last glacial period. *Nature*, **389**, 929–935.

FREUDENTHAL, T., NEUER, S., MEGGERS, H., DAVENPORT, R. & WEFER, G. 2001a. Influence of lateral particle advection and organic matter degradation on sediment accumulation and stable nitrogen isotope ratios along a productivity gradient in the Canary Island region. *Marine Geology*, **177**, 93–109.

FREUDENTHAL, T., WAGNER, T., WENZHÖFER, F., ZABEL, M. & WEFER, G. 2001b. Early diagenesis of organic matter from sediments of the eastern tropical Atlantic: evidence from stable nitrogen and carbon. *Geochimica et Cosmochimica Acta*, **65**, 1795–1808.

FRY, B. 1988. Food web structure on Georges Bank from stable C, N, and S isotopic composition. *Limnology and Oceanography*, **33**, 1182–1190.

HELDER, W., KLOOSTERHUIS, R.T. & NOLTING, R.F. 1995. Early diagenesis in sediments from the oxygen minimum zone off Yemen. *In*: VAN HINTE, J.E., VAN WEERING, T.C.E. & TROELSTRA, S.R. (eds) *Tracing a Seasonal Upwelling. Cruise Reports Netherlands Indian Ocean Programme*. National Museum of Natural History, Leiden, **4**, 67–77.

HERBLAND, A. & VOITURIEZ, B. 1979. Hydrological structure analysis for estimating the primary production in the tropical Atlantic Ocean. *Journal of Marine Research*, **37**, 87–101.

HOLMES, M.E., MUELLER, P.J., SCHNEIDER, R.R., SEGL, M., PAETZOLD, J. & WEFER, G. 1996. Stable nitrogen isotopes in Angola Basin surface sediments. *Marine Geology*, **134**, 1–12.

HOLMES, M.E., SCHNEIDER, R.R., MUELLER, P.J., SEGL, M. & WEFER, G. 1997. Reconstruction of past nutrient utilization in the eastern Angola Basin based on sedimentary $^{15}N/^{14}N$ ratios. *Paleoceanography*, **12**, 604–614.

HOLMES, M.E., MUELLER, P.J., SCHNEIDER, R.R., SEGL, M. & WEFER, G. 1998. Spatial variations in euphotic zone nitrate utilisation based on $\delta^{15}N$ in surface sediments. *Geo-Marine Letters*, **18**, 58–65.

HOLMES, M.E., EICHNER, C., STRUCK, U. & WEFER, G. 1999. Reconstruction of surface ocean nutrient utilization using stable nitrogen isotopes in single particles and sediments. *In*: FISCHER, G. & WEFER, G. (eds) *The Use of Proxies in Paleoceanography: Examples from the South Atlantic*. Springer, Berlin, 447–468.

IVANOVA, E. 1999. *Late Quaternary monsoon history*

and paleoproductivity of the western Arabian Sea. PhD thesis, Free University, Amsterdam.

KERHERVÉ, P., MINAGAWA, M., HEUSSNER, S. & MONACO, A. 2001. Stable isotopes ($^{13}C/^{12}C$ and $^{15}N/^{14}N$) in settling organic matter of the northwestern Mediterranean Sea: biogeochemical implications. *Oceanologica Acta*, **24**, S77–S85.

KNAPPERTSBUSCH, M. & BRUMMER, G.-J.A. 1995. A sediment trap investigation of sinking coccolithophorids in the North Atlantic. *Deep-Sea Research*, **42**, 1083–1109.

KONING, E., BRUMMER, G.-J.A., VAN RAAPHORST, W., VAN BENNEKOM, A.J., HELDER, W. & VAN IPEREN, J.M. 1997. Settling, dissolution and burial of biogenic silica in the sediments off Somalia (northwestern Indian Ocean). *Deep-Sea Research*, **44**, 1341–1360.

KONING, E., VAN IPEREN, J.M., VAN RAAPHORST, W., HELDER, W., BRUMMER, G.-J.A. & VAN WEERING, T.C.E. 2001. Selective preservation of upwelling-indicating diatoms in sediments off Somalia, NW Indian Ocean. *Deep-Sea Research*, **48**, 2473–2495.

LEVITUS, S. 1994. *Climatological Atlas of the World Ocean*. NOAA Professional Paper, **13**.

MONTOYA, J.P. 1994. Nitrogen isotope fractionation in the modern ocean: implications for the sedimentary record. *In:* ZAHN, R., PEDERSEN, T.F., KAMINSKI, M.A. & LABEYRIE, L. (eds) *Carbon Cycling in the Glacial Ocean: Constraints on the Ocean's Role in Global Change*. NATO ASI Series, **117**, 257–279.

MONTOYA, J.P., HORRIGAN, S.G. & MCCARTHY, J.J. 1991. Rapid, storm-induced changes in the natural abundance of ^{15}N in a planktonic ecosystem. *Geochimica et Cosmochimica Acta*, **55**, 3627–3638.

NAQVI, S.W.A., YOSHINARI, T., JAYAKUMAR, D.A. & 5 OTHERS 1998. Budgetary and biogeochemical implications of N_2O isotope signatures in the Arabian Sea. *Nature*, **394**, 462–464.

OSTROM, N.E., MACKO, S.A., DEIBEL, D. & THOMPSON, R.J. 1997. Seasonal variation in the stable carbon and nitrogen isotope biogeochemistry of a coastal cold ocean environment. *Geochimica et Cosmochimica Acta*, **61**, 2929–2942.

OWENS, N.J.P. 1987. Natural variations in ^{15}N in the marine environment. *Advances in Marine Biology*, **24**, 390–451.

OWENS, N.J.P. & REES, A.P. 1989. Determination of nitrogen-15 at sub-microgram levels of nitrogen using automated continuous-flow isotope ratio mass spectrometry. *Analyst*, **114**, 1655–1657.

SCHAEFER, P. & ITTEKKOT, V. 1993. Seasonal variability of $\delta^{15}N$ in settling particles in the Arabian Sea and its paleogeochemical significance. *Naturwissenschaften*, **80**, 511–513.

SCHAEFER, P. & ITTEKKOT, V. 1995. Isotopic biogeochemistry of nitrogen in the northern Indian Ocean. *Mitteilungen des Geologisch-Paläontologisches Institut der Universität Hamburg*, **78**, 67–93.

SCHOTT, F.A. & FISCHER, J. 2000. Winter monsoon circulation of the northern Arabian Sea and Somali Current. *Journal of Geophysical Research*, **C105**, 6359–6376.

SCHOTT, F., SWALLOW, J.C. & FIEUX, M. 1990. The Somali Current at the equator: annual cycle of currents and transports in the upper 1000 m and connection to neighbouring latitudes. *Deep-Sea Research*, **12**, 1825–1848.

SCHUBERT, C.J. & CALVERT, S.E. 2001. Nitrogen and carbon isotope composition of marine and terrestrial organic matter in Arctic Ocean sediments: implications for nutrient utilization and organic matter composition. *Deep-Sea Research I*, **48**, 789–810.

SHEMESH, A., MACKO, S.A., CHARLES, C.D. & RAU, G.H. 1993. Isotopic evidence for reduced productivity in the glacial Southern Ocean. *Science*, **262**, 407–410.

SIGMAN, D.M., ALTABET, M.A., MICHENER, R., MCCORCLE, D.C., FRY, B. & HOLMES, R.M. 1997. Natural abundance-level measurements on the nitrogen isotopic composition of oceanic nitrate: an adaptation of the ammonia diffusion method. *Marine Chemistry*, **57**, 227–242.

VAN COUWELAAR, M., ANGEL, M.V. & MADIN, L.P. 1997. The distribution and biology of the swimming crab *Charybdis smithii* McLeay, 1838 (Crustacea; Brachyura; Portunidae) in the NW Indian Ocean. *Deep-Sea Research II*, **44**, 1251–1280.

VAN HINTE, J. E., VAN WEERING, T. C. E. & TROELSTRA, S. R. (eds) 1995. *Tracing a Seasonal Upwelling. Cruise Reports Netherlands Indian Ocean Programme, 4*. National Museum of Natural History, Leiden.

VAN WEERING, T.C.E., HELDER, W. & SCHALK, P. 1997. Netherlands Indian Ocean Program, first results and an introduction. *Deep-Sea Research*, **44**, 1177–1195.

VELDHUIS, M.J.W., KRAAY, G.W., VAN BLEIJSWIJK, J.D.L. & BAARS, M.A. 1997. Seasonal and spatial variability in phytoplankton biomass, productivity and growth in the northwestern Indian Ocean (the NW- and SE-Monsoon, 1992–1993. *Deep-Sea Research*, **44**, 425–449.

VON GYLDENFELDT, A.-B., CARSTENS, J. & MEINECKE, J. 2000. Estimation of the catchment area of a sediment trap by means of current meters and foraminiferal tests. *Deep-Sea Research II*, **47**, 1701–1718.

VOSS, M., DIPPNER, J.W. & MONTOYA, J.P. 2001. Nitrogen isotope patterns in the oxygen-deficient waters of the Eastern Tropical North Pacific Ocean. *Deep-Sea Research I*, **48**, 1905–1921.

WADA, E. 1980. Nitrogen isotope fractionation and its significance in biogeochemical processes occurring in marine environments. *In:* GOLDBERG, E., HORIBE, Y. & SARUHASHI, K. (eds) *Isotope Marine Chemistry*. Uchida Rokakuho, Tokyo, 375–398.

WADA, E. & HATTORI, A. 1990. *Nitrogen in the Sea: Forms, Abundances and Rate Processes*. CRC Press, Boca Raton, FL.

WANIEK, J., KOEVE, W. & PRIEN, R.D. 2000. Trajectories of sinking particles and the catchment areas above sediment traps in the northeast Atlantic. *Journal of Marine Research*, **58**, 983–1006.

ZONNEVELD, K.A.F. & BRUMMER, G.-J.A. 2000. (Palaeo-)ecological significance, transport and preservation of organic-walled dinoflagellate cysts in the Somali Basin, NW Arabian Sea. *Deep-Sea Research II*, **47**, 2229–2256.

Late Quaternary highstand deposits of the southern Arabian Gulf: a record of sea-level and climate change

ALUN H. WILLIAMS[1] & GORDON M. WALKDEN[2]

[1] *Oolithica Geoscience Ltd, 489 Union Street, Aberdeen AB11 6DB, UK*
[2] *Department of Geology and Petroleum Geology, University of Aberdeen, King's College, Aberdeen AB24 3UE, UK (e-mail: g.walkden@abdn.ac.uk)*

Abstract: The southern Arabian (Persian) Gulf is at present the site of extensive carbonate sedimentation, as was the case during Pleistocene interglacial marine highstands. During glacial lowstands the basin was subaerially exposed, and aeolian sedimentation predominated. Most of the southern Arabian Gulf floor is underlain by Quaternary carbonates, and scattered outcrops may be found onshore. These belong to three formations: the aeolian Ghayathi Formation, the continental Aradah Formation and the marine Fuwayrit Formation. The Fuwayrit Formation consists of three members, separated by subaerial exposure surfaces. These are, from the base upwards, the shallow marine Futaisi and Dabb'iya Members, and the aeolian Al Wusayl Member. Offshore, at least six Quaternary sequences are present within the uppermost 50 m of sediment. No reliable direct age dates have been acquired from Pleistocene shallow marine or coastal deposits in the southern Arabian Gulf. It has therefore been necessary to infer the ages of these sediments by a comparison of their stratigraphy and elevation with deposits known from other parts of the world. We regard this approach as valid because the southern Gulf coastline lacks evidence for significant widespread neotectonic uplift, and halotectonic effects are localized. This comparison indicates that the Fuwayrit Formation was deposited during the last interglacial (oxygen isotope substage 5e), as (1) these sediments represent the youngest pre-Holocene marine deposits, and (2) they are found at an elevation correlative with many substage 5e deposits from other parts of the globe. Sedimentary evidence reveals two highstands during this period, peaking at around 1.5 m and 6 m above present sea level, respectively. Offshore sediments indicate that sea level did not fall as far as 24 m below present level in the intervening regression. Following the second highstand, sea level fell to more than 23 m below present level, before briefly rising once again (late isotope stage 5). This later highstand probably peaked between 17 and 7 m below present level. The sequence underlying the Fuwayrit Formation was probably deposited during the penultimate interglacial (late oxygen isotope stage 7). It is also likely that the Ghayathi Formation aeolianites were largely sourced from this sequence. Facies analysis of offshore core sediments indicates that sea level reached at least 15 m below present level during this period. Widespread evidence exists for a Holocene sea level higher than at present in the southern Arabian Gulf, indicating that it peaked at 1–2 m above present level, *c.* 5.5 ka bp. Pleistocene deposits preserved in the southern Arabian Gulf provide a record of changing palaeowinds and palaeoclimates. Currently, the region experiences a hyper-arid to arid climate, with facies patterns dominated by the northwesterly shamal wind. The Ghayathi Formation was originally deposited under an arid climatic regime, which allowed the sediments to remain unconsolidated. The dunefield was later remodelled under conditions of increasing wind speed, with a change in wind direction from NNW to WNW. These changes are thought to reflect the onset of glaciation. Palaeocurrent directions from the Al Wusayl Member, combined with sedimentary evidence from the Futaisi and Dabb'iya Members, indicate that during the peak of the last interglacial the prevailing wind (the 'palaeo-shamal') blew from the NE. Compelling evidence for a pluvial episode during this period is provided by abundant and widespread dissolution (palaeokarstic) pits found in the top surface of the Futaisi Member, believed to represent the former positions of abundant trees or large plants.

The Arabian Gulf is the marine flooded portion of a more extensive foreland basin that forms a northwestern extension of the Arabian Sea, to which it is connected via the Straits of Hormuz. It is shallow, with a maximum depth of around 90 m. The southern Gulf, from Saudi Arabia to the United Arab Emirates (UAE), is generally very shallow, with a gentle northwards-sloping floor (Fig. 1). This area is at present the site of extensive carbonate sedimentation, with the prevailing northwesterly 'shamal' wind acting as one of the major controls on facies patterns. The southern Arabian Gulf has been frequently cited in the literature as a modern example of a carbonate ramp (e.g. Tucker & Wright 1990), although the validity of this model has been questioned (Walkden & Williams 1998). During Pleistocene glacial maxima sea level fell by as much as 120 m below present level, leaving the entire Gulf subaerially exposed. Melting of polar ice masses was associated with repeated flooding of the basin, leading to the deposition of carbonate sediments. During lowstands aeolian sedimentation predominated, under the influence of a strengthened glacial Shamal. Most of the southern Arabian Gulf sea floor is underlain by Quaternary marine carbonates, and scattered outcrops may be found onshore. Quaternary aeolianite deposits are fairly widespread in Abu Dhabi.

As a result of the physical isolation imposed by the Musandam Peninsula, oceanic influences on the Arabian Gulf are minimal. However, the Gulf exerts a significant influence on the waters of the Arabian Sea, most markedly in the Gulf of Oman. Located in a hyper-arid to semi-arid climate, and with only one permanent inflowing river, the Arabian Gulf suffers a net loss of water through evaporation. This increases the salinity of the water, which sinks and flows back into the Arabian Sea through the Straits of Hormuz. The outflow increases during the winter, when Gulf waters are often colder than the neighbouring ocean. The bottom current has a high salinity, oxygen content and nutrient content, and its effects can be traced well beyond the continental shelf (Purser & Seibold 1973; Bower et al. 2000). During glacial periods, when the Gulf is empty, this current, and its influence on the oceanography of the Arabian Sea, is turned off.

Despite the proximity to an active plate margin, we see no unequivocal evidence for late Pleistocene uplift of the coastline of the southern Arabian Gulf. Values of up to 190 cm ka^{-1} were reported for Qatar and Saudi Arabia (Vita-Finzi 1978; Ridley & Seeley 1979), but were later disputed (McClure & Vita-Finzi 1982). We believe that these results can be explained by failure to account for a Holocene highstand of 1–2 m above present sea level. By the same token we propose that onshore outcrops of upper Pleistocene marine deposits are also mostly unaffected by tectonism. Study of offshore sediments shows that there is no need to invoke tectonic movements for much of the sea floor in the southern Gulf, such as Umm Shaif shoal (Williams 1999). However, active halotectonics means that some offshore shoals and islands do show signs of vertical movements.

This paper is based on work carried out in the UAE and Qatar. It gives an overview of the Quaternary deposits found in the region, and of

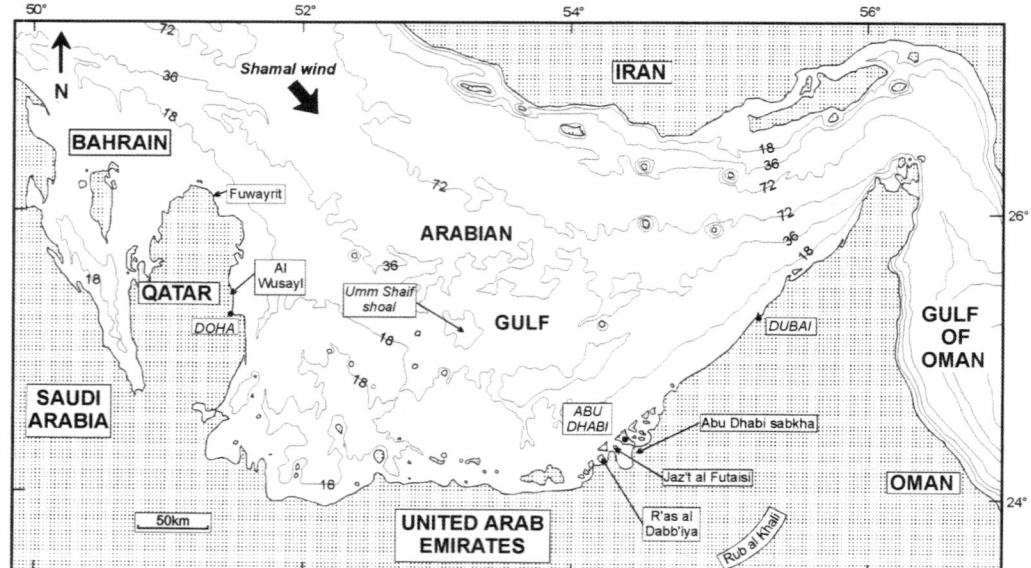

Fig. 1. Map showing the location of the southern Arabian Gulf, including bathymetry, localities mentioned in the text, political geography and position of major cities. Depths are in metres.

the information they can provide on changing sea levels and climate. The coastline of the southern Arabian Gulf provides a far more reliable indicator of palaeo-sea levels than the coastlines of the Arabian Sea, which have been largely modified by neotectonics. In addition, the deposits provide evidence for palaeoclimatic changes that are likely to have affected the whole region. Our database includes coastal and inland outcrops, man-made exposures and continuously cored boreholes up to 60 m deep in the floor of the southern Arabian Gulf (Williams 1999).

Pleistocene deposits in the southern Arabian Gulf

Scattered outcrops of Pleistocene sediments are found onshore in the southern Arabian Gulf. They belong to three formations: the Ghayathi Formation, the Aradah Formation and the Fuwayrit Formation. The stratigraphy of these deposits is shown in Fig. 2. Although outcrops are rare, they have a widespread distribution, and are found in Abu Dhabi, Qatar, Saudi Arabia, Bahrain and Kuwait. Both the Ghayathi Formation and the Fuwayrit Formation crop out in coastal regions, whereas the Aradah Formation is found inland.

Pleistocene deposits are much more widespread beneath the sea floor of the Arabian Gulf, but have received little study.

Ghayathi Formation

Hadley et al. (1998) proposed the name 'Ghayathi Formation' to cover 'remnant paleodune deposits composed of carbonate and siliciclastic material exposed slightly inland and along the Arabian Gulf coast of Abu Dhabi Emirate'. They are also found in Saudi Arabia. The Ghayathi Formation consists of mixed carbonate–clastic aeolianites, which occur up to 80 km inland in Abu Dhabi, where they are by far the most voluminous Pleistocene sediments (Glennie 1998; Hadley et al. 1998). Near the coast the aeolianites are composed almost entirely of marine-derived carbonate grains, whereas inland they contain up to 89% quartz. Deposits of the Ghayathi Formation are morphologically and chronostratigraphically complex, and have been discussed by Williams (1999) and Williams & Walkden (2001).

Aradah Formation

The Aradah Formation consists of continental sabkha-type deposits overlying the Ghayathi For-

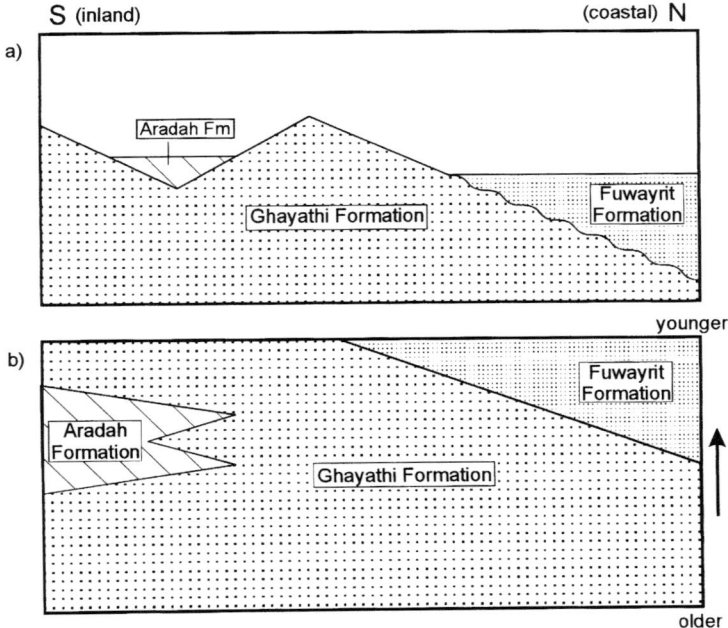

Fig. 2. Schematic stratigraphy of the Pleistocene deposits in Abu Dhabi, showing (**a**) morphological relationship and (**b**) chronological relationship.

mation in southern UAE (Hadley *et al.* 1998). These deposits were not observed in this study, and will not be discussed.

Fuwayrit Formation

The Fuwayrit Formation consists of shallow marine and marine-derived aeolian carbonates, preserved in coastal localities (Williams 1999; Williams & Walkden 2001). In Abu Dhabi they mostly overlie Ghayathi Formation aeolianites, and are often preserved as 'zeugen' (erosional remnants; Fig. 3). In Qatar they are found overlying Eocene sediments. Three members have been recognized within the Fuwayrit Formation, separated by subaerial exposure surfaces (Fig. 4). These are, from base upwards: the Futaisi Member and the Dabb'iya Member (shallow marine deposits of up to 6 m thickness in the UAE and Qatar); and the Al Wusayl Member (marine-derived aeolianites of up to 20 m thickness found only in Qatar).

Offshore deposits

Study of sea-floor cores from six localities in the southern Arabian Gulf indicates that at least six Quaternary sequences, consisting of marine carbonates and sabkha-derived evaporites, are present within the uppermost 50 m of sediment (Williams 1999). The uppermost Pleistocene sequence represents the offshore equivalent of the Fuwayrit Formation, and has been seen to a maximum thickness of 29 m. Otherwise, there are no correlative submarine deposits onshore, and the stratigraphy of these sequences has not been properly formalized. Only the two uppermost Pleistocene sequences (sequences 2 and 3 of Williams (1999)) will be considered here.

Deposition of the Pleistocene sediments

Ghayathi Formation

The Ghayathi Formation provides a good example of 'regressive aeolianites', deposited through reworking of coastal and shallow marine sediments during and following sea-level fall (Williams & Walkden 2001). The large-scale cross-bedding that is characteristic of these deposits extends below present-day sea level (Fig. 3).

Some of the best exposures of the Ghayathi Formation palaeodunes are where they have been dissected by sea and/or wind, leaving flat-topped

Fig. 3. Zeugen exposure showing Ghayathi Formation aeolianites (pre-isotope stage 5) capped by shelly beach facies of the Dabb'iya Member, Fuwayrit Formation (isotope substage 5e), Futaisi Island, Abu Dhabi.

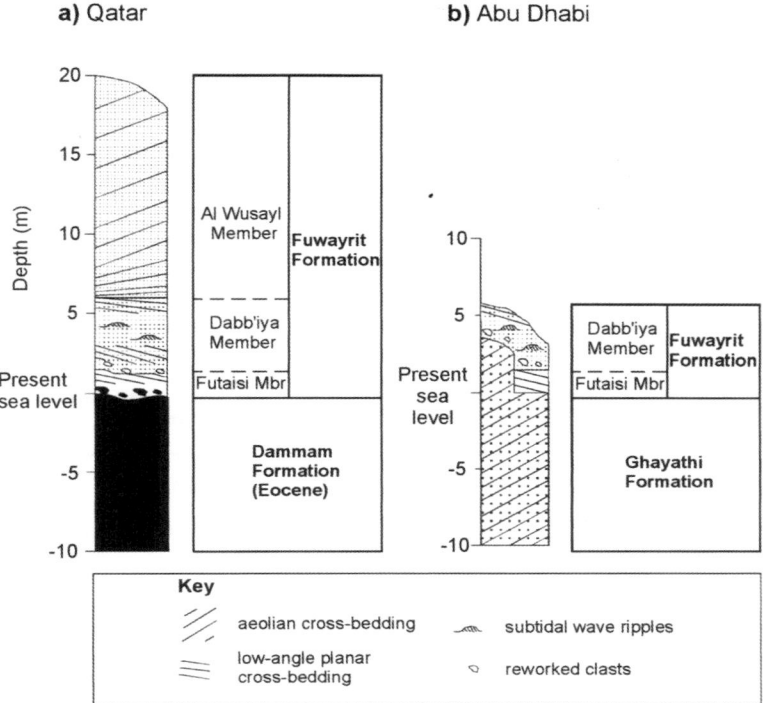

Fig. 4. Generalized stratigraphy of the Pleistocene deposits at outcrop in (a) Qatar and (b) Abu Dhabi, showing subdivisions of the Fuwayrit Formation.

and steep-sided zeugen (Fig. 3). Kirkham (1998) noticed that many of these features are aligned as parallel discontinuous ridges about 150 m apart. Mapping revealed these to be up to 7 km long, and they are visible on satellite images. Comparison of these ridges with features in the neighbouring siliciclastic sand sea (the Rub al Khali) led Kirkham (1998) to conclude that they represent ancient seif dunes. These have an orientation of c. 100° (ESE), parallel to the extensive siliciclastic seif dunes that currently dominate the desert of the eastern Emirates. This implies that they belong to the same dune system. The inland seifs are out of phase with the modern shamal winds, and are believed to be fairly inactive today (K. W. Glennie, pers. comm.).

Despite their overall geometry, the primary bedding within the Ghayathi Formation palaeodunes in Abu Dhabi is barchanoid, with dips predominantly to the SE. It is therefore apparent that the original dune morphology was remodelled before lithification (Kirkham 1998). Transverse dunes such as barchans become unstable with increasing wind velocities, especially when accompanied by changes in wind direction. This often leads to the formation of longitudinal dunes, such as seifs (Glennie 1993). Sedimentological evidence from the Ghayathi Formation indicates that an originally barchan dunefield was reworked into seif dunes under conditions of increasing wind speed, accompanied by a change in direction from NNW to WNW. This is illustrated in Fig. 5.

Fuwayrit Formation: onshore

The lowermost sediments of the Fuwayrit Formation found onshore belong to the Futaisi Member. These unconformably overlie the Ghayathi Formation in Abu Dhabi and the Eocene Dammam Formation in Qatar (Fig. 4). Reworked blocks of aeolianite indicate that the Ghayathi Formation was indurated by the time it was transgressed. The Futaisi Member consists of littoral and shallow subtidal deposits, which are found to an elevation of 1.5 m above present sea level (Williams 1999). Mangrove root traces (Fig. 6) and abundant crab burrows are found within these deposits, and provide a good indicator of palaeo-sea level.

The Futaisi Member is overlain by the Dabb'iya Member. The contact between the two is commonly indistinct, but there is abundant evidence in both Qatar and Abu Dhabi that the Futaisi Mem-

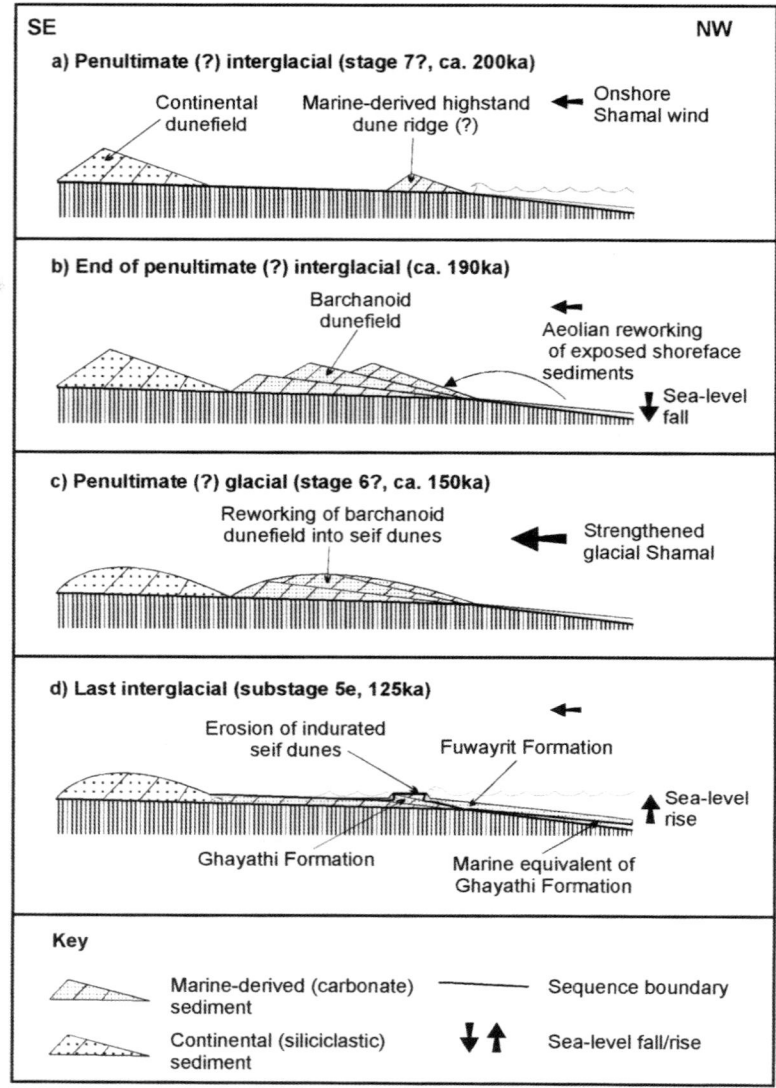

Fig. 5. Schematic illustration of the depositional model for the Ghayathi Formation in Abu Dhabi (from Williams & Walkden 2001).

ber underwent subaerial exposure and lithification before deposition of the Dabb'iya Member. This evidence includes a lag-covered, pot-holed contact, neptunean dykes, and reworked blocks of marine limestone found within the base of the Dabb'iya Member (Fig. 7). The most striking exposure-related features are found in Abu Dhabi. These are abundant circular pits, up to 1 m in diameter, and up to 1 m deep (Fig. 8a). We believe that they formed through karstic dissolution, linked to the position of large plants. Similar pits are created where trees grow in Pleistocene limestone in Florida, and the formation of 'solution pipes' in Bermuda has been attributed to the stemflow influence of trees (Herwitz 1993). Comparable features are fairly common in cyclic successions in other parts of the geological record (e.g. Walkden & De Matos, 2000).

Where the Dabb'iya Member directly overlies the Ghayathi Formation, the contact is highly erosive, and overlain by a lag of aeolianite clasts. Channels with a vertical relief of up to 2 m have elevation are cross-bedded carbonate sands, deposited in a foreshore setting. These provide a good indication of the highest sea level attained. In Abu Dhabi the Dabb'iya Member is dominated by skeletal grainstones, whereas in northern and eastern Qatar oolites predominate.

The lowermost Dabb'iya Member sediments preserved in Abu Dhabi consist of a rather perplexing facies, which is found infilling the pits within the Futaisi Member (Fig. 8a). It comprises a poorly sorted, structureless conglomerate of marine pebbles and bioclasts, floating in coarse bioclastic sand. The larger clasts are dominated by lithified burrow-fragments, with echinoids, bivalve and gastropod shells, corals and rhodoliths also abundant. Barnacles and serpulids encrust some clasts. Two interpretations for the origins of this conglomerate have been considered (Williams 1999): it represents either a high-energy transgressive lag, formed through a reworking of the nearshore sea bed, or possibly a tsunami deposit. Both hypotheses will be discussed more fully elsewhere.

The Al Wusayl Member is found only in Qatar. It consists of carbonate aeolianites composed mainly of ooids, conformably overlying the uppermost foreshore deposits of the Dabb'iya Member. In places a beach-backshore dune profile can be recognized. The dunes are preserved as coastal ridges up to 20 m high, and form the most voluminous Pleistocene deposits in Qatar. They provide a good example of 'highstand aeolianites' (Fig. 9), and seem to have become rapidly lithified shortly after deposition (Williams & Walkden 2001). The

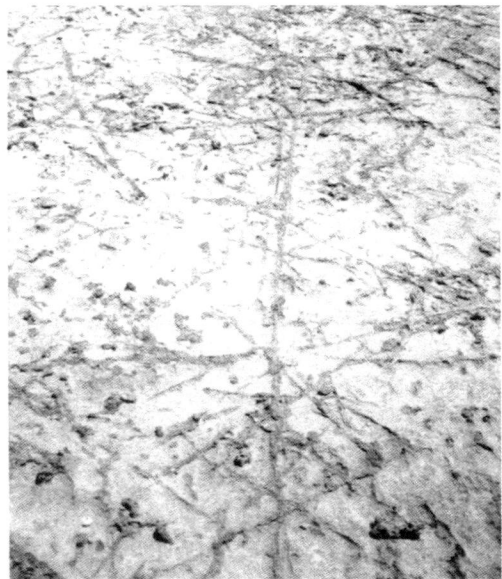

Fig. 6. Mangrove traces, Futaisi Member, Fuwayrit Formation (isotope substage 5e), Futaisi Island, Abu Dhabi. Field of view along base of frame c. 2.5 m.

been eroded into the aeolianite, and angular blocks of up to 2 m across, interpreted as collapsed cliff scree, may be found.

The Dabb'iya Member reaches a maximum elevation of 6 m above present sea level, and consists of shallow subtidal and littoral deposits (Williams 1999). The deposits with the highest

Fig. 7. Block of reworked Futaisi Member at base of Dabb'iya Member (both isotope substage 5e), Dabb'iya peninsula, Abu Dhabi.

Fig. 8. (a) Circular pits in the top surface of the Futaisi Member, filled with conglomerate of the Dabb'iya Member (both isotope substage 5e), Futaisi Island, Abu Dhabi. (b) Date palm eroded from beach sands, Abu Dhabi beach. The bowl-shaped depression formed by the roots is of very similar dimensions to many of the pits shown in (a). Note hose for irrigation.

aeolianites of the Al Wusayl Member have undergone extensive meteoric diagenesis, leading to the creation of up to 35% secondary porosity.

Correlation with offshore sediments

Offshore, the Fuwayrit Formation (sequence 2 of Williams (1999)) consists of a diverse range of submarine carbonate sediments. On Umm Shaif shoal oolitic and coral reef deposits predominate. They have been found in core at a minimum depth of 15 m below present sea level, and reach a maximum thickness of 29 m. A subaerial exposure surface has been recognized near the top of the formation. Facies patterns indicate that this surface does not correspond to the boundary separat-

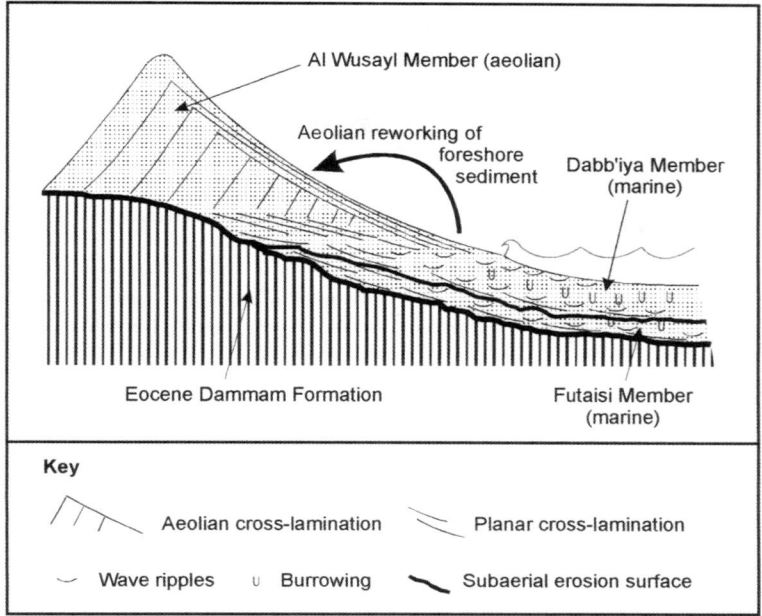

Fig. 9. Schematic model for the deposition of the Fuwayrit Formation in Qatar during the peak of the last interglacial (Sangamon; oxygen isotope substage 5e; from Williams & Walkden 2001).

ing the Futaisi and Dabb'iya Members, but occurred later. There is no evidence for subaerial exposure during the peak of the highstand.

Although there are no onshore submarine deposits equating to the sequences underlying the Fuwayrit Formation, the coastal aeolianites of the Ghayathi Formation are dominated by marine-derived sediments. These are likely to have been sourced from the equivalent of offshore sequence 3 of Williams (1999), consisting of submarine carbonates and sabkha evaporites. This sequence has been seen in core to a minimum depth of 27 m below present sea level, and to a maximum thickness of 9 m.

Age of Pleistocene deposits

Only four absolute ages have been reported from Pleistocene sediments in the southern Arabian Gulf (although more dates have been obtained from the unconsolidated clastic dune sands). These were all obtained by luminescence dating of Ghayathi Formation aeolianites. Two of these dates (45 and 130 ka) are from unspecified localities in the Emirates, reported by Glennie & Gokdag (1998). The others (64 and 99 ka) (Juyal et al. 1998) are from a site c. 25 km from the coast of western Abu Dhabi. Attempts to date deposits from nearby sites in Abu Dhabi (Hadley et al. 1998) have indicated that they were depos-

ited before 160 ka. The variability of these dates gives an indication of the chronostratigraphic complexity of the Ghayathi Formation.

Dating of sediments from the Fuwayrit Formation is hampered by the extensive diagenetic alteration they have suffered. As a result of this, most of the original carbonate material has been leached, or has been contaminated by younger material. Luminescence dating may provide possibilities, but this is inhibited by the paucity of clastic material present. So far, little work has been published on luminescence dating of marine carbonates. Attempts to gain optically stimulated luminescence (OSL) ages from the Fuwayrit Formation of Qatar, Abu Dhabi and offshore proved unsuccessful (Williams 1999).

Five samples were chosen from the Dabb'iya Member for radiocarbon dating (SRR-6248, SRR-6249, AA-30840, AA-30841, AA-30842). Samples SRR-6248 and SRR-6249 were dated at the NERC Radiocarbon Laboratory in East Kilbride, Scotland. The other samples were prepared to graphite in East Kilbride and then analysed by ^{14}C AMS at the University of Arizona NSF facility. All gave ages of between $29\,250 \pm 340$ and $33\,420 \pm 550$ radiocarbon years (effectively years before 1950) (Table 1). We do not believe that these ages represent the true age of the samples, but are a result of diagenetic contamination and alteration. It has been shown that only a few per

Table 1. *Radiocarbon age dates of Arabian Gulf samples*

Sample	Nature	Locality	Member	Age (a bp)
SRR-6248	Coral	Futaisi Island	Dabb'iya	30 670 ± 225
SRR-6249	Red algae	Futaisi Island	Dabb'iya	30 490 ± 260
AA-30840	Barnacle	Futaisi Island	Dabb'iya	31 570 ± 470
AA-30841	Barnacle	Near Doha	Dabb'iya	29 250 ± 340
AA-30842	Barnacle	Near Doha	Dabb'iya	33 420 ± 550

cent contamination by young material can give a very old sample a radiocarbon age in the range of 20–45 ka (Olsson 1968; Burr *et al.* 1992). The ages do, however, prove that the samples are not of Holocene age.

Correlation with other stable platform deposits

In the absence of direct age dates, it has been necessary to infer the ages of the Pleistocene sediments in the southern Arabian Gulf, by a comparison of the stratigraphy with deposits described from other parts of the world (Williams 1999).

The Fuwayrit Formation, stratigraphically the youngest pre-Holocene sediments preserved in the region, can be directly correlated with deposits that date from the last interglacial (oxygen isotope substage 5e). In Bermuda and the Bahamas, two sequences were deposited during this time, reaching respective elevations of 4 m and *c.* 6 m above present sea level (Hearty & Kindler 1995). In the Mediterranean, substage 5e deposits contain two sequences of between 5 and 7 m above present sea level (Hearty 1986; Kindler *et al.* 1997). Coral reefs dated to substage 5e are found at 2 m and 7.5 m above sea level in the Loyalty Islands (Marshall & Launay 1978). On Hawaii, substage 5e deposits have been correlated with shoreline notches cut in lithified dunes at 6.7 and 8.2 m elevation (Ku *et al.* 1974; Sherman *et al.* 1993). Substage 5e deposits from the eastern coast of the USA are found up to 14 m above present sea level, although these may have been uplifted somewhat (Hollin & Hearty 1990; Toscano & York 1992). In Australia, deposits dated to the last interglacial indicate that sea levels were in the range of 2–8 m above present level during substage 5e (Chappell 1987; Nott 1996).

If it is accepted that the Fuwayrit Formation was deposited during the last interglacial, then logic suggests that the underlying sequence (sequence 3 of Williams (1999)) was deposited during the penultimate interglacial (late oxygen isotope stage 7). It is very likely that the Ghayathi Formation aeolianites were largely sourced from these sediments. In Bermuda and the Bahamas, deposits from the late stage 7 interglacial are found up to 2.5 m above present sea level (Hearty & Kindler 1995).

Pleistocene sea levels

All of the above regions show evidence for maximum sea levels of between +2 and +14 m elevation during substage 5e, dating between 130 and 117 ka. With the exception of Australia, they all record unequivocal evidence for two highstands, separated by a regression, during which sea level fell below the present level. In all of the localities the second sea-level peak was higher than the first. If the Fuwayrit Formation is correlated with these deposits, sedimentary evidence shows that sea levels in the Arabian Gulf were higher than at present on at least two occasions during the last interglacial, peaking at *c.* 1.5 and 6 m above present sea level. This is consistent with other low-latitude sites around the world. Umm Shaif shoal remained submerged during this time, indicating that sea level remained higher than 24 m below the present level during exposure of the Futaisi Member.

A large body of evidence exists to suggest that two later highstands occurred in isotope stage 5, during substages 5c and 5a. Of these, the later (substage 5a) is believed to have had the higher sea level, with deposits from Bermuda and the Bahamas (Hearty & Kindler 1995) and Florida (Ludwig *et al.* 1996) indicating sea levels close to or higher than at present. There is no evidence for sea levels higher than at present during late stage 5 in the southern Arabian Gulf. However, at least one late highstand is recorded in core sediments from offshore Abu Dhabi. An exposure surface near the top of the Fuwayrit Formation indicates that sea level fell to lower than 23 m below the present level on Umm Shaif shoal, before rising to at least 17 m below the present level. Tentative interpretation of sedimentary facies indicates that sea levels may have reached 7 m below present level during this transgression (Williams 1999). It is not known whether this highstand corresponds to substage 5c or 5a.

No evidence for stage 7 sea levels higher than at present has been seen in the Arabian Gulf. It is possible, however, that such deposits could have been removed by deflation or erosion. Facies analysis of offshore core sediments indicates that sea level reached at least 15 m below present level during the deposition of sequence 3, and may have been higher.

Evidence for a Holocene highstand in the southern Arabian Gulf

Evidence for Holocene sea levels higher than at present is widespread and compelling in the southern Arabian Gulf (Table 2).

In Qatar, Holocene highstand deposits are generally preserved as beachrock. Taylor & Illing (1969) described cemented beach sands from the northwestern coast, standing between 1.5 and 2.5 m above present sea level, which they dated to between 3930 ± 130 and 4340 ± 180 a bp. Vita-Finzi (1978) described raised beachrock deposits of up to 2 m elevation from several localities in Qatar. These date from between 4690 ± 80 and 5830 ± 70 a bp.

In the Abu Dhabi sabkha, beach ridges are found 3–4 km inland from the intertidal algal flats. They stand c. 1–2 m above the adjacent sabkha, and are easily picked out on satellite photographs. These features have been described by several workers (e.g. Evans et al. 1969, 1973; Kendall & Skipwith 1969; Kirkham 1997), and have been dated to between 3465 ± 173 and 4191 ± 193 a bp (Evans et al. 1973). Kenig (1991) studied a canal cutting in the sabkha SW of Abu Dhabi island, and was able to map the sediments deposited at the end of the post-glacial transgression. He recorded a transgressive and a regressive phase, bracketed by intertidal algal sediments. At the peak of the transgression 'washover fans' of coarse shelly sand were deposited 2 m above present sea level. These were dated at 5400 ± 126 and 5110 ± 167 a bp. A shell sample from lagoonal sediments gave a date of 4280 ± 186 a bp, whereas another shell sample collected from the regressive algal layer, at approximately present sea level, was dated at 1580 ± 186 a bp. On Zabbut Island, in the west of Abu Dhabi, Williams (1999) observed a stranded mussel bed at 1 m above high-tide level. Two *in situ* specimens of the lower intertidal bivalve *Barbatia* were dated at 2485 ± 50 and 2315 ± 55 a bp.

The above evidence indicates that following the post-glacial (Flandrian) transgression, Holocene sea level in the southern Arabian Gulf peaked at 1–2 m above present level at c. 5500 a bp. It did not return to its present position until after 2300 a bp. As previously argued with regard to the late Pleistocene (oxygen isotope substage 5e) highstand, we see no reason to attribute this Holocene highstand record to local neotectonic change. Amongst a range of variable data, Pirazzoli (1991, 1996) recorded numerous sites in the Arabian Sea, Indian Ocean and NW Pacific where a sea-level high of similar age, magnitude and pattern has been recorded. Elsewhere, well-documented comparable case studies in relatively stable areas include those by Angulo et al. (1999) for southern Brazil (a high of up to 2.1 m c. 5410 a bp) and Beaman et al. (1994) for NE Australia (a high of up to 1.7 m c. 5660 a bp). These data support a

Table 2. *Reported ages and elevations of Holocene sediments found above present sea level in the southern Arabian Gulf*

Locality	Elevation (m)	Age (a bp)	Reference
NW Qatar	1.5–2.5	3930 ± 130	Taylor & Illing (1969)
NW Qatar	1.5–2.5	4200 ± 200	Taylor & Illing (1969)
NW Qatar	1.5–2.5	4340 ± 180	Taylor & Illing (1969)
Western Qatar	1.5	5370 ± 80	Vita-Finzi (1978)
NE Qatar	1.7	4690 ± 80	Vita-Finzi (1978)
Eastern Qatar	2	5830 ± 70	Vita-Finzi (1978)
Abu Dhabi sabkha	1–2	3465 ± 173	Evans et al. (1973)
Abu Dhabi sabkha	1–2	3948 ± 185	Evans et al. (1973)
Abu Dhabi sabkha	1–2	4191 ± 193	Evans et al. (1973)
Mussafah	2	5400 ± 126	Kenig (1991)
Mussafah	2	5110 ± 167	Kenig (1991)
Mussafah	1–2	4280 ± 186	Kenig (1991)
Mussafah	0	1580 ± 186	Kenig (1991)
Zabbut Island	1	2485 ± 50	Williams (1999)
Zabbut Island	1	2315 ± 55	Williams (1999)

case that the mid-Holocene highstand in the southern Arabian Gulf represents a widespread eustatic effect and that the area is tectonically stable.

Palaeowinds and palaeoclimates in the southern Arabian Gulf

The Arabian Gulf region is located within the subtropical trade wind belt, and is at present dominated by the northwesterly 'shamal' wind, strongest during the winter months (Fig. 1). This traces a clockwise path across Arabia, a pattern that is reflected in the dune morphologies of the peninsula's desert regions (Glennie 1998). During glacial periods, an increase in the size of the polar climate cells results in an intensification in the strength and duration of the trade winds. This also has the effect of squeezing the lower-latitude climate cells towards the equator, leading to a shift in the direction of the shamal wind. Most of the southern Gulf currently falls within the hyperarid zone in the UNESCO (1979) classification scheme of the arid lands of the world. This has not always been the case, however. Evidence of pluvial episodes has been reported from the Arabian desert (McClure 1976; Wood & Imes 1994) and from the northern Gulf (Uchupi et al. 1999). It has been hypothesized that increased influence of the SW monsoon during warm interglacials may have led to wetter climates in the Arabian Gulf region (Glennie 1998). These palaeoclimatic changes are reflected in the Pleistocene deposits preserved in the southern Arabian Gulf.

The Ghayathi Formation was deposited during falling sea level, following the (?)penultimate interglacial (late oxygen isotope stage 7). As the sea floor was exposed, carbonate sediments were reworked into a barchan dunefield by winds from the NNW (Fig. 5). An arid climatic regime allowed these dunes to remain unconsolidated, and they were later remodelled into seif dunes. The changing dune morphologies indicate a change in direction of the prevailing wind, from NNW to WNW, accompanied by an increase in maximum wind speeds. These changes are thought to reflect the onset of glaciation (Glennie 1998). The Ghayathi Formation was later lithified before transgression.

The Fuwayrit Formation, exposed on the present-day coastline, consists of coastal deposits from the last interglacial. Thus a useful comparison can be made between the sedimentary environments that existed during the last interglacial and those that are currently found in the region. In general, these are found to be very similar; however, a few notable features indicate that both the wind direction and the amount of precipitation may have varied. Both of these are well illustrated by the deposits found in Qatar.

One of the most notable differences between deposits of the Fuwayrit Formation and present-day sediments on the coast of Qatar is that the Fuwayrit Formation consists largely of ooids. On the present-day Arabian Gulf coastline ooid build-ups are primarily located in tidally dominated settings on windward shorelines. The most striking examples of these are the eastern Abu Dhabi tidal deltas, described by Loreau & Purser (1973). This is in marked contrast to the eastern coast of Qatar today, where the dominant current regime is one of longshore drift (Shinn 1973; Fig. 10), and ooids are not actively forming. Indeed, in many areas sediment production is so low that extensive hardgrounds have formed on the sea floor (Shinn 1969; Williams 1999). During the peak of the last interglacial, ooids were so abundant in eastern Qatar that aeolian dune ridges of 20 m height built up along the coast (the Al Wusayl Member) (Fig. 9), comparable with those found today on barrier islands adjacent to the Abu Dhabi tidal deltas (Evans et al. 1973).

These dune ridges provide another contrast with the present-day sedimentary regime. The only significant present-day aeolian accumulations found in Qatar are siliciclastic dunes located in the SE of the peninsula (Fig. 10). These are a remnant of the sand sea that covered southern Qatar during the last glacial, and they are at present being deflated into the sea by the northwesterly shamal (Shinn 1973). No significant aeolianites are accumulating today in Qatar. As well as being almost exclusively located on the eastern coast of Qatar (although Vita-Finzi (1978) reported deposits from the eastern side of Ras Abaruk peninsula in the west; Fig. 10), palaeocurrent directions from the Al Wusayl Member indicate that they were deposited under the influence of northeasterly winds (Fig. 11). The prevalence of winds from the NE would also account for the abundant ooid production, at present associated with tidal, windward coastlines. Further evidence for a tidally dominated coastline is found near Doha (Fig. 1), where coarse tidal bars are found within the Fuwayrit Formation (Williams 1999).

Another interesting feature of the Fuwayrit Formation in Qatar is the *in situ* preservation of the Al Wusayl Member palaeodunes. In contrast to the aeolianites of the Ghayathi Formation, deposits of the Al Wusayl Member were sufficiently well lithified to avoid remobilization and deflation when sea level dropped. This may indicate that they were deposited under more humid conditions than the Ghayathi Formation.

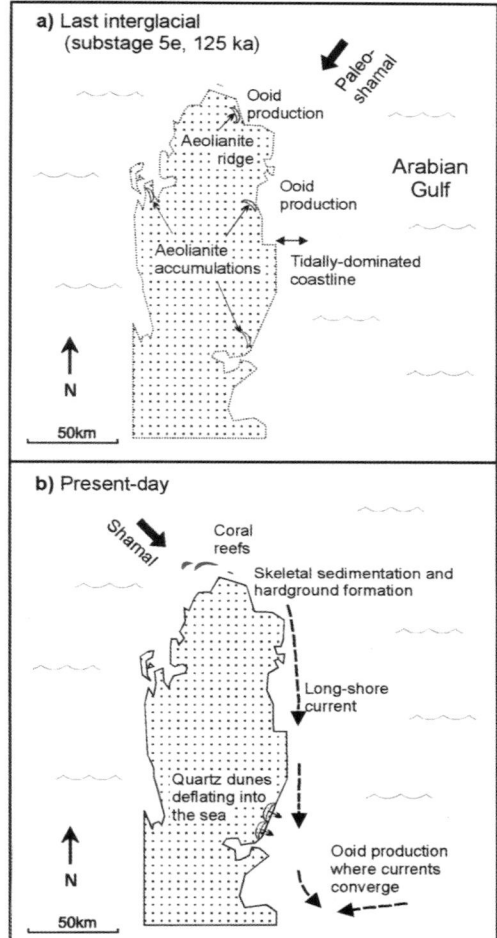

Fig. 10. Schematic illustration of the principal differences in the sedimentary environments of eastern Qatar during the last interglacial (substage 5e) compared with those of the present day.

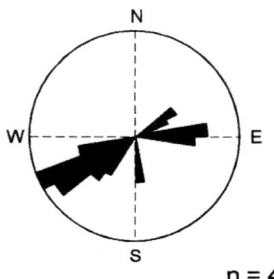

Fig. 11. Foreset measurements from the aeolian Al Wusayl Member, Fuwayrit Formation, eastern Qatar. Only surfaces of >20° dip have been plotted.

However, the paucity of rootlets within sediments of the Al Wusayl Member indicates that the climate was semi-arid at best (Williams & Walkden 2001).

Compelling evidence for a more pluvial episode during the last interglacial is provided by the pits found in the top surface of the Futaisi Member in Abu Dhabi (Fig. 8a). If these pits were formed through the influence of plant stems, they indicate a far higher density of plant life than is supportable under the current climate. The regular spacing of the pits is reminiscent of a plantation or mature forest. Similar stands of date palms are found at present in the region (Fig. 8b), but these are artificially irrigated. Such a density of plants would indicate a significantly more humid climate than is found today in the southern Arabian Gulf.

Conclusions

(1) The southern Arabian Gulf is at present the site of extensive carbonate sedimentation, as was the case during Pleistocene interglacial highstands. During glacial lowstands the basin was subaerially exposed, and aeolian sedimentation predominated. Most of the southern Arabian Gulf floor is underlain by Quaternary marine carbonates, and scattered outcrops may be found onshore. Quaternary aeolianite deposits are fairly widespread in Abu Dhabi.

(2) Pleistocene sediments found onshore in the southern Arabian Gulf belong to three formations: the Ghayathi Formation, the Aradah Formation, and the Fuwayrit Formation. Offshore, at least six Quaternary sequences are present within the uppermost 50 m of sediment. The uppermost Pleistocene sequence represents the offshore equivalent of the Fuwayrit Formation. Otherwise, there are no correlative submarine deposits preserved onshore.

(3) The Ghayathi Formation consists of mixed carbonate–clastic aeolianites, found in Abu Dhabi and Saudi Arabia. They occur up to 80 km inland in Abu Dhabi, where they are by far the most voluminous Pleistocene sediments. The Ghayathi Formation provides a good example of 'regressive aeolianites', deposited through reworking of coastal and shallow marine sediments during and after sea-level fall. They were originally deposited as barchan dunes, but were later reworked into seif dunes under conditions of increasing wind speed, accompanied by a change in wind direction from NNW to WNW. The Aradah Formation consists of continental sabkha-type deposits found overlying the Ghayathi Formation in southern Abu Dhabi.

(4) The Fuwayrit Formation consists of shallow marine and marine-derived aeolian carbonates, preserved in coastal localities in Abu Dhabi,

Qatar, Bahrain, Saudi Arabian and Kuwait. Onshore, it is composed of three members, separated by subaerial exposure surfaces. These are, from base upwards: the Futaisi Member, shallow marine deposits preserved to a maximum elevation of 1.5 m above present sea level; the Dabb'iya Member, shallow marine deposits preserved to 6 m above present sea level; and the Al Wusayl Member, oolitic aeolianites found only in Qatar, to a maximum elevation of 20 m above present sea level. Offshore, on Umm Shaif shoal, the Fuwayrit Formation has a thickness of 29 m. Here it is dominated by oolitic and coral reef sediments.

(5) No direct age dates have been acquired from Pleistocene marine or coastal deposits in the southern Arabian Gulf. It has therefore been necessary to infer the ages of these sediments by a comparison of their elevation and stratigraphy with deposits described from other parts of the world. This approach is valid in the absence of unequivocal or independent evidence for significant late Quaternary neotectonic change along the coastline, with the exception of local halotectonics. We propose that late Quaternary onshore deposits, and much of the offshore deposits, are in place and largely unaffected by tectonism.

(6) Comparison with deposits from other apparently stable platforms indicates that the Fuwayrit Formation was deposited during the last interglacial (oxygen isotope substage 5e). We believe that this is a valid assumption as (a) these represent the youngest pre-Holocene marine deposits, and (b) they are found at an elevation correlative with substage 5e deposits from other parts of the globe. Sedimentary evidence reveals that there were two highstands in the southern Gulf during this period. The first peaked at around 1.5 m above present sea level, with the second reaching 6 m. Offshore sediments indicate that sea level did not fall as far as 24 m below present level in the intervening regression. Following the second highstand, sea level fell to below 23 m below present level, before briefly rising again (late oxygen isotope stage 5). This later highstand probably peaked between 17 and 7 m below present level.

(7) The sequence underlying the Fuwayrit Formation was probably deposited during the penultimate interglacial (late oxygen isotope stage 7). It is also likely that the Ghayathi Formation aeolianites were largely sourced from this sequence. No evidence for stage 7 sea levels higher than present level has been seen in the Arabian Gulf. Facies analysis of offshore core sediments indicates that sea level reached at least 15 m below present level during this period.

(8) Given our case for minimal neotectonic change after c. 120 ka, we further invoke a mid-Holocene sea level higher than at present in the southern Arabian Gulf. Elevated Holocene sediments indicate that sea level peaked at 1–2 m above present level at c. 5500 a bp. It did not return to its current position until after 2300 a bp.

(9) Pleistocene deposits preserved in the southern Arabian Gulf provide a record of changing palaeowinds and palaeoclimates. Currently the region experiences a hyper-arid to arid climate, with facies patterns dominated by the northwesterly shamal wind. The Ghayathi Formation was originally deposited under an arid climatic regime, which allowed the sediments to remain unconsolidated. The dunefield was later remodelled under conditions of increasing wind speed, with a change in wind direction from NNW to WNW. These changes are thought to reflect the onset of glaciation.

(10) Palaeocurrent directions from the Al Wusayl Member, combined with sedimentary evidence from the Futaisi and Dabb'iya Members, indicate that during the peak of the last interglacial the prevailing wind (the 'palaeo-shamal') blew from the NE. Compelling evidence for a pluvial episode during this period is provided by palaeokarstic pits found in the top surface of the Futaisi Member. These pits probably correspond to the former positions of trees or other large plants.

The authors would like to thank K. Glennie, A. Kirkham and M. Simmons for guidance in the field and stimulating discussion. A. Williams was in receipt of grant GT4/95/2/E from the Natural Environment Research Council. G. Walkden thanks the NERC Radiocarbon Laboratory under allocation 719.1297 for undertaking the dating of samples. We also thank ADMA for the generous availability of Pleistocene cores and data.

References

ANGULO, J.R., GIANNINI, P.C.F., SUGUIO, K. & PESSENDA, L.C.R. 1999. Relative sea-level changes in the last 5500 years in southern Brazil (Laguna–Imbituba region, Santa Catarina State) based on vermetid ^{14}C ages. *Marine Geology*, **159**, 323–339.

BEAMAN, R., LARCOMBE, P. & CARTER, R.M. 1994. New evidence for the Holocene sea-level high from the inner shelf, central Great Barrier Reef, Australia. *Journal of Sedimentary Research*, **64**, 881–885.

BOWER, A.S., HUNT, H.D. & PRICE, J.F. 2000. Character and dynamics of the Red Sea and Persian Gulf. *Journal of Geophysical Research*, **105**, 6387–6414.

BURR, G.S., EDWARDS, R.L., DONAHUE, D.J., DRUFFEL, E.R.M. & TAYLOR, F.W. 1992. Mass spectrometric ^{14}C and U–Th measurements in coral. *Radiocarbon*, **34**, 611–618.

CHAPPELL, J.M.A. 1987. Late Quaternary sea-level changes in the Australian region. *In:* TOOLEY, M.J. & SHENNAN, I. (eds) *Sea-level Changes*. Blackwell, Oxford, 296–328.

Evans, G., Schmidt, V., Bush, P. & Nelson, H. 1969. Stratigraphy and geologic history of the sabkha, Abu Dhabi, Persian Gulf. *Sedimentology*, **12**, 145–159.

Evans, G., Murray, J.W., Biggs, H.E., Bate, R. & Bush, P.R. 1973. The oceanography, ecology, sedimentology and geomorphology of parts of the Trucial Coast barrier island complex, Persian Gulf. *In:* Purser, B.H. (ed.) *The Persian Gulf: Holocene Carbonate Sedimentation and Diagenesis in a Shallow Epicontinental Sea*. Springer, Berlin, 233–278.

Glennie, K.W. 1993. Wind action and desert landscapes (Revision of Chapter 22). *In:* Duff, D. (ed.) *Holmes' Principles of Physical Geology*. Chapman & Hall, London, 407–504.

Glennie, K.W. 1998. The desert of southeast Arabia: a product of Quaternary climatic change. *In:* Alsharhan, A.S., Glennie, K.W., Whittle, G.L. & Kendall, C.G.St.C. (eds) *Quaternary Deserts and Climatic Change*. Balkema, Rotterdam, 279–292.

Glennie, K.W. & Gokdag, H. 1998. Cemented Quaternary dune sands, Ras Al Hamra housing area, Muscat, Sultanate of Oman. *In:* Alsharhan, A.S., Glennie, K.W., Whittle, G.L. & Kendall, C.G.St.C. (eds) *Quaternary Deserts and Climatic Change*. Balkema, Rotterdam, 109–116.

Hadley, D.G., Brouwers, E.M. & Brown, T.M. 1998. Quaternary paleodunes, Arabian Gulf coast: age and palaeoenvironmental evolution. *In:* Alsharhan, A.S., Glennie, K.W., Whittle, G.L. & Kendall, C.G.St.C. (eds) *Quaternary Deserts and Climatic Change*. Balkema, Rotterdam, 123–140.

Hearty, P.J. 1986. An inventory of last interglacial (*sensu lato*) age deposits from the Mediterranean Basin: a study of isoleucine epimerization and U-series dating. *Zeitschrift für Geomorphologie N.F, Supplementband*, **62**, 51–69.

Hearty, P.J. & Kindler, P. 1995. Sea-level highstand chronology from stable carbonate platforms (Bermuda and the Bahamas). *Journal of Coastal Research*, **11**, 675–689.

Herwitz, S.R. 1993. Stemflow influences on the formation of solution pipes in Bermuda eolianite. *Geomorphology*, **6**, 253–271.

Hollin, H. & Hearty, P.J. 1990. South Carolina interglacial sites and stage 5 sea levels. *Quaternary Research*, **33**, 1–17.

Juyal, N., Singhvi, A.K. & Glennie, K.W. 1998. Chronology and paleoenvironmental significance of Quaternary desert sediment in southeastern Arabia. *In:* Alsharhan, A.S., Glennie, K.W., Whittle, G.L. & Kendall, C.G.St.C. (eds) *Quaternary Deserts and Climatic Change*. Balkema, Rotterdam, 315–325.

Kendall, C.G.St.C. & Skipwith, P.A.D'E. 1969. Geomorphology of a recent shallow-water carbonate province: Khor Al Bazam, Trucial Coast, southwest Persian Gulf. *Geological Society of America Bulletin*, **80**, 865–892.

Kenig, F. 1991. *Sédimentation, distribution et diagenèse de la matière organique dans un environnement carbonaté hypersalin: le système lagune–sabkha d'Abu Dhabi*. PhD thesis, Université d'Orléans.

Kindler, P., Davaud, E. & Strasser, A. 1997. Tyrrhenian coastal deposits from Sardinia (Italy): a petrographic record of high sea levels and shifting climate belts during the last interglacial (isotopic substage 5e). *Palaeogeography, Palaeoclimatology, Palaeoecology*, **133**, 1–26.

Kirkham, A. 1997. Shoreline evolution, aeolian deflation and anhydrite distribution of the Holocene, Abu Dhabi. *GeoArabia*, **2**, 403–416.

Kirkham, A. 1998. Pleistocene carbonate seif dunes and their role in the development of complex past and present coastlines of the U.A.E. *GeoArabia*, **3**, 19–32.

Ku, T.L., Kimmel, M.A., Easton, W.H. & O'Neill, T.J. 1974. Eustatic sea level 120,000 years age on Oahu Hawaii. *Science*, **183**, 959–961.

Loreau, J.-P. & Purser, B.H. 1973. Distribution and ultrastructure of Holocene ooids in the Persian Gulf. *In:* Purser, B.H. (eds) *The Persian Gulf: Holocene Carbonate Sedimentation and Diagenesis in a Shallow Epicontinental Sea*. Springer, Berlin, 279–329.

Ludwig, K.R., Muhs, D.R., Simmons, K.R., Halley, R.B. & Shinn, E.A. 1996. Sea-level records at ~80 ka from tectonically stable platforms: Florida and Bermuda. *Geology*, **24**, 211–214.

Marshall, J.F. & Launay, J. 1978. Uplift rates of the Loyalty Islands as determined by ^{230}Th/^{234}U dating of raised coral terraces. *Quaternary Research*, **9**, 186–192.

McClure, H.A. 1976. Radiocarbon chronology of late Quaternary lakes in the Arabian desert. *Nature*, **263**, 755–756.

McClure, H.A. & Vita-Finzi, C. 1982. Holocene shorelines and tectonic movements in eastern Saudi Arabia. *Tectonophysics*, **85**, T37–T43.

Nott, J. 1996. Late Pleistocene and Holocene sea-level highstands in northern Australia. *Journal of Coastal Research*, **12**, 907–910.

Olsson, I.U. 1968. Modern aspects of radiocarbon dating. *Earth-Science Reviews*, **4**, 203–218.

Pirazzoli, P.A. 1991. *World Atlas of Holocene Sea-level Changes*. Elsevier, Amsterdam.

Pirazzoli, P.A. 1996. *Sea-level Changes: the Last 20,000 Years*. Wiley, Chichester.

Purser, B.H. & Seibold, E. 1973. The principal environmental factors influencing Holocene sedimentation and diagenesis in the Persian Gulf. *In:* Purser, B.H. (eds) *The Persian Gulf: Holocene Carbonate Sedimentation and Diagenesis in a Shallow Epicontinental Sea*. Springer, Berlin, 1–11.

Ridley, A.P. & Seeley, M.W. 1979. Evidence for Recent coastal uplift near Al Jubail, Saudi Arabia. *Tectonophysics*, **52**, 319–327.

Sherman, C.E., Glenn, C.R., Jones, A.T., Burnett, W.C. & Schwarcz, H.P. 1993. New evidence for two highstands of the sea during the last interglacial, oxygen isotope substage 5e. *Geology*, **21**, 1079–1082.

Shinn, E.A. 1969. Submarine lithification of Holocene carbonate sediments in the Persian Gulf. *Sedimentology*, **12**, 109–144.

Shinn, E.A. 1973. Sedimentary accretion along the leeward, SE coast of Qatar Peninsula, Persian Gulf.

In: PURSER, B.H. (eds) *The Persian Gulf: Holocene Carbonate Sedimentation and Diagenesis in a Shallow Epicontinental Sea.* Springer, Berlin, 179–191.

TAYLOR, J.C.M. & ILLING, L.V. 1969. Holocene intertidal calcium carbonate cementation, Qatar, Persian Gulf. *Sedimentology*, **12**, 69–107.

TOSCANO, M.A. & YORK, L.L. 1992. Quaternary stratigraphy and sea-level history of the U.S. middle Atlantic coastal plain. *Quaternary Science Reviews*, **11**, 301–328.

TUCKER, M.E. & WRIGHT, V.P. 1990. *Carbonate Sedimentology.* Blackwell Scientific, Oxford.

UCHUPI, E., SWIFT, S.A. & ROSS, D.A. 1999. Late Quaternary stratigraphy, paleoclimate and neotectonism of the Persian (Arabian) Gulf region. *Marine Geology*, **160**, 1–23.

UNESCO 1979. *Map of the world distribution of arid regions.* MAB Technical Note, 7.

VITA-FINZI, C. 1978. Environmental history. *In:* DE CARDI, B. (ed.) *Qatar Archaeological Report: Excavations 1973.* Oxford University Press, Oxford, 9–25.

WALKDEN, G.M. & DE MATOS, J.E. 2000. 'Tuning' high frequency cyclic platform successions using omission surfaces: Lower Jurassic of the U.A.E. and Oman. *In:* ALSHARHAN, A.S. & SCOTT, R.W. (eds) *Middle East Models of Jurassic/Cretaceous Carbonate Systems.* SEPM (Society for Sedimentary Geology) Special Publication, **69**, 37–52.

WALKDEN, G.M. & WILLIAMS, A.H. 1998. Carbonate ramps and the Pleistocene–Recent depositional systems of the Arabian Gulf. *In:* WRIGHT, V.P. & BURCHETTE, T.P. (eds) *Carbonate Ramps.* Geological Society, London, Special Publications, **149**, 43–53.

WILLIAMS, A.H. 1999. *Glacioeustatic cyclicity in Quaternary carbonates of the southern Arabian Gulf: sedimentology, sequence stratigraphy, paleoenvironments and climatic record.* PhD thesis, Aberdeen University.

WILLIAMS, A.H. & WALKDEN, G.M. 2001. Carbonate eolianites from a eustatically-influenced ramp-like setting: the Quaternary of the southern Arabian Gulf. *In:* ABEGG, F. (ed.) *Modern and Ancient Carbonate Eolianites: Sedimentology, Sequence Stratigraphy, and Diagenesis.* SEPM (Society for Sedimentary Geology) Special Publication, **71**, 77–92.

WOOD, W.W. & IMES, J.L. 1994. How wet is wet? Precipitation constraints on late Quaternary climate in the southern Arabian Peninsula. *Journal of Hydrology*, **164**, 263–268.

Varves, turbidites and cycles in upper Holocene sediments (Makran slope, northern Arabian Sea)

U. VON RAD[1], A. ALI KHAN[2], W. H. BERGER[3], D. RAMMLMAIR[1] & U. TREPPKE[4]

[1]*Bundesanstalt für Geowissenschaften und Rohstoffe (BGR), Stilleweg-2, D-30655 Hannover, Germany (e-mail: u.vonrad@t-online.de)*
[2]*National Institute of Oceanography, St-47, Block-1, Clifton, Karachi, Pakistan*
[3]*University of California San Diego, Scripps Institution of Oceanography, La Jolla, CA 92093-0524, USA*
[4]*Institut für Ostseekunde, PF 301161, D-16112 Rostock-Warnemünde, Germany*

Abstract: We have analysed two Late Holocene records, each about 5 ka long, consisting of varved sediments deposited in the oxygen minimum zone off Pakistan (upper continental slope off Ormara and west of Karachi). Varve counting was checked by accelerator mass spectrometry (AMS) ^{14}C dating. Detailed lithofacies analysis, ultra-high-resolution X-ray fluorescence scanning, flux rates from sediment traps and the lamina-by-lamina-analysis of a 5 year record (1993–1998) support our interpretation of the annual character of the laminae. Although the pelagic material is deposited throughout the year, most of it is apparently laid down during the high-productivity period of the late summer monsoon. During the winter (mainly mid-December to February–March) mainly light-coloured detrital material (siliciclastic material and reworked carbonate flour) is deposited, probably by river flood events. Event deposits include turbid-plume or suspensate deposits (light grey homogeneous to graded silty clay layers), which are being laid down at decadal or shorter intervals; they are explained by episodically strong river floods after heavy rains transporting mud-charged waters to the narrow shelf and onto the steep continental slope. Medium grey or reddish grey, graded and laminated silt turbidites originated from less frequent, unchannelized, low-density turbidity currents. In general, periods with thin varves (generally correlated with rare turbidites) are correlated with minima of detrital element ratios (TiO_2/Al_2O_3, Zr/Al_2O_3, K_2O/Al_2O_3), especially in Period I (c. 5600–4700 a bp), suggesting a climate with reduced precipitation and river runoff (possibly winter monsoon dominated). Period II (4700–2600 a bp) is characterized by comparatively thick varves documenting a generally wet period (with possibly summer monsoon domination). During Period III (2600–1000 a bp) a gradual thinning of varves and a decrease of turbidite abundance (thickness) per century is interpreted as a gradual decline of precipitation and river runoff leading finally to dry conditions from 1600 to 1000 a bp. The sequence of cycles detected by autocorrelation and standard Fourier analysis seems to contain a large proportion of multiples of the lunar perigee cycle (4.425 a, 8.85 a) and the lunar (half) nodal cycle (9.3 a, 18.6 a). Our test for cyclicity in the series of varve thickness (varve cycles) and of abundance of turbidites (turbidite cycles) detected prominent high-frequency cycles. Some cycles of varve thickness match the cyclicity of turbidite frequency. We also detected the presence of a 1470 a cycle, previously reported from the glacial-age Greenland ice record.

The continental slope off Pakistan is characterized by a conspicuous, stable and expanded oxygen minimum zone (OMZ) which influenced sedimentation by preventing bioturbation, especially during late Holocene times (von Rad *et al.* 1995, 1999b). Hence the upper slope sediments off Pakistan between 250 and 1000 m water depth show well-preserved lamination (von Stackelberg 1972; Schulz *et al.* 1996). Annually laminated or 'varved' marine sediments are excellent archives of past climate record (Hughen *et al.* 1996; Kemp 1996; Overpeck 1996; Schimmelmann & Lange 1996; von Rad *et al.* 1999a). During SONNE cruises SO 90 (1993) and SO 130 (1998) several varved sediment cores were collected along the Pakistan margin (Schulz *et al.* 1996; von Rad *et al.* 1998). The sediments recovered from the OMZ west of Karachi (core SO 90-56KA; see Fig. 1) are well laminated and >5000 varves have been counted and measured by von Rad *et al.* (1999a). This continuous Late Holocene record shows: (1) seasonal variability of dark and light laminations, which

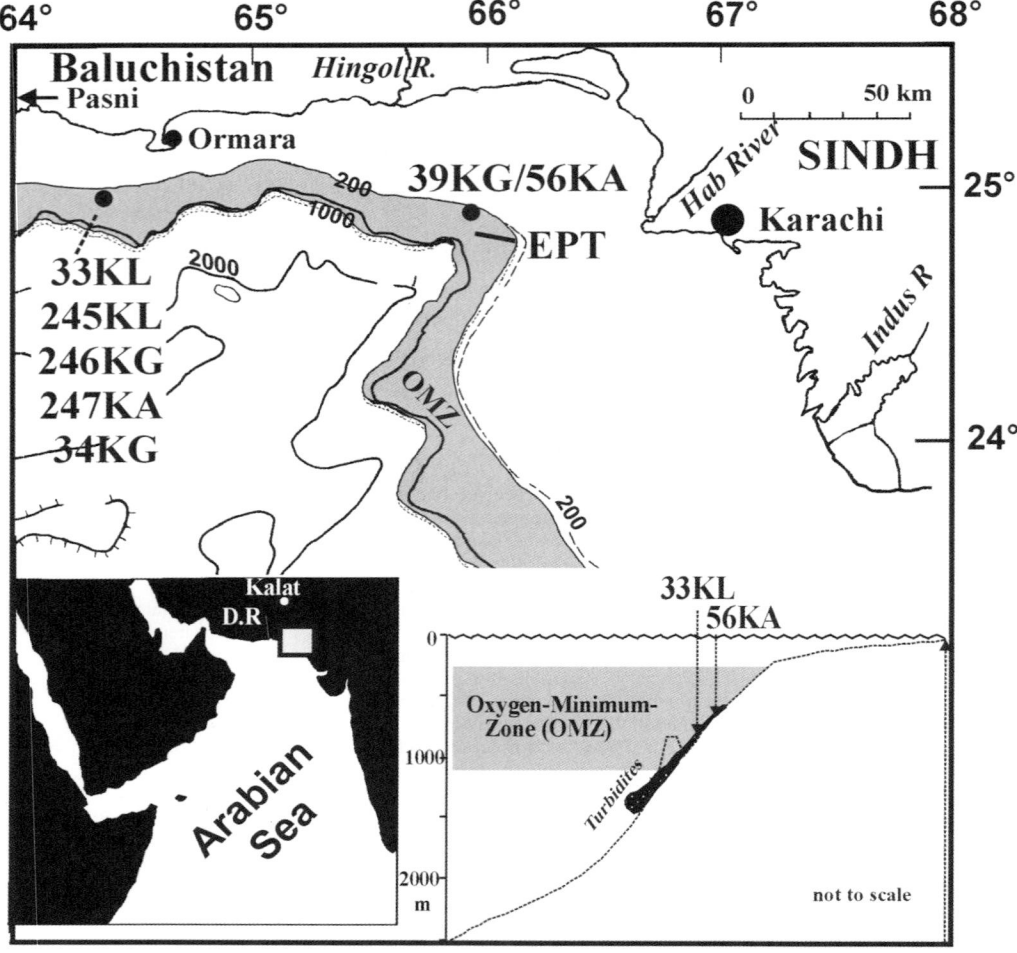

Fig. 1. Area of investigation off Pakistan with location of cores studied. Insets show location in Arabian Sea and a generalized profile across the OMZ off Karachi. D.R., Dasht River; EPT, Eastern PAKOMIN Trap.

is due to variations of sediment supply and/or productivity during the annual monsoon cycle; (2) decadal–centennial (–millennial) climatic changes within the Late Holocene period, which are shown by variable terrigenous silt and clay supply to the slope, explained by variations in precipitation and river runoff in the Makran hinterland. Lückge et al. (2001) studied the same high-resolution core (SO 90-56KA) to infer the monsoon-driven 'moisture history' in the northeastern Arabian Sea during the past 5 ka using three types of independent proxies: (1) varve thickness; (2) stable isotope ratios; (3) inorganic geochemical composition.

The main objectives of this study are: (1) to describe the sedimentological and geochemical variability within the varved hemipelagic sediments and event deposits and discuss their origin; (2) to reconstruct monsoon-controlled climatic changes during late Holocene time (the last 5600 a) from high-resolution time series of laminated sediments deposited on the upper continental slope off Ormara and Karachi (Fig. 1); (3) to document the presence and origin (forcing) of cycles in the sediments, using spectra of varve thickness and turbidites.

Geological and oceanographic–meteorological setting

The on- and offshore Makran accretionary wedge belongs to one of most dynamic and rapidly changing active continental margins of the world. The Makran slope forms the seaward part of the

>500 km wide, folded accretionary prism, which extends inland into Baluchistan (Arthurton et al. 1982; White 1982; von Rad et al. 2000). The semi-arid hinterland of the Makran Ranges is characterized by tectonic uplift, earthquakes and tsunamis, as well as by rapid denudation, especially during episodic river floods after heavy winter rains and beach erosion during the summer monsoon (Snead 1993).

Tectonics

For the past million years, the accretionary process has caused dramatic vertical faulting and warping: marine terraces uplifted in the order of tens of metres to 100 m during the past 2 ka (near Pasni, Ormara or west of the Hingol River mouth) document dramatic uplift, which takes place mainly during strong destructive earthquakes, such as the 1945 earthquake near Pasni (Snead 1993). Often the earthquakes were followed by tsunamis causing extensive flooding and rapid coastal erosion. The coastal Makran of Iran and Pakistan is characterized by several raised terraces leading to coastal plain emergence and progradation, seaward shift of the coastline and to upward shoaling (regressive) slope to inner shelf to beach cycles (Harms et al. 1984).

Morphology, climate, precipitation and erosion

As a result of the extreme insolational heating in desert conditions, the mechanical and chemical weathering of the sediments is accentuated and erosional processes are enhanced. During the strong swells of the SW monsoon season, especially during cyclones (commonly in April–June), the uplifted coastal terraces are rapidly eroded (Snead 1993). Torrential downpours during storms or cyclones cause excessive precipitation and flash floods. The recent erosion and redeposition of the un- or semiconsolidated clastic sediments is evidenced by the deep V-shaped canyons (e.g. that of the Hingol River), which have formed by rapid headward-cutting streams (Snead 1967). The studied cores are located on the upper Makran slope SSW of the Ormara peninsula (off the Basul River) and west of Karachi (SSE of the Hingol River). The Basul River (catchment c. 15 000 km^2) and the

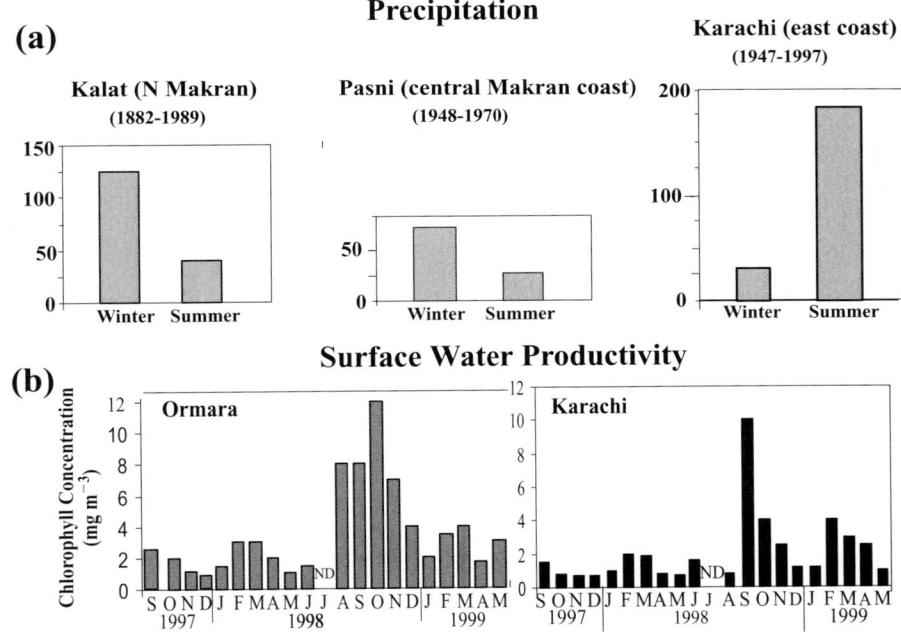

Fig. 2. (a) Seasonal precipitation record (mm) of Kalat (northern Makran), Pasni (central Makran coast) and Karachi in the east (winter: December–March; summer: May–October). Westward increasing importance of winter monsoon rains in the coastal Makran region, as compared with the strong prevalence of summer monsoon rains in Karachi, should be noted (data by Deutscher Wetterdienst 1999). (b) Surface water productivity (calculated in chlorophyll concentration), estimated from monthly averages from high-resolution ocean colour (pigment) maps from the SeaWifs satellite (September 1997–May 1999). Source: NASA Goddard Space Flight Center (1999).

>600 km long Hingol River (catchment >60 000 km^2) are perennial rivers transporting large amounts of suspended sediments episodically onto the continental shelf off Ormara. The active margin-type shelf off Ormara is exceptionally narrow (<20–30 km) and the shelf edge exceptionally shallow (only 45 m), as compared with a shelf edge of 125 m west of Karachi. The passive margin off the Indus delta is characterized by a 90 km wide shelf and a shelfbreak located at a water depth of 135 m. This indicates that the shelf off the convergent Makran margin is uplifted and almost eliminated by the accretionary process. Because most sediments bypass this narrow shelf, they reach the upper slope as turbid plumes or turbidity currents.

The depositional area of the uppermost Makran slope provides a complete record of past variations in monsoonal precipitation and fluvial discharge (von Rad et al. 1999a). The precipitation pattern in southern Pakistan is very different for (1) the Karachi–Indus delta area, with a strong maximum during the summer (June–September) associated with the SW monsoon (Fig. 2a), and (2) the semi-arid central Makran coast (Pasni), with rainfall in both seasons, dominated by winter rainfall (especially January–March; Fig. 2). West of the Pakistani–Iranian border and north of Quetta precipitation is mainly in the winter (e.g. in Kalat, Fig. 2a). The winter rains in the Makran coastal area are related to western depressions or Mediterranean cyclonic storms crossing Mesopotomia and Baluchistan. The Hingol and Basul rivers, which drain the eastern and northern Makran and are mainly characterized by winter rains (Snead 1993), are responsible for most fluvial sediments deposited off the eastern Makran coast. Although annual mean precipitation is relatively higher in Sindh than in the coastal areas of Baluchistan, the episodic heavy winter rains and associated river floods in the Makran can be catastrophic. During high floods turbid plumes have been observed along the Makran coast by satellite imagery. Shortly before our last expedition in April 1998 (SO 130) there was a tremendous flash flood in the Dasht River area (close to the Pakistani–Iranian border) and further east up to the Hingol River, costing the lives of >200 people. This flash flood occurred close to the end of the major El Niño year 1997–1998.

Oceanography

Primary productivity in waters above the Makran shelf and upper slope (Banse 1994) is due to coastal upwelling and/or advection of nutrient-rich waters from offshore Oman during the SW monsoon, with a second smaller maximum during February and March (Fig. 2b). Monthly averages of high-resolution ocean colour data from the Sea-Wifs Satellite (giving estimates of chlorophyll concentrations of the surface waters) show a strong maximum in September–October west of Karachi (location of core 56KA) and a similar maximum from August to September (SW monsoon) at the location of the Ormara cores (NASA Goddard Space Flight Center 1999). In both areas a smaller productivity peak is observed during February–March (Fig. 2b), indicating the influence of convective winter mixing during the NE monsoon.

Samples and methods

Samples and data

We studied two cores from the upper Makran slope between Ormara and Karachi (Fig. 1). As a result of their location at this tectonically active margin with a very narrow shelf and poorly vegetated mountain ranges in the hinterland, the Ormara cores are characterized by a higher rate of terrigenous deposition and much higher frequency of turbidites than core 56KA from the Karachi slope, 200 km further east (von Rad et al. 1999a). To obtain a continuous record of the sediments deposited during the past 5500 a, we used a stacked record of four cores, all taken from the same position (24°54'N, 64°17'E) and a water depth of about 660 m (Fig. 3): piston core SO 90-33KL, box core SO 130-246KG, piston core SO 130-245KL and a 5.6 m long Kastenlot (30 cm × 30 cm diameter) core SO 130-247KA). During the SO 130 cruise (1998) a new large-diameter (125 mm) piston core was used to obtain longer and less disturbed cores than the equivalent SO 90-33KL core from the previous cruise (90 mm diameter), which showed more compaction. Because of the very sticky nature of the cohesive carbonate-poor silty clays, the sediment recovery in core SO 130-247KA that is completely undisturbed was only 5.24 m. The varves in core 247KA are slightly thicker than in the piston core 245KL, indicating some compaction in the piston core. As the top 25 cm of the core 247KA are disturbed, we used the undisturbed box core 246KG for the near-surface sediment.

We were able to correlate the four Ormara cores exactly by identifying about 340 very distinct marker turbidites or event deposits (> 5.0 mm thick). For example, from the base of the box core 246KG (c. 37 cm bsf (cm below sea floor)) a marker layer (number 52) was identified in core 247KA (38 cm bsf); marker layer 143 in core 247KA (294 cm bsf) can be discovered in core 33KL at a core depth of 144 cm. In core 33KL the deepest marker layer, number 300 (about 830 cm bsf) was matched with the same marker layer in 245KL at 709 cm bsf.

All varve thickness and ^{14}C age data of core SO 90-56KA and the Ormara cores (33KL,

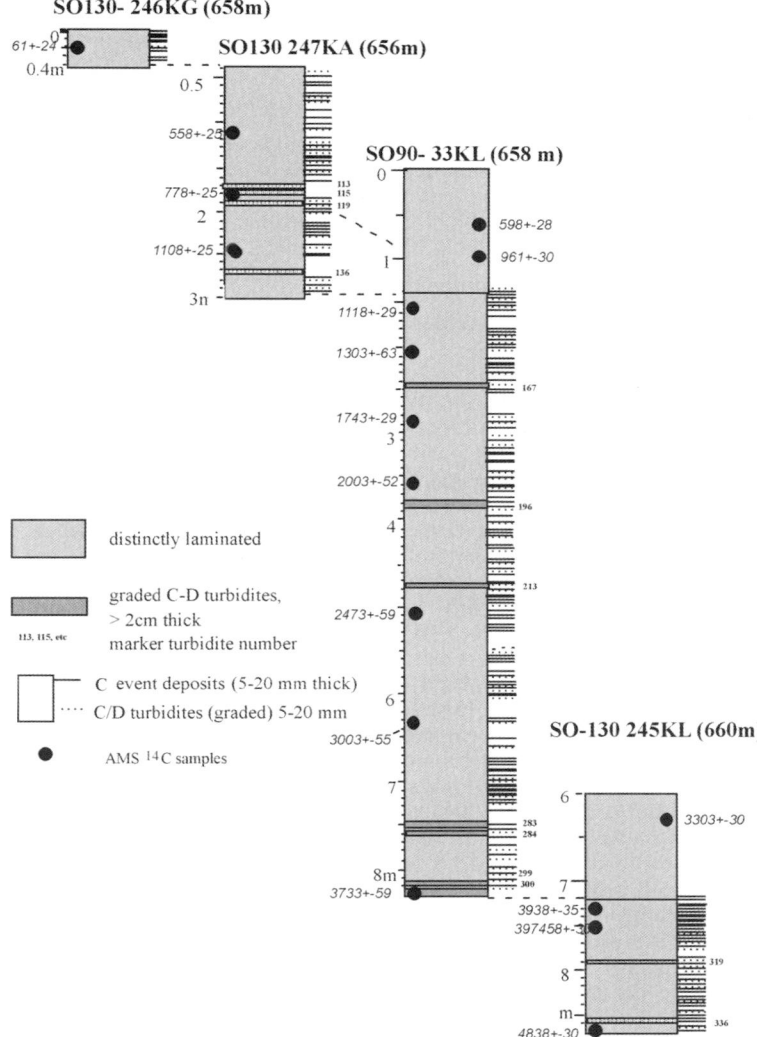

Fig. 3. Lithological profiles of the investigated sections of laminated cores 246KG, 247KA, 33KL and 245KL used for a stacked varve record of the upper slope sediments SW of Ormara (water depth c. 655 m). As a result of the abundance of turbidites not all individual turbidites >2 cm are indicated. ^{14}C age data calibrated to calendar years (Table 1) are shown in italics.

245KL, 246KG, 247KA) discussed in this paper were deposited in the PANGAEA data bank (http://www.pangaea.de/PangaVista?query=vonradu)

X-radiography, varve counting and varve thickness measurements

The cores were described, photographed and slabbed (8–10 mm thick) for X-ray study performed at the Bundesanstalt für Geowissenschaften und Rohstoffe using 5 mA, 50 kV and 2 min exposure time. Contact prints were then made for detailed studies and varve counting. The varves were counted and their thickness was measured with a tree-ring measuring device (Aniol 1983; von Rad et al. 1999a) under a binocular microscope at the Tree-ring Laboratory of the University of Göttingen. From cores SO 130-246KG, SO 130-247KA, SO 90-33KL and SO 130-245KL we counted the thickness of 5600 varve couplets (corresponding to 5600 'varve years before 1998'). We estimate the counting error as maximally 2–3% for the integrated count, that is, smaller than errors of accelerator mass spectrometry (AMS) ^{14}C age data. We allow for a probable error of 1-2% when discussing the significance of spectral lines. As all event deposits were measured together with one underlying varve, we reinserted the thickness of the arithmetic mean of the underlying plus the overlying hemipelagic varve, where an event deposit was eliminated.

High-resolution X-ray fluorescence scanning

To study the sub-millimetre geochemical variability of the individual laminae and check their seasonal nature, we scanned an undisturbed, Araldite-impregnated, polished section (originally used for thin-sectioning) of box core SO 90-39KG (30-40 cm; same position as core 56KA) with BGR's Itrax X-ray fluorescence (XRF) microscope (Cox Analytical Systems, Sweden) using scanning steps of 100 μm parallel and perpendicular to the lamination. The instrument is based on an Iso-Debeyeflex 3003 HV-Generator (Richard Seifert & Co.) and a long fine focus Philips X-ray diffraction (XRD) Mo tube (3 kW, 60 kV, Type PW 2275/20). The beam is focused by a glass monocapillary to 100 μm diameter. The energy-dispersive Si(Li) detector (Röntec X flash) has a 139 eV resolution at Mn K_α. It is mounted at 45° and has a nozzle that can be evacuated to reduce the air absorption for light elements. The elemental maps were run simultaneously at 45 kV at 30 mA, boosted by the 100 μm mono-capillary, 0.5 s detector accumulation time and transferred to a processing unit for image optimization. This was achieved by repeated multiplication and application of a mean filter. The elemental maps were rotated parallel to the optical scan and a profile with relative intensity of the elements Ca and Fe was obtained for 100 points per lamina. In addition, a Cox geoscanner prototype was used to obtain chemical patterns. This instrument is based on the same type of generation as the Itrax system, but the beam is focused by a slit capillary to 1000 μm × 50 μm and the step size used was 50 μm. We used an energy-dispersive, peltien-cooled drift chamber Röntec detector with 149 eV resolution at Mn K_α and an exposure time of 20 s per 50 μm. Scans of pure Araldite resin (without sediment) indicated zero counts for the elements Ca and Fe, proving that the resin is not influencing our compositional results. The noise at the scale of individual laminae is due to the fact that the integration of elemental counts (during scans parallel to the somewhat wavy laminae) might measure parts of the over- or underlying lamina, whereas scans perpendicular to the lamination might give blurred results, because lamina boundaries are not always strictly perpendicular to the surface of the thin-section.

Chronology of varved sediments

The Ormara cores were dated by varve counting (von Rad et al. 1999a), which was checked by 17 AMS ^{14}C dated samples. As there were only very few planktonic foraminifera in our samples, even if large amounts of sediment were washed, we did not pick G. ruber or G. sacculifer but used about 10 mg of mixed planktonic foraminifera for AMS ^{14}C dating. The samples were freeze-dried and cleaned in an ultrasonic bath. AMS ^{14}C ages were measured at the Leibniz Laboratory of Kiel University following standard procedures (Nadeau et al. 1997). Dates were calibrated and converted to calendar ages using assumed reservoir ages of 400 a (following Stuiver & Braziunas 1993, fig. 5B) and adding 48 a to reach calendar years before 1998, the 'year zero' of our varve counting (instead of 1950 used by standard radiocarbon analysis; see Table 1). For only the near-surface sample (at 20.4 cm) we used a reservoir age of 640 a determined at the varve-calibrated core 56KA (von Rad et al. 1999a).

Frequency analysis

To generate a 'varve series' for frequency analysis we used the following procedure (for more detail, see Berger & von Rad 2002).

(1) The raw varve-thickness data were cleaned by eliminating the event deposits ('suspensate deposits' and turbidites). We patched occasional gaps or disturbances by averaging the thickness of adjacent varves. The resulting thickness series was log-transformed, which had the effect of centring the mean within the range of variation. A few values greater or smaller than 2.5 SD from the mean (considered to be erroneous measurements) were replaced and averaged with neighbouring values. The resulting series were standardized to a mean of unity and an SD of 0.25. The overall trend was determined by smoothing the record using a 10-point boxcar 400 times. The difference between the original record and the trend is the residual, which was taken as the base for further analysis. A record with 4 a resolution was generated by using a five-point boxcar and resampling at 4 a intervals. Data are available for more exacting analysis from the PANGAEA data bank.

(2) A 'turbidite series' was generated by retaining all previously rejected layers (when cleaning the varve-thickness record) and assigning the value of unity to each (in the fashion of a bar code).

(3) Analysis of the time series was by autocorrelation and Fourier expansion (see Berger & von Rad 2002).

Table 1. AMS-^{14}C data of the stacked Ormara cores

Lab no.	Core	Depth (cm)	Cumulative depth (cm)	Forams	Size (μm)	Conventional ^{14}C age	Calibrated age (Stuiver & Braziunas 1993) calendar years before 1950	Reservoir age	Calendar years before 1998	Varve age (years before 1998)	Difference Varve/calendar years (% varve years)
KIA 7965	246 KG	22-26	20.40	mixed plankton	315-400	653±24		640	61	75±10	c. 0
KIA 7966	247KA	106-112	106	G menardi dominant	315->400	923±25	510	400	558	485±35	15
KIA 769	33KL	55-66	155	mixed plankton	>200	994±28	550	400	598	695±40	14
KIA 7968	247KA	183.5-188	184	mixed plankton	315-400	1180±25	730	400	778	825±25	6
KIA 770	33KL	95-106	205	mixed plankton	>200	1346±30	895	400	943	872±40	8
KIA 7969	247KA	248-254	251	mixed plankton	315->400	1515±25	1060	400	1108	1115±35	c. 0
KIA 771	33KL	146-156	305	mixed plankton	>200	1531±29	1070	400	1118	1286±40	13
KIA 121	33KL	201.5-211.5	358	mixed plankton	>150	1702±63	1255	400	1303	1555±50	16
KIA 772	33KL	295-305	457	mixed plankton	>200	2122±29	1695	400	1743	2050±50	15
KIA 122	33KL	364-374	531	mixed plankton	>150	2355±52	1955	400	2003	2531+-50	15
KIA 123	33KL	500-510	673	mixed plankton	>150	2751±59	2425	400	2473	2940±50	16
KIA 124	33KL	635-645	815	mixed plankton	>150	3179±55	2955	400	3003	3650±50	18
KIA 7971	245KL	626-634	915	G.sacculifer dominant	315->400	3388±30	3255	400	3303	4100±25	19
KIA 125	33KL	835-840	1028	mixed plankton	>150	3768±59	3685	400	3733	4590±50	19
KIA 7973	245KL	731-738	1051	G.sacculifer dominant	315->400	3935±35	3890	400	3938	4700±25	16
KIA 7975	245KL	750-760	1067	mixed plankton	315-400	3950±30	3910	400	3958	4800±50	18
KIA 7976	245KL	865-871	1181	mixed plankton	315-400	4575±30	4790	400	4838	5600±25	14

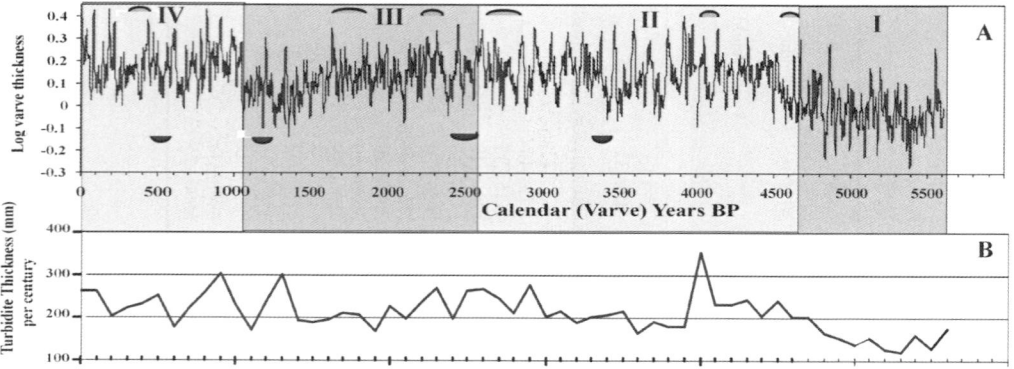

Fig. 4. Varve stratigraphy and turbidite frequency. (**a**) Varve thickness (despiked, turbidite-free, log-transformed; see 'Samples and methods' section) v. age in calendar years ('varve years'). Convex-upward symbols (grey) indicate maxima of TiO_2/Al_2O_3, Zr/Al_2O_3 and K_2O/Al_2O_3 (suggesting enhanced terrigenous input as a result of increased precipitation and fluvial input), convex-downward symbols (black) indicate minima of the same element ratios. (**b**) Turbidite thickness (mm) per century plotted v. varve years.

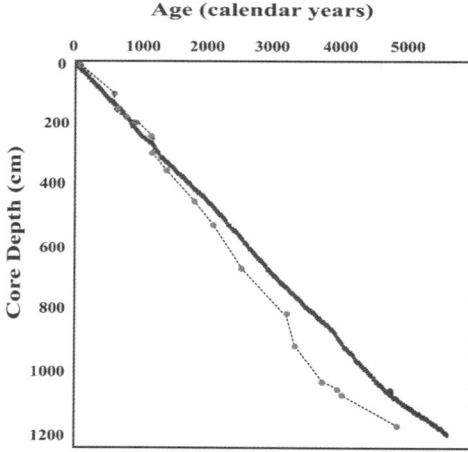

Fig. 5. Depth–age diagram for Ormara stack: continuous line indicates cumulative thickness of hemipelagic varves plus turbidites (varve curve) plotted v. age in varve years. Calibrated AMS ^{14}C ages (●; see Table 1) are plotted v. core depth in the same diagram. Varve data indicate a very uniform, high sedimentation rate of 2.3 mm a^{-1} (2.3 m ka^{-1}). The divergence of ^{14}C age data below c. 300 cm core depth should be noted (see text).

Sedimentological and geochemical results (hemipelagic sediments)

Varve chronology

Using the stacked record of the Ormara cores (Fig. 3) we counted individual varve couplets and digitized the varve thickness to establish a continuous (non-floating) varve chronology back to about 5600 varve years BP (Fig. 4). The cumulative depth of these varve couplets was plotted against the number of years, assuming that one dark–light couplet was deposited within 1 year ('varve years'). In constructing the varve curve, event deposits (which are deposited quasi-instantaneously) add only thickness, but no age. The visual stratigraphic record provided us with an excellent age–depth scale and a good indication of accumulation rate changes (Fig. 5). The sedimentation rate resulting from this varve thickness curve is c. 2.3 mm a^{-1}. Below a core depth of 400 cm the cumulative varve curve and the ^{14}C age v. depth curves diverge slightly (the varve ages are between 14 and 19% older than the ^{14}C ages, calibrated to calendar years). This indicates either a somewhat overcounted varve age or too young ^{14}C ages. This observation is in contrast to that for core SO 90-56KA, where we found the opposite: lower varve ages than ^{14}C ages toward the base of the core, suggesting slight undercounting (von Rad et al. 1999a), possibly caused by amalgamation of varves owing to compaction (or local erosion by turbidites?). For the Ormara cores, which have far more event deposits than core 56KA off Karachi, we can only guess that maybe more than one light lamina was deposited during some years. Also, there are inherent potential errors in the calibration of ^{14}C data (e.g. unknown reservoir ages, especially as the 'mixed' planktonic foraminiferal assemblage might result in a mixture of different reservoir ages). In summary, we maintain that our interpretation that the dark–light couplets are indeed varves is confirmed also for the Ormara cores, at least for the

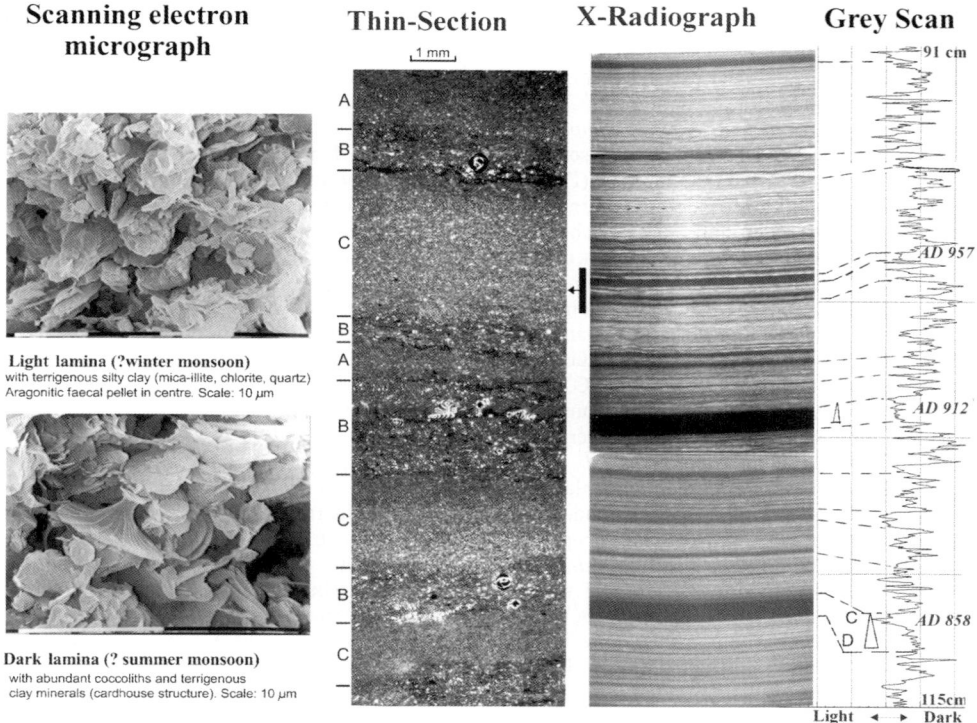

Fig. 6. Grey-scan (digitized lightness values), X-radiograph photo, thin-section photomicrograph (see black bar for location in core) and scanning electron micrographs of varved sediment sequence in core SO 90-33KL (91–115 cm; for X-radiograph, equivalent section of core SO 130-245KL was inserted).

past 2000 a, where ^{14}C ages and varve ages agree fairly well.

Hemipelagic laminated sediments: are they really varves?

On the basis of the varve counting from the X-radiographs, the depositional history of c. 5600 a can be reconstructed (Fig. 5). Available AMS ^{14}C dates from core 33KL support the varve chronology for the upper section of the composite varve record with ±500 a. Unfortunately, we were unable to correlate the Ormara varve record with the published varve chronology (von Rad et al. 1999a) from site 56KA west of Karachi, 200 km further east.

Individual hemipelagic 'varve' couplets are about 0.8–1.5 (up to 2.5) mm thick (Fig. 6). The significant redistribution of river-transported shelf material onto the slope adds complexity to the 'normal' seasonal cycle of downslope deposition of terrigenous material and the pelagic rain of biogenic material and organic matter (Staubwasser & Sirocko 2001). Light terrigenous laminae (A), which were deposited by turbid layer suspensions, are well sorted and rich in chlorite and quartz, but almost devoid of nanno- and microplankton; they contain about 0.3% organic carbon (mainly terrigenous organic macerals). The dark laminae (B) are poorly sorted, highly porous and rich in organic matter (about 1%; derived from marine phytoplankton), coccoliths, OMZ-specific benthic foraminifera and fish debris, but also diluted by clastic terrigenous material (Fig. 6).

We offer the following five arguments in favour of a seasonal origin of the varve couplets in our cores:

(1) *Good correlation of varve counts with ^{14}C dates (as discussed above)*.

(2) *Lamina-by-lamina analysis of a 5 year record (1993–1998)*. By analysing a 5 year record, deposited between 1993 in core 34KG (taken in 1993) and 1998 in core 246KG, Lückge et al. (2002) found five olive–grey and four light grey layers, proving their origin as seasonal varve couplets. The 'normal' (hemi)pelagic sedimentation active during the main part of the year produces thin olive–grey laminae with an elevated input of marine organic matter and biogenic

calcareous and opaline material. The thick light grey layers consist of terrigenous silty clay and are interpreted as 'plumites' derived from short-term heavy rainfalls leading to flood events.

(3) *The mineralogical and biogenic composition of individual laminae.* Thin-section and scanning electron microscopy (SEM) analyses (Fig. 6) indicate that the dark (olive–grey) laminae are poorly sorted silty clay and contain abundant benthic and planktonic foraminifera, coccoliths, diatoms (Lückge et al. 2002) and minute brownish organic spherules (faecal pellets?). Their total organic carbon (TOC) content is comparatively high (0.7–1.2%) and of marine origin (Lückge et al. 2002). Light laminae (type A) or C layers contain essentially well-sorted terrigenous silty clay (mica–illite, chlorite, quartz), with very little terrestrial organic matter (Lückge et al. 2002). They are barren of marine phyto- or zooplankton. C layers that are >1 cm thick may be graded (see Fig. 6), as is typical for gravitative event deposits.

(4) *High-resolution geochemical variability of varved sequence.* The high-resolution XRF scan of a varved sequence in core 39KG (see Fig. 9, below) shows clearly that the compositional variability (indicated by the marker elements Ca and Fe) co-varies with the dark- and light-coloured laminae, i.e. at a semiannual (seasonal) pattern. In Fig. 9 we show an X-radiograph and a grey-value scan (digitized imaging, resolution 5 μm), as well as the elemental distribution of Ca and Fe of the analysed area of about 20 mm × 1000 mm. After checking the elemental distributions for Ca, Fe, Zr, Si, Ti, K and Mn, we found that only the distribution spectra of Ca and Fe showed a significant variability that changes in step with the colour of the dark and light sub-laminae (see Fig. 9). The light ('winter') detrital laminae are rich in Fe (mainly from heavy minerals, Fe oxides and pyrite) and poor in Ca. The dark ('summer', mixed detrital–organic laminae are anti-correlated, being enriched in Ca (mainly from coccoliths, foraminiferal fragments and calcite flour, and poor in Fe. The suspensate event deposits (e.g. AD 1869) show the same geochemical pattern as the detrital 'winter' laminae. The negative correlation is also shown by the scatter plot of Ca v. Fe (see Fig. 10, below).

(5) *Comparison of lamina composition with sediment trap results (see Fig. 11, below).* Although we realize that a straightforward comparison of sediment trap results (which document mainly vertical settling, and no lateral near-bottom sediment transport) with the composition of sea-floor sediments is impossible, we will discuss some of the recent results of Treppke (1999) and Andruleit et al. (2000) for a better understanding. The total sediment flux rates, as measured in the Eastern PAKOMIN Trap (EPT) sediment traps (1166 m water depth, 562 m trap depth) west of Karachi (see Fig. 11), are dominated by lithogenic particles, followed by carbonates. The flux of biogenic opal and total organic matter is usually <6% of the total flux. During the NE monsoon (especially January–February) the maximum lithogenic and carbonate fluxes (Treppke 1999), as well as coccolith fluxes (Andruleit et al. 2000) were observed in sediment trap EPT-2 west of Karachi (c. 15 km SW of site 56KA, Fig. 1). The opal and especially the diatom flux, however, show two maxima, a major one in September, and a smaller one in January–February (Treppke 1999). Similar to the coccolith flux, the silicoflagellate flux has its maximum during the winter monsoon (January–February) with a minor peak in May. We conclude that, off the eastern Makran (west of 66°E) there is periodic, but at times major precipitation followed by river floods, during the winter months (mid-December–March). The main surface productivity of diatoms (and organic matter) is during the summer monsoon (mainly September), whereas the fluxes of coccoliths were within the same range for the summer monsoon (peak in June–July) as for the winter monsoon (maximum in January–February). The silicoflagellate flux resembles the coccolith flux, with a strong maximum in December–February. The organic matter flux (not shown) resembles that of opal, with three maxima: a strong one in December–February, a moderate maximum in September and a minor one in May. Andruleit et al. (2000) explained the summer monsoon productivity by the presence of nutrient-rich surface waters advected from the upwelling areas off Oman, whereas during the winter monsoon the locally high productivity appears to respond to local injections of nutrient-rich deep water into the surface water as a result of sea-surface cooling leading to convective overturn.

Hence we postulate that, in general, one couplet consisting of one dark plus one light lamina actually documents one annual monsoon cycle, although the significant dilution by river-derived and shelf-slope redistributed terrigenous material (with peaks in January–March, and July–August) and the fact that we have to consider two productivity seasons (see Fig. 11) complicates the picture. Dark (olive–grey) laminae consist of opaline and calcareous phyto- and zooplankton, organic matter (plus bacterial mats) and detrital material. Although the pelagic material is deposited throughout the year, most of it is apparently deposited during the high-productivity period of the late summer monsoon. A minor productivity maximum during the winter (especially in January and February) is responsible for the deposition of most coccolith-derived calcite, as well as some opal,

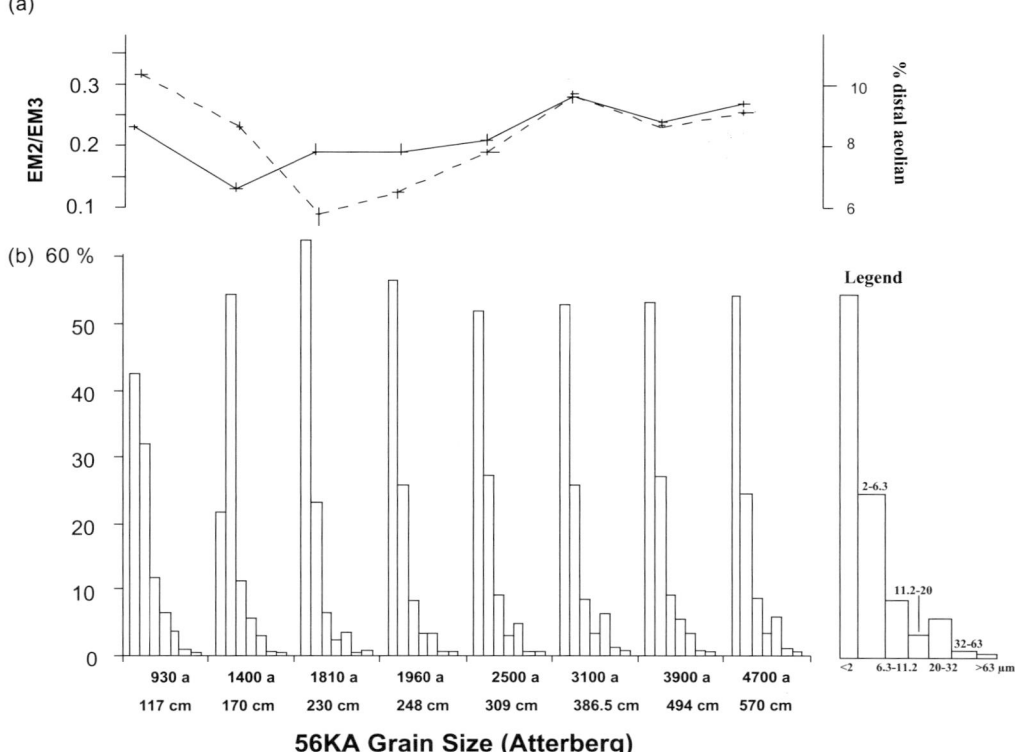

Fig. 7. Grain-size variability of eight selected sediment samples of core SO 90-56KA, fractionated by Atterberg size analysis. Grain-size distribution: EM 2 (11.2–32 μm fraction; distal aeolian dust); EM 3 (2–11.2 μm fraction; fluvial mud).

especially silicoflagellates (see Fig. 11). During the winter (mainly mid-December to February–March) mainly light detrital material (siliciclastic material and reworked carbonate flour) is deposited, probably by short-term flood events (Lückge et al. 2002), although detrital deposition takes place also during the enhanced wave activity of the summer monsoon (Staubwasser & Sirocko 2001).

Temporal variability of the texture and composition of the lithogenic fraction and their interpretation

The hemipelagic sediments, studied in some detail in core 56KA, are fine-silty clays with a median around 1.8 μm. In general, the textural sediment properties of the Upper Holocene sediments are very uniform (Fig. 7a): except for the sample at 170 cm (1400 a BP), which has a strong mode around 2–6.3 μm, the clay (<2 μm) fraction is dominant (40–60%) in all samples. Prins & Weltje (1999) found that the grain-size distributions of the sediments from the Pakistan continental margin consist of mixtures of three end-members: (1) 'fluvial mud' (mode around 9 μm, their end-member EM 3); (2) 'distal aeolian dust' (mode around 22 μm, their end-member EM 2); (3) 'proximal aeolian dust' (mode around 50 μm, their end-member EM 1). In our sediments from the upper Makran continental slope (Fig. 7a) we discovered only the very dominant fluvial mud end-member (EM 3; we used the 2–11.2 μm fraction) and a small contribution of distal aeolian mud (EM 2; we used the 11.2–32 μm fraction).

The mineralogical composition of the four analysed size fractions (Fig. 6b) shows within the fluvial mud fraction a strong increase between the <2 μm and the 2–11.2 μm fractions of the quartz content (from 11 to 43%) and concomitant decrease of the mica–illite content. The distal aeolian (medium silt; 11.2–32 μm) fraction has very high quartz contents, some feldspar and about one-quarter each of mica–illite and chlorite. For unknown reasons, the quartz content of the two oldest samples (3900 and 4700 a BP) is somewhat smaller

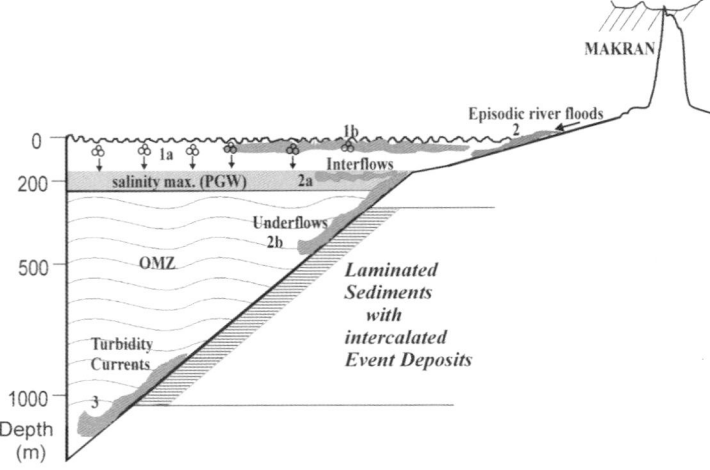

Fig. 8. Sketch of transport and deposition processes of individual laminae of the hemipelagic facies and event deposits at the Pakistani continental slope. (1a) Origin of dark, mixed detrital–biogenic–organic laminae: they consist of terrigenous detrital silty clay plus organic matter and plankton from productive surface waters, and are deposited throughout the year (especially during the late summer monsoon). (1b) Origin of light, detrital laminae derived from suspensions of 'dirty water' after river floods (mainly during the winter monsoon). Event deposits include light grey suspensate deposits or plumites, deposited either from interflows (2a) or from underflows (2b), as well as medium grey (or reddish grey) silt turbidites (3). PGW, Persian Gulf Water.

than the quartz content in the remaining samples and there is a considerable proportion (40–45%) of fine silt (2–20 μm). The grain-size distribution shows a strong mode in the <2 to 11.2 μm range (Fig. 6a), interpreted as fluvial mud, whereas there are minor modes between 11.2 and 32 μm, interpreted as distal aeolian dust (Prins & Weltje 1999; von Rad et al. 1999b). Our grain size and mineralogical data suggest that during the past 5000 a fluvial mud is by far the dominant sediment source for the upper Makran slope sediments. However, there is a subtle, but distinctly noticeable admixture of distal aeolian silt, especially from 5000 to 3000 a BP, and again c. 1000 a ago.

Event deposits (suspensate deposits and turbidites)

Turbid-plume or suspensate deposits

Event deposits (Fig. 8) include silt turbidites and turbid-plume or suspensate deposits (C layers; Behl 1995). The last two are more common and deposited at decadal or shorter (sometimes even annual) intervals. The light grey 'plumites' or 'suspensate' event deposits (von Rad et al. 1999a) are clayey fine silts with a median around 3–4 μm and silt contents of 50–60%. Similar sediment layers >5 mm thickness have been interpreted from the Santa Barbara Basin and the northern California margin as 'suspensate event deposits' or 'plumites' (Behl 1995; Weatcroft et al. 1997). From the shelf these clouds of dirty water are being dispersed either as interflows in the dense salinity maximum zone between 200 and 300 m water depth (Persian Gulf Outflow Water), which tops the OMZ, or as underflows of nepheloid or turbid layer flows (Fig. 8). The result is frequent thin, light grey, homogeneous to graded silty clay layers within the varved sequence. This very fine-grained, but proximal, allochthonous slope facies appears to be very frequent along steep slopes of active continental margins, but has up to now not been thoroughly studied. The event deposits were ultimately derived from the continent by rivers, which deposit this material first on the shelf. The process of cross-shelf and down-slope transport of mud-charged waters might have been triggered by cyclonic storms leading to very thin low-density turbidity currents, or, more likely, near-bottom and mid-water nepheloid suspensions followed by suspension settling (Fig. 8; von Stackelberg 1972). These poorly studied mechanisms are probably very active and frequent at the upper slopes of active continental margins, especially as the hinterland of the Makran region is characterized by high rates of coastal uplift, semi-arid desert–prairie conditions, and rare, but episodically strong, monsoonal floods.

Recently, Staubwasser & Sirocko (2001) postulated that the white–light grey layers (our C layers) were formed after periods of large-scale mud expulsion by mud volcanoes on the shelf, subsequent erosion by waves and redeposition onto the upper slope. They based this assumption on a comparison of various radiogenic Nd and Pb

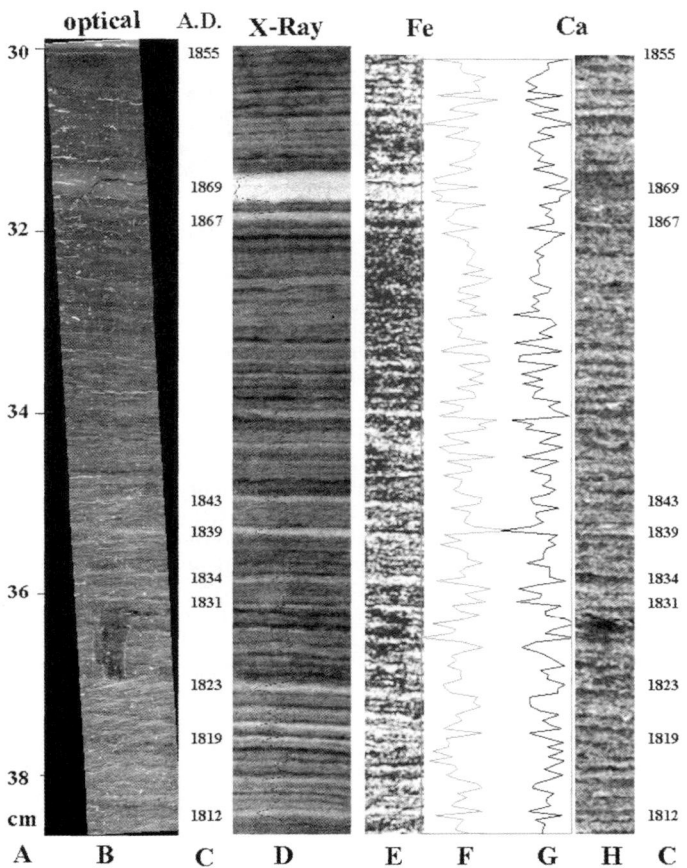

Fig. 9. High-resolution geochemical scan of varved series of core SO 90-39KG (30–38.5 cm). (**a**) Core depth (cm); (**b**) digital (grey-value) scan of investigated, impregnated specimen (note thick light, type-C event deposit at ad 1869); (**c**) age (ad) from varve counting; (**d**) X-radiograph (inverted colours); (**e**) Fe distribution map converted into grey-scan picture (ITRAX: 100 μm diameter mono-capillary, 0.5 s per point); (**f**) elemental concentration profile for Fe (relative intensity, Geoscanner: slit capillary 1 cm by 50 μm, time 20 s per point, i.e. 50 μm values added up for 1 mm intervals); (**g**) elemental concentration profile for Ca (as in F); (**h**) Ca distribution map (100 μm diameter mono-capillary, 0.5 s per point). (Note excellent negative correlation of Ca with Fe within individual laminae.)

isotope ratios of one mud volcano sample (Chandragup), which are similar to those from 'white layers', but different from the olive–grey and black laminae. However, our data suggest strongly that the C layers are suspensate deposits or plume deposits originating from mud-charged waters after heavy river floods. In a separate paper (von Rad *et al.* 2002), we discuss our arguments against the 'mud volcano origin hypothesis'.

Medium grey or reddish grey, graded and laminated silt turbidites

The medium grey (D layers) or reddish grey, graded and laminated silt turbidites (F layers) originated from less frequent, unchannelized, low-density, low-velocity turbidity currents. Type-D turbidites have a thickness >5 mm and consist of medium to dark grey, homogeneous silt. The thicker layers are commonly graded and internally laminated. Graded reddish brown silt turbidites (F) with finely dispersed hematite (Staubwasser & Sirocko 2001) were identified only off Karachi; they are rare and form valuable marker horizons that can be correlated for up to 50 km. Staubwasser & Sirocko (2001) suggested an origin from the Indus River (which indeed has deposited reddish grey laminated clays during the Preboreal period; von Rad *et al.* 1999*b*). However, the lack of any reddish silt turbidites (or indeed of any turbidites) in the slope cores off the Indus delta, the long

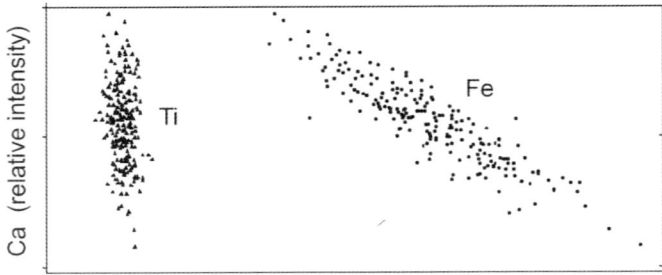

Fig. 10. Scatter plot of relative intensity of Ca v. Fe (values of Fig. 9f and g) showing clearly negative correlation of both elements whereas Ti does not show any variability in ratio to Ca.

distance between the mouth of the Indus River and the location of core SO 90-56KA west of Karachi for such a thick (>10 cm) silt turbidite, the dominant west–east-directed longshore currents, and morphological barriers to cross-slope transport make an Indus source highly unlikely. Hence we suggest that the reddish grey F turbidites were derived from a local source in the nearby Makran.

Turbidite frequency and varve thickness variability

There are about 340 marker layers (C layers and D turbidites) making up 46% of the total composite section (Fig. 3) of the Ormara cores. The variation of the cumulative thickness of event deposits and turbidites (> 50 mm) per century is shown in Fig. 4b. The frequency of turbidites shows a weak correlation with the thickness of hemipelagic varves (calculated 'turbidite-free'), especially in the period between 5600 and 4500 a BP, where turbidites are rare and thin, and varve thickness is significantly reduced. A maximum in turbidite thickness per century is observed near 4000 a BP, coinciding with generally thick varves and maxima of Zr, Ti and K input. During the past 3000 a the turbidite thickness per century is rather variable (a general turbidite minimum between 2200 and 1500 a BP coincides roughly with the thinning of varves). A minimum in the turbidite thickness per century coincides roughly with the varve thickness minimum around 1100 a BP. At site 56KA west of Karachi the frequency and cumulative thickness of the event deposits per century correlates positively with the varve thickness of the hemipelagic sediments, suggesting that both depositional processes are mainly controlled by precipitation, i.e. by monsoon-influenced climate variability (von Rad et al. 1999a).

Looking at the varve thickness variability (Fig. 4a), we detect the following general trends.

During Period I (about 5600–4700 a BP) varves are generally thin (and turbidites rare), indicating a generally dry climate with reduced precipitation and river runoff (possibly winter monsoon dominated, following Lückge et al. 2001).

Period II (4700–2600 a BP) is characterized by comparatively thick varves, and a few maxima of TiO_2/Al_2O_3, Zr/Al_2O_3 and K_2O/Al_2O_3 (at 4600, 4100 and 2700 a BP) interrupted by a minimum of these ratios around 3600 a BP. These data suggest a generally wet period (with possibly summer monsoon domination).

Period III (2600–1000 a BP) is characterized by gradually thinning varves associated with a decrease of turbidite thickness per century. There are two maxima of TiO_2/Al_2O_3, Zr/Al_2O_3 and K_2O/Al_2O_3 (at 2300 and c. 1700 a BP) and a minimum of these ratios around 1200 a BP. This is interpreted as a gradual aridification of the hinterland (decline of precipitation and river runoff) leading finally to dry conditions from 1600 to 1000 a BP.

Period IV (1000 a BP to present) shows variable, but generally thick varves, suggesting alternately wet to rather dry conditions (e.g. at the Zr/Al_2O_3, K_2O/Al_2O_3 minimum around 500 a BP).

Cyclicity

Core 56KA

Decadal to millennial cyclicity of varves and turbidites has been studied in our 'standard varve section' at site 56KA west of Karachi (von Rad et al. 1999b; Berger & von Rad 2002). A first inspection of the cleaned varve-thickness series of core 56KA (Fig. 12a) shows broad maxima around 100, 1600, 3000 and 3750 a BP and minima around 500, 1950 and 3450 a BP (Fig. 12a). If we consider the period between 1950 a BP ('Roman Warm Period') and 500 a BP ('Medieval Climatic Optimum'), we detect a low-frequency cycle of roughly 1450 a. Autocorrelation of this series identified a 1460 a cycle (Berger

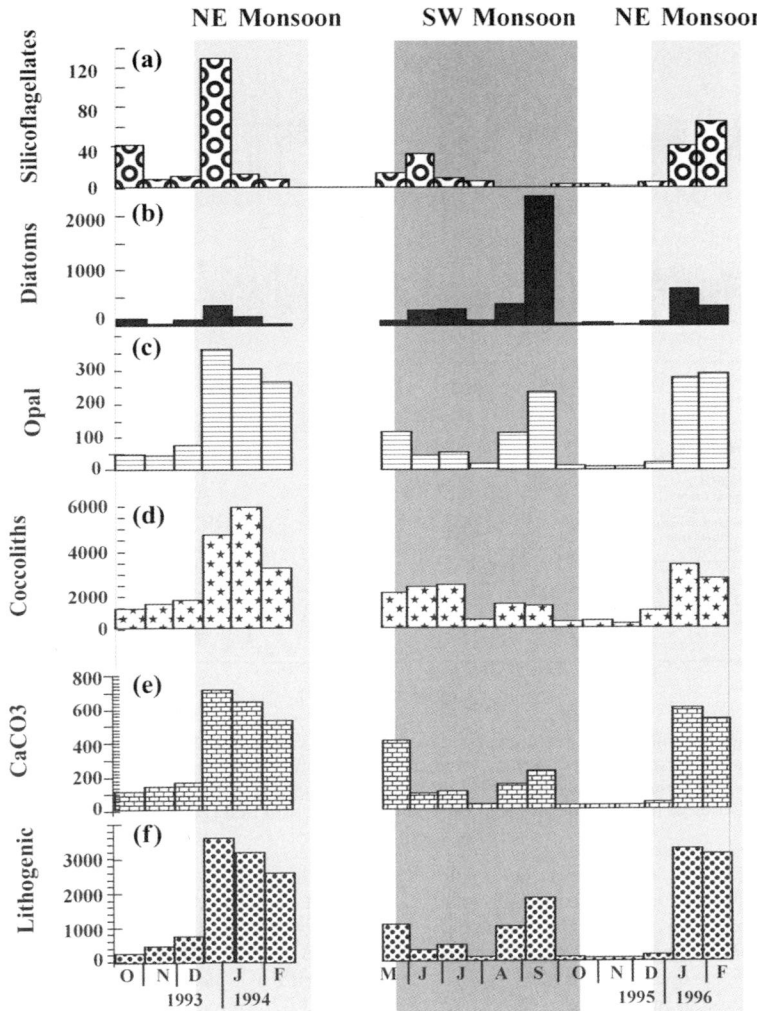

Fig. 11. Flux rates derived from Eastern PAKOMIN Trap (EPT-1: 15 October 1993–24 February 1994; EPT-2: May 1995–28 February 1996). (**a**) and (**b**) from Treppke (Treppke 1999); (**c**), (**e**) and (**f**) from Ittekkot (in Treppke 1999; Andruleit et al. 2000); (**d**) from Andruleit (Andruleit et al. 2000). Lithogenic, $CaCO_3$ and opal flux rates in mg m^{-2} d^{-1}. Flux rates of coccoliths (**c**) include all coccolithophore species (except coccospheres and calcisperes): coccoliths $\times$ 10^{-6} $\times$ m^{-2} $\times$ d^{-1} per count. Diatom flux: 10^5 valves $\times$ m^{-2} $\times$ d^{-1}; silicoflagellate flux: 10^4 individuals $\times$ m^{-2} $\times$ d^{-1}.

& von Rad 2002, fig. 2b), especially for latest Holocene time (the past 2000 a). The most prominent peak in the Fourier spectrum of Fig. 2b is a 125 a cycle. Other noteworthy peaks are at 280, 95, 56, 39, 26 and 23 a, with minor peaks at 245, 202, 164, 29.5, 18.7, 14 and 12.2 a (Fig. 12b). The significance of the peaks can be appreciated by considering that most of the cycles mentioned are close to or beyond 2 SD from the overall mean (mean 1.4; 1 SD is 0.6).

Previous work (von Rad et al. 1999b) detected cycles of 750, 250, 125, 95, 45, 39, 29–31 and 14 a length, in good agreement with the present analysis. Also, strong variability in the 55 a band and a 31 a cycle were found in the series of grey values of core 56KA (von Rad et al. 1999b).

The most prominent cycles in the 56KA varve series (Fig. 12d) have periods of 26, 39, 56, 95 and 125 a length. We note that each of these values is close to a whole-number multiple of the lunar tidal cycle at 4.425 a (lunar perigee cycle) or 9.3 a (lunar nodal half-cycle). The period near 26 a is close to six times the perigee cycle

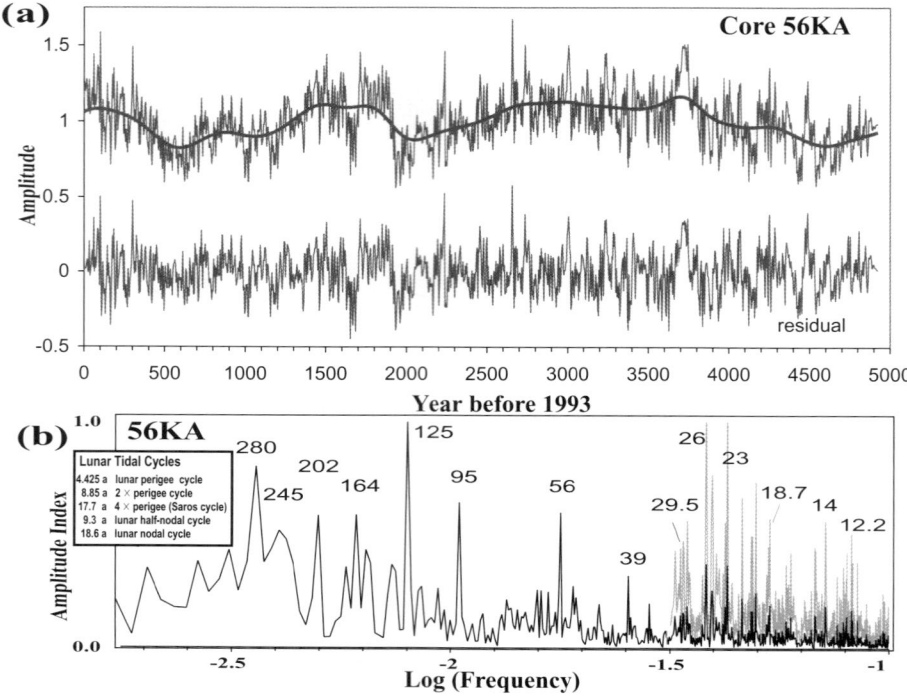

Fig. 12. Detection of cycles within the varve thickness series of the stacked record of core SO 90-56KA west of Karachi (see Fig. 1). (**a**) Despiked thickness series (turbidite-free, log normalized), resampled for 4 a intervals and smoothed to find long-term variations (bold line). The difference is the residual (lower curve). (**b**) Fourier spectrum of the autocorrelation series of the residual shown in (**a**). From Berger & von Rad (2002).

(26.55 a); 39 a is close to nine times that cycle (39.83 a); 56 a is close to six times the nodal cycle (55.8 a); 95 a is close to five times the full nodal cycle ($5 \times 18.6 = 93.0$) and 125 a is near seven times the large perigee cycle (4×4.425), which emerges from the nodal modulation (18.6 a) of perigee ($7 \times 17.7 = 123.9$). This apparent coincidence suggests that tidal forcing may play a role in producing the cycles.

Ormara cores

The stacked record of the extremely turbidite-rich varved sediments of the Ormara cores represents about 5500 varve years (Fig. 13a). The autocorrelation series is much less distinct than in 56KA, but a cycle near 1490 emerges upon repeating the procedure on the mirrored autocorrelation series, with zero lag at the centre.

Fourier analysis of the mirrored autocorrelation series (Berger & von Rad, unpublished) shows remarkable simplicity: there is much less information than in core 56 (Fig. 12b).

The cycles seen in the varve record in the Ormara cores (Berger & von Rad, unpublished) are as follows: 178, 167, 107 (103), 74, 57, 43 (42.6) and 32 a. Of these, 178 a may be assigned to 5×35.4 (177), 167 a to 19×8.85 (168.15), 107 a to 3×35.4 (106.2), and 43 a to 5×8.85 (44.25), with a typical error of 1%. The periods 74, 57 and 32 a remain unassigned. The period of 57 a may be taken as identical with that of 56 a of core 56. Other similarities stem from the tidal staircase phenomenon.

The turbidite record of the Ormara cores was treated the same as that of core 56KA (Berger & von Rad 2002). This core is close to the mouths of the very active Makran rivers (Basul and Hingol); thus, there are many more turbidites in this record than in core 56KA west of Karachi. The record has cycles with dominant periods at 460 and 183 a, and lesser periods at 288, 155, 94.5 and 47.5 a. In the third rank there are the periods near 74, 54, 42.5, 29.6, 25.3 and 23.3 a. A number of these are identical or close to periods found in core 56KA.

Combining all periods found in the Ormara cores, we obtain (from the top down): 460, 288, 183, 178, 167, 155, 107 (103), 94.5, 74, 57 (54), 47.5, 43 (42.5), 32, 29.6, 25.3 and 23.3 a. Several of these

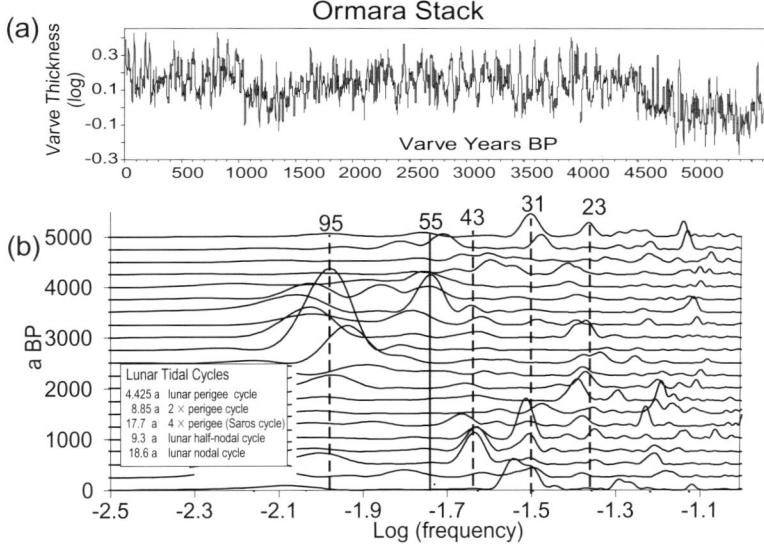

Fig. 13. Detection of cycles within the varve thickness series of stacked 5000 a section of cores off Ormara (see Figs. 3 and 4). (**a**) Despiked thickness series, resampled for 4 a intervals, residual after subtracting smoothed version (as in Fig. 12a). (**b**) Fourier spectra of the autocorrelation series of 20 non-overlapping 'gliding' 500 a windows, overlapping by 250 a. Periods >100 a are greatly attenuated. Amplitudes were further adjusted by dividing each position by the square of log(F), to emphasize the shorter periods. It should be noted that there are preferred periods of oscillation, but they change with time (the power seems to jump from longer cycles to shorter ones).

are readily assigned to multiples of tidal periods: 460 a (13 × 35.4 = 460.2), 288 a (8 × 35.4 = 283.2), 178 a (5 × 35.4 = 177), 107 a (3 × 35.4 = 106.2), 43 a (5 × 8.85 = 44.25) and 32 a (7 × 4.425 = 30.975). Others do not readily find a tidal slot, and if one has to be found, arguments are more involved. Suffice it to say that six of the 13 periods found between 460 and 30 a are assignable to whole-number multiples of tidal periods, in the fashion demonstrated for core 56KA.

Figure 13 shows the periodicity of varve thickness of the Ormara stack applying a combination autocorrelation–Fourier analysis to 20 'gliding' 500 a windows, overlapping by 250 a. By this method the variability of cyclicity through time is demonstrated, but periods greater than 100 a are greatly attenuated. As discovered in other land sections and ocean areas (e.g. in the Santa Barbara Basin), we note preferred periods of climate oscillation around 95 a, near 55 a, at 43 a, near 31 a and 23 a. Apparently, the power seems to jump from longer cycles to shorter ones. But this is not a gradual process; instead, it appears that shorter cycles (e.g. the 43 a cycle) become more prevalent in successively younger time series (e.g. it is present, but small in the 3600–4500 a series; but becomes stronger in the 2700–3600 a record). The important result is that there are preferred periods of (climate) oscillations, but they change with time.

Significance of cycles

The significance of the cycles found in Ormara cores, as evidence for tidal influence on both varve thickness and turbidite frequency, has been discussed by Berger & von Rad (2002). The lunar perigee cycle of 4.425 a, the 'Saros Cycle' of 17.7 a (4 × 4.425; Hoyt & Schatten 1997, p. 167) and the lunar half-nodal cycle of 9.3 a result in variations in the amplitudes of tides. These cycles may influence climate by controlling the intensity of tidal mixing on shelves, which can affect surface water temperature. Also, and perhaps more importantly in present circumstances, cyclic changes in tidal amplitudes should modulate the resedimentation of soft clay- and silt-sized sediment piled onto the shelf by river input. Brückner (1890) described a 35 a cycle in weather and climate and we ascribe this 'Brückner cycle' to the first harmonic of the Saros cycle of 17.7 a (Berger et al. 2000). An important cycle in our records occurs around 29–31 a. It might be a beat frequency of the lunar tide (17.7 a) and the sunspot cycle (11 a). Alternatively, it might be ascribed to a subharmonic of the lunar cycle (7 × 4.425).

A prominent period in our varves and turbidites is around 55–56 a, a cycle that could be ascribed either to solar forcing (5 × 11 a) or lunar forcing

(3 × 17.7 a). A similar cycle (58 a) has been described from the Santa Barbara Basin from fish-scale populations by Baumgartner et al. (1992), and from varve thickness data by Lange et al. (1996; 56–59 a) and by Biondi et al. (1997; 58 a). Evolutive spectral analysis of sunspot data over the past 300 a revealed a dominant 55 a cycle besides an 11 a and a 100 a cycle (Berger et al. 1990; Loutre et al. 1992).

A cycle around 95 a is very prominent in the 56KA varves and Ormara turbidite cycle. It might tentatively be assigned to a multiple of the Saros cycle (11 × 8.85). Going to low-frequency cycles, a cycle around 107 a is very prominent (3 × 35.4). Similar cycles around 100 a were also observed in the Santa Barbara Basin (Baumgartne et al. 1992; Biondi et al. 1997; Lange et al. 1996). Other common low-frequency cycles centre around 125, 250, 280, 350 and 460 a.

Also, the 1470 a cycle is clearly shown in the Late Holocene varve record of core 56KA (less clearly in the Ormara cores; 1490 a). This cycle has been observed in various records: Sirocko et al. (1996) stressed a 775 a cycle in Arabian Sea deep-sea cores, a subharmonic of a '1500 a cycle', which was also described from the South China Sea by Wang et al. (1999). The 1470 a cycle was described from the amplitude variations in the $\delta^{18}O$ record of the GISP2 ice core from Greenland (Grootes & Stuiver 1997), where it is strong during glacial conditions: Schulz et al. (1999) presented evidence that its amplitude depends on continental ice mass (as recorded in sea-level fluctuations) and also on whether the system is in transition (ready to respond) or stationary (stable). Bond et al. (1997) described a pervasive millennial-scale cycle with a period of 1470 ± 500 a in the North Atlantic as 'pacemaker for rapid climate change', not only during glacial times ('Heinrich events' or 'Bond cycles'), but also during the Holocene period (e.g. the 'Little Ice Age'). Opposite opinions on the significance of this millennial cycle have been discussed by Wunsch (2000) and Meeker et al. (2001). We take the presence of the 1470 a cycle in our Holocene record as evidence that the varves here analysed contain information of global significance. For a brief discussion of the significance of the 1470 a cycle and its correlation with the ice-rafting cycle, the reader is referred to Berger & von Rad (2002).

Conclusions

(1) We offer the following arguments in favour of an annual (seasonal) character of the couplets of dark grey organic-rich mixed biogenic–detrital and light grey detrital laminae: good to fair correlation of the varve curve with ^{14}C datings; lamina-by-lamina analysis of a 5 year record in box cores obtained in 1993 and 1998; specific mineralogical and biogenic composition of individual laminae; ultra-high-resolution geochemical variability of a varved series measured by an XRF scanner; comparison of lamina composition with sediment trap results spanning (almost) a full monsoonal cycle.

(2) The varve chronology of the stacked Ormara core record is based on varve counting and AMS ^{14}C dating. We established an uninterrupted time series of varve thickness values for the past 5600 a, comparable with the 5000 a standard varve core 56KA west of Karachi.

(3) Detailed grain-size analysis of the fine silty clays demonstrated that they are dominated by fluvial mud with a small contribution of distal aeolian silt.

(4) Turbid-plume or suspensate deposits (light grey homogeneous to graded silty clay layers) are being deposited at decadal or shorter intervals and are explained by episodically strong river floods after heavy rains transporting mud-charged waters across the narrow shelf and onto the steep continental slope.

(5) Medium grey or reddish grey, graded and laminated silt turbidites originated from less frequent, unchannelized, low-density turbidity currents. The variation of the cumulative thickness of event deposits and turbidites per century is weakly correlated with the thickness of hemipelagic varves.

(6) In general, periods with thin varves (often associated with rare turbidites) are correlated with minima of detrital element ratios (TiO_2/Al_2O_3, Zr/Al_2O_3, K_2O/Al_2O_3), especially in Period I (c. 5600–4700 BP). This period is characterized by generally thin varves (and rare turbidites), suggesting a climate with reduced precipitation and river runoff (possibly winter monsoon dominated). Period II (4700–2600 a BP) is characterized by comparatively thick varves documenting a generally wet period (with possibly summer monsoon domination). During Period III (2600–1000 a BP) a gradual thinning of varves and decrease of turbidite thickness per century is interpreted as a gradual decline of precipitation and river runoff, leading finally to dry conditions from 1600 to 1000 a BP. Period IV (1000 a BP to present) shows variable, but generally thick varves, suggesting alternately wet to rather dry conditions.

(7) The sequence of cycles detected by autocorrelation and standard Fourier analysis seems to contain a large proportion of multiples of the lunar perigee cycle (4.425 a) and the lunar half-nodal cycle (9.3 a). Some cycles of varve thickness match the cyclicity of turbidite frequency. Our record also contains evidence for the presence

of the 1470 a cycle, previously reported from the glacial-age Greenland ice record and North Atlantic deep-sea cores (ice-rafting or 'Bond cycles'). As proposed by Berger & von Rad (2002), we hypothesize that a large proportion of the detected cyclicity (including the 1470 a cycle) can be ascribed to lunar tidal action.

We are grateful to P. Grootes (Leibniz-Labor, Kiel), who supplied AMS ^{14}C data, and to H. Leuschner (Göttingen) for his support during the use of the tree-ring measuring device (Aniol 1983) at the Institute of Palaeobotany and Quaternary Research of the University of Göttingen. A. Kemp (Southampton), J. Pike (Cardiff) and A. Lückge (Hannover) kindly reviewed the manuscript and gave many helpful comments. A.A.K. acknowledges a grant by BGR/BEO for a 6 month post-cruise study visit at Hannover. Thanks go also to the Federal Ministry of Education and Research (BMBF, Bonn) for funding the cruises SO 90 and SO 130 (Projects 03G0090 A/130 A).

References

ANDRULEIT, H., VON RAD, U., BRUNS, A. & ITTEKKOT, V. 2000. Coccolithophore fluxes from sediment traps in the northeastern Arabian Sea off Pakistan. *Marine Micropaleontology*, **38**, 285–308.

ANIOL, R.W. 1983. Treering analysis CATRAS. *Dendrochronologica*, **1**, 45–53.

ARTHURTON, R.S., FARAH, A. & AHMAD, W. 1982. The Late Cretaceous–Cenozoic history of western Baluchistan, Pakistan. *In:* LEGGETT, J.K. (ed.) *Trench and Fore-arc Geology: Sedimentation and Tectonics on Modern and Ancient Active Plate Margins*. Geological Society, London, Special Publications, **10**, 373–385.

BANSE, K. 1994. On the coupling of hydrography, phytoplankton, zooplankton, and settling organic particles offshore in the Arabian Sea. *Proceedings of the Indian Academy of Sciences (Earth and Planetary Sciences)*, **103**(83), 125–161.

BAUMGARTNER, T.R., SOUTAR, A. & FERREIRA-BARTRINA, V. 1992. Reconstructions of the history of Pacific sardine and northern anchovy populations over the past two millennia from sediments of the Santa Barbara Basin. *California Cooperation Fishery Investigations, Report*, **33**, 24–40.

BEHL, R.J. 1995. Sedimentary facies and sedimentology of the late Quaternary Santa Barbara Basin, Site 893. *In:* KENNETT, J.P., BALDAUF, J.G. & LYLE, M. (eds) *Proceedings of the Ocean Drilling Program, Scientific Results, 146 (Part 2)*. Ocean Drilling Program, College Station, TX, 295–308.

BERGER, W.H. & VON RAD, U. 2002. Decadal to millennial cyclicity in varves and turbidites from the Arabian Sea: hypothesis of tidal origin. *Global and Planetary Change*, **34**, in press.

BERGER, A., MÉLICE, J.L. & VAN DER MERSCH, I. 1990. Evolutive spectral analysis of sunspot data over the past 300 years. *Philosophical Transactions of the Royal Society of London, Part A*, **330**, 529–541.

BERGER, W.H., VON RAD, U. & LANGE, C. B. 2000. Brückner cycles in marine varves—detection and possible significance. Geophysical Research. *Abstracts, European Geophysical Society Annual Meeting, March 2000, Nice*.

BIONDI, F., LANGE, C.B., HUGHES, M.K. & BERGER, W.H. 1997. Inter-decadal signals during the last millennium (AD 1117–1992) in the varve record of Santa Barbara Basin, California. *Geophysical Research Letters*, **24**, 193–196.

BOND, G., SHOWERS, W., CHESEBY, M. & 7 OTHERS 1997. A pervasive millennial-scale cycle in North Atlantic Holocene and Glacial climate. *Science*, **278**, 1257–1266.

BRÜCKNER, A. 1890. Klimaschwankungen seit 1700 (nebst Bemerkungen über die Klimaschwankungen der Diluvialzeit). *In:* PENCK, A. (ed.) *Geographische Abhandlungen, Band IV, Heft 2*. E.D. Hölzel, Vienna.

GROOTES, P.M. & STUIVER, M. 1997. Oxygen $^{18/16}$ variability in Greenland snow and ice with 10^3–10^5–year time resolution. *Journal of Geophysical Research*, **C102**, 26455–26470.

HARMS, J.C., CHAPPEL, H.N. & FRANCIS, D.C. 1984. The Makran coast of Pakistan. Its stratigraphy and hydrocarbon potential. *In:* HAQ, B.U. & MILLIMAN, J.D. (eds) *Marine Geology and Oceanography of Arabian Sea and Coastal Pakistan*. Van Nostrand Reinhold, New York, 3–26.

HOYT, D.V. & SCHATTEN, K.H. 1997. *The Role of the Sun in Climate Change*. Oxford University Press, Oxford.

HUGHEN, K.A., OVERPECK, J.T., PETERSON, L.C. & ANDERSON, R.F. 1996. The nature of varved sedimentation in the Cariaco Basin, Venezuela, and its palaeoclimatic significance. *In:* KEMP, A.E.S. (eds) *Palaeoclimatology and Palaeoceanography from Laminated Sediments*. Geological Society, London, Special Publications, **116**, 171–183.

KEMP, A.E.S. (ed.) 1996. *Palaeoclimatology and Palaeoceanography from Laminated Sediments*. Geological Society, London, Special Publications, **116**.

LANGE, C.B., SCHIMMELMANN, A., YASUDA, M.L. & BERGER, W.H. 1996. Paleoclimatic significance of marine varves off southern California. *In:* WIGAND, P. & ROSE, M. (eds) *Southern California Climate: Trends and Extremes of Past 2000 Years*. Los Angeles County Natural History Museum, Los Angeles, CA, 45–100.

LOUTRE, M.F., BERGER, A., BRETAGNON, P. & BLANC, P.-L. 1992. Astronomical frequencies for climate research at the decadal to century time scale. *Climate Dynamics*, **7**, 181–194.

LÜCKGE, A., DOOSE-ROLINSKI, H., ALI KHAN, A., SCHULZ, H. & VON RAD, U. 2001. Monsoonal variability in the northeastern Arabian Sea during the past 5000 years: geochemical evidence from laminated sediments. *Palaeogeography, Palaeoclimatology, Palaeoecology*, **167**, 273–286.

LÜCKGE, A., REINHARDT, L., ANDRULEIT, H., DOOSE-ROLINSKI, H., VON RAD, U., SCHULZ, H. & TREPPKE, U. 2002. Decoding five years of sedimentation: are the laminated sediments off Pakistan (NE Arabian Sea) really varves? *In:* CLIFT, P.D., KROON, D., GAEDICKE, C. & CRAIG, J. (eds) *The Tectonic and Climatic Evolution of the Arabian Sea*

Region. Geological Society, London, Special Publications, **195**, 421–431.

MEEKER, D., MAYEWSKI, P.A., GROOTES, P.M., ALLEY, R.A. & BOND, G.C. 2001. Comment on 'On sharp spectral lines in the climate record and the millennial peak' by Carl Wunsch. *Paleoceanography*, **16**(5), 544–547.

NADEAU, M.J., SCHLEICHER, M., GROOTES, P.M. & 5 OTHERS 1997. The Leibniz-Labor AMS facility at the Christian-Albrechts-University, Kiel, Germany. *Nuclear Instruments and Methods in Physics Research*, **B123**, 22–30.

NASA GODDARD SPACE FLIGHT CENTER 1999. *Goddard DAAC Ocean Color Data* (Support Team Ch.R. McClain and J. Acker), Greenbelt, MD.

OVERPECK, J.T. 1996. Varved sediments records of recent seasonal to millennial scale environmental variability. *In:* JONES, P.D., BRADLEY, R.S. & JOUZEL, J. (eds) *Climatic Variations and Forcing Mechanisms of the Last 2000 Years*. NATO ASI Series, **131**, 479–498.

PRINS, M. & WELTJE, G.J. ET AL. 1999. End-member modeling of siliciclastic grain-size distributions: the Late Quaternary record of eolian and fluvial sediment supply to the Arabian Sea and its paleoclimatic significance. *In:* HARBAUGH, J. (eds) *Numerical Experiments in Stratigraphy: Recent Advances in Stratigraphic and Sedimentologic Computer Simulations*. SEPM (Society for Sedimentary Geology) Special Publication, **62**, 91–111.

SCHIMMELMANN, A. & LANGE, C.B. 1996. The tale of 1001 varves: a review of Santa Barbara Basin sediment studies. *In:* KEMP, A.E. (ed.) *Palaeoclimatology and Palaeoceanography from Laminated Sediments*. Geological Society, London, Special Publications, **116**, 121–142.

SCHULZ, M., BERGER, W.H., SARNTHEIN, M. & GROOTES, P.M. 1999. Amplitude variations of 1470-year climate oscillations during the last 100,000 years linked to fluctuations of continental ice mass. *Geophysical Research Letters*, **26**(22), 3385–3388.

SCHULZ, H., VON RAD, U. & VON STACKELBERG, U. 1996. Laminated sediments from the Oxygen-Minimum Zone of the Northeastern Arabian Sea. *In:* KEMP, A.E.S. (ed.) *Paleoclimatology and Palaeoceanography from Laminated Sediments*. Geological Society, London, Special Publications, **116**, 185–207.

SIROCKO, F., GARBE-SCHÖNBERG, D., MCINTYRE, A. & MOLFINO, B. 1996. Teleconnections between the subtropical monsoons and high-latitude climates during the last deglaciation. *Science*, **272**, 526–529.

SNEAD, R.E. 1967. Recent morphological changes along the coast of West Pakistan. *Annals of the Association of American Geographers*, **57**, 550–565.

SNEAD, R.E. 1993. Geography, geomorphologic process and effects on archeological sites on the Makran coast. *In:* SHRODER, J.F. (ed.) *Himalaya to the Sea: Geology, Geomorphology and the Quaternary*. Routledge, London, 363–378.

STAUBWASSER, M. & SIROCKO, F. 2001. On the formation of laminated sediments on the continental margin off Pakistan: the effects of sediment provenance and sediment redistribution. *Marine Geology*, **172**, 43–56.

STUIVER, M. & BRAZIUNAS, T.F. 1993. Modeling atmospheric ^{14}C ages of marine samples to 10,000 BC. *Radiocarbon*, **35**(1), 137–189.

TREPPKE, U. 1999. *Diatomeen-Analyse an den Sinkstoff-Fallen EPT/WPT aus dem Arabischen Meer vor Pakistan*. Bundesanstalt für Geowissenschaften und Rohstoffe, Hannover, Archive No. **0119194**.

VON RAD, U., SCHULZ, H. & SONNE 90 SCIENTIFIC PARTY 1995. Sampling the Oxygen Minimum Zone off Pakistan: Glacial/Interglacial variations of anoxia and productivity. *Marine Geology*, **125**, 7–19.

VON RAD, U., DOOSE, H. & CRUISE PARTICIPANTS 1998. *SONNE Cruise SO 130 Cruise Report MAKRAN II*. Bundesanstalt für Geowissenschaften und Rohstoffe, Hannover, Archive No. **117368**.

VON RAD, U., SCHAAF, M., MICHELS, K.H., SCHULZ, H., BERGER, W.H. & SIROCKO, F. 1999a. A 5,000-year record of climate change in varved sediments from the oxygen minimum zone off Pakistan (Northeastern Arabian Sea). *Quaternary Research*, **51**, 39–51.

VON RAD, U., SCHULZ, H., RIECH, V., DEN DULK, M., BERNER, U. & SIROCKO, F. 1999b. Multiple monsoon-controlled breakdown of oxygen-minimum conditions during the past 30,000 years documented in laminated sediments off Pakistan. *Palaeogeography, Palaeoclimatology, Palaeoecology*, **152**, 129–161.

VON RAD, U., BERNER, U., DELISLE, G. & 8 OTHERS 2000. Gas and fluid venting at the Makran Accretionary Wedge off Pakistan: initial results. *Geo-Marine Letters*, **20**, 10–19.

VON RAD, U., LÜCKGE, A., DELISLE, G. & ALI KHAN, A. 2002. On the formation of laminated sediments on the continental margin of Pakistan—Comment. *Marine Geology*, in press.

VON STACKELBERG, U. 1972. Faziesverteilung in Sedimenten des indisch–pakistanischen Kontinentalrandes (Arabisches Meer). *'Meteor' Forschungs-Ergebnisse, Reihe C*, **9**, 1–73.

WANG, L., SARNTHEIN, M., ERLENKEUSER, H., GROOTES, P.M., GRIMALT, J.O., PELEJERO, C. & LINK, G. 1999. Holocene variations in Asian monsoon moisture: a bidecadal sediment record from the South China Sea. *Geophysical Research Letters*, **26**(18), 2889–2892.

WEATCROFT, R.A., SOMMERFIELD, C.K., DARKE, D.E., BORGELD, J.C. & NITTROUER, C.A. 1997. Rapid and widespread dispersal of flood sediment on the northern California margin. *Geology*, **25**(2), 163–166.

WHITE, R.S. 1982. Deformation of the Makran accretionary prism in the Gulf of Oman (north-west Indian Ocean). *In:* LEGGETT, J.K. (ed.) *Trench and Fore-arc Geology: Sedimentation and Tectonics on Modern and Ancient Active Plate Margins*. Geological Society, London, Special Publications, **10**, 373–385.

WUNSCH, C. 2000. On sharp spectral lines in the climate record and the millennial peak. *Paleoceanography*, **15**(4), 417–424.

Periodical breakdown of the Arabian Sea oxygen minimum zone caused by deep convective mixing

GERT JAN REICHART[1], JURIAAN NORTIER[1], GERARD VERSTEEGH[2] & WILLEM JAN ZACHARIASSE[1]

[1]*Utrecht University, Institute of Earth Sciences, PO Box 80.021, 3508TA Utrecht, The Netherlands (e-mail: reichart@geo.uu.nl)*
[2]*Netherlands Institute for Sea Research, Department of Marine Biogeochemistry and Toxicology, PO Box 59, 1790AB Den Burg (Texel), The Netherlands*

Abstract: The northern Arabian Sea is at present characterized by a pronounced oxygen minimum zone (OMZ) with oxygen concentrations reaching values as low as 2 µM between 150 and 1250 m. This intense mid-water OMZ results from high annual organic particle fluxes and a moderate rate of thermocline ventilation. Sediment studies have shown that the intensity of the northern Arabian Sea OMZ has fluctuated on Milankovitch and sub-Milankovitch time scales, in conjunction with changes in either surface water productivity or thermocline ventilation. Here we evaluate the role of convective mixing in the periodical breakdown of the OMZ by reconstructing the density gradient for periods showing a well-ventilated water column. For this reason we reconstructed sea surface temperatures and salinities for the last 70 ka based on alkenone thermometry and $\delta^{18}O$ analyses on planktic and benthic foraminifers. For the studied time span thermocline ventilation by intermediate water formation in the northern Arabian Sea is a viable mechanism to explain observed fluctuations in the intensity of the OMZ. We postulate that the necessary decrease in the vertical density gradient during well-ventilated periods resulted from intensified winter monsoonal winds in combination with effects caused by glacio-eustacy.

Today, the northern Arabian Sea is characterized by an intense oxygen minimum zone (OMZ), but the thickness and intensity of this OMZ have varied significantly over geological time (Hermelin 1991; Altabet et al. 1995; Reichart et al. 1998; Schulz et al. 1998; Schulte et al. 1999; Den Dulk et al. 2000). Modern low oxygen values ($[O_2]$ <0.05 ml l^{-1}) between 150 and 1200 m are the result of high biological productivity and a poorly ventilated thermocline. Past variability in OMZ intensity has been attributed to changes in productivity (Altabet et al. 1995; Schulz et al. 1998), or the advection of intermediate waters from the south (Boyle et al. 1995; Schulte et al. 1999). A third possibility invokes changes in the depth of local overturning in the northern Arabian Sea (Reichart et al. 1998). This hypothesis is based on the observation that well-ventilated water column conditions are often accompanied by substantial numbers of *Globorotalia truncatulinoides* and/or *Globorotalia crassaformis*. These species, at present absent in the Arabian Sea (e.g. Cullen & Prell 1984; Anderson 1991), require seasonal deep mixing to complete their annual life cycle.

Here we will test whether the increase in the depth of convective mixing from 100 to >600 m during certain intervals as predicted by the presence of *G. truncatulinoides* and/or *G. crassaformis* is confirmed by reconstructions of the vertical density gradient for these periods. For this reason we reconstructed sea surface temperature (SST) and salinity (SSS) for the past 70 ka by combining planktic and benthic oxygen isotopes and alkenone unsaturation index ($U_{37}^{K'}$) values.

Climate and hydrography of the northern Arabian Sea

During the northern summer, heating of the Tibetan Plateau is at a maximum, resulting in a strong pressure gradient between the Tibetan low-pressure cell and a belt of high pressure over the southern Indian Ocean. This pressure gradient drives warm and humid southwesterly winds, causing coastal and open-ocean upwelling off Oman and Yemen. Under these conditions surface water productivity rises to values that are amongst the highest known for the open ocean (Wyrtki 1973; Smith & Bottero 1977;

Swallow 1984; Brock et al. 1992). During winter cold and dry northeasterly winds blow from the high-pressure cell above central Asia to the region of low pressure associated with the inter-tropical convergence zone (ITCZ) at c. 10°S. These winds cause onshore Ekman transport along the eastern coast of the Arabian Peninsula, suppressing coastal upwelling and productivity (Slater & Kroopnick 1984). Open ocean upwelling during winter ceases as a result of minimum wind-stress curl. The cold and dry northeasterly winds on the other hand result in a deepening of the mixed layer off Pakistan down to 100–125 m (Banse 1984), causing a second productivity maximum (Madhupratap et al. 1996), which is smaller than that in summer.

Annually integrated productivity values amount to 200–400 g C m^{-2} a^{-1} (Kabanova 1968; Qasim 1982). These high values combined with a moderate rate of thermocline ventilation (You & Tomczak 1993) result in the intense OMZ between 150 and 1200 m (Wyrtki 1973; Deuser et al. 1978; Olson et al. 1993) with oxygen concentrations below 0.05 ml l^{-1} (e.g. Van Bennekom et al. 1995). Such low bottom water oxygen concentrations have been shown to strongly affect biological, geochemical and sedimentary processes at the sediment–water interface (Schulz et al. 1996; Jannink et al. 1998; Van der Weijden et al. 1999; Staubwasser & Sirocko 2001).

Surface water temperatures in the northern Arabian Sea show strong seasonal and spatial variation over the year. In the NW, minimum temperatures of 22–23°C are reached during summer as a result of upwelling of cold water (Wyrtki 1971). There is no discernible effect of upwelling upon sea surface temperatures in the northeastern Arabian Sea. Here minimum temperatures of c. 23°C occur during winter whereas maximum temperatures of >28°C are measured during early summer (Levitus & Boyer 1994).

Surface water salinities in the northern Arabian Sea are high (average 36.4 PSU, reaching values up to 37 PSU) as a result of excess evaporation (Wyrtki 1973). Arabian Sea surface water flowing into the Persian Gulf has salinities up to 42 PSU. This high-salinity water flows back into the Arabian Sea, where it is found as a salinity maximum between 150 and 300 m (Wyrtki 1971). The contribution of Red Sea outflow water to the northern Arabian Sea is evident as a small salinity maximum between 700 and 800 m (Wyrtki 1971).

Material and methods

Analyses

Sedimentary organic carbon content was measured on a CNS analyser (Fison NA 1500) after removal of carbonates. Carbonates were removed by extraction using 1M HCl and shaking for 24 h. Samples were then centrifuged for 15 min at 2800 rev min^{-1} and decanted. This was repeated but with time for shaking reduced to 4 h. By rinsing with demineralized water $CaCl_2$ was removed. Analytical precision and accuracy were determined by replicate analyses of samples, and by comparison with an international standard (BCR-71) and in-house standards (F-TURB and MM-91). The relative standard deviations, analytical precision and accuracy were better than 3%.

Element concentrations were measured by inductively coupled plasma atomic emission spectrometry (ICP-AES) (Perkin Elmer Optima 3000) after part of the sample was dried at 60°C for 4 days, thoroughly ground in an agate mortar before $HClO_4$, HNO_3 and HF digestion, and the final residue was taken up in 1 M HCl. Comparison with an international standard (SO-1) and in-house standards (MM-91) and the analyses of duplicate samples revealed that the relative standard deviations, analytical precision and accuracy were better than 3%.

About 2.5 g dry sediment per sample was used for ultrasonic extraction of alkenones with methanol (CH_3OH), methanol and dichloromethane (DCM; CH_2Cl_2) (1:1), and dichloromethane, respectively. The extracts were combined and evaporotated to dryness. Extracts were dried under N_2. Then the extracts were dissolved in DCM–hexane (C_6H_{14}) 1:1, brought on a precombusted aluminium oxide (Al_2O_3) column and eluted with DCM–hexane (1:1). The eluent containing the apolar fraction was dried under N_2 and concentrated in 1 ml ethyl acetate ($C_2H_8O_2$), of which 1 μl was injected into the gas chromatograph. The Hewlett–Packard 6890 series gas chromatograph was equipped with a CPSil5CB column (length 25 m, diameter 0.32 mm, film thickness 0.12 μm), flame ionization detector, and He as carrier gas. The temperature was programmed from 70 to 130°C at 20° min^{-1} and from 130 to 320°C at 4° min^{-1}. The final temperature was maintained for 15 min. C_{37} alkenones were identified on the basis of gas chromatography–mass spectrometry (GC–MS) analysis and the specific retention times. Conditions were: MS operation energy 70 eV; mass range m/z 50–800; cycle time 1.7 ms (resolution 1000). Duplicate measurements on 30 samples have shown a precision for the temperature estimate of about 0.6°C.

For the oxygen isotope measurements about 20 specimens of the selected species were roasted for 30 min at 380°C under vacuum to remove organic remains. Samples were transferred to an automated carbonate preparation unit (IsoCarb) in which the foraminifera were dissolved in 100%

phosphoric acid at 90.0°C. From the released CO_2 the isotopes were measured on an isotope ratio mass spectrometer (VG SIRA 24). Values are reported relative to the Peedee belemnite (PDB) in standard δ notation; calibration was achieved through analyses of National Bureau of Standards (NBS) 18 and 19 reference materials. Precision for $\delta^{18}O$ measurements was better than 0.1‰.

Planktonic foraminiferal and pteropod counts for NIOP478 were carried out on the 150–595 μm fraction. Because relative abundances of *G. truncatulinoides* and *G. crassaformis* are low (<3%) they were quantified using the complete sample, counting their numbers in 27 (of 45) fields of a rectangular picking tray, to a maximum of 30 specimens (Reichart *et al.* 1998). For NIOP458 *G. truncatulinoides* and *G. crassaformis* counts were performed by counting all specimens of both species present in splits (using an Otto micro splitter) from the 150–595 μm fraction.

Sediments and age model

Sediment cores NIOP458 and -478 were taken during the Netherlands Indian Ocean Programme (1992–1993). NIOP478 was taken from an area where the OMZ impinges on the Pakistan margin (24°12.7′N, 065°39.7′E, depth 565 m) (Fig. 1). Sediments at this location consist of dark greenish to light greenish–grey hemipelagic muds, which are mostly homogeneous (bioturbated), but sometimes laminated (non-bioturbated) (Fig. 2). NIOP458 is also located on the Pakistan margin but is from a depth well below the present day OMZ (201°59.4′N, 063°48.7′E, depth 3001 m, Fig. 1). Sediments at station 458 consist of light greenish–grey hemipelagic homogeneous mud. Only a few turbiditic layers were observed in this core, which were avoided during sampling.

The age model for NIOP458 and -478 is based on correlating the *Neogloboquadrina dutertrei* oxygen isotope record (Fig. 3) with that of NIOP464 from the Murray Ridge (Reichart *et al.* 1998). The age model for NIOP464 is based on correlation with the oxygen isotope record of Mediterranean core MD84641, which was tuned by correlating the sapropel pattern with astronomical target curves. This time scale is independent of the SPECMAP time scale. The age model for the upper part of NIOP478 was further improved by including three accelerator mass spectrometry (AMS) ^{14}C datings (Fig. 2). The ^{14}C ages were corrected for a reservoir age of 400 a (Bard 1988), and converted to calendar ages using Stuiver *et al.* (1986) and Bard *et al.* (1990). The base of NIOP478 has an age of about 70 ka. The base of NIOP458 is much older (c. 230 ka), but for this paper we use the time interval also represented by

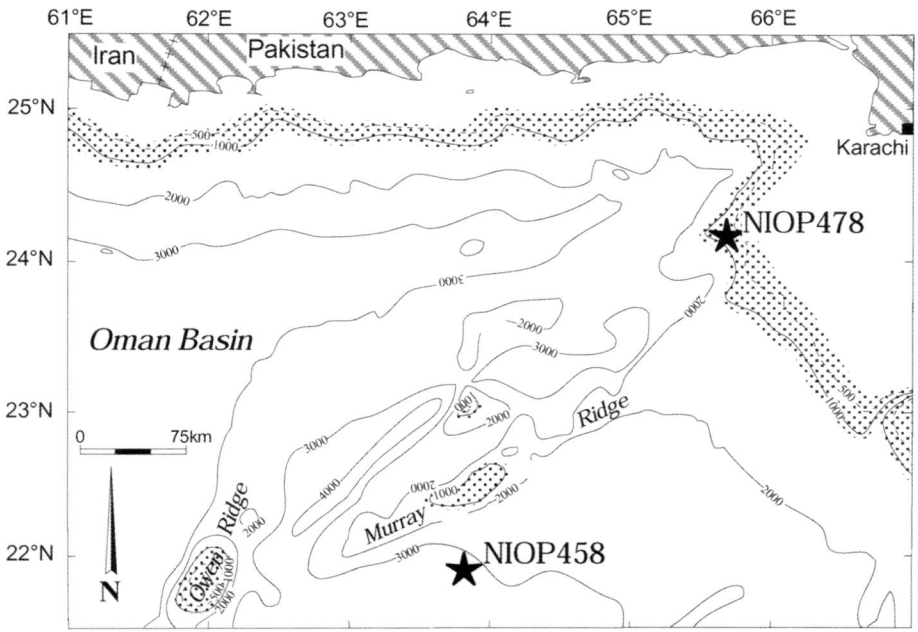

Fig. 1. Location map showing sites NIOP458 and -478 in the northeastern Arabian Sea. Stippled areas indicate intersection of OMZ with sea-floor topography.

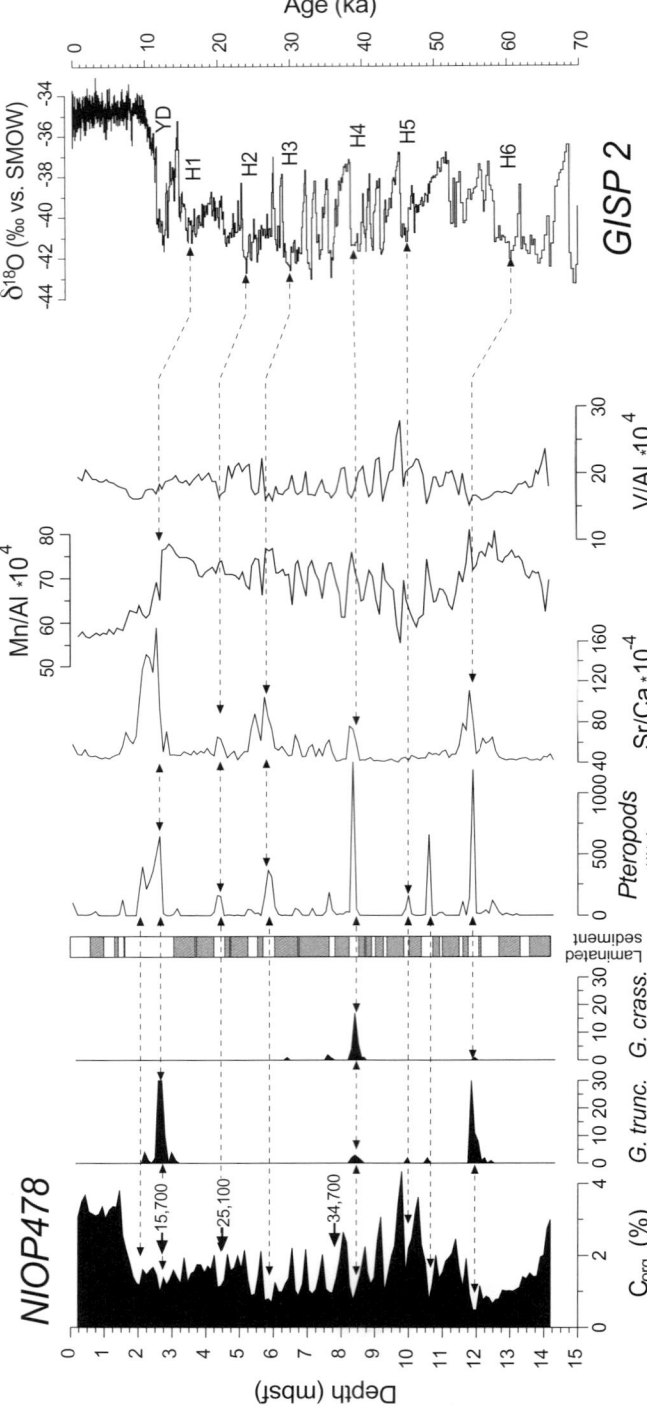

Fig. 2. Records of C$_{org}$, occurrences of *Globorotalia truncatulinoides* and *Globorotalia crassaformis*, laminated intervals, pteropods, and ratios of Sr/Ca, Mn/Al and V/Al (weight ratios) for NIOP478. Proxy records are correlated with the Greenland Ice Sheet Project (GISP2) δ^{18}O$_{ice}$ record (Grootes *et al.* 1993; Sowers *et al.* 1993). Interstadials from the GISP 2 ice core can be correlated with the C$_{org}$ record in great detail. Positions of the Heinrich events and the Younger Dryas (YD) are according to Bond *et al.* (1993).

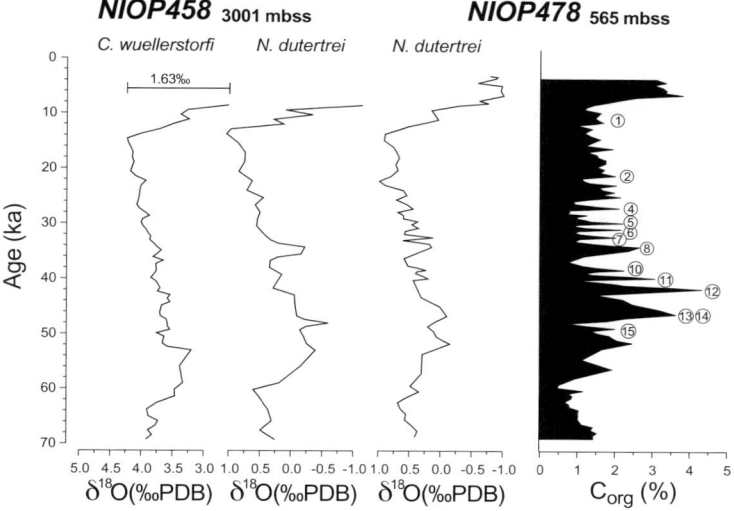

Fig. 3. Correlation between the *N. dutertrei* oxygen isotope records of NIOP458 and -478. Also indicated is the benthic (*C. wuellerstorfi*) oxygen isotope record of NIOP458 and the organic carbon (C_{org}) record of NIOP478. Numbers refer to oxygen isotope maxima of the Dansgaard–Oeschger cycles from the isotope record of the Greenland ice cores, and are transferred from Fig. 2, C_{org} maxima corresponding to relatively high temperatures over Greenland.

NIOP478. Sample resolution for NIOP478 is c. 500 a, whereas sample spacing and lower sedimentation rate at NIOP458 resulted in a c. 1000 a resolution.

Results

Alkenone-derived sea surface temperatures

To derive sea surface temperatures from alkenone unsaturation ratios we used the original equation by Prahl & Wakeham (1987) based on cultures by *Emiliani huxleyi*, which has been confirmed by a global core top calibration (Müller et al. 1998):

$$\text{SST} \;(°C) = 29.412 U_{37}^{K'} - 1.147$$

with $U_{37}^{K'} = C_{37:2}/(C_{37:2} + C_{37:3})$.

As there might be slight differences in calibration between alkenone-producing species it is important to note that the dominant alkenone producers in the Arabian Sea are *E. huxleyi* and/or strains of *Gephyrocapsa oceanica* that behave similarly to *E. huxleyi* (Sonzogni et al. 1997). These algae have been responsible for the alkenone production for at least the last 240 ka (Rostek et al. 1997). For the Holocene period we reconstruct SST of 26.3°C (Fig. 4), which corresponds well to modern annual mean SST of 26.6°C (Levitus & Boyer 1994). As productivity in the study area is highly seasonal, reconstructed temperatures might be biased towards the SST of the productive season. The productivity weighed mean temperature of 26.9°C calculated from pigment data (Banse 1994) is within the analytical range of the alkenone-derived temperature (0.6°C). As the seasonal SST amplitude is about 6°C it is believed that the alkenone-derived temperatures represent annual mean instead of seasonal temperatures (Rostek et al. 1997).

Glacial ice volume effect

The oxygen isotope ratio in the calcitic test of foraminifera is a function not only of sea-water temperature but also of global ice volume, ambient sea-water salinity and (for some species) a biological fractionation (vital effect). Global ice volume through the preferential fixation of ^{16}O in ice sheets, and salinity through the preferential removal of ^{16}O during evaporation, together determine the $\delta^{18}O$ of sea water. The $\delta^{18}O$ composition of sea water can thus be reconstructed from calcitic $\delta^{18}O$ when temperature, vital effects and ice volume are known. By using the relationship between modern sea surface water $\delta^{18}O$ and SSS, past surface water salinities in the northern Arabian Sea can be calculated.

The temperature-dependent isotopic fractionation between sea water and calcite was derived by Epstein et al. (1953) and subsequently modified by Shackleton (1974):

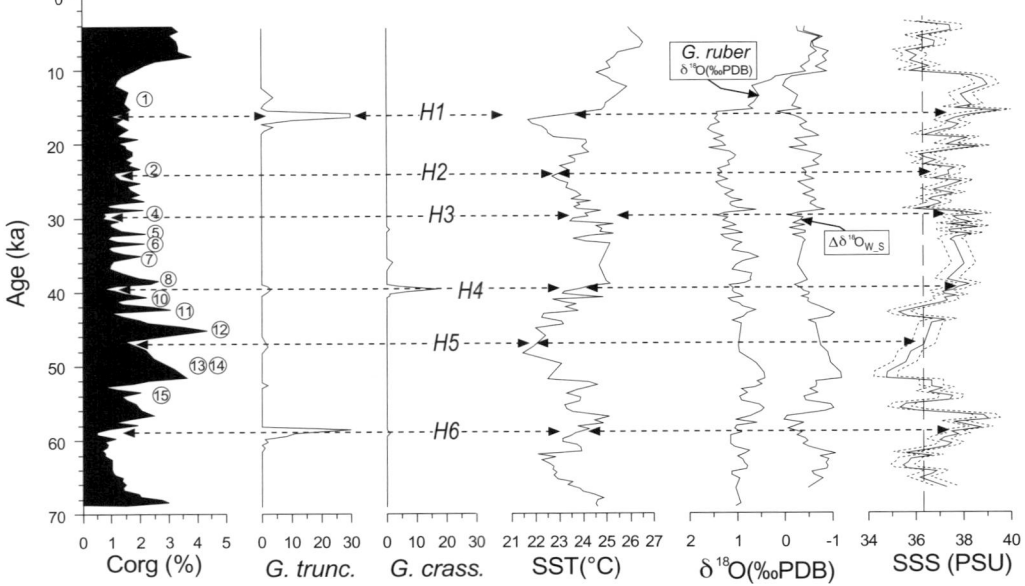

Fig. 4. Time series for C_{org}, occurrences of *G. truncatulinoides* and *G. crassaformis*, and SST calculated from $U_{37}^{K'}$ values; for *G. ruber* both $\delta^{18}O$ and $\Delta\delta^{18}O_{w\_S}$ are plotted. Reconstructed salinity is plotted at the right. Numbers refer to the warm parts of the Dansgaard–Oeschger cycles; H1–H6 refer to the North Atlantic Heinrich events.

$$SST(°C) = 16.9 - 4.38(\delta_c - \delta_w) + 0.10(\delta_c - \delta_w)^2$$

where δ_c is the $\delta^{18}O$ of foraminiferal calcite and δ_w is the $\delta^{18}O$ of sea water (both expressed relative to the same standard).

The $\delta^{18}O$ of deep water benthic foraminifera primarily records global ice volume because temperature changes in the deep sea are generally small. Changes in $\delta^{18}O$ of benthic foraminifera over longer time spans are, therefore, mainly caused by changing ice volume. The $\delta^{18}O$ record based on *Cibicides wuellerstorfi* in NIOP458 is considered to approximate the global ice volume record over the past 70 ka well because present-day water depth at this site is 3000 m. The global glacial–interglacial ice volume change is estimated to have resulted in a 1.2‰ change in $\delta^{18}O$ (Shackleton 1987; Fairbanks 1989). The $\Delta\delta^{18}O$ change for *C. wuellerstorfi* between the last glacial maximum and Recent time in NIOP458 is 1.63‰, implying that 0.43‰ accounts for a glacial bottom water temperature of 1.9°C lower. A previous estimate for the Arabian Sea showed a bottom water temperature rise of *c.* 1.1°C from the last glacial maximum to the Holocene period (Kallel *et al.* 1988). The slight deep water temperature variability can be corrected for, assuming deep water temperature changes proportional to and in phase with ice volume changes (Labeyrie *et al.* 1996):

$$\Delta\delta^{18}O_{(ice\ volume)} = 1.2(\Delta\delta^{18}O_{(C.\ wuellerstorfi)}/1.63).$$

The deep water $\delta^{18}O$ record of NIOP458, corrected for bottom water temperature, was subsequently transferred to NIOP478.

Sea surface temperature and salinity effects on $\delta^{18}O$

The surface water $\delta^{18}O$ record is based on *Globigerinoides ruber* in NIOP478 (Fig. 4). This species inhabits the relatively warm oligotrophic surface mixed layer, and is abundant throughout the year (Conan & Brummer 2000), although highest shell fluxes are recorded during summer (Curry *et al.* 1992). *Globigerinoides ruber* $\delta^{18}O$ values from box core tops, corresponding to present SST, are 2.15 ± 0.1‰. Measured $\delta^{18}O$ values of *G. ruber* were raised by +0.2‰ to correct for the vital effect of this species (e.g. Fairbanks *et al.* 1980, 1982). The corrected values result in a modern isotope temperature of 29.0°C, corresponding to the average temperature in the intermonsoonal periods. Correction for a 0.85‰ offset found in the Indian Ocean by Duplessy *et al.* (1981) results in reconstructed modern SST of 26.0°C, corresponding to the mean annual SST (26.6°C, Levitus & Boyer 1994) and close to the reconstructed SST for Holocene time with $U_{37}^{K'}$ (26.3°C; see above). As we use for the reconstructions the difference relative to the Holocene period, $\Delta\delta^{18}O$, the small offset in Holocene estimates has no effect on the SSS reconstructions. Correction of $\delta^{18}O$ values for *C. wuellerstorfi* for vital effects is not necessary, as values are already normalized to 1.2‰ using the method of Labeyrie *et al.* (1996).

Changes in surface water $\delta^{18}O_w$ over time are defined by changes in global ice volume ($\Delta\delta^{18}O_{w\_Ice}$) and local salinity ($\Delta\delta^{18}O_{w\_S}$)

$$\Delta T_{dt} = -4.38[\Delta\delta_{ruber(dt)} - (\Delta\delta_{w\_Ice} + \Delta\delta_{w\_S})_{(dt)}]$$
$$+ 0.10[\Delta\delta_{ruber(dt)} - (\Delta\delta_{w\_Ice} + \Delta\delta_{w\_S})_{(dt)}]^2$$

with ΔT_{dt} being the change in SST over time relative to Holocene SST.

The alkenone-derived ΔT_{dt} and $\Delta\delta_w$ related to changes in ice volume $(\Delta\delta_{w\_Ice})_{(dt)}$ are known and can be inserted in the formula so that $\Delta\delta^{18}O_{w\_S}$ can be calculated for any interval over the past 70 ka. Subsequently, surface water salinities can be calculated from the $\Delta\delta^{18}O_{w\_S}$ values by using the published relationships between surface water $\delta^{18}O$ and SSS in the Arabian Sea: $\delta^{18}O_w(VSMOW) = -8.98 + 0.26 \times SSS$ (Peeters 2000) or $\delta^{18}O_w(SMOW) = -9.24 + 0.28 \times SSS$ (Rostek *et al.* 1993). The two equations are similar, but we prefer here to use the relationship published by Rostek *et al.* (1993) because this relationship was derived from sea-water samples between 20°S and 21°N covering a larger area than the samples used by Peeters (2000), which were limited to the western Arabian Sea.

The relationship between $\delta^{18}O_w$ and salinity may have been different in the past, when other moisture sources may have provided fresh water to the Arabian Sea. Considering the geographical position of the Arabian Sea this seems not very likely. We therefore assume that this $\delta^{18}O$-salinity relationship remains applicable over the studied period.

The modern mean salinity of 36.4‰ in the northeastern Arabian Sea (Levitus & Boyer 1994) is assigned to the mean Holocene $\Delta\delta^{18}O_{w\_S}$. The error for $\Delta\delta^{18}O_{w\_S}$ calculations based on the propagation of errors for alkenone (±0.6°C, corresponding to c. 0.12‰) and isotope (0.10‰ for *G. ruber*) measurements into the SST calculations is about 0.22‰, which results in an error of about 1 PSU for the salinity calculations. Uncertainties in splicing the deep water $\delta^{18}O$ record of NIOP458 to the surface water $\delta^{18}O$ record of NIOP478 may have introduced an additional small error in the reconstructed salinity record.

Discussion

Short-term changes in OMZ intensity in the northern Arabian Sea are reflected by an alternation of bioturbated and non-bioturbated (laminated) sediments (Fig. 2). The laminated sediments are enriched in organic matter and trace metals (e.g. vanadium) characteristic of water column dysoxia. The bioturbated sediments, on the other hand, show high concentrations of Mn, the deposition of which is favoured under oxygenated bottom water conditions (Calvert *et al.* 1996), and high Sr to Ca ratios as a result of deepening of the local aragonite compensation depth (ACD) (Reichart *et al.* 1997, 1998).

Comparing northern Arabian Sea proxy records with the $\delta^{18}O$ record of the Greenland ice cores shows that well-oxygenated conditions in the Arabian Sea correspond to cold climatic conditions in Greenland and North Atlantic Heinrich events (Reichart *et al.* 1998, 2002; Schulz *et al.* 1998; Von Rad *et al.* 1999).

Well-ventilated conditions in the northern Arabian Sea are mostly associated with the occurrence of substantial numbers of *G. truncatulinoides* and/or *G. crassaformis* (Fig. 2). In particular, the well-ventilated conditions during periods equivalent to Heinrich events H1, H4 and H6 are characterized by relatively high abundances of one of the two species. Both species are strikingly absent, however, during periods with a pronounced OMZ. A perusal of the literature suggests that this presence–absence pattern of the two species is controlled somehow by changes in the depth of convective overturning in the northern Arabian Sea. Both of these species have an annual life cycle and individuals sink to deeper water as they increase in size (Hemleben *et al.* 1985; Lohmann & Schweitzer 1990). After reproduction at depth juveniles are returned to the surface.

Sediment trap and tow data suggest that *G. truncatulinoides* and *G. crassaformis* reproduce at c. 600 and c. 300 m, respectively (see discussion by Reichart *et al.* 1998). Maintaining living populations thus requires a mechanism to return juveniles to the surface. A viable mechanism would be seasonal (winter) overturning. A strong permanent pycnocline would prevent juvenile shells from reaching surface waters. With Stokes' law it can be calculated that without physical mixing it is impossible for juveniles to reach the sea surface, even with unrealistic foraminiferal specific densities of less than 0.5 g cm^{-3}. (Considering viscous fluid resistance forces only, the ascending velocity (v) of an ideally spherical ($\emptyset$,

d 25 µm) foraminifer with specific mass ρ_{foram} in sea water with density ρ_{sw} (1.03 g cm^{-2}) and viscosity υ (1.5×10^{-2} cm^2 s^{-1}), and gravity g (980 cm s^{-2}) can be predicted with Stokes' law: $v = [gd^2(\rho_{sw} - \rho_{foram})]/(18\upsilon)$.

The reconstructed salinity and temperature record (Fig. 4) allows us to test whether the depth of convective mixing indeed increased during periods corresponding to Heinrich events as predicted by the substantial numbers of *G. truncatulinoides* and/or *G. crassaformis* present during such periods. In quantifying the depth of overturning in the northern Arabian Sea during these periods we should note that the alkenone-derived SST values represent mean annual temperatures, and that the derived surface water densities are thus underestimated. Reconstructed salinities also represent mean annual values, but are less likely to have varied significantly over the year.

During glacial times, fresh water is stored in continental ice sheets, and thus global salinity increased during these periods. This salinity increase can be approximated from $Sp = Sm + [\Delta SL(35/3800)]$, where Sp is past salinity, Sm is modern salinity, ΔSL is sea-level change, 35 represents modern average salinity and 3800 average ocean depth in meters (Rostek et al. 1993). This implies that during the last glacial maximum, when sea level was 120 m lower, salinity should have been c. 1.1 PSU higher on average. For the average SSS in the northern Arabian Sea this would mean c. 37.5 PSU. The reconstructed 39 PSU (Fig. 4) value indicates that other, regional factors must have influenced surface water salinity substantially. Increased evaporation as a result of intensified cold and dry winter monsoons may be one of these factors. A possible other factor is a dried-up Persian Gulf and a virtually isolated Red Sea as a result of glacio-eustasy. Today surface water from the Arabian Sea flows into the Persian Gulf and Red Sea and some of this water is returned at depth. Excess surface water salt is thus exported to subsurface depth. No such export mechanism existed during glacial lowstands through which overall surface water salinity in the northern Arabian Sea could have been increased.

Previous temperature reconstructions for the upwelling area off Arabia show glacial cooling of somewhat more than 3°C, but in this area temperature changes are dampened by reduced upwelling during glacial times (Emeis et al. 1995). A similar change in glacial to Holocene SST has been reconstructed for the eastern part of the Arabian Sea; Rostek et al. (1993) reconstructed for offshore India a minimum glacial SST of 25°C. For the mouth of the Gulf of Aden glacial temperatures between 23 and 24°C were calculated, implying a ±4°C glacial to Holocene difference. Reconstructed temperatures of less than 22°C for the last glacial, based on the alkenone record of NIOP478, indicate a more severe cooling in the northernmost part of the Arabian Sea (Fig. 4). Glacial SST reconstructions generally suggest only a moderate cooling (<3°C) for the (sub)tropics (Billups & Spero 1996; Bard et al. 1997; Sonzogni et al. 1998). Both the exceptionally low temperatures in the northern Arabian Sea and the rapid fall in temperatures towards the north suggest that cold and intensified winter monsoon winds were responsible for low SST values offshore Pakistan during the last glacial. A dominant role of the winter monsoon in causing a fall in glacial temperatures is also indicated by concentrations of noble gases in ground waters from northern Oman, which indicate a lowering of glacial winter temperatures of ±6.5°C (Weyhenmeyer et al. 2000).

Calculated modern sea surface water density (SSD) using mean annual temperature and salinity data is about 1024 kg m^{-3} (UNESCO 1981; Gill 1982). This value is in agreement with the density values calculated by Wyrtki (1971). At the salinity maximum between 200 and 300 m (representing Persian Gulf outflow water) density has increased to 1026.5 kg m^{-3}. Density increases further to 1027.5 kg m^{-3} at the small salinity maximum between 700 and 800 m (representing Red Sea outflow water).

To reconstruct the depth of convective mixing for periods corresponding to the Heinrich events, deep water salinities were corrected for the ice volume effect, resulting in salinities at 700–800 m of about 36.5‰, and densities of about 1028–1028.5 kg m^{-3} during the last glacial. Surface water densities calculated using the reconstructed mean annual temperatures and salinities for the intervals in the record corresponding to North Atlantic Heinrich events never reach values over c. 1027 kg m^{-3}. This would not allow convective mixing to depths predicted by the presence of substantial numbers of *G. truncatulinoides* during H1 and H6. However, deep convective mixing does not depend on mean annual temperature, but rather on the minimum temperature reached during the winter. Today, winter temperatures are about 4°C colder than the mean annual temperature, and this value, if applied to the periods corresponding to Heinrich events 1 and 6, would have resulted in surface water densities of c. 1028 kg m^{-3} (Table 1) and winter mixing to depths of 700–800 m. It is likely that actual overturning during winter extended even deeper, as the subsurface salinity maxima related to the outflow of Persian Gulf and Red Sea water were absent during the last glacial. Sea level was

Table 1. *Reconstructed mean annual and winter sea surface temperatures, salinities and densities in the northern Arabian Sea for time slices corresponding to the North Atlantic Heinrich events 1–6*

Event	Salinity (PSU)	Mean annual temperature (°C)	Winter temperature (°C)	Density (kg m^{-3})
H1	38.5	21.5	17.5	1028.1
H2	37.5	22.5	18.5	1027.1
H3	38.2	23.5	19.5	1027.3
H4	37.6	23.0	19.0	1027.0
H5	36.4	22.0	18.0	1026.3
H6	38.5	23.0	19.0	1027.7

already more than 60 m lower at 70 ka bp and did not rise substantially until Holocene time (Lambeck & Chappell 2001), which resulted in the drying up of the Persian Gulf and isolation of the Red Sea,. The reconstructed sea surface densities for the winter (Table 1) indicate high values for all six periods, with maxima reached during H1 and H6, corresponding to the *G. truncatulinoides* peaks. However, the *G. crassaformis* peak during H4 is not matched by a similar sea-water density maximum; densities were higher during H2 and H3. Although convective overturn was deep enough to allow *G. crassaformis* to complete its annual life cycle, during H1, H2 and H5 other factors prevented this species from reaching substantial numbers.

Minimum sea surface temperatures are related to distinct cold surges within the winter monsoon flow. These cold surges are primarily known from SE Asia, where they cause abnormal low temperatures of more than 15°C below the average winter temperature, but are also known from other Asian monsoon areas and even occur over the South American continent (McGregor & Nieuwolt 1997). An intensification of these cold surges could explain the 7°C colder winter temperatures during the last glacial found in the South China Sea (Thunnell & Miao 1996).

Calibration of the northern Arabian Sea TOC record to the Greenland Ice Sheet Project (GISP) $\delta^{18}O$ record and the North Atlantic Heinrich massive ice rafting events (Fig. 2) shows that low-productivity conditions in the northern Arabian Sea prevailed during the cold phases of the Dansgaard–Oeschger cycles and thus Heinrich events. Low-productivity conditions during Heinrich events are compatible with an increase in the depth of winter overturning, as the local ventilation of the thermocline should have reduced subsurface nutrient concentration and thus lowered summer productivity, even under conditions of strong SW monsoonal winds. High-productivity conditions, on the other hand, are in phase with the warm phases of the Dansgaard–Oeschger cycles, when upwelling of nutrient-rich waters from the pronounced OMZ enhanced upwelling.

During Holocene time the OMZ in the northern Arabian Sea has been a persistent feature (Schulz et al. 1998; Von Rad et al. 1999), whereas during the last glacial rapid changes between a situation with a well-oxygenated water column and a situation comparable with the present occurred. Figure 5 gives a schematic outline of the switches in hydrography in the northern Arabian Sea as a response to cooling during Heinrich events. The surface waters flowing into the marginal basins (Red Sea and Persian Gulf), which are returned at depth, strengthen stratification during Holocene time (Fig. 5a). High surface water productivity and ventilation of the intermediate water with relatively old water are responsible for a pronounced OMZ. When sea level is lowered during glacials the marginal basins are dry or effectively isolated, which indirectly increases surface water salinity and reduces water column stratification. Without deep convective overturn, however, a stable OMZ can still develop, nutrients from the subsurface are easily recycled and high productivity persists (Fig. 5b). Because of the reduced stability of the water column, an increased cooling and evaporation at the sea surface by an intensified winter monsoon will result in deep convective overturn and the local formation of intermediate water (Fig. 5c). The OMZ is ventilated and the subsurface waters become relatively nutrient poor. The oxygen demand of the subsurface waters further decreases as the upwelling of nutrient-depleted waters no longer supports a massive summer bloom. When the winter climate becomes milder again, between the Heinrich events or after the flooding of the marginal basins, a stable OMZ develops again.

Conclusions

A reconstruction of surface water temperatures and salinities in the northern Arabian Sea over the past 70 ka indicates that the depth of convective winter mixing increased substantially during periods characterized by a well-ventilated water column and low-productivity conditions. These periods correspond to the coldest phases in the Dansgaard–Oeschger cycles, the Heinrich events. This breakdown of the OMZ thus seems related to local intermediate water formation and confirms earlier conclusions based on the periodical presence of substantial numbers of *G. truncatulinoides* and/or *G. crassaformis* in the northern Arabian Sea (Reichart et al. 1998). Subsurface nutrient concentrations during periods of inter-

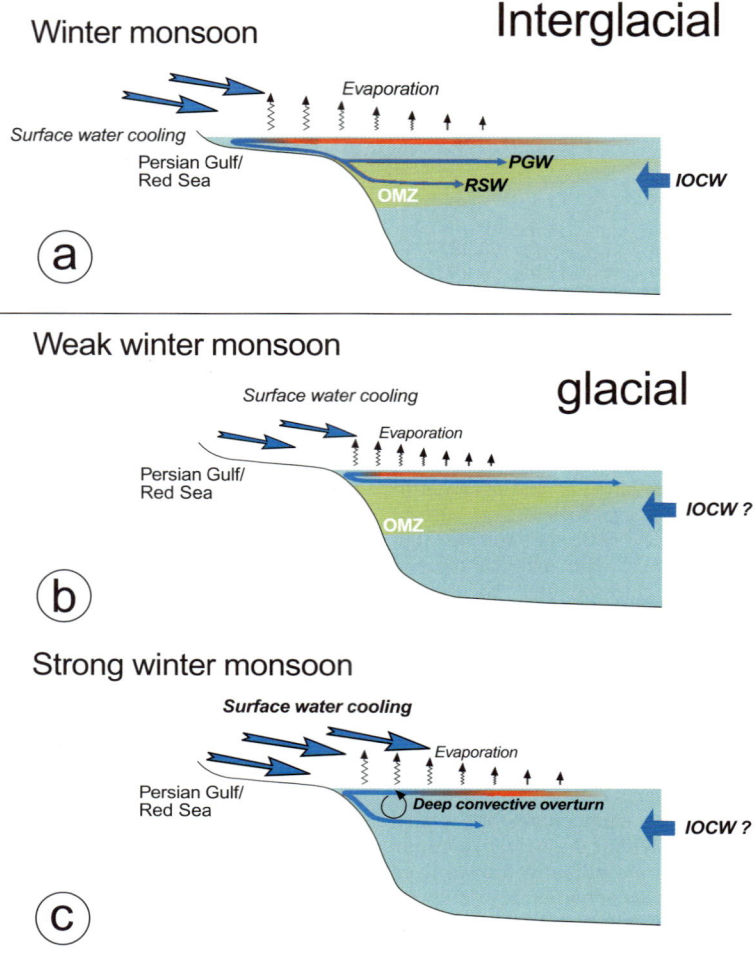

Fig. 5. Schematic representation of surface and intermediate water circulation during present (**a**), and glacial times (**b, c**). During glacials the marginal basins dry up or become effectively isolated. An intensified winter monsoon enhances both evaporation and surface water cooling, triggering deep convective overturn (**c**). Superimposed on the winter monsoon variability the inflow of Indian Ocean Central Water (IOCW) may have varied. PGW, Persian Gulf water; RSW, Red Sea water.

mediate water formation must have been low, resulting in reduced (summer) productivity conditions and a further decrease in subsurface oxygen consumption. The physical cause for intermediate water formation in the northern Arabian Sea during the most pronounced cold phases of the Dansgaard–Oeschger cycles is surface water cooling and salinification as a result of intensified winter monsoonal winds and a lower sea level.

This research was carried out as part of the Netherlands Indian Ocean Programme 1992–1993, supported by the Netherlands Marine Research Foundation (SOZ) of the Netherlands Organization of Scientific Research (NWO).

R. Schneider and F. Jansen are acknowledged for their constructive reviews. We thank the officers and crew of the R.V. *Tyro*, and acknowledge the technical support of the technicians from NIOZ. We also thank A. van Dijk, G. Itman, G. Nobbe, H. de Waard and G. van het Veld for analytical assistance. This research was funded partly by NWO, Grant 800.09.120, to G.J.R.

References

ALTABET, M.A., FRANCOIS, R., MURRAY, D.W. & PRELL, W.L. 1995. Climate-related variations in denitrification in the Arabian Sea from sediment $^{15}N/^{14}N$ ratios. *Nature*, **373**, 506–509.

ANDERSON, D.M. 1991. *Foraminifer evidence of monsoon upwelling off Oman during the late Quaternary*. PhD thesis, Brown University, Providence, RI.

BANSE, K. 1984. Overview of the hydrography and associated phenomena in the Arabian Sea, off Pakistan. *In:* HAQ, B.U. & MILLIMAN, J.D. (eds) *Marine Geology and Oceanography of Arabian Sea and Coastal Pakistan*. Van Nostrand Reinhold, New York, 271–303.

BANSE, K. 1994. On the coupling of hydrography, phytoplankton, zooplankton, and settling organic particles offshore in the Arabian Sea. *In:* LAL, D. (ed.) *Biogeochemistry of the Arabian Sea*. Indian Academy of Sciences, Bangalore, 27–63.

BARD, E. 1988. Correction of accelerator mass spectrometry ^{14}C ages measured in planktonic foraminifers: paleoceanographic implications. *Paleoceanography*, 3, 635–645.

BARD, E., HAMELIN, B., FAIRBANKS, R.G. & ZINDER, A. 1990. Calibration of the ^{14}C timescale over the past 30,000 years using mass spectrometric U–Th ages from Barbados corals. *Nature*, 345, 405–410.

BARD, E., ROSTEK, F. & SONZOGNI, C. 1997. Interhemispheric synchrony of the last deglaciation inferred from alkenone palaeothermometry. *Nature*, 385, 707–710.

BILLUPS, K. & SPERO, H.J. 1996. Reconstructing the stable isotope geochemistry and paleotemperatures of the equatorial Atlantic during the last 150,000 years: results from individual foraminifera. *Paleoceanography*, 11, 217–238.

BOND, G., BROECKER, W., JOHNSEN, S., MCMANUS, J., LABEYRIE, L., JOUZEL, J. & BONANI, G. 1993. Correlations between climate records from the North Atlantic sediments and Greenland ice. *Nature*, 365, 143–147.

BOYLE, E.A., LABEYRIE, L. & DUPLESSY, J. 1995. Calcite foraminiferal data confirmed by cadmium in aragonite *Hoeglundina*: application to the last glacial maximum in the northern Indian Ocean. *Paleoceanography*, 10, 881–900.

BROCK, J.C., MCCLAIN, C.R., ANDERSON, D.M., PRELL, W.L. & HAY, W.W. 1992. Southwest monsoon circulation and environments of recent planktonic foraminifera in the northwest Arabian Sea. *Paleoceanography*, 7, 799–813.

CALVERT, S.E., BUSTIN, R.M. & INGALL, E.D. 1996. Influence of water column anoxia and sediment supply on the burial and preservation of organic carbon in marine shales. *Geochimica et Cosmochimica Acta*, 60, 1577–1593.

CONAN, S.M.H. & BRUMMER, G.J.A. 2000. Fluxes of planktic foraminifera in response to monsoonal upwelling on the Somalia Basin margin. *Deep-Sea Research*, 47, 2207–2227.

CULLEN, J.L. & PRELL, W.L. 1984. Planktonic foraminifera of the northern Indian Ocean: distribution and preservation in surface sediments. *Marine Micropaleontology*, 9, 1–52.

CURRY, W.B., OSTERMAN, D.R., GUPTHA, M.V.S. & ITTEKKOT, V. 1992. Foraminiferal production and monsoonal upwelling in the Arabian Sea: evidence from sediment traps. *In:* SUMMERHAYES, C.P., PRELL, W.L. & EMEIS, K.C. (eds) *Upwelling Systems: Evolution Since the Early Miocene*. Geological Society, London, Special Publications, 64, 93–106.

DEN DULK, M., REICHART, G.J., VAN HEYST, S., ZACHARIASSE, W.J. & VAN DER ZWAAN, G.J. 2000. Benthic foraminifera as proxies of organic matter flux and bottom water oxygenation? A case history from the northern Arabian Sea. *Palaeogeography, Palaeoclimatology, Palaeoecology*, 161, 337–359.

DEUSER, W.G., ROSS, E.H. & MLODZINSKA, C. 1978. Evidence for and rate of denitrification in the Arabian Sea. *Deep-Sea Research*, 25, 431–445.

DUPLESSY, J.C., BLANC, P.L. & BÉ, A.W.H. 1981. Oxygen-18 enrichment of planktonic foraminifera due to gametogenic calcification below the euphotic zone. *Science*, 213, 1247–1250.

EMEIS, K.-C., ANDERSON, D.M., DOOSE, H., KROON, D. & SCHULZ-BULL, D. 1995. Sea-surface temperatures and the history of monsoon upwelling in the northwestern Arabian Sea during the last 500,000 years. *Quaternary Research*, 43, 355–361.

EPSTEIN, S., BUCHSBAUM, R., LOWENSTAM, H.A. & UREY, H.C. 1953. Revised carbonate–water isotopic temperature scale. *Geological Society of America Bulletin*, 64, 1315–1326.

FAIRBANKS, R.G. 1989. A 17000-year glacio-eustatic sea level record: influence of glacial melting rates on the Younger Dryas event and deep ocean circulation. *Nature*, 342, 637–642.

FAIRBANKS, R.G., WIEBE, P.H. & BÉ, A.W.H. 1980. Vertical distribution and isotopic composition of living planktonic foraminifera in the western North Atlantic. *Science*, 207, 61–63.

FAIRBANKS, R.G., SVERDLOVE, M., FREE, R., WIEBE, P.H. & BÉ, A.W.H. 1982. Vertical distribution and isotopic fractionation of living planktonic foraminifera from the Panama Basin. *Nature*, 298, 841–844.

GILL, A. 1982. *Atmosphere–Ocean Dynamics*. International Geophysical Series, 30, 599–600.

GROOTES, P.M., STUIVER, M., WHITE, J.W.C., JOHNSEN, S. & JOUZEL, J. 1993. Comparison of oxygen isotope records from GISP2 and GRIP Greenland ice cores. *Nature*, 366, 552–554.

HEMLEBEN, C., SPINDLER, M., BREITINGER, I. & DEUSER, W.G. 1985. Field and laboratory studies on the ontogeny and ecology of some globorotaliid species from the Sargasso Sea off Bermuda. *Journal of Foraminiferal Research*, 15, 254–272.

HERMELIN, J.O.R. *et al.* 1991. The benthic foraminiferal faunas of sites 725, 726, and 728 (Oman Margin, northwestern Arabian Sea). *In:* PRELL, W.L. & NIITSUMA, N. (eds) *Proceedings of the Ocean Drilling Program, Scientific Results, 117*. Ocean Drilling Program, College Station, TX, 55–87.

JANNINK, N.T., ZACHARIASSE, W.J. & VAN DER ZWAAN, G.J. 1998. Living (Rose Bengal stained) benthic foraminifera from the Pakistan continental margin (northern Arabian Sea). *Deep-Sea Research I*, 45, 1483–1513.

KABANOVA, Y.G. 1968. Primary production in the northern part of the Indian Ocean. *Oceanology*, 8, 214–225.

KALLEL, N., LABEYRIE, L.D., JULLIER-LECLERC, A. & DUPLESSY, J.-C. 1988. A deep hyrological front between intermediate and deep-water masses in the glacial Indian Ocean. *Nature*, **333**, 651–655.

LABEYRIE, L., LABRACHERIE, M., GORFTI, N. & 9 OTHERS 1996. Hydrographic changes of the Southern Ocean (southeast Indian sector) over the last 230 kyr. *Paleoceanography*, **11**, 57–76.

LAMBECK, K. & CHAPPELL, J. 2001. Sea level change through the last glacial cycle. *Science*, **292**, 679–685.

LEVITUS, S. & BOYER, T.P. 1994. *World Ocean Atlas 1994, Vol. 4: Temperature, NOAA Atlas NESDIS 4.* US Government Printing Office, Washington, DC.

LOHMANN, G.P. & SCHWEITZER, P.N. 1990. *Globorotalia truncatulinoides* growth and chemistry as probes of the past thermocline, 1, shell size. *Paleoceanography*, **5**, 55–75.

MADHUPRATAP, M., PRASANNA KUMAR, S., BHATTATHIRI, P.M.A., DILEEP KUMAR, M., RAGHUKUMAR, S., NAIR, K.K.C. & RAMAIAH, N. 1996. Mechanism of the biological response to winter cooling in the northeastern Arabian Sea. *Nature*, **384**, 549–552.

MCGREGOR, G.R. & NIEUWOLT, S. 1997. *Tropical Climatology: an Introduction to the Climates of the Low Latitudes*. Wiley, Chichester.

MÜLLER, P.J., KIRST, G., RUHLAND, G., VON STORCH, I. & ROSSELL-MELÉ, A. 1998. Calibration of the alkenone paleotemperature index Uk'37 based on core-tops from the eastern South Atlantic and the global ocean (60°N–60°S). *Geochimica et Cosmochimica Acta*, **62**, 1757–1772.

OLSON, D.B., HITCHCOCK, G.L., FINE, R.A. & WARREN, B.A. 1993. Maintenance of the low-oxygen layer in the Central Arabian Sea. *Deep-Sea Research II*, **40**, 473–685.

PEETERS, F. 2000. *The distribution and stable isotope composition of living foraminifera in relation to seasonal changes in the Arabian Sea*. PhD thesis, Vrije Universiteit Amsterdam.

PRAHL, F.G. & WAKEHAM, S.G. 1987. Calibration of unsaturation patterns in long-chain ketone compositions for palaeotemperature assessment. *Nature*, **330**, 367–369.

QASIM, S.Z. 1982. Oceanography of the northern Arabian Sea. *Deep-Sea Research*, **29**, 1041–1068.

REICHART, G.J., DEN DULK, M., VISSER, H.J., VAN DER WEIJDEN, C.H. & ZACHARIASSE, W.J. 1997. A 225 kyr record of dust supply, paleoproductivity and the oxygen minimum zone from the Murray Ridge (northern Arabian Sea). *Palaeogeography, Palaeoclimatology, Palaeoecology*, **134**, 149–169.

REICHART, G.J., LOURENS, L.J. & ZACHARIASSE, W.J. 1998. Temporal variability in the northern Arabian Sea Oxygen Minimum Zone (OMZ) during the last 225,000 years. *Paleoceanography*, **13**, 607–621.

REICHART, G.J., SCHENAU, S.J., DE LANGE, G.J. & ZACHARIASSE, W.J. 2002. Synchroneity of oxygen minimum zone intensity on the Oman and Pakistan margins at sub-Milankovitch time scales. *Marine Geology*, **185**, 403–415.

ROSTEK, F., BARD, E., BEAUFORT, L., SONZOGNI, C. & GANSSEN, G. 1997. Sea surface temperature and productivity records for the past 240 kyr in the Arabian Sea. *Deep-Sea Research II*, **44**, 1461–1480.

ROSTEK, F., RUHLAND, G., BASSINOT, F.C., MÜLLER, P.J., LABEYRIE, L.D., LANCELOT, Y. & BARD, E. 1993. Reconstructing sea surface temperature and salinity using $\delta^{18}O$ and alkenone records. *Nature*, **364**, 319–321.

SCHULTE, S., ROSTEK, F., BARD, E., RULLKÖTTER, J. & MARCHAL, O. 1999. Variations of oxygen-minimum and primary productivity recorded in sediments of the Arabian Sea. *Earth and Planetary Science Letters*, **173**, 205–221.

SCHULZ, H., VON RAD, U. & VON STACKELBERG, U. 1996. Laminated sediments from the oxygen-minimum zone of the northeastern Arabian Sea. *In:* KEMP, A.E.S. (ed.) *Paleoclimatology and Paleoceanography from Laminated Sediments*. Geological Society, London, Special Publications, **116**, 185–207.

SCHULZ, H., VON RAD, U. & ERLENKEUSER, H. 1998. Correlation between Arabian Sea and Greenland climate oscillations of the past 110,000 years. *Nature*, **393**, 54–57.

SHACKLETON, N.J. 1974. Attainment of isotopic equilibrium between ocean water and the benthonic foraminifera genus *Uvigerina*: isotopic changes in the ocean during the last glacial. *In:* LABEYRIE, J. (ed.) *Les Méthodes Quantitatives d'Étude des Variations du Climat au Cours du Pleistocene*. Centre National de la Recherche Scientifique, Paris, 203–210.

SHACKLETON, N.J. 1987. Oxygen isotopes, ice volume and sea level. *Quaternary Science Reviews*, **6**, 183–190.

SLATER, R.D. & KROOPNICK, P. 1984. Controls on dissolved oxygen distribution and organic carbon distribution in the Arabian Sea. *In:* HAQ, B.U. & MILLIMAN, J.D. (eds) *Marine Geology and Oceanography of Arabian Sea and Coastal Pakistan*. Van Nostrand Reinhold, New York, 305–313.

SMITH, R.L. & BOTTERO, J.S. 1977. Nutrients as tracers of water mass structure in the coastal upwelling off northwest Africa. *In:* ANGEL, M. (ed.) *A Voyage of Discovery*. Pergamon, Oxford, 291–304.

SONZOGNI, C., BARD, E., ROSTEK, F., DOLLFUS, D., ROSELL-MELÉ, A. & EGLINTON, G. 1997. Temperature and salinity effects on alkenone ratios measured in surface sediments from the Indian Ocean. *Quaternary Research*, **47**, 344–355.

SONZOGNI, C., BARD, E. & ROSTEK, F. 1998. Tropical sea-surface temperatures during the last glacial period: a view based on alkenones in Indian Ocean sediments. *Quarterly Science Reviews*, **17**, 1185–1201.

SOWERS, T., BENDER, M., LABEYRIE, L. & 5 OTHERS 1993. 135,000 year Vostok-SPECMAP common temporal framework. *Paleoceanography*, **8**, 737–766.

STAUBWASSER, M. & SIROCKO, F. 2001. On the formation of laminated sediments on the continental margin off Pakistan: the effects of sediment provenance and sediment redistribution. *Marine Geology*, **172**, 43–56.

STUIVER, M., PEARSON, G.W. & BRAZIUNAS, T.F. 1986. Radiocarbon age calibration of marine samples back to 9000 cal yr BP. *Radiocarbon*, **28**, 980–1021.

SWALLOW, J.C. 1984. Some aspects of the physical oceanography of the Indian Ocean. *Deep-Sea Research Part A*, **31**, 639–650.

THUNNELL, R.C. & MIAO, Q. 1996. Sea surface temperature of the Western Equatorial Pacific Ocean during the Younger Dryas. *Quaternary Research*, **46**, 72–77.

UNESCO 1981. *Tenth Report of the Joint Panel on Oceanographic Tables and Standards*. UNESCO Technical Papers in Marine Science, **36**.

VAN BENNEKOM, A.J., HIEHLE, M., VAN OOYEN, J., VAN WEERLEE, E. & VAN KOUTRIK, M. 1995. CTD and hydrography. *In:* VAN HINTE, J.E., VAN WEERING, T.C.E. & TROELSTRA, S.R. (eds) *Tracing a Seasonal Upwelling, Report on Two Cruises of R.V. Tyro to the NW Indian Ocean in 1992 and 1993*. National Museum of Natural History, Leiden, 41–54.

VAN DER WEIJDEN, C.H., REICHART, G.J. & VISSER, H.J. 1999. Enhanced preservation of organic matter in sediments underlying the Oxygen Minimum Zone in the northeastern Arabian Sea. *Deep-Sea Research I*, **46**, 807–830.

VON RAD, U., SCHULZ, H., RIECH, V., DEN DULK, M., BERNER, U. & SIROCKO, F. 1999. Multiple monsoon-controlled breakdown of oxygen-minimum conditions during the past 30,000 years documented in laminated sediments off Pakistan. *Palaeogeography, Palaeoclimatology, Palaeoecology*, **152**, 129–161.

WEYHENMEYER, C.E., BURNS, S.J., WABER, H.N., AESCHBACH-HERTIG, W., KIPFER, R., LOOSLI, H.H. & MATTER, A. 2000. Cool glacial temperatures and changes in moisture source recorded in Oman groundwaters. *Science*, **287**, 842–845.

WYRTKI, K. 1971. *Oceanographic Atlas of the International Indian Ocean Expedition*. Publisher, Town.

WYRTKI, K. 1973. Physical oceanography of the Indian Ocean. *In:* ZEITSCHEL, B. (ed.) *The Biology of the Indian Ocean*. Springer, Berlin.

YOU, Y. & TOMCZAK, M. 1993. Thermocline circulation and ventilation in the Indian Ocean derived from water mass analysis. *Deep-Sea Research*, **40**, 13–56.

Formation of varve-like laminae off Pakistan: decoding 5 years of sedimentation

A. LÜCKGE[1], L. REINHARDT[1], H. ANDRULEIT[1], H. DOOSE-ROLINSKI[1], U. VON RAD[1], H. SCHULZ[2] & U. TREPPKE[2]

[1]*Bundesanstalt für Geowissenschaften und Rohstoffe (BGR), Stilleweg 2, 30655 Hannover, Germany (e-mail: a.lueckge@bgr.de)*
[2]*Institut für Ostseeforschung (IOW), Seestraße 15, 18119 Warnemünde, Germany*

Abstract: We studied Holocene sediments from the northeastern Arabian Sea near Pakistan, which were obtained from the same location in 1993 and 1998, to determine the composition and origin of laminated sediments for this 5 year time interval. Methods included geochemical, sedimentological and palaeontological analyses. We then compared our results with meteorological records, and satellite and sediment trap data. We suggest that short-term (few days) heavy rainfall periods in the hinterland and at the coast lead finally to flood events causing the deposition of light-coloured layers as event deposits on the continental slope. These layers are characterized by low percentages of biogenic compounds (i.e. organic matter, coccoliths and diatoms) and interpreted to have been deposited mainly during the winter season, when heavy rainfall can be expected. The thickness of the light layers seems to be related to the intensity of precipitation during a single flood event. In the 1997–1998 El Niño year, which was characterized by the strongest anomalies for the last 20 years, the thickest layer was deposited. The dark layers accumulate over the remaining larger part of the year and are characterized by an elevated input of biogenic material (marine organic matter, skeletal opal, foraminifera and coccoliths).

Annually laminated or 'varved' marine sediments occurring world-wide in marine and lacustrine environments are an excellent archive to decipher climatic change with a temporal resolution down to years or even seasons (Kemp 1996; Overpeck et al. 1996; von Rad et al. 1999). The principal prerequisite for lamination of marine sediments is seasonal variation in sediment input, which creates changes of the sediment colour, texture, structure and composition (Kemp 1996, and references therein; Schimmelmann & Lange 1996). Moreover, anoxic or suboxic conditions are the most important factors to prevent bioturbation and to preserve lamination. Laminated sediments occur mainly under zones of high primary productivity, where rapid oxygen consumption in the water column leads to the formation of an oxygen-minimum zone (OMZ). One prominent area where laminated sediments can be studied is the continental margin off Pakistan in the northeastern Arabian Sea. Sediments from the continental slope off Pakistan recovered during three cruises, in 1969 (METEOR-Indian Ocean Expedition; von Stackelberg 1972), in 1993 (SONNE 90; von Rad et al. 1995; Schulz et al. 1996) and in 1998 (SONNE 130; von Rad et al. 1998), show these millimetric laminations. The observed alternating dark and light sediments were interpreted as seasonal varves, which were deposited during summer and winter monsoon, respectively (von Rad et al. 1999). Recently, Staubwasser & Sirocko (2001) presented an alternative scenario for the deposition of the laminated sediments. From inorganic geochemical data they inferred that the light layers were derived from the expulsion of sediment from mud volcanoes on the shelf. A detailed discussion of the results presented by Staubwasser & Sirocko (2001) has been given by von Rad et al. (2002).

We have undertaken this study to understand very recent sedimentation patterns. To check whether the observed lamination represents varves, we compared sediments from a box core (34KG; 24°53.95'N, 64°17.23'E; water depth 667 m) taken in August 1993 during cruise SO-90 and a box core (246KG; 24°53.96'N, 64°17.24'E; water depth 658 m) and a multicore (236MC; 24°53.96'N, 64°17.25'E; water depth 658 m) taken at nearly the same position 5 years later during SO-130 cruise in April 1998. Until this study, such a straightforward test of the addition of annual laminae by correlating and counting marine varves obtained from the same location and sampled at known time intervals was successful only in the

Santa Basin (Berger, pers. comm.). According to Bates & Jackson (1987), varves can be defined as a sequence of laminae deposited within 1 year and produced by seasonal changes in sediment supply. In this context, we subsampled individual dark and light sediments, which were studied by a variety of sedimentological, geochemical and palaeontological analyses.

Environmental setting

The Arabian Sea hydrography is controlled by the Indian Ocean monsoon. The monsoon influences climate seasonally between eastern Africa and southeastern Asia. During the winter period of the annual monsoon cycle the Arabian Sea is dominated by dry and cold northeasterly winds originating from the snow-covered Asian mountains. During summer, heating of the Tibetan Plateau causes a strong pressure gradient between the Tibetan low-pressure cell and the high-pressure zone over the Indian Ocean. This causes warm and moist winds blowing from the SW. The India–Pakistan region is generally characterized by precipitation maxima during the summer monsoon, particularly from June to September. Almost no rain falls between December and April. In September reduced solar insolation marks the onset of the dry winter monsoon season. In contrast to this 'normal' pattern on the Indian subcontinent, precipitation records for the Makran coast show rainfall maxima during the winter season (Snead 1993a). Between November and March the area is under the influence of (north)-westerly winds. Precipitation in winter is linked to storms in the eastern Mediterranean Sea (Roberts & Wright 1993).

The studied cores 34KG, 246KG and 236MC are located on the uppermost continental slope about 30 km SW of Omara on the eastern Makran coast (Fig. 1). Here, the margin is characterized by a narrow shelf and an extremely steep upper slope (inclination up to 3–4°). The main oceanographic feature of this margin is the expanded, well-developed OMZ between 200 and 1200 m water depth with suboxic or anoxic conditions at the sea floor (von Rad et al. 1995; Schulz et al. 1996). This environment restricts benthic faunal activity and facilitates preservation of a primary laminated sediment fabric.

Samples and methods

Subsamples of individual alternating dark and light laminae from multicore 236MC and box core 246KG were carefully scraped from the sediments with a scalpel. Altogether nine samples, five dark layers and four light layers, were separated. These

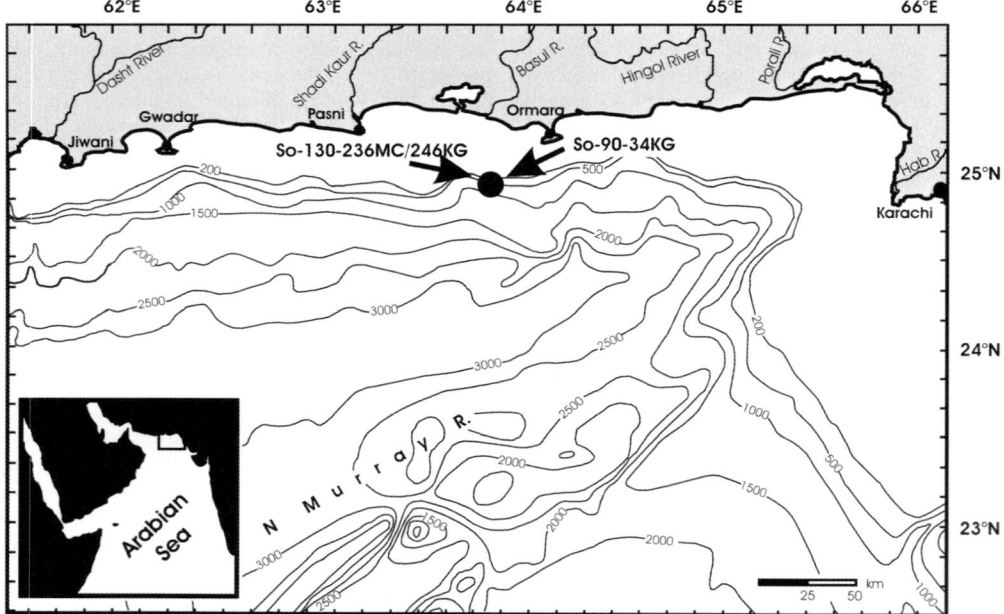

Fig. 1. Location map indicating the position of box core SO-90-34KG, box core SO-130-246KG and multicore SO-130-236MC in the northeastern Arabian Sea off the Makran coast (Pakistan).

samples make up the time interval between August 1993 and April 1998. The samples were analysed by X-ray fluorescence (XRF). Before analysis the sediment was dried and then ground and homogenized in an agate ball mill. XRF analyses were performed on Philips PW 1400 and PW 1480 instruments. Analytical precision, checked with 106 international standards, combined with 24 synthetic standards for elements or ranges of concentrations that are not covered by international standards, was for major elements (Ca, K, Al, Ti, Si) generally better than 1%. $CaCO_3$ contents were calculated from CaO percentages obtained by XRF. Using a polarizing microscope, we semiquantitatively estimated the proportion of foraminifera, quartz and clay minerals in smear slides. Coccoliths were counted quantitatively by scanning electron microscopy (SEM) according to the method proposed by Andruleit (1996). The total organic carbon (TOC) content of dried sediments was determined by the combustion of acid-treated (10% HCl at 80°C) samples in a LECO CS-444 instrument. Standard deviation was ±0.02 wt %. A detailed description of the method used for the alkenone measurements has been given by Doose-Rolinski et al. (2001). For reconstruction of sea-surface temperatures (SST) based on the alkenone unsaturation index ($U^{k'}_{37}$; Brassell et al. 1986), we used the temperature calibration equation of Sonzogni et al. (1998) for the Arabian Sea. The diatoms and silicoflagellates were enriched by preparing the dry sediment subsamples with hydrochloric acid and hydrogen peroxide to remove carbonate and organic material according to Schrader & Gersonde (1978). The clean samples were prepared on 18 mm cover slips, which were mounted on slides for light microscopy with Mountex as mounting medium. Quantification was done on 4–12 traverses at 1000× magnification. Counting procedure and definition of counting units follows the procedure described by Schrader & Gersonde (1978). Diatoms were identified down to species level whereas silicoflagellates were counted as a group.

Results

Lithological description

Visual inspection demonstrated the excellent correlation of individual laminae between the two cores 34KG and 246KG (Fig. 2). Using thick light deposits as marker beds, the two cores can be matched layer by layer over the entire length of the box cores. In fact, it was even possible to make a layer-by-layer correlation between up to >10 m long piston and kasten cores taken in 1993 and 1998 (von Rad et al. 2002). As shown in Fig. 2, a sequence of dark and light laminae of about 30 mm was deposited between August 1993 and April 1998 (on average 6 mm a^{-1}). Altogether nine layers (five dark and four light) were deposited within this 5 year interval. The top layers of both cores consist of a dark layer, which implies similar deposition conditions during the 'pre'-summer monsoon in spring in April 1998 and the 'real' summer monsoon period in August 1993. The light grey layers are, with the exception of layer 3 (L-3; Fig. 2), distinctly thicker than the dark ones. As shown by geochemical, as well as palaeontological results (see below) layer L-3 is probably a mixture of a dark and a light layer (because of the thickness of only 0.3 mm, it was difficult to separate this light lamina from the surrounding dark layers). The light laminae are 0.3–20 mm thick whereas the dark olive laminae are 0.3–0.8 mm thick. In general, the light laminae consist of well-sorted, terrigenous silty clays. Smear slide analyses of the individual laminae show that the light layers are dominated by terrigenous material such as quartz and (undifferentiated) clay minerals, and reworked, finely dispersed calcite. In contrast, the dark layers are poorly sorted silty clays, which are comparatively rich in organic matter and other biogenic material (mainly coccoliths, planktic foraminifera and subordinate benthic foraminifera and diatoms).

Geochemical analyses

The difference between dark and light layers can also be inferred from the elemental XRF data. The clastic compounds such as Al_2O_3, SiO_2 and TiO_2 are relatively enriched in the light layers (Fig. 3). However, inorganic variations of light and dark laminae are caused by dilution with biogenic compounds (e.g. $CaCO_3$; Fig. 3).

An obvious difference between dark and light layers exists in the organic matter content. The organic carbon content (TOC) of the light layers is below 0.7%, whereas that of the dark layers have TOC contents up to 2%. Only layer 3 has an elevated TOC value of 1.2%.

The difference of dark and light laminae can also be seen in the alkenone record of the samples. Long-chain, unsaturated methyl and ethyl ketones, which are biosynthesized by marine coccolithophores (belonging to the class Pyrmnesiophyceae), are strongly enriched in the dark laminae. Moreover, the alkenone unsaturation index is due to the growth temperature of coccolithophores. Depending on water temperature, coccolithophores produce different ratios of di- and tri-unsaturated C_{37}-alkenones. These ratios can be used to calculate variations in SST. The

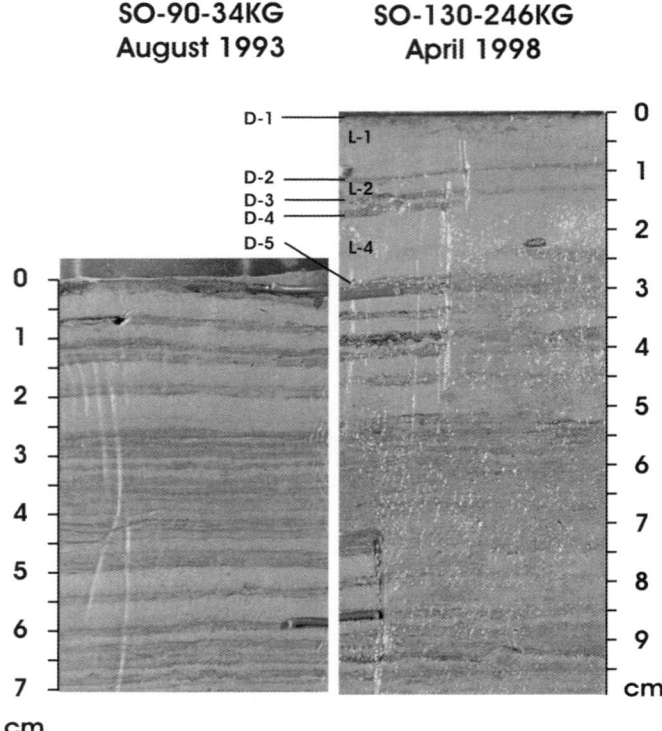

Fig. 2. Photographs showing core SO-90-34KG recovered in August 1993 (left) and core SO-130-246KG recovered in April 1998 (right). The layer-by-layer correlation and the 1993–1998 sedimentation interval can be seen.

SST in the dark layers varies from 26.5 to 27.5°C, which is in accordance with temperature data obtained from water stations during cruise SO-130 (von Rad et al. 1998). The temperature for at least the two thick light layers is somewhat lower (down to 24°C), suggesting a cooling during these events (Fig. 3). However, the alkenone concentrations in light layers are very low and an analytical error cannot be excluded.

Micropalaeontological analyses

Total abundances of coccoliths strongly vary between the dark and light layers and are up to 90% lower in the latter (Fig. 3). The light samples are also characterized by the occurrence of reworked and redeposited coccoliths. Absolute abundances of individual species vary parallel to the total abundances. Diversity was rather low with 21 identified taxa. Dominant species are *Gephyrocapsa oceanica*, *Florisphaera profunda* and *Emiliania huxleyi* (Fig. 4a). Relative species abundances show completely different patterns and seem to be uninfluenced by the laminations (Fig. 4b).

The diatoms were preserved as single valves. Almost 100 taxa were identified. The abundance of diatoms is distinctly different in the dark and light layers (Figs. 3 and 5). The concentration in the light layers varies between 25×10^3 and 547×10^3 valves, and in the dark layers between 420×10^3 and 1786×10^3 in 1 g dry sediment, respectively. With the exception of light layer 3, the diatom concentration is much higher in the dark layers. The diversity as well as the number of species tend to be higher in the dark layers (Fig. 5a). The most abundant taxa are *Thalassionema nitzschioides* var. *nitzschioides* and resting spores of the genus *Chaetoceros* (Fig. 5b). Only the top layer contains a high amount of the *Nitzschia bicapitata* group. Abundant oceanic species are *Planktoniella sol* and *Rhizosolenia bergonii* with up to 8% of the total diatom assemblage. More frequent pelagic species that prefer the vicinity of the coast are *Actinocylus octonarius* and *Actinoptychus senarius*, which have their highest abundances in the dark layers. Sporadically, freshwater taxa such as *Stephanodiscus astreae*, *Aulacoseira granulata* and *Hantschia amphioxys* were ob-

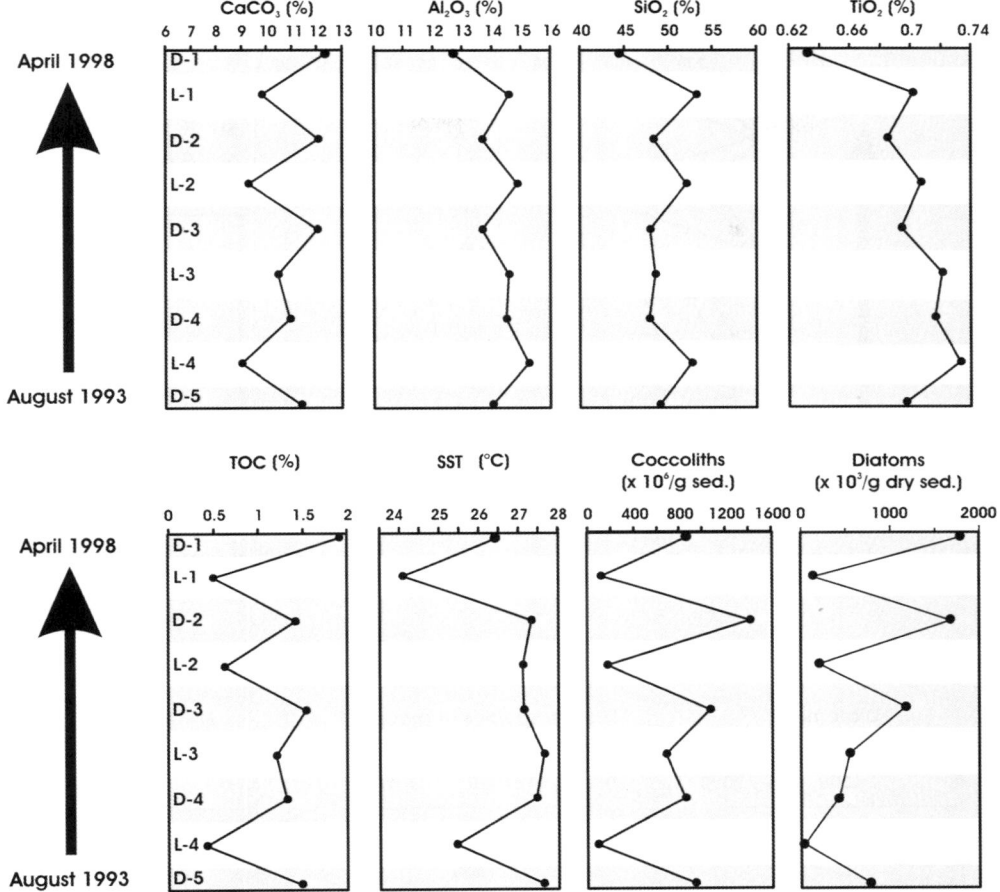

Fig. 3. Chemical composition (CaCO$_3$, Al$_2$O$_3$, SiO$_2$, TiO$_2$ and total organic carbon), alkenone-derived sea surface temperatures (SST), and total abundances of coccoliths as well as diatoms in dark and light layers of multicore SO-130-236MC.

served. Generally, the amount of weakly silicified diatom taxa decreases with depth (Fig. 5c). The abundance of the silicoflagellates parallels that of the diatoms, but with no silicoflagellate preservation in the light layers (Fig. 5c).

Discussion

According to the interpretation of von Rad et al. (1999), light terrigenous laminae were deposited during the winter monsoon season, whereas the dark intervals were deposited during summer monsoonal high-productivity phases from August to October. In this case we would expect one more light layer in the studied interval of the last 5 years. However, the top layers in August 1993 and in April 1998, respectively, are dark. This suggests similar sedimentation patterns for the intermonsoon and the summer monsoon period.

At the studied location, which is situated close to the coast, pigment concentrations in surface waters obtained from satellite observation data (SeaWifs deployed by NOAA) show only moderate differences in primary productivity throughout the year (see also http://daac.gsfc.nasa.gov/CAMPAIGN_DOCS/OCDST/classic_scenes/04_classics_arabian.html (QT-Animation)). Madhupratap et al. (1996) showed that, at least for mesozooplankton assemblages, biomass remains nearly constant throughout the year. However, sediment trap studies in the same region on coccolithophores and diatoms revealed clear seasonal cycles (Treppke et al. 2002).

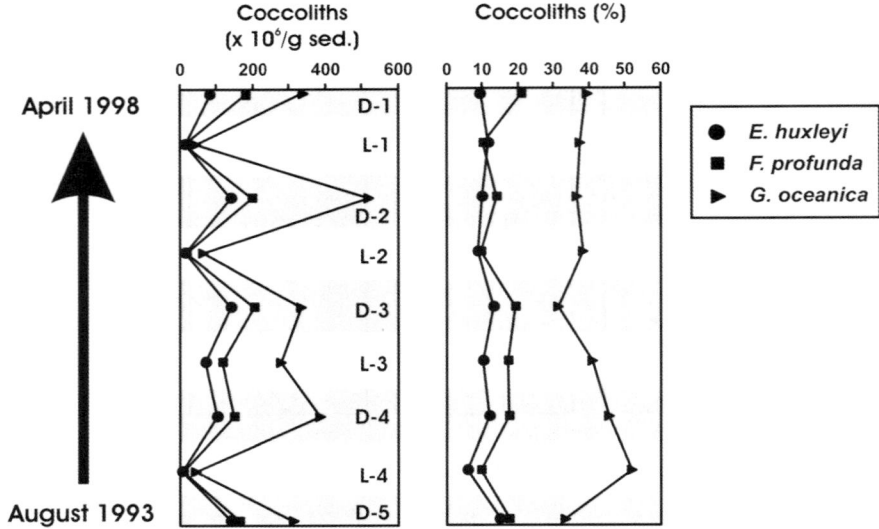

Fig. 4. Total and relative abundances of selected coccolithophore species in dark and light layers. The samples were taken from multicore SO-130-236MC.

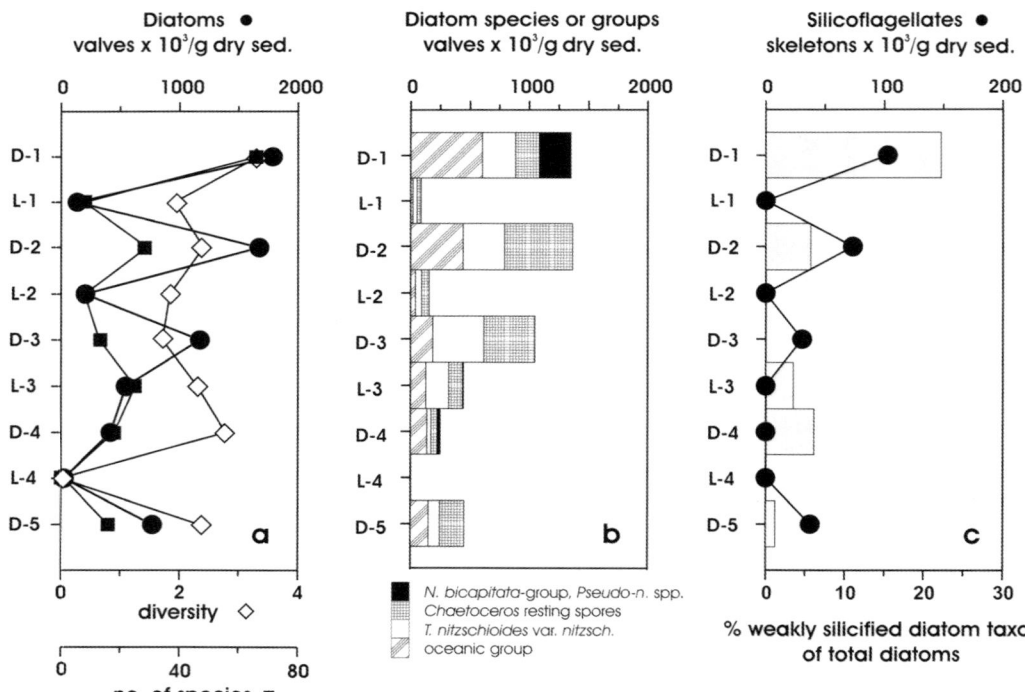

Fig. 5. Total diatom concentrations, diversity and number of species (**a**), concentrations of species or species groups (**b**) and silicoflagellate concentrations as well as percentage of weakly silicified diatoms of total diatoms (**c**) in samples from multicore SO-130-236MC.

Coccolithophores show highest fluxes at the beginning of the SW (summer) and the NE (winter) monsoons (Andruleit et al. 2000). The comparison of coccolithophore and diatom assemblages in the sediments with sediment trap data (Andruleit et al. 2000; Treppke et al. in press) reveals that the dark layer assemblages represent the average of the trapped assemblage in the water column. However, the variations in the relative abundances of distinct species or species groups within the dark layers, as well as the diversity and species number, may indicate different production conditions in recent years. The relatively high amount of T. nitzschioides var. nitzschioides and of the resting spores of Chaetoceros together with a lower diversity and species number in dark layer 3 (D-3) may indicate higher productivity during the SW monsoon. These taxa (T. nitzschioides var. nitzschioides and Chaetoceros spp.) represent generally high-nutrient conditions (Cupp 1943; Pitcher et al. 1991; Treppke et al. 1996). The rather low concentrations of these taxa and the increase of oceanic species in D-4 point to reduced productivity. This is corroborated by the occurrence of N. bicapitata, indicating oceanic upwelling and good opal preservation (Lange et al. 1994) because of its small size and weak silicification. The decreasing amount of rather weakly silicified diatom taxa with core depth (Fig. 5c) represents dissolution of their delicate valves after sedimentation and the contemporary enrichment of strongly silicified taxa. Surprisingly, L-3 and D-4 show an increase of the delicate species. All other light layers contain none of these taxa, indicating either different original diatom assemblages or stronger dissolution.

As shown above, the composition of the trap material resembled the sediment of the dark layers, with a high amount of biogenic particles and organic matter. The organic matter in the dark, TOC-rich laminae is dominated by unstructured, amorphous compounds, which do not show any similarity to biogenic structures. Lückge et al. (1999) demonstrated that these amorphous aggregates were derived from marine organic matter, which is supported by $\delta^{13}C$ values of -20 to $-21‰$ (Hemmann et al. 1997). One striking feature within the organic matter fraction is the occurrence of small (<5 μm) dark brown spherules, which may be faecal pellets from nanno- or picoplanktic organisms (Lückge 1997). Regrettably, all traps were clogged in winter at the same time and therefore no full seasonal cycle could be recorded. The clogging was interpreted as a result of an extreme sedimentation event (Andruleit et al. 2000), which eventually might correspond to the formation of a light sediment layer. As revealed by microscopical studies (Lückge 1997), the organic matter pool in the light layers is dominated by reworked terrigenous material (i.e. vitrinite, inertinite) that is derived from higher land plants. Marine-derived organic matter was observed only in traces. This is supported by isotopic measurements of bulk organic matter from similar sediments that were obtained during cruise SO-90 (Hemmann et al. 1997). The analyses have shown excursions to more negative $\delta^{13}C$ values in the light intervals. In this context, we propose that the accumulation of organic matter as well as other biogenic components occurs within one seasonal cycle and produces the dark olive layers. This means that the dark olive layers characterize the 'normal' hemipelagic, annual sedimentation pattern (Fig. 6a). The light grey layers represent the displaced (terrigenous) sediments, which can be interpreted as suspensate event deposits or plumites (Fig. 6b; Behl 1995; Wheatcroft et al. 1997; Wright et al. 1988). Recently, the dynamics of such sediment transport from river mouths to the sea floor were monitored in great detail off northern California, applying a multidisciplinary approach (Nittrouer 1999). Considering virtually all aspects of the process including the discharge of sediment-laden fresh water to the coastal waters (Syvitski & Morehead 1999), the formation of the river-induced plume and subsequent sediment settling (taking into account, for example, overall sediment load, flocculation of particles, their settling rate, oceanic currents, oceanic density structure, etc.), the formation of centimetre-thick sediment layers on the upper slope within only a few days after the flood event was proven (Morehead & Syvitski 1999; Nittrouer 1999, and references therein). Von Rad et al. (1999) concluded that detrital material is transported episodically from rivers of the nearby Makran coast (i.e. about 50 km) into the sea (Fig. 6b). The silty clays are transported across the shelf and down the upper slope by plumes of nepheloid suspensions (von Rad et al. 1999, and references therein). This process could be favoured by sinking high-salinity surface waters during the winter monsoon (Kumar & Prassad 1999). Between November and February, latent heat release, driven by cool, dry continental winds, results in an increased density of surface waters (Kumar & Prassad 1999). This cooling may be indicated by the calculated SST, which is lower for two of the light layers. Finally, suspension settling leads to the deposition of the light grey layers without any indication of erosion of underlying laminae. However, the depositional process may temporarily affect or even destroy the benthic fluffy layer that is usually found on top of the sediment surface (observed in August 1993 and April 1998). Generally, the fluffy layer acts as an important interface where hydraulic sorting of particles, as well

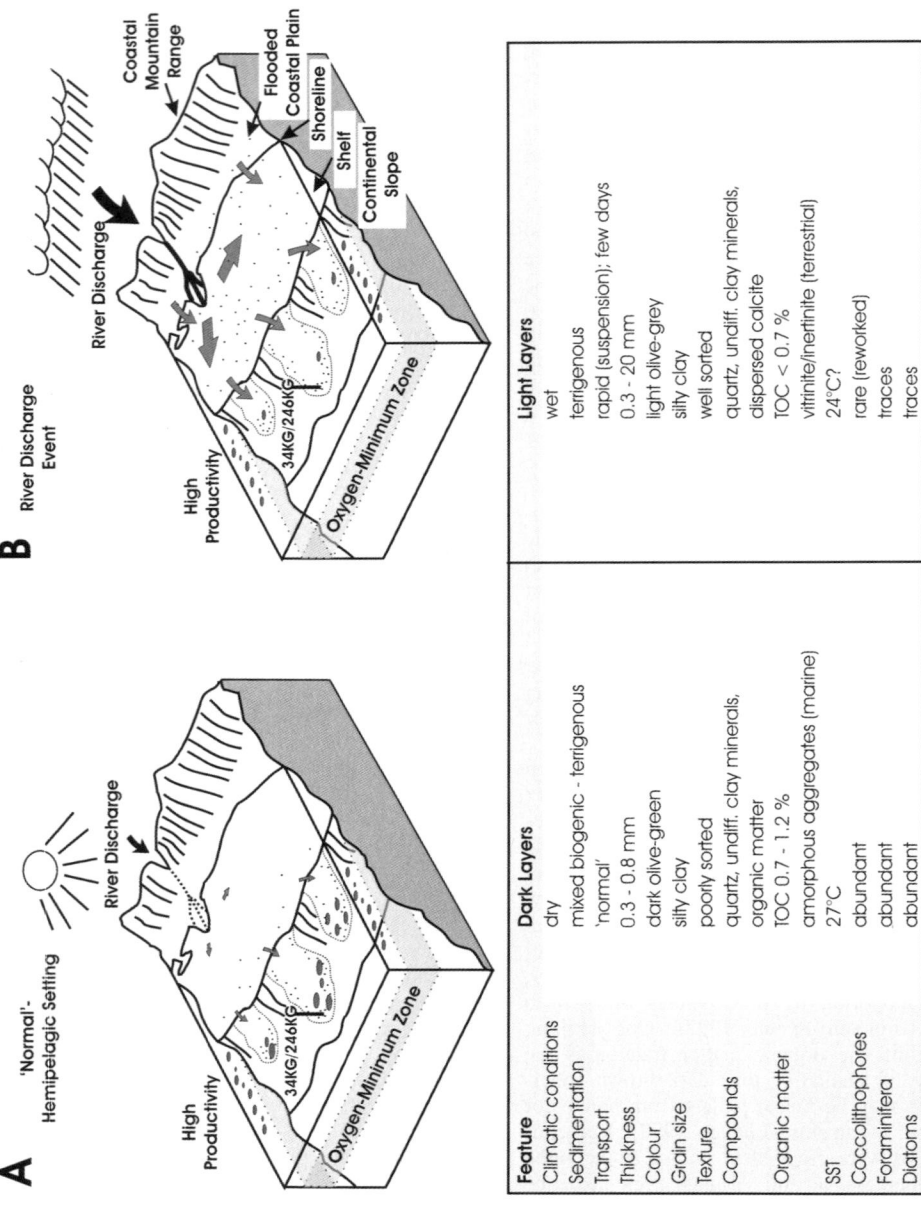

Fig. 6. Schematic model and typical features of laminae formation in the northeastern Arabian off Pakistan. The 'normal' hemipelagic sedimentation pattern (**a**) prevails throughout the year whereas (**b**) characterizes single short-term rainfall events of a few days, which mainly occur during the winter season in December and/or February–March. The grey shaded arrows indicate sediment transport into the Arabian Sea. The size of these arrows indicates the relative sediment flux rates.

as various chemical alterations of material, takes place before incorporation into the sediment occurs (Pilskaln & Pike 2001).

Geochemical studies by Lückge *et al.* (2001), Staubwasser & Sirocko (2001) and von Rad *et al.* (2002) suggest a genetic relationship between the marine deposits and recently sedimented fluvial silty clays of the lower flood terraces and river bed of the Hingol River (Fig. 1) at the Makran coast. Moreover, observations on the coastal plain, where recently built bridges, dams and small gravel roads were destroyed or damaged, imply high-energy flood events (see also Snead 1993*a*). The flood events are supported by the steep topography (up to 460 m; Snead 1993*b*) between coastal plain and coastal mountain range (consisting of weakly consolidated strata), which is drained by several (partly episodic) rivers along the Makran coast (Fig. 6). The more or less complete absence of vegetation in the semiarid hinterland also favours rapid erosion. The catastrophic flood events are probably due to heavy rainfall of short duration. Meteorological data for the Makran coast show that precipitation occurs mainly during the winter season with maxima in December and/or in February–March. Therefore, we propose in accordance with the trap results that the light grey layers represent single short-term events (few days), which occur mostly during the winter season. The year 1997–1998 was one of the strongest El Niño years in the last century (e.g. Yu & Rienecker 1999). On 2–5 March 1998 catastrophic rainfall and flood events occurred along the coast of SE Iran and SW Pakistan as reported in the '1998 Global Register of Large River Flood Events' by the Dartmouth Flood Observatory (http://www.dartmouth.edu/artsci/geog/floods/1998sum.html). Thus, it seems to be likely that the uppermost light layer L-1 (Fig. 2) of about 10 mm thickness, which is covered only by a thin dark layer, was deposited during this flood event shortly before we took the core (in April 1998).

Sedimentation and deposition on the upper slope must be a very rapid process, or the light layers would contain higher amounts of biogenic material as a result of mixing with the normal marine signal. As shown above, this can be excluded, because the light layers contain terrigenous organic matter and reworked fossils. Also, because relative species abundances of coccolithophores seem to be uninfluenced by the laminations, we think that the light layers do not bear a true seasonal signal.

The fact that during the last 5 years only four light greyish layers were deposited suggests that at least during 1 year there was no heavy rainfall. Apart from this, referring to the precipitation data, it is possible that more than one rain event and, consequently, more than one light layer could be deposited within 1 year. Nevertheless, the calibration of laminae countings by accelerator mass spectrometry (AMS) ^{14}C datings confirms the deposition of one light and one dark layer within 1 year on average (von Rad *et al.* 1999).

We cannot exclude a time lag between rainfall event and deposition on the slope. Soutar & Crill (1977) and Schimmelmann & Lange (1996) pointed out the existence of a delay between rainfall and final deposition for the Santa Barbara Basin sediments. They inferred that much of the delay is due to storage of sedimentation on the shelf for a short period. However, in the case of the Makran area, this seems to be less likely because of the narrow and relatively steep shelf and upper slope. This is supported by the sediment trap studies, which show highest lithogenic flux during the winter season, when much of the rain can be expected, and that the sediment traps were clogged by the end of February within a very short period probably by an extreme sedimentation event (Andruleit *et al.* 2000). If deposition or at least reworking of sediments on the shelf, caused by tropical storms as proposed by Staubwasser & Sirocko (2001), were important, the transport and final deposition of clastic detritus should take place during the summer season when the magnitude of winds is much higher than in winter.

Summary

By analysing a 5 year record (1993–1998) of alternating dark olive and light grey laminae deposited within the oxygen-minimum zone of the northeastern Arabian Sea we proved the varve-like character of the seasonal sediment layer couplets and developed a simple model to describe the sedimentation patterns, as follows.

(1) The 'normal' hemipelagic, dark layers accumulate over a large part of the year and are characterized by an elevated amount of marine organic matter and biogenic calcareous and opaline material.

(2) The light grey layers consist of terrigenous silty clay and are interpreted as 'plumites' derived from short-term heavy rainfalls leading to flood events. The last event deposit found in our cores documents the disastrous river flood in March 1998 connected with the very strong 1997–1998 El Niño event.

(3) The investigated 5 year record proves the perfect suitability of laminated sediments for high-resolution (i.e. down to individual years) studies of monsoon-induced climate change through time. However, one has to keep in mind

that varve-like couplets can sometimes be the result of more than one year of sedimentation (i.e. exceptional drought) or vice versa of several rainfall events during a single year.

J. Lodziak (BGR) kindly performed the XRF analyses. We are indebted to J. J. G. Reijmer (Kiel) and C. Betzler (Hamburg) for their constructive reviews. We are also grateful to the Bundesministerium für Bildung und Forschung (BMBF, Bonn) for funding cruises SO-90 and SO-130 (Projects 03G0090A and 03G0130A) and post-cruise research of H. Doose-Rolinski.

References

ANDRULEIT, H. 1996. A filtration technique for quantitative studies of coccoliths. *Micropaleontology*, **42**, 403–406.

ANDRULEIT, H., VON RAD, U., BRUNS, A. & ITTEKOT, V. 2000. Coccolithophore fluxes from sediment traps in the northeastern Arabian Sea off Pakistan. *Marine Micropaleontology*, **38**, 285–308.

BATES, R.L. & JACKSON, J.A. 1987. *Glossary of Geology*. American Geological Institute, Alexandria, VA.

BEHL, R.J. 1995. Sedimentary facies and sedimentology of the late Quaternary Santa Barbara Basin, Site 893. *In:* KENNETT, J.P., BALDAUF, J. & LYLE, M. (eds) *Proceedings of the Ocean Drilling Program, Scientific Results, 146, Part 2.* Ocean Drilling Program, College Station, TX, 295–308.

BRASSELL, S.C., EGLINTON, G., MARLOWE, I.T., PFLAUMANN, U. & SARNTHEIN, M. 1986. Molecular stratigraphy: a new tool for climatic assessment. *Nature*, **320**, 129–133.

CUPP, E.F. 1943. *Marine Plankton Diatoms of the West Coast of North America*. Bulletin of the Scripps Institution of Oceanography, Technical Series, **5**.

DOOSE-ROLINSKI, H., ROGALLA, U., SCHEEDER, G., LÜCKGE, A. & VON RAD, U. 2001. High-resolution sea-surface temperatures and evaporation changes during the Late Holocene in the Northeastern Arabian Sea. *Paleoceanography*, **16**, 358–367.

HEMMANN, A.G., LITTKE, R., LÜCKGE, A., SCHULZ, H. & VON RAD, U. 1997. Isotopic evidence for organic sources in laminated sediments in the northeastern Arabian Sea. *Extended Abstracts, 18th International Meeting on Organic Geochemistry, 22-26 September, 1997*, 465–466.

KEMP, A.E.S. (ed.) 1996. *Palaeoclimatology and Palaeoceanography from Laminated Sediments*. Geological Society, London, Special Publications, **116**.

KUMAR, S.P. & PRASAD, T.G. 1999. Formation and spreading of Arabian Sea high-salinity water mass. *Journal of Geophysical Research*, **104**(C1), 1455–1464.

LANGE, C.B., TREPPKE, U.F. & FISCHER, G. 1994. Seasonal diatom fluxes in the Guinea Basin and their relationship to trade winds, hydrography and upwelling events. *Deep-Sea Research*, **41**, 859–878.

LÜCKGE, A. 1997. Ablagerung und Frühdiagenese organischen Materials in marinen Hochproduktivitätsgebieten. *Berichte Forschungszentrum Jülich*, 3413.

LÜCKGE, A., ERCEGOVAC, M., STRAUSS, H. & LITTKE, R. 1999. Early diagenetic alteration of organic matter by sulfate reduction in Quaternary sediments from the northeastern Arabian Sea. *Marine Geology*, **158**, 1–13.

LÜCKGE, A., DOOSE-ROLINSKI, H., KHAN, A.A., SCHULZ, H. & VON RAD, U. 2001. Monsoonal variability in the northeastern Arabian Sea during the past 5,000 years: Geochemical evidence from laminated sediments. *Palaeogeography, Palaeoclimatology, Palaeoecology*, **167**, 273–286.

MADHUPRATAP, M., PRASANNA KUMAR, S., BHATTATHIRI, P.M.A., DILEEP KUMAR, M., RAGHUKUMAR, S., NAIR, K.K.C. & RAMAIAH, N. 1996. Mechanism of the biological response to winter cooling in the northeastern Arabian Sea. *Nature*, **384**, 549–552.

MOREHEAD, M.D. & SYVITSKI, J.P. 1999. River-plume sedimentation modeling for sequence stratigraphy: application to the Eel margin, northern California. *Marine Geology*, **154**, 29–41.

NITTROUER, C.A. 1999. STRATAFORM: overview of its design and synthesis of its results. *Marine Geology*, **154**, 3–12.

OVERPECK, J.T., ANDERSON, D., TRUMBORE, S. & PRELL, W. 1996. The southwest Indian monsoon over the last 18000 years. *Climate Dynamics*, **12**, 213–225.

PILSKALN, C.H. & PIKE, J. 2001. Formation of Holocene sedimentary laminae in the Black Sea and the role of the benthic flocculent layer. *Paleoceanography*, **16**, 1–19.

PITCHER, G.C., WALKER, D.R., MITCHELL-INNES, B.A. & MOLONEY, C.L. 1991. Short-term variability during an anchor station study in the southern Benguela upwelling system: phytoplankton dynamics. *Progress in Oceanography*, **28**, 39–64.

ROBERTS, N. & WRIGHT, H.E. ET AL. 1993. Vegetational, lake-level, and climatic history of the Near East and Southwest Asia. *In:* WRIGHT, H.E. (ed.) *Global Climates since the Last Glacial Maximum*. University of Minnesota Press, Town, 194–220.

SCHIMMELMANN, A. & LANGE, C.B. 1996. The tale of 1001 varves: a review of Santa Barbara Basin sediment studies. *In:* KEMP, A.E.S. (ed.) *Palaeoclimatology and Palaeoceanography from Laminated Sediments*. Geological Society, London, Special Publications, **116**, 121–142.

SCHRADER, H. & GERSONDE, R. *et al.*1978. Diatoms and silicoflagellates. *In:* ZACHARIASSE, W.J. (ed.) *Micropaleontological Counting Methods and Techniques—an Exercise on an Eight Metres Section of the Lower Pliocene of Capo Rosello, Silicy*. Micropaleontological Bulletin, 129–176.

SCHULZ, H., VON RAD, U. & VON STACKELBERG, U. 1996. Laminated sediments from the oxygen-minimum zone of the northeastern Arabian Sea. *In:* KEMP, A.E.S. (eds) *Palaeoclimatology and Palaeoceanography from Laminated Sediments*. Geological Society, London, Special Publications, **116**, 185–207.

SNEAD, R.J. 1993a. Geography, geomorphic process and effects on archaelogical sites on the Makran coast. *In:* SCHRODER, J.F. (ed.) *Himalaya to the Sea. Geology, Geomorphology and the Quaternary*. Routledge, London, 363–378.

SNEAD, R.J. 1993b. Uplifted marine terraces along the Makran coast of Pakistan and Iran. *In:* SCHRODER, J.F. (ed.) *Himalaya to the Sea. Geology, Geomorphology and the Quaternary.* Routledge, London, 327–363.

SONZOGNI, C., BARD, E. & ROSTEK, F. 1998. Tropical sea-surface temperatures during the last glacial period: a view based on alkenones in Indian Ocean sediments. *Quaternary Science Reviews*, **17**, 1185–1201.

SOUTAR, A. & CRILL, P.A. 1977. Sedimentation and climatic patterns in the Santa Barbara Basin during 19th and 20th centuries. *Geological Society of America Bulletin*, **88**, 1161–1172.

STAUBWASSER, M. & SIROCKO, F. 2001. On the formation of laminated sediments on the continental margin off Pakistan: the effects of sediment provenance and sediment distribution. *Marine Geology*, **172**, 43–56.

SYVITSKI, J.P. & MOREHEAD, M.D. 1999. Estimating river-sediment discharge to the ocean: application to the Eel margin, northern California. *Marine Geology*, **154**, 13–28.

TREPPKE, U.F., LANGE, C.B., DONNER, B., FISCHER, G., RUHLAND, G. & WEFER, G. 1996. Diatom and silicoflagellate fluxes at the Walvis Ridge: an environment influenced by coastal upwelling in the Benguela System. *Journal of Marine Research*, **54**, 991–1016.

VON RAD, U., DOOSE-ROLINSKI, H. & Shipboard Scientific Party SO130 1998. *MAKRAN II: the Makran accretionary wedge of Pakistan: Tectonic Evolution and Fluid Migration (Part 2). Research Cruise SO 139 with R.V. SONNE.* Cruise Report, Bundesanstalt für Geowissenschaften und Rohstoffe, Hannover, Archive No. 117368.

VON RAD, U., SCHAAF, M., MICHELS, K.H., SCHULZ, H., BERGER, W.H. & SIROCKO, F. 1999. A 5000-yr record of climate change in varved sediments from the oxygen minimum zone off Pakistan, northeastern Arabian Sea. *Quaternary Research*, **51**, 39–53.

VON RAD, U., SCHULZ, H., ALI KHAN, A. & 17 OTHERS 1995. Sampling the oxygen minimum zone off Pakistan: glacial–interglacial variations of anoxia and productivity (preliminary results, SONNE 90 cruise). *Marine Geology*, **125**, 7–19

VON RAD, U., KHAN, A.A., BERGER, W.H., RAMMLMAIR, D. & TREPPKE, U. 2002. Varves, turbidites and cycles in upper Holocene sediments (Makran Slope, Northern Arabian Sea). *In:* CLIFT, P.D., KROON, D., GAEDICKE, J. & CRAIG, J. (eds) *The Tectonic and Climatic Evolution of the Arabian Sea Region.* Geological Society, London, Special Publications, **195**, 387–406.

VON STACKELBERG, U. 1972. Fazies verteilung in Sedimenten des indisch-pakistanischen Kontinentalrandes (Arabisches Meer). *Meteor-Forschungs-Ergebnisse, Reihe C*, **9**, 1–73.

WHEATCROFT, R.A., SOMMERFIELD, C.K., DRAKE, D.E., BORGELD, J.C. & NITTROUER, C.A. 1997. Rapid and widespread dispersal of flood sediment on the northern California margin. *Geology*, **25**, 163–166.

WRIGHT, L.D., WISEMAN, W.J., BORNHOLD, B.D. & 5 OTHERS 1988. Marine dispersal and deposition of Yellow River silts by gravity-driven underflows. *Nature*, **332**, 629–632.

YU, L. & RIENECKER, M.M. 1999. Mechanisms for the Indian warming during the 1997–98 El Niño. *Geophysical Research Letters*, **26**, 735–738.

On the evolution of the oxygen minimum zone in the Arabian Sea during Holocene time and its relation to the South Asian monsoon

MICHAEL STAUBWASSER[1,2] & PETER DULSKI[2]

[1]*Department of Earth Sciences, University of Oxford, Parks Road, Oxford OX1 3PR, UK*
(e-mail: michaels@earth.ox.ac.uk)
[2]*GeoForschungsZentrum Potsdam, Telegrafenberg, 14473 Potsdam, Germany*

Abstract: The evolution of the oxygen minimum zone within the permanent thermocline of the Arabian Sea (AS) during early and mid-Holocene time was reconstructed from a laminated sediment core taken from the Pakistani continental margin (316 m water depth). A trace metal proxy for water column ventilation (authigenic U) was extracted by principal component analysis from a large dataset of inorganic and total organic carbon (TOC) measurements. This proxy is compared with preservation of lamination and paired benthic–planktonic ^{14}C data. The latter record the $\Delta^{14}C$ depth gradient in the AS and may provide a sensitive indicator for ventilation by enhanced surface convection. Laminated sediments were preserved between 10 and 7.5 ka bp on the Pakistani continental margin and accumulated authigenic U independently from TOC accumulation. The inferred reducing conditions in the AS thermocline are in agreement with high palaeoproductivity in the western AS upwelling region. Century-scale variability in northern AS surface hydrography (recorded as $\delta^{18}O$ in planktonic foraminifera) is reflected in the accumulation of authigenic U on the Pakistani margin. The agreement of AS surface conditions, which generally reflect the South Asian monsoon (SAM), with ventilation of the OMZ confirms a dominant influence of the SAM and summer monsoon upwelling in particular on AS thermocline ventilation during early Holocene time. However, the preservation of laminated sediments off Pakistan and palaeoproductivity in the western AS disagree before 10 ka cal. bp, and between 7.5 and 5.5 cal. ka bp. Here, the absence of lamination indicates better ventilation of the thermocline, whereas palaeoproductivity in the upwelling region was high. This suggests that other factors may also have contributed in variable proportions to AS thermocline ventilation. At present, these factors include lateral advection of oxygenated Central Indian Water and ventilation by winter surface convection in the northern AS.

Arabian Sea oceanography is strongly influenced by the South Asian monsoon, with coastal and open ocean upwelling taking place along its western boundary under the sustained southwesterly winds during the summer monsoon (Tomczak & Godfrey 1994). A pronounced oxygen minimum zone (OMZ) can be found throughout the Arabian Sea (AS) within the permanent thermocline between 200 and 1000 m water depth (Fig. 1). It is maintained by a balance of enhanced oxygen consumption underneath the highly productive upwelling region during the summer monsoon season and advection of Central Indian Water naturally low in oxygen (Swallow 1984). In the eastern AS the core of the OMZ between 200 and 500 m is permanently nitrate reducing (Morrison *et al.* 1999). Central Indian Water, which forms at the Southern Subtropical Front, is the largest oxygen supplier to the AS thermocline despite its moderate ventilation, whereas Persian Gulf Water and Red Sea Water at present are only minor contributors to the oxygen budget (You & Tomczak 1993; You 1998). In the northern AS the top of the OMZ is additionally ventilated from above by surface convection as a result of cooling during the winter monsoon (Banse 1984; Morrison *et al.* 1999).

In this study we seek to reconstruct the evolution of the OMZ in the AS during early and mid-Holocene time. Previous studies have attempted to infer past thermocline ventilation in the AS from bulk sediment parameters such as trace metal concentration (von Rad *et al.* 1999b) and by a combination of palaeoproductivity proxies and stable N-isotope analysis (Reichart *et al.* 1998). We initially followed the classical, but also problematic approach of bulk sedimentary trace metal geochemistry (Calvert & Pedersen 1993) and applied it to the laminated sediment core 63KA, which was recovered from the Pakistani continental margin at a depth of 316 m, well within the core of the OMZ. This approach is based on the observation that a number of trace metals may become relatively enriched in sediments beneath a suboxic or anoxic water column, but is hampered

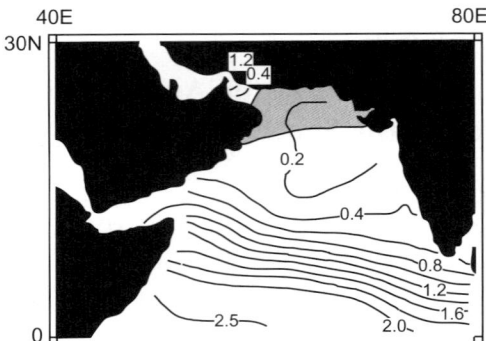

Fig. 1. Arabian Sea thermocline oxygen concentration (ml l^{-1}) at 300 m water depth; annual average according to Levitus & Boyer (1994). Shaded area: approximate area where winter monsoon surface convection takes place (after Banse & McClain 1986).

by the fact that trace metal enrichment can also be caused by a number of other independent factors, such as shoaling of redox zones at high organic carbon flux to the sediment surface, variable input of Al–Si detritus, and variable sediment source lithology (Shaw et al. 1990; Calvert & Pedersen 1993; Lapp & Balzer 1993; Crusius et al. 1996; Staubwasser & Sirocko 2001). During the course of this study we single out redox-sensitive trace elements potentially suitable as proxies for water column ventilation. By using additional information derived from the degree of lamination preservation and paired benthic–planktonic radiocarbon measurements we also demonstrate the limitations of bulk sediment chemical parameters as proxies for OMZ ventilation in the AS. Finally, we infer the evolution of the OMZ from a combination of all parameters with respect to the different factors controlling the OMZ at present.

Methods

Bulk inorganic analysis was performed on 250 mg of dried sediment weighed into Teflon vials and digested in a HNO_3–HF–$HClO_4$ solution. The samples were dissolved in 0.7 M HNO_3 and measured for major, minor and trace metals by inductively coupled plasma atomic emission spectrometry (ICP-AES) and ICP mass spectrometry (ICP-MS). Mg, Ca, Ti, V, Mn and Fe were measured by ICP-AES on an ARL 35000C system, whereas Al and Sc were measured by ICP-AES on a Varian Liberty 2000 system. To minimize the effect of spectral interference, the ICP-AES measurements were calibrated with matrix-matched multi-element synthetic standard solutions. Rb, Sr, Y, Cs, Ba, the rare earth elements (REE), Pb, Th and U were analysed by ICP-MS on a Perkin–Elmer Elan 5000 system following the analytical protocol of Dulski (2001). The determination of Cr, Ni, Cu, Zn, Ga and Cd was performed in a separate ICP-MS analysis, using Sc, Ar_2 and Ru as internal standards for drift correction. Isobaric interference of ^{112}Sn and ^{114}Sn on the respective Cd isotopes were routinely corrected for. Potential interference from molecular Na and Mn, and from double-charged Ba, was tested with standard solutions. As a result, a correction for interference of double-charged Ba species on ^{65}Cu, ^{66}Zn, ^{68}Zn and ^{69}Ga was applied.

Trace element recovery of the above open vessel digestion was cross-checked on selected samples with pressurized dissolution and ICP-MS analysis, as well as additional X-ray fluorescence measurements (Cr, Zr). On these grounds Zr and the heavy REE values were rejected from the dataset. External reproducibility was determined from a total of 10 replicates of one sample in addition to several 2- to 3-fold replicates. Values ranged between 2 and 4% in most cases (1 SD, Table 1). Total organic carbon (TOC) was measured with an Eltra Metalyt-CS-1000 S IR spectrometer on decarbonated samples with an external reproducibility of 7.7%.

The laminated sediments show an alternation of black and olive–green hemipelagic layers caused by seasonal variability of organic matter deposition, with occasional thin white layers of 0.5–2 mm thickness in sharp visual contrast to the hemipelagic lamination (von Rad et al. 1999a; Staubwasser & Sirocko 2001). These white layers probably originate from mud expulsion on the tectonically active Makran shelf (Staubwasser & Sirocko 2001). An arbitrarily defined classification scheme similar to that of Behl & Kennett (1996) based on visual inspection of radiographs was used to describe the degree of lamination preservation. Four subdivisions were defined: 0, homogeneous; 1, only white layers traceable; 2, indistinct olive–green and black layers; 3, clearly visible olive green and black lamination.

The stratigraphy of core 63KA is based on 46 accelerator mass spectrometry (AMS) ^{14}C dates of planktonic foraminifera Globigerinoides sacculifer (Staubwasser 1999; Staubwasser et al. 2003). On five samples both planktonic and benthic foraminifera (Uvigerina sp.) were ^{14}C dated. All AMS measurements were made at the Leibniz-Labor facility at the University of Kiel (Nadeau et al. 1998; Schleicher et al. 1998).

Inorganic geochemical indicators of water column ventilation

Under reducing conditions in the sediment column a number of trace metals such as V, Cr, Cu, Zn,

Table 1. *Analytical uncertainties, principal components and relative variance explained in core 63KA*

Element	External reproducibility (RSD %)	PC1	PC2	PC3	PC4	PC5	PC6
Al	3.6	0.57	0.18	0.00	0.61	0.36	0.02
Sc	6.8	0.35	0.01	0.05	0.83	0.26	0.04
Ca	1.8	−0.76	−0.10	−0.07	−0.26	−0.50	−0.05
Mg	1.3	0.41	−0.16	0.01	0.14	0.75	0.11
Ti	1.2	0.68	−0.11	0.12	0.24	0.61	0.08
Fe	1.4	0.57	0.02	0.13	0.29	0.65	0.13
Mn	2.1	0.64	0.57	0.19	0.20	0.39	0.07
V	2.0	0.35	−0.56	0.02	0.24	0.61	0.14
TOC	7.7	−0.60	−0.65	−0.15	−0.12	−0.25	−0.03
Ni	2.3	−0.34	−0.90	−0.05	−0.09	0.04	0.04
Cr	5.7	−0.10	−0.88	0.00	−0.07	0.13	−0.06
Cu	2.7	−0.33	−0.87	−0.08	0.00	0.01	0.05
Zn	3.4	0.47	−0.04	0.04	0.14	0.68	−0.09
Cd	7.6	−0.15	−0.66	−0.54	0.00	−0.05	−0.28
Pb	1.9	0.49	−0.03	0.19	0.43	0.28	0.26
U	2.6	−0.08	−0.11	−0.95	−0.07	−0.06	0.04
Ba	2.1	0.69	0.61	0.02	0.16	0.27	0.02
Ga	2.9	0.66	0.14	0.05	0.26	0.59	0.17
Rb	3.1	0.74	0.29	0.00	0.22	0.38	0.34
Sr	2.1	−0.37	0.76	−0.15	−0.25	−0.32	−0.02
Y	3.4	0.72	−0.08	−0.04	0.17	0.24	0.56
Cs	2.1	0.77	0.22	0.04	0.28	0.47	0.12
La	2.7	0.90	0.20	0.07	0.16	0.29	0.09
Ce	2.8	0.89	0.22	0.07	0.17	0.30	0.07
Pr	2.6	0.90	0.24	0.07	0.18	0.28	−0.01
Nd	2.8	0.90	0.25	0.07	0.17	0.27	0.03
Sm	3.5	0.89	0.28	0.07	0.18	0.23	0.03
Eu	3.1	0.89	0.12	0.11	0.20	0.29	0.03
Gd	2.7	0.85	0.26	0.08	0.19	0.28	0.09
Th	6.4	0.66	0.36	0.03	0.22	0.41	0.35
	% of total variance explained	40	20	4.7	7.2	15	2.8
	% explained by PC 1–5	87					

Cd, Pb and U are removed from pore water at their respective redox boundaries (Shaw et al. 1990; Klinkhammer & Palmer 1991; Calvert & Pedersen 1993; Lapp & Balzer 1993). This may lead to continued metal diffusion from the overlying water column along a pore-water concentration gradient and further precipitation at the redox boundary. When water column ventilation decreases, the respective redox boundaries shoal (Shaw et al. 1990; Lapp & Balzer 1993). The resulting larger concentration gradients in the pore water enhance diffusion and precipitation, which can lead to enrichment of redox-sensitive elements in the solid phase. However, the same effect is achieved by an increase in TOC flux without invoking a change in water column ventilation (Calvert & Pedersen 1993). In addition, variable degrees of clastic input along the Pakistani margin and changes in sediment provenance may blur any authigenic signal in the bulk sediment record (Staubwasser & Sirocko 2001). Consequently, the relative contribution of these factors to bulk sediment geochemical composition has to be assessed to identify which element expresses variability predominantly because of authigenic trace metal enrichment as a consequence of past changes in water column ventilation.

Principal component analysis of 63KA inorganic sedimentary chemistry

The requirement for any redox-sensitive trace element to meet to be considered a potential indicator of water column ventilation is independence from variability caused by detrital or biogenic components such as TOC. To extract the best suitable trace metal proxy for past water column ventilation change from the dataset, a principal component (PC) analysis was performed with the commercially available Systat software. The PC analysis included (varimax) rotation of the PCs calculated from the correlation matrix (Rayment & Jöreskog 1993). The results are presented with a focus on the potentially redox-sensitive trace metals mentioned above.

Of the total variance in the record 87% is explained by the first five PCs, whereas the variance explained by higher PCs is within the range of analytical uncertainty (Table 1). Of these five significant PCs, the last two appear to reflect specific analytical procedures rather than true variation in the sediment core. PC4 (7.2% of total variance explained) is dominated by those elements measured on one of the ICP-AES systems (Al, Sc), and PC5 (15% of total variance explained) by those measured on the other ICP-AES system (Ca, Fe, Ti, Mg, V). This is probably due to different non-linear drift between the three analytical techniques used. Only the first three PCs (Fig. 2) appear to have a geochemical meaning. PC1 comprises major and minor elements such as Al, Ti and Fe, and trace elements such as Ga (Fig. 2a), Rb, Cs and Th (Table 1), which are commonly associated with Al–Si detritus (e.g. Krauskopf & Bird 1995). Also associated with PC1 are the potentially redox-sensitive elements V, Mn, Zn, Ba and Pb, which consequently seem to be significantly influenced by Al–Si detritus. Ca and TOC are anti-correlated with the above elements, indicating that dilution by Al–Si detritus is a major factor in determining the content of biological material preserved in these sediments.

Associated with PC2 are TOC (Fig. 2b), Cr, Ni, Cu and, to a lesser extent, Cd and V (Table 1). This could indicate a direct dependence of Cr, Ni, Cu, V and Cd on TOC flux or authigenic enrichment as a result of enhanced early diagenesis fuelled by high TOC. However, a number of additional observations suggest that the relation is not as straightforward for Cr, Ni and Cu (those elements with the highest correlation with PC2). On three transects of surface cores across the OMZ off the Pakistani margin, these elements showed little relation to TOC or the OMZ (Staubwasser & Sirocko 2001). Instead, Ni and Cr were found to reflect sediments with abundant mafic minerals discharged from the Makran (western Pakistan), whereas sediments discharged from the river Indus (eastern Pakistan) were enriched in Cu. The sediments deposited in the vicinity of core 63KA were found to have originated from a local river source of intermediate Cu concentration, but with a variable contribution from the Makran. By contrast, Cd was found to be enriched in TOC-rich cores from the OMZ at levels that can be explained by direct association with TOC (Staubwasser & Sirocko 2001).

Mn and Ba are anti-correlated with PC2 (Table 1). These elements can be remobilized from the solid phase under reducing conditions, which may result in depletion from the solid phase as they are

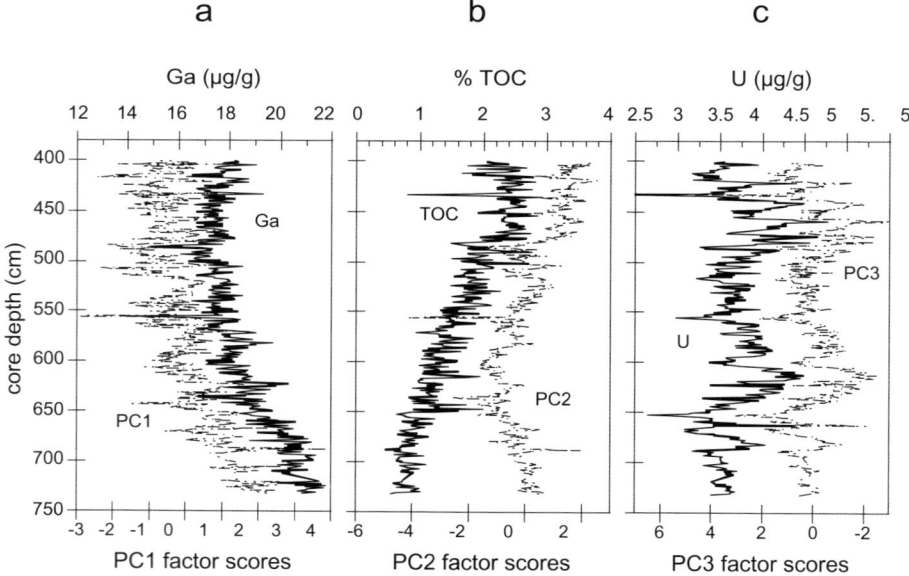

Fig. 2. Geochemically significant principal components (PC) and typical element or biogenic component associated with them: (**a**) PC1 and Al–Si detritus (Al shows the same distribution with depth as Ga, but is not shown here for reasons explained later in the text); (**b**) PC2 and TOC; (**c**) PC3 and U.

subsequently lost from the sediment column by upward diffusion into the water column (Calvert & Pedersen 1993; Torres et al. 1996). The association of Mn and Ba with PC2 could therefore indicate an early diagenetic depletion in the sediments along the Pakistani continental margin. However, an interpretation of any downcore variability of these elements is complicated by the fact that they are also associated with PC1 (Al–Si detritus).

PC3 comprises only U and Cd, with U (Fig. 2c) being the dominant contributor (Table 1). Both elements are also slightly enriched compared with average shale (Table 2). Cd and U concentrations are commonly elevated in the anoxic sediments from the Arabian Sea (Nath et al. 1997; Staubwasser & Sirocko 2001). These elements have a low abundance in average crustal material (Taylor & McLennan 1985) and may therefore become enriched during early diagenesis in the Al–Si detritus-dominated sediments of the Pakistani margin. In contrast to Cd, which is also associated with PC2, U varies independently from other components in core 63KA. Its zero correlation with PC1 shows that most of the U does not rest within Al–Si detritic phases, and that it is not diluted by variable input of Al–Si detritus. Enrichment as a result of early diagenesis fuelled by high flux of organic carbon does not seem to have happened either, perhaps with the exception of the uppermost part of the core (Fig. 2b and c).

On the basis of the PC analysis, ratios of U over a detrital denominator should describe authigenic U variability that is likely to be related to changes in water column ventilation. Commonly used detrital denominators are Al, Ti and Fe, but because a significant part of the downcore variability of these elements is explained by the variety of analytical methods used in this study (association of PC4 and PC5 with individual analytical techniques; see above), other elements that are exclusively associated with PC1 may be better suited in this case. Ga does not form an individual mineral phase, but is always present at trace level in minerals containing Al (Gottardi et al. 1969; Krauskopf & Bird 1995). Therefore, Ga was chosen as the detrital denominator for the following discussion of the 63KA U record. For comparison, the U/Ga ratios can be converted to U/Al ($\times 10^4$) ratios by dividing by 0.38 (derived from a correlation of Ga and Al in core 63KA with $r = 0.8$).

Lamination preservation as an indicator of water column ventilation

At present, dissolved oxygen concentrations remain below 5 $\mu M\ l^{-1}$ all year in the OMZ of the northern AS, and transient secondary nitrite maxima even indicate continuing denitrification in the water column (Morrison et al. 1999). Sediments deposited under such conditions along the Pakistani margin are preserved undisturbed by bioturbation as macrobenthos is absent (von Stackelberg 1972). Consequently, the preservation of lamination in a downcore record is a direct indicator of suboxic–anoxic conditions in the water column. On the basis of the qualitative lamination preservation index defined above (see Methods), the section of core 63KA discussed in this study can be divided into a bioturbated sub-section below 630 cm, where in situ mollusc shells can be observed. This section is separated by a short transition (630–595 cm) from a laminated section between 595 and 480 cm, which indicates a decrease in water column ventilation to suboxic conditions (Fig. 3a). This section is in turn followed by an indistinctly laminated section (480–400 cm) with intercalated breaks in the lamination sequence, but in situ mollusc shells cannot be observed.

Comparison of proxies for past OMZ ventilation

The U/Ga record agrees reasonably well with the lamination preservation index in the lower sections of the core (Fig. 3a and c), if a small downcore offset as a result of subsurface authigenic U precipitation is accounted for. However, in the mid-Holocene part of the record (480–400 cm) the absence or poor preservation of lamination is in contrast to a maximum in authigenic U. This suggests that the relation between water column ventilation, authigenic U accumulation and preservation of lamination observed in the lower parts of the core is more complex in the upper part.

Table 2. *Some redox-sensitive trace elements in core 63KA and in average shale (Turekian & Wedepohl 1961)*

	V	Cr	Ni	Cu	Cd	U
[63KA]/Al × 10^4	16–19	13–18	9–14	4–6	0.05–0.14	0.51–0.80
[av. shale]/Al × 10^4	16	11	8.5	5.6	0.04	0.46

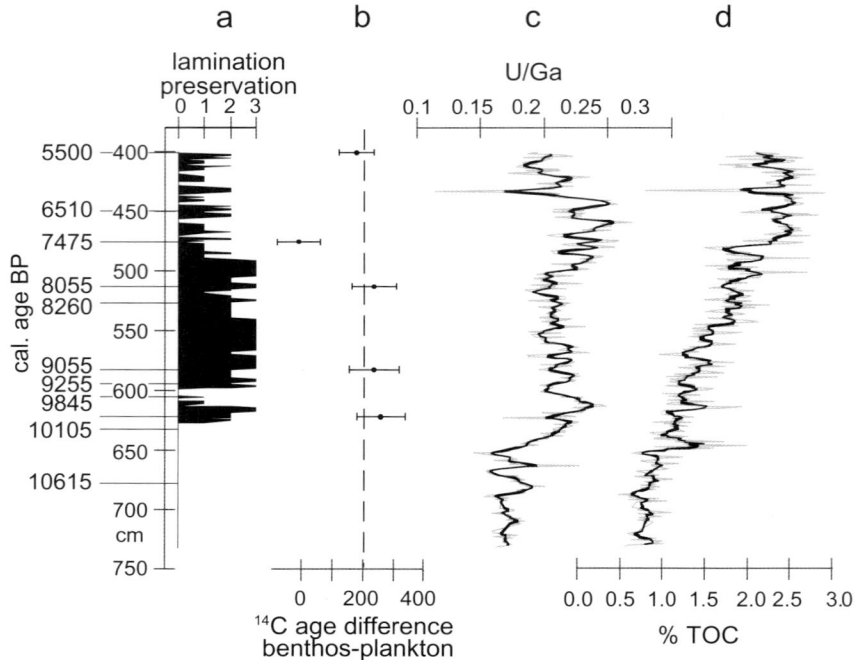

Fig. 3. 63KA (400–732 cm) records: (**a**) arbitrary lamination preservation index (0, homogeneous; 1, only white layers traceable; 2, indistinct olive–green and black layers; 3, clearly visible olive green and black lamination) (see Methods section for details on lamination); (**b**) ^{14}C age difference between *Uvigerina* sp. (benthos) and *Globigerinoides sacculifer* (plankton) with uncertainty (1σ); (**c**) U/Ga ratios (see Principal component analysis section for explanation of Ga denominator); (**d**) % TOC. Bold lines show the five-point smoothed record.

This apparent contradiction can be explained in several ways. The first is that the OMZ became better ventilated, but a higher accumulation of TOC on the Pakistani margin (Fig. 3d) fuelled early diagenesis and enhanced authigenic U enrichment. Another scenario is that seasonality in sediment flux may have been reduced, so that no lamination was formed at the first place. Although lake records from NW India indicate that mid-Holocene time on annual average was wetter than any other period recorded so far (Enzel *et al.* 1999), there are no data available that suggest changes in sediment flux seasonality at that time.

The third possible explanation takes into account mechanisms of OMZ ventilation observed at present. In contrast to the annually sustained suboxic OMZ in the eastern AS, the top of the OMZ in the northern AS is ventilated from above by surface convection during the winter monsoon (Banse 1984; Morrison *et al.* 1999). Only transient true suboxic time intervals (with nitrate reduction) are observed here. In the following discussion, we utilize a small number of paired benthic–planktonic ^{14}C ages to elaborate on the possibility of seasonally enhanced OMZ ventilation caused by deeper convection during the winter monsoon.

^{14}C in paired benthic–planktonic foraminifera samples as potential palaeoconvection proxy

Reconstructed vertical pre-bomb ^{14}C profiles show that Δ^{14}C decreases with depth in the upper ocean (Broecker *et al.* 1995), equivalent to increasing ^{14}C reservoir ages. In the North Atlantic this depth gradient is eroded during surface convection in winter, and the resulting approximately 200 m deep mixed layer in early spring has a 30‰ lower Δ^{14}C value than the shallow mixed layer in autumn (Broecker & Peng 1980). As living benthic foraminifera occur only within the upper 1–1.5 cm (Jannink *et al.* 1998), which is equivalent to 10–20 a in core 63KA, paired ^{14}C measurements of benthic and planktonic foraminifera from the laminated sediments should record the ^{14}C depth gradient in the northern AS. Enhanced surface convection should reduce this gradient in paired benthic–planktonic foram samples.

A reconstructed ^{14}C depth profile for the Arabian Sea for pre-bomb time (Bhushan et al. 2000) shows a Δ^{14}C value of just above $-75‰$ in the surface mixed layer (equal to a conventional ^{14}C age of 625 a), and just below $-100‰$ at 300 m depth (the depth of the 63KA core site and equal to a ^{14}C age of 850 a). The ^{14}C age difference between the surface and the depth of core 63KA is thus just over 200 a. The paired benthic–planktonic ^{14}C measurements reproduce this age difference within uncertainties apart from one pair from a non-laminated mid-Holocene interval, which shows no difference (Fig. 3b). This particular value is significant at the 3σ level in comparison with the others, and could suggest deeper surface convection and better ventilation of the OMZ. Exposure of anoxic sediments to oxic conditions causes remobilization of authigenic U, which will partly reprecipitate deeper in the sediment column and enhance authigenic U accumulation there (Thomson et al. 1995). As a result of simultaneous burn-down of TOC, any present lamination may be diminished. This scenario would make high U/Ga ratios and poor preservation of lamination consistent with each other, but unfortunately the limited number of paired benthic–planktonic ^{14}C data does not allow us to confirm enhanced convection during the winter monsoon. We cannot derive a satisfactory explanation for the discrepancy between the ventilation proxies, but prefer the lamination index because of the many potential factors that may contribute to authigenic trace metal enrichment (see above).

Early and mid-Holocene evolution of the OMZ

Because of the different and spatially separated processes involved in thermocline ventilation at present (see Introduction), and because of the occasional discrepancy of proxies for past water column ventilation outlined above, it is unlikely that a single core will capture the entire picture of Holocene OMZ evolution. To overcome the spatial limitation, the 63KA records are compared with a previously published palaeoproductivity record from the upwelling region in the western Arabian Sea (core 74KL, Sirocko et al. 1996).

11–10 ka BP and the transition period from 10 to 9.3 ka BP

Before 10 ka BP, the absence of lamination but presence of mollusc shells, and the low accumulation of authigenic U indicate that the OMZ was ventilated in the northern AS (Fig. 3a and c). Rising palaeoproductivity in western AS was apparently not sufficient to cause the OMZ to become suboxic (Fig. 4). Because of lack of data, neither surface convection nor the ventilation of Central Indian Water can be inferred.

The transition period from 10 to 9.3 ka cal. BP begins with a brief laminated sequence and increased accumulation of authigenic U, indicating a suboxic OMZ. The difference between benthic and planktonic forams of 260 ^{14}C a in the laminated interval is within error of the pre-bomb value, suggesting that winter convection in the northern AS was similar to that at present (Fig. 3b). This interval is followed by a brief period of oxic conditions (in situ mollusc shells), which is not well resolved in the U/Ga record. The palaeoproductivity record from the western AS does not show a significant reduction (Fig. 4). Consequently, a past change in oxygen supply either by lateral advection of oxygen or surface convection seems likely. A point can be made for the case of better ventilation of Central Indian Water, as during the above transition interval significantly lower δ^{18}O values were recorded in the Antarctic Byrd ice core (Sowers & Bender 1995), which may indicate a climate change in the formation region of Central Indian Water.

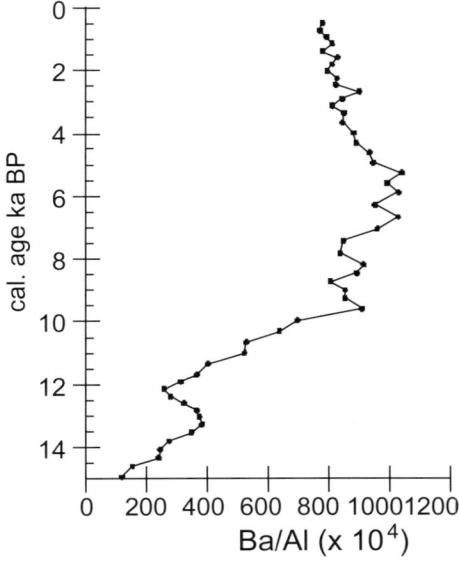

Fig. 4. Ba/Al ratios for core 74KL, western Arabian Sea, after Sirocko et al. (1996), with revised stratigraphy using ^{14}C reservoir ages of 850 a (Uerpmann 1991).

9.4 to 7.5 ka cal. BP

A well-preserved lamination and high accumulation of authigenic U indicate suboxic conditions in the OMZ during early Holocene time (Fig. 3a and c). This is consistent with stronger than present summer monsoon upwelling in the western AS (Luther et al. 1990; Anderson & Prell 1992; Overpeck et al. 1996; Sirocko et al. 1996) (Fig. 4). As the benthic–planktonic ^{14}C age difference (240 a) was comparable with modern values, winter convection must have been similar to that at present. A comparison of the U/Ga record with a δ^{18}O (G. ruber) record from the same core shows a lagged anti-correlation (U/Ga lag: 19 cm, Fig. 5). The δ^{18}O (G. ruber) record reflects surface hydrographic conditions, mostly salinity and temperature (e.g. Duplessy 1982). AS surface temperature and salinity are dominated by the atmospheric circulation pattern of the South Asian monsoon (Levitus & Boyer 1994; Webster et al. 1998). At the site of core 63KA, discharge from the nearby river Indus during the rainy summer monsoon season is thought to dominate δ^{18}O in surface-dwelling G. ruber (Staubwasser et al. 2003). This correlation suggests a close coupling

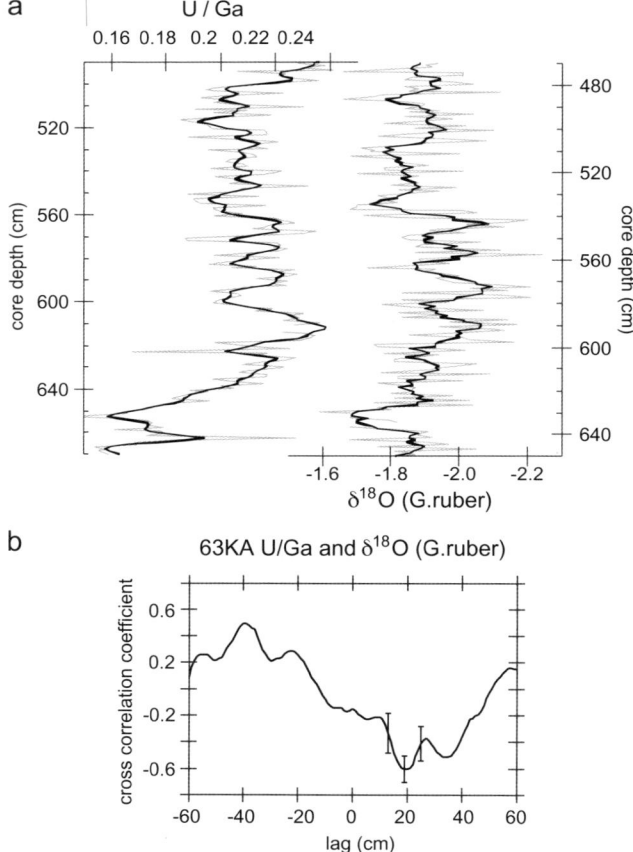

Fig. 5. (a) 63KA U/Ga record plotted with a 19 cm downcore offset to a δ^{18}O of the planktonic foraminifera *Globigerinoides ruber* record from the same core (fine line, raw data; bold line, five-point smoothed record). (b) Cross-correlation function of δ^{18}O (G. ruber) and U/Ga (raw data). The minimum of the function at lag 19 cm is significant at the 95% level (Davis 1987) and the uncertainty interval (2σ) is given for it and surrounding representative points. Auto-correlation has been taken into account for both records (Schönwiese 2000). The 19 cm lag of U/Ga with respect to the planktonic δ^{18}O record suggests that the redox front at which U precipitated in the past was at that depth relative to the sediment surface. This would be still within the present-day zone of Fe reduction (Lückge et al. 1999).

of the monsoon-controlled AS surface hydrography and the ventilation of the OMZ, most probably through small changes in upwelling during the summer monsoon.

7.5 to 5.5 ka cal. BP

The interpretation of the records from this section is hampered by the inconsistency of the lamination preservation and U/Ga palaeo-ventilation proxies. Palaeoproductivity in the western AS appears to have been at its highest (Fig. 4), as was TOC accumulation on the Pakistani margin (Fig. 3d). However, in contrast to the early Holocene period, only indistinctly laminated or non-laminated sediments were preserved on the Pakistani continental margin, which suggest better ventilation of the OMZ (Fig. 3a). This scenario, however, does not agree with high palaeoproductivity in the western AS nor with the reduced mid-Holocene thermocline ventilation proposed by Reichart et al. (1998). These discrepancies suggest that palaeoproductivity in the AS might not have been the single controlling factor of OMZ ventilation. Seasonal ventilation as a result of enhanced winter convection is suggested in this study (see above), but the scarcity of paired benthic–planktonic ^{14}C data allows no final conclusion to be drawn.

Summary

Authigenic enrichment of U seems to be a sensitive indicator for the intensity of past water column suboxic conditions in the Arabian Sea for early Holocene time in laminated sediment cores from the Pakistani continental margin. Other factors potentially contributing to such an enrichment, such as TOC flux, can be ruled out for this time interval on the basis of principal component analysis. A continuous reconstruction of the evolution of Arabian Sea thermocline ventilation from laminated core 63KA is limited, because of an inconsistency between authigenic U accumulation and the preservation of lamination during mid-Holocene time. Redistribution of U within the sedimentary column under enhanced winter convection and ventilation may be the cause of this discrepancy.

In general, explaining the evolution of the OMZ in the Arabian Sea and the causes of its variability is hampered by the complex interplay of a number of controlling processes observed at present, such as oxygen consumption, lateral oxygen advection and surface convection. Because these ventilating mechanisms at present are spatially separated in the Arabian Sea, it is unlikely that one sediment core is sufficient for this purpose. Variable upwelling in the western AS is a major controlling factor of OMZ ventilation in the Arabian Sea during early Holocene time. However, high palaeoproductivity does not always coincide with the preservation of laminated sediments (and reducing conditions). Other factors, which directly control OMZ ventilation at present, such as the ventilation of Central Indian Water on advection into the AS and the intensity of surface convection in the northern AS during the winter monsoon, may have contributed in variable proportions to the evolution of the OMZ. Therefore, the presence or absence of a reducing OMZ does not allow us to infer the evolution of the South Asian monsoon in general.

F. Sirocko is thanked for scientific guidance during the early stages of this work. We also thank J. Negendank for his support during the generation of the data at the GeoForschungsZentrum (GFZ), Potsdam. G. Schettler and E. Kramer kindly provided laboratory space and support on the ICP-AES analysis, and R. Naumann is thanked for X-ray diffraction measurements. This study was funded by the Deutsche Forschungsgemeinschaft.

References

ANDERSON, D.M. & PRELL, W.L. 1992. The structure of the southwest monsoon winds over the Arabian Sea during the late Quaternary: observations, simulations, and marine geologic evidence. *Journal of Geophysical Research*, **97**, 15481–15487.

BANSE, K. 1984. Overview of the hydrography and associated biological phenomena in the Arabian Sea off Pakistan. *In*: HAQ, B.U. & MILLIMAN, J.D. (eds) *Marine Geology and Oceanography of the Arabian Sea and Coastal Pakistan*. Van Nostrand Reinhold, New York, 273–301.

BANSE, K. & MCCLAIN, C.R. 1986. Winter blooms of phytoplankton in the Arabian Sea as observed by the Coastal Zone Color Scanner. *Marine Ecology Progress Series*, **34**, 201–211.

BEHL, R.J. & KENNETT, J.P. 1996. Brief interstadial events in the Santa Barbara Basin, NE Pacific, during the past 60 kyr. *Nature*, **379**, 243–246.

BROECKER, W.S. & PENG, T.-H. 1980. Seasonal variability in the ^{14}C/^{12}C ratio for surface ocean water. *Geophysical Research Letters*, **7**, 1020–1022.

BROECKER, W., SUTHERLAND, S., SMETHIE, W., PENG, T.-H. & OSTLUND, H.G. 1995. Oceanic radiocarbon: separation of the natural and bomb components. *Global Biogeochemical Cycles*, **9**, 263–288.

BHUSHAN, R., SOMAYAJULU, S., CHAKRABORTY, S. & KRISHNASWAMI, S. 2000. Radiocarbon in the Arabian Sea water column: temporal variations in bomb ^{14}C inventory since the GEOSECS and CO$_2$ air–sea exchange rates. *Journal of Geophysical Research*, **105**, 14273–14282.

CALVERT, S.E. & PEDERSEN, T.F. 1993. Geochemistry of recent oxic and anoxic marine sediments: implications for the geological record. *Marine Geology*, **113**, 67–88.

CRUSIUS, J., CALVERT, S., PEDERSEN, T. & SAGE, D. 1996. Rhenium and molybdenum enrichments in sediments as indicators of oxic, suboxic and sulfidic conditions of deposition. *Earth and Planetary Science Letters*, **145**, 65-78.

DAVIS, J.C. 1987. *Statistics and Data Analysis in Geology*. Wiley, New York.

DULSKI, P. 2001. Reference materials for geochemical studies: new analytical data by ICP-MS and critical discussion of reference values. *Geostandards Newsletter*, **25**, 87-125.

DUPLESSY, J.C. 1982. Glacial to inter-glacial contrasts in the northern Indian ocean. *Nature*, **295**, 494-498.

ENZEL, Y., ELY, Y.Y., MISHRA, S. & 6 OTHERS 1999. High resolution Holocene environmental changes in the Thar desert, Northwestern India. *Science*, **284**, 125-128.

GOTTARDI, G., BURTON, J.D. & CULKIN, F. 1969. Gallium. *In:* WEDEPOHL, K. (ed.) *Handbook of Geochemistry*. Springer, Berlin, **II/3**.

JANNINK, N.T., ZACHARIASSE, W.J. & VAN DER ZWAAN, G.J. 1998. Living (Rose Bengal stained) benthic foraminifera from the Pakistan continental margin (northern Arabian Sea). *Deep-Sea Research I*, **45**, 1483-1513.

KLINKHAMMER, G.P. & PALMER, M.R. 1991. Uranium in the oceans: where it goes and why. *Geochimica et Cosmochimica Acta*, **55**, 1799-1806.

KRAUSKOPF, K.B. & BIRD, D.K. 1995. *Introduction to Geochemistry*. McGraw-Hill, New York.

LAPP, B. & BALZER, W. 1993. Early diagenesis of trace metals used as an indicator of past productivity changes in coastal sediments. *Geochimica et Cosmochimica Acta*, **57**, 4639-4652.

LEVITUS, S. & BOYER, T.P. 1994. *World Ocean Atlas 1994, Volume 2: Oxygen*. NOAA Atlas NESDIS 2. US Department of Commerce, Washington, DC.

LÜCKGE, A., ERCEGOVAC, M., STRAUSS, H. & LITTKE, R. 1999. Early diagenetic alteration of organic matter by sulfate reduction in Quaternary sediments from the northeastern Arabian Sea. *Marine Geology*, **158**, 1-13.

LUTHER, M.E., O'BRIEN, J.J. & PRELL, W.L. 1990. Variability in upwelling fields in the northwestern Indian Ocean; 1. Model experiments for the past 18,000 years. *Paleoceanography*, **5**, 433-445.

MORRISON, J.M., CODISPOTI, L.A., SMITH, S.L. & 8 OTHERS 1999. The oxygen minimum zone in the Arabian Sea during 1995. *Deep-Sea Research II*, **46**, 1903-1931.

NADEAU, M.-J., GROOTES, P.M., SCHLEICHER, M., HASSELBERG, P., RIECK, A. & BITTERLING, M. 1998. Sample throughput and data quality at the Leibniz-Labor AMS facility. *Radiocarbon*, **40**, 239-245.

NATH, B.N., BAU, M., RAO, B.R. & RAO, C.M. 1997. Trace and rare earth elemental variation in Arabian Sea sediments through a transect across the oxygen minimum zone. *Geochimica et Cosmochimica Acta*, **61**, 2375-2388.

OVERPECK, J., ANDERSON, D., TRUMBORE, S. & PRELL, W. 1996. The southwest Indian Monsoon over the last 18,000 yrs. *Climate Dynamics*, **12**, 213-225.

REICHART, G.J., LOURENS, L.J. & ZACHARIASSE, W.J. 1998. Temporal variability in the northern Arabian Sea Oxygen Minimum Zone (OMZ) during the last 225,000 years. *Paleoceanography*, **13**, 607-621.

RAYMENT, R.A. & JÖRESKOG, K.G. 1993. *Applied Factor Analysis in the Natural Sciences*. Cambridge University Press, Cambridge.

SCHLEICHER, M., GROOTES, P.M., NADEAU, M.-J. & SCHOON, A. 1998. The carbonate C-14 background and its components at the Leibniz AMS facility. *Radiocarbon*, **40**, 85-93.

SCHÖNWIESE, C.D. 2000. *Praktische Statistik*. Borntraeger, Stuttgart.

SHAW, T.J., GIESKES, J.M. & JAHNKE, R.A. 1990. Early diagenesis in differing depositional environments: the response of transition metals in pore water. *Geochimica et Cosmochimica Acta*, **54**, 1233-1246.

SIROCKO, F., GARBE-SCHÖNBERG, C.-D., MCINTYRE, A. & MOLFINO, B. 1996. Teleconnections between the subtropical monsoons and high-latitude climates during the last deglaciation. *Science*, **272**, 526-529.

SOWERS, T. & BENDER, M. 1995. Climate records covering the last deglaciation. *Science*, **269**, 210-214.

STAUBWASSER, M. 1999. *Early Holocene variability of the Indian monsoon and Arabian Sea thermocline ventilation*. PhD thesis, Christian-Albrechts-Universität zu Kiel.

STAUBWASSER, M. & SIROCKO, F. 2001. On the formation of laminated sediments on the continental margin off Pakistan: the effects of sediment provenance and sediment redistribution. *Marine Geology*, **172**, 43-56.

STAUBWASSER, M., SIROCKO, F., GROOTES, P.M. & ERLENKEUSER, H. 2003. South Asian monsoon climate change and radiocarbon in the Arabian Sea during early and mid Holocene *Paleoceanography*, in press.

SWALLOW, J.C. 1984. Some aspects of the physical oceanography of the Indian Ocean. *Deep-Sea Research*, **31**, 639-650.

TAYLOR, S.R. & MCLENNAN, S.M. 1985. *The Continental Crust: its Composition and Evolution*. Blackwell, Oxford.

THOMSON, J., HIGGS, N.C., WILSON, T.R.S., CROUDACE, I.W., DELANGE, G.J. & VAN SANTVOORT, P.J.M. 1995. Redistribution and geochemical behaviour of trace elements around S1, the most recent Mediterranean sapropel. *Geochimica et Cosmochimica Acta*, **59**, 3487-3501.

TOMCZAK, M. & GODFREY, J.S. 1994. *Regional Oceanography: an Introduction*. Pergamon, Oxford.

TORRES, M.E., BOHRMANN, G. & SUESS, E. 1996. Authigenic barites and fluxes of barium associated with fluid seeps in the Peru subduction zone. *Earth and Planetary Science Letters*, **144**, 469-481.

TUREKIAN, K.K. & WEDEPOHL, K.H. 1961. Distribution of the elements in some major units of the Earth's crust. *Geological Society of America Bulletin*, **72**, 175-192.

UERPMANN, H.-P. 1991. Radiocarbon dating of shell middens in the Sultanate of Oman. *PACT*, **29**(IV.5), 335-347.

VON RAD, U., SCHAAF, M., MICHELS, K.H., SCHULZ, H., BERGER, W.H. & SIROCKO, F. 1999a. A 5000-yr record of climate change in varved sediments from the oxygen minimum zone off Pakistan, northeastern Arabian Sea. *Quaternary Research*, **51**, 39-53.

VON RAD, U., SCHULZ, H., RIECH, V., DEN DULK, M., BERNER, U. & SIROCKO, F. 1999b. Multiple monsoon-controlled breakdown of oxygen-minimum conditions during the past 30,000 years documented in laminated sediments off Pakistan. *Palaeogeography, Palaeoclimatology, Palaeoecology*, **152**, 129–161.

VON STACKELBERG, U. 1972. Faziesverteilung in Sedimenten des indisch-pakistanischen Kontinentalrandes (Arabisches Meer). *'Meteor'-Forschungserg., Reihe C*, **9**, 1–73.

WEBSTER, P.J., MAGANA, V.O., PALMER, T.N., SHUKLA, J., TOMAS, R.A., YANAI, M. & YASUNARI, T. 1998. Monsoons: processes, predictability, and the prospects for prediction. *Journal of Geophysical Research*, **103**, 14451–14510.

YOU, Y. & TOMCZAK, M. 1993. Thermocline circulation and ventilation in the Indian Ocean derived from water mass analysis. *Deep-Sea Research I*, **40**, 13–56.

YOU, Y. 1998. Seasonal variations of thermocline circulation and ventilation in the Indian Ocean. *Journal of Geophysical Research*, **102**, 10391–10422.

Discovery of the Toba Ash (c. 70 ka) in a high-resolution core recovering millennial monsoonal variability off Pakistan

ULRICH VON RAD[1], KLAUS-PETER BURGATH[1], MUHAMMAD PERVAZ[2] & HARTMUT SCHULZ[3]

[1] *Bundesanstalt für Geowissenschaften und Rohstoffe, D-30631 Hannover, Germany (e-mail: u.vonrad@t-online.de)*
[2] *Hydrocarbon Development Institute of Pakistan, Islamabad, Pakistan*
[3] *Institut für Ostseeforschung, PF 301161, D-16112 Rostock-Warnemünde, Germany*

Abstract: A discrete Toba Ash layer in the northeastern Arabian Sea was detected near the base of a 20.2 m long piston core (289KL) recovered from the oxygen minimum zone off the Indus delta. In addition to the Toba Ash, we discovered two highly disseminated, vitreous, rhyolitic 'ash layers' in two annually laminated box cores: a 'Younger Ash' (about AD 1885–1900), and an 'Older Ash' (about AD 1815–1830). The glass shards were probably derived from eruptions of Indonesian volcanoes, although it was not possible to correlate these two ashes with well-known historical eruptions. We discuss source, transport and deposition of distal ash-fall layers in the Arabian Sea, which are derived from violent ultra-Plinian eruptions on the Indonesian volcanic archipelago, as well as their use for palaeoclimatic correlation. Core 289KL has a complete, high-resolution stratigraphic record of the past 75 ka with 21 interstadials (IS) or Dansgaard–Oeschger (D–O) cycles and equivalents of Heinrich events H1–H6. The high-frequency record of this core shows rapid climate oscillations with periods around 1.5 ka and can be tuned precisely to the $\delta^{18}O$ record of a Bay of Bengal core and to the GISP-2 ice core from Greenland. The Toba event (70 ± 4 ka BP), which is well documented in the Arabian Sea and Bay of Bengal records at the end of IS-20, as well as in the Greenland ice, is an excellent stratigraphic marker horizon to validate this correlation. The apparent synchronous appearance of the various D–O oscillations and Heinrich events, which has been documented for many northern hemisphere localities, can be explained only by fairly rapid atmospheric circulation changes. Changes in the intensity of the Indian summer monsoon are tightly coupled with suborbital climate oscillations in the northern hemisphere via atmospheric moisture and heat circulation.

The northeastern Arabian Sea is characterized by an exceptionally stable oxygen minimum zone (OMZ) between 200 and 1200 m water depth (Fig. 1) and a strong, seasonal, monsoon-controlled variability of primary productivity and terrigenous supply (Reichart *et al.* 1997; Schulz *et al.* 1998; von Rad *et al.* 1999b). This results in annually laminated sediments that can be analysed by varve counting to construct ultra-high-resolution time series (von Rad *et al.* 1999a).

A new long piston core recovered from the OMZ at the continental slope off the Indus delta (SO 130-289KL, 23°07.3′N, 66°29.8′E, 571 m water depth; Figs 1 and 2) documents millennial-scale cycles of interstadials and stadials (represented by laminated or bioturbated intervals, respectively) in great detail. The new wide-diameter (125 mm) piston core has the advantage of recovering an undisturbed and complete, >75 ka stratigraphic record with a moderately high sedimentation rate (c. 25 cm ka^{-1}). As the core is from the OMZ, all the interstadials (Dansgaard–Oeschger event equivalents) are laminated (Schulz *et al.* 1998, 2002; von Rad *et al.* 1999b). Near the base of this core we discovered within a laminated interval a distinct volcanic ash layer, which is correlated with the Youngest Toba Tuff (YTT) in Sumatra (Indonesia), the first discovery of a discrete Toba Ash layer in the northeastern Arabian Sea. Because ash-fall layers document virtually instantaneous events, they form ideal chronostratigraphic event markers, if the composition of the glass shards can be traced to source areas of known and dated historical or prehistoric eruptions. The oxygen-depleted environment of the upper continental slope off Pakistan is particularly well suited for tephrochronological studies, as benthic activity (bioturbation) is absent, which in other environments significantly disturbs or destroys thin ash layers or 'disseminated ashes'

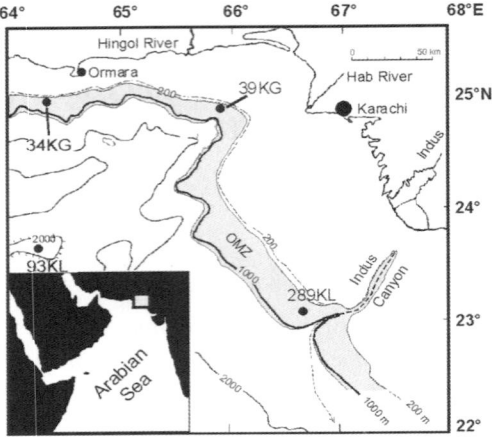

Fig. 1. Area of investigation with location of investigated cores SO 90-34KG, 39KG, 93KL, SO 130-289KL (SO90–136 KL at same location). Oxygen minimum zone is stippled.

after their deposition. In annually laminated box cores we also discovered historical, disseminated rhyolitic ash layers of unknown origin that can be dated by varve stratigraphy (AD 1885–1900: 'Younger Ash'; AD 1815–1830: 'Older Ash').

In this paper we will present the lithology and chemical composition of discrete and disseminated tephra in the northeastern Arabian Sea in the context of the chronostratigraphy of three OMZ cores off Pakistan. We will discuss the source, transport and deposition of these distal ash-fall layers, which are probably derived from eruption centres in the Indonesian volcanic archipelago.

In addition, we will also consider the millennial-scale oscillations of the monsoonal climate in the high-resolution core 289KL during the past 75 ka and discuss their cyclicity and origin, as well as the use of a global tephra event for palaeoclimatic correlation with other monsoon-controlled areas in southeastern Asia (e.g. the Bay of Bengal) and the Greenland ice record.

Material and methods

Material

The varve stratigraphy and varve thickness of a Holocene 'kasten core' SO 90-56 KA (30 cm × 30 cm diameter, 10 m long box core) and box core SO 90-39KG taken from the centre of the OMZ at the same location west of Karachi (24°50′N, 65°55′E; water depth 695 m; Fig. 1) were studied by von Rad et al. (1999a). Piston cores (e.g. core SO 90-93KL from the Northern Murray Ridge;

about 1800 m water depth; Fig. 1), which have a much lower sedimentation rate (about 7 cm ka^{-1}), document a palaeoclimatic record of >100 ka. In this core, Schulz et al. (1998, 2002) discovered a concentration of disseminated ash in the upper part of Interstadial (IS)-20 (about 70–72 ka BP); those workers studied this core by isotope stratigraphy, sonic velocity logs and colour stratigraphy (reflecting total organic content (TOC) variability) and correlated the chemical composition of this ash with the Toba Ash, erupted at c. 70 ± 4 ka BP from the Toba caldera in Sumatra. Because this ash is also preserved as sulphate aerosol in the GISP-2 ice core from Greenland, it is an excellent time marker to correlate the monsoonal climate records from the Arabian Sea with the northern hemisphere ice core record (Schulz et al. 1998, 2002).

A 20.2 m long piston core (SO 130-289KL) with a complete high-resolution palaeoceanographic record of the past 75 ka was obtained in 1998 during the SO 130 cruise (von Rad & Doose, 1998). This core contains a discrete volcanic ash layer with an age of c. 70 ka that can be correlated with the YTT.

Analytical methods

Unopened 1 m sections of the 20.2 m long core 289KL were analysed on board ship with a Geotek multisensor core scanner for sonic velocity, gamma-ray attenuation and magnetic susceptibility. After opening, the cores were photographed, slabbed for X-radiographs, and sampled for organic and inorganic geochemical analyses. The colour and lightness were digitized using a hand-held Minolta Chromameter (CR-2002) with an opening of 1 cm. Measurements were spaced at 2 cm corresponding to a temporal resolution ranging from 15 to 200 a. The lightness values were used to estimate TOC % after a careful calibration of lightness data from various SO 90 cores with TOC % measured by the Leco carbon combustion method (Schulz et al. 1998; see Fig. 2).

The varves of box cores 34KG and 39KG were counted and their thickness was measured with a tree-ring measuring device under a binocular microscope at the Tree-ring Laboratory of the University of Göttingen (see von Rad et al. 1999b). We estimate the counting error for the last 200 a to be c. ± 5 a.

Using the varve stratigraphy we detected highly diluted, dispersed ash-fall deposits in two box cores (39KG and 34KG; Fig. 1) at the general position where the Krakatau (AD 1883) and Tambora ash (AD 1815) were expected. To determine the presence and semi-quantitative concentration of volcanic glass, we took 50 samples from box

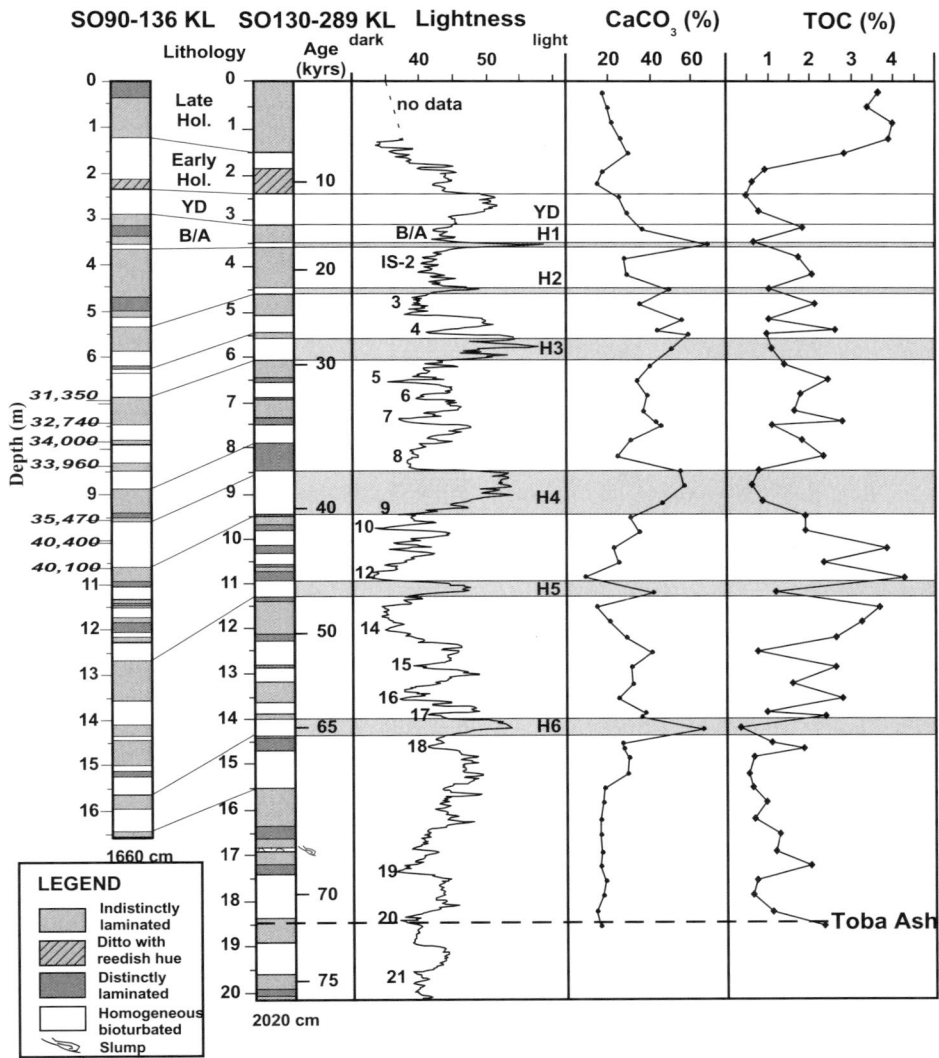

Fig. 2. Lithostratigraphy of cores SO 90-136KL and SO 130-289KL obtained at same position from the centre of the OMZ off the Indus delta (water depth c. 570 m), with location of discrete Toba Ash layer, grey-scan (lightness) values (reflecting TOC), measured TOC and $CaCO_3$ concentrations. YD, Younger Dryas; B/A, Bölling–Alleröd; H1 to H6, Heinrich event equivalents, IS-2 to IS-30, interstadials (Dansgaard–Oeschger event equivalents). Italics indicate calibrated ^{14}C ages (core 136KL).

cores 34KG (between 17.4 and 20.5 cm, and from 31.5 to 34 cm) and 39KG (15.1–16.7, 27.3–29.1 cm, 36.6–37.8 cm) spaced about 3–5 mm apart. The samples were treated with H_2O_2 to destroy organic matter and de-carbonatized with hydrochloric acid. The 20–63 μm fraction was then sieved and weighed. Grain mounts were made from this fraction for optical studies under the petrographic microscope. The slide was searched for volcanic glass shards using shape, colour, refractive index and the lack of birefringence. The identified glass shards were counted, and the proportion of glass shards calculated as per mil of an estimate of all grains on the slide (Fig. 3). Grain mounts were carefully ground with a precision grinding machine (15 μm grain) and polished. The polished surfaces of the glass shards were exposed at the surface of the slide and photographed under reflected light, noting the exact coordinates of the shards on the thin-section holder.

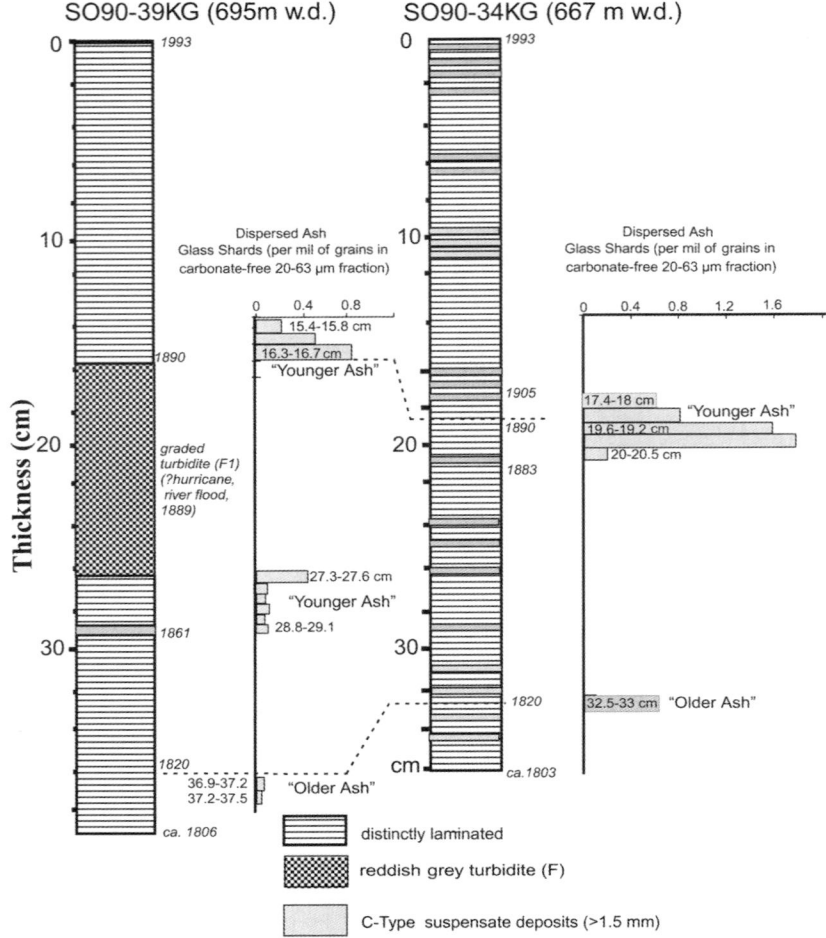

Fig. 3. Box cores SO 90-39KG and -34KG showing lithology, varve stratigraphy and frequency of ash shards. *1993, 1890*, etc. indicate approximate location of years AD from varve counts; 15.4–15.8 cm, position of sub-samples for ash counts in centimetres below sea floor. Suspensate deposits (C layers) are indicated only approximately.

From the distinct ash layer at the base of core 289KL (correlated with the YTT) we separated the 32–160 μm fraction and concentrated the volcanic glass by dissolving the carbonate with formic acid and by magnetic separation at different Ampere settings to eliminate the paramagnetic 'heavy minerals' and concentrate the light-coloured glass fraction. This fraction was impregnated by Araldite resin to make polished sections of glass grain mounts for microprobe analysis (Fig. 4). The fraction was also used for standard inductively coupled plasma mass spectrometry (ICP-MS) analysis of rare and trace elements analysis (see Table 2, below).

A 5 cm × 2 cm section of the sediment including the Toba Ash layer was carefully cut from the core for making a thin-section (Fig. 5). Before impregnation the sample was desalinized (by treatment with distilled water) and the pore water replaced by repeated steps of acetone treatments; the sample was carefully impregnated with the epoxy resin Araldite 2020, taking special care to suck out all the acetone and to slow down the polymerization process by cooling the sample in a refrigerator.

Selected glass shards could be analysed from polished and carbon-coated thin-sections with the BGR Cameca Camebax electron microprobe (EMP) system using an acceleration voltage of 15 kV and a current of 15 nA. The glass shards were identified by an optical (reflected light) system, as well as viewed and printed from back-

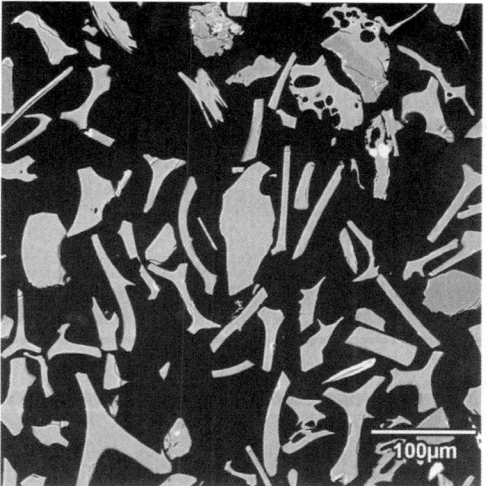

Fig. 4. Rhyolitic bubble-wall and pumice shards of Toba Ash (back-scattered electron (BSE) image of microprobe sample).

scattered electron images. We analysed 11 elements from small areas, using for calibration the standards of three synthetic glasses (for Si, Al, Fe, Ca, Mg) and other standards (for Cr, Na, K, Mg, Ti, P). For the Toba Ash, where enough material was available for analysis, the data from EMP analysis were combined with the X-ray fluorescence (XRF) and ICP-MS data from bulk samples and trace element analysis.

Toba Ash in core SO 130-289KL

Chronostratigraphy and millennial-scale climate cycles

The litho- and chronostratigraphy of the 20.2 m long core SO130-289KL taken at the same position in the centre of the OMZ off the Indus delta as its precursor core SO90-136KL (16.6 m long; Schulz et al. 1998) is shown in Fig. 2. Core 289KL extends the record of 136KL considerably, documenting the complete palaeoclimate record of the past 75 ka, down to IS-21. As proxies for palaeoclimate variability, a number of variables proved useful (Schulz et al. 1998; von Rad et al. 1999b; Leuschner & Sirocko 2000): in addition to $\delta^{18}O$ values of planktonic foraminifera, inorganic geochemical parameters, and the abundance of planktonic and benthic foraminifera, we have used gamma-ray attenuation and magnetic susceptibility logs, lightness (grey-scale) measurements (which can be positively correlated with TOC), as well as $CaCO_3$ and TOC percentages. For the study of core 289KL we used the gamma-ray attenuation (wet bulk density; not shown), lightness values, TOC and carbonate (Fig. 2). The laminated intervals of the Late Holocene period, Preboreal period, and IS-1 to IS-21 (including the 'last glacial maximum') are characterized by dark (olive–grey) colours (lightness <45), high TOC (>2%, with maxima around 4% during Holocene time, IS-10 and IS-12), and low $CaCO_3$ values (<40%), whereas the bioturbated intervals of the Younger Dryas and the stadials or Heinrich event (H) equivalents (H1–H6) are light coloured (lightness >45), with low TOC (<1.5%, with minima during H1 and H2) and high $CaCO_3$ (> 40–65%) contents (Fig. 2). The linear sedimentation rates are significantly reduced during the dark laminated interstadials (about 5–30 cm ka^{-1}), and enhanced during the light stadials: H1, H3, H4 and H7 attain sedimentation rates of 70–80 cm ka^{-1}; during the Younger Dryas, H2, H5 and H6, we observe moderate (20–40 cm ka^{-1}) sedimentation rates.

Lithology, fabric and chemical composition of the Toba Ash in core 289KL

A distinct, about 1–1.5 mm thick white layer (Fig. 5a), correlated with the Youngest Toba Tephra (YTT), was discovered in core 289KL at a depth of 1843 cm below sea floor (bsf) within an indistinctly laminated interval (IS-20, Fig. 6). At first glance, there appear to be two ash layers in this core (Fig. 3a). However, detailed analysis of the apparent 'upper ash layer' in a thin-section (Fig. 3b) showed that it is identical in thickness, fabric and chemical composition to the 'lower ash layer', but shows inverse grading. This suggests strongly that the 'upper ash layer' is in an upside-down position and that this 'false upper ash layer' is the sheared upper limb of an *en miniature* slump fold, repeating this ash layer in our section. This is supported by the fact that minor slump folding and microfaulting was observed in the X-radiograph of this section and that in core 287KL, taken from the northwestern levee of Indus Canyon, about 20 km SW of core 289KL, we discovered only one, 0.5–0.8 mm thick ash layer at 1461.5 cm bsf that can also be correlated with the YTT.

The ash layer in core 289KL shows a very distinct lower boundary overlain by size-graded glass shards, and a rather indistinct upper boundary (Fig. 5b and c). Whereas no glass shards were observed below the lower ash boundary, shards are present up to c. 1.5 mm above the upper ash boundary. This indicates an instantaneous fallout and settling event, followed by weak bioturbation after the ash layer was deposited.

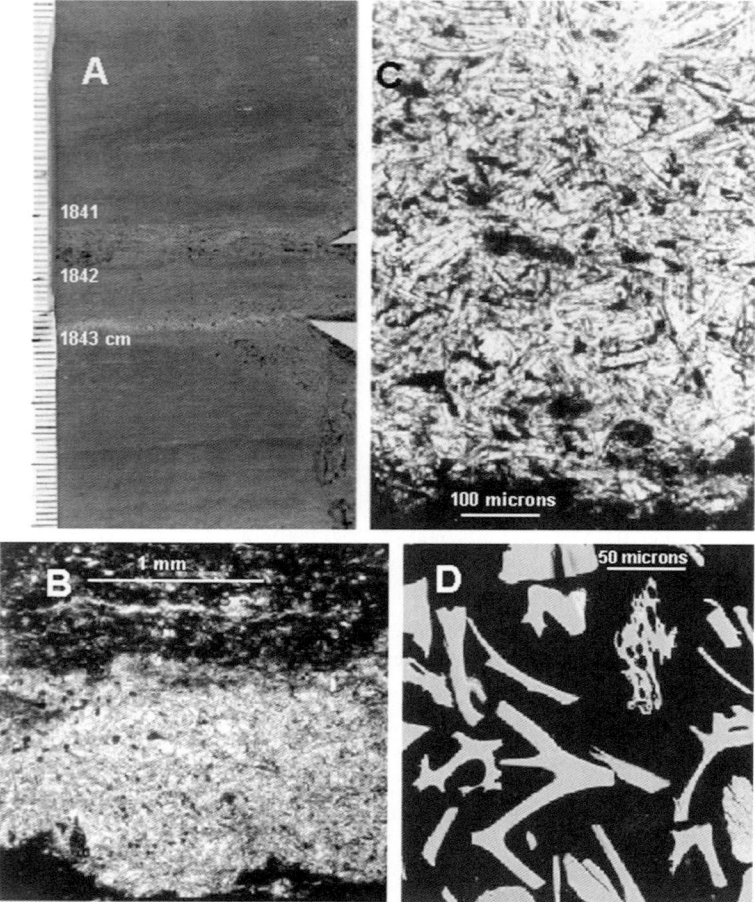

Fig. 5. Toba Ash (c. 70 ka BP) in core 289KL (c. 1843 cm). (**a**) Photograph of core 289KL showing the two apparent ash layers (see arrows; see text for explanation). (**b**) Thin-section photograph of impregnated part of core showing texture and structure of 'lower' ash layer of (**a**) (note distinct lower and indistinct upper boundary of 1 mm thick ash layer). (**c**) Detail of (**b**) showing indistinct size grading of shards within lower part of ash layer. (**d**) BSE image of microprobe sample of lower layer of Toba Ash in core 289KL, c. 1843 cm (MS208-BTA_L1b).

The vitreous ash layer consists of >90% colourless, rhyolitic glass shards ranging from 25 to 180 μm (mean c. 70 μm), traces of plagioclase and minor amounts of admixed terrigenous quartz, clay aggregates and biogenous carbonate. Typically, the coarse-silt to fine-sand-size glass shards have a refractive index of c. 1.495 and a plain or cuspate shape, with bubble-wall shards and junction shards being less frequent (Figs 4 and 5d). Pumice shards (Fig. 5d) are rare.

The glass shards of the ash in core 289KL have a calc-alkaline character and a remarkably homogeneous, rhyolitic composition (Table 1, Fig. 7): 77.81% SiO_2, 0.05% TiO_2, 12.66% Al_2O_3, 0.9% FeO, 0.07% MnO, 0.06% MgO, 0.72% CaO, 2.74% Na_2O, 5.05% K_2O, 0.01% P_2O_5, 0.05% TiO_2, 0.03% Cr_2O_3 (mean of 40 microprobe analyses of individual glass shards, recalculated water-free). The position of the ash layer at about the 70 ka level in core 289KL and the high silica composition of the shards indicate that it correlates with the YTT eruption in Sumatra (Rose & Chesner 1987; Dehn et al. 1991; Pattan et al. 1999). Major element (especially alkali) composition of the shards is also identical to glass compositions from the YTT on Sumatra and the widespread ash layer in the Gulf of Bengal (Gasparotti et al. 2000) and India (Westgate et al. 1998; Fig. 7). In addition, the trace element composition (Table 2) is very similar to compositional data from the Toba caldera (Sumatra) and from Toba Ash occurrences in the Bay of Bengal

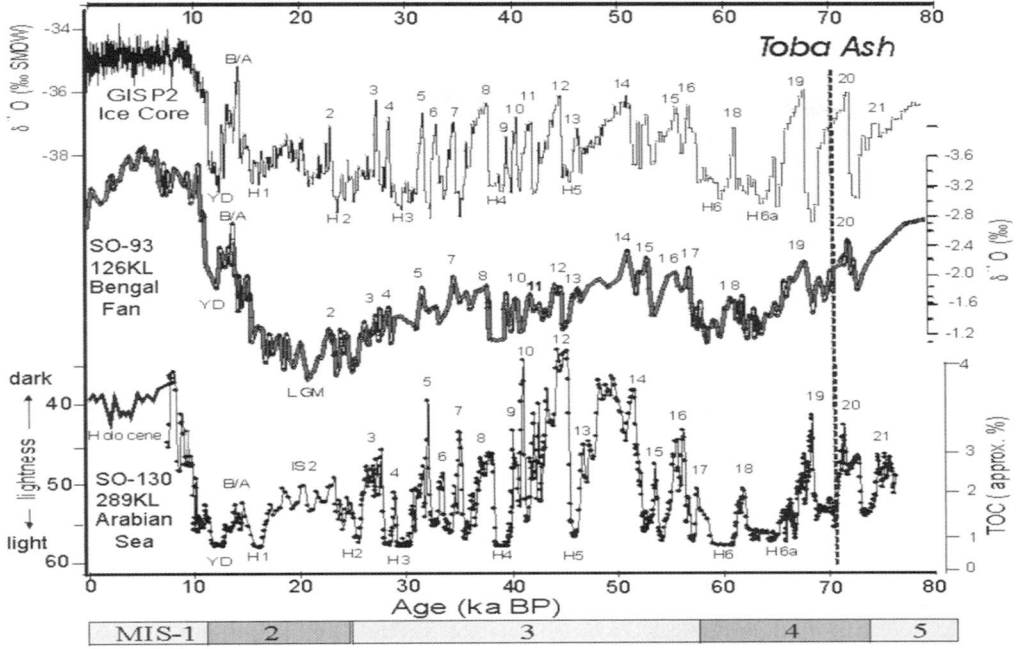

Fig. 6. Correlation of TOC concentrations–lightness values of core SO 130-289KL (for mid- and late Holocene time, data from core SO 90-136KL were inserted) v. age, with marine $\delta^{18}O$ values from core SO 93-126KL from the Bengal Fan (modified from Kudrass *et al.* 2001), and with Greenland GISP-2 ice core record (Dansgaard *et al.* 1993; Stuiver & Grootes 2000). MIS, marine isotope stage. Note the presence of Toba Ash (or aerosols) and excellent correlation of all interstadials and stadials (Heinrich-equivalent events) in all three records.

and India. Although the glass shards appear to be rather fresh, some hydration during subaqueous alteration must have taken place, considering the contents of *c.* 5% H_2O, similar to the Toba Ash studied from Ocean Drilling Program (ODP) Site 758 NW of Sumatra (Dehn *et al.* 1991).

Records of historical ash falls during the past 200 a

'Younger Ash' (*c.* AD 1885–1900)

The concentration of felsic glass shards in box cores 34KG and 39KG is only about 0.2–0.8‰ of the carbonate-free 20–63 μm fraction (see Fig. 3). However, a few unequivocal colourless, elongate, blocky or platy glass shards (with a longest diameter of *c.* 50–160 μm) were identified (Fig. 8a and b). We did not observe any microlites in the glass, but foliated shards with a central, very straight and narrow 'ridge', apparently formed by a bubble-wall junction (Fig. 8b). Microprobe analysis (Fig. 8a) shows that all these glass shards have a K-rich rhyolitic composition with about 75% SiO_2, 2.2% Na_2O and 3.9% K_2O (Table 1, Fig. 7).

'Older Ash' (*c.* AD 1815–1830)

The concentration of colourless, silicic glass shards of this age range was even smaller in both investigated box cores than that of the 'Younger Ash' (<0.1‰; see Fig. 3). However, Fig. 8c–f shows examples of elongate rhyolitic glass shards, again with an elongate Y-shaped bubble-wall junction ridge (Fig. 8f) and typical bubble-wall shards. The glass shards of box core 34KG that could be analysed by the microprobe (Fig. 8d) indicated a somewhat more Mg-rich and Na-poor rhyolitic composition than the 'Younger Ash', with about 74% SiO_2, 0.6% Na_2O and 3.8% K_2O. The glass from core 39KG has somewhat less SiO_2, less MgO, but more Na_2O, and especially K_2O (5.2%). Although original H_2O contents up to 7 wt % have been reported for rhyolitic magmas (Schmincke 1986), rhyolitic deep-sea tephra layers around Indonesia have an original H_2O content of only 1.3% (subrecent glass), increasing progressively with age by secondary alteration to about 5% in

Table 1. *Microprobe geochemical data (major and minor elements) of 'Younger Ash' and 'Older Ash' from cores SO 90-34 and 39KG, and Toba Ash samples (SO 130-289KL)*

Core no.	Depth (cm)	Microprobe no.	SiO_2	TiO_2	Al_2O_3	FeO	MnO	MgO
Younger Ash (AD 1900–1910)								
SO 90-34KG	18.6–19.2	G34_1	74.45	0.21	12.18	1.06	0.06	0.23
SO 90-34KG	18.6–19.2	G34_2	76.01	0.22	12.36	1.04	0.05	0.23
SO 90-34KG	18.6–19.2	G34_6	74.61	0.07	12.19	0.87		0.07
Mean Younger Ash			75.02	0.17	12.24	0.99	0.06	0.18
SD			0.70	0.07	0.08	0.09	0.01	0.08
Older Ash A (AD 1815–1830)								
SO 90-34KG	32.5–33	T34_2	77.08	0.11	11.93	0.71	0.01	1.62
SO 90-34KG	32.5–33	T34_3	76.69	0.13	11.83	0.73	0.01	1.63
SO 90-34KG	32.5–33	T34_4	75.14	0.04	12.45	0.69	0.01	1.70
SO 90-34KG	32.5–33	T34_5	75.03	0.07	12.98	0.93	0.04	1.81
SO 90-34KG	32.5–33	F34_1	78.02	0.08	12.52	0.62	0.01	1.72
Mean Older Ash A			76.39	0.09	12.34	0.74	0.02	1.70
SD (2–5)			1.15	0.03	0.42	0.10	0.01	0.07
Older Ash B (AD 1815–1839)								
SO 90-39KG	37.2–37.5	D39_1	73.96	0.08	14.20	0.70	0.04	1.02
SO 90-39KG	37.2–37.5	D39_2	73.33	0.07	14.21	0.75	0.09	1.10
SO 90-39KG	37.2–37.5	D39_3	74.94	0.06	13.75	0.67	0.01	1.01
SO 90-39KG	37.2–37.5	D39_4	75.83	0.12	13.30	0.65	0.01	1.01
SO 90-39KG	37.2–37.5	D39_5	73.38	0.11	14.36	0.80	0.02	1.18
Mean Older Ash B			74.29	0.09	13.96	0.71	0.03	1.06
SD			0.96	0.02	0.39	0.05	0.03	0.07
Upper Toba Ash (71.3 ± 3.5 ka BP)								
SO 130-289KL	1842	BTA_U1	74.33	0.05	11.81	0.86	0.07	0.04
SO 130-289KL	1842	BTA_U2	74.92	0.04	12.12	0.85	0.05	0.05
SO 130-289KL	1842	BTA_U3	74.87	0.05	12.09	0.88	0.07	0.03
SO 130-289KL	1842	BTA_U4	74.02	0.07	12.32	0.92	0.09	0.08
SO 130-289KL	1842	BTA_U5	74.33	0.05	11.93	0.87	0.09	0.06
SO 130-289KL	1842	BTA_U6	74.05	0.05	12.15	0.85	0.07	0.05
SO 130-289KL	1842	BTA_U7	74.24	0.08	12.19	0.86	0.09	0.08
SO 130-289KL	1842	BTA_U8	73.39	0.06	12.03	0.89	0.11	0.04
SO 130-289KL	1842	BTA_U9	74.09	0.05	12.13	0.82	0.09	0.06
SO 130-289KL	1842	BTA_U10	74.22	0.04	11.98	0.84	0.03	0.07
SO 130-289KL	1842	BTA_U11	74.62	0.04	12.00	0.83	0.07	0.06
SO 130-289KL	1842	BTA_U12	73.99	0.05	11.94	0.80	0.12	0.05
SO 130-289KL	1842	BTA_U13	74.34	0.07	11.93	0.81	0.07	0.06
SO 130-289KL	1842	BTA_U14	73.99	0.05	11.97	0.84	0.09	0.05
SO 130-289KL	1842	BTA_U15	74.37	0.07	12.40	0.96	0.06	0.09
SO 130-289KL	1842	BTA_U16	74.56	0.06	12.23	0.82	0.07	0.09
SO 130-289KL	1843.5	BTA_L17	73.85	0.05	12.00	0.80	0.05	0.03
SO 130-289KL	1843.5	BTA_L18	74.78	0.03	11.99	0.90	0.04	0.04
SO 130-289KL	1843.5	BTA_L19	75.63	0.05	12.23	0.85	0.06	0.06
SO 130-289KL	1843.5	BTA_L20	74.92	0.02	12.01	0.89	0.01	0.05
SO 130-289KL	1843.5	BTA_L21	74.57	0.05	12.14	0.92	0.07	0.07
SO 130-289KL	1843.5	BTA_L22	74.40	0.02	12.04	0.81	0.09	0.04
SO 130-289KL	1843.5	BTA_L23	75.55	0.08	12.43	0.80	0.07	0.08
SO 130-289KL	1843.5	BTA_L24	75.27	0.06	12.34	0.84	0.14	0.07
SO 130-289KL	1843.5	BTA_L25	75.16	0.07	12.47	0.95	0.03	0.06
SO 130-289KL	1843.5	BTA_L26	74.70	0.07	12.28	0.92	0.09	0.07
SO 130-289KL	1843.5	BTA_L27	74.24	0.07	12.26	0.91	0.06	0.08
SO 130-289KL	1843.5	BTA_L28	74.34	0.07	12.39	1.08	0.05	0.10
SO 130-289KL	1843.5	BTA_L29	74.04	0.04	12.33	0.91	0.04	0.07
SO 130-289KL	1843.5	BTA_L30	74.69	0.04	11.88	0.80	0.09	0.03
SO 130-289KL	1843.5	BTA_L31	74.56	0.08	11.98	0.97	0.12	0.08
SO 130-289KL	1843.5	BTA_L32	74.70	0.02	11.83	0.78	0.07	0.06
SO 130-289KL	1843.5	BTA_L33	74.87	0.05	12.00	0.92	0.07	0.05
SO 130-289KL	1843.5	BTA_L34	73.67	0.07	12.25	0.99	0.02	0.07
SO 130-289KL	1843.5	BTA_L35	73.75	0.06	12.15	0.91	0.05	0.06
SO 130-289KL	1843.5	BTA_L36	74.88	0.06	12.00	0.88	0.05	0.05
SO 130-289KL	1843.5	BTA_L37	74.73	0.05	12.08	0.76	0.04	0.07
SO 130-289KL	1843.5	BTA_L38	74.12	0.06	12.18	0.91	0.05	0.06
SO 130-289KL	1843.5	BTA_L39	74.33	0.03	12.16	0.79		0.04
SO 130-289KL	1843.5	BTA_L40	74.18	0.05	12.14	0.89	0.05	0.07
Mean Toba Ash			74.46	0.05	12.12	0.87	0.07	0.06
SD			0.480	0.0158	0.166	0.064	0.027	0.0170

CaO	Na$_2$O	K$_2$O	P$_2$O$_5$	Cr$_2$O$_3$	Total	n	Remarks	Petrology
1.26	2.51	3.21	0.12		95.29		Defoc. (6 μm)	Rhyolite
1.17	2.05	3.04	0.04		96.21		Defoc. (6 μm)	Rhyolite
0.57	2.12	5.33	0.02	0.03	95.85		Defoc. (10 μm)	Rhyolite
1.00	2.23	3.86	0.06	0.03	95.78	3		
0.31	0.20	1.04	0.04	0.00	0.38			
1.05	0.69	3.85	0.14		97.19		Defoc. (<10 μm)	Rhyolite
1.09	0.46	3.58	0.15		96.30		Foc. (2 μm)	Rhyolite
1.19	0.74	3.89	0.19		96.04		Defoc. (<10 μm)	Rhyolite
1.24	0.50	3.49	0.19		96.28		Foc. (2 μm)	Rhyolite
1.25	0.52	4.01	0.17		98.92			Rhyolite
1.16	0.58	3.76	0.17	0.00	96.95	4		
0.08	0.11	0.20	0.02	0.00	1.06			
2.89	0.88	5.30	0.21		99.28		Defoc. (10 μm)	Rhyolite
2.89	0.89	5.25	0.21		98.79		Defoc. (10 μm)	Rhyolite
2.79	0.81	5.32	0.16		99.52		Defoc. (10 μm)	Rhyolite
2.56	0.89	5.26	0.11	0.01	99.74		Defoc. (15 μm)	Rhyolite
2.99	0.81	5.07	0.17	0.06	98.89		Defoc. (10 μm)	Rhyolite
2.82	0.86	5.24	0.17	0.04	99.24	5		
0.15	0.04	0.09	0.04	0.03	0.36			
0.65	2.95	4.71			95.47		Defoc. (5 μm)	Rhyolite
0.58	2.37	4.72		0.05	95.70		Defoc. (5 μm)	Rhyolite
0.63	2.21	4.68		0.01	95.51		Defoc. (5 μm)	Rhyolite
0.74	2.66	4.85	0.01	0.02	95.76		Defoc. (5 μm)	Rhyolite
0.60	2.25	4.61			94.79		Defoc. (5 μm)	Rhyolite
0.58	2.36	4.41			94.57		Defoc. (5 μm)	Rhyolite
0.66	2.27	4.55	0.01	0.01	95.03		Defoc. (5 μm)	Rhyolite
0.68	3.06	5.03	0.01	0.02	95.30		Defoc. (5 μm)	Rhyolite
0.64	2.73	4.68		0.01	95.29		Defoc. (5 μm)	Rhyolite
0.61	3.09	4.88	0.02	0.01	95.78		Defoc. (5 μm)	Rhyolite
0.60	3.09	4.97	0.01		96.29		Defoc. (5 μm)	Rhyolite
0.64	3.18	4.87			95.64		Defoc. (15–20 μm)	Rhyolite
0.65	2.79	5.04	0.01	0.13	95.77		Defoc. (5 μm)	Rhyolite
0.61	2.79	4.91		0.03	95.30		Defoc. (5 μm)	Rhyolite
0.80	1.62	4.92			95.29		Defoc. (<5 μm)	Rhyolite
0.75	2.07	4.92			95.57		Defoc. (5 μm)	Rhyolite
0.59	3.18	4.95		0.04	95.50		Defoc. (5 μm)	Rhyolite
0.68	3.07	4.96			96.49		Defoc. (5 μm)	Rhyolite
0.68	2.04	4.97		0.01	96.57		Defoc. (5 μm)	Rhyolite
0.65	2.76	4.98			96.29		Defoc. (5 μm)	Rhyolite
0.76	2.52	5.10	0.01		96.21		Defoc. (5 μm)	Rhyolite
0.65	2.91	4.96	0.01		95.93		Defoc. (5 μm)	Rhyolite
0.75	1.35	4.68			95.79		Defoc. (5 μm)	Rhyolite
0.78	1.82	5.14			96.46		Defoc. (5 μm)	Rhyolite
0.78	1.83	5.04			96.39		Defoc. (5 μm)	Rhyolite
0.78	1.82	4.93	0.03		95.69		Defoc. (5 μm)	Rhyolite
0.76	2.32	4.80	0.01		95.51		Defoc. (5 μm)	Rhyolite
0.89	2.33	4.77			96.02		Defoc. (5 μm)	Rhyolite
0.82	2.69	5.04	0.01	0.01	95.99		Defoc. (5 μm)	Rhyolite
0.69	3.08	4.99	0.02		96.31		Defoc. (5 μm)	Rhyolite
0.64	3.31	4.65		0.02	96.39		Defoc. (20 μm)	Rhyolite
0.68	3.09	4.85		0.01	96.08		Defoc. (20 μm)	Rhyolite
0.58	3.09	4.77			96.40		Defoc. (20 μm)	Rhyolite
0.76	2.80	4.83	0.01	0.04	95.47		Defoc. (20 μm)	Rhyolite
0.78	2.95	4.76			95.47		Defoc. (20 μm)	Rhyolite
0.60	2.97	4.65	0.03		96.17		Defoc. (20 μm)	Rhyolite
0.71	2.91	4.82			96.17		Defoc. (20 μm)	Rhyolite
0.68	2.89	4.74			95.69		Defoc. (20 μm)	Rhyolite
0.63	3.03	4.56			95.57		Defoc. (20 μm)	Rhyolite
0.78	2.87	4.62			95.65		Defoc. (20 μm)	Rhyolite
0.69	2.63	4.83	0.01	0.03	95.78	40		
0.077	0.485	0.166	0.007	0.030	0.469			

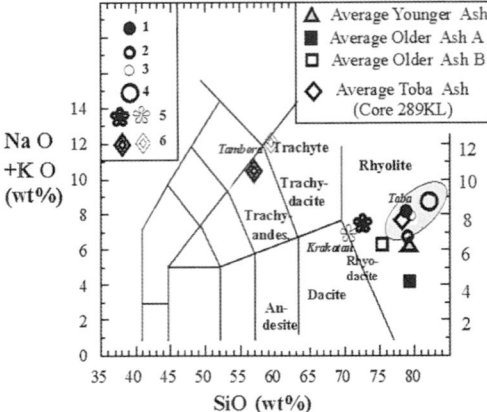

Fig. 7. Chemical composition of glass shards in investigated ash layers in Arabian Sea cores in the total alkali–silica diagram after Le Bas et al. (1986). Composition of investigated glass shards (Table 1), recalculated LOI (H_2O)-free. For comparison, published data are included for Toba Ash (1–4), Krakatau (1883) eruption (5: Sigurdsson & Carey 1989) and Tambora (1815) eruption (6: Mandeville et al. 1996) are shown (1: Toba Caldera, Sumatra; 2: data from Rose & Chesner 1987; 3: ODP Site 758, Layer A; Dehn et al. 1991; 4: Bay of Bengal, Gasparotti et al. 2000). Field of all Toba data is stippled.

the case of a tephra of 250 ka age (Ninkovich 1979).

Discussion

Toba Ash (70 ka BP ± 4 ka)

The mega-eruption of the 'Youngest Toba Tephra' (YTT) from the Toba crater in western Sumatra around 70 ka BP was the largest explosive eruption known from Quaternary times (Ninkovich et al. 1978a, 1978b; Rose & Chesner 1987, 1990; Chesner et al. 1991; Dehn et al. 1991). The rhyolitic ash ejected during this extremely voluminous ultra-Plinian co-ignimbrite explosion was thrown up into the upper stratosphere and troposphere and produced a tephra fall layer covering an area of $>4 \times 10^6$ km^2 (Rose & Chesner 1987, 1990; Zielinski et al. 1996, 1997). The YTT covers a huge area ENE of the Toba caldera (Sumatra) toward the South China Sea (Bühring et al. 2000), whereas the main dispersal was directed NW to the Bay of Bengal, India, and finally to the Arabian Sea (Ninkovich et al. 1978b; Fig. 9), and SW from Sumatra to the central Indian Basin. The Toba Ash has been discovered in cores from the Bengal Fan (Dehn et al. 1991; Gasparotti et al. 2000), in India (Acharyya & Basu 1993; Westgate et al. 1998), in the Central Indian Basin (Pattan et al. 1999), in the NW Arabian Sea (Leuschner & Sirocko 2000), and in the NE Arabian Sea (Schulz et al 1998, 2002). The volume of erupted tephra during the Toba eruption was estimated as 800–840 km^3 dense rock equivalent (Rose & Chesner 1987), which, because of the newly discovered occurrences further to the NE and NW, should be a minimum value. The Toba Ash discovered in South China Sea cores by Bühring et al. (2000) indicates that the tephra was also transported >1800 km to the east. Those workers explained this bi-directional dispersal pattern by the fact that during the northern hemisphere summer, mid- to low-level tropospheric SW monsoonal winds transport ash from the lower parts of the eruption column and co-ignimbrite plumes to the east–NE, whereas upper tropospheric and stratospheric easterly winds ('jet streams') are active throughout the year and are responsible for the mainly westward dispersal of the Toba Ash over the Indian Ocean (Fig. 9).

The progressive loss of (more mafic) crystals (pyroxene, clinopyroxene, magnetite), plagioclase and of higher-density glass shards (as a result of crystal inclusions) can cause a downwind aeolian fractionation of ashes, leading to a downwind increase of the bulk SiO_2 and K_2O content (Fisher & Schmincke 1984). Surprisingly, the silt-sized ash from the Bay of Bengal shows no linear trends of fining grain size over a distance of 1500 km from Toba, Sumatra, to the Bay of Bengal (Gasparotti et al. 2000). The silt-sized vitreous ash from core 289KL contains hardly any crystals and appears to have a very similar grain size to the Bay of Bengal ash. Hence we exclude any remarkable downwind aeolian fractionation of the Toba Ash between Sumatra and the Arabian Sea.

Published radiometric (K/Ar and Ar/Ar) ages of the Toba Ash vary from 73.5 ± 3.5 ka BP (Ninkovich et al. 1978a) to 73 ± 4 or 74 ± 2 ka BP (weighted mean; Chesner et al. 1991). Recently, Schulz et al. (2002) pointed out that the very detailed correlation of the ash occurrence (upper part of IS-20) in our Arabian Sea cores with the GISP-2 core suggests a somewhat younger age (about 70 ± 5 ka). For this paper we estimate the YTT age to be c. 70 ka with a standard deviation of 4 ka.

Ash layers in deep-sea sediments are mostly fallout tephra formed during highly explosive eruptions, which resulted in high fragmentation and wide dispersal of the glass shards. The ash is injected into the stratosphere and travels thousands of kilometres before being deposited to the ocean surface and settling to the sea floor (Fisher & Schmincke 1984; Dehn et al. 1991). This can explain why we found ash from Indonesian volca-

Table 2. Trace and minor element analysis (ICP-MS) of two samples from the Toba Ash in core 289KL (Hans Lorenz, BGR, analyst; bsf, below sea floor) compared with data from the Toba caldera (Sumatra) and the Toba Ash in India (Westgate et al. 1998) (all values in mg kg^{-1})

Sample: Location: Analysis no.:	Toba Caldera (Sumatra) UT-1298	Kukdi River (India) UT-1068	SO 130-289KL 1842 cm bsf IM-23411	SO 130-289KL 1843.5 cm bsf IM-23412
Ag			0.28	0.08
As			13.3	24.3
Ba	117	439	387	391
Bi			0.4	0.45
Ce	40	53	26.3	30.7
Cr			12.8	173
Cs	9.52	7.6	8.06	7.93
Hf	3.42	3.34	3.11	7.93
Li			34.7	39.2
La	20.35	28.4	13.7	17.2
Nb	12.2	14.9	15.3	15.4
Ni			10.5	52.5
Pb			33.1	30.6
Rb	265	213	172	193
Sr	28	48	79	91
Ta	1.15	1.28	1.81	1.76
Th	30.3	28.1	16.9	15.7
U	6.04	4.99	3.17	4.04
Y	39	29	17.0	16.8
Zn			154	92.2
Zr	73	77	88.4	294
Pr	4.92	5.9	3.07	3.70
Nd	18.2	19.6	10.7	12.9
Sm	4.21	4.16	2.24	2.47
Eu	0.3	0.41	0.32	0.31
Gd	4.92	3.84	2.31	2.42
Tb	0.92	0.67	0.39	0.39
Dy	6.25	4.56	2.60	2.38
Ho	1.35	0.98	0.56	0.50
Er	3.9	2.83	1.82	1.69
Tm	0.77	0.49	0.31	0.28
Yb	4.88	3.34	2.10	1.98
Lu	0.88	9.57	0.37	0.36

noes in the Arabian Sea, about 5200 km away from its volcanic origin.

Historical ash layers

When looking for potential source areas and type of volcanism for any Arabian Sea ash, it is necessary to focus on the most violent Plinian or ultra-Plinian explosive eruptions in Indonesia, with a volcanic explosivity index (VEI) >4 (Simkin & Siebert 1994). In equatorial latitudes, the easterly tropical jetstream, active in the upper troposphere and stratosphere between 10 and 15 km altitude, is responsible for the west to NW transport of most ash plumes from Indonesia towards India and Africa. The ash layers derived from volcanoes in Sumatra have a high silica content, as they are associated with ignimbrite eruptions from a comparatively thick pre-Cenozoic crust, whereas ashes derived from the Sunda Strait and western Java (e.g. Krakatau) are dacitic, and those of eastern Java and the Lesser Sunda Islands (e.g. Tambora) are generally more andesitic in nature. Therefore it is most likely that our Holocene high-silica rhyolitic tephra layers in the Arabian Sea were derived from Sumatra. Up to now, no widely distributed rhyolitic tephra layer of post-Toba age has been described from the deep-sea sediments around Indonesia, not even from ODP Site 758 NW of Sumatra (Fig. 9), which was carefully studied by Dehn et al. (1991), who described seven Quaternary ash layers, all of rhyolitic composition.

Obviously, we considered the explosive eruption of Krakatau between Java and Sumatra (with an VEI of 6) in 1883 as a potential source for our

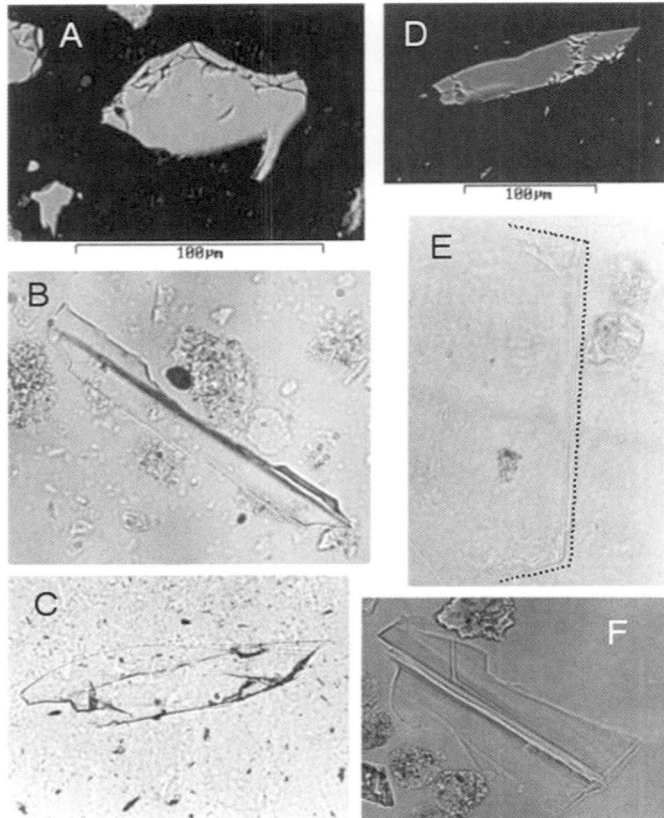

Fig. 8. Photomicrographs and microprobe BSE images of glass shards of 'Younger Ash' ((**a**) and (**b**) AD 1890–1920) and of 'Older Ash' ((**c**)–(**f**) AD 1815–1830). (a) 34KG, 18.6–19.2 cm (BSE image of microprobe C34_3b); (**b**) 34KG, 19.2–19.8 cm (length of shard 160 μm); (**c**) 39KG, 37.2–37.5 cm (length of shard 180 μm); (**d**) same shard as C (microprobe BSE image); (**e**) 34KG, 32.5–33 cm (length of shard 170 μm, central part lost during polishing); (**f**) 34KG, 32.5–33 cm (length of shard 180 μm).

'Younger Ash'. About 18–21 km³ of tephra of rhyodacitic composition (Fig. 7) were erupted as co-ignimbrite ash and widely distributed (Mandeville et al. 1996). According to Verbeek (1885, fig. 55) the maximum reported dispersal of the Krakatau ash was about 900 km to the NNW (NW of Singapore) and 1200 km to the SW (to Cocos–Keeling Island). In the central Indian Basin (south of Sri Lanka) fresh pumice with a composition similar to the Krakatau ash was discovered by Mudholkar & Fuji (1995). However, the rhyodacitic composition of the Krakatau tephra (Fig. 7) is too different from our 'Younger Ash' to be a likely source.

For the 'Older Ash' we first considered the great 1815 eruption of Tambora as a potential source. Although aerosols from the vigorous Plinian co-ignimbrite eruption of the Tambora volcano in 1815 (Self et al. 1984) were detected in the Greenland ice cores (94 ppb SO_4^{2-}) by Zielinski et al. (1997), the maximum dispersal of the trachyandesitic ash (Fig. 7), apparently mainly by current-transported pumice, was only about 800 km. Hence it is unlikely that coarse-silt-sized Tambora glass was transported as far as to the Arabian Sea (Sparks, pers. comm.). Also, the eruptions of Kelut (1826), Galunggung (1822) and Raung (1817) had andesitic compositions and can be excluded as possible sources.

Millennial monsoonal climate variability and northern hemispheric teleconnections

The high-frequency record of core 289KL shows rapid climate oscillations, which can be tuned precisely to the $\delta^{18}O$ record of a Bay of Bengal core and to the GISP-2 ice core from Greenland

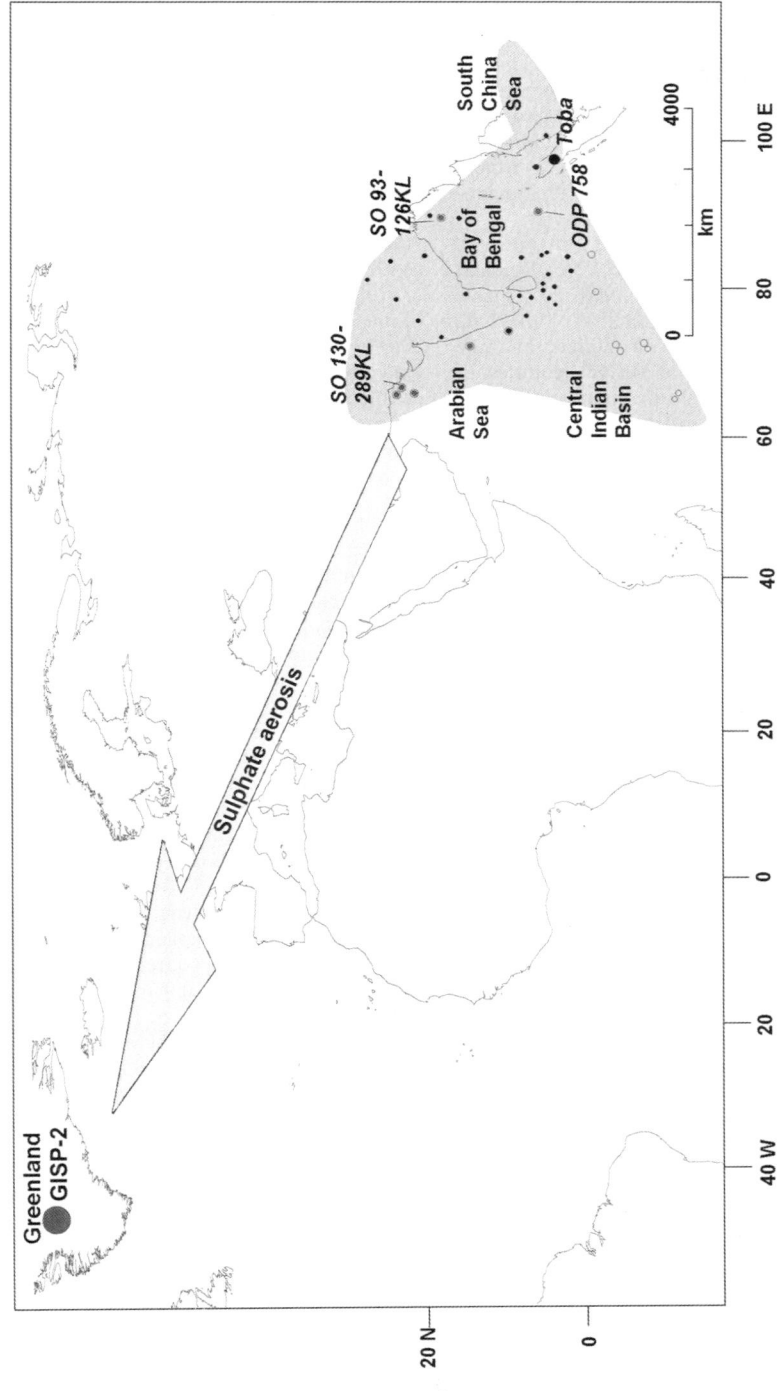

Fig. 9. Distribution of Toba Ash between the source area (Sumatra), the Arabian Sea and Greenland (modified after Schulz et al. 2002).

(Fig. 6). The Toba event (70 ± 4 ka BP), which is well documented in the Arabian Sea and Bay of Bengal records at the end of IS-20, as well as in the Greenland ice (Fig. 6), is an excellent chronostratigraphic marker horizon. Hence the discovery of the Toba Ash is of paramount importance to validate the correlation of the high-frequency cyclicity from the monsoon-controlled subtropical latitudes to the high latitudes of the Northern Atlantic and Iceland by independent tephrochronology.

Previous high-resolution studies of northern Arabian Sea cores (Sirocko et al. 1996; Reichart et al. 1997, 1998; Schulz et al. 1998, 2002) have shown the high-frequency instability of the Late Pleistocene monsoon-controlled climate on time scales of a few millennia down to centuries or decades. In general, the palaeoclimate and palaeocirculation in the Northern Arabian Sea fluctuated between two extremes: (1) periods of intensified summer monsoon leading to high primary productivity, reduced terrigenous input and sedimentation rates, as well as to oxygen-deficient bottom water conditions (OMZ) during warm periods (interstadials); (2) periods of weakened summer monsoon leading to low primary productivity, increased terrigenous (aeolian) input, enhanced sedimentation rates, deep winter mixing, fully oxygenated bottom waters and breakdown of the OMZ conditions during stadials (Reichart et al. 1997; von Rad et al. 1999b; Schulz et al. 2002).

Especially between c. 60 and 25 ka, core 289KL (Fig. 6) shows well-developed interstadials (IS-17 to IS-3) or 'Dansgaard–Oeschger (D–O) event equivalents'. This agrees with results from the Greenland ice record (Alley et al. 1999) that D–O oscillations were largest at times of intermediate temperature and ice extent during marine isotope stage (MIS) 3 (60–25 ka), rather than during coldest or warmest times near insolation and CO_2 maxima and minima. Especially at IS-19, -16, -12 and -3 we note that the warming trends of the interstadials are very rapid (within a few decades) with subsequent gradual cooling, a phenomenon well known from the D–O cycles in the Greenland ice cores (Fig. 6). The interstadials are separated by the cool rebounds of the D–O cycles and by stadials, equivalent to the Heinrich events (H6a to H1, Younger Dryas) which precede D–O interstadials 1, 2, 4, 8, 12, and 17 (Figs 2 and 6). In the North Atlantic, the Heinrich events are caused by large meltwater inputs from the Laurentide ice sheet leading to freshwater discharge by 'iceberg armadas', cooling and freshening of the North Atlantic surface waters, and a turnoff of the Atlantic thermohaline deep-water circulation. In the Arabian Sea, the Heinrich events are characterized by more arid conditions in the Arabian desert and Pakistan, resulting in stronger dust flux to the ocean by northwesterly winds, strongly reduced productivity and improved ventilation by deep winter mixing (Madhupratap et al. 1996; Schulz et al. 1998, 2002; Reichart et al. 1998; Sirocko et al. 1999; von Rad et al. 1999b).

The D–O events show a quasi-periodic spacing of c. 1.5 ka, which has been described from the Greenland ice cores (Dansgaard et al. 1993; Stuiver & Grootes 2000), the subpolar North Atlantic cores (ice-rafting cycles: Bond et al. 1997, 1999), and the South China Sea (Wang et al. 1999). A 1470 a cycle was recently discovered in the GISP-2 oxygen isotope record, which is closely tied to the D–O cycles and has been explained as being tied to threshold levels of continental ice mass (Schulz et al. 1999); the internal dynamics of the East Greenland ice sheet may have been the ultimate pacemaker of the D–O cycles (van Kreveld et al. 2000). Berger & von Rad (2002) reported the same 1470 a cycle in varved Holocene Arabian Sea sediments and suggested that the detected cyclicity can be ascribed to lunar tidal action.

In Fig. 6 we have correlated the millennial-scale climate cycles of core 289KL tuned precisely to the $\delta^{18}O$ core record of the GISP-2 ice core from Greenland (Dansgaard et al. 1993), as well as to core SO 93-126 (Fig. 6) from the Bay of Bengal (Kudrass et al. 2001). The apparent synchronous appearance of the various D–O oscillations and Heinrich events in the Indian Ocean and the Greenland ice (Schulz et al. 1998), which was also discovered in the Santa Barbara Basin off southern California (Hendy & Kennett 1999) and many other marine and continental areas over the whole northern hemisphere, is a striking phenomenon, which can be explained only by decadal–centennial atmospheric circulation changes. The Toba event (70 ± 4 ka BP) is well documented in our Arabian Sea and Bay of Bengal records at the end of IS-20, as well as (by sulphate aerosols) in the Greenland ice (Zielinski et al. 1996, 1997). Apparently, the (sub)tropical monsoon circulation is tightly coupled with the west wind belt of the mid- to high-latitude northern hemisphere (Porter & An 1995; An 2000). As in the Arabian Sea, the palaeoclimate of the Gulf of Bengal region is dominated by strong millennial variations in the intensity of the summer monsoon. Apparently, the high-frequency suborbital climate variations are caused by feedback processes involving snow and dust of the Tibetan Plateau changing the summer monsoon capacity to transport moisture and heat to central Asia (Kudrass et al. 2001).

During the stadials, central Asia is dominated by dry westerlies; this leads to intensified dust transport, lowering the albedo of the (dust-covered) ice, and resulting in a cooler and more arid climate. The same oscillations can be studied also in the aridity–humidity cycles of the Chinese loess region (Fang et al. 1999; An 2000) and in the sediments of the South China Sea (Wang et al. 1999). Sirocko et al. (1999), Wang et al. (1999) and Kudrass et al. (2001) have postulated that changes in the intensity of the Indian summer monsoon are tightly coupled with suborbital climate oscillations in the northern hemisphere via atmospheric moisture and heat circulation.

Conclusions

(1) We discovered one of the most distal western occurrences of the famous 'Youngest Toba Tephra' (YTT). The discrete rhyolitic ash layer was found in OMZ core 289KL off Pakistan at the base of IS-20 (c. 70–71 ka). The age and chemical composition match exactly published data from several occurrences of the YTT.

(2) Two highly disseminated vitreous 'ash layers' of rhyolitic composition were detected in two annually laminated box cores: a 'Younger Ash' (c. AD 1885–1900), and an 'Older Ash' (c. AD 1815–1830). The glass shards were probably derived from Indonesian volcanoes (possibly from Sumatra). So far, we are unable to correlate them with well-known historical eruptions.

(3) The litho- and chronostratigraphy of the 20.2 m long core SO 130-289KL documents the complete palaeoclimate record of the past 75 ka, to IS-21. The laminated intervals of the Late Holocene and Preboreal periods, and IS-1 to IS-21 (including the 'last glacial maximum') are characterized by dark colours, high TOC and low $CaCO_3$ values, as well as reduced sedimentation rates, whereas the bioturbated intervals of the Younger Dryas and the stadials or Heinrich event equivalents (H1–H6) are light, with low TOC %, high $CaCO_3$ contents and enhanced sedimentation rates (Fig. 2).

(4) We correlated the millennial-scale climate cycles of core 289KL (Fig. 9) precisely with the $\delta^{18}O$ core record of the GISP-2 ice core from Greenland, as well as with core SO 93-126 from the Bay of Bengal (Kudrass et al. 2001). The Toba event (70 ± 4 ka BP), which is well documented during the end of IS-20 in the Arabian Sea and Bay of Bengal records, as well as in the Greenland ice (Zielinski et al. 1996), is an excellent stratigraphic marker horizon to validate this correlation. The apparent synchronous appearance of the various D–O oscillations and Heinrich events, which now has been documented for many marine and continental locations in the northern hemisphere, can be explained only by fairly rapid (decadal–centennial) atmospheric circulation changes that are tightly coupled with suborbital climate oscillations in the northern hemisphere via atmospheric moisture and heat circulation. Tentatively, we suggest that the variability of the 'greenhouse conditions' of the Asian monsoon drives the 'icehouse conditions' of the high latitudes by triggering, amplifying and terminating the interstadials and stadials in the northern hemisphere (Kudrass et al. 2001).

We are grateful to M. Schmidtke (BGR) for making grain mounts, to T. Malarski and P. Rendschmidt (both BGR) for preparing accurate thin-sections and polished sections of grain mounts for the following microprobe investigation, and to J. Lodziak (BGR) for expert assistance at the BGR microprobe. At an early stage of this study, M. Kraml (Freiburg) and E. Lazzaretti (Marseille) helped with the identification and selection of glass shards. J. Merkt and A. Kleinmann (both Niedersächsisches Landesamt für Bodenforschung, Hannover) helped with the preparation of a thin-section. A. Lückge (BGR) assisted with the digital presentation of figures. H. R. Kudrass (BGR) and the anonymous reviewers kindly read the manuscript and gave helpful comments. The SONNE cruises were funded by the German Federal Ministry of Education, Science, Research and Technology (BMBF; grants 03G 090A and 0130A). During this study, M.P. was supported at BGR by a 6 month post-doctoral research fellowship, also funded by BMBF.

References

ACHARYYA, S.K. & BASU, P.K. 1993. Toba ash on the Indian subcontinent and its implications for correlation of Late Pleistocene alluvium. *Quaternary Research*, **40**, 10–19.

ALLEY, R.B., CLARK, P.U., KEIGWIN, L.D. & WEBB, R.S. 1999. Making sense of millennial-scale climate change. *In:* CLARK, P.U., WEBB, R.S. & KEIGWIN, L.D. (eds) *Mechanisms of Global Change at Millennial Time Scales.* Geophysical Monograph, American Geophysical Union, **112**, 385–394.

AN, Z. 2000. The history and variability of the East Asian paleomonsoon climate. *Quaternary Science Reviews*, **19**, 171–187.

BERGER, W.H. & VON RAD, U. 2002. Decadal to millennial cyclicity in varves and turbidites from the Arabian Sea: hypothesis of tidal origin. *Global and Planetary Change*, **729**, in press.

BOND, G., SHOWERS, W., CHESEBY, M. & 7 OTHERS 1997. A pervasive millennial-scale cycle in North Atlantic Holocene and Glacial climate. *Science*, **278**, 1257–1266.

BOND, G., SHOWERS, W., ELLIOT, M. & 5 OTHERS 1999. The North Atlantic 1–2 kyr climate rhythm: relation to Heinrich events, Dansgaard/Oeschger cycles and the Little Ice Age. *In:* CLARK, P.U., WEBB, R.S. & KEIGWIN, L.D. (eds) *Mechanisms of Global Change at Millennial Time Scales.* Geophysical Monograph, American Geophysical Union, **112**, 35–58.

BÜHRING, C., SARNTHEIN, M. & Leg 184 Shipboard Scientific Party 2000. Toba ash layers in the South China Sea: evidence of contrasting wind directions during eruption ca. 74 ka. *Geology*, **28**, 275–278.

CHESNER, C.A., ROSE, W.I., DEINO, A., DRAKE, R. & WESTGATE, J.A. 1991. Eruptive history of the Earth's largest Quaternary caldera (Toba, Indonesia) clarified. *Geology*, **19**, 200–203.

DANSGAARD, W., JOHNSON, S.J., CLAUSEN, H.B. & 8 OTHERS 1993. Evidence for great instability of past climate in a 250 kyr ice-core record. *Nature*, **374**, 218–220.

DEHN, J., FARRELL, J.W. & SCHMINCKE, H.U. ET AL. 1991. Neogene tephrochronology from Site 758 on northern Ninetyeast Ridge: Indonesian arc volcanism of the past 5 Ma. *In:* WEISSEL, J., PEIRCE, J., TAYLOR, E. & ALT, J. (eds) *Proceedings of the Ocean Drilling Program, Scientific Results, 121.* Ocean Drilling Program, College Station, TX, 273–295.

FANG, X.-M., ONO, Y., FUKUSAWA, H. & 6 OTHERS 1999. Asian summer monsoon instability during the past 60,000 years: magnetic susceptibility and pedogenic evidence from the western Chinese Loess Plateau. *Earth and Planetary Science Letters*, **168**, 219–232.

FISHER, R.V. & SCHMINCKE, H.-U. 1984. *Pyroclastic Rocks.* Springer, Berlin.

GASPAROTTI, G., SPADAFORA, E., SUMMA, V. & TATEO, F. 2000. Contribution of grain size and compositional data from the Bengal Fan sediment to the understanding of Toba volcanic event. *Marine Geology*, **162**, 561–572.

HENDY, I.L. & KENNETT, J.P. 1999. Latest Quaternary North Pacific surface water responses imply atmosphere-driven climate instability. *Geology*, **27**, 291–294.

KUDRASS, H.R., HOFMANN, A., DOOSE, H., EMEIS, K. & ERLENKEUSER, H. 2001. Modulation and amplification of climatic changes in the Northern Hemisphere by the Indian summer monsoon during the past 80 k.y. *Geology*, **29**, 63–66.

LE BAS, M.J., LE MAITRE, R.W., STRECKEISEN, A. & ZANETTIN, B. 1986. A chemical classification of volcanic rocks based on the total alkali–silica diagram. *Journal of Petrology*, **27**, 745–750.

LEUSCHNER, D.C. & SIROCKO, F. 2000. The low-latitude monsoon climate during Dansgaard–Oeschger cycles and Heinrich events. *Quaternary Science Reviews*, **19**, 243–254.

MADHUPRATAP, M., KUMAR, S.P., BHATTATHIRI, P.M.A., KUMAR, M.D., RAGHUKUMAR, S., NAIR, K.K.C. & RAMAIAH, N. 1996. Mechanism of the biological response to winter cooling in the northeastern Arabian Sea. *Nature*, **384**, 549–552.

MANDEVILLE, C.W., CAREY, S. & SIGURDSSON, H. 1996. Magma mixing, fractional crystallization and volatile degassing during the 1883 eruption of Krakatau volcano, Indonesia. *Journal of Volcanology and Geothermal Research*, **74**, 243–274.

MUDHOLKAR, A. & FUJI, T. 1983. Fresh pumice from the central Indian Basin: a Krakatau 1883 signature. *Marine Geology*, **125**, 143–151.

NINKOVICH, D. 1979. Distribution, age and chemical composition of tephra layers in deep-sea sediments off western Indonesia. *Journal of Volcanology and Geothermal Research*, **5**, 67–86.

NINKOVICH, D., SHACKLETON, N.J., ABDEL-MONEM, A.A., OBRADOVICH, J.D. & IZETT, G. 1978a. K–Ar age of the Late Pleistocene eruption of Toba. *Nature*, **276**, 574–577.

NINKOVICH, D., SPARKS, R.S. & LEDBETTER, M.T. 1978b. The exceptional magnitude and intensity of the Toba eruption Sumatra: an example of the use of deep-sea tephra layers as a geological tool. *Bulletin of Volcanology*, **41**, 286–298.

PATTAN, J.N., SHANE, P. & BANAKAR, V.K. 1999. New occurrences of Youngest Toba Tuff in abyssal sediments of the Central Indian Basin. *Marine Geology*, **155**, 243–248.

PORTER, S. & AN, Z. 1995. Correlation between climate events in the North Atlantic and China during the last glaciation. *Nature*, **375**, 305–308.

REICHART, G.J., DEN DULK, M., VISSER, H.J., VAN DER WEIJDEN, C.H. & ZACHARIASSE, W.J. 1997. A 225 kyr record of dust supply, paleoproductivity and the oxygen minimum zone from the Murray Ridge (northern Arabian Sea). *Palaeogeography, Palaeoclimatology, Palaeoecology*, **134**, 146–169.

REICHART, G.J., LOURENS, L.J. & ZACHARIASSE, W.J. 1998. Temporal variability in the Northern Arabian Sea oxygen minimum zone (OMZ) during the last 225,000 years. *Paleoceanography*, **13**, 607(621.

ROSE, W.I. & CHESNER, C.A. 1987. Dispersal of ash in the great Toba eruption, 75 ka. *Geology*, **15**, 913–917.

ROSE, W.I. & CHESNER, C.A. 1990. Worldwide dispersal of ash and gases from the Earth's largest known eruption: Toba, Sumatra, 75 ka. *Palaeogeography, Palaeoclimatology, Palaeoecology*, **89**, 269–275.

SCHMINCKE, H.U. 1986. *Vulkanismus.* Wissenschaftliche Buchgesellschaft, Darmstadt.

SCHULZ, H., EMEIS, K., ERLENKEUSER, H., VON RAD, U. & ROLF, C. 2002. The Toba volcanic event and interstadial/stadial climates at the Marine Isotope Stage 5/4 transition in the northern Indian Ocean. *Quaternary Research*, **57**, 22–31.

SCHULZ, H., VON RAD, U. & ERLENKEUSER, H. 1998. Correlation between Arabian Sea and Greenland climate oscillations of the past 110,000 years. *Nature*, **393**, 54–57.

SCHULZ, M., BERGER, W.H., SARNTHEIN, M. & GROOTES, P.M. 1999. Amplitude variations of 1470-year climate oscillations during the last 100,000 years linked to fluctuations of continental ice mass. *Geophysical Research Letters*, **26**, 3385–3388.

SELF, S., RAMPINO, M.R. & NEWTON, M.S. 1984. Volcanological study of the Great Tambora Eruption of 1815. *Geology*, **12**, 659–663.

SIGURDSSON, H. & CAREY, S. 1989. Plinian and coignimbrite tephra fall from the 1815 eruption of Tambora volcano. *Bulletin of Volcanology*, **51**, 243–270.

SIMKIN, T. & SIEBERT, L. 1994. *Volcanoes of the World.* Geoscience Press, Tucson, AZ.

SIROCKO, F., GARBE-SCHÖNBERG, D., MCINTYRE, A. & MOLFINO, B. 1996. Teleconnections between the subtropical monsoons and high-latitude climates during the last deglaciation. *Science*, **272**, 526–529.

SIROCKO, F., LEUSCHNER, D., STAUBWASSER, M., MALEY, J. & HEUSER, L. 1999. High-frequency oscillations of the last 70,000 years in the tropical/subtropical and polar climates. *In:* CLARK, P.U., WEBB, R.S. & KEIGWIN, L.D. (eds) *Mechanisms of Global Change at Millennial Time Scales.* Geophysical Monograph, American Geophysical Union, **112**, 113–126.

STUIVER, M. & GROOTES, P.M. 2000. GISP2 oxygen isotope ratios. *Quaternary Research*, **53**, 277–284.

VAN KREVELD, S., SARNTHEIN M., ERLENKEUSER, H. & 5 OTHERS 2000. Potential links between surging ice sheets, circulation changes, and the Dansgaard–Oeschger cycles in the Irminger Sea, 60–18 kyr. *Paleoceanography*, **15**, 425–445.

VERBEEK, R.D.M. 1885. *Krakatau.* Landsdrukkerij, Batavia.

VON RAD, U., DOOSE, H. & CRUISE PARTICIPANTS 1998. *SONNE Cruise SO 130 Cruise Report MAKRAN II.* Bundesanstalt für Geowissenschaften und Rohstoffe, Archive No. **117368**.

VON RAD, U., SCHAAF, M., MICHELS, K.H., SCHULZ, H., BERGER, W.H. & SIROCKO, F. 1999*a*. A 5,000-year record of climate change in varved sediments from the Oxygen Minimum Zone off Pakistan (Northeastern Arabian Sea). *Quaternary Research*, **51**, 39–51.

VON RAD, U., SCHULZ, H., RIECH, V., DEN DULK, M., BERNER, U. & SIROCKO, F. 1999*b*. Multiple monsoon-controlled breakdown of oxygen-minimum conditions during the past 30,000 years documented in laminated sediments off Pakistan. *Palaeogeography, Palaeoclimatology, Palaeoecology*, **152**, 129–161.

WANG, L., SARNTHEIN, M., ERLENKEUSER, H., GROOTES, P.M., GRIMALT, J.O., PELEJERO, C. & LINK, G. 1999. Holocene variations in Asian monsoon moisture: a bidecadal sediment record from the South China Sea. *Geophysical Research Letters*, **26**(18), 2889–2892.

WESTGATE, J.A., SHANE, P.A.R., PEARCE, N.J.G. & 5 OTHERS 1998. All Toba Tephra occurrences across Peninsular India belong to the 75,000 yr B.P. eruption. *Quaternary Research*, **50**, 107–112.

ZIELINSKI, G.A., MAYEWSKI, P.A., GRÖNVOLD, K., GERMANI, M.S., WHITLOW, S., TWICKLER, M.S. & TAYLOR, K. 1997. Volcanic aerosol records and tephrochronology of the Summit, Greenland ice cores. *Journal of Geophysical Research*, **102**, 26625–26640.

ZIELINSKI, G.A., MAYEWSKI, P.A., MEEKER, L.D., WHITLOW, S., TWICKLER, M.S. & TAYLOR, K. 1996. Potential atmospheric impact of the Toba mega-eruption. *Geophysical Research Letters*, **23**, 837–840.

The seasonal and vertical distribution of living planktic foraminifera in the NW Arabian Sea

FRANK J. C. PEETERS & GEERT-JAN A. BRUMMER

Department of Marine Chemistry and Geology (MCG), Netherlands Institute for Sea Research (NIOZ), PO Box 59, 1790 Den Burgh, Texel, The Netherlands

Abstract: The NW Arabian Sea is characterized by a strong seasonal contrast in surface water hydrography. During the SW monsoon of 1992, we encountered strong coastal upwelling characterized by low sea surface temperatures (SST), high nutrient concentrations, a shallow thermocline and a near-surface chlorophyll maximum. By contrast, the hydrography during the NE monsoon of 1993 was characterized by a relatively warm nutrient-depleted surface mixed layer and a deep chlorophyll maximum. We show that the faunal composition, depth habitat and abundance of living planktic foraminifera respond to the hydrographic changes controlled by the seasonally reversing monsoon system. Total shell concentrations (>125 μm) ranged from 4 to 332 individuals (ind.) m^{-3} during upwelling and from 3 to 85 ind. m^{-3} during the non-upwelling season. During upwelling, the fauna was dominated by *Globigerina bulloides*. During non-upwelling the fauna was characterized by relatively high concentrations of tropical symbiont-bearing species such as *Globigerinoides ruber*, *Globigerinoides sacculifer* and *Globigerinella siphonifera*, whereas concentrations of *Globigerina bulloides* were an order of magnitude lower. Factor analysis on 15 species yields an upwelling assemblage (UA), a tropical assemblage (TA) and a subsurface assemblage (SA). A fourth factor represents the distribution of the species *Globigerina falconensis*, which is mainly found in subsurface waters during the non-upwelling period (NE monsoon). A model is presented to calculate the base of the productive zone from the vertical shell concentration profile of a given species. The model is validated by comparing the range in calcification temperatures of *G. bulloides*, derived from its $\delta^{18}O$, with the *in situ* sea-water temperature range of the productive zone as predicted from the model. It appears that shell growth (calcite precipitation) is restricted to the productive zone as defined by this method. The average calcification temperature of *G. bulloides* corresponds to the point of maximum change in the shell concentration profile (i.e. the inflection point). For most shallow-dwelling species, the inflection point is found at or below the depth of the chlorophyll maximum, although above the main thermocline. This study indicates that the depth habitat and abundance of different species varies seasonally. Consequently, the abundance and stable isotope composition of specimens in the fossil record reflects a mixture of specimens that were produced at various depths during the different seasons.

In the NW Arabian Sea, coastal upwelling and biological productivity are strongly coupled and mainly controlled by the seasonally reversing monsoon system (Prell & Curry 1981; Anderson & Prell 1991; Prell *et al.* 1992, 1993; Banse 1994; McCreary *et al.* 1996; Rostek *et al.* 1997). The stable isotope composition of planktic foraminiferal shells ($\delta^{13}C$, $\delta^{18}O$) and the abundance of species have been used to reconstruct changes in monsoon history, upwelling intensity, biological productivity and the sea surface temperature of the Arabian Sea (Hutson & Prell 1980; Kroon *et al.* 1990; Prell *et al.* 1992; Anderson & Prell 1991, 1993; Anderson *et al.* 1992; Brock *et al.* 1992; Steens *et al.* 1992; Anderson & Thunell 1993; Naidu & Malmgren 1996a, b; Rostek *et al.* 1997; Ivanova *et al.* 1999; Ivanova 2000). Paradoxically, much of our knowledge on the environmental preferences of planktic foraminifera has been derived from fossil shells on the ocean floor. The characteristics of the faunal assemblages and the stable isotope composition of individual species were, by analogy, related to characteristics of the present-day upper ocean (Hutson & Prell 1980; Kroon *et al.* 1990; Steens *et al.* 1992; Ivanova 2000). In the sediment, however, the faunal assemblages are mixed by bioturbation and thus may represent a time span from tens to hundreds of years, thus covering many annual cycles.

Investigations on field collected foraminifera have proven to be very useful to characterize the ecology of the species, hence their usage for palaeoenvironmental reconstructions (e.g. Bé & Tolderlund, 1971; Fairbanks & Wiebe 1980; Deuser *et al.* 1981; Kroon & Ganssen 1989;

Ravelo et al. 1990; Ottens 1992; Ravelo & Fairbanks 1992; Kemle-von Mücke 1994; Watkins et al. 1996, 1998; Ufkes et al. 1998; Watkins & Mix 1998). In the Arabian Sea, however, the number of systematic studies on living foraminifera is rather limited and often lacks a seasonal coverage. The Arabian Sea region is most interesting for the study of ecology of living planktic foraminifera because of the strong monsoon-controlled seasonal variation in sea surface hydrography.

In this paper we discuss the temporal and spatial distribution of planktic foraminifera in relation to seasonal upwelling in the northwestern Arabian Sea. The seasonal contrast is covered by two sample-sets of depth-stratified plankton tows that were collected along a transect perpendicular to the coast of Yemen–Oman in August 1992 (SW monsoon) and February 1993 (NE monsoon), respectively (Fig. 1). Our objectives are to determine the seasonal and depth distribution of various species, to identify foraminiferal variables that represent a certain state of the ocean and to assess the relationship between the living standing stock and the export flux of foraminiferal shells settling to greater depth.

Material

Sample collection

During the C cruises of R.V. *Tyro* to the Arabian Sea (Van Hinte et al. 1995), depth-stratified

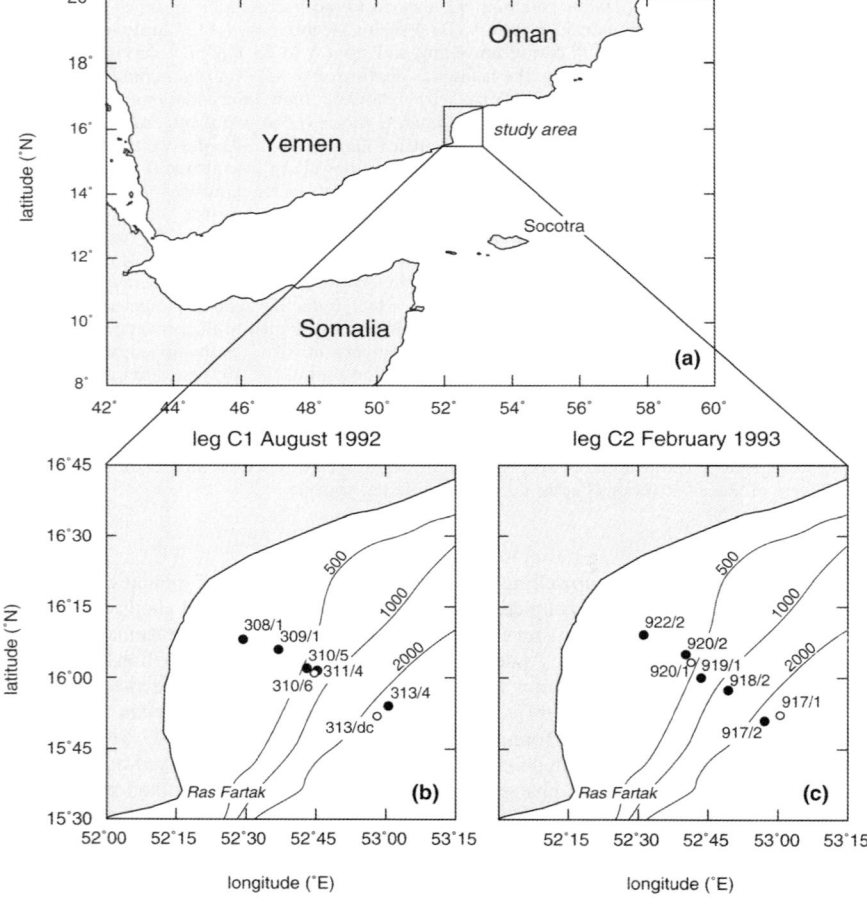

Fig. 1. (**a**) Map showing location of (**b**) and (**c**). (**b**) Map of the Oman–Yemen transect with locations of multinet casts taken during leg C1 of the Netherlands Indian Ocean Program of R.V. *Tyro*, in August 1992. ●, shallow casts (100–0 m); ○, deep casts (500–0 m). Isobaths are given in metres. (**c**) same as (**b**) for leg C2 in February 1993. It should be noted that the same locations have different station numbers; stations 313 and 310 are identical to stations 917 and 920.

plankton tows were collected off the Arabian Peninsula in August 1992 (leg C1) and February 1993 (leg C2) (Fig. 1). The transect was sampled at five stations (Table 1). The upper water column was sampled at five standard depth intervals, 100–75, 75–50, 50–25, 25–10 and 10–0 m. A Hydrobios Multinet system with depth sensor and flow-meter was equipped with five 100 µm plankton nets, which are opened at depth and closed consecutively during the upcast while towed behind the ship. During each leg, an additional deep cast was taken at two stations to a depth of 500 m, covering intervals of 500–300, 300–200, 200–150, 150–100 and 100–0 m. The shallowest net of the deep cast covers the entire shallow cast and has not been studied. To avoid clogging of the net, varying amounts of water were filtered depending on the expected amount of particulate matter. For the deep cast characteristically 100–500 m^3 were filtered per interval, and for shallow casts generally 20–100 m^3 per interval. Samples were stored in a borax-buffered formalin–sea-water solution (pH >8.0), in 500 ml LDPE bottles, kept under dark and cool conditions and checked for their pH. Further details have been given by Brummer (1995).

Conductivity–temperature–depth (CTD) profiles

To compare the faunal distribution patterns and depth habitat of living planktic foraminifera with water column properties, 11 CTD-rosette casts were taken along the transect (Figs. 2 and 3, Table 2). The following selection of variables is used in this study: temperature, salinity, density, fluorescence and nitrate concentration. Technical procedures and calibration of the data have been given by Van Hinte et al. (1995). On the basis of the temperature profile the upper water column is subdivided into the surface mixed layer, the upper thermocline and the lower thermocline. In this study the depth of the surface mixed layer is found at SST − 0.5 °C, whereas the depth of the thermocline is defined as the largest minimum in the first derivative of temperature with depth.

Sample preparation

In the laboratory, the plankton tow samples were split into eight aliquots using a Fritsch Laborette 27 Rotary Cone Sample Divider. Six aliquots (three-quarters of the total sample) were used for foraminiferal analysis and one aliquot for chemical analysis (not discussed herein). The remaining aliquot is stored in a sea-water solution with a pH of 8.0–8.5 and kept under cool and dark conditions. The sample aliquots for foraminiferal analysis were frozen in liquid nitrogen using preweighed aluminium cups, freeze-dried, stored in an desiccator with silica gel for at least 3 h and weighed on a Sartorius Analytic balance with a precision of ±0.1 mg. The dry weight of this residue, representing the total biomass between 100 and 1000 µm (Table 1; Figs. 2f and 3f), is weighed at 30 s intervals over a period of 3 min to correct for moisture uptake. The organic matter was removed using a low-temperature asher. The ashed residue was wet sieved over mesh sizes of 500, 355, 250, 150 and 125 µm and rinsed with ethanol (Peeters et al. 1999). The separate size fractions were dried on a hot plate at 80°C for c. 30 min and analysed for their foraminiferal composition. At least 200 specimens per fraction were counted. If fewer than 200 specimens were present, which was the case in most of the larger fractions, the entire fraction was counted. Total standing stock (individuals (ind.) m^{-3}) and planktic foraminiferal census counts are given in Appendices A and B. The average size of foraminiferal shells (>125 µm) is calculated using a weighted averaged mean of number of shells per size fraction following the method discussed by Peeters et al. (1999). In total, 64 samples have been used in this study. Planktic foraminifera were identified using a WILD M5A binocular microscope following the taxonomy of Bé (1967) and Hemleben et al. (1989).

Stable isotope analysis

Whole shells of G. bulloides were hand picked from the size fractions 150–250 and 250–355 µm. The stable isotope composition of the shells was measured on a Finnigan 251 gas source mass spectrometer equipped with an automated carbonate extraction line. All analyses were carried out at the Faculty of Earth and Life Sciences at the Vrije Universiteit in Amsterdam, The Netherlands. Samples were dissolved in concentrated orthophosphoric acid at 80 °C. Approximately 30–80 µg of sample was used per analysis. Results are reported in the conventional δ-notation as per mil deviation from the Vienna-Pee Dee Belemnite standard (V-PDB). Calibration to the V-PDB standard was performed via NBS-18, NBS-19 and NBS-20 international standards (Coplen et al. 1983; Hut 1987). The external reproducibility for $\delta^{18}O$ is ±0.08‰ and for $\delta^{13}C$ is ±0.04‰.

Table 1. Location and sample details of multi-net stations 308–313 of the C1 cruise in August 1992, and of stations 917–922 of the C2 cruise in February 1993

Station-cast-net	Date, local time	Decimal longitude (°N)	Decimal latitude (°E)	Distance from coast (km)	Depth start (m)	Depth end (m)	Volume filtered (m³)	Biomass (mg m⁻³)
308-1-5	19 Aug. 1992, 09:25	52.50	16.14	30.28	22.8	0.0	103	51.7
308-1-4					47.5	22.8	55	13.2
308-1-3					72.1	47.5	48	8.0
308-1-2					98.1	72.1	66	3.2
309-1-5	19 Aug. 1992, 17:31	52.64	16.09	45.84	8.0	0.0	36	60.5
309-1-4					18.2	8.0	26	25.9
309-1-3					28.3	18.2	25	17.8
309-1-2					49.5	28.3	42	11.7
310-5-5	20 Aug. 1992, 15:37	52.73	16.02	59.12	8.1	0.0	41	62.9
310-5-4					23.3	8.1	34	23.9
310-5-3					48.5	23.3	44	12.7
310-5-2					74.3	48.5	40	5.4
310-5-1					99.6	74.3	33	4.2
310-6-4	20 Aug. 1992, 17:08	52.75	16.00	61.72	148.1	97.6	35	6.5
310-6-3					200.2	148.1	53	4.9
310-6-2					298.3	200.2	104	4.9
310-6-1					498.4	298.3	270	2.9
311-1-5	21 Aug. 1992, 09:50	52.75	16.00	61.42	7.6	0.0	52	60.2
311-1-4					23.3	8.1	50	29.1
311-1-3					49.5	23.3	132	11.4
311-1-2					73.8	49.5	124	7.3
311-1-1					96.1	74.8	36	4.2
313-4-5	21 Aug. 1992, 12:20	53.02	15.91	93.42	8.1	0.0	25	259.9
313-4-4					24.3	10.1	41	171.9
313-4-3					48.5	24.3	87	34.8
313-4-2					73.8	49.5	87	18.8
313-4-1					99.6	74.8	60	9.3
313-dc-4	22 Aug. 1992, 12:10	52.94	15.88	86.43	147.6	99.1	94	6.2
313-dc-3					196.1	149.6	98	9.8
313-dc-2					300.3	199.2	278	14.2

PLANKTIC FORAMINIFERA IN NW ARABIAN SEA

Sample	Date/Time							
313-dc-1					498.0	300.3	478	10.9
917-2-5	25 Feb. 1993, 09:15	52.92	15.90	83.68	12.1	0.0	21	40.7
917-2-4					27.3	12.1	22	66.2
917-2-3					51.6	27.3	43	80.8
917-2-2					77.9	51.6	79	19.6
917-2-1					102.1	77.9	68	12.1
917-1-4	25 Feb. 1993, 08:15	52.97	15.89	88.71	153.0	101.0	75	10.1
917-1-3					203.0	153.0	75	10.7
917-1-2					302.0	203.0	184	2.8
917-1-1					502.0	302.0	240	6.6
918-2-5	26 Feb. 1993, 09:40	52.84	15.96	72.74	12.1	0.0	35	43.2
918-2-4					27.3	12.1	47	71.1
918-2-3					51.6	27.3	97	76.1
918-2-2					77.9	51.6	132	23.9
918-2-1					103.1	77.9	101	14.1
919-1-5	26 Feb. 1993, 16:47	52.73	16.00	58.99	10.1	0.0	12	61.0
919-1-4					26.3	10.1	19	22.3
919-1-3					51.6	26.3	50	89.4
919-1-2					76.8	51.6	91	23.6
919-1-1					102.1	76.8	70	9.5
920-2-5	27 Feb. 1993, 09:15	52.64	16.08	47.08	11.1	0.0	16	54.0
920-2-4					27.3	11.1	46	36.3
920-2-3					52.6	27.3	79	52.1
920-2-2					77.9	52.6	135	15.9
920-2-1					102.1	77.9	105	12.2
920-1-4	27 Feb. 1993, 08:15	52.70	16.09	52.53	153.0	103.0	168	7.0
920-1-3					202.0	153.0	146	7.7
920-1-2					303.0	202.0	269	27.9
920-1-1					501.0	303.0	282	15.1
922-2-5	27 Feb. 1993, 08:00	52.52	16.17	30.99	12.1	0.0	48	77.2
922-2-4					26.3	12.1	58	67.2
922-2-3					51.6	26.3	106	45.7
922-2-2					77.0	51.6	149	16.3
922-2-1					102.1	77.0	169	4.7

Biomass represents the dry weight of particulate matter between 100 and 1000 μm. 'Swimmers' were excluded from these measurements.

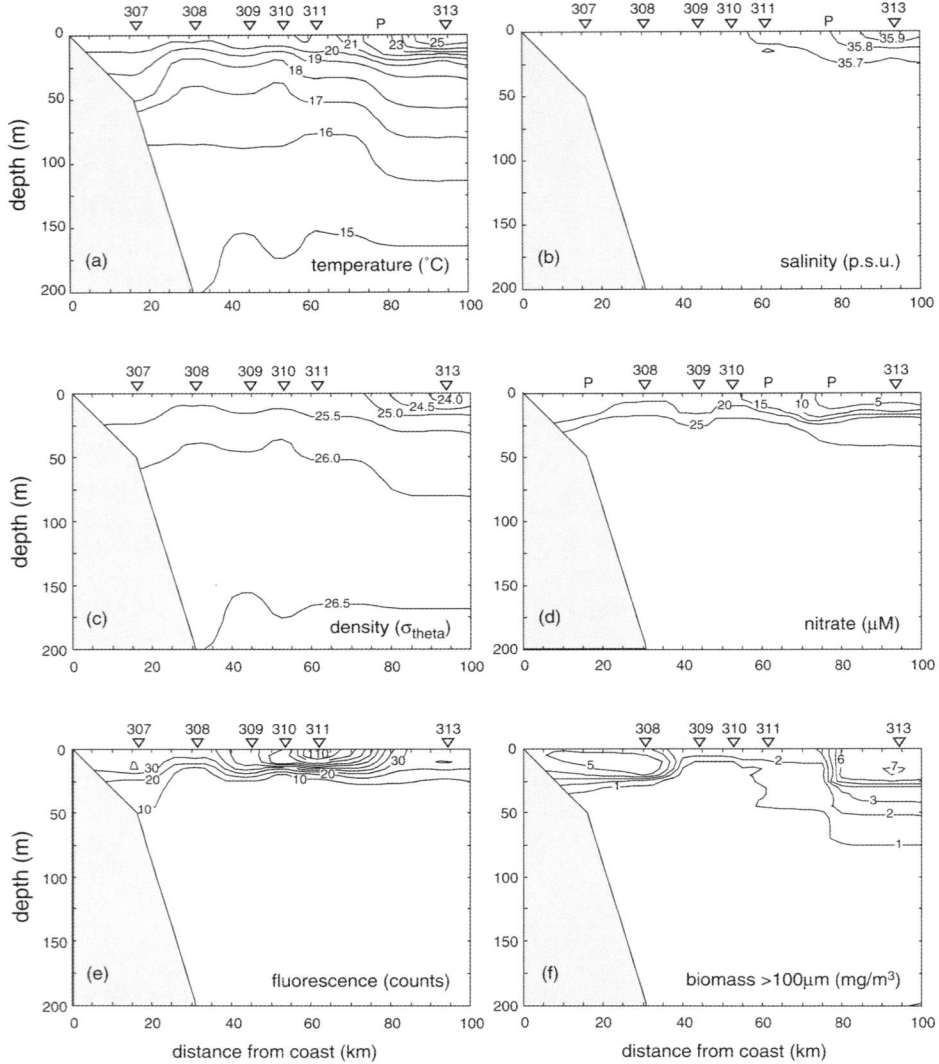

Fig. 2. (**a**) Temperature, (**b**) salinity, (**c**) density, (**d**) nitrate concentration, (**e**) fluorescence and (**f**) biomass distribution along the Oman–Yemen transect in August 1992 (SW monsoon season). At stations 308–311, cold and nutrient-rich water is found near the sea surface, indicating freshly upwelled waters. At the end of the transect (station 313), a relatively warm-water eddy is encountered, as indicated by a higher temperature and salinity and lower nitrate concentrations near the sea surface. It should be noted that fluorescence (chlorophyll) maxima are found near the sea surface. At top of figures: triangles indicate, station numbers; P, surface water observation from 3 m water depth, either from aqua-flow system or plankton pump sample.

Methods

Contouring data

To assess the spatial distribution pattern of foraminifera, CTD data and $\log_4$ transformed species concentrations were plotted and computer contoured using an inverse distance squared variable radius search algorithm (Figs 2 and 3; see also Figs 5 and 6). Because the largest changes in the distribution (absolute abundance) pattern of planktic foraminifera occur in the upper part of the water column, contour plots and shell concentration profiles are drawn for the upper 200 m.

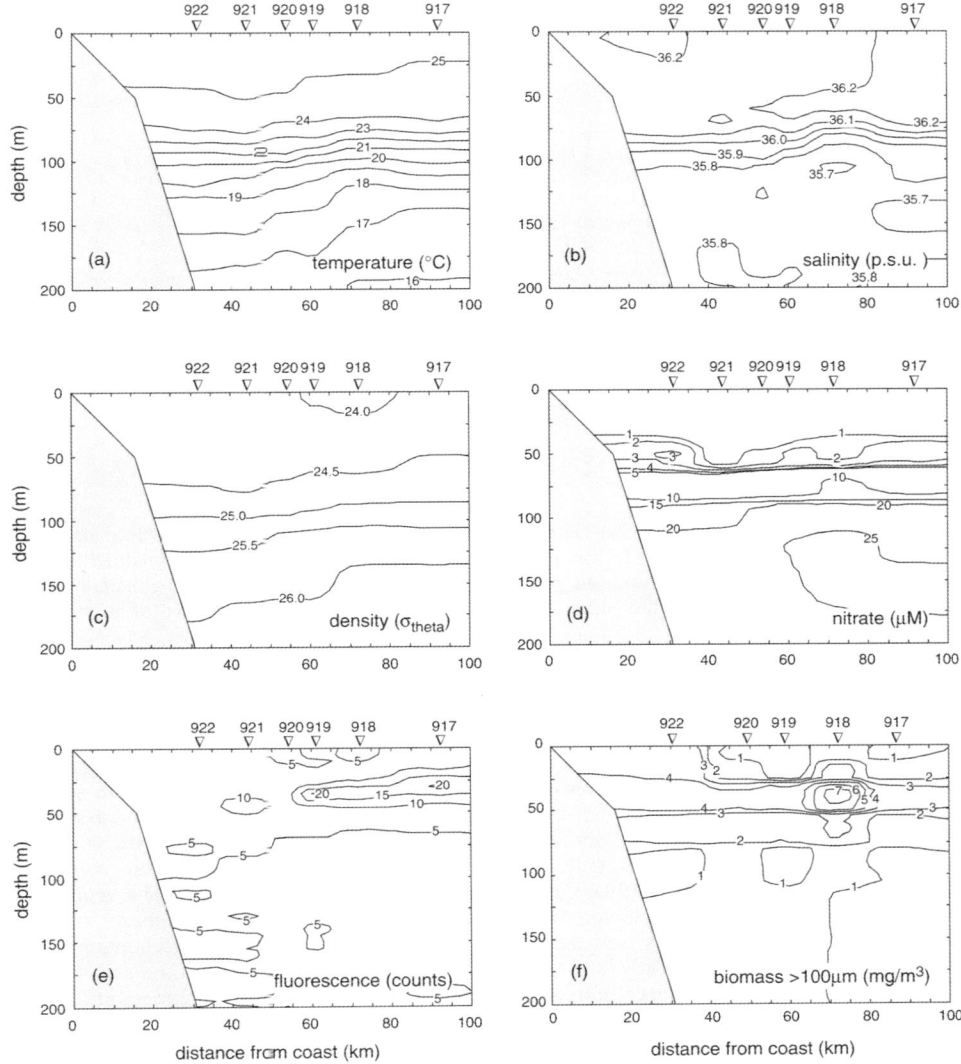

Fig. 3. (a) Temperature, (b) salinity, (c) density, (d) nitrate concentration, (e) fluorescence and (f) biomass distribution in February 1993 (NE monsoon season). A warm and oligotrophic uniform surface layer characterizes the transect in winter. A deep fluorescence (chlorophyll) maximum is found at c. 40 m and the main thermocline at c. 100 m.

Dead or alive?

Living and dead specimens in plankton nets may be distinguished by the presence or absence of cytoplasm and sometimes by its colour (e.g. Berger 1971; Hemleben et al. 1989; Kemle-von Mücke 1994; Ortiz et al. 1995; Schiebel et al. 1995). Separating living from dead specimens in this way is time consuming and not very precise considering the slow cytoplasm decomposition after death, during the descent to the sea floor.

Often, plasma-filled tests of foraminifera are found far below the zone in which shell growth (productive zone) takes place (Schiebel et al. 1995). By contrast, empty tests have been found within the zone where production takes place (Berger 1971; Schiebel et al. 1995). The presence/absence of colour of the cytoplasm therefore does not allow identification of the exact position between the productive zone and the settling flux zone below, i.e. the base of the productive zone

Table 2. Locality details of CTD stations taken during C cruises leg 1 (August 1992) and leg 2 (February 1993)

Station-cast	Date, local time	Dec. longitude (°N)	Dec. latitude (°E)
307-1	19 Aug. 1992, 6:15	52.38	16.18
308-2	19 Aug. 1992, 9:40	52.52	16.13
309-2	19 Aug. 1992, 18:21	52.63	16.10
310-1	20 Aug. 1992, 11:35	52.70	16.07
313-4	21 Aug 1992, 14:47	53.00	15.88
917-1	25 Feb. 1993, 12:02	53.00	15.87
918-1	25 Feb. 1993, 19:15	52.83	15.98
919-1	26 Feb. 1993, 15:44	52.75	16.02
920-1	27 Feb. 1993, 10:40	52.70	16.07
921-1	27 Feb. 1993, 17:32	52.60	16.07
922-1	28 Feb. 1993, 8:31	52.53	16.18

(Z_{BPZ}). Consequently, one needs to define Z_{BPZ} for the various species on other criteria. In this paper, we propose to use the shell concentration profile to define Z_{BPZ} and to estimate the size of the standing stock.

Standing stock and the base of the productive zone

Previous studies have shown that shell concentration profiles of planktic foraminifera often have a characteristic shape (see, e.g. the shell concentration profiles given by Berger 1969, 1971; Kahn & Williams 1981; Bijma & Hemleben 1994; Ortiz et al. 1995; Schiebel et al. 1995; Kemle-von Mücke & Oberhänsli 1999). These profiles, and the data presented in this paper, consistently show that shell densities are high in the upper part of the water column and low at greater depths in the ocean. Obviously, the high shell densities in the upper part of the water column mostly represent specimens that are alive and capable of adjusting their buoyancy by producing low-density lipids or gases to counter gravitational settling (Furbish & Arnold 1997). At greater depth in the water column, the shell densities are low and relatively constant. Specimens found in this part of the water column represent the 'pelagic rain' of exported shells that settle to the sea floor. We propose to use the shape of the shell concentration profile to identify a depth level that separates the water column into two zones: (1) the productive zone (PZ), in which most specimens are alive and precipitate their primary calcite; (2) the settling flux zone (SFZ), in which the specimens mostly represent the exported shells that settle to the sea floor. The boundary between the PZ and the SFZ is characterized by a rapid decline in the shell concentration with depth and is named the base of the productive zone (Z_{BPZ}). To calculate the depth of the base of the productive zone we use the following approach.

The SFZ is characterized by a constant and relatively low shell concentration, which we will call C_{SF}. The depth at which the concentration of shells approximately equals C_{SF} is considered as the base of the productive zone. In practice, the base of the productive zone will be somewhere within a tow interval and cannot be assigned to a specific depth. We therefore use a curve fit procedure to describe the shell concentration profile by a continuous curve. The geometry of a Gaussian distribution curve appears to be very similar to most shell concentration profiles. Although we realize that a Gaussian distribution function does not describe the mechanistic process of shell production and the loss of foraminifer shells as a result of grazing and mortality and subsequent settling, it does describe the most important phenomena that are frequently observed in the concentration profiles of different species. In Fig. 4, for example, we show that the maximum in the shell concentration profile is described by the maximum of the Gaussian curve, whereas the lower shell concentrations that are found at greater depths in the ocean are fitted by the 'tail' of the curve. Obviously, modelling the characteristics of the shell concentration profile of a given species is very informative and permits a detailed comparison of the vertical distribution of living planktic foraminifera to hydrographic features (e.g. mixed-layer depth, thermocline depth or depth of chlorophyll maximum). It also may be used to extract environmental information from the living planktic foraminifera that can be used for the reconstruction of SST, thermocline depth, biological productivity, mixed-layer depth, etc. (e.g. Ravelo & Fairbanks 1992; Watkins & Mix, 1998).

Collecting foraminifera using depth-stratified tows results in a total number of foraminifera caught per depth interval, N_i (number of specimens). The concentration of foraminifera per unit volume of water, C_i (ind. m^{-3}), is calculated using

$$C_i = \frac{N_i}{V_i} \quad (1)$$

in which V_i (m^3) is the volume of sea water filtered per net as measured by a flow meter. It can be expected that most of the specimens caught in the upper part of the water column are alive (e.g. Berger 1969), whereas deeper in the water column an increasing proportion of dead specimens is found. The total concentration of specimens as a function of depth ($C(z)$) is given by the sum of living ($C_L(z)$) and dead specimens ($C_D(z)$), so that

$$C(z) = C_L(z) + C_D(z). \quad (2)$$

Below the base of the productive zone, we must assume that specimens are dead, and that their concentration remains approximately constant and equals C_{SF}:

$$C(z) = C_D(z) = C_{SF} \quad (\text{for} \quad z > z_{BPZ}). \quad (3)$$

With z representing the depth of the water column in metres. The concentration of dead specimens at any depth above z_{BPZ} is assumed to linearly increase with depth from zero at the sea surface to a value of C_{SF} at the base of the productive zone (Fig. 4). The concentration of dead specimens at any depth above Z_{BPZ} is thus given by

$$C_D(z) = C_{SF} \frac{z}{z_{BPZ}} \quad (\text{for} \quad z \leq z_{BPZ}). \quad (4)$$

To fit the concentration profile of living specimens, $C_L(z)$, we use a Gaussian distribution:

$$C_L(z) = \frac{a}{\sigma \sqrt{2\pi}} \exp\left[-0.5\left(\frac{z-c}{\sigma}\right)^2\right] \quad (5)$$

in which a, c and σ are parameters related to the total area under the curve, the centre of distribution and standard deviation, respectively (Fig. 4). Consequently, the total concentration as a function of depth, including both living and dead specimens, is described by the sum of equations (4) and (5), yielding

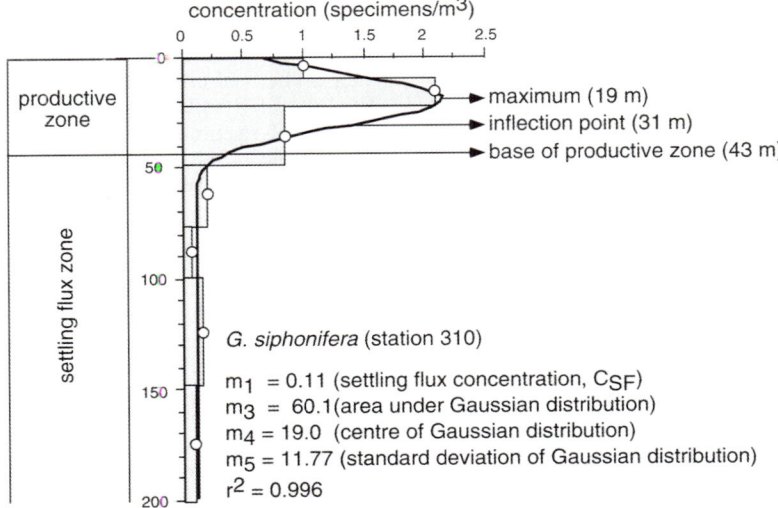

Fig. 4. Schematic representation of the vertical distribution of *Globigerinella siphonifera* at station 310 (upwelling station during the SW monsoon season, August 1992), illustrating position of concentration maximum, inflection point and the base of the productive zone. Parameter m_1 reflects the export flux shell concentration (C_{SF}) and is assumed to be constant (i.e. the shell concentration below the productive zone). The inflection point, representing the point of maximum change of the concentration with depth, is found by adding one standard deviation (m_5) to the depth at which the maximum concentration (m_4) is found, which in this case equals 30.8 m (i.e. 19.0 m + 11.8 m). The base of the productive zone is defined at two standard deviations below the depth at which the maximum is found ($m_4 + 2m_5$). Parameter m_3 represents the total area under the total Gaussian distribution, and can be used in standing stock calculations. It should be noted that m_3 does not represent the standing stock, as it needs to be corrected for that part of the curve that is present between the sea surface and the base of the productive zone. (For further explanation, see text.)

$$C(z) = C_{SF}\frac{z}{z_{bpz}} + \frac{a}{\sigma\sqrt{2\pi}}\exp\left[-0.5\left(\frac{z-c}{\sigma}\right)^2\right]$$
(for $z \leq z_{BPZ}$). (6)

If two maxima are present in the shell concentration profile (see, for example, the shell concentration profile of G. menardii at station 310 in Fig. 8, below), we use an additional Gaussian distribution to fit the data. In such cases the parameters of the first and second distribution are numbered (e.g. the centres of the first and second Gaussian distribution are named c_1 and c_2, respectively). The depth at which the total shell concentration $C(z)$ approaches C_{SF} must be defined as the base of the productive zone. However, one might argue about the criteria that should be used to place this boundary. We must therefore find a practical measure to identify the base of the productive zone; that is, a depth level that separates the majority of the living specimens in the surface layers from a zone below that is dominated by dead specimens. Assuming that living specimens in the water column are described by the Gaussian distribution function of equation (5), one might expect that about 95% of the living population is present within two times the standard deviation (SD) from the centre of the distribution; that is, 2 SD below the depth at which the maximum shell concentration is found. Consequently, the base of the productive zone is found at

$$z_{BPZ} = c + 2\sigma.$$ (7)

At this depth, the concentration of shells approximately equals the settling flux concentration (Fig. 4) and: $C(z_{BPZ}) \approx C_{SF}$. If the shell concentration profile shows a maximum near the sea surface (in the uppermost plankton net), then $c = 0$ m. It should be noted that the maximum change of the shell concentration with depth (i.e. the inflection point, z_i) is found at 1 SD from the maximum of the Gaussian curve:

$$z_i = c + \sigma.$$ (8)

In this paper we define the number of foraminifera that is present within the productive zone as standing stock (SS):

$$SS = \int_{z_0}^{z_B} C(z)\,dz.$$ (9)

However, because a small part of the specimens above z_{BPZ} will be dead and should not be included in the standing stock estimates, we also define the corrected standing stock (SS_{COR}), by subtracting the concentration of dead specimens from the total concentration:

$$SS_{COR} = \int_{z_0}^{z_B} C_L(z)\,dz = \int_{z_0}^{z_B} [C(z) - C_D(z)]\,dz$$
(10)

which thus represents the expected number of living specimens (ind. m^{-2}) found in the productive zone; that is, in a vertical section of the water column ranging from the sea surface to z_{BPZ}. In practice, the number of dead specimens in the productive zone appears to be very small compared with the number of living specimens, so that $SS \approx SS_{COR}$.

Results

Hydrographic conditions

During leg C1, in August 1992 (SW monsoon), the Oman transect was characterized by coastal upwelling (Figs 1 and 2). Near the coast, surface water temperatures (at 5 m) were low (20.5 °C) and nitrate concentrations at the surface were high (17.2–18.5 μM) (Fig. 2a and d). At the offshore end of the transect (at station 313), we encountered a warm mesoscale eddy with a surface temperature of 25.7 °C and 2.5 μM of nitrate. Depth profiles indicate a shallow surface mixed layer depth of c. 10 m at all stations.

In contrast, hydrographic conditions during the leg C2 cruise in February 1993 (NE monsoon) were rather uniform (Figs 1 and 3). Along the transect, SST varied around 26 °C and a deep chlorophyll maximum (DCM), associated with the base of the mixed layer, had developed at about 40–50 m (Fig. 3e). The main thermocline is situated at about 100 m. Surface water nutrient concentrations were two orders of magnitude lower than measured during the SW monsoon (Brummer 1995).

The hydrography, as sampled during the two expeditions, characteristically represents the seasonal hydrographic variability in this area, in that during the summer SW monsoon cold and nutrient-rich water reaches the surface, enhancing biological productivity, whereas during the winter NE monsoon the ocean stratifies and a relatively thick, warm and oligotrophic surface layer develops. Consequently, the hydrography during leg C1 and leg C2 may be considered as hydrographic end-members.

Fauna composition and species distribution patterns

In total, 26 species of planktic foraminifera (>125 μm) are identified in the 64 plankton net samples (Appendices A and B). The 10 most abundant species (Table 3) account for more than 95% of the fauna. The concentration of foraminifera (>125 μm) in the water column ranges from 4 to 332 ind. m^{-3} in August 1992 during upwelling and from 3 to 85 ind. m^{-3} in February 1993 when no upwelling occurred. Contour plots of log$_4$ transformed shell concentrations of the 15 most abundant species >125 μm are given in Figs. 5 and 6. It should be noted that we use a log$_4$ transformation on the shell concentrations only for the purpose of contouring. This transformation results in contour intervals that appeared to be most practical for visualization of the data.

A factor analysis with Varimax Rotation using SYSTAT was performed on 15 species, using all 64 plankton tows. The remaining 11 species, with a per cent abundance of lower than 0.4%, had low squared multiple correlation with all other variables and were removed from the original data before analysis (Tables 3 and 4). To increase normality of the data (Tabachnick & Fidell 1996), the species concentrations were log$_e$ transformed.

In case a species was absent in a sample, the sample is excluded from the analysis (i.e. pairwise deletion). Considering only eigenvectors greater than 1.0, there appear to be four important factors, which together explain 85% of the total variance (Table 4). Species factor scores are given in Table 4 and represent correlations between species (variables) and factors. The higher the score, the better the species is a pure measure of the factor.

Factor 1: the upwelling assemblage (UA, explaining 29.6% of total variance). Species that show high scores on the first factor are *D. anfracta*, *N. pachyderma* (sin), *T. quinqueloba*, *G. bulloides*, *N. dutertrei* and *G. menardii*. Highest shell concentrations of these species are mostly found during upwelling as sampled during leg C1 (Figs 5 and 6). According to Bé & Tolderlund (1971), the species in the upwelling assemblage have their maximum occurrence in different faunal provinces: *N. pachyderma* (sin) in the polar province, *G. bulloides* and *T. quiqueloba* in the subtropical province, *N. dutertrei* in the transitional to subtropical provice and *G. menardii* in the subtropical to tropical provice. *D. anfracta* is probably restricted to transitional–temperate to tropical provinces. Because of its small adult size it is routinely underrepresented in conventional collections >150 μm (Brummer & Kroon 1988; Peeters *et al.* 1999). Despite their maximum abundance in different faunal provinces, the concomitant occurrence of these species indicates high nutrient availability. Judging from the factor pattern, two sub-assemblages may be recognized within the upwelling assemblage.

(1) Upwelling assemblage-A (UA-A). The scores of *G. bulloides*, *N. dutertrei* and *G. menardii* on the first factor are slightly lower than those of the other species involved and may show an increased score on the second and/or third factor to account for presence during non-upwelling (Factor 2, *N. dutertrei*) or suggesting a deeper habitat (*G. menardii*). Although more abundant during upwelling than during non-upwelling, these species are consistently present during both monsoon seasons. It is suggested that they reflect fertile regions in a more general sense; that is, as also observed in central parts of the Arabian Sea (Curry *et al.* 1992).

(2) Upwelling assemblage-B (UA-B). *D. anfracta*, *N. pachyderma* (sin) and *T. quinqueloba* have highest positive scores only on Factor 1. These (small) species were abundant only during upwelling, and were virtually absent during the non-upwelling period. As *N. pachyderma* (sin) and *T. quinqueloba* are known to represent cold waters at higher latitudes (Bé & Tolderlund, 1971), these species may be close to their upper thermal

Table 3. *Per cent abundance of planktic foraminifera (>125 μm) for leg 1 (August 1992) and leg 2 (February 1993) samples, as well as for the total dataset*

Species	% in C1 dataset	% in C2 dataset	% in total dataset
G. bulloides	62.04	4.25	43.22
G. ruber	2.51	26.63	10.37
G. glutinata	8.99	11.98	9.96
G. calida	10.37	8.35	9.71
G. siphonifera	1.63	16.74	6.55
G. sacculifer	0.54	16.87	5.86
N. dutertrei	3.55	2.61	3.25
G. tenellus	0.63	6.85	2.65
G. menardii	2.45	1.64	2.18
T. quinqueloba	1.89	0.03	1.29
P.F. indet.	0.67	1.12	0.82
N. pachyderma (sin)	1.19	0.01	0.80
G. falconensis	0.35	1.46	0.71
H. parapelagica	0.63	0.32	0.53
P.F. aberrant	0.66	0.13	0.49
D. anfracta	0.68	0.06	0.48
T. iota	0.63	0.01	0.43
G. uvula	0.24	0.02	0.17
G. hexagona	0.10	0.27	0.15
O. universa	0.01	0.38	0.13
G. scitula	0.15	0.03	0.11
B. digitata	0.02	0.15	0.06
P. obliquiloculata	0.05	0.06	0.05
G. theyeri	0.03	0.01	0.02
G. rubescens	0.00	0.01	0.00
G. tumida	0.00	0.00	0.00
G. adamsi	0.00	0.00	0.00
T. humilis	0.00	0.00	0.00

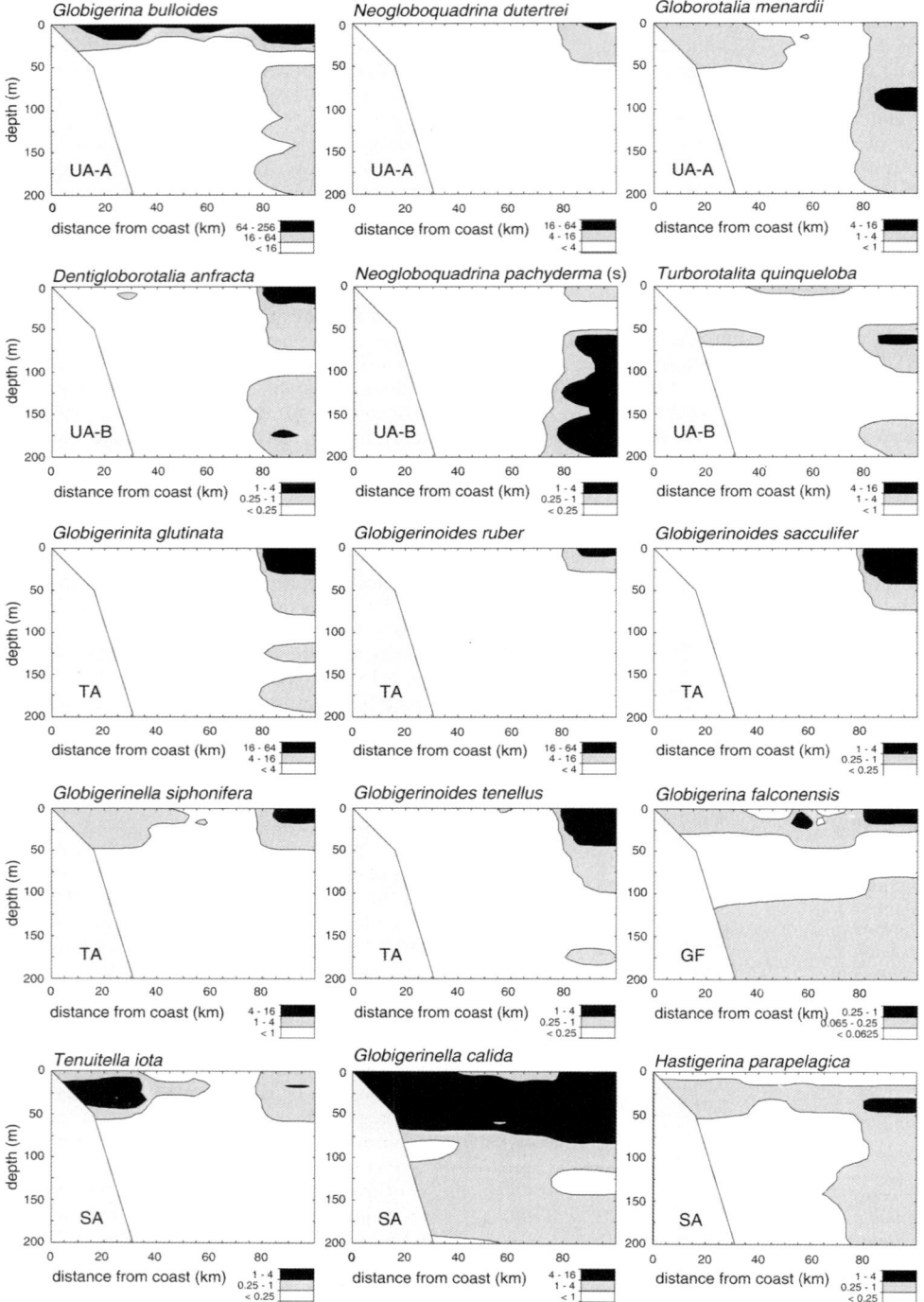

Fig. 5. Contoured distribution patterns of the most abundant species of planktic foraminifera during the SW monsoon leg 1 survey in August 1992. It should be noted that the shell concentrations have been $\log_4$ transformed and that the contour intervals are adjusted to the maximum concentration.

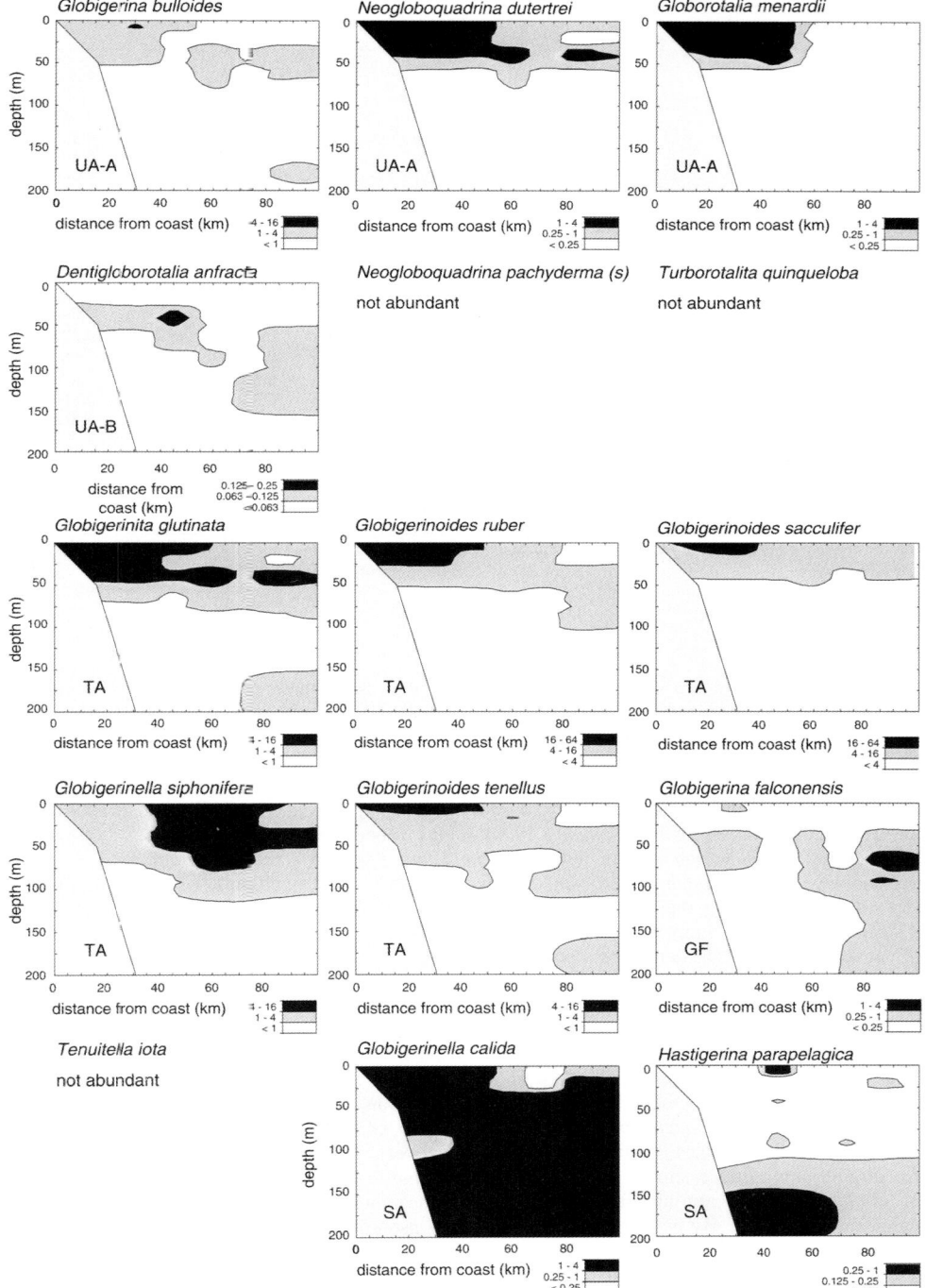

Fig. 6. Contoured distribution patterns of the most abundant species of foraminifera during the NE monsoon leg 2 survey in February 1993. It should be noted that the shell concentrations of the species have been $\log_4$ transformed and that the contour intervals are adjusted to the maximum concentration.

Table 4. Factor scores of $\log_e$ transformed species concentrations

	Rotated factor pattern			
	Factor 1	Factor 2	Factor 3	Factor 4
D. anfracta	**0.933**	0.141	0.258	−0.143
N. pachyderma (sin)	**0.903**	0.170	−0.258	0.299
T. quinqueloba	**0.885**	−0.179	0.248	−0.034
G. bulloides	**0.845**	−0.191	0.464	0.059
N. dutertrei	**0.656**	0.422	0.479	−0.016
G. menardii	**0.562**	0.052	0.418	0.009
G. glutinata	0.559	**0.731**	−0.002	0.157
G. ruber	0.044	**0.965**	−0.081	0.247
G. sacculifer	−0.014	**0.898**	0.107	−0.245
G. siphonifera	−0.036	**0.871**	0.272	0.047
G. tenellus	−0.021	**0.860**	0.037	0.314
T. iota	0.346	0.157	**0.907**	0.181
G. calida	0.047	−0.036	**0.801**	0.072
H. parapelagica	0.273	0.204	**0.576**	−0.176
G. falconensis	0.052	0.239	0.100	**0.964**
% of total variance explained	29.6	28.0	18.1	9.1
Cumulative % of total variance explained	29.6	57.6	75.7	84.7

All 64 plankton tow samples were used for the analysis. Three assemblages are found (see text for discussion). The fourth factor reflects the distribution pattern of G. falconensis, explaining an additional 9.1% of the total variance. Bold indicates highest score for each species.

tolerance limit in the Arabian Sea, explaining their deeper habitat and absence during non-upwelling (Figs. 5 and 6). Upwelling of deep cold water during the SW monsoon will bring these species into surface waters indicating the areas of most intense coastal upwelling (Thiede 1975; Giraudeau raudeau & Rogers 1994; Ivanova et al. 1999). Interestingly, T. quinqueloba occurs not only in the assemblage with N. pachyderma (sin) in the colder waters of the lower thermocline but also in the surface waters at temperatures as high as 21 °C. It is possible that these may be a population of the warm-water T. quinqueloba, similar to those found in the Coral Sea by Darling et al. (2000).

Factor 2: the tropical assemblage (TA, explaining 28.0% of total variance). The species that show high positive scores on the second factor include G. glutinata, G. ruber, G. sacculifer, G. siphonifera and G. tenellus. The species inhabit the relatively warm oligotrophic surface mixed layer and were most abundant in the non-upwelling period as sampled during leg C2. By having symbionts, the species are adapted to the nutrient-depleted levels of the photic zone. G. glutinata also shows some affinity with the upwelling assemblage.

Factor 3: the subsurface assemblage (SA, explaining 18.1% of total variance). The species present in this assemblage are G. calida, H. parapelagica and T. iota. Although highest concentrations are found during upwelling, these species are also found during the non-upwelling period. Highest concentrations are mostly found in the subsurface. During upwelling, the shell concentration maxima are found shallower than during non-upwelling (Figs 5 and 6). H. parapelagica and T. iota are known to bear symbionts, which would imply that species distributions are limited to the photic zone.

Factor 4: Globigerina falconensis (GF, explaining 9.1% of total variance). The distribution pattern of G. falconensis accounts for the presence of a fourth factor. The species inhabits subsurface waters and, in contrast to the SA species, is most abundant during non-upwelling conditions. Maximum concentrations of this species are found between 50 and 100 m, i.e. below the DCM (Figs 5 and 6). G. falconensis is approximately four times more abundant during non-upwelling (NE monsoon) than during upwelling (SW monsoon). Probably, this species is the best faunal indicator for NE monsoon conditions. Its relative abundance in the surface sediments of the Arabian Sea increases towards the NE, indicating a stronger influence of NE monsoon conditions (see also Schulz et al. 2002).

Depth habitat and hydrography

In this section the depth habitat of various species will be discussed in relation to the hydrography. First we will discuss the results on the two 'upwelling stations', 310 and 313, followed by the 'non-upwelling stations' 920 and 917. Only species with a minimum concentration of 1 ind. m^{-3} in at least one of the nine plankton nets of a given station will be discussed.

Station 310 was located in the centre of upwelling during the SW monsoon of August 1992, characterized by low SST and high nitrate concentrations, indicating freshy upwelled waters (Figs 2 and 7). *G. bulloides* is the most abundant species at this station and inhabits the uppermost part of the water column. The base of the productive zone is found at 17 m. Its inflection point (z_i), which indicates the point of maximum change of the shell concentration with depth, is found at 8 m and coincides with the thermocline and the chlorophyll maximum. *G. glutinata* and *N. dutertrei* show a similar concentration profile: a maximum shell concentration at the sea surface and a shallow base of the productive zone at 21 m and 26 m, respectively. *G. siphonifera* and *G. calida*, however, have subsurface maxima, with the base of their productive zone at 43 m and 41 m, respectively. The latter species thus reflect waters from slightly below the main thermocline. The inflection point of these species appears to coincide with a second, although less pronounced thermocline at a depth of 30 m. The concentration profile of *G. menardii* shows two maxima: a shallow maximum at 17 m and a deeper maximum at 65 m. At this station, the average test size of *G. menardii* increases to a depth of about 85 m (Fig. 8). The shallow maximum apparently represents small or immature specimens, whereas the deeper maximum is associated with relatively large mature specimens. The base of the productive zone is therefore situated below the deepest maximum, close to where *G. menardii* reaches its maximum test size (Fig. 8). This observation confirms that the base of the productive zone indeed reflects the maximum depth at which *G. menardii* lives. The decreasing average test size below the productive zone may be explained by the fact that the specimens are dead, and have lost their ability to counter gravitational settling. The average test size decreases here, because larger shells will settle to the sea floor faster than smaller ones.

Station 313 was located in the core of a relatively warm-water eddy during the SW monsoon of August 1992 (Figs 2 and 9). The chlorophyll maximum, mixed layer and thermocline at this station were all present at a depth of 8–12 m. Lower nitrate concentrations and higher temperature of the surface mixed layer indicate more 'aged' upwelled waters than recorded at station 310 (centre of upwelling). Shell concentrations are generally high, and species show a shell concentration maximum at or near the sea surface, except for *G. calida* and *G. menardii*. The inflection point of *G. bulloides* is found at 12 m and coincides with the thermocline and chlorophyll maximum. The base of its productive zone is found at 23 m. Again, the concentration profile of *G. glutinata* and *N. dutertrei* is similar to that of *G. bulloides*. However, the base of their productive zone is found slightly deeper, at 43 m and 38 m, respectively. Species from the tropical assemblage (TA) (i.e. *G. ruber*, *G. sacculifer*, *G. siphonifera* and *G. tenellus*) are abundant as well at this station, possibly because of the presence of a thin but relatively warm and nutrient-depleted mixed layer. Concentration maxima of these species were found at, or slightly below, the sea surface, whereas inflection points coincide with, or are found just below, the thermocline.

The base of the productive zone of these species varies between 25 m (*G. ruber*) and 51 m (*G. siphonifera*). *G. calida* and *G. menardii* show a subsurface maximum with a slight increase of the concentration near the sea surface. Again, the base of the productive zone is found near the depth of test size of *G. menardii* (Fig. 8). Remarkably, neither the inflection points (for *G. calida* at 57 m and for *G. menardii* at 111 m) nor the bases of the productive zone (for *G. calida* at 72 m and for *G. menardii* at 132 m) appear to coincide with any particular hydrographic feature in the water column. It may be anticipated that temperature, light or food availability may limit the depth distribution of these species rather than the position of the thermocline or the fluorescence maximum.

Station 920 (at the same location as 310) reflects non-upwelling NE monsoon conditions of February 1993 (Figs 3 and 10). Contrasting the hydrography during upwelling, the DCM and thermocline were found at different depths in the water column. The surface mixed layer is c. 45 m thick and its base coincides with the DCM. Nitrate rapidly increases to 10–15 µM below the depleted surface mixed layer. At this station the main thermocline was found at 99 m. All species show a maximum of the shell concentration close to the sea surface. Although less pronounced, the concentration profiles consistently indicate a subsurface maximum too. Surface maxima are associated with small, immature specimens, whereas the deeper maxima, associated with the DCM, represent the larger (adult) specimens. The base of the productive zone is consistently found in the zone between the mixed layer (DCM) and

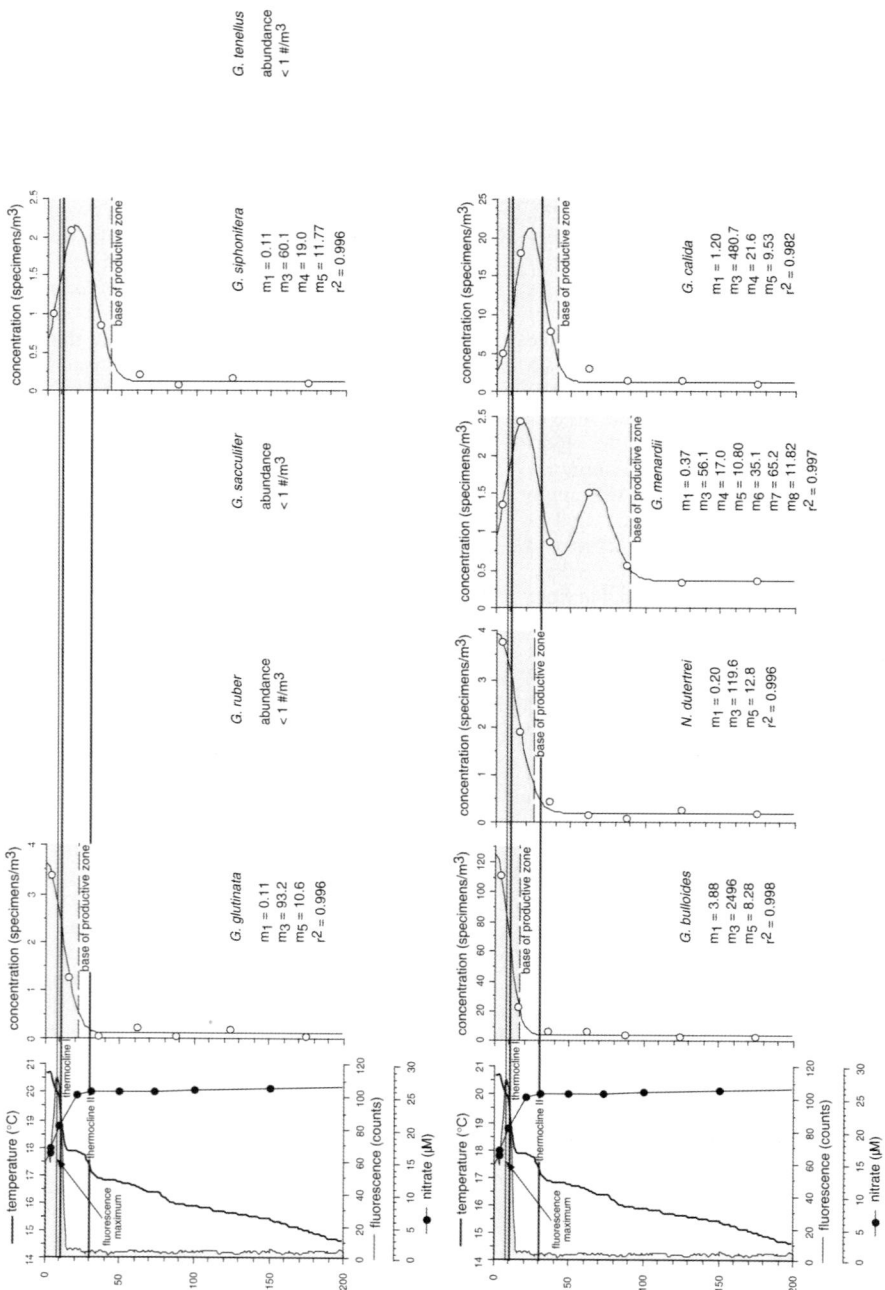

Fig. 7. CTD profile of station 310 in the centre of upwelling during the SW monsoon in August 1992, showing temperature, fluorescence, nitrate concentration and the shell concentration profiles of the most abundant species. The low temperatures and very high nitrate availability at the surface, indicating freshly upwelled waters, should be noted. m_1 represents the value of the settling flux concentration (C_{SF}), m_3 represents the total surface-area under the Gaussian distribution, m_4 represents the centre of the Gaussian distribution (c) and indicates the depth of the maximum in the shell concentration profile, whereas m_5 represents the width of the Gaussian distribution (σ). Note that m_6, m_7 and m_8 have similar meaning as m_3, m_4 and m_5 but now refer to the second Gaussian distribution that is used to fit a second (deepest) maximum in the shell concentration profile.

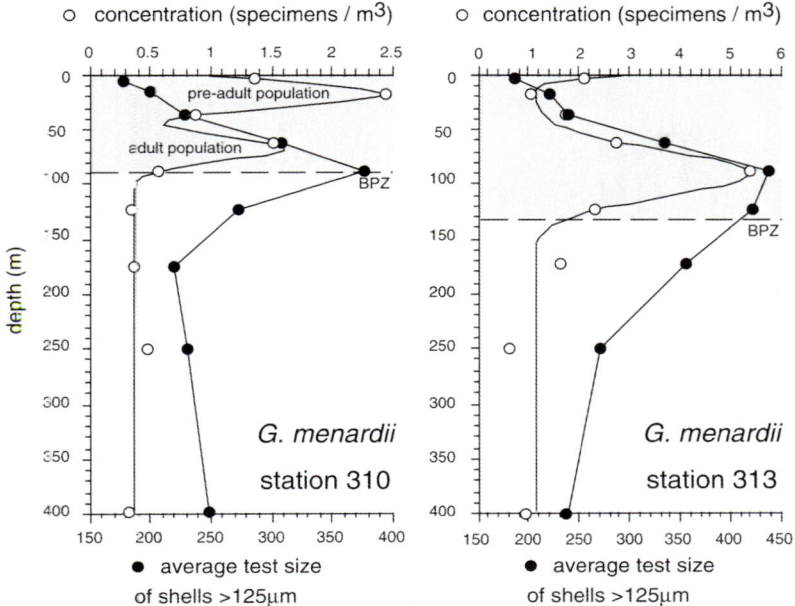

Fig. 8. Vertical shell concentration profiles and average test size of *G. menardii* at stations 310 and 313 (SW monsoon in August 1992). The average test size of *G. menardii* increases with depth down to the base of the productive zone, as defined by fitting the shell concentration profile, and indicates that specimens are growing down to the base of the productive zone. Below the productive zone the specimens are dead and have lost their ability to counter gravitational settling. Because larger (heavier) specimens settle faster to the sea floor the average test size decreases below the productive zone.

the thermocline, for both upwelling as well as non-upwelling species. However, concentrations of UA species (*G. bulloides*, *N. dutertrei* and *G. menardii*) are low compared with those of TA species (*G. glutinata*, *G. ruber*, *G. sacculifer*, *G. siphonifera* and *G. tenellus*). The shell concentration profile of *G. calida* is very different from that of the other species and our curve fitting procedure suggests that the base of the productive zone is at c. 280 m (i.e. far below the thermocline).

Station 917 (at the same location as 313), reflects non-upwelling conditions of February 1993 (Figs 3 and 11). At this station the mixed layer is about 15 m thick, whereas the DCM is present at 29 m and appeared to coincide with the first measurable increase in nitrate. The thermocline was found at 96 m. Most species show a subsurface maximum, between the DCM and the thermocline, with strikingly similar concentration profiles of *G. bulloides*, *G. glutinata*, *N. dutertrei* and *G. ruber*. *G. sacculifer*, however, shows a shell concentration maximum near the sea surface. It is known that *G. sacculifer* migrates vertically during ontogeny between the chlorophyll maximum and the sea surface (Bijma et al. 1990; Bijma et al. 1994; Hemleben & Bijma 1994) and we

suggest that the *G. sacculifer* population at this station is relatively young. The base of the productive zone is consistently found between 59 and 74 m. Although no curve fit for the shell concentration profile of *G. calida* could be established by our model, the shell concentration profile suggests that for this species the base of the productive zone is below 200 m.

Depth habitat and $\delta^{18}O_{shell}$

The shell concentration profiles suggest that living specimens may be found in a zone ranging from the sea surface to the base of the productive zone. It can thus be expected that, within the productive zone, the oxygen isotope composition of the shells increases with depth because of continuing shell growth at lower temperatures. Below a certain depth level, however, the oxygen isotope composition should remain constant because shell growth has ceased. This, of course, is true only if upper ocean conditions are more or less constant for the period of the observations. It is interesting to know what depth level of the upper water column is represented by the oxygen isotope composition of the shells. To establish a relationship between

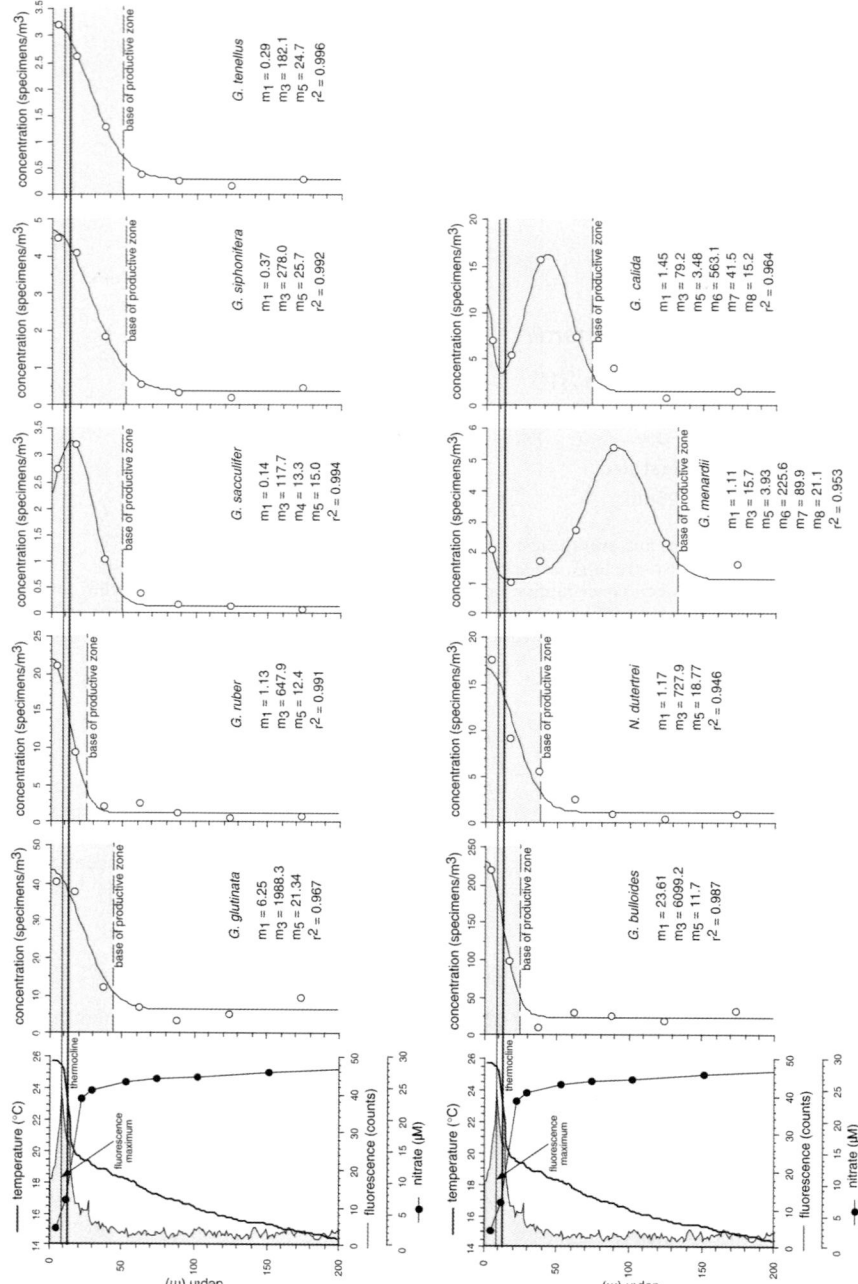

Fig. 9. CTD profile of station 313 in a warm core eddy during the SW monsoon in August 1992, showing temperature, fluorescence, nitrate concentration profile and the shell concentration profiles of the most abundant species at this station. The lower nitrate availability and higher temperature of the mixed layer, indicating more aged surface waters relative to station 310, should be noted. m_1 represents the value of the settling flux concentration (C_{SF}), m_3 represents the total surface-area under the Gaussian distribution, m_4 represents the centre of the Gaussian distribution (c) and indicates the depth of the maximum in the shell concentration profile, whereas m_5 represents the width of the Gaussian distribution (σ). Note that m_6, m_7 and m_8 have similar meaning as m_3, m_4 and m_5 but now refer to the second Gaussian distribution that is used to fit a second (deepest) maximum in the shell concentration profile.

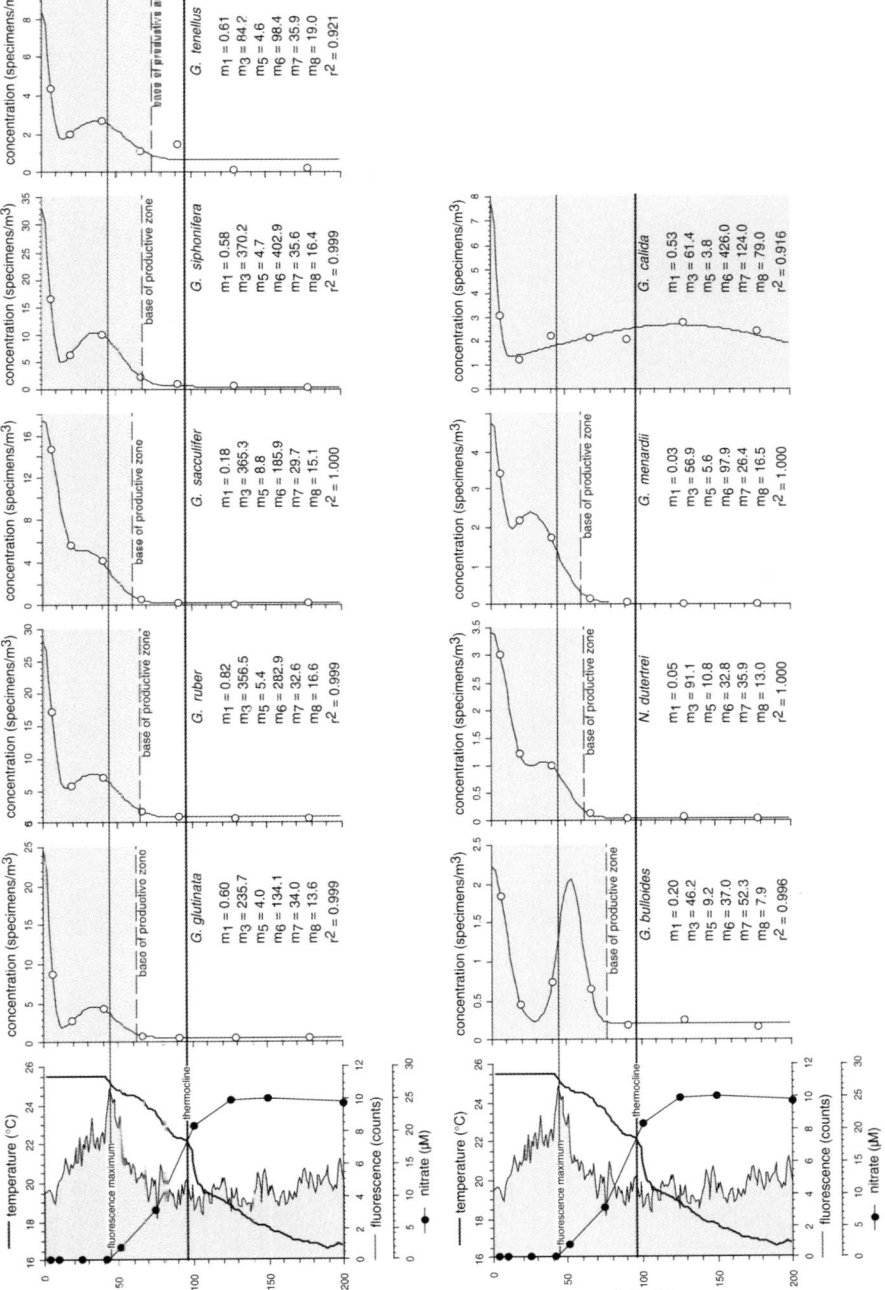

Fig. 10. CTD profile of station 920 during the NE monsoon in February 1993, showing temperature, fluorescence, nitrate concentration and the shell concentration profiles of the most abundant species at this station. In contrast to upwelling conditions the fluorescence maximum and thermocline are found at different depths in the water column. m_1 represents the value of the settling flux concentration (C_{SF}), m_3 represents the total surface-area under the Gaussian distribution, m_4 represents the centre of the Gaussian distribution (c) and indicates the depth of the maximum in the shell concentration profile, whereas m_5 represents the width of the Gaussian distribution (σ). Note that m_6, m_7 and m_8 have similar meaning as m_3, m_4 and m_5 but now refer to the second Gaussian distribution that is used to fit a second (deepest) maximum in the shell concentration profile.

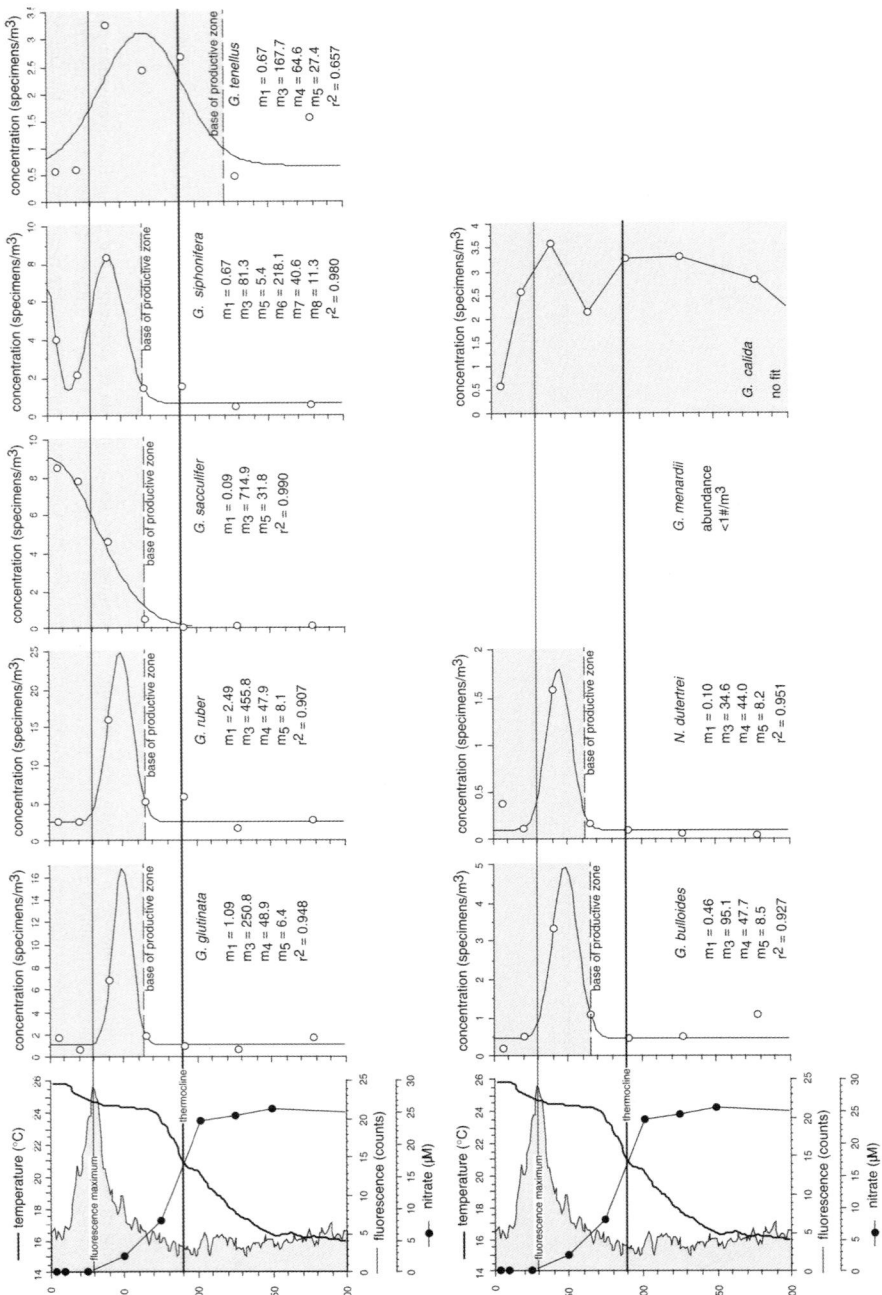

Fig. 11. CTD profile of station 917 during the NE monsoon in February 1993, showing temperature, fluorescence, nitrate concentration and the shell concentration profiles of the most abundant species at this station. m_1 represents the value of the settling flux concentration (C_{SF}), m_3 represents the total surface-area under the Gaussian distribution, m_4 represents the centre of the Gaussian distribution (c) and indicates the depth of the maximum in the shell concentration profile, whereas m_5 represents the width of the Gaussian distribution (σ). Note that m_6, m_7 and m_8 have similar meaning as m_3, m_4 and m_5 but now refer to the second Gaussian distribution that is used to fit a second (deepest) maximum in the shell concentration profile.

the shell concentration profile and calcification temperatures of specimens in the water column (T_c), the oxygen isotope composition of *G. bulloides* was measured at stations 310, 313, 917 and 920. All $\delta^{18}O_{shell}$ measurements are converted into calcification temperature following the methods discussed by Peeters *et al.* (2002). As we are interested in the calcification temperatures of the exported specimens, we excluded the measurements from the uppermost two plankton nets representing mostly living specimens. The obtained calcification temperatures (see Table 6) clearly indicate a shallower calcification depth during upwelling (stations 310 and 313) than during non-upwelling (stations 917 and 920), when the ocean is thermally stratified. The average calcification temperatures suggest that during upwelling *G. bulloides* calcifies in the upper 10–15 m and during non-upwelling between 50 and 60 m. These temperatures exactly correspond to the depth level in the shell concentration profile that is characterized by a rapid decline in the shell concentration (i.e. the inflection point). In addition, the total range in calcification temperatures corresponds to the expected *in situ* temperature range from the sea surface to the base of the productive zone.

Depth habitat and ambient temperatures

Above, we have shown that the average calcification temperature of *G. bulloides* corresponds to the temperature found at the inflection point in the concentration profile. It also has been shown that the total range of calcification temperatures corresponds to the temperature range from the sea surface to the base of the productive zone. However, it is known that the $\delta^{18}O$ of foraminifera from plankton tows have lower values than the $\delta^{18}O$ of the shells on the sea floor (Bé 1980; Curry & Matthews 1981; Duplessy *et al.* 1981). Secondary calcification (e.g. during gametogenesis at the end of the foraminiferal life cycle) and vital effects related to ontogenetic processes may explain this difference (e.g. Duplessy *et al.* 1981; Spero & Lea 1996; Bemis *et al.* 1998). Among others, we observe that the average test size of species is relatively small near the sea surface and increases with depth in the water column. This suggests that during their life cycle the specimens descent to deeper water while they grow larger. The $\delta^{18}O_{shell}$ thus represents an integrated signal over the productive zone.

If we assume that the relationship between the oxygen isotope composition and the shell concentration profile can be used for other species as well, we may use the depth of the inflection point to estimate the expected average $\delta^{18}O_{shell}$. In Figs 12 and 13 we plotted, for various species, the inflection point of the shell concentration profile on the temperature profile. We have limited the ambient temperature ranges for the species to those temperatures found in a range of 1 SD above and below the inflection point. This represents the range from the maximum in the shell concentration profile to the base of the productive zone.

Upwelling conditions. Figure 12 shows the expected average calcification temperature for various species of planktic foraminifera at upwelling stations 310 and 313. For each species the inflection point is plotted on the temperature profile (see also Fig. 4). The vertical bars indicate the inferred depth habitat from the concentration profile, ranging from the maximum in the concentration profile to the base of the productive zone. The horizontal bars indicate the range in ambient temperatures that will be recorded in the oxygen isotope composition of the shells. Most of the species inhabit the upper 35 m of the water column. Inflection points are consistently found below the fluorescence maximum. Species such as *G. bulloides*, *G. glutinata*, *N. dutertrei* and *G. ruber* show the shallowest depth habitats. At station 310 ambient temperatures range from 20.7 °C at the sea surface, to as low as 17.7 °C. At station 313, these species live at temperatures ranging from 25.8 °C to 18.5 °C. The inflection points of *G. siphonifera*, *G. calida* and *G. sacculifer* are found deeper in the water column between 20 and 57 m, reflecting temperatures down to 16.8 °C. At both stations *G. menardii* shows the deepest habitat. Its oxygen isotope composition will reflect temperatures below the main thermocline, i.e. between 16.7 and 15.5 °C.

Non-upwelling conditions. Figure 13 illustrates the depth ranges and associated temperatures for various species of planktic foraminifera at non-upwelling stations 920 and 917. At station 920, the fluorescence maximum coincides with the base of the mixed layer. Although there is a temperature gradient starting directly below the surface mixed layer at 40 m, the main thermocline is present at *c.* 100 m. In contrast to the upwelling conditions, the inflection points of species are consistently found between 40 and 60 m, except for the inflection point of *G. sacculifer* at station 917, which is found slightly shallower. The species thus represent temperatures below the surface mixed layer and above the main thermocline. The total temperature range in the upper 70 m at station 920 is only 1.7 °C, from 25.4 °C at the surface to 23.7 °C at 70 m. At station 917 the fluorescence maximum and mixed layer are decoupled. The onset of a small thermocline, di-

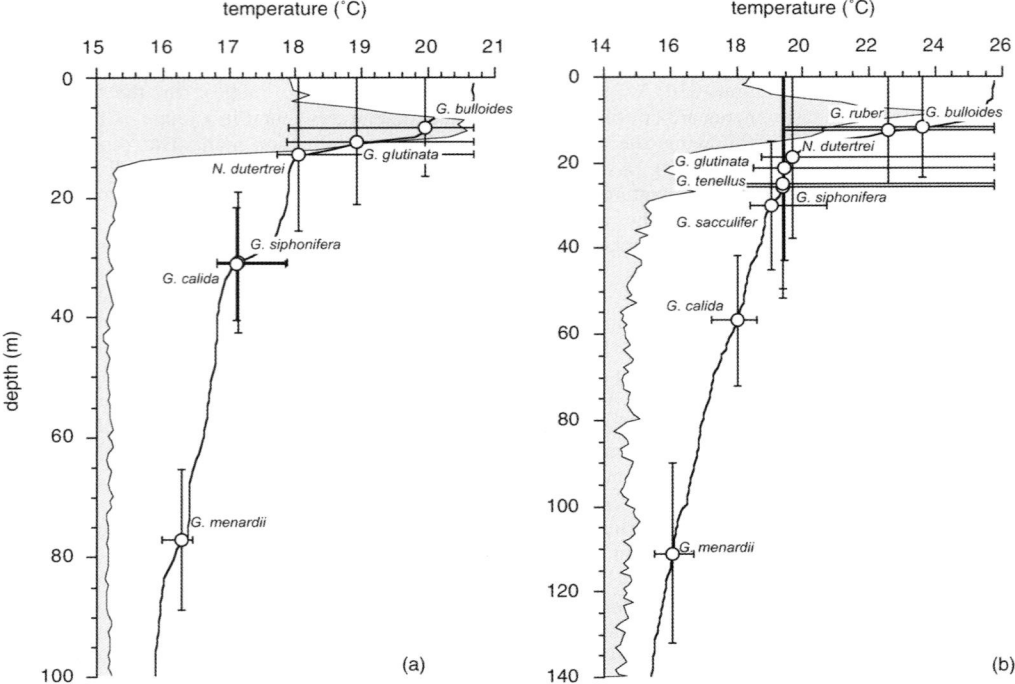

Fig. 12. Depth habitat and ambient temperatures for various species of planktic foraminifera at upwelling stations: (**a**) 310; (**b**) 313. For each species, the inflection point of the shell concentration profile is plotted on the temperature profile. This temperature will be reflected in the mean oxygen isotope composition of the species. Horizontal bars indicate the ambient temperature for each species. The vertical bars indicate the depth habitat of adult specimens as inferred from curve fitting, and range from the level at which the maximum concentration is found to the base of the productive zone. The shaded area shows the fluorescence profile.

rectly below the mixed layer, is found above the DCM. Because inflection points of species are present between 50 and 60 m, all species reflect the same ambient temperature of 24.2 °C. Only *G. sacculifer* shows a shallower inflection point at the depth of the DCM and will reflect slightly higher temperatures than the other species. The base of the productive zone of all species is consistently found at about 60–65 m.

Standing stock

The size of the standing stock (SS_{COR}) may be interpreted as a measure of the success of a given species in response to environmental conditions. It can be expected that a high standing stock will produce a high shell flux (i.e. C_{SF}), whereas a low standing stock will produce a low shell flux. If shell production did not change with time, the standing stock would produce a constant 'rain' of shells to the sea floor (steady-state conditions). Consequently, one would expect to find a species-specific relationship between the integrated standing stock and the concentration of shells below the productive zone (i.e. C_{SF}). To test this hypothesis, we have plotted the standing stocks of various species (SS_{COR}) v. C_{SF} (Fig. 14b). Because both the size of the standing stock and C_{SF} vary over several orders of magnitude, we used logarithmic scales to visualize the data. For this figure, we used the data tabulated in Table 5. We found a positive relationship between SS_{COR} and C_{SF} for all species. Apparently, higher standing stocks indeed correspond to a higher export flux. However, the relationship between standing stock and export flux may differ with species. At a given size of the standing stock, species such as *G. bulloides*, *G. glutinata*, *G. ruber* and possibly *N. dutertrei* show a higher settling flux concentration compared with *G. menardii*, *G. sacculifer*, *G. calida* and *G. siphonifera*. This suggests that turnover rates are higher or settling speeds of these species are lower. Although the relationship between the standing stock and the settling flux might be affected by different processes, it will provide the answer to the turnover rates of differ-

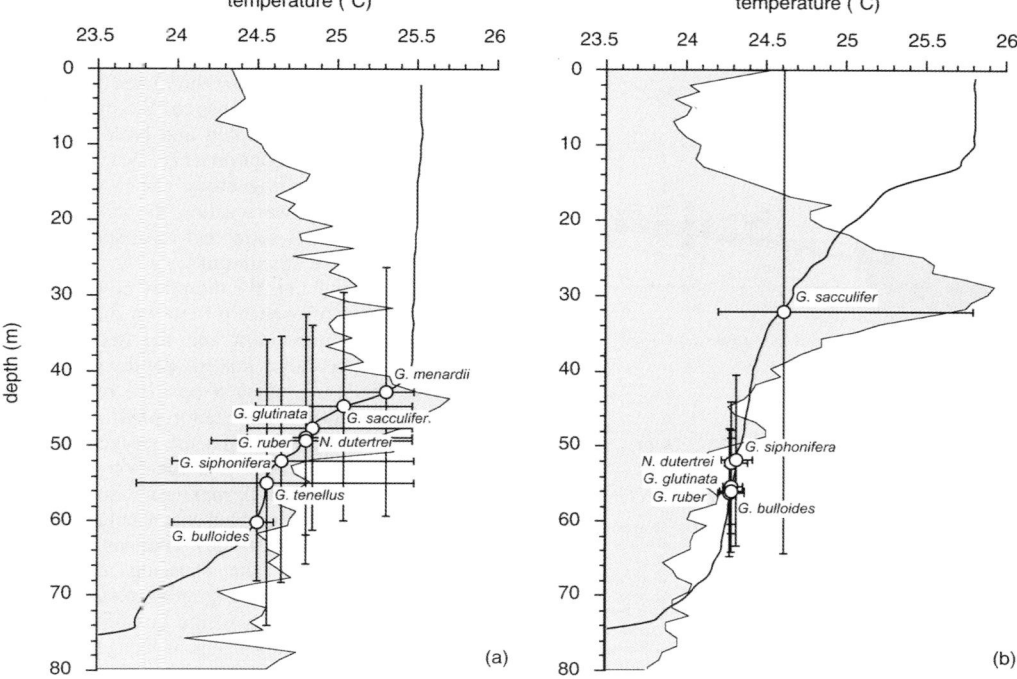

Fig. 13. Depth habitat and ambient temperatures for various species of planktic foraminifera at non-upwelling stations: (a) 920 (b) 917. For each species, the inflection point of the distribution profile is plotted on the temperature profile. The vertical bars indicate the depth habitat of adult specimens as inferred by Gaussian fitting, from the level at which the maximum concentration is found to the base of the productive zone. Horizontal bars indicate the ambient temperature for each species. The shaded area shows the fluorescence profile.

ent species. For such an approach, however, one would need plankton tow data and sediment trap observations from one location.

Discussion

Planktic foraminifera respond to changing environmental conditions of the upper water column. From a palaeoceanographic point of view it is important to know how biological and physical features of the upper water column (e.g. depth of chlorophyll maximum, mixed layer depth, thermocline depth) control the abundance and the calcification depths of various species. Bé & Hutson (1977) noted that: 'The net effect of depth stratification of species has important implications for paleoecological analysis in that species living at different depths (with different environmental conditions) are mixed together in the same dead assemblage on the sea floor'.

Plankton tow studies in the tropical Atlantic and Pacific Ocean indicate that mixed layer depth, thermocline depth, integrated primary productivity and light attenuation levels may control the absolute abundance and depth habitat of planktic foraminifera (Ravelo & Fairbanks 1992; Watkins et al. 1998). The definition of the base of the productive zone, introduced in this paper, provides an independent tool to estimate the depth of calcification for various species. Estimates of the standing stock and base of the productive zone may help to identify the environmental conditions that control the abundance and vertical distribution of a given species.

Our plankton tow data show that planktic foraminifera, collected during the two monsoon seasons, do not have an absolute depth habitat but rather live at depths directly related to the local hydrography. This observation is in agreement with previous work in other areas (e.g. Ravelo & Fairbanks 1992; Thunell & Reynolds-Sautter 1992). In general, most (non-globorotaloid) species appear to have a shallower habitat during upwelling (SW monsoon season) than during non-upwelling (NE monsoon season). The environmental parameters controlling the depth habitat, however, may vary depending on the species. Symbiont-bearing species such as G. sacculifer

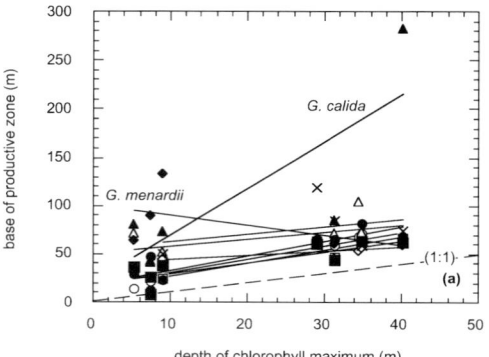

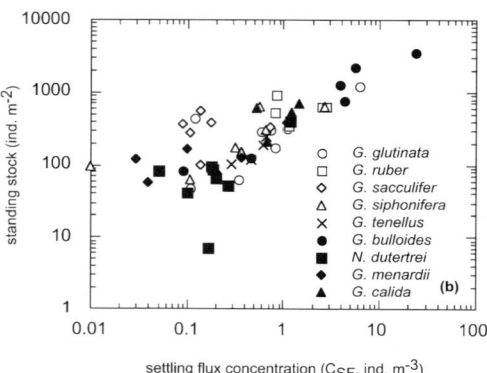

Fig. 14. (**a**) Scatter plot showing that the depth of the chlorophyll maximum and the base of the productive zone are positively correlated for most species. The species *G. menardii*, however, shows a negative relationship. (**b**) Relationship between the standing stock (SS_{COR}) and the shell concentration below the productive zone (C_{SF}). Higher standing stocks correspond to higher shell concentrations below the productive zone.

and *G. ruber* may obtain nutrition from their symbionts and proliferate in oligotrophic environments such as found in the mixed layer. For these species, light availability or mixed layer temperature may control their distribution (Mulitza *et al.* 1998). On the other hand, the availability of particulate food may be more important for other species such as, for example, *G. bulloides* (Watkins *et al.* 1996). To define whether a certain hydrographic state of the upper ocean is favourable for a given species, one should consider the size of the standing stock. Fairbanks & Wiebe (1980) showed that the DCM plays an important role in the vertical distribution patterns of planktic foraminifera in the upper water column. It provides the primary food source for most foraminiferal species, and highest concentrations are often associated with the DCM. Off the Arabian Peninsula, the DCM appears to be close to the sea surface during upwelling and coincides with the main thermocline just below the mixed layer (Fig. 2e). During winter (NE monsoon) the thermocline and DCM may be decoupled and both are found deeper in the water column (Fig. 3e). Although the size of the standing stock for most species differs between the two seasons, the vertical shell concentration profiles show that most species have their concentration maximum close to the DCM (Figs. 7, 9, 10 and 11). We therefore conclude that there is a direct relationship between the depth of the chlorophyll maximum and the depth of the BPZ. Figure 14a shows that all dominant species, except *G. menardii*, show a positive relationship between the depth of the chlorophyll maximum and the BPZ, suggesting that the oxygen isotopic composition of the shells is a function of (1) the depth of the chlorophyll maximum and (2) the thermal structure of the upper water column.

As well as the (seasonal) changes in depth habitat, the size of the standing stock also responds to the local hydrography, resulting in a strong seasonal variability of the flux of foraminiferal shells to the sea floor (e.g. Curry *et al.* 1992; Conan & Brummer, 2000). The most striking seasonal change in the size of the standing stock is observed in *G. bulloides*. Previous studies have shown that this species is the best indicator for upwelling (Anderson & Prell 1991; Brock *et al.* 1992; Curry *et al.* 1992; Anderson & Prell 1993; Naidu & Malmgren 1996*a*, *b*). During upwelling conditions, the standing stock varied between 760 and 3500 ind. m^{-2}, whereas during the non-upwelling period, the standing stock did not exceed 140 ind. m^{-2}. Remarkably, the size of the standing stock of *G. bulloides* and the depth of the DCM are negatively correlated (Fig. 15). During upwelling conditions this species inhabits the uppermost part of the water column and its inflection point coincides with the DCM, suggesting that food availability is the most important environmental parameter controlling the size of the standing stock. Although the reproductive cycle and thus the standing stock of *G. bulloides* may be related to lunar periodicity in regions with relatively stable upper ocean conditions (Schiebel *et al.* 1997), in the Arabian Sea the size of its standing stock is mainly controlled by the process of upwelling and related phyto- and zooplankton blooms, which serve as food for this species. Other abundant species in the Arabian Sea, such as *G. siphonifera*, *G. ruber* and *G. sacculifer*, show a positive correlation with the depth of the DCM. These are all symbiont-bearing species, which are most abundant in the surface mixed layer (Watkins *et al.* 1996).

As a result of the seasonal changes in hydrography, the faunal assemblages found during the two monsoon seasons clearly differ from one another. During the SW monsoon, species from the upwelling assemblage, which are also characteristic for cold-water regions at high latitudes (Bé & Tolderlund 1971) dominate in the faunal assemblage, whereas during the NE monsoon the faunal assemblage is dominated by species from the tropical assemblage. It is known, however, that during the SW monsoon, fluxes of most species increase (Curry et al. 1992; Conan & Brummer, 2000). Our data indicate that, during upwelling, species from the tropical assemblage are not present in the centre of upwelling, but are found in 'aged' upwelled waters, with lower nitrate concentrations and higher sea surface temperatures (i.e. station 313, Figs 2a and 5). The tropical assemblages may thus be found predominantly in warm-water eddies. These eddies mostly have a shallow chlorophyll maximum and therefore during upwelling species from the tropical assemblage also reflect shallower water column properties than during non-upwelling.

A small sample from the sedimentary record often represents a time span from tens to hundreds of years. Here, the shells of foraminifera that were produced during various seasons are mixed together in one fossil assemblage. This study shows that a seasonal decomposition of the fossil assemblage is feasible. The abundance and stable isotopic composition of species that were produced during different times of the year may be used to reconstruct the palaeoceanographic conditions of the two monsoon seasons. Potentially, such an approach would allow for a reconstruction of sea surface temperature, biological productivity and water column stratification of both monsoon seasons. However, for such an approach, it is essential to know the turnover rates of each of the species. Only when shell fluxes of the various species are coupled to standing stock estimates will it be possible to improve palaeoceanographic interpretations of monsoon history. We suggest that future work should concentrate on the relationship between living planktic foraminifera and their export flux, to quantify turnover rates for the various species.

Conclusions

In this study, we compared the depth habitat and seasonal distribution of living foraminifera in the upwelling area off the Arabian Peninsula in response to monsoon-controlled hydrographic changes. The following conclusions are drawn from this study.

(1) In the area off Oman–Yemen, the faunal assemblage, shell concentrations, standing stock and depth habitat of planktic foraminifera change seasonally. Because hydrographic changes in this area are directly coupled to the monsoon system, the changes in the faunal assemblage and isotopic composition of individual species allow reconstruction of surface water conditions and monsoon history over longer time scales.

(2) Factor analysis on 15 species yields three faunal assemblages: an upwelling assemblage, a tropical assemblage and a subsurface assemblage. A fourth factor represents the distribution pattern of *Globigerina falconensis*. This species is mainly found during NE monsoon conditions and inhabits subsurface waters. Consequently, this species is the best faunal indicator of NE monsoon conditions.

(3) Living planktic foraminifera do not have an absolute depth habitat (calcification depth), but rather live at depths related to the local hydrography. Most shallow-dwelling species have a shallower habitat during upwelling (SW monsoon) than during non-upwelling (NE monsoon). The depth of the chlorophyll maximum appears to be an important environmental parameter controlling the depth habitat of species.

(4) The shell concentration profile of most planktic foraminifera may be described by a Gaussian distribution model that can be used to calculate the base of the productive zone, reflecting the boundary between the productive and settling flux zone. For non-globorotaloid species, the base of the productive zone depends on the depth of the DCM. Below the productive zone the shell concentration is low and represents dead specimens settling to the sea floor.

(5) The standing stock varies seasonally and is related to the shell concentration below the productive zone, representing the shells of dead specimens settling to the sea floor. Higher standing stocks correspond to higher settling flux concentrations, suggesting a direct relationship between shell production and export flux.

(6) During upwelling, tropical species do not proliferate in the centre of upwelling. However, high abundances of these species may be found in relatively warm-water eddies or upwelling filaments.

(7) The standing stock of *G. bulloides* correlates negatively with mixed layer depth and the depth of the chlorophyll maximum. This suggests that food availability may be the most important environmental parameter controlling the size of the standing stock. The standing stocks of species that possess photosynthetic symbionts, such as *G. sacculifer*, *G. siphonifera* and *G. ruber*, correlate

Table 5. *Results of calculations*

	Upwelling stations						Non-upwelling stations					
	SS	SS_{COR}	C_{SF}	z_{MAX}	z_i	z_{BPZ}		SS	SS_{COR}	C_{SF}	z_{MAX}	z_i	z_{BPZ}
Station 313 (D)							**Station 917 (D)**						
G. glutinata	1215	1082	6.25	0	21	43	*G. glutinata*	312	279	1.09	49	55	62
G. ruber	337	323	1.13	0	12	25	*G. ruber*	615	535	2.49	48	56	64
G. sacculifer	99	96	0.14	13	28	43	*G. sacculifer*	363	360	0.09	0	32	64
G. siphonifera	152	142	0.37	0	26	51	*G. siphonifera*	296	275	0.67	41	52	63
G. tenellus	101	94	0.29	0	25	49	*G. tenellus*	244	204	0.67	65	92	119
G. bulloides	3463	3187	23.61	0	12	23	*G. bulloides*	123	108	0.46	48	56	65
N. dutertrei	391	369	1.17	0	19	38	*N. dutertrei*	40	37	0.10	44	52	60
G. menardii	375	302	1.11	90	111	132	*G. menardii*						
G. calida	691	638	1.45	42	57	72	*G. calida*						
Station 310 (D)							**Station 920 (D)**						
G. glutinata	47	46	0.11	0	11	21	*G. glutinata*	286	267	0.60	0	48	61
G. ruber							*G. ruber*	509	482	0.82	0	49	66
G. sacculifer							*G. sacculifer*	370	365	0.18	0	45	60
G. siphonifera	60	58	0.11	19	31	43	*G. siphonifera*	612	592	0.58	0	52	68
G. tenellus							*G. tenellus*	183	161	0.61	0	55	74
G. bulloides	1255	1223	3.88	0	8	17	*G. bulloides*	73	66	0.20	0	60	68
N. dutertrei	65	62	0.20	0	13	26	*N. dutertrei*	81	79	0.05	0	49	62
G. menardii	123	107	0.37	17	77	89	*G. menardii*	120	119	0.03	0	43	59
G. calida	513	489	1.20	22	31	41	*G. calida*	596	522	0.53	0	203	282
Station 311 (S)							**Station 918 (S)**						
G. glutinata							*G. glutinata*[1]	172	144	0.82	33	51	68
G. ruber							*G. ruber*[2]						
G. sacculifer							*G. sacculifer*	266	263	0.11	0	26	52
G. siphonifera							*G. siphonifera*[1]	529	n.d	n.d	.37	70	104
G. tenellus							*G. tenellus*[2]						
G. bulloides	758	733	4.31	0	6	12	*G. bulloides*						
N. dutertrei	7	6	0.17	0	4	7	*N. dutertrei*						
G. menardii							*G. menardii*						
G. calida							*G. calida*						

Station 308 (S)

Species						
G. glutinata	63	60	0.35	0	7	14
G. ruber						
G. succulifer						
G. siphonifera[1]	95	95	0.01	28	49	70
G. tenellus						
G. bulloides[1]	2173	2094	5.52	0	14	29
N. dutertrei	84	81	0.19	0	18	36
G. menardii[1]	165	162	0.10	0	32	63
G. calida[1]	667	667	0.00	0	40	79

Station 919 (S)

Species						
G. glutinata[1]	292	266	0.74	43	57	71
G. ruber[1]	615	530	2.76	42	52	61
G. succulifer[1]	320	299	0.74	39	47	55
G. siphonifera[1]	620	521	2.72	46	60	73
G. tenellus[1]	119	105	0.46	38	49	60
G. bulloides[1]	140	n.d	n.d	46	63	81
N. dutertrei[1]	52	44	0.27	40	52	63
G. menardii						
G. calida						

Station 922 (S)

Species						
G. glutinata	438	434	0.12	0	30	59
G. ruber	899	879	0.85	0	23	46
G. sacculifer	550	547	0.14	0	23	46
G. siphonifera[1]	172	160	0.32	35	53	71
G. tenellus[1]	205	n.d	n.d	47	66	85
G. bulloides[1]	81	78	0.09	0	32	65
N. dutertrei	93	89	0.18	0	22	43
G. menardii	57	55	0.04	30	45	61
G. calida[1]	203	174	0.69	42	63	84

[1] C_{SF} and C_{COR} not well defined.
[2] Shell concentration profile does not allow a curve fit.

Calculations were carried out on: standing stock (SS) following equation (9), corrected standing stock (SS_{COR}) following equation (10), settling shell concentration (C_{SF}) following equation (6), depth of maximum shell concentration (z_{MAX}) following equation (6), depth of shell concentration inflection point (z_1) following equation (8) and depth of the productive zone (z_{BPZ}) following equation (7). D, station with deep cast (0–500 m); S, station with shallow cast (0–100 m). Stations with a deep multi-net in general yield better estimates for C_{SF}, SS and SS_{COR}.

Table 6. *Calcification temperature statistics of* G. bulloides *and corresponding depth levels*

Station	T_c average (°C)	SD (°C)	T_c min. (°C)	T_c max. (°C)	n	Average calcif. depth (m)	Maximum depth (m)	Shell concentration inflection point (m)
310	19.8	0.6	18.6	21.1	12	10	12	8
313	23.6	0.3	23.0	24.3	15	12	13	12
917	24.3	1.0	23.3	25.5	4	53	76	56
920	24.5	0.1	24.3	24.6	3	61	64	60

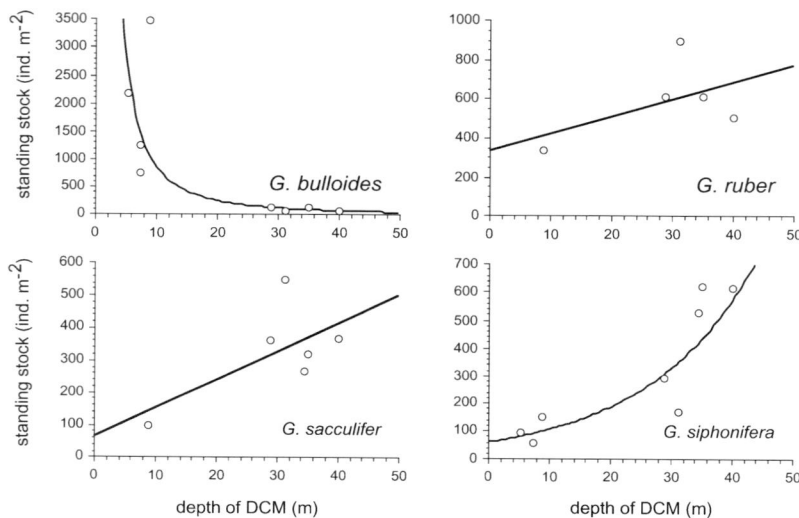

Fig. 15. The integrated standing stock of living *Globigerina bulloides* correlates inversely with the depth of the chlorophyll maximum. The standing stock of symbiont-bearing species such as *Globigerinella siphonifera* and *Globigerinoides sacculifer* correlates positively with the depth of the chlorophyll maximum.

positively with mixed layer depth and the depth of the chlorophyll maximum.

The Netherlands Indian Ocean Programme was funded and coordinated by the Netherlands Marine Research Foundation (SOZ) of the Netherlands Organization for Scientific Research (NWO). We thank Captain J. de Jong, technicians and crews of R.V. *Tyro.* J. Kranenborg and W. van Raaphorst are thanked for helpful suggestions and discussion. We are grateful to J. van Hinte for his critical comments on an earlier version of this work. We much appreciate the constructive reviews of K. Darling and L. Sautter. This is Publication 3649 of the Netherlands Institute for Sea Research (NIOZ).

References

ANDERSON, D.M. & PRELL, W.L. 1991. Coastal upwelling gradient during the Late Pleistocene. *In:* PRELL, W.L. & NITSUMA, N. (eds) *Proceedings of the Ocean Drilling Program, Scientific Results, 117.* Ocean Drilling Program, College Station, TX, 265–275.

ANDERSON, D.M. & PRELL, W.L. 1993. A 300 kyr record of upwelling off Oman during the Late Quaternary: evidence of the Asian southwest monsoon. *Paleoceanography,* **8**, 193–208.

ANDERSON, D.M. & THUNELL, R.C. 1993. The oxygen-isotope composition of tropical ocean surface water during the last deglaciation. *Quaternary Science Reviews,* **12**, 465–473.

ANDERSON, D.M., BROCK, J.C. & PRELL, W.L. 1992. Physical upwelling processes, upper ocean environment and the sediment record of the southwest monsoon. *In:* SUMMERHAYS, C.P., PRELL, W.L. & EMEIS, K.C. (eds) *Upwelling Systems: Evolution since the Early Miocene.* Geological Society, London, Special Publications, **64**, 121–129.

BANSE, K. 1994. On the coupling of hydrography, phytoplankton, zooplankton, and settling organic particles offshore in the Arabian Sea. *In:* LAL, D. (eds) *Biogeochemistry of the Arabian Sea.* Indian Academy of Sciences, Bangalore, 27–63.

BÉ, A.W.H. 1967. Foraminifera families: Globigerinidae and Globorotaliidae. In: FRASER, J. H. (ed.) Fiches d'Identification du Zooplankton, sheet 108. Conseil Permanent International pour l'Exploration de la Mer, Charlottenlund, Denmark, 1–3.

BÉ, A.W.H. 1980. Gametogenetic calcification in a spinose planktonic foraminifer. Globigerinoides sacculifer (Brady) Marine Micropaleontology, 5, 283–310.

BÉ, A.W.H. & HUTSON, W.H. 1977. Ecology of planktonic foraminifera and biogeographic patterns of life and fossil assemblages in the Indian Ocean. Micropaleontology, 23, 369–414.

BÉ, A.W.H. & TOLDERLUND, D.S. 1971. Distribution and ecology of living planktonic foraminifera in surface waters of the Atlantic and Indian Oceans. In: FUNNELL, B.M. & RIEDEL, W.R. (eds) Micropaleontology of Oceans. Cambridge University Press, Cambridge, 105–149.

BEMIS, B.E., SPERO, H.J., BIJMA, J. & LEA, D. 1998. Reevaluation of the oxygen isotopic composition of planktonic foraminifera: experimental results and revised paleotemperature equations. Paleoceanography, 13, 150–160.

BERGER, W.H. 1969. Ecologic patterns of living planktonic foraminifera. Deep-Sea Research, 169, 1–24.

BERGER, W.H. 1971. Planktonic foraminifera: sediment production in an oceanic front. Journal of Foraminiferal Research, 1, 95–118.

BIJMA, J. & HEMLEBEN, C. 1994. Population dynamics of the planktic foraminifer Globigerinoides sacculifer (Brady) from the central Red Sea. Deep-Sea Research I, 41, 485–510.

BIJMA, J., EREZ, J. & HEMLEBEN, C. 1990. Lunar and semi-lunar reproductive cycles in some spinose planktonic foraminifers. Journal of Foraminiferal Research, 20, 117–127.

BIJMA, J., HEMLEBEN, C. & WELLNITZ, K. 1994. Lunar-influenced carbonate flux of the planktic foraminifer Globigerinoides sacculifer (Brady) from the central Red Sea. Deep-Sea Research, I, 41, 511–530.

BROCK, J.C., MCCLAIN, C.R., ANDERSON, D.M., PRELL, W.L. & HAY, W.W. 1992. Southwest monsoon circulation and environments of recent planktonic foraminifera in the northwestern Arabian sea. Paleoceanography, 7, 799–813.

BRUMMER G.-J.A. 1995. Plankton pump and multinet. In: VAN HINTE, J.E., VAN WEERING, T.J.C. & TROELSTRA, S.R. (eds) Tracing a Seasonal Upwelling: Report on Two Cruises of RV Tyro to the NW Indian Ocean in 1992 and 1993. National Museum of Natural History, Leiden, 4, 31–40.

BRUMMER, G.-J.A. & KROON, D., 1988. Planktonic foraminifers as tracers of ocean–climate history. Ph.D. thesis, Free University Press Amsterdam.

CONAN, S.M.-H. & BRUMMER, G.-J.A. 2000. Fluxes of planktic foraminifera in response to monsoonal upwelling on the Somalia Basin margin. Deep-Sea Research II, 47, 2207–2227.

COPLEN, T.B., KENDALL, C. & HOPPLE, J. 1983. Comparison of stable isotope reference sample. Nature, 302, 236–238.

CURRY, W.B. & MATTHEWS, R.K. 1981. Equilibrium ^{18}O fractionation in small size fraction planktic foraminifera: evidence from Recent Indian Ocean sediments. Marine Micropaleontology, 6, 327–337.

CURRY, W.B., OSTERMANN, D.R., GUPTHA, M.V.S. & ITTEKOT, V. 1992. Foraminiferal production and monsoonal upwelling in the Arabian Sea: evidence from sediment traps. In: SUMMERHAYS, C.P., PRELL, W.L., & EMEIS, K.C. (eds) Upwelling Systems: Evolution since the Early Miocene. Geological Society, London, Special Publications, 64, 93–106.

DARLING, K.F., WADE, C.M., STEWART, I.A., KROON, D., DINGLE, R. & LEIGH-BROWN, A.J. 2000. Molecular evidence for genetic mixing of Arctic and Antarctic subpolar populations of planktonic foraminifers. Nature, 405, 43–47.

DEUSER, W.G., ROSS, E.H., HEMLEBEN, C. & SPINDLER, M. 1981. Seasonal changes in species composition, numbers, mass, size, and isotopic composition of planktonic foraminifera settling into the deep Sargasso Sea. Palaeogeography, Palaeoclimatology, Palaeoecology, 33, 103–127.

DUPLESSY, J.-C., BLANC, P.-L. & BÉ, A.W.H. 1981. Oxygen-18 enrichment of planktonic foraminifera due to gametogenetic calcification below the euphotic zone. Science, 213, 1247–1250.

FAIRBANKS, R.G. & WIEBE, P.H. 1980. Foraminifera and chlorophyll maximum: vertical distribution, seasonal succession, and paleoceanographic significance. Science, 209, 1524–1526.

FURBISH, D.J. & ARNOLD, A.J. 1997. Hydrodynamic strategies in the morphological evolution of spinose planktonic foraminifera. Geological Society of America Bulletin, 109, 1055–1072.

GIRAUDEAU, J. & ROGERS, J. 1994. Phytoplankton biomass and sea-surface temperature estimates from sea-bed distribution of nannofossils and planktic foraminifera in the Benguela upwelling system. Micropaleontology, 40, 275–285.

HEMLEBEN, C. & BIJMA, J. 1994. Foraminiferal population dynamics and stable carbon isotopes. In: ZAHN, R., PEDERSEN, T.F., KAMINSKI, M.A. & LABEYRIE, L. (eds) Carbon Cycling in the Glacial Ocean: Constraints on the Ocean's Role in the Global Change. NATO ASI Series I, 17, 145–166.

HEMLEBEN, C., SPINDLER, M. & ANDERSON, O.R. 1989. Modern Planktonic Foraminifera. Springer, Berlin.

HUT, G. 1987. Stable Isotope Reference Samples for Geochemical and Hydrological Investigations. Consultants Group Meeting, 16-18.09.1985. International Atomic Energy Agency, Vienna.

HUTSON, W.H. & PRELL, W.L. 1980. A paleoecological transfer function, FI-2, for Indian Ocean planktonic foraminifera. Journal of Paleontology, 54, 381–399.

IVANOVA, E.M., 2000. Late Quaternary monsoon history and paleoproductivity of the western Arabian Sea. PhD thesis, Free University Amsterdam.

IVANOVA, E.M., CONAN, S.M-H., PEETERS, F.J.C. & TROELSTRA, S.R. 1999. Living Neogloboquadrina pachyderma sin and its distribution in the sediments from Oman and Somalia upwelling areas. Marine Micropaleontology, 36, 91–107.

KAHN, M.I. & WILLIAMS, D.F. 1981. Oxygen and carbon isotopic composition of living planktonic foraminifera from the Northern Pacific Ocean. Palaeogeography, Palaeoclimatology, Palaeoecology, 33, 47–69.

KEMLE-VON MÜCKE, S. 1994. *Oberflächenwasserstruktur und -zirkulation des Südostatlantiks im Spätquartär*. Berichte, Fachbereich Geowissenschaften, Universität Bremen, **55**.

KEMLE-VON MÜCKE, S. & OBERHÄNSLI, H. 1999. The distribution of living planktic foraminifera in relation to southeast Atlantic oceanography. *In:* FISCHER, G & WEFER, G. (eds) *Use of Proxies in Paleoceanography: Examples from the South Atlantic*. Springer, Berlin, 91–115.

KROON, D. & GANSSEN, G. 1989. Northern Indian Ocean upwelling cells and the stable isotope composition of living planktonic foraminifers. *Deep-Sea Research*, **36**, 1219–1236.

KROON, D., BEETS, K., MOWBRAY, S., SHIMMIELD, G. & STEENS, T. 1990. Changes in northern Indian Ocean monsoonal wind activity during the last 500 ka. *Memorie della Societa Geologica Italiana*, **44**, 189–207.

MCCREARY, J.P., KOHLER, K.E., HOOD, R.R. & OLSON, D.B. 1996. A four-component ecosystem model of biological activity in the Arabian Sea. *Progres in Oceanography*, **37**, 193–240.

MULITZA, S., WOLFF, T., PÄTZOLD, J., HALE, H. & WEFER, G. 1998. Temperature sensitivity of planktic foraminifera and its influence on the oxygen isotope record. *Marine Micropaleontology*, **33**, 223–240.

NAIDU, P.D. & MALMGREN, B.A. 1996a. Relationship between Late Quaternary upwelling history and coiling properties of *N. pachyderma* and *G. bulloides* in the Arabian Sea. *Journal of Foraminiferal Research*, **26**, 64–70.

NAIDU, P.D. & MALMGREN, B.A. 1996b. A high-resolution record of late Quaternary upwelling along the Oman Margin, Arabian Sea based on planktonic foraminifera. *Paleoceanography*, **11**, 129–140.

ORTIZ, J.D., MIX, A.C. & COLLIER, R.W. 1995. Environmental control of living symbiotic and asymbiotic foraminifera of the California Current. *Paleoceanography*, **10**, 987–1009.

OTTENS, J.J., 1992. *Planktic Foraminifera as indicators of ocean environments in the Northeast Atlantic*. PhD thesis, Free University Amsterdam.

PEETERS, F.J.C., BRUMMER, G.-J.A. & GANSSEN, G.M. 2002. The effect of upwelling on the distribution and stable isotope composition of *Globigerina bulloides* and *Globigerinoides ruber* (planktic foraminifera) in modern surface waters of the NW Arabian Sea. *Global and Planetary Change*, **35**, in press.

PEETERS, F.J.C., IVANOVA, E.M., CONAN, S.M.-H., BRUMMER, G.-J.A., GANSSEN, G.M., TROELSTRA, S.R. & VAN HINTE, J.E. 1999. A size analysis of planktic foraminifera from the Arabian Sea. *Marine Micropaleontology*, **36**, 31–63.

PRELL, W.L. & CURRY, W.B. 1981. Faunal and isotopic indices of monsoonal upwelling, Western Arabian Sea. *Oceanologica Acta*, **4**, 91–98.

PRELL, W.L., MURRAY, D.W., CLEMENS, S.C. & ANDERSON, D.M. 1992. Evolution and variability of the Indian Ocean summer monsoon: evidence from the Western Arabian Sea Drilling Programme. *In:* DUNCAN, R.A. ET AL. (eds) *Syntheses of Results from Scientific Drilling in the Indian Ocean*. Geophysical Monograph, American Geophysical Union, **70**, 447–469.

RAVELO, A.C. & FAIRBANKS, R.G. 1992. Oxygen isotopic composition of multiple species of planktonic foraminifera: recorders of the modern photic zone temperature gradient. *Paleoceanography*, **7**, 815–831.

RAVELO, A.C., FAIRBANKS, R.G. & PHILANDER, S.G.H. 1990. Reconstructing tropical Atlantic hydrography using planktonic foraminifera and an ocean model. *Paleoceanography*, **5**, 409–431.

ROSTEK, F., BARD, E., BEAUFORT, L., SONZOGNI, C. & GANSSEN, G. 1997. Sea surface temperature and productivity records for the past 240 kyr in the Arabian Sea. *Deep-Sea Research II*, **44**, 1461–1480.

SCHIEBEL, R., BIJMA, J. & HEMLEBEN, C. 1997. Population dynamics of the planktic foraminifer *Globigerina bulloides* from the eastern North Atlantic. *Deep-Sea Research*, **44**, 1701–1713.

SCHIEBEL, R., HILLER, B. & HEMLEBEN, C. 1995. Impacts of storms on recent planktic foraminiferal test production and $CaCO_3$ flux in the north Atlantic at 47°N, 20°W (JGOFS). *Marine Micropaleontology*, **26**, 115–129.

SCHULZ, H., VON RAD, U. & ITTEKKOT, V. 2002. Planktonic foraminifera, particle flux and oceanic productivity off Pakistan, NE Arabian Sea: modern analogues and application to the palaeoclimatic record. *In:* CLIFT, P.D., KROON, D., GAEDICKE, C. & CRAIG, J. (eds) *The Tectonic and Climatic Evolution of the Arabian Sea Region*. **195**, 499–516.

SPERO, H.J. & LEA, D. 1996. Experimental determination of stable isotope variability in *Globigerina bulloides*: implications for paleoceanographic reconstructions. *Marine Micropaleontology*, **28**, 231–246.

STEENS, T.N.F., GANSSEN, G. & KROON, D. 1992. Oxygen and carbon isotopes in planktic foraminifera as indicators of upwelling intensity and upwelling-induced high productivity in sediments from the north-western Arabian Sea. *In:* SUMMERHAYS, C.P., PRELL, W.C. & EMEIS, K.C. (eds) *Upwelling Systems: Evolution since the Early Miocene*. Geological Society, London, Special Publications, **64**, 107–119.

TABACHNICK, B.G. & FIDELL, L.S. 1996. *Using Multivariate Statistics*. HarperCollins, New York.

THIEDE, J. 1975. Distribution of foraminifera in surface waters of a coastal upwelling area. *Nature*, **253**, 712–714.

THUNELL, R. & REYNOLDS-SAUTTER, L. 1992. Planktonic foraminiferal faunal and stable isotopic indices of upwelling: asediment trap study in the San Pedro Basin, Southern California Bight. *In:* SUMMERHAYS, C.P., PRELL, W.C. & EMEIS, K.C. (eds) *Upwelling Systems: Evolution since the Early Miocene*. Geological Society, London, Special Publications, **64**, 77–91.

UFKES, E., JANSEN, J.H.F. & BRUMMER, G.-J.A. 1998. Living planktonic foraminifera in the eastern South

Atlantic during spring: indicators of water masses, upwelling and Congo (Zaire) River plume. *Marine Micropaleontology*, **33**, 27–53.

VAN HINTE J.E., VAN WEERING, T.J.C. & TROELSTRA, S.R. 1995. *Tracing a Seasonal Upwelling, Report on Two Cruises of RV* Tyro *to the NW Indian Ocean in 1992 and 1993*. National Museum of Natural History, Leiden.

WATKINS, J.M. & MIX, A.C. 1998. Testing the effects of tropical temperature, productivity, and mixed-layer depth on foraminiferal transfer functions. *Paleoceanography*, **13**, 96–105.

WATKINS, J.M., MIX, A.C. & WILSON, J. 1996. Living planktic foraminifera: tracers of circulation and productivity regimes in the central equatorial Pacific. *Deep-Sea Research II*, **43**, 1257–1282.

WATKINS, J.M., MIX, A.C. & WILSON, J. 1998. Living planktic foraminifera in the central tropical Pacific Ocean: articulating the equatorial 'cold tongue' during La Niña, 1992. *Marine Micropaleontology*, **33**, 157–174.

Appendix A. *Total planktic foraminifer shell concentration >125 μm (ind. m^{-3}) and species percentage data for C1 cruise, August 1992*

Station-cast-net	Date, time (GMT +3)	Dec. longitude (°E)	Dec. latitude (°N)	Depth start (m)	Depth end (m)	T_{start} (°C)	T_{end} (°C)	Vol. filt. (m$_3$)	Biomass (mg m^{-3})	Tot. shell conc. (ind. m^{-3})	B. digitata	D. anfracta	G. bulloides	G. falconensis	G. rubescens	G. adamsi	G. aequilateralis	G. calida	G. glutinata
308-1-5	19.08.92	52.50	16.14	23	0	17.53	20.49	103	51.7	120.4	0.0	0.3	75.8	0.1	0.0	0.0	1.2	11.5	1.7
308-1-4	9:05	52.50	16.14	48	23	16.89	17.53	55	13.2	28.6	0.0	0.0	39.2	0.2	0.2	0.0	6.7	30.0	0.8
308-1-3	9:25	52.50	16.14	72	48	16.38	16.89	48	8.0	18.5	0.0	0.3	44.7	0.0	0.0	0.1	3.6	29.7	1.5
308-1-2		52.50	16.14	98	72	15.83	16.38	66	3.2	5.2	0.0	0.0	53.7	1.2	0.0	0.0	1.2	15.8	10.0
309-1-5	20.08.92	52.64	16.09	8	0	20.64	20.70	36	60.5	107.2	0.0	0.0	87.9	0.0	0.0	0.0	0.9	3.0	1.9
309-1-4	9:39	52.64	16.09	18	8	18.49	20.64	26	25.9	25.3	0.0	0.0	45.1	0.6	0.0	0.0	4.5	29.4	3.0
309-1-3	9:50	52.64	16.09	28	18	17.70	18.49	25	17.8	20.7	0.0	0.0	26.5	0.5	0.0	0.0	3.6	48.5	2.1
309-1-2		52.64	16.09	50	28	16.71	17.70	42	11.7	9.4	0.0	0.0	18.6	0.3	0.0	0.0	2.7	51.7	2.7
310-5-5	20.08.92	52.73	16.02	8	0	19.96	20.68	41	62.9	129.5	0.0	0.2	85.8	0.8	0.0	0.0	0.8	3.7	2.6
310-5-4	15:19	52.73	16.02	23	8	17.77	19.96	34	23.9	52.7	0.0	0.5	43.3	2.5	0.0	0.0	3.9	34.3	2.4
310-5-3	15:37	52.73	16.02	49	23	16.78	17.77	44	12.7	17.9	0.2	0.0	33.4	0.5	0.0	0.0	4.7	43.6	0.3
310-5-2		52.73	16.02	74	49	16.38	16.78	40	5.4	12.2	0.0	0.0	54.0	0.3	0.0	0.0	1.6	24.0	1.9
310-5-1		52.73	16.02	100	74	15.88	16.38	33	4.2	6.5	0.6	0.0	55.0	1.3	0.0	0.0	1.3	21.3	0.6
310-6-4	20.08.92	52.75	16.00	148	98	15.43	15.88	35	6.5	6.1	1.9	1.3	47.5	1.9	0.0	0.0	2.5	21.9	3.1
310-6-3	17:12	52.75	16.00	200	148	14.65	15.43	53	4.9	5.3	0.9	0.0	50.0	1.4	0.0	0.0	1.9	18.9	0.9
310-6-2	17:08	52.75	16.00	298	200	13.60	14.65	104	4.9	4.6	0.0	0.8	58.6	0.0	0.0	0.0	1.7	9.3	3.1
310-6-1		52.75	16.00	498	298	11.98	13.60	270	2.9	4.4	0.3	3.6	63.0	2.7	0.0	0.0	1.3	3.4	1.8
311-1-5	21.08.92	52.75	16.00	8	0	19.96	20.68	52	60.2	93.1	0.0	0.0	93.1	0.0	0.0	0.0	0.3	1.3	0.7
311-1-4	9:30	52.75	16.00	23	8	17.77	19.96	50	29.1	14.6	0.0	0.0	47.5	0.2	0.0	0.0	0.5	44.3	0.0
311-1-3	9:50	52.75	16.00	50	23	16.74	17.77	132	11.4	16.6	0.0	0.0	38.3	1.0	0.0	0.0	1.6	48.9	0.0
311-1-2		52.75	16.00	74	50	16.39	16.74	124	7.3	14.4	0.0	0.0	35.9	0.0	0.0	0.0	1.8	54.5	0.0
311-1-1		52.75	16.00	96	75	15.88	16.39	36	4.2	6.7	0.0	0.0	20.9	0.0	0.0	0.0	0.5	67.6	1.6
313-4-5	21.08.92	53.02	15.91	8	0	25.33	25.75	25	259.9	331.7	0.0	1.0	66.0	0.2	0.0	0.0	1.4	2.1	12.2
313-4-4	11:54	53.02	15.91	24	10	19.38	25.33	41	171.9	176.5	0.0	0.6	55.9	0.1	0.0	0.0	2.3	3.0	21.3
313-4-3	12:10	53.02	15.91	49	24	18.26	19.38	87	34.8	54.5	0.0	0.8	18.0	0.0	0.0	0.0	3.4	28.6	21.9
313-4-2		53.02	15.91	74	50	17.18	18.26	87	18.8	63.3	0.0	0.8	46.0	0.0	0.0	0.0	0.9	11.7	11.1
313-4-1		53.02	15.91	100	75	16.52	17.18	60	9.3	46.8	0.0	0.3	55.7	0.2	0.0	0.0	0.7	8.3	7.4
313-dc-4	22.08.92	52.94	15.88	148	99	15.37	16.52	94	6.2	33.7	0.0	2.5	55.6	0.3	0.0	0.0	0.5	2.5	14.5
313-dc-3	12:45	52.94	15.88	196	150	14.45	15.37	98	9.8	52.8	0.0	2.1	58.7	0.4	0.0	0.0	0.9	2.7	17.7
313-dc-2	13:49	52.94	15.88	300	199	13.42	14.45	278	14.2	37.3	0.0	3.7	48.8	1.3	0.0	0.0	1.6	2.5	17.0
313-dc-1		52.94	15.88	498	300	11.75	13.42	478	10.9	58.7	0.0	1.5	55.9	0.3	0.0	0.0	0.6	0.8	13.7

G. uvula	G. conglobatus	G. ruber	G. tenellus	G. sacculifer (−)	G. sacculifer (+)	G. menardii	G. scitula	G. theyeri	G. tumida	G. hexagona	H. parapelagica	N. dutertrei	N. pach (sin)	O. universa	P. obliquiloculata	T. iota	T. humillis	T. quinqueloba	Aberrant	P.F. indet.
0.3	0.0	0.1	0.0	0.0	0.0	3.3	0.0	0.0	0.0	0.0	0.2	2.6	0.0	0.0	0.0	1.0	0.0	0.3	0.3	1.3
0.2	0.0	0.0	0.0	0.0	0.0	8.4	0.0	0.0	0.0	0.0	1.8	2.5	0.0	0.0	0.0	7.0	0.0	0.8	1.1	1.1
0.3	0.0	0.1	0.1	0.0	0.0	4.0	0.4	0.0	0.0	0.0	1.2	1.5	0.0	0.0	0.0	1.2	0.0	9.9	0.7	0.4
0.4	0.0	3.9	0.0	0.0	0.0	4.6	1.5	0.0	0.0	0.8	0.8	2.3	0.0	0.0	0.0	0.4	0.0	2.3	0.4	0.8
0.3	0.0	0.0	0.0	0.0	0.0	0.2	0.0	0.0	0.0	0.0	0.0	2.4	0.0	0.0	0.0	0.0	0.0	2.8	0.7	0.0
0.0	0.0	0.4	0.4	0.2	0.0	4.5	0.0	0.0	0.0	0.0	1.4	4.7	0.2	0.0	0.0	4.0	0.0	1.0	0.4	0.2
0.5	0.0	0.0	0.0	0.3	0.0	6.4	0.3	0.0	0.0	0.0	3.1	3.6	0.0	0.0	0.0	2.8	0.0	1.5	0.0	0.3
0.0	0.0	0.7	0.3	0.3	0.0	16.2	0.0	0.0	0.0	0.3	1.4	2.7	0.3	0.0	0.0	0.0	0.0	1.4	0.0	0.3
0.0	0.0	0.0	0.4	0.0	0.0	1.1	0.0	0.0	0.0	0.0	0.1	2.9	0.0	0.0	0.0	0.4	0.0	1.0	0.0	0.2
0.0	0.0	0.0	0.0	0.0	0.0	4.6	0.0	0.0	0.0	0.0	1.6	3.6	0.1	0.0	0.1	1.3	0.0	1.0	0.4	0.5
0.0	0.0	0.0	0.0	0.0	0.0	4.9	0.3	0.0	0.0	0.0	3.0	2.4	0.2	0.0	0.0	2.7	0.0	2.9	0.5	0.3
0.0	0.0	0.5	0.0	0.0	0.0	12.3	0.5	0.0	0.0	0.0	0.3	1.4	0.3	0.0	0.0	0.0	0.0	2.5	0.5	0.0
0.0	0.0	0.6	0.6	0.0	0.0	8.8	0.0	0.0	0.0	0.0	4.4	1.3	1.3	0.0	0.0	1.3	0.0	1.3	0.0	0.6
0.0	0.0	0.6	0.0	0.6	0.0	5.6	0.0	0.6	0.0	0.6	3.8	4.4	0.0	0.0	0.0	0.0	0.0	3.1	0.0	0.6
0.0	0.0	0.9	0.5	0.0	0.0	6.6	0.0	1.4	0.0	0.9	1.9	3.8	0.9	0.0	0.0	0.0	0.0	9.0	0.0	0.0
0.0	0.0	0.8	0.0	0.0	0.0	10.1	0.0	0.6	0.0	2.8	0.3	3.1	1.4	0.0	0.3	0.0	0.0	5.6	0.3	1.1
0.0	0.0	1.1	0.4	0.2	0.0	7.0	0.4	0.0	0.0	3.4	0.7	4.2	1.6	0.0	0.0	0.9	0.0	2.9	0.1	0.9
0.2	0.0	0.2	0.2	0.3	0.0	0.2	0.0	0.0	0.0	0.0	0.0	1.0	0.0	0.0	0.0	0.0	0.0	2.6	0.0	0.0
0.0	0.0	0.0	0.0	0.0	0.0	0.2	0.0	0.0	0.0	0.0	2.4	1.1	0.0	0.0	0.0	0.4	0.0	1.8	0.9	0.7
0.0	0.0	0.0	0.0	0.0	0.0	0.0	0.0	0.2	0.0	0.0	2.7	1.2	0.0	0.0	0.0	0.0	0.0	3.9	0.4	1.9
0.1	0.0	0.1	0.0	0.0	0.0	0.4	0.0	0.0	0.0	0.0	1.1	0.6	0.0	0.0	0.0	0.7	0.0	2.7	0.4	1.5
0.0	0.0	1.1	0.0	0.0	0.0	2.2	0.0	0.0	0.0	0.0	1.1	3.3	0.0	0.0	0.0	0.0	0.0	0.5	0.0	1.1
0.6	0.0	6.3	1.0	0.8	0.0	0.6	0.0	0.0	0.0	0.0	0.0	5.3	0.2	0.0	0.0	0.2	0.0	0.0	1.6	0.6
0.0	0.0	5.3	1.5	1.8	0.0	0.6	0.0	0.0	0.0	0.0	0.4	5.2	0.1	0.0	0.2	0.6	0.0	0.4	0.5	0.3
0.1	0.0	3.7	2.4	1.9	0.1	3.1	0.0	0.0	0.0	0.0	2.6	10.1	0.0	0.1	0.0	1.0	0.0	1.1	0.1	1.0
0.2	0.0	3.9	0.6	0.6	0.0	4.3	0.8	0.0	0.0	0.0	0.8	4.0	2.3	0.0	0.2	0.4	0.0	7.6	1.6	2.4
1.0	0.0	2.7	0.6	0.3	0.0	11.5	0.8	0.0	0.0	0.1	0.6	2.1	2.2	0.0	0.1	0.3	0.0	2.4	1.1	1.7
0.0	0.0	1.7	0.5	0.4	0.0	6.8	1.6	0.4	0.0	0.9	1.3	1.4	5.9	0.0	0.0	0.1	0.0	2.0	0.3	0.7
0.1	0.0	1.3	0.6	0.2	0.0	3.1	0.6	0.1	0.0	0.0	1.6	1.7	5.2	0.0	0.2	0.2	0.0	2.4	0.1	0.5
0.2	0.0	1.4	0.8	0.1	0.0	1.6	0.6	0.1	0.0	0.4	1.1	1.1	8.7	0.0	0.0	0.0	0.0	8.1	0.0	0.9
0.3	0.0	1.2	0.8	0.2	0.0	1.6	0.2	0.2	0.0	1.0	0.0	1.3	11.7	0.0	0.3	0.3	0.0	6.5	0.2	1.4

Appendix B. *Planktic foraminifer shell concentration >125 μm (ind. m^{-3}) and species percentage data for C2 cruise, August 1992*

Station-cast-et	Date, time (GMT +3)	Dec. longitude (°E)	Dec. latitude (°N)	Depth start (m)	Depth end (m)	T_{start} (°C)	T_{end} (°C)	Vol. filt. (m$_3$)	Biomass (mg m^{-3})	Tot. shell conc. (ind. m^{-3})	B. digitata	D. anfracta	G. bulloides	G. falconensis	G. rubescens	G. adamsi	G. aequilateralis	G. calida	G. glutinata
917-2-5	25.02.93	52.92	15.90	12	0	25.71	25.80	21	40.7	19.2	0.0	0.0	1.0	0.7	0.0	0.0	20.8	3.0	8.6
917-2-4	8:50	52.92	15.90	27	12	24.89	25.71	22	66.2	17.5	0.0	0.0	2.8	0.7	0.0	0.0	12.2	14.6	4.2
917-2-3	9:15	52.92	15.90	52	27	24.32	24.89	43	80.8	49.1	0.0	0.0	6.8	1.1	0.0	0.0	17.0	7.3	13.8
917-2-2		52.92	15.90	78	52	23.09	24.32	79	19.6	17.1	0.0	0.2	6.3	9.6	0.0	0.0	8.6	12.5	10.8
917-2-1		52.92	15.90	102	78	19.96	23.09	68	12.1	16.7	0.0	0.1	2.7	6.4	0.0	0.0	9.4	19.5	5.9
917-1-4	25.02.93	52.97	15.89	153	101	16.26	19.96	75	10.1	8.0	0.0	0.7	6.4	4.2	0.0	0.0	6.4	41.3	8.4
917-1-3	7:23	52.97	15.89	203	153	15.67	16.26	75	10.7	12.5	2.2	0.0	8.7	7.4	0.2	0.0	4.2	22.4	13.7
917-1-2	8:15	52.97	15.89	302	203	13.75	15.67	184	2.8	3.8	0.6	0.2	3.6	26.0	0.0	0.0	5.6	20.9	11.9
917-1-1		52.97	15.89	502	302	11.88	13.75	240	6.6	7.7	1.2	0.6	5.1	15.1	0.0	0.0	6.5	10.3	18.9
918-2-5	26.02.93	52.84	15.96	12	0	25.86	25.95	35	43.2	25.3	0.0	0.0	1.2	0.2	0.0	0.0	20.8	0.2	8.6
918-2-4	7:05	52.84	15.96	27	12	25.22	25.86	47	71.1	21.9	0.0	0.0	0.6	0.0	0.0	0.0	26.7	0.7	10.9
918-2-3	8:30	52.84	15.96	52	27	24.60	25.22	97	76.1	26.5	0.0	0.0	3.4	0.5	0.0	0.0	29.6	10.3	14.1
918-2-2		52.84	15.96	78	52	22.56	24.60	132	23.9	11.6	0.0	0.0	5.5	1.2	0.0	0.0	45.1	15.8	8.8
918-2-1		52.84	15.96	103	78	19.25	22.56	101	14.1	11.5	0.0	0.0	7.2	3.4	0.0	0.0	17.4	17.5	7.0
919-1-5	26.02.93	52.73	16.00	10	0	25.76	25.90	12	61.0	32.6	0.0	0.0	2.0	0.0	0.0	0.0	16.7	1.4	12.3
919-1-4	16:47	52.73	16.00	26	10	25.59	25.76	19	22.3	24.1	0.0	0.0	2.6	0.3	0.0	0.0	19.2	1.2	12.8
919-1-3	—	52.73	16.00	52	26	24.36	25.59	50	89.4	55.4	0.0	0.0	5.9	0.6	0.0	0.0	23.7	4.4	11.2
919-1-2		52.73	16.00	77	52	23.52	24.36	91	23.6	21.6	0.0	0.0	7.8	1.8	0.0	0.0	35.5	13.5	11.7
919-1-1		52.73	16.00	102	77	19.40	23.52	70	9.5	12.3	0.0	0.2	7.3	4.0	0.0	0.0	22.6	20.5	6.2
920-2-5		52.64	16.08	11	0	25.52	25.52	16	54.0	75.2	0.0	0.0	2.4	0.0	0.0	0.0	22.0	4.1	11.5
920-2-4	27.02.93	52.64	16.08	27	11	25.49	25.52	46	36.3	27.9	0.0	0.0	1.6	0.0	0.0	0.0	23.0	4.5	9.8
920-2-3	8:45	52.64	16.08	53	27	24.60	25.49	79	52.1	34.6	0.0	0.2	2.1	0.6	0.0	0.0	29.1	6.4	12.1
920-2-2	9:15	52.64	16.08	78	53	23.11	24.60	135	15.9	10.4	0.0	0.4	6.3	0.9	0.2	0.0	23.1	20.5	8.2
920-2-1		52.64	16.08	102	78	19.93	23.11	105	12.2	7.0	0.0	0.0	2.5	1.0	0.0	0.0	14.0	30.0	8.9
920-1-4	27.02.93	52.70	16.09	153	103	17.48	19.93	168	7.0	5.7	0.0	0.0	4.3	0.4	0.0	0.0	12.4	49.4	10.7
920-1-3	7:15	52.70	16.09	202	153	16.53	17.48	146	7.7	5.6	0.0	0.2	2.9	1.4	0.0	0.0	9.2	42.9	10.4
920-1-2	8:15	52.70	16.09	303	202	13.98	16.53	269	27.9	4.0	1.3	0.9	4.4	1.9	0.0	0.0	10.9	22.7	10.0
920-1-1		52.70	16.09	501	303	12.56	13.98	282	15.1	6.1	11.7	0.6	4.3	3.3	0.0	0.0	5.6	10.0	12.9
922-2-5		52.52	16.17	12	0	25.54	25.55	48	77.2	85.0	0.0	0.0	4.9	0.4	0.0	0.0	3.9	3.1	14.5
922-2-4	27.02.93	52.52	16.17	26	12	25.53	25.54	58	67.2	59.5	0.0	0.0	4.8	0.2	0.0	0.0	3.6	2.8	15.8
922-2-3	7:00	52.52	16.17	52	26	24.56	25.53	106	45.7	33.1	0.0	0.2	6.8	1.0	0.0	0.0	9.1	12.0	16.7
922-2-2	—	52.52	16.17	77	52	23.75	24.56	149	16.3	9.5	0.0	0.0	5.3	4.8	0.0	0.1	11.1	19.0	12.6
922-2-1		52.52	16.17	102	77	21.00	23.75	169	4.7	2.9	0.0	0.0	7.2	6.4	0.0	0.0	11.8	24.1	8.3

G. uvula	G. conglobatus	G. ruber	G. tenellus	G. sacculifer (−)	G. sacculifer (+)	G. menardii	G. scitula	G. theyeri	G. tumida	G. hexagona	H. parapelagica	N. dutertrei	N. pach (sin)	O. universa	P. obliquiloculata	T. iota	T. humillis	T. quinqueloba	Aberrant	P.F. indet.
0.0	0.0	13.5	3.0	44.2	0.0	1.0	0.0	0.0	0.0	0.0	0.0	2.0	0.0	0.3	0.0	0.0	0.0	0.0	0.0	2.0
0.0	0.0	14.6	3.5	44.8	0.0	0.0	0.0	0.0	0.0	0.0	0.7	0.7	0.0	0.3	0.0	0.0	0.0	0.0	0.7	0.3
0.0	0.0	32.6	6.7	9.4	0.4	0.2	0.0	0.0	0.0	0.0	0.0	3.2	0.0	0.4	0.1	0.0	0.0	0.0	0.0	1.1
0.2	0.0	30.4	14.2	3.0	0.2	0.0	0.0	0.0	0.0	0.0	0.0	1.0	0.0	0.7	0.0	0.0	0.0	0.0	0.2	2.2
0.0	0.0	34.9	16.1	0.1	0.0	0.0	0.1	0.1	0.0	0.6	0.0	0.6	0.0	0.0	0.0	0.0	0.1	0.0	1.5	1.9
0.0	0.0	19.1	6.0	2.0	0.0	0.0	0.0	0.0	0.0	0.2	2.4	0.7	0.0	0.7	0.2	0.0	0.0	0.0	0.4	0.7
0.0	0.0	22.6	12.5	1.0	0.1	0.0	0.0	0.0	0.0	1.3	1.1	0.3	0.0	0.0	0.4	0.0	0.0	0.0	0.2	1.5
0.0	0.0	18.6	5.8	2.3	0.0	0.0	0.0	0.0	0.0	0.0	1.3	0.8	0.0	0.0	0.2	0.0	0.0	0.2	0.2	1.9
0.0	0.0	18.4	11.8	0.7	0.0	0.1	0.2	0.0	0.0	9.1	0.3	0.0	0.0	0.1	0.0	0.0	0.0	0.0	0.1	1.4
0.0	0.0	27.5	6.6	30.8	0.2	0.6	0.0	0.0	0.0	0.0	0.0	2.1	0.0	0.6	0.0	0.0	0.0	0.0	0.0	0.6
0.0	0.0	20.3	5.0	32.2	0.1	1.0	0.0	0.0	0.0	0.0	0.0	1.7	0.0	0.4	0.0	0.0	0.0	0.0	0.0	0.3
0.0	0.0	22.3	6.2	9.2	0.4	0.1	0.0	0.0	0.0	0.1	0.0	2.5	0.0	0.8	0.0	0.0	0.0	0.0	0.2	0.4
0.0	0.0	13.6	3.6	3.4	0.0	0.0	0.0	0.0	0.0	0.0	0.5	1.3	0.0	0.4	0.1	0.0	0.0	0.0	0.0	0.6
0.3	0.0	27.2	13.7	2.5	0.1	0.1	0.0	0.1	0.0	0.3	0.7	1.3	0.0	0.2	0.0	0.0	0.0	0.2	0.0	0.7
0.0	0.0	35.2	4.1	24.9	0.0	0.7	0.0	0.0	0.0	0.0	0.0	1.0	0.0	0.3	0.0	0.0	0.0	0.0	0.0	1.4
0.0	0.0	34.7	4.1	20.4	0.0	0.9	0.0	0.0	0.0	0.0	0.0	2.0	0.0	0.3	0.0	0.0	0.0	0.0	0.0	1.5
0.0	0.0	28.3	6.4	14.4	0.3	0.4	0.0	0.0	0.0	0.0	0.0	2.7	0.0	0.7	0.0	0.0	0.0	0.0	0.0	1.0
0.1	0.0	17.2	2.9	3.7	0.6	0.0	0.0	0.0	0.0	0.0	0.0	1.9	0.0	1.2	0.1	0.1	0.0	0.1	0.1	1.6
0.0	0.0	22.5	3.7	6.0	0.5	0.0	0.2	0.0	0.0	0.3	0.0	1.9	0.0	1.1	0.2	0.0	0.0	0.2	0.0	2.8
0.0	0.2	22.6	5.8	19.6	0.1	4.5	0.0	0.0	0.0	0.0	1.1	4.0	0.0	0.2	0.0	0.0	0.0	0.0	0.0	1.8
0.0	0.0	20.5	7.1	20.2	0.0	7.8	0.0	0.0	0.0	0.0	0.0	4.3	0.0	0.6	0.3	0.0	0.0	0.0	0.2	0.2
0.0	0.0	20.2	7.7	11.9	0.2	5.0	0.0	0.0	0.0	0.0	0.2	2.9	0.0	0.5	0.0	0.0	0.0	0.0	0.0	0.6
0.0	0.0	16.8	10.4	4.5	0.7	1.6	0.0	0.0	0.0	0.0	0.6	1.2	0.5	0.6	0.4	0.2	0.0	1.3	0.1	1.5
0.0	0.0	14.1	21.0	2.7	0.2	1.1	0.2	0.0	0.0	0.4	1.3	0.7	0.2	0.0	0.2	0.0	0.0	0.0	0.4	1.1
0.0	0.0	9.1	1.8	2.4	0.0	0.4	0.6	0.3	0.0	0.1	4.2	1.3	0.0	0.6	0.3	0.0	0.0	0.0	0.1	1.7
0.2	0.0	13.2	4.4	4.9	0.0	0.2	0.5	0.4	0.0	0.2	6.6	0.6	0.2	0.2	0.0	0.3	0.0	0.2	0.0	1.0
1.5	0.0	17.4	13.8	4.7	0.6	0.4	0.8	0.0	0.0	2.6	2.2	1.4	0.0	0.3	0.0	0.2	0.2	0.2	0.3	1.0
0.0	0.0	19.4	14.2	1.8	0.0	0.0	0.8	0.0	0.0	12.7	0.5	0.4	0.0	0.0	0.0	0.0	0.0	0.2	0.1	1.3
0.0	0.0	36.4	7.7	22.3	0.1	1.4	0.0	0.0	0.0	0.0	0.0	4.2	0.0	0.1	0.0	0.0	0.0	0.0	0.1	0.8
0.0	0.0	38.1	4.0	24.6	0.0	1.8	0.0	0.0	0.0	0.0	0.0	3.3	0.0	0.2	0.1	0.0	0.0	0.0	0.3	0.5
0.0	0.0	25.4	6.3	13.3	0.1	3.3	0.0	0.0	0.0	0.0	0.0	3.5	0.0	0.2	0.2	0.0	0.0	0.0	0.0	2.0
0.0	0.0	17.7	16.0	6.9	0.0	1.5	0.0	0.0	0.0	0.2	0.0	2.0	0.0	0.0	0.0	0.0	0.0	0.0	0.4	2.3
0.0	0.0	23.3	6.2	5.9	0.0	1.3	0.0	0.0	0.0	0.3	0.0	2.7	0.0	0.3	0.3	0.0	0.0	0.0	0.3	1.6

Planktic foraminifera, particle flux and oceanic productivity off Pakistan, NE Arabian Sea: modern analogues and application to the palaeoclimatic record

HARTMUT SCHULZ[1], ULRICH VON RAD[2] & VENUGOPALAN ITTEKKOT[3]

[1] *IOW Institut für Ostseeforschung Warnemünde, PF 301161, D-18119 Rostock, Germany*
[2] *Bundesanstalt für Geowissenschaften und Rohstoffe BGR, PF 510153, D-30631 Hannover, Germany*
[3] *Zentrum für Marine Tropenökologie ZMT, Fahrenheitstraße 1, D-28 359 Bremen, Germany*

Abstract: We use the flux of bulk sediment ($CaCO_3$, biogenic opal, organic carbon, lithogenic material), and of planktic foraminifera (PF) and other shell-bearing plankton from sediment trap EPT-2 off Pakistan to (1) constrain the seasonal pattern of regional productivity and (2) search for indications of the NE monsoon winter situation that may serve as a modern analogue to better reveal the seasonal climatic signals preserved in the sedimentary record of the Arabian Sea. Our trap data show a clear seasonality of fluxes that can also be traced in the composition of non-bioturbated (varved) summer and winter sediment laminae preserved within the oxygen minimum zone. In EPT-2, the flux of PF is low during summer, but during winter and late spring it is higher, as at trap station WAST, in the upwelling area of the western Arabian Sea. *Globigerina bulloides*, a PF species linked to summer upwelling and high productivity, is of minor importance off Pakistan. In contrast, *Globigerina falconensis* dominates in flux and relative abundance, and is indicative of winter mixing, when NE monsoonal winds cool the highly saline surface waters and break up stratification. An enhanced horizontal flux of suspended sediments stirred up on the shelf and upper slope is clearly shown by the peak in occurrence of small benthic foraminifera during winter. Altogether, our data suggest that the particle flux in the northeastern Arabian Sea is determined by local sediment resuspension and winter productivity rather than by summer monsoonal upwelling, representing a 'non-upwelling' environment, in contrast to the 'summer upwelling' regime off Oman, Somalia and southern India. We used this evidence to reconstruct the seasonal intensity of both monsoons for the past 25 ka: the SW and NE monsoon both were weak during the last glacial period. The NE monsoon peaked during the cool phases of the glacial to interglacial climatic transition (i.e. during the Younger Dryas (YD) and Heinrich Event H1). The SW monsoon was reinforced after the YD. Both monsoons were enhanced during early Holocene time, when summer insolation and hence atmospheric forcing was at a maximum.

Various studies from the northwestern Indian Ocean have documented the high biological productivity of coastal and open ocean upwelling off Oman and Somalia, driven by the strong SW monsoonal winds during summer that evolve as a result of the strong pressure gradient between the continental low-pressure cell over central Asia and the high-pressure cell over the subtropical SW Indian Ocean (Zeitschel & Gerlach 1973; Bauer *et al.* 1991; Clemens *et al.* 1991; Brock *et al.* 1992; Rixen *et al.* 2000). Winds over the western Arabian Sea reach velocities of 30 m s^{-1} during the SW monsoon, but are much weaker during the winter NE monsoon, at 2–4 m s^{-1} (Ramage *et al.* 1972). Sediment trap data (Fig. 1) confirm that up to 58% of the total particle flux and 78% of planktic foraminiferal flux is confined to the summer SW monsoon i.e. June–October (Nair *et al.* 1989; Curry *et al.* 1992; Haake *et al.* 1993).

As a result of high monsoonal productivity, the Arabian Sea is characterized by a pronounced mid-depth oxygen minimum zone (OMZ). The primary reason for the very low oxygen concentration at intermediate depths is the high oxygen consumption by microbial decomposition of sinking organic matter from the photic zone. Other factors, such as the advection of low-oxygenated waters from the south, 'preformed' outside the Arabian Sea basin, and stratification by highly saline (but relatively well-oxygenated) water masses entering from the Red Sea and Persian Gulf, may lead to a reduced vertical mixing and sluggish circulation, and thus contribute to the stability of the OMZ (Olson et al. 1993). The OMZ is strongest in the northeastern part of the Arabian Sea, off Pakistan and northern India, where extremely low oxygen concentrations of only 0.2 ml l^{-1} between 200 and 1000 m water depth result in large parts of the continental slope and topographic heights having dysaerobic to nearly anoxic conditions (Fig. 1). In the western and central Arabian Basin, oxygen concentrations drop to comparable low values. However, the OMZ is much reduced to a core layer between 300–350 and 700 m water depth (Wyrtki 1971).

The question arises: what factors drive the OMZ to be so pronounced in the northeastern Arabian Sea, apart from the western upwelling areas of high summer productivity? On the one hand, Schulz et al. (1996) and Andruleit et al. (2000) suggested that the eastward drift of highly productive surface and subsurface water cells, which follow the strong anticyclonic surface gyre during summer, could lead to high fluxes in the northeastern Arabian Sea and hence fuel the OMZ. On the other hand, Banse & McClain (1986) and Madhupratab et al. (1996), stressed the importance of the NE monsoon for the regional productivity, and hence for the maintenance of the strong OMZ in the northeastern Arabian Sea. On the basis of water column temperature and biological data, those workers showed that surface cooling during winter reduces thermal stratification. Mixing to a depth of >120 m favours the injection of nutrients into the photic zone, and allows high productivity in winter. These observations agree well with the Coastal Zone Color Scanner satellite data: a winter field of high chlorophyll concentrations sets up in the open waters in the northernmost Arabian Sea in December and culminates in March, at the end of the NE monsoon.

Additional evidence for the importance of the NE monsoon for the extreme oxygen minimum comes from the study of sediments from the Pakistan margin, which are characterized by a distinct lamination. At least for the past 5 ka there is a continuous deposition of 'varved' couplets of thin, dark, organic carbon-rich and several times thicker, light, organic carbon-poor laminae (von Rad et al. 1999a; Doose-Rolinski et al. 2001; Lückge et al. 2001, 2002). Von Stackelberg (1972) first concluded that a major part of these sediments deposited along the Indian and Pakistan continental margin may be of allochthonous origin, and hence might be the product of high lateral flux of resuspended matter from the shelf and upper continental slope areas. Schulz et al. (1996) and von Rad et al. (1999a) presented a more detailed model of varve formation in the OMZ. These authors pointed out that major flux of lithogenic matter during winter may be reflected in the light, organic carbon-poor suspension layers of up to several millimetres thickness, whereas the

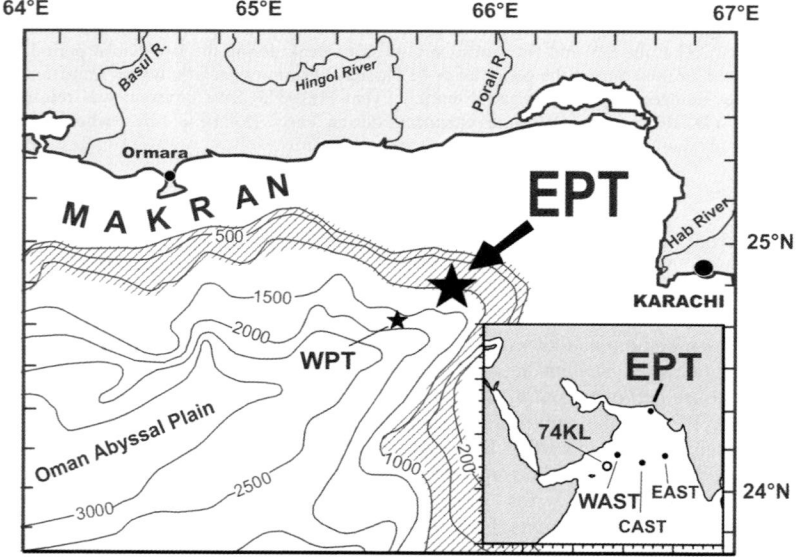

Fig. 1. Area and location of traps EPT-2 and WPT (Eastern and Western Pakomin Trap) in the oxygen minimum zone (stippled area) off Pakistan. Inset map: locations of Arabian Sea traps WAST, CAST and EAST (Curry et al. 1992; Haake et al. 1993) and of sediment core SONNE42-74KL (Sirocko et al. 1993; Schulz 1995).

flux of organic matter during summer and autumn may result in the deposition of thinner, dark, organic carbon-rich summer laminae. This observation is corroborated by microscopic analyses of thin-sections of laminated sediment, which clearly show a cyclic pattern, presumably associated with changes in the amount and composition of the flux during summer and winter months (Schulz et al. 1996).

Although the strong contrast between these two types of layers is evident from numerous sediment profiles, the origin of the large amount of allochthonous matter, and hence the mechanism of varve formation, is still under debate. Schulz et al. (1996) argued that during winter, large amounts of sediments deposited on the shelf are stirred up by winter storms and are transported as suspensions across the shelf edge, finally deposited as suspensates in the deeper waters. However, Staubwasser & Sirocko (2001) and Lückge et al. (2001, 2002) invoked mud volcanoes or heavy rainfall during winter as possible triggers of these high-sedimentation events along the steep Makran continental margin.

Recently, Andruleit et al. (2000) first presented flux data from two trap moorings EPT and WPT (Western and Eastern Pakomin Trap), deployed in 1993–1994 and 1995–1996 about 70 km offshore Pakistan (Fig. 1). A strong seasonality in the flux of organic and lithogenic matter was observed, being highest during the months January and February (up to >4000 mg m^{-2} day^{-1}), and also during August and September. Lithogenic matter dominated the flux, at 63–77%. Andruleit et al. (2000) suggested that a significant part of the trapped material may originate from lateral advection. However, the presence of well-preserved coccospheres and consistent abundance patterns of individual species from a total of four deployments show that the flux of biogenic particles is not only the result of local resuspension, but must be also attributed to changes in regional productivity and related oceanographic and biological processes.

In this study, we further investigate the role of the NE and SW monsoons and their imprint on the flux and sedimentary record in the northeastern Arabian Sea. Data sets from the longest sediment trap record available (EPT 2; May 1995–February 1996) are combined with new evidence from oceanographic and sedimentary data (Table 1). Emphasis is on the flux and relative abundance changes of planktic foraminiferal (PF) species. Finally, our results are used to shed light on the palaeointensity of the NE and SW monsoons in the Arabian Sea for the past 25 ka, during the last major shift in global climate from glacial to interglacial conditions.

Material and methods

CTD and sediment trap logistics

Water temperature profiles of the upper water column were assembled to assess the seasonal hydrographic dynamics off Pakistan and the physical process of deep mixing during winter. Most of the conductivity–temperature–depth (CTD) data were taken between the months of September and May during recent cruises of R.V.s *Tyro*, *Sonne* and *Meteor* (Table 1). Onboard R.V. *Sonne* a SEABIRD SEACAT SBE and a CTD-probe with an oxygen sensor by ME Meerestechnik were used. Further details of cruises IIOE-METEOR and R.V. *Tyro* have been given by Dietrich et al. (1966) and van der Linden et al. (1993), respectively.

A cone-shaped trap of type MARK 7G-2 (Honjo & Doherty 1988) was deployed within the OMZ at 590 m water depth. The trap was equipped with 21 cups, and the sampling interval

Table 1. *Overview of sample material used in this study*

Site	Latitude (N)	Longitude (E)	Water depth (m)	Sampling time	Reference
CTD stations					
IIOE Station 236	23°25′	065°53′	1225	12.03.65	Dietrich et al. 1966
TYRO92-474	23°49.7′	065°00.2′	1472	22.10.92	van der Linden et al. 1993
SONNE90-147MS	24°36.3′	065°35.3′	1958	24.09.93	T. Jennerjahn, pers comm.
SO119-EPT	24°44.6′	065°48.6′	1237	14.05.97	This study
SO123-CTD 01	24°03.0′	062°20.0′	3288	09.09.97	E. Flüh, pers. comm.
SO124-CTD 01	24°08.4′	062°28.6′	3300	08.10.97	H. Villinger, pers. comm
SO130-290MS	24°45.4′	065°35.8′	1393	22.04.98	U. Berner, pers. comm.
Trap station					
EPT-2	24°45.6′	065°48.7′	1166	05.05.95–28.02.96	Andruleit et al. 2000; this study
Sediment core					
SONNE 42-74KL	14°19.3′	057°20.8′	3212	0–25 ka	Sirocko et al. 1993; Schulz 1995

was 22 days. The sampling period for EPT-2 was from 5 May 1995 to 11 October 1996. This trap revealed the longest record of 264 days, ending on 28 February 1996 with two cups filled nearly to the top. Like the three Pakistan traps deployed about 18 months before, EPT-2 was clogged, presumably as a result of an extremely high flux in winter (Andruleit et al. 2000). Therefore, no record is available for the transition from the late winter to spring intermonsoon. Further details of trap and sample preparation have been given by Haake et al. (1993). For methods used for bulk sediment analysis of lithogenic and biogenic fluxes (flux of total particles, of calcium carbonate, of opal, of particulate organic carbon (POC), and of lithogenic matter), readers are referred to Haake et al. (1993). Parts of these flux data have been recently published by Andruleit et al. (2000).

Coarse fraction and foraminiferal analysis

We used a 1 mm sieve to remove the large swimmers in the 12 samples from EPT-2. Aliquots of 1/16 to 1/64 were obtained by a rotary splitter and filtered through a Nucleopore filter. The residue together with the filter was oven-dried immediately at 40 °C, weighed, and soaked overnight in distilled water. The residue was carefully washed from the filter and wet-sieved over a 63 μm mesh. The fraction >63 μm was dried again at 40 °C.

The components from this fraction were identified under a binocular microscope as seven categories: intact planktic foraminifera and planktic foraminiferal fragments, pteropods (including fragments), fish debris, mollusc larvae, benthic foraminifera, siliceous plankton and detrital remains. If necessary, aliquots were obtained with an Otto microsplitter. Between 290 and 2350 grains in the fraction >63 μm were counted.

For PF species analysis, the fraction >125 μm was separated from the fraction >63 μm by dry-sieving. The whole fraction was used and between 66 and 1073 intact PF specimens were encountered. Species identification follows the taxonomy of Be (1967), and the illustrations of Zobel (1973), Thunnell & Reynolds (1984) and Hemleben et al. (1989). Identification and flux estimates of distinct PF species are restricted to the fraction >125 μm, according to the observations of Zobel (1973) from plankton tows off Pakistan, that 40–60% of the specimens <125 μm may not be determined to the species level.

The flux of components or foraminiferal species was calculated as the product of the number of specimens, the split factor, the opening of the trap (0.5 m^2) and the collecting time (in days) and is given in mg or specimens m^{-2} day^{-1}. Assuming a settling velocity of about 100–200 m day^{-1}, for-aminiferal tests and other sand-sized particles of the fraction >63 μm should reach trap depth (590 m) within 1–2 weeks at most (Berger & Piper 1972; Takahashi & Be 1984). However, the much lower sinking speed of smaller particles is difficult to estimate. No attempt was made to correct for the temporal offset between oceanographic changes and biological response in shell production, which should be in the range of several days to a few weeks for most of the small and juvenile PF species encountered (Hemleben et al. 1989).

For PF analysis of surface sediment and core samples, we used the size fraction >150 μm, and the standards set up by the CLIMAP Project (CLIMAP Project Members 1976). To avoid bias as a result of dissolution, our comparison between the trap and the sediment foraminiferal assemblages presented at the end of this study is based on sites from above the present level of the foraminiferal lysocline of 3200 m (Cullen & Prell 1984). Altogether, 216 genuine core-top samples, mostly collected from box- and multicorers of recent cruises of R.V.s Tyro, Sonne and Meteor and from sample sets of Zobel (1973) and Paropkari (pers. comm), were used. A Holocene age of most of these samples is verified by stable oxygen isotope data on benthic and/or planktic foraminifera species. These sediment data will be published elsewhere. Additional PF surface sediment data were used from the Brown University Foraminiferal Data Base, NOAA Paleoclimatology Program, Boulder, Colorado (Prell et al. 1999). The PF census data and lithological and stratigraphic details of sediment core SONNE42-74KL have been presented by Schulz (1995), and by Sirocko (1989) and Sirocko et al. (1993), respectively.

Weekly composites of sea-surface temperatures (SST) with 18 km resolution from the Physical Oceanography DAAC at Jet Propulsion Laboratory–California Institute of Technology for the years 1985–1995 were used to obtain the local SST pattern at the trap site. These data were not available for the months of January–May 1996. We used the estimates of the year 1985 for that interval as the analysis of the 10-year-long record shows that estimates of this year represented best the average SST pattern during winter and spring. The final record of weekly data also agrees well with the long-term monthly averages adopted from Hastenrath & Lamb (1979).

Results and discussion

CTD–temperature profiles of the winter season

Following the long-term and weekly averages of SST (Fig. 2a), temperatures for the year 1995 off

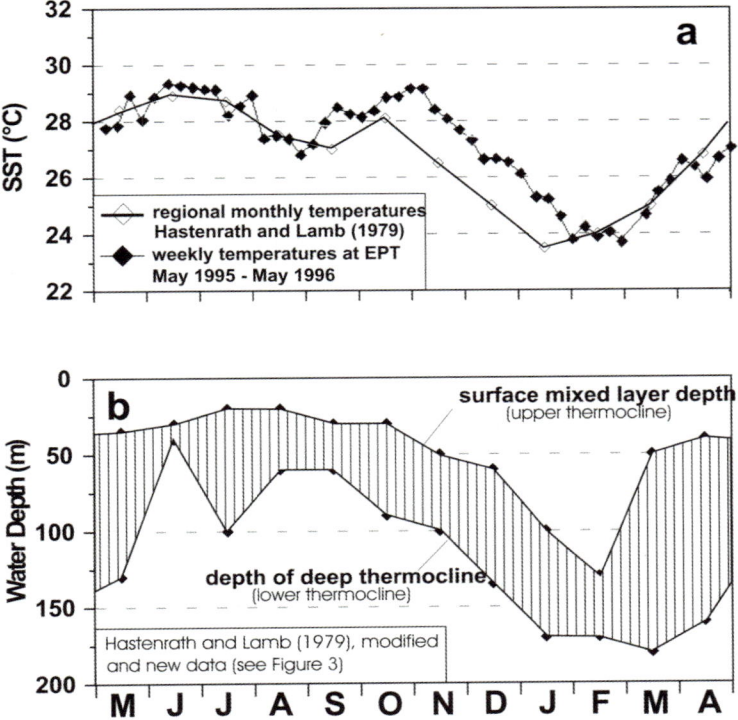

Fig. 2. (a) Annual pattern of sea surface temperatures from monthly and weekly datasets at trap station EPT-2 from May (M) to April (A). (b) Depth of surface mixed layer and deep thermocline at trap station EPT-2.

Pakistan are highest in the months of June and July, in the range of 28–30 °C. A second maximum is observed from September to October. Between these two periods, we note a significant cooling of 1–2 °C in August–September. Analysis of the local record for 1985–1995 confirms that the temperature drop during late summer is a recurrent feature in the summer SST pattern of the northwestern Arabian Sea.

In October, with the onset of the autumn intermonsoon phase, SST starts to drop towards a clear winter minimum of about 23 °C in the months of January and February. The thermocline during winter deepens from about 50 m during summer to about 150 m depth in these months (Fig. 2b).

Our compilation of recent CTD data shows the process of surface cooling and winter mixing in detail (Fig. 3). Taking the 100 m isobath for reference, warming of deeper waters as a result of sinking of highly saline surface waters starts at the surface during autumn. The upper thermocline is still pronounced in late October. Most intense deep mixing can be deduced from the temperature profile of March, when a nearly homogeneous body of warm waters at 23 °C extends from the surface to a depth of more than 150 m. The presence of these warm waters is still feasible in the CTD–temperature record of April, when a new thermocline and shallow mixed-layer form as a result of SST rise caused by increasing insolation. Pollehne *et al.* (1993) observed extreme oligotrophic conditions in the region during early summer (June 1987). Those workers interpreted the unexpected low biological production and intense regeneration of nutrients in the surface waters on the Pakistan shelf as the result of onshore transport of warm oligotrophic open oceanic waters during the early stage of the SW monsoon. Our detailed records may offer an alternative explanation for the low productivity: because of the re-formation of a shallow thermocline in springtime, the photic zone is cut off from the nutrient-rich deeper waters, leading to intense regeneration, and to oligotrophic surface waters during that season. The deep winter thermocline, still marked in the record for mid-May, is progressively eroded by thermal diffusion.

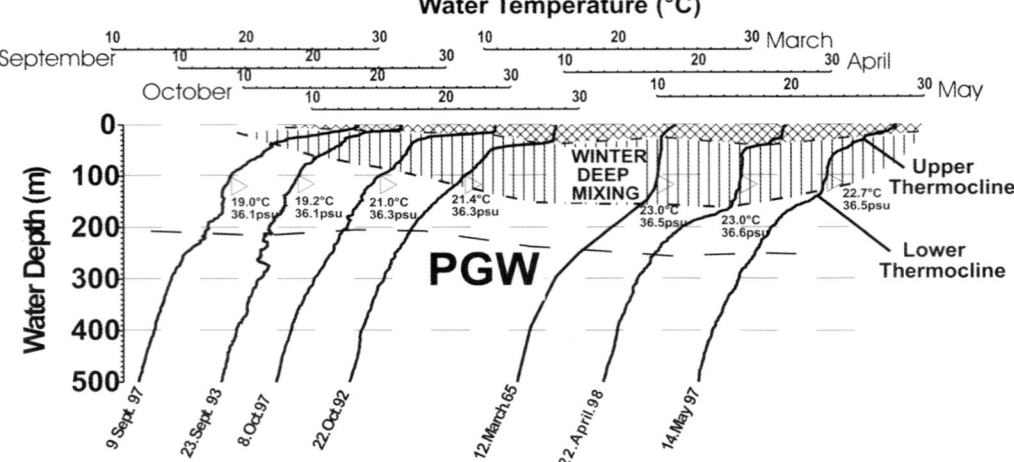

Fig. 3. CTD–temperature profiles for the upper 500 m in the northeastern Arabian Sea, off Pakistan (9 September–14 May; see Table 1), showing the evolution of surface and subsurface water masses during the winter season. Deep mixing leads to a nearly homogeneous water mass (profile of March) with temperatures of c. 23 °C, linked to sinking and mixing of saline surface water to more than 150 m depth. The process of deeper warming and mixing from the surface starts in October, as can be seen in the higher values of temperature and salinity at 100 m depth (triangles). PGW, underlying Persian Gulf Water (for details see text).

Particle flux of EPT-2

The patterns of total particle flux (Fig. 4a), as well as the flux of individual components (Fig. 4b–e), show three maxima: (1) during late spring (May); (2) near the end of the summer monsoon (August and September); (3) the highest, during January and February. During these months, total particle flux is 4300–4100 mg m^{-2} day^{-1}, about two times higher than during the summer maximum of August–September (1300 and 2400 mg m^{-2} day^{-1}) and the maximum in May (1700 mg m^{-2} day^{-1}). Total particle flux is low during July–August (sample 4) at only 180 mg m^{-2} day^{-1}, and from October to December (samples 7–10) with values of 140–250 mg m^{-2} day^{-1}. As shown by Andruleit et al. (2000), the total flux and the flux of calcium carbonate are positively correlated ($R^2 = 0.96$), with a rather stable ratio of c. 6:1. The fluxes of POC and opal are also well correlated with the total flux, suggesting a small variability in the composition along with the trap record (Fig. 4f), except for September (samples 5 and 6), when the fluxes of opal and POC are nearly of the same magnitude as during January–February.

The major components of the fraction >63 μm (Fig. 5) are the shells of pteropods, of PF, and the remains of large diatoms and mostly spongy radiolaria. An extremely large flux of limacid pteropod shells of >100 000 specimens m^{-2} day^{-1} is found at the start of the record for May 1995 (Fig. 5a). The flux of pteropods of this spike is more than 20 times higher than the flux of PF at the same time. After a rapid decline in June and July, pteropods are nearly absent throughout most of the summer and autumn. Pteropods usually prefer the warm and highly saline mesotrophic to oligotrophic regions, where intense regeneration of nutrients occurs. The extreme flux of pteropods may indicate the decline of nutrients at the end of the winter season of high productivity. As discussed above, shallow surface waters off Pakistan may become progressively nutrient depleted in spring, as a result of the cut-off from deeper waters when surface waters heat up again (see Figs. 2 and 3). A second, but minor flux of pteropods is noticed in the winter months.

In contrast to the pteropod abundance, the flux of PF (and fragments) (Fig. 5b), of siliceous plankton (diatoms, radiolarians) (Fig. 5c), and of bivalve larvae (Fig. 5e), displays a clear bimodal distribution in the fraction >63 μm, with maxima in late spring (11 600, 1000 and 700 specimens m^{-2} day^{-1}, respectively) and winter (9100, 2800 and 350 specimens m^{-2} day^{-1}, respectively). During summer and autumn, flux values for these groups are distinctly lower. Little is known about the lifestyle and ecological significance of bivalve larvae. Zobel (1973) reported increased abundances of swimming bivalves in water depths between 20 and 50 m even in the size fraction

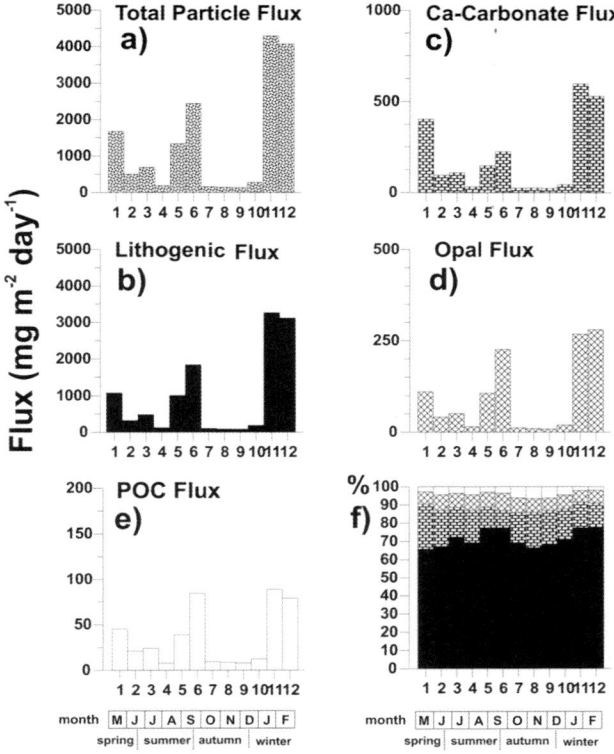

Fig. 4. Flux (in mg m^{-2} day^{-1}) of (**a**) total particles, (**b**) lithogenic material, (**c**) calcium carbonate, (**d**) opal, and (**e**) particulate organic carbon (POC) for the period May 1995–February 1996 in trap EPT-2 off Pakistan. (**f**) shows the relative abundance of the four major components in per cent.

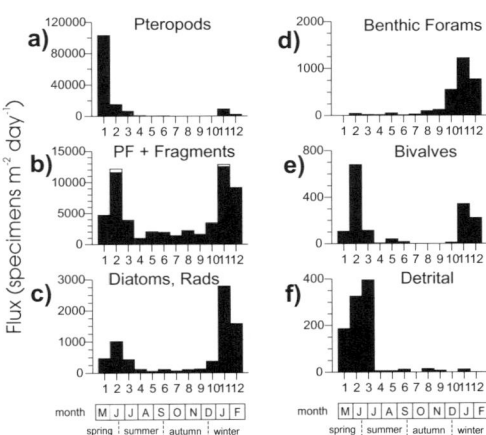

Fig. 5. Flux of particles > 63 μm (in specimens m^{-2} day^{-1}) in trap EPT-2 for (**a**) pteropods (and fragments), (**b**) planktic foraminifera (black bars) and fragments (white bars), (**c**) large diatoms and radiolaria, (**d**) benthic foraminifera, (**e**) marine bivalves, and (**f**) detrital grains (quartz, large clay mineral particles).

>160 μm off the Indus delta. Thiede (1975) suggested that the high flux of bivalve shells in the NE Atlantic may be linked to increased productivity of the nearshore waters.

The flux of PF fragments is well correlated with the flux of intact PF tests, displaying rather uniform values of fragmentation. The presence of well-preserved pteropods and small PF specimens (often bearing spines) further supports the suggestion that foraminiferal dissolution in the surface waters and upper OMZ was small during the period under observation. Lowest fragmentation and hence best preservation is observed for samples 11 and 12 during winter.

Benthic foraminiferal (BF) tests were found in all samples, except sample 1 (Fig. 5d). In contrast to the good preservation of pteropods and PF, a strong corrosion of these tests must indicate a long transport path or remobilization from older sediment. A clear flux maximum of BF occurs from December and February. In January the flux of BF reaches more than 1200 specimens m^{-2} day^{-1}. No BF were observed in the size fraction >125 μm.

Lithogenic material accounts for 63–77% and thus clearly dominates the total mass flux (Fig. 4f). However, only small numbers of detrital grains of quartz and large clay mineral particles were found in the fraction >63 μm, with a clear maximum from May to July and a sharp decline in July (Fig. 5f). A pronounced summer maximum is hardly seen in any of the records of organisms >63 μm (Fig. 5), in contrast to the flux patterns of the bulk components (Fig. 4).

Flux and relative abundance of planktic foraminiferal species

The flux of PF of the size fraction >125 μm displays the same bimodal pattern as in the fraction >63 μm (Fig. 5b), with maxima in spring and winter (Table 2). Twenty-three PF species were encountered in the trap record off Pakistan. Ten species occurred with an average relative abundance of more than 2%. Among these, the eight species *Globigerinita rubescens* (Hofker, 1956), *Globigerinoides ruber* (white) (d'Orbigny, 1839), *Globigerinoides tenellus* (Parker, 1958), *Neogloboquadrina dutertrei* (d'Orbigny, 1839), *Globigerinita glutinata* (Egger, 1895), *Globigerina falconensis* (Blow, 1959), *Globigerina bulloides* (d'Orbigny, 1829) and *Turborotalia quinqueloba* (Natland, 1938) account for more than 85% of the total assemblage. Because of the small sample volume available from samples 4–7 and 9, the total number of intact specimens was below 300 (Table 2) which may affect the statistical significance of their relative species abundances (van der Plas & Tobi 1965). Therefore, the abundance data from these samples should be interpreted in terms of general trends rather than in terms of individual quantities.

Up to 10% of PF specimens of the assemblages could not be identified to the species level even in the large fraction >125 μm. This group of specimens may consist mostly of juvenile *Turborotalia quinqueloba* and of *Globigerinita iota* (Parker, 1962) or *Globorotalia anfracta* (Parker, 1967).

The flux and relative abundance of the eight frequent species are presented in Fig. 6a–h. According to Be & Tolderlund (1971), these PF can be grouped as typical for tropical–subtropical (*Globigerinita rubescens, Globigerinoides ruber* (white), *Globigerinoides tenellus*), for subtropical to transitional (*Neogloboquadrina dutertrei, Globigerina falconensis),* and for temperate to subpolar climatic conditions (*Turborotalia quinqueloba, Globigerina bulloides*). *Globigerinita glutinata* may be regarded as a cosmopolitan species with a broad tolerance with respect to water temperature and salinity.

Highest flux of individual species of the size fraction >125 μm is observed for *G. rubescens*, with 670 specimens m^{-2} day^{-1} during June–July (Fig. 6a), and during January for *G. falconensis* (Fig. 6e) and *G. glutinata* (Fig. 6f), with 540 and 520 specimens m^{-2} day^{-1}, respectively. During these periods, the three species also make up a large proportion (20–40%) of the total assemblage.

Globigerinita rubescens displays the highest flux of any species in the record. Little is known about the ecological preferences of this small species, which seems to be a shallow dweller from warm subtropical to temperate waters (Fairbanks *et al.* 1982; Hemleben *et al.* 1989). We interpret the high flux of *G. rubescens* in May–June together with the flux maximum of pteropods in May to reflect relatively low levels of productivity in the early and late stages of winter mixing (compare Fig. 3, profiles 1 and 7).

Globigerinoides ruber, G. tenellus and *N. dutertrei* (Fig. 6b–d) also show highest flux during spring to early summer, with maximum values of 240, 230 and 68 specimens m^{-2} day^{-1}, respectively. Together with *G. rubescens*, or the group of tropical–subtropical species (*G. rubescens, G. ruber, G. tenellus* and also for *N. dutertrei*) (Fig. 6a–d), only a small flux maximum is noticed in January and February.

The species *G. falconensis, G. glutinata, G. bulloides* and perhaps also *T. quinqueloba* (Fig. 6e–g) show highest flux during winter. However, the relative abundance patterns of the frequent three species are different: *G. falconensis* shows highest relative abundance (and flux) from December to February, *G. glutinata* in autumn, from September to November, whereas highest abundances of *G. bulloides* are seen during August to October. *Globigerina falconensis* (Fig. 6e) clearly is the dominant species during winter with regard to its flux and also relative abundance, making up as much as 35% of the total fauna in January and February.

The relative abundance of *G. bulloides* is >10% during summer and autumn (Fig. 6g). High frequencies of this species have been reported to be centred in subpolar latitudes in three oceans (Be 1977; Vincent & Berger 1981). However, in the western Arabian Sea, *G. bulloides* is also dominant in the areas of summer upwelling and is considered as an indicator of cool conditions as a result of upwelling of subsurface waters, and hence of high monsoonal productivity (Prell & Curry 1981; Curry *et al.* 1992; Guptha & Mohan 1996). Moreover, Naidu & Malmgren (1995) showed that the test size of *G. bulloides* is positively correlated with upwelling strength in the waters off Oman. However, the high relative

Table 2. *EPT-2 planktic foraminiferal flux data (specimens $m^{-2} d^{-1}$)*

Sample	Interval	No. of PF counted	Flux PF	Flux PF fragments	Flux G. rubescens	Flux G. ruber	Flux G. tenellus	Flux N. dutertrei	Flux G. falconensis	Flux G. glutinata	Flux G. bulloides	Flux T. quinqueloba
Size fraction >125 µm												
EPT-II-s1	5.5.95–29.5.95	487	687		266	240	87	0	26	20	15	4
EPT-II-s2	30.5.95–23.6.95	1073	1501		669	189	228	68	154	79	61	25
EPT-II-s3	24.6.95–18.7.95	349	477		159	113	52	15	12	77	25	9
EPT-II-s4	19.7.95–12.8.95	90	106		16	33	9	13	7	13	7	1
EPT-II-s5	13.8.95–6.9.95	66	343		47	122	35	0	17	58	41	6
EPT-II-s6	7.9.95–1.10.95	87	495		93	99	41	12	41	116	64	0
EPT-II-s7	2.10.95–26.10.95	179	255		32	55	12	4	3	100	31	1
EPT-II-s8	27.10.95–20.11.95	370	529		54	106	25	10	23	214	60	10
EPT-II-s9	21.11.95–15.12.95	163	231		17	45	7	3	7	105	16	17
EPT-II-s10	16.12.95–9.1.96	241	320		28	54	20	1	60	112	28	10
EPT-II-s11	10.1.96–3.2.96	700	1911		265	233	113	17	544	518	137	20
EPT-II-s12	4.2.96–28.2.96	383	1094		125	111	67	15	375	241	111	23

Sample	Flux pteropods	Flux PF	Flux PF fragments	Flux Diat/Rads	Flux BF	Flux bivalves	Flux detrital
Size fraction >63 µm							
EPT-II-d1	103075	4618	105	473	0	106	186
EPT-II-d2	15325	11615	573	1012	47	679	326
EPT-II-d3	6458	3766	150	436	23	113	396
EPT-II-d4	934	975	19	122	17	6	6
EPT-II-d5	308	1903	122	58	52	41	6
EPT-II-d6	436	1815	122	116	12	17	12
EPT-II-d7	6	1345	35	67	29	0	0
EPT-II-d8	58	2092	105	112	99	0	15
EPT-II-d9	31	1572	28	128	125	0	7
EPT-II-d10	141	3306	141	375	556	12	0
EPT-II-d11	9088	12509	340	2790	1222	343	12
EPT-II-d12	2225	9015	116	1577	774	221	0

PF, planktic foraminifera; Diat/Rads, diatoms and radiolaria; BF, benthic foraminifera.

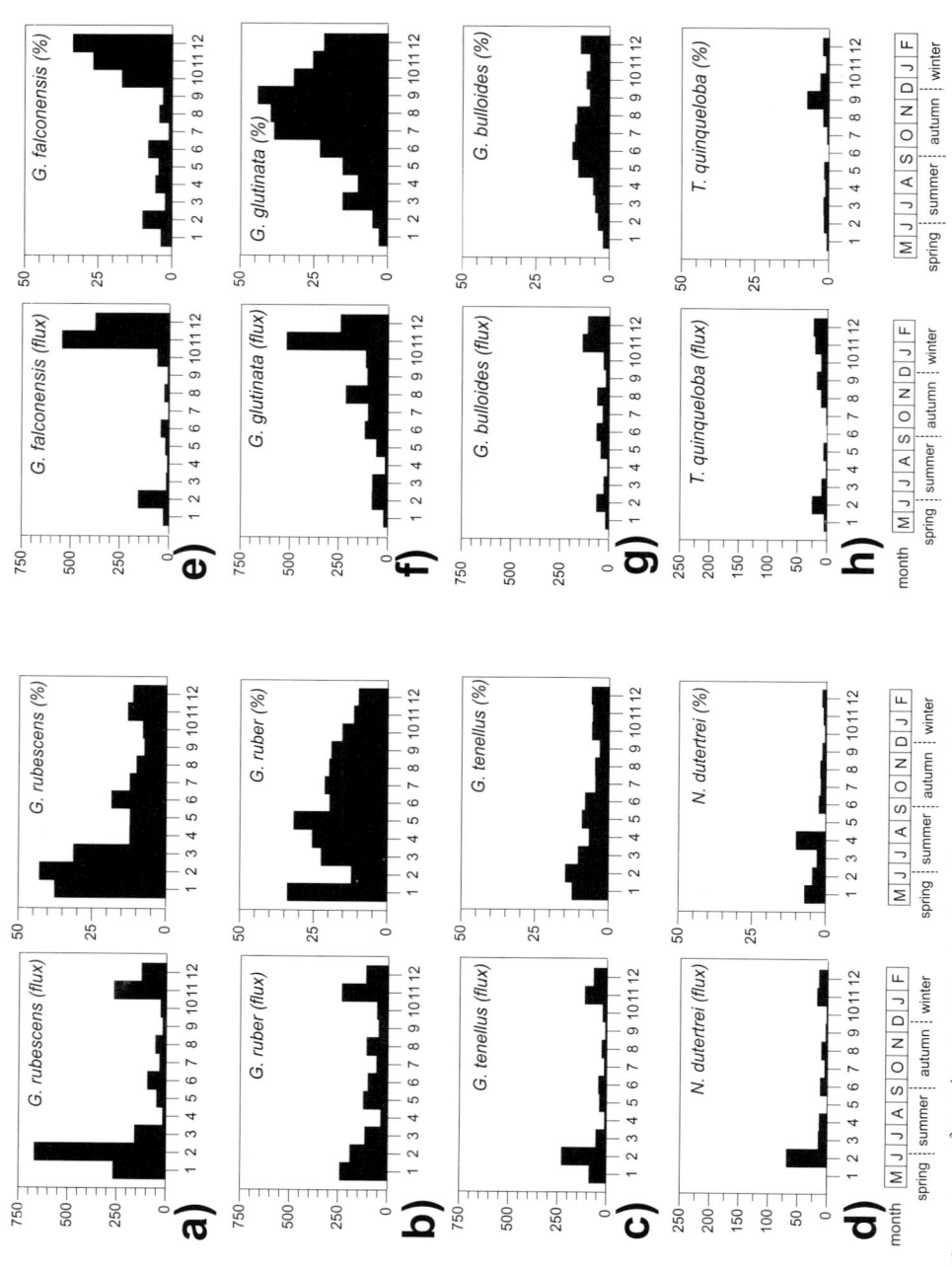

Fig. 6. Flux (in specimens^{-2} day^{-1}) and relative abundance (in per cent) of eight most frequent planktic foraminiferal species in the size fraction >125 μm in trap EPT-2.

abundances of *G. bulloides* off western India (Naidu *et al.* 1992) indicate that this species also thrives in the warm, nutrient-rich nearshore waters of the eastern part of the basin. This observation is further corroborated by the rather low, but significant flux and relative abundance of this species in our trap record from the northeastern Arabian Sea.

Globigerinita glutinata (Fig. 6f) occurs with significant abundances, often >20%, in subpolar (for example, in northern North Atlantic sediments; Kipp 1976), but also in tropical waters (for example, in Arabian Sea sediments; Cullen & Prell 1984). This species displays a distinct flux maximum during winter. It also shows a broad maximum in its relative abundance during late autumn to winter, making it the most important species for that season in our trap record. In the surface sediments of the Arabian Sea (fraction >150 μm), *G. glutinata* is most frequent, with more than 30% in the northwestern part of the basin, although outside the areas of intense monsoonal summer upwelling (Cullen & Prell 1984). Therefore, the broad maximum and also the increased flux of *G. glutinata* during late autumn and winter may reflect the specific nutrient conditions after the summer upwelling season, or marginal to the western upwelling areas, where *G. bulloides* dominates.

Comparing the patterns of relative abundances of the PF species displayed in Fig. 6, the overall impression is the presence of a seasonal succession in PT-2: *G. rubescens*, together with *G. tenellus*, is most frequent during late spring to early summer, whereas *G. ruber* dominates during summer to autumn. Then follows *G. bulloides* with increased frequencies during late summer–autumn and *G. glutinata* during late autumn to winter. Finally, *G. falconensis* dominates during winter. Unfortunately, the incompleteness of our trap record prevents us extending this observation to a full annual cycle.

Origin of flux and sediment resuspension in winter

The trap record shows strongly increased total and lithogenic fluxes during May, August–September and January–February. The onset of the SW monsoon, progressing from south to north along the west coast of India, can be expected off Pakistan for the period after mid-July at the earliest. We therefore exclude that the high flux of the various compounds during spring (see Figs. 4–6) may be linked to the SW monsoon, but must be related to winter to springtime processes. If we take into account that shell production and the flux of particles lags physical monsoonal forcing by several weeks (Rixen *et al.* 2000), the time offset between SW monsoon forcing and the high flux events would be even greater.

The total particle flux during winter–spring and during late summer is 7–20 times higher than that found in the open Arabian Sea (Nair *et al.* 1989; Haake *et al.* 1993). The origin of the high flux of particles, however, is poorly constrained by direct observations. The sedimentary origin of at least parts of the fluxes found in the trap is indicated by the presence of the small BF in the fraction >63 μm (Fig. 5d) in nearly all samples. BF usually form a major fraction of the biogenic shells in the sediments within the OMZ (Zobel 1973; den Dulk *et al.* 1998; von Rad *et al.* 1999b). The maximum of BF flux in December–February coincides well with the highest flux of lithogenic, but also other bulk particles in the record (Fig. 4). Therefore, we suggest that the lateral flux of allochthonous sediment is highest during winter and must have a significant effect on the trap record.

In contrast, large detrital grains in the fraction >63 μm, such as quartz and large clay mineral particles, display their maximum during May–July, with a sharp decline in mid-July (Fig. 5f). Unexpectedly, we did not find an increased flux of these particles during winter. It can be speculated that these grains have been transported to the trap location by the mid-tropospheric northwesterly winds, which carry dust from sources in Arabia and Mesopotamia to the northern Arabian Sea. Satellite images show the tracks of dust plumes that change in position and dust load during the summer season: in spring to early summer (March–May) a northeastern branch of the main trajectory is situated over the Persian Gulf and carries dust plumes to the Pakistan margin (Sirocko & Sarnthein 1989). During July–August, the main branch reaches further south and traverses the Arabian Sea over the Gulf of Aden. Assuming the aeolian origin of this detrital fraction, the sudden drop in the record to very low dust abundance in the middle of the summer monsoon season could be explained in terms of these changes in the wind trajectories. There is no indication in our record for the input of dust carried by the winter winds of the NE monsoon from the northeastern dry areas and the Thar desert to the trap site. Sirocko (1995) pointed out that the atmospheric winter dust load is generally small and could not be observed on satellite imagery.

The seasonality of sedimentation seen in the trap record (Figs. 4 and 5) is confirmed by the studies of laminated sediments. Schulz *et al.* (1996) and von Rad *et al.* (1999a) suggested that high flux of particles during winter may origin from the shelf, where sediments are stirred up by turbulent mixing as a result of cold winter winds blowing offshore. In the light of our new oceanographic and trap data, a different process may also be important in causing the extremely high winter and spring flux in our trap record: the cooling and sinking of highly saline surface waters as a result of the NE monsoon leads to a seasonal densification and oxygenation of bottom waters, which are able to entrain small particles and remineralize nutrients from the sea floor, and to carry a suspension load over longer distances. Similar to the effects of wind-driven turbulent mixing, the uptake of fine-grained detritus would be most intense in the nearshore waters, where winter cooling is strongest.

The slightly reduced SST values seen in the temperature record during summer (Fig. 2a) may also have their counterpart in the distinct summer maximum of the flux of small particles (Fig. 4). The monthly data of Hastenrath & Lamb (1979) show a spiked increase in the thickness of the well-mixed surface waters for the month of July (Fig. 2b), coeval with the slight drop in SST during the months of July and August and the summer flux maximum. We therefore may speculate that during mid-summer, sediment remobilization on much smaller scales may be operating, similar to the winter situation. The forcing agent would then be the SW monsoonal winds, which cool the surface and lead to wind-driven deeper mixing.

Conversely, the surface to subsurface waters are most stratified for a short period in July, and in October–December (Fig. 2). These periods roughly correlate with the intervals of lowest total particle flux of our trap record, with a time offset of a few weeks (Fig. 4).

Increased winter productivity

A number of studies have previously emphasized the importance of deep winter mixing for the local productivity budget of the northeastern Arabian Sea (Banse & McClain 1986, and references therein; Madhupratap et al. 1996). They showed that the high concentrations of surface water phytoplankton chlorophyll during winter could not be explained alone by upwelling processes related to changes in the surface water temperatures and winds. In their study of coccolithophorids, Andruleit et al. (2000) found consistent patterns of relative and absolute species abundances of coccospheres and coccoliths in four traps deployed at two sites off Pakistan. The good fit of the abundance patterns at sites WPT and EPT (Fig. 1) and at various depths suggested that biological processes, rather than sediment resuspension, are responsible for the increased flux of coccoliths and other organic particles (Fig. 4) during the early summer monsoon (May–July), the early winter monsoon (January and February) and possibly also during late summer. However, as outlined by Andruleit et al. (2000), lateral advection, for example from the western upwelling fields during summer but also during winter and spring, could not be excluded as important factors.

The size of allochthonous particles that can be transported from the continental margin to the trap site, more than 500 m above the sea floor and c. 70 km offshore, is limited to grain sizes smaller than 125 µm, as previously shown by the presence of benthic foraminifera (Fig. 5d). Therefore, we suggest that the flux of particles in the size fraction >63 µm and in the fraction >125 µm may be used to trace more reliably the productivity patterns, without minor effects of remobilized matter contained in the smaller size fractions.

PF together with the group of limacid pteropods, large diatoms–radiolarians and bivalve shells form the dominant fraction in the size class >63 µm (Fig. 5). The groups of pteropods and PF clearly dominate the size fraction >125 µm. However, the interpretation of their flux patterns may be complicated by the observation that different planktic groups closely interact and may respond differently to oceanographic processes in their abundance and temporal distribution. For example, a trap study of species interrelationships of diatoms and PF from the San Pedro Basin off Southern California showed that total flux of these groups varied independently (Sautter & Sancetta 1992).

Off Pakistan, the fluxes of PF, but also of diatoms–radiolarians and marine bivalves in the record >63 µm show very similar patterns (Fig. 5). The patterns of these three groups correlate well with those of the total flux and of individual bulk components in spring and winter (Fig. 4), when highest fluxes are observed. In contrast, the flux maximum of lithogenic and also of biogenic opal and of POC observed in summer is not reflected in the flux of large particles. It may be suggested that favourable conditions during summer off Pakistan either were too short to result in a large flux of large shell-bearing plankton, but not for small siliceous and coccolithophorid plankton, causing the high flux of opal and POC during that season (Fig. 4d and e).

Various studies confirm that PF species respond positively to seasonally enhanced surface water productivity. Curry et al. (1992) and Conan & Brummer (2000) demonstrated this many-fold increase in the flux of all species as a result of coastal and open ocean upwelling in the Arabian

Sea during the summer months. If we use the flux of PF to trace primary production, local productivity is highest during the winter and spring off Pakistan (Fig. 6). The importance of winter productivity is also evident from a comparison of PF fluxes in the fraction >125 μm between traps EPT-2 off Pakistan and WAST from the western summer upwelling area (Fig. 7): the winter PF flux off Pakistan is similar to or even higher than at WAST. In the summer months, however, the PF flux off Pakistan is far below those values off Oman.

An index of non-upwelling conditions: Globigerina falconensis

Globigerina falconensis dominates the assemblage in its flux and relative abundance during winter (Fig. 6e). In many surface sediment and trap studies, *G. falconensis* has been lumped together with *G. bulloides* because of its close morphological resemblance and its sporadic occurrence at <20%, and in most cases <5% (Zobel 1973; Pflaumann 1985; Curry et al. 1992). However, a detailed study by Malmgren & Kennett (1977) on the southern Indian Ocean revealed a clear biometric differentiation between the two species: *G. falconensis* is distinctly different from *G. bulloides* in the thickness of the test and smaller size of the final chamber, in the apertural dimensions, in the presence of an apertural lip, and in its more elongate chambers and smaller size. *Globigerina falconensis* bears symbionts, whereas *G. bulloides* is barren.

Apart from their possibly different preferences in prey, it has been consistently reported from the North Atlantic (Kipp 1976; Ottens 1992) and the SW Pacific (Thiede et al. 1997) that *G. falconensis* lives at the warm 'edge' of the distribution of *G. bulloides*, thus indicating a preference for slightly warmer waters than *G. bulloides*. If we compare the flux patterns of the two species in the northern Arabian Sea (Figs. 6 and 7), *G. bulloides* appears to dominate under the highly productive summer conditions of the western and southeastern Arabian Sea. During upwelling, local SST may drop to <20 °C and a extremely shallow thermocline develops. *Globigerina bulloides* reaches summer flux maxima of >12 000 specimens m^{-2} day^{-1} in the size fraction >125 μm (Curry et al. 1992). In contrast, *G. falconensis* occurs only with very low frequencies under these conditions (Ostermann, pers. comm.). Our trap record of *G. falconensis* off Pakistan indicates that high abundances of this species may also reflect enhanced productivity and cooled subsurface waters, but clearly outside the summer season (Fig. 6e). Hence, *G. falconensis* may represent an end-member of deep winter mixing and a deep thermocline, and hence, of most distinct non-upwelling conditions in the Arabian Sea as a result of the NE monsoon.

Application to the sediment record of the past 25 ka

To support this hypothesis, we mapped the relative abundance patterns of *G. bulloides* and *G. falconensis* on the basis of about 200 sediment surface samples from the northeastern Arabian Sea, combined with published datasets (Fig. 8). The distribution of *G. bulloides* clearly identifies the areas of high coastal and open oceanic upwelling driven by the SW monsoon: high abundances of *G. bulloides* are centred off northern Somalia (maximum of 57% near Socotra), off Oman (43%), and off southern India (67%). In contrast, the relative abundance of *G. falconensis* shows very low frequencies in these regions. A considerable amount of specimens of *G. falconensis* of <10% was found only in the Somali upwelling area. In contrast, a clear maximum of >20% is situated along the coastal areas off northern India

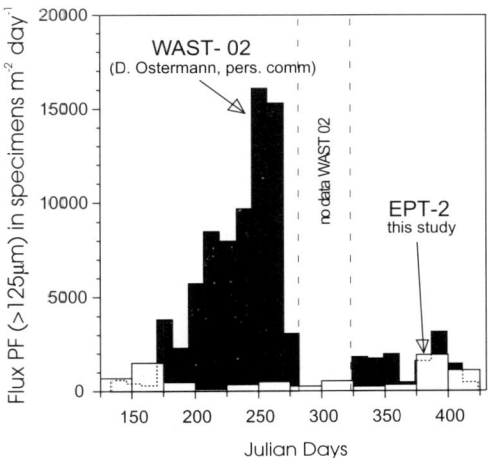

Fig. 7. Comparison of flux of planktic foraminifera in the size fraction >125 μm for the period of May–February in Julian days between Western Arabian Sea Trap WAST 02 (1040 m water depth) and EPT-2. No record from WAST 02 is available for autumn (September–October); unpublished WAST PF data for the fraction > 125 μm were provided by D. Ostermann, Woods Hole Oceanographic Institution. A complete description of the PF fluxes of the size fraction >150 μm from this trap was published by Curry et al. (1992).

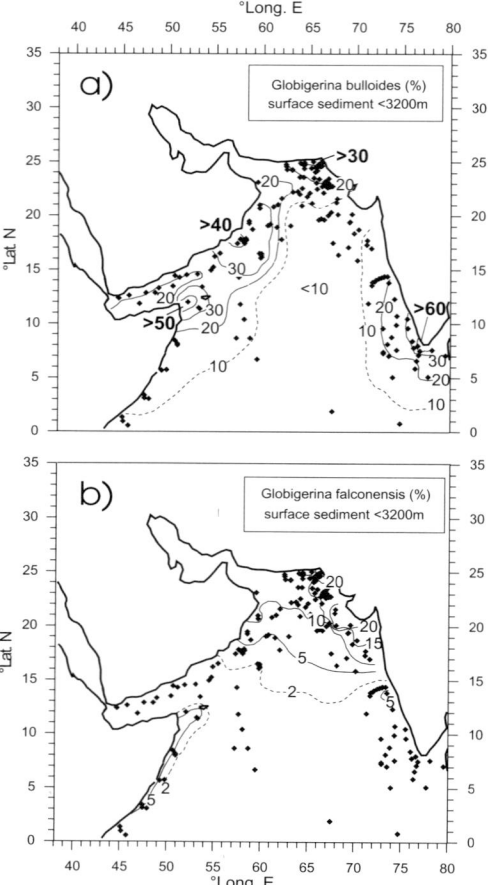

Fig. 8. Maps showing the relative abundance of the two planktic foraminiferal species (**a**) *Globigerina bulloides* and (**b**) *Globigerina falconensis* in the size fraction >150 μm in surface sediments from the Arabian Sea. To avoid possible bias of selective dissolution on individual species, only samples from above the lysocline (*c.* 3200 m water depth) were used.

and Pakistan in the northeastern part of the Arabian Sea. These different distributions suggest that the intensity of the SW and NE monsoons, i.e. summer upwelling and winter deep-mixing conditions, can be mirrored by the relative abundances of these two species in the deep-sea sediment. In particular, the high percentages of *G. falconensis* in the northeastern part of the Arabian Sea seem to reflect the areas of most intensive winter cooling, where winter temperatures fall below 23 °C and winter mixing occurs.

On the basis of stable isotope, alkenone–SST and foraminiferal evidence, it has been suggested that the NE monsoon played a major role in the primary productivity of the northern Indian Ocean during the last glacial period (Emeis *et al.* 1995; Reichart *et al.* 1998; Cayre & Bard 1999). The SW monsoon was weakened and the Arabian Sea surface temperatures were considerably lower, as a result of the cool glacial conditions (Rosteck *et al.* 1993; Bard *et al.* 1997; Sonzogni *et al.* 1998). From the higher glacial abundances of *G. falconensis* in a sediment record from the northeastern Arabian Sea, den Dulk *et al.* (1998) and Reichart *et al.* (1998) concluded that the NE monsoon was enhanced during the cool periods of Marine Isotope Stages (MIS) 2–4, and also during the cool stadial periods of interglacial MIS 5. In a detailed record of SONNE core 74KL from the western Arabian Sea upwelling area, we examine the role of the NE monsoon in more detail in relation to a number of well-defined sedimentary and isotopic events (Sirocko *et al.* 1993), including the periods of the last glacial maximum, the last glacial to interglacial transition and the Holocene period (Fig. 9a).

Our records from *G. bulloides* and *G. falconensis*, scaled to the modern abundance of these two species in the area (Fig. 9b), show coherent changes in both the SW and the NE monsoon. The SW monsoonal intensity was relatively weak during the last glacial period. During the early part of the last glacial termination, from 18 to 11.5 cal. ka BP, the SW monsoon was slightly stronger. Present-day intensity was reached after about 11–10 cal. ka BP. In contrast, the intensity of the NE monsoon during the last glacial maximum was comparable with that today, presumably because of low insolation forcing. During the glacial–interglacial termination, the NE monsoon peaked, and it shows two maxima during the Heinrich Meltwater Event 1 (H1) and the Younger Dryas (YD) cool periods. Only for the period of 10–7 cal. ka BP, i.e. for the maximum of summer insolation forcing and the following 3 ka, were both the NE and SW monsoons strong, suggesting a pronounced seasonality.

For most of the past 25 ka, the SW and NE monsoons seem to be rather anti-correlated on glacial–interglacial but also millenial scale, or may have even varied independently. Rapid millenial- to centennial-scale fluctuations are also clearly recognized in our foraminiferal data. The timing of these monsoonal shifts agrees well with the series of $CaCO_3$ and isotopic events towards lighter $\delta^{18}O$ of the planktic species *G. ruber* from the same core. They indicate the stepwise re-intensification of the summer monsoon and can be attributed to rapid rise in SST and/or decreases in surface salinity (Sirocko *et al.* 1993) after the last glacial period. However, most of these rapid shifts during the last glacial termination before 10–

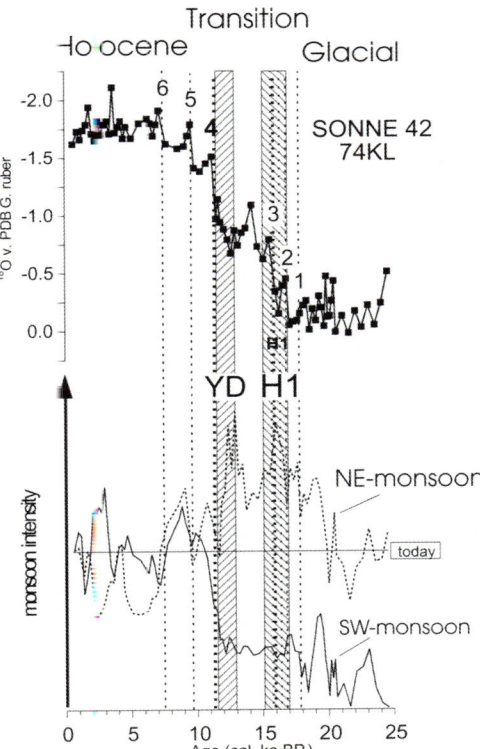

Fig. 9. Reconstruction of seasonal monsoon intensity from sediment core SO42-74KL in the western Arabian Sea (see Fig. 1) using the monsoon indices of *Globigerina bulloides* and *Globigerina falconensis*, and comparison with the rapid monsoonal events (1–6) from the same core (Sirocko et al. 1993). H1 and YD correspond to cool northern hemisphere climatic events first reported from the North Atlantic and the Greenland ice cores. The foraminiferal data have been smoothed by a three-point running average and are scaled to the percentages of the two species found today at the position of core 74KL.

11 cal ka BP seem to be triggered by decreases in the NE monsoonal intensity rather than by increased SW monsoons. This hypothesis further highlights the importance of the NE monsoon for palaeoclimatic reconstructions in the Arabian Sea.

Summary and conclusions

(1) Our trap record from the northeastern Arabian Sea off Pakistan shows a clear flux maximum in winter (January–May) and a second, but minor maximum in summer (late August–September). The summer maximum has not been recognized in the flux of PF and other coarse shells. Total particle flux of >4300 mg m^{-2} day^{-1} is highest during winter, up to 10 times higher than further out in the Arabian Sea during the summer monsoon maximum. The bulk composition of flux is rather uniform throughout the record, with about 70% of lithogenic matter, 24–9% calcium carbonate, 3–6% POC and 6–8% opal. As also revealed from the study of laminated sediments, a major part of the flux at the trap site consists of laterally advected sediment, previously deposited on the nearby shelf and upper slope.

(2) Sediment resuspension and the peak in fallout during winter (as also seen in the peak in flux of small benthic foraminifera far offshore), in combination with mixed layer deepening and the injection of nutrients, favour high productivity during winter in the northeastern Arabian Sea and may maintain the pronounced oxygen minimum there. The maximum flux of shell-bearing plankton during winter and spring further supports increased winter productivity off Pakistan. The imprint of the summer monsoon on the productivity in this region is small and poorly constrained by our flux data.

(3) Off Pakistan, the dominant PF species are *G. rubescens*, *G. tenellus* and *G. ruber* in spring and *G. falconensis*, *G. glutinata* and *G. bulloides* in winter. *Globigerina falconensis* shows the strongest flux and abundance in January–February. We suggest that *G. falconensis* is an indicator of deep winter mixing. This species reflects the specific non-upwelling conditions during the NE monsoon, in an oceanic basin that today is dominated by the large-scale upwelling phenomena linked to the summer SW monsoon. The summer upwelling species *Globigerina bulloides* is of minor importance and has only a gentle relative abundance maximum off Pakistan.

(4) These indications from the trap record are further corroborated by the distribution of *G. bulloides* and *G. falconensis* in the sediment. The areas of high summer SW monsoonal upwelling are identified by high abundances of *G. bulloides*. The relative abundance of *G. falconensis* shows a contrasting distribution maximum in the northeastern Arabian Sea, under the region of most intense winter mixing driven by the NE monsoon. We used the abundance records of these two species to reconstruct the intensity of the two monsoonal systems for the past 25 ka, including a number of independently defined monsoonal events. Distinct changes are observed in the intensity of both monsoons. It becomes feasible to reconstruct the history of the Arabian monsoon on seasonal scales.

We thank A. Lückge for sampling during cruise SONNE 130, and A. Suthhof for handling the trap samples. Numerous colleagues have provided surface sediment samples and helped to improve the CTD and foraminiferal database. D. Ostermann (Woods Hole Oceanographic Institution) kindly provided unpublished census data of trap WAST 02. R. Bahlo carried out extensive scanning electron microscope analyses. We are indebted to the officers and crews of R.V. *Sonne* and R.V. *Meteor* for good collaboration. Constructive reviews by A. D. Singh and an anonymous reviewer improved the manuscript. This study was carried out under the German research programme 'JGOFS-Indik' and was made possible by BMFT–BMBF grants 03G0090A and 03F0137C.

References

ANDRULEIT, H.A., VON RAD, U., BRUNS, A. & ITTEKKOT, V. 2000. Coccolithophore fluxes from sediment traps in the northeastern Arabian Sea off Pakistan. *Marine Micropaleontology*, **38**, 285–308.

BANSE, K. & MCCLAIN, C.R. 1986. Winter blooms of phytoplankton in the Arabian Sea as observed by the Coastal Zone Color Scanner. *Marine Ecology Progress Series*, **34**, 201–211.

BARD, E., ROSTECK, F. & SONZOGNI, C. 1997. Interhemispheric synchrony of the last deglaciation inferred from alkenone paleothermometry. *Nature*, **385**, 707–710.

BAUER, S., HITCHCOCK, G.L. & OLSEN, D.B. 1991. Influence of monsoonally-forced Ekman dynamics upon surface layer depth and plankton biomass distribution in the Arabian Sea. *Deep-Sea Research I*, **38**, 531–553.

BE, A.W.H. 1967. Foraminifera, families: Globigerinidae and Globorotaliidae, Fiche no. 108. *In:* FRASER, J.H. (ed.) *Fiches d'identification du zooplancton.* Conseil International pour l'Exploitation de la Mer, Charlottenlund, Denmark, sheet 108.

BE, A.W.H. 1977. An ecological, zoogeographic and taxonomic review of recent planktonic foraminifera. *In:* RAMSAY, A.T.S. (ed.) *Oceanic Micropaleontology.* Academic Press, London, **1**, 1–100.

BE, A.W.H. & TOLDERLUND, D.S. 1971. Distribution and ecology of living planktonic foraminifera in surface waters of the Atlantic and Indian Oceans. *In:* FUNNEL, B.M. & RIEDEL, W.R. (eds) *The Micropaleontology of the Oceans.* Publisher, Town, 105–149.

BERGER, W.H. & PIPER, D.J.W. 1972. Planktonic foraminifera: differential settling, dissolution and redeposition. *Limnology and Oceanography*, **17**, 275–287.

BROCK, J.C., MCCLAIN, C.R., ANDERSON, D.M., PRELL, W.L. & HAY, W.W. 1992. Southwest monsoon of recent planktonic foraminifera in the northwestern Arabian Sea. *Paleoceanography*, **7**, 799–815.

CAYRE, O. & BARD, E. 1999. Planktonic foraminiferal and alkenone records of the Last Deglaciation from the Eastern Arabian Sea. *Quaternary Research*, **52**, 337–342.

CLEMENS, S., PRELL, W., MURRAY, D., SHIMMIELD, G. & WEEDON, G. 1991. Forcing mechanisms of the Indian Ocean monsoon. *Nature*, **353**, 720–725.

CLIMAP Project Members 1976. The surface of the Ice-Age-Earth. *Science*, **191**, 1131–1137.

CONAN, S.M.-H. & BRUMMER, G.-J.A. 2000. Fluxes of planktic foraminifera in response to monsoonal upwelling on the Somalia Basin margin. *Deep-Sea Research II*, **47**, 2207–2227.

CULLEN, J.L. & PRELL, W.L. 1984. Planktonic foraminifera of the Northern Indian Ocean: distribution and preservation in surface sediments. *Marine Micropaleontology*, **9**, 1–52.

CURRY, W.B., OSTERMANN, D.R., GUPTHA, M.V.S. & ITTEKKOT, V. 1992. Foraminiferal production and monsoonal upwelling in the Arabian Sea: evidence from sediment traps. *In:* SUMMERHAYES, C.P., PRELL, W.L. & EMEIS, K.C. (eds) *Upwelling Systems: Evolution Since the Early Miocene.* Geological Society, London, Special Publications, **64**, 93–106.

DEN DULK, M., REICHART, G.J., MEMON, G.M., ROELOFS, E.M.P., ZACHARIASSE, W.J. & VAN DER ZWAAN, G.J. 1998. Benthic foraminiferal response to variations in surface water productivity and oxygenation in the Northern Arabian Sea. *Marine Micropaleontology*, **35**, 43–66.

DIETRICH, G., DÜING, W., GRASSHOFF, K. & KOSKE, P.H. 1966. Physikalische und chemische Daten nach Beobachtungen des Forschungsschiffes 'Meteor' im Indischen Ozean 1964/1965. *'Meteor' Forschungsergebnisse, Reihe A*, **2**, 1–5.

DOOSE-ROLINSKI, H., ROGALLA, U., SCHEEDER, G., LÜCKGE, A. & VON RAD, U. 2001. High-resolution temperature and evaporation changes during the late Holocene in the northeastern Arabian Sea. *Paleoceanography*, **16**, 358–367.

EMEIS, K.-C., ANDERSON, D., DOOSE, H., SCHULZ-BULL, D. & KROON, D. 1995. Sea-surface temperatures and the history of monsoon upwelling in the NW Arabian Sea during the last 500,000 yr. *Quaternary Research*, **43**, 355–361.

FAIRBANKS, R.G., SVERDLOVE, M., FREE, R., WIEBE, P. & BE, A.W.H. 1982. Vertical distribution and isotopic fractionation of living planktonic foraminifera from the Panama Basin. *Nature*, **298**, 841–844.

GUPTHA, M.V.S. & MOHAN, R. 1996. Seasonal variability of the vertical fluxes of *Globigerina bulloides* (D'ORBIGNY) in the northern Indian Ocean. *Mitteilungen Geologisch-Paläontologisches Institut Universität Hamburg*, **79**, 1–17.

HAAKE, B., ITTEKKOT, V., RAMASWAMY, V., NAIR, R.R. & CURRY, W.B. 1993. Seasonality and interannual variability of particle fluxes to the deep Arabian Sea. *Deep-Sea Research I*, **40**, 1323–1344.

HASTENRATH, S. & LAMB, P.J. 1979. *Climatic Atlas of the Indian Ocean, Part 1: Surface Climate and Atmospheric Circulation.* University of Wisconsin Press, WI.

HEMLEBEN, C., SPINDLER, M. & ANDERSON, O.R. 1989. *Modern Planktonic Foraminifera.* Springer, New York.

HONJO, S. & DOHERTY, K.W. 1988. Large aperture time-series oceanic sediment traps; design objectives, construction, and application. *Deep-Sea Research I*, **35**, 133–149.

KIPP, N.G. 1976. New transfer function for estimating past sea-surface conditions from sea-bed distribution of planktonic foraminiferal assemblages in the North Atlantic. In: CLINE, R.M. & HAYS, J. (eds) *Investigation of Late Quaternary Paleoceanography and Paleoclimatology.* Geological Society of America Memoir, **145**, 3–41.

LÜCKGE, A. DOOSE-ROLINSKI, H., KHAN, A.A., SCHULZ, H. & VON RAD, U. 2001. Monsoonal variability in the northeastern Arabian Sea during the past 5,000 years: geochemical evidence from laminated sediments. *Palaeogeography, Palaeoclimatology Palaeoecology,* **167**, 273–286.

LÜCKGE, A., REINHARDT, L., ANDRULEIT, H., DOOSE-ROLINSKI, H., VON RAD, U., SCHULZ, H. & TREPPKE, U. 2002. Formation of varve-like laminae off Pakistan: decoding 5 years of sedimentation. In: CLIFT, P.D., KROON, D., GAEDICKE, C. & CRAIG, J. (eds) *The Tectonic and Climatic Evolution of the Arabian Sea Region.* Geological Society, London, Special Publications, **195**, 421–432.

MADHUPRATAP, M., PRASANNA KUMAR, S., BHATTATHIRI, P.M.A., DILEEP KUMAR, M., RAGHUKUMAR, S., NAIR, K.K.C. & RAMAIAH, N. 1996. Mechanism of the biological response to winter cooling in the northeastern Arabian Sea. *Nature,* **384**, 549–552.

MALMGREN, B.A. & KENNETT, J.P. 1977. Biometric differentiation between Recent *Globigerina bulloides* and *Globigerina falconensis* in the Southern Indian Ocean. *Journal of Foraminiferal Research,* **7**, 130–148.

NAIDU P.D. & MALMGREN, B.A. 1995. Monsoon upwelling effects on test size of some planktonic foraminiferal species from the Oman Margin, Arabian Sea. *Paleoceanography,* **10**, 117–122.

NAIDU P.D., BABU, C.P. & RAO, CH.M. 1992. Upwelling record in the sediments of the western continental margin of India. *Deep-Sea Research I,* **39**, 715–723.

NAIR, R.R., ITTEKKOT, V., MANGANINI, S.J. & 5 OTHERS 1989. Increased particle flux to the deep ocean related to monsoons. *Nature,* **338**, 749–751.

OLSON, D.B., HITCHCOCK, G.L., FINE, R.A. & WARREN, B.A. 1993. Maintenance of the low-oxygen layer in the central Arabian Sea. *Deep-Sea Research II,* **40**, 673–685.

OTTENS, J.J. 1992. April and August northeast Atlantic surface water masses reflected in planktic foraminifera. *Netherlands Journal of Sea Research,* **28**, 261–283.

PFLAUMANN, U. 1985. Transfer-Function 134/6—a new approach to estimate sea-surface temperatures and salinities of the Eastern North Atlantic from planktonic foraminifera in the sediment. '*Meteor*' *Forschungsergebnisse, Reihe C,* **39**, 37–71.

POLLEHNE, F., ZEITSCHEL, B. & PEINERT, R. 1993. Short-term sedimentation patterns in the northern Indian Ocean. *Deep-Sea Research II,* **40**, 737–752.

PRELL W.L. & CURRY, W.B. 1981. Faunal and isotopic indices of monsoonal upwelling: Western Arabian Sea. *Oceanologica Acta,* **4**, 91–98.

PRELL W., MARTIN, A., CULLEN, J. & TREND, M. 1999. *The Brown University Foraminiferal Data Base. IGBP PAGES—World Data Center-A for Paleoclimatology, Data Contribution Series No. 1999-027.* NOAA–NGDC Paleoclimatology Program, Boulder, CO.

RAMAGE, C.S., MILLER, F.R. & JEFFRIES, C. 1972. *Meteorological Atlas of the International Indian Ocean Expedition. The Surface Climate of 1963, 1964.* US National Science Foundation and India Meteorological Department.

REICHART, G.J., LOURENS, L.J. & ZACHARIASSE, J.W. 1998. Temporal variability in the Northern Arabian Sea oxygen minimum zone (OMZ) during the last 225,000 years. *Paleoceanography,* **13**, 607–621.

RIXEN, T., HAAKE, B. & ITTEKKOT, V. 2000. Sedimentation in the western Arabian Sea. The role of coastal and open-ocean upwelling. *Deep-Sea Research II,* **47**, 2155–2178.

ROSTECK, F., RUHLAND, G., BASSINOT, F.C., MÜLLER, P.J., LABEYRIE, L.D., LANCELOT, Y. & BARD, E. 1993. Reconstructing sea-surface temperature and salinity using $\delta^{18}O$ and alkenone records. *Nature,* **364**, 319–321.

SAUTTER, L.R. & SANCETTA, C. 1992. Seasonal associations of phytoplankton and planktic foraminifera in an upwelling region and their contribution to the seafloor. *Marine Micropaleontology,* **18**, 263–278.

SCHULZ, H. 1995. *Sea-surface Temperatures 10,000 years B.P.—Consequences of the Early Holocene Insolation Maximum.* Berichte–Reports, Geologisch-Paläontolologisches Institute Universität Kiel, **73**.

SCHULZ, H., VON RAD, U. & STACKELBERG, U. 1996. Laminated sediments from the oxygen-minimum zone of the northeastern Arabian Sea. In: KEMP, A.E.S. (ed.) *Paleoclimatology and Paleoceanography from Laminated Sediments.* Geological Society, London, Special Publications, **116**, 185–207.

SIROCKO, F. 1989. *Zur Akkumulation von Staubsedimenten im nördlichen Indischen Ozean: Anzeiger der Kimageschichte Arabiens und Indiens.* Berichte–Reports, Geologisch-Paläontologisches Institut Universität Kiel, **27**.

SIROCKO, F. 1995. *Abrupt change in monsoonal climate: evidence from the geochemical composition of Arabian Sea sediments.* Habilitation thesis, University of Kiel.

SIROCKO, F. & SARNTHEIN, M. 1989. Wind-borne deposits in the Northwestern Indian Ocean: record of Holocene sediment versus modern satellite data. In: Leinen, M. & Sarnthein M. (eds) *NATO ASI Series C,* **282**, 401–433.

SIROCKO, F., SARNTHEIN, M., ERLENKEUSER, H., LANGE, H., ARNOLD, M. & DUPLESSY, J.C. 1993. Century-scale events in monsoonal climate over the past 24,000 years. *Nature,* **364**, 322–324.

SONZOGNI, C., BARD, E. & ROSTECK, F. 1998. Tropical sea-surface temperatures during the Last Glacial Period: a view based on alkenones in Indian Ocean sediments. *Quaternary Science Reviews,* **17**, 1185–2101.

STAUBWASSER, M. & SIROCKO, F. 2001. On the formation of laminated sediments on the continental margin off Pakistan: the effects of sediment provenance and sediment redistribution. *Marine Geology,* **172**, 43–56.

TAKAHASHI, K. & BE, A.W.H. 1984. Planktonic foraminifera: factors controlling sinking speeds. *Deep-Sea Research I,* **31**, 1477–1500.

THIEDE, J. 1975. Shell- and skeleton-producing plankton and nekton in the eastern North Atlantic Ocean. *'Meteor' Forschungsergebnisse, Reihe C*, **20**, 33–79.

THIEDE, J., NEES, S., SCHULZ, H. & DE DEKKER, P. 1997. Oceanic surface conditions recorded on the sea floor of the Southwest Pacific Ocean through the distribution of foraminifers and biogenic silica. *Palaeogeography, Palaeoclimatology, Palaeoecology*, **131**, 207–239.

THUNNELL, R.C. & REYNOLDS, L.A. 1984. Sedimentation of planktonic foraminifera: seasonal changes in species flux in the Panama Basin. *Micropaleontology*, **30**, 243–262.

VAN DER LINDEN, W.J.M., ZACHARIASSE, J.W., VAN DER WEIJDEN, K.H. & SHIPBOARD PARTY 1993. *Late Quaternary productivity and the dynamics of the Oxygen Minimum Zone in the northeast Arabian Sea, Part 2*. Unpublished Shipboard Report of NIOP Cruise D2, Karachi–Karachi, University of Utrecht.

VAN DER PLAS, L. & TOBI, A.C. 1965. A chart for judging the reliability of point counting results. *American Journal of Science*, **263**, 157–182.

VINCENT, E. & BERGER, W.H. 1981. Planktonic foraminifera and their use in paleoceanography. *In:* EMILIANI, C. (ed.) *The Oceanic Lithosphere: The Sea*. Wiley, New York, **7**, 1025–1119.

VON RAD, U., SCHAAF, M., MICHELS, K.H., SCHULZ, H., BERGER, W.H. & SIROCKO, F. 1999*a*. A 5,000 year record of climate change in varved sediments from the Oxygen Minimum Zone off Pakistan, Northeastern Arabian Sea. *Quaternary Research*, **51**, 39–53.

VON RAD, U., SCHULZ, H., RIECH, V., DEN DULK, M., BERNER, U. & SIROCKO, F. 1999*b*. Multiple monsoon-controlled breakdown of oxygen-minimum conditions during the past 30,000 years documented in laminated sediments off Pakistan. *Palaeogeography, Palaeoclimatology, Palaeoecology*, **152**, 129–161.

VON RAD, U., SCHULZ, H. & SONNE 90 SCIENTIFIC PARTY 1995. Sampling the Oxygen Minimum Zone off Pakistan: Glacial/Interglacial variations of anoxia and productivity. *Marine Geology*, **125**, 7–19.

VON STACKELBERG, U. 1972. Faziesverteilung in Sedimenten des indisch-pakistanischen Kontinentalrandes (Arabisches Meer). *'Meteor' Forschungsergebnisse, Reihe C*, **9**, 1–73.

WYRTKI, K. 1971. *Oceanographic Atlas of the International Indian Ocean Expedition*. National Science Foundation, Washington, DC.

ZEITSCHEL, B. & GERLACH, S.A. (eds) 1973. *The Biology of the Indian Ocean*. Springer, Berlin.

ZOBEL, B. 1973. Biostratigraphische Untersuchungen an Sedimenten des indisch-pakistanischen Kontinentalrandes (Arabisches Meer). *'Meteor' Forschungsergebnisse, Reihe C*, **12**, 9–73.

Index

Page numbers in italic, e.g. *208*, refer to figures. Page numbers in bold, e.g. **171**, signify entries in tables.

Ab-e Shahr Unit **171**
Ab-i Lashkar *208*
Abnama Fault *160*
Abu Dhabi *305*, *372*
Aden, Gulf of *8*, *88*
Ahmedabad *312*
Ajmer *312*
Al Ain *306*
Al Ghabbi *312*
Al Mintrib *312*
Al Wasil *312*
Al Wusayl *372*
Al Wusayl Member 374, *375*, *379*
Angohran Unit *166*, *167*, *169*, **171**, *172*
Arabia 91–92
Arabia, Southern 301–302
    Arabian Desert 302, 313–314
        aeolian sequences 308–310, *311*
        ages **305**
        fluvial sequences 302–308
        OSL sand dates *304*
        SW Monsoon winds *303*, *308*
Arabian Gulf *41*, 371–373, *372*, 383–384
    age of Pleistocene deposits 379–380, **380**
        correlation with other deposits 380
    deposition of Pleistocene sediments
        correlation between onshore and offshore sediments 378–379
        Fuwayrit Formation 375–378, *375*, *377*
        Ghayathi Formation 374–375, *374*, *376*
    Holocene highstand 381–382, **381**
    palaeowinds and palaeoclimates 382–383, *383*
    Pleistocene deposits 373, *373*
        Aradah Formation 373–374, *373*
        Fuwayrit Formation *373*, 374
        Ghayathi Formation 373, *373*
        offshore 374
    Pleistocene sea levels 380–381
Arabian Plate *26*, *40*
Arabian Sea and Basin *72*, 317–320
    bathymetry *318*
    climate and hydrography 407–408
    coastal neotectonics 87–88, *88*, 92–93
        Arabia 91–92
        India 92
        Makran 88–91
    dinoflagellate abundance *319*
    distribution of foraminifera 463–464, *464*, 485–490
        calcification temperature **490**
        calculation results **488–489**
        conductivity–temperature–depth (CTD) profiles 465, *468*, *469*
        contouring data 468
        depth habitat and ambient temperatures 483–484, *484*, *485*
        depth habitat and hydrography 477–479, *478*, *479*, *480*, *481*, *482*
        depth habitat and oxygen balance 479–483
        fauna composition and species distribution 472–476, **473**, *474*, *475*, **476**, **494–495**, **496–497**
        hydrographic conditions 472
        living and dead specimens 469–470
        productive zone (PZ) 470–472, *471*
        sample collection 464–465, **466–467**, 470
        sample preparation 465
        stable isotope analysis 465
        standing stock 484–485, *486*, *490*
gravity chart *8*
marine data coverage *72*, *74*
orogeny and climate change 1
oxygen minimum zone (OMZ) breakdown study 407, *409*, 413–416, *416*
    alkenone-derived sea surface temperatures 411, *412*
    glacial ice volume effect 411–412
    sea surface temperature and salinity effects 412–413
    sediments and age model 409–411, *410*, *411*
oxygen minimum zone (OMZ) evolution study 433–434, *434*, 439
    comparison of proxies 437–439
    Early and mid-Holocene evolution 439–441, *439*, *440*
    inorganic geochemical indicators of water column ventilation 434–437
    lamination preservation as indicator of water column ventilation 437, *438*
    methods 434, **435**
    principle component analysis 435–437, *436*, **437**
palaeogeography *54*
Paleogene magnetic isochrons and palaeo-propagators 71–72, 82–83
    ages of magnetic Chrons **73**
    analytical signal 73–76
    data 72–73
    identification of magnetic anomalies 73
    magnetic anomaly profiles *75*, *78*, *79*
    magnetic lineation pattern 78–80
    method of analysis 73–77

propagating ridge segments 80, *81*
spreading and propagation rates 80–82, *81*, *82*, **82**, *83*
validation of interpreted isochrons 76–77
Paleogene plate tectonic evolution 7, 18–21
    Chron 27 (61Ma) 11–12, *12*
    Chrons 26 (58Ma) and 25 (56Ma) 12–13, *12*
    Chron 24 (52Ma) to 22 (49Ma) 13–18, *14*, *15*, *16*, *17*
    Chron 21 (46Ma) *15*, *17*, 18
    plate reconstructions, problems and method 10–11
    review of previous models 7–9
    revised tectonic chart for Paleogene time 9, *10*
    rotation parameters **11**
particle flux and productivity of foraminifera 499–501, *500*, **501**, 513–514
    coarse fraction and analysis 502
    conductivity–temperature–depth profiles and sediment traps 501–502
    conductivity–temperature–depth profiles for winter 502–503, *503*, *504*
    flux and relative abundance of species 506–509, **507**, *508*
    increased winter productivity 510–511, *511*
    index of non-upwelling conditions 511
    origin of flux and sediment resuspension 509–510
    particle flux of EPT-2 504–506, *505*
    sediment record 511–513, *512*, *513*
regional stratigraphy and tectonic evolution 31, *32*
    Cretaceous time 31–33
    Late Miocene to Recent time 33–34
    Late Oligocene to Miocene time 33
    Paleocene to Early Oligocene time 33
salinity and temperature variations **328**
seismic stratigraphy and major unconformities 25–27, *26*, *27*, *28*, 34–35
    methods, data acquisition and interpretation 27
    pre-drift sequence P 30–31, *31*
    sequence M1 29–30, *30*, *31*
    sequence M2 28–29, *30*, *31*
    sequence M3 and M-unconformity 27–28
Aradah Formation 373–374, *373*
Aravalli Range *312*
$^{40}Ar/^{39}Ar$ geochronology 109–111
Az Zahir *311*

Badamu-Siahan Unit *162*, **180**, 183
Baft *149*
Baghraband syncline *167*
Bahrain *372*
Baiban Shelf *38*, 55–56
Bajgan *149*, *152*
Bajgan Complex 151–153
Bakhtiari Fold Zone *206*, 210, *211*, *213*
Bam *187*
Bamposht Unit *169*, **171**
Bandar-e-Abbas *150*, *196*
Band-e Chaker Unit **171**, 175–176, *175*

Band-e Zeyarat Complex *152*, **153**, 156, *157*
Bangalore *88*
Bard-e Marz Limestone 188, *188*
Bashakerd Fault *197*
Bazman *197*, *199*
Bhuj *88*
Bidak Unit *152*, 187–188
Birjand *149*
Birk Fault *162*
Birk Unit **153**, 161–163, *162*
Bombay *88*
Burma Basin *252*

Calcigonellum infula **340**
    taxonomy 336
Calciodinellum operosum 320, *321*, *337*
    abundances **338–339**
    distribution 323, *323*
    ecology 331–332
    taxonomy 336
Calciperidinium asymmetricum **340**
    taxonomy 336
Cambay, Gulf of *312*
Carlsberg Ridge *8*, *72*, *197*
Caspian Sea *149*
Central Indian Ridge *72*
Chagos–Laccadive Ridge *54*, *72*, *88*
Chah Bahar *89*
Chah Mirak *152*
Chah Mirak Unit *152*, **153**, *159*
Chahbahar *150*
Chain Fracture Zone *8*, *10*
Chain Ridge *8*, *10*
Chaman Fault *88*, *238*
Chambal River *312*
Chandragup *38*
Chang-La *98*
Chilling *98*, *99*, *100*, *110*
Chogdo Formation *100*, *102*, 105–106
    palaeo-currents *106*
Choksti *99*
climate–tectonic models, testing 2–3
Cochin *88*
Coloured Mélange Complex *152*, 153–156, **153**, *154*, *155*, *166*

Dabb'iya Member 374, *375*, *377*, *378*, *379*
Dalrymple Fault *46*
Dalrymple Trough *30*, *32*, *38*, *118*, *197*
    magnetic anomalies *120*
    seismic stratigraphy 63–64
Dar Anar Complex **153**, 156–158
Dar Anar Fault *152*, *160*
Dar Anar Unit *152*
Dar Pahn Unit *166*, **171**, *177*
Darban *169*
Darban Unit *169*, **171**
Darkhunish shale Unit **171**, *175*

INDEX

Jinnah Trough *119*
Jiroft *187*
Jiroft Fault *152*
Jiwani *138*
Jodhpur *312*
Jumna River *312*
Jurutze *98*
Jurutze Formation *109*, 101–105, *102*
    palaeo-currents *106*

Kabr Kuh anticline *206*
Kahnuj *152*
Kam Sefid *151*
Kam Sefid Sandstone 188
Kam Sefid Unit *152*
Karachi *138*, *260*, *312*
Karakoram *238*
Karanj Fault *166*
Kargil *98*, *239*
Karvandar Sequence *187*
Katawaz Basin *238*
Kazerun Line *206*
Kenar Fault *169*
Kenar Unit *169*, **171**
Kerguelen Hotspot *54*
Kerguelen Plateau *54*
Kermanshah *149*
Khalsi *98*, *239*
Khalsi Flysch *100*, *102*, 105
    palaeo-currents *106*
    sedimentary log *103*
Khodar Fault *167*
Khuwaymah *89*
Kirthar Fold Belt 273–275, *274*, 281–286, **287**, 292
    collision 291–292
    drift phase 1 (Late Jurassic to Santonian time) 289
    drift phase 2 (Santonian to Maastrichtian time) 289–291
    drift phase 3 (Paleocene to Eocene time) 291
    rift period (Early to Late Jurassic) 286–289, *288*
    sequence stratigraphy
        key outcrops *285*
        methods of determination 276–279, *277*, *278*
        results 279–281, *280*, **282–284**
    tectonic setting 275–276
Kirthar Range *312*
Kohat Plateau *238*
Koh-i Soltan *197*, *199*
Kohkiluyeh Fold Zone *206*, *211*, *213*
Konarak *38*, *89*
Konashamir Sandstone 189
Kordestan Fold Zone *206*
Kuhak syncline *162*
Kuh-e Birk *162*
Kuh-e Birk Massif *163*
Kuh-e Murdan *162*
Kuh-e Tahtun *166*
Kuh-i Asmari *208*

Kutch *312*

Laccadive Ridge *8*, *10*
Ladakh *238*, *239*
Ladakh Batholith
    biotite and hornblende ages *110*
    diagenetic grade versus distance *110*
    thermal evolution of suture zone 111–112, *112*, 113
Ladakh Himalaya *98*, *99*
Lahabari *208*
Lamayuru Group turbidites *100*
Las Bela Valley *40*, 58–59, *59*
Las Bela–Chaman Structural Axis *40*, *59*
Laxmi Basin *72*
Laxmi Ridge *8*, *10*, 18–19, *46*, *72*, *238*
Leh *98*, *239*
Little Murray Ridge *26*, *38*, *46*, *89*, *118*
Liwa *303*, *306*
Lorestan Fold Zone *206*, *211*, *213*
Lut Block *149*
Lut Depression *38*
Lut microcontinent *40*

Madagascar *54*
    separation from India 9
Madagascar Ridge *54*
Mahdaneh Fault *162*
Makran 88–91, *88*
Makran Accretionary Prism/Wedge 25, *26*, *34*, 137, 144–145, *238*, 387–388, *388*
    bottom simulating reflector (BSR) 137, *138*
    heat flow density 138
    Eocene–Early Oligocene flysch sequences 165–171, *166*, *167*
    geological and oceanographic–meteorological setting 388
        morphology, climate, precipitation and erosion 388–389, *389*
        oceanography 389–390
        tectonics 388
    Holocene sediment study
        chronology of varved sediments 392
        chronology of varved sediments **393**
        frequency analysis 392
        high-resolution X-ray fluorescence 390–391
        lithological profiles *391*
        samples and data 390
        X-radiography, varve counting and varve thickness 390
    Holocene sediment study results
        cyclicity 400–404, *402*, *403*
        depth–age diagram *394*
        graded and laminated silt turbidites 398–399
        hemipelagic laminated sediments 394–396, *395*
        texture and composition variations 396–397, *397*
        turbidite frequency and varve thickness variability 399–400
        turbid-plume or suspensate deposits 397–398, *398*

varve chronology 392, *394*
Late Miocene–Pliocene deposits 180–182
M-unconformity 27–28, 33
Miocene neritic sequences 174–179
numerical simulation of geothermal state 139
    Model 1 139–141, *141*, **141**, *142*
    Model 2 141–143, *143*
    Model 3 144, *144*
Oligocene–Miocene flysch sequences 171–174
stratigraphic units **171**
trace gases in water column 138–139, *140*
Makran Coast Range 38
Makran Escarpment *40*, *46*
Makran front 195–199
Makran margin, Quaternary sedimentation 219–220, *221*
  morphotectonic framework 220
  sediment facies
    bed thickness and vertical sequences 224–225, **225**, *226*, *227*
    characteristics and distribution 223–224, **223**
    grain size 228–231, *231*, *232*
    structures and sequence 225–228, *228*, **229**, *230*
  sediment source, distribution and controls 232–234, *233*
  stratigraphic framework 220–223, *222*
    sedimentation rates 223
  turbidite statistics 231–232
Makran Unit **171**
Makran Wedge *32*, 39–42, *59*
Malay Basin 252
Mangalore 88
Manujan *150*, *152*
Marich Unit *152*, 188
Mascarine Basin 9
Mascarine Plateau *8*, *10*, *54*
Mashkel depression *162*, *199*
Mashkid Unit *162*, **180**, 185–186
Masirah Line *88*
Massirah Island *306*
Mekong Basin 252
Melodomuncula berlinensis **340**
  taxonomy 336
Middle Indus Basin 273–275, *274*, 281–286, **287**, 292
  see also Indus Basin
  collision 291–292
  drift phase 1 (Late Jurassic to Santonian time) 289
  drift phase 2 (Santonian to Maastrichtian time) 289–291
  drift phase 3 (Paleocene to Eocene time) 291
  rift period (Early to Late Jurassic) 286–289, *288*
  sequence stratigraphy
    methods of determination 276–279, *277*, *278*
    results 279–281, *280*, **282–284**
  tectonic setting 275–276
Minab *150*, *152*, *196*
Minab anticline *206*
Minab Conglomerate **171**, 181–182, *182*
Minab Unit *166*

Miru *98*, *100*
Mokhtarabad Complex **153**, 159–161
monsoon 8
  influence of Indus River 247–248
monsoon variability evidence in Toba ash layer 445–446, *446*, 454–459, *457*
  analytical methods 446–449, *449*
  chronostratigraphy and climate cycles 445
  historical records of ash falls
    older ash 451–454
    younger ash 451, *456*
  lithology 448
  lithology, fabric and chemical composition 449–451, *450*, *451*, **452–453**, *454*, **455**
  lithostratigraphy *447*
  materials 446
Morghak Unit **153**, 163
Morton Fault *167*, *169*
Mosri Unit *162*, *169*, **171**
Mudayrib 311
Murray Ridge *8*, 25, *26*, *30*, *32*, *38*, *46*, *72*, *88*, *97*
  age 65
  seismic stratigraphy 63
  topography 64–65
Murray Ridge complex 117–119, *118*, 128–29
  evolution 131–132
  Little Murray Ridge *118*
  magnetic investigations 127–128
  metabasalt composition *129*, *130*
  North Murray Ridge *118*
  petrography and geochemistry of igneous rocks
    Harzburgite of clinopyroxene-poor lherzolite, completely serpentinized 126–127
    microporphyric tholeiitic metabasalt 121–126, *121*, **122**, **123**, *124*, *125*
  possible supra-subduction origin 128–131, *131*
  sedimentary rocks
    carbonate crusts 127
    quartz siltstone 127
  South Murray Ridge *118*, *119*
    magnetic anomalies *120*
  study materials and methods
    geophysical surveys 119
    petrographic, geochemical and palaeontological analyses 120–121
    sampling 119–120
Musandam 88
Musandam Peninsula 38
Musandam Valley 38
Muscat 306

Nahang Unit *167*, *169*, **171**, *192*
Nain 149
Nam Con Son Basin 252
Nanga Parbat *98*, *238*
Nargakan Unit *169*, **171**
Narreh-Now *199*
neotectonics 4–5

Neyriz *149*
Nikshahr *150*
Nimu *99, 102*
Nindam Formation *100, 102*
Ninetyeast Ridge *54*
nitrogen isotope mass balance *363*
Nokju Fault *162*
Nummulitic Limestone *100*
Nurabad *151, 152*
Nurla Formation *102*

Oman *372*
Oman Abyssal Plain *25, 26, 32*
Oman Basin *18*
Oman, Gulf of *149*
Oman Basin, Gulf of *37, 65–66*
 central area *53–54*
 eastern side of Makran subduction complex *58–59*
 eastern transform margin
  crustal structure *61*
  morphology and structural setting *59–61*
  proto-Indus and Indus fan *61–62*
  seismic stratigraphy of Murray Ridge complex *63–64*
  topography of Murray Ridge region *64–65*
 geological cross-section *41*
 gravity map *46*
 morpho-tectonic setting of the convergent northern margin
  coastal Makran wedge *39–42*
  Eocene–Holocene southern accretionary wedge *38–39*
  offshore accretionary wedge *42–52, 43*
  pre-Eocene northern accretionary wedge *37–38*
 paleogeography *54*
 seismic reflection profiles *39, 44, 45, 47, 48, 49, 50, 51, 56, 57, 60, 62*
 southern passive rifted margin *55*
 topographic map *38*
 western translation margin *55*
  Baiban Shelf *55–56*
  Strait of Hormuz *56–58, 56*
Ormara *89, 138*
Ormara microplate *40, 46, 59*
Ornach-Nal Fault *40, 59*
Orthopithonella granifera *320, 321, 337*
 abundances **338–339**
 distribution *322, 322*
 ecology *330–331*
 taxonomy *336*
Owen Basin *8, 10, 26*
Owen Fracture Zone *8, 10, 26, 72, 88, 197*
Owen Ridge *2, 10, 46*

Padgan Fault *169*
Pakistan, offshore varve-like laminae *421–422, 422, 425–430, 428*
 environmental setting *422*

results
 geochemical analyses *423–424, 425*
 lithological description *423, 424*
 micropalaeontological analyses *424–425, 426*
 samples and methods *422–423*
Palami Conglomerate **171**
partial annealing zone (PAZ) *107–108*
Pasni *89, 138*
Patkon Conglomerate *189*
Pattani Trough *252*
Pearl River Mouth Basin *252*
Persian Gulf *149*
 palaeogeography *54*
Pishamag Fault *169*
Pishamag Unit *167*
Pishin *150*
Pishin Unit *151*, **171**
Pleistocene–Holocene sedimentation *3–4*
productive zone (PZ) of foraminifera *470–472, 471*
Proximity Index for fine-grained turbidites *228*

Q'al eh Shaikan *166, 182*
Qatar *372*
Qeshm Island *38, 206*
Qiang Tang Block *98*
Qishm Island *89*
Qualhat Seamount *38*

Ram Hormuz syncline *208*
Ra's al Hadd *89*
Ra's al Hamra *89*
Ras al Khaimah *306*
R'ass al Dabb'iya *372*
Ravi River *312*
Remeshk Complex **153**, *158–160*
Remesk *160*
Reunion Hotspot *54*
Rig Unit *151*, **171**
Roksha Unit *167*, **171**, *176*
Rub al Khali *303, 306*
Rudan Fault *152*
Ruk Unit *165–168, 167, 169*, **171**

Sabarmati River *312*
Sabkha Fuwat Ash Sham *306*
Sabkha Matti *303*
Sabz Unit *166*, **171**, *174–175*
Sabzevar *149*
Sahan Tang Unit *151*, **171**
Sanandaj *149*
Sanandaj–Sirjan–Bajdan–Dur Kan sliver *40, 149*
Saravan *149, 150, 162, 199*
Saravan accretionary prism **180**
 Cenozoic sediments *190–191*
 Eocene flysch sequences *182–186*
 Eocene–Oligocene outliers *186–187*, **187**
 geotectonic overview *199–200*
 Oligocene–Miocene conglomerates *186, 187*

Pliocene–Pleistocene fanglomerates 191, *192*
  superficial deposits 191–193
    mud volcano *193*
Saravan Fault *162*
Saravan Unit *162*, **180**, 183–184
Sarbaz *151*
Sardasht *151*
Saspol *239*
Saudi Arabia *372*
Saurastra *312*
Scrippsiella trochoidea (spiny cysts) 320, *321*, *337*
  abundances **338**–**339**
  distribution 323, *324*
  ecology 332
  taxonomy 336
settling flux zone (SFZ) of foraminifera 470, *471*
Seychelles Bank *8*, *10*
Seychelles–Mascarine Plateau *72*
Shah Kuh *162*, 186, *199*
Shahiq *311*
Shahr Pum Unit *166*, *173*
Shahr Unit **171**
Shamal wind 3, 302, *303*, *372*
Sheba Ridge *88*, *197*
Shey *98*
Shirinzad Unit *169*, *170*, **171**
Simply Folded Belt *206*, 208–209, *211*
Sirjan *149*, *187*
Sistan Suture *40*
Somali Plate, Indian Plate relative movement *20*
Somalian monsoon 341–342, 351, 353–355, 367–368
  controls on nitrogen isotopes of export production 364–365
  diagenesis and burial 365–367, *367*
  foraminifera study methods 342–344
    proxies 344–345, *345*, *346*
  foraminifera study results 343–344
    atmospheric variations and foraminiferal productivity 350–351
    long-term foraminiferal productivity variations 347–349, *349*
    short-term foraminiferal productivity history 349–350
  lateral transport and annual export 363–364
  modern annual dynamics 342, *342*
  particle flux study materials and methods *354*, 355–358, *355*, **356**, *357*
  particle flux study results 361–362
    autumn intermonsoon and NE monsoon 362
    core-top sediments 362–363, *363*
    settling nitrogen flux *361*
    SW monsoon 362
  patterns of change 345–346
    atmospheric variations 347
    foraminiferal productivity variations 347
  regional hydrography and pelagic system 358
    autumn intermonsoon and NE monsoon 361
    sea surface temperatures *360*

SW monsoon 358–361, *359*
Sonne Fault *46*, *59*, *89*, *197*
Sphaerodinella albatrosiana 320, *321* *337*
  abundances **338**–**339**
  distribution 322–323, *323*
  ecology 331–332
  taxonomy 336
Sphaerodinella tuberosa (var. 1) **340**
Sphaerodinella tuberosa (var. 2) 320, *321*, *337*
  abundances **338**–**339**
  distribution 322, *322*
  ecology 329–330
  taxonomy 336
Spontang *98*, *239*
Sulaiman Ranges *238*, *312*
Sumda Formation, sedimentary log *103*
Sumda-Do *99*, *100*, *110*
Suran *162*
Surjan Fault *40*, *59*, *162*
Sutlej River *312*

Tabas Block *149*
Taftan *197*, *199*
Tahrule *150*, *166*
Tahtun Unit **171**
Talkhab Fault *162*
Talkhab Mélange Complex **153**, 163–164
Talkhab Unit *162*, **180**, 186
Tehran *150*
Thar Desert 301–302, 310–312, *312*, 313–314
  supporting data from Arabian Sea 312–313
Thok *38*
Thoracosphaera heimii 320, *321*, *337*
  abundances **338**–**339**
  accumulation rates *329*
  distribution 321–322, *321*
  ecology 329–330
  fragmentation *324*
  ratio of other species *324*
  taxonomy 336
Tiab anticline *166*
Tibetan Plateau *239*
Toba 3
Toba Ash and monsoon variability 445–446, *446*, 454–459, *457*
  analytical methods 446–449, *449*
  chronostratigraphy and climate cycles 449
  historical records of ash falls
    older ash 451–454
    younger ash 451, *456*
  lithology *448*
  lithology, fabric and chemical composition 449–451, *450*, *451*, **452**–**453**, *454*, **455**
  lithostratigraphy *447*
  materials 446
Trans Himalaya *98*, *239*

Ulug Muztagh *239*

Umm as Samim 303, 306
United Arab Emirates (UAE) 372
Upsi 98
Urtsi 98
Uruq Al Mutaridan 303, 306

Vaziri Unit 167, 178–179, 179, 166
Veranj Fault 160

Wadi Batha 311
Wadi Du Taymat 311
Wadi Qabit 311
Wahibah Sands 303, 306
Wanlah 98
West Natuna Basin 252

Yazd Block 149

Zaboli 162
Zaboli Fault 169
Zaboli Unit 162, 180, 184–185
Zagros 38
Zagros collision 194–195, 196

Zagros crush zone 197
Zagros Foldbelt 40
Zagros Mountains 55, 205–208, 206
   evolution 213–215
   fold geometry 209, 210
      elevation and inclination 209–212
      half-wavelength/amplitude ratio 209, 211
      shortening of sedimentary cover 212, 213
   fold structures 208–209
   neotectonic deformation 216
   salt deposits 208
      influence of thickness on deformation 215
      stratigraphic control 215
   Simply Folded Belt 206, 208–209, 211, 213, 214
   stratigraphy and divisions 207
   synchronous thrust and wrench faulting 215–216
Zagros Thrust 206
Zahedan 149, 199
Zanskar Platform 98, 100, 102
Zanskar River 99, 239
Zendan Fault 46, 152, 166, 196, 197
   Miocene neritic sequences 174–179
Zendan Fault–Oman Line 40, 56